GROUP 24
Physical and Mathematical Aspects of Symmetries

Other titles in the series

The Institute of Physics Conference Series regularly features papers presented at important conferences and symposia highlighting new developments in physics and related fields. Previous publications include:

GROUP 24
Physical and Mathematical Aspects of Symmetries

Proceedings of the Twenty-Fourth International Colloquium on Group Theoretical Methods in Physics held in Paris, France, 15–20 July 2002

Edited by
J-P Gazeau, R Kerner, J-P Antoine, S Métens and J-Y Thibon

CRC Press
Taylor & Francis Group
Boca Raton London New York

CRC Press is an imprint of the
Taylor & Francis Group, an **informa** business

CRC Press
Taylor & Francis Group
6000 Broken Sound Parkway NW, Suite 300
Boca Raton, FL 33487-2742

First issued in paperback 2019

ISBN-13: 978-0-7503-0933-2 (hbk)
ISBN-13: 978-0-367-39478-3 (pbk)

British Library Cataloguing in Publication Data

A catalogue record for this book is available from the British Library.

Library of Congress Cataloging-in-Publication Data are available

Visit the Taylor & Francis Web site at
http://www.taylorandfrancis.com

and the CRC Press Web site at
http://www.crcpress.com

Foreword

The 24th International Colloquium on Group Theoretical Methods in Physics, also known as the "Group-24" Conference, was held from the 15th July to the 20th July 2002 in Paris, France. Although one digit short of the genuine silver jubilee, it was in many ways a remarkable and special event.

About 30 years ago a group of enthusiasts, headed by H. Bacry of Marseille and A. Janner of Nijmegen, initiated a series of annual meetings intended to provide a common forum for scientists interested in group theoretical methods, represented at that time mostly by two important communities: on the one hand, elementary particle theorists and phenomenologists, and solid state specialists, on the other, mathematicians eager to apply newly-discovered group and algebraic structures.

First four meetings took place alternatively in Marseille and Nijmegen. Soon after, the workshop acquired an international flavour, especially following the 1976 Colloquium in Montreal, Canada, and has since been held in many places around the world. It has become a bi-annual Colloquium since 1990, the year it was organized in Moscow.

This has been the first time since the foundation of this series of meetings that the Colloquium was organized once more in France. Moreover, it was held in the city of Paris, one of the cradles of group theory. Here Galois wrote his notes the night before the fatal duel; here Elie Cartan introduced the classification of semi-simple Lie algebras; one can continue *ad libitum* the list of famous mathematicians connected with this exceptional place. The very names of the streets in the Latin Quarter gave a unique flavour to our enterprise: Monge, Legendre, Lagrange, Laplace, – and many others, physicists, chemists, botanists, – the city of Paris hosted them all, and dwells on its glorious past.

Organizing a prestigious scientific meeting in a great capital is a great challenge. On the other hand, it gave the organizers a unique possibility to use its extremely rich scientific and educational infrastructure. Three Universities were associated with the organization of the Colloquium: Université Pierre et Marie Curie-Paris 6, Université Paris 7-Denis Diderot and the suburban Université de Marne-la-Vallée; they provided us with important financial support and material help. All the sessions took place in the Institutes of the Montagne Sainte Geneviève campus, just behind the Panthéon: Institut Océanographique, Institut Henri Poincaré, Institut de Chimie-Physique and Institut de Géographie. We address our special thanks to the directors of the Institut Henri Poincaré, M. Broué and A. Comtet, and to the director of Institut de Chimie-Physique Alfred Maquet, and their staff, for exceptional quality of hospitality extended to us during the Colloquium.

Important financial help has been provided by the Ministry of Research and Technology, the Ministry of Foreign Affairs and the Ministry of Education. The colloquium was also supported by French Embassies in Algeria, in Bulgaria, in Czech Republic, in Hungary, in Poland, and in Ukraine. The Wigner Medal and Hermann Weyl prize ceremonies were held at the Mairie of the 5ème arrondissement (the City Hall of the Latin Quarter in Paris) whose

Mayor Jean Tibéri extended his hospitality to the Colloquium. In addition, the meeting was sponsored by IUPAP and the CNRS, the French National Research Council. We express our thanks to all these Institutions, above all to our Universities, their Presidents, the Directors of Physics Departments and the research groups Laboratoire de Physique des Liquides (Paris 6) and Laboratoire de Physique théorique de la Matière condensée (Paris 7), who helped us financially and morally.

Several aspects of the "Group-24" meeting may be considered as original and innovative. They concern the organization of our sessions as well as their scientific content. A new form of contribution was introduced: *rapid plenary presentations of poster*s, called *poster clips* . Besides the regular poster display in the halls of the Institutes in which parallel sessions took place, accessible to all during numerous coffee breaks and other, those contributing posters were offered the opportunity to present their work on a couple of slides during two to three minute talk in special *plenary poster session*. Moreover, poster contributions were offered equal space in the proceedings as the regular parallel session contributions.

Another important innovation came from the deliberate choice of novel domains of research. These new applications were represented in special parallel sessions, devoted to such subjects as mathematical biology, chemistry and biochemistry, quantum control, cryptography and other subjects until recently considered as "exotic". A special "round table" was organized to discuss some historical and philosophical aspects of group theory in physics, and mathematical physics in general. Another scientific event was a mini-symposium on *Quantum Groups and Integrable Systems* held on July 17.

An important *Satellite Workshop* devoted to *Coherent States, Wavelets, and Applications* was organized by J.-P. Antoine and his group in Louvain-la-Neuve (Belgium). The proceedings of this conference are included in this volume.

Traditionally, during the ICGTMP Colloquium, the prestigious Wigner Medal is awarded by an International Committee headed by Arno R. Bohm to a scientist who has made an outstanding contribution to the development of group theoretical methods in physics and their applications. In 2002, the *Wigner Medal* was awarded to *Harry Zvi Lipkin* from the Weizmann Institute, Israel, for his life-long work in physics using novel group-theoretical methods.

For the first time, the newly established *Hermann Weyl Prize* for a young scientist who has made an important contribution to the domain of symmetries in physics, was awarded in conjunction with the Wigner Medal. The recipient was *Edward Frenkel* from the University of California at Berkeley. Both ceremonies were held in the Mairie of the 5ème arrondissement, place du Panthéon. A beautiful musical program performed by a group of Professor Davy Erlih's students of the Paris Chamber Music Conservatory conveyed to these ceremonies an unforgettable artistic touch.

It takes about two years to prepare a Colloquium of this importance. The scientific preparation owes a lot to the International Standing Committee, headed by Professor H.D. Doebner, as well as to the International Advisory Committee.

The organization of the Colloquium could never have been accomplished without such a devoted and efficient team. The logistics of the Conference was taken care of by the *Congrès Scientifiques Services* agency, headed by Chantal Iannarelli. The secretarial work was diligently taken over by Evelyne Authier. The accompanying persons' programme was conceived and carried out talentedly by Grazyna Kerner and Paulette Solomon.

The Local Organizing Committee was constantly helped by the group of young and devoted assistants and graduate students: Miroslav Andrle, Cécile Barbachoux, Avi El Kharrat, Sara Franceschelli, Tarik Garidi, Christophe Tollu and Frédéric Toumazet. We express our deep gratitude to all our collaborators and friends who contributed to the our endeavour, and we hope that the participants will keep in memory the nice week spent together in the middle of July 2002 in Paris.

Jean-Pierre Gazeau
Richard Kerner
Chairmen of the Local Organizing Committee

Introduction

This volume contains contributions to the 24th International Colloquium on Group Theoretical Methods in Physics. All the contributed papers have undergone an independent refereeing process; as a result of this and of the editorial decisions most, but not all, of the contributions have been included.

This volume's organization follows the general scheme of the Colloquium. It starts with the contributions related to special events: the *Wigner Medal* and the *Hermann Weyl prize*, preceded by short descriptions and the texts of *Laudatios* by Yuval Ne'eman and H.D. Doebner. Special session dedicated to Allan Solomon (Session 7, on Methods and applications of Quantum Theory) is preceded by Karol Penson's *Laudatio*.

Next, the invited talks at plenary sessions and round table follows Then all other contributions, including poster clips, are displayed, following organization in parallel sessions, in alphabetical order inside each session, as follows:

* Session 1: *Discrete Groups and Geometry*
* Session 2: *Strings, Cosmology, and Gravitation*
* Session 3: *Conformal Theories and Integrable Systems*
* Session 4: *Particle and Nuclear Physics*
* Session 5: *Atomic and Molecular Physics, Theoretical Chemistry, Condensed Matter*
* Session 6: *Noncommutative Geometry*
* Session 7: *Group Theoretical Methods and Applications in Quantum Theory*
* Session 8: *Algebras, Groups, and their Representations*
* Session 9: *Nonlinear Systems and their Symmetries*

The contributions to the Mini-Symposium on *Quantum Groups and Integrable Systems* form special section that follows after the above 9. The contributions to the Satellite Colloquium on *Coherent States, Wavelets, and Applications* also form a special section in the volume.

The Editors:

J-P Gazeau, R Kerner, J-P Antoine, S Métens, J-Y Thibon

and the Local Organization Committee

International Standing Committee

Akito Arima	University of Tokyo, Tokyo
Joseph Birman	City University of New York, New York
Arno Böhm	University of Texas, Austin
Luis Boya	University of Zaragoza, Zaragoza
Laurence Boyle	University of Kent, Canterbury
Vladimir Dobrev	INRNE, Sofia
Heinz-Dietrich Doebner	University of Clausthal, Clausthal
Jean-Pierre Gazeau	University Paris 7-Denis Diderot, Paris
Mo-Lin Ge	Nankai Institute of Mathematics, Tianjin
Robert Gilmore	Drexel University, Philadelphia
Gerald Goldin	Rutgers University, Piscataway
Bruno Gruber	University of Illinois, Carbondale
Francesco Iachello	Yale University, New Haven
Richard Kerner	University Pierre et Marie Curie-Paris 6, Paris
Young Kim	University of Maryland, College Park
Allan Kostelecki	Indiana University, Bloomington
Vladimir Man'ko	Lebedev Physical Institute, Moscow
Mariano A. del Olmo	University of Valladolid, Valladolid
Tchavdar Palev	INRNE, Sofia
George Pogosyan	JINR, Dubna and ITP, Erevan
Allan Solomon	Open University, Milton Keynes
Luc Vinet	University of Montréal, Mc Gill University, Montréal
Kurt Bernardo Wolf	CCF-UNAM, Cuernavaca

International Advisory Committee

R. Aldrovandi	IFT, São Paulo
S.T. Ali	Concordia University, Montréal
T. Arecchi	University of Florence, Florence
P. Binétruy	University Paris-Sud, Orsay
A. Chamseddine	American University, Beirut
F. Calogero	INFN, Roma
Y.-M. Cho	Seoul National University, Seoul
A. Comtet	University Pierre et Marie Curie-Paris 6, Paris
M. Dubois-Violette	University Paris-Sud, Orsay
J. Fröhlich	ITH, Zurich
J. Faraut	University Pierre et Marie Curie-Paris 6, Paris
D. Gal'tsov	Moscow State University, Moscow
G. Gibbons	DAMTP, Cambridge
M. Henneaux	Université Libre, Bruxelles

V. Kac	MIT, Boston
E. Knobloch	University of California, Berkeley
P. Kulish	POMI, St. Petersburg
N. Manton	DAMTP, Cambridge
A. Morozov	ITEP, Moscow
Y. Ne'eman	Tel-Aviv University, Tel-Aviv
S. Okubo	University of Rochester, Rochester
S. Randjbar-Daemi	ICTP, Trieste
S. Shnider	Bar-Ilan University, Bar-Ilan
M. Senechal	Smith College Northampton, Northampton
E. Stanley	Boston University, Boston
P. Steinhardt	University of Princeton, Princeton
O. Suzuki	Nihon University, Tokyo
I. Todorov	INRNE, Sofia
J. Wess	MPI, Munchen
P. Winternitz	University of Montréal, Montréal
S. Woronowicz	University of Warsaw, Warsaw
B. Wybourne	Nicolas Copernicus University, Toruń

Local Organizing Committee

Chairmen:

Jean-Pierre Gazeau	University Paris 7-Denis Diderot, Paris
Richard Kerner	University Pierre et Marie Curie-Paris 6, Paris

Members:

Jean-Pierre Françoise	University Pierre et Marie Curie-Paris 6, Paris
Yannick Giraud-Héraud	Collège de France
Eric Huguet	University Paris 7-Denis Diderot, Paris
Ernest Ilisca	University Paris 7-Denis Diderot, Paris
Jean-Claude Lévy	University Paris 7-Denis Diderot, Paris
Stéphane Métens	University Paris 7-Denis Diderot, Paris
André Rassat	École Normale Supérieure, Paris
Jean-Yves Thibon	Marne-la-Vallée University
Michel Rausch de Traubenberg	University Louis Pasteur, Strasbourg
Jean-Louis Verger-Gaugry	University Joseph Fourier, Grenoble

Obituary: Lochlainn O'Raifeartaigh   1933-2000.

Dear Treasa O'Raifeartaigh, colleagues and friends.

It is a great honour for me to present to you today a memorial tribute to one of our colleagues and friends, Lochlainn O'Raifeartaigh, who died on November 18th, 2000 aged 67 after a short illness. He was one of the world leaders in the theory of elementary particles, and was far better known outside his native Ireland than within. From 1960 to his death, he was a professor, later Senior Professor, at the School of Theoretical Physics in the Dublin Institute for Advanced Studies.

In 1958, Lochlainn O'Raifeartaigh married Treasa Donnelly, a Celtic studies graduate of Queen's University, Belfast. They had met in an Irish Language-speaking area called the Donegal Gaeltacht and they both shared a life-long interest in the Irish language and culture.

Most of Lochlainn's scientific career was centred around the city of Dublin, where he graduated with first-class honours in mathematical physics from University College Dublin in 1953. As a student, he had attended lectures by the Nobel Laureate Erwin Schrödinger, the first director of the School of Theoretical Physics. When Schrödinger returned to Vienna in 1956 Lochlainn O'Raifeartaigh worked with the famous Irish relativitist John L. Synge on the Theory of Relativity.

He entered the Dublin Institute for Advanced Studies in 1956 and a year later was awarded a studentship to study under Walter Heitler, one of the pioneers of Quantum Field Theory, at the University of Zurich where he was awarded his doctorate in 1960. He returned to DIAS in 1961 as assistant professor and was elected to membership of the Royal Irish Academy at the age of 29.

Lochlainn O'Raifeartaigh specialised in the application of group theory to physics. His research on the symmetries of physical theories attracted much attention - particularly in the US. He spent the winter of 1963/1964 at the Madras Institute for Mathematical Sciences and in the autumn of 1964 went on extended leave from DIAS to Syracuse University in New York State.

It was during his stay at Syracuse University (1964-8) that he made the discovery that established his reputation. This result, which became known as *O'Raifeartaigh's No-Go Theorem*, showed that it was impossible to combine internal and relativistic symmetries in other than a trivial fashion, thus ending a widespread quest by the particle physics community to achieve this fusion.

Its announcement made him famous overnight. It also brought to an abrupt end the research programmes of many of his colleagues. Although his future career in the US was now assured, he chose instead to return to DIAS in 1968 as a senior professor, after spending one year at the Institute for Advanced Study at Princeton.

O'Raifeartaigh's prolific career in theoretical physics was manifested by many fundamental contributions to the application of symmetries in particle physics. In the 1970's he showed that the new supersymmetries could provide a mechanism (*O'Raifeartaigh's mechanism*) for circumventing his No-Go theorem which had assumed only classical Lie group symmetries. In the 80's he applied non-abelian gauge theory to the analysis of magnetic monopoles. His interests encompassed the Spin-Statistics theorem, Kac-Moody and W-algebras, and included early contributions to the theory of non-invariance (dynamical) groups, among much else. This long and productive career in the application of symmetry methods to theoretical physics was acknowledged by much international recognition including the von Humbolt Research Award in 1998 and the Wigner Medal in August 2000, the latter for his "pioneering contributions to particle physics". His book, *The Dawning of Gauge Theory,* showed him to be also an accomplished historian of physics.

Lochlainn O'Raifeartaigh's interests were not confined to Physics. He was an enthusiastic theatre-goer and keen hill-walker. He put his interest in international politics to good use in the cause of nuclear disarmament; along with the Irish Nobel Laureate Ernest Walton, he helped found the Irish Pugwash group, bringing together physicists and experts on international affairs. He was also President of the Ireland-Israel Friendship league for many years.

Lochlainn O'Raifeartaigh had a profound understanding of theoretical physics, and the even rarer ability to impart this understanding in class and in seminar. His remarkable talent for clear exposition ensured that he was always in demand as a lecturer, and his gentle manner and patience made him the first port of call for colleagues and students who wanted clarification of some aspect of their subject. He was a man of faith and humanity, devoted to his family and his country, and to the wider community of mankind beyond the borders of his profession. His passing leaves a great void in Irish theoretical physics in particular, and in the world of physics in general.

He will be remembered by his colleagues as much for his humility, patience, kindness and humour as for his academic excellence.

Lochlainn O'Raifeartaigh is survived by his wife Treasa, his sons, Conor, Finbar and Cormac, and his daughters Una and Aoife.

He will be sorely missed, not only by his family, but also by all of us who were lucky enough to have come into contact with this wonderful human being.

Allan Solomon,   Paris, July 2002.

Contents

Special Plenary Session & Round Table:
Historical and Epistemological Aspects of Group Theory in Physics   169

Section 1: Discrete Groups and Geometry   183

The 2002 Wigner Medal

The Wigner Medal 2002

Introduction
Arno R. Bohm
Group Theory and Fundamental Physics Foundation
Physics Department, University of Texas, Austin

As chairman of the Group Theory and Fundamental Physics Foundation, I would like to welcome you to the presentation of the 2002 Wigner Medal. The Wigner Medal is unique because it is a truly international award. It is not affiliated with any state or national society.

The Wigner Medal is awarded for "outstanding contributions to the understanding of physics through group theory". The awardee is chosen by an international selection committee whose members are elected by the Board of Trustees of the Foundation and by the Standing Committee of the International Group Theory Colloquium.

The previous recipients of the Wigner Medal are:

> 1978 E. P. Wigner
> 1978 V. Bargmann
> 1980 I. M. Gel'fand
> 1982 L. Michel
> 1984 Y. Ne'eman
> 1986 F. Gürsey
> 1988 I. M. Singer
> 1990 F. Iachello
> 1992 J. Wess and B. Zumino
> 1996 V. Kac and R. V. Moody
> 1998 M. Moshinsky
> 2000 L. O'Raifeartaigh

The 2002 International Selection Committee consisted of A. I. Solomon (UK), Chairman; Members: J. P. Gazeau (France), R. Moody (Canada), S. Weinberg (USA) and A. Bohm (USA).

The International Selection Committee has awarded the 2002 Wigner Medal to:
Harry Jeannot Lipkin
For contributions to the algebraic method in nuclear and particle physics and its extension to describe the spectra of nuclei.

We have also come together to remember another of our distinguished colleagues who received the Wigner Medal in 2000, Lochlainn O'Raifeartaigh. I am happy to welcome Mrs. Treasa O'Raifeartaigh who has come to Paris to be with us tonight. Lochlainn has been a very dear and admired friend of mine since my first days in the US and when I gave a lecture at Syracuse University on the O'Raifeartaigh theorem with Lochlainn in the audience. It was probably the first of many lectures he heard on this subject, for which he then finally received the Wigner Medal.

Professor Allan Solomon will next deliver the O'Raifeartaigh Memorial Lecture. Then Yuval Ne'eman will give the Laudatio and Professor Lipkin the Response to the Award of the Wigner Medal.

4

Harry J. Lipkin and his wife Malka, with the Wigner medal

Awarding of the Wigner Medal for 2002 to Prof. H.J. Lipkin, of the Weizmann Institute of Science Rehovoth, Israel

Laudatio,
composed and read by Y. Ne'eman on behalf of the
Group Theory Foundation

Ladies and Gentlemen,

ICGTMP 24 is the first conference in the ICGTMP series in the XXIth Century. To a large extent - although not yet entirely so - XXth-century physics has fulfilled the *cloud-clearing* task assigned to it by Lord Kelvin in his April 1900 speech before the British Association in Brighton. Assessing the means used in the process of reconstruction, in laying new foundations, in the making of the new physics - we realize *the importance* and *the power of the algebraic methodology* to which we have dedicated this series. The XIXth Century gave us Group theory, invented by two pairs of youthful figures - the tragic pioneers - Evariste Galois and Hendrik Abel – and the luckier pair, Felix Klein and Sophus Lie. Emmy Noether then provided a link to the *variational* approach in Physics and thereby readapted and repowered that ancient tool, connecting it as the harbinger of Symmetry, to the sometimes (though not always) dynamically perceived Conservation Laws. The variational method had been launched by Aristotle, used by Heron, reintroduced by Fermat and by Young, mathematically sharpened by Leibniz, Bernoulli, Euler, Lagrange and Maupertuis, broadened by Hamilton and Jacobi. It was now used in clearing both of Kelvins clouds, though much more so in the Quantum case. Pauli applied it to the hydrogen spectrum and the algebraically-inspired Matrix Mechanics appeared as an alternative to Schroedinger's calculus-based approach. Moreover, it thereby included from the start the extension of the concept of symmetry beyond that which is explicit in the input data - namely to Dynamical Degeneracy Groups (DDG), such as the SO(4) arising in hydrogen as the quantized Kepler problem.

Hermann Weyl strengthened the new algebraic foundations and Eugene Wigner became the Cyrano de Bergerac who first used that sword massively for good causes. One after the other, the various *levels of matter in the micro* were conquered by the spectroscopic approach in this drive as initiated by Wigner - *atomic, molecular* - followed first by *nuclear* spectroscopy and then, in the early Sixties, *hadron* spectroscopy in particle physics.

The symmetries were those of the Hamiltonian, until 1959 when there occurred a great leap forward, namely the discovery of *spectrum-generating algebras* (SGA) or *spectrum generating groups* (SGG), by **Prof. Harry J. Lipkin of the Weizmann Institute of Science in Rehovot, Israel**. Professor Lipkin, at that time already considered as a leader in the derivation of symmetries of the Hamiltonian, had now generalized his treatment and discovered, in an article with his then student S. Goshen, the novel algebraic method of a SGG which they applied to nuclear and many-body physics. The context was that of the *harmonic oscillator* in $d = 1 \cdots 3$ dimensions and for 1 to N oscillators. Aside from the conservation of angular momentum - denoting an *explicit* SO(3) symmetry - and the U(3) *DDG* displayed by the oscillator solutions - it was now shown that *the entire set of solutions forms one infinite-dimensional representation* of Sp(3,R), an SGG whose generators are given by the algebra of double-quanta such as $a_i^+ a_j^+; a_i^- a_j^-; a_i^+ a_j^-$, describing giant monopole vibrations or breathing modes in a many-body system. It has been applied in nuclear physics and in the study of Bose-Einstein condensates.

Goshen and Lipkin had introduced a method which, after it had been rediscovered five years later by Dothan, Gell-Mann, Neeman and by Barut and Bohm, was extended to relativistic particle physics. This was in the nineteen-sixties, during the apparent crisis in

Quantum Field Theory; in view of the on-mass-shell and highly phenomenological methods of Dispersion Relations, the SGG concept was grabbed as a lifebuoy, in the expectation that it might provide a framework for dynamical considerations. This need vanished in 1971, with the successful renormalization of Yang-Mills gauge theories (broken and unbroken). Yet there remained a prominent survivor from the Dispersion era, namely the Superstring. Although this relates to an off-mass-shell methodology, in a sense going beyond Quantum Field Theory, the study of the relevant spectrum is still managed with SGG methodology.

In Particle Physics, Lipkin was the initiator of a very successful drive aiming at extracting the maximal amount of predictive results from both candidate SU(3) symmetries of the early Sixties, the Sakata triplet and the Octet with its structural quark triplet. The triplet Levinson-Lipkin-Meshkov displayed skill and thoroughness in its assault on these models and were the first (1962) to point to an experimental test in the annihilation of $p + \bar{p}$ into two mesons which could clearly distinguish between the two models - a test in which the Sakata triplet failed and which the octet definitely passed. In 1965, Lipkin (with F. Scheck) pushed through a thorough exploration of the Quark Model.

In nuclear and many-body physics, Lipkin developed the methodology for a separation of the *collective* from the *individual* particle degrees of freedom, overcoming in a creative and imaginative way the difficulties generated by the Pauli Principle. Another class of contributions, such as the Lipkin-Meshkov-Glick two-shell SU(2) model, is used nowadays in nuclear and condensed matter physics because of the light it sheds over the dynamical set-up.

The Wigner Medal is awarded for "outstanding contributions to the understanding of physics through group theory", and is certainly fully warranted here. But in this case, it is impossible not to mention two other very special aspects, namely (1) Lipkin's didactic contributions, popularizing difficult topics, in his for "pedestrian" series of books - and (2) his humoristic contributions, mostly via the *Journal of Irreproducible Results*, of which he was, with A. Cohen, one of the two central pillars.

Response to the Award of the Wigner Medal

Harry J. Lipkin

Department of Particle Physics Weizmann Institute of Science, Rehovot 76100, Israel
School of Physics and Astronomy, Raymond and Beverly Sackler Faculty of Exact Sciences,
Tel Aviv University, Tel Aviv, Israel
High Energy Physics Division, Argonne National Laboratory, Argonne, IL 60439-4815, USA

Abstract. An exciting journey through frontier science against obstacles now forgotten. Electrical engineering professors told us that there was no future in electronics. Niels Bohr told us that there was no future for quantum mechanics which was a theory for the atomic scale; the nuclear scale would need a new theory as different from quantum mechanics as quantum mechanics was from Newtonian mechanics. Fermi's theory of beta decay was wrong; a new theory was needed. The first experiment showing that relativistic positrons obeyed the Dirac equation. The nuclear shell model was nonsense. Parity was conserved, the Mossbauer effect was nonsense, group theory was useless and quarks were nonsense.

There were side trips like nuclear reactor dynamics, where this postdoc was allowed to test a theory of reactor stability by turning on a reactor and watching to see whether it would blow up like Chernobyl or stabilize according to his theory.

1. Introduction - What I learned from Wigner

I am very pleased to be honored at this meeting by the award of the Wigner Medal. I learned many things from my former teacher Eugene Wigner[1].

1.1. My first published paper in theoretical physics

When I was a graduate student at Princeton I suggested to Wigner one day at tea time that a particular radiative correction might explain a discrepancy between some new beta decay experiments (which were later shown to be wrong) and the Fermi theory. Wigner said "It sounds like a crazy idea, but why don't you calculate it?" I began the calculation, using the old quantum electrodynamics which I was just learning from Heitler's book before the new QED of Feynman, Schwinger and Tomonaga. If you don't know what you are doing and know the answer you want, you get it. But when I showed it to Wigner, he said "This is very interesting. But what is this? And how did you get that?" I learned a great deal by making all possible mistakes as he then guided me through the calculation. Although Wigner guessed correctly that this idea was useless he encouraged me to calculate the effect, to show that it would never explain anything, and then to publish it. I did not want to write the work up. It had nothing to do with my experimental thesis work, and it didn't solve the problem. But Wigner insisted. "You have done the work. You must publish it."

The publication[2] attracted the attention of a Harvard student, Eugene Merzbacher, whose thesis problem was a proper calculation of the same effect with the new QED. He confirmed that the effect was too small to ever be seen in any beta decay experiment.

1.2. An important lesson learned from Wigner

I always remember Wigner's remark: "I believe that this theory is wrong. But you know, the old quantum theory of Bohr and Sommerfeld was wrong, too. And it is hard to see how we could ever have reached the correct quantum theory without first going through this stage." I have been following this advice throughout my career and pursued approaches believed to be wrong by conventional wisdom; e.g. electronics, the future of quantum mechanics, the Mossbauer effect, group theory in physics, SU(3) symmetry and the quark model. In the remainder of this talk I shall try to describe how I became a theoretical particle physicist.

2. From Electrical Engineering to Physics

In high school in Rochester, New York in 1934 I liked building small radios and never thought of becoming a physicist. Before Hiroshima, most people had no idea what physicists did. I remember how a local draft board during World War II refused to defer physicists from army service. Their instructions were to defer only "chemists and scientists, but not physicists".

2.1. Cornell E. E. students told electronics impractical - no jobs, no future

When I entered the Electrical Engineering School at Cornell University in 1938 the curriculum included only one semester of electronics. The professors assured us that there were no jobs and no future in electronics. We had better study our machinery and power transmission courses and forget this useless electronics. Fortunately the university was flexible and allowed students to choose additional courses. Having heard that there were very interesting physics courses, some of us went over to the physics building and listened to two new refugee professors from Hitler's Europe, Hans Bethe and Bruno Rossi.

In the engineering school we learned that electrical energy traveled through wires. The engineers also knew that radio existed and that electrical energy also traveled through the air. But they didn't really understand it and it wasn't practical. In the physics department we learned about the basic properties of matter and energy without any pretense that this was practical. We also learned how electrical energy traveled through the air, as described by the famous equations of Maxwell, which engineering students did not study in those days.

2.2. Microwave radar at MIT

I graduated in 1942, after the U.S. had entered World War II, and joined the Radiation Laboratory at M.I.T. in the development of microwave radar. This was all based on electronics and electrical energy traveling through air, both considered impractical by engineers.

A radar receiver that I developed was produced by a well known industrial company in Chicago. I had to make a special trip from Boston to tell their engineers why the first model built according to my original design didn't work. The small changes introduced to facilitate mass production allowed electrical energy to travel through the air in peculiar ways that completely ruined the receiver performance. They looked at me as if I were crazy when I told them to move a wire soldered at one point on the chassis back to another point a centimeter away where I had originally put it. Every electrical engineer knew that when a chassis was grounded it made no difference where you soldered a wire to it. They had never worked with such high frequency, high gain amplifiers before. They moved the connection to humor this young fool and were amazed when all their troubles went away and the receiver worked. To them it seemed like black magic.

Electrical engineers with the standard training aimed at specializing in "practical" directions were unable to cope with the new phenomena of high frequencies and wave guides. They had learned how to solve the problems that were practical today, but were unprepared for the completely new problems that become practical and even urgent tomorrow. They knew all about how electrical energy traveled in wires but could not understand how it could travel through the air. They did not know how to design radar equipment and make it work.

The microwave radar program was an outstanding success. One of its major achievements, detecting German submarines from the air was possible because the German establishment had made a high level decision that radar at microwave frequencies was not practical. Their submarines were not equipped with microwave receivers which would warn them of an approaching airplane carrying microwave radar.

The key people at the Radiation Laboratory were all physicists, not engineers. The staff included Rabi, Alvarez, Bloch, Purcell, Schwinger, Dicke and many others close to the Nobel prize level.

2.3. Why I moved to physics

After working four years at M.I.T. as an electronic engineer, I decided to go to graduate school in physics and study the basic properties of matter and energy, rather than more "practical" subjects, in order to be better prepared for future developments.

3. From Experiment to Theory

3.1. Niels Bohr on Quantum Mechanics in 1946

Princeton in the late 1940's was very much influenced by the Copenhagen school and by the great giants who had carried through the remarkable revolution of the 1920's which completely transformed our ideas of space, time and continuity.

Niels Bohr and his associates told us that their revolution had succeeded in explaining all phenomena on the atomic scale, but was completely useless for the smaller nuclear scale. A new revolution was needed, leading to a theory as different from Copenhagen quantum theory as quantum mechanics was different from Newtonian mechanics. New revolutionary ideas like quantization of space-time might be needed. The road to the new physics would be paved by new exciting experiments whose results defied explanation by conventional quantum theory. I left Princeton as an experimentalist who had performed as a Ph.D. thesis the first experiment showing that relativistic positrons obeyed the Dirac Equation. But I never found any experiments showing that the old Bohr-Heisenberg-Schroedinger-Pauli-Dirac quantum theory was inadequate for the description of small distance phenomena.

Nobody has yet found such experiments. One of my quantum mechanics teachers at Princeton, David Bohm, tried very hard to find the keys to new physics in investigating the foundations of quantum mechanics. The Aharonov-Bohm effect, which caused great controversy when it was first proposed, is now one of the pillars of conventional Copenhagen quantum mechanics.

3.2. Experiment to test Dirac theory for relativistic positrons

My thesis work in experimental physics, Mott scattering of 1 MeV electrons and positrons, might be called in today's language "An experimental test of the standard model". It showed that the relativistic corrections to the Rutherford scattering formula which had opposite signs for electrons and positrons were correctly predicted by the Dirac equation. This was not easily achieved because positrons were only available from radioactive sources which had to be produced in cyclotrons and lived only a few hours.

The way I got my positrons would make anyone worried about nuclear safety and radioactivity shudder today. A copper target was bombarded all day in a cyclotron in Washington, D.C. Then a member of Princeton's administrative staff who knew no nuclear physics and nothing about nuclear safety flew down to Washington in a private plane, took the radioactive copper target from the cyclotron in a car to the airport, then flew it to Princeton in the private plane, and then took it by car to the Princeton physics building. A radiochemist then separated out the positron-emitting radioactive gallium and deposited it on my source holder. I then put it into my apparatus at about midnight and took data all by myself, day and night until the nine hour half-life radioactivity had decayed to the point where I could no longer get useful data. I repeated this several times to get enough data to confirm Dirac's theory.

4. The Beginnings of Nuclear Physics and Nuclear Energy in Israel

4.1. A Year in France

In 1950 when I came to Israel nobody in the country understood what a nuclear reactor was and how it worked, and it was impossible for students to get a Ph. D. in nuclear physics in Israel. The Israeli government sent three physicists including me to Paris in 1953 to learn about nuclear reactors. There I was asked to read an article by Alvin Weinberg about the nonlinear kinetics and stability of nuclear reactors. I generalized his treatment for a homogeneous reactor to the case of a heterogeneous reactor containing uranium and heavy water. I was then asked to test my theory on the French reactor ZOE at the Chatillon Nuclear Center. When a reactor is turned on, its power level increases. When it reaches the desired power level control rods are adjusted to keep the desired power. But suppose the reactor is left alone after it is turned on. Will the power stabilize itself at some level or will it blow up like Chernobyl? Lipkin's extension of Weinberg's theory said that the ZOE reactor would be stable. One shudders today to think that the French Atomic Energy Commission allowed this young postdoc who was only learning about nuclear reactors to test his theory one evening alone with two technicians at the reactor control room. Since we were working in the evening, our dinners were brought from the laboratory restaurant. Since this was France, the dinner included a carafe of wine. And since the technicians were working overtime, they received an extra carafe of wine with their dinner. We turned the reactor on and the technicians read the power level and uranium and heavy water temperatures at regular time intervals. It fit my theory, I wrote the paper and everyone was happy. The reactor was stable and did not blow up.

Some time later Weinberg invited me to visit Oak Ridge and introduced me to a reactor engineer and a mathematician who told me that I had found an application of Liapounov's second method for stability. We then wrote a paper about this mathematics.

4.2. Back in Israel

In 1954 the three of us came back to Israel along with three other Israeli nuclear physicists who had just returned from study abroad. The first stage toward developing any realistic nuclear energy program was to enable students to learn basic nuclear physics at home instead of abroad. This was achieved by founding a new Nuclear Physics Department at the Weizmann Institute and establishing a center for basic research with the installation of the country's first nuclear accelerator and a graduate training program. I became the country's leading expert in nuclear reactor physics and our group played key roles in the initial stages of the Atomic Energy program which led to the building and operation of Israel's two nuclear reactors at Nahal Sorek and Dimona.

The theoretical work in the department began with work on nuclear structure using the then very new nuclear shell model. The collective model and the applications of group theory and the new BCS theory of superconductivity to nuclear physics were also investigated. Rehovot rapidly became an international center for nuclear physics. The 1957 International Conference on Nuclear Structure was the first such physics conference held in Israel. I was on the organizing committee, edited the proceedings and prepared a list of humorous daily bulletins and jocular physics articles. These jokes are probably remembered much more than my physics and led to the founding of the "Journal of Irreproducible Results."

The applications of the mathematical techniques of group theory to collective motion and nuclear many-body problems led to the development of simple models which have by now become classics[3] and of the spectrum-generating algebra[4] for which I have now been awarded the Wigner Medal.

5. Parity and the Mössbauer Effect

In the late 1950's two opportunities arose for exciting research requiring only radioactive sources and simple detectors. Experimental breakthroughs in two areas of nuclear physics called "parity nonconservation" and the "Mössbauer effect" opened up possibilities for us to get in at the very beginning of these rapidly developing areas of frontier research.

In 1957, after the experimental discovery of parity nonconservation, I developed a "double-scattering method for measuring beta ray polarization". This simple extension of my Ph. D. scattering experiment became a classic described in textbooks. I spent the academic year 1958-59 at the University of Illinois in Urbana directing Hans Frauenfelder's group doing beta-ray polarization experiments, while Hans was on sabbatical at CERN.

During the summer of 1958 I lectured at the Ecole d'Eté at Les Houches and included the Spectrum-Generating Algebra[4]. I also visited Princeton and told Wigner about this work but did not do a very good job at explaining it. The name "spectrum generating group" had not yet been invented, and Wigner could not understand the point of a group that was not a symmetry of the Hamiltonian of the harmonic oscillator.

During that year in Urbana, we heard about Rudolf Mössbauer's discovery of the effect which now bears his name and won him a Nobel Prize. His original experiments were misunderstood and greeted with skepticism by the physics community because its understanding required the combination of languages of nuclear and solid state physicists who did not talk to one another. I learned enough of both at Urbana to become a Mössbauer expert, was the first to suggest that the effect was important enough to be called the Mössbauer Effect[5], and began to work on it with the Frauenfelder group..

6. From Groups and Many-body Physics to Particle Physics and Quarks

6.1. Lie Groups for Pedestrians

In his recollections Wigner[6] refers to Pauli's derisive popular label "Die Gruppenpest" and attributes this resistance to group theory to the absence of a first-rate textbook. Wigner's "little book" published in 1931 helped but did not solve this problem. In 1950 the top young particle theorists at Princeton including at least four future Nobel prize winners were sure that group theory was completely useless. None attended Giulio Racah's now famous Princeton lectures on "Group Theory and Spectroscopy". They thought that isospin was a rotation in some abstract three-dimensional space and did not realize that the combination of isospin and strangeness symmetries was $SU(2) \times U(1)$. Knowing nothing about unitary groups they spent eight years looking fruitlessly for higher symmetries by considering only rotations in higher and higher dimensions until Gell-Mann found SU(3) by accident..

Racah's remarkable insight into the relevancy of group theory for physics was evident in his lectures, but not in his papers and unavailable in the existing literature. My ability to translate what I had learned from Racah into a language understandable to physicists produced a series of lecture notes that eventually appeared in books "for pedestrians". "Beta Decay for Pedestrians" gave physicists the tools to calculate angular distributions of the new parity experiments. "Lie Groups for Pedestrians"[7] enabled the nuclear and particle physicists to understand the group theory they could use. It also included my own original development and classification of the algebra of bilinear products of second-quantized creation and annihilation operators. Lie Groups for Pedestrians has now been reprinted by Dover and is available at a price students can afford.

In 1958 Victor Weisskopf told me about one such new algebra found by Arthur Kerman for nuclear physics with three operators satisfying angular momentum commutation rules. In Urbana I heard the same algebra described by Phil Anderson who had found them independently for the electron gas[8]. As an interpreter between these two groups who did not talk to one another, I told each about the other and noted that they had really discovered a two dimensional symplectic algebra that was isomorphic to the algebra of three dimensional rotations.

In September 1967 I attended an international conference in Warsaw celebrating the 100th anniversary of the birth of Marie Sklodowska Curie. My invitation to the conference had been sent in April, 1967. But in June 1967 after the six-day war Poland broke off diplomatic relations with Israel. I had no idea whether I would be welcome in communist Poland. The conference organizers did everything possible to make me feel welcome. They sent a young Polish student Richard Kerner to meet me at Warsaw airport and reassure me that everything was OK. I very much appreciated Richard's hospitality and gave him an autographed copy of my book "Lie Groups for Pedestrians." I lost track of Richard until July, 2002, when he re-introduced himself at this conference and asked me to add another note to the book that he had kept all these years.

6.2. Unitary Symmetry

The 1960's brought a sudden realization that group theory could be useful for particle physics and brought us to frontier particle physics. There was no particle theory at all in Israel when Yuval Ne'eman returned from London and gave a seminar in Rehovot about his "Eightfold Way" Unitary Symmetry. But in the nuclear physics group at Weizmann Carl Levinson and Sydney Meshkov were using the group $SU(3)$ to study nuclear structure and realized that they had all the mathematical tools needed to calculate experimental predictions from the new theory. Since particle physicists knew no group theory at that time we were able to get into the lead in this activity.

In the spring of 1961 at a small meeting on unitary symmetry at Imperial College I told Abdus Salam about our $SU(3)$ calculations for proton - antiproton annihilation into two mesons. Salam said new experimental results from CERN were now available to be compared with our predictions. We went up to his office after the session and found that the data did not fit the predictions from the Eightfold Way, but favored the currently competing Sakata model. Salam was flying to Pakistan the next day and suggested that I write up the paper together with my colleagues in Israel.

Back at Weizmann Carl, Syd and I immediately saw that Salam and I had looked in the wrong column of a table of $SU(3)$ Clebsch-Gordan coefficients. The results from the right column were even more exciting but opposite. They killed the Sakata model and left the Eightfold Way in agreement with experiment. We immediately wrote up the paper. But we couldn't leave Salam out of it, because he had been in on the original idea. We couldn't put his name on the paper because the conclusions were now reversed and there was no communication between Israel and Pakistan.

I sent the paper to my good friend Gerry Brown who was then starting a new journal "Physics Letters". I explained the situation and left further processing to his discretion. He found that a Pakistani student of Salam's, Munir Ahmed Rashid, had independently discovered the error. Gerry accepted the suggestion from Imperial College that Salam's and Rashid's names be added to the list of authors. So a paper appeared as a collaboration of three Israelis and two Pakistanis[9].

6.3. Beyond SU(3) symmetry to quarks

The work on symmetries continued as a group of very talented students (now all professors) joined in the effort, which culminated in the development of a symmetry called $SU(6)_W$ with the W for Weizmann. The group continued the tradition of plunging into new areas before they became fashionable by developing the quark model seriously while the particle physics establishment rejected quarks as nonsense. Combining the quark picture with dispersion relations led to a new approach called duality, the "Veneziano dual resonance model"[10] and "duality diagrams".

With this background I eagerly embraced the quark model as the key to new physics. Experiments told us that quarks were real objects. Establishment theorists insisted that quarks were nonsense. Their arguments recalled the arguments against Bohr's atom in which electrons moved in orbits without losing energy by radiation. Nobody found free quarks, but more and more experiments showed that hadrons were made out of quarks. Was this the key to the new revolution? Unfortunately no. Bohr, Heisenberg, Schroedinger, Pauli and Dirac can still explain everything with appropriate mathematical techniques.

When SLAC found scaling, Bjorken and Feynman explained it all with the quark-parton model, and the theory establishment insisted that this was all nonsense, high energy physics looked exciting again. Bjorken and Feynman were clearly describing the physics of the real world. Was this an opening to the new physics which theorists could not explain? Again disappointment. Someone found how to explain the experiments with the old theories.

7. Conclusions

The history of high-energy physics in the second half of the twentieth century has been the carrying of 1920 quantum mechanics into higher and higher energies and smaller and smaller distances finding very interesting physics and many new phenomena completely unexpected in 1950. But no new revolution. The quantum mechanics of Bohr, Heisenberg, Schroedinger, Pauli and Dirac not only stood the test of time. They were the pillars of the new knowledge accumulated that completely changed the quality of life of the ordinary citizen in ways that were completely unimaginable a half century ago when radio and television were in their infancy. Transistors, lasers, personal computers, cellular telephones and the internet did not exist, and all required for their development the application of quantum mechanics in ways that the creators of quantum mechanics could never imagine.

All this makes one wonder how to direct promising young scientists toward fruitful applied work. What is most practical today will probably be out of date tomorrow. We cannot tell researchers to concentrate on directions which will be important in the future.

Who can predict the future? My professors at the university could not foresee the importance of electronics. Niels Bohr could not foresee the future consequences of his own quantum mechanics. When I hear some older people trying to tell younger people what they should be doing, I am reminded of the words my father used to say to me when I thought I had been very clever. "If you knew what you don't know, you would know more than you know."

Graduate study in a good university in a pure science provides the training necessary for work in new areas which cannot possibly be anticipated at the time the student begins his studies. The student learns to solve new problems by developing new techniques and discovering new things. Exactly what he develops and what he discovers at this stage is not so important. It is learning the approach to search and discovery and gaining experience in attacking new problems, where one cannot find the techniques for solution in any text book or hand book, and one has to work it out all alone.

For this conference it is appropriate to note that the developments of the past half century have completely overturned our understanding of the fundamental building blocks of matter and the forces that bind them together. Instead of neutrons and protons bound into nuclei we have quarks and gluons bound into new families of hadrons unknown in 1950. And a crucial ingredient of our new understanding is group theory. We have come a long way from the days when particle physicists had never heard of SU(n), thought it irrelevant to physics, and discovered the algebra of SU(3) by examining the commutators of weak currents without knowing group theory.

8. Epilogue - The Source of Wigner's Encouragement

I return to Wigner and find in his book of recollections[1] some insight into his encouragement to me. In discussing the exciting colloquia where he heard great physicists as a student in Berlin, he writes: "One element missing from the colloquia was concrete encouragement. Einstein was very kind to young physicists, but even he did not push us along as he might have done. He never said, 'Look here, this idea of yours is quite promising. Why don't you work it out and publish it?' I waited in vain to hear such words."

Wigner evidently kept such words in mind for young students like me after he became a world famous professor.

References

[1] Andrew Szanton, "The Recollections of Eugene P. Wigner as told to Andrew Szanton" Plenum Press, New York (1992). See in particular pp.74-75.
[2] H. J. Lipkin, Phys. Rev. 76, 567 (1949)
[3] Harry J. Lipkin, N. Meshkov and A.J. Glick, Nucl. Phys. 62, 188, 199 and 211 (1965)
[4] S. Goshen and H. J. Lipkin, Ann. Phys 6, 301 (1959)
[5] Harry J. Lipkin, Ann. Phys. 9 332 (1960)
[6] ref[1] pp. 116-119
[7] Harry J. Lipkin, Lie Groups for Pedestrians, Second Edition. North-Holland Publishing Co. Amsterdam (1966), unabridged republication, Dover Publications, New York (2002)
[8] A. K. Kerman, Ann. Phys. 12, 300 (1961); P. W. Anderson, Phys. Rev. 112, 164 (1958).
[9] C. A. Levinson, H. J. Lipkin, S. Meshkov, A. Salam and R. Munir, Physics Letters 1, 44 (1962).
[10] G. Veneziano, Phys. Reports 9C (1974) 199

The 2002 Hermann Weyl Prize

During several meetings of Standing Committee preceding the Paris Colloquium, in 2000 and 2001, H. D. Doebner, Chairman, put forward the idea of creating a new international prize intended to award a young mathematician or physicist for her/his important contribution to develoment of concepts related to symmetry in physics. The award was given the name "*The Hermann Weyl Prize*". The first Selection Committee was elected, comprising the following persons: I. T. Todorov (Chairman), H. D. Doebner (Clausthal-Zellerfeld), V. Kac (Boston), R. Kerner (Paris) and J. Wess (München). The last version of the by-laws follows below

The Hermann Weyl Prize
awarded by the
Standing Committee of the International Group Theory Colloquium
The bylaws

- **Article 1**

The Hermann Weyl prize consists of a certificate citing the accomplishments of the recipient, a prize money of 500 USD and some allowance to attend the biannual International Group Theory Colloquium at which the award is presented.

- **Article 2**

The purpose of the prize shall be to provide recognition for young scientists who have performed original work of significant scientific quality in the area of the understanding of physics through symmetries. Their contributions shall be in one or more of the following fields: group theory; development of new mathematical tools for the description of physical phenomena; interpretation of experimental results; formulation of general laws based on symmetries.

- **Article 3**

A In selecting recipients of the prize, there shall be no discrimination based on sex, race or religion.

B As a rule the age of the awardees should be less than 35-years.

- **Article 4**

Members of the Selection Committee and of the Standing committee shall be ineligible to receive the prize.

- **Article 5**

The prize shall be awarded every two years.
The ceremony for the awarding of the prize shall be held at a Colloquium on Group Theoretical Methods in Physics. The Chairperson of the Standing Committee shall organize the ceremony in collaboration with the organizers of the conference.

- **Article 6**

The Selection Committee shall be constituted as follows:

A There shall be five members including the chairperson.

B The members and the chairperson shall be appointed by the Standing Committee of the International Group Theory Colloquium. Only two members of the Standing Committee are eligible for the selection Committee.

C Initially, two members (including one member of the Standing Committee) shall be appointed for two years and three members for four years. Thereafter, all appointments shall be for four years. It is being understood that after a term served as a member of the Selection Committee, four years must elapse before a person can be reappointed to this committee.

- **Article 7**

Duties of the Selection Committee:

The chairperson of the Selection Committee shall organize the work of the Committee according to the present rules:

A The chairperson shall organise together with the chairperson of the Standing Committee the publicity for the awarding of the Hermann Weyl prize. The chairperson calls for nominations through senior scientists at least ten months before the planned date for the awarding. He informs the chairperson of the Standing Committee on the nominations.

B The nominations should include

    1 Five copies of the PhD thesis, publications and reports of the nominee together with a summary up to 4 pages of her/his work.

    2 A brief curriculum vitae of the nominee.

    3 A letter from the nominating senior scientist introducing the nominee and explaining the nomination.

    4 At least two letters seconding the nomination.

D The Selection Committee decides on the awardee through a secret majority vote four months before the planned date for the awarding. The chairperson informs the members of the Standing Committee in detail on this decision.

Although the existence of this newly established prize was certainly not publicized widely enough during half a year preceding the Colloquium, many excellent candidates were proposed. After long and vivid discussions, the first *Hermann Weyl Prize* has been awarded to *Edward Frenkel* from the University of California at Berkeley for his outstanding results in the development of new algebraic tools enriching our knowledge of symmetries.

Edward Frenkel is Professor of Mathematics at the University of California in Berkeley. Born in Russia, he has got his B. A. degree in Moscow's Gubkin Institute in 1989, then completed his studies by acquiring a Ph. D. in Mathematics in 1991 at Harvard University in Boston. His expertise field is in mathematical physics and string theory. His major contributions concern infinite-dimensional algebras, integrable systems, conformal field theory and supersymmetry. He has published, jointly with David Ben-Zvi, an excellent book on vertex algebras and algebraic curves.

In the following Hermann Weyl Prize lecture, Edward Frenkel gives an overview of his latest contributions to his area of interest.

Edward Frenkel awaiting the Weyl Prize ceremony

Inst. Phys. Conf. Ser. No 173
Paper presented at 24th Int. Coll. Group Theoretical Methods in Physics, Paris, France, 15–20 July 2002
©2003 IOP Publishing Ltd

Affine Kac-Moody algebras, integrable systems and their deformations

Edward Frenkel

Department of Mathematics, University of California, Berkeley, CA 94720, USA

Abstract. Representation theory of affine Kac-Moody algebras at the critical level contains many intricate structures, in particular, the hamiltonian structures of the KdV and modified KdV hierarchies and the Miura transformation between them. In this talk I will describe these structures and their deformations which will lead us to the deformed Virasoro and $\mathcal{W}$–algebras and the integrable hierarchies associated to them. I will also discuss briefly the relation of these matters to the geometric Langlands correspondence.

It is a great honor for me to give this talk as the first recipient of the Hermann Weyl Prize. Weyl was a pioneer of applications of symmetry in quantum physics, a scientist who truly appreciated the beauty of mathematics. He once said: "My work has always tried to unite the true with the beautiful and when I had to choose one or the other, I usually chose the beautiful."

1. Affine Kac-Moody algebras

Let $\mathfrak{g}$ be a simple Lie algebra over $\mathbb{C}$. Fix an invariant inner product κ on $\mathfrak{g}$ and let $\widehat{\mathfrak{g}}_\kappa$ denote the one-dimensional central extension of $\mathfrak{g} \otimes \mathbb{C}((t))$ with the commutation relations

$$[A \otimes f(t), B \otimes g(t)] = [A, B] \otimes f(t)g(t) - (\kappa(A, B)\operatorname{Res} f dg)K, \qquad (1)$$

where K is the central element. The Lie algebra $\widehat{\mathfrak{g}}_\kappa$ is the *affine Kac-Moody algebra* associated to $\mathfrak{g}$.

A representation of $\widehat{\mathfrak{g}}_\kappa$ on a complex vector space V is called *smooth* if for any vector $v \in V$ there exists $N \in \mathbb{Z}_+$ such that $\mathfrak{g} \otimes t^N \mathbb{C}[[t]] \cdot v = 0$. We also require that K acts on V as the identity (this is equivalent to a more traditional approach whereby one fixes the inner product κ but allows K to act as the identity times a scalar, called the level).

Let $U_\kappa(\widehat{\mathfrak{g}})$ be the quotient of the universal enveloping algebra $U(\widehat{\mathfrak{g}}_\kappa)$ of $\widehat{\mathfrak{g}}_\kappa$ by the ideal generated by $(K - 1)$. Define its completion $\widetilde{U}_\kappa(\widehat{\mathfrak{g}})$ as follows:

$$\widetilde{U}_\kappa(\widehat{\mathfrak{g}}) = \varprojlim U_\kappa(\widehat{\mathfrak{g}})/U_\kappa(\widehat{\mathfrak{g}}) \cdot (\mathfrak{g} \otimes t^N \mathbb{C}[[t]]).$$

It is clear that $\widetilde{U}_\kappa(\widehat{\mathfrak{g}})$ is a topological algebra which acts on all smooth representations of $\widehat{\mathfrak{g}}_\kappa$, on which K acts as the identity. We shall recall the description of the center $Z(\widehat{\mathfrak{g}})$ of $\widetilde{U}_{\kappa_c}(\widehat{\mathfrak{g}})$ from [FF3, F3].

First we need to introduce the notion of an oper.

2. Opers

We start with the definition of an $\mathfrak{sl}_n$–oper. Let X be a smooth algebraic curve and Ω the line bundle of holomorphic differentials on X. We will fix a square root $\Omega^{1/2}$ of Ω. An $\mathfrak{sl}_n$–oper on X is an nth order differential operator acting from the holomorphic sections of $\Omega^{-(n-1)/2}$ to those of $\Omega^{(n+1)/2}$ whose principal symbol is equal to 1 and subprincipal symbol is equal to 0 (note that for these conditions to be coordinate-independent, this operator must act from $\Omega^{-(n-1)/2}$ to $\Omega^{(n+1)/2}$). Locally, we can choose a coordinate t and write this operator as

$$L = \partial_t^n + v_1(t)\partial_t^{n-2} + \ldots + v_{n-1}(t). \tag{2}$$

It is not difficult to obtain the transformation formulas for the coefficients $v_1(t), \ldots, v_{n-1}(t)$ of this operator under changes of coordinates.

For example, an $\mathfrak{sl}_2$–oper is nothing but a Sturm-Liouville operator of the form $\partial_t^2 + v(t)$ acting from $\Omega^{-1/2}$ to $\Omega^{3/2}$. Under the change of coordinates $t = \varphi(s)$ we have the following transformation formula

$$v \mapsto \widetilde{v}, \qquad \widetilde{v}(s) = v(\varphi(s))\left(\varphi'(s)\right)^2 + \frac{1}{2}\{\varphi, s\},$$

where

$$\{\varphi, s\} = \frac{\varphi'''}{\varphi'} - \frac{3}{2}\left(\frac{\varphi''}{\varphi'}\right)^2$$

is the Schwarzian derivative. Operators of this form are known as *projective connections* (see, e.g., [FB], Sect. 7.2).

Drinfeld and V. Sokolov [DS] have introduced an analogue of operators (2) for a general simple Lie algebra $\mathfrak{g}$. Their idea was to replace the operator (2) by the first order matrix differential operator

$$\partial_t + \begin{pmatrix} 0 & v_1 & v_2 & \cdots & v_{n-1} \\ -1 & 0 & 0 & \cdots & 0 \\ 0 & -1 & 0 & \cdots & 0 \\ \vdots & \ddots & \ddots & \cdots & \vdots \\ 0 & 0 & \cdots & -1 & 0 \end{pmatrix}. \tag{3}$$

Now consider the space of more general operators of the form

$$\partial_t + \begin{pmatrix} * & * & * & \cdots & * \\ -1 & * & * & \cdots & * \\ 0 & -1 & * & \cdots & * \\ \vdots & \ddots & \ddots & \ddots & \vdots \\ 0 & 0 & \cdots & -1 & * \end{pmatrix} \tag{4}$$

The group of upper triangular matrices with 1's on the diagonal acts on this space by gauge transformations

$$\partial_t + A(t) \mapsto \partial_t + gA(t)g^{-1} - g^{-1}\partial_t g(t).$$

It is not difficult to show that this action is free and each orbit contains a unique operator of the form (3). Therefore, locally, over a sufficiently small neighborhood U of X, equipped with a coordinate t, the space of $\mathfrak{sl}_n$–opers on U may be identified with the quotient of the space of operators of the form (4) by the gauge action of the group of upper triangular matrices.

Now let us generalize the latter definition to the case of an arbitrary simple Lie algebra $\mathfrak{g}$. Let $\mathfrak{g} = \mathfrak{n}_+ \oplus \mathfrak{h} \oplus \mathfrak{n}_-$ be the Cartan decomposition of $\mathfrak{g}$ and e_i, h_i and $f_i, i = 1, \ldots, \ell$, be the Chevalley generators of $\mathfrak{n}_+, \mathfrak{h}$ and $\mathfrak{n}_-$, respectively. Then the analogue of the space of operators of the form (4) is the space of operators

$$\partial_t - \sum_{i=1}^{\ell} f_i + \mathbf{v}(t), \qquad \mathbf{v}(t) \in \mathfrak{b}_+, \tag{5}$$

where $\mathfrak{b}_+ = \mathfrak{h} \oplus \mathfrak{n}_+$. This space is preserved by the action of the group of N_+–valued gauge transformations, where N_+ is the Lie group corresponding to $\mathfrak{n}_+$. Following [DS], we define the space of $\mathfrak{g}$–opers over a sufficiently small neighborhood U of X, equipped with a coordinate t, as the quotient of the space of all operators of the form (5) by the N_+–valued gauge transformations. One shows that these gauge transformations act freely, and one can find canonical representatives of each orbit labeled by ℓ–tuples of functions $v_1(t), \ldots, v_\ell(t)$. The first of them, $v_1(t)$, transforms as a projective connection, and $v_i(t), i > 1$, transforms as a k_i–differential (i.e., a section of Ω^{k_i}), where $\{k_i\}_{i=1,\ldots,\ell}$ are the orders of the Casimir elements of $\mathfrak{g}$, i.e., the generators of the center of $U(\mathfrak{g})$. Using these transformations, one glues together the spaces of $\mathfrak{g}$–opers on different open subsets of X and thus obtains the notion of a $\mathfrak{g}$–oper on X. A. Beilinson and V. Drinfeld have given a more conceptual definition of opers on X as G–bundles with a connection and a reduction to the Borel subgroup B_+ (the Lie group of $\mathfrak{b}_+$) satisfying a certain transversality condition, see [BD], Sect. 3.

Denote by $\mathrm{Op}_{\mathfrak{g}}(D^\times)$ the space of $\mathfrak{g}$–opers on the (formal) punctured disc. This is the quotient of the space of operators of the form (5), where $\mathbf{v}(t) \in \mathfrak{b}_+((t))$, by the gauge action of $N_+((t))$.

Drinfeld and Sokolov have obtained $\mathrm{Op}_{\mathfrak{g}}(D^\times)$ as the result of the hamiltonian reduction of the space of all operators of the form $\partial_t + A(t), A(t) \in \mathfrak{g}((t))$. The latter space may be identified with a hyperplane in the dual space to the affine Lie algebra $\widehat{\mathfrak{g}}_{\nu_0}, \nu_0 \neq 0$, which consists of all linear functionals taking value 1 on K. It carries the Kirillov-Kostant Poisson structure; the non-zero invariant inner product ν_0 on $\mathfrak{g}$ appears as a parameter of this Poisson structure.

Applying the Drinfeld-Sokolov reduction, we obtain a Poisson structure on the algebra $\mathrm{Fun}(\mathrm{Op}_{\mathfrak{g}}(D^\times))$ of functions on $\mathrm{Op}_{\mathfrak{g}}(D^\times)$. This Poisson algebra is called the *classical W–algebra* associated to $\mathfrak{g}$. In the case when $\mathfrak{g} = \mathfrak{sl}_n$, this Poisson structure is the (second) *Adler–Gelfand–Dickey* Poisson structure. Actually, it is a member of a two-dimensional family of Poisson structures on $\mathrm{Op}_{\mathfrak{sl}_n}(D^\times)$ with respect to which the flows of the nth KdV hierarchy are hamiltonian. We recall that in terms of the first description of $\mathfrak{sl}_n$–opers, i.e., as operators L of the form (2), the KdV equations may be written in the Lax form

$$\partial_{t_m} L = [(L^{m/n})_+, L], \qquad m > 0, m \nmid n, \tag{6}$$

where $L^{1/n} = \partial_t + \ldots$ is the pseudodifferential operator obtained by extracting the nth root of L and $+$ indicates the differential part of a pseudodifferential operator.

Drinfeld and Sokolov have defined an analogue of the KdV hierarchy on the space of $\mathfrak{g}$–opers for an arbitrary $\mathfrak{g}$. The equations of this hierarchy are hamiltonian with respect to the above Poisson structure (in fact, they are bihamiltonian, but we will not discuss here the other hamiltonian structure).

3. The center

Let us go back to the completed universal enveloping algebra $\widetilde{U}_\kappa(\widehat{\mathfrak{g}})$ and its center $Z_\kappa(\widehat{\mathfrak{g}})$. Note that $Z_\kappa(\widehat{\mathfrak{g}})$ is a Poisson algebra. Indeed, choosing a non-zero invariant inner product κ_0, we may write a one-parameter deformation of κ as $\kappa + \epsilon\kappa_0$. Given two elements, $A, B \in Z_\kappa(\widehat{\mathfrak{g}})$, we consider their arbitrary ϵ–deformations, $A(\epsilon), B(\epsilon) \in \widetilde{U}_{\kappa+\epsilon\kappa_0}(\widehat{\mathfrak{g}})$. Then the ϵ–expansion of the commutator $[A(\epsilon), B(\epsilon)]$ will not have a constant term, and its ϵ–linear term, specialized at $\epsilon = 0$, will again be in $Z_\kappa(\widehat{\mathfrak{g}})$ and will be independent of the deformations of A and B. Thus, we obtain a bilinear operation on $Z_\kappa(\widehat{\mathfrak{g}})$, and one checks that it satisfies all properties of a Poisson bracket.

Now we can describe the Poisson algebra $Z_\kappa(\widehat{\mathfrak{g}})$. For a simple Lie algebra $\mathfrak{g}$ we denote by $^L\mathfrak{g}$ its Langlands dual Lie algebra, whose Cartan matrix is the transpose of that of $\mathfrak{g}$ (note that this duality only affects the Lie algebras of series B and C, which get interchanged). We have a canonical identification $^L\mathfrak{h} = \mathfrak{h}^*$.

Let κ_c be the critical inner product on $\mathfrak{g}$ defined by the formula $\kappa_c(x,y) = -\frac{1}{2}\mathrm{Tr}_\mathfrak{g}\mathrm{ad}x\mathrm{ad}y$. In the standard normalization of [K], the modules over $\widetilde{U}_{\kappa_c}(\widehat{\mathfrak{g}})$ on which K acts as the identity are the $\widehat{\mathfrak{g}}$–modules of *critical level* $-h^\vee$, where $h^\vee$ is the dual Coxeter number.

Theorem 1 ([FF3, F3]). (1) *If $\kappa \neq \kappa_c$, then $Z_\kappa(\widehat{\mathfrak{g}}) = \mathbb{C}$.*
(2) $Z_{\kappa_c}(\widehat{\mathfrak{g}})$ *is isomorphic, as a Poisson algebra, to the classical W–algebra* $\mathrm{Fun}(\mathrm{Op}_{^L\mathfrak{g}}(D^\times))$.

Thus, we recover the (second) Poisson structure of the $^L\mathfrak{g}$–KdV hierarchy from the center of the completed universal enveloping algebra $\widetilde{U}_{\kappa_c}(\widehat{\mathfrak{g}})$. Note that the two Poisson structures appearing in the theorem depend on parameters: the inner products κ_0 on $\mathfrak{g}$ and ν_0 on $^L\mathfrak{g}$. In the above isomorphism they have to agree in the obvious sense, namely, that the restriction of κ_0 to $\mathfrak{h}$ is dual to the restriction of ν_0 to $^L\mathfrak{h} = \mathfrak{h}^*$.

For example, $\mathrm{Op}_{\mathfrak{sl}_2}(D^\times) = \{\partial_t^2 - v(t)\}$, where $v(t) = \sum_{n\in\mathbb{Z}} v_n t^{-n-2}$ is a formal Laurent series. Therefore $\mathrm{Fun}(\mathrm{Op}_{\mathfrak{sl}_2}(D^\times))$ is a completion of the polynomial algebra $\mathbb{C}[v_n]_{n\in\mathbb{Z}}$. The Poisson structure is that of the classical Virasoro algebra; it is uniquely determined by the Poisson brackets between the generators

$$\{v_n, v_m\} = (n-m)v_{n+m} - \frac{1}{2}(n^3 - n)\delta_{n,-m}.$$

Under the above isomorphism, the generators v_n are mapped to the *Segal-Sugawara operators* S_n. Those are defined (for an arbitrary $\mathfrak{g}$) by the formula

$$S(z) = \sum_{n\in\mathbb{Z}} S_n\, z^{-n-2} = \sum_a :J^a(z)^2:,$$

where

$$J^a(z) = \sum_{n\in\mathbb{Z}} J_n^a\, z^{-n-1}, \qquad J_n^a = J^a \otimes t^n,$$

and $\{J^a\}$ is an orthonormal basis of $\mathfrak{g}$ with respect to κ_0.

For $\mathfrak{g} = \mathfrak{sl}_2$, the center $Z_{\kappa_c}(\widehat{\mathfrak{sl}}_2)$ is a completion of the polynomial algebra generated by $S_n, n \in \mathbb{Z}$. For general $\mathfrak{g}$, we also have $\ell - 1$ "higher" Segal–Sugawara operators $S_n^{(i)}, i = 2, \ldots, \ell, n \in \mathbb{Z}$, of orders equal to the orders of the Casimirs of $\mathfrak{g}$, and the center $Z_{\kappa_c}(\widehat{\mathfrak{g}})$ is a completion of the algebra of polynomials in these operators. However, explicit formulas for $S_n^{(i)}$ with $i > 1$ are unknown in general.

4. Miura transformation

In addition to operators of the form (3), it is useful to consider the operators

$$
\partial_t + \begin{pmatrix} u_1 & 0 & 0 & \cdots & 0 \\ -1 & u_2 & 0 & \cdots & 0 \\ 0 & -1 & u_3 & \cdots & 0 \\ \vdots & \ddots & \ddots & \cdots & \vdots \\ 0 & 0 & \cdots & -1 & u_n \end{pmatrix}, \qquad \sum_{i=1}^n u_i = 0. \tag{7}
$$

It is easy to see that the operator (7) defines the same oper as the operator (3) (i.e., that they are gauge equivalent under the action of the group of upper triangular matrices) if and only if we have the following identity:

$$
\partial_t^n + v_1(t)\partial_t^{n-2} + \ldots + v_{n-1}(t) = (\partial_t + u_1(t)) \ldots (\partial_t + u_n(t)), \tag{8}
$$

This equation expresses $v_1, \ldots, v_{n-1}$ as differential polynomials in $u_1, \ldots, u_\ell$. For example, for $n = 2$ we have

$$
\partial_t^2 - v = (\partial_t - u)(\partial_t + u), \qquad \text{i.e.,} \qquad v = u^2 - u'. \tag{9}
$$

The latter formula is known as the *Miura transformation*. R. Miura had found that this formula relates solutions of the KdV equation to solutions of another soliton equation, called the modified KdV, or mKdV, equation. This suggested that the KdV equation has something to do with the second order operators $\partial_t^2 - v$, because this formula appears in the splitting of this operator into two operators of order one. This observation has subsequently led to the discovery of the inverse scattering method (in a subsequent work by Gardner, Green, Kruskal and Miura).

The Miura transformation may be viewed as a map from the space of the first order operators $\{\partial_t + u(t)\}$ to the space $\mathrm{Op}_{\mathfrak{sl}_2}(D^\times) = \{\partial_t^2 - v(t)\}$ of $\mathfrak{sl}_2$–opers (or projective connections) on $D^\times$. In order to make it coordinate-independent, we must view the operator $\partial_t + u(t)$ as acting from $\Omega^{-1/2}$ to $\Omega^{1/2}$, i.e., consider it as a connection on the line bundle $\Omega^{-1/2}$. Denote the space of such connections on $D^\times$ by $\mathrm{Conn}_{\mathfrak{sl}_2}(D^\times)$. Then this map is actually a Poisson map $\mathrm{Conn}_{\mathfrak{sl}_2}(D^\times) \to \mathrm{Op}_{\mathfrak{sl}_2}(D^\times)$ if we introduce the Poisson structure $\mathrm{Conn}_{\mathfrak{sl}_2}(D^\times)$ by the formula

$$
\{v_n, v_m\} = \frac{1}{2}n\delta_{n,-m}, \tag{10}
$$

where $u(t) = \sum_{n \in \mathbb{Z}} u_n t^{-n-1}$. Thus, the algebra of functions on $\mathrm{Conn}_{\mathfrak{sl}_2}(D^\times)$, which is a completion of the polynomial algebra $\mathbb{C}[u_n]_{n \in \mathbb{Z}}$, is a Heisenberg–Poisson algebra.

Drinfeld and Sokolov have defined an analogue of the Miura transformation for an arbitrary simple Lie algebra $\mathfrak{g}$. The role of the operator $\partial_t + u(t)$ is now played by the operator $\partial_t + \mathbf{u}(t)$, where $\mathbf{u}(t)$ takes values in $\mathfrak{h}((t))$, considered as a connection on the H–bundle $\Omega^{\rho^\vee}$. In other words, under the change of variables $t = \varphi(s)$ it transforms as follows:

$$u \mapsto \widetilde{u}, \qquad \widetilde{u}(s) = v(\varphi(s))\varphi'(s) - \rho^\vee \left(\frac{\varphi''(s)}{\varphi'(s)} \right).$$

Denote the space of such operators on the punctured disc by $\mathrm{Conn}_\mathfrak{g}(D^\times)$. Then we have a natural map

$$\mu : \mathrm{Conn}_\mathfrak{g}(D^\times) \to \mathrm{Op}_\mathfrak{g}(D^\times)$$

which sends $\partial_t + \mathbf{u}(t)$ to the oper which is the gauge class of the operator $\partial_t - \sum_{i=1}^\ell f_i + \mathbf{u}(t)$. The is the Miura transformation corresponding to the Lie algebra $\mathfrak{g}$.

Define a Poisson structure on $\mathrm{Conn}_\mathfrak{g}(D^\times)$ as follows. Write $u_i(t) = \langle \alpha_i, \mathbf{u}(t) \rangle$ and $u_i(t) = \sum_{n \in \mathbb{Z}} u_{i,n} t^{-n-1}$. Then set

$$\{u_{i,n}, u_{j,m}\} = n\nu_0^{-1}(\alpha_i, \alpha_j)\delta_{n,-m},$$

where ν_0^{-1} is the inner product on $\mathfrak{h}^*$ induced by $\nu_0|_\mathfrak{h}$. Then the Miura transformation $\mathrm{Conn}_\mathfrak{g}(D^\times) \to \mathrm{Op}_\mathfrak{g}(D^\times)$ is a Poisson map (if we take the Poisson structure on $\mathrm{Op}_\mathfrak{g}(D^\times)$ corresponding to ν_0).

Drinfeld and Sokolov [DS] have defined the modified KdV hierarchy corresponding to $\mathfrak{g}$ on $\mathrm{Conn}_\mathfrak{g}(D^\times)$. The equations of this $\mathfrak{g}$–mKdV hierarchy are hamiltonian with respect to the above Poisson structure. The Miura transformation intertwines the $\mathfrak{g}$–mKdV and $\mathfrak{g}$–KdV hierarchies.

5. Wakimoto modules

In Theorem 1 we identified the Poisson structure of the $^L\mathfrak{g}$–KdV hierarchy with the Poisson structure on the center $Z_{\kappa_c}(\widehat{\mathfrak{g}})$. It is natural to ask whether one can interpret in a similar way the Poisson structure of the $^L\mathfrak{g}$–mKdV hierarchy and the Miura transformation. This can indeed be done using the Wakimoto modules of critical level [W, FF1, FF2, F3].

Let us briefly explain the idea of the construction of the Wakimoto modules (see [FB], Ch. 10-11, and [F3] for more details). Set $\mathfrak{b}_- = \mathfrak{h} \oplus \mathfrak{n}_-$. Given a linear functional $\chi : \mathfrak{h}((t)) \to \mathbb{C}$, we extend it trivially to $\mathfrak{n}_-((t))$ and obtain a linear functional on $\mathfrak{b}_-((t))$, also denoted by χ. Let $\mathbb{C}_\chi$ be the corresponding one-dimensional representation of $\mathfrak{b}_-((t))$. We would like to associate to it a smooth representation of $\widehat{\mathfrak{g}}$. It is clear that the induced module $\mathrm{Ind}_{\mathfrak{b}_-((t))}^{\mathfrak{g}((t))} \mathbb{C}_\chi$ is not smooth. Therefore we need to modify the construction of induction corresponding to a different choice of vacuum. In the induced module the vacuum is annihilated by the Lie subalgebra $\mathfrak{n}_-((t))$, while in the Wakimoto module obtained by the "semi-infinite" induction the vacuum is annihilated by $t\mathfrak{g}[t] \oplus \mathfrak{n}_+$.

However, when one applies the "semi-infinite" induction procedure one has to deal with certain "quantum corrections". The effect of these corrections is two-fold: first of all, the resulting module is a module over the central extension of $\mathfrak{g}((t))$, i.e., the affine algebra $\widehat{\mathfrak{g}}$, of critical level. Second, the parameters of the module no longer behave as linear functionals on $\mathfrak{h}((t))$, or equivalently, as elements of the space $\mathfrak{h}^*((t))dt = {}^L\mathfrak{h}((t))dt$ of $\mathfrak{h}^*$–valued one-forms on $D^\times$, but as connections on the ${}^LH$–bundle Ω^ρ. They are precisely the elements of the space $\mathrm{Conn}_{L\mathfrak{g}}(D^\times)$ which is a principal homogeneous space over ${}^L\mathfrak{h}((t))dt$.

Thus we obtain a family of smooth representations of $\widetilde{U}_{\kappa_c}(\widehat{\mathfrak{g}})$ (on which the central element K acts as the identity) parameterized by points of $\mathrm{Conn}_{L\mathfrak{g}}(D^\times)$. These are the *Wakimoto modules* of critical level. For $\chi \in \mathrm{Conn}_{L\mathfrak{g}}(D^\times)$ we denote the corresponding module by W_χ.

Example. Let $\mathfrak{g} = \mathfrak{sl}_2$ with the standard basis $\{e, h, f\}$. Consider the Weyl algebra with generators $a_n, a_n^*, n \in \mathbb{Z}$, and relations $[a_n, a_m^*] = \delta_{n,-m}$. Let M be the Fock representation generated by a vector $|0\rangle$ such that $a_n|0\rangle = 0, n \geq 0$ and $a_n^*|0\rangle = 0, n > 0$. Set $a(z) = \sum_{n\in\mathbb{Z}} a_n z^{-n-1}$, etc. Then for any Laurent series $u(t)$ the formulas

$$e(z) = a(z),$$
$$h(z) = -2{:}a(z)a^*(z){:} + u(z),$$
$$f(z) = -{:}a(z)a^*(z)^2{:} + u(z)a^*(z) - 2\partial_z a^*(z)$$

define an $\widehat{\mathfrak{sl}}_2$–module structure on M. This is the Wakimoto module attached to $\partial_t - u(t)$. One checks easily that in order for the h_n's to transform as the functions t^n (or equivalently, for $h(z)dz$ to transform as a one-form), $\partial_t - u(t)$ needs to transform as a connection on $\Omega^{-1/2}$.

One also checks that the Segal-Sugawara operator $S(z)$ acts on this module as $u^2 - u'$, i.e., through the Miura transformation. This statement has the following generalization for an arbitrary $\mathfrak{g}$.

Theorem 2 ([FF3, F3]). *The center $Z_{\kappa_c}(\widehat{\mathfrak{g}})$ acts on $W_\chi, \chi \in \mathrm{Conn}_{L\mathfrak{g}}(D^\times)$, according to a character. The corresponding point in $\mathrm{Spec}\,Z_{\kappa_c}(\widehat{\mathfrak{g}}) = \mathrm{Op}_{L\mathfrak{g}}(D^\times)$ is $\mu(\chi)$, where $\mu : \mathrm{Conn}_{L\mathfrak{g}}(D^\times) \to \mathrm{Op}_{L\mathfrak{g}}(D^\times)$ is the Miura transformation.*

Thus, we obtain an interpretation of the Miura transformation as an affine analogue of the Harish-Chandra homomorphism $Z(\mathfrak{g}) \to \mathbb{C}[\mathfrak{h}^*]^W$. We remark that the map $\mathrm{Conn}_{L\mathfrak{g}}(D^\times) \to \mathrm{Spec}\,Z_{\kappa_c}(\widehat{\mathfrak{g}})$ that we obtain this way is manifestly Poisson because the Wakimoto modules may be deformed away from the critical level (this deformation gives rise to a Poisson structure on $\mathrm{Conn}_{L\mathfrak{g}}(D^\times)$ which coincides with the one introduced above).

In summary, we have now described the phase spaces of both the generalized KdV and mKdV hierarchies, their Poisson structures and a map between them in terms of representations of affine Kac-Moody algebras of critical level.

6. Local Langlands correspondence for affine algebras

The classical local Langlands correspondence aims to describe the isomorphism classes of smooth representations of $G(F)$, where G is a reductive algebraic group and $F = \mathbb{Q}_p$ or $\mathbb{F}_q((t))$, in terms of homomorphisms from the Galois group of F to the Langlands dual group ${}^LG$ (this is the group for which the sets of characters and cocharacters of the maximal torus are those of G, interchanged).

Let us replace $\mathbb{F}_q((t))$ by $\mathbb{C}((t))$ and G by its Lie algebra $\mathfrak{g}$. Then we try to describe smooth representations of the central extension of the loop algebra $\mathfrak{g}((t))$ in terms of some Galois data. But in the geometric context the Galois group should be thought of as a sort of fundamental group. Hence we replace the notion of Galois representations by the notion of a ^LG–local system, or equivalently, a ^LG–bundle with connection on $D^\times$.

The local Langlands correspondence in this context should be a statement that to each ^LG–bundle with connection on $D^\times$ corresponds a category of $\widehat{\mathfrak{g}}$–modules. Here is an example of such a statement in which the Wakimoto modules of critical level introduced in the previous section play an important role (this is part of an ongoing joint project with D. Gaitsgory).

First we introduce the notion of *nilpotent opers*. Those are roughly those opers on $D^\times$ which have regular singularity at the origin and unipotent monodromy around 0. We denote the space of nilpotent $\mathfrak{g}$–opers by $\mathrm{nOp}_\mathfrak{g}$. For $\mathfrak{g} = \mathfrak{sl}_2$, its points are the projective connections of the form $\partial_z^2 - v(z)$, where $v(t) = \sum_{n \le -1} v_n t^{-n-2}$. We have a residue map $\mathrm{Res} : \mathrm{nOp}_\mathfrak{g} \to \mathfrak{n}$ which for $\mathfrak{g} = \mathfrak{sl}_2$ takes the form $\partial_t^2 - v(t) \mapsto v_{-1}$.

Recall that for a nilpotent element $x \in {}^L\mathfrak{g}$, the *Springer fiber* of x is the variety of all Borel subalgebras of $^L\mathfrak{g}$ containing x. For example, the Springer fiber at 0 is just the flag variety of $^L\mathfrak{g}$.

Lemma 1. *The set of points of the fiber $\mu^{-1}(\rho)$ of the Miura transformation over a nilpotent $^L\mathfrak{g}$–oper ρ is in bijection with the set of points of the Springer fiber of $\mathrm{Res}(\rho) \in {}^L\mathfrak{n}$.*

For example, the set of points of the fiber $\mu^{-1}(\rho)$ of the Miura transformation over a regular $^L\mathfrak{g}$–oper is the set of points of the flag variety of $^L\mathfrak{g}$.

Now fix $\rho \in \mathrm{nOp}_{L\mathfrak{g}}$ and consider the category $\mathcal{C}_\rho$ of $\widehat{\mathfrak{g}}$–modules of critical level on which $Z_{\kappa_c}(\widehat{\mathfrak{g}}) \simeq \mathrm{Fun}(\mathrm{Op}_{L\mathfrak{g}}(D^\times))$ acts through the central character $Z_{\kappa_c}(\widehat{\mathfrak{g}}) \to \mathbb{C}$ corresponding to ρ and such that the Lie subalgebra $(t\mathfrak{g}[[t]] \oplus \mathfrak{n}_+) \subset \widehat{\mathfrak{g}}$ acts locally nilpotently and the Lie subalgebra $\mathfrak{h}$ acts with integral generalized eigenvalues. Then, according to a conjecture of Gaitsgory and myself, the derived category of $\mathcal{C}_\rho$ is equivalent to the derived category of quasicoherent sheaves on the Springer fiber of $\mathrm{Res}(\rho)$ (more precisely, the corresponding DG-scheme).

In particular, under this equivalence the skyscraper sheaf at a point of the Springer fiber of $\mathrm{Res}(\rho)$, which is the same as a point χ of $\mathrm{Conn}_{L\mathfrak{g}}(D^\times)$ projecting onto ρ under the Miura transformation, should correspond to the Wakimoto module W_χ. Thus, the above conjecture means that, loosely speaking, any object of the category $\mathcal{C}_\rho$ may be "decomposed" into a "direct integral" of Wakimoto modules.

7. A q–deformation

Now we wish to define q–deformations of the structures described in the previous sections. In particular, we wish to introduce q–analogues of opers and connections (together with their Poisson structures) and of the Miura transformation between them. We also wish to define q–analogues of the KdV hierarchies. For that we replace the universal enveloping algebra of the affine Lie algebra $\widehat{\mathfrak{g}}$ by the corresponding *quantized enveloping algebra* $U_q(\widehat{\mathfrak{g}})$.

Then the center of $U_q(\hat{\mathfrak{g}})$ at the critical level (with its Poisson structure defined in the same way as in the undeformed case) should be viewed as q–analogue of the algebra of functions on opers (i.e., the classical $\mathcal{W}$–algebra). One the other hand, parameters of Wakimoto modules should be viewed as q–analogues of connections, and the action of the center of $U_q(\hat{\mathfrak{g}})$ on Wakimoto modules should give us a q–analogue of the Miura transformation.

In [FR1], N. Reshetikhin and I have computed these structures in the case when $\mathfrak{g} = \mathfrak{sl}_n$ (we used the Wakimoto modules over $U_q(\hat{\mathfrak{sl}}_n)$ constructed in [AOS]). Let us describe the results. The q–analogues of $\mathfrak{sl}_n$–opers are q–difference operators of the form

$$L_q = D^n + t_1(z)D^{n-1} + \ldots + t_{n-1}(z)D + 1, \tag{11}$$

where $(Df)(z) = f(zq^2)$. The q–analogues of connections are operators $D + \Lambda(z)$, where $\Lambda = (\Lambda_1, \ldots, \Lambda_n)$ and $\prod_{i=1}^n \Lambda_i(z) = 1$. The q–analogue of the Miura transformation is the formula expressing the splitting of the operator (11) into a product of first order operators

$$D^n + t_1(z)D^{n-1} + \ldots + t_{n-1}(z)D + 1 = (D + \Lambda_1(z)) \ldots (D + \Lambda_n(z)).$$

For example, for $\mathfrak{g} = \mathfrak{sl}_2$ the q–Miura transformation is

$$t(z) = \Lambda(z) + \Lambda(zq^2)^{-1}. \tag{12}$$

Note that in the limit $q = e^h, h \to 0$ we have $t(z) = 2 + 4h^2 v(z)z^2 + \ldots$ and $\Lambda(z) = e^{2hu(z)z}$ so that we obtain the ordinary Miura transformation $v = u^2 - u'$.

The Poisson structures with respect to which this map is Poisson are given by the formulas [FR1]

$$\{\Lambda(z), \Lambda(w)\} = (q - q^{-1})f(w/z)\,\Lambda(z)\Lambda(w), \qquad f(x) = \sum_{n \in \mathbb{Z}} \frac{q^n - q^{-n}}{q^n + q^{-n}}\, x^n,$$

$$\{t(z), t(w)\} = (q - q^{-1}) \left(f(w/z)\, t(z)t(w) + \delta\left(\frac{w}{zq^2}\right) - \delta\left(\frac{wq^2}{z}\right) \right),$$

where $\delta(x) = \sum_{n \in \mathbb{Z}} x^n$. Analogous formulas for $\mathfrak{sl}_n, n > 2$, may be found in [FR1].

The equations of the q–analogue of the nth KdV hierarchy are given by the formulas

$$\partial_{t_m} L_q = \left[(L_q^{m/n})_+, L_q \right], \qquad m > 0, m \nmid n.$$

These equations are hamiltonian with respect to the above Poisson structure [F2].

For simple Lie algebras other than $\mathfrak{sl}_n$, Wakimoto modules have not yet been constructed. Nevertheless, in [FR2] we have generalized the above formulas to the case of an arbitrary $\mathfrak{g}$.

Another approach is to define q–analogues of the classical $\mathcal{W}$–algebras by means of a q–analogue of the Drinfeld-Sokolov reduction. This has been done in [FRS, SS].

If we write $\Lambda(z) = Q(zq^{-2})/Q(z)$, then formula (12) becomes

$$t(z) = \frac{Q(zq^2)}{Q(z)} + \frac{Q(zq^{-2})}{Q(z)},$$

which is the Baxter formula for the Bethe Ansatz eigenvalues of the transfer-matrix in the XXZ spin chain model. This is not coincidental. The transfer-matrices of spin models may be obtained from central elements of $U_q(\widehat{\mathfrak{g}})$ of critical level and the Bethe eigenvectors can be constructed using Wakimoto modules. Then the formula for the eigenvalues becomes precisely the formula for the q–Miura transformation (see [FR2]). In the quasi-classical limit we obtain that the formula for the Bethe Ansatz eigenvalues of the Gaudin model (the quasi-classical limit of the XXZ model) may be expressed via the (ordinary) Miura transformation formula. This was explained (in the case of an arbitrary simple Lie algebra $\mathfrak{g}$) in [FFR, F1].

8. The q–characters

Let $\operatorname{Rep} U_q(\widehat{\mathfrak{g}})$ be the Grothendieck ring of finite-dimensional representations of $U_q(\widehat{\mathfrak{g}})$. N.Reshetikhin and M.Semenov-Tian-Shansky [RS] have given an explicit construction of central elements of $U_q(\widehat{\mathfrak{g}})$ of critical level. It amounts to a homomorphism from $\operatorname{Rep} U_q(\widehat{\mathfrak{g}})$ to $Z_q((z))$, where Z_q is the center of $U_q(\widehat{\mathfrak{g}})$ at the critical level. Combining this homomorphism with the q–Miura transformation defined in [FR2], we obtain an injective homomorphism

$$\chi_q : \operatorname{Rep} U_q(\widehat{\mathfrak{g}}) \to \mathbb{Z}[Y_{i,a}^{\pm 1}]_{i=1,\dots,\ell; a \in \mathbb{C}^\times}$$

which we call the q–character homomorphism (see [FR3]). It should be viewed as a q–analogue of the ordinary character homomorphism

$$\chi : \operatorname{Rep} U(\mathfrak{g}) \to \mathbb{Z}[y_i^{\pm 1}]_{i=1,\dots,\ell},$$

where the y_i's are the fundamental coweights of $\mathfrak{g}$. Under the forgetful homomorphism $Y_{i,a} \mapsto y_i$, the q–character of a $U_q(\widehat{\mathfrak{g}})$–module V becomes the ordinary character of the restriction of V to $U_q(\mathfrak{g})$ (specialized at $q = 1$). For instance, if $V(a)$ is the two-dimensional representation of $U_q(\widehat{\mathfrak{sl}}_2)$ with the evaluation parameter a, then $\chi_q(V(a)) = Y_a + Y_{aq^2}^{-1}$, which under the forgetful homomorphism become the character $y + y^{-1}$ of the two-dimensional representation of $U_q(\mathfrak{sl}_2)$ (or $U(\mathfrak{sl}_2)$). In [FM2], the notion of q–characters was extended to the case when q is a root of unity.

When $\mathfrak{g} = \mathfrak{sl}_n$, the variables $Y_{i,a}$ correspond to the series $\Lambda_i(z)$ introduced above by the following rule: $\Lambda_i(za) \mapsto Y_{i,aq^{-i+1}} Y_{i-1,aq^{-i+2}}^{-1}$, where $Y_0 = Y_n = 1$.

One gains a lot of insight into the structure of finite-dimensional representations of quantum affine algebras by analyzing the q–characters. For instance, it was conjectured in [FR3] and proved in [FM2] that the image of the q–character homomorphism is equal to the intersection of the kernels of certain *screening operators*, which come from the hamiltonian interpretation of the q–character homomorphism as the q–Miura transformation. This enabled us to give an algorithm for the computation of the q–characters of the fundamental representations of $U_q(\widehat{\mathfrak{g}})$ [FM2].

H. Nakajima [N] has interpreted the q–characters in terms of the cohomologies of certain quiver varieties. Using this interpretation, he was able to describe the multiplicities of irreducible representations of $U_q(\widehat{\mathfrak{g}})$ inside tensor products of the fundamental representations.

9. Deformed $\mathcal{W}$–algebras

As we mentioned above, the q–character homomorphism expresses the eigenvalues of the transfer-matrices in spin models obtained via the Bethe Ansatz (see [FR2]). The transfer-matrices form a commutative algebra in a quantum object, the quantized enveloping algebra $U_q(\widehat{\mathfrak{g}})$. But now we know that the algebra of transfer-matrices carries a Poisson structure and that the q–character homomorphism defines a Poisson map, i.e., the q–Miura transformation. This immediately raises the question as to whether the algebra of transfer-matrices (already a quantum, albeit commutative, algebra) may be further quantized. This "second quantization" was defined in [SKAO, FF5, AKOS, FR2], and it leads us to deformations of the $\mathcal{W}$–algebras.

The deformed $\mathcal{W}$–algebra $\mathcal{W}_{q,t}(\mathfrak{g})$ is a two-parameter family of associative algebras. It becomes commutative in the limit $t \to 1$, where it coincides with the q–deformed classical $\mathcal{W}$–algebra discussed above. In another limit, when $t = q^\beta$ and $q \to 1$, one obtains the conformal $\mathcal{W}$–algebras which first appeared in conformal field theory (see [FF4]).

For $\mathfrak{g} = \mathfrak{sl}_2$ we have the deformed Virasoro algebra $\mathcal{W}_{q,t}(\mathfrak{sl}_2)$. It has generators $T_n, n \in \mathbb{Z}$, satisfying the relations

$$f\left(\frac{w}{z}\right)T(z)T(w) - f\left(\frac{z}{w}\right)T(w)T(z) = (q - q^{-1})(t - t^{-1})\left(\delta\left(\frac{w}{zq^2t^2}\right) - \delta\left(\frac{wq^2t^2}{z}\right)\right)$$

where $T(z) = \sum_{n \in \mathbb{Z}} T_n z^{-n}$ and

$$f(z) = \frac{1}{1 - z}\frac{(zq^2; q^4t^4)_\infty(zt^2; q^4t^4)_\infty}{(zq^4t^2; q^4t^4)_\infty(zq^2t^4; q^4t^4)_\infty}, \qquad (a; b)_\infty = \prod_{n=0}^{\infty}(1 - ab^n).$$

There is also an analogue of the Miura transformation (free field realization) given by the formula

$$T(z) = :\Lambda(z): + :\Lambda(zq^2t^2)^{-1}: ,$$

where $\Lambda(z)$ is the exponential of a generating function of generators of a Heisenberg algebra.

Just like the Virasoro and other conformal $\mathcal{W}$–algebras, which are symmetries of CFT, the deformed $\mathcal{W}$–algebras appear as dynamical symmetry algebras of various models of statistical mechanics (see [LP, AJMP]).

References

[AJMP] Y. Asai, M. Jimbo, T. Miwa, Ya. Pugai, *Bosonization of vertex operators for the $A_{n-1}^{(1)}$ face model*, J. Phys. **A29** (1996) 6595–6616.

[AKOS] H. Awata, H. Kubo, S. Odake, J. Shiraishi, *Quantum $\mathcal{W}_N$ algebras and Macdonald polynomials*, Comm. Math. Phys. **179** (1996) 401–416.

[AOS] H. Awata, S. Odake, J. Shiraishi, *Free boson realization of $U_q(\widehat{\mathfrak{sl}}_N)$*, Comm. Math. Phys. **162** (1994) 61–83.

[BD] A. Beilinson and V. Drinfeld, *Quantization of Hitchin's integrable system and Hecke eigensheaves*, Preprint, available at www.math.uchicago.edu/~benzvi.

[DS] V. Drinfeld and V. Sokolov, *Lie algebras and KdV type equations*, J. Sov. Math. **30** (1985) 1975–2036.

[FF1] B. Feigin, E. Frenkel, *A family of representations of affine Lie algebras*, Russ. Math. Surv. **43**, N 5 (1988) 221–222.

[FF2] B. Feigin, E. Frenkel, *Affine Kac-Moody Algebras and semi-infinite flag manifolds*, Comm. Math. Phys. **128**, 161–189 (1990).

32

[FF3]   B. Feigin and E. Frenkel, *Affine Kac–Moody algebras at the critical level and Gelfand–Dikii algebras*, Int. Jour. Mod. Phys. **A7**, Supplement 1A (1992) 197–215.

[FF4]   B. Feigin, E. Frenkel, *Integrals of motion and quantum groups*, in Proceedings of the C.I.M.E. School *Integrable Systems and Quantum Groups*, Italy, June 1993, Lect. Notes in Math. **1620**, pp. 349–418, Springer, 1995.

[FF5]   B. Feigin, E. Frenkel, *Quantum W–algebras and elliptic algebras*, Comm. Math. Phys. **178** (1996) 653–678.

[FFR]   B. Feigin, E. Frenkel and N. Reshetikhin, *Gaudin model, Bethe ansatz and critical level*, Comm. Math. Phys. **166** (1994) 27–62.

[F1]   E. Frenkel, *Affine algebras, Langlands duality and Bethe ansatz*, in Proceedings of the International Congress of Mathematical Physics, Paris, 1994, ed. D. Iagolnitzer, pp. 606–642, International Press, 1995.

[F2]   E. Frenkel, *Deformations of the KdV hierarchies and related soliton equations*, Int. Math. Res. Notices. **2**, 55–76 (1996)

[F3]   E. Frenkel, *Lectures on Wakimoto modules, opers and the center at the critical level*, Preprint math.QA/0210029.

[FB]   E. Frenkel, D. Ben-Zvi, *Vertex algebras and algebraic curves*, Mathematical Surveys and Monographs, vol. 88. AMS 2001.

[FM1]   E. Frenkel, E. Mukhin, *Combinatorics of q-characters of finite-dimensional representations of quantum affine algebras*, Comm. Math. Phys. **216** (2001) 23–57.

[FM2]   E. Frenkel, E. Mukhin, *The q-characters at roots of unity*, Adv. Math. **171** (2002) 139–167.

[FR1]   E. Frenkel, N. Reshetikhin, *Quantum affine algebras and deformations of the Virasoro and W–algebras*, Comm. Math. Phys. **178**, 237–264 (1996).

[FR2]   E. Frenkel, N. Reshetikhin, *Deformations of W–algebras associated to simple Lie algebras*, Comm. Math. Phys. **197** (1998), no. 1, 1–32.

[FR3]   E. Frenkel, N. Reshetikhin, *The q–characters of representations of quantum affine agebras and deformations of W–algebras*, Preprint math.QA/9810055; in Contemporary Math **248**, 163–205, AMS 2000.

[FRS]   E. Frenkel, N. Reshetikhin, M.A. Semenov-Tian-Shansky, *Drinfeld-Sokolov reduction for difference operators and deformations of W–algebras I*, Comm. Math. Phys. **192** (1998) 605–629.

[K]   V.G. Kac, *Infinite-dimensional Lie Algebras*, 3rd Edition, Cambridge University Press, 1990.

[LP]   S. Lukyanov, Ya. Pugai, *Multi-point local height probabilities in the integrable RSOS model*, Nuclear Phys. **B473** (1996) 631–658.

[N]   H. Nakajima, *Quiver varieties and finite-dimensional representations of quantum affine algebras*, J. Amer. Math. Soc. **14** (2001) 145–238

[RS]   N.Yu. Reshetikhin, M.A. Semenov-Tian-Shansky, *Central extensions of quantum current groups*, Lett. Math. Phys. **19** (1990) 133–142.

[SS]   M.A. Semenov-Tian-Shansky, A.V. Sevostyanov, *Drinfeld-Sokolov reduction for difference operators and deformations of W-algebras. II. The general semisimple case*, Comm. Math. Phys. **192** (1998) 631–647.

[SKAO]   J. Shiraishi, H. Kubo, H. Awata, S. Odake, *A quantum deformation of the Virasoro algebra and the Macdonald symmetric functions*, Lett. Math. Phys. **38** (1996) 33-51.

[W]   M. Wakimoto, *Fock representations of affine Lie algebra $A_1^{(1)}$*, Comm. Math. Phys. **104** (1986) 605–609.

Plenary Sessions

Inst. Phys. Conf. Ser. No 173: Plenary Sessions
Paper presented at 24th Int. Coll. Group Theoretical Methods in Physics, Paris, France, 15–20 July 2002
©2003 IOP Publishing Ltd

Quantum computing using dissipation

A Beige

Blackett Laboratory, Imperial College London, Prince Consort Road, London, SW7 2BW, UK

Abstract. The principal obstacle to quantum information processing with many qubits is decoherence. One source of decoherence is spontaneous emission which causes loss of energy and information. Inability to control system parameters with high precision is another possible source of error. As a solution we propose quantum computing experiments *using* dissipation based on an environment-induced quantum Zeno effect. As an example we present a simple scheme for quantum gate implementations with cold trapped ions in the presence of cooling.

1. Introduction

Following the theoretical formulation of quantum computing [1] and the first algorithms for problems which can be solved more easily on a quantum computer than on a classical computer [2, 3] the practical implementation of such a device has become a challenging task. Initial steps have already been taken. Quantum bits (qubits) can be realised for instance by storing the information in a superposition of the internal states of two-level atoms. However building systems with many coupled qubits remains a huge challenge. Many demands must be met: reliable qubit storage, preparation and measurement, gate operations with high fidelity and low failure rate and scalability of the system to many qubits. The biggest problem is posed by decoherence which can cause the loss of information.

In this paper we review the idea of quantum computing using dissipation [4, 5] which might help to overcome the decoherence problem in some systems. On the contrary, it might even be useful to introduce an additional spontaneous decay channel into a system in order to create a realm of new possibilities to implement gate operations between qubits. To illustrate this we present a scheme for the realisation of quantum computing between cold trapped ions in the presence of cooling. The ions are stored inside a linear trap and each qubit is obtained from two different ground states of the same ion, as in [6, 7].

In the quantum computing scheme proposed here, the role of cooling during gate operations is twofold. On one hand it decreases the sensitivity of the scheme with respect to heating [8]. On the other hand, the presence of the cooling lasers introduces a decay channel into the system whose presence leads to a restriction of the time evolution of the system onto the computational subspace and facilitates the possibility for quantum gate implementations within one step. The probability for photon emission is relatively small since the system remains to a good approximation within a decoherence-free subspace [9, 10, 11]. If heating nevertheless populates the vibrational mode or gate failure moves the system out of the decoherence-free subspace, then photons are emitted at a high rate [12, 8, 13]. This can be detected and the computation can be restarted.

Many schemes for quantum computing with trapped ions have already been proposed. Some of them require cooling of the ions into the ground state of a common vibrational mode. That this is possible has been demonstrated recently in Innsbruck [14], where the Cirac-Zoller controlled-NOT quantum gate [15] has been implemented with the help of six concatenated laser pulses individually addressing each of the two ions. At the same time, the group in Boulder demonstrated a robust, high-fidelity geometric two ion-qubit phase gate in the laboratory. This was achieved with a sequence of laser pulses and without individual addressing of the ions [16]. But are these schemes really suitable for quantum computation with many qubits? Finding reliable ways to scale present schemes to many qubits might require further simplifications of the experimental setup without decreasing the precision of gate operations.

2. Non-Hermitian Hamiltonians and no-photon time evolutions

Before we discuss the coherent control of an open quantum system with the help of dissipation in more detail, let us first introduce the theoretical model for describing the time evolution of the system under the condition of no photon emission. We give a short review of the quantum jump method [17] which is equivalent to the Monte Carlo wave-function [18] and the quantum trajectory [19] approaches [20].

2.1. The quantum jump approach

In the last decades, several quantum optics experiments have been performed studying the statistics of photons emitted by *single* quantum mechanical systems, like one or two laser-driven trapped atoms or ions. Effects have been found that would be averaged out in the statistics of photons emitted by a whole *ensemble* and which cannot be predicted with the help of expectation values calculated for a statistical ensemble [21]. New formalisms describing single systems interacting with the environment had to be developed.

A typical example for such an experiment is electron shelving [22], i.e. the occurance of stochastic macroscopic light and dark periods in the resonance fluorescence of a laser-driven atom with a metastable state. Another example is the two-atom double-slit experiment by Eichmann *et al.* [23] which demonstrated that the photons emitted by two atoms at a fixed distance can create an interference pattern on a distant screen. These experiments suggest that the effect of the environment on the state of the atoms is the same as the effect of rapidly repeated measurements of whether a photon has been emitted or not [24, 25]. From this assumption the quantum jump approach has been derived.

Assume that a measurement is performed on a quantum optical system surrounded by a free radiation field initially in its vaccum state $|0_{ph}\rangle$ and prepared in $|\psi\rangle$ determining after a time Δt whether or not a photon has been created. If H is the Hamiltonian including the interaction of the system with its environment, the state of the system equals

$$|0_{ph}\rangle\, U_{cond}(\Delta t, 0)|\psi\rangle \equiv |0_{ph}\rangle\langle 0_{ph}|\, U(\Delta t, 0)\, |0_{ph}\rangle|\psi\rangle \qquad (1)$$

under the condition that the free radiation field is still in the vacuum state. For quantum optical experiments, it has been shown that the dynamics under the conditional time evolution operator $U_{cond}(\Delta t, 0)$, defined by the right hand side of (1), can be summarised by a Hamiltonian H_{cond} that is largely independent of the choice of Δt. The conditional Hamiltonian H_{cond} is non-Hermitian and the norm of a state vector developing with H_{cond} decreases in general in time. Under the condition of no photon emission, the state of the system equals at time t

$$|\psi^0(t)\rangle = U_{cond}(t, 0)\, |\psi\rangle / \|\cdot\| . \qquad (2)$$

For convenience, H_{cond} has been defined such that

$$P_0(t, \psi) = \|\, U_{cond}(t, 0)\, |\psi\rangle\, \|^2 \qquad (3)$$

is the probability for no photon emission in $(0, t)$.

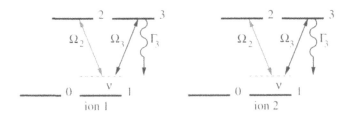

Figure 1. Level scheme of the two ions involved in the gate operation. Each qubit is obtained from the ground states $|0\rangle$ and $|1\rangle$ of one ion. In addition a metastable state $|2\rangle$, a rapidly decaying state $|3\rangle$ with decay rate Γ_3 and two strong laser fields with coupling strength $g_j = \frac{1}{2}\eta_j\Omega_j$ and detuning ν are required.

The non-Hermiticity of the conditional Hamiltonian H_{cond} and the continuous decrease of the amplitude of the unnormalised state vector $U_{\mathrm{cond}}(t,0)\,|\psi\rangle$ reflect that the observation of no photons reveals information about the system. The longer no photon is emitted the more unlikely it becomes that there is excitation that might cause an emission and the amplitudes of states with spontaneous decay rates decrease exponentially [26].

2.2. Example: Two four-level ions inside a linear ion trap

The level scheme of the two cold trapped ions that we consider in this paper in order to discuss the role of dissipation in quantum gate realisations is shown in Figure 1. Each qubit is obtained from two different atomic ground states $|0\rangle$ and $|1\rangle$ of the same ion. In addition, a metastable state $|2\rangle$ and a rapidly decaying level 3 are required. First, the ions have to be cooled into the ground state of a common vibrational mode. Two strong laser fields detuned by the frequency ν of a common vibrational mode should be applied. The laser field coupling to the 1-2 transition establishes a coupling between the two qubits involved in the gate operation. The laser field driving the 1-3 transition represents the laser cooling setup and can be replaced by any other laser cooling configuration without changing the effective time evolution of the system.

In the following, we denote the spontaneous decay rate of level 3 by Γ_3 while b and $b^\dagger$ are the annihilation and creation operator of a phonon in the common vibrational mode. The coupling constant of this mode to the atomic 1-j transition equals $g_j \equiv \frac{1}{2}\eta_j\Omega_j$ where Ω_j is the Rabi frequency of the applied laser field and η_j is the Lamb Dicke parameter depending on the characteristics of the ion trap. Proceeding as in [17], one finds that the conditional Hamiltonian within the dipole and the rotating wave approximation and in the interaction picture with respect to the free Hamiltonian equals

$$H_{\mathrm{cond}} = \sum_{i=1}^{2} \mathrm{i}\hbar\left[\, g_2\,|1\rangle_i\langle 2|\, b^\dagger + g_3\,|1\rangle_i\langle 3|\, b^\dagger - \mathrm{h.c.}\,\right] - \sum_{i=1}^{2} \tfrac{\mathrm{i}}{2}\hbar\Gamma_3\,|3\rangle_i\langle 3|\,. \qquad (4)$$

Here the Lamb-Dicke regime and the condition $\nu \gg \Omega_2, \Omega_3$ has been assumed, as in [15].

3. Quantum computing using dissipation

The basic idea of *quantum computing using dissipation* is to utilise the *coherent* no-photon time evolution given by H_{cond} for the implementation of gate operations. Whenever a photon emission occurs the computation fails and has to be repeated. Nevertheless, we show that fidelities F and success rates P_0 close to one can be achieved for a wide range of experimental parameters. The reason that spontaneous emission from the ions is in the following negligible is that the system remains during the whole computation in a decoherence-free (DF) state [9, 10, 11]. In this section we define DF states and describe a mechanism which restricts the time evolution of a system onto its decoherence-free subspace (DFS).

3.1. Decoherence-free states

The DFS of a system is a subspace of states whose population does not lead to decoherence. Using the quantum jump approach, a state $|\psi\rangle$ is DF if $P_0(t, \psi) = 1$ for all times t [11]. Hence, the DFS is spanned by the eigenvectors of the conditional Hamiltonian with real eigenvalues λ_i. The eigenvectors $|\lambda_i\rangle$ of H_{cond} are in general non-orthogonal. It is therefore useful to introduce the reciprocal basis vectors $|\lambda^j\rangle$ with $\langle\lambda^j|\lambda_i\rangle = \delta_{ij}$ and to write the conditional Hamiltonian as

$$H_{\text{cond}} = \sum_i \lambda_i |\lambda_i\rangle\langle\lambda^i| . \tag{5}$$

Suppose that all non-DF states couple strongly to the environment and populating them leads typically to a photon emission within a time Δt. The time Δt is greater than a certain minimal size which can be determined from the quantum jump approach. Provided that the eigenvalues λ_k corresponding to non-DF states fulfil the condition

$$e^{-i\lambda_k \Delta t/\hbar} = 0 , \tag{6}$$

the no-photon time evolution operator becomes [27]

$$U_{\text{cond}}(\Delta t, 0) = \sum_{i:|\lambda_i\rangle\in\text{DFS}} e^{-i\lambda_i \Delta t/\hbar} |\lambda_i\rangle\langle\lambda_i| . \tag{7}$$

This operator projects every initial state onto the DFS. The action of the environment over a time Δt can therefore be interpreted as a measurement whether the system is DF or not. The probability for no emission in Δt equals the probability to be in a DF state.

3.2. An environment-induced quantum Zeno effect

These continuous measurements caused by the environment lead to a realm of possibilities to manipulate a system within the DFS. Actually, any arbitrary interaction can be used as long as its typical time scale is much longer than Δt defined by condition (6) [4]. This restriction of the system onto the DFS can intuitively be understood with the help of the quantum Zeno effect [28]. Within Δt, a weak interaction can only transfer population proportional Δt out off the DFS. During the next measurement the system is found in a non-DF state with a probability proportional Δt^2. Otherwise it is projected back onto the DFS by the time evolution operator (7). Important is that the probability to find the system always in a DF state, i.e. $T/\Delta t$ times if T is the gate operation time, goes in the limit of weak interactions, i.e. for $\Delta t/T \rightarrow 0$, to one.

However, the time evolution of the system inside the DFS is not inhibited. Using first order perturbation theory with respect to the weak interaction, (7) and the assumption that the system is initially in a DF state, one can show that the conditional time evolution in Δt is the same as the one that one obtains from the Hamiltonian

$$H_{\text{eff}} = I\!P_{\text{DFS}} H_{\text{cond}} I\!P_{\text{DFS}} \tag{8}$$

using the same approximations. Here $I\!P_{\text{DFS}} = \sum_{i:|\lambda_i\rangle\in\text{DFS}} |\lambda_i\rangle\langle\lambda_i|$ is the projector onto the DFS. The effective Hamiltonian H_{eff} is used in the following to find the appropriate laser configuration for the realisation of certain gate operations.

One motivation among others to use an environment-induced quantum Zeno effect for quantum computing is the simplicity of the resulting schemes. As we see in the next section, the DFS of the cold-ion system contains in addition to all ground states, highly entangled states. Through populating these states entanglement between qubits can be created during the effective time evolution even if this is in general not possible using only a single laser pulse.

3.3. Scaling of probabilistic quantum computing schemes

Let us now estimate the probability of finding the result of a whole computation assuming that each gate can be performed with maximum fidelity (as it applies as long as the corrections to the desired effective time evolution are not too big [7]) but only with a finite success rate P_0. The probability of implementing an algorithm of N gates faultlessly is P_0^N and grows exponentially with N. On the other hand, if one always knows whether an algorithm has failed or not, the computation can be repeated until a result is obtained. The probability for not having a result after M runs equals

$$P_{\text{no result}} = (1 - P_0^N)^M . \tag{9}$$

For large N this is approximately $\exp(-M P_0^N)$. Many repetitions might be necessary to implement a computation. However, for smaller numbers of N and if P_0 is sufficiently close to unity, the failure probability is already nearly negligible for $M \approx N$. For example, if $P_0 = 95\%$ and an algorithm with $N = 50$ gates is performed, then repeating the computation 50 times yields a success rate above 98%.

4. Quantum computing with cold trapped ions in the presence of cooling

We now consider again the concrete setup described in subsection 2.2 and describe the realisation of two-qubit quantum gates between cold trapped ions in the presence of cooling with the help of additional weak laser fields.

4.1. Decoherence-free states of the ions

To predict the effect of these fields we first determine the decoherence-free (DF) states of the system in the absence of any additional interaction. As shown in subsection 3.1, they can be calculated by finding the eigenvectors of the Hamiltonian (4) with real eigenvalues. This yields that the decoherence-free subspace (DFS) of the system contains only superpositions of states with the ions either in the state $|00\rangle$ combined with an arbitrary state of the vibrational mode or both ions in $|01\rangle$, $|10\rangle$, $|11\rangle$ or in the antisymmetric and maximally entangled state

$$|a\rangle \equiv \tfrac{1}{\sqrt{2}} [|12\rangle - |21\rangle] \tag{10}$$

while the vibrational mode is not populated. Here all real eigenvalues are zero and the system does not evolve as long as no additional interaction is applied.

4.2. Single laser pulse quantum gates

To realise gate operations between the two qubits formed by the ground states of the ions weak laser fields are applied in addition to the strong lasers shown in Figure 1. Let us denote the Rabi frequency of the laser with respect to the j-2 transition in ion i by $\Omega_j^{(i)}$ and assume that

$$\Omega_j^{(i)} \ll g_2, \ g_3 \text{ and } \Gamma_3 . \tag{11}$$

40

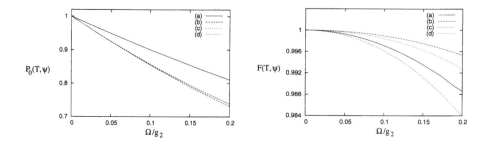

Figure 2. Success rate and fidelity under the condition of no photon emission of a single CNOT gate as a function of the Rabi frequency Ω for $\Gamma_3 = 2\sqrt{37}\,g_2$, $g_3 = \sqrt{2}\,g_2$ and for the initial qubit states $|00\rangle$ (a), $|10\rangle$ (b), $|11\rangle$ (c) and $[\,|10\rangle - |11\rangle\,]/\sqrt{2}$ (d). The gate success rate and fidelity are always maximal if the ions are initially prepared in $|01\rangle$.

The conditional Hamiltonian H_{cond} of the system then becomes the sum of the Hamiltonian (4) and the laser Hamiltonian

$$H_{\text{laser}} = \sum_{i=1,2}\sum_{j=0,1} \tfrac{1}{2}\hbar\Omega_j^{(i)}\,|j\rangle_i\langle 2| + \text{h.c.} \tag{12}$$

If the time evolution of the system is restricted onto the DFS, as predicted in the previous section, then it is effectively given by the Hamiltonian (8) which equals

$$H_{\text{eff}} = \tfrac{1}{2\sqrt{2}}\hbar\,[\,-\Omega_0^{(1)}\,|01\rangle + \Omega_0^{(2)}\,|10\rangle + (\Omega_1^{(2)} - \Omega_1^{(1)})\,|11\rangle\,]\,\langle a| + \text{h.c.} \tag{13}$$

This Hamiltonian can be used to implement quantum gate operations if the operation time T is chosen such that at the end of each gate again only qubit states are populated. A more detailed analysis of the no-photon time evolution of the ions is given in [7].

4.3. The controlled-NOT gate

As a concrete example we consider the controlled NOT (CNOT) gate. If ion 1 contains the target qubit and ion 2 provides the control qubit this gate corresponds to the time evolution operator

$$U_{\text{gate}} = |00\rangle\langle 00| + |01\rangle\langle 01| + |10\rangle\langle 11| + |11\rangle\langle 10|\,. \tag{14}$$

The easiest way to realise a CNOT gate is to couple one laser with the (real) Rabi frequency $\Omega \equiv \Omega_1^{(1)}$ to the 1-2 transition of ion 1 and another one with the same Rabi frequency to the 0-2 transition of ion 2 which yields the effective Hamiltonian

$$H_{\text{eff}} = \tfrac{1}{2\sqrt{2}}\hbar\Omega\,[\,|10\rangle - |11\rangle\,]\,\langle a| + \text{h.c.} \tag{15}$$

If the duration of the laser pulse equals $T = 2\pi/\Omega$ the resulting time evolution can be shown to be exactly the desired operation [7].

Figure 2 results from a numerical solution of the no-photon time evolution given by the sum of the Hamiltonians (4) and (12). For very small Rabi frequencies, the gate success rate and fidelity is for all initial states close to one. For larger values of Ω, the $P_0(T, \psi)$ decreases and is for $\Omega = 0.2\,g_2$ as low as 73 %. The gate fidelity is in this case still above 98.4 %. The smallest gate success rate is found when the atoms are initially in $|10\rangle$, $|11\rangle$ or in a superposition of these two states. In general, success rates $P_0 > 90$ % are achieved as long as $\Omega < 0.07\,g_2$.

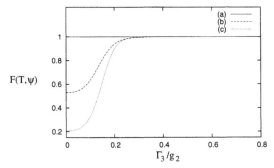

Figure 3. Fidelity for a single CNOT gate under the condition of no photon emission as a function of the spontaneous decay rate Γ_3/g_2 for $g_3 = \Gamma_3$, $\Omega = 0.01\,g_2$ and for the initial qubit states $|00\rangle$ (a), $|10\rangle$ (b) and $[\,|10\rangle - |11\rangle\,]/\sqrt{2}$ (c). If the ions are initially prepared in $|01\rangle$, the gate success rate equals one. For $|\psi\rangle = |10\rangle$ the fidelity is about the same as in graph (c).

Finally, we would like to comment on the role of the spontaneous decay rate of level 3 in the scheme. Figure 3 shows the fidelity of a single CNOT gate as a function of Γ_3 and for $g_3 = \Gamma_3$. In the chosen parameter regime, the effective damping rate of unwanted population in non-DF states can be shown to equal $g_3^2/\Gamma_3 = \Gamma_3$ to a very good approximation and the effective decay rate of non-DF states increases linearly in Γ_3. Figure 3 confirms that the presence of the auxiliary dissipation channels is crucial for the scheme to work! For small damping rates the mechanism which restricts the time evolution of the system onto the computational subspace fails and the minimum gate fidelity is well below 50 %.

5. Conclusions

The paper reviews the idea of quantum computing using dissipation which offers a great variety of possibilities to implement gate operations in the presence of only one dissipation channel in a system. As an example we described the possibility to implement precise quantum gates between cold trapped ions in the presence of cooling of a common vibrational mode and within one step. Because of its simplicity and robustness against parameter fluctuations [7], the proposed scheme might help to increase the number of qubits in present quantum computing experiments.

Acknowledgement. This work was supported by the Royal Society in form of a University Research Fellowship and by the EPSRC and the European Union in part.

References

[1] D. Deutsch, Proc. R. Soc. A **400**, 97 (1985); *ibid.* **425**, 73 (1989).
[2] P. W. Shor, *Algorithms for Quantum Computation: Discrete Log and Factoring*, eds by S. Goldwasser, Proceedings of the 35th Annual Symposium on the Foundations of Computer Science, IEEE Computer Society, Los Alamitos, CA (1994), p. 124.
[3] L. K. Grover, Phys. Rev. Lett. **79**, 325 (1997).
[4] A. Beige, D. Braun, B. Tregenna, and P. L. Knight, Phys. Rev. Lett. **85**, 1762 (2000).
[5] B. Tregenna, A. Beige, and P. L. Knight, Phys. Rev. A **65**, 032305 (2002).
[6] A. Beige, Phys. Rev. A **67**, 020301(R) (2003).

42

[7] A. Beige, *Quantum computing with cold trapped ions in the presence of cooling*, quant-ph/quant-ph/0304168.

[8] J. Eschner, B. Appasamy, and P. E. Toschek, Phys. Rev. Lett. **74**, 2435 (1995).

[9] G. M. Palma, K. A. Suominen, and A. K. Ekert, Proc. Roy. Soc. London Ser. A **452**, 567 (1996).

[10] P. Zanardi and M. Rasetti, Phys. Rev. Lett. **79**, 3306 (1997).

[11] A. Beige, D. Braun, and P. L. Knight, New J. Phys. **2**, 22 (2000).

[12] D. J. Wineland and W. M. Itano, Phys. Rev. A **20**, 1521 (1979).

[13] D. Leibfried, R. Blatt, C. Monroe, and D. J. Wineland, Rev. Mod. Phys. **75**, 281 (2003).

[14] F. Schmidt-Kaler, H. Häffner, M. Riebe, S. Gulde, G. P. T. Lancaster, T. Deuschle, C. Becher, C. F. Roos, J. Eschner, and R. Blatt, Nature **422**, 408 (2003).

[15] J. I. Cirac and P. Zoller, Phys. Rev. Lett. **74**, 4091 (1995).

[16] D. Leibfried, B. DeMarco, V. Meyer, D. Lucas, M. Barrett, J. Britton, W. M. Itano, B. Jelenkovic, C. Langer, T. Rosenband, and D. J. Wineland, Nature **422**, 412 (2003).

[17] G. C. Hegerfeldt, Phys. Rev. A **47**, 449 (1993).

[18] J. Dalibard, Y. Castin, and K. Mølmer, Phys. Rev. Lett. **68**, 580 (1992).

[19] H. Carmichael, *An Open Systems Approach to Quantum Optics*, Lecture Notes in Physics, Vol. **18** (Springer, Berlin, 1993).

[20] For a recent review see M. B. Plenio and P. L. Knight, Rev. Mod. Phys. **70**, 101 (1998).

[21] G. C. Hegerfeldt, Fortschr. Phys. **46**, 595 (1998).

[22] H. G. Dehmelt, Bull. Am. Phys. Soc. **20**, 60 (1975).

[23] U. Eichmann, J. C. Berquist, J. J. Bollinger, J. M. Gilligan, W. M. Itano, and D. J. Wineland, Phys. Rev. Lett. **70**, 2359 (1993).

[24] A. Beige and G. C. Hegerfeldt, Phys. Rev. A **53**, 53 (1996).

[25] C. Schön and A. Beige, Phys. Rev. A **64**, 023806 (2001).

[26] R. J. Cook, Phys. Scr. **T21**, 49 (1988).

[27] The eigenvectors $|\lambda_i\rangle$ of H_{cond} which correspond to DF states are orthogonal to each other and one has $|\lambda^i\rangle = |\lambda_i\rangle$.

[28] B. Misra and E. C. G. Sudarshan, J. Math. Phys. **18**, 756 (1977).

Inst. Phys. Conf. Ser. No 173: Plenary Sessions
Paper presented at 24th Int. Coll. Group Theoretical Methods in Physics, Paris, France, 15–20 July 2002
©*2003 IOP Publishing Ltd*

Properties of Non-Hermitian Quantum Field Theories

Carl M Bender

Department of Physics, Washington University, St. Louis, MO 63130 USA
E-mail: cmb@wuphys.wustl.edu

Abstract. In this talk I discuss quantum systems whose Hamiltonians are non-Hermitian but whose energy levels are all real and positive. Such theories are required to be symmetric under $\mathcal{CPT}$, but not symmetric under $\mathcal{P}$ and $\mathcal{T}$ separately. Recently, quantum mechanical systems having such properties have been investigated in detail. In this talk I extend the results to quantum field theories. Among the systems that I discuss are $-\phi^4$ and $i\phi^3$ theories. These theories all have unexpected and remarkable properties. I discuss the Green's functions for these theories and present new results regarding bound states, renormalization, and nonperturbative calculations.

1. Introduction

The Hamiltonian of a physical system must embody the continuous symmetries and discrete symmetries of that system. There is little doubt that the continuous symmetry group of the universe is the proper Lorentz group. (The *proper* Lorentz group contains all those elements of the Lorentz group that are continuously connected to the identity.) This talk addresses the question of what the discrete symmetry of the universe is.

It is clear that the universe is *not* invariant under the full Lorentz group. Recall that the full Lorentz group is in four distinct parts: (1) the proper Lorentz group; (2) the elements of the proper Lorentz group multiplied by the parity reflection operator $\mathcal{P}$; (3) the elements of the proper Lorentz group multiplied by the time reflection operator $\mathcal{T}$; (4) the elements of the proper Lorentz group multiplied by the parity reflection and time reflection operators $\mathcal{PT}$. It has been experimentally verified that the universe does not exhibit parity reflection symmetry and it also does not exhibit time reversal symmetry. However, a famous theorem in the subject of quantum field theory (one of the few rigorous theorems in quantum field theory!) known as the $\mathcal{PCT}$ theorem, states that the universe *is* invariant under combined space-time reflection and particle-antiparticle interchange [1].

The proof of the $\mathcal{PCT}$ theorem rests on several crucial assumptions, namely, that the Hamiltonian is Hermitian (so that the spectrum is real) and that the spectrum is bounded below. The existence of a real positive spectrum allows one to extend the Lorentz group to the *complex* Lorentz group. The complex Lorentz group consists of *two*, and not four, disconnected parts because in the complex Lorentz group there is a continuous path from the identity to the element $\mathcal{PT}$ that reflects space-time. Note that $\mathcal{PCT}$ symmetry is a much weaker condition than Hermiticity; one must assume that $H = H^\dagger$ in order to prove the $\mathcal{PCT}$ theorem, but we cannot conclude that $H = H^\dagger$ from $\mathcal{PCT}$ symmetry.

The hypothesis made in this talk is that the discrete symmetry of the universe is $\mathcal{PCT}$ symmetry. In this talk we only consider quantum theories in which particles are their own antiparticles; thus, we will assume that the symmetry of the universe is space-time reflection, or $\mathcal{PT}$ symmetry. We argue that space-time reflection symmetry ($\mathcal{PT}$ symmetry) is a simple and natural physical constraint on the Hamiltonian. Hermiticity symmetry $H = H^\dagger$ is a convenient mathematical condition, but one whose physical justification is remote and obscure. We will see that in many (but not all) cases the assumption of $\mathcal{PT}$ symmetry leads to a spectrum that is real and positive.

2. Origin of the Idea

In the late 1980s I coauthored a series of papers in which a technique was developed for solving nonlinear problems in classical and quantum mechanics and quantum field theory by expanding perturbatively in powers of a parameter that measures the nonlinearity of the problem. To illustrate, let us consider the Thomas-Fermi differential-equation boundary-value problem [2]

$$y''(x) = y^{3/2}/\sqrt{x}, \qquad y(0) = 1, \ y(\infty) = 0. \tag{1}$$

This is a difficult problem to solve numerically because there are instabilities and there is no analytical solution. Our approach to this problem is to introduce a small parameter ϵ in the *exponent*:

$$y''(x) = y\,(y/x)^\epsilon, \qquad y(0) = 1, \ y(\infty) = 0, \tag{2}$$

and to solve for $y(x)$ as a series in powers of ϵ:

$$y(x) = e^{-x} + \epsilon Y_1(x) + \epsilon^2 Y_2(x) + \epsilon^3 Y_3(x) + \cdots. \tag{3}$$

The advantage of this procedure is that, unlike many perturbation expansions, the perturbation expansion (3) has a nonzero radius of convergence. The solution to the original Thomas-Fermi boundary-value problem is obtained by setting $\epsilon = 1/2$ in (1) [2].

While I was visiting Saclay, Bessis told to me that he and Zinn-Justin had come across the complex non-Hermitian Hamiltonian

$$H = p^2 + ix^3, \tag{4}$$

whose spectrum appeared to be real and positive. To examine this surprising conjecture we used the perturbation method described above to calculate the eigenvalues of the class of quantum mechanical Hamiltonians

$$H = p^2 + x^2(ix)^\epsilon, \tag{5}$$

where ϵ is a real parameter. Using a variety of analytical and numerical methods we were able to establish with confidence [3, 4] that for the infinite class of Hamiltonians for which $0 \leq \epsilon$ the entire spectrum of H in (5) is real and positive (see Fig. 1). The Hamiltonian (4) considered by Bessis and Zinn-Justin is a special case corresponding to $\epsilon = 1$. This class of Hamiltonians includes the interesting special case $\epsilon = 2$ for which $H = p^2 - x^4$. It is most surprising that the spectrum of this Hamiltonian is real and positive even though it contains a wrong-sign potential.

These quantum mechanical models can be immediately extended to quantum field theory. For example, in a $-x^4$ theory the expectation value $\langle x \rangle$ is *not* zero. The corresponding result for a $-g\phi^4$ quantum field theory in D-dimensional Euclidean space is that the one-point Green's function $G_1 = \langle \phi \rangle$ is also nonzero. This finding may allow us to construct new models for the Higgs boson. We also examine bound states in a $-g\phi^4$ quantum field theory.

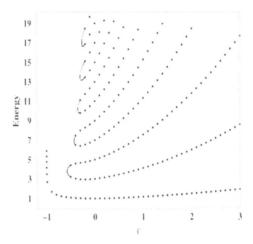

Figure 1. Energy levels of the Hamiltonian $H = p^2 + x^2(ix)^\epsilon$ as a function of the parameter ϵ. There are three regions: When $\epsilon \geq 0$, the spectrum is real and positive and the energy levels rise with increasing ϵ. The lower bound of this region, $\epsilon = 0$, corresponds to the harmonic oscillator, whose energy levels are $E_n = 2n + 1$. When $-1 < \epsilon < 0$, there are a finite number of real positive eigenvalues and an infinite number of complex conjugate pairs of eigenvalues. As ϵ decreases from 0 to -1, the number of real eigenvalues decreases; when $\epsilon \leq -0.57793$, the only real eigenvalue is the ground-state energy. As ϵ approaches -1^+, the ground-state energy diverges. For $\epsilon \leq -1$ there are no real eigenvalues.

3. Positive Spectrum for Wrong Sign Potentials

The spectrum shown in Fig. 1 depends crucially on the boundary conditions for the Hamiltonian in (5). For example, let us consider the case $\epsilon = 2$ for which the Hamiltonian is

$$H = p^2 - gx^4, \tag{6}$$

where we have inserted a coupling constant $g > 0$. There are several ways to obtain the Hamiltonian H in (6). One way is to substitute $g = |g|e^{i\theta}$ into the Hamiltonian $H = p^2 + gx^4$ and to rotate from $\theta = 0$ to $\theta = \pi$. Under this rotation, the ground-state energy $E_0(g)$ becomes complex. Evidently, $E_0(g)$ is real and positive when $g > 0$ and complex when $g < 0$.‡ One can also obtain (6) as the limit of the Hamiltonian $H = p^2 + gx^2(ix)^\epsilon$ as $\epsilon : 0 \to 2$. Having studied Hamiltonians like that in (6) in great detail, we and others have shown that for $\epsilon \geq 0$ the spectra of such Hamiltonians are real, positive, and discrete. The spectrum of the limiting Hamiltonian (6) obtained in this manner is similar to that of the Hamiltonian $H = p^2 + gx^4$ $(g > 0)$; it is entirely real, positive, and discrete. Very recently, the reality and positivity of the spectra have been established rigorously[5].

How can one Hamiltonian (6) possess two different spectra? The answer lies in the boundary conditions satisfied by the eigenfunctions $\psi_n(x)$. In the first case, in which $\theta = \arg g$ is rotated from 0 to π, $\psi_n(x)$ vanishes in the complex-x plane as $|x| \to \infty$

‡ Rotating from $\theta = 0$ to $\theta = -\pi$, we obtain the same Hamiltonian as in (6) but the spectrum is the complex conjugate of the spectrum obtained when we rotate from $\theta = 0$ to $\theta = \pi$.

inside the wedges $-\pi/3 < \arg x < 0$ and $-4\pi/3 < \arg x < -\pi$. In the second case, in which ϵ runs from 0 to 2, $\psi_n(x)$ vanishes in the complex-x plane as $|x| \to \infty$ inside the wedges $-\pi/3 < \arg x < 0$ and $-\pi < \arg x < -2\pi/3$. In this case the boundary conditions hold in wedges that are symmetric with respect to the imaginary axis; these boundary conditions enforce the $\mathcal{PT}$ symmetry of H and account for the reality of the spectrum.

4. One-Point Green's Function G_1

There is another striking difference between the two theories corresponding to H in (6). The one-point Green's function $G_1(g)$ is given by

$$G_1(g) = \langle 0|x|0\rangle/\langle 0|0\rangle \equiv \int_C dx\, x\psi_0^2(x) \left/ \int_C dx\, \psi_0^2(x), \right. \tag{7}$$

where C is a contour that lies in the asymptotic wedges described above. The value of $G_1(g)$ for H in (6) depends on the limiting process by which we obtain H. If we substitute $g = g_0 e^{i\theta}$ into the Hamiltonian $H = p^2 + gx^4$ and rotate from $\theta = 0$ to $\theta = \pi$, we get $G_1(g) = 0$ for all g on the semicircle in the complex-g plane. Thus, this rotation in the complex-g plane preserves parity symmetry ($x \to -x$). However, if we define H in (6) by using the Hamiltonian $H = p^2 + gx^2(ix)^\epsilon$ and allowing ϵ run from 0 to 2, we find that $G_1(g) \neq 0$. Indeed, $G_1(g) \neq 0$ for *all* values of $\epsilon > 0$. Thus, in this theory $\mathcal{PT}$ symmetry (reflection about the imaginary axis, $x \to -x^*$) is preserved, but parity symmetry is permanently broken.

These two different results for $G_1(g)$ emphasize the importance of the boundary conditions in the integrals in (7) for determining the one-point Green's function. We are concerned in this talk with the theory that preserves $\mathcal{PT}$ symmetry. In this theory the energy spectrum is real and positive and $G_1(g)$ is nonzero.

We have extended these quantum-mechanical arguments to the quantum field theory whose D-dimensional Euclidean space Lagrangian is

$$\mathcal{L} = (\nabla\phi)^2/2 + m^2\phi^2/2 - g\phi^4/4. \tag{8}$$

What is remarkable about this "wrong-sign" field theory is that, when it is obtained using the $\mathcal{PT}$-symmetric limit, the energy spectrum is real and positive, and the one-point Green's function is nonzero. Furthermore, the field theory is renormalizable, and in four dimensions is asymptotically free (and thus nontrivial). Based on these features of the theory, we believe that the theory may provide a useful setting to describe the Higgs particle.

The one-point Green's function G_1 is a *complex* functional integral in Euclidean space: $G_1 = \int_C \mathcal{D}\phi\, \phi(0)e^{-L[\phi]}/\int_C \mathcal{D}\phi\, e^{-L[\phi]}$. Here, $L[\phi] = \int d^D x\, \mathcal{L}$ and C is a contour in the complex-ϕ plane defined as follows: Functional integrals are infinite products of ordinary integrals, one integral for each lattice point in Euclidean space. For these ordinary integrals the contour of integration must lie within 45° wedges that lie in the lower-half plane and are centered about $-45°$ and $-135°$. In D-dimensional space we use $\epsilon = gm^{D-4}/4$ to represent the dimensionless coupling constant. The small-ϵ asymptotic behavior of G_1 is determined by a *soliton* (not an instanton). In general, G_1 has a *negative imaginary* value:

$$D = 0: \ G_1 \sim -\frac{i}{m}2^{-1/2}\epsilon^{-1/2}e^{-1/\epsilon} \quad (\epsilon \to 0^+);$$

$$D = 1: \ G_1 \sim -\frac{i}{\sqrt{m}}16\sqrt{\pi}\epsilon(2/\epsilon)^{2/3}e^{-16/(3\epsilon)}3^{-1/6}/\Gamma^2(1/3) \quad (\epsilon \to 0^+). \tag{9}$$

In dimension D, $G_1 \sim e^{-4\Lambda[D]/\epsilon}$ as $\epsilon \to 0^+$, where $\Lambda[D]$ is determined by a spherically symmetric soliton. Numerical values of $\Lambda[D]$ for $0 \leq D \leq 4$ are given in Ref. [6].

5. Bound States

A significant difference between the conventional Lagrangian

$$\mathcal{L} = (\nabla\phi)^2/2 + m^2\phi^2/2 + g\phi^4/4 \tag{10}$$

and the $\mathcal{PT}$-symmetric Lagrangian (8) is that when g is sufficiently small, the $\mathcal{PT}$-symmetric theory possesses bound states while the conventional theory does not. These bound states persist in the non-Hermitian $\mathcal{PT}$-symmetric $-g\phi^4$ quantum field theory for all dimensions $0 \leq D < 3$ but are not present in the conventional Hermitian $g\phi^4$ field theory.

We calculate the bound-state energies perturbatively. For the conventional Lagrangian (10) in one dimension (the anharmonic oscillator) the perturbation series for the kth energy level E_k begins

$$E_k \sim m[k + 1/2 + 3(2k^2 + 2k + 1)\epsilon/4 + O(\epsilon^2)] \quad (\epsilon \to 0^+), \tag{11}$$

where $\epsilon = g/(4m^3)$. The *renormalized mass* M is the first excitation above the ground state:

$$M \equiv E_1 - E_0 \sim m[1 + 3\epsilon + O(\epsilon^2)] \text{ as } \epsilon \to 0^+. \tag{12}$$

To determine if the two-particle state is bound, we examine the *second* excitation above the ground state. We define $B_2 \equiv E_2 - E_0 \sim m[2 + 9\epsilon + O(\epsilon^2)]$ as $\epsilon \to 0^+$. If $B_2 < 2M$, then a two-particle bound state exists and the (negative) binding energy is $B_2 - 2M$. If $B_2 > 2M$, then the second excitation above the vacuum is interpreted as an unbound two-particle state. In the small-coupling regime, where perturbation theory is valid, the conventional anharmonic oscillator does not possess a bound state. Indeed, using WKB, variational methods, or numerical calculations one can show that there is no two-particle bound state for any $g > 0$. Thus, the gx^4 interaction represents a repulsive force.§

We obtain the perturbation series for L in (8) from the perturbation series for the conventional theory by replacing ϵ with $-\epsilon$. Thus, while the conventional anharmonic oscillator does not possess a two-particle bound state, the $\mathcal{PT}$-symmetric oscillator does indeed possess such a state. We give the binding energy of this state in units of the renormalized mass M and we define the *dimensionless* binding energy Δ_2 by

$$\Delta_2 \equiv (B_2 - 2M)/M \sim -3\epsilon + O(\epsilon^2) \quad (\epsilon \to 0^+). \tag{13}$$

This bound state evaporates when ϵ increases beyond $\epsilon = 0.0465\ldots$. As ϵ continues to grow, Δ_2 reaches a maximum of 0.427 at $\epsilon = 0.13$ and then approaches 0.28 as $\epsilon \to \infty$.

In the $\mathcal{PT}$-symmetric anharmonic oscillator, there are not only two-particle bound states for small coupling constant but also k-particle bound states for all $k \geq 2$. The dimensionless binding energies are $\Delta_k \equiv (B_k - kM)/M \sim -3k(k-1)\epsilon/2 + O(\epsilon^2)$ as $\epsilon \to 0+$. Since the coefficient of ϵ is negative, the dimensionless binding energy becomes

§ In general, a repulsive force in a quantum field theory is represented by an energy dependence in which the energy of a two-particle state decreases with separation. The conventional anharmonic oscillator Hamiltonian corresponds to a field theory in one space-time dimension where there cannot be any spatial dependence. The repulsive nature of the force is understood to mean that the energy B_2 needed to create two particles at a given time is more than twice the energy M needed to create one particle.

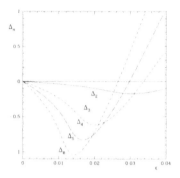

Figure 2. Dimensionless binding energies Δ_2, Δ_3, Δ_4, Δ_5, and Δ_6 for the two-particle, three-particle, four-particle, five-particle, and six-particle bound states plotted as functions of the dimensionless coupling constant ϵ. Note that the multiparticle bound states cease to be bound as ϵ increases past 0.0465, 0.039, 0.034, 0.030, and 0.027.

negative as ϵ increases from 0, and there is a k-particle bound state. The higher k-particle bound states cease to be bound for smaller values of ϵ; the binding energies Δ_3, Δ_4, Δ_5, and Δ_6 become positive as ϵ increases past 0.039, 0.034, 0.030, and 0.027 [7].

Figure 2 shows that for any value of ϵ there are always a finite number of bound states and an infinite number of unbound states. The number of bound states decreases with increasing ϵ until there are no bound states at all. Observe that there is a range of ϵ for which there are only two- and three-particle bound states. This situation is analogous to the physical world in which one observes only states of two and three bound quarks. In this range of ϵ if one has an initial state containing a number of particles (renormalized masses), these particles will clump together into bound states, releasing energy in the process. Depending on the value of ϵ, the final state will consist either of two- or of three-particle bound states, whichever is energetically favored. Note also that there is a special value of ϵ for which two- and three-particle bound states can exist in thermodynamic equilibrium.

These results generalize from quantum mechanics to the D-dimensional $\mathcal{PT}$-symmetric $-g\phi^4$ quantum field theory. There exists a bound state because *to leading order in the dimensionless coupling constant ϵ* the binding energy becomes negative as ϵ increases from 0. We calculate the bound-state energy by summing all "sausage-link" graphs and identifying the bound-state pole. The dimensionless binding energy to leading order in ϵ is

$$\Delta_2 \sim -(4\pi)^{(D-1)/(D-3)}[3\Gamma(3/2 - D/2)]^{2/(3-D)}\epsilon^{2/(3-D)}, \tag{14}$$

which reduces to (13) at $D = 1$. Equation (14) holds for $0 \le D < 3$ because we have performed mass renormalization (but not wave function or coupling-constant renormalization).

Let us compare a $g\phi^3$ theory with a $g\phi^4$ theory: A $g\phi^3$ theory represents an attractive force. The bound states that arise as a consequence of this force can be found by using the Bethe-Salpeter equation. However, the $g\phi^3$ field theory is unacceptable because the spectrum is not bounded below. If we replace g by ig, the spectrum becomes real and positive, but the force becomes repulsive and there are no bound states. The same is true for a two-scalar theory with interaction of the form $ig\phi^2\chi$ [8]. This latter theory is an acceptable model of scalar electrodynamics, but has no analog of positronium.

6. Two-Point Green's Function

There are many other quantum field-theoretic results. For example, we have just completed a study of the two-point Green's function in quantum field theory [9]. A byproduct of this research shows that in a $\mathcal{PT}$-symmetric quantum theory the eigenstates of the Hamiltonian are *complete*.

7. New Results

I conclude by reporting a major breakthrough in $\mathcal{PT}$-symmetric quantum theory [10]. Subsequent to the presentation of this talk we have now been able to establish that $\mathcal{PT}$-symmetric quantum mechanics has an inner product that is associated with a *positive-definite* norm. Thus, a $\mathcal{PT}$-symmetric quantum theory is a fully consistent, unitary, probabilistic, physical quantum mechanical theory. We have found that every $\mathcal{PT}$-symmetric Hamiltonian has a symmetry $\mathcal{C}$ that has until now not been discovered. The linear operator $\mathcal{C}$ commutes with the $\mathcal{PT}$ operator and also with the Hamiltonian H. Also, $\mathcal{C}^2 = 1$, so the eigenvalues of $\mathcal{C}$ are ± 1. The positive-definite inner product is taken with respect to the $\mathcal{CPT}$ operator.

In summary, we have generalized the condition of Hermiticity in quantum mechanics to the statement of $\mathcal{CPT}$ invariance. In effect we have established the converse of the $\mathcal{CPT}$ theorem. If we assume that the Hamiltonian is symmetric and possesses space-time reflection symmetry, and that this symmetry is not spontaneously broken, then the Hamiltonian is Hermitian with respect to $\mathcal{CPT}$ conjugation. In effect, we are replacing the usual mathematical condition of Hermiticity, whose physical content is questionable, by the physical condition of space-time symmetry. This symmetry ensures the reality of the spectrum of the Hamiltonian in complex quantum theories.

Conventional Hermitian Hamiltonians and $\mathcal{PT}$-symmetric Hamiltonians have two important features in common, namely, symmetry and even-dimensionality. When a conventional Hermitian theory is formulated in real Hilbert space, Hamiltonians are required to be symmetric because they represent physical observables. In $\mathcal{PT}$-symmetric quantum theory we are extending this real formulation of quantum mechanics into the complex domain. However, we must retain the symmetry of Hamiltonians for the same reason as in conventional theory. Also, in the real formulation of quantum theory the dimensionality of the Hilbert space must be even. This is necessary in order to introduce a complex structure in the real Hilbert space. In the present theory we require the introduction of the $\mathcal{CPT}$ structure. From a physical point of view this is because half of the eigenstates — those having negative $\mathcal{PT}$ norm — might be interpreted as states representing antiparticles. Therefore, for each particle state there is a corresponding antiparticle state. These two states are always formed pairwise, in the sense that when $\mathcal{PT}$ symmetry is spontaneously broken, corresponding pairs of eigenstates and eigenvalues become complex conjugates of one another. This is because the secular equation for a $\mathcal{PT}$ symmetric Hamiltonian is always real [11].

In a conventional Hermitian quantum field theory the operators $\mathcal{C}$ and $\mathcal{P}$ commute, but in a $\mathcal{PT}$-symmetric quantum field theory these operators do not commute. As a consequence, it is not necessarily true that particles and antiparticles have the same energy eigenvalues. Recall that the condition of space-time reflection symmetry is weaker than the condition of Hermiticity, and therefore it is possible to consider new kinds of quantum field theories, whose self-interaction potential are, for example, $ig\phi^3$ or $-g\phi^4$, that have previously been thought to be unacceptable. A plausible signal of one of these new theories would be the observation of a particle and its corresponding antiparticle having different masses.

References

[1] Streater R F and Wightman A S 1964 PCT, Spin & Statistics, and all that (New York: Benjamin)
[2] Bender C M, Milton K A, Pinsky S S, and Simmons, Jr. L M (1989) J. Math. Phys. 30 1447 - 1455
[3] Bender C M and Boettcher S (1998) Phys. Rev. Lett. 80 5243 - 5246
[4] Bender C M, Boettcher S, and Meisinger P N (1999) J. Math. Phys. 40 2201 - 2229
[5] Dorey P, Dunning C, and Tatao R (2001) J. Phys. A: Math. Gen. 34 L391 - L400 and 34 5679 - 5704
[6] Bender C M, Meisinger P N, and Yang H (2001) Phys. Rev. D 63 45001-1 - 45001-10
[7] Bender C M, Boettcher S, Jones H F, Meisinger P N, and Şimşek M (2001) Phys. Lett. A 291 197 - 202.
[8] Bender C M, Dunne G V, Meisinger P N, and M. Şimşek (2001) Phys. Lett. A 281 311 - 316
[9] Bender C M, Boettcher S, Meisinger P N, and Wang Q (2002) hep-th/0208136 Phys. Lett. A, to appear
[10] Bender C M, Brody D C, and Jones H F (2002) quant-ph/0208076
[11] Bender C M, Berry M V, and Mandilara A (2002) J. Phys. A: Math. Gen. 35 L467 - L471

Inst. Phys. Conf. Ser. No 173: Plenary Sessions
Paper presented at 24th Int. Coll. Group Theoretical Methods in Physics, Paris, France, 15–20 July 2002
©2003 IOP Publishing Ltd

51

Cooperativity and symmetry at biological scales

A. Carbone ‡

Institut des Hautes Études Scientifiques, 35 route de Chartres, 91440 Bures-sur-Yvette, France

Abstract. Facts and ideas presented in this review paper have been written mostly as a guideline to orient the reader through some references in the field. Biological scales are intended in a broad sense and refer to bio-molecular structures as well as supra-molecular and cellular organisations.

1. Two examples.

The world of biology is a world of organisation: spatial and temporal. In general these two aspects are intricate and cannot be dissociated, with the consequent effect that the mechanisms underlying certain biological phenomena are hard to guess. In the following two examples, I still want to emphasize each aspect separately: temporal organisation is illustrated by protein networks in chemotaxis, and spatial organisation, by the structural regularity of protein coats of viruses.

Chemotaxis. Cells employ protein networks to control their behavior. In bacterial chemotaxis, for instance, the bacterium runs, tumbles, senses the average levels of nutrient. When the bacteria (too small in size to sense gradient differences in its receptors) senses *over time* that the attractant concentration is high, it keeps on swimming in the same direction, whereas when the time average concentration of the attractant decreases, it switches into its tumbling mode and changes direction, with the global result to swim towards high nutrient regions. The paradigm followed can be roughly described as a sequence of three actions: sense, compute, respond.

- *Sense:* peptide signals bind to receptors and they are read out in tiny amounts.

- *Compute:* the receptor, by various mechanisms, transduces this event into a change in the chemical state of a messenger molecule; this pathway involves thousands of bio-molecules, transport into the membrane and the cytoplasm, and polarization and migration of chemical gradients.

- *Respond:* molecules interact with others that affect the cell physiology, mechanics, or gene expression in the cell.

This rough schema is present in neutrophil chemotaxis as well, where the neutrophil (much bigger in size than a bacterium) locally senses the bacteria and by contracting its actin filaments accordingly, it chases the bacteria until phagocyting it [1, 2]. Neutrophil chemotaxis is a different and much more complicated process (at the molecular level) than bacterial chemotaxis, due to the feed-forward and feed-back coupling with the *mechanical* properties of the cytoskeleton (induced actin contraction). In both bacterial and neutrophil chemotaxis, the integration of multiple signals among different pathways is poorly understood.

‡ Electronic address: carbone@ihes.fr

Viruses: the example of bacteriophage HK97. The crystal structure of the mature empty capsid of the double-stranded DNA bacteriophage HK97 was recently determined at 3.6Å resolution [3]. It is an icosahedral capsid of 660Å in diameter formed by the assembly of 420 copies of a single 385-residue protein. This virus remarkably displays topologically linked proteins in its capsid, and subunits rotations and local refolding mechanisms in the mature capsid formation.

Each capsid subunit, a flat-folded protein, is joined to its neighbor by ligation on the side chain. This generates pentameric and hexamerics topologically linked rings of covalently joined subunits (recall that a covalent phosphate bond between nucleotides corresponds to $50kcal/mol$ while the energy of an hydrogen bond is $1 - 5kcal/mol$!). These catenanes have not been previously observed in viral capsid and they provide here a stabilisation mechanism for this very thin virus envelop [3].

Besides the beauty of the geometrical arrangements of proteins, an amazing discovery has been associated to bacteriophage HK97: it undergoes several capsid large-scale conformational changes which transform precursors into infectious virions [4]. Both the first and the last capsid of this series of conformations, have been experimentally identified (the first by cryoelectron microscopy and the last by crystallography), and the intermediate ones have been deduced by computational modeling, which turned out to be a crucial tool to understand the various steps of the transformation. It was detected that a rigid body rotation (≈ 40 degrees) caused switching to an entirely different set of interactions during folding and that in this occasion two motifs undergo refolding. These conformational changes stabilize the capsid by increasing the surface area buried at interfaces and bringing the cross-linking forming residues close together. The inner surface of one of the intermediate capsid is negatively charged and this suggests that the transition is triggered electrostatically by DNA packaging. These transitions are irreversible and their energetic basis lies in exothermic switching from one local minimum of the conformational free energy to another lower-energy state.

2. Genomes

The life of a cell is orchestrated by the information encoded in its *genome*, a double-stranded DNA molecule consisting of four nucleotides A, T, C, G. We shall point out some basic facts on cells and genomes of various organisms.

The bacterium *E. coli*, of volume $\approx 1\mu m^3$, is a rod-shaped cell with a thick cell wall composed of two concentric layers. Extending from the cell surface there are a few rotating flagella used by the cell as propellers. In the center of the cell, there is a region where the $4 \cdot 10^6$ base-pair DNA is packed. Between the inner cell wall and the region containing the DNA, the cell is filled in by the cytoplasm, a dense crowd of large and small molecules in thermal motion. There are about 300,000 (non-ribosomal) proteins of $\approx 10nm$ in diameter, 20,000 ribosomes (making the 25% of the cytoplasmic volume), 300,000 tRNAs, a couple of thousand mRNA molecules, 50,000,000 small organic molecules, including amino-acids, nucleotides, sugars, ATPs, and various ions. Yet, 70% of the cell volume is filled by $2 \cdot 10^{10}$ water molecules. Eukaryotic cells, which are typically 10 times bigger than bacteria (1,000 more voluminous), have also a cytoplasm but they are architecturally organized into several relatively well defined compartments and structural networks: nucleus, cytoskeleton, Golgi apparatus, endoplasmic reticulum, vacuole, other organelles, etc.

Molecules in the cytoplasm "communicate" with each other through *physical interactions*. This molecular communication represents the central interest of ongoing research: data gathered with different techniques are stored in databases and combined with DNA sequence information to infer, through intermediate conjectures and hypothesis, possible mechanisms regulating the cellular machinery.

What is coded in the genome? It is tempting to think that a genome encodes the full description of a given organism. But if the genome does encode the full description it is only by a series of indirect implications. The genome is expressed in the cell environment, and the cell carries a long history in its structure and molecular content. Initial conditions for a living cell are determined by the mother cell, and "the rest", such as the architecture, the consistency of the initial conditions, and the cell behavior, can be thought as being coded in its DNA. Still non-deterministic features are also omnipresent due to the importance of thermal agitation at sub-micrometric scales. For instance, the neuronal architecture is only roughly coded in DNA while many synaptic junctions between neurons can be seen as randomly distributed. More precisely, there are organized regions in the brain which are connected to each other in a pre-determined way and this is due to the cooperation of protein networks that induce neurons to grow in specific directions. Another basic observation which is important to keep in mind is that only a fraction of the genomic information is used at different stages of a cell life. There are genes whose role is crucial under only extreme conditions, and such events might never appear in a cell lifetime. Also, in multicellular organisms, differentiated cells employ distinct parts of the genome corresponding to their specialized function within the organism as a whole.

Genome sizes. There is a huge dispersion of genome sizes, going from $\approx 10^6$ basepairs (bps) for bacteria, to $\approx 10^{11}$ bps for certain plants, like lily. *H. sapiens* has a genome of $\approx 10^9$ bps, like sharks and frogs, or certain mollusks, reptiles, birds, plants, etc. In most multicellular eukaryotes, the actual DNA *coding* regions, where genomic information is stored, cover a small fraction of the genome sequence. *H.sapiens* has a 1% of coding DNA, the lungfish has 0.02% over its 140Gbps, the fruit-fly has 20%, etc... while the unicellular yeast has 70% and the bacteria *E.coli* nearly 100% of coding genome.

Genomes are highly constraint sequences. Genomes might be very long but they are far from being random; in fact, their emergence along evolution imposed constraints which are very limiting as can be seen easily with some simple observations. Take, for instance, the number 10^8 of existing viable genomes (that is, 10^8 different species on earth, including the large variety of living bacteria) against the 4^{10^6} possible genome sequences, where 4 is the number of bases A, T, C, G, and 10^6 is the average size of bacteria (since most existing species are bacterial species, it is fair to consider their genome size). Another example concerns proteins. If we assume that the "meaningful" part of a genome corresponds to proteins, then the space of all 4^{10^6} sequences is reducible to a space of 20^{300} potential proteins, where 300aa is the average length of a protein. Even if we would consider a larger size for proteins, as 2000aa for instance, the two numbers 4^{10^6} and 20^{2000} would still be far from being comparable. Of all potential protein sequences, only 5^{300} are (potentially) *functional* proteins (this rough estimate is easily computable from the secondary structure of a protein and the potential disposition of α-helices and β-sheets in the 3-dimensional space). If we look at proteins with *two functions*, this number decreases again to 2^{300}, and so on.

What are the factors that imposed these restrictions along Evolution? To give a definite answer is impossible, but two aspects concerning the structure of a cell can be noticed. The first is the presence of *symmetries* at the different levels of organisation in the cell. Syntactic symmetry (seen in genomic repetitions, and in the use of codons), spatial symmetry (seen in viral coats), temporal symmetry (seen in the cyclic behavior of bio-chemical processes), combinatorial symmetry (seen in the homology of protein sub-complexes), functional symmetry (seen in the function similarity of homologous complexes).

The second aspect concerns the presence of *universal machines* in the cell. These universal machines have the property to interact with a large class of molecules of a given type. For instance the ribosome which translates essentially any kind of RNA sequences into proteins, the RNA polymerase which transcribes essentially any DNA sequence into RNA sequences, the splicing machinery which cuts out introns from RNA sequences transforming them in mature messenger RNAs, the ubiquitin machinery that helps protein degradation, chaperons that help protein folding, etc.

What is the difference between organisms. One would like to understand where the difference between different living organisms lies. Remember that a man and a gorilla share more than the 99% of their coding sequences. To try to answer to the question we need to go down to a molecular level, but let us first fix some numbers:

organism	Number of genes	genome size
Homo sapiens (human)	$3 - 5 \cdot 10^4$	$3.3 \cdot 10^9$
Fugu rebripes (fish)	$3 \cdot 10^4$	$4 \cdot 10^8$
Drosophila melanogaster (fly)	$1.4 \cdot 10^4$	$1.8 \cdot 10^8$
Caenorhabditis elegans (worm)	$1.8 \cdot 10^4$	$1 \cdot 10^8$

The size of a genome is measured by its number of base pairs (remember that a stretched human DNA sequence is about 2 meters long). The table above shows that living organisms have genome sizes which are often very different, while the number of their genes is rather close. Actually, different organisms not only share the same number of genes, but those are almost identical (*homologous*), as shown by the recent sequencing of the mouse, rat, drosophila genomes, etc. In fact, less and less genes remain unknown as new genomes are sequenced: in 1996, with the completion of the *S. cerevisiae* genome sequence, a third of its genes did not match any previously detected sequence; in 1998, a sixth of the genes had no known homologues in *C. elegans*, and in 2000 only a tenth of the genes of *D. melanogaster* were defined as "unknown". Even if genes are detected, it does not mean that their functions are known. The recently sequenced micro-organism *Plasmodium falciparum*, the malaria eukaryotic parasite, is $22 \cdot 10^6$bps long and contains (an estimation of) 5,200 genes whose 60.9% have been identified as hypothetical proteins, that is coding sequences with unknown function [5].

Families of homologous genes. When we look closer to organisms with bilateral symmetry, like humans, flies, rats, sea urchins, we see that

- their genome contains at least one specific gene from each family of homologous genes, for each process of exchange of information in the cell, and that

- it is the number of genes coding for regulatory factors that varies from organism to organism (see Section 3).

This suggests that these organisms seem to be "equivalent" on the basis of genetic contents, and that we need to go down to the molecular level and study how their genes interact (through regulatory factors) to understand their actual differences. In practice, one looks for a model that is able to generate many different organisms from a combination of very few signals and very few genes! The proposed model is based on the idea of *cooperativity*: many different proteins interact in *concert* and with the DNA sequence to give a rich combinatorics of cellular behaviours.

3. Regulatory networks

A simplified model of transcription, translation and regulation. Before discussing the interactions between genes, let us briefly review basic mechanisms of gene expression. A gene along DNA is preceded by a so-called *promoter* region, where "regulatory" proteins interact with each other via direct or indirect binding to the DNA to "decide" whether the gene should be transcribed or not by an RNA-polymerase, the basic component of the transcriptional machinery in the cell. Hence, regulatory proteins *regulate the expression* of the gene.

Such proteins bind to promoter regions to either prevent transcription or enhance it. Therefore the level of gene expression decreases in presence of proteins inhibiting transcription, while it increases with a high concentration of proteins enhancing it. In between, the status of a gene is often tuned between on and off, by a relative concentration of inhibitors and enhancers. (A realistic picture should also take into account the continuous reduction in concentration due to protein degradation.)

If the combination of proteins on the promoter region enhances gene translation, the RNA-polymerase produces an RNA sequence, which is an exact copy of the gene. This sequence, after being often edited and spliced by the machinery of the cell (in case of eukaryotic cells), is ready to be translated into a protein by the ribosomal complex. The ribosome, which is the basic component of the translational machinery of the cell, gets the messenger RNA as input and synthesizes the corresponding protein from individual amino-acids.

Since regulatory proteins are themselves produced by the same transcription/translation process, their expression is also controlled by other regulatory proteins and even sometimes auto-regulated by their own concentration.

The complexity of regulation. The unfolding of development, from fertilization to the formation of a differentiated adult organism follows a blueprint encoded in its DNA both as transcriptional units and *cis*-regulatory elements. The DNA information is interpreted by transcription factors which respond to inputs from upstream stimuli and transfer these signals downstream to other regulatory genes and to structural genes controlling the proper development of the organism. Linear regulatory pathways have been studied from many years, but recent studies also point towards more complex transcriptional networks, reminiscent of electrical engineering diagrams. Such networks involve a vast number of genes and rely on positive and negative feed-back loops, amplification mechanisms, combinatorial interactions, cross-talks, signal thresholds and gradients of regulatory factor concentrations.

A logic behind regulation. Behind the combination of numerous regulatory proteins on promoter regions, there is a logic that biologists try to understand and which can be "rigorously" formalized. The regulation of a gene can be approximated by some boolean function of several variables describing the absence or presence of regulatory proteins. Suppose, for simplicity, to have just two sites of interaction in the promoter region and two proteins, interacting each one with one site. The presence or absence of these proteins at their respective sites, determines the on/off state of the gene. For two variables there are already 16 different possible boolean functions that can describe the gene expression. In general the story is much more complicated. Instead of 2 sites, several and organized subsets of sites are involved. Such regulatory modules in the promoter region are often called cis-*regulatory sequences*. Different parts of the control mechanism are under selection in different regimes. We shall see this in what follows.

Modules and the regulation network in early development: the case study of Sea Urchin. The Sea Urchin genome (complete sequence expected by 2004) is 800 Mega bases long, about $1/4$

of the human genome. Regulatory networks and *cis*-regulatory elements of some of the genes involved in endo-mesodermal specification during the early development of the Sea Urchin have been identified and thoroughly studied [6, 7, 8]. Let us outline some of the results and the complexity behind this regulation network by considering the *endo 16* gene, which plays a role in the development of the gastric apparatus.

The promoter region of *endo 16* is 2300bps long. A map of the regulatory proteins interacting with this region has been reconstructed: there are 7 modules consisting of 12 binding sites for *specific* proteins, as well as 23 other sites where *several different* proteins can bind. The regulatory circuit involving seven modules and 40 proteins was identified and can be compared to a computer program with 40 different variables. This logic model allows testable predictions to be emitted.

The data to reconstruct this circuit have been obtained experimentally. The protein associated to *endo 16* has been fluorescently labelled, and optical slices were recorded. Each slice corresponding to a different state of the "boolean circuit", corresponds to a different state of development of the Sea Urchin. Davidson and his group are about to reconstruct the *full* "logical" circuit of endo-mesodermal specification during early development (*endo 16* is just one player amongst more than 50 involved genes!). Their approach will remain based on careful analysis of expression data, perturbation data, *cis*-regulatory analysis of several genes, and other kinds of experiments [8].

Another example: the Drosophila development. By contrast with the previous example where a single mutation in one of the genes is usually sufficient to produce a new phenotype, John Reinitz showed how the "eve" gene of the Drosophila fly required a combination of mutations of its binding sites to display a phenotypic change [9]. Here the combinatorics becomes much more complicated. Reinitz developed a refined three-tiered model of transcriptional regulation. In particular, above direct DNA-binding by factors, the second tier involves adapter molecules binding to DNA-bound factors, while the third tier includes a physical model of how adapters initiate transcription.

4. A second level of organisation: network connectivity

The intertwined signaling cascades, superimposed to the transcriptional circuitry, together with the communication among structural genes producing the main building blocks of the organism, make matters even more complex. The most elementary kind of circuits that one can think of consists of two expression cascades. The first produces a protein that catalyzes a reaction providing some metabolite for the organism; the second leads to a regulatory protein with no catalytic function. The small molecular weight signaling molecule obtained from the first cascade, and the regulatory factor obtained from the second cascade, interact and mutually affect each other's production by interfering with the transcription of the two genes which are at the origin of the two cascades. This mutual connectivity can be thought as an effective wiring between the two cascades.

The connectivity of several signaling cascades easily generates a complicated network of interactions. Macromolecular networks come in three major flavours: *metabolic* networks, where interactions mostly involve enzymatic proteins and small molecules or metabolites, *protein-protein interaction* networks, and *genetic* networks, where regulatory proteins interact with promoter regions on DNA. These networks have been analyzed and are still studied from physiological, comparative and informational angles. Real progresses have been made in the study of metabolic flow charts, while methodologies for the other two kinds of networks are still rather poorly developed.

Genetic network homology does not imply gene homology: the example of circadian clocks. Different organisms are found to share the same networks. Circadian rhythms, for instance, are used by a wide range of organisms to provide an internal daily periodicity, and the associated clock networks share common features from cyanobacteria to mammals. Namely, all networks seem to include an interaction between two kinds of components: *positive* elements and *negative* elements [10]. In the simplest case, two proteins P and N, which play the role of the positive and negative elements, interact with each other as follows: P increases its own expression and that of N by binding to the promoter regions P_P of itself and P_N of N; on the other hand, a strong binding of N to P inhibits P activity, and represses the expression of both elements by preventing P to bind to the promoters P_P and P_N.

Circadian circuits show that the topology of a circuit is preserved from species to species but that this is not related to gene homology. For instance, the positive elements KaiA in *Synachococcus*, Clc and Cyc in *Drosophila* and Clock and Bmal in mice are not homologous, and this is also not the case for the negative elements KaiB and KaiC in *Synachococcus*, Tim and Per in *Drosophila*, Tim, Per1 and Per2 in mice.

An example of topologically equivalent genetic circuits that perform the same function and that display homologous genes across species, is bacterial chemotaxis [11].

Large networks: the protein-protein interaction networks. Protein-protein interaction networks are constructed from data on pairwise interactions between proteins (detected with two-hybrid systems, for instance) or on protein complexes (detected by mass spectrometry). The networks derived from these kinds of data are usually very different. This is due to two reasons: first, different experimental techniques cover different fractions of the complete set of proteins of a given genome, and second, data are strongly dependent on the experimental protocol.

A graph of pairwise protein interactions has been constructed for the yeast genome: nodes are proteins and edges are interactions. It contains a giant component of 466 proteins and 163 small disjoint subgraphs. (This graph has to be seen more as a statistical estimate of the actual protein-protein interaction in yeast because of the high number of false positive and false negative that the yeast-two-hybrid experiments might detect.) The mathematical analysis of its structure identifies it as a "small world" graph, which means that any two points in the graph can be connected by relatively short paths along existing links. In particular, these graphs are sparse graphs, cliquish, they are "scale-free", that is they satisfy a power law distribution of the node degrees $P(k) \approx k^{-\gamma}$, where $k \gg K$ for some threshold K, they contain few highly connected nodes, and they are insensitive to random errors, that is, if we take away a randomly chosen node, the average distance between nodes does not change [12, 13].

Visualization of protein complexes and a comparative study between complexes and pairwise protein interactions has been recently proposed [14].

Another example of large networks: metabolic networks. Metabolic networks represent the bio-chemistry of an organism, that is those processes that generate mass, energy, information transfer and cell-fate. Nodes are molecules and edges are reactions. As for protein-protein interaction networks, metabolic networks are scale-free graphs. A basic difference between the two types of graph is that protein-protein interaction networks have disjoint clusters while metabolic networks are connected. In particular, it has been noticed that metabolic networks of *different organisms* present the same topological scaling properties [15, 17, 16].

Networks and evolution. What is the dynamical process generating large networks? How metabolic and protein interaction networks have been shaped? Does their present structure provides robustness against mutations? What are the evolutionary models compatible with their growth? One way to *generate* a small world network is to take a regular graph and randomly reassign some of the connections. Another way is by adding new nodes to a graph, and preferentially connect them to existing nodes that have already large connection degrees. The first method displays a peaked distribution of connections, while the second displays a power law distribution for the connection degrees of the nodes. As mentioned above, large biological networks appear to be scale-free, and it was suggested that the most connected nodes in large networks might also be the most ancient [16].

By taking into consideration biologically relevant hypothesis in the network construction such as gene duplications, and both the possibility of adding and deleting nodes from the graph, Andreas Wagner [18] showed that theoretical models of network growth bore little relevance to evolutionary processes leading to real-life networks: protein connectivity does not reflect evolutionary constraints.

Networks and biological functions. What kind of biological information can be deduced from the structure of biological networks? Does the large scale structure of metabolic and protein-protein interaction networks reflect any biological function? Does it reflect chemistry or physics? Are the common combinatorial features of regulatory networks, as for instance circadian networks, consequences of some underlying "design principle"?

Network design. The ability to predict, control and design biological complex networks depends on having detailed quantitative descriptions of their dynamics, but absolute descriptions of mechanistic accuracy seem hopeless. Hence, one way to proceed might be to apply *physical/chemical* knowledge to develop abstract statistical models of these processes and use them to refine or disprove particular physical hypothesis.

Le us consider a concrete example. Regulatory networks that produce a regular oscillatory behavior over a 24-hour period and that are driven by light or temperature might be many. Biological circadian networks have, however, to satisfy some extra constraints such as a circadian period remaining relatively constant over a wide range of temperatures, and functioning reliably in the presence of internal noise, or under global changes in transcription and translation levels under certain nutrition or growth conditions. It seems plausible that a large scale design of "stable" *physical* components might induce a relatively "robust" behavior of such regulatory networks [10].

The study of cellular signal-processing pathways reveals recurring regulatory motifs. Small graph structures (where nodes are molecules and edges are *chemical* reactions) appear multiple times within an organism and among organisms. In this direction, Adam Arkin emphasizes the interest of breaking down some of the regulatory networks into recognizable engineering functions, irrespective of the identities of their constituent genes. Autoregulatory feedback loops in gene expression and cross-regulatory systems as well as amplifier functions, frequency-filtering functions and biphasic response motifs are identified in several concrete biological examples [19, 20].

A combinatorial approach to network design, has been taken in [21], where the design principles of transcriptional regulatory networks is delineated both through time-resolved gene expression on living bacteria and through a graph-theoretic approach to identify basic building blocks within the whole network. Relations between these basic blocks and their information processing function are established.

Conclusion. Networks should be viewed as a beginning, and not as an end to our understanding of the developmental, differentiation, growth, apoptosis programs of an organism. Equally complex mechanisms involving alterations in chromatin configurations, epigenetic events such as DNA methylation, associations of transcriptional activators and repressors that do not directly contact DNA, post-translational modifications of the transcriptional machinery impose additional layers of regulation on the hierarchy of mechanisms relating sequences to phenotypes.

Acknowledgment

We apologize for the many omitted and missed references. For more references and an introduction to some mathematical models in molecular biology the reader may consult [22].

References

[1] Ch.V. Rao, J.R. Kirby, A.P. Arkin. Design and diversity in bacterial chemotaxis, *Science*, 2002.
[2] "Neutrophil crawling", http://expmed.bwh.harvard.edu/projects/motility/neutrophil.html. Movie on a neutrophil chasing a bacteria which was turned in 1950 by David Rogers.
[3] W.R. Wikoff, L. Liljas, R.L. Duda, H. Tsuruta, R.W. Hendrix, J.Johnson, Topologically linked protein rings in the bacteriophage HK97 capsid, *Science*, 289:2129-2133, 2000.
[4] J.F. Conway, W.R. Wikoff, N. Chen, R.L. Duda, R.W. Hendrix, J.E. Johnson and A.C. Steven, Virus maturation involving large subunit rotations and local refolding, *Science*, 292:744-748, 2001.
[5] Plasmodium genomics, *Nature*, 419:489–542, 2002.
[6] C.H. Yuh, H. Bolouri, E.H. Davidson, Genomic cis-regulatory logic: Experimental and computational analysis of a sea urchin gene. *Science* 279: 1896-902, 1998.
[7] E.H. Davidson, *Genomic Regulatory Systems*, Academic Press, USA, 2001.
[8] E.N. Olson, Ed. *cis Network*, a special issue of *Developmental Biology*, Academic Press, USA, 2002. The issue collects papers on the work done in Davidson Laboratory around the deconstruction of the transcriptional circuitry of development.
[9] J. Reinitz, D.H. Sharp, Mechanism of eve stripe formation. *Mech. Dev.*, 49: 133-58, 1995.
[10] N. Barkai, S. Leibler, Circadian clocks limited by noise, *Nature*, 403:267-268, 1999.
[11] N. Barkai, S. Leibler, United we sense..., *Nature*, 393:18-21, 1998.
[12] D.J. Watts, S.H. Strogatz, Collective dynamics of 'small world' networks, *Nature*, 393:440-442, 1998.
[13] R. Albert, H. Jeong, A.L. Barabási, Error and attacks tolerance of complex networks, *Nature*, 406:378-382, 2000.
[14] G.D. Bader, Ch.W.V. Hogue, Analyzing yeast protein-protein interaction data obtained from different sources, *Nature Biotechnology*, 20:991–997, 2002.
[15] D.A. Fell, A. Wagner, The small world of metabolism, *Nature Biotechnology*, 18:1121-1122, 2000.
[16] H. Jeong, B. Tombor, R. Albert, Z.N. Oltvai, A.L. Barabási, The large scale organisation of metabolic networks, *Nature*, 407:651-654, 2000.
[17] A.L. Barabási, R. Albert, Emergence of scaling in random networks *Science*, 286(5439):509-12, 1999.
[18] A. Wagner, The yeast protein interaction network evolves rapidly and contains few redundant duplicate genes. *Molecular Biology and Evolution*, 18:1283-1292, 2002.
[19] A. Gilman, A.P. Arkin, Genetic "code": Representations and dynamical models of genetic components and networks, *Annu. Rev. Genomics Hum. Genet.*, 3:341-69, 2002.
[20] H.H. McAdams, A.P. Arkin, Towards a circuit engineering discipline. *Curr. Biol.*, 10: R318-20, 2000.
[21] S.S Shen-Orr, R. Milo, S. Mangan, U. Alon, Network motifs in the transcriptional regulation network of *Escherichia coli*, *Nature Genetics*, 2002.

[22] A. Carbone, M. Gromov. Mathematical slices of molecular biology, *La Gazette des Mathématiciens*, Société Mathématique de France, numéro spécial, 11–80, avril 2001.

Inst. Phys. Conf. Ser. No 173: Plenary Sessions
Paper presented at 24th Int. Coll. Group Theoretical Methods in Physics, Paris, France, 15–20 July 2002
©2003 IOP Publishing Ltd

Quantum geometry of ADE diagrams and generalized Coxeter-Dynkin systems

R. Coquereaux

Centre de Physique Théorique & CIRM, CNRS, Luminy, Marseille

Abstract. We describe the quantum geometry of ADE diagrams (or generalizations of them), and show how to relate this geometry, in particular the Ocneanu quantum symmetries, to the partition functions that appear in conformal field theory.

Introduction

The purpose of this talk‡ is to present several ideas and results on the quantum geometry of special families of graphs (ADE diagrams or their generalizations). We hope to convince the reader of the beauty and richness of the subject.

This paper in a nutshell :

- **Classical situation**: Representation theory of Lie groups ($SU(2)$, $SU(3)$, etc) and their subgroups can be encoded by graphs. These graphs tell us how to decompose the representations obtained by tensor multiplying irreducible representations (irreps); actually it is enough to know what happens when one tensor multiplies by the fundamental representations. Such a graph defines an associative algebra (the "graph algebra") which is the the Grothendieck ring spanned by characters of the group. Notice that the graph algebra of a subgroup is a module over the graph algebra of the group.

- **Quantum situation**:

  - By truncating the diagrams of tensorisation for $SU(2)$, $SU(3)$,...$SU(h)$, ..., one obtains the usual A_r Dynkin diagrams (for $SU(2)$) or their higher dimensional analogues (Di Francesco – Zuber diagrams of $\mathcal{A}$ type for $SU(3)$). All these graphs have self - fusion (an associative multiplication law with positive integral structure constants), but they are not the only ones to enjoy this property.

  - For a given h (the choice of $SU(h)$), the first task is to find all the graphs that have self-fusion. For the $SU(2)$ family, and besides A_r diagrams, one discovers in this way the D_{even}, E_6 and E_8 Dynkin diagrams.

  - The next task is to identify all those diagrams which do not necessarily enjoy self-fusion, but which generate a module over one of the algebras defined by the previous family. In this way, and for the $SU(2)$ family, one discovers the D_{odd} and E_7 diagrams.

  - All the diagrams obtain so far (with or without self - fusion) can be labelled by an integer k, called the level of the diagram (not equal to the level h of $SU(h)$ characterizing a chosen family of diagrams). A diagram of level k is always a module over the member $\mathcal{A}_k$ of the $\mathcal{A}$ family with the same level§.

‡ Conference given at ICGTMP ("Group24") June 15-21, 2002, Paris

§ Warning, in the $SU(2)$ case, we have two notations for the same objects since the subindex of A_r refers usually to the number of vertices (the rank), but in this case $k = r - 1$, so that $\mathcal{A}_{r-1} = A_r$

- We then move from the geometry of the "space" G to the geometry of the paths on G (a procedure that is quite common in quantum physics!) Paths on G build a vector space $Paths$ which comes with a grading: paths of definite grade are associated with Young Frames of SU(h). In the case of $SU(2)$ this grading is just an integer (a length, or a point on $\mathcal{A}_k$). What turns out to be most interesting is a particular vector subspace of $Paths$ whose elements are called "essential paths". The space of essential paths $EssPaths$ is itself graded in the same way as $Paths$. By using the possibility of concatenating paths on the chosen diagram, one may define another multiplicative (associative) structure on the algebra of graded endomorphisms of essential paths. This leads to the definition of a bi-algebra $\mathcal{B}G$; actually this is a di-algebra which is semi-simple for both structures but existence of a scalar product allows one to transmute one of the multiplications into a co-multiplication compatible with the other structure.
- There are two — usually distinct — block decompositions for $\mathcal{B}G$ (ideals corresponding to simple blocks). One type of blocks corresponds to the grading associated with points of $\mathcal{A}_k$, another type of blocks corresponds to points of another graph called $Oc(G)$, the Ocneanu graph of G (or graph of quantum symmetries). The algebra spanned by the associated minimal central projectors is also called $Oc(G)$ and is a bimodule over the graph algebra of $\mathcal{A}_k$; the bimodule structure is encoded by a set of matrices $W_{x,y} = (W_{x,y})^i_j$, where x and y refer to points of the graph $Oc(G)$ and where i and j run on the set of vertices of $\mathcal{A}_k$.
- These matrices $W_{x,y}$ can be interpreted as partition functions of a conformal field theory with a boundary and defects lines (also labelled by x and y). When both x and y coincide with the identity element of $Oc(G)$, the matrix $W0 \doteq W_{0,0}$ is a modular invariant: it commutes with the generators S and T, representing $SL(2, \mathbb{Z})$ in the vector space spanned by the vertices of the graph $\mathcal{A}_k$ (in conformal field theory, these vertices label the characters of an affine Lie algebra). The sesquilinear form associated with $W0$ is the modular invariant partition function. Other matrices $W_{x,y}$ are associated with partition functions that are not modular invariant but they can be given a nice interpretation in terms of conformal field theory (see our historical section for references).

- **An example: the $\mathcal{E}_5$ diagram of the $SU(3)$ system.**
The Di Francesco – Zuber $G = \mathcal{E}_5$ diagram is displayed on Figure 1, it is a module over the $\mathcal{A}_5$ diagram (the generator corresponding to the given orientation is the vertex $(1,0)$). The dimension of the space of paths on $\mathcal{E}_5$ is infinite, but when we restrict our attention to essential paths (one type of essential path for every vertex of $\mathcal{A}_5$), we find 21 possibilities i.e., 21 blocks of dimensions (d_p, d_p) for the first algebra structure of $\mathcal{B}G$. The integers d_p are given by the list:
(12), $(24, 24)$, $(36, 48, 36)$, $(36, 60, 60, 36)$, $(24, 48, 60, 48, 24)$, $(12, 24, 36, 36, 24, 12)$
For its other multiplicative structure, $\mathcal{B}G$ has 24 blocks. Its dimensions d_x are as follows: six blocks with $d_x = 12$, twelve blocks with $d_x = 24$ and six blocks with $d_x = 60$.
Notice that $\sum_p d_p^2 = 29376$ and $\sum_x d_x^2 = 29376$; moreover $\sum_p d_p = 720$ and $\sum_x d_x = 720$. The first equality (quadratic sum rule) is non trivial but nevertheless expected since we have a bi-algebra structure. The other equality (linear sum rule) is unexpected but seems to hold in all cases explicitly studied so far (in some cases one has to introduce a natural correction factor). The indexing set for x, i.e., the Ocneanu graph of $\mathcal{E}_5$, has 24 points; it was obtained in [5] and is displayed on Figure 2.
One obtain in this way 24 toric matrices (and partition functions) of type $W_{x,0}$, and 24^2 matrices of type $W_{x,y}$. Many of them happen to coïncide. The modular invariant partition

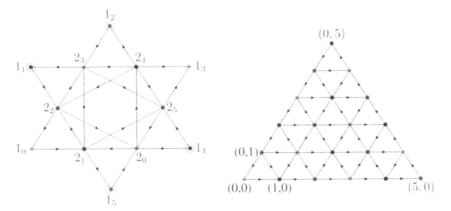

Figure 1. The $\mathcal{E}_5$ and $\mathcal{A}_5$ generalized Dynkin diagrams

Figure 2. Ocneanu graph for $\mathcal{E}_5$

function is associated with $W_{0,0}$ and is given by

$$
\begin{aligned}
\mathcal{Z}_{\mathcal{E}_5} \doteq \mathcal{Z}_{1_0 \otimes 1_0} = \ & |\chi_{(0,0)} + \chi_{(2,2)}|^2 + |\chi_{(0,2)} + \chi_{(3,2)}|^2 + |\chi_{(2,0)} + \chi_{(2,3)}|^2 \\
+ \ & |\chi_{(2,1)} + \chi_{(0,5)}|^2 + |\chi_{(3,0)} + \chi_{(0,3)}|^2 + |\chi_{(1,2)} + \chi_{(5,0)}|^2
\end{aligned}
$$

It agrees $\|$ with the expression first obtained by [11], using entirely different techniques.

Brief historical section:
Here we make a long story short and gather only a few references. Many others can be found by looking at the quoted material or appear at the end of the present article. Apologies for omissions.

$\|$ There is a global shift of $(1,1)$ because of our conventions.

The study of quantum geometry of ADE graphs was, at the beginning, only a nice example illustrating the general theory of paragroups and "Ocneanu cells" ([14]). This example and its generalizations turned out to be so rich that it became, by itself, a subject of study. Most of the theory was developed by A. Ocneanu himself and described (sometimes in a rather allusive way) at several meetings and conferences in the years 95 − 2000 (for instance [15]). As far as we know, the first published material describing this theory is [16].

From the physics side, many relations existing between ADE graphs and physics (models of statistical mechanics) had been already observed and investigated by V. Pasquier in his thesis (see [18]). Classification of modular invariant partition functions for conformal field theories of $SU(2)$ type was obtained at the same time, i.e., at the end of the eighties, by [1] in a celebrated paper. Later T. Gannon (and collaborators) could obtain ([11] similar results for conformal field theories based on other affine Kac − Moody algebras (of type $SU(3)$, etc.).

Di Francesco and Zuber made the crucial observation [10] that the $SU(3)$ classification could be related to a family of particular graphs (that we call the Di Francesco − Zuber graphs), in a way similar to the relation existing between the $SU(2)$ classification and the ADE Dynkin diagrams. Several precisions concerning this classification were brought by A. Ocneanu at the Bariloche school ([17], see also the lectures of J.-B. Zuber and D. Evans at the same school).

After the (unpublished) work by Ocneanu concerning the ADE themselves, it was more or less clear that the existence of a modular invariant partition functions associated with these diagrams (or their generalizations) was only the tip of a theoretical iceberg. For instance, from the existence of a toric structure on ADE diagrams (several toric matrices implementing the bimodule structure of the Ocneanu graph $Oc(G)$ associated with the graph G), it was clear that the modular invariant partition function was only describing a particular point of $Oc(G)$, and that other "interesting" partition functions claiming for a physical interpretation existed in the theory. A simple algorithm allowing one to obtain these toric matrices $W_{x,y}$ was explained in [2], following the example of E_6, and a physical interpretation in terms of conformal field theory with a boundary and defects lines was given in [20]. Using the techniques explained in [2], a systematic study of all ADE cases was performed in [4] and several interesting cases belonging to the $SU(3)$ family were analyzed in [5].

1. The diagrams

1.1. The classical $SU(2)$ system of diagrams: a classical analogy

1.1.1. Representation theory of $SU(2)$: the A_∞ diagram In order to decompose a product of irreducible representations of $SU(2)$ (composition of spins), it is enough to know how to decompose the product $\tau_p \otimes \tau_1$ where τ_p is the irreducible representation of dimension $p + 1 = 2j + 1$ and τ_1 is the fundamental (spin $1/2$): $\tau_p \otimes \tau_1 = \tau_{p-1} \oplus \tau_{p+1}$. The "rule of multiplication by τ_1" can be encoded by a diagram ("diagram of tensorisation by the fundamental") whose vertices are labelled by p and where the neighbors of p label those irreps that appear on the right hand side of the tensor multiplication of τ_p by τ_1 : these are the vertices with labels $p - 1$ and $p + 1$. The diagram (see Fig. 3) is an infinite half line (called A_∞).

Equivalently, the rule of multiplication by τ_1 is encoded by the adjacency matrix of this graph; it is an infinite matrix since its row or column indices ($p + 1$) range from 1 to ∞.

This diagram has self-fusion, in the sense that we have an associative multiplication $\tau_p \otimes \tau_q$ with positive integral structure constants (the multiplicity coefficients that appear on the r.h.s. of the reduction of such a tensor product). Notice that the corresponding table of multiplication (composition of arbitrary spins) is fully determined as soon as we impose associativity, the fact that τ_0 acts as the unit, and the rule of multiplication by τ_1 (i.e., the diagram A_∞).

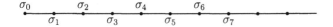

$$\sigma_0 \quad\quad \sigma_2 \quad\quad \sigma_4 \quad\quad \sigma_6$$
$$\sigma_1 \quad\quad \sigma_3 \quad\quad \sigma_5 \quad\quad \sigma_7$$

Figure 3. A_∞, the fusion diagram of $SU(2)$

1.1.2. Representation theory of finite subgroups of $SU(2)$: the affine $A^{(1)}, D^{(1)}, E^{(1)}$ diagrams diagram

- Choose a (finite) subgroup of $SU(2)$, i.e., one of the so-called binary polyhedral groups.
- The fundamental representation is again 2 dimensional and the multiplication of any of its irreps by the fundamental is encoded by the corresponding diagram of tensorisation, which, for the binary groups of symmetries of platonic bodies coïncide with the affine exceptional Dynkin diagrams $E_6^{(1)}, E_7^{(1)}, E_8^{(1)}$ (McKay correspondence).
- The vector space generated by the set of irreducible representations of such a subgroup is a module over the set of irreps of $SU(2)$ (reduce irreps from the group $SU(2)$ to its subgroup and use tensor multiplication of representations). In diagrammatic parlance, we may say that affine ADE diagrams are modules over the A_∞ diagram.
- Irreps of a binary polyhedral group can also be tensor multiplied and decomposed into irreps (with positive integral structure constants). In other words : affine ADE diagrams have self fusion. In particular one of its vertices σ_p acts as the unit; we call it σ_0.
- Call G_1 the adjacency matrix of each of these diagrams. Its highest eigenvalue is $\beta = 2$ (in all cases) and is called the Perron - Frobenius norm of the diagram. It happens to coïncide with the dimension of the fundamental. For a given diagram, dimensions of the irreps are given by components of the (unique)normalized eigenvector corresponding to β (it is normalized to 1 at the unit point σ_0).
- The table of characters happens to be equal to the matrix of eigenvectors (properly normalized) of G_1. This is a way to express the general McKay correspondence in the case of $SU(2)$.

1.2. The quantum $SU(2)$ system of diagrams

- Replace the A_∞ diagram by A_n diagrams (truncated A_∞ diagrams).
- A_n diagrams have self - fusion.
- Construct those diagrams that generate modules over A_n : get the A, D and E diagrams. For example, E_6 is an A_{11} module, E_7 an A_{17} module, and E_8 an A_{29} module.
- Some of them have self-fusion (A_n, D_{eve}, E_6, E_8), other don't (D_{odd}, E_7).
- Dimensions obtained from the components of the Perron – Frobenius eigenvector are not integers (quantum dimensions).

1.3. The classical $SU(3)$ system of diagrams

- Representation theory of $SU(3)$ is characterized by two generalized $\mathcal{A}_\infty$ diagrams differing only by orientation (multiplication by the fundamentals $3 = (1,0)$ of $\overline{3} = (0,1)$).
- Representation theory for finite subgroups of $SU(3)$ is fully characterized by a family of diagrams that have self – fusion and generate modules over the graph algebra of the generalized $\mathcal{A}_\infty$ diagram of $SU(3)$.

1.4. The quantum $SU(3)$ system of diagrams

- Replace the $\mathcal{A}_\infty$ diagram by $\mathcal{A}_k$ diagrams (truncated $\mathcal{A}_\infty$ diagrams)
- $\mathcal{A}_k$ have self-fusion.
- Construct those diagrams that are modules over the $\mathcal{A}_k$: get the Di Francesco – Zuber diagrams.
- Some of them have self-fusion and other don't.
- The system contains a "principal" series (the $\mathcal{A}$ series) and a finite number of "genuine exceptional" cases ($\mathcal{E}_5$, $\mathcal{E}_9$ and $\mathcal{E}_{21}$). The other diagrams of the system are obtained as orbifolds of the genuine diagrams (exceptional or not) and as twists or conjugates (sometimes both) of the genuine diagrams and of their orbifolds.

1.5. Classical and quantum examples

The fundamental diagram of tensorisation for the binary tetrahedral group $E_6^{(1)}$ and its quantum counterpart is the diagram E_6.

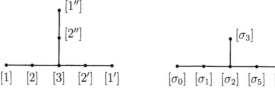

Figure 4. The diagrams $E_6^{(1)}$ and E_6

2. Paths and essential paths

2.1. Notations

Start from a Dynkin diagram G of type ADE. Call $\beta = 2\cos\pi/\kappa$ its Perron Frobenius norm (highest eigenvalue of the adjacency matrix) and μ the normalized Perron Frobenius vector (eigenvector of β). For affine ADE diagrams, $\beta = 2$ and the Perron Frobenius vector gives the dimensions of irreps for the corresponding binary polyhedral group. Call κ the (dual) Coxeter number of G , obtained from the value of β, define $k = \kappa - 2$ as the level of the diagram. For $SU(h)$ generalized Coxeter - Dynkin systems, the expression of β is not the same (it is equal to $1 + 2\cos(2\pi/\kappa)$ for $SU(3)$) but still allows one to define a generalized Coxeter number; the integer h itself coïncide with the Coxeter number of $SU(h)$ and the level of a such a generalized diagram is defined to be $k \doteq \kappa - h$.

2.2. Definitions (case of ADE diagrams)

We now describe essential paths for the case of the usual $SU(2)$ system.

- Consider elementary paths on G (elementary in the sense of "naive" i.e., a succession of contiguous edges ξ_p). Now build the vector space $Paths$ spanned by the elementary paths. This vector space is graded by n, the length, or equivalently, by horizontal Young frames with n boxes.

- Declare that elementary paths are orthonormal. This defines a Hilbert space structure on Paths.

- Define operators C_p from $Paths^n$ to $Paths^{n-2}$, as follows: if $p \geq n$, the result is 0; for a path $\gamma = \xi_1\xi_2\ldots\xi_p\xi_{p+1}\xi_{p+2}\ldots\xi_n$ of length n, if ξ_{p+1} is *not* equal to the opposite of ξ_p (same edge with opposite orientation), the result is again the null vector; otherwise the result is a path of length $n-2$ obtained by removing from γ the round trip $\xi_p\xi_{p+1}$ and multiplying what remains by $\sqrt{\mu(r(\xi_p))/\mu(s(\xi_p))}$. Here, r and s denote the range and source of the oriented edge ξ_p.

- Define $EssPath$ (the subspace of essential paths) as the intersection of the kernels of all annihilation operators C_p, or, equivalently, as the intersection of the kernels of all Jones operators $e_p = C_p^\dagger C_p / \beta$. This vector subspace is also graded by the length n.

The notion of essential paths can be generalized to higher Coxeter – Dynkin systems (classical or quantum) associated with $SU(h)$, like the Di Francesco – Zuber diagrams, which is associated with $SU(3)$. The length grading becomes a grading by Young frames.

2.3. The bigebra $\mathcal{B}G \doteq End_{\#}(EssPath(G))$

The (first) algebra structure of $\mathcal{B}G$ comes from its definition in term of direct sum of algebras of endomorphisms.

$$\mathcal{B}G = \oplus_p End(EssPath^{(p)}G)$$

- Facts
    - Path is not only a Hilbert space but an associative algebra (use concatenation).
    - $\mathcal{B}G$ is (obviously) an algebra for ∘ (composition of endomorphisms)
    - $\mathcal{B}G$ is (not obviously¶) an algebra for × (convolution)[+]
    - The di-algebra $\mathcal{B}G$ can be transformed into a bi-algebra by using the existence of a scalar product (the convolution law × is traded against a convolution coproduct). The composition coproduct is found to be compatible with the first associative multiplication ∘. The situation is self - dual.

- $\mathcal{B}G$ is semi-simple for ∘ and we can write it as a direct sum of blocks (of dimensions d_p)
    - For a graph of level k, blocks are labelled[*] by vertices of the diagram $\mathcal{A}_k$
    - Example. For $G = E_6$, the blocks are labelled by the diagram A_{11} ($\equiv \mathcal{A}_{10}$).

- $\mathcal{B}G$ is semi-simple for × and we can write it as a direct sum of blocks (of dimensions d_x)
    - Blocks are labelled by vertices of the Ocneanu graph of G. Actually this defines what the Ocneanu graph is.

¶ This is not obvious because the concatenation product of two essential paths is not an essential path
+ The definition of this product uses the existence of a law of concatenation for paths.
* Warning (again): for ADE diagrams, $\mathcal{A}_{k=\kappa-2} = A_{\kappa-1}$

– The multiplication table (for ∘) of the minimal central projectors associated with the blocks labelled by x (law $\times$) defines the so – called "algebra of quantum symmetries", also denoted $Oc(G)$. The Ocneanu graph of G is the Cayley graph of the algebra $Oc(G)$.

BG has been claimed to possess a structure of weak Hopf algebra, but this property, as far as I know, was never shown explicitly (see the forthcoming article [6]).

2.4. Classical and quantum examples

Here we display the paths and essential paths starting from the origin of the diagram $E_6^{(1)}$ (describing the binary tetrahedral group $\widetilde{Tet}$). The number of paths of length n starting from the origin and reaching a given vertex gives the multiplicity of the corresponding irreducible representation of the binary tetrahedral group in the n-th tensor power of the fundamental representation. Essential paths of length n starting from the origin describe the restriction of the irreducible symmetric representation of dimension $n+1$ of $SU(2)$ to the chosen subgroup (in this case, taking for example $n = 6$, one gets the branching rule [7] $\rightarrow 1[1] + 2[3]$). Paths and essential paths on affine ADE diagrams may have an arbitrary length.

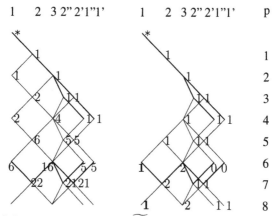

Figure 5. Paths and essential paths for $\widetilde{Tet}$

Paths on the E_6 diagram (or on ADE) may have an arbitrary length but essential paths have a length smaller or equal to 10 (in general, if the Coxeter number is κ, the bound is $\kappa - 2$). The structure of the space of essential paths starting from any vertex a is described by the rectangular matrix E_a (see its definition in the next section). Here is, for example, E_0, with the vertex ordering $0, 1, 2, 5, 4, 3$:

$$
E_0 = \begin{pmatrix}
1 & . & . & . & . & . \\
. & 1 & . & . & . & . \\
. & . & 1 & . & . & . \\
. & . & . & 1 & 1 & . \\
. & . & . & 1 & 1 & 1 \\
. & . & 1 & 1 & . & . \\
. & 1 & 1 & . & . & . \\
. & 1 & . & . & . & 1 \\
. & . & 1 & . & . & . \\
. & . & . & 1 & . & . \\
. & . & . & . & 1 & . \\
\end{pmatrix}
$$

3. A zoo of interesting matrices

3.1. Notations

The representation theory of $SU(h)$ can be encoded by a set of $h-1$ diagrams (with oriented edges and infinitely many vertices) generalizing the A_∞ diagram of $SU(2)$; there is one such oriented diagram for each fundamental representation.

Members of the corresponding A family (call them $\mathcal{A}_k$, k standing for the level) are obtained as truncated♯ A_∞ diagrams; all these quantum diagrams have self-fusion (associative algebra structure with positive integral structure constants). They can be related to a particular category of representations of quantum groups (at roots of unity) but we shall not discuss this aspect here.

The others members of a given generalized $\mathcal{A}, \mathcal{D}, \mathcal{E}$ family are such that the vector space spanned by the set of their r vertices is a module over the associative algebra defined by the corresponding $\mathcal{A}_k$ diagram (the diagram of $\mathcal{A}$ type with the same level). The "rank" (number of vertices) of this diagram $\mathcal{A}_k$ is called s, so that $s = k+1$ for usual ADE diagrams (and the usual notation is A_s) but $s = (k+1)(k+2)/2$ for Di Francesco – Zuber diagrams (generalized ADE diagrams of type $SU(3)$).

3.2. Fusion matrices: the N_i's

Fusion matrices are associated with $\mathcal{A}_k$ diagrams. They are square matrices of dimension $s \times s$ called N_i. Here i is a multi-index referring to a Young frame of $SU(h)$ and the cardinality of the indexing set is also s. When the Young frame refers to a fundamental representation (only one column), this fusion matrix is the adjacency matrix of the corresponding oriented diagram. Other matrices N_i are obtained from the fundamental ones by applying the particular recurrence relation specific to $SU(h)$. Example: In the case of $SU(2)$, each Young frame is an horizontal string of boxes and is characterized by its length (a positive or null integer which also coïncides with the length of essential paths). The matrix N_1 is the adjacency matrix of $\mathcal{A}_k = A_{r=k+1}$ and N_0 is the unit. The recurrence relation (coupling of spins) is $N_{i+1} = N_1 N_i - N_{i-1}$. Matrices N_i have indices (j, k) referring to vertices of $\mathcal{A}_k$. These matrices generate a (commutative) associative algebra isomorphic with the algebra of the given $\mathcal{A}$ diagram. The index i runs from 1 to s.

3.3. Fused adjacency matrices: the F_i's

Let G be a given diagram with given level k. It has r vertices. Call F_0 the unit matrix of dimension $r \times r$. For usual ADE diagrams, each edge carries both orientations and we call F_1 the adjacency matrix; it is symmetric. For generalized diagrams, there is more than one adjacency matrix. Other matrices F_i are then obtained by imposing the same recurrence relation as for fusion matrices. Matrices F_i have indices (a, b) referring to vertices of G; they characterize G as a module over the corresponding $\mathcal{A}$ graph. They are also in one to one correspondence with the minimal central projectors diagonalizing one of the two associative structures of the di-algebra $\mathcal{B}G$, in other words they characterize the corresponding blocks and give their dimensions $d_i = \sum_{a,b}(F_i)_{a,b}$. Here i runs from 1 to s (Young frames) and a, b from 1 to r.

♯ Truncation is made by removing the parts of the diagram with level bigger than k

3.4. Graph matrices: the G_a's

Sometimes the diagram G admits self-fusion. In those cases we call $G_0 \doteq F_0$, $G_1 \doteq F_1$ and more generally G_a the set of matrices (one for each vertex of G) representing faithfully the multiplication of vertices. Warning: with the exception of F_0 and F_1, the matrices F_i and G_a are distinct (in the case of $\mathcal{A}$ diagrams, of course, they are of course the same).

3.5. Essential matrices: the E_a's

For every vertex a of G, we set $(E_a)_{i,b} \doteq (F_i)_{a,b}$. These are rectangular matrices of dimension (s, r). They are sometimes called intertwining matrices. They display "visually" the structure of essential paths emanating from a vertex a on the diagram G.

3.6. Matrices for $Oc(G)$

Since we have a di-algebra $\mathcal{B}G$ we have also a set of matrices S_x which characterize the blocks of the other associative structure (one for each point of the Ocneanu graph). The corresponding dimension d_x is obtained by summing the non zero matrix elements of S_x.

3.7. Toric matrices and generalized toric matrices: the W_x and $W_{x,y}$

We know that $\mathcal{A}_k$ acts on G, therefore $\mathcal{A}_k$ also acts (from both sides) on $Oc(G)$. Example : A_{11} acts on E_6, hence A_{11} also acts on the Ocneanu algebra†† of quantum symmetries $Oc(E_6) = E_6 \otimes_{A_3} E_6$, from the left and from the right. In general $Oc(G)$ is an $\mathcal{A}_k$ bimodule and the action is encoded as follows: $p \, x \, q = \sum_{y \in Oc(G)} (W_{xy})_q^p \ y$ with $x, y \in Oc(G)$ and $p, q \in \mathcal{A}_k$.

The partition function associated with W_{xy} is $Z_{xy} = \bar{\chi} W_{xy} \chi$ where χ_i label a particular basis of the vector space $\mathbb{C}^s$ (they can be interpreted as characters of an affine Lie algebra). W_0 defines the usual partition function (it is modular invariant). W_x defines a twisted partition function with one defect line. W_{xy} defines a twisted partition function with two defect lines. In the case of E_6, one obtains $12 \times 12 = 144$ matrices W_{xy} of dimension 11×11, (many of them happen to be equal), the 12 matrices $W_x \doteq W_{x0}$ and the matrix $W_0 \doteq 0 \otimes_{A_3} 0$ associated with the origin of the Ocneanu graph.

3.8. Modular aspects: S, T and $SL(2, \mathbb{Z})$

Modular invariance of Z_{00} can be proven either by checking that it is invariant when we replace the modular parameter τ by $\tau + 1$ or $-1/\tau$ in the functions χ_p (they are generalized Jacobi's theta functions) or, much more simply, by showing that the matrix W_0 commutes with the generators S and T of the modular group in this representation (they obey the relations $S^4 = (ST)^3 = 1$). Notice that the matrix S in this Hurwitz - Verlinde representation can be obtained as a properly normalized table of eigenvectors for the adjacency matrix of the diagram $\mathcal{A}_k$: this kind of quantum Fourier transform implements the quantum McKay correspondence.

[1] A. Cappelli, C. Itzykson et J.B. Zuber; *The ADE classification of minimal and $A_1^{(1)}$ conformal invariant theories*. Commu. Math. Phys. 13,1 (1987).
[2] R. Coquereaux, Notes on the quantum tetrahedron, Moscow Math. J. vol2, n1, Jan.-March 2002, 1-40, hep-th/0011006.

†† This tensor product is taken above the subalgebra A_3 generated by vertices $0, 4, 3$, so that $a \otimes ub = au \otimes b$ when $u \in A_3$.

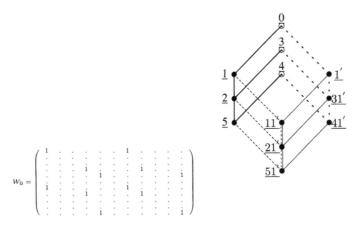

Figure 6. The E_6 Ocneanu graph and its modular invariant

[3] R. Coquereaux; *Classical and quantum polyhedra: A fusion graph algebra point of view* AIP Conference Proceedings 589 (2001), 37th Karpacz Winter School of Theor. Phys., J. Lukierski and J. Rembielinski eds, 181-203, hep-th/0105239.

[4] R. Coquereaux, G. Schieber, Twisted partition functions for ADE boundary conformal field theories and Ocneanu algebras of quantum symmetries, J. of Geom. and Phys. 781 (2002), 1-43, hep-th/0107001.

[5] R. Coquereaux, G. Schieber, Determination of quantum symmetries for higher ADE systems from the modular T matrix. hep-th/0203242.

[6] R. Coquereaux, A. Garcia, R. Trinchero, Ocneanu bi-algebras for ADE diagrams and quantum groupoïds. To appear.

[7] J. Böckenhauer and D. Evans; *Modular invariants, graphs and α induction for nets of subfactors II.* Commun. Math. Phys. 200, 57-103 (1999).

[8] J. Böckenhauer and D. Evans; *Modular invariants from subfactors* D. Evans lectures at Bariloche Summer School, Argentina, Jan 2000. AMS Contemporary Mathematics 294, R. Coquereaux, A. García and R. Trinchero eds.

[9] P. Di Francesco and J.-B. Zuber, in *Recent developments in Conformal Field Theory*, Trieste Conference, 1989, S. Randjbar-Daemi, E. Sezgin and J.-B. Zuber eds, World Scientific, 1990; P. Di Francesco, Int. J. of Mod. Phys. A7, 92, 407.

[10] P. Di Francesco and J.-B. Zuber; $SU(N)$ *lattice integrable models associated with graphs* Nucl. Phys. B338, 602 (1990).

[11] T. Gannon, The Classification of affine su(3) modular invariants, Comm. Math. Phys. 161, 233-263 (1994).

[12] A. Hurwitz; *Uber endliche Gruppen, welche in der Theorie der elliptschen Transzendenten auftreten* Math. Annalen 27, 183-233 (1886).

[13] J. McKay; *Graphs, singularities and finite groups* Proc. Symp. Pure Math. 37, 183 (1980).

[14] A. Ocneanu; *Quantized groups, string algebras and Galois theory for algebras.* Warwick (1987), in Operator algebras and applications, Lond Math Soc Lectures note Series 136, CUP (1988).

[15] A. Ocneanu; *Paths on Coxeter diagrams: from Platonic solids and singularities to minimal models and subfactors.* Talks given at the Centre de Physique Théorique, Luminy, Marseille, 1995.

[16] A. Ocneanu; *Paths on Coxeter diagrams: from Platonic solids and singularities to minimal models and subfactors.* Notes taken by S. Goto, Fields Institute Monographs, AMS 1999, Rajarama Bhat et al, eds.

[17] A. Ocneanu; *Quantum symmetries for $SU(3)$ CFT-models* A. Ocneanu, The Classification of subgroups of quantum SU(N), Lectures at Bariloche Summer School, Argentina, Jan.2000, AMS Contemporary Mathematics 294, R. Coquereaux, A. García and R. Trinchero eds.

[18] V. Pasquier; *Two-dimensional critical systems labelled by Dynkin diagrams* Nucl.Phys. B285, 162 (1987).

[19] P.A. Pearce and Yk Zhou; *Intertwiners and A-D-E lattice models*, IJMPhys B, Vol 7, Nos 20-21 (1993) 3469.

[20] V.B. Petkova and J.B. Zuber; *The many faces of Ocneanu cells*, hep-th/0101151, Nucl. Phys. B 603(2001) 449.

[21] J.B. Zuber; Lectures at Bariloche Summer School, Argentina, Jan 2000. hep-th/0006151. AMS Contemporary Mathematics 294, R. Coquereaux, A. García and R. Trinchero eds.

Inst. Phys. Conf. Ser. No 173: Plenary Sessions
Paper presented at 24th Int. Coll. Group Theoretical Methods in Physics, Paris, France, 15–20 July 2002
©2003 IOP Publishing Ltd

73

A review of Berezin - and other - quantization methods

M. Engliš ‡

MÚ AV ČR, Žitná 25, 11567 Prague 1, Czech Republic

Abstract. This is a survey (somewhat biased in favour of results obtained by the author) of recent progress, new ideas and some open problems in several approaches to quantization from a mathematician's viewpoint. In particular, we discuss current developments in Berezin, Berezin-Toeplitz and prime quantization, and indicate related questions from the theory of Fourier integral operators, several complex variables, group theory and complex geometry.

1. Introduction

The original concept of quantization on $\mathbf{R}^n$, going back to Dirac, Weyl, and von Neumann, consists in assigning to the classical observables (= real-valued functions f of $(p,q) \in \mathbf{R}^n \times \mathbf{R}^n$) the quantum observables (= self-adjoint operators $\mathrm{Op}(f)$ on a separable Hilbert space) in such a way that:

(Q1) *(linearity)* the correspondence $f \mapsto \mathrm{Op}(f)$ is linear;

(Q2) *(von Neumann rule)* for any C^∞ function $\phi : \mathbf{R} \to \mathbf{R}$, $\mathrm{Op}(\phi \circ f) = \phi(\mathrm{Op}(f))$;

   (in particular: $\mathrm{Op}(1) = I$)

(Q3) *(Poisson brackets)* $[\mathrm{Op}(f), \mathrm{Op}(g)] = -\frac{ih}{2\pi} \mathrm{Op}(\{f,g\})$, where

$$\{f,g\} = \sum_{j=1}^{n} \left(\frac{\partial f}{\partial p_j} \frac{\partial g}{\partial q_j} - \frac{\partial f}{\partial q_j} \frac{\partial g}{\partial p_j} \right)$$

is the Poisson bracket of f and g; in particular:

$$[\mathrm{Op}(p_j), \mathrm{Op}(p_k)] = [\mathrm{Op}(q_j), \mathrm{Op}(q_k)] = 0, \qquad [\mathrm{Op}(p_j), \mathrm{Op}(q_k)] = -\frac{ih}{2\pi} \delta_{jk} I ; \quad (1)$$

(Q4) *(Schrödinger representation)* the operators corresponding to the coordinate functions p_j, q_j $(j = 1, \ldots, n)$ are unitarily equivalent to the operators

$$\mathrm{Op}(q_j) : f(q) \mapsto q_j f(q), \qquad \mathrm{Op}(p_j) : f(q) \mapsto -\frac{ih}{2\pi} \frac{\partial f(q)}{\partial q_j}$$

on $L^2(\mathbf{R}^n)$.

By the Stone-von Neumann theorem, if (1) is satisfied then (Q4) is equivalent to the condition that $\mathrm{Op}(p_j), \mathrm{Op}(q_j)$ act irreducibly (i.e. have no common invariant subspace).

   Unfortunately, it is well known that these axioms are not consistent. First of all, using (Q1), (Q2) and (1) it is possible to express Q_f for $f(p,q) = p_1^2 q_1^2 = (p_1 q_1)^2$ in two ways with two different results; hence, (Q1)+(Q2)+(Q4), as well as (Q1)+(Q2)+(Q3), lead to

‡ The author was supported by GA ČR grant no. 201/00/0208 and GA AV ČR grant no. A1019005.

a contradiction. Second, a theorem of Groenewold [16] and van Hove [18] asserts that (Q1)+(Q3)+(Q4) leads to a contradiction as well. Finally, it was shown in [12] that also (Q2)+(Q3)+(Q4) yields a contradiction. Thus even <u>any three</u> of the axioms (Q1)–(Q4) are inconsistent. (It can be shown that <u>any two</u> of these axioms, with the exception of (Q1)+(Q3) — which are fulfilled for the prequantization representation of van Hove, see below — and, possibly, (Q2)+(Q4) and (Q3)+(Q4), are either inconsistent or lead to something trivial. See the introduction in [1] for the details.)

The story is, of course, similar in the case of quantization on a general symplectic manifold (M, ω). The latter consists, again, in seeking an assignment $f \mapsto \mathrm{Op}(f)$ satisfying the same axioms (Q1)–(Q3) (with the Poisson brackets given now by

$$\{f, g\} = \omega(X_f, X_g),$$

where X_f, the Hamiltonian vector field of $f \in C^\infty(M)$, is defined by $\omega(\cdot, X_f) = df(\cdot))$ and a suitable substitute of the "irreducibility" axiom (Q4): for instance, one can require that the procedure be *functorial* in the sense that for any symplectomorphism $\phi : (M^{(1)}, \omega^{(1)}) \to (M^{(2)}, \omega^{(2)})$, there should exist a unitary operator U_ϕ such that

$$\mathrm{Op}^{(1)}(f \circ \phi) = U_\phi^* \mathrm{Op}^{(2)}(f) U_\phi, \qquad \forall f,$$

and that for $(M, \omega) = \mathbf{R}^{2n}$ with the standard symplectic form we should recover the operators $\mathrm{Op}(q_j)$, $\mathrm{Op}(p_j)$ given by the Schrödinger representation (Q4).

Again, of course, this set of axioms is inconsistent, and the furthest one can get is to fulfill the two axioms (Q1)+(Q3) by the so-called <u>prequantization</u> representation of van Hove:

$$\mathrm{Op}(f) = -\tfrac{ih}{2\pi} \nabla_{X_f} + f,$$

operating on sections of a certain line bundle over M, which exists if and only if $h^{-1}[\omega] \in H^2(M, \mathbf{R})$ is an integral cohomology class.

There are two traditional approaches to handle this disappointing situation. In both of them, one first of all discards the von Neumann rule, except for $\phi = \mathbf{1}$, i.e.

$$\mathrm{Op}(\mathbf{1}) = I.$$

The first approach is to insist on the linearity (Q1), the Poisson brackets (Q3), and the Schrödinger representation (Q4), but restrict the space of quantizable observables (the domain of the map $f \mapsto \mathrm{Op}(f)$). For instance, for the quantization on $\mathbf{R}^n$, if we allow only functions f at most linear in the p_j variables, and let $\mathrm{Op}(f)$ be the restriction of the prequantization operator

$$\psi \longmapsto -\frac{ih}{2\pi} X_f \psi - \theta(X_f)\psi + f\psi, \qquad \psi = \psi(p, q),$$

to the functions ψ depending only on q, then everything works. For general symplectic manifolds, restricting to "functions depending on only half of the variables" requires the use of polarizations. This approach is the <u>geometric quantization</u> of Kostant and Souriau.

The second approach is to insist only on the linearity (Q1) and the Schrödinger representation (Q4), but relax (Q3) to hold only asymptotically as $h \to 0$:

$$[\mathrm{Op}^{(h)}(f), \mathrm{Op}^{(h)}(g)] = -\frac{ih}{2\pi} \mathrm{Op}^{(h)}(\{f, g\}) + O(h^2). \tag{2}$$

The simplest example of this is the celebrated Weyl calculus, $f \mapsto W_f$ on $L^2(\mathbf{R}^n)$, where W_f is given by the oscillatory integral

$$W_f \psi(x) = h^{-n} \iint f\left(p, \frac{x+y}{2}\right) e^{2\pi i(x-y)p/h} \psi(y)\, dy\, dp.$$

It can be shown that for nice f and g (for instance — if $f, g \in \mathscr{S}(\mathbf{R}^{2n})$), $W_f W_g = W_{f \natural g}$ where $f \natural g$, called the *Moyal* product of f and g, has an asymptotic expansion

$$f \natural g = \sum_{j=0}^{\infty} h^j \rho_j(f, g) \quad \text{as } h \to 0,$$

with $\rho_0(f, g) = fg$ and $\rho_1(f, g) - \rho_1(g, f) = -\frac{i}{2\pi}\{f, g\}$. Hence

$$f \natural g - g \natural f = -\frac{ih}{2\pi}\{f, g\} + O(h^2)$$

and so (2) holds for $\mathrm{Op}(f) = W_f$.

It turns out that, in fact, it is often not really necessary to have the operators $\mathrm{Op}(f)$, but suffices to have the noncommutative product like $\natural$. This is the idea underlying the deformation quantization.

2. Deformation quantization

Let $C^\infty(M)[[h]]$ denote the ring of all formal power series in h over $C^\infty(M)$. A star-product is an associative $\mathbf{C}[[h]]$-bilinear mapping $*$ such that $\forall f, g \in C^\infty(M)$,

$$f * g = \sum_{j=0}^{\infty} h^j C_j(f, g),$$

where the bilinear operators C_j satisfy

$$C_0(f, g) = fg, \qquad C_1(f, g) - C_1(g, f) = -\frac{i}{2\pi}\{f, g\},$$
$$C_j(f, 1) = C_j(1, f) = 0 \qquad \forall j \geq 1.$$

Two star-products $*, *'$ are called equivalent if there exists a formal power series of operators $M = I + M_1 h + M_2 h^2 + \dots$ such that $Mf *' Mg = M(f * g)$. A star-product is called differential if C_j are differential operators; and if M has complex structure, a star-product is said to have the property of separation of variables if $f * g = fg$ whenever either f or $\bar{g}$ is holomorphic. See the recent survey of Gutt [17] for more information on star-products and deformation quantization.

On the basis of a star-product, one can define the spectrum of $f \in C^\infty(M)[[h]]$ using the so-called $*$-exponential, and use it as the substitute for the ordinary spectrum of the quantum operators $\mathrm{Op}(f)$ in the conventional quantum mechanics. It turns out, however, that equivalent (hence, mathematically isomorphic) star-products may lead to different spectra for the same observable [8]. Hence, it is of interest to have some "natural" or "distinguished" star-products (for instance — defined in terms of some natural geometric objects associated to the manifold, etc.). Also, the star product is defined to be just a formal power series — it is of interest to know if there are some which would really converge for a given value of the Planck constant h.

For Kähler manifolds, examples (of both resp. of the former) are furnished by the Berezin and the Berezin-Toeplitz star-products, which we now discuss.

3. Bergman spaces and the Berezin(-Toeplitz) quantization

Consider a bounded domain $\Omega \subset \mathbf{C}^n$, a weight function $\rho > 0$ on Ω (assumed continuous and integrable), and denote by $L^2_{\text{hol}}(\Omega, \rho)$ the subspace of holomorphic functions in $L^2(\Omega, \rho)$ (called the <u>weighted Bergman space</u>). Then the point evaluations $f \mapsto f(y)$ are continuous on L^2_{hol}, hence there exists for each $y \in \Omega$ a function $K_y \in L^2_{\text{hol}}$ such that

$$f(y) = \langle f, K_y \rangle \qquad \forall f \in L^2_{\text{hol}}.$$

The function $K(x, y) := K_y(x) = \langle K_y, K_x \rangle$, which is holomorphic in x and $\bar{y}$, is called the <u>reproducing kernel</u> of $L^2_{\text{hol}}(\Omega, \rho)$. One has $K(x, y) = \overline{K(y, x)}$ and $K(x, x) > 0$.

For any (bounded linear) operator A on L^2_{hol},

$$Af(x) = \langle Af, K_x \rangle = \langle f, A^* K_x \rangle = \int_\Omega f(y) \langle AK_y, K_x \rangle \rho(y)\, dy.$$

Hence A is uniquely determined by the function

$$\widetilde{A}(x, y) := \frac{\langle AK_y, K_x \rangle}{\langle K_y, K_x \rangle},$$

which (being holomorphic in $x, \bar{y}$) is in turn uniquely determined by its values on the diagonal

$$\widetilde{A}(x) := \widetilde{A}(x, x).$$

The function $\widetilde{A}(x)$ is called the <u>symbol</u> of the operator A. It is a bounded C^ω (= real-analytic) function on Ω, depending linearly on A; further, $\widetilde{A^*} = \overline{\widetilde{A}}$ and $\widetilde{I} = 1$.

For $\phi \in L^\infty(\Omega)$, the <u>Toeplitz operator</u> T_ϕ is defined on L^2_{hol} by $T_\phi f := P(\phi f)$ where P is the orthogonal projection of L^2 onto L^2_{hol}. Clearly, T_ϕ depends linearly on ϕ, $T_\phi^* = T_{\bar{\phi}}$, $T_1 = I$, $T_{\phi f} = T_\phi T_f$ and $T_{\bar{f}\phi} = T_{\bar{f}} T_\phi$ is f is holomorphic (but $T_\phi T_\psi \neq T_\psi T_\phi$ and $\neq T_{\phi\psi}$ in general), and $\|T_\phi\| \leq \|\phi\|_\infty$.

The <u>Berezin transform</u> is the composition $B\phi := \widetilde{T_\phi}$, i.e. for $\phi \in L^\infty(\Omega)$:

$$B\phi(x) = \frac{\langle \phi K_x, K_x \rangle}{\langle K_x, K_x \rangle} = \int_\Omega \phi(y) \frac{|K(x, y)|^2}{K(x, x)} \rho(y)\, dy.$$

Again, B is linear, $Bf = f$ if either f or $\bar{f}$ is holomorphic, $B1 = 1$, and $\|B\phi\|_\infty \leq \|\phi\|_\infty$.

Now denote

$$\mathscr{A}_\rho := \{\widetilde{A} : A \text{ a bounded linear operator on } L^2_{\text{hol}}(\Omega, \rho)\}$$

(this is a space of bounded real-analytic functions on Ω) and define

$$\widetilde{A} *_\rho \widetilde{B} := \widetilde{AB}.$$

(Note that this makes sense since $A \leftrightarrow \widetilde{A}$ is 1-to-1.) Then $(\mathscr{A}_\rho, *_\rho)$ is a (noncommutative, associative) algebra of smooth functions on Ω. (It is isomorphic to the algebra of bounded linear operators on $L^2_{\text{hol}}(\Omega, \rho)$.)

I. The idea of the <u>Berezin quantization</u> is to find a family $\rho = \rho_h$ of weights depending on h such that the corresponding products $*_{\rho_h} \equiv *_h$ yield a star product.

More precisely, consider the direct sum $(\mathbb{A}, *)$ of all the algebras $(\mathscr{A}_{\rho_h}, *_h)$, and let $\mathbb{A}_0$ be a linear subset of $\mathbb{A}$ such that each $f = \{f_h(x)\}_h \in \mathbb{A}_0$ has an asymptotic expansion

$$f_h(x) = \sum_{j=0}^{\infty} h^j f_j(x) \qquad \text{as } h \to 0$$

with some $f_j \in C^\omega(\Omega)$. Suppose that we can show that

- $\mathbb{A}_0$ is <u>total</u> in the sense that for any $m > 0$, $x \in \Omega$ and $F_0, F_1, \ldots, F_m$ in $C^\omega(\Omega)$ there exists $f \in \mathbb{A}_0$ such that $\partial^\alpha f_j(x) = \partial^\alpha F_j(x)$ for all multiindices $|\alpha| \le m$ and $j = 0, 1, \ldots, m$;
- for any $f, g \in \mathbb{A}_0$, one has $f * g \in \mathbb{A}_0$ and

$$(f * g)_h(x) = \sum_{i,j,k \ge 0} h^{i+j+k} C_k(f_i, g_j)(x) \qquad \text{as } h \to 0,$$

where $C_k : C^\omega(\Omega) \times C^\omega(\Omega) \to C^\omega(\Omega)$ are bilinear differential operators such that

$$C_0(\phi, \psi) = \phi\psi, \qquad C_1(\phi, \psi) - C_1(\psi, \phi) = -\frac{i}{2\pi}\{\phi, \psi\}.$$

Then the recipe

$$\left(\sum_{i \ge 0} f_i h^i \right) * \left(\sum_{j \ge 0} g_j h^j \right) := \left(\sum_{i,j,k \ge 0} C_k(f_i, g_j) h^{i+j+k} \right)$$

defines a star-product on Ω with separation of variables. (Berezin [2], Engliš [11].) Moreover, this time it is not just a formal star product, since for functions in the total set $\mathbb{A}_0$ it really exists as an element of $C^\infty(\Omega)$.

II. There is the following <u>equivalent approach</u> to the Berezin quantization, due to Karabegov [19]. Let B_h be the Berezin transform corresponding to $L^2(\Omega, \rho_h)$. Assume that we can establish the asymptotic expansion

$$B_h = Q_0 + Q_1 h + Q_2 h^2 + \ldots \qquad \text{as } h \to 0,$$

where Q_j are differential operators such that

$$Q_0 = I, \qquad Q_1 = \Delta \quad \text{(the Laplace-Beltrami operator)}.$$

Denote by $c_{j\alpha\beta}$ the coefficients of Q_j:

$$Q_j f = \sum_{\alpha, \beta \text{ multiindices}} c_{j\alpha\beta} \, \partial^\alpha \overline{\partial}^\beta f,$$

and set

$$C_j(f, g) := \sum_{\alpha, \beta} c_{j\alpha\beta} \, (\overline{\partial}^\beta f)(\partial^\alpha g).$$

The $C_j(\cdot, \cdot)$ define a differential star-product with separation of variables (the same one as in the preceding paragraph).

III. The idea of the <u>Berezin-Toeplitz quantization</u> is, again, to try to find a family of weights ρ_h such that the corresponding Toeplitz operators $T_\phi^{(h)}$ now satisfy:

$$T_f^{(h)} T_g^{(h)} = \sum_{j=0}^{\infty} h^j T_{C_j(f,g)}^{(h)} \qquad \text{as } h \to 0, \qquad \forall f, g \in C^\infty(\Omega), \tag{3}$$

where $C_j(\cdot,\cdot)$ are bidifferential operators such that

$$C_0(f,g) = fg, \qquad C_1(f,g) - C_1(g,f) = -\tfrac{i}{2\pi}\{f,g\}. \tag{4}$$

Then C_j define a star-product with separation of variables.

Here (3) is to be understood in the sense of operator norms, i.e.

$$\left\| T_f^{(h)} T_g^{(h)} - \sum_{j=1}^{m} h^j T_{C_j}^{(h)} \right\| = O(h^{m+1}), \qquad \forall m = 1,2,\ldots.$$

Remark. Note that the three procedures I, II, III above use, in turn, only the products $*_h$, the Berezin transforms B_h, and the Toeplitz operators $T^{(h)}$, respectively. $\square$

Of course, the first question in implementing any of the above three procedures is how to choose the weights ρ_h. For general symplectic manifolds, the answer seems to be unknown; however, for Kähler manifolds — that is, for complex manifolds for which $\omega = \partial\bar{\partial}\Phi$ (locally) for some (real-valued) function Φ — it can be shown by group-invariance considerations (see [1]) that the only sensible choice is

$$\rho_h(x)\,dx = e^{c\Phi(x)}\,\omega^n(x),$$

where $c = c(h)$ depends only on h.

We have thus arrived at the final recipe for the Berezin (and the Berezin-Toeplitz) quantization on Kähler manifolds: namely, choose a potential Φ for ω (assumed, for simplicity, to exist globally); take the Bergman space $L^2_{\mathrm{hol}}(\Omega, e^{-\nu\Phi}\omega^n)$ ($\nu \in \mathbf{R}$); let $K_\nu(x,y)$ be its reproducing kernel, B_ν its Berezin transform, and $T_\phi^{(\nu)}$ its Toeplitz operators; and see if any of the three approaches I, II, III above work.

The answer is: yes they do, with $\nu = 1/h$.

Theorem 1 (Berezin quantization) [11] *Assume that Ω is a strictly pseudoconvex domain in $\mathbf{C}^n$, with real-analytic boundary and a Kähler form ω whose potential Φ is such that $e^{-\Phi} \asymp \mathrm{dist}(\cdot,\partial\Omega)$ near the boundary. Then for $\nu = 1/h$,*

$$K_\nu(x,y) = \frac{1}{\pi^n h^n} \sum_{j=0}^{\infty} b_j(x,y)\,h^j \qquad \text{as } h \to 0$$

for (x,y) close to the diagonal; and

$$B_\nu = I + h\Delta + h^2 Q_2 + h^3 Q_3 + \ldots,$$

where Q_j are differential operators whose coefficients involve only the curvature tensor of the Riemannian metric induced by ω, and its covariant derivatives.

Remarks. 1) The assumptions can be weakened (weakly pseudoconvex, smooth boundary, Φ such that $\{(x,t) \in \Omega \times \mathbf{C} : |t|^2 < e^{-\Phi(x)}\}$ is a domain of finite type), see [11], [10]. There are still, however, some interesting situations which the theorem misses (e.g. the Cheng-Yau metric on Ω.)

2) Pseudoconvexity is a kind of "holomorphic completeness": Ω is not pseudoconvex if $\exists \widetilde{\Omega} \supsetneq \Omega$ such that any $f \in \mathrm{Hol}(\Omega)$ extends holomorphically to $\widetilde{\Omega}$.

3) If the potential Φ exists only locally, things still work, but one has to pass to sections of a certain line bundle (which exists if and only if $\nu[\omega]$ is an integral cohomology class).

4) The simplest situation (not covered by the last theorem) is that of bounded symmetric domains in $\mathbf{C}^n$, which is the original situation considered by Berezin [2]. $\square$

Theorem 2 (Berezin-Toeplitz quantization) *Let* Ω *be compact* [3], *or a planar domain with the Poincaré Kähler structure* [20], *or a bounded symmetric domain with the invariant Kähler form* [5], *or* $\mathbf{C}^n$ [9] [4], *or as in the preceding theorem* [11]. *Then with* $\nu = 1/h$, *the Berezin-Toeplitz quantization works.*

Remark. For Ω compact, we again need to pass to sections of line bundles. $\quad\square$

The proof for the Berezin case uses the asymptotics of $K(x,y)$ obtained by Fefferman [15] and Boutet de Monvel and Sjöstrand [7]; the one for the Berezin-Toeplitz case uses the generalized Toeplitz operators of Boutet de Monvel and Guillemin [6]. Both depend heavily on the theory of Fourier integral operators, or on the results on the $\bar{\partial}$-operator. In particular, the various restrictions to smoothly bounded (or compact) domains above are dictated by the fact that there is no satisfactory theory of FIOs (or $\bar{\partial}$-operator) on more general domains, hence, so far, ad hoc methods had to be applied in such cases.

4. Possible extensions of the Berezin/Berezin-Toeplitz quantization

We have seen that both the Berezin and the Berezin-Toeplitz quantizations originate from the Bergman spaces (spaces of square-integrable holomorphic functions); hence, in particular, they require complex structure, and work only on Kähler manifolds. It seems natural to ask if other function spaces with reproducing kernels could be used instead.

For instance, one such space is the space $L^2_{\text{harm}}(\Omega, \rho)$ of the <u>harmonic</u> functions in $L^2(\Omega, \rho)$, or more generally, the space $L^2_A(\Omega, \rho)$ of the functions in $L^2(\Omega, \rho)$ annihilated by a given hypoelliptic partial differential operator $\mathbf{A}$. Another example are the Sobolev spaces $H^s(\Omega)$: by the Sobolev theorem, if $s > \frac{n}{2}$ then $H^s(\Omega)$ has a reproducing kernel (however, since the scalar product is no longer defined by integration, it is not possible to define Toeplitz operators and the Berezin transform). One can also combine the last two ideas and consider the subspaces H^s_A of the functions in $H^s(\Omega)$ annihilated by a given (hypo)elliptic differential operator $\mathbf{A}$ (thus, for instance, the Sobolev spaces of holomorphic or harmonic functions). These spaces have reproducing kernels even for all real s [21].

Can one still make things work in these settings?

Another way of extending the Berezin-Toeplitz quantization is to look, quite generally, at "operator calculi" of the form

$$f \mapsto \mathbf{Q}_f := \int_{\Omega} f(y)\, \mathbf{Q}_y\, dy$$

with some field $y \mapsto \mathbf{Q}_y$ of operators ("quantizers") on Ω, and try to establish an asymptotic formula

$$\mathbf{Q}_f \mathbf{Q}_g \simeq \sum_{j=0}^{\infty} h^j \mathbf{Q}_{C_j(f,g)} \qquad \text{as } h \to 0.$$

(The Toeplitz operators correspond to $\mathbf{Q}_y = \|K_y\|^{-2} \langle \cdot, K_y \rangle K_y$.)

Can one get an interesting extension of the Berezin-Toeplitz quantization by using other operator calculi then the Toeplitz operators?

Due to lack of space we only sketch the answers to the questions just posed; the details can be found in the forthcoming papers [13] and [14].

First of all, neither the Berezin nor the Berezin-Toeplitz quantization based on the harmonic Bergman space work. (For the former, the reason is that the correspondence $A \leftrightarrow \tilde{A}$ is not 1-to-1 anymore.) The same is true for the Sobolev spaces $H^s(\Omega)$, and for their harmonic subspaces H^s_{harm} (and, hence, very probably also for L^2_A and H^s_A with more fancy hypoelliptic operators **A**). This means that the only ones of our suggested extensions that have a chance to work are, essentially, only the Berezin and the Berezin-Toeplitz quantizations based on the holomorphic Sobolev spaces, and the Berezin-Toeplitz quantization based on operator calculi. It turns out that these, at least (the last — in the presence of a transitive Lie group action) indeed lead to some nontrivial results. See [13] and [14].

References

[1] S.T. Ali, M. Engliš: *Quantization methods: a rapid survey,* preprint (2002).

[2] F.A. Berezin: *Quantization,* Math. USSR Izvestiya **8** (1974), 1109–1163.

[3] M. Bordemann, E. Meinrenken, M. Schlichenmaier: *Toeplitz quantization of Kähler manifolds and $\mathfrak{gl}(N)$, $N \to \infty$ limits,* Comm. Math. Phys. **165** (1994), 281–296.

[4] D. Borthwick: *Microlocal techniques for semiclassical problems in geometric quantization,* Perspectives on quantization (South Hadley, MA, 1996), pp. 23–37, Contemp. Math. **214**, AMS, Providence, 1998.

[5] D. Borthwick, A. Lesniewski, H. Upmeier: *Nonperturbative deformation quantization of Cartan domains,* J. Funct. Anal. **113** (1993), 153–176.

[6] L. Boutet de Monvel, V. Guillemin: *The spectral theory of Toeplitz operators,* Ann. Math. Studies **99**, Princeton University Press, Princeton, 1981.

[7] L. Boutet de Monvel, J. Sjöstrand: *Sur la singularité des noyaux de Bergman et de Szegö,* Astérisque **34–35** (1976), 123–164.

[8] M. Cahen, M. Flato, S. Gutt, D. Sternheimer: *Do different deformations lead to the same spectrum?,* J. Geom. Phys. **2** (1985), 35–48.

[9] L.A. Coburn: *Deformation estimates for the Berezin-Toeplitz quantization,* Comm. Math. Phys. **149** (1992), 415–424; *Berezin-Toeplitz quantization,* Algebraic methods in operator theory, pp. 101–108, Birkhäuser, Boston, 1994.

[10] M. Engliš: *Pseudolocal estimates for $\bar\partial$ on general pseudoconvex domains,* Indiana Univ. Math. J. **50** (2001), 1593–1607.

[11] M. Engliš: *Weighted Bergman kernels and quantization,* Comm. Math. Phys. **227** (2002), 211–241.

[12] M. Engliš: *A no-go theorem for nonlinear canonical quantization,* Comm. Theor. Phys. **37** (2002), 287–288.

[13] M. Engliš: *Berezin and Berezin-Toeplitz quantizations for general spaces of holomorphic functions,* in preparation.

[14] M. Engliš: *Berezin-Toeplitz quantization and invariant symbolic calculi,* in preparation.

[15] C. Fefferman: *The Bergman kernel and biholomorphic mappings of pseudoconvex domains,* Inv. Math. **26** (1974), 1–65.

[16] H.J. Groenewold: *On the principles of elementary quantum mechanics,* Physica **12** (1946), 405–460.

[17] S. Gutt: *Variations on deformation quantization,* preprint math.QA/0003107.

[18] L. van Hove: *Sur certaines représentations unitaires d'un groupe infini de transformations,* Mem. Acad. Roy. de Belgique, Classe des Sci. **26** (1951), no. 6.

[19] A.V. Karabegov: *Deformation quantization with separation of variables on a Kähler manifold,* Comm. Math. Phys. **180** (1996), 745–755.

[20] S. Klimek, A. Lesniewski: *Quantum Riemann surfaces, I: The unit disc,* Comm. Math. Phys. **146** (1992), 103–122; *II: The discrete series,* Lett. Math. Phys. **24** (1992), 125–139; *III: The exceptional cases,* Lett. Math. Phys. **32** (1994), 45–61.

[21] J.-L. Lions: *Remarks on reproducing kernels of some function spaces,* Function spaces, interpolation theory, and related topics (Lund, 2000) (A. Kufner, M. Cwikel, M. Engliš, L.-E. Persson, and G. Sparr, editors), pp. 49–59, Walter de Gruyter, Berlin, 2002.

Inst. Phys. Conf. Ser. No 173: Plenary Sessions
Paper presented at 24th Int. Coll. Group Theoretical Methods in Physics, Paris, France, 15–20 July 2002
©*2003 IOP Publishing Ltd*

Symmetry and graph theory as tools for the treatment of π-systems

P.W. Fowler

School of Chemistry, University of Exeter, Stocker Road, Exeter EX4 4QD, UK

Abstract. The intertwined roles of graph theory and group theory in the theoretical treatment of π systems are highlighted with reference to three applications in molecular chemical physics. In the first, eigenvectors of the adjacency matrix of a fullerene are used to define the embedding of the graph in three dimensions, from which point-group symmetry assignment and prediction of NMR spectra, crucial for experimental assignment of isomers, follow. In the second, a simple Ansatz is generalised to pick out the most efficient symmetry-breaking modes of small-gap planar and non-planar π systems subject to second-order Jahn-Teller effects. In the third, analysis of the symmetries of products of occupied and unoccupied frontier orbitals is used to predict the existence and sense of the ring currents of cyclic π systems that are conventionally used to define their aromaticity or antiaromaticity.

1. Introduction

Unsaturated systems offer a natural arena for the application of point-group symmetry in chemistry. In the classical planar π systems exemplified by benzene, the planarity of the nuclear framework allows rigorous symmetry separation between the mobile π electrons that define many of the chemical and physical properties of the molecule, and occupy orbitals that are antisymmetric with respect to reflection in the plane, and the more sluggish σ electrons that provide the background potential against which the π electrons move, and occupy orbitals that have maximal density in the plane. The subject has been reinvigorated by the discovery of curved π systems such as the fullerenes and nanotubes where σ/π separation is no longer a rigorous symmetry property, but still gives a useful first picture of electronic structure and other properties of the system. The present contribution reviews briefly three aspects of the theory of curved and flat π systems in which group theory and graph theory combine to give informative descriptions of the electronic and geometric structure. More details are given in the references.

2. Fullerenes and topological coordinates

Fullerenes are discrete molecular forms of elemental carbon. The prototype is C_{60}, the discovery of which [1] was recognised by the award of the 1996 Nobel Prize in Chemistry. The general member of the class is a molecule C_n in which n carbon atoms are arranged on a pseudospherical surface, each joined by localised two-electron 'σ' bonds to three neighbours, and in which the set of such bonds forms a closed cage of exactly 12 pentagonal and $n/2 - 10$ hexagonal rings. In the simplest qualitative picture of the chemical bonding in these molecules, the remaining valence of each carbon atom is used to form a global 'π' system over the surface of the cage, each atom contributing one radial p orbital to a system of n molecular orbitals. So far, about a dozen structures of fullerenes prepared by the Krätschmer-Huffman [2] synthesis are fully characterised. Symmetry has played an important role in their characterisation in that the main experimental evidence of structure, the ^{13}C NMR signature, is a direct reflection of the number and size of the orbits (sets of equivalent atoms) of the point group of the molecular structure.

Symmetry and graph theory also play major roles in the general theoretical account of the electronic and geometric structures of fullerenes [3]. The deceptively simple recipe of {12 faces pentagonal *plus* all other faces hexagonal} is compatible with multiple isomeric possibilities, starting with the combinatorially unique dodecahedral C_{20}, and soon reaching thousands of isomers (e.g 1812 for C_{60}, 8149 for C_{70}), all representing local minima on the potential energy surface for n carbon atoms. The restriction to exactly 12 pentagons follows from Euler's theorem $v + e = f + 2$ connecting the numbers of *vertices*, *edges* and *faces* of a spherical polyhedron without holes or handles. A trivalent polyhedron with $v = n$ vertices has $e = 3n/2$ edges, and hence $n/2 + 2$ faces; reconciliation of the edge count determined from the sum of vertex degrees with that determined from the sum of face sizes shows that any trivalent polyhedron has a constant sum $\sum_r (6 - r)f_r = 12$, where f_r is the number of faces of size r. The constant 12 expresses the intrinsic curvature of the sphere, and would take different values for a carbon network on a surface of higher genus. The simplest solution for a vector of face sizes on the sphere that is compatible with the carbon atom's predilection for obtuse bond angles is $f_5 = 12, f_6 = n/2 - 10 = h$. At least one fullerene structure with h hexagonal faces is possible for $n = 20 + 2h$ vertices for every h except $h = 1$. The experimentally observed isomer of C_{60} emerges as the smallest fullerene where the pentagonal faces are disjoint, as required on grounds of curvature and π stability and the second most abundant fullerene, C_{70}, follows as the second smallest isolated-pentagon structure.

Enumeration and construction of the isomers compatible with the combinatorial definition of the fullerene polyhedra is a prerequisite for systematic theoretical study, and it has been carried out in at least two ways in the fullerene literature. One powerful method, on which the IUPAC nomenclature for fullerenes [4] is based, and which is easily implemented, is the face-spiral algorithm [5]. Fullerenes are represented as one-dimensional sequences of pairwise adjacent faces, obtained by a notional unwrapping or 'peeling' of the pseudospherical polyhedron into a single, continuous strip, exactly as one might hope to peel an orange. As the face sizes in a fullerene are all either 5 or 6, there is the possibility of coding each isomer by a single $n/2 + 2$-digit number, and hence ordering the set of isomers lexicographically and non-redundantly. Computation of codes for all fullerene isomers of a given size by this method is an expensive but once-and-for-all task [3]. At very large vertex count ($n \geq 380$), well beyond the present reach of chemical characterisation, there are known problems with a minority of unspirallable fullerenes [3,6], and there are now available some faster modes of construction that avoid these problems and capture all isomers [7], but the transparency of the spiral algorithm and its in-built ordering criterion have made it the basis of much of the theoretical work on the chemistry and physics of fullerenes.

A problem related to the symmetry theme of the present contribution is: given the spiral code (or any other graph-theoretical representation) of some fullerene, how do we assign a symmetry and realistic three-dimensional structure to this candidate for realisation as a molecule? An answer is given by the 'topological coordinates' [8], based on the adjacency matrix of the fullerene graph. From the face spiral, it is straightforward to determine the list of connections between vertices of the molecular graph. From this list, the adjacency matrix of the graph follows. The adjacency matrix $\mathbf{A}$ has entries

$A_{ij} = 1$ for pairs i, j of atoms joined by a σ bond, and $A_{ij} = 0$ otherwise.

Eigenvectors and eigenvalues can be calculated for this square symmetric matrix. The eigenvectors span the permutation representation of the vertices of the graph in the point symmetry group G of graph (i.e. in the 3D embedding that we intend to determine). As fullerene graphs are trivalent, their adjacency eigenvalues lie in the range $+3 \geq \lambda^A \geq -3$, and as they are non-bipartite graphs, the limit -3 is not attained. (The vertices of a bipartite graph fall into two disjoint sets such that each vertex of each set has nearest neighbours only in the other. Bipartite graphs are either acyclic or have only even cycles.) For fullerenes, therefore, the adjacency eigenvalues in non-increasing order are

$$+3 = \lambda_1^A > \lambda_2^A \geq \lambda_3^A \geq \ldots \geq \lambda_{n-2}^A \geq \lambda_n^A > -3. \tag{1}$$

The eigenvector with the unique maximum eigenvalue $+3$ has equal coefficients $1/\sqrt{n}$ on all vertices, and can be seen as a uniform spherical S wave spread over the fullerene, carrying no geometric information. This vector spans Γ_0, the totally symmetric representation of the point group G.

The spherical-wave analogy suggests that the other eigenvectors of $\mathbf{A}$ should resemble waves that are cut by more and more nodal surfaces as λ^A decreases. In particular it should be possible to find three independent vectors with the overall appearance of P spherical harmonics. Once such vectors are found, since the three P functions on the unit sphere are $\sim x$, y and z, the set of their coefficients on each fullerene vertex can be taken in scaled form as Cartesian coordinates [8]. In this way, the graph, a purely combinatorial object, has been shown to define its own embedding in 3D space.

The analogy has a graph-theoretical foundation, since for a regular graph (one where all vertex degrees are equal) the eigenvalue equation

$$\mathbf{A} | \lambda_i^A > = \lambda_i^A | \lambda_i^A > \qquad (i = 1, n) \tag{2}$$

shares its set of eigenvectors with the discrete Laplace equation defined on the same graph

$$\mathbf{L} | \lambda_i^L > = (\mathbf{D} - \mathbf{A}) | \lambda_i^L > = \lambda_i^L | \lambda_i^L >, \qquad (i = 1, n) \tag{3}$$

where $\mathbf{D}$ is the $n \times n$ diagonal matrix of vertex degrees, equal for a fullerene to three times the identity matrix $\mathbf{I}$. For fullerenes, the eigenvalues and eigenspaces of the two matrices are related by

$$\lambda_i^L = 3 - \lambda_i^A, \quad |\lambda_i^L > = |\lambda_i^A > . \qquad (i = 1, n) \tag{4}$$

The solutions of the Laplace equation are adaptations of the spherical harmonics to the polyhedral graph and include P-like functions, though with departures from the simple 'more-nodes-higher-energy' spherical order when the underlying polyhedron is highly non-spherical. A long cylindrical fullerene, for example, will typically have a low-energy P function $|\lambda_2^A >$ cut by a nodal plane across the long axis, and two functions of higher energy λ^L corresponding to cuts along this axis; D, F, ... vectors with two or more cuts across the axis may intervene before the second and third P vectors are reached in the sequence of decreasing eigenvalues of $\mathbf{A}$/increasing eigenvalues of $\mathbf{L}$.

There is also a group-theoretical justification of the wave analogy, in that the permutation representation (see below) of the vertices of a polyhedral graph contains at least the totally symmetric and the translational representations, Γ_0 and Γ_{XYZ}. The simplest polyhedron, the tetrahedron, has four vertices spanning $A_1 + T_2 \equiv \Gamma_0 + \Gamma_{XYZ}$

in T_d; three independent cartesian basis functions can be constructed by taking opposite signs on vertices belonging to opposite edges. Other polyhedra consist of larger single orbits of vertices or multiple orbits of vertices, but similar single-noded combinations can always be found for them too.

The single-noded eigenvectors of $\mathbf{A}$ and $\mathbf{L}$ are readily identified by a disconnection test: for each candidate vector, a modified graph is constructed by deleting all vertices bearing zero coefficients, all edges incident on them, and all edges between coefficients of opposite sign; for vectors of the required P type, this breaks the graph into exactly two components. Extra solutions can arise for small general polyhedra [9], but at least three suitable vectors exist for all fullerenes. For all cases examined, the topological coordinates are faithful to the adjacency information in that the three neighbours of a vertex in the graph turn out also to be nearest in space.

Once identified, the three P vectors $|\lambda_x>, |\lambda_y>, |\lambda_z>$ lead to a geometry which has the maximum symmetry attainable for the fullerene graph realised as a single spherical shell. Analysis of the topological coordinates is one of several ways to assign this symmetry. The possible maximum point groups of fullerenes can be shown to be 28 in number [10]: I_h, I, T_h, T_d, T, D_{6h}, D_{6d}, D_6, D_{5h}, D_{5d}, D_5, D_{3h}, D_{3d}, D_3, D_{2h}, D_{2d}, D_2, S_6, S_4, C_{3h}, C_{2h}, C_{3v}, C_3, C_{2v}, C_2, C_s, C_i, C_1. Any symmetry is usually apparent on viewing the molecule along the three orthogonal directions of the topological coordinate frame, as these are also principal inertial axes.

The connection to NMR spectra is that in ideal circumstances the ^{13}C NMR spectrum of a fullerene isomer will be very simple. It will consist of a number of peaks, one for each occupied orbit of atomic sites, with intensities proportional to the number of sites in each set, and hence inversely proportional to the order of the site group (i.e., with relative heights of 1 for C_{3v}, 2 for C_3, 3 for C_s, and 6 for C_1 sites). In fact, the idealized stick spectrum of a pure fullerene isomer contains at most three different peak heights, as the C_{3v} and C_3 site groups are mutually exclusive. Arguments based on the composition of these NMR stick spectra, catalogued in [3], along with energy considerations, have been used, for example, to assign the experimental isomers of C_{84}.

A novel use of topological coordinates has been proposed in connection with description and quantification of chirality of molecules such as the fullerenes [11]. The idea is to use adjacency eigenvalues and eigenvectors to define a model π electronic structure of the fullerene in the normal Hückel approach, and compute model rotational strengths of the $\pi \rightarrow \pi^*$ transitions in order to associate a unique sign, and hence configurational descriptor, with each enantiomer of a chiral fullerene. The intermediate quantities are electric and magnetic dipole transition integrals, which in turn require specification of a geometry. If the topological coordinates are used to supply this geometry, then the whole procedure can be carried out automatically from adjacency information alone, independently of vertex numbering and avoiding the complexities of schemes based on vertex labels, which would have to be devised painfully and individually for every new fullerene isomer [4].

3. Distortivity of π systems

The previous section showed that the adjacency matrix of a graph defines a geometric structure. The more conventional use of adjacency information in π systems is in modelling *electronic* structure. Since the Hückel Hamiltonian of a π system is

$$\mathbf{H} = \alpha\mathbf{I} + \beta\mathbf{A}, \tag{5}$$

where α and β are Coulomb and resonance parameters [12], the π molecular orbitals in simple Hückel theory are combinations $|\lambda_i^A>$ of basis functions, with energies

$$\epsilon_i = \alpha + \lambda_i^A \beta. \tag{6}$$

Bonding orbitals have $\lambda_i^A > 0$, antibonding $\lambda_i^A < 0$. Aufbau, Pauli and Hund's rules generate π electronic configurations and lead to magic electron counts as in the famous Hückel $4n + 2$ rule for monocycles, and the leapfrog rule $n = 60 + 6k$ for fullerenes [13].

In contrast to the topological-coordinate application, here it is the behaviour of the eigenvalue spectrum $\{\lambda_i^A\}$ around $i = n/2$ that is of interest. A properly closed-shell configuration for neutral C_n fullerenes has $\lambda_{n/2}^A > 0$, $\lambda_{n/2+1}^A < 0$; such configurations occur for leapfrog fullerenes, and for some others [3], but in fullerenes the overwhelmingly dominant π pattern is the *pseudo-closed* shell, where $\lambda_{n/2}^A > \lambda_{n/2+1}^A > 0$, as a natural consequence of the electron deficiency of the pentagonal ring.

A pseudo-closed shell in a symmetrical molecule, fullerene or not, has the possibility of improving its π energy by adopting a geometrical distortion such that previously different occupied and unoccupied orbitals become equisymmetric and by their interaction produce a splitting of orbital energies. This is a second-order Jahn-Teller effect, discussed in chemistry under the title of π-distortivity [14]. The idea is that the π electrons have a preference for strong localised double bonds over larger numbers of weaker delocalised bonds but are prevented from achieving this π-optimal situation by the stronger preference of the σ electrons for uniformity.

Symmetry appears in this area on reconsidering a long-standing tutorial model for π distortivity introduced by Heilbronner [15]. In this model, geometric distortion of the benzene ring is represented by variation in the resonance integrals, taking higher and lower values $(1 \pm \delta)\beta$ on alternate edges. The total π energy is then

$$E_\pi = 6\alpha + 4(1 + \sqrt{1 + 3\delta^2})\beta \tag{7}$$

and hence has a minimum stabilisation at $\delta = 0$ (α and β are both negative quantities), demonstrating the preference of the π electrons for a localised Kekulé over a delocalised Clar structure. The pattern of integrals used in the model follows the b_{2u} distortion mode that would, if not blocked by σ destabilisation, drive benzene to the D_{3h} cyclo-hexatriene structure. Significantly, this pattern also has the property that the sum of β integrals around each vertex remains constant. The generalisation [16] then defines as a *Heilbronner mode* any pattern of distortion that maintains a constant sum around each (non-pendant) vertex. Intuitively, it is clear that some such modes will represent efficient downhill directions for localisation of the π system. If the curvature of the π energy is sufficiently large, the resistance of the σ electrons will be overcome, and actual geometric distortion will take place; the predicted distortion threshold is $(\partial^2 E_\pi/\partial\beta^2) \sim 1.7\beta^{-1}$ [17].

A counting rule for the Heilbronner modes available to a given π system without pendant vertices follows from the balance of (edge-related) freedoms and (vertex-related) constraints [16]:

$$n(\delta) = e - v \ (+1) \tag{8}$$

where the $+1$ is to be included on the RHS for bipartite graphs, such as the benzene ring or the cube, but omitted for non-bipartite graphs such as fullerenes. This dependence on the nature of the graph arises as follows: in any bipartite graph it is possible to construct a fully alternating combination of vertices, in which every positive coefficient is surrounded by negative and *vice-versa*; this carries zero weight on every edge and so the corresponding combination of vertex constraints has no effect on Heilbronner modes. Pendant vertices (i.e. unsaturated carbon atoms with one unsaturated neighbour) introduce extra freedoms. The counting rule connecting numbers of structural components can be generalised as a relationship between their symmetries. The symmetry form of the rule is [16]

$$\Gamma(\delta) = \Gamma(e) - \Gamma(v) \ (+\Gamma_\star) \tag{9}$$

where $\Gamma(x)$ is the permutation representation of the set of objects x in the symmetry group G of the spatially embedded graph, with character under each operation equal to the number of objects left unshifted, and $\Gamma_\star$ is the one-dimensional representation of the alternating combination of vertices. Again, the bracketed term is included only for bipartite graphs. $\Gamma_\star$ appears in the Hückel theory of bipartite (alternant) π systems, where it links conjugate bonding and anti-bonding molecular orbitals $\mid \pm \lambda >$.

The utility of these two rules emerges when the Heilbronner approach is compared with the traditional [17] approach to π distortivity. The bond-polarisability matrix, confusingly also known as π, is a second-order perturbation property whose elements $\pi_{ij,kl}$ define the derivative of the π bond order of ij with respect to the resonance integral of kl [12]. Bond order is defined from the product of the orbital coefficients for the two ends of the bond, summed over all occupied spin orbitals. Elements of π are functions of the energies and coefficients of occupied and unoccupied orbitals, and so follow directly from the matrix $\mathbf{A}$. With additional approximations, the eigenvectors that result from diagonalisation of the $e \times e$ bond-polarisability matrix correspond to the principal directions of distortion by which the π system lowers its energy. The eigenvectors of π span the full permutation representation of the edges, $\Gamma(e)$.

Thus one way to decide whether Hückel theory predicts distortion for a π system such as pentalene (the C_8H_6 molecule with a framework of two edge-fused pentagons) is to evaluate the π matrix (in this case 9×9), diagonalise it, and check the resulting eigenvalues against the threshold of $\sim 1.7\beta^{-1}$. It will be found that π for the graph has one large eigenvalue $(2.357\beta^{-1})$, corresponding to a vector that leads to the twisted C_{2h} structure that is found by full ab $initio$ optimisation. However, the two rules give immediately the number $(9 - 8 = 1)$, symmetry $(\Gamma(e) = 3A_{2g} + 2B_{1u} + 2B_{3u} + 2B_{2g}$, $\Gamma(v) = 3A_{2g} + 2B_{1u} + 2B_{3u} + B_{2g}$, $\Gamma(\delta) = B_{2g})$ and the almost exact composition of the maximal eigenvector of π [17], by inspection alone. A single-point energy calculation confirms that this vector gives a large second derivative of E_π.

More generally, the Heilbronner vectors have a strong tendency to pick out the most distortive modes of the π system [16]. For dodecahedral C_{20}^{2+}, the symmetry rule gives the Heilbronner modes as five-fold degenerate sets $H_g + H_u$, and of these H_u shadows exactly the most distortive bond-polarisability eigenvector. Symmetry not only reduces the effort of calculation, it gives a pictorial explanation of the results - in this case, showing the cylindrically symmetric component of the distortion mode as an alternating expansion and contraction of the equator of the dodecahedron. Likewise for neutral C_{60}, the Heilbronner modes span the permutation representation of the 30 formal double bonds of the fully symmetric Kekulé structure, and the 7 distinct non-symmetric Heilbronner modes all cluster around the most distortive of the eigenvectors of the 90×90 π matrix. In neither fullerene is the distortion threshold reached, but the key advantage of this symmetry based approach is that, when distortive fullerenes are identified, the results will have a ready rationalisation.

4. An orbital model of ring currents

Chemists use the term aromaticity in many ways, implying significant similarity between a given π system and the benzene archetype. An aromatic molecule may be that is more stable, less reactive in some directions, has more nearly equal bond lengths, or greater magnetisability anisotropy than expected. The *definition* of aromatic π systems as exactly those that support a diamagnetic ring current [18] (circulation of π electrons induced by a perpendicular magnetic field) has the advantage that the theoretical determination of whether a system supports a current with a given sense should be a yes/no decision. It turns out that symmetry and graph-theoretical arguments are informative here too.

Ring currents are inferred in experiment from ^{1}H NMR chemical shifts at sites attached to the carbon skeleton, and by analogy, are often inferred in calculation from calculated chemical shifts. However, as it is now possible to calculate the pattern of current density induced in a molecule by an external magnetic field, it seems preferable to make a more direct attack, and assign aromaticity on the basis of the current-density map itself. A key advance in *ab initio* methods was the proposal by Keith and Bader [19] to use a distributed-gauge method for computation of current density, and in one variant to make the choice that current density at a point should be calculated with the point itself taken as the origin of vector potential. This choice yields maps of excellent quality, even with basis sets of only moderate size, and the maps can be used routinely to discuss the sense, intensity and physical location of ring currents. However there also a conceptual advantage that arises from the choice of each point as its own origin.

It turns out that this *ipsocentric* feature of the method leads to a natural partition of the perturbed wavefunction into two infinite sums of orbital terms [20]. These replace the traditional common-origin division into ground-state diamagnetic and excited-state paramagnetic contributions. The orbital contributions to the perturbed wavefunction have several key features: (i) each term is a transition integral divided by an orbital energy difference; (ii) the integrals define virtual excitations from occupied to unoccupied molecular orbitals - there is no remixing of occupied orbitals; (iii) the terms leading to diamagnetic circulations in the presence of a magnetic field perpendicular to the molecular plane obey *in-plane translational* selection rules; (iv) the terms leading to paramagnetic circulations obey *in-plane rotational* selection rules.

Given (i), the current map is, other things being equal, governed by transitions between frontier orbitals. Given (ii), the orbital contributions have a unique physical status, and allow first-order prediction of the changes in maps caused by electron gain or loss, or chemically modification of substituents on the ring. Given (iii), the sense of a current can often be deduced from the symmetry characteristics of the HOMO, LUMO and nearby orbitals, and even simple Hückel calculations may be sufficient to give these symmetries. An example shows the power of the symmetry argument [21].

The Hückel rule for monocycles predicts aromaticity for $4n + 2$ π systems, and anti-aromaticity for $4n$ systems. If energetic and magnetic criteria of aromaticity are consistent, this implies diamagnetic ring currents in the first case and paramagnetic ring currents in the second. Using the characterisation of the π orbitals of an N-monocycle as sine and cosine combinations associated with an angular momentum quantum number $k = 0, 1 \ldots (N/2)$ (N even), $(N-1)/2$ (N odd), the result for $N = 4n + 2$ follows immediately. In a closed-shell configuration, either all orbitals with given k are occupied, or all are empty. Rotation about the out-of-plane axis mixes only orbitals of equal k, and cannot produce occupied-to-unoccupied virtual transitions. In-plane translational operators mix orbitals that differ in k by ± 1. The sole occupied-to-unoccupied transitions are therefore HOMO to LUMO. Hence in a (4n+2)-electron monocycle, the ring current is wholly diamagnetic, i.e. in the classical Lenz's Law direction, and attributable to the mobility of just the four HOMO electrons (two in 2-electron systems).

The result for $4n$ depends on the fact that singlet states of these monocycles are formed by distortion from the D_{Nh} regular polygon to a bond-alternated $D_{(N/2)h}$ polygon, which induces a splitting in the angular-momentum pairs at the HOMO-LUMO divide. The transition across the HOMO-LUMO gap is between $+k$ and $-k$, is rotationally allowed, and favoured by a small energy denominator, leading to a strong two-electron paramagnetic π ring current, running in the opposite direction to the Lenz's Law prediction. Overall the molecule will remain diamagnetic however, because of local atomic and σ bond circulations.

Both predictions are borne out in *ab initio* calculations on planar and planarised cycles. Extension of the argument shows why the benzene current can be switched on and off by fusing various rings to its perimeter; unsaturated clamping rings that introduce new π orbitals into the HOMO-LUMO gap switch off the current; saturated clamping rings leave the π frontier orbitals essentially undisturbed and the current intact [22].

References

[1] Kroto H W, Heath J R, O'Brien S C, Curl R F, and Smalley R E 1985 Nature 318 162-163
[2] Krätschmer W, Lamb L D, Fostiropoulos, K and Huffman D R 1990 Nature 347 354-358
[3] Fowler P W and Manolopoulos D E 1995 An Atlas of Fullerenes (Oxford: OUP)
[4] Wudl F, Smalley R E, Smith A B, Taylor R, Wasserman E and Godly E W 1997 Pure Applied Chem. 69, 1412-1434
[5] Manolopoulos D E , May J C and Down S E, 1991 Chem. Phys. Lett. 181 105-111
[6] Manolopoulos D E and Fowler P W 1993 Chem. Phys. Lett. 204 1-7
[7] Brinkmann G and Dress A W M 1997 J. Algorithms 23 345-358
[8] Manolopoulos D E and Fowler P W 1992 J. Chem. Phys. 96 7603-7614
[9] Graovac A, Plavsic D, Kaufman M, Pisanski T and Kirby E C 2000 J. Chem. Phys. 113 1925-1931
[10] Fowler P W, Manolopoulos D E, Redmond D B and Ryan R P 1993 Chem. Phys. Lett. 202 371-378
[11] Rassat A, Làszló I and Fowler P W 2002 Chemistry - A Eur. J. (in press)
[12] Streitwieser A 1961 Molecular Orbital Theory for Organic Chemists (New York: Wiley)
[13] Fowler P W and Steer J I 1987 J. Chem. Soc. Chem. Comm. 1987 1403-1405
[14] S. Shaik S, A. Shurki A, D. Danovich D and P. Hiberty P 2001 Chem. Rev. 101, 1501-1539.
[15] Heilbronner E 1989 J. Chem. Ed. 66 471-478
[16] Rassat A, and Fowler P W 2002 Phys. Chem. Chem. Phys. 4 1105-1113
[17] Binsch G, Heilbronner E and Murrell J N 1966 Mol. Phys. 11 305-320
[18] Schleyer P von R, H. Jiao H 1996 Pure. Appl. Chem. 68 209-218
[19] Keith T A and Bader R F W 1993 J. Chem. Phys. 99 3669-3682
[20] Steiner E and Fowler PW 2001 J. Phys. Chem. A 105 9553-9562
[21] Steiner E and Fowler PW 2001 J. Chem. Soc. Chem. Comm. 2001 2220-2221
[22] Fowler P W, Havenith R W A, Jenneskens L W, Soncini A and Steiner E 2001 J. Chem. Soc. Chem. Comm. 2001 2386-2387

Inst. Phys. Conf. Ser. No 173: Plenary Sessions
Paper presented at 24th Int. Coll. Group Theoretical Methods in Physics, Paris, France, 15–20 July 2002
©*2003 IOP Publishing Ltd*

Noncommutative Geometry and Physics: The First 10 Years

José M. Gracia-Bondía

Departamento de Física, Universidad de Costa Rica, 2060 San Pedro, Costa Rica

1. Introduction

As a branch of mathematics, Noncommutative Geometry (NCG) is 20 years old. At the Oberwolfach meeting in September 1981 [1], Connes unveiled an approach to *differential geometry on noncommutative algebras*, already worked out in detail for the noncommutative torus, then known as the "irrational rotation C^*-algebra". The differential calculus for the torus had been embryonically developed in 1980 by Connes himself. The canonical trace on the torus algebra yields a "noncommutative integral". The related calculus, reformulated in the new language of cyclic cohomology, was developed in detail in the basic paper on "Noncommutative Differential Geometry" [2]; which started to circulate in preprint form around Christmas, 1982. Noncommutative geometry is an operator algebraic, variational reformulation of the foundations of geometry, extending to noncommutative spaces. NCG allows consideration of "singular spaces", erasing the distinction between the continuous and the discrete. Its main specific tools are Dirac operators, C^*-modules, Fredholm modules, the noncommutative integral, Hochschild and cyclic homology of algebras... and Hopf algebras. On the mathematical side, NCG has had a vigorous development. Current topics of interest include index theory, the Baum–Connes and Novikov conjectures, mathematical quantization, the Riemann hypothesis; and locally compact quantum groups. In the mainstream of mathematics, *noncommutative geometry is here to stay*.

2. The Interface with Group Theory and Hopf Algebras in renormalization

If G is a compact topological group, then G can be recovered from the algebra of real representative functions $R(G)$, which turns out to be a commutative Hopf algebra. The reconstruction theorem of Tanaka and Kreimer builds up an equivalence between the categories of compact groups and of these Hopf algebras. (A satisfactory correspondence for locally compact groups, including an extension of Pontryagin duality, has only recently been found; it requires a considerable amount of C^* technology.)

The industry of quantum group theory is based on relinquishing commutativity of the Hopf algebras. But I want to call your attention here to the appearance of commutative Hopf algebras in renormalization theory. Bogoliubov's combinatorial renormalization scheme in dimensional regularization can be summarized thus. If Γ is 1PI and *primitive* (has no subdivergences), set

$$C(\Gamma) := -T(f(\Gamma)), \quad \text{and then} \quad R(\Gamma) := f(\Gamma) + C(\Gamma),$$

where $C(\Gamma)$ is the *counterterm* and $R(\Gamma)$ is the desired finite value. Next, recursively define Bogoliubov's $\bar{R}$-operation by setting

$$\bar{R}(\Gamma) = f(\Gamma) + \sum_{0 \subsetneq \gamma \subsetneq \Gamma} C(\gamma) f(\Gamma/\gamma),$$

with the proviso that $C(\gamma_1 \ldots \gamma_r) := C(\gamma_1) \ldots C(\gamma_r)$, whenever $\gamma = \gamma_1 \ldots \gamma_r$ is a disjoint union of several components. Finally, remove the pole part of the previous expression: $C(\Gamma) := -T(\bar{R}(\Gamma))$ and $R(\Gamma) := \bar{R}(\Gamma) + C(\Gamma)$. In summary,

$$C(\Gamma) := -T\left[f(\Gamma) + \sum_{\emptyset \subsetneq \gamma \subsetneq \Gamma} C(\gamma) f(\Gamma/\gamma) \right], \tag{1}$$

$$R(\Gamma) := f(\Gamma) + C(\Gamma) + \sum_{\emptyset \subsetneq \gamma \subsetneq \Gamma} C(\gamma) f(\Gamma/\gamma). \tag{2}$$

Let Φ stand for any particular QFT. The *Hopf algebra* H_Φ is a commutative algebra generated by 1PI graphs: i.e., connected graphs with at least 2 vertices which cannot be disconnected by removing a single line. The product $\Gamma_1 \Gamma_2$ means the disjoint union of the graphs Γ_1 and Γ_2. The counit is given by $\varepsilon(\Gamma) := 0$ on any generator, with $\varepsilon(\emptyset) := 1$ (empty graph $\leftrightarrow$ identity element). The *coproduct* Δ is given, on any 1PI graph Γ, by

$$\Delta\Gamma := \sum_{\emptyset \subseteq \gamma \subseteq \Gamma} \gamma \otimes \Gamma/\gamma.$$

As before, the sum ranges over all subgraphs which are divergent and proper (i.e., removing one internal line cannot increase the number of connected components); γ may be either connected or a disjoint union of several pieces. The notation Γ/γ denotes the (connected, 1PI) graph obtained from Γ by replacing each component of γ by a single vertex. Δ is coassociative, because if $\gamma \subseteq \gamma' \subseteq \Gamma$, then γ'/γ can be regarded as a subgraph of Γ/γ; and it is obvious that $(\Gamma/\gamma)/(\gamma'/\gamma) \simeq \Gamma/\gamma'$. The coassociativity relation $(\Delta \otimes \mathrm{id})(\Delta\Gamma) = (\mathrm{id} \otimes \Delta)(\Delta\Gamma)$ can now be expressed as

$$\sum_{\emptyset \subseteq \gamma \subseteq \gamma' \subseteq \Gamma} \gamma \otimes \gamma'/\gamma \otimes \Gamma/\gamma' = \sum_{\substack{\emptyset \subseteq \gamma \subseteq \Gamma \\ \emptyset \subseteq \gamma'' \subseteq \Gamma/\gamma}} \gamma \otimes \gamma'' \otimes (\Gamma/\gamma)/\gamma'',$$

which is verified directly almost at once.

That defines H_Φ as a bialgebra. A grading is provided by *depth*. A graph Γ in the space H_Φ has depth k (or is k-primitive) if

$$P^{\otimes k+1}(\Delta^k \Gamma) = 0 \quad \text{and} \quad P^{\otimes k}(\Delta^{k-1})\Gamma \neq 0,$$

where P is the projection $u \circ \varepsilon - \mathrm{id}$, whose importance was "experimentally" discovered by Figueroa and myself [3] in the context of the Hopf algebra of rooted trees. Here, depth measures the maximal length of inclusion *chains* of subgraphs appearing in the Bogoliubov recursion. If $\gamma \subsetneq \Gamma$ and Γ has depth l, then γ has depth $\leq l - 1$. Elements of H_Φ without subdivergences have depth 1, as is the case of the "fish" graph in the φ_4^4 model. The "ice-cream" graph has depth 2. Of the eight four-vertex (3-loop) graphs relevant for the 4-point structure of that model, five have depth 3, two have depth 2 and one (the tetrahedron graph) has depth 1. The antipode S can then be defined as the inverse of $\mathrm{id} = u \circ \varepsilon - P$ for the convolution, and so (after some calculation)

$$S(\Gamma_l) := \sum_{k=1}^{l} P^{*k} \Gamma_l = -\Gamma + \sum_{\emptyset \subsetneq \gamma \subsetneq \Gamma} S(\gamma) \Gamma/\gamma, \tag{3}$$

for a graph Γ_l of depth l. The formula gives in a nutshell the equivalence of Zimmermann's and Bogoliubov's renormalization procedures.

The foregoing is my own summary of the Hopf algebra of Feynman diagrams discovered by Connes and Kreimer [4]. As it stands, the Hopf algebra H_Φ corresponds to a formal manipulation of graphs. These formulas can then be matched to expressions for numerical values. Firstly, the Feynman rules for the unrenormalized theory prescribe a linear map

$$f : H_\Phi \to A$$

Some *coproducts* for φ_4^4 diagrams

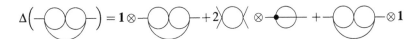

Figure 1. The "setting sun": a primitive diagram

Figure 2. The "double ice cream in a cup"

Figure 3. The "triple sweet"

$$\Delta\left(\begin{array}{c}\end{array} \right) = 1\otimes \quad + \quad \otimes \quad + \quad \otimes \quad + \quad \otimes \quad + \quad \otimes \quad + \quad \otimes 1$$

Figure 4. The "rag-doll"

into some commutative algebra A, which is multiplicative on disjoint unions: $f(\Gamma_1\Gamma_2) = f(\Gamma_1)f(\Gamma_2)$. In other words, f is actually a homomorphism of algebras. In dimensional regularization, A is an algebra of Laurent series in a complex parameter ε, and A is the direct sum of two *subalgebras*:

$$A = A_+ \oplus A_-.$$

Let $T: A \to A_-$ be the projection on the second subalgebra, with $\ker T - A_+$. Λ_+ is the holomorphic subalgebra of Taylor series and A_- is the subalgebra of polynomials in $1/\varepsilon$ without constant term; the projection T picks out the pole part, in a minimal subtraction scheme. Now T is not a homomorphism, but the property that both its kernel and image are subalgebras is reflected in a "multiplicativity constraint": $T(ab) + T(a)T(b) = T(T(a)b) + T(aT(b))$.

Equation (1) means that "the antipode delivers the counterterm": one replaces S in the calculation (3) by C to obtain the right hand side, before projection with T. From the definition of the coproduct in H_Φ, (2) is a *convolution* in $\mathrm{Hom}(H_\Phi, A)$, namely, $R = C * f$. To show that R is multiplicative, it is enough to verify that the counterterm map C is multiplicative, since the convolution of homomorphisms is a homomorphism, because A is commutative. The Hopf algebra approach to renormalization arose in parallel with the Connes–Moscovici NC theory of foliations [5]. A foliation is described by a noncommutative algebra of functions twisted by local diffeomorphisms, $A = C_c^\infty(F) \rtimes \Gamma$; horizontal and vertical vector fields on the frame bundle $F \to M$ are represented on A by the action of a Hopf algebra H_{CM} which simplifies computing a local index formula in NCG. One can map H_{CM} into (an extension of) the Hopf algebra of rooted trees, a precursor of the graphical Hopf algebras which is described in Connes and Kreimer [6] and our book [7]. On extending the Hopf algebra H_Φ of graphs with operations of insertion of subgraphs, one obtains a NC Hopf algebra of the H_{CM} type, which gives a handle on the combinatorial structure of H_Φ. For H_{CM} and the algebra of trees, the corresponding groups are known; for the H_Φ's the mystery remains.

In the last few years, Epstein–Glaser (EG) renormalization has been revived by the Hamburg group around Fredenhagen. It has also been sharpened into a more versatile tool ("T-renormalization") in my [8]. The concept of *optimal rescaling* of the T-amplitudes is the soul of this improvement. When handled through the Hopf algebra of Feynman diagrams H_Φ, EG renormalization provides a most direct route to the core results of renormalization theory. A similar tack was recently taken by Hollands and Wald [9] and Brunetti, Fredenhagen and Verch [10] for the precise definition of local, covariant quantum field. The basic idea (very much in the spirit of NCG) is that, if a quantum field is thought of as a distribution on a globally hyperbolic spacetime, with values in the *algebra of the Wick products*, then isometric embeddings from one spacetime into another are realized as suitable homomorphisms of the fields. The key property in the existence proof by Hollands and Wald is the postulated, and then recursively proved, optimal rescaling of the metric g, up to the logarithmic terms:

$$\lambda^a f_n[\lambda^2 g] = f_n[g] + \log \lambda \, f_{n-1}[g].$$

3. Other applications to Physics

Indeed, there is no question of "application" of NCG to physics, but of mutual intercourse. The original approach of noncommutative geometry to physics was a model of humility. Far from trying to *dictate*, from NCG principles, what the entrails of the subatomic world should be, we were trying to *learn* from mainstream physics the noncommutative geometry of the world. By "mainstream physics" we mainly understand the Standard Model (SM) of fundamental interactions. The Standard Model is dramatically elegant and beautiful in that its crucial concepts are among the deepest and most powerful in modern mathematics. I am talking about the concepts of gauge field and of chiral fermion. Gauge fields are identical with connections, perhaps the most important objects in the modern formulation of geometry. Thinking seriously about the space of all connections has been a very fruitful idea that mathematicians have picked up from physicists. Chiral fermions are acted on by Dirac operators. Perhaps it is not so well known is that Dirac operators are a source of NCG, where they play the role of "fundamental classes". It is precisely the role of the Dirac operator both in the Standard Model and in NCG that made the rapprochement between this branch of mathematics and fundamental physics natural and unavoidable.

There have been two main lines for making sense of the SM in terms of NCG: one developed by Connes himself (and followers), the other related to the Lie superalgebra $su(2|1)$, mainly exploited at the same time by the Mainz–Marseille group (Scheck, Häußling, Papadopoulos, Coquereaux, Esposito-Farèse). The first stage of the process of linking NCG and fundamental physics finished around the winter of 1995–96, when the very structure of the SM gave Connes the inspiration for the construction of noncommutative spin manifolds. Since that time we know how to put fermion fields on a noncommutative manifold. This knowledge is backed by a fundamental mathematical result, whose proof you can find in our book [7]. Afterwards, there has been little activity on the SM front. The unimodularity condition, imposed on the noncommutative gauge field in order to reproduce the correct hypercharges, was "explained" in terms of anomaly cancellation by Alvarez, Martín and myself [11]; but not understood on a native NCG basis. This loose thread has been tied up by Schücker and

Lazzarini [12].

4. Arrival of the Top-down Approach

The first cohabitation of NCG and the physics of fundamental interactions followed a *bottom-up* pattern. The *top-down* approach goes back to the classic "Quantized space-time" by Snyder (January 1947, [13]). In this charming paper, it was suggested that coordinates are noncommuting operators; throughout, Lorentz covariance is maintained. Then as now, motivations for using NC coordinates of spacetime were the hope of improving the renormalizability of QFT and of grappling with the attendant nonlocality of physics at the Planck scale. This top-down approach has been resurrected recently by string theorists. In the most popular model, the commutation relations are simply of the form

$$[x^j, x^k] = i\theta^{jk}, \tag{4}$$

breaking Lorentz invariance. As anticipated by Douglas and Hull [14], by Sheikh-Jabbari [15], and by Schomerus [16], and plausibly argued by Seiberg and Witten in their renowned 100-page paper [17], open strings with allowed endpoints on 2Dirichlet-branes in a B-field background act as electric dipoles of the abelian gauge field of the brane; the endpoints live on the noncommutative space determined by (4). Thus it was unavoidable that string theory and NCG recognize their points of contact. Slightly before, Douglas, Schwarz and Connes [18] had arrived at the conclusion that compactification of M-theory, in the context of dimensionally reduced gauge theory actions, leads ineluctably to NC spaces.

A very important construction in [17] is the so-called Seiberg–Witten map in gauge theory. This map relates the gauge fields and the gauge variations in a noncommutative theory with commutative counterparts. Let $\hat{A}(A)$ and $\hat{\lambda}(A,\lambda)$, where

$$\hat{\delta}\hat{A}_j = \partial_j\hat{\lambda} + i(\hat{A}_j \star \hat{\lambda} - \hat{\lambda} \star \hat{A}_j),$$

be, respectively, the NC gauge potentials and gauge variation in terms of the commutative ones; the $\star$ denotes the Moyal product, of which more below. Then necessarily, in order to have

$$\hat{A}(A) + \hat{\delta}\hat{A}(A) = \hat{A}(A + \delta A)$$

for θ small,

$$\hat{A}_j(A) = A_j - \tfrac{1}{4}\theta^{kl}\{A_k, \partial_l A_j + F_{lj}\} + O(\theta^2),$$
$$\hat{\lambda}(A,\lambda) = \lambda + \tfrac{1}{4}\theta^{kl}\{\partial_k\lambda, A_l\} + O(\theta^2), \tag{5}$$

where $F(A)$ is the ordinary gauge field for A. These equations have been seen by Jurčo, Schupp and Wess [19] and also by Jackiw and Pi [20] to correspond to an infinitesimal 1-cocycle for a projective representation of the underlying gauge group in the Moyal algebra. The equations (5) are valid for arbitrary θ, and so, when using the Seiberg–Witten map, the components of θ can be regarded as variables. We point out that Seiberg–Witten maps between noncommutative gauge theories have ambiguities when coupled to matter fields.

Before continuing, we need a mathematical interlude on the Moyal product. It is correctly (nonperturbatively) defined, for nondegenerate θ, as

$$f \star g(u) := (\pi\theta)^{-4} \int_{\mathbb{R}^4} \int_{\mathbb{R}^4} f(u+s)g(u+t) e^{2is\theta^{-1}t} \, ds \, dt, \tag{6}$$

and this gives rise to the commutation relations (4) Mathematically, those are the commutation relations of Quantum Mechanics, when the reduced Planck constant replaces θ! Formula (6) is then the basis of the "fourth" formalism of Quantum Mechanics, in which observables, states and transitions are described by functions (or distributions) on phase space. This formalism, which goes back to Wigner, Weyl and Moyal, has had already a long and proud history when (a version of) the Moyal product was rediscovered by string theorists. An even happier circumstance was that the QFT framework for making sense of the noncommutative limit of string theory also *preexisted* the paper by Seiberg and Witten. This story deserves its own chapter.

5. Noncommutative Field Theory (NCFT)

NCFT can be and was developed *independently of its string theory motivation* and background; and I shall do here. (I do not discuss quantum fields on ad-hoc discrete spaces.)

Let us explain why NC field theory predated the Seiberg–Witten paper. Connes' NC spin manifold theorem and Fredholm module theory, referred to above, comes in handy here. QFT has an algebraic core independent of the nature of space-time. From the representation theory of the ∞-diml orthogonal group, plus an appropriate 1-particle space, one can derive all quantities of interest. Nothing changes if the "matter field" evolves on a NC space. In a nutshell: the canonical quantization machinery may be applied to a NC 1-particle space. We recall the long-standing hope, that giving up *locality*, one of the basic tenets of quantum field theory (indeed, one of the main selling points by the forefathers) would be rewarded with a better UV behaviour. This hope was now amenable to rigorous scrutiny, and it is not borne out. Thus, QFT on noncommutative manifolds *also requires renormalization*. This, in some sense the first result of NCFT, was proved in general by Várilly and myself [21].

Of course, one can prove the same in the context of a *particular* NCG model, by writing down the integral corresponding to a Feynman diagram and finding it to be divergent.

What, in my opinion, are the main issues in NCFT? There has been an enormous outpouring of literature concerning these issues. The quality of many of those papers is substandard; they should come with an attached notice of *Warning*. To begin with, the mathematics of the Moyal product, as already said, antedates its current use by string theorists. As Fedele Lizzi put it [22], "*in the string community we can happily go on rediscovering quantum mechanics...*" There has been no shortage of mathematical mistakes, attesting that many people did not bother to study the Weyl–Wigner–Moyal formalism in any depth. Some papers read as if they were conceived by randomly opening a book on QFT and trying to figure out the noncommutative analog of the section which appeared. Particularly obnoxious is the pest of the so-called "no-go theorems", whereby trivial observations have been published again and again. The whole field shows the traces of having grown too quickly, tending to obscure the lasting contributions.

Planar versus nonplanar Feynman diagrams: Consider, e.g., the theory given by the action functional

$$S = \int d^4x \left(\frac{1}{2} \frac{\partial \phi}{\partial x^\mu} \frac{\partial \phi}{\partial x_\mu} + \frac{1}{2} m^2 \phi^2 + \frac{g}{4!} \phi \star \phi \star \phi \star \phi \right)$$

The propagators do not change with respect to the commutative theory, but the vertices get in momentum space a factor proportional to

$$\exp\left(-i/2 \sum_{1 \leq k < l \leq 4} p_{k\alpha} \theta_{\alpha\beta} p_{l\beta} \right);$$

we suppose the momenta are all incoming on the vertex.

Consider then a simple diagram of the model like the tadpole diagram; let p denote the incoming momentum and k the loop variable. Depending on the order of the momenta, the previous factor is equal to 1 or to $e^{-ip_k\theta_{kl}k_l}$. In general, planar diagrams get phase factors depending only on external momenta; for nonplanar diagrams there are phase factors which depend on loop variables, and the corresponding integrals become convergent. For the tadpole diagram, we get amplitudes of the form

$$\Gamma_{\mathrm{pl}} \propto \int \frac{d^4k}{k^2+m^2}, \qquad \Gamma_{\mathrm{npl}} \propto \int \frac{d^4k}{k^2+m^2} e^{-ip\theta k}.$$

The second integral is finite.

UV/IR mixing, supersymmetry: For more complicated diagrams with subdivergences, however, the dependence on p of the amplitudes corresponding to nonplanar diagrams spells trouble. This is because these diagrams may become divergent again for particular values of the momenta. Here rears its ugly head the notorious UV/IR mixing, which is thought to reflect the latent stringy degrees of freedom, and tends to spoil renormalizability. It is less of a monster in the broken phase: the pion selfenergy does not have noncommutative IR singularities at all [23]. And, perhaps needless to say, it has been argued that actually RG equations are finite in the IR, and *only perturbative approximations* of them are plagued by IR divergences. Supersymmetric theories have advantages in regard to renormalizability and the UV/IR trouble. In this context, it has been proved by Paban et al [24] that the deformation leading to NC Yang–Mills theory is about the only one compatible with supersymmetry.

Renormalizability in context of Seiberg–Witten map: Another possibility in order to circumvent the UV/IR problem and obtain renormalizable theories is to use the SW map. The trouble with this approach is that many new vertices appear, at different orders in θ and $\hbar$. Extensive calculations [25] have shown that in θ-expanded QED, many miraculous cancellations occur. However, the fermion 4-point function becomes divergent, so, strictly speaking, that model is nonrenormalizable.

Unitarity of theories on spaces with timelike noncommutativity: In connection with the UV/IR problem, it has been asserted that theories with timelike noncommutativity suffer from violation of unitarity. Nonplanar contributions to the "square" of the Feynman propagator contain the unitarity-violating terms. Even so, Dorothea Bahns et al [26] have cogently argued that, when using perturbative Hamiltonian or Yang–Feldman approaches to QFT, this problem can be exorcised.

Anomalies in NCFT: The part of the nonabelian anomaly which is quadratic in the gauge potentials was found by Carmelo P. Martín and myself [27] to be

$$-\frac{1}{96\pi^2} \int d^4x \, \varepsilon_{\mu_1\mu_2\mu_3\mu_4} \operatorname{Tr} T^a \, [T^b, T^c] \partial_{\mu_1} \theta^a \left[\hat{A}^b_{\mu_2} \star \partial_{\mu_3} \hat{A}^c_{\mu_4} - \partial_{\mu_3} \hat{A}^c_{\mu_4} \star \hat{A}^b_{\mu_2} \right]$$

$$-\frac{1}{96\pi^2} \int d^4x \, \varepsilon_{\mu_1\mu_2\mu_3\mu_4} \operatorname{Tr} T^a \, \{T^b, T^c\} \partial_{\mu_1} \theta^a \left[\hat{A}^b_{\mu_2} \star \partial_{\mu_3} \hat{A}^c_{\mu_4} + \partial_{\mu_3} \hat{A}^c_{\mu_4} \star \hat{A}^b_{\mu_2} \right].$$

The first term of this innocent-looking formula is a new contribution that does not vanish in the noncommutative case. It makes it very difficult to construct nonanomalous chiral theories in NCG. Further work by C. P. Martín has established that the anomalies are related only to the (divergent) planar diagrams: nonplanar diagrams do not break the Ward identities.

Construction of gauge-covariant observables: It will have been clear to the listener that the space coordinates themselves are not gauge-covariant. On the other hand,

$$x^\mu + \theta^{\mu\nu} A_\nu$$

is gauge-covariant, as was well known to the practitioners of Weyl–Wigner–Moyal theory. This noncovariance of the space coordinates brings NC theories closer with gravity. The construction of gauge-invariant observables has been taken up by Harald Dorn and coworkers in several interesting papers.

In summary, NCFT offers a brand-new laboratory for QFT. However, when eating this fish, one should beware of the fishbones.

6. Spectral Action and Other Developments

Meanwhile, Alain Connes has not remained idle. Dissatisfaction with absence of gravity in the old Connes–Lott model led him to develop the *spectral action principle*. With Chamseddine, he proposed a universal formula for an action associated with a NC spin geometry [28]. It is based on the spectrum of the Dirac operator, and is a geometric invariant. Automorphisms of the underlying manifold combine both diffeomorphisms and internal symmetries. The Yang–Mills action functional is in general replaced by a "universal" bosonic functional of the form

$$B_\phi[D] = \operatorname{Tr}\phi(D^2),$$

with ϕ being an "arbitrary" positive function of the Dirac operator D. Chamseddine and Connes argue that B_ϕ has the following asymptotic development:

$$B_\phi[D/\Lambda] \sim \sum_{n=0}^{\infty} f_n \Lambda^{4-2n} a_n(D^2) \quad \text{as} \quad \Lambda \to \infty,$$

the a_n being heat kernel coefficients, $f_0 = \int_0^\infty x\phi(x)\,dx$, $f_1 = \int_0^\infty \phi(x)\,dx$, $f_2 = \phi(0)$, $f_3 = -\phi'(0)$, and so on.

They compute the development for the Dirac–Yukawa operator associated to the SM, obtaining all terms in the bosonic part of the action for the Standard Model, plus gravity couplings. Thus, the spectral action for the SM unifies with gravity at a very high energy scale. Wulkenhaar has conjectured that the spectral action has the necessary additional symmetries for gauge theories on θ-deformed spacetime to become renormalizable. This is indeed one of the burning questions of the hour.

Somewhat related to the above, Connes and also Landi, Várilly, Sitarz and Dubois-Violette, have been working on systematically obtaining *new noncommutative spaces*. One motivation is to do quantum gravity in the noncommutative Euclidean context. For that, commutative and NC manifolds alike should be "counted" by means of an algebraic K-theory invariant. Here quantum groups come into their own in NCG, as symmetry groups of homogeneous noncommutative spaces [29, 30].

7. Outlook

Will NCG live up to its promise with regard to fundamental physics?

One should not confuse learning a new language with expansion of physical knowledge. A word of caution about the changing relationships between physics and mathematics is always in order. For instance, Yang and Mills found a family of theories pertinent to describe the observed phenomena. Mathematicians have translated into fibre bundle theory, and now we teach them to our students with the fixity of mathematics, so they seem unavoidable. However, they are suspect (at least to Mills and this reviewer) for the good reason that quantizing them is a nightmare. It may come to pass that eventually Yang–Mills theories will be seen as an incomplete preliminary stage of the "true" physical theory.

By the standard of fashion, the NCG conquest of physics is now ebbing. This ebb will surely prove temporary, since the concepts and tools of noncommutative geometry are indeed powerful and helpful in QFT. On the other hand, our ability to reformulate the SM as a noncommutative geometry is not, in itself, all that meaningful. A truer test for the NCG paradigm would be the understanding of the fermion mass and mixing matrices.

As for NC field theory, the models available so far, while they amount to a useful mathematical laboratory for the fusion of gravity and the other fundamental interactions, are still too rough. Despite the ideology of effective field theory, the fantastic successes of the renormalizability program, and of renormalizability itself as a heuristic principle in selecting physical models, tell us that the *violation of locality* which we expect at some level is *much more subtle* than we have been able to dream of until now. Some crucial ingredient is missing.

References

[1] A. Connes, Tagungsbericht 42/81, Mathematisches Forschungszentrum Oberwolfach, 1981
[2] A. Connes, Publ. Math. IHES **39** (1985) 257
[3] H. Figueroa and J. M. Gracia-Bondía, Mod. Phys. Lett. **A16** (2001) 1427
[4] A. Connes and D. Kreimer, Commun. Math. Phys. **210** (2000) 249
[5] A. Connes and H. Moscovici, Commun. Math. Phys. **198** (1998) 198
[6] A. Connes and D. Kreimer, Commun. Math. Phys. **199** (1998) 203
[7] J. M. Gracia-Bondía, J. C. Várilly and H. Figueroa, Elements of Noncommutative Geometry, Birkhäuser, Boston, 2001
[8] J. M. Gracia-Bondía, hep-th/0202023
[9] S. Hollands and R. M. Wald, gr-qc/0111108
[10] R. Brunetti, K. Fredenhagen and R. Verch, math-ph/0112041
[11] E. Alvarez, J. M. Gracia-Bondía and C. P. Martín, Phys. Lett. **B364** (1995) 33
[12] S. Lazzarini and T. Schücker, Phys. Lett. **B510** (2001) 277
[13] H. S. Snyder, Phys. Rev. **71** (1947) 38
[14] M. R. Douglas and C. M. Hull, JHEP **9802** (1998) 008
[15] M. M. Sheikh-Jabbari, Phys. Lett. **B455** (1999) 129
[16] V. Schomerus, J. High Energy Phys. **9906** (1999) 030
[17] N. Seiberg and E. Witten, J. High Energy Phys. **9909** (1999) 032
[18] A. Connes, M. R. Douglas and A. Schwartz, J. High Energy Phys. **9802** (1998) 003
[19] B. Jurčo, P. Schupp and J. Wess, hep-th/0106110
[20] R. Jackiw and S.-Y. Pi, Phys. Lett. B **534** (2002) 181
[21] J. C. Várilly and J. M. Gracia-Bondía, Int. J. Mod. Phys. **A14** (1999) 1305
[22] F. Lizzi, Nucl. Phys. B (Proc. Suppl.) **104** (2002) 143
[23] F. Ruiz-Ruiz, hep-th/0202011
[24] S. Paban, S. Sethi and M. Stern, hep-th/0201259
[25] R. Wulkenhaar, J. High Energy Phys. **0203** (2002) 024
[26] D. Bahns, S. Doplicher, K. Fredenhagen and G. Piacitelli, Phys. Lett. **B533** (2002) 178
[27] J. M. Gracia-Bondía and C. P. Martín, Phys. Lett. **B479** (2000) 321
[28] A. H. Chamseddine and A. Connes, Commun. Math. Phys. **186** (1997) 731
[29] J. C. Várilly, Commun. Math. Phys. **221** (2001) 511
[30] A. Connes, math.QA/0209142

Inst. Phys. Conf. Ser. No 173: Plenary Sessions
Paper presented at 24th Int. Coll. Group Theoretical Methods in Physics, Paris, France, 15–20 July 2002
©2003 IOP Publishing Ltd

Delone Set Models for Quasicrystals

Jeffrey C. Lagarias

AT&T Labs- Research, Florham Park, New Jersey 07932 USA

Abstract. This work surveys recent mathematical developments motivated by quasicrystalline materials, i.e. materials whose atomic structure has (statistical) long-range translational order evidenced by X-ray diffraction patterns having sharp spots, which exhibit symmetries forbidden to crystals. Such materials cannot have a periodic arrangement of atoms.

A *Delone set* or (r, R)-set in Euclidean n-space is a discrete set which has a positive packing radius r and covering radius R by equal spheres. It may be thought of as modelling positions of atoms of an infinite idealized structure. The talk describes a taxonomy of such sets. A *Delone set of finite type* is a Delone set X such that its set of interpoint distance vectors $X - X$ is a discrete closed set, see [1]. This class of point sets includes Meyer sets [2, 3, 4], cut-and-project sets and most proposed models for quasicrystalline structures, including random tiling models [5]. Such sets naturally have associated a finite number of extra "internal dimensions" in a way similar to cut-and-project sets, see [1].

More precisely, a Delone set of finite type is always contained in a *quasilattice*, which is a finitely generated subset of Euclidean n-space. The dimension of this set may exceed that of the ambient space, thus providing extra dimensions. The possible statistical symmetries of such sets are restricted by the possible symmetries of this quasilattice. It is well known that for interesting examples, such as the Penrose tiling, that these allowed symmetries may include some forbidden for crystals. The structure of permitted symmetries of quasilattices has been extensively studied mathematically [6].

One can attempt to specify structure of Delone sets by restricting the local patterns of points that may occur, with local patterns specified up to a general Euclidean motion. An interesting problem is to explain the appearance of sets with regularity under translations forced by such "local rules under isometries". The existence of such local rules for regular point systems (crystal with one atom in the unit cell) was shown by Delone et. al [7] in 1976, and their bounds were extended to arbitrary crystals [8]. It was recently shown in [9] that such "local rules" exist for a large class of Delone sets of finite type.

Repetitive Delone sets are ones in which every local pattern repeats infinitely often with a bounded spacing between their appearances, the bound depending on the pattern. This condition is roughly analogous to the pattern being a "ground state" though possibly permitting enough disorder for positive entropy to occur. Linearly repetitive Delone sets are one in which this bound grows linearly with the size of the pattern. These have zero entropy in a combinatorial sense, and provide possible models for perfect quasicrystals, that is, ones which are (believed to be) thermodyamically stable. Linearly repetitive Delone sets are always diffractive in a weak sense [10].

One large class of models for quasicrystalline materials are based on self-similar constructions. It is natural to do this using tiling models, where there are many interesting examples of self-similar tilings, some having pure point diffraction. It is less natural to use Delone sets, but it turns out to be possible to build self-similar constructions in the framework of Delone sets [11]. Mathematically the (uniform) discreteness of Delone sets is a replacement for the compactness properties used in studying self-similar tilings.

The discovery of quasicrystalline materials raises again the question of what conditions on the local structure of a Delone set are needed to enforce crystallinity. There are now quite sharp criteria in terms of the combinatorial complexity of how many inequivalent local regions of a given size are permitted [12].

Quasicrystalline materials were discovered based on their diffraction spectra exhibiting point peaks exhibiting non-crystalline symmetries. There has been much effort spent on finding models exhibiting pure point diffractivity with such symmetries. There are many subtle mathematical issues in modelling diffractivity, not all of them solved [13, 14]. Attached to any Delone set is a dynamical system induced by the $\mathbf{R}^n$ action of translations, see [15]. J.-Y. Lee, R. V. Moody and B. Solomyak [16] recently showed under mild hypotheses, that the Delone set has pure point diffraction spectrum if and only if the dynamical system is uniquely ergodic and has pure point spectrum. Their methods permit pure point diffractivity to be proved for interesting examples.

References

[1] Lagarias J C 1999, Geometric Models for Quasicrystals I. Delone Sets of Finite Type, Discrete & Computational Geometry 21 161-191.

[2] Meyer Y 1995, Quasicrystals, Diophantine Approximation and Algebraic Numbers, in: *Beyond Quasicrystals* (F. Axel and D. Gratias, Ed), Les Editions de Physique, Springer-Verlag 3-16.

[3] Lagarias J C 1996, Meyer's concept of quasicrystal and quasiregular sets, Comm. Math. Phys. **179** 365-376.

[4] Moody R V 1997, Meyer Sets and their Duals, in: *The mathematis of long-range aperiodic order (Waterloo, ON 1995)*, NATO ASI Series C Math. Phys. Sci. 489, Kluwer Acad. Publ. Dordrecht 403-441.

[5] Henley C L 1999, Random tiling models, in: *Quasicrystals: The state of the art*, Second Edition (D..P. DiVencenzo and P. J. Steinhardt, Eds.), Directions in Condensed Matter Physics 16, world Scientific: River Edge, NJ. 459-560.

[6] Le T Q T, Piunikhin S A, and Sadov V A 1993, The geometry of quasicrystals, Russian Math. Surveys **48**, No. 1 37-100.

[7] Delone B N, Dolbinin N P, Stogrin M I, and Galiulin R V 1976, A local criterion for regularity of a system of points, Soviet Math. Dokl. **17**, No. 2 319-322.

[8] Dolbilin N I, Lagarias J C, and Senechal M 1998, Multiregular point systems, Discrete & Computational Geometry **20** 477-498.

[9] Lagarias J C 1999, Geometric Models for Quasicrystals II. Local Rules Under Isometries, Discrete & Computational Geometry 21 345-372.

[10] Lagarias J C and Pleasants P A B 2003, Repetitive Delone Sets and Quasicrystals, Ergodic Theory and Dynamical Systems, **23** to appear. eprint: `arXiv math.DS/9909033`

[11] Lagarias J C and Yang Wang 2003, Substitution Delone Sets, Discrete & Computational Geometry **29** 175-209. eprint: `arXiv math.MG/0110222`, Oct. 2001.

[12] Lagarias J C and Pleasants P A B 2002, Local complexity of Delone sets and crystallinity, Caand. Math. Bull. **45** 634-652.

[13] Hof A 1997, Diffraction by aperiodic structures, in: *The mathematis of long-range aperiodic order (Waterloo, ON 1995)*, NATO ASI Series C Math. Phys. Sci. 489, Kluwer Acad. Publ. Dordrecht 239-268.

[14] Lagarias J C 2000, Mathematical quasicrystals and the problem of diffraction, in: *Directions in mathematical quasicrystals*, CRM Monographs Ser. 13, Amer. Math. Soc.: Providence, RI 61-93.

[15] Solomyak B 1998, Spectrum of dynamical systems arising from Delone sets, in: *Quasicrystals and Discrete Geometry*, Fields Inst. Monograph 10, Amer. Math. Soc., Providence, RI 265-275.

[16] Lee J Y, Moody R V, and Solomyak B 2002, Pure point dynamical and diffraction spectrum, Ann. Henri Poincaré **3** 1003-1018.

Inst. Phys. Conf. Ser. No 173: Plenary Sessions
Paper presented at 24th Int. Coll. Group Theoretical Methods in Physics, Paris, France, 15–20 July 2002
©2003 IOP Publishing Ltd

Lie Symmetries for Lattice Equations

D. Levi
Dipartimento di Fisica "E. Amaldi",
Universitá degli Studi Roma Tre and Sezione INFN, Roma Tre,
Via della Vasca Navale 84, 00146 Roma, Italy

Abstract. Lie symmetries have been introduced by Sophus Lie to study differential equations. They have been one of the most efficient way for obtaining exact analytic solution of differential equations. We show how one can extend this technique to the case of differential difference and difference equations.

1. Introduction

Lie groups have long been used to study differential equations. As a matter of fact, they originated in that context [1,2]. They have been put to good use to solve differential equations, to classify them, and to establish properties of their solution spaces [3–8]. In particular, it provides one of the most efficient method for obtaining exact analytic solutions for partial differential equations, i.e. symmetry reduction. This method consists of a sequence of algoritmic steps, the first of which is finding the Lie group G of local point transformations:

$$\tilde{\mathbf{x}} = F_g(\mathbf{x}, u) = \mathbf{x} + g\xi(\mathbf{x}, u) + \mathcal{O}(g^2)$$
$$\tilde{u} = H_g(\mathbf{x}, u) = u + g\phi(\mathbf{x}, u) + \mathcal{O}(g^2) \tag{1}$$

where g is the set of group parameters. Given a partial differential equation of order k $E_k(\mathbf{x}, u, u_{x_i}, u_{x_{i,j}}, \ldots, u_{x_{i_1,\ldots,i_k}}) = 0$, G is obtained requiring that the transformation (1) leaves the set of solutions invariant, i.e. defining the infinitesimal generator of the Lie point symmetry $\hat{X} = \xi_i \partial_{x_i} + \phi \partial_u$,

$$pr\hat{X} E_k|_{E_k=0} - 0. \tag{2}$$

Then we look for solutions which are invariant under a subgroup G_0 of G. The finite group transformations, obtained by integrating the infinitesimal generators or by exponentiation, transform solutions of E_k into solutions of the same equation. Other classes of exact solutions can be obtained by considering *conditional symmetries* [9]. We can introduce an equivalent (for partial differential equations) representation of the infinitesimal generator of the symmetry, often called *evolutionary* [3],

$$\hat{X}_e = Q(\mathbf{x}, u, u_{x_1}, \ldots, u_{x_k})\partial_u \quad Q = \phi - \xi_i u_{x_i}, \quad pr\hat{X}_e E_k|_{E_k=0} = 0. \tag{3}$$

The existence of the evolutionary symmetry (3) implies that an invariant solution of E_k must satisfy

$$u_g = Q(\mathbf{x}, u, u_{x_1}, \ldots, u_{x_k}) \tag{4}$$

and the request that (3) be a symmetry is equivalent to require that (4) commutes with E_k. As Q is linear in the first derivatives of u eq.(4) is integrable on the characteristics, being a quasilinear partial differential equation of first order. The notion of point symmetries can be extended [3] to the case of *generalized symmetries* by requiring that $Q = Q(\mathbf{x}, u, u_{x_{i_1}}, \ldots, u_{x_{i_1}, \ldots, x_{i_s}})$. In this case eq.(4) is a partial differential equation of order s and thus it cannot be solved explicitly to get the group transformation. However, by Noether theorem [3], they still can provide conservation laws and are usually associated to *exactly integrable equations*.

Applications of Lie group theory to discrete equations, like difference equations, differential-difference equations, or q-difference equations are much more recent [10–17]. By a difference equation we mean a functional relation, linear or non-linear, between a function calculated at different points of a lattice. Why are these systems important? They appear in many applications. They can be written down as a discretization of a differential equation and in such a case the differential equation is reduced to a recurrence relation:

$$\frac{du}{dx} = f(x, u) \quad \Rightarrow \quad v(n+1) = g(n, v(n)).$$

We can consider dynamical systems defined on a lattice, i.e. systems where the dependent fields depend on a set of independents variables which vary partly on the integers and partly on the reals. For example we can have:

$$\frac{d^2 u(n, t)}{dt^2} = F(t, u(n, t), u(n-1, t), \ldots, u(n-a, t), u(n+1, t), \ldots, u(n+b, t))$$

These kind of equations can appear in many different setting. Those which are more interesting from a scientific point of view, are associated to the evolution of many body problems, to the study of crystals, to biological systems, etc. .

To conclude I would like to comment on the richness of the world of the differential difference equations with respect to that of partial differential equations by presenting a result obtained by MacKay and Aubry [18]. They showed that dynamical chains are much richer than their continuum conterpart as almost any Hamiltonian network of weakly coupled oscillators has a 'breather' solution while the existance of breathers for a nonlinear wave equation is rare. This implies that the discrete world can be richer of interesting solutions and thus worthwhile studying by itself; symmetries are a simple and efficient way to do so.

In Section 2, after a brief description of discrete systems, we introduce the Lie point symmetries for discrete equations while in Section 3 we consider some of their possible extensions. Section 4 is devoted to some conclusions.

2. Lie point symmetries for discrete equations

The first steps in the construction of Lie symmetries for difference equations were taken by Shiguro Maeda in 1980 [10] and later extended by many authors [11–17,21]. For semplicity in the following I will consider just a scalar equation in two independent variables but equivalent results can be obtained in the case of N independent and M dependent variables.

A discrete equation in $\Re^2$ is a functional relation for a field $u(P)$ at a finite number, say L, of different points P_i in R^2, i.e. $E = E(x, t, u(P_1), \ldots, u(P_L)) = 0$. A differential difference equation is obtained by considering the points P_i uniformly spaces in one direction, say t, with spacing h_t, in such a way that we are allowed to consider the continuous limit when h_t goes to zero.

When we embed the points in a lattice, the points P_i in $\Re^2$ can be labelled by two discrete indexes, (n, m), which characterize them with respect to two independent directions, $P_{n,m}$. For example, in the Cartesian plane (x, t), we have:

$$P_{n,m} = (x_{n,m}, t_{n,m}), \tag{5}$$

and the function $u(P)$ reads

$$u(P_{n,m}) = u(x_{n,m}, t_{n,m}) = u_{n,m}. \tag{6}$$

A difference scheme will be a set of relations (equations) among the values of $\{x, t, u(x, t)\}$ at a finite number, say L, of points in $\Re^2$ $\{P_1, \dots, P_L\}$ around a reference point, say P_1. Some of these relations will define where the points are in $\Re^2$ and others how $u(P)$ transforms in $\Re^2$. In our case, as we have one only dependent variable and two independent variables, we expect to have five equations, four which define the two independent variables in the two independent directions in $\Re^2$, and one the dependent variable in terms of the lattice points:

$$E_a(\{x_{n+j,m+i}, t_{n+j,m+i}, u_{n+j,m+i}\}) = 0 \tag{7}$$
$$1 \le a \le 5; \qquad -i_1 \le i \le i_2, \quad -j_1 \le j \le j_2 \quad (i_1, i_2, j_1, j_2) \in Z$$
$$i_1 + i_2 = N, \qquad j_1 + j_2 = M$$

System (7) must be such that, starting from L points we are able to calculate $\{x, t, u\}$ in all points. If a continuous limit of (7) exists, than one of the equations will go over to a partial differential equation and the others will be identically satisfied (generically $0 = 0$). We can also do partial continuous limits when only one of the independent variables become continuous while the other is still discrete. In this case only part of the lattice equations are identically satisfied and we obtain a differential difference equation for the dependent variable and an equation for the lattice variable. To clarify the ideas, let us present some examples of difference scheme.

Let as consider the case of the discrete heat equation at first on a uniform orthogonal lattice:

$$\frac{u_{n+1,m} - u_{n,m}}{t_{n+1,m} - t_{n,m}} = \frac{u_{n,m+2} - 2u_{n,m+1} + u_{n,m}}{(x_{n,m+1} - x_{n,m})^2} \tag{8}$$
$$x_{n,m+1} - x_{n,m} = h_x; \qquad t_{n,m+1} - t_{n,m} = 0 \tag{9}$$
$$x_{n+1,m} - x_{n,m} = 0; \qquad t_{n+1,m} - t_{n,m} = h_t,$$

where h_x, h_t are two a priory fixed constants which define the spacing between two neighbouring points in the two directions of the orthogonal lattice. The lattice equations (9) could be substituted by different ones which will provide different Lie point symmetries for the difference scheme while keeping the continuous limit. As examples let us consider

$$x_{n,m+2} - 2x_{n,m+1} + x_{n,m} = 0; \qquad t_{n,m+1} - t_{n,m} = 0 \tag{10}$$
$$x_{n+1,m} - x_{n,m} = 0; \qquad t_{n+2,m} - 2t_{n+1,m} + t_{n,m} = 0,$$

or

$$x_{n,m+2} - 2x_{n,m+1} + x_{n,m} = 0; \qquad t_{n,m+1} - t_{n,m} = 0 \tag{11}$$
$$x_{n+1,m} - (1+c)x_{n,m} = 0; \qquad t_{n+1,m} - t_{n,m} = h.$$

Eq.(10) corresponds to a five point lattice scheme when the lattice spacings can vary, while the case of eq.(11) to a five points exponential lattice.

2.1. Symmetries of a difference scheme

As we are interested in Lie point symmetries, we look for transformations of the form:

$$\tilde{x} = F_g(x, t, u) = x + g\,\xi(x, t, u) + \dots \tag{12}$$
$$\tilde{t} = G_g(x, t, u) = t + g\,\tau(x, t, u) + \dots$$
$$\tilde{u} = H_g(x, t, u) = u + g\,\phi(x, t, u) + \dots$$

where g, as before, is the group parameter, which leave (7) invariant.

Such a transformation acts on the whole space of the independent and dependent variables $\{x, t, u\}$, at least in some neighborhoud of P_1 including all L points. This means that the same set of functions F, G and H will determine the transformation in all points of the scheme. In the point P_1 we define the infinitesimal generator as:

$$\hat{X}_{P_1} = \xi(x, t, u)\partial_x + \tau(x, t, u)\partial_t + \phi(x, t, u)\partial_u \tag{13}$$

and than we prolong it to all other $L - 1$ points of the scheme. Since the transformation is given by the same set of functions $\{F, G, H\}$ at all points, the *prolongation* of $\hat{X}_{P_1}$ is obtained simply by evaluating $\hat{X}_{P_1}$ at the corresponding points involved in the scheme. So

$$pr\hat{X} = \sum_{i=1}^{L} \hat{X}_{P_i} \tag{14}$$

and consequently the invariance condition for the difference scheme is:

$$pr\hat{X}E_a|_{E_a=0} = 0. \tag{15}$$

Eq.(15) is a set of functional equations whose solution is obtained, as usual, by turning them into differential equations by successive derivation with respect to the independent variables $\{x, t, u\}$.

The solution of (15) provide us with the function $\xi(x, t, u)$, $\tau(x, t, u)$ and $\phi(x, t, u)$, the infinitesimal coefficients of the local Lie point symmetry group. The transformation is obtained by integrating the vector field, i.e. by solving the following system of differential equations:

$$\frac{d\tilde{x}}{dg} = \xi(\tilde{x}, \tilde{t}, \tilde{u}), \qquad \tilde{x}|_{g=0} = x,$$

$$\frac{d\tilde{t}}{dg} = \tau(\tilde{x}, \tilde{t}, \tilde{u}), \qquad \tilde{t}|_{g=0} = t, \tag{16}$$

$$\frac{d\tilde{u}}{dg} = \phi(\tilde{x}, \tilde{t}, \tilde{u}), \qquad \tilde{u}|_{g=0} = u.$$

In general we expect the infinitesimal coefficients ξ and τ to be determined by the lattice equations. So according to the form of the lattice, different symmetries can appear.

In fact, in the case of eq.(9), by applying the infinitesimal generator (13) to the lattice equations we get:

$$\xi(x_{n,m+1}, t_{n,m+1}, u_{n,m+1}) = \xi(x_{n,m}, t_{n,m}, u_{n,m});$$
$$\xi(x_{n+1,m}, t_{n+1,m}, u_{n+1,m}) = \xi(x_{n,m}, t_{n,m}, u_{n,m}).$$

$u_{n,m+1}$, $u_{n+1,m}$ and $u_{n,m}$ are independent functions so we get $\xi = \xi(x, t)$. $t_{n,m+1} = t_{n,m}$ but $x_{n,m+1} \neq x_{n,m}$ and consequently $\xi = \xi(t)$. As $x_{n+1,m} = x_{n,m}$ but $t_{n+1,m} \neq t_{n,m}$ we get that the only possible value for ξ with this lattice scheme is ξ=costant. In a similar fashion we derive that also τ must be a constant and that $\phi=u + s(x, t)$, where $s(x, t)$ is a solution of the heat equation, the linear superposition formula. Summarizing we get that the infinitesimal generators of the symmetries for the heat eqaution are given by

$$\hat{P}_0 = \partial_t; \quad \hat{P}_1 = \partial_x; \quad \hat{W} = u\partial_u; \quad \hat{S} = s(x, t)\partial_u. \tag{17}$$

If we associate to the heat equation (8) the lattice (10) we get that we have the generators (17) and an extra generator, $\hat{D} = x\partial_x + 2t\partial_t$. See [19, 20] for other examples.

3. Extensions

As one can see in the examples considered so far, in the case of discrete equations we will, in general, get less symmetries than in the corresponding continuous system. We are often laking symmetries which involve both dependent and independent variables. The possible extension to a variable lattice considered in Section 2 is not always sufficient to obtain them [21].

Research for more general symmetries requires the extension of the intrinsic Lie point ansatz. We have various possibilities.

Let us consider the fields $u(P_j)$ in different lattice points P_j as different independent fields related by the difference equation. Under such an hypotesis the difference equation is just an algebraic relation between the fields. The discrete variables are just indeces of the dependent fields and the only independent variables are the continuous ones. In this case the infinitesimal generator reads:

$$\hat{X} = \tau(t, \{u(P_i)\})\partial_t + \sum_{i=1}^{L} \phi_{P_i}(t, \{u(P_j)\})\partial_{u(P_i)} \tag{18}$$

As, by Taylor expansion, $u_{n,m+1}$ can be expressed in term of $u_{n,m}$ and all of its derivatives, the presence of other points in the definition of the infinitesimal generator of the symmetries is equivalent to consider generalized symmetries. This explains the difficulties in obtaining nontrivial results in this case [11].

On the other hand, as was shown in [15], by allowing symmetry generators which depends on different lattice points, linear difference equations admit a symmetry group which is isomorph to the one of the continuous equation. As an example, let us consider the discrete heath equation:

$$\Delta_t u = \Delta_{xx} u \tag{19}$$

where the discrete derivative $\Delta = \Delta^+$ is defined as the incremental ratio,

$$\Delta_z^+ = \frac{T_z - 1}{h_z}, \qquad T_z f(z) = f(z + h_z), \tag{20}$$

with h_z the lattice spacing in the z direction. In [15] it was showned that eq.(19) has the following symmetry group:

$$\hat{P}_0 = \Delta_t^+ u \partial_u, \tag{21}$$
$$\hat{P}_1 = \Delta_x^+ u \partial_u,$$
$$\hat{W} = u \partial_u,$$
$$\hat{B} = (2t T_t^{-1}\Delta_x^+ u + x T_x^{-1} u)\partial_u,$$
$$\hat{D} = (2t T_t^{-1}\Delta_t^+ u + x T_x^{-1}\Delta_x^+ u + \frac{1}{2}u)\partial_u,$$
$$\hat{K} = \{t^2 T_t^{-1}\Delta_t^+ u - h_t t T_t^{-2}\Delta_t^+ u + tx T_t^{-1} T_x^{-1}\Delta_x^+ u +$$
$$+ \frac{1}{4}x^2 T_x^{-2} u - \frac{1}{4}h_x x T_x^{-2} u + \frac{1}{2}t T_t^{-1} u\}\partial_u.$$

In a sequent work [22] one was able to prove that the previous result can be extended to any discretization for any linear partial differential equation by using the following prescriptions:

- we write down the vector fields in evolutionary form

- we substitute

$$u_{,t} \to \Delta_t u \tag{22}$$
$$u_{,x} \to \Delta_x u$$
$$x \to x\beta_x$$
$$t \to t\beta_t$$

where β_x, β_t are functions of the same shift operators as Δ_x and Δ_t such that the following commutation relation is satisfied:

$$[\Delta_z, z\beta_z] = 1. \tag{23}$$

Equation (23) define completely the function β in terms of Δ. If by Δ we consider the continuous derivative than $\beta = 1$, for Δ given by eq.(20) than $\beta_z = T_z^{-1}$, completely in agreement with the results presented in (21), while if we choose $\Delta_z^s = \frac{T_z - T_z^{-1}}{2h_z}$ than $\beta_z = 2(T_z + T_z^{-1})^{-1}$. Moreover, from the correspondence (22) it follows that for any analytic solution of the linear partial differential equation we find a solution of the discrete counterpart. In the case of Δ^+, or the completely symmetric case of $\Delta_z^- = \frac{1 - T_z^{-1}}{h_z}$, the continuous case will give all solutions of the discrete one. However this will not be the case for a more general Δ like Δ^s when the set of solutions so obtain is a subset of the possible ones. The result contained in (22, 23) is a generalization of the *umbral calculus* [23].

As pointed out before, if the symmetries depend on more points on the lattice, we are in a situation similar to generalized symmetries and we are no more able to get the corresponding group transformations. However we can still use the symmetries to do symmetry reduction and obtain explicit solutions. For the heat equation (18) in the case of a symmetry generator $\hat{P}_1 - a\hat{P}_0$, Eq. (4) reads:

$$\frac{du_{n,m}}{dg} = -\left(\frac{1}{h_x} - a\frac{1}{h_t}\right)u_{n,m} + \frac{1}{h_x}u_{n,m+1} - \frac{a}{h_t}u_{n+1,m}. \tag{24}$$

Eq. (24) is a linear differential difference equation and to get the group transformations we need the explicit solution of its initial problem. The symmetry reduction is obtained by setting $\frac{du_{n,m}}{dg} = 0$. In this case we get:

$$u_{n,m} = c_0 + c_1(1 + a^2 h_t)^n (1 + ah_x)^m \tag{25}$$

which is the discrete representation of the continuous solution.

What happens in the case of nonlinear difference equations? In general we are no more able to obtain a symmetry group isomorphic to the continuous one, the application of the *umbral* corrispondence to a nonlinear equation is questionable and reaserch on it is in progress [24, 25].

In all cases, both for linear and nonlinear difference equations,

- we can still compute intrinsic Lie point symmetries
- generalized symmetries, i.e. symmetries depending on a finite number of points of the lattice, are usually associated to integrable equations and their form is usually very complicate and very difficult to predict apriori without any further information on the structure of the system at study. As an example of this let us show few examples of symmetries of the discrete Burgers equation [12]

$$\Delta_t u = \frac{1 + h_x u}{1 + h_t(\Delta_x u + uT_x u)}\Delta_x(\Delta_x u + uT_x u) \tag{26}$$

i.e.

$$u_{,\lambda_1} = [1 + h_t(\Delta_x u + uT_x u)]\Delta_t u \tag{27}$$

a time translation and

$$u_{,\lambda_2} = [1 + h_x u] \Delta_x \{2t T_t^{-1} \frac{u}{1 + h_t(\Delta_x u + u T_x u)} + \tag{28}$$

$$+ (x + \frac{h_x}{2}) T_x^{-1} \frac{1}{1 + h_x u}\}$$

a boost. Let us notice that both symmetries are nonlinear and depend on a finite number of lattice points but, while eq.(27) is polynomial in u, eq.(28) is rational.

As a further exemplification of the complicate structure of symmetries for nonlinear discrete equations let us present the simplest symmetries one obtains for the discrete Nonlinear Schrödinger equation [26–28] :

$$i\dot{Q}_n + \frac{1}{h_x^2}[2Q_n - (1 - \epsilon|Q_n|^2)(Q_{n+1} + Q_{n-1})] = 0 \tag{29}$$

where $\epsilon = \pm 1$ and h_x the spacing between two lattice consecutive points, i.e.

$$\hat{X}_1 = Q_n \partial_{Q_n} - Q_n^* \partial_{Q_n^*} \tag{30}$$

$$\hat{X}_2 = (1 - \epsilon|Q_n|^2)[Q_{n+1}\partial_{Q_n} - Q_{n-1}^* \partial_{Q_n^*}] \tag{31}$$

$$\hat{X}_3 = (1 - \epsilon|Q_n|^2)Q_{n-1}\partial_{Q_n} - (1 - \epsilon|Q_n|^2)Q_{n+1}^* \partial_{Q_n^*} \tag{32}$$

$$\hat{Y} = [-\frac{2t}{h_x^2}(1 - \epsilon|Q_n|^2)(Q_{n+1} - Q_{n+1}) + i(2n + 1)Q_n]\partial_{Q_n} + \tag{33}$$

$$- [\frac{2t}{h_x^2}(1 - \epsilon|Q_n|^2)(Q_{n+1}^* - Q_{n+1}^*) + i(2n + 1)Q_n^*]\partial_{Q_n^*}.$$

$\hat{X}_1$ and $\hat{X} = \hat{X}_2 + \hat{X}_3$ are intrinsic Lie point symmetries, $\{\hat{X}_1, \hat{X}_2 \text{ and } \hat{X}_3\}$ are isospectral symmetries while $\hat{Y}$ is a nonisospectral symmetry. The $\hat{X}$ commute among themselves while the commutator of $\hat{Y}$ with the $\hat{X}$'s give $\hat{X}$'s. .

We can use $\frac{\hat{Y} - i\hat{X}_1}{2}$ to do a symmetry reduction for the Nonlinear Schrödinger equation (29). In the continuous limit this would correspond to a dilation symmetry which would give rise to elliptic functions or Painlevé solutions. This reduces to solving the Nonlinear Schrödinger equation (29) together with

$$-\frac{t}{h_x}(1 - \epsilon|Q_n|^2)(Q_{n+1} - Q_{n+1}) + inQ_n = 0. \tag{34}$$

Defining $Q_n = \rho_n e^{i\theta_n}$, under the assumption that $\rho_n^2 \neq \epsilon$ we get the following nonlinear reduced equation:

$$\sqrt{\rho_n^2 \rho_{n+1}^2 - h_x^4 c_0^2} + \sqrt{\rho_n^2 \rho_{n-1}^2 - h_x^4 c_0^2} = \frac{nh_x^2}{t}\frac{\rho_n^2}{1 - \epsilon\rho_n^2}. \tag{35}$$

Up to now we have shown that we can find more general symmetries than the intrinsic Lie point ones in the case of linear and integrable difference equations and that we can use them to get solutions. In the following we want to suggest an answer for the generic case. Let's consider the class of differential difference equations:

$$\dot{u}_n = F_n(u_{n+1}, u_n, u_{n-1}) \tag{36}$$

on a uniform unchangable orthogonal lattice and look for symmetries depending linearly on t, as it is the case of the symmetries $\hat{Y}$ (33) for the discrete Nonlinear Schrödinger equations. To exclude the case of integrable equations, see (33), we look for symmetries of the form:

$$u_{n,\lambda} = t\dot{u}_n + H_n(u_{n+1}, u_n, u_{n-1}) \tag{37}$$

Under the assumption that the differential difference equation (36) is really nonlinear

$$\left(\frac{\partial F_n}{\partial u_{n+1}}, \frac{\partial F_n}{\partial u_{n-1}}\right) \neq (0,0) \quad \forall n \tag{38}$$

we can, by a simple transformation reduce in all generality the function H_n to the form

$$H_n = H_n(u_{n+1}, u_n), \tag{39}$$

with $\frac{\partial H_n}{\partial u_{n+1}} \neq 0$ as we are not interested in intrinsic Lie point symmetries. We can state the following theorem, proved in [14];

Theorem 1 *If a nonlinear equation (36) has a symmetry of the form (38, 39) than it is equivalent, up to a Lie point transformation*

$$\tilde{t} = \omega t, \quad \tilde{u}_n = \phi_n(u_n); \quad \omega \neq 0, \quad \phi_n' \neq 0 \quad \forall n \tag{40}$$

to an equation of the form

$$\dot{u}_n = A_n + B_n \tag{41}$$
$$A_n = a_{n+1}e^{u_{n+1}} - a_n e^{u_n} - 1$$
$$B_n = a_n e^{-u_n} - a_{n-1}e^{-u_{n-1}} - 1$$
$$a_n^2 = n^2 + \alpha n + \beta \quad \forall n$$

where α and β are arbitrary constants. Than the symmetry is given by

$$u_{n,\lambda} = t\dot{u}_n + A_n \tag{42}$$

and the equation is linearizable.

Consequently an equation of the form (36) can have dilation symmetries of the form (37) only if it is equivalent to a linear equation.

4. Conclusions

We have shown how one can construct in a coherent way Lie point symmetries for discrete equations.

If we want to extend the symmetries to the case when they depend on more point of the lattice than this can be done only in the case of linear, linearizable or integrable equations. This statement is very plausible but no complete proof of this statement has been given up to now. In the case of linear, linearizable or integrable equations we can find generalized symmetries from which we are not able to construct group transformations but we can use them to get explicit solutions via symmetry reduction.

For a generic equation we can at least construct intrinsic Lie point symmetries and we can use them, as for partial differential equations, to construct group transformations, to do symmetry reduction and obtain explicit solutions, to classify the discrete equations according to their symmetries, etc..

References

[1] Lie S 1888 Math. Ann. 32 213–
[2] Lie S 1893 Theorie der Transformationgruppen (Leipzig: B.G. Teubner)
[3] Olver P J 1993 Applications of Lie Groups to Differential Equations (New York: Springer)
[4] Ibragimov N H 1985 Transformation Groups Applied to Mathematical Physics (Boston: Reidel)

[5] Ovsiannikov L V 1982 Group Analysis of Differential Equations (New York: Academic)
[6] Bluman G W and Kumei S 1989 Symmetries and Differential Equations (Berlin: Springer)
[7] Gaeta G 1994 Nonlinear Symmetries and Nonlinear Equations (Dordrecht: Kluwer)
[8] Winternitz P 1993 Integrable Systems, Quantum Groups and Quantum Field Theories (Dordrecht: Kluwer) 429–495
[9] Levi D and Winternitz P 1989 J. Phys. A: Math. Gen. 22 2915–2924
[10] Maeda S 1980 Math. Japan 25 405–420
[11] Levi D and Winternitz P 1993 J. Math. Phys. 34 3713–3730
[12] Hernandez Heredero R, Levi D and Winternitz P 1999 J. Phys. A Math. Gen. 32 2685–2695
[13] Hernandez Heredero R, Levi D, Rodriguez M A and Winternitz P 2000 J. Phys. A: Math. Gen. 33 5025–5040
[14] Levi D and Yamilov R 1999 J. Phys. A: Math. Gen. 32 8317–8323
[15] Floreanini R, Negro J, Nieto L M and Vinet L 1996 Lett. Math. Phys. 36 351–355
[16] Quispel G R W, Capel H W and Sahadevan R 1992 Phys. Lett. A 170 379–383
[17] Ames W F, Anderson R L, Dorodnitsyn V A, Ferapontov E V, Gazizov R K, Ibragimov N H and Svirshchevskii S R 1994 CRC Hand-book of Lie Group Analysis of Differential Equations Volume I: Symmetries, Exact Solutions and Conservation Laws (Boca Raton: CRC Press)
[18] MacKay R S and Aubry S 1994 Nonlinearity 7 1623–1643
[19] Levi D, Tremblay S and Winternitz P 2000 J. Phys. A : Math Gen. 33 8507–8523
[20] Levi D, Tremblay S and Winternitz P 2001 J. Phys. A: Math. Gen. 34 9507–9524
[21] Dorodnitsyn V A, Kozlov R and Winternitz P 2000 J. Math. Phys. 41 480–504
[22] Levi D, Negro J and dell'Olmo M 2001 J. Phys. A: Math. Gen. 34 2023–2030
[23] Roman S and Rota G C 1978 Adv. Math. 27 95–188
[24] Levi D, Ragnisco O and Tempesta P, work in progress
[25] Carl B and Schiebold C 2000 Jber. d. Dt. Math.-Verein. 120 102–148
[26] Ablowitz M J and Ladik J F 1975 J. Math. Phys. 16 598–603
[27] Levi D and Quispel G R W 2000 Symmetries and integrability of difference equations (Providence: AMS/CRM) 363–366
[28] Hernandez Heredero R and Levi D, The discrete Nonlinear Schroedinger equation and its Lie symmetries, *J. Phys. A: Math. Gen.* submitted to.

Inst. Phys. Conf. Ser. No 173: Plenary Sessions
Paper presented at 24th Int. Coll. Group Theoretical Methods in Physics, Paris, France, 15–20 July 2002
©*2003 IOP Publishing Ltd*

Path Group in gauge theory and gravity

Michael B. Mensky

P.N. Lebedev Physics Institute, Moscow,Russia

Abstract. Applications of the Path Group (consisting of classes of continuous curves in Minkowski space-time) to gauge theory and gravity are reviewed. Covariant derivatives are interpreted as generators of an induced representation of Path Group. Non-Abelian generalization of Stokes theorem is naturally formulated and proved in terms of paths. Quantum analogue of Equivalence Principle is formulated in terms of Path Group and Feynman path integrals.

1. Introduction

One of the most important concepts in the modern quantum field theory is gauge field. This concept was introduced [1] on the basis of gauge symmetry, i.e. invariance under localized (depending on space-time points) symmetry groups. Later it became clear that gauge theory may be naturally formulated on the basis of such mathematical formalism as connections in fiber bundles and their curvatures. Gravity was from the very beginning formulated as theory of curved (pseudo-)Riemannian spaces. Fiber bundles turned out to be also efficient mathematical formalism for gravitational fields.

Here we shall give a short survey of an alternative mathematical background for both gauge theory and gravity, namely Path Group ([2, 3, 4, 5], see [6, 7] for reviews). Path Group (PG) is a generalization of translation group differing from the latter in that it may be applied to particles in external gauge and gravitational fields.

Formally Path Group may be defined as a set of certain classes of continuous curves in Minkowski space (generalization on the case of paths in an arbitrary group space is possible). Concept of PG came up as development of the Suvegesh's groupoid of parallel transports [8, 9] (a groupoid differs from a group in that not any pair of its elements may be multiplied). The goal was to find a universal group such that parallel transports in various space-times be its representations.

The concept of PG may be obtained also in the attempt to globalize infinitesimal translations of a tangent space to a curved space-time. PG arises then instead of the usual translation group since curvature makes translations in different directions not commutative (see [10, 11, 12, 13] for other types of non-commutative translations).

PG reduces geometry to algebra: various geometries (gauge and gravitational fields) are nothing else than representations of the universal PG. In case of gauge fields the representation is simpler in that Lorentz group may be factorized out. In case of gravitational fields both Lorentz and PG (united to give the generalized Poincaré group) essentially participate in the constructions.

The natural character of PG is seen from the facts that it allows 1) to give a group-theoretical interpretation of covariant derivatives (both for gravitational or/and gauge fields) as generators of relevant representations of PG, 2) to formulate and prove a non-Abelian version of Stokes theorem and 3) to reduce the path integral in a curved space-time to the path integral in the flat space (a quantum version of Equivalence Principle)

2. Path Group

An element of Path Group (PG) is defined as a *class of curves* in Minkowski space constructed in such a way that the classes form a group. For curves in Minkowski space, $\{\xi\} = \{\xi(\tau) \in \mathcal{M}\}$, operation of multiplication $\{\xi'\}\{\xi\}$ may be naturally defined as passing of two curves one after another and inversion $\{\xi\}^{-1}$ as passing the same curve in the opposite direction. However, the set of all continuous curves is not, in respect to these operation, a group: 1) not all pairs of curves may be multiplied to give a continuous curve, 2) multiplication (when defined) is not assotiative and 3) product of a curve by its inverse does not yield a unit element (i.e. such one that multiplication by it does not change an arbitrary curve).

To correct the second defect, one may consider differently parametrized curves to be equivalent (and include them in the same class). To correct the third defect, one may consider equivalent those curves which differ by inclusion of 'appendices' of the form $\{\xi\}^{-1}\{\xi\}$ (and go over to else wider classes of curves). At last, the first defect is overcome if we include in the same class those curves which differ by general shift: $\xi'(\tau) = \xi(\tau) + a$ (see Figure).

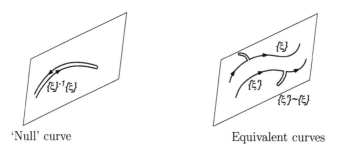

'Null' curve   Equivalent curves

The resulting class of curves is called a *path* and denoted by p or $[\xi]$. All paths form *Path Group P*. A path may be presented by any curve from the corresponding class.

If the curves which differ by general shift are not considered equivalent, we have more narrow classes. The end points are the same for all curves in such a class. It may be therefore called a *pinned path* and denoted as $\hat{p} = p_x^{x'} = [\xi]_x^{x'}$ (or p_x) where x is the initial and x' final point of any curve presenting the given pinned path.

All pinned paths form a groupoid $\hat{P}$ since not any pair of pinned paths may be multiplied. Pinned paths are often convenient for constructing representations of the group P of free paths. It is important that the pair p, x of a free path and a point unambiguously determines the pinned path p_x starting in x and having the same shape as p.

Paths may be defined [6, 7] for an arbitrary group space G leading to the group of paths $P(G)$ in G (the role of general shift is played in this case by right shift in the group). We shall restrict ourselves by considering only the path group in Minkowski space, $P = P(\mathcal{M})$. This will prove to be sufficient for applications to gauge theory and gravity.

3. Gauge fields as representations of Path Group

Theory of free elementary particles is governed by the group of translations. Starting from Path Group instead of the translation group we obtain theory of particles in an external gauge field. Gauge fields thus arise independently of the idea of gauge symmetry.

The key point for constructing theory of particles is that Path Group P (generalized translations) acts on Minkowski space $\mathcal{M}$ transitively. An important consequence is that particles have to be described by the induced representations of Path Group.

1. Theory of elementary particles may be constructed starting from a certain group and implying the requirement of locality [14, 15, 6] (see [16] for an analogous construction). In our case Path Group P will play the role of the governing group and Minkowski space $\mathcal{M}$ the role of the localization space. Note that the action of the group P on the space $\mathcal{M}$ is naturally defined as shifting a point $x \in \mathcal{M}$ along the path $p \in P$ (see Figure). An arbitrary point in $\mathcal{M}$ is invariant under the action of closed paths

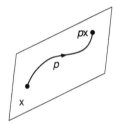   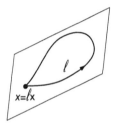

Path acting on a point   Invariance of points under action of loops

(loops). Therefore, the subgroup of loops $L \subset P$ is a stabilizer of an arbitrary point. This means that the space $\mathcal{M}$ may be presented as a quotient space $\mathcal{M} = P/L$.

If some point $\mathcal{O} \in \mathcal{M}$ is chosen as an origin in $\mathcal{M}$, then the point $x \in \mathcal{M}$ is identified with the coset $p'L \in P/L$ where $x = p'\mathcal{O}$. The space $\mathcal{M} = P/L$ will serve in our case as a localization space.

2.The general scheme of constructing theory of particles starting from the group and localization space is following [14, 15, 6].

Let theory of particles is governed by the group P (which may be symmetry or kinematic group) and characterized by localization in the space $\mathcal{M}$ (i.e. the subspace of states $\mathcal{H}_x$ is defined in which the particle is localized in an arbitrary point x of $\mathcal{M}$). Then the group P must act on the space $\mathcal{M}$ (so that $p: \; x \to px$ for $x \in \mathcal{M}$, $p \in P$), and this action must be in accord with the representation U of P acting in the space of the particle's states:

$$U(p)\mathcal{H}_x = \mathcal{H}_{px}.$$

A representation $U(P)$ possessing this property is called imprimitive.

Assume that the action of P on $\mathcal{M}$ is transitive (this may be done without loss of generality, because otherwise we may divide $\mathcal{M}$ in imprimitive subspaces). Then $\mathcal{M}$ is a homogeneous space and can be presented as a quotient space, $\mathcal{M} = P/L$, with an appropriate subgroup $L \subset P$.

The representation $U(P)$ acts then transitively on the set of subspaces $\{\mathcal{H}_x | x \in \mathcal{M} = P/L\}$. According to the imprimitivity theorem [17], such a representation (transitive but imprimitive) is equivalent to the representation induced from some representation $\alpha(L)$ of the subgroup L:

$$U(P) = \alpha(L) \uparrow P.$$

This gives a receipt for constructing theory of local particles. Knowing the group P and the space $\mathcal{M} = P/L$ of localization we can restore the representation $U(P) = \alpha(L) \uparrow P$ acting in the space of states $\mathcal{H}$ of the particle (and therefore can restore the space $\mathcal{H}$ itself). To do this, we have 1) to choose arbitrarily a representation $\alpha(L)$ of the subgroup L and 2) to induce it onto the whole group P.

Inducing a representation $\alpha(L)$ of a subgroup onto the whole group P may be achieved in the following way [17]. Vectors of the carrier space $\mathcal{H}$ of the induced representation $U(P) = \alpha(L) \uparrow P$ are presented by functions $\Psi(p)$ on P with values in the carrier space $\mathcal{L}_\alpha$ of $\alpha(L)$ with the additional *structure condition* imposed on the functions:

$$\Psi(pl) = \alpha(l^{-1})\Psi(p), \quad l \in L, p \in P$$

The induced representation acts on these functions by left shifts:

$$(U(p)\Psi)(p') = \Psi(p^{-1}p'), \quad p, p' \in P.$$

The representation $\alpha(L)$ is arbitrary in this construction. It describes *internal degrees of freedom* of the particle which it possesses even if its localization in $\mathcal{M}$ is fixed. States of the particle are described by the functions $\Psi(p)$ depending on elements of the group P. In our case these are paths, so we arrive to the non-local formalism of path-dependent wave functions. It is a group-theoretical version of Mandelstam's path-dependent fields [18, 19, 20].

However, despite of the non-local form, the theory must be essentially local because the requirement of locality was imposed from the very beginning. Therefore, *explicitly local form* of the representation must exist.

For constructing this form we need an extension $\alpha(P)$ of the representation $\alpha(L)$ onto the group P such that $\alpha(pl) = \alpha(p)\alpha(l)$ for arbitrary $p \in P$ and $l \in L$. Local wave functions $\psi(x)$ may be defined then as

$$\psi(x) = \alpha(p')\Psi(p')$$

where the point $x \in \mathcal{M}$ corresponds to the coset $p'L \in P/L$. Now the action of the induced representation is given by

$$(U(p)\psi)(x) = \alpha(p')[\alpha(p^{-1}p')]^{-1}\psi(p^{-1}x)$$

(for simplicity we denote this representation by the same letter although it is only equivalent but not identical to the preceding one).

3. Applying the general scheme to the Path Group P, its subgroup of loops L and Minkowski space $\mathcal{M} = P/L$, we obtain that the representation $\alpha(L)$ describes a gauge field and the induced representation $U(P) = \alpha(L) \uparrow P$ presents a particle in this field. We shall illustrate this approach starting from the usual description of a gauge field by vector-potential.

Making use of a vector-potential $A_\mu(x)$, introduce a representation of the groupoid of pinned paths by ordered exponentials

$$\hat{\alpha}(\hat{p}) = \mathcal{P} \exp\left\{i \int_{\hat{p}} A_\mu(x)dx^\mu\right\}$$

(integration here is performed along any of the curves from the class $\hat{p}$). Fixing an (arbitrarily) point $\mathcal{O} \in \mathcal{M}$ as an origin of $\mathcal{M}$, we may associate a pinned paths $p_{\mathcal{O}}$ (starting in $\mathcal{O}$) with any free path $p \in P$. In these notations, the representation of loops may be expressed as $\alpha(l) = \hat{\alpha}(l_{\mathcal{O}})$ and the expansion of this representation onto the whole group P (with the properties specified above) as $\alpha(p) = \hat{\alpha}(p_{\mathcal{O}})$.

This determines a local form of the representation $U(P) = \alpha(L) \uparrow P$ which is given by the following elegant formula:

$$(U(p)\psi)(x) = \hat{\alpha}(p^x_{x'})\psi(x').$$

Another very convenient form of the same (or rather equivalent) representation may be expressed in terms of only free paths:

$$U(p) = \mathcal{P}\exp\left\{-\int_p d\xi^\mu \nabla_\mu\right\} \quad \text{where} \quad (\nabla_\mu\psi)(x) = \left(\frac{\partial}{\partial x^\mu} - iA_\mu(x)\right).$$

The covariant derivatives are therefore generators of the representation $U(P)$ of Path Group (just as ordinary derivatives are generators of translations). By this the *group-theoretical interpretation of covariant derivatives* is given.

The representation of the group of loops, $\alpha(L)$, provides in this scheme a non-local description of a gauge field. This description is better than local descriptions by vector-potential and by field stregth. Indeed, the first of these descriptions is redundant since various vector-potentials may correspond to the same physical situation, and the second is insufficient because non-local topological effects (such as Aharonov-Bohm effect) cannot be described by field strength. The 'intermediate' non-local description by $\alpha(L)$ is adequate.

The scheme presented here leads to gauge fields without usage of gauge transformations and the idea of gauge invariance. *Gauge transformations* arise in this scheme as presentation of *arbitrariness in the local description* of gauge fields by vector-potentials. Let us change the vector-potential A_μ to A'_μ in such a way that the representation of L does not change, $\alpha'(l) = \alpha(l)$, i.e. the non-local path-dependent description of the gauge field is the same. Then the new and old vector-potentials may be shown to be connected by a gauge transformation.

4.   One more evidence of natural character of the non-local path-dependent presentation of gauge fields is the non-Abelian version of Stokes theorem [5, 6].

Any loop may be presented as a *product of small 'lasso'* of the form $p^{-1}\delta l\, p$ with δl being a very small loop and p finite path (see Figure). The product is then of the

form $l = \mathcal{P}\prod_j p_j^{-1}\,\delta l_j\, p_j$. ¿From the multiplicative properties of the representation $\alpha(L)$ we have

$$\alpha(l) = \mathcal{P}\prod_j[\alpha(p_j)]^{-1}\hat{\alpha}\left((\delta l_j)_{x_j}\right)\alpha(p_j) = \mathcal{P}\prod_j\exp\left(\frac{i}{2}\mathcal{F}_{\mu\nu}(p_j)\sigma_j^{\mu\nu}\right)$$

where $x = p\mathcal{O}$ is a point which the path p brings the origin $\mathcal{O}$ to, δl_x is a pinned loop having the shape δl and starting in x, and $\mathcal{F}_{\mu\nu}(p) = [\alpha(p)]^{-1} F_{\mu\nu}(p\mathcal{O})\alpha(p)$. Symbolically this may be presented in the form

$$\mathcal{P}\exp\left(i\int_{\partial\Sigma} A_\mu(x)\,dx^\mu\right) = \mathcal{P}\exp\left(\frac{i}{2}\int_\Sigma F_{\mu\nu}(x)\,dx^\mu \wedge dx^\nu\right)$$

which is an ordered-exponential form of the non-Abelian Stokes theorem. Other forms of this theorem are in [21, 22].

4. Paths in gravity

Path Group may be used for description of a gravitational field (curved space-time) and particles in a gravitational or gauge + gravitational field [6, 7]. Minkowski space plays the role of a standard tangent space, connected with the curved space-time in a nonholonomic way.

1. There is a natural (but non-holonomic) mapping [2, 23] of curves (paths) in the tangent space to the curves in the curved space-time (see Figure). Instead of an explicit

form of this mapping, the fiber bundle $\mathcal{B}$ (over the curved space-time $\mathcal{X}$) of local frames $b = \{b_\alpha^\mu\}$ may be used to express the necessary relations.

The key instrument for this end are (horizontal) *basis vector fields* in $\mathcal{B}$:

$$B_\alpha = b_\alpha^\mu \left(\frac{\partial}{\partial x^\mu} - \Gamma_{\mu\nu}^\lambda(x)\, b_\beta^\nu \frac{\partial}{\partial b_\beta^\lambda}\right)$$

The *representation of Path Group* by operators acting on functions in $\mathcal{B}$ as well as the *action of P on $\mathcal{B}$* are readily defined:

$$U(p) = \mathcal{P}\exp\left(\int_p d\xi^\alpha\, B_\alpha\right), \quad (U(p)\psi)(b) = \psi(bp).$$

Thus defined mapping $p : b \to bp$ is a parallel transport along the curve in $\mathcal{X}$ which corresponds to the path p in $\mathcal{M}$ in the sense of the above mentioned mapping.

If the connection conserves the metric, these operatins may be restricted on the *subbundle $\mathcal{N} \in \mathcal{B}$ of orthonormal frames* yielding operators $U(p)$ acting on functions in $\mathcal{N}$ and the mapping $p : n \to np$ such that

$$(U(p)\psi)(n) = \psi(np).$$

Thus, the action of Path Group P by *parallel transports* is defined on $\mathcal{N}$. Lorentz group Λ acts on $\mathcal{N}$ as the *structure group* of the fiber bundle, i.e. $\lambda : n \to n\lambda$ where $(n\lambda)_\alpha^\mu = n_\beta^\mu \lambda_\alpha^\beta$. Therefore the action of their semidirect product (generalized Poincaré group $Q = \Lambda \otimes P$) is defined.

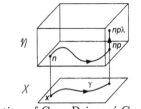

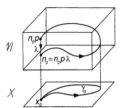

Action of Gen. Poincaré Group   Element of Holonomy Subgroup

More precisely, the elements of Q are of the form $q = p\lambda$ (where $p \in P$ and $\lambda \in \Lambda$), multiplication of them is determined by the relation $\lambda[\xi]\lambda^{-1} = [\lambda\xi]$, while the action of Q on fiber bundle $\mathcal{N}$ is defined as $nq = np\lambda = (np)\lambda$ ($p : n \to np$ being a parallel transport and $\lambda : n \to n\lambda$ the action of the structure group). The resulting action of $p\lambda$ (Figure, left diagram) is in accord with the multiplication law in the group Q.

2. Holonomy Subgroup H of the generalized Poincaré group Q is defined in respect to the (arbitrarily chosen and fixed) local frame $n_0 \in \mathcal{N}$ as the subset leaving n_0 invariant (a stationary subgroup of n_0) so that $h = p\lambda \in H$ if $n_0 p\lambda = n_0$ (see Figure, right diagram).

Holonomy Subgroup H in the generalized Poincaré group Q plays the role (for gravity) analogous to the subgroup of loops L in the Path Group P (for gauge theory). This is why a representation $\alpha(H)$ represents a *gauge + gravitational field* while the induced representation $U(Q) = \alpha(H) \uparrow Q$ describes particles in this combined field. If the representation α is trivial, $\alpha(h) \equiv 1$, the induced representation describes particles in a pure gravitational field (corresponding to the Holonomy Subgroup H).

Holonomy Subgroup $H \subset Q$ determines geometry. The geometry (including topology) may be restored if the subgroup H is given. This reconstruction procedure may be applied [24] to explore geometry of hyperbolic, or Lorentzian, cones, starting from $H = \{l\nu^k | l \in L, k \in Z\}$ where ν is a fixed element of the Lorentz group.

3. Various quantum analogues of the Einstein's Equivalence Principle may be defined. Validity of the quantum equivalence principle (QEP) depends of course on its definition. It is advantageous however to define QEP in such a way that it be valid. Path Group (together with the natural non-holonomic mapping of the curves in Minkowsky space onto the curves in the curved space-time) makes this possible [2, 3, 23] (see also [25, 26]).

A very brief formulation of QEP may be given as follows: evolution of a quantum particle in a curved space-time $\mathcal{X}$ must be described by the *same Feynman path integral* as in the flat space-time, if the paths in the standard tangent space $\mathcal{M}$ are used in the integral instead of the corresponding curves in the curved space-time $\mathcal{X}$.

Technically it is convenient to express QEP in terms of the representation $U(P)$ introduced above, putting $U(P)$ in the integrand of (flat) Feynman path integral instead of the ordinary translation. This may be done not only for purely gravitational but also for gravitational + gauge field (see [23] for details).

5. Conclusion

In the present paper we considered Path Group P (which generalizes translation group) and demonstrated the following issues:

1) Gauge fields are described by representations $\alpha(L)$ of the subgroup $L \subset P$ of loops. Particles in such a field are presented by the induced representation $U(Q) = \alpha(L) \uparrow Q$. Non-Abelian Stokes theorem is naturally formulated and proved in terms of the representation $\alpha(L)$.

2) For application to gravity the generalized Poincaré group Q is necessary which is a semidirect product of Path Group by Lorentz group.

3) Geometry of a curved space-time (including non-trivial topology) is presented by the Holonomy Subgroup $H \in Q$.

4) Gauge + gravitational field is presented by the Holonomy Subgroup H together with its representation $\alpha(H)$ while particles in this field are described by the representation $\alpha(H) \uparrow Q$.

5) Quantum Equivalence Principle is naturally formulated in terms of Feynman path integral and natural non-holonomic mapping of curves in Minkowski space onto the curves in the curved space-time.

References

[1] Yang C N and Mills R L 1954 Phys. Rev. 96 191
[2] Mensky M B 1972 The equivalence principle and symmetry of Riemannian space, in: "Gravitation: Problems and Prospects", the memorial volume dedicated to A.Z.Petrov, (Naukova Dumka: Kiev), 157-167 [in Russian]
[3] Mensky M B 1974 Theor. Math. Phys. 18 136
[4] Mensky M B 1978 Letters in Math. Phys. 2 175-180
[5] Mensky M B 1979 Letters in Math. Phys. 3 513-520
[6] Mensky M B 1983 Path Group: Measurements, Fields, Particles (Nauka: Moscow) [in Russian; Japanese extended translation: 1988 (Yoshioka: Kyoto)]
[7] Mensky M B 1990 Applications of the path group in gauge theory, gravitation and string theory, in: Pawlowski M and Raczka R (eds.) Gauge Theories of Fundamental Interactions (World Scientific: Singapore etc.) 395-422
[8] Süveges M 1966 Acta Phys. Acad. Sci. Hung. 20 41, 51, 274
[9] Süveges M 1969 Acta Phys. Acad. Sci. Hung. 27 261
[10] Jackiw R 1978 Phys. Rev. Lett. 41 1635
[11] Jackiw R 1985 Phys. Rev. Lett. 54 159
[12] Jackiw R 1985 Phys. Lett. B 154 404
[13] Jackiw R 2002 Phys. Rev. Lett. 88 1116031
[14] Mensky M B 1976 Induced Representations Method: Space-Time and Concept of Particles (Nauka: Moscow) [in Russian]
[15] Mensky M B 1976 Commun. Math. Phys. **47**, 97-108
[16] Rowe D 2002 Journal of Physics 35 5599, 5625
[17] Coleman A J 1968 Induced and subduced representations, in: Loeble E M (ed.) "Group Theory and Its Applications" (Academic Press: New York and London) 57-118
[18] Mandelstam S 1962 Ann. Phys. (USA) 19 1, 25
[19] Mandelstam S 1968 Phys. Rev. 175 1580, 1604
[20] Bialynicki-Birula I 1963 Bull. Acad. polon. sci. Sér. sci. math. astron. et Phys. 11, 135
[21] Bralic N 1980 Phys. Rev. D 22 3090
[22] Aref'eva I 1980 Teor. Mat. Fiz. B 43 111
[23] Mensky M B 1996 Helvetica Physica Acta 69 301-304
[24] Mensky M B 1985 The group of paths in gravitation and gauge theory, in: "Quantum Gravity (Proceed. 2nd Intern. Sem. on Quant. Gravity, Moscow, October 13-15, 1981)" (Plenum: New York) 527-546
[25] Pažma V and Prešnaider P 1988 Czechosl. J. Phys. B 38 968
[26] Kleinert H 1995 Path Integrals in Quantum Mechanics, Statistics, and Polymer Physics (World Scientific: Singapore)

Inst. Phys. Conf. Ser. No 173: Plenary Sessions
Paper presented at 24th Int. Coll. Group Theoretical Methods in Physics, Paris, France, 15–20 July 2002
©2003 IOP Publishing Ltd

Recent applications of Topology and Group Theory in Condensed Matter.

Michael I.Monastyrsky

ITEP, Moscow, Russia and MPIPKS Dresden, Germany

1. Introduction.

Interactions of Topology and Group Theory with Physics have a long and fruitful history. Investigation of new physical systems like membranes, nanotubes, fulerens and so on, lead to nonstandard mathematical problems at the cross-roads of topology, differential geometry and group theory.

In my talk I briefly discuss some mathematical problems concerning the theory of phase transitions in two media: fluid membranes and lattice spin systems.

The first is concerned with topology and geometry and the second with group theory.

2. Fluid membranes.

The interest in membranes grew substantially in the last years. There are some specific reasons for this.

First of all membranes play an important role in many physical and biological processes. For example, red blood cells are membranes, and the flickering of red blood cells relates to thermal fluctuations of it is shapes.

Secondly, the study of membranes has a lot in common with string theory and two dimensional conformal models. This allows the application of methods of field theory to membranes.

Here interesting and unsolved mathematical problems appear in junction of the turn of the theory of Riemann surfaces, minimal surfaces, the calculus of variations and others.

Most of the problems we disscuss here were obtained in collaboration with E.Kats and were published in [1]. We refere to [1] for more detailed explanations and references.

3. Physics of Membranes.

The membranes that we consider here are thin and flexible sheets of amphilic molecules. Such a construction distinguishes this object from similar systems like liquid interfaces or shells. At the interface of liquids the exchange of particles between both volume phases occurs. It corresponds to a finite surface tension $\approx 100 \; erg/sm^2$. For membranes which are in equilibrium with the solvent of amphilic molecules the number of molecules is fixed. That means zero surface tension. So, the energy of membranes is characterised mainly by bending energy rather than surface tension. For liquid membranes which we are discussing here the characteristic bending module is of the order of ambient temperature, $\kappa \simeq 10^{-14}$ erg. This is its principal distinction from shells. So, for membranes we must a use statistical mechanics approach , whereas for shells we apply elasticity theory. In solvents lipids or surfactants bilayers typically form closed surfaces or vesicles. The characteristic length scale of a vesicle is of the order $1 - 10\ \mu m$. There are a diversity of forms of membranes. Better known membranes with a simple topology are the sphere and the torus but in experiments one observs membranes with higher genuses, e.g. with $g = 2, 3$. Here g is the number of handles.

There exist also non-compact membranes like a stack of layers which possess cylindrical or plane geometry. Such super-molecular aggregates can be self-organising on a large scale and exhibit phases with different types of orientational and translational order.

For example, spherical vesicles can be organised in three-dimensional cubic lattices, and infinite cylinders can be packed in two-dimensional hexagonal lattices. In this a large-scale crystalline order exists (in typical examples of order 10^{-4} cm, whereas on smaller scales the system remains liquid). There exist also other structures built by infinite layers. The simplest example is the one-dimensional lattice of almost plane layers. Such a structure L_α, called laminar, is the analog of a smectic liquid crystal. There exists also the phase Q_α with a space cubic symmetry and the bloc structure generated by infinite periodic surfaces with complicated topology. Schematically such a structure can be determined in thermodynamic limit $g \to \infty$, as a three-dimensional cubic set of "handles". This structure is in principal different from a topologically trivial cubic lattice of spherical vesicles. The melting process of periodically distributed phase Q_α leads to the isotropic phase, the so called sponge phase L_3. The structural bloc of the phase Q_α is a non-periodic minimal surface with a finite density of handles g. However the space distribution of g is determined by the nearest order. In our talk we consider the new method of classification of different types of membranes and their phase transitions.

4. Mathematics of membranes.

Here we discuss some mathematics which we use to describe phase transitions in membranes.

Let M^2 be a surface (closed, with boundary, or non-compact) embedded in R^3.

Definition 1 ‡ A surface M^2 is called a Willmore surface if it is an extremum of the Helfrich-Willmore (HW) functional:

$$F = \int_{M^2} H^2\, dA \qquad (1.1)$$

where H is the mean curvature, and dA is an area element.

By the Gauss-Bonnet theorem the HW functional is equivalent to:

$$F = \int_{M^2} (k_1^2 + k_2^2)\, dA$$

where k_1 and k_2 are the principal curvatures of M and

$$H = \frac{k_1 + k_2}{2}$$

Therefore the extrema of the functional (1.1) determine most plane surfaces of a fixed topology. The following result due to J.Weiner is important for future use.

Proposition 1. Let M^2 be a minimal surface in S^3 and γ a stereographic projection $S^3 \to R^3$. Then

$$\gamma(\hat{M}^2) = M^2 \qquad \text{and} \qquad F(M^2) = \sigma(\hat{M}^2) \qquad (1.2)$$

where $\sigma(\hat{M}^2)$ is the area of the minimal surface.

‡ We use here standard modern terminology but such functionals were considered by S.Germain, S.Poisson in the XIX century, but of course in a different context.

This result is valid both for closed surfaces and for one with a boundary. Willmore surfaces are in not general exhausted by the projection of minimal surfaces. For example, there exists an infinite set of Willmore tori which are not equivalent to minimal tori in S^3. These tori are extrema of (1.1) but not minimal. This result, apparently, is true for a surfaces of a higher genus. Willmore surfaces of higher genus generated by minimal surfaces in S^3 have been constructed. It is important to emphasise that Willmore surfaces embedded in R^3 satisfy the following estimation:

$$\int_{M^2} H^2 \, dA \geq 4\pi n$$

where n is the number of intersection points (n–different pre-images under projection). From this, and the square estimation of minimal surfaces embedded in S^3, it follows that all Willmore surfaces for $g \geq 2$ have at least one point of intersections §

5. Phase transitions with a change of membrane topology

In the previous section we have shown that Willmore surfaces providing minimum Helfrich energy in the fixed topological class $g = const$ are related to minimal surfaces in S^3 and provide the minima of the functional F. Therefore instead of calculating the partition function of the Hamiltonian (1.1) we are able to calculate the partition function determined by the Gibbs factor $e^{-\int \sigma(\hat{M}^2) d\hat{A}}$ for minimal surfaces in S^3. This approach is especially useful in determining the structure of the topology (which can be complicated), like cubic or sponge phases.Both of these structures are phases with a finite density of handles. In the terminology of minimal surfaces in S^3 such structures, according (1.2), relate to minimal surfaces with a finite density of holes. Refering the reader to a detailed discussion of this correspondence in the paper [1] we add some words about the to physical properties of the cubic and sponge phases.

The cubic phase of complex liquids is constructed by means of triple-periodic surfaces. The elementary cell of such a structure at a scale of 10 μm to 150 μm consists of a large number of molecules which diffuse freely along the membrane. In this case the cubic structure is determined by the periodic distribution of the mass density $\rho(r)$ and other characteristics, including a $g(r)$ related to this one. The sponge phase L_3 is determined by the short-range order of distribution $\rho(r)$ or $g(r)$. In the language of a dual system distribution of holes in S^3 we can speak about periodic in the cubic case or amorphous in the sponge phase distribution of holes with density $n(r)$.

Here we point out some similarities with the fluctuation theory of holes in ordinary liquid, developed by Yakov Frenkel in 1926.

If the energy E_h of the hole created in a minimal surface is finite, then the mean density of holes can be estimated by

$$n \approx \frac{1}{\xi^2} \exp\left(-\frac{E_h}{T}\right) \tag{1.3}$$

where ξ^2 is the mean size of an elementary cell created by the holes on a minimal surface. This size is related to the mean value of the physical cell of a cubic structure by transformation (1.2). The exponential factor in Formula (1.3) has the meaning of the probability of creation

---

§ The equivalence between special Willmore surfaces and minimal surfaces which follows from Proposition 1 leads to a deep analogy with string theory. Minimal surfaces $M^2 \subset S^3$ are the "world sheets" of the string determined in S^3 by Nambu-Goto action. The surfaces M^2 play the role of instantons in the Yang-Mills theory defined in the space S^3.

of a hole. To estimate ξ we use the following observation. Let us write the full HW functional (1.1) with the Gaussian curvature term:

$$F = \frac{\kappa}{2} \int H^2 \, dA + \bar{\kappa} \int K \, dA \tag{1.4}$$

where two bending modules κ and $\bar{\kappa}$ (saddle-splay) determine a deformation of the membrane with respect to the mean and the Gaussian curvature correspondingly. The creation of a sponge and a cubic phase depends on a spontaneous rise of the complexity of the topology of the surface, that is, the creation of handles or of passages between layers. It happens when $\bar{\kappa} > 0$ because according to the Gauss-Bonnet theorem the second summand in (1.4) is equal $4\pi\bar{\kappa}(1 - N_h)$, where N_h is the number of handles. The process of a spontaneous creation of the handles leads to the appearance of an increasing number of handles, but of smaller size. The process stops at the equilibrium scale ξ for which we search. It happens when the positive terms of higher order of the mean curvature equal the negative term of the Gaussian curvature. For finding ξ we might estimate the summands of order H^{-4} and K^{-2} in the expansion (1.4). The minimisation of the energy developed in such a way, determine the equilibrium scale

$$\xi \simeq \sqrt{\frac{\kappa}{\bar{\kappa}}} a$$

We used the natural estimate κa^2 (a the molecular size) for the fourth order elasticity moduli.

The process of the spontaneous creation of holes and the consequent changing of the topology of membranes remind us of some constructions of a more complicated nature, namely in quantum gravity, the theory of space-time foams developed by J.Wheeler, S.Hawking and many others. It is not out of place to mention their famous predecessor W.K.Clifford with his space-theory of matter.

6. Unsolved problems and discussion.

In this section we discuss some open questions which seem important and interesting both for physics and mathematics.

1. The structure of Willmore surfaces.

As we mentioned above the construction of an infinite family of tori (the so called Hopf tori, by U.Pinkal) embedded in S^3, which are extremals but not minimal surfaces in S^3. Is it possible to construct such manifolds for surfaces of higher genus? The estimation of the dimension of the space of parameters yields the number $6g - 10(g > 2)$. This question is interesting for physics from the following point of view. We develop our theory of phase transitions as the theory of holes in minimal surfaces embedded in S^3. Does the picture change if we sum to the partition function over all extremals. A similar question exists in considering the of space-time foam of to quantum gravity.

2. Defects in Membranes.

Up to now we considered membranes with no internal order. It is interesting to study defects in membranes. The most investigated phase is the so called hexatic phase of point disclinations which is constructed by adding to the model (1.1) a local order parameter $\vec{n}$ which is a tangent vector to the surface. See the detailed discription in [2]. It is possible to obtain membranes with a more general lattice structure of defects. Our method uses a minimal periodic surfaces generated by a period lattice which is the discrete uniform subgroup of the group of motions, of R^3.

7. Generalized Kramers-Wannier Duality in Spin Systems.

In 1941 H.A.Kramers and G.H.Wannier discovered a particular symmetry, connecting low temperature and high temperature phases in the planar Ising model.

The corresponding transformation, the so called Kramers-Wannier (KW) transformation, is a special nonlocal substitution of variables in the partition function:

$$\sum_{x,x+\hat\alpha} exp\beta H(\sigma_x, \sigma_{x+\hat\alpha}) = \sum_{x',x'+\hat\alpha} exp\beta^* \tilde H(\mu_{x'}, \mu_{x'+\hat\alpha'}) \qquad (2.1)$$

Here the σ_x are "spin" variables determined on a two-dimensional lattice, the $\mu_{x'}$ are "spins" on the dual lattices; $(\sigma_x, \mu_x \in Z_2)$, H and $\tilde H$ are Hamiltonians of the system expressed in terms of σ and μ correspondingly and the parameter of the inverse temperature $\beta = (kT)^{-1}$ is related to β^* by the well-known formula

$$\beta^* = arth\ exp(-2\beta) \qquad (2.2)$$

The existence of such type of transformations is a very general property of lattice statistical systems that possess a discrete (and not only discrete) symmetry. The KW transformation enables us to find in many important cases the point of phase transition when the exact analytic expression for the partition function is unknown. Analogs of the KW transformation were constructed for many systems important from a physical point of view, e.g. XY-models, Z_N Ising models and some others (see [3] and [4] for more detailed references). These systems possess a commutative symmetry group.

The systems with "spin" variables taking values in a non-commutative group are also very interesting for physics. But this case is more difficult and up to now only very partial results have been obtained.

To the best of my knowledge there have been no serious achievements since the end of the 1970s, when a lot of attempts were made to generalize the KW transformation to non-Abelian case.

In the paper by A.Zamolodchikov and myself [4] was solved this problem in certain special cases, related to some subgroups of O(3). Our method was based on reduction to the commutative case.

The KW-transformation for systems with commutative symmetry group K (such as the Ising model) can be carried out by general methods. In this case the KW-transformation is a Fourier transformation of a spin system on the lattice L to the spin system on the dual lattice $\tilde L$ with spin variables taking values in the group $\hat K$, the group of characters of K.

The main difficulty in the construction of the KW transformation for a non-Abelian group G concerns the description of the dual object $\hat G$ of G. The space $\hat G$ is the space of all irreducible representations and, unlike the commutative case, is not a group. In this section I annonce the new results obtained by V.Buchstaber and myself which solve this problem for finite non-commutative groups. This method was inspired by some recent achievements in the theory of multivalued groups [5] and also has its roots in the classical work of G.Frobenius. A detailed exposition of our method will be published soon in [6].

We use standard mathematical notations and definition which can be found in any complete manual for the group theory for example in [7].

8. Generalized KW-duality.

Let $G = \{g_0 = e, ..., g_{n-1}\}$ be a finite non-commutative group.

We denote by $C(G)$ the group algebra of G and by $C[G]$ the ring of complex valued functions on G.

We define the canonical spliting $< \cdot, \cdot >$:

$$C[G] \bigotimes C[G] \to \mathbf{C} :< f, x >= f(\bar{x})$$

which determines two types of duality:

$$C(G) = Hom(C[G], \mathbf{C}), \quad C[G] = Hom(C(G), \mathbf{C})$$

where $\mathbf{C}$ is the field of complex numbers.

Let us consider the representation ad of the group G in the space C(G):

$$g_l : g_k \to g_l g_k g_l^{-1}$$

Denote by g_k^G the orbit of the action of ad on $g_k \in G$ and by $\delta_k \in C[G]$ characteristic function on G:

$$\delta_k(g_s) = \begin{cases} 1 \ if \ g_s \in g_k^G \\ 0 \ if \ g_s \notin g_k^G \end{cases}$$

Let m be the number of conjugacy classes under (the action of) the ad. Let us chooce representatives of these classes:

$$g_0, g_{k_1}, \dots g_{k_m}.$$

Proposition 1. A linear map:

$$W : C(G) \to \mathbf{C}$$

satisfies the condition:

$$W(g_k) = W(g_l g_k g_l^{-1})$$

iff

$$W = \sum \gamma_j \delta_{k_j} \in C[G] = Hom(C(G), \mathbf{C})$$

i.e.

$$W(g_s) = \sum_{j=1}^{m} \gamma_j \delta_{k_j}(g_s).$$

If we choose $\gamma = (\gamma_1, \dots, \gamma_m)$ as a vector of free parameters we obtain the general form of an ad-invariant linear mapping.

It is well known that on the space $C[G]$ there exists scalar product

$$< \varphi, \psi >= 1/n \sum_{g \in G} \varphi(g) \bar{\psi}(g).$$

The characters of the irreducible unitary representations of G; $\chi^1, \dots, \chi^m$, are ortonormal functions on G.

Let us consider the linear map:

$$\hat{W} : C[G] \to \mathbf{C}, \quad \hat{W} = \sum_{j=1}^{m} \hat{\gamma}_j \chi^j$$

$$\hat{W}(\psi) = \sum_{j=1}^{m} \hat{\gamma}_j < \chi^j, \psi >$$

Since the characters χ^j are ad-invariant functions, by the proposition 1, there exists, for each irreducible character χ^l the expansion:

$$\chi^l = \sum_{j=1}^{m} \gamma_j^l \delta_{k_j}.$$

Proposition 2. If we set

$$\gamma_j = \sum_{l=1}^{m} \gamma_j^l \hat{\gamma}_l$$

then linear mappings of dual spaces $C(G)$ and $C[G]$

$$W : C(G) \to \mathbf{C} \qquad W(g) = \sum_{j=1}^{m} \gamma_j \delta_{k_j}(g)$$

$$W : C[G] \to \mathbf{C} \qquad \hat{W}(\psi) = \sum_{j=1}^{m} \hat{\gamma}_j < \chi^j, \psi >$$

will be determined by the same function, or more precisely:

$$\sum \gamma_j \delta_{k_j} = n \sum_{j=1}^{m} \hat{\gamma}_j \chi^j$$

Proposition 3. The transformation

$$\gamma_j = \sum_{l=1}^{m} \gamma_j^l \hat{\gamma}_l \tag{2.3}$$

is valid for all finite groups and coinsides in all known cases with the ordinary Kramers-Wannier transforms.

The transformation (2.3) we shall call the Kramers-Wannier transformation for a non-commutative group G. Using this approach we obtain very easily the KW-transformation for the case of the group I_5 the group of symmetries of the icosahedron. This case was not considered in our earlier paper [4] and is important for studying phase transitions in quasicrystals.

The main idea of our approach is to study duality transformations of two spaces related to group G: the group algebra $C(G)$ and the space of regular functions $C[G]$. Our method can be generalized to more complicated cases of compact groups and has interesting consequences in the theory of quantum groups.

We hope to study these very intriguing possibilities in the near future.

Acknowledgements.

Most of the results presented in this talk were developed with my colleagues and friends E.Kats (part I) and V.Buchstaber (part II). I am very grateful to them for very fruitful collaboration. Part of the work was prepared during my visit to the Max-Planck Institut für Physik Komplexer Systeme, Dresden, Germany. I thank this Institute for its very favorable environment and financial support. This work was also partially supported by Grant 02-01-00734 of RFBR.

[1] E.I.Kats, M.I.Monastyrsky, Minimal surfaces and fluctuation of membranes with non-trivial topology, JETP **91**, 1279-1285 (2000).

[2] F.David, Introduction to the statistical mechanics of random surfaces and membranes in two dimensional guantum gravity and random sufaces, ed. by D.J.Gross, T.Piran, S.Weinberg, 81-123, World Scientific, Singapore, 1992.

[3] J.B.Kogut, An Introduction to lattice gauge theory and spin systems, Reviews of Modern Phys. **51**, 659-713 (1979).

[4] A.B.Zamolodchikov,M.I.Monastyrsky, The Kramers-Wannier transformation for Spin Systems, JETP, **50**, 167-172 (1979).

[5] V.M.Buchstaber, E.G.Rees, Multivalued group, their representations and Hopfs-alqebras transformation groups, **2**, 325-349 (1997)

[6] V.M.Buchstaber, M.I.Monastyrsky, Generalised Kramers- Wannier duality for finite groups, Submitted for Publ in J.Phys A.

[7] G. James, M.Liebeck, Representations and Characters of Groups, 2nd ed., Cambridge University Press, Cambridge (2001)

Inst. Phys. Conf. Ser. No 173: Plenary Sessions
Paper presented at 24th Int. Coll. Group Theoretical Methods in Physics, Paris, France, 15–20 July 2002
©*2003 IOP Publishing Ltd*

Exceptional Groups and Physics

Pierre Ramond

Institute for Fundamental Theory, Physics Department, University of Florida,
Gainesville, FL 32611, USA

Abstract. Quarks and leptons charges and interactions are derived from gauge theories associated with symmetries. Their space-time labels come from representations of the non-compact algebra of Special Relativity. Common to these descriptions are the Lie groups stemming from their invariances. Does Nature use Exceptional Groups, the most distinctive among them? We examine the case for and against their use. They do indeed appear in charge space, as the Standard Model fits naturally inside the exceptional group E_6. Further, the advent of the $E_8 \times E_8$ Heterotic Superstring theory adds credibility to this venue. On the other hand, their use as space-time labels has not been as evident as they link spinors and tensors under space rotations, which flies in the face of the spin-statistics connection. We discuss a way to circumvent this difficulty in trying to generalize eleven-dimensional supergravity.

1. Introduction

With the advent of Quantum Mechanics, Lie algebras and the groups they generate have found widespread uses in the description of physical systems. The quantum-mechanical state of a particle is determined by labels. Some, like the particle's momentum (or position) are continuous, others like its spin, and charges assume discrete values. All stem from irreducible unitary representations of Lie algebras. The continuous ones pertain to irreps of non-compact groups, and the discrete ones to compact groups. Mass and spin label the representations of the non-compact group of special relativity the Poincaré group, and the color of a quark roams inside a representation of the compact color group $SU(3)$. Moreover, their interactions are determined by dynamical structures based on these invariance groups. Although Nature does not use *all* mathematical structures created by our mathematical friends, it seems to favor some particularly unique and beautiful ones for the description of its inner secrets. Alas, they often appear in disguised broken-down form, so it is up to us to divine their existence from incomplete evidence: awareness of these structures is an important research tool.

There are four infinite families of simple Lie algebras, the garden variety algebras: $A_n \sim SU(n+1)$, $B_n \sim SO(2n+1)$, $C_n \sim Sp(2n)$, and $D_n \sim SO(2n)$, all with n extending to ∞. They describe spacetime rotation, quark and lepton charges, and their associated Yang-Mills gauge structures. Today, $SU(N)$ gauge theories with N large are intensely studied.

In the Lie garden, one also finds five rare flowers, the exceptional algebras: G_2, F_4, E_6, E_7 and E_8, their rank indicated by the subscripts. In view of Nature's fascination with unique structures, they merit further study.

2. A Short Course on Exceptional Algebras

The smallest exceptional algebra [1] is G_2. It has 14 parameters and its smallest representation is seven-dimensional, the seven imaginary directions of octonions (Cayley numbers). It is in fact the automorphism group of the octonion algebra. An octonion ω is written as

$$\omega = a_0 + a_\alpha e_\alpha , \qquad \alpha = 1, 2, \ldots, 7 , \qquad e_\alpha^2 = -1, \qquad e_\alpha e_\beta = \Psi_{\alpha\beta\gamma} e_\gamma, \quad (1)$$

for $\alpha \neq \beta$, where $\Psi_{\alpha\beta\gamma}$ are totally antisymmetric and equal to $+1$ for the combinations $(\alpha\beta\gamma) = (123) , (246) , (435) , (651) , (572) , (714) , (367)$, and zero otherwise. This algebra is non-associative as their associator does not vanish:

$$[e_\alpha, e_\beta, e_\gamma] \equiv (e_\alpha e_\beta) e_\gamma - e_\alpha (e_\beta e_\gamma) = 2\tilde{\Psi}_{\alpha\beta\gamma\delta} e_\delta, \qquad (2)$$

where $\tilde{\Psi}_{\alpha\beta\gamma\delta}$ is the dual of the structure constants. G_2 acts on the seven imaginary units. There are four Hurwitz algebras, the real numbers R, the complex numbers C, the quaternions Q, and the octonions Ω. The three quaternion imaginary units are the Pauli spin matrices (multiplied by i), and their automorphism group is $SU(2)$. All have the property that the norm of their product is the product of their norms.

All other exceptional algebra can be constructed terms of (3×3) antihermitian traceless matrices with elements over products of two sets of Hurwitz algebras. This leads to the "magic square" of Tits and Freudenthal.

Apply the construction to a matrix with elements over $\Omega \times \Omega'$. An "octonionic octonion" has $8 \times 8 = 64$ elements, while an imaginary one has $7 + 7 = 14$ elements. In an antihermitian traceless matrix, this accounts for $3 \times 64 + 2 \times 14 = 220$ parameters. Adding to them the two automorphism groups, we get the 248 parameters of E_8, the largest exceptional Lie algebra . If Lie algebras can be associated with cars, surely E_8 is the Delahaye of Lie algebras!

3. Charge Spaces

The state of an elementary particle is labelled at a given time as

$$| m , s , x^i , x^- , \{s_a\} ; \xi_1, \xi_2, \ldots \xi_N >_t , \qquad (3)$$

where the first set are the space-time labels given in light-cone coordinates: the continuous transverse positions x^i, where i runs over the transverse dimensions of space, the spins s (more than one in higher dimensions). The second labels are the internal charges ξ_α which are described by irreps of compact Lie algebras. The space-time is thus written in terms of orthogonal group of rotations in the transverse space, subgroup of the semi-simple non-compact Poincaré group. On the other hand, the discrete internal charges belong to representations of compact simple Lie groups.

Quarks and lepton charges span representations of the Standard Model group $SU(3) \times SU(2) \times U(1)$. Remarkably they fit snuggly into two representations of the larger $SU(5)$ [2].

$$SU(5) \supset SU(3) \times SU(2) \times U(1) , \qquad (4)$$

with three families transforming as $\bar{5} \oplus 10$. With the discovery of neutrino masses it is almost certain that each neutrino has a Dirac partner, the right-handed neutrino. With it, each family fits in the fundamental spinor representation of $SO(10)$ [3]:

$$SO(10) \supset SU(5) \times U(1) ; \qquad 16 = \bar{5} \oplus 10 \oplus 1 . \qquad (5)$$

It is amazing that the natural algebra with one rank higher is that of the exceptional E_6 [4], with

$$E_6 \supset SO(10) \times U(1) ; \qquad \mathbf{27} = \mathbf{16} \oplus \mathbf{10} \oplus \mathbf{1} , \qquad (6)$$

which is a complex representation. E_6 and the spin representations of orthogonal groups $SO(4n+2)$, $n \geq 2$ are the only fundamental complex representations with no anomalies. This of course opens the road to E_8:

$$E_8 \supset E_7 \times SU(2) \supset E_6 \times U(1) . \qquad (7)$$

This ladder to exceptional algebras is even more apparent through their Dynkin diagrams.

$\mathbf{E_8}$

By chopping-off one dot at a time, one arrives at the Dynkin of the Standard Model

$\mathbf{SU(3) \times SU(2)}$

Exceptional groups make their appearance in superstring theory. The gauge group of the most promising heterotic string [5] in ten space-time dimensions is nothing but $E_8 \times E_8$ with 496 gauge parameters (496 is a perfect number: can anyone doubt string theory [6]?). There, one compactifies over a six-dimensional manifold to get to four space-time dimensions. To preserve supersymmetry, the manifold must have $SU(3)$ holonomy. A trip in the extra dimensions gets you back where you started modulo $SU(3)$, and this is compensated by the $SU(3)$ obtained from $E_8 \supset E_6 \times SU(3)$. Thus E_6 is naturally obtained! The number of families is the number of holes in the six-dimensional Calabi-Yau manifold [7].

4. Space Charges

Exceptional groups naturally contain orthogonal groups as subgroups

$$E_8 \supset SO(16) ; \qquad E_7 \supset SO(12) \times SO(3) ;$$
$$E_6 \supset SO(10) \times SO(2) ; \qquad F_4 \supset SO(9) , \qquad G_2 \supset SO(3) \times SO(3) . (8)$$

They could therefore contain (in their non-compact form) the conformal group in D spacetime dimensions $SO(D,2)$ and its Poincaré subgroup or else as contracted form of the above. However any role that exceptional groups may play in the description of space charges (position, mass, spin,...) has to be quite subtle.

The reason is that their representations contain both spinorial and tensorial representations of their orthogonal subgroups. For instance, the fundamental irrep of F_4

$$26 = 16 \oplus 9 \oplus 1 \ ,$$

contains both the $SO(9)$ spinor and the vector representations. Thus F_4 transformations naturally mix these, but in quantum theory this is like mixing apples and oranges: space spinors obey Fermi-Dirac statistics while the vectors are Bose-Einstein. This simple fact makes their relevance to space charges indirect to say the least. On the other hand, fermions and bosons do coexist in Nature and there must be some symmetry which links them. It is well at this point to examine the difference between bosons and fermions

4.1. Fermion-Boson Confusion

In four dimensions, fermions and bosons are naturally differentiated, as fermions have half-odd integer helicities while the boson helicities are integers. In $d + 1$ spacetime dimensions, fermions transform as spin representations of the transverse little group $SO(d-1)$, while bosons are transverse tensors. As a result in most dimensions, fermions and bosons have different dimensionalities, but there are exceptions: in $1+1$ dimensions, there is no transverse little group and both fermions and bosons are uni-dimensional. This makes it easier to confuse them and there is the well-know phenomenon of bosonisation or fermionisation.

In $9 + 1$ dimensions, the little group is $SO(8)$, with its unique triality property according to which bosons and fermions are group-theoretically equivalent, and this is the domain of superstring theories where this triality is put to excellent use. The Dynkin diagram of $SO(8)$ displays this triality

SO(8)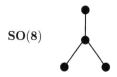

and it is the Mercedes of Lie groups. This triality is explicit in the $F_4 \supset SO(8)$ decomposition. One of the great surprises in string theories has been the emergence of a new theory which contains all string theories; it is called M-theory and it is not a string theory and lives in one more space dimension than the superstrings. The heterotic string theory can be obtained by compactifying M-theory over the line S^1/Z_2. The infrared limit of M-theory is $N = 1$ supergravity in eleven dimensions [8]. When compactified on a d-torus, one finds a non-compact exceptional group, $E_{d(d)}$, where the number in parenthesis is the number of non-compact generators minus the number of compact ones. In particular for $d = 8$ one obtains a theory in $2 + 1$ dimensions with a non-compact version of E_8 [9].

It seems that there is also a special arrangement between fermions and bosons in eleven spacetime dimensions. Yet there is nothing remarkable about the transverse little group $SO(9)$. Its Dynkin diagram

SO(9)

does not display any symmetries; it is more like a Trabant than a Mercedes or a Delahaye! How can there be any confusion between bosons and fermions? Yet it describes the space in which M-theory roams! Finally we note two interesting anomalous Dynkin embeddings which might have hitherto unsuspected applications. The first is

$$SO(16) \supset SO(9) ,$$

in which the sixteen-dimensional *spinor* representation of $SO(9)$ fits snuggly into the sixteen-dimensional *vector* representation of $SO(16)$. We will come back to it later. The second is

$$SO(26) \supset F_4 ,$$

which equates the 26-dimensional vector irrep of $SO(26)$ to that of F_4. Its real form

$$SO(25,1) \supset F_{4(-20)} \supset SO(9) ,$$

may provide a heterotic path to eleven dimensions starting from the original bosonic theory.

5. $N = 1$ Supergravity in 11 Dimensions

Supergravity in eleven spacetime dimensions is the infrared limit of M-theory. It is a local field theory, believed to diverge at three loops. On the light-cone, the theory is described by a chiral superfield with 256 components

$$\Phi(y^-, \vec{x}, \theta^\alpha) = \phi(y^-, \vec{x}) + \theta^\alpha \, \psi_\alpha(y^-, \vec{x}) + \cdots \theta^1\theta^2 \cdots \theta^8 \lambda(y^-, \vec{x}) ,$$

expanded in terms of eight complex anticommuting Grassman variables, and where y^- is the displaced chiral coordinate

$$y^- = x^- - i\bar{\theta}\,\theta/\sqrt{2} .$$

Introduce the sixteen (256×256) Dirac matrices

$$\{ \Gamma^a, \Gamma^b \} = 2\delta^{ab} , \qquad a, b = 1, 2 \ldots 16 ,$$

with vector indices transforming as the $SO(9)$ spinor(recall the anomalous Dynkin embedding). These are not to be confused with the (16×16) nine Dirac matrices which transform as $SO(9)$ vectors

$$\{ \gamma^i, \gamma^j \} = 2\delta^{ij} , \qquad i, j = 1, 2 \ldots, 9 .$$

Together they allow for a neat way of writing the $SO(9)$ generators acting on this superfield

$$S^{ij} = -\frac{i}{4}\left(\gamma^{ij}\right)_{ab}\Gamma^{ab} \ ,$$

where in the usual notation $\gamma^{ij} = \gamma^i\gamma^j$, $i \neq j$, $\Gamma^{ab} = \Gamma^a\Gamma^b$, $a \neq b$. The 52 F_4 parameters i split into the 36 S^{ij} which generate $SO(9)$, and sixteen $SO(9)$ spinors, T^a. Algebraic closure is given by

$$[T^a, T^b] = \frac{i}{2}\left(\gamma^{ij}\right)^a S^{ij} \ ,$$

so there is a whiff of F_4 in the light-cone description of $N = 1$ SUGRA in eleven dimensions. We shall see later this is the tip of a beautiful algebraic structure. Their action on the superfield show it to split in three $SO(8)$ representations, the **44** of the symmetric second rank traceless tensor, the **84** the antisymmetric third rank tensor, and **128** the Rarita-Schwinger spinor-vector field. It is convenient to write the superfield in terms of the three highest weight components of each representation (in our basis)

$$\Phi(y^-, \vec{x}, \theta^\alpha) = \theta^1\theta^8\left(h(y^-, \vec{x}) + \theta^4\,\psi(y^-, \vec{x}) + \theta^4\theta^5\,A(y^-, \vec{x})\right) \ ,$$

where h, ψ, A are the highest weights of the **44**, **128**, and **84**, respectively. Define the Dynkin indices as

$$I^{(k)} = \sum_{\text{irrep}} w^k \ ,$$

in terms of the length of each weight w. They satisfy the composition law

$$I^{(k)}_{\mathbf{r}\times\mathbf{s}} = I^{(0)}_{\mathbf{r}} I^{(k)}_{\mathbf{s}} + I^{(0)}_{\mathbf{s}} I^{(k)}_{\mathbf{r}} \ ,$$

where $\mathbf{r}$ and $\mathbf{s}$ are any two irreps. $I^{(0)}$ is the dimension of the representation. Since $SO(9)$ has rank four it has four such independent indices. The three SUGRA irreps have much in common, as the following table shows

irrep	(1001)	(2000)	(0010)
D	128	44	84
I_2	256	88	168
I_4	640	232	408
I_6	1792	712	1080
I_8	5248	2440	3000

It shows that

$$I^{(k)}_{128} = I^{(k)}_{44} + I^{(k)}_{84} \ , \qquad k = 0, 2, 4, 6 \ ,$$

but the sum rule fails for the higher invariant $k = 8$. It has been conjectured [10] that it is this failure that is responsible for the non-renormalizability of $N = 1$ SUGRA.

Amazingly, this pattern of equalities is repeated for an infinite number of sets of three $SO(9)$ representations, which describe higher spin massless particles [11]. It has to do with the fact that there are three equivalent ways to embed $SO(9)$ inside F_4 [12]. This is the octonionic equivalent I-spin, U-spin and V-spin which label three equivalent ways to embed $SU(2)$ inside $SU(3)$. The F_4 Weyl chamber is $1/3$ that of $SO(9)$. Take a highest weight in the F_4 Weyl chamber, λ. Let ρ be the sum of the fundamental weights. There exist two Weyl reflections C, which map λ outside the F_4 Weyl chamber, but stay inside that of $SO(9)$. Hence there is a unique way to associate one F_4 representation to three $SO(9)$ irreps. The mapping is

$$C \bullet \lambda \ = \ C \left(\lambda + \rho_{F_4} \right) - \rho_{SO(9)} \ .$$

This mapping associates with each F_4 irrep, a set of three $SO(9)$ representations called Euler triplets. Equality betwen its Dynkin indices is guaranteed by the character formula

$$V_\lambda \otimes S^+ - V_\lambda \otimes S^- \ = \ \sum_C sgn(C) \mathcal{U}_{C \bullet \lambda} \ ,$$

where V_λ is any F_4 representation written in terms of its $SO(9)$ content, $S^\pm$ are the two spinor irreps of $SO(16)$ also written in terms of $SO(9)$ through the anomalous Dynkin embedding

$$\mathbf{128} \ = \ \mathbf{44} \oplus \mathbf{84} \ ; \qquad \mathbf{128} \ = \ \mathbf{128} \ .$$

The failure of the equality for the eigth order invariant is linked to the fact that S^+ and S^- have different Pfaffian invariants [13]. One recognizes the "trivial" Euler triplet as the three fields of $N = 1$ SUGRA in eleven dimensions associated with $\lambda = 0$. This character formula is akin to an index formula for Kostant's operator associated with the coset $F_4/SO(9)$, the sixteen-dimensional projective Cayley-Moufang plane. Euler triplets are solutions of Kostant's equation [14]

$$\mathcal{K} \, \Psi \ \equiv \ \Gamma^a \, T^a \, \Psi \ = \ 0 \ ,$$

where the T^a generate the $F_4/SO(9)$ tranformations.

$$[\, T^a \, , T^b \,] \ = \ i \, f^{[ij] \, ab} \, T^{ij} \ .$$

Kostant's operator commutes with the generalized $SO(9)$ generator made up of an "orbital" and the previously defined "spin" part

$$L^{ij} \ \equiv \ T^{ij} + S^{ij} \ .$$

The solutions to Kostant's equation are the Euler triplets, and the trivial solution is the SUGRA triplet. The number of representations in each Euler set is the ratio of the order of the F_4 and $SO(9)$ Weyl groups. It is also the Euler number of the coset manifold, hence the name.

It is convenient [15] to express the F_4 in terms of three sets of 26 real coordinates: u_i which transform as transverse space vectors, u_0 as scalars, and ζ_a as space spinors. This enables us to write the Euler triplets as chiral superfields of the form [16]

$$\Phi(y^-, \vec{x}, \theta^\alpha) \ = \ \theta^1 \theta^8 \Big(h(y^-, \vec{x}, u_i, \zeta_a) + \theta^4 \, \psi(y^-, \vec{x}) + \theta^4 \theta^5 \, A(y^-, \vec{x}) \Big) \ ,$$

where now the components h, ψ and A are the highest weight components of the three irreps with definite polynomial dependence on the new coordinates. For the proper spin-statistics interpretation, the twistor-like variables ζ_a must appear quadratically. It turns out that the ζ's appear in even powers only for those Euler triplets that have the same number of bosons and fermions!

The physical interpretation of these triplets is still unclear. Their quantum numbers suggest that they can be related by the emission of fields with specified quantum number, in analogy with the transition within a gauge multiplet by emission of a W-boson. In particular one recognizes the two-form field, so it is possible that emission of a two-form potential from the superparticle in eleven dimension might generate the other triplets, but supersymmetry is broken in the process.

Poincaré invariance requires the Euler triplets to be massless as there are not enough fields among them to complete into massive $SO(10)$ little group multiplets. Furthermore there is no supersymmetry relating members of an Euler triplet except, of course, the first one. There are grave difficulties [17] when coupling to gravity a massless particle with spin greater than two in flat space-time [18]: either one gets a relativistic theory ghosts or else a theory that does not satisfy Lorentz invariance. There are no such objections with an infinite number of such particles, which would correspond to a highly non-local theory.

There are indications that one may need in fact an infinite number of Euler triplets. If the divergences of supergravity are linked to the lack of cancellation in $I^{(8)}$, the same would be true for any Euler triplet contribution to a loop amplitude, but the sign of the deficit is the same for all triplets as it is proportional to the dimension of the F_4 representation from which it originates. To get a cancellation with manifestly positive quantities, an infinite number are required, in the sense of ζ-function regulariçation. The dimension of any F_4 representation is a 24th order polynomial in the Dynkin integer indices, and the ζ-function of even order vanish, so there might be hope. However it is clear that this is not the language to address this issue as there are formidable technical difficulties to overcome before being able to carry out this program.

6. Exceptional Jordan Algebra

This tour ofExceptional groups in Physics would not be complete without a mention of the Exceptional Jordan Algebra [19]. Jordan algebras provide an alternate way of describing Quantum Mechanics in terms of its observables. Let J_a be any observable, we introduce the commutative product

$$J_a \circ J_b = J_b \circ J_a ,$$

(9)

which maps observables into observables. Since matrices do not commute, the Jordan associator

$$(J_a, J_b, J_c) \equiv J_a \circ (J_b \circ J_c) - (J_a \circ J_b) \circ J_c .$$

(10)

is not zero, but satisfies the Jordan identity

$$(J_a, J_b, J_a^2) = 0 .$$

(11)

These equations serve as the postulates of the commutative but non-associative Jordan Algebras.By writing the Hamiltonian in terms of two hermitian matrices

$$H = \frac{i}{4}[A, B] ,$$

(12)

we can express time evolution in terms of the Jordan associator

$$i\hbar \frac{\partial J}{\partial t} = (A, J, B) .$$

(13)

There would be nothing new in this rewriting of Quantum Mechanics, if it were not for Jordan, von Neumann and Wigner [20] who noticed that the Jordan axioms were satisfied by (3×3) hermitian matrices over octonions. The non-associativity of the octonion forbids a Hilbert space interpretation, and this is what makes it special. It is known as the exceptional Jordan algebra (EJA). Its group of derivations of the EJA is nothing but the exceptional F_4!

Gürsey suggested the EJA as label for internal charges, especially since $F_4 \supset SU(3) \times SU(3)$. Our analysis with Euler triplets suggest rather that the $SO(9)$ subgroup be interpreted as the light-cone little group in eleven dimensions [21]. If $SO(9)$ is the light-cone little group in eleven dimensions, we want time evolution to preserve it. To that effect, we need to couple the EJA to an external field that transforms non-trivially under $SO(9)$. Otherwise, time evolution with a fixed external potential preserves at most $SO(7)$. If the $SO(9)$ subgroup of EJA automorphism group F_4 can indeed be identified with the light-cone little group in eleven space-time dimensions, it will suggest the EJA as the charge space of a very special system.

References

[1] For references and more details, see P. Ramond, Introduction to Exceptional Lie Groups and Algebras, CALTECH-68-577 (Dec 1976), unpublished.

[2] H. Georgi and S.L. Glashow, *Phys.Rev.Lett.* 32:438-441,1974

[3] H. Fritzsch and P. Minkowski, *Annals Phys.* 93:193-266,1975 ; H. Georgi, Invited Talk at Williamsburg Conference, 1975.

[4] F. Gürsey, P. Ramond and P. Sikivie, *Phys.Lett.* B60:177,1976

[5] David J. Gross, Jeffrey A. Harvey, Emil J. Martinec, Ryan Rohm, *Phys.Rev.Lett.* 54:502-505,1985

[6] Also remarked by L. Boya, an aficionado of exceptional groups, these Proceedings.

[7] P. Candelas, Gary T. Horowitz, Andrew Strominger, Edward Witten, *Nucl.Phys.* B258:46-74,1985

[8] E. Cremmer, B. Julia, J. Scherk , *Phys. Lett* **B76**, 409(1978)

[9] for a modern treatment and references therein, see M. Günaydin, K. Koepsell, H. Nicolai . Adv.Theor.Math.Phys.5:923-946,2002, hep-th/0109005

[10] T. Curtright, *Phys. Rev. Lett.* **48**, 1704(1982)

[11] T. Pengpan and P. Ramond, *Phys. Rep.* **315**. 137(1999)

[12] B. Gross, B. Kostant, P. Ramond, and S. Sternberg, *Proc. Natl. Acad. Scien.*, 8441 (1998)

[13] I thank E. Witten for this remark

[14] B. Kostant, *Duke J. of Mathematics* **100**, 447(1999).

[15] T. Fulton, *J. Phys. A:Math. Gen.* **18**, 2863(1985).

[16] Lars Brink, P. Ramond, and X.Xiong, Supersymmetry and Euler Multiplets, JHEP 0210:058,(2002); HEP-TH 0207253.

[17] Loyal Durand III, *Phys. Rev.* **128**, 434 (1962); K. Case and S. Gasiorowicz, *Phys. Rev.* **125**, 1055 (1962); M. Grisaru, R. Pendleton, P. Van Nieuwhenhuizen, *Phys. Rev.* **D15**, 496(1977); B. DeWit and D.Z. Freedman, *Phys. Rev.* **D 21**, 358(1980); C. Aragone and S. Deser, *Nuovo Cim.* **57B**, 33 (1980). E. Witten and S. Weinberg, *Phys. Lett.* **B96**, 59 (1980).

[18] In curved space the no-go theorems can be evaded, see M. A. Vasiliev, hep-th/0104246, and references therein.

[19] See F. Gürsey, Invited Talk given at the Conference on Non-Associative Algebras at the University of Virginia in Charlottesville, March 1977.

[20] P. Jordan, J. Von Neumann, and E. Wigner, *Ann. Math.* **35**, 65(1933)

[21] P. Ramond, Algebraic Dreams, UFIFT-HET-01-27 (Dec 2001) HEP-TH 0112261; L. Smolin, hep-th/0104050.

Inst. Phys. Conf. Ser. No 173: Plenary Sessions
Paper presented at 24th Int. Coll. Group Theoretical Methods in Physics, Paris, France, 15–20 July 2002
©2003 IOP Publishing Ltd

Group projection method in statistical systems

Krzysztof Redlich [a,b], **Frithjof Karsch** [a] **and Ahmed Tounsi** [c]

[a] Fakultät für Physik, Universität Bielefeld, Postfach 100 131, D-33501 Bielefeld, Germany
[b] Institute of Theoretical Physics, University of Wrocław, PL-50204 Wrocław, Poland
[c] Laboratoire de Physique Théorique et Hautes Energies, Université Paris 7, Paris, France

Abstract. We discuss an application of group theoretical methods to the formulation of the thermodynamics of systems constrained by the conservation laws described by a semi–simple compact Lie group. A general projection method that allows to construct a partition function for a given irreducible representation of the Lie group is outlined. Applications of the method in Lattice Gauge Theory (LGT) for non–zero baryon number and in the phenomenological description of particle production in ultrarelativistic heavy ion collisions are also indicated.

1. Introduction

In the formulation of the thermodynamics of a strongly interacting medium one needs in general to implement the constraints imposed by the conservation laws that are governed by an internal symmetry of the Hamiltonian [1, 2, 3, 4, 5, 6, 7, 8]. In this paper, we discuss how applying the basic properties of the Lie groups and their representations one can derive the partition function that accounts for these constraints. We present examples of the application of the method for different statistical systems and phenomenological models.

2. Projected partition function

The usual way of treating the problem of quantum number conservation in statistical physics is by introducing the grand canonical partition function,

$$Z(\mu_S, T, V) = \text{Tr}[e^{-\beta(\hat{H} - \mu_S \hat{S})}], \tag{1}$$

with $\hat{H}$ being the Hamiltonian, $\hat{S}$ the charge operator, β the inverse temperature, V the volume of the system and μ_S the chemical potential associated with the conserved charge S. The chemical potential in Eq. (1) plays the role of the Lagrange multiplier which is fixed by the condition that the charge is conserved on the average and has the required value $\langle S \rangle$ such that:

$$\langle S \rangle = T \frac{\partial \ln Z(\mu_S, T)}{\partial \mu_S}. \tag{2}$$

The grand canonical partition function (1) provides an adequate description of the statistical properties of the system only if the number of particles carrying charge S is asymptotically large and if its fluctuations can be neglected.

In order to derive a more general statistical operator that is free from the above requirements one usually replaces the function (1) by the canonical partition function Z_S that accounts for an exact charge conservation

$$Z_S(T, V) := \text{Tr}_S[e^{-\beta \hat{H}}]. \tag{3}$$

The subscript S under the trace indicates that it is restricted to the states that carry an exact value S of the conserved charge. Obviously $Z_S(T, V)$ from Eq. (3) and $Z(\mu_S, T, V)$ from Eq. (1) are connected via cluster decomposition

$$Z(\mu_S, T, V) = \sum_{s=-\infty}^{s=+\infty} Z_S \lambda_S^s, \tag{4}$$

with the fugacity parameter $\lambda_S := e^{\beta \mu_S}$. Thus, Z_S is viewed as a coefficient in the Laurent series of Z in the fugacity. Applying the Cauchy formula in Eq. (4) one calculates Z_S by taking an inverse transformation:

$$Z_S(T, V) = \int_{-\pi}^{+\pi} \frac{d\phi}{2\pi} e^{-is\phi} \tilde{Z}(\phi, T, V). \tag{5}$$

The generating function $\tilde{Z}(\phi, T, V) := Z(\lambda_S = e^{i\phi}, T, V)$ is obtained from the grand canonical partition function (1) by "Wick rotation" of the chemical potential, $\mu_S \to i\phi$. The function $\tilde{Z}$ is unique for all canonical partition functions that are formulated for a fixed value of the conserved charge.

The integral (5) describes the projection onto the canonical partition function that accounts for an exact conservation of an abelian charge. This is the projection procedure as Z_S is also obtained from

$$Z_S(T, V) = \text{Tr}[e^{-\beta \hat{H}} \hat{P}_S] \tag{6}$$

where $\hat{P}_S$ is the projection operator on the states with a given value of the conserved charge S. For an additive quantum numbers, $\hat{P}_S$ is just a delta function $P_S = \delta_{S', S} \hat{I}$. Using the Fourier expression of $\delta_{S', S}$ in Eq. (6) one reproduces the projected formula (5).

The conservation of an additive quantum number is usually related with an invariance of the Hamiltonian under the abelian U(1) internal symmetry group. In many physics applications it is of importance to generalize the projection method to symmetries that are related with a non-abelian Lie groups G. An example is a special unitary group SU(N) that plays an essential role in the strong interaction. A generalization of the projection method would require to specify the projection operator or the generating function. Consequently, the partition function obtained with a specific eigenvalue of the Casimir operator that fixes the multiplet of the irreducible representation of the symmetry group G can be determined.

To find the generating function for the canonical partition function with respect to the symmetry group G, one introduces $\tilde{Z}(g)$ [1]

$$\tilde{Z}(g) := \text{Tr}[U(g)e^{-\beta \hat{H}}] \tag{7}$$

as the function on the group G with U(g) being a unitary representation of the group and $g \in G$. The U(g) can be decomposed into irreducible representations $U_\alpha(g)$,

$$U(g) = \sum_{\alpha}^{\oplus} U_\alpha(g), \tag{8}$$

and thus, from Eqs. (7) and (8) one can write explicitly

$$\begin{aligned} \tilde{Z}(g) : &= \sum_{\alpha} \text{Tr}_\alpha[U_\alpha(g)e^{-\beta \hat{H}}] \\ &= \sum_{\alpha} \sum_{\nu_\alpha, \xi_\alpha} \langle \nu_\alpha, \xi_\alpha \mid U_\alpha(g)e^{-\beta \hat{H}} \mid \nu_\alpha, \xi_\alpha \rangle, \end{aligned} \tag{9}$$

where ν_α labels the states within the representation α and ξ_α are degeneracy parameters of a given representation. Due to the requirement of an exact symmetry the only non–vanishing matrix elements of the evolution operator $e^{-\beta \hat{H}}$ are those diagonal in ν_α. The matrix elements of $U_\alpha(g)$ are non-zero if they are diagonal in ξ_α. Finally, the matrix elements of the Hamiltonian are independent of the states within representation and those of $U(g)$ of degeneracy factors. Consequently, the matrix elements in Eq. (9) factorize and the generating function is

$$\tilde{Z}(g) = \sum_\alpha \frac{\chi_\alpha(g)}{d(\alpha)} Z_\alpha(T, V), \tag{10}$$

where $d(\alpha)$ is the dimension and χ_α is the character of the irreducible representation α. In the above expression Z_α is introduced as a *canonical partition function* with respect to G symmetry and is defined as

$$Z_\alpha(T, V) := \mathrm{Tr}_\alpha e^{-\beta \hat{H}} = d(\alpha) \sum_{\xi_\alpha} \langle \xi_\alpha \mid e^{-\beta \hat{H}} \mid \xi_\alpha \rangle. \tag{11}$$

Eq. (10) connects a canonical partition function Z_α to the generating function on the group. Thus, Z_α is the coefficient in the cluster decomposition of the generating function with respect to the characters of the representations associated with the symmetry group.
The orthogonality relation of the characters,

$$\frac{1}{d(\alpha)} \int d\mu(g) \chi_\alpha^*(g) \chi_\gamma(g) = \delta_{\alpha,\gamma} \tag{12}$$

allows to find the canonical partition function. From (10) and (12) one gets

$$Z_\alpha(T, V) = d(\alpha) \int d\mu(g) \chi_\alpha^*(g) \tilde{Z}(g). \tag{13}$$

This result is a generalization of (5) to an arbitrary internal symmetry group that is a compact Lie group. The formula holds for any dynamical system as it is independent of the specific form of the Hamiltonian.

To find the canonical partition function one needs to determine first the generating function $\tilde{Z}(g)$ defined on the symmetry group G. If the group is of rank r, then the character of any irreducible representation is a function of r variables $\{\gamma_1, \ldots, \gamma_r\}$, thus also is the generating function (10). Diagonalizing the unitary operators $U(g)$ under the trace in Eq.(7) and denoting by J_k ($k = 1, \ldots, r$) the commuting generators of G, the generating function can be formulated on the maximal abelian subgroup of G as

$$\tilde{Z}(\gamma_1, \ldots, \gamma_r) = \mathrm{Tr}[e^{-\beta \hat{H} + i \sum_{i=1}^{r} \gamma_i J_i}]. \tag{14}$$

Thus $\tilde{Z}(\vec{\gamma})$ is just the GC partition function with complex chemical potentials that are associated with all generators of the Cartan sub-algebra.

Equations (13) and (14) provide a complete description of the statistical operators that account for the constraints imposed by the conservation laws of internal symmetries of the Hamiltonian. The simplicity of the projection formula (13) is that the operators that appear in the generating function are additive. One sees that the problem of extracting the canonical partition function with respect to an arbitrary compact Lie group G is reduced to the projection onto a maximal abelian subgroup of G.

The generating function (14) can be calculated by applying standard perturbative diagrammatic methods or a mean field approach. However, if the interactions in the

Hamiltonian can be omitted or can effectively be described as a modification of the particle dispersion relations by implementing an effective particle mass, then the trace in Eq. (14) can be done exactly. In this particular situation the generating function can be written [1] as

$$\tilde{Z}(\vec{\gamma}) = \exp[\sum_\alpha \frac{\chi_\alpha(\vec{\gamma})}{d(\alpha)} Z_\alpha^1] \tag{15}$$

where the one-particle partition function

$$Z_\alpha^1 = V \int \frac{d^3 p}{(2\pi)^3} (1 \pm \exp(-\beta\sqrt{p^2 + m_\alpha^2}))^{-1} \tag{16}$$

is just a thermal phase–space that is available to all particles of mass m_α that belong to a given irreducible multiplet. The sum is taken over all particle representations that are constituents of a thermodynamical system.

2.1. Phenomenological model of colour confinement

To illustrate how the projection method results in the partition function, we discuss a statistical model that accounts for the conservation of a non-abelian charge related with the global $G = SU_c(N) \times U_B(1)$ internal symmetry of the Hamiltonian. As a physical system one considers a thermal fireball that is composed of quarks and gluons that carry the colour degrees of freedom related with $SU_c(N)$ symmetry [3, 4] and baryon number related with $U_B(1)$ subgroup. The system has temperature T and volume V. The interactions between quarks and gluons are implemented effectively as resulting in dynamical particle masses that are T dependent, e.g. through $m_{q,(g)} \sim gT$. Thus, since under these conditions the free–particle dispersion relations are preserved, Eq. (15) provides a correct description of the generating function on the symmetry group G. The sum in the exponent in (15) gets the contributions from quarks, anti–quarks and gluons that transform respectively under the fundamental (0,1), its conjugate (1,0) and adjoint (1,1) representation of the $SU_c(N)$ symmetry group. Thus [3],

$$\ln \tilde{Z}(T, V, \vec{\gamma}, \gamma_B) = \frac{\chi_Q}{d_Q} Z_Q^1 + \frac{\chi_Q^*}{d_Q} Z_{\bar{Q}}^1 + \frac{\chi_G}{d_G} Z_G^1 \tag{17}$$

where $\vec{\gamma} = (\gamma_1, .., \gamma_{N-1})$ are the parameters of the $SU_c(N)$ and γ_B of the $U_B(1)$ symmetry groups. Through an explicit calculation of one-particle partition functions for massive quarks the corresponding generating functions read:

$$\ln \tilde{Z}_Q(T, V, \vec{\gamma}, \gamma_B) = \frac{g_Q}{d_Q} \frac{m_Q^2 VT}{2\pi^2} \sum_{n=0}^\infty \frac{(-1)^{n+1}}{n^2} K_2(nm_Q/T)$$
$$[\ e^{i\gamma_B n/T} \chi_Q(n\vec{\gamma}) + e^{-i\gamma_B n/T} \chi_Q^*(n\vec{\gamma})] \tag{18}$$

where the two terms in the brackets represent the contribution of quarks and antiquarks respectively. The corresponding contribution for massive gluons is obtained as

$$\ln \tilde{Z}_G(T, V, \vec{\gamma}, \gamma_B) = \frac{g_G}{d_G} \frac{m_G^2 VT}{2\pi^2} \sum_{n=0}^\infty \frac{1}{n^2} K_2(nm_G/T)[\chi_G(n\vec{\gamma}) + \chi_Q^*(n\vec{\gamma})] \tag{19}$$

The coefficients g_G, g_Q and $d_Q = N, d_G = N^2 - 1$, are, respectively, the quark and gluon spin–isospin degeneracy factors and dimensions of the representations.

The statistical operator is now obtained from (13) through the projection of the generating functions (18) and (19) onto a given sector of quantum numbers. Here we quote the results for the $SU_C(3)$ colour singlet partition function that represents a global colour neutrality condition (phenomenological confinement) of the quark–gluon plasma droplet [3]:

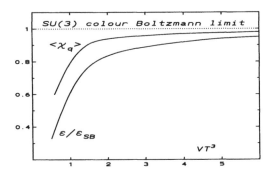

Figure 1. Thermodynamical functions obtained from the projected partition function (20) are compared to their Stefan–Boltzmann limit as a function of VT^3. The results for the energy density ϵ/ϵ_{SB} and the expectation value of the character in the fundamental representation of $SU(3)$ are indicated in the figure. The results are obtained for a vanishing baryon–chemical potential μ_B in (20).

$$Z_{\alpha=0}(\mu_B, T, V) = \int d\mu(\gamma_1, \gamma_2) \exp\{c_1\chi_G +$$
$$c_2[L_R\cosh(\beta\mu_B) + iL_I\sinh(\beta\mu_B)]\}. \quad (20)$$

where a finite average value of the baryon number is controlled by the chemical potential μ_B. The constants c_1 and c_2 can be extracted from Eqs. (18–19) and taking only the first terms in the series. The parameters L_I, L_R denotes the imaginary and real part of the character in the fundamental representation of the $SU_C(3)$ group whereas χ_G is the character of the adjoint representation. The integration is done on the $SU_C(3)$ group with an appropriate Haar measure $\mu(\vec{\gamma})$.

To illustrate how the group projection influence the thermodynamics of the system we display in Fig. 1 the behavior of the energy density and the thermal average of the characters as obtained from the colour singlet statistical operator from Eq. (20). The results are shown as a function of dimensionless parameter VT^3. In the absence of group projection and for a vanishing value of μ_B the partition function would have a simple form; $Z(T, V) = \exp[c_1 + c_2]$. The resulting energy density ϵ_{SB} describes a well known Stefan-Boltzmann limit and $\langle\chi_q\rangle = 1$ in this case. For an asymptotically large value of $VT^3 \to \infty$, that corresponds to the thermodynamical limit, the colour projected results coincide with their Boltzmann values. However, for small values of VT^3 a large suppression of thermodynamical quantities is seen in Fig. 1. This is a generic feature of the partition function restricted to a fixed representation. The requirement of an exact conservation of the quantum numbers impose a strong constraint on the particle thermal phase–space that results in the suppression seen in Fig. 1. In the thermodynamical limit, the number of particles that carry a quantum number related with a given representation is so large that the above constraints are irrelevant.

In finite temperature gauge theory the zero component of the gauge field A_0 takes on the role of the Lagrange multiplier which guaranties that all states satisfy the Gauss law [3, 10, 11]. In the Euclidean space one can choose a gauge in such a way that $A_0^\nu(x, \tau)\lambda_\nu$ is a constant in Euclidean time, so that $A_0^{ab} = g^{-1}\alpha^{ab}\delta_{ab}$. In such a gauge the Wilson loop defined as

$$L(x) = \frac{1}{N}\text{Tr}P\exp[ig\int_0^\beta A_0(x, \tau)d\tau] \quad (21)$$

represents the character of the fundamental representation of the $SU_c(N)$ group [3]. Thus, the projected partition function could be viewed as an effective model that connects the coloured quasi-particles [9] with the Wilson loop [11]. This partition function can also be related to

144

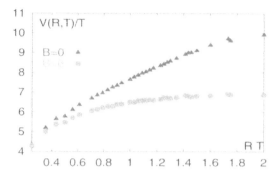

Figure 2. Heavy quark potential from Ref. [16] as a function of quark anti–quark separation RT. The results were obtained from Monte–Carlo simulations of quenched QCD at fixed value of the coupling constant $\beta = 5.62$ and baryon number $B = 0$ and $B = 6$.

the strong coupling effective free energy of the lattice gauge theory with a finite chemical potential [3] and to the effective potential of the $SU(N)$ spin model for the Wilson loop.

2.2. Projected partition function in Lattice Gauge Theory

The formulation of a finite baryon number density in QCD on the lattice leads to a sever numerical problem of complex probability. The QCD partition function formulated on the Euclidean lattice and after integrating out the Wilson fermions, becomes [12, 13]

$$Z(\mu_B, T, V) = \int \prod dU e^{S_G} \det[\aleph(\mu_B, U)] \tag{22}$$

where S_G is the gluon action, $\aleph := 1 - \kappa M(U, \mu_B)$ is the fermion matrix and κ is the hopping parameter. In order to perform the Monte-Carlo simulations with this statistical operator it is necessary that the measure be real and positive [13, 14, 15]. One situation where $\det[\aleph]$ is real is when there exist an invertible operator P such that $\aleph^+ = P\aleph P^{-1}$. For the Wilson fermions at zero chemical potential and for complex μ_B this relation hold with $P = \gamma_5$. However, for real and non-vanishing $\mu_B \neq 0$ this is not anymore valid and the probabilistic interpretation of the path integral representation of the QCD partition function is violated [13, 16]. Consequently the Monte-Carlo method is not valid and the numerical solution of the QCD thermodynamics is not accessible anymore.‡ One way to partly overcome the above problems in finite density QCD is to use a projection method [3, 10, 16]. Rather than introducing a non–vanishing chemical potential [14], that is formulate QCD in the grand canonical ensemble with respect to U(1)-baryon symmetry, one my go over to a canonical formulation of the thermodynamics and fix directly the baryon number [3, 10, 16]. Following the general discussion of Section 1 the projected partition function onto a given sector of baryon number could be obtained from [3, 16]:

$$Z_B(T, V) = \int_0^{2\pi} \frac{d\phi}{2\pi} e^{-iB\phi} Z(\mu_B = i\phi, T, V) \tag{23}$$

where the function Z under the integral represents the grand canonical partition function in Eq.(22) calculated with a complex μ_B, and consequently is a real and positive quantity. Nevertheless, due to a Fourier integration the projected partition function still suffers a numerical problem of oscillating functions. However, in the quenched limit of QCD and

---

‡ The complex structure of the fermionic contribution to the partition function is very transparent in the model discussed in the last section. From Eq. (20) it is clear that the complex structure appears for $\mu_B \neq 0$ since $L_I \neq 0$ for the $SU(N)$ system with $N > 2$. On the other hand the quark generating function (18) is indeed real for the complex baryon chemical potential.

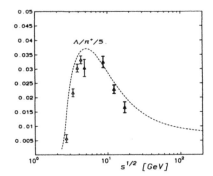

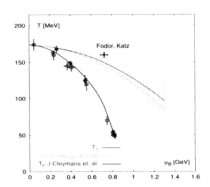

Figure 3. The left hand figure: the statistical model results (broken line) on the energy dependence of Λ/π ratio. The data are from [21]. The right hand figure: a comparisons of the chemical freeze-out parameters with the phase boundary line. The upper thin line represents the LGT results obtained in Ref. [22] and the lower thin line describes the conditions of constant energy density that was fixed at $\mu = 0$ [22]. The full (Fodor-Katz) point is the end-point of the crossover transition from Ref. [23]. The lower line is the unified freeze-out curve from [20].

for moderate values of baryon number the group integration was done analytically and the first results on QCD thermodynamics at finite baryon density were established [16].

The presence of a finite net baryon number modifies non-trivially the properties of QCD medium. As an example of Monte–Carlo study we show in Fig. 2 a heavy quark anti–quark potential obtained from the Polyakov loop correlations [16]. For zero baryon number it shows the usual linearly rising behavior for the quenched case. For finite $B = 6$ the potential remains finite at large distances due to screening of the static quark anti–quark pair by the already present net baryon charge in the system.

2.3. Canonical projection in heavy ion collisions

Central heavy ion collisions at relativistic incident energies represent an ideal tool to study nuclear matter at high temperatures and densities. Particle production is – at all incident energies – a key quantity to extract information on the properties of nuclear matter under these extreme conditions. In this context a particular role has been attributed to particles carrying strangeness that is related with U(1) invariance of the strong interactions [2, 6]. The production of secondaries measured in heavy ion collisions was shown in the literature to be very satisfactory described in the context of statistical thermal models [6, 7, 17, 18, 19, 20]. However, already the first attempts to describe strange particle production in low energy central and high energy peripheral heavy ion collisions have shown that the conservation of strangeness should be implemented exactly [6].

An exact formulation of strangeness conservation is implemented through the projection of the partition function onto states of fixed representation of the U(1) group [1]. For the hadron resonance gas the strangeness neutral partition function is obtained from [7]:

$$Z_{S=0}^{C}(T,V) = \frac{1}{2\pi} \int_{-\pi}^{\pi} d\phi \, \exp\left(\sum_{n=-3}^{3} S_n e^{in\phi} \right) \tag{24}$$

where S_n is a thermal phase space available to all particles and resonances that carry strangeness n with $n \in (-3, 3)$. The density n_s^i of particle i carrying strangeness s is derived

146

form (24) as [7]

$$(n_s^i)^C = (n_s^i)^{GC} F_s(T, V) \tag{25}$$

where $(n_s^i)^{GC}$ is the grand canonical density and $F_s(T, V)$ is the suppression factor that measures a deviation of $(n_s^i)^C$ from its asymptotic, grand canonical value. For large V and/or T the factor $F_s(T, V) \rightarrow 1$. The group projection implies that $F \leq 1$, thus it suppresses a particle densities that carries the U(1) charge. This suppression was found to increase with decreasing collision energy, increasing strangeness content of the particles and decreasing centrality of the collisions [7]. These properties are well observed in experimental data. As an illustration for the model comparison with experimental data [21] we show in Fig. (3–left) the results on the ratio of lambda to pion multiplicities [7].

The statistical model that accounts for an exact conservation of U(1) charges was found in the literature to reproduce the basic features of particle yields obtained in heavy ion and hadron–hadron collisions. The yields were found to be well reproduced with thermal parameters (the temperature and baryon chemical potential) that follow a universal freeze-out line of *fixed energy per particle* of 1 GeV [20]. The freeze-out line is shown in Fig. (3–right) together with the most recent LGT results on the position of the critical curve in the $(T-\mu_B)$ plane [22, 23].

Acknowledgements: We wish to thank P. Braun-Munzinger and L. Turko for interesting discussions. K.R acknowledges the support of the Alexander von Humboldt Foundation.

[1] K. Redlich and L. Turko, Z. Phys. B97 (1980) 279; L. Turko, Phys. Lett. B104 (1981).
[2] 153; R. Hagedorn and K. Redlich, Z. Phys. C27 (1985) 541; Rafelski and M. Danos, Phys. Lett. B97 (1980) 279; L. Turko and J. Rafelski, Eur. Phys. J. C18 (2001) 587.
[3] D. Miller and K. Redlich, Phys. Rev. D37 (1988) 3716; Phys. Rev. D35 (1987) 2524; H. Th. Elze, D. Miller and K. Redlich, Phys. Rev. D35 (1987) 748.
[4] D.H. Rischke, M.I. Gorenstein, A. Schäfer, H. Stöcker and W. Greiner, Phys. Let. B278 (1992) 19; Z. Phys. C56 (1992) 325.
[5] A.B. Balantekin, Phys. Rev. E64 (2001) 066105; P.N. Meisinger, et al., Phys. Rev. D65 (2002) 034009; M.I. Gorenstein, et al., Phys. Let. B524 (2002) 265; H. Th. Elze, et. al., Phys. Let. B506 (2001) 123; A.G. Michael, et. al., Phys. Rev. D59 (1999) 034009; M.G. Mustafa, et al., Euro Phys. J. C5 (1988) 711; C.D. Fosco, Phys. Rev. D57 (1988) 6554; C. Spieles, et. al., Phys. Rev. C57 (1998) 908; L.D. Mc Lerran and A. Sen, Phys. Rev. D32 (1985) 2794.
[6] P. Braun-Munzinger, et al., Nucl. Phys. A697 (2002) 902 and references therein.
[7] A. Tounsi, et al., J. Phys G28 (2002) 2095; Eur. Phys. J. C24 (2002) 35; J. S. Hamieh, K. Redlich, and A. Tounsi, Phys. Lett. B486 (2000) 61
[8] C.M. Ko, et al., Phys. Rev. Lett. 86 (2001) 5438; S. Jeon, et al., Nuc. Phys. A697 (2002) 546.
[9] J. Engels, et al., Z. Phys. C42 (1989) 341; A. Peshier, et al., Phys. Rev. C61 (2000) 045203.
[10] A. Roberge and N. Weiss, Nucl. Phys. B275 [FS17] (1986) 734.
[11] A. Gocksch, R.D. Pisarski, Nucl. Phys. B402 (1993) 657; R. Pisarki, hep-ph/0203271; Nucl. Phys. A702 (2002) 151; A. Dumitru and R. Pisarski, Phys. Lett. B525 (2002) 95.
[12] For a recent review see: I.M. Barbour, S.E. Morrison, E.G. Klepfish, J.B. Kogut and M.P. Lombardo, Nucl. Phys. B (proc. Suppl.) 60 (1998) 220.
[13] F. Karsch, Lect. Notes. Phys. 583 (2002) 209; I. Barbour, et al., Nucl. Phys. Proc. Supp. 60A (1998) 220.
[14] P. Hasenfratz and F. Karsch, Phys. Lett. B125 (1983); J. Kogut et al., Nucl. Phys. B225 [FS9] (1983) 93.
[15] B. Berg, J. Engels, E. Kehl, B. Waltl and H. Satz, Z. Phys. C31 (1986) 167.
[16] F. Karsch, Nucl. Phys. Proc. Suppl. 83 (2000) 14; J. Engels et al., Nucl. Phys. Proc. Suppl. 83 (2000) 366; Nucl. Phys. B558 (1999) 307; O. Kaczmarek et al., Phys. Rev. D62 (2000) 034021.
[17] P. Braun-Munzinger, I. Heppe, and J. Stachel, Phys. Lett. B465 (1999) 15; P. Braun-Munzinger and J. Stachel, J. Phys. G28 (2002) 1971; Phys. Lett. B490 (2000)196; Nucl. Phys. A690 (2001) 119c; A. Keränen, and F. Becatini, J. Phys. G28 (2002) 2041; Becatini, et al., Phys. Rev. C64 (2001) 024901.
[18] J. Letessier, and J. Rafelski, Int. J. Mod. Phys. E9 (2000) 107.
[19] P. Braun-Munzinger, et al., Phys. Lett. B518 (2001) 41; D. Magestro, J. Phys. G28 (2002) 1745.
[20] J. Cleymans, et al., Phys. Rev. C60 (1999) 0544908; Phys. Rev. Lett. 81 (1998) 5284; Phys. Rev. C59 (1999) 1663; Phys. Lett. B485 (2001) 27.
[21] A. Mischke, for NA49 Collaboration, nucl-ex/0209002.
[22] S. Ejiri, et al., hep-lat/0209012; C. Schmidt, et al., hep-lat/0209009.
[23] Z. Fodor and S.D. Katz, Phys. Lett. B534 (2002) 87.

Inst. Phys. Conf. Ser. No 173: Plenary Sessions
Paper presented at 24th Int. Coll. Group Theoretical Methods in Physics, Paris, France, 15–20 July 2002
©2003 IOP Publishing Ltd

Symmetric polynomials in physics

H J Schmidt and J Schnack

Department of Physics, University of Osnabrück, Germany

Abstract. We give two examples where symmetric polynomials play an important rôle in physics: First, the partition functions of ideal quantum gases are closely related to certain symmetric polynomials, and a part of the corresponding theory has a thermodynamical interpretation. Further, the same symmetric polynomials also occur in Berezin's theory of quantization of phase spaces with constant curvature.

1. Introduction

It often happens that mathematical theories have unexpected applications in physics. In the present case of the theory of symmetric polynomials (SP) we have, additionally, the remarkable situation, that physicists have re-discovered certain fragments of the theory of SP in order to solve problems in few-particle quantum statistical mechanics [1],[2],[3],[4],[5],[6]. Actually, it turns out that a part of the physical theory of ideal quantum gases is equivalent to a part of the theory of SP if a certain translation scheme is applied, see below. The central idea of this scheme, namely that partition functions can be considered as evaluations of certain SP, is not novel, but appeared at various places in the literature, see [7],[8],[9],[10], often in the context of generalized statistics. Nevertheless, the relevance for the problems treated in the above-mentioned articles and the consequences of this observation seem to have remained largely unnoticed.

A second field where SP might be important tools is the theory of quantization in the form suggested by F. A. Berezin [11] and subsequently further developed, see e. g. [12]. Here the same SP as in quantum statistical mechanics occur in the expansion of the quantization operator for two-dimensional phase spaces with constant curvature. One could speculate about the underlying reasons and possible extensions of this connection.

2. SP and ideal quantum gases

We will only explain the basic idea of the connection between SP and quantum statistical theory of ideal gases. Further details may be found in [13] and [14] and in the literature quoted there.

It is well-known that the eigenstates of the N-particle Hamiltonian without interactions can be characterized by "occupation number sequences" $i \mapsto n_i$. Here n_i is the occupation number of the i-th energy level E_i of the 1-particle Hamiltonian. Hence

$$\sum_i n_i = N, \tag{1}$$

and, for fermions, additionally

$$n_i \in \{0, 1\}. \tag{2}$$

Equivalently, each eigenstate can be characterized by a monomial of degree N

$$x^n \equiv \prod_i x_i^{n_i}, \tag{3}$$

where the x_i are abstract, commuting variables corresponding to the energy levels and n denotes the whole occupation number sequence. The energy eigenvalues corresponding to these eigenstates are

$$E = \sum_i E_i n_i. \tag{4}$$

Now we can express the N-particle partition function as

$$Z_N^{\pm}(\beta) = \sum_E e^{-\beta E} = \sum_n \exp\left(-\beta \sum_i E_i n_i\right) \tag{5}$$

$$= \sum_n \prod_i \left(e^{-\beta E_i}\right)^{n_i} = \left.\sum_n \prod_i x_i^{n_i}\right|_{x_i = \exp(-\beta E_i)} \tag{6}$$

$$\equiv \begin{cases} b_N(x_1, x_2, \ldots) & : \text{Bosons}(+) \\ f_N(x_1, x_2, \ldots) & : \text{Fermions}(-) \end{cases} \tag{7}$$

Here the sum over n is subject to the constraint (1) for bosons, and to (1), (2) for fermions. Since these constraints are invariant under permutations of the variables $x_1, x_2, \ldots$, the resulting sum of the monomials x^n in (6) will be a symmetric polynomial of the $x_1, x_2, \ldots$. We call these SP "fermi polynomials" f_N or "Bose polynomials" b_N, respectively. In the theory of SP the f_N are called "elementary SP" and the b_N "complete SP". However, in this article we will stick to our more physical nomenclature.

The partition function $Z_N(\beta)$ of a particular system is obtained by evaluation of the corresponding SP along the curve $\beta \mapsto x_i(\beta) = \exp(-\beta E_i)$, see figure 1. Hence we have a $1 : 1$ correspondence between certain SP and certain "partition types" of ideal gases. Here the "partition type" of a system is given by the number N of particles, the number $L \in \{1, 2, \ldots, \infty\}$ of abstract energy levels (or the dimension of the 1-particle Hilbert space) and the type of the statistics, Bose or Fermi. The values E_i of the energy levels, including their degeneracy, only determine the system and its particular partition function. Hence a SP corresponds not to a single system but to a large class of systems. It is then obvious, that mathematical relations between the f_N and the b_N can be translated into physical relations between the corresponding partition functions, irrespective of the values of the E_i.

There is a third kind of SP with a physical meaning in the theory of ideal gases, the "power sums"

$$p_n \equiv \sum_i x_i^n. \tag{8}$$

Evaluation at $x_i = \exp(-\beta E_i)$ gives

$$p_n|_{x_i = \exp(-\beta E_i)} = \sum_i \exp(-n\beta E_i) = Z_1(n\beta). \tag{9}$$

One of the central results of the elementary theory of SP is that each of the above families of SP, the f_N, the b_N, and the p_n, can be used as a "basis" of SP, in the sense that any SP can be expressed as a polynomial of the f_N (resp. b_N or p_n). This implies that the fermionic partition functions can be expressed by means of the bosonic ones and vice versa. Moreover,

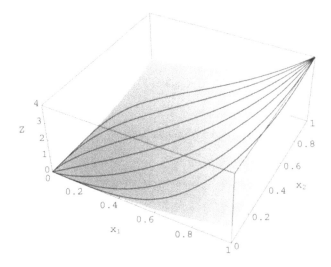

Figure 1. This figure shows the graph of the Bose polynomial $b_3 = x_1^3 + x_1^2 x_2 + x_1 x_2^2 + x_2^3$ with $N = 3$ and $L = 2$. A selected number of curves parametrized by the inverse temperature β is shown which illustrate the partition functions for special systems obtained by the evaluation of b_3 at the values $x_1(\beta) = \exp(-\beta E_1)$ and $x_2(\beta) = \exp(-\beta E_2)$.

both in turn can be expressed by means of the 1-particle partition function at different inverse temperatures $n\beta$. These relations are more or less known in the physical literature but their origin in the theory of SP has only recently be disclosed [13].

SP can be defined through their generating functions. In statistical mechanics the corresponding generating functions are called "grand canonical partition functions" and will be denoted by $B(z)$ for bosons and $F(z)$ for fermions. The formal parameter z is physically interpreted as the fugacity $z = \exp(-\beta\mu)$, where μ is the chemical potential. The physical domain of z is $(0, 1)$ for bosons and $(0, \infty)$ for fermions.

There exists a fundamental symmetry $\omega : \Lambda \longrightarrow \Lambda$ of the ring Λ of SP, which maps the f_N onto the b_N, see [15]. It is connected with the equation $F(z)B(-z) = 1$. The physical interpretation of this relation is that the fermionic grand canonical partition function is related to the analytical continuation of the bosonic one to negative z and vice versa. Similar relations for other thermodynamic functions are implied. Whether analytical continuation is possible depends on the system under consideration. For the system of particles in a box it has been shown that analytical continuation of the partition functions is possible in the thermodynamic limit [3] and the resulting Bose-Fermi symmetry has been discussed. Another Bose-Fermi symmetry has been observed for particles in odd space dimensions confined by a common harmonic oscillator potential [5]. Here $Z_1(\beta) = -Z_1(-\beta)$ implies $Z_N^+(\beta) = (-1)^N Z_N^-(-\beta)$. Since here analytical continuation is involved w. r. t. the β-plane, it is not yet clear how this symmetry is related to that given by ω

These and further relations and translation schemes will be sketched in the table below. For details see [13] and, as a standard reference for the theory of SP, [15].

Physics	Mathematics
Abstract energy levels	Variables $x_1, x_2, \ldots$
Occupation by N particles	Monomials $\mathbf{x^n}$
Partition types	Symmetric polynomials $p(\mathbf{x}) = \sum_{\mathbf{n}} \mathbf{x^n}$
-	Ring of symmetric polynomials Λ
Partition function $Z_N(\beta)$	Evaluation of $p(\mathbf{x})$ at $x_i = e^{-\beta E_i}$, $i = 1, \ldots, L$
$\left.\begin{array}{l} F(z) \\ B(z) \end{array}\right\}$ Grand canonical partition functions	$\left.\begin{array}{l} F(z) \\ B(z) \end{array}\right\}$ Generating functions for f_n, b_n
Fugacity $z = e^{-\beta\mu}$	$F(z) \equiv \prod_{i=1}^{L} (1 + x_i z) = \sum_{n=0}^{L} f_n z^n$ $B(z) \equiv \prod_{i=1}^{L} \frac{1}{1-x_i z} = \sum_{n=0}^{L} b_n z^n$
Bose- Fermi symmetry by analytical continuation in the z-plane [3]	$B(z)F(-z) = 1$
	$\left.\begin{array}{l} \omega : \Lambda \longrightarrow \Lambda \\ \omega : f_n \mapsto b_n \end{array}\right\}$ involutive automorphism of graded rings
$\sum_i \left(e^{-\beta E_i} \right)^n = Z_1(n\beta)$	p_n, evaluated
-	Generating function: $P(z) = \frac{d}{dz} \log B(z) = \sum_{n=1}^{L} p_n z^n$
$\langle N \rangle$	$zP(z)$
Landsberg's identities (Appendix E of his 1961 textbook, rediscovered several times)	Newton's identity: $n f_n = \sum_{r=1}^{n} (-1)^{r-1} p_r f_{n-r}$

3. SP and Berezin quantization

The main idea of Berezin's approach to quantization is the use of "generalized coherent states" (GCS) in order to establish an approximate equivalence between classical and quantum observables and states. GCS will be denoted by $|\alpha\rangle$ where the "parameter" α runs through some phase space M. The most important property of GCS is the completeness relation

$$1 = \int_M |\alpha\rangle\langle\alpha| \, d\alpha. \tag{10}$$

CGS are used to map operators A onto functions on M by means of

$$j(A)(\alpha) \equiv \frac{\langle\alpha|A|\alpha\rangle}{\langle\alpha|\alpha\rangle}. \tag{11}$$

The adjoint map is given by

$$j^*(f) = \int_M f(\alpha)|\alpha\rangle\langle\alpha| \, d\alpha, \tag{12}$$

where we ignore all questions about the exact domains of definition of (11) and (12). The "quantization operator" $j \circ j^*$ maps functions onto functions. It is used for semiclassical expansions of physical quantities w. r. t. powers of h, where h is a formal parameter ("Planck's constant") on which the Hilbert space and the GCS depend.

In the case of a flat phase space $M = \mathbf{R}^2$, the GCS are chosen as the usual coherent states discovered by E. Schrödinger. Then

$$j \circ j^* = \exp(-h^2\Delta), \tag{13}$$

where Δ is the Laplace operator in $\mathbf{R}^2$, see e. g. [16]. There are at least two other cases where $j \circ j^*$ can be calculated: The Lobachevskij plane L and the 2-sphere S^2. Here $j \circ j^*$ is given in the form of an infinite product, see [11], which resembles the grand canonical partition functions considered in the previous section. Hence $j \circ j^*$ can be expanded into power series w. r. t. Δ, where the coefficients are Bose polynomials for L and Fermi polynomial for S^2. Δ is the corresponding Laplace-Beltrami operator of L, resp. S^2. More explicitely:

Case 1

$$M \quad = \text{LOBACHEVSKIJ plane } L$$
$$|\alpha\rangle \quad : \text{BEREZIN'S coherent states [11]}$$

$$j \circ j^* = \prod_{n=0}^{\infty} \left(1 - \frac{h^2}{(1+nh)(1-(n-1)h)}\Delta\right)^{-1} \tag{14}$$

$$= \sum_{N=0}^{\infty} b_N(x_1, x_2, \ldots)\Delta^N \tag{15}$$

$$\text{where} \quad x_n = \frac{h^2}{(1+nh)(1+(n-1)h)} \tag{16}$$

Case 2

$$M \quad = \text{2-sphere } S^2$$
$$|\alpha\rangle \quad : \text{BLOCH'S coherent states [16]}$$
$$h \quad = \frac{1}{2s} \quad (s : \text{ spin quantum number}) \tag{17}$$

152

$$j \circ j^* = \prod_{n=0}^{\infty} \left(1 + \frac{h^2}{(1+nh)(1+(n+1)h)} \Delta \right) \tag{18}$$

$$= \sum_{N=0}^{\infty} f_N(x_1, x_2, \ldots) \Delta^N \tag{19}$$

$$\text{where} \quad x_n = \frac{1}{(2s+n)(2s+n+1)} \tag{20}$$

It is remarkable, that the bose and fermi polynomials occur for the two simplest cases of 2-dimensional phase spaces with constant negative and positive curvature. Both kind of polynomials can be considered as special cases of the so-called Schur polynomials s_λ, where λ denotes a Young diagram or, equivalently, a partition of N, see [15]. In fact, $b_N = s_{(N)}$ and $f_N = s_{(1^N)}$, where (N) (resp. (1^N)) is the Young diagram consisting of a single row (resp. column). One might wonder whether also other Schur polynomials occur in the expansion of the quantization operator for other choices of phase spaces and GCS.

References

[1] Landsberg P T 1961 Thermodynamics with Quantum Statistical Illustrations (New York: Interscience)
[2] Borrmann P and Franke G 1993 J. Chem. Phys. **98** 2484-2485
[3] Lee N H 1997 Phys. Rev. E **55** 1518-1520
[4] Schmidt H J and Schnack J 1998 Physica A **260** 479-489
[5] Schmidt H J and Schnack J 1999 Physica A **265** 584-589
[6] Arnaud J Chusseau L and Philippe F 2000 Phys. Rev. B **62** 13482-13489
[7] Balentekin A B 1992 Proceedings of the International Symposium on Group Theory and Special Symmetries in Nuclear Physics, Ann Arbor, 1991, edited by Draayer J (Singapore: World Scientific) 16-24
[8] Meljanac S, Stojić M and Svrtan D 1997 Phys. Lett. A **224** 319-325
[9] Aldrovandi R 2001 Special Matrices of Mathematical Physics (Singapore: World Scientific)
[10] Jellal A, Palev T D, and Van der Jeugt J 2001 J. Phys. A **34** 10179-10199
[11] Berezin F A 1975 Commun. Math. Phys. **40** 153-174
[12] English 2002 this volume
[13] Schmidt H J and Schnack J 2001 Am. J. Phys. **69** (12) 53-57
[14] Balentekin A B 2001 Phys. Rev. E **64** 066105
[15] Macdonald I G 1979 Symmetric Functions and Hall Polynomials (Oxford: Clarendon)
[16] Klauder J R and Skagerstam B S 1985 Coherent states (Singapore: World Scientific)

Inst. Phys. Conf. Ser. No 173: Plenary Sessions
Paper presented at 24th Int. Coll. Group Theoretical Methods in Physics, Paris, France, 15–20 July 2002
©2003 IOP Publishing Ltd

Strings, symmetries and representations

C. Schweigert, I. Runkel
LPTHE, Université Paris VI, 4 place Jussieu, F – 75 252 Paris Cedex 05

J. Fuchs
Institutionen för fysik, Universitetsgatan 5, S – 651 88 Karlstad

Abstract. Several aspects of symmetries in string theory are reviewed. We discuss the rôle of symmetries both of the string world sheet and of the target space. We also show how to obtain string scattering amplitudes with the help of structures familiar from the representation theory of quantum groups.

1. Strings and conformal field theory

Symmetries of various types, and consequently representation theoretic tools, play an important rôle in string theory and conformal field theory. The present contribution aims at reviewing some of their aspects, the choice of topics being influenced by our personal taste. After a brief overview of string theory and conformal field theory, we first discuss orbifolds and duality symmetries. We then turn to D-branes and theories of open strings, which we investigate in the final section using Frobenius algebras in representation categories.

A minimalistic point of view on (perturbative) string theory is to regard it as a perturbative quantization of a field theory, with the perturbation expansion being organized not in terms of graphs, i.e. one-dimensional objects, but rather in terms of surfaces. What makes this perturbation expansion particularly interesting is that it even covers the quantization of theories that include a gravitational sector.

The configuration space of a classical string is given by the embeddings of its two-dimensional world sheet Σ, with local coordinates τ and σ, into a target space M with coordinates X^μ. Both Σ and M are supposed to be endowed with a metric and possibly other background fields. It is convenient to change the perspective and regard the target space coordinate X^μ, via the embedding of Σ into M, as a function of the world sheet coordinates τ, σ, and thus as a classical field $X^\mu(\tau, \sigma)$ on Σ. This classical field theory is governed by an action S, the sigma-model action, which depends on the embedding X and on the metric h on the world sheet. S must in particular be invariant under local rescalings of the metric h of Σ

$$S[e^{\Phi(\sigma,\tau)}h, X] = S[h, X] \tag{1}$$

so that we deal with a classical 2D conformal field theory (CFT). Such theories carry an action of an infinite-dimensional symmetry algebra, the Witt algebra, respectively (after quantization) of its central extension, the Virasoro algebra. There can also be further symmetries, so that the Virasoro algebra is generically only a subalgebra of the so-called *chiral algebra*, which is provided by all holomorphic fields in the theory (the field affiliated with the Virasoro algebra is the stress-energy tensor T). In simple cases, like a flat background

$M = \mathbb{R}^{p,q}$, the solutions of the classical field equations split into left moving and right moving modes,

$$X^\mu(\sigma, \tau) = X_L^\mu(\sigma{+}\tau) + X_R^\mu(\sigma{-}\tau). \tag{2}$$

When the classical field theory can be quantized, one obtains a quantization of the space-time coordinates as fields in a two-dimensional theory. As a result there is a rich interplay between world sheet physics and space-time physics. For instance, the vanishing of the β-functions in the two-dimensional field theory implies field equations of a theory on target space that include a gravitational sector. To describe string theory, the BRST procedure must be applied to the conformal symmetry. Nilpotency of the BRST charge restricts the target space M, in the case of superstrings and of a flat metric on M, to 9+1 space-time dimensions. To give another example, modular invariance of the torus partition function

$$Z(\beta) = \mathrm{tr}_{\mathcal{H}_{\mathrm{string}}} e^{-\beta H_{\mathrm{string}}} \tag{3}$$

of the world sheet theory, i.e. invariance of $Z(\beta)$ under transformations

$$i\beta \mapsto \frac{a\,i\beta + b}{c\,i\beta + d} \tag{4}$$

with $a, b, c, d \in \mathbb{Z}$, $ad - bc = 1$, implies absence of anomalies [28]. One should be aware, though, that the converse is not true – it is easy to write down anomaly-free spectra not coming from a modular invariant partition function. Thus, at least as closed strings are concerned, string theory is a strictly more selective theoretical framework than particle field theory.

Bosonic string theory has a serious drawback: its spectrum contains tachyons, i.e. states of negative mass squared. This problem can be circumvented by enlarging the symmetries of the world sheet to a Lie *super*algebra containing the super-Virasoro algebra. Then among the holomorphic fields one finds, along with the stress-energy tensor T, its superpartner, the supercurrent G. Correspondingly, the world sheet must now carry the structure of a super-Riemann surface, rather than a Riemann surface. (In practice, one often works in a bosonic setting, employing the equivalence between super-Riemann surfaces and Riemann surfaces with a spin structure. Special care is then required, in particular when studying such theories on surfaces with boundaries [16].) The extension to world sheet supersymmetry can be implemented either for both left moving and right moving degrees of freedom, leading to superstring theories of type II, or for one chirality only, leading to heterotic strings.

If *space-time* supersymmetry is imposed as an additional requirement, say for 'phenomenological' reasons in compactifications to four space-time dimensions, then a further extension of the world sheet symmetry is required. One needs $N = 2$ supersymmetry on the world sheet (of which an $N = 1$ subalgebra is gauged), which means the presence of two supercurrents $G^{(\pm)}$ as well as an abelian current J. In addition a projection, the so-called GSO projection, must be imposed. This can be done in two different ways, depending on a relative sign for left movers and right movers, which leads to superstring theories of type IIA and IIB, respectively.

As mentioned above, for superstrings BRST invariance requires space-time to be 9+1-dimensional. But the space-time we experience has only 3+1 dimensions. Therefore one considers targets M for which six dimensions are compactified, i.e., in the simplest case,

$$M = \mathbb{R}^{3,1} \times M_6 \tag{5}$$

with some compact Euclidean six-dimensional internal space M_6.

There are two major approaches for arriving at such a string background:

■ A geometric setting, in which one studies a sigma model with target space of the form (5). Space-time supersymmetry then requires the compact six-dimensional space M_6 to be a Calabi-Yau manifold, i.e. to admit a covariantly constant spinor.

- An entirely different approach describes the internal space through exactly solvable CFTs built from infinite-dimensional symmetry algebras, in particular affine Lie algebras. (Affine Lie algebras also provide the gauge sector in heterotic theories.) This purely algebraic construction allows one to investigate string theory in the regime of strong curvature of the target space in which geometric methods break down. More specifically, for every homogeneous space G/H, the so-called coset construction supplies an exactly solvable CFT model. It can be proven [23, 30] that such a coset model has $N = 2$ superconformal symmetry if and only if G/H is Kähler. These models, called Kazama-Suzuki models (including $N = 2$ minimal models), are the building blocks for the Gepner construction of string vacua.

2. Orbifolds

It is a general experience that physical systems with symmetries admit reduction procedures which can result in interesting new systems. String theory is no exception; here reduction appears in the form of orbifold methods, which constitute a powerful tool for obtaining new string backgrounds. The orbifold construction has an algebraic and a geometric aspect, too.

On the algebraic side one considers the situation that the chiral algebra $\mathcal{A}$ has a non-trivial group G of automorphisms. The fixed point subalgebra $\mathcal{A}^G$ of $\mathcal{A}$ is again a chiral algebra – the chiral algebra of the *orbifold theory* [5]. Results similar to the theory of dual pairs (cf. [19] for a review) give a good control on the representations of $\mathcal{A}^G$ and thus on the superselection sectors of the (chiral) orbifold theory. The chiral data of orbifold theories have been worked out explicitly for large classes of examples, like orbifolds of affine Lie algebras by inner [22] and outer ([4], compare also [15]) automorphisms and permutation orbifolds [2]. The formalization of chiral algebras as vertex algebras allows one to prove many statements about the representation theory of orbifolds. There is in particular a Galois theory for chiral algebras [7].

In the geometric setting [6], one considers the action of a group G on a target space M by isometries. (This action must also preserve other background fields than the metric.) In case the action is not free, this gives rise to a singular space $M//G$. However, as a consequence of including so-called twisted sectors, which is necessary to achieve modular invariance, string propagation in such a background is still well-defined, i.e. the CFT does not feel the singularity. It is worth emphasizing that for those models which possess both a geometric and an algebraic description, the geometric orbifold does not always yield the same theory as an algebraic orbifold. For example, when M is a compact Lie group and G a subgroup of its center, the geometric orbifold construction can lead to an enhancement, rather than a reduction, of the chiral algebra. (This happens e.g. for $M = SU(2)$ and $G = \mathbb{Z}_2$, i.e. for the WZW model based on the group $SO(3)$. If the level is divisible by four, then the corresponding modular invariant is of extension type.)

As a technical aside, we mention that the computation of string amplitudes that include twist fields is based on the theory of Riemann surfaces with holomorphic symmetries [18].

3. Duality symmetries

Another task that symmetries can fulfill is to relate different backgrounds. In the simplest case of a free background, along with the combination (2) obviously also the function

$$\tilde{X}^\mu(\sigma, \tau) = X_L^\mu(\sigma + \tau) - X_R^\mu(\sigma - \tau) \tag{6}$$

is a classical solution. A more complete analysis shows that this duality continues to be realized in the quantized theory; it connects the theory of a free boson compactified on a circle of radius R with the same theory compactified on a circle of radius $2/R$.

This operation is known as *T-duality*. It typically connects two *different* backgrounds; self-dual theories possess an enhanced symmetry. For example, at radius $R^2 = 2$ the free boson on a circle acquires a non-abelian current algebra of type $\widehat{su}(2)$ at level 1 as an additional symmetry. This is a generic feature of duality symmetries. It already appears in the Kramers-Wannier duality of lattice systems that relates the high- and low-temperature regime; the self-dual theory is at the critical point (the reader may guess herself what the enhanced symmetry is in this example). A direct lesson from T-duality for string theory is that geometric interpretations of string theory backgrounds typically require to make arbitrary choices – e.g. in the case of a compactified free boson none of the two possible interpretations, i.e. either using the radius R or using $2/R$, is preferred.

T-duality possesses the following important generalization. The $N = 2$ superconformal algebra has an automorphism of order two, acting as

$$\omega(J) = -J, \qquad \omega(G^{(\pm)}) = G^{(\mp)}, \qquad \omega(T) = T, \tag{7}$$

which just like T-duality flips the sign of a chiral abelian current, and as a consequence can also be used to relate different string backgrounds. In the geometric setting, it relates different Calabi-Yau manifolds. This relationship, called *mirror symmetry*, has turned out to be both mathematically deep and technically extremely useful (for a review, see e.g. [17]).

The dualities discussed so far are so-called perturbative dualities – they relate terms at the same order in string perturbation theory. ‡ In contrast, dualities like the Olive-Montonen duality are non-perturbative. Such dualities have found many generalizations in string theory in the last eight years. In particular, so-called S-dualities act also on the coupling constant and thereby allow us to relate different backgrounds in the regime of strong and weak coupling. For example, it has been found that the type IIA superstring on K3 surfaces is connected via such a duality with the heterotic string on a four-dimensional torus [20].

4. D-branes and open strings

So far we have restricted our attention to symmetries acting either on the target space or on the chiral algebra. There are also symmetries that act on the world sheet as well. These have been termed *orientifolds*; they result in yet another type of string theories, of type I, in which the perturbation theory is organized in terms of orientable *and* unorientable surfaces that are also allowed to have boundaries. In these theories the torus amplitude $Z(\tau)$, constituting the partition function of bulk fields, is projected by a contribution given by the amplitude $K(t)$ of the CFT on a Klein bottle, which leads to a total partition function $\frac{1}{2}(Z + K)$. In terms of the characters $\chi_i(\tau)$ of the chiral CFT, the torus amplitude reads

$$Z(\tau) = \sum_{i,j} Z_{ij} \chi_i(\tau) \chi_j(\tau)^*, \tag{8}$$

with non-negative integers Z_{ij}, while the Klein bottle amplitude is

$$K(t) = \sum_i K_i \chi_i(2it). \tag{9}$$

Again there is a twisted sector – we must include open strings. For these, the partition function

$$A(t) = \sum_{a,b} n_a n_b A_{ab,i} \chi_i(\tfrac{it}{2}) \tag{10}$$

‡ Notice, though, that these dualities are non-perturbative as far as perturbation theory in the sigma model is concerned.

on the annulus gets projected by the contribution from the Möbius strip, i.e.

$$M(t) = \sum_a n_a \, M_{a,i} \, \chi_i\left(\tfrac{\mathrm{i}t+1}{2}\right), \tag{11}$$

leading to the partition function $\frac{1}{2}(A + M)$ for unoriented open strings. (The index a labels boundary conditions for open strings and the non-negative integers n_a are so-called Chan-Paton multiplicities. They give rise to gauge symmetries in theories of open strings [24].)

There are in fact many other motivations to study boundary conditions in conformal field theories. They arise e.g. in the description of defects in quasi one-dimensional condensed matter systems, in percolation problems, and in the analysis of string propagation in the background of certain solitonic solutions, so-called D-banes. D-branes are distinguished submanifolds occurring in these solutions, similar to the position of a black hole in the Schwarzschild solution. Open strings are restricted to start and end on these submanifolds [26].

String scattering amplitudes are obtained as a perturbation series, with each term an integral over an associated CFT correlation function. In the presence of boundaries, the latter can be analyzed with methods based on the orientifold projection, which we will not describe here (for a recent review see [1]). Another approach, to be reviewed below, is based on representation theoretic structures.

5. String amplitudes and Frobenius algebras

It is an old insight that a group § G contains as much information as its category $\mathcal{C}(G)$ of finite-dimensional complex representations. (Much of the usefulness of groups in physics can be traced to this equivalence.) The category $\mathcal{C}(G)$ has the following crucial properties:
- It has a tensor product $\otimes \colon \mathcal{C}(G) \times \mathcal{C}(G) \to \mathcal{C}(G)$, the tensor product of representations, with the trivial representation as its tensor unit.
- It has a braiding, i.e. a family of isomorphisms $c_{V,W} \in \mathrm{Hom}(V \otimes W, W \otimes V)$ satisfying various axioms like functoriality and tensoriality; the braiding $c_{V,W}$ is simply the permutation of the two factors in the tensor product.
- The notion of a dual, or contragredient, representation endows $\mathcal{C}(G)$ with a duality.

There are also two more technical properties:
- There exists an object V such that every object W of $\mathcal{C}(G)$ appears as the subobject of $V^{\otimes N}$ for sufficiently large N.
- The forgetful functor $\omega \colon \mathcal{C}(G) \to Vect_{\mathrm{fin}}(\mathbb{C})$, assigning to a representation the underlying vector space, is a fiber functor. This means in particular that it respects the tensor structure. It is the central insight of Tannaka theory (see e.g. [21] for a review) that G can be reconstructed from the abstract category $\mathcal{C}(G)$ as the group of automorphisms of the functor ω.

The approach to conformal field theory that we describe in the sequel is set up in a similar spirit – given the chiral algebra, formalized e.g. as a vertex algebra, we can determine its category $\mathcal{C}$ of representations and the associated conformal blocks. Rather than directly with the chiral algebra, we then work with the representation category $\mathcal{C}$; its objects are representations, while its morphisms are built from the conformal blocks.

For a rational CFT, the category $\mathcal{C}$ is a *modular tensor category*. (For an early review of this relationship see [25].) $\mathcal{C}$ should be thought of as a basis-independent version of the chiral data like braiding and fusing matrices and the (fractional part of) conformal weights. A modular tensor category is an abelian, semi-simple, braided tensor category with a duality, obeying also a few further conditions: it must be $\mathbb{C}$-linear; the tensor unit $\mathbf{1}$ must

§ We omit here technically important qualifiers which, depending on the chosen framework, could be "locally compact" or "algebraic".

be (absolutely) simple, i.e. $\operatorname{End}(1) = \mathbb{C}\,\mathrm{id}_1$; there must be only finitely many (isomorphism classes of) simple objects; and finally, the quadratic matrix

$$s_{ij} := \operatorname{tr} c_{X_i, X_j}\, c_{X_j, X_i} \qquad (12)$$

must be invertible. According to the Verlinde conjecture, this matrix s also appears in the modular transformation of the characters of the vertex algebra. Modular tensor categories also arise as (truncations of) categories of representations of quantum groups and as categories of representations of weak Hopf algebras.

As an important consequence of the axioms, each modular tensor category allows for the construction of a topological field theory in three dimensions. The latter assigns to every closed oriented two-manifold $\hat{\Sigma}$ a finite-dimensional vector space $\mathcal{H}(\hat{\Sigma})$, the space of conformal blocks, on which the mapping class group $Map(\hat{\Sigma})$ acts projectively. The conformal blocks should be thought of as multivalued functions; accordingly, they are the *pre*-correlators of the local conformal field theory. Moreover, to every three-manifold M with boundary $\hat{\Sigma}$ that contains a Wilson graph, a TFT assigns a vector in $\mathcal{H}(\hat{\Sigma})$. The axioms of a TFT formalize the well-known relation between a TFT on a three-dimensional manifold and a *chiral* CFT on its boundary, a structure that is e.g. central to the description of universality classes of quantum Hall fluids (for a review, see e.g. [11]).

Our goal is now to describe CFT correlators on a surface Σ that possibly has a boundary. A major input in our construction are tools from topological field theory. To be able to apply them, we need a two-manifold $\hat{\Sigma}$ without boundaries. A natural candidate for $\hat{\Sigma}$ is the complex double of Σ. It comes with an orientation reversing involution σ such that Σ is identified with the quotient of $\hat{\Sigma}$ by the action of σ. This orientifold map is rather different from the symmetries arising in the description of twist fields of orbifolds, which are holomorphic. Indeed, it is natural [31] to view the world sheet Σ as a real scheme and $\hat{\Sigma}$ as its complexification; the orientifold map is then just the action of the Galois group $Gal(\mathbb{C}/\mathbb{R}) \cong \mathbb{Z}_2$.

With the help of $\hat{\Sigma}$ we can in particular give a concise version of the principle of *holomorphic factorization*: The correlators of the CFT on Σ are specific vectors in $\mathcal{H}(\hat{\Sigma})$, which are subject to two types of constraints:
(1) They must be invariant under the action of $Map(\Sigma) \cong Map(\hat{\Sigma})^\sigma$.
(2) They must satisfy factorization rules.

The second input in our construction is algebraic – a symmetric special Frobenius algebra in the modular tensor category $\mathcal{C}$. An algebra in $\mathcal{C}$ is an object A of $\mathcal{C}$ together with morphisms $m \in \operatorname{Hom}(A \otimes A, A)$ and $\eta \in \operatorname{Hom}(1, A)$ that turn it into an algebra, i.e. m is an (associative) multiplication and η is a unit for m. A Frobenius algebra has in addition the structure of a co-algebra, i.e. there are a (co-associative) co-product $\Delta \in \operatorname{Hom}(A, A \otimes A)$ and a co-unit $\epsilon \in \operatorname{Hom}(A, 1)$, and product and co-product are connected by the condition that

$$(m \otimes \mathrm{id}_A) \circ (\mathrm{id}_A \otimes \Delta) = \Delta \circ m = (\mathrm{id}_A \otimes m) \circ (\Delta \otimes \mathrm{id}_A). \qquad (13)$$

The additional requirements that the algebra must also be symmetric and special are of a somewhat more technical nature.

In (unitary) conformal field theories, such a Frobenius algebra object is supplied by the algebra of open string states for any single given boundary condition [13]. The product comes from the operator product of boundary fields, and associativity of the OPE implies associativity of the algebra A. The Frobenius structure follows from the non-degeneracy of two-point functions of boundary fields on the disk. It is worthwhile to remark that this algebra is not necessarily (braided-) commutative.

Next we combine non-commutative algebra and 3D TFT. According to the principle of holomorphic factorization we need to select an element of $\mathcal{H}(\hat{\Sigma})$; by the principles of TFT, such a vector is determined by a three-manifold M_Σ whose boundary is the double $\hat{\Sigma}$ and a ribbon graph in M_Σ. For M_Σ we take the so-called connecting manifold [9], defined as the quotient of the interval bundle $\hat{\Sigma} \times [-1,1]$ by the $\mathbb{Z}_2$ that acts as σ on $\hat{\Sigma}$ and as $t \mapsto -t$ on the interval. The points with $t = 0$ provide a distinguished embedding of Σ into M_Σ; in fact, Σ is a retract of M_Σ, or in more intuitive terms, M_Σ is just a fattening of the world sheet Σ. (To give one example, the double of a disk is a sphere, and the orientifold map is the reflection at the equatorial plane; the connecting manifold is in this case a full 3-ball.) Concerning the prescription for the ribbon graph in M_Σ we refer to [12]. Here we merely recall that it involves a (dual) triangulation of Σ with ribbons labelled by the Frobenius algebra A.

In this framework, one can prove the consistency requirements of modular invariance and factorization. We also recover the combinatorial data (partition functions, NIM-reps [27], classifying algebras [14], ...) that have arisen as necessary conditions in earlier work. Moreover, Morita equivalence, combined with orbifold technology, allows for an elegant proof of T-dualities for arbitrary topologies of the world sheet.

The following table, associating algebraic structures to physical concepts, can serve as a succinct summary of our results [13]:

Physical concept	Algebraic structure
Boundary condition	A-left module
Boundary field Ψ_i^{MN}	$\mathrm{Hom}_A(M \otimes i, N)$
Defect line	A-bimodule
Bulk field Φ_{ij}	$\mathrm{Hom}_{A,A}((A \otimes i)^-, (A \otimes j)^+)$
Disorder field $\Phi_{ij}^{B_1 B_2}$	$\mathrm{Hom}_{A,A}((B_1 \otimes i)^-, (B_2 \otimes j)^+)$

Our algebraization of physical concepts leads to rigorous proofs. It also allows for powerful algorithms. In particular, for constructing a full local CFT only a single non-linear constraint needs to be solved: the one encoding associativity of the Frobenius algebra A. Moreover, old physical questions amount to standard problems in algebra and representation theory:

- The classification of CFTs with given chiral data $\mathcal{C}$ amounts to classifying Morita classes of symmetric special Frobenius algebras in the category $\mathcal{C}$. In particular, modular invariants of automorphism type are classified by the Brauer group of $\mathcal{C}$.
- The classification of boundary conditions and defects is reduced to the standard representation theoretic problem of classifying modules and bi-modules. As a consequence, powerful methods like induced modules and reciprocity theorems are at our disposal.
- The problem of deforming CFTs is related to the problem of deforming algebras, which is a cohomological question. For the moment, the only known results in this direction are rigidity theorems [8]: a rational CFT cannot be deformed within the class of rational CFTs.

Let us finally collect a few aspects of string theory for which symmetries play a rôle that for lack of space were omitted in the present contribution:

- The derivation of the standard model and its symmetries from string theory. Using compactifications of the type (5), it is rather difficult to come close to the standard model. More recently, so-called brane-world models have attracted much attention (see e.g. [32]).
- Ideas about an underlying unifying symmetry of string theory. Recently, speculations that the underlying symmetry can be described in terms of a real form of the hyperbolic Kac-Moody algebra E_{10} have been upgraded [29] to E_{11}, a Kac-Moody algebra of rank 11.
- The construction of cosmological backgrounds using non-compact Lie groups.
- The description of the chiral symmetry of conformal field theories. The structure of a

160

vertex algebra originally arose as one mathematical formalization of these symmetries. It has by now evolved in a useful mathematical structure of independent interest, with a rich structural theory. (For a recent review see [10].)

Acknowledgements: C.S. thanks the organizers for their invitation to present these results.

[1] C. Angelantonj and A. Sagnotti, *Open strings*, Phys. Rep. (to appear)
[2] P. Bantay, *Characters and modular properties of permutation orbifolds*, Phys. Lett. B 419 (1998) 175
[3] R.E. Behrend, P.A. Pearce, V.B. Petkova, and J.-B. Zuber, *Boundary conditions in rational conformal field theories*, Nucl. Phys. B 579 (2000) 707
[4] L. Birke, J. Fuchs, and C. Schweigert, *Symmetry breaking boundary conditions and WZW orbifolds*, Adv. Theor. Math. Phys. 3 (1999) 671
[5] R. Dijkgraaf, C. Vafa, E. Verlinde, and H. Verlinde, *The operator algebra of orbifold models*, Commun. Math. Phys. 123 (1989) 485
[6] L.J. Dixon, J.A. Harvey, C. Vafa, and E. Witten, *Strings on orbifolds*, Nucl. Phys. B 261 (1985) 678
[7] C. Dong, H. Li, and G. Mason, *Twisted representations of vertex operator algebras*, Math. Annal. 310 (1998) 571
[8] P. Etingof, D. Nikshych, and V. Ostrik, *On fusion categories*, preprint math.QA/0203060
[9] G. Felder, J. Fröhlich, J. Fuchs, and C. Schweigert, *Conformal boundary conditions and three-dimensional topological field theory*, Phys. Rev. Lett. 84 (2000) 1659; *Correlation functions and boundary conditions in RCFT and three-dimensional topology*, Compos. Math. 131 (2002) 189
[10] E. Frenkel, *Vertex algebras and algebraic curves*, Séminaire Bourbaki No. 875 (1999-2000)
[11] J. Fröhlich, B. Pedrini, C. Schweigert, and J. Walcher, *Universality in quantum Hall systems: coset construction of incompressible states*, J. Stat. Phys. 103 (2001) 527
[12] J. Fuchs, I. Runkel, and C. Schweigert, *Conformal correlation functions, Frobenius algebras and triangulations*, Nucl. Phys. B 624 (2002) 452
[13] J. Fuchs, I. Runkel, and C. Schweigert, *TFT construction of RCFT correlators I: Partition functions*, preprint hep-th/0204148, to appear in Nucl. Phys. B
[14] J. Fuchs and C. Schweigert, *A classifying algebra for boundary conditions*, Phys. Lett. B 414 (1997) 251
[15] J. Fuchs and C. Schweigert, *Lie algebra automorphisms in conformal field theory*, Contemp. Math. (2002)
[16] J. Fuchs, C. Schweigert, and J. Walcher, *Projections in string theory and boundary states for Gepner models*, Nucl. Phys. B 588 (2000) 110
[17] B.R. Greene and M.R. Plesser, *Introduction to mirror manifolds*, in: *Mirror Symmetry I*, S.-T. Yau, ed. (American Mathematical Society, Providence 1998), p. 1
[18] S. Hamidi and C. Vafa, *Interactions on orbifolds*, Nucl. Phys. B 279 (1987) 465
[19] R. Howe, *Perspectives on invariant theory: Schur duality, multiplicity-free actions and beyond*, in: *The Schur Lectures*, I. Piatetski-Shapiro and S. Gelbart, eds. (Isr. Math. Conf. Proc., Ramat-Gan 1995), p. 1
[20] C.M. Hull and P.K. Townsend, *Unity of superstring dualities*, Nucl. Phys. B 436 (1995) 507
[21] A. Joyal and R. Street, *An introduction to Tannaka duality and quantum groups*, Springer Lecture Notes in Mathematics 1488 (1991) 413
[22] V.G. Kac and I.T. Todorov, *Affine orbifolds and rational conformal field theory extensions of $W_{1+\infty}$*, Commun. Math. Phys. 190 (1997) 57
[23] Y. Kazama and H. Suzuki, *Characterization of $N=2$ superconformal models generated by the coset space method*, Phys. Lett. B 216 (1989) 112
[24] N. Marcus and A. Sagnotti, *Group theory from "quarks" at the ends of strings*, Phys. Lett. B 188 (1987) 58
[25] G. Moore and N. Seiberg, *Lectures on RCFT*, in: *Physics, Geometry, and Topology*, H.C. Lee, ed. (Plenum Press, New York 1990), p. 263
[26] J. Polchinski, *Dirichlet-branes and Ramond-Ramond charges*, Phys. Rev. Lett. 75 (1995) 4724
[27] G. Pradisi, A. Sagnotti, and Ya.S. Stanev, *Completeness conditions for boundary operators in 2D conformal field theory*, Phys. Lett. B 381 (1996) 97; see also [3]
[28] A.N. Schellekens and N.P. Warner, *Anomalies and modular invariance in string theory*, Phys. Lett. B 177 (1986) 317
[29] I. Schnakenburg and P. West, *Kac-Moody symmetries of IIB supergravity*, Phys. Lett. B 517 (2001) 421
[30] C. Schweigert, *On the classification of $N=2$ superconformal coset theories*, Commun. Math. Phys. 149 (1992) 425
[31] C. Schweigert and J. Fuchs, *The world sheet revisited*, preprint hep-th/0105266, to appear in Fields Institute Commun.
[32] A. Uranga, *From quiver diagrams to particle physics*, in: *European Congress of Mathematics*, C. Casacuberta et al., eds. (Birkhäuser, Progress in Mathematics Vol. 202, Basel 2001), p. 499

Inst. Phys. Conf. Ser. No 173: Plenary Sessions
Paper presented at 24th Int. Coll. Group Theoretical Methods in Physics, Paris, France, 15–20 July 2002
©2003 IOP Publishing Ltd

Solitons, Platonic Symmetry and Fullerenes

Paul M. Sutcliffe

Institute of Mathematics, University of Kent at Canterbury, Canterbury, CT2 7NF, U.K.

Abstract. Skyrmions are topological solitons in three space dimensions which are candidates for an effective description of nuclei. It is of interest to determine the structure and symmetry of the classical Skyrmion solution with minimal energy for a given soliton number. Numerical solutions of the relevant nonlinear PDE yield interesting results, with the Skyrmions often having unexpected Platonic symmetry. For large soliton numbers the solutions have a structure similar to Fullerenes in carbon chemistry. These results are reviewed, together with an analytical perspective which involves an approximation based on rational maps between Riemann spheres.

1. Introduction

The Skyrme model [16] was first proposed in the early sixties as a model for the strong interactions of hadrons, but it was set aside after the advent of quantum chromodynamics (QCD). Much later Witten [18] showed that it could arise as an effective description at low energies in the limit where the number of quark colours is large. Subsequent work [1] demonstrated that the single soliton solution (known as a Skyrmion) reproduced the properties of a nucleon to within an accuracy of around 30%; quite an achievement, given that there is, at present, no practical way of calculating the properties of nuclei from QCD via, for example, lattice gauge theory.

In order to study nuclei of larger atomic number one first needs to compute the minimal energy configurations of multi-solitons, since in the Skyrme model there is an identification between the numbers of solitons and nucleons. In this review the focus is on the classical static Skyrmion solutions, their symmetry and structure. This problem involves solving a highly nonlinear elliptic PDE for a matrix-valued function in three space dimensions. A numerical solution of this problem yields interesting results in which the Skyrmions sometimes have unexpected Platonic symmetry. The energy and baryon densities are localized around the edges of polyhedra, which for large soliton numbers have a structure similar to Fullerenes in carbon chemistry. First we shall describe these results and then discuss how they can be further understood through the use of an analytic approach in which Skyrmions are approximated by an ansatz that uses rational maps between Riemann spheres.

2. Skyrmions

A static Skyrme field, $U(\mathbf{x})$, is an $SU(2)$ matrix defined throughout $\mathbb{R}^3$ and satisfying the boundary condition that $U \to 1$ as $|\mathbf{x}| \to \infty$. This boundary condition implies a compactification of space so that the Skyrme field becomes a mapping $U : S^3 \mapsto SU(2)$, and so can be classified by an integer valued winding number

$$B = \frac{1}{24\pi^2} \int \varepsilon_{ijk} \operatorname{Tr} \left(\partial_i U\, U^{-1} \partial_j U\, U^{-1} \partial_k U\, U^{-1} \right) d^3x, \tag{1}$$

Table 1. The symmetry group G of the minimal energy Skyrmion with baryon number B.

B	1	2	3	4	5	6	7	8	9	10	11	12	13	14	15	16	17	18	19	20	21	22
G	$O(3)$	$D_{\infty h}$	T_d	O_h	D_{2d}	D_{4d}	Y_h	D_{6d}	D_{4d}	D_3	D_{3h}	T_d	O	C_2	T	D_2	Y_h	D_2	D_3	D_{6d}	T_d	D_3

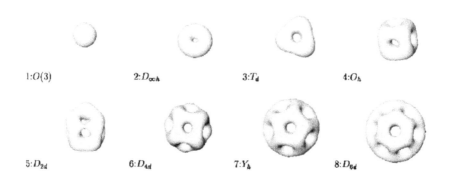

1:$O(3)$ 2:$D_{\infty h}$ 3:T_d 4:O_h

5:D_{2d} 6:D_{4d} 7:Y_h 8:D_{6d}

Figure 1. Baryon density isosurfaces for the minimal energy Skyrmions with $1 \le B \le 8$.

which is a topological invariant; it gives the number of times that the target 3-sphere is covered when the domain 3-sphere is covered once. This winding number counts the number of solitons in a given field configuration and is identified with baryon number in the application to modelling nuclei.

The energy of a static Skyrme field is given by

$$E = \frac{1}{12\pi^2} \int \mathrm{Tr}\left(-\frac{1}{2}\left(\partial_i U\, U^{-1}\right)^2 - \frac{1}{16}\left[\partial_i U\, U^{-1}, \, \partial_j U\, U^{-1}\right]^2\right) d^3x \qquad (2)$$

and for each integer B, the problem is to minimize this energy within the class of fields with baryon number B in order to find the minimal energy Skyrmion. Critical points of this energy functional are static solutions of the nonlinear Skyrme equation, which for static fields is precisely the equation obtained from the variation of this energy. It can be shown that the Skyrme energy satisfies the Faddeev-Bogomolny lower bound [10, 7]

$$E \ge |B| \qquad (3)$$

but this bound cannot be saturated for any non-trivial (i.e. $B \ne 0$) finite energy configuration.

Using numerical methods and a powerful parallel machine, minimal energy Skyrmions have been computed [3, 4] for all $B \le 22$ yielding the results presented in Table 1 for their symmetries. Note the appearance of Platonic symmetry groups T, O, Y, (and their extension by reflection symmetries) corresponding to the rotational symmetries of the tetrahedron, octahedron/cube, and icosahedron/dodecahedron.

When we refer to the symmetry of a Skyrmion we do not mean that the Skyrme field itself is invariant under particular spatial rotations, but rather that the effect of a spatial rotation can be undone by the application of the global $SO(3)$ symmetry of the Skyrme model, which acts through the conjugation $U \mapsto \mathcal{O}U\mathcal{O}^\dagger$, where $\mathcal{O} \in SU(2)$ is a constant matrix. In particular this means that the baryon and energy densities (the integrands in (1) and (2)) are strictly invariant. In fig. 1 we display baryon density isosurfaces for the minimal energy Skyrmions with $1 \le B \le 8$. Energy density isosurfaces are qualitatively similar.

The $B = 1$ Skyrmion is spherically symmetric [16] but for $B = 2$ the Skyrmion has only an axial symmetry [13, 15, 17], with a toroidal structure for the baryon density isosurface. For $B > 2$ the baryon density is localized around the edges of a polyhedron, which for $B = 3$ and $B = 4$ is a tetrahedron and cube respectively [8]. From fig. 1 it can be seen that each of the polyhedra is trivalent and contains $2B - 2$ faces. This led to a conjecture that this pattern would continue to larger values of B and hence that the baryon density of minimal energy Skyrmions for $B \geq 7$ would be localized around the polyhedra which arise in the study of Fullerenes with $4B - 8$ carbon atoms [5]. These trivalent polyhedra are composed of 12 pentagons and $2B - 14$ hexagons (see [11] and the article by Prof. Fowler in these proceedings). In fig. 2 we display baryon density isosurfaces, together with models of the associated polyhedra, for minimal energy Skyrmions with $7 \leq B \leq 22$. From this figure it can be seen that the polyhedron is indeed of the Fullerene type, being trivalent and consisting of 12 pentagons and $2B - 14$ hexagons, in all cases except $B = 9, 13$. These two anomalous cases contain quadvalent vertices but can also be understood in terms of Fullerenes through the application of a symmetry enhancement process in which certain edges are shrunk to zero length [4].

In the following section we describe how several aspects of the numerical results just described can be understood by studying rational maps between Riemann spheres.

3. Rational maps

In the rational map ansatz [12], a Skyrme field with baryon number B is constructed from a degree B rational map between Riemann spheres. Although the rational map ansatz does not give exact solutions of the static Skyrme equations, it produces approximations which have energies only a couple of percent above the numerically computed solutions. Briefly, use spherical coordinates in $\mathbb{R}^3$, so that a point $\mathbf{x} \in \mathbb{R}^3$ is given by a pair (r, z), where $r = |\mathbf{x}|$ is the distance from the origin, and z is a Riemann sphere coordinate giving the point on the unit two-sphere which intersects the half-line through the origin and the point $\mathbf{x}$. Now, let $R(z)$ be a degree B rational map between Riemann spheres, that is, $R = p/q$ where p and q are polynomials in z such that $\max[\deg(p), \deg(q)] = B$, and p and q have no common roots. Given such a rational map the ansatz for the Skyrme field is

$$U(r, z) = \exp \left[\frac{i f(r)}{1 + |R|^2} \begin{pmatrix} 1 - |R|^2 & 2\bar{R} \\ 2R & |R|^2 - 1 \end{pmatrix} \right],$$ (4)

where $f(r)$ is a real profile function satisfying the boundary conditions $f(0) = \pi$ and $f(\infty) = 0$, which is determined by minimization of the Skyrme energy of the field (4) given a particular rational map R.

Substitution of the rational map ansatz (4) into the Skyrme energy functional results in the following expression for the energy

$$E = \frac{1}{3\pi} \int \left(r^2 f'^2 + 2B(f'^2 + 1)\sin^2 f + \mathcal{I} \frac{\sin^4 f}{r^2} \right) dr,$$ (5)

where $\mathcal{I}$ denotes the integral

$$\mathcal{I} = \frac{1}{4\pi} \int \left(\frac{1 + |z|^2}{1 + |R|^2} \left| \frac{dR}{dz} \right| \right)^4 \frac{2i \, dz d\bar{z}}{(1 + |z|^2)^2}.$$ (6)

To minimize the energy (5) one first determines the rational map which minimizes $\mathcal{I}$, then given the minimum value of $\mathcal{I}$ it is a simple exercise to find the minimizing profile function. Thus, within the rational map ansatz, the problem of finding the minimal energy Skyrmion

164

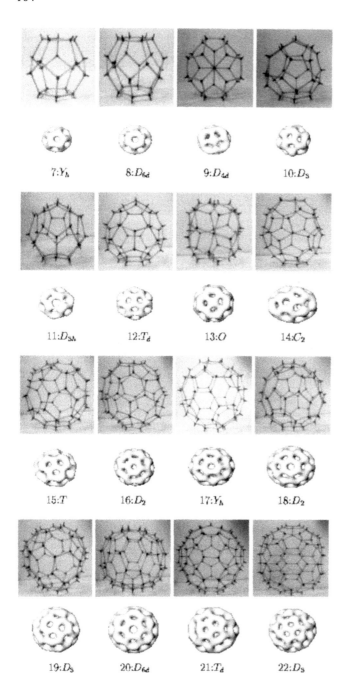

Figure 2. Baryon density isosurfaces (to scale) for minimal energy Skyrmions with $7 \leq B \leq$ 22, and the associated polyhedral models (not to scale).

reduces to the simpler problem of calculating the rational map which minimizes the function $\mathcal{I}$.

The baryon density of the rational map $R = p/q$ is proportional to the Wronskian

$$w(p,q) = p'q - q'p \tag{7}$$

which has $2B - 2$ roots, giving the points on the Riemann sphere for which the baryon density vanishes along the corresponding half-lines through the origin. These $2B - 2$ points on the sphere give the face-centres of the Skyrmion polyhedron, thus explaining the relationship between the number of faces and the number of solitons.

Using a simulated annealing algorithm the $\mathcal{I}$ minimizing rational maps for $1 \leq B \leq 22$ have been computed [3, 4] and found to be in good agreement with the results of full Skyrme field minimization. As an example, the $\mathcal{I}$ minimizing map of degree three is given by

$$R = \frac{\sqrt{3}iz^2 - 1}{z^3 - \sqrt{3}iz}. \tag{8}$$

The Wronskian of this map is

$$w = -\sqrt{3}i(z^4 + 2\sqrt{3}iz^2 + 1) \tag{9}$$

whose roots are the four points on the Riemann sphere located at the face centres of an appropriately oriented and scaled tetrahedron, in agreement with the structure of the baryon density isosurface of the $B = 3$ Skyrmion displayed in fig. 1. The fact that the Skyrme field constructed from this rational map has tetrahedral symmetry is a consequence of the tetrahedral symmetry of the rational map, as follows.

In terms of the rational map approach a spatial rotation acts on the Riemann sphere coordinate z as an $SU(2)$ Möbius transformation. Similarly the $SO(3)$ global symmetry of the Skyrme model acts on the Riemann sphere coordinate R of the target two-sphere also as an $SU(2)$ Möbius transformation. Hence a map is G-symmetric if, for each $g \in G$, there exists a target space rotation D_g so that $R(g(z)) = D_g(R(z))$. In the case of the tetrahedral group the generators are $z \mapsto -z$ and $z \mapsto (iz + 1)/(-iz + 1)$, which correspond to a $180°$ rotation about the x_3-axis and a $120°$ rotation about the line $x_1 = x_2 = x_3$ respectively. The rational map (8) satisfies the relations

$$R(-z) = -R(z) \quad \text{and} \quad R\left(\frac{iz+1}{-iz+1}\right) = \frac{iR(z)+1}{-iR(z)+1} \tag{10}$$

and hence is tetrahedrally symmetric. In fact, upto fixing orientations, this is the unique tetrahedrally symmetric map of degree three.

The construction of symmetric rational maps for arbitrary degree and any finite point group is a matter of classical group theory, and explicit formulae can be found in [12]. The symmetry of each $\mathcal{I}$ minimizing rational map is analyzed in [4] for all $B \leq 22$. In most cases the symmetry of the minimizing map agrees with that of the minimal energy Skyrmion, and the Skyrme field produced from the map is a good approximation to the minimal energy Skyrmion. Occasionally this is not the case and the symmetry of the $\mathcal{I}$ minimizing map is different from that of the minimal energy Skyrmion. In these situations a different map, with a very slightly larger value of $\mathcal{I}$, matches the symmetry of the minimal energy Skyrmion and provides a good approximation to it through the rational map ansatz. Thus, in situations where there are rational maps which are local minima of the function $\mathcal{I}$, with values which are very close to the global minimum, only numerical computations of the full Skyrme energy can be certain to correctly identify the symmetry of the minimal energy Skyrmion.

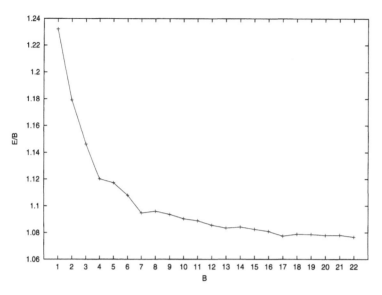

Figure 3. The energy per baryon E/B for minimal energy Skyrmions with $1 \leq B \leq 22$.

4. Shells, lattices and crystals

In fig. 3 we plot the energy per baryon E/B for the minimal energy Skyrmions with $1 \leq B \leq 22$. Note the dips in this function at $B = 7$ and $B = 17$, highlighting the fact that these Skyrmions have unusually low energies. A glance at Table 1 reveals that these are precisely the values at which the minimal energy Skyrmion is particularly symmetric, having icosahedral symmetry. (There are also less pronounced dips at $B = 4$ and $B = 13$, where the Skyrmion again has Platonic symmetry, but this time octahedral.) This raises the issue of determining the values of B at which the minimal energy Skyrmion has icosahedral symmetry and a resulting unusually low energy. This has recently been addressed in [2], where evidence is presented that the values which continue the sequence $B = 7, 17$ are $B = 37, 67, 97$.

Fig. 3 shows that the energy per baryon of the shell-like Skyrmions is, generally, decreasing as B increases, and appears to be approaching an asymptotic value. This asymptotic value can be investigated by studying a periodic arrangement of Skyrmions in the form a two-dimensional lattice, as follows. In very large Fullerene-type polyhedra, where hexagons are dominant, the twelve pentagons may be viewed as defects inserted into a flat structure, to generate the curvature necessary to close the shell. Energetically, the optimum infinite structure is a hexagonal lattice (the analogue of a graphite sheet). This configuration has infinite energy, since it has infinite extent in two directions, but its energy per baryon should be lower than that of any of the known finite energy Skyrmions, and will be the asymptotic value approached by large Fullerene-like Skyrmions. Such a hexagonal Skyrme lattice can be constructed using a variant of the rational map ansatz and has energy per baryon $E/B = 1.061$ [6], consistent with an asymptotic value in fig. 3.

In fig. 4 we display a surface of constant baryon density for this hexagonal Skyrme lattice. The structure is clearly visible, the baryon density having a hole in the centre of each of the hexagonal faces. Note that the displayed region contains exactly eight full hexagons and has baryon number four, so each hexagon may be thought of as having baryon number $1/2$. This is the expected limit of the polyhedral structures discussed earlier, where a Skyrmion with

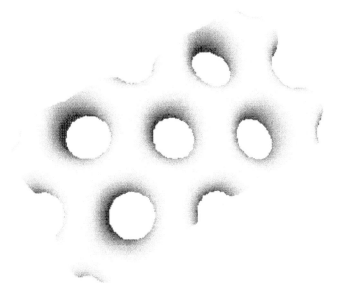

Figure 4. A baryon density isosurface for the Skyrme lattice.

soliton number B has $2B - 2$ faces.

Although the energy per baryon of the Skyrme lattice is low it is not the lowest known value for E/B. This occurs for an infinite three-dimensional cubic crystal of Skyrmions [9, 14] which has $E/B = 1.036$. In fig. 5 we plot a baryon density isosurface for the Skyrme crystal. Each lump represents a half-Skyrmion and the total baryon number shown is four. Since the energy per baryon of the Skyrme lattice exceeds that of the Skyrme crystal it is reasonable to expect that above some critical soliton number, the minimal energy Skyrmion will resemble a portion of the crystal rather than a shell constructed from the planar lattice by inserting pentagonal defects. However, even the order of magnitude of this critical value is difficult to estimate since it relies on a delicate comparison of the surface to volume energy of a finite portion of the crystal and this is very sensitive to the way in which the portion of the crystal is smoothed off at its boundary.

5. Conclusion

In conclusion, we see that the topological solitons of the Skyrme model have a rich and interesting structure, often with a large amount of symmetry. Some of these aspects can be explained by studying rational maps between Riemann spheres, but there are still many open questions. Finally, in the difficult task of Skyrmion quantization and comparison with experimental data for nuclei there is still much work to do, but hopefully the improved understanding of the classical Skyrmion solutions and their symmetries will help with this aspect.

Acknowledgements

The material reviewed here is work I have done over the last few years in various collaborative combinations with Richard Battye, Conor Houghton and Nick Manton. I acknowledge the

168

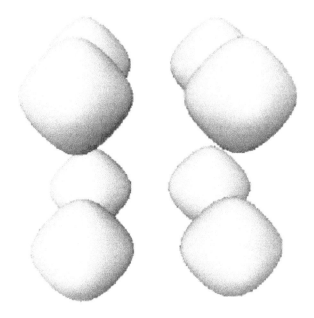

Figure 5. A baryon density isosurface for the Skyrme crystal.

EPSRC for an Advanced Fellowship.

References

[1] Adkins G S, Nappi C R and Witten E 1983 Nucl. Phys. B228 552
[2] Battye R A, Houghton C J and Sutcliffe P M 2002 *Icosahedral Skyrmions*, in preparation.
[3] Battye R A and Sutcliffe P M 2001 Phys. Rev. Lett. 86 3989
[4] Battye R A and Sutcliffe P M 2002 Rev. Math. Phys. 14 29
[5] Battye R A and Sutcliffe P M 1997 Phys. Rev. Lett. 79 363
[6] Battye R A and Sutcliffe P M 1998 Phys. Lett. B416 385
[7] Bogomolny E B 1976 Sov. J. Nucl. Phys. 24 449
[8] Braaten E, Townsend S and Carson L 1990 Phys. Lett. B235 147
[9] Castillejo L, Jones P S J, Jackson A D, Verbaarschot J J M and Jackson A 1989 Nucl. Phys. A501 801
[10] Faddeev L D 1976 Lett. Math. Phys. 1 289
[11] Fowler P W and Manolopoulos D E 1995 *An Atlas of Fullerenes*, Clarendon Press
[12] Houghton C J, Manton N S and Sutcliffe P M 1998 Nucl. Phys. B510 507
[13] Kopeliovich V B and Stern B E 1987 JETP Lett. 45 203
[14] Kugler M and Shtrikman S 1988 Phys. Lett. B208 491
[15] Manton N S 1987 Phys. Lett. B192 177
[16] Skyrme T H R 1961 Proc. Roy. Soc. A260 127
[17] Verbaarschot J J M 1987 Phys. Lett. B195 235
[18] Witten E 1983 Nucl. Phys. B223 422

Special Plenary Session and Round Table

Historical and Epistemological Aspects of Group Theory in Physics

Inst. Phys. Conf. Ser. No 173: Special Plenary Session and Round Table
Paper presented at 24th Int. Coll. Group Theoretical Methods in Physics, Paris, France, 15–20 July 2002
©2003 IOP Publishing Ltd

Historical and epistemological aspects of Group Theory in Physics

An interlude

S. Twareque Ali, Gerald A. Goldin, and Richard Kerner

As we climb the steep, vertiginous slopes of the Mountain of Science, we are tempted to cast our eyes back on the path already traversed. The Peak of the Mountain is hidden by dense clouds, and some doubt if it will ever be accessible—or even that it really exists! In contrast, the way behind is visible to all, filled with the traces of giants who opened it up for the crowd that followed.

However the rear view is sometimes also blurred, shrouded by rapidly accumulating mists of uncertainty. Powerful myth-creating forces are at work, and myths quickly replace historical reality in our perception of the past. A better understanding of the past should help us forecast the future. But many believe the sad observation that the only lesson history can teach us is that no history lessons have ever been learned.

An important scientific meeting is an ideal place to exchange news of our latest achievements in mathematical physics, in group theory and its applications, and to comment on the most recent developments in our branches of science. But it is also an occasion to share our views on the past history and current events of mathematics and physics—on the roles played by our famous predecessors, on the possible pathways that could have been explored but were left aside, on the present-day possibilities that are perhaps being overlooked. In a word, it is an opportunity to indulge ourselves in activities closely related to our work that actually belong to other disciplines: the history and philosophy of science.

Many renowned mathematicians and physicists, as their work matured, came to take a deep interest in the process of development of scientific ideas, in epistemology, ontology, and the history of science. Profound and fascinating texts were written by Planck, Einstein, Heisenberg and Schrödinger (to cite but a few). Thus it was proposed, on the occasion of our largely interdisciplinary meeting, that we also invite contributions addressing epistemological, historical and philosophical issues related to the development of symmetries in physics, especially group theory and its applications.

Hence a special Plenary Session devoted to these subjects was held on Saturday, the 20th of July. It was the last day of our Colloquium. A Round Table discussion, attended by many participants, followed the plenary talks. Pierre Cartier of the École Normale Supérieure, and Jean-Marc Lévy-Leblond from the University of Nice gave the main plenary contributions. The Round Table was chaired by Sara Franceschelli, and the additional participants included S. Twareque Ali, Gerald Goldin, and Rémi Mosseri.

As a matter of fact, both Harry J. Lipkin's laudatio (authored by Yuval Ne'eman) and Lipkin's response during the Wigner Medal ceremony also provide excellent examples of historical-epistemological analyses. For obvious reasons, these contributions are placed at the beginning of this volume as a separate section.

Pierre Cartier, who was himself involved in the elaboration of the modern approach to mathematics, and particularly to group theory, brilliantly analyzed the contributions of three undisputedly dominant figures of the XXth century, the three "giants" Hermann Weyl, Elie Cartan and Claude Chevalley. To paraphrase remarks that Cartier made during the discussion, there are few dark rooms remaining today in the castle of group theory. Now we must go out from the stronghold to explore the surrounding landscape. As we do so, we can retreat when we wish to the castle, using the methods of group theory—in which we are secure—in our new explorations.

Jean-Marc Lévy-Leblond contributed an amusing lecture, verging on science fiction, analyzing the recent history of scientific ideas that led to the Theory of Relativity in a parallel world, where no physicist named Albert Einstein ever existed. In such a world, Lévy-Leblond imagines a parallel physicist, Albrecht Zweistein, less lucky and not recognized by his peers. Nevertheless soemeone else—with a French-sounding name, this time—would produce an even better theory with the same mathematical content, called chronogeometry. A plausible corollary to this hypothesis is that in the same parallel world, a scientist named Jean-Luc Vély-Lebrun would contribute a paper to a parallel Colloquium, arguing that the history of physics could have been different if instead of Albrecht Zweistein, an obscure physicist named Anton Dreistein had written an unnoticed article much better than the one authored by Zweistein.

The Round Table took up as a discussion question, "What are the obstacles that keep us from 'breakthrough' insights?" Participants spoke of the processes of mathematical and scientific analysis, of psychological or cognitive obstacles (in individuals), of cultural obstacles (in the prevailing scientific ambience, where financial support for research is increasingly contingent on short-term success in the marketplace), and of epistemological obstacles (stemming from how the structure of knowledge itself develops).

One example for discussion was that of quasicrystals and quasiperiodicity, where for a time the 'knowledge' stemming from the complete classification of the crystallographic point and space groups may perhaps have been a barrier to recognizing new possibilities. Additional examples included wavelet transforms and wavelet analysis, nonlinear dynamical systems (both classical and quantum), and gauge field theory. The latter example invites the distinction between aspects of our descriptions intended to correspond to 'real' entities, and aspects that are arbitrary or conventional—a distinction that it may not be possible to maintain sharply. The case of wavelet analysis was recognized as a subject whose theoretical underpinnings began to be understood only as a consequence of a chance meeting of people working in widely separated disciplines (in this instance, group theory and seismic analysis). This was a perfect example of cross-fertilization of knowledge, a time-tested historical path for the development of science. The session concluded with lively questions from the floor, with seeming unanimity that it is important to pursue theoretical disciplines for their creative and intellectual merit, not necessarily to follow the dictates of fashion or the financial incentives of industry and the marketplace.

We believe this special session deserves a special place, as an interlude between the Invited Plenary Talks and the remainder of the volume containing Contributed Talks and Posters. We hope that the diversity and the touch of humour will be appreciated.

Inst. Phys. Conf. Ser. No 173: Special Plenary Session and Round Table
Paper presented at 24th Int. Coll. Group Theoretical Methods in Physics, Paris, France, 15–20 July 2002
©2003 IOP Publishing Ltd

What if Einstein had not been there?
A Gedankenexperiment in Science History

Jean-Marc Lévy-Leblond

Laboratoire de Physique Théorique
Université de Nice, Parc Valrose, F-06108 Nice Cedex 2, France
jmll@unice.fr

Abstract. Suppose Einstein had not existed. How, in the early twentieth century, might our understanding of space-time physics have developed? This paper proposes a reconstruction of history as it could have evolved, drawing on attested pre-Einsteinian works, and some little known post-Einsteinian ones, as well as introducing a few imaginary characters. The virtual rise of Minkowskian chronogeometry will be reviewed, focussing on:

 – the discovery of the inertia of energy and the subsequent construction of velocity-dependent formulas for energy and momentum (Von Dida 1909, Hunepierre 1909),
 – the application of the Erlanger program to the construction of possible space-time structures and classification of chronogeometrical groups, from von Ignatowsky (1910, 1911), Franck and Rothe (1911,1912) to more recent work by Blondeville and Prostov.

Some secondary sources will also be discussed (Zweistein, 1905). Finally, the relevance of such an approach, on educational, epistemological and cultural grounds, will be highlighted.

This paper is dedicated to the memory of Stephen J. Gould
who so eloquently highlighted the contingency of natural
history , notwithstanding its rationality .

1. Introduction

Gedankenexperiments consist in asking contrafactual questions - "What if . . .?". It all started when Galileo wondered "What if air resistance did not exist?" and discovered the law of free fall [1] . This strategy was later fruitfully extended and popularised in physics, in particular by Einstein. I propose to use it here, not *in* physics, but *on* physics. Understanding the unfolding of any specific stretch of history greatly benefits from imagining how it could have developed otherwise, if such and such crucial circumstance had been different [2]. I will advance what I believe to be a plausible scenario of the history of space-time physics at the beginning of the twentieth century had Einstein not been there to make his justly celebrated 1905 breakthrough. I will rely heavily on the thorough study by Arthur Miller of pre-Einsteinian developments [3], while inventing a few characters and contributions (the latter usually corresponding to real later work). May I suggest that, when first perusing this paper, the reader should not consult the references, so as to make up his/her own mind regarding the veracity of the facts provided?

2. The Inertia of Energy

By the end of the XIXth century, the development of electromagnetic theory was leading physicists to a fully electromagnetic worldview, according to which all phenomena could ultimately be reduced to the interactions of electrical charges through the ether. All properties of matter would be explainable in electromagnetic terms. Mass itself should then have its origin in the electromagnetic field, as first suggested by Thomson (1881). Drawing on a hydrodynamical analogy, Thomson demonstrated that by electrically charging a conducting sphere, one would change its inertia. The physical reason was that the motion of body through the ether generated an electrical current that gave rise to a magnetic field acting back upon the charge; this self-induction effect would modify the inertia of the moving body (a primitive example of renormalization). Thomson found that the additional inertia was proportional to the energy of the electrostatic field, with a coefficient of proportionality $8/15c^2$. Heaviside (1889) corrected some errors of Thomson and found a coefficient $4/3c^2$ (a correct derivation of the exact factor $1/c^2$ would have to wait for Fermi, in 1922).

At all events, whatever the value of the coefficient, due to the velocity dependence of the field and its energy, the total kinetic energy of the body could no longer show the usual quadratic dependence on velocity. However, obtaining the exact formula proved rather difficult, as the calculations depended upon ad hoc assumptions concerning the cohesive forces holding the body against the internal repulsive Coulomb forces, as well as on its form and its possible deformations when set into motion. Thus ensued a decade of confusion, with various proposals as to the dependence of mass on the velocity, by Searle (1897), Wien (1900), Abraham (1903), Lorentz (1904), Langevin and Bucherer (1905); the trouble was compounded by the distinction between a "longitudinal mass" m_L and a "transversal mass" m_T. Here is a small sample of the variety of the results obtained:

$$Abraham(1903) \begin{cases} m_L = m_e\,\beta^{-3}[\beta(1-\beta^2)^{-1} - \tanh^{-1}\beta] \\ m_T = m_e\,\beta^{-3}[\tfrac{1}{2}(1+\beta^2) - \tanh^{-1}\beta - \tfrac{1}{2}\beta] \end{cases} \quad (1)$$

$$Lorentz(1903) \begin{cases} m_L = m_e\,\tfrac{2}{3}(1-\beta^2)^{-3/2} \\ m_T = m_e\,\tfrac{2}{3}(1-\beta^2)^{-1/2} \end{cases} \quad (2)$$

$$Langevin-Bucherer(1904) \begin{cases} m_L = m_e\,\tfrac{2}{3}(1-\tfrac{1}{3}\beta^2)(1-\beta^2)^{-4/3} \\ m_T = m_e\,\tfrac{2}{3}(1-\beta^2)^{-1/3} \end{cases} \quad (3)$$

(with the definitions $\beta = v/c$ and $m_e = e^2/Rc^2$ in the usual notations, R being the radius of the spherical charged body).

Meanwhile, Kaufmann's experiments (1902-1903) confirmed the reality of the variation of mass with velocity. However the confusion was exacerbated by the question of the agreement betwen the theoretical and the experimental results, the precision of which was clearly overrated at the time (Kaufmann claimed a 1% precision for $0.6 < \beta < 0.9$).

Many discussions on the nature and significance of this velocity dependence took place between Abraham, Lorentz, Kaufmann, Planck and Poincaré, who had written as early as 1900 that "Electromagnetic energy can be viewed as a fluid with inertia". Finally, Planck, at a famous Köln conference (1908) concluded to a general "law of inertia of energy" by showing that the flow of any sort of energy, whether thermal, chemical, elastic, gravitational, etc., could be associated with a momentum density. At the same period, in one of the first important American contributions, Lewis and Tolman (1909) stated that the variation of mass with velocity transcended electromagnetic theory, and must be a universal feature, independent of the electric charge of the body [4]. Then, by the end of 1909, came the breakthrough. Von Dida, a student of Planck, asked himself what could be the simplest and most natural form for the inertia r of a body as a function of its velocity [5]. Here is how he proceeded, starting from first principles. Von Dida first defined the (possibly variable) inertia N of a body as the coefficient of the velocity in the expression of the momentum:

$$p = Nv. \quad (4)$$

He then *assumed* the inertia of energy in its most elementary form, asking that any variation in the energy E of the body would entail a proportional variation of its inertia N:

$$dN = \chi dE \tag{5}$$

where χ should be a universal constant. Finally, by recalling Hamilton's first equation,

$$v = \frac{dE}{dp}, \tag{6}$$

he had three relationships between the four magnitudes v, p, N, E, enabling him to express for example the last three of them in terms of the first one. One now computes quite simply $dN = \chi v dp = \chi v(dNv + Ndv)$, hence

$$\frac{dN}{N} = \frac{\chi v dv}{1 - \chi v^2}, \tag{7}$$

so that

$$N = \frac{N_0}{\sqrt{1 - \chi v^2}}. \tag{8}$$

Of course, considering the low-velocity limit leads one to identify N_0 with the mass m of the body. Since the constant χ has the dimensions of the inverse-square of a velocity, it is natural to set $\chi = c^{-2}$, defining a *universal* constant c with the dimension of a velocity, acting as a limit velocity for any massive body. Note that c is not necessarily the velocity of any specific physical agent, although of course it was ultimately to be identified with the velocity of electromagnetic waves. Von Dida finally obtained the following expressions for the dynamical properties of a body :

$$N = \frac{m}{\sqrt{1 - v^2/c^2}}, \quad p = \frac{mv}{\sqrt{1 - v^2/c^2}}, \quad E_{kin} = \frac{mc^2}{\sqrt{1 - v^2/c^2}} - mc^2. \tag{9}$$

(the last expression results from the fact that, because of (3), $E = \chi^{-1}N + cst$, and the requirement that for the purely kinetic energy, one should have $E_{kin}(v)|_{v=0} = 0$).

A puzzling consequence of these formulas was that they put into question the very notion of velocity. Indeed, since the momentum is no longer proportional to the velocity, the conservation of the total momentum becomes inconsistent with the usual law of addition of velocities. But the trouble with velocity was still to deepen, as we are going to see.

3. The New Conservation Laws

At the very same time, a young Swiss physicist, Albert Hunepierre, hit on a different but related approach. Attending the 1908 Köln conference already alluded to, he was stimulated by conversations with Paul Langevin on the nature of conservation laws, and by a comment of Minkowski at the end of Planck's lecture: "In my view, the law of momentum is obtained from the energy law; namely, in Lorentz's theory, the energy law depends on the reference system. We write the energy law for every possible reference system, so that we have many equations and in those are contained the law of momentum" [6]. Hunepierre's idea was to exploit this clue in order to characterize the functions $E(u)$ and $p(u)$ giving the energy and momentum of a body in terms of its velocity u (the change in notation with respect to the preceding section is purposeful). In fact, he simply revived, in very general terms, considerations going back to Huyghens. For the sake of simplicity, let us confine ourselves to the one-dimensional case, so that E and p are an even and an odd function respectively [7]. Consider a system of interacting particles, with velocities u_k $(k = 1, 2, \ldots)$; it is ruled by the conservation laws for the total energy E_{tot} and total momentum p_{tot}:

$$\left\{ \begin{array}{rcl} E_{tot} & = & \sum_k E(u_k) = Cst \\ p_{tot} & = & \sum_k p(u_k) = Cst \end{array} \right. \tag{10}$$

expressed in a particular inertial reference frame. Let us now describe the system in another inertial reference frame, moving with respect to the first one with velocity U. The crux of the argument is to *assume the additive law of combination for velocities*, so that the particles now have velocities

$$u'_k = u_k + U \quad (k = 1, 2, \ldots). \tag{11}$$

The conservation laws in this new reference frame then read

$$\left\{ \begin{array}{rcl} E'_{tot} & = & \sum_k E(u_k + U) = Cst' \\ p'_{tot} & = & \sum_k p(u_k + U) = Cst'. \end{array} \right. \tag{12}$$

If we now require the two conservation laws to hold in all equivalent inertial frames, we see, by developing the expressions (10) in power series of U, that not only are the total energy and momentum E and p conserved, but also all the quantities formed by adding up the successive derivatives of the individual energies and momenta,

$$\sum_k \frac{dE}{du}(u_k), \sum_k \frac{d^2E}{du^2}(u_k), \ldots, \sum_k \frac{dp}{du}(u_k), \sum_k \frac{d^2p}{du^2}(u_k) \ldots \tag{13}$$

But we cannot have more than two independent conservation laws; otherwise the collision process would be overdetermined (as the case of two particle makes clear). The "new" conservation laws must therefore be offshoots of the "old" ones. Since the individual velocities are arbitrary, the derivatives of the energy and momentum functions must depend on the energy and momentum for each particle separately. And the additivity of the conservation laws requires this functional dependence to be linear. Finally, the even and odd nature respectively of E and p severely restricts the possibilities. For the first derivatives, indeed, the most general expressions are:

$$\left\{ \begin{array}{l} \dfrac{dE}{du} = p \\[2ex] \dfrac{dp}{du} = m + \chi E \end{array} \right. \tag{14}$$

where we have chosen a coefficient unity in the first equation according to the usual dimensional convention, and where the constant term in the right-hand-side of the second equation is identified with the mass of the particle in order to recover the standard low-velocity Newtonian expressions. By so doing, we automatically take E to be the purely kinetic energy of the particle, since, in the case $\chi = 0$, one recovers from (11) the expressions $p = mu$ and $E = mu^2/2$. Of course, we know, with hindsight, that the coefficient χ is the same as in the previous Section, and we put $\chi = c^{-2}$. The solution of the system (11) of differential equations, given the even and odd nature of E and p respectively, is unique:

$$\left\{ \begin{array}{rcl} E_{kin} & = & mc^2[\cosh(u/c) - 1] \\ p & = & mc \sinh(u/c) \end{array} \right. \tag{15}$$

Obviously, the comparison of the expressions (12) with those given by (7) compounded the problem concerning the very notion of velocity, to which we will return anon. But

Hunepierre's major accomplishment, in a second paper of the same year 1909, was to realize that the new form (12) of the conservation laws was inconsistent with the hitherto unquestioned additivity of mass. It suffices to consider two particles, with the same mass m. In the reference frame of their center of inertia, they have respective velocities $(u, -u)$; the internal energy of the system, that is, the kinetic energy of the particles in this very reference frame thus is $E_0 = 2mc^2 \left[cosh(u/c) - 1 \right]$. In another reference frame, moving with respect to the first one with velocity U, the particles have velocities $(u + U, -u + U)$, so that the total momentum is $p = mc \sinh(u/c + U/c) + mc \sinh(-u/c + U/c) = 2mc \cosh(u/c) \sinh(U/c)$. However, on the other hand, the momentum of the system, moving with global velocity U, should read, according to the second of equations (12), $p = Mc \sinh(U/c)$, where M is its (total) mass. We are then led to recognize that the total mass is given by $M = 2m \cosh(u/c) = 2m + E_0/c^2$. Mass is no longer additive, and any change in the internal energy of the system entails a proportional change in its mass, $\Delta m = \Delta E_0/c^2$. Hunepierre then proposed to normalize the (hitherto arbitrary) zero of energies so as to include, for any M, its "mass energy" Mc^2 in its internal energy, which enabled to write in full generality the now famous Hunepierre's equation

$$M = E_0/c^2 \tag{16}$$

which he jokingly transcribed as

$$\text{Mass} \propto \text{Innergy.} \tag{17}$$

4. The New Space-Time

By that time, that is, around 1910, it was becoming obvious that the changes in dynamics (i.e. in the expressions for the energy and momentum) required a parallel modification in kinematics (i.e. in the structure of space-time), as illustrated by the debate around the notion of velocity.

It had been known for several years that Maxwell equations were invariant under a particular set of transformations, as noted by Lorentz (1904) - who left his name to them - , and by Poincaré (1905), who emphasized the group structure of this set. Minkowski had given in 1908 a neat mathematical description of this pseudo-euclidean group. Still, this invariance property was thought to be specific to electromagnetic theory, and its interpretation was unclear. Poincaré, for instance, up to the end of his life (1912), clung to an epistemology which required Lorentz's ether, and hardly allowed for a physical significance of the Lorentz transformations [8]. However, Sommerfeld, who had been a student of Felix Klein, was familiar with the Erlanger program and the geometrical interpretation of transformation groups [9]. He thus suggested to investigate the possible transformation groups in space-time, depending only on general and abstract requirements, irrespective of the specific physical phenomena (electromagnetic, gravitational or others) taking place on a supposedly universal spatio-temporal stage.

In the early 1910s, several independent works appeared resolving this question - with, it must be said, rather awkward approaches and complicated calculations, due to a lack of familiarity with group theory (and with infinitesimal Lie algebraic methods in particular). Various contributions by von Ignatowsky (1910, 1911), Franck and Rothe (1911,1912), van Rijn (1912), Hahn (1913), etc., finally converged towards the following conclusion [10]. Under the very general assumptions of:

- homogeneity of space-time
- isotropy of space

- existence of causal relationships,

the Lorentz-type transformations (written here in the one-dimensional case)

$$\begin{cases} x' = \dfrac{x - vt}{\sqrt{1 - \chi v^2}} \\[3mm] t' = \dfrac{t - \chi vx}{\sqrt{1 - \chi v^2}} \end{cases} \tag{18}$$

with a constant $\chi \geq 0$, are the *only* possible ones. Of course, if these transformations are to be universal, and if Maxwell equations are indeed correct, one has again to identify $\chi = c^{-2}$. Under the influence of the Kleinian point of view, the theory of space-time soon became known under the very apt name of "chronogeometry", or, more precisely, "Minkowskian chronogeometry", when necessary to distinguish it from the classical conception of space-time or "Galilean chronogeometry" (which, it must be stressed, indeed is a space-with-time theory, as it already mixes space and time - or, rather, time with space, if not space with time as in the true Minkowskian case). Once this generalized geometrical perspective was adopted, it was a simple matter to understand the puzzling phenomena of so-called "length contraction" and "time dilatation" as *parallax* effects in space-time, quite analogous to the customary parallax effects in space. Langevin, in particular, was especially keen on this interpretation [11] .

5. Speeds

The confusion around the notion of speed was due to the existence of two equally natural derivations for the energy and momentum, yielding the two expressions stemming from (7) and (12) respectively, which we rewrite in the natural system of units already advocated by Planck, where $c = 1$:

$$\begin{cases} E = \dfrac{m}{\sqrt{1 - v^2}} = m \cosh u \\[3mm] p = \dfrac{mv}{\sqrt{1 - v^2}} = m \sinh u \end{cases} \tag{19}$$

It was immediately clear that the two competing notions were related through

$$v = \tanh u, \tag{20}$$

but it was generally held that the 'true' velocity was v , since, after all it entered Hamilton's equation and was consistent with the old Galilean definition of velocity as the ratio of the span of space traveled by a mobile object to the time taken, i.e. $v = \Delta x / \Delta t$. While most physicists maintained that u was merely a formal quantity, a purely mathematical device, others insisted that the limitation $v \leq c$ and the non-additivity of the velocity parameter v required u to be given a primary physical meaning - as well as a name of its own, for which the term 'rapidity' was chosen [12] . Agreement was finally reached that a single notion in Galilean chronogeometry split into two different albeit related ones in Minkowskian chronogeometry. The situation, it was understood, is quite similar to that in ordinary geometry where slope, when small, is characterized by a single parameter, while one has to distinguish angle and tangent for larger values. Finally, rapidity was given a direct physical meaning by Blondeville who discussed the operational definitions of speed; he stressed that an observer, if isolated within his own reference frame without any external clue, cannot measure the distance he covers (so that the Galilean definition is useless to him). However, he can measure his (proper) acceleration as a function of his (proper) time, and integrate it, which results, as a simple calculation shows, in the increment of his rapidity:

$$\int_A^B dt_0 \gamma_0(t_0) = u_B - u_A \tag{21}$$

The most natural *intrinsic* measurement of speed changes thus yields directly the variation of rapidity [13]. Once the status of rapidity as a *bona fide* physical magnitude was established, it was used by Blondeville and Prostov for a much simpler derivation of the Lorentz group as the only possible chronogeometrical group (along with its singular Galilean limit) by relying on the elementary but deep lemma that any one-parameter continuous group can be additively parametrized [14].

6. Another Route to Spacetime?

No account of the birth of modern chronogeometry would be complete without mentioning a recent historical discovery, which shows that another path could have been taken. Recently, in the archives of the famous journal *Annalen der Physik*, the manuscript was uncovered of a paper submitted in 1905 by a young physicist named Albrecht Zweistein. In this paper, which was rejected for reasons we will shortly discuss, Zweistein studied the electrodynamics of moving bodies. While sticking to the customary (at the time) electromagnetic worldview of Lorentz and Poincaré, he consistently pursued an earlier argument by Cohn (1901,1902), who, following a Machian train of thought stressing 'scientific economy', had shown the superfluousness of the ether. Zweistein then completely did away with the ether and its privileged frame [15]. Instead, he proposed an interpretation of the Lorentz invariance of Maxwell equations based on two principles:

1) the classical equivalence of reference frames in uniform motion (which he called the "Principle of relativity", borrowing the terminology from Poincaré),

2) the apparently weird idea that the velocity of light was the same in all these reference frame - his 'second postulate'.

Zweistein then proceeded to analyse time and space measurements, through detailed gedankenexperiments using the exchange of light signals to synchronize distant clocks. He was able to show that the time and space coordinates defined by such operational procedures, when compared in two equivalent reference frames, were linked by the standard Lorentz transformations.

Although this was quite a clever paper, it raised strong objections and was refused for publication; it must be said that the author held no academic position, which certainly did not help. A first referee sternly condemned the dismissal of the ether as "unphysical", in line with Poincaré's position. A second one pointed to the *ad hoc* and fragile nature of the operational point of view taken by Zweistein: what, he asked, if non-electromagnetic signals, of a hitherto unknown nature, were discovered and used? Would we not have to face the possibility of a different description of space-time for these new phenomena, negating the idea of a universal spatio-temporal arena? And what if Maxwell's equations were only approximate? [In modern terms, think of the situation if the photon finally had a nonzero mass, however small; light would not travel with the invariant velocity, and any derivation of chronogeometry based on the exchange of light signals would become invalid.] The paper was resubmitted by the author a few years later, claiming priority after the development of chronogeometry as sketched above. It was then rebutted by Sommerfeld on more epistemological grounds. Here is what he wrote: "[The theory of space-time] is an *Invariantentheorie* of the Lorentz group. The name 'relativity theory' is an unfortunate choice: the relativity of space and time is not the essential thing, which is the independence of the laws of nature from the point of view of the observer" [16].

One may only wonder what the dominant interpretation and terminology of our theory of space-time would have been if Zweistein's paper had been published as early as 1905 . . .

7. Conclusions

I will now abandon fiction and propose some conclusions.

But first, in order to ward off any accusation of blasphemy or lese-majesty, let me stress that Einstein himself was quite aware that his path towards what we now know as 'special relativity theory' was not the only possible one, and that others could have originated the new space-time physics. Here is, for instance, what he wrote in his obituary to Langevin (1947): "It appears to me as a foregone conclusion that he would have developed the special relativity theory, had not that been done elsewhere" [17].

I leave it to specialists in science history to assess the degree of plausibility of the alternate account narrated here. But what is the purpose of this fictitious reconstruction? In fact, I see three areas of relevance for such considerations:

a) *educational*. The presentation of scientific notions as they unfolded historically is not the only one, nor even the best one. Alternative arguments and novel derivations should be pursued and developed, not necessarily to replace, but at least to supplement the standard ones.

b) *epistemological*. As I have tried to show, if Einstein had not existed, we still would have 'his' theory - or would it really be the same? Apart from a few details perhaps, the formalism, that is, the notations and equations, would be very similar to ours today. But the language and, underneath, the words and the ideas we would use could be rather different. Many familiar terms, beginning with 'relativity' itself, might be absent from our vocabulary. My point here is to stress the polysemy of science, even for such a highly formalized science as physics. We need to actualise the variety of potential meanings behind our symbols - would it be only to keep open impredictable future paths of development.

c) *cultural*. Science is too often perceived by lay people as a mechanistic and inhuman endeavour. We should try to offer a less absolute and less deterministic view of its development, so that science clearly appears for what it is, that is, a human venture, the solidity of which is no doubt based on its collective structure, where the role of the individuals nevertheless cannot be neglected. Think for instance of the whole mythology of the twentieth century without the figure of Einstein (it is on purpose that in the above fiction the emergence of chronogeometry has been attributed to a largely collective effort).

In more general terms, my aim here was to advocate a view of the history of science which gives due credit to the notion of contingency. This is the reason of my dedicating this paper to the memory of Stephen J. Gould who passed a few weeks ago. In his work, and especially in his book *Full House*, he stressed the importance of the notion of contingency for the history of life [18] . If you 'rewind the film', as he was fond of saying, and let it unfold again starting a billion years ago, there is no reason the story would follow the same course. Chance events, it is now accepted, have played a crucial role in the history of life, events such as the crashing of a meteorite in Yucatan 65 million years ago, which tolled the knell of dinosaurs and paved the way for mammals. Of course, as Gould repeatedly emphasised, this view is not akin to an irrationalistic one: the general features of history, would it be natural, social or scientific, must and can be explained. But historical phenomena are so complex as to be extremely sensitive to a host of apparently secondary conditions. Of course, this should not come as a surprise to us physicists who today are familiar with deterministic chaos, sensitivity to initial conditions, etc.

Might it not be claimed, then, that Einstein was a Loren(t)z butterfly?

It is a pleasure to thank Françoise Balibar and Bruno Latour for their comments on a preliminary version of this paper and George Morgan for his most helpful linguistic advice, as well as Stéphane Métens for typesetting the text.

References

[1] See J.-M. Lévy-Leblond, "Science's fiction", *Nature* **413**, 573 (2001).

[2] As a remarkable example, think of the 1972 novel by Philip K. Dick, *The Man in High Castle* (Vintage Books, 1992), describing America in 1962, twenty years after it has lost the war and is occupied by Nazi Germany and imperial Japan. On a more scientific side, a recent paper by E. B. Davies uses the fictional device of an imaginary Earth permanently covered with a cloud hiding the sky to investigate the role of astronomy (non-existent on this world) in the history of science (arXiv: physics/0207043).

[3] Arthur I. Miller, *Albert Einstein's Special History of Relativity, Emergence (1905) and Early Interpretation (1905-1911)*, Addison-Wesley, 1981.

[4] All of the preceding is historically correct, as can be checked from the seminal book by Arthur I. Miller (ref. 3). In particular, for the discussions of the mass variation with velocity within electromagnetic theory, see Sections 1.8 to 1.14. For Planck's law of inertia of energy, see Sections 12.5.7. For Tolman and Lewis contribution, see Section 12.2, note 4.

[5] Here we depart from history as it unfolded. The following derivation in fact is due to W.C. Davidon, *Foundations of Physics* **5**,525 (1975), who also give an account of the historical developments sketched above.

[6] While Hunepierre's character is a fiction, the remark by Minkowski is true to the facts; see A. Miller, ref.3, p. 367.

[7] The following derivation has been proposed by J.-M. Lévy-Leblond, "What is So Special in Relativity?", in *"Group-Theoretical Methods in Physics"*, A. Janner & al., Lecture Notes in Physics n^{0}50, Springer-Verlag, 1976.

[8] See, for instance, the comments by O. Darrigol in his introduction to Albert Einstein, *Œuvres choisies*, vol.2 (*Relativités I*), Seuil-CNRS, 1993,P.23.

[9] See A. Miller, ref.3, p. 181 (note 41).

[10] None of these names or contributions are invented. Here are the exact references: W.I. Ignatowsky, *Archiv der Math. und Phys. III*, **17**, 1 (1910), *Phys. Z.* **11**, 972 (1910), &**12**, 776 & 779 (1911), P. Franck & H. Rothe, *Ann.Phys.* **34**, 825 (1911), *Phys.Z* **13**, 750 (1912), A.C. van Rijn van Alkemade,*Ann.Phys.* **38**, 1033 (1912), E. Hahn, *Archiv Math.Phys.* III, **21**,1 (1913). These papers were immediately neglected and forgotten and their conclusions soon rediscovered by L. A. Pars, *Phil. Mag.* **42**, 249 (1921), E. Esclangon, *C.R.Acad. Sci.* **202**, 1492 (1936), E. Le Roy, *C.R. Acad Sci.* **202**, 794 (1936), V. Lalan, *C.R. Acad. Sci.* **203**, 794 (1936)& *Bull. Sci. Math.France*, **65**,83 (1937), which were no more successful. A later independent proof, using simpler and more modern arguments, was proposed by J.-M. Lévy-Leblond,*Am. J. Phys.* **44**, 271 (1976). A general review of this long but occult line of arguments has been made by J.-P. Lecardonnel, in his Thesis, University Paris VI (Pierre et Marie Curie), 1979, and in a paper, *Bull. U. Phys.*, **615**,1171 (1979).

[11] Although Langevin did not explicitly use the idea of parallax, he nevertheless put forward a very modern view of space-time; see P. Langevin, "L'évolution de l'espace et du temps", *Scientia* **10**, (1911).

[12] Here is a serious historical question: when and by whom was the term 'rapidity' used for the first time-certainly rather late in the historical development? I would indulge in conjecturing that J.A. Wheeler could be at the origin of the word in the fifties.

[13] While this simple fact may have been known to quite a number of people, I do not know of any published mention before J.-M. Lévy-Leblond, *Am. J. Phys.* **48**, 345 (1980).

[14] It remains surprising that such a simple mathematical result was recognized so lately as a solid founding stone for chronogeometry. See J.-M. Lévy-Leblond & J.-P. Provost, *Am. J. Phys.* **47**, 1045 (1979). It may also be mentioned here that a still more general derivation of the (very few) possible chronogeometries, relaxing the linearity of the action of the group on space-time, has been given by H. Bacry & J.-M. Lévy-Leblond, *J. Math. Phys.* **9**, 1605 (1968).

[15] The important contributions of Emil Cohn and his epistemological stand are barely known today. They

probably exerted a definite influence on Einstein who thought highly of Cohn (as did the usually overcritical Pauli); see A. Miller, ref.3, p.181-182, note 42.

[16] This statement(as well as other similar ones) by Sommerfeld is well documented although it came historically much later, that is, in a 1948 review paper on philosophy and physics after 1900; see A. Miller, ref. 3, p.181, note 41.

[17] See A. Miller, ref. 3, p. 388.

[18] Stephen J. Gould, *Full House*, Harmony Books (1966).

Section 1

Discrete Groups and Geometry

Inst. Phys. Conf. Ser. No 173: Section 1
Paper presented at 24th Int. Coll. Group Theoretical Methods in Physics, Paris, France, 15–20 July 2002
©*2003 IOP Publishing Ltd*

Wavelet Basis for Quasicrystal diffraction

Miroslav Andrle[1,2] † and Avi Elkharrat[1] ‡

[1]LPTMC, Université Paris 7 - Denis Diderot, 2 Place Jussieu 75251, Paris Cedex 05, France
[2]Dep. of Mathematics, FNSPE-CTU, Trojanova 13, 120 00 Prague 2, Czech Republic

Abstract. In this article we present (1) a scale dependent partitioning procedure of the Fourier space based on beta-integers, with β a quadratic Pisot number, for computing quasicrystals diffraction patterns, (2) a Haar wavelet analysis with beta-adic wavelets. These technics enhance the precision of numerical computation and analysis of the intensity function arising from diffraction patterns of one-dimensional model sets.

1. Introduction

Quasicrystals have startled physicists because they contradicted preconceived ideas about symmetries in Condensed Matter. In the field of crystals, discrete group theory allowed to derive physical properties from electronic states to elasticity [6]. For quasicrystals, a great breakthrough was achieved with the 'Cut and Project Method', but more needs to be made. As a complementary alternative of using higher-dimensional representations of quasiperiodic point sets, it is interesting to use tools adapted to quasicrystalline symmetries in physically relevant dimensions. Here, the tool we use is a class of irrational numbers, the so-called set of beta-integers, denoted by $\mathbb{Z}_\beta$, with β a quadratic Pisot-Vijayaraghavan unit. We suggest that the set of beta-integers is a relevant numerical frame in which we could think of the structural properties of quasicrystals, similarly to what crystallographers do with integers. The set of beta-integers is the starting point of more complex constructions. In the present article study we use beta-integers to construct (1) a scale dependent partitioning procedure of Fourier space for numerical computation of aperiodic diffraction patterns (Section 3); (2) an aperiodic multiresolution analysis of $L^2(\mathbb{R})$ and the construction of a Haar beta-adic wavelet basis (Section 4). Both, computation of the diffraction pattern and wavelet analysis of the later, are enhanced, since these technics allow greater numerical precision.

2. Tau-integers

Let $\tau = (1 + \sqrt{5})/2$ denote the *Golden Ratio*, solution greater than 1 of the algebraic equation $X^2 = X + 1$. Recall that the extension ring of τ is $\mathbb{Z}[\tau] = \{m + n\tau \mid m, n \in \mathbb{Z}\}$. The set of tau-integers $\mathbb{Z}_\tau$ is a subset of $\mathbb{Z}[\tau]$. Positive tau-integers can be constructed by the following substitution algorithm: $L \to LS$, $S \to L$, starting from letter L, and by associating to letters L and S tiles of length 1 and $1/\tau$ respectively. The set of tau-integers is a selfsimilar set, symmetrical with respect to the origin.

$$\mathbb{Z}_\tau = -\mathbb{Z}_\tau, \quad \cdots \subset \mathbb{Z}_\tau/\tau^{j-1} \subset \mathbb{Z}_\tau/\tau^j \subset \mathbb{Z}_\tau/\tau^{j+1} \subset \cdots, \ j \in \mathbb{Z}. \tag{1}$$

For a mathematical review on tau-integers see [3].

† andrle@ccr.jussieu.fr
‡ kharrat@ccr.jussieu.fr

186

Figure 1. Diffraction pattern of the Fibonacci chain in the internal Fourier space. On the left are the peaks supported by scale 0, in the middle scale 1, and on the right scale 2.

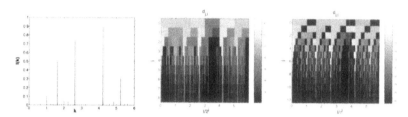

Figure 2. Diffraction pattern of the Fibonacci chain in physical Fourier space (left), dyadic wavelet transform in $\log_2$ scale (middle) and tau-adic wavelet transform in $\log_\tau$ scale (right).

3. Numerical Diffraction

The starting point of our study is the numerical calculation of diffraction patterns of one-dimensional cut and project sets (Fig. 2). Throughout this article we shall only discuss the Fibonacci chain, although our results are not restricted to this set. The issue is how to take a calculation in order to have the best approximation of the diffraction pattern, knowing that it is made of sharp Bragg peaks of various intensity that are supported by the normalized dense ring $\mathbb{Z}[\tau]$.

When we calculate this diffraction pattern on an evenly discretized support the positions of Bragg peaks are poorly approximated. Consequently to the intrinsic quasiperiodic nature of the Fibonacci chain, each peak is computed with a different accuracy and it becomes difficult to determine the meaning and the relevance of the information we collect. A wiser move is to take a calculation on points of the type $m + n\tau$, with $m, n \in \mathbb{Z}$. Obviously we now have a better approximation of the positions of Bragg peaks, and therefore a better approximation of their intensities.

Tau-integers allow us to go one step further, providing us, from Equation (1), with a scale dependent partitioning procedure of the **k**-space — a discretization scheme that emphasizes on the selfsimilar structure of the spectrum. We shall refer to *scaling* as the branch of measurement that involves the construction of an instrument that associates qualitative constructs with quantitative metric units [7], and we shall define a scale in the diffraction pattern in the following way:

Definition 1 *We say that a Bragg peak belongs to a certain scale j if j is the smallest integer such that the Bragg peak is located on the set $\mathbb{Z}_\tau/\tau^j \setminus \{\mathbb{Z}_\tau/\tau^k, k < j\}$.*

Thus, by using this definition for *scale*, we can replace qualitative sentences like "Bragg peaks of low intensity" by quantitative sentences like "Bragg peaks supported by scales j, for $j > N_0 \in \mathbb{N}$". The diffraction pattern is decomposed along the scales $\mathbb{Z}_\tau/\tau^j$ for $j = 0, 1, \ldots, N$, see Figure 1, the integer N characterizing working precision.

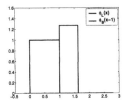 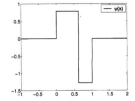

Figure 3. Haar scaling functions (left) and Haar wavelet (right) for tau-integer support.

4. Wavelet Analysis

Our next step is to take the wavelet transform of the diffraction pattern. As in the case of numerical diffraction, the issue here is how to discretize the support of the signal, and how to sample the signal, in order to render its intrinsic symmetries and properties accurately. The wavelet analysis is based on a multiresolution analysis of $L^2(\mathbb{R})$ [1]. Taking the wavelet transform of the diffraction pattern of the Fibonacci chain allows us to visualize all peaks at once (see Fig. 2).

4.1. Mathematical Insights on Haar Tau-adic Wavelet Basis

Haar scaling functions and Haar wavelets are both piecewise constant non-continuous functions (Fig. 3). The tau-adic wavelet transform is similar to the dyadic transform, except that we need two scaling functions, one for the tile L and one for the tile S, $\phi_L(x) = \mathbf{1}_{[0,1)}(x)$ and $\phi_S(x) = \tau^{1/2}\mathbf{1}_{[1,\tau)}(x)$ respectively. Moreover instead of cutting each tile into two smaller tiles of equal length we cut the tiles with respect to the substitution of tau-integers, shown in Section 2.

We construct the Haar wavelet basis on non-negative tau-integers $\mathbb{Z}_\tau^+$. Denote by Λ_L^+ (resp. Λ_S^+) the subset of $\mathbb{Z}_\tau^+$ of left-hand ends of all tiles L (resp. tiles S) [2]. Define V_0 as the linear span $V_0 = \mathrm{span}\{\phi_L(x-l), \phi_S(x-s)\}$, with $l \in \Lambda_L^+$ and $s \in \Lambda_S^+$. The latter forms an orthonormal basis of V_0. Consequently for all $p \in \mathbb{Z}$ we have

$$V_j = \mathrm{span}\{\phi_{Lj,l}(x), \phi_{Sj,s}(x)\}, \quad l \in \Lambda_L^+, \ s \in \Lambda_S^+, \tag{2}$$

with $\phi_{Lj,l}(x) = \tau^{j/2}\phi_L(\tau^j x - l)$ and $\phi_{Sj,s}(x) = \tau^{j/2}\phi_S(\tau^j x - s)$. The nested sequence of spaces

$$\cdots \subset V_{-1} \subset V_0 \subset V_1 \subset \cdots \subset V_j \subset \cdots, \tag{3}$$

forms a multiresolution analysis of $L^2(\mathbb{R}^+)$, of piecewise constant functions embedded in the nested sequence $\cdots \subset \tau\mathbb{Z}_\tau^+ \subset \mathbb{Z}_\tau^+ \subset \mathbb{Z}_\tau^+/\tau \subset \cdots \subset \mathbb{Z}_\tau^+/\tau^j \subset \cdots$ respectively.

We introduce the space W_0 as the orthogonal complement of V_0 in V_1, $V_1 = V_0 \oplus W_0$. We define the Haar wavelet

$$\psi(x) = \tau^{-1/2}(\mathbf{1}_{[0,1/\tau)}(x) - \tau\mathbf{1}_{[1/\tau,1)}(x)), \tag{4}$$

such that $W_0 = \mathrm{span}\{\psi(x-l)\}_{l \in \Lambda_L^+}$, with $\{\psi(x-l)\}_{l \in \Lambda_L^+}$ an orthonormal basis. Then we define the space W_j for all $j \in \mathbb{Z}$ as $W_j = \mathrm{span}\{\tau^{j/2}\psi(\tau^j x - l)\}_{l \in \Lambda_L^+}$. Once again we have $V_{j+1} = V_j \oplus W_j$. We say that W_j is the space of details in space V_{j+1} while V_j is the space of approximations of order j. The set of functions $\{\psi_{j,l} = \tau^{j/2}\psi(\tau^j x - l)\}_{j \in \mathbb{Z}, l \in \Lambda_L^+}$ forms

---

(1) A sequence of closed nested subspaces of $L^2(\mathbb{R})$ verifying precise properties. For instance, see [8].
(2) We obviously have $\mathbb{Z}_\tau^+ = \Lambda_L^+ \cup \Lambda_S^+$, and it is a partition.

188

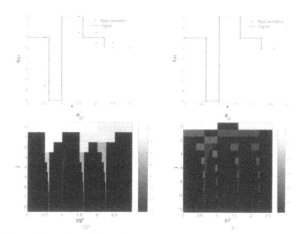

Figure 4. Dyadic (up-left) and tau-adic (up-right) approximation of an aperiodic signal; dyadic (bottom-left) and tau-adic (bottom-right) wavelet transform of the signal in $\log_2$ resp. $\log_\tau$ scale.

an orthonormal basis of $L^2(\mathbb{R}^+)$, therefore all $f(x) \in L^2(\mathbb{R}^+)$ has a unique decomposition as $f(x) = \sum_{j,l} d_{j,l} \psi_{j,l}(x)$, where $d_{j,l} = \langle f, \psi_{j,l} \rangle = \int_{\mathbb{R}} f(x) \psi_{j,l}(x) \mathrm{d}x$ are the wavelet coefficients of $f(x)$. The extension of the basis to $L^2(\mathbb{R})$ is carried out by symmetry of the support, the scaling functions and the wavelet, with respect to the origin. Thus we construct a relevant basis of $L^2(\mathbb{R})$ for wavelet analysis adapted to quasiperiodic sets. For a full mathematical development of tau-adic wavelets see [4], [1] and [2].

4.2. Relevance of Tau-adic Wavelet Transform

At this point we would like to give some insight on the interpretation of tau-adic wavelet transform. We perform dyadic and tau-adic wavelet transform of the signal displayed on Fig. 4. This piecewise constant mono-scale function is in V_0. Since the signal is aperiodic, the coefficients of the dyadic wavelet transform are scattered throughout all scales. On the other hand tau-adic wavelet transform of the signal renders information at the scale it belongs.

Acknowledgements

Authors would like to express their gratitude to J. Le Bourlot and N. Décamp for insightful discussions on numerical methods and wavelets, and referee for helpful comments about redaction.

References

[1] M. Andrle (2002), Ph.D. Thesis, Univ. Paris 7, UFR de Physique et Physique Théorique.
[2] M. Andrle, Č. Burdík, J.-P. Gazeau, J. Fourier Anal. and Application, to be published.
[3] Ch. Frougny, J.-P. Gazeau and R. Krejcar (2002), *Theor. Comp. Sci.*, to appear.
[4] J.P. Gazeau, J. Patera (1996), J. Phys. A: Math. Gen. **29** p. 4549.
[5] R. Krejcar (2000), Ph.D. Thesis, Univ. Paris 7, UFR de Physique et Physique Théorique.
[6] J.F. Nye (1957), *Physical Properties of Crystals*, Oxford.
[7] W. M. Trochim, *Knowledge Base*, http://trochim.omni.cornell.edu/kb
[8] P. Wojtaszczyk (1997), *A Mathematical Introduction to Wavelets*, Cambridge Univ. Press.

Inst. Phys. Conf. Ser. No 173: Section 1
Paper presented at 24th Int. Coll. Group Theoretical Methods in Physics, Paris, France, 15–20 July 2002
©2003 IOP Publishing Ltd

Haar wavelets for the quasiperiodic icosahedral Danzer tiling.

Miroslav Andrle[1,2] **and Peter Kramer**[3]

[1]LPTMC, Université Paris 7 - Denis Diderot, 2 Place Jussieu 75251, Paris Cedex 05, France
[2]Dep. of Mathematics, FNSPE-CTU, Trojanova 13, 120 00 Prague 2, Czech Republic
[3]Institut für Theoretische Physik der Universität, D 72076 Tübingen, Germany

Abstract. Orthogonal wavelet bases on the quasiperiodic icosahedral Danzer tiling are constructed. The stone inflation is written in terms of discrete Euclidean operations on the prototiles. By geometric inflation we represent any fixed prototile as the union of scaled copies of prototiles. Application of the unitary representations of the Euclidean group for the inflation to the Haar characteristic functions on the prototiles yields a wavelet basis. This basis is constructed on any tile of the tiling and orthogonalized in closed form.

1. Introduction

We are interested in the construction of wavelets on quasiperiodic tilings. Wavelets can be used to analyse properties of quasicrystals. The simplest tilings are the ones with stone-inflation. For these, any inflated tile can be packed from a finite set of prototiles. For functions with domains on the tiles this leads to very simple decompositions. The explicit construction of orthogonal wavelet bases for the Penrose-Robinson and the triangle tiling was given in [5]. We extend this construction to the 3D icosahedral Danzer tiling. A basic step is to write the stone inflation of the tiling in explicit form by use of operations from the Euclidean group $IO(3, \mathbb{R})$ in E^3. These Euclidean operations are lifted into linear unitary representations acting on the function space of wavelets. We sketch the steps whose explicit algebraic form is given in [1].

2. Icosahedral basis vectors and point group

The basis of the icosahedral primitive P-module in E^3 are six vectors $\langle e_1, \ldots, e_6 \rangle$ of length $|e_i| = \sqrt{1/2}$. These vectors are icosahedral projections of a lattice basis for the hypercubic lattice in 6D. They are shown in Fig. 1 in relation to the icosahedron.

In an appropriate system of coordinates the six vectors are

$$\langle e_1, e_2, e_3, e_4, e_5, e_6 \rangle = \sqrt{1/2} \begin{bmatrix} 0 & s & \bar{s} & \bar{c} & 0 & c \\ s & c & c & 0 & \bar{s} & 0 \\ c & 0 & 0 & s & c & s \end{bmatrix} \quad (1)$$

where $s = \sin \beta, c = \cos \beta, \tan \beta = \tau^{-1}$ where $\tau = (1 + \sqrt{5})/2$ is the golden mean. We use a bar overlining to denote a minus sign in front of an expression. Any translation t in E^3 is expressed in the P-module by a linear combination of the vectors Eq. 1.

The point group elements g are given in terms of the icosahedral Coxeter group H_3. The three generators and the relations of H_3 are

$$R_1 = (23)(46), R_2 = (45)(36), R_3 = (15)(2\bar{3}), \quad (2)$$
$$(R_1 R_2)^5 = (R_2 R_3)^3 = (R_3 R_1)^2 = e.$$

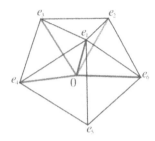

Figure 1. Icosahedron and basis vectors $e_1, \ldots, e_6$.

We write all operations of the icosahedral Coxeter group as signed permutations of the vectors $e_i := i, i = 1, \ldots, 6$ in cycle notation.

3. The Danzer tiles.

The Danzer tiling [3] has four tetrahedral prototiles named (A, B, C, K). All edges of the prototiles run along 5, 2 and 3fold icosahedral axes. All faces are perpendicular to 2fold axes. Each prototile has a mirror image which we denote for short by A', B', C', K'. The pairs of tiles X, X' are shown in Fig. 2.

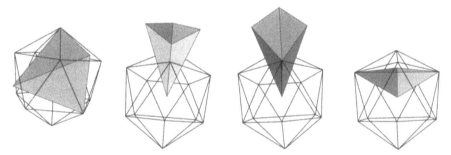

Figure 2. Tiles A, A', B, B', C, C', K, K' in their position inside the icosahedron.

The Danzer tiling can be constructed [3] by inflation, compare the next section. It can also be projected and derived by local rules from a canonical icosahedral tiling [8]$(\mathcal{T}, D_6)$. This tiling has 6 tiles, projections of 3dimensional faces from the Voronoï domains of the D_6 root lattice, whose points in turn are the even lattice points of P [6, 7]. The derivative relation of the tilings allows to use the projection and window technique for the Danzer tiling. The Danzer tiling is also equivalent [4] to the Socolar-Steinhardt tiling [9].

4. Inflation of the Danzer tiles.

The Danzer tiling admits a stone inflation with the linear factor τ. This means that any tile $X = A, A', B, B', C, C', K, K'$ when linearly scaled by τ can be packed from copies of the unscaled prototiles. The Danzer tiling can be generated by repeated inflation of a fixed prototile. We pass to an algebraic description of the inflation.

For an explicit algebraic description we choose [1] for each prototile a reference vertex 0 and a standard orientation. By writing $X_j \to (t_j, g_j)X_j$ we denote an Euclidean operation on the prototile X_j. Here (t_j, g_j) is an element of the Euclidean group $IO(3, \mathbb{R})$ acting on E^3, a rotation $g_j \in H_3$ followed by a translation t_j. In the inflation of a fixed Danzer prototile, a given prototile may participate in the packing more than once. In this case we need a second subscript to distinguish the corresponding occurrences. The inflation of the tile X then takes the general form of a sum of Euclidean operators applied to the prototiles,

$$\tau X = \sum_j \left(\sum_{l_j} (t_{j,l_j}, g_{j,l_j}) X_j \right). \tag{3}$$

For each prototile τX the pairs (t_{j,l_j}, g_{j,l_j}) are listed in [1]. Extension of the Euclidean group by linear scalings with τ yields elements of the affine group. The scaling by τ as an affine group element reads $(0, \tau e)$. Using $(0, \tau e)(t, g) = (\tau t, g)(0, \tau e)$, relations similar to Eq. 3 can be written to decompose $\tau^n X$, $n = 0, \pm 1, \pm 2, \ldots$ in terms of subtiles $\tau^{n-1} X_j$.

5. Haar wavelets on the Danzer tiling.

To any operation of (t, g) of $IO(3, \mathbb{R})$ or its affine extension there corresponds a unitary operation $U(t, g)$ acting on functions with domain E^3. For Haar scaling functions we consider the normalized characteristic functions $\chi(X_k)$ on the tiles X_k. When the inflated tile τX_k is packed according to Eq. 3 from say m_k tiles, the characteristic function on τX_k becomes the sum of m_k orthogonal characteristic functions on these m_k tiles. A corresponding subdivision into m_k subtiles applies to X_k. The operators $U(t, g)$ relate the characteristic functions to their standard positions. We denote the space spanned by normalized characteristic functions living on all Danzer tiles V_0. By n-fold repetition of inflation steps on each tile X_k of the Danzer tiling one can recursively construct an orthonormal basis to any level of precision V_n [1, 2]. The space of wavelet W_0 will be find as an orthogonal complement V_0 in V_1, $V_1 = V_0 \oplus_\perp W_0$. Then for each tile we find $m_k - 1$ orthonormal functions (wavelet) forming with all its admissible unitary operations (translation and rotations) an orthonormal basis of W_0 [1, 2].

References

[1] Andrle M and Kramer P, *Haar wavelets for the icosahedral Danzer tiling*. In preparation
[2] Andrle M 2002, *Model Sets and Adapted Wavelet Transform*, PhD Thesis, Université Paris 7-Denis Diderot, and CTU-FNSPE, Dept. of Mathematics, Prague, *http://bimbo.fjfi.cvut.cz/~andrle*
[3] Danzer L 1991 *Quasiperiodicity: Local and Global Aspects*, in: Lecture Notes in Physics 382 Berlin: Springer, 561-572
[4] Danzer L Papadopolos Z and Talis A 1993, *Full equivalence between Socolar's tilings and the (A, B, C, K)-tilings leading to a rather natural decoration* J. Mod. Phys. B 7 1379-1386
[5] Gazeau J-P and Kramer P 2000, *From quasiperiodic tilings with τ-inflation to τ-wavelets*, Mat. Science and Eng. 294-296 425-428
[6] Kramer P Papadopolos Z and Zeidler D 1992, *Concepts of symmetry in quasicrystals*, in: AIP Conf Proc 266, Eds. A Frank, T H Seligman, K B Wolf, New York: AIP 179-200
[7] Kramer P and Papadopolos Z 1995, *Symmetry concepts for quasicrystals and noncommutative crystallography*, in: Proc. ASI Aperiodic Long Range Order, Waterloo 1995, ed. R.V. Moody, New York: Kluwer, 307-330
[8] Kramer P Papadopolos Z Schlottmann M and Zeidler D 1994, *Projection of the Danzer tiling* J. Phys. A 27, 4505-4517
[9] Socolar J E S and Steinhardt P J 1986, *Quasicrystals, I: Definition and structure, II: Unit-cell configurations* Phys. Rev. B34 596-616, 617-647

Inst. Phys. Conf. Ser. No 173: Section 1
Paper presented at 24th Int. Coll. Group Theoretical Methods in Physics, Paris, France, 15–20 July 2002
©*2003 IOP Publishing Ltd*

193

Quasicrystalline Combinatorics

Michael Baake

Institut für Mathematik, Universität Greifswald, Jahnstr. 15a, 17487 Greifswald, Germany

Uwe Grimm

Applied Mathematics Department, Faculty of Mathematics and Computing,
The Open University, Walton Hall, Milton Keynes MK7 6AA, UK

Abstract. Several combinatorial problems of (quasi)crystallography are reviewed with special emphasis on a unified approach, valid for both crystals and quasicrystals. In particular, we consider planar sublattices, similarity sublattices, coincidence sublattices and their module counterparts. The corresponding counting functions are encapsulated in Dirichlet series generating functions, with worked out results for the square lattice and the Tübingen triangle tiling. Finally, we discuss a novel approach to central and averaged shelling for these examples, also involving Dirichlet series.

1. Introduction

The discovery of non-periodic solids has motivated the construction of numerous examples of aperiodic tiling models, and led to the systematic theory of model sets, see [15, 4, 19] for details and further references. Most combinatorial questions of crystallography, such as sublattice or shelling structures, have a natural analogue for aperiodic systems. However, the traditional methods of crystallography often do not apply. Fortunately, the most important systems possess a high degree of symmetry, which manifests itself in an intimate relation to algebraic number theory [17]. This relationship has been exploited successfully to tackle combinatorial questions for quasicrystals [2, 7, 16, 18, 21].

Interestingly, this approach also simplifies the treatment of crystals. One common feature is the use of Dirichlet series generating functions, which emerges from the observation that the counting functions, when properly normalised, can be expressed in terms of multiplicative arithmetic functions, see [1] for background material. This leads to a systematic and unified approach which will be summarised in this article. Besides explicit results, the generating functions also allow a precise calculation of asymptotic properties.

We shall use the square lattice, written as $\mathbb{Z}^2 = \mathbb{Z}[i]$, the set (ring) of Gaussian integers, and the vertex set of the Tübingen triangle tiling (TTT) for explaining and illustrating our concepts, see figure 1. We concentrate on explicit results for these examples, and refer to original sources for a more general exposition and for details on the asymptotic behaviour.

2. Counting general sublattices

Let us consider $\mathbb{Z}[i]$. Our first question is for the number of *sublattices* of given index m, called $\ell_2(m)$. This is a multiplicative function, i.e. $\ell_2(mn) = \ell_2(m)\ell_2(n)$ for m, n coprime.

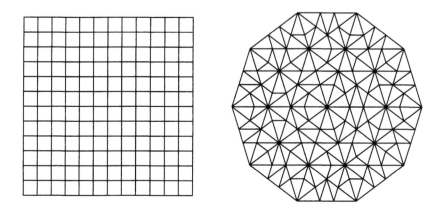

Figure 1. Finite patches of the square lattice and the Tübingen triangle tiling.

This suggests a Dirichlet series generating function, and one obtains [2, Eq. (3.9)]

$$F_2(s) = \sum_{m=1}^{\infty} \frac{\ell_2(m)}{m^s} = \zeta(s)\zeta(s-1) = \prod_p \frac{1}{1-p^{-s}} \frac{1}{1-p^{1-s}} \tag{1}$$

$$= 1 + \frac{3}{2^s} + \frac{4}{3^s} + \frac{7}{4^s} + \frac{6}{5^s} + \frac{12}{6^s} + \frac{8}{7^s} + \frac{15}{8^s} + \frac{13}{9^s} + \frac{18}{10^s} + \frac{12}{11^s} + \frac{28}{12^s} + \frac{14}{13^s} + \dots$$

where $\zeta(s) = \sum_{m=1}^{\infty} m^{-s}$ is Riemann's zeta function, p in the product runs over all (rational) primes, and $\ell_2(m) = \sum_{d|m} d$ is the ordinary divisor function, compare [1] for details.

The corresponding question is well defined also for $L = \mathbb{Z}[\xi]$, $\xi = \exp(2\pi i/5)$, when considered as a (free) $\mathbb{Z}$-module of rank 4. The number of full rank submodules of index m, $\ell_4(m)$, is again multiplicative and has the generating function [2, App. A]

$$F_4(s) = \sum_{m=1}^{\infty} \frac{\ell_4(m)}{m^s} = \zeta(s)\zeta(s-1)\zeta(s-2)\zeta(s-3) \tag{2}$$

$$= 1 + \frac{15}{2^s} + \frac{40}{3^s} + \frac{155}{4^s} + \frac{156}{5^s} + \frac{600}{6^s} + \frac{400}{7^s} + \frac{1395}{8^s} + \frac{1210}{9^s} + \frac{2340}{10^s} + \frac{1464}{11^s} + \dots$$

where $\ell_4(m) = \sum_{d_1 \cdots d_4 = m} d_1^0 d_2^1 d_3^2 d_4^3$ with the sum running over positive d_i only.

Clearly, this combinatorial question is purely algebraic, and the method also applies to counting finite index subgroups of free Abelian groups, see [2, App. A] for details. The generating function for counting index m subgroups of $\mathbb{Z}^n$ reads

$$F_n(s) = \sum_{m=1}^{\infty} \frac{\ell_n(m)}{m^s} = \zeta(s)\zeta(s-1) \cdot \ldots \cdot \zeta(s-n+1) \tag{3}$$

and gives the formula $\ell_n(m) = \sum_{d_1 \cdots d_n = m} d_1^0 d_2^1 \cdot \ldots \cdot d_n^{n-1}$.

3. Counting similarity sublattices

Let us turn to (geo)metric properties and ask for the number $a_4(m)$ of *square* sublattices of $\mathbb{Z}^2$ of a given index m, which coincides with the number of ideals of $\mathbb{Z}[i]$ of norm m. The answer is given by the Dedekind zeta function of the quadratic (or cyclotomic) field $\mathbb{Q}(i)$, i.e., one obtains [3, Eq. (7)]

$$\zeta_{\mathbb{Q}(i)}(s) = \sum_{m=1}^{\infty} \frac{a_4(m)}{m^s} = \frac{1}{1-2^{-s}} \prod_{p\equiv1\,(4)} \frac{1}{(1-p^{-s})^2} \prod_{p\equiv3\,(4)} \frac{1}{1-p^{-2s}} \tag{4}$$

$$= 1 + \tfrac{1}{2^s} + \tfrac{1}{4^s} + \tfrac{2}{5^s} + \tfrac{1}{8^s} + \tfrac{1}{9^s} + \tfrac{2}{10^s} + \tfrac{2}{13^s} + \tfrac{1}{16^s} + \tfrac{2}{17^s} + \tfrac{1}{18^s} + \tfrac{2}{20^s} + \cdots$$

The corresponding question for the module $\mathbb{Z}[\xi]$ is answered by the Dedekind zeta function [20] of the cyclotomic field $\mathbb{Q}(\xi)$ which reads [3, Eq. (10)]

$$\zeta_{\mathbb{Q}(\xi)}(s) = \sum_{m=1}^{\infty} \frac{a_{10}(m)}{m^s} \tag{5}$$

$$= \frac{1}{1-5^{-s}} \prod_{p\equiv1\,(5)} \frac{1}{(1-p^{-s})^4} \prod_{p\equiv-1\,(5)} \frac{1}{(1-p^{-2s})^2} \prod_{p\equiv\pm2\,(5)} \frac{1}{1-p^{-4s}}$$

$$= 1 + \tfrac{1}{5^s} + \tfrac{4}{11^s} + \tfrac{1}{16^s} + \tfrac{1}{25^s} + \tfrac{4}{31^s} + \tfrac{4}{41^s} + \tfrac{4}{55^s} + \tfrac{4}{61^s} + \tfrac{4}{71^s} + \tfrac{1}{80^s} + \cdots$$

These two generating functions also have an interpretation in terms of colourings: $a_4(m)$ is (up to permutation) the number of colourings of the square lattice where one colour occupies a sublattice of $\mathbb{Z}^2$ of index m and the remaining colours occupy its cosets. The function $a_{10}(m)$, in turn, counts the colourings of $\mathbb{Z}[\xi]$ via submodules of index m, all of which are similarity submodules because $\mathbb{Z}[\xi]$, as $\mathbb{Z}[i]$, is a principal ideal domain, see [20, 7] for details. For a discussion of the corresponding colour groups, see [14].

This approach can be extended to all cyclotomic fields with class number one (the corresponding rings of integers are then principal ideal domains), see [20, 3, 5], and references given there, for details.

4. Counting coincidence sublattices

Another geometric problem, with interesting applications to grain boundaries and twinning phenomena in (quasi)crystals, is the classification of sublattices of $\mathbb{Z}^2$ that can be seen as an intersection of $\mathbb{Z}^2$ with a rotated copy of itself. These are the *coincidence sublattices*, see [2] for details, and the corresponding coincidence index is called Σ-factor in materials science. Once more, the generating function (for the number of coincidence sublattices of index m) is best written as a Dirichlet series, where (again using $\mathbb{Z}^2 = \mathbb{Z}[i]$) one finds [18]

$$\Phi_{\mathbb{Z}[i]}(s) = \prod_{p\equiv1\,(4)} \frac{1+p^{-s}}{1-p^{-s}} = \frac{1}{1+2^{-s}} \frac{\zeta_{\mathbb{Q}(i)}(s)}{\zeta(2s)} \tag{6}$$

$$= 1 + \tfrac{2}{5^s} + \tfrac{2}{13^s} + \tfrac{2}{17^s} + \tfrac{2}{25^s} + \tfrac{2}{29^s} + \tfrac{2}{37^s} + \tfrac{2}{41^s} + \tfrac{2}{53^s} + \tfrac{2}{61^s} + \tfrac{4}{65^s} + \cdots$$

which is derived in [2, Prop. 3.1].

Behind this combinatorial problem is a group structure. It turns out that the set of rotations R such that $\mathbb{Z}^2 \cap R\mathbb{Z}^2$ is a coincidence sublattice of $\mathbb{Z}^2$ forms a group, called $\mathrm{SOC}(\mathbb{Z}^2)$, which equals $\mathrm{SO}(2,\mathbb{Q})$, see [2, Thm. 3.3] for details. Also, the possible coincidence indices $[\mathbb{Z}^2 : (\mathbb{Z}^2 \cap R\mathbb{Z}^2)] =: \Sigma(R)$ (excluding 0 and ∞) form a monoid (a semigroup with unit), generated by the (rational) primes $p \equiv 1\,(4)$.

The formulation with cyclotomic integers admits the corresponding result for $\mathbb{Z}[\xi]$, and the generating function for the number of coincidence submodules of index m is [18]

$$\Phi_{\mathbb{Z}[\xi]}(s) = \prod_{p\equiv1\,(5)} \left(\frac{1+p^{-s}}{1-p^{-s}}\right)^2 = \frac{1}{1+5^{-s}} \frac{\zeta_{\mathbb{Q}(\xi)}(s)}{\zeta_{\mathbb{Q}(\tau)}(2s)} \tag{7}$$

$$= 1 + \tfrac{4}{11^s} + \tfrac{4}{31^s} + \tfrac{4}{41^s} + \tfrac{4}{61^s} + \tfrac{4}{71^s} + \tfrac{4}{101^s} + \tfrac{8}{121^s} + \tfrac{4}{131^s} + \tfrac{4}{151^s} + \cdots$$

where $\tau = (1+\sqrt{5})/2$ is the golden ratio and $\zeta_{\mathbb{Q}(\tau)}(s)$ is the Dedekind zeta function of the quadratic field $\mathbb{Q}(\tau)$, i.e.

$$\zeta_{\mathbb{Q}(\tau)}(s) = \frac{1}{1-5^{-s}} \prod_{p \equiv \pm 1\,(5)} \frac{1}{(1-p^{-s})^2} \prod_{p \equiv \pm 2\,(5)} \frac{1}{1-p^{-2s}}. \tag{8}$$

There is one subtlety in the application to the *discrete* point set of the TTT — a small acceptance correction factor is needed, which can be calculated explicitly from the cut and project scheme. This, together with a general discussion of the n-fold symmetry case, can be found in [18].

5. Central shelling

Let us come to the shelling problem, first in its version for the *central* shelling. Here, one asks for the number of points of $\mathbb{Z}[i]$ on circles of radius r around the origin. The result is usually given in terms of lattice theta functions, see [9, Ch. 4] for an extensive exposition, but, in our situation, it can also be encapsulated in a Dirichlet series. To this end, one considers only radii $r > 0$ and divides the corresponding shelling number, $c(r^2)$, by 4, which is the trivial symmetry factor. What remains, since $r^2 = m$ is an integer, is the multiplicative function $a_4(m)$ whose generating function was given above in equation (5). Further details and references are given in [6].

In the case of $\mathbb{Z}[\xi]$, the central shelling function c depends on $r^2 \in \mathbb{Z}[\xi + \bar{\xi}] = \mathbb{Z}[\tau]$, and one needs the primes of $\mathbb{Z}[\tau]$ (see [10, Appendix B] for details) to derive a formula for $c(r^2)$, compare [5]. Let us write $\tilde{p}$ for a prime in $\mathbb{Z}[\tau]$, and let $t(\tilde{p})$ be the highest power t such that $\tilde{p}^t | r^2$. Then, whenever $t(\tilde{p})$ is odd for a $\tilde{p}$ that is also a prime in $\mathbb{Z}[\xi]$ (i.e., an inert prime), one has $c(r^2) = 0$. Otherwise, one finds the formula

$$c(r^2) = 10 \prod_{\substack{\tilde{p}|r^2 \\ \tilde{p}\text{ splits}}} (t(\tilde{p})+1) \tag{9}$$

where the product runs only over those primes $\tilde{p}$ of $\mathbb{Z}[\tau]$ which split in the extension to $\mathbb{Z}[\xi]$. Here, these are precisely the primes that originate from rational primes $p \equiv 1$ (5).

Can the result be given in a simpler way? Suppose, for a moment, that the function $c(r^2)$ would only depend on the *norm* of r^2, i.e., on $m = r^2 \sigma(r^2)$, where σ is the Galois automorphism of $\mathbb{Q}(\tau)$ defined by $\sqrt{5} \mapsto -\sqrt{5}$. Consequently, m is a rational integer. This would mean that $c(r^2) = 10f(m)$ with $f(m)$ a multiplicative arithmetic function, whose Dirichlet series turns out to be

$$C_{10}(s) = \sum_{m=1}^{\infty} \frac{f(m)}{m^s} \tag{10}$$

$$= \frac{1}{1-5^{-s}} \prod_{p \equiv 1\,(5)} \frac{1}{(1-p^{-s})^2} \prod_{p \equiv -1\,(5)} \frac{1}{1-p^{-2s}} \prod_{p \equiv \pm 2\,(5)} \frac{1}{1-p^{-4s}}$$

$$= 1 + \frac{1}{5^s} + \frac{2}{11^s} + \frac{1}{16^s} + \frac{1}{25^s} + \frac{2}{31^s} + \frac{2}{41^s} + \frac{2}{55^s} + \frac{2}{61^s} + \frac{2}{71^s} + \frac{1}{80^s} + \frac{1}{81^s} + \ldots$$

As can be seen from the exact shelling formula (9), this cannot be true in general. However, it gives the correct answer for many cases, and is then by far the simplest way to calculate $c(r^2)$. To be more precise, consider $m = r^2 \sigma(r^2)$. Whenever m is not divisible by the square of *any* rational prime $p \equiv 1$ (5), we have $c(r^2) = 10f(m)$, where $f(m)$ is then 0, 1 or a power of 2.

The first value of m where this fails is $m = 121 = 11^2$. Since $11 = (3+\tau)(4-\tau) = \tilde{p}\tilde{p}'$ in $\mathbb{Z}[\tau]$, where we write $\tilde{p}'$ for $\sigma(\tilde{p})$, the possible values of r^2 with norm m are $\tilde{p}^2$, $\tilde{p}\tilde{p}'$ and

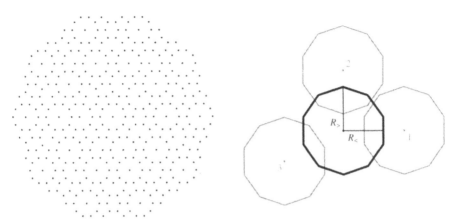

Figure 2. A TTT vertex set and its decagonal window (black). Also shown are three shifted copies of the window (grey), see text for details.

$(\tilde{p}')^2$, with shelling numbers 30, 40 and 30, respectively. Since $f(121) = 3$, we would get a wrong answer for the middle case. Nevertheless, provided that r is the radius of a non-empty shell (i.e., $t(\tilde{p})$ is even for all inert primes $\tilde{p}$ of $\mathbb{Z}[\tau]$, as explained above), $10f(m)$ still gives the correct shelling number if, whenever m is divisible by a rational prime $p \equiv 1$ (5) (whence $p = \tilde{p}\tilde{p}'$ is a complex splitting prime, and $\tilde{p}$ splits in the extension to $\mathbb{Z}[\xi]$), r^2 is divisible by either $\tilde{p}$ or $\tilde{p}'$, but never by both. Beyond this situation, one has to use formula (9).

Equation (9), together with the partial simplification of equation (10), gives the result for the central shelling of the full module; an alternative, and more systematic, formulation is also possible by means of a primitive character of $\mathbb{Q}(\tau)$. When passing to a model set [15], which is a discrete subset of the full module, an obvious selection takes place for the possible radii. Furthermore, a correction factor may become necessary which depends on the window chosen, see [5] for details.

6. Averaged shelling

In this last application, we are interested in the *average* number of points on shells of radius r, where the average is taken over all points of our point set as possible centres. In the lattice situation, the averaged and central shelling numbers coincide, but this is not so for model sets [6, 15]. Note that the averages needed do exist for model sets, so that the question posed is well defined. To be specific, we explain the averaged shelling for the vertex set of the TTT.

Let us first describe the TTT vertex set in algebraic terms. We use $L = \mathbb{Z}[\xi]$ as above, with $\xi = \exp(2\pi i/5)$, together with the Galois automorphism $\star$ of $\mathbb{Q}(\xi)$ given by $\xi \mapsto \xi^2$. This is the star map of the standard cut and project or model set construction [15], which gives

$$\Lambda = \{x \in L \mid x^\star \in W\} \tag{11}$$

where the window W is a relatively compact set with nonempty interior. For the TTT, we choose W as a regular decagon of edge length $\tau/\sqrt{\tau+2}$, hence of inradius $R_< = \tau^2/2$ and circumradius $R_> = \tau^2/\sqrt{\tau+2}$, see figure 2 for the correct orientation.

This construction gives a triangle tiling with long (short) edges of lengths 1 $(1/\tau)$. If the window is centred at 0, the model set is singular, while generic examples are obtained by

Table 1. Averaged shelling numbers for the TTT for all possible distances $0 < r \leq 2$.

r^2	representative	orbit length	$t^2 = \sigma(r^2)$	shift type	shelling number
$2 - \tau$	$\xi + \bar{\xi}$	10	$1 + \tau$	1	$6 - 2\tau$
1	1	10	1	1	$28 - 14\tau$
$3 - \tau$	$\xi^2 - \bar{\xi}^2$	10	$2 + \tau$	2	$-50 + 32\tau$
$4 - \tau$	$1 - \xi - \bar{\xi}^2$	20	$3 + \tau$	3	$112 - 68\tau$
$1 + \tau$	$\xi^2 + \bar{\xi}^2$	10	$2 - \tau$	1	$20 - 8\tau$
$2 + \tau$	$\xi - \bar{\xi}$	10	$3 - \tau$	2	$-44 + 30\tau$
4	2	10	4	1	$-18 + 12\tau$

shifting the window, i.e. $\Lambda^u = \{x \in L \mid x^* \in W + u\}$, for almost all $u \in \mathbb{R}^2$.

Table 1 lists all distances $0 < r \leq 2$ between points of Λ, together with a representing point in L. The corresponding orbit length is given w.r.t. the point symmetry group D_{10}. Here, it coincides with the value of the central shelling number $c(r^2) = 10f(m)$, as calculated by means of equation (10). For larger values of r^2, $c(r^2)$ can comprise several D_{10}-orbits, but only complete ones because the window is tenfold and reflection symmetric. For the averaged shelling, we need to calculate the relative overlap area of the window W with a shifted copy (the so-called covariogram of W), where the shift is the star image of the representing vector [5]. For our examples, only the three relative orientations indicated in figure 2 emerge, with midpoint distance t, where t^2 is the algebraic conjugate of r^2 in $\mathbb{Q}(\tau)$. The resulting averaged shelling numbers (last column) are elements of $\mathbb{Z}[\tau]$ in all cases shown here. In general, one can easily see that they are elements of $\mathbb{Q}(\tau)$, e.g., as a result of the inflation structure via a standard Perron-Frobenius argument. The topological structure [13, 11] of the tiling, however, further restricts the possible frequencies of finite patches, so that the frequency module is $\frac{1}{10}\mathbb{Z}[\tau]$ in this case [11]. This, in turn, can be used to show that the averaged shelling numbers of the TTT are always in $2\mathbb{Z}[\tau]$, i.e., of the form $m + n\tau$ with m, n even.

For the overlap scenarios 1 and 2 of figure 2, where the shifts are along the two principal reflection axes of the decagon, the covariogram has a relatively simple form. Measuring the distance between the decagons in units of the circumradius, with $q = t/R_>$, we find

$$
h_1(q) = \begin{cases}
\frac{1}{5}[5 - 2\tau\frac{q}{\eta} + (\frac{q}{2\eta})^2], & 0 \leq q < \frac{2\eta}{\tau} \\
\frac{1}{5}[2 - \frac{q}{\eta} + \tau^3(1 - \frac{q}{2\eta})^2], & \frac{2\eta}{\tau} \leq q < 2\eta \\
0, & 2\eta \leq q
\end{cases}
\tag{12}
$$

where $\eta = \sqrt{2 + \tau}/2$, and

$$
h_2(q) = \begin{cases}
\frac{1}{5}(5 - 2\tau q), & 0 \leq q < \frac{1}{\tau} \\
\frac{1}{5}(2\tau^2 - 4q + \frac{q^2}{\tau}), & \frac{1}{\tau} \leq q < \tau \\
\frac{\tau^2}{5}(2 - q)^2, & \tau \leq q < 2 \\
0, & 2 \leq q.
\end{cases}
\tag{13}
$$

These functions have to be multiplied by the orbit length to give the averaged shelling number, resp. the corresponding contribution to it.

The averaged shelling number for the example of type 3 given in table 1 was calculated separately (and exactly). We do not give the general formula for the covariogram of the regular decagon, but mention that

$$
h(q) = \frac{2}{\pi} \arccos\left(\frac{q}{2\rho}\right) - \frac{q}{\pi\rho}\sqrt{1 - \left(\frac{q}{2\rho}\right)^2}, \tag{14}
$$

with $\rho = \sqrt{\frac{5}{2\pi\tau}\sqrt{2+\tau}} \approx 0.9672$, is a very good approximation. It is obtained by replacing the decagon W with a disk of equal area, hence of radius $\rho R_>$. This function is also known as Euclid's hat, see [12, p. 100] and references given there for details.

7. Open problems

The above examples demonstrate that a systematic and unified approach to combinatorial problems of (quasi)crystallography is possible, at least in the planar case. The situation is more involved in dimensions $d \geq 3$, where satisfactory results so far exist only for the most symmetric cases, i.e. those with (hyper)cubic or (hyper)icosahedral symmetry [2, 7, 8, 16, 21]. Still, even some of these cases leave room for improvement and simplification, e.g., along the lines mentioned around equation (10).

On the other hand, model sets with high symmetry are closely related to lattices in higher dimensions, where many of these questions are still open, compare [9]. One may expect some progress at least for the class of root lattices, hence also for quasicrystals derived from them. Further progress is also needed in the investigation of colour symmetry groups, see [14] for a summary of the present state of affairs.

Perhaps the biggest mystery is the meaning of the averaged shelling function. For the standard tilings with all magic properties (inflation rule, perfect matching rules, pure point diffraction etc.), the averaged shelling numbers always seem to be "nice" ($\mathbb{Z}[\tau]$ numbers in the TTT example), while this is not the case for model sets with generic windows. This phenomenon points towards another function defined by these numbers that admits similarly nice analytic and topological interpretations, but we do not know how to substantiate this at present.

Acknowledgment

It is a pleasure to thank Robert V. Moody and Alfred Weiss for helpful discussions and suggestions. This work was partially supported by the German Research Council (DFG).

References

[1] T. M. Apostol, *Introduction to Analytic Number Theory*, Springer, New York (1976).

[2] M. Baake, Solution of the coincidence problem in dimensions $d \leq 4$, in: *The Mathematics of Long-Range Aperiodic Order*, ed. R. V. Moody, Kluwer, Dordrecht 1997, pp. 9–44.

[3] M. Baake, Combinatorial aspects of colour symmetries, *J. Phys. A: Math. Gen.* **30** (1997) 2687–2698; mp_arc/02-323.

[4] M. Baake, A guide to mathematical quasicrystals, in: *Quasicrystals*, eds. J.-B. Suck, M. Schreiber and P. Häussler, Springer, Berlin (2002), pp. 17–48; math-ph/9901014.

[5] M. Baake and U. Grimm, A note on shelling, preprint math.MG/0203025.

200

[6] M. Baake, U. Grimm, D. Joseph and P. Repetowicz, Averaged shelling for quasicrystals, *Mat. Sci. Eng.* **A 294–296** (2000) 441–445; math.MG/9907156.

[7] M. Baake and R. V. Moody, Similarity submodules and semigroups, in: *Quasicrystals and Discrete Geometry*, ed. J. Patera, Fields Institute Monographs, vol. 10, AMS, Providence, RI (1998), pp. 1–13.

[8] M. Baake and R. V. Moody, Similarity submodules and root systems in four dimensions, *Can. J. Math.* **51** (1999) 1258–1276; math.MG/9904028.

[9] J. H. Conway and N. J. A. Sloane, *Sphere Packings, Lattices and Groups*, 3rd ed. Springer, New York (1999).

[10] F. W. Dodd, *Number Theory in the Quadratic Field with Golden Section Unit*, Polygonal Publishing House, Passaic, NJ (1983).

[11] F. Gähler, private communication (2002).

[12] T. Gneiting, Radial positive definite functions generated by Euclid's hat, *J. Multivariate Anal.* **69** (1999) 88–119.

[13] J. Kellendonk, The local structure of tilings and their integer group of coinvariants, *Commun. Math. Phys.* **187** (1997) 115–157.

[14] R. Lifshitz, Theory of color symmetry for periodic and quasiperiodic crystals, *Rev. Mod. Phys.* **69** (1997) 1181–1218.

[15] R. V. Moody, Model sets: A Survey, in: *From Quasicrystals to More Complex Systems*, eds. F. Axel, F. Dénoyer and J. P. Gazeau, EDP Sciences, Les Ulis, and Springer, Berlin (2000), pp. 145–166; math.MG/0002020.

[16] R. V. Moody and A. Weiss, On shelling E_8 quasicrystals, *J. Number Theory* **47** (1994) 405–412.

[17] P. A. B. Pleasants, Designer quasicrystals: Cut-and-project sets with pre-assigned properties, in: *Directions in Mathematical Quasicrystals*, eds. M. Baake and R. V. Moody, CRM Monograph Series, vol. 13, AMS, Providence, RI (2000), pp. 95–141.

[18] P. A. B. Pleasants, M. Baake and J. Roth, Planar coincidences for N-fold symmetry, *J. Math. Phys.* **37** (1996) 1029–1058.

[19] M. Schlottmann, Cut-and-project sets in locally compact Abelian groups, in: *Quasicrystals and Discrete Geometry*, ed. J. Patera, Fields Institute Monographs, vol. 10, AMS, Providence, RI (1998), pp. 247–264.

[20] L. C. Washington, *Introduction to Cyclotomic Fields*, 2nd ed., Springer, New York (1997).

[21] A. Weiss, On shelling icosahedral quasicrystals, in: *Directions in Mathematical Quasicrystals*, eds. M. Baake and R. V. Moody, CRM Monograph Series, vol. 13, AMS, Providence, RI (2000), pp. 161–176.

Inst. Phys. Conf. Ser. No 173: Section 1
Paper presented at 24th Int. Coll. Group Theoretical Methods in Physics, Paris, France, 15–20 July 2002
©2003 IOP Publishing Ltd

Self-similar Cut-and-project Sequences

Peter Baláži and Edita Pelantová

Department of Mathematics, Faculty of Nuclear Sciences and Physical Engineering, Czech Technical University, Trojanova 13, 120 00 Prague, Czech Republic

Abstract. We consider one-dimensional aperiodic sequences Σ arising from a two-dimensional cut-and-project scheme. Such sequences are given by two irrational numbers η, ε and by a bounded interval Ω. We give a necessary and sufficient condition on the parameters η, ε and Ω under which Σ is selfsimilar, i.e. there exists a real number $\gamma > 1$ such that $\gamma\Sigma \subset \Sigma$.

1. Introduction

Cut-and-project sequences have often been considered in physics literature in connection with modelling quasicrystals [1]. Mathematicians study infinite words associated to cut-and-project sequences as they are generalizations of the Sturmian words, particularly of the famous Fibonacci chain. Recently a nice application of the cut-and-project sequences in the theory of wavelets has appeared [2]. For modelling of the structure with long range aperiodic order, the existence of non-trivial Bragg spectrum is important; to generate an aperiodic infinite word is easy, if there exists a substitution rules under which the word is invariant; the construction of wavelets based on an aperiodic structure exploits the presence of a scaling factor. For all those required properties the selfsimilarity of cut-and-project sequences is crucial.

We consider a strip in $\mathbf{R}^2$ defined by two parallel lines with an irrational slope ε. Let us project all lattice points $(a, b) \in \mathbf{Z}^2$ from this strip onto the line $y = \varepsilon x$. The direction of projection need not be orthogonal and in generally it is given by the line $y = \eta x$, where η is an irrational slope, $\eta \neq \varepsilon$. Projected points form a so called cut-and-project set, which can be explicitly written as

$$\Sigma_{\varepsilon,\eta}(\Omega) = \{a + b\eta \mid a, b \in \mathbf{Z}, \ a + b\varepsilon \in \Omega\}$$

where Ω is an bounded interval corresponding to the chosen strip. To abbreviate notation we denote the Abelian groups $\{a + b\varepsilon \mid a, b \in \mathbf{Z}\}$ and $\{a + b\eta \mid a, b \in \mathbf{Z}\}$ by $\mathbf{Z}[\varepsilon]$ and $\mathbf{Z}[\eta]$ respectively. The bijection between them defined by

$$x := a + b\eta \mapsto x^* := a + b\varepsilon$$

will be called the starmap. Using this notation we can rewrite the cut and project set as

$$\Sigma_{\varepsilon,\eta}(\Omega) = \{x \in \mathbf{Z}[\eta] \mid x^* \in \Omega\}$$

It is known [4], that for any $\Sigma_{\varepsilon,\eta}(\Omega)$ there exist an increasing sequence $(x_n)_{n\in\mathbf{Z}}$ and two positive values Δ_1 and Δ_2 such that $\Sigma_{\varepsilon,\eta}(\Omega) = \{x_n \mid n \in \mathbf{Z}\}$ and distances $x_{n+1} - x_n$ between consecutive points take at most three values among $\{\Delta_1, \Delta_2, \Delta_1 + \Delta_2\}$. The aperiodicity of $\Sigma_{\varepsilon,\eta}(\Omega)$ implies that values Δ_1 and Δ_2 are linearly independent over $\mathbf{Q}$.

Since the set $\{x_n^* \mid n \in \mathbf{Z}\} \subset \Omega$ is bounded, the starmap images Δ_1^* and Δ_2^* must take opposite signs. Without loss of generality we assume that $\Delta_1^* > 0$, $\Delta_2^* < 0$ and a semiclosed window $\Omega = [c, d)$. There exists a simple rule how for any $x_n \in \Sigma_{\varepsilon,\eta}[c, d)$ to determine its right neighbour x_{n+1}:

$$x_{n+1} = \begin{cases} x_n + \Delta_1 & \text{if} & x_n^* \in [c, d - \Delta_1^*) \\ x_n + \Delta_1 + \Delta_2 & \text{if} & x_n^* \in [d - \Delta_1^*, c - \Delta_2^*) \\ x_n + \Delta_2 & \text{if} & x_n^* \in [c - \Delta_2^*, d) \end{cases}$$

Note that if $d - c = \Delta_1^* - \Delta_2^*$ then only two distances Δ_1 and Δ_2 occur in the cut-and-project sequence. The function f which assign to the starmap image of a point $x \in \Sigma_{\varepsilon,\eta}[c, d)$ the starmap image of its neighbour is therefore piecewise linear bijection on $[c, d)$ with at most two points of discontinuity.

Since distances $x_{n+1} - x_n$ take at most three values, we can code the sequence of distances by a 3-letter alphabet. The ternary infinite bidirectional word $(u_n)_{n \in \mathbf{Z}}$ which we obtain is uniquely given by the function f and in fact it is the coding of three interval exchange with permutation (3,2,1), for more details see [3].

2. Selfsimilarity of Cut-and-project Sequences

We say that a set $M \subset \mathbf{R}$ is selfsimilar if there exists $\gamma > 1$ such that $\gamma M \subset M$. Many selfsimilar cut-and-project sequences are used in different applications. For all those sequences, ε and η are different roots of one quadratic equation with integer coefficients and the closure $\overline{\Omega}$ of the window Ω contains 0. It is easy to verify that such a setting gives selfsimilar cut-and-project sequence. We shall see that such setting is also necessary for the selfsimilarity. To deduce our main result we will use some facts and notion of the number theory.

$\mathbf{Q}[\alpha]$ denotes the minimal number field containing $\mathbf{Q}$ and α. If α is an irrational solution of a quadratic equation with integer coefficients then $\mathbf{Q}[\alpha] = \{a + b\alpha \mid a, b \in \mathbf{Q}\}$. The second root α' is called the algebraic conjugate of α and clearly $\alpha' \in \mathbf{Q}[\alpha]$. The mapping on $\mathbf{Q}[\alpha]$ defined by $x = a + b\alpha \mapsto x' = a + b\alpha'$ is an automorphisms on $\mathbf{Q}[\alpha]$, i.e. $(x + y)' = x' + y'$ and $(x.y)' = x'.y'$ for all $x, y \in \mathbf{Q}[\alpha]$. This automorphism is called the Galois automorphism. If $\gamma > 1$ is a root of a monic quadratic polynomial with integer coefficients and if its conjugate γ' is in modulus less then 1, then γ is called the quadratic Pisot number.

Theorem: *Let $\gamma > 1$ be a selfsimilarity factor of the cut-and-project set $\Sigma_{\varepsilon,\eta}(\Omega)$. Then γ is a quadratic Pisot number, $\varepsilon, \eta \in \mathbf{Q}[\gamma]$, $\varepsilon = \eta'$ and $0 \in \overline{\Omega}$.*

Proof: If $x = a + b\eta$ belongs to $\Sigma_{\varepsilon,\eta}(\Omega)$, then $\gamma x \in \Sigma_{\varepsilon,\eta}(\Omega) \subset \mathbf{Z}[\eta]$, i.e. there exist integers $\tilde{a}, \tilde{b}$ such that $\gamma(a + b\eta) = \tilde{a} + \tilde{b}\eta$. It means $\gamma = \frac{\tilde{a} + \tilde{b}\eta}{a + b\eta}$ and $\eta = \frac{-\tilde{a} + a\gamma}{\tilde{b} - \gamma b}$. Therefore $\mathbf{Q}[\gamma] = \mathbf{Q}[\eta]$.

Let us note that for points $x, y \in \Sigma_{\varepsilon,\eta}(\Omega)$, $x < y$, their distance $y - x \in \Sigma_{\varepsilon,\eta}(\Omega) - \Sigma_{\varepsilon,\eta}(\Omega) = \Sigma_{\varepsilon,\eta}(\Omega - \Omega)$ and $y - x$ is a non-negative integer combination of the lengths Δ_1 and Δ_2. This obvious fact combined with selfsimilarity gives

$$\gamma\Delta_1 = \gamma(x_{n+1} - x_n) = \gamma x_{n+1} - \gamma x_n = k_{11}\Delta_1 + k_{12}\Delta_2$$

for some integer $n \in \mathbf{Z}$ and non-negative integers k_{11} and k_{12}. Analogously, $\gamma\Delta_2 = k_{21}\Delta_1 + k_{22}\Delta_2$, for $k_{21}, k_{22} \in \mathbf{N}_0$. If we denote by $\mathbf{K}$ the 2×2-matrix with the elements k_{ij}, we can write

$$\mathbf{K}\begin{pmatrix} \Delta_1 \\ \Delta_2 \end{pmatrix} = \gamma \begin{pmatrix} \Delta_1 \\ \Delta_2 \end{pmatrix}, \tag{1}$$

i.e. γ is an eigenvalue of the integer matrix $\mathbf{K}$ and therefore γ is a root of a quadratic monic polynomial; the corresponding eigenvector is (Δ_1, Δ_2). As $\Delta_1, \Delta_2 \in \mathbf{Q}[\eta] = \mathbf{Q}[\gamma]$, we can apply the Galois automorphism on (1) to obtain the second eigenvector and eigenvalue:

$$\mathbf{K} \begin{pmatrix} \Delta_1' \\ \Delta_2' \end{pmatrix} = \gamma' \begin{pmatrix} \Delta_1' \\ \Delta_2' \end{pmatrix}.$$

From the Perron-Frobenius theorem we have $|\gamma'| < \gamma$. Moreover, $\gamma^m \Delta_1$ and $\gamma^m \Delta_2$ reprezent distances between some points from $\Sigma_{\varepsilon,\eta}(\Omega)$ for any positive integer $m \in \mathbf{N}$. It forces $(\gamma^m \Delta_1)^*, (\gamma^m \Delta_2)^* \in \Omega - \Omega$. The definition of the starmap gives $(k.x)^* = k.x^*$ for any $k \in \mathbf{Z}$ and $x \in \mathbf{Z}[\eta]$. We therefore obtain

$$\mathbf{K}^m \begin{pmatrix} \Delta_1^* \\ \Delta_2^* \end{pmatrix} = \left(\mathbf{K}^m \begin{pmatrix} \Delta_1 \\ \Delta_2 \end{pmatrix} \right)^* = \left(\gamma^m \begin{pmatrix} \Delta_1 \\ \Delta_2 \end{pmatrix} \right)^* \in (\Omega - \Omega)^2 \tag{2}$$

This says that the sequence of vectors $\mathbf{K}^m \begin{pmatrix} \Delta_1^* \\ \Delta_2^* \end{pmatrix}$ is bounded.

Since the eigenvectors of the matrix $\mathbf{K}$ form a basis of $\mathbf{R}^2$, we can write for some real coefficients α_1 and α_2:

$$\begin{pmatrix} \Delta_1^* \\ \Delta_2^* \end{pmatrix} = \alpha_1 \begin{pmatrix} \Delta_1 \\ \Delta_2 \end{pmatrix} + \alpha_2 \begin{pmatrix} \Delta_1' \\ \Delta_2' \end{pmatrix}.$$

Substituting this expression into (2) we have that the sequence

$$\alpha_1 \gamma^m \begin{pmatrix} \Delta_1 \\ \Delta_2 \end{pmatrix} + \alpha_2 (\gamma')^m \begin{pmatrix} \Delta_1' \\ \Delta_2' \end{pmatrix}$$

is bounded. The fact that γ is a quadratic irrationality, $\gamma > 1$ implies $\alpha_1 = 0$ and $|\gamma'| < 1$, i.e. γ is a quadratic Pisot number.

We know that $\Delta_1, \Delta_2 \in \mathbf{Z}[\eta]$ and thus they can be written as $\Delta_1 = a_1 + b_1 \eta$, $\Delta_2 = a_2 + b_2 \eta$ for some integers a_1, a_2, b_1, b_2. The equality of vectors $(\Delta_1^*, \Delta_2^*) = \alpha_2 (\Delta_1', \Delta_2')$ means

$$(a_1 + b_1 \varepsilon)/(a_2 + b_2 \varepsilon) = (a_1 + b_1 \eta')/(a_2 + b_2 \eta')$$

It gives $(a_1 b_2 - a_2 b_1)(\varepsilon - \eta') = 0$. As we have already mentioned, Δ_1 and Δ_2 are independent over $\mathbf{Q}$, i.e. $(a_1 b_2 - a_2 b_1) \neq 0$. Therefore $\varepsilon = \eta'$ as the theorem claims. It also means that the starmap coincides with the Galois automorphism.

The property of selfsimilarity requires that $\gamma^m x$ belongs to $\Sigma_{\varepsilon,\eta}(\Omega)$ for any $m \in \mathbf{N}$ and $x \in \Sigma_{\varepsilon,\eta}(\Omega)$. This is equivalent to $(\gamma x)^* = (\gamma')^m x' \in \Omega$. Since $\lim_{m \to \infty} (\gamma')^m x' = 0$, we have $0 \in \overline{\Omega}$.

3. Comments on Substitution Rules

A substitution rule φ on a 3-letter alphabet $\mathcal{A} = \{A, B, C\}$ assigns to each letter A, B and C a finite non-empty word in the same alphabet $\varphi(A)$, $\varphi(B)$, and $\varphi(C)$ respectively. We say that a infinite word $\ldots u_{-3} u_{-2} u_{-1} | u_0 u_1 u_2 \ldots$ is invariant under substitution rule φ if

$$\ldots u_{-3} u_{-2} u_{-1} | u_0 u_1 u_2 \ldots = \ldots \varphi(u_{-2}) \circ \varphi(u_{-1}) | \varphi(u_0) \circ \varphi(u_1) \circ \varphi(u_2) \ldots$$

where the symbol $\circ$ stands for the concatenation of words.

Suppose that the ternary word $(u_n)_{n \in \mathbf{Z}}$ corresponding to a cut and project sequence $\Sigma_{\varepsilon,\varepsilon'}(\Omega)$ is invariant under a substitution φ. Geometrically it means that there exists a selfsimilarity factor γ such that the strings of letters corresponding to the segment $[\gamma x_q, \gamma x_{q+1}] \cap \Sigma_{\varepsilon,\varepsilon'}(\Omega)$ and to the segment $[\gamma x_p, \gamma x_{p+1}] \cap \Sigma_{\varepsilon,\varepsilon'}(\Omega)$ coincide if distances $x_{p+1} - x_p$ and $x_{q+1} - x_q$ corresponds to the same letter. The consequence of this property is that a selfsimilarity factor γ corresponding to a substitution φ must satisfy the condition $\mathbf{Z}[\varepsilon] = \gamma \mathbf{Z}[\varepsilon]$ which is equivalent to the condition $\gamma.\gamma' = \pm 1$.

The existence of substitution under which cut-and-project sequence $\Sigma_{\varepsilon,\varepsilon'}(\Omega)$ is invariant requires that the both boundary points of the acceptance interval Ω belong to the field $\mathbf{Q}[\varepsilon]$, see [5]. This is only the necessary but not a sufficient condition. Our recent study [6] of the cut-and-project sequences associated to the golden mean $\tau = \frac{1+\sqrt{5}}{2}$ shows that the complexity of the infinite word plays an important role for the form of the sufficient condition. Recall that the complexity of an infinite word $(u_n)_{n \in \mathbf{Z}}$ is a mapping $\mathcal{C} : \mathbf{N} \mapsto \mathbf{N}$ defined by

$$\mathcal{C}(k) = \#\{u_{i+1}u_{i+2}\ldots u_{i+k} \mid i \in \mathbf{N}\},$$

i.e. $\mathcal{C}(k)$ is the number of different subwords of length k in the word $(u_n)_{n \in \mathbf{Z}}$. Any cut-and-project sequence has the complexity either $\mathcal{C}(k) = 2k+1$ or $\mathcal{C}(k) = k + const$. We have found out that the existence of substitution rule for a sequence with the complexity $\mathcal{C}(k) = 2k+1$ is possible only if the Galois image $(d-c)'$ of length of acceptance interval $\Omega = [c, d)$ belongs to a certain set M whereas the complexity $\mathcal{C}(k) = k + const$ enables the existence of a substitution rule independently of the magnitude of $(d-c)'$.

References

[1] The Physics of Quasicrystals, eds. P.J.Steinhardt and S. Ostlund, World Scientific, 1987

[2] J.-P. Gazeau, J. Patera *Tau wavelets of Haar*, J. Phys. A: Math. Gen **29**, (1996) 4549-4559

[3] S. Ferenczi, Ch. Holton, L. Zamboni, *Combinatorics of three-interval exchanges* ICALP 2001, 567-578.

[4] L. S. Guimond, Z. Masáková and E. Pelantová *Combinatorial properties of cut-and-project sequences*, to appear in J. of Théorie des Nombres de Bordeaux (2002)

[5] Z. Masáková, J. Patera and E. Pelantová *Substitution rules for aperiodic sequneces of cut-and-project type*, J. of Phys. A: Math.Gen., **33** (2000) 8867–8886.

[6] P. Baláži, *Substitution Matrices with Complexity 2n+1*, Preprint FNSPE, Czech Technical Univ. (2002)

Inst. Phys. Conf. Ser. No 173: Section 1
Paper presented at 24th Int. Coll. Group Theoretical Methods in Physics, Paris, France, 15–20 July 2002
©2003 *IOP Publishing Ltd*

Closed geodesics in cosmology and discrete group actions on hyperbolic manifolds

Peter Kramer

Institut fuer Theoretische Physik, U. 72076 Tuebingen, Germany

Abstract. Hyperbolic spaces H^n [7] are models of an infinite cosmos with constant negative curvature. Moreover they provide universal covering spaces for compact finite manifolds M of nontrivial topology with the same metric and curvature. The double torus $T \cup T$ and the Weber-Seifert manifold are hyperbolic manifolds M covered by H^2, H^3 respectively. Both display closed geodesics with observable effects on the distribution of matter. The preimage and image points under any homotopy $g \in \pi(M)$ when acting on H^n are identified on M. Therefore any closed geodesic corresponds to a homotopy g. It is shown that the closed geodesics on M associated with g can be compared and classified in length and relative direction by orbits under the continuous normalizers N_g, $g \in \pi(M)$.

1. The hyperbolic spaces H^n and B^n

The hyperbolic space [7] is a coset space $H^n = SO(1, n, R)/SO(n, R)$. Its metric is inherited by restricting the Lorentz scalar product $\langle , \rangle$ from the embedding Minkowski space $M(1, n)$ to pairs of vectors pointing to the unit hyperboloid H^n. This restricted scalar product is the hyperbolic cosine of the hyperbolic angle between points on H^n. H^n has negative constant curvature. The hyperbolic ball B^n is a conformal image of H^n. The intersections of hyperplanes in $M(1, n)$, orthogonal to space-like vectors, with the unit hyperboloid H^n when mapped to B^n become Moebius spheres $S^{(n-1)}$ intersecting B^n.

2. The double torus $T \cup T$ and its homotopies

We follow the sources, in particular Magnus [6], the set-up and analysis given in [2], and then describe results elaborated in [3]. H^2 is the unit hyperboloid and B^2 the unit disc. The proper time-preserving Lorentz group $S_\uparrow^+(1, 2, R)$ has the universal covering group $SU(1, 1)$. To the Lorentz action on H^2 there correspond linear fractional transforms of $SU(1, 1)$ acting on a complex variable within the unit disc B^2. The double torus $T \cup T$ unfolds [1] into an octagon. The condition that 8 octagons share any vertex enforces H^2, tesselated by congruent octagons of fixed edge length, as the universal covering. The homotopy group $\pi(T \cup T)$ has 4 generators and one relation. It is closely related to a hyperbolic Coxeter group [6] with Dynkin diagram $\circ \overset{2}{-} \circ \overset{8}{-} \circ \overset{2}{-} \circ$. Any homotopy $g \in \pi(T \cup T)$ maps equivalent points of the tesselation into one another. The geodesics on B^2 are Moebius circles S^1.

3. The normalizer N_g and its orbit lines

A homotopy corresponds to a fixed hyperbolic element $g \in SU(1, 1)$. This element can be written as

$$g = g(\theta, \omega) \tag{1}$$

$$= \left[\begin{array}{cc} \cosh(\theta) + i\sinh(\theta)\omega_0 & \sinh(\theta)(\omega_2 - i\omega_1) \\ \sinh(\theta)(\omega_2 + i\omega_1) & \cosh(\theta) - i\sinh(\theta)\omega_0 \end{array} \right].$$

Here θ fixes the hyperbolic class of $g \in SU(1,1)$ and the space-like vector $\omega = (\omega_0, \omega_1, \omega_2)$ in $M(1,2)$ fixes g within its class. We shall assume $\omega_1 = 0$.

The continuous normalizer $N_g < SU(1,1)$ consists of all elements which commute with g. With eq. 1 it can be written as the commutative one-parameter subgroup

$$N_g = \langle g(\lambda\theta/2, \omega), \quad -\infty < \lambda < \infty \rangle. \tag{2}$$

On the complex unit disc B^2 we choose as orbit representatives under N_g the points on the imaginary axis. Each of these points determines an orbit line parametrized by λ. In general the orbit lines are not geodesics! The transforms of the representative points under $g(-\theta/2, \omega)$ and $g(\theta/2, \omega)$ respectively are on two Moebius circles which from eqs. 1, 2 are preimages and images under the homotopy g. In Fig. 1 the orbit lines are shown for the generator $g = C_5 \in \pi(T \cup T)$. The orbit lines are orthogonal to a family of Moebius circles. There is a single orbit line which coincides with the shortest geodesic under g.

4. Classification of closed geodesics

Closed geodesics for a fixed homotopy g of $T \cup T$ run between pairs of preimage and image points under g. The geodesics between the pairs close on $T \cup T$. The variety of closed geodesics associated with a fixed g is organised by the orbits of the normalizer. From the properties of the normalizer N_g and its orbits we find the following classification for closed geodesics:

(1) Any closed geodesic on $T \cup T$ is uniquely associated with an element $g \in \pi(M)$ and with a geodesic on B^2. Properties (2-5) hold for these geodesics.

(2) A geodesic with preimage on an orbit line under N_g has its image on the same orbit line.

(3) The geodesic length depends only on the orbit line of its preimage point.

(4) The angle between an orbit line and a geodesic starting on it depends only on the chosen orbit line.

(5) For given homotopy g, there is a single orbit line coincident with a shortest geodesic. The length of this shortest geodesic is determined by the character $\chi(g)$,

$$\cosh(2\theta) = \frac{1}{2}(\chi(g))^2 - 1. \tag{3}$$

This length is unchanged under conjugation with elements of $\pi(T \cup T)$ and moreover under conjugations with symmetries from $SU(1,1)$ like those from the Coxeter group.

(6) Shortest closed geodesics for different homotopies $g \neq g'$ can be compared in terms of their characters.

In Fig. 2 we illustrate geodesics associated with the homotopy C_5 between pairs of edges of an octagon. These edges are part of the Moebius circles from Fig. 1. Two geodesics connect vertices of the edges. The shortest geodesic in between connects two general points on the edges.

The analysis of closed geodesics by use of the normalizers of homotopies and their orbits is general. For its application to the Weber-Seifert hyperbolic dodecahedral manifold compare [4]. A review of cosmic topology and its test from astronomical observations is given in [5]. The most effective way of detecting closed geodesics is the analysis of the pair correlation function of the matter density.

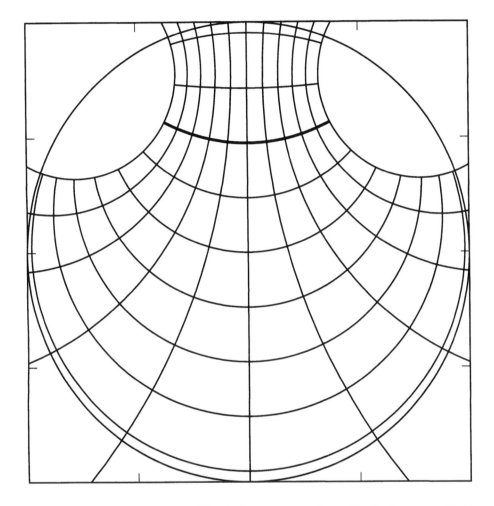

Figure 1. Orbit lines on B^2 under the generator $g = C_5 \in \pi(T \cup T)$. C_5 maps two Moebius circles into one another. The orbit lines between them are orthogonal to a family of Moebius circles. A unique single heavy orbit line coincides with the shortest closed geodesic associated with C_5.

References

[1] Hilbert D and Cohn-Vossen S 1932
   Anschauliche Geometrie, Berlin
[2] Kramer P and Lorente M 2002
   The double torus as a 2D cosmos: Groups, geometry, and closed geodesics
   J. Phys. A 35 1-21
[3] Kramer P 2002
   General closed geodesics on the double torus manifold
   submitted for publication
[4] Kramer P 2002

208

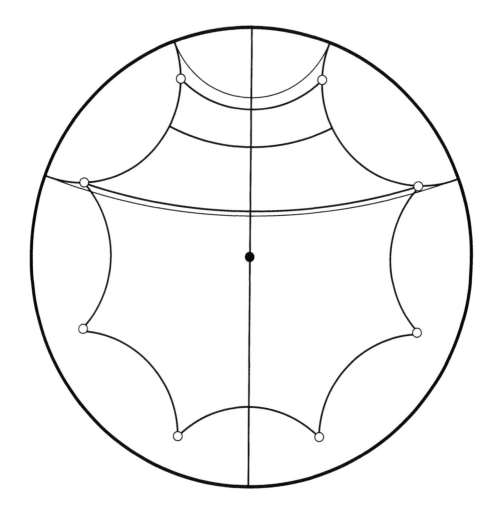

Figure 2. The generator C_5 of $\pi(T \cup T)$ maps 2 edges of the fundamental octagon into one another. Two geodesics close on vertices of these edges. The shortest geodesic associated with C_5 joins a general pair of edge points.

Closed geodesics on the Weber-Seifert hyperbolic dodecahedral manifold
submitted for publication
[5] Lachieze-Rey M and Luminet J-P 1995,
Cosmic topology, Physics Reports 254 135-214
[6] Magnus J W 1974,
Non-Euclidean Tesselations and Their Groups, New York:Academic
[7] Ratcliffe J G 1994,
Foundations of Hyperbolic Manifolds, New York:Springer

Inst. Phys. Conf. Ser. No 173: Section 1
Paper presented at 24th Int. Coll. Group Theoretical Methods in Physics, Paris, France, 15–20 July 2002
©2003 IOP Publishing Ltd

Using the renormalization group for the evaluation of electronic localization

Gerardo G. Naumis

Instituto de Fisica.Universidad Nacional Autónoma de México. Apdo. Postal 20-364, 01000, México D.F., México.

E-mail: naumis@fisica.unam.mx

Abstract. A method to evaluate electronic localization properties is proposed. The method is based in the scaling of bands, and for one dimension, it is shown that the localization nature is related with the stability of the renormalization group around the fixed points of the transfer matrix trace map. The Lyapunov exponents of the trace map determine the localization length and scaling exponents of critical states. As an example, the case of the Fibonacci chain is analyzed.

PACS numbers: 71.23.Ft, 71.23.An, 72.10.Bg, 72.15.Rn

The renormalization group has been a very useful tool in order to study the phonon and electronic spectrum of quasiperiodic systems, due to the self-similar nature of the potential that makes this formalism particularly involved. However, in the study of electronic and phonon localization, the renormalization group has not been fully exploited. The reason is that localization is evaluated through the Lyapunov exponents[1,2](LE), but most of the calculations of these exponents do not consider that the definition involves a limit from the right and from the left in the localization center[3]. Instead, the norm of a transfer matrix is taken, and the problem of the match between the solutions from the left and from the right is not taken into account (this well known problem is called the Borland paradox [3]). Furthermore, in this article we show that the LE of an eigenenergy is zero, whatever the shape of the wave function. To avoid all these problems, here we propose an alternative approach that uses the scaling of the bands as a measure of localization. In 1D, the final result depends on the LE of the renormalization group of the transfer matrix trace map, but now taken in the sense of stability. This method also leads to a classification of the localization nature in terms of the flow of the renormalization group of the trace map. According to this, the nature of the fixed points (hyperbolic, repulsive or hyperbolic fixed points of the map cycles) determine the type of the spectrum and the localization properties. The method is very appropriate for quasiperiodic systems, since Kohmoto, Kadanoff and Tao [4] found a recurrence relation for the trace of a Fibonacci chain [4], that defines a map, from which the spectrum can be found by successive iterations. In this work, we will use this renormalization group of the trace in order to obtain the scaling exponents of the critical states for the electronic problem with a Fibonacci potential. The method can be extended to other quasiperiodic [5,6] and disordered potentials[7].

As a model, we use a very simple hamiltonian that contains a lot of the physics involved in the localization problem: a 1D s-band tight-binding hamiltonian. In more dimensions, the method of evaluating localization using the scaling of bands works in a similar way. The model hamiltonian is defined on a chain of n sites, with an on-site potential V_n at site n, and hopping integral t between sites n and $n+1$. The corresponding Schröedinger equation for this model is,

$$t\psi_{n-1} + t\psi_{n+1} + V_n\psi_n = E\psi_n, \tag{1}$$

where ψ_n is the value of the wavefunction at site n. Eq.(1) can be rewritten in terms of the transfer matrix $M(n)$ and a vector Ψ_n with components (ψ_n, ψ_{n-1}),

$$\Psi_n \equiv \begin{pmatrix} \psi_n \\ \psi_{n-1} \end{pmatrix} = \begin{pmatrix} (E-V_n)/t & -1 \\ 1 & 0 \end{pmatrix} \Psi_{n-1} \equiv M(n)\Psi_{n-1}. \tag{2}$$

The wave-function at site n in the end of the chain, as a function of the value at the beginning, is given by a successive application of Eq. (2),

$$\Psi_n = M(n)M(n-1)M(n-2)M(n-3)...M(2)\Psi_1 \equiv T(n)\Psi_1. \tag{3}$$

The eigenenergies are those for which the trace norm of $T(n)$ ($\tau_n \equiv trT(n)$) is less than two [8]. Thus the trace determines the spectrum, and in many cases there are recurrence relations, obtained from the renormalization group, where the trace of a certain length is a function of the trace of previous lengths, as follows

$$\tau_n(E) = f(\tau_{n-1}(E), \tau_{n-2}(E), ..., \tau_1(E)). \tag{4}$$

Here is natural to ask if the trace can also give information about localization. To answer this question, we observe that usually, the localization length of an eigenstate with energy E is obtained through the Lyapunov exponent, defined as[8]

$$\gamma(E) = \lim_{n\to\infty} \frac{1}{n} \ln \| T(n) \| = \lim_{n\to\infty} \frac{1}{n} \ln | \lambda_{\max} |, \tag{5}$$

where $\lambda_{\max}$ is the greatest eigenvalue of $T(n)$. λ_{max} can be found by using the characteristic equation of the transfer matrix, and that the trace is an invariant under unitary transformations. The two eigenvalues of $T(n)$ are,

$$\lambda_{\pm} = \frac{\tau_n \pm \sqrt{\tau_n^2 - 4}}{2}. \tag{6}$$

¿From here, is clear that for energies inside the spectrum ($\| \tau_n \| \le 2$), $\lambda_{\pm}$ are complex, both with unitary norm, and the LE is zero, since $\| T(n) \|$ is always 1. Thus, they do not give information about localization for energies that are eigenvalues. A solution for this problem, is to relate localization with the scaling of bands. The idea is to construct a supercrystal, by considering as unit cell a piece of the chain of size n, that can be amorphous, quasiperiodic or periodic. Using this cell, we can construct the infinite supercrystal by joining the cells side by side. The resultant supercrystal has Bloch solutions and a band spectrum, and the bandwidths are determined by the wave-function overlap at the border of the cells. Next we study how the bandwidths scale with the size of the unit cell. If a cell of size n has exponentially localized states, the bandwidth of the crystal is $W_n \sim < \psi_1 \mid H \mid \psi_2 > \sim t e^{-\frac{2n}{\xi}}$, where $< \psi_1 \mid$ and $\psi_2 >$ are two similar eigenstates in contiguous cells, and ξ is the localization length. A similar calculation for extended states shows that the bandwidth do not depend on n, since the states do not decay with the distance. For quasiperiodic cells, the corresponding states decay as a power law, and $W_n \sim n^{2\beta(E)}$ where $\beta(E)$ is a scaling exponent, that can be obtained from W_n,

$$\beta_n(E) = \ln W_n / 2 \ln n. \tag{7}$$

The bandwidth depends on the trace, since the band edges (energies that we denote by E_s) are the points where $\tau(E_s) = \pm 2,$. From this, we observe that the properties of localization are determined by the stability of the trace map around the point $\tau_n = \pm 2$. For a localized state, the band shrinks in an exponential way as the system grows, due to a decreasing overlap between neighbouring cells. If for a certain generation we have $\tau_n(E_s) = 2$, after N generations, the trace evaluated at the same energy must be outside the band, $\tau_{n+N}(E_s) > 2$. Thus, a localized state corresponds to a repulsive fixed point of the trace map. For extended states, the bands edges do not change with the system size, and $\tau_n(E_s) = \pm 2$ for any n and fixed E_s. Thus, ± 2 is a fixed point of the trace map, hyperbolic in nature since the trace for energies inside the band remains bounded, while for other energies goes to infinity. In quasiperiodic systems, usually the bands are subdivided as n grows since the spectrum is a type of Cantor-set. Thus, the number of points E_s grows with the system size, and as a conclusion one can say that the trace map must be non-linear, as we will see for the case of the Fibonacci chain.

In order to make more quantitative the relationship between stability and localization, we proceed as follows: let us write the trace as a polynom,

$$\tau_n(E) - 2 = \prod_{i=1}(E - E_i), \qquad (8)$$

where E_i are the roots of $\tau_n(E) - 2 = 0$. After taking the logaritm of the derivative evaluated in one of the band edges (E_s) we get,

$$\ln\left(\frac{d\tau_n(E)}{dE}\right)_{E=E_s} = \sum_{i \neq s} \ln(E_s - E_i). \qquad (9)$$

Using that E_i are the only values that satisfies a cyclic boundary condition[9] for the cell of size n (in other words, E_i is an eigenvalue of the *finite* system), and dividing by n, we can use the density of states ($\rho_n(E)$) of the unit cell to perform the sum. This leads to the Thouless formula[3] for the inverse of the localization length $\xi(E_s)$, but this time we get an expression that depends on the trace,

$$\frac{1}{n}\ln\left(\frac{d\tau_n(E)}{dE}\right) = \int_{-\infty}^{\infty} \rho_n(E')\ln(E - E')dE' \equiv \frac{1}{\xi(E)}. \qquad (10)$$

Notice that we drop the s since it is understood that the derivative must be evaluated at a band edge. Last equation shows that the Lyapunov exponent, used in the sense that it gives how much differs two trajectories under the action of the map, gives the inverse of the localization length. For quasiperiodic systems, instead of dividing by n, we divide by $\ln n$ in order to obtain a value different from zero in the inverse of $\xi(E)$, and thus we get a modified Thouless formula that is appropriate for quasiperiodic systems since it gives the scaling exponent of the wave-function,

$$\beta_n(E) = \frac{1}{\ln n}\ln\left(\frac{d\tau_n(E)}{dE}\right) = \frac{n}{\ln n}\int_{-\infty}^{\infty} \rho_n(E')\ln(E - E')dE' \qquad (11)$$

As an example of this technique, we will obtain the scaling exponents of the wave-function in the most simple quasicrystal: the Fibonacci chain. Here the potential $V(n)$ is taken from a sequence generated by two letters, A and B. This sequence follows the recursive rule: $A \to B$ and $B \to BA$. It has been proved, that for this case, the trace for a chain of length $F(l)$ is given by[4],

$$x_{F(l)}(E) = x_{F(l-1)}(E)x_{F(l-2)}(E) - x_{F(l-2)}(E),$$

where x_n is half the trace ($x_n = \tau(E)/2$), $F(l)$ is the l-esim Fibonacci number with the initial conditions, $x_{-1}(E) = 1$, $x_0(E) = (E + \lambda)/2$ and $x_1(E) = (E + \lambda)/2$, where $\lambda = (V_A - V_B)/2$. The nature of the states is easy obtained by using the approach presented in this work. The fixed points of the map in this case are 2 and 0. However, 0 is not consistent with the initial conditions (since the map has the invariant[4]: $x_{l+1}^2(E) + x_l^2(E) + x_{l-1}^2(E) - 2x_{l+1}(E)x_l(E)x_{l-1}(E) = \lambda^2 - 1$). From the initial conditions, it is easy to show the other fixed point ($\tau = 2$) occurs only when we have a periodic chain, *i.e.*, when all the $V(n)$ are equal. Since $\tau(E) = 2$ is an hyperbolic fixed point (because the eigenvalues of the gradient matrix around the fixed point has an eigenvalue less than one, and the other is greater), we conclude that there are extended states, but they are only observed for a periodic chain. Localized states are not observed since there are not repulsive fixed points. The only remaining possibility is to have critical states, as is reveal by the fact that the map contains two cycles: one of period two and the other with period six[4]. For the last cycle, which corresponds to energies at the center of the spectrum, a linear stability analysis gives that the trace scales as $\tau_{l+6}(E) = \tau_l^\alpha(E)$, where

$$\alpha = \ln \sigma^3 / \ln \left(\left(1 + 4 \left(1 + \lambda^2 \right)^2 \right)^{\frac{1}{2}} \pm 2 \left(1 + \lambda^2 \right) \right) \tag{12}$$

(σ is the golden mean $(\sqrt{5} + 1)/2$). Using eq.(7), we obtain that the scaling exponent of the wave-function is $\beta = \alpha/2$. The result for the off-diagonal problem (where $V(n) = 0$ and t_n is given by a Fibonacci sequence of t_A and t_b) is similar, but λ is replaced by $\lambda = |y - (1/y)|/2$, where $y = t_A/t_B$. This result can be compared for example, with the scaling exponent for the special case of the state with $E = 0$, for which an analytical expression is know[10],

$$\beta = \ln y / \ln \sigma^3 \tag{13}$$

When $y << 1$ or $y >> 1$, the method presented here gives exactly the same result. For $y \approx 1$ the results are a little bit different, due to the fact that in the limit of the linear chain, we need to consider more than one eigenvalue of the gradient matrix of the map around the cycles, and two scaling indices are present. However, the difference is minimal. In conclusion, this work shows that the renormalization group is a very useful tool for studying localization and gives a natural classification for the type of localization.

I would like to thank DGAPA-UNAM project IN-108199 for the financial help and R. Kerner for useful comments.

References

[1] R.B. Capaz, B. Koiller and S.L.A. de Queiroz 1990 *Phys. Rev.* **B42** 6402
[2] M.T. Velhinho and I.R. Pimentel 2000 *Phys. Rev.* **B61** 1043.
[3] J.M. Ziman Models of disorder,Cambridge University Press, Cambridge, (1979).
[4] M. Kohmoto, L.P. Kadanoff and Ch. Tao 1983 *Phys. Rev. Lett.* **50** 540
[5] J.M. Luck 1989 *Phys. Rev.* **B39** 5834
[6] A. Ghosh and S.N. Karmakar 1998 *Phys. Rev.* **B58** 2586
[7] P. Le Doussal and K.J. Wiese 2002 *Phys. Rev. Lett.* **89** 125702
[8] A. Sütö, in *Beyond Quasicrystals*, ed. by F. Axel and D. Gratias, Les Editions de Physique, France, (1994)
[9] G.G. Naumis 1999 Phys. Rev. **B59** 11 315
[10] M. Kohmoto, B. Sutherland and Ch. Tang 1987 *Phys. Rev.***B35** 1023.

Inst. Phys. Conf. Ser. No 173: Section 1
Paper presented at 24th Int. Coll. Group Theoretical Methods in Physics, Paris, France, 15–20 July 2002
©*2003 IOP Publishing Ltd*

Tilings and coverings embedded into the canonical decagonal tiling $\mathcal{T}^{*(A_4)}$

Z Papadopolos[1], **G Kasner**[2]

[1]Institut für Theoretische Physik, Universität Tübingen,
[2]Institut für Theoretische Physik, Universität Magdeburg

Abstract. This paper considers some tilings and coverings embedded in a quasilattice of the canonical tiling $\mathcal{T}^{*(A_4)}$.

1. Introduction

The icosahedral canonical tiling $\mathcal{T}^{*(D_6)} \equiv \mathcal{T}^{*(2F)}$ [1] is a skeleton of the icosahedral (I_h) model $\mathcal{M}[2, 3, 4, 5]$ of F-phases. It describes a bulk of the alloys i-AlPdMn [6] and i-AlCuFe [7]. The model $\mathcal{M}$ we obtained through a decoration of a quasilattice $\mathcal{T}^{*(D_6)}$ by Bergman polytopes [2]. The resulting atomic positions of the model $\mathcal{M}$ are coded by three windows, W_q, W_b and W_a [2, 4]. Each (polytopal) window defines a quasilattice in an icosahedrally (I_h) projected copy of a D_6 root lattice, which becomes, after the projection, a 3-dimensional $\mathbb{Z}(\tau)$ ($\tau = (\sqrt{5} + 2)/2$) module of type $2F$ (see it's definition over a basis in Ref [8]).

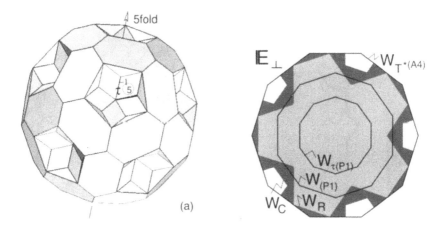

Figure 1. Left (a) is the window W_q. Right (b) are two sections of W_q by two 5fold planes labeled "C" and "R", coding the corresponding planes in $\mathbb{E}_{\parallel}$. W_{P1} is coding the Penrose P1 tiling [9] with maximal possible edge length contained in W_C and $W_{\tau(P1)}$ is coding the P1 tiling with maximal edge length contained in W_R, which is by factor τ bigger then the previous.

We studied the most stable surfaces of i-AlPdMn, and these are surfaces orthogonal to a 5fold symmetry axis of an icosahedron [10]. These (terrace like) surfaces appear in the

Fibonacci sequence with intervals $m = 4.08$ Å and $l = \tau m = 6.60$ Å [11]. But also an interval of length $s = \tau^{-1}m$ was observed [12]. All these intervals may appear together in a "decorated" Fibonacci sequence, to be defined later. The codings of quasilattices in all 5fold terraces are derived from the single window W_q shown in Fig 1(a). More accurately, each 5fold surface plane $P_\parallel^{5-}$ (in $\mathbb{E}_\parallel$) is coded by an intersection of the 3-dimensional window W_q by the corresponding plane in $\mathbb{E}_\perp$, $P_\perp^{5-}$, see Fig 1(b)

$$W_q \bigcap P_\perp^{5-}, \tag{1}$$

where $P_\perp^{5-} = {}^* P_\parallel^{5-}$, and "*" is a symbol for the *-map: $\tau \longrightarrow -1/\tau$.

The "decorated" Fibonacci sequence of planes along a 5fold symmetry axis of an icosahedron with appropriate intervals is coded by the window 2⑤ [5] along a 5fold axes in $\mathbb{E}_\perp$ in (3-dimensional icosahedral) $2F$ module. The standard distance $\text{⑤} = 1/\sqrt{2}$ is obtained by the icosahedral (I_h) projection of a vector e_i from an orthonormal basis of the $\mathbb{Z}^6$ lattice. The standard distance ⑤ in i-AlPdMn is $\text{⑤} = 4.56$. The decorated sequence is defined through the Fibonacci sequence coded by the window $(2\tau/(\tau + 2))\text{⑤}$ and realized in $\mathbb{E}_\parallel$ by intervals $S = (2\tau^2/(\tau + 2))\text{⑤}$ and $L = \tau S$. A decoration defined by $S \longrightarrow l$ and $L \longrightarrow m \bigcup s \bigcup m$ (where $l = \tau m = \tau^2 s$) leads to the decorated Fibonacci sequence coded by the window 2⑤, mentioned above. This window fits into W_q with it's maximal size $2\tau\text{⑤}$ [2, 4] along the 5fold direction and is moreover shifted by the vector $(2\tau/(\tau + 2))\text{⑤}$ from the center of W_q along the 5fold direction [5].

The decagonal (2-dimensional) submodule in each 5fold plane of a (3-dimensional) $2F$ module is a C_5 projected A_4 root lattice ($A_4 \subset D_6$) into the considered plane. Hence, in a 5fold plane (in $\mathbb{E}_\parallel$ and $\mathbb{E}_\perp$) in which 2-dimensional $\mathbb{Z}(\tau)$ module is embedded (C_5 projected A_4 module), a decagon of an appropriate size and orientation in $\mathbb{E}_\perp$ [5] defines the $T^{*(A_4)}$ tiling [13].

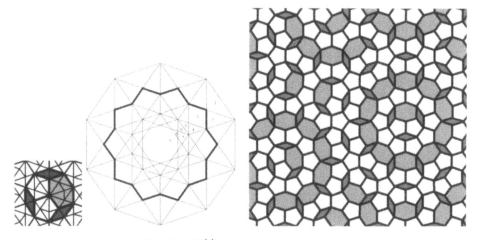

Figure 2. (a) The tiling $T^{*(z)}$ of the plane by the acute rhombus, pentagon and hexagon as the prototiles, locally derived from $T^{*(A_4)}$. (b) The window of the tiling $T^{*(z)}$, without the content of golden triangles, inscribed in the window of the tiling $T^{*(A_4)}$. (c) The only 4 existing isolated local configurations of hexagons in $T^{*(z)}$.

If one intends to put on a 5fold plane of a monograin imaged by STM any tiling, only the tilings derivable out of the quasilattice are coming in question. Because the $T^{*(A_4)}$

tiling is the most rich known structure in a C_5 projected A_4 module, we look for embedded tilings/coverings with vertices from a sub-quasilattice of the tiling $T^{*(A_4)}$.

2. Local derivation of the tiling $T^{*(z)}$ from the canonical tiling $T^{*(A_4)}$

From the canonical tiling [8] $T^{*(A_4)}$ with (two) golden triangles as prototiles [13] we locally derive a tiling of a plane by pentagons, hexagons and acute rhombuses as shown in Fig 2 (a). The window of the tiling $T^{*(z)}$ is shown in Fig 2 (b). And the result of the local derivation in Fig 2 (c). In Fig 2 (c) are preseted the only four existing local non overlapping configurations of hexagons: (i) a full circle of ten hexagons, (ii) seven hexagons forming part of a circle, (iii) four hexagons forming part of a circle (all circles are of the same radius) and (iv) two isolated hexagons along a twofold line. In each of the configurations (i)-(iii) there is also an additional isolated hexagon inside the corresponding circle.

3. Coverings derived locally from the tiling $T^{*(z)}$

If we try to place in each hexagon a pentagon, we see that there are two possibilities by each hexagon. Let us put on each hexagon both pentagons that will overlap and denote a pentagon by $D_{\parallel}^y$. Instead of an acute rhombus let us place a pair of, by factor τ^{-1} smaller pentagons $D_{\parallel}^x$. Like this we get a covering (embedded in $T^{*(A_4)}$) of both tilings $T^{*(A_4)}$ and $T^{*(z)}$, see Fig 3. It is a subcovering $C_{T^{*(A_4)}}^s$ [14] of the Kramer's Delone covering $C_{T^{*(A_4)}}^k$ by into $\mathbb{E}_{\parallel}$ projected Delone cells D^y and D^x, see [15].

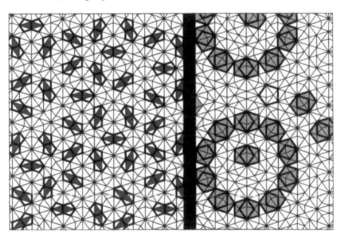

Figure 3. Local derivation of the covering $C_{T^{*(A_4)}}^s$ from the tiling $T^{*(z)}$: $T^{*(z)} \longrightarrow C_{T^{*(A_4)}}^s$. Left: the local derivation of the pairs of small pentagons $D_{\parallel}^x$ from the acute rhombus of the tiling $T^{*(z)}$. Right: the local derivation of the overlapping pair of big pentagons $D_{\parallel}^y$ from the hexagon of the tiling $T^{*(z)}$.

Whereas thickness of the Delone covering $C_{T^{*(A_4)}}^k$ is

$$C^k = -\tau + 3 \approx 1.382 \tag{2}$$

the thickness of the subcovering $C_{T^{*(A_4)}}^s$ is

$$C^s = 2\tau - 2 \approx 1.236 < 1.382. \tag{3}$$

4. Tilings derived from $T^{*(z)}$ and embedded into $T^{*(A_4)}$

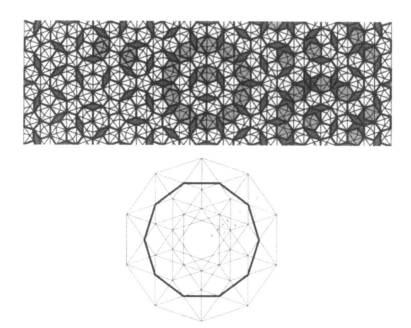

Figure 4. Top (a) from left to right is the local derivation of the partly random tiling $T^{*(p1)_r}$, $T^{*(z)} \longrightarrow T^{*(p1)_r}$. Bottom (b) the window of the tiling $T^{*(p1)}$ inside the window of the tiling $T^{*(A_4)}$.

In $T^{*(z)}$ (with the raster of the tiling $T^{*(A_4)}$) we can embed a tiling of a plane by 4 Penrose prototiles of the tiling P1[9]: pentagon, pentagonal star, crown and acute rhombus. This procedure reminds to the ancient problem of placing the pentagons without overlap in the plane. If we randomly choose in each hexagon of $T^{*(z)}$ a single pentagon, and the rest unify with a neighboring acute rhombus, there appears a partly random version of the Penrose tiling P1, $T^{*(p1)_r}$, see Fig 4 (a). From the window of the Penrose tiling P1 embedded into the window of the tiling $T^{*(A_4)}$ (Fig 4 (b) it is clear that it is impossible to derive locally the Penrose tiling P1 out of the tiling $T^{*(A_4)}$ (or $T^{*(z)}$). It is so, because the shape of the window of the tiling P1 can not be obtained by intersections or unions of the codings of golden tetrahedra in the window of $T^{*(A_4)}$, see Fig 4 (b) and Ref [13]. But, the fraction of the plane covered by pentagons in a partly random variant of P1 is the same as by the P1 tiling. The tiling $T^{*(z)}$ with it's 4 local configurations of hexagons marks the positions in a plane on which the pentagons, that can be randomly chosen, are to be placed.

The Niizeki star tiling $T^{*(n)}$ [16] is also a candidate to be seen on a 5fold surface (is to be locally derived out of the atomic positions). In [14] we show from the relation of the tiling $T^{*(n)}$ to the (by factor τ) inflated tiling $\tau\, T^{*(z)}$, that the Niizeki tiling can be 1-1 related to the tiling P1.

References

[1] Kramer P et al. 1991 in *Symmetries in Science V: Algebraic structures, their representations, realizations and physical applications*, edited by B. Gruber and L. C. Biedenharn (Plenum, New York) p 395

[2] Papadopolos Z et al. 1998 in *Proc. of the Int. Conf. on Aperiodic Crystals, Aperiodic 1997* edited by Marc de Boissieu et al. (World Scientific, Singapore) p 173

[3] Kasner G et al. 1999 Phys. Rev. B **60** 3899

[4] Papadopolos Z et al. 1999 in *Quasicrystals* edited by J. M. Dubois et al. Mater. Res. Soc. Symp. Proc. No. **553** p 231

[5] Papadopolos Z et al. 2002 preprint No.: cond-mat/0111479 to appear in Phys. Rev. B **66** (1. November 2002)

[6] de Boissieu M et al. 1994 *J. Phys.: Condens. Matter* **6** 10725

[7] Katz A and Gratias D 1995 in *Proceedings of the 5th International Conference on Quasicrystals*, edited by Janot C and Mosseri R (World Scientific, Singapore) p 194

[8] Papadopolos Z and Kramer P 1995 in *Proc. of Aperiodic '94*, ed. by Chapuis G and Paciorek W (World Scientific, Singapore) p 70

[9] Grünbaum B and Shepard G C 1987 *Tilings and Patterns* W.H. Freeman, San Francisco

[10] Shen Z et al. 2000 Surf. Sci. 450 1

[11] Schaub T M et al. 1994 Phys. Rev. Lett. **73** 1255; Schaub T M et al. 1995 Appl. Phys. A **61** 491

[12] Shen Z et al. 1999 Phys. Rev. B **60** 14688

[13] Baake M et al. 1990 in Int. J. Mod. Phys. **B4** 2217

[14] Papadopolos Z and Kasner G 2002 in "Coverings of Discrete Quasiperiodic Sets - Theory and Applications to Quasicrystals" Springer Tracts in Modern Physics Vol 180, Editors Kramer P and Papadopolos Z p 165

[15] Kramer P 2000 J. Phys. A: Math. Gen. **34** 1885

[16] Niizeki K 1989 J. Phys. A: Math. Gen. **22** 4281

Inst. Phys. Conf. Ser. No 173: Section 1
Paper presented at 24th Int. Coll. Group Theoretical Methods in Physics, Paris, France, 15–20 July 2002
©*2003 IOP Publishing Ltd*

Delone sets of beta-integers, Rauzy fractals and adeles

J-L Verger-Gaugry

Institut Fourier, CNRS, University of Grenoble I, BP 74 - Domaine Universitaire, 38402
Saint Martin d'Hères, France

Abstract. We recall the two canonical cut-and-project schemes which lie over the set $\mathbb{Z}_\beta$ of β-integers when β is a Perron number of degree $m \geq 2$. If β is Pisot only, the canonical windows in the two internal spaces are fractals, called Rauzy fractal in one of them, and $\mathbb{Z}_\beta$ is a Meyer set. The existence of supplementary p-adic internal spaces with p dividing the norm $N(\beta)$, shown recently by A. Siegel in the study of the dynamics of susbstitution systems of Pisot type, in which Rauzy fractals are extended, suggests to consider cut-and-project schemes over the ring of adeles of $\mathbb{Q}(\beta)$ as a complete geometrical framework over $\mathbb{Z}_\beta$. This indicates a deep link with the works of A. Weil on adeles, already partially discovered by J.-P. Schreiber.

1. Introduction

The sets of β-integers, called $\mathbb{Z}_\beta$, reveal to be extremely pertinent and quite appropriate for investigating quasicrystals and quasicrystalline models [8] [9] [10] [23]. Initially studied as uniformly discrete sets on the line with β a Pisot number such as $(1+\sqrt{5})/2, 1+\sqrt{2}, 2+\sqrt{3}$, they can be spread vectorially to form so-called β-grids in $\mathbb{R}^n$ and are relevant for quasicrystallography in arbitrary dimension with arbitrary Pisot numbers β. The sets $\mathbb{Z}_\beta$ (when uniformly discrete) form a large class in the set of Delone sets, this one being classified by Lagarias [13] [14], in particular in the set of Meyer sets ([16] [17]) if β is a Pisot number. By construction the sets $\mathbb{Z}_\beta$ are self-similar since $\beta\mathbb{Z}_\beta \subset \mathbb{Z}_\beta$ always holds.

The algebraic description of such objects (on the line) was done by Frougny and Gazeau [8] [10] when β is a general Pisot number. The geometrical counterpart of this algebraic approach is also extremely rich and was developed by Verger-Gaugry and Gazeau [23] even in the case where β is a Perron number. We will recall in section 2 the two canonical cut-and-project schemes over $\mathbb{Z}_\beta$.

But the geometrical representations of the set $\mathbb{Z}_\beta$ of beta-integers were also studied independently in another field, namely in dynamical systems of substitutive systems [3] [4] [6] where β is the dominant eigenvalue of the incidence matrix, and several related domains such as the coding of toroidal automorphisms [3]; the dynamics of substitutive systems leads to representations by the so-called Rauzy fractals when β is a Pisot number [6] [15] [19] (section 3). The story of beta-integers is then double, at least, and merits to be viewed from both sides leading to a mutual enrichment in each domain.

It is the subject of this note to explain briefly this exchange. In this spirit, it is also the purpose of this note to show that from some recent results obtained by Siegel [22] there is an indication that another geometrical framework of cut-and-project schemes, more suitable and more general, lies over the set of β-integers, that provided by using the ring of adeles of $\mathbb{Q}(\beta)$ (Weil [24]) (section 4). Cut-and-project schemes with adeles of number fields were investigated by Schreiber in 1973 in an inaugural work [21]. We will briefly suggest how Rauzy fractals can be extended in a natural way (with p-adic components) in this new cut-and-project scheme over the ring of adeles of $\mathbb{Q}(\beta)$.

2. beta numeration and beta-integers

Let $\beta \in (1, +\infty)$ be a Perron number of degree $m \geq 2$. Assume $P(X) = X^m - a_{m-1}X^{m-1} - a_{m-2}X^{m-2} - \ldots - a_1X - a_0$, $a_i \in \mathbb{Z}$ its minimal polynomial and $\beta^{(j)}$, $1 \leq j \leq m-1$, its conjugates with $\beta^{(0)} := \beta$ and $|\beta^{(j)}| < \beta$ for all $1 \leq j \leq m-1$. Let $A = \{0, 1, \ldots, \lfloor\beta\rfloor\}$ be the alphabet associated with the β-shift [18] [20]. Let us denote by $d_\beta(1)$ the Renyi-expansion of 1 in base β. We will denote

$$c_1c_2c_3\ldots = \begin{cases} t_1t_2t_3\ldots & \text{if } d_\beta = 0.t_1t_2\ldots \text{ is infinite} \\ (t_1t_2\ldots t_{m-1}(t_m - 1))^\omega & \text{if } d_\beta \text{ is finite and equal to } 0.t_1t_2\ldots t_m \end{cases}$$

where $(\ )^\omega$ means that the word within $(\ )$ is infinitely repeated. Let us define the companion matrix ${}^tQ$ of $P(X)$ as the transposed matrix of Q and

$$Q := \begin{pmatrix} 0 & 1 & 0 & \ldots & 0 \\ 0 & 0 & 1 & \ldots & 0 \\ \vdots & \vdots & & \ddots & 0 \\ 0 & 0 & & \ldots & 1 \\ a_0 & a_1 & & \ldots & a_{m-1} \end{pmatrix}$$

We deduce the sequence $(Z_k)_k$ of vectors of $\mathbb{Z}^m$: for $k \in \mathbb{Z}, Z_k = ({}^tQ)^k Z_0$ where $Z_0 = {}^t(100\ldots0)$ is the first vector of the standard basis $\{Z_0, Z_1, \ldots, Z_{m-1}\}$ of $\mathbb{R}^m$. Then let us define (the coefficients x_i are said to satisfy the conditions of Parry)

$$\mathcal{L}_\beta := \{x_kZ_k + x_{k-1}Z_{k-1} + \ldots + x_1Z_1 + x_0Z_0 \mid x_i \in A, k \geq 0 \text{ and}$$

$$(x_j, x_{j-1}, \ldots, x_1, x_0, 0, 0, \ldots) < (c_1, c_2, \ldots) \text{ for all } j, 0 \leq j \leq k\}$$

where the symbol " $<$ " means "lexicographically smaller than"; and

$$\mathbb{Z}_\beta^+ := \{x_k\beta^k + x_{k-1}\beta^{k-1} + \ldots + x_1\beta^1 + x_0\beta^0 \mid x_i \in A, k \geq 0 \text{ and}$$

$$(x_j, x_{j-1}, \ldots, x_1, x_0, 0, 0, \ldots) < (c_1, c_2, \ldots) \text{ for all } j, 0 \leq j \leq k\}$$

By definition, the set of beta-integers is $\mathbb{Z}_\beta := \mathbb{Z}_\beta^+ \cup (-\mathbb{Z}_\beta^+)$. This set is always relatively dense. It is uniformly discrete when β is a Pisot number. It is conjectured to be uniformly discrete when β is a Perron number. This occurs iff the β-shift is specified [7].

There exist two canonical cut-and-project schemes [17] $E \longleftarrow (E \times D = \mathbb{R}^m, L) \longrightarrow D$ over the beta-integers which lie on the line E of $\mathbb{R}^m$. In both cases the lattice L is $\mathbb{Z}^m \subset \mathbb{R}^m$, E is the line generated by the vector $B = B^{(0)} := {}^t(1\ \beta\ \beta^2\ \ldots\ \beta^{m-1})$ and the projection map $\pi_B : (E \times D) = \mathbb{R}^m \to E$ is orthogonal and satisfies

$$\pi_B\left(\sum_{i=0}^k \alpha_iZ_i\right) = \frac{\sum_{i=0}^k \alpha_i\beta^i}{\|B\|^2} B, \quad \alpha_i \in \mathbb{Z}, k \geq 0, i.e. \ \pi_B(\mathcal{L}_\beta \cup -\mathcal{L}_\beta) = \mathbb{Z}_\beta \frac{1}{\|B\|^2} B$$

All the points of the subset $\mathcal{L}_\beta \subset \mathbb{Z}^m$ gather about the line generated by the vector W of coordinates say (ξ_j) in the standard basis of $\mathbb{R}^m$ with $\xi_j = (a_{j-1}\beta^{j-1} + a_{j-2}\beta^{j-2} + \ldots + a_1\beta + a_0)\beta^{-j}(P'(\beta))^{-1}$. In the first case, the internal space D is the hyperplane of $\mathbb{R}^m$ formed by the direct sum of all ${}^tQ$-invariant subspaces F of $\mathbb{R}^m$ except $\mathbb{R}W$ and $\pi_D : \mathbb{R}^m \to D$ is equal to $\oplus\pi_F$ with $\pi_F : \mathbb{R}^m \to F$ the projection map along the complementary direct sum. In the second case, the internal space D is

$$D = (\oplus_{i=1}^{s-1}\mathbb{R}B^{(i)}) \oplus (\oplus_{i=s,i-s \text{ even}}^{m-1} (\mathbb{R}\,\text{Im}(B^{(i)}) + \mathbb{R}\,\text{Re}(B^{(i)})))$$

where $B^{(q)} = {}^t(1\,\beta^{(q)}\,\beta^{(q)^2}\ldots\beta^{(q)^{(m-1)}})$, $0 \le q \le m-1$, and s, $t = (m-s)/2$ are respectively the number of real and complex embeddings of $\mathbb{Q}(\beta)$. Let us denote by $\pi_{B,i}$ the respective orthogonal projection maps on the lines $\mathbb{R}\,B^{(i)}, 1 \le i \le s-1$, and planes $\mathbb{R}\,\mathrm{Im}(B^{(i)}) + \mathbb{R}\,\mathrm{Re}(B^{(i)})$, $i \ge s$, $i - s$ even. We have $\pi_D = \oplus_{i \ge 1}\pi_{B,i}$. The dominant eigenspace $\mathbb{R}W$ of ${}^tQ$ is orthogonal to D. Componentwise we have $\pi_{B,i}\left(\sum_{j=0}^k \alpha_j Z_j\right) = \left(\sum_{j=0}^k \alpha_j \beta^{(i)^j}\right) B^{(i)}/\|B^{(i)}\|^2$, $\alpha_j \in \mathbb{Z}, i = 1, 2, \ldots, s-1$ and similarly in a vectorial way by 2×2 blocks when $i = s, s+1, \ldots, m$ (see [23]). The numeration in base $\beta^{(j)}, j \ge 1$, appears in each component of D.

3. Rauzy fractals

Let us denote by $* := \pi_D \circ (\pi_B)^{-1}$ the star-operation, restricted to the subset $\mathbb{Z}_\beta\,B/\|B\|^2$ of E, then injective, of the (first or second) cut-and-project scheme $(E \times D, \mathbb{Z}^m)$. In both cases, the set $*(\mathbb{Z}_\beta\,B/\|B\|^2)$ is bounded in D when β is a Pisot number [23]. When β has conjugates $\beta^{(j)}$ which lie outside the closed unit disc of $\mathbb{C}$, then this set is not bounded. When β is a Salem number, this set may be bounded or not [1]. When β is a Pisot number, the adherence of the set $*(\mathbb{Z}_\beta\,B/\|B\|^2)$ is fractal in D. This fractal set is called Rauzy fractal when D is the internal space orthogonal to the line $\mathbb{R}W$ (second case). The Rauzy fractal is the subset of D whose components are convergent formal series in $\beta^{(j)}, 1 \le j \le m-1$, obtained by the operation $*$ from the beta-integers. Some basic questions questions about it are: is it connected ?, is its interior simply connected? What is the Hausdorff dimension of its boundary ? Is it the fundamental region of a lattice in D ? Which one ? In [19] Rauzy takes for β the dominant eigenvalue of the incidence matrix of the substitution system over 3 letters $0 \to 01, 1 \to 02, 2 \to 0$. The above questions about this Rauzy fractal were completely answered by Messaoudi in [15]. The interested reader may refer to [4] for general considerations about Rauzy fractals and their relations with other domains.

4. Cut-and-project schemes over the ring of adeles of $\mathbb{Q}(\beta)$

Let us consider the second cut-and-project scheme and β a Perron number of degree m. We will extend the internal space in order to have the formal series in $\beta^{(j)}, 1 \le j \le m-1$, obtained by the operation $*$ from $\mathbb{Z}_\beta$ still convergent in it (with coefficients satisfying the conditions of Parry). The set thus formed will be still called Rauzy fractal. Let us observe that the m coordinates of the elements of $\mathbb{Z}_\beta\,B/\|B\|^2$, resp. of $\mathbb{Z}_{\beta^{(j)}}\,B^{(j)}/\|B^{(j)}\|^2$, $j \ge 1$, in the standard basis of $\mathbb{C}^m$ are in $K := \mathbb{Q}(\beta)$, resp. in the conjugate field $K_j := \mathbb{Q}(\beta^{(j)})$; therefore $\mathbb{Z}_\beta\,B/\|B\|^2$, resp. $\mathbb{Z}_{\beta^{(j)}}\,B^{(j)}/\|B^{(j)}\|^2$, $j \ge 1$, lies on the line generated by $u_0 := B/\|B\|^2$ over K, resp. by $u_j := B^{(j)}/\|B^{(j)}\|^2$ over K_j, in the vector space K^m, resp. K_j^m. The space $E := Ku_0 + K_1 u_1 + \ldots K_{m-1}u_{m-1}$ is a vector space from which we will make the adele space $E_{\mathbb{A}} := K_{\mathbb{A}}u_0 \oplus \sum_{j=1}^{m-1} K_{\mathbb{A}}u_j$, Ku_0 remaining the 'representation space' containing $\mathbb{Z}_\beta$, where $K_{\mathbb{A}}$ denotes the ring of adeles of K; in $E_{\mathbb{A}}$ we will define a suitable internal space. First component: if ν is a place of K, denote by K_ν its completion for ν. With the notations of Schreiber [21] p 315-319, we have the isomorphism $K_{\mathbb{A}} \simeq K_{\mathbb{A}}'(\nu) \times K_\nu$. Let us denote by π_ν, resp. π_ν', the canonical projection of $K_{\mathbb{A}}$ to K_ν, resp. to $K_{\mathbb{A}}'(\nu)$, and $\phi(K)$ the image of K in $K_{\mathbb{A}}$. The subring $\phi(K)$ is known to be discrete and relatively dense in $K_{\mathbb{A}}$, the restriction of π_ν to $\phi(K)$ is injective and $\pi_\nu'(\phi(K))$ is dense in $K_{\mathbb{A}}'(\nu)$. In other terms, $K_{\mathbb{A}}'(\nu) \leftarrow (K_{\mathbb{A}}, \phi(K)) \to K_\nu$ is a cut-and-project scheme. Take for $\nu : K \hookrightarrow \mathbb{R}$ the embedding $\mathbb{Q}(\beta) \subset \mathbb{R}$. Consider that any real number is the sum of a beta-integer and a fractional part (formal series in $\beta^j, j \le 0$).

Since, for all $j \leq 0$, we have $Z_j \in \mathbb{Z}[N(\beta)^{-1}]^m$ we have to retain the finite places ν of K which lie over $\mathbb{Q}_p$ with p dividing the norm $N(\beta)$ for a (p-adic) space contribution to the internal space. Other components: the fields K_j correspond to the other infinite places of K. For $1 \leq j \leq m-1$, take the places ν (finite or infinite) such that the above formal series are converging in K_{j_ν} The sum of these spaces K_{j_ν} gives the other (space) contribution to the internal space. From [24] theorem 2 p 64, E is discrete in $E_\mathbb{A}$ and $E_\mathbb{A}/E$ is compact. When beta is Pisot, the internal space is the sum of the infinite completions $\mathbb{R} \oplus K_j u_j, 1 \leq j \leq m-1$ and of the ν-adic ones K_ν with ν above p and $p|N(\beta)$ (this last condition was discovered by Siegel [22]). The existence of additive and multiplicative laws on $\mathbb{Z}_\beta$ [9] [23] and problems of algebraicity of β [2] can be reformulated in this scheme.

References

[1] Adamczewski B 2002 Symbolic discrepancy and self-similar dynamics, preprint.

[2] Akiyama S 2000 in Algebraic Number Theory and Diophantine Analysis (Graz 1998) (Berlin: de Gruyter) 11-26

[3] Arnoux P, Berthé V, Ferenczi S, Ito S, Mauduit C, Mori M, Peyrière J, Siegel A, Tamura J-I and Wen Z-Y 2002 Introduction to finite automata and substitution dynamical systems, (Berlin: Springer)

[4] Arnoux P, Berthé V, Ei H and Ito S 2001 Tilings, quasicrystals, discrete planes, generalized substitutions and multidimensional continued fractions, preprint.

[5] Arnoux P, Berthé V, Siegel A and Verger-Gaugry J-L 2002 Rauzy fractals, Delone sets of beta-integers and adeles, in preparation.

[6] Arnoux P and Ito S 2001 Bull. Belg. Math. Soc. 8 181-207

[7] Blanchard F 1989 Theor. Comput. Sci. 65 131-141

[8] Burdik Č, Frougny Ch, Gazeau J-P and Krejcar R 1998 J. Phys. A: Math. Gen. 31 6449-6472

[9] Burdik Č, Frougny Ch, Gazeau J-P and Krejcar R 2000 Dynamical Systems (Luminy Marseille 1998) (River Edge: World Scientific) 125-136

[10] Gazeau J-P 1997 The Mathematics of Aperiodic Long Range Order (NATO-ASI Proceedings, Waterloo 1995) ed R V Moody (Dordrecht: Kluwer) 175-198

[11] Ito S and Kimura M 1991 Japan J. Indus. Appl. Math. 8 461-486

[12] Ito S and Sano Y 2001 Osaka J. Math. 38 349-368

[13] Lagarias J 1999 Disc. Comput. Geom. 21 161-191 and 345-372

[14] Lagarias J 2000 in Directions in Mathematical Physics CRM Monograph Series 13 Ed. M. Baake and R.V. Moody 61-94

[15] Messaoudi A 2000 Acta Arithmetica XCV.3 195-224

[16] Meyer Y 1972 Algebraic Numbers and Harmonic Analysis (Amsterdam: North-Holland)

[17] Moody R V 1997 in The Mathematics of Long-Range Aperiodic Order Ed R V Moody (Dordrecht: Kluwer) 403-441

[18] Parry W 1960 Acta Math. Acad. Sci. Hung. 11 401-416

[19] Rauzy G 1982 Bull. Soc. Math. France 110 147-178

[20] Renyi A 1957 Acta Math. Acad. Sci. Hung. 8 477-493

[21] Schreiber J-P 1973 Bull. Soc. Math. France 101 297-332

[22] Siegel A 2001 Représentation des systèmes dynamiques substitutifs non unimodulaires, preprint.

[23] Verger-Gaugry J-L and Gazeau J-P 2000 Geometric study of the beta-integers for a Perron number and mathematical quasicrystals, preprint.

[24] Weil A 1974 Basic Number Theory, Third Edition, (Berlin: Springer-Verlag)

Inst. Phys. Conf. Ser. No 173: Section 1
Paper presented at 24th Int. Coll. Group Theoretical Methods in Physics, Paris, France, 15–20 July 2002
©2003 IOP Publishing Ltd

Superspace groups for incommensurate composite systems

P. Zeiner
Institute for Theoretical Physics & CMS, TU Wien, Wiedner Hauptstraße 8–10, 1040 Vienna, Austria

T. Janssen
Institute for theoretical Physics, University of Nijmegen, Toernooiveld, 6525 ED Nijmegen, The Netherlands

Abstract. We discuss the notion of superspace groups for incommensurate composite structures. We illustrate their main properties for the special case of a composite structure consisting of two subsystems.

1. Introduction

The concept of superspace groups was introduced a few decades ago for the needs of (in)commensurately modulated systems (for a detailed description of superspace groups and their properties see [1]). It is based on the idea that modulated systems have a periodic average structure which is (in)commensurately modulated. In the diffraction pattern one can distinguish clearly between a set of main reflections corresponding to the n–dimensional average structure and the much weaker satellite reflections which correspond to the d–dimensional modulation.

In the case of composite structures there does not exist a unique average structure anymore. In fact every subsystem will be a modulated system with its own periodic average structure and these average structures will be mutually incommensurate. One could choose one of these average structures as periodic basic structure and then apply the usual theory of superspace groups. However, the choice of the periodic average structure is arbitrary and by no means unique. In general, different choices of the periodic average structure will lead to different (and inequivalent) superspace groups. This is of course not acceptable for the crystallographer, since each physical system should be characterized by a unique symmetry group.

In order to remedy this situation we generalize the notion of superspace groups for the needs of composite systems. The important feature is that all subsystems are treated in the same way and no one plays a distinguished role. This generalization will provide us in a natural way with a definition of equivalence of superspace groups for composite systems.

2. Superspace groups for (in)commensurately modulated systems

Since the subsystems of incommensurate composite systems are modulated systems, we recall the ordinary definition of superspace groups here and summarize some of the most important facts. For more details we refer to [2].

Let $V_S = V_E \oplus V_I$ be the $n + d$–dimensional superspace, where V_E and V_I denote the n–dimensional physical space and the d–dimensional internal space, respectively. Let us denote the Euclidean groups of V_S, V_E and V_I by $E(n+d)$, $E(n)$ and $E(d)$, respectively and let $T(n+d)$, $T(n)$ and $T(d)$ be the corresponding subgroups of translations and $O(n+d)$, $O(n)$ and $O(d)$ the corresponding orthogonal groups. Then superspace groups for modulated systems are defined as follows:

Definition 2.1 (Superspace group for modulated systems) *An* (n,d)*–dimensional superspace group* G *is a subgroup of* $E(n) \times E(d)$ *such that*

'S1) $G \cap T(n+d) =: T_\Sigma$ *is an* $(n+d)$*–dimensional lattice group and*

'S2) $G \cap T(d) =: T_D$ *is a* d*–dimensional lattice group.*

Thus superspace groups are $n + d$–dimensional space groups where the distinguished translational subgroup T_D plays a special role. The notion of equivalence of superspace groups has to take into account this special role of T_D and thus the ordinary definition of space group equivalence has to be modified for superspace groups:

Definition 2.2 *Two* (n,d)*–dimensional superspace groups* G *and* G' *are equivalent if and only if there exists a group isomorphism* $\chi : G \to G'$ *such that* $\chi(G \cap T(d)) = G' \cap T(d)$.

As a consequence the point group of a superspace group is a finite subgroup of $O(n) \times O(d)$, furthermore it is a subdirect product of a finite subgroup of $O(n)$ with a finite subgroup of $O(d)$. If we choose a so called standard basis as lattice basis, i.e. a basis $a_i \in \Sigma, i = 1, \ldots, n+d$ such that $a_i \in V_I$ for $i = n+1, \ldots, n+d$, then the rotation matrices take the form

$$\Gamma(R) = \begin{pmatrix} \Gamma_E(R) & 0 \\ \Gamma_M(R) & \Gamma_I(R) \end{pmatrix}, \qquad \Gamma_M(R) = \sigma\Gamma_E(R) - \Gamma_I(R)\sigma, \qquad (1)$$

where $\Gamma_E(R)$, $\Gamma_I(R)$, $\Gamma_M(R)$ are $n \times n$, $d \times d$ and $d \times n$ integer matrices and σ is the (ir)rational connecting matrix. Thus one can assign to each arithmetic class of superspace groups a finite subgroup of $GL(n,d,\mathbb{Z})$, where the latter is defined as the group of all integer matrices of the form (1). In fact there is a one-to-one correspondence between the arithmetic classes of superspace groups and the conjugation classes of finite subgroups of $GL(n,d,\mathbb{Z})$. Moreover any superspace group is equivalent to an extension of $\mathbb{Z}^{n+d}$ by a finite subgroup of $GL(n,d,\mathbb{Z})$ and vice versa.

3. Superspace groups for incommensurate composite systems

If we embed a modulated structure or a composite structure into an $n + d$–dimensional superspace then to each atom in physical space there corresponds a d–dimensional manifold, the so called atomic surface [2]. We assume that all these atomic surfaces are d–dimensional hyperplanes (the subsystems are strictly periodic) or continuous deformations of hyperplanes (the modulations of the subsystems are continuous). Otherwise the atomic surfaces are not unique and it might be impossible to make a distinction between modulated systems and composite systems [3]. This assumption rules out (amongst others) systems with discontinuous modulations and commensurate composite systems. Physically this assumption means that we can distinguish clearly between the different subsystems.

According to our assumption we can assign to each subsystem a d–dimensional subspace V_ν, such that $V_E \oplus V_\nu = V_S$. If we denote the group of translations corresponding to V_ν by $T_\nu(d)$ then we can define a superspace group for incommensurate composite systems:

Definition 3.1 (Superspace group for composite systems, m inequivalent subsystems) *An* (n,d)*–dimensional superspace group* G *is a subgroup of* $E(n) \times E(d)$ *such that*

C1) $G \cap T(n+d) =: T_\Sigma$ *is an* $(n+d)$*–dimensional lattice group,*

(C2) $G \cap T_\nu(d) =: T_{D_\nu}$ *are d–dimensional lattice groups for all* $\nu = 1, \ldots, m$ *and*

(C3) $G \cap T_\nu(d) = T_{D_\nu}$ *are invariant subgroups of G for all* $\nu = 1, \ldots, m$.

Condition (C1) just guarantees that $G \subset E(n+d)$ is a space group, whereas the condition $G \subset E(n) \times E(d)$ implies that the physical space is mapped onto itself under every rotation. The last condition imposes that each subsystem is mapped onto itself under each symmetry operation. This definition is too restrictive in general since it does not allow the possibility of mapping different subsystems on each other. We will call these groups superspace groups of type one. They are applicable if all subsystems consist of different chemical compounds and thus cannot be mapped onto each other by a symmetry transformation. A more general definition is the following one, let us call them superspace groups of type two if not all π_g in (C3') are trivial, otherwise it reduces to definition 3.1:

Definition 3.2 (Superspace group for composite systems, equivalent subsystems) *An* (n, d)*–dimensional superspace group G is a subgroup of* $E(n) \times E(d)$ *such that*

(C1) $G \cap T(n + d) =: T_\Sigma$ *is an* $(n + d)$*–dimensional lattice group,*

(C2) $G \cap T_\nu(d) =: T_{D_\nu}$ *are d–dimensional lattice groups for all* $\nu = 1, \ldots, m$ *and*

(C3') $g T_{D_\nu} g^{-1} = T_{D_{\pi_g(\nu)}}$, *where* π_g *is a permutation of* $\nu = 1, \ldots, m$.

The last condition takes into account that subsystems may be permutated under a symmetry operation. It also assures that main reflections are mapped onto main reflections. Note that any superspace group has a unique maximal normal subgroup of type one, i.e. the subgroup of all symmetry operations that leave all subsystems fixed.

In case of a single subsystem these definitions reduce to the conventional definition of superspace groups. One can then always choose $V_1 = V_I$ and it can be proved that the conditions (C3) and (C3'), respectively are satisfied automatically. Of course a choice $V_1 \neq V_I$ is also possible, which corresponds to a non standard embedding.

Some composite systems may show hidden symmetries (mixing symmetries) that mix the physical space with the internal space [2]. One could take such symmetries into account if one replaces $E(n) \times E(d)$ by $E(n + d)$, but we do not want to pursue this subject here.

In this setting the following definition of equivalence emerges naturally (of course only composite systems with the same number of subsystems can be equivalent):

Definition 3.3 *Two* (n, d)*–dimensional superspace groups G and G' are equivalent if and only if there exists a group isomorphism* $\chi : G \to G'$ *such that* $\chi(G \cap T_\nu(d)) = G' \cap T_{\pi(\nu)}(d)$, *where* π *is an appropriate permutation.*

4. Composite systems with two subsystems

We want to illustrate the consequences of the above definitions for composite systems consisting of two subsystems. In addition we assume that $T_1(d)$ and $T_2(d)$ have only the unit element in common, i.e. the modulation vectors of one system are the reciprocal lattice vectors of the periodic average structure of the other system.

4.1. Inequivalent subsystems

Since we are free to choose any lattice basis we want we could choose a standard basis for one of the two subsystems. In this case the rotation matrices would take the form (1) with matrices $\Gamma_E^\nu(R)$, $\Gamma_I^\nu(R)$, $\Gamma_M^\nu(R)$ and σ^ν, $\nu = 1, 2$. However, such a choice would attribute a distinct position to one of the two subsystems, which would only make sense if one is interested only in the symmetry of this subsystem. However, we can choose a lattice basis such that $a_i \in D_1$

for $i = n - d + 1, \ldots, n$ and $a_i \in D_2$ for $i = n + 1, \ldots, n + d$. With respect to this basis the rotation matrices take the form

$$\Gamma(R) = \begin{pmatrix} \Gamma_0(R) & 0 & 0 \\ \Gamma_{M1}(R) & \Gamma_1(R) & 0 \\ \Gamma_{M2}(R) & 0 & \Gamma_2(R) \end{pmatrix}, \qquad \Gamma_{M\nu}(R) = \sigma_\nu \Gamma_0(R) - \Gamma_\nu(R)\sigma_\nu, \quad (2)$$

where $\Gamma_0(R)$, $\Gamma_{M\nu}(R)$ and $\Gamma_\nu(R)$ are integer matrices. Moreover Γ_1 and Γ_2 are equivalent representations of the point group P. However, note that the corresponding intertwining matrices need not be integer matrices. From this we can infer that the superspace groups of the two subsystems (seen as conventional superspace groups) need not be equivalent. In fact they may belong to different arithmetic classes of superspace groups, but they have to be in the same geometric class (in the conventional sense). One can show that to each arithmetic class of superspace groups there exists a finite group of integer matrices of the form (2). On the other hand, there exists a unique arithmetic class of superspace groups to each finite group of integer matrices of the form (2), if Γ_1 and Γ_2 are equivalent representations and an additional compatibility relation for $\Gamma_{M\nu}$ is satisfied. Thus we can prove a one-to-one correspondence between arithmetic crystal classes and equivalence classes of certain finite groups of integer matrices [3].

4.2. Equivalent subsystems

Let us consider now a superspace group of type II, i.e. there is a symmetry operation that maps the two different subsystems onto each other. This superspace group G contains a normal subgroup H of index 2 which is a superspace group of type I, whose point group elements can thus be written in the form (2). Let us denote the point groups of these superspace groups with P_G and P_H, respectively. Then we can write the rotation matrices as follows

$$\Gamma(R) = \begin{pmatrix} \Gamma_0(R) & 0 \\ \Gamma_M(R) & \Gamma_{12}(R) \end{pmatrix}, \qquad \Gamma_M(R) = \sigma \Gamma_0(R) - \Gamma_{12}(R)\sigma, \quad (3)$$

where all matrices involved are integer matrices (except σ) and the representation Γ_{12} of P_G is obtained from the integer representation Γ_1 of P_H by induction (or alternatively from Γ_2). This representation looks similar to (1), but let us point out an important difference: $\Gamma_E(R)$ is an $n \times n$ matrix whereas $\Gamma_0(R)$ is a $(n - d) \times (n - d)$ matrix. Correspondingly $\Gamma_I(R)$ and $\Gamma_{12}(R)$ have different dimensions, too.

Again one can prove that there is a one-to-one correspondence of the arithmetic classes of superspace groups of type II with certain equivalence classes of finite groups of integer matrices. This can be generalized easily for the case that $T_1(d)$ and $T_2(d)$ have more than the zero translation in common, too. The case of more than two subsystems is more difficult and shall be discussed elsewhere [3].

Acknowledgements

One of the authors (P. Z.) acknowledges financial support by the Austrian Academy of Sciences (APART-program).

References

[1] A. Janner and T. Janssen, Physica **99**, 47–76 (1979)
[2] T. Janssen, Acta Cryst. A **47**, 243–255 (1991)
[3] P. Zeiner, T. Janssen, to be published

Section 2

Strings, Cosmology and Gravitation

Inst. Phys. Conf. Ser. No 173: Section 2
Paper presented at 24th Int. Coll. Group Theoretical Methods in Physics, Paris, France, 15–20 July 2002
©2003 IOP Publishing Ltd

Spacetime algebraic skeleton

R. Aldrovandi and A. L. Barbosa

Instituto de Física Teórica
State University of São Paulo - UNESP
Rua Pamplona 145
01405 - 900 São Paulo Brazil

Abstract.
The cosmological constant is shown to have an algebraic meaning: it is essentially an eigenvalue of a Casimir invariant of the Lorentz group acting on the spaces tangent to every spacetime. This is found in the context of de Sitter spacetimes, for which the Einstein equation is a relation between operators. Nevertheless, the result brings to the foreground the "skeleton" algebraic structure underlying the geometry of general physical spacetimes, which differ from one another by the "fleshening" of that structure by different tetrad fields.

1. Introduction

In a very simplified picture, the Universe has begun as an inflationary de Sitter spacetime which changed to the solution describing the present-day state — a Friedmann solution [1, 2, 3] with a significant de Sitter remnant [4]. A spacetime S is a four-dimensional differentiable manifold whose tangent spaces are copies of the Minkowski space M, the typical fiber in the tangent bundle TS. A tetrad frame–field solders a copy of M at each point $p \in S$ [5]. TS is associated to the bundle BM of frames and S is the quotient of BM by the Lorentz group. For de Sitter spacetimes the bundles BM are Lie groups — just the corresponding de Sitter groups.

A tetrad field is a set of four vectors e_a constituting a local vector basis on S. Their dual forms ω^b, with $\omega^b(e_a) = \delta_a^b$, constitute a covector basis. We shall be using latin letters $(a, b, \ldots = 0, 1, 2, 3)$ to label components on M, greek letters for spacetime indices and $\eta = (\eta_{ab}) = \text{diag}(1, -1, -1, -1)$ for the Lorentz metric. A coordinate system $\{x^\mu\}$ defines a natural pair of vector/covector basis $\{\partial_\mu = \frac{\partial}{\partial x^\mu}, dx^\nu\}$, in terms of which [6]

$$e_a = h_a{}^\mu \partial_\mu \text{ and } \omega^b = h^b{}_\nu dx^\nu, \tag{1}$$

with the conditions $h^b{}_\mu h_a{}^\mu = \delta_a^b$ and $h^a{}_\mu h_a{}^\nu = \delta_\mu^\nu$. Under a Lorentz transformation with parameters α^{cd}, the tetrads change according to

$$h^{a'}{}_\mu(x) = \Lambda^{a'}{}_b(x) \ h^b{}_\mu(x) = \left(exp[\tfrac{1}{2}\alpha^{cd} J_{cd}]\right)^{a'}{}_b \ h^b{}_\mu(x). \tag{2}$$

Each J_{cd} is a 4×4 matrix representing one of the Lorentz group generators:

$$[J_{cd}]^{a'}{}_b = \eta_{db} \ \delta_c^{a'} - \eta_{cb} \ \delta_d^{a'} . \tag{3}$$

A tetrad field transmutes tensors on M into tensors on S. For a vector, for instance, $\phi^\mu(x) = h_a{}^\mu(x)\phi^a$. The Lorentz metric, in particular, is taken into a Riemannian metric,

$$g_{\mu\nu}(x) = \eta_{ab} \ h^a{}_\mu(x) \ h^b{}_\nu(x). \tag{4}$$

The Minkowski indices are contracted, so that $\phi^\mu(x)$ and $g_{\mu\nu}(x)$ are Lorentz-invariant. The role of the Lorentz group is concealed by the transmutations, and remains out of sight in the usual metric formalism. Connections have a special behavior under transmutation. A general Lorentz connection

$$\Gamma = \tfrac{1}{2}J_{ab}\,\Gamma^{ab}{}_\mu\,dx^\mu = \tfrac{1}{2}J_{ab}\Gamma^{ab}{}_c\,\omega^c \tag{5}$$

is changed into the Lorentz-invariant form

$$\Gamma^\lambda{}_{\nu\mu} = h_b{}^\lambda\partial_\mu h^b{}_\nu + h_a{}^\lambda\Gamma^a{}_{b\mu}h^b{}_\nu. \tag{6}$$

2. de Sitter spacetimes

There are two kinds (de Sitter proper, and the anti-de Sitter) of such spacetimes [2], with groups of motions $SO(4,1)$ and $SO(3,2)$. These groups are the principal bundles BM of Lorentz frames on the corresponding spacetimes, which are the quotients $SO(4,1)/SO(3,1)$ and $SO(3,2)/SO(3,1)$ and can be seen as hypersurfaces of pseudo-radii L in the pseudo–Euclidean spaces $\mathbb{E}^{4,1}$ and $\mathbb{E}^{3,2}$ with metrics $\eta = (1,-1,-1,-1,s = \mp 1)$, whose points in Cartesian coordinates $(\xi) = (\xi^0,\xi^1,\xi^2,\xi^3,\xi^4)$ satisfy

$$s\,\eta_{ab}\,\xi^a\xi^b + \left(\xi^4\right)^2 = L^2\ .$$

Introducing $\sigma^2 = \eta_{ab}\,\delta^a{}_\mu\delta^b{}_\nu\,x^\mu x^\nu$ and $n = \tfrac{1}{2}\left(1 - \tfrac{\xi^4}{L}\right) = \tfrac{1}{1+s\sigma^2/4L^2}$, a simple tetrad field $h^a{}_\mu = n\delta^a_\mu$ turns up. The $\{x^\mu\}$'s are stereographic coordinates satisfying $x^\mu = h_a{}^\mu\xi^a$, and the line element on the hypersurfaces is $ds^2 = g_{\mu\nu}\,dx^\mu dx^\nu$, with

$$g_{\mu\nu} = h^a{}_\mu h^b{}_\nu\eta_{ab} = n^2\delta^a_\mu\delta^b_\nu\eta_{ab}\ . \tag{7}$$

This metric leads to the Riemann and Ricci tensors

$$R^\alpha{}_{\beta\rho\sigma} = -\tfrac{s}{L^2}\left[\delta^\alpha_\sigma g_{\beta\rho} - \delta^\alpha_\rho g_{\beta\sigma}\right],\quad R_{\mu\nu} = \tfrac{3s}{L^2}\,g_{\mu\nu}, \tag{8}$$

and to the scalar curvature

$$R = 12\,\tfrac{s}{L^2}\ . \tag{9}$$

If the curvature tends to zero by a Inonü-Wigner contraction $L \to \infty$, both groups reduce to the Poincaré group and both spacetimes to Minkowski space [7, 8].

As the bundles BM coincide with the groups, their tangent vectors constitute just the de Sitter Lie algebras. The structure coefficients reduce to constants, the vertical fields are directly related to the Lorentz generators and the horizontal fields will be the generators $\{T_c\}$ of spacetime translations. Consider the three sets of structure constants:

$$f^{(ef)}{}_{(ab)(cd)} = \eta_{bc}\delta^e{}_a\delta^f{}_d + \eta_{ad}\delta^e{}_b\delta^f{}_c - \eta_{bd}\delta^e{}_a\delta^f{}_c - \eta_{ac}\delta^e{}_b\delta^f{}_d \tag{10}$$

$$f^{(e)}{}_{(ab)(c)} = \eta_{cb}\delta^e{}_a - \eta_{ca}\delta^e{}_b \tag{11}$$

$$f^{(ef)}{}_{(a)(b)} = \left(\delta^f{}_a\delta^e{}_b - \delta^e{}_a\delta^f{}_b\right) = -f^e{}_{(ab)}{}^f\ . \tag{12}$$

The Lie algebras of the de Sitter groups are then given by

$$[J_{cd},J_{ef}] = \tfrac{1}{2}f^{(ab)}{}_{(cd)(ef)}\,J_{ab} \tag{13}$$

$$[J_{cd},T_e] = f^{(a)}{}_{(cd)(e)}\,T_a \tag{14}$$

$$[T_c,T_e] = \tfrac{s}{2L^2}\,f^{(ab)}{}_{(c)(e)}\,J_{ab}\ . \tag{15}$$

The geometry of any spacetime is encapsulated in its bundle of Lorentzian frames. Different spacetimes will differ in Eq.(15), but will keep Eqs.(13–14) in their algebraic skeletons. The direct-product character of BM appears in terms of ∂_μ or $\nabla_\mu = h^a{}_\mu T_a$, for common (Poincaré) or covariant derivatives (de Sitter translations).

The Jacobi identity $[J_{ab}, [J_{cd}, T_e]] + [T_e, [J_{ab}, J_{cd}]] + [J_{cd}, [T_e, J_{ab}]] = 0$ is written

$$f^{(h)}{}_{(ab)(f)} f^{(f)}{}_{(cd)(e)} - f^{(h)}{}_{(cd)(f)} f^{(f)}{}_{(ab)(e)} = \tfrac{1}{2} f^{(f)}{}_{(ab)(cd)} f^{(h)}{}_{(gf)(e)}.$$

From Eq.(13) it is seen that each J_{ab} can be represented by a 4×4 matrix with elements

$$(J_{ab})^c{}_e = f^{(c)}{}_{(ab)(e)}. \tag{16}$$

This corresponds to the representation (3) of the Lorentz group and gives the meaning of Eq.(14): the T_a's are Lorentz vectors. The action of J_{ab} on them is given by a commutation in the left-hand side and by matrix multiplication in the right-hand side.

3. Curvature and algebra

The Jacobi identity $[T_a, [T_b, T_c]] + [T_c, [T_a, T_b]] + [T_b, [T_c, T_a]] = 0$ gives

$$\tfrac{s}{2L^2} f^{(h)}{}_{(ef)[(a)} f^{(ef)}{}_{(b)(c)]} = 0, \tag{17}$$

where $[abc]$ stands for the summation over all cyclic permutations of the included indices. The left factor is a matrix element $(J_{ef})^h{}_a$ as in (16). Each right factor is the component, along J_{ef}, of a matrix $F_{bc} = - \tfrac{s}{2L^2} J_{ef} f^{(ef)}{}_{(b)(c)}$ whose entries $R^c{}_{dab} = (F_{ab})^c{}_d$ are

$$R^c{}_{dab} = - \tfrac{s}{2L^2} f^{(c)}{}_{(ef)(d)} f^{(ef)}{}_{(a)(b)}. \tag{18}$$

This is actually the Riemann curvature tensor seen from a tetrad frame. Adding a connection corresponds to a non-trivial extension [9] of the usual abelian translation algebra, a passage from $s = 0$ to $s \neq 0$ in Eq.(15). We can change the notation $T_a \Rightarrow D_a$, as the "extended" translation is the covariant derivative

$$D_c = e_c + \tfrac{1}{2}\Gamma^{ab}{}_c J_{ab}. \tag{19}$$

The $\Gamma^{ab}{}_c$'s are the connection components in base $\{e_a\}$ and the J_{ab}'s are the generators in the due representation. Applied to a vector field their commutator is, by Eq.(15),

$$[D_a, D_b]V^e = \tfrac{s}{2L^2} f^{(cd)}{}_{(a)(b)} (J_{cd})^e{}_f V^f = \tfrac{s}{2L^2} f^{(cd)}{}_{(a)(b)} f^{(e)}{}_{(cd)(f)} V^f. \tag{20}$$

Comparison of (20) with the standard formula

$$V^c{}_{;a;b} - V^c{}_{;b;a} = R^c{}_{dab} V^d$$

shows that $R^c{}_{dab}$ as introduced in Eq.(18) is indeed the curvature tensor. The Jacobi identity (17) will then be the vanishing–torsion Bianchi identity,

$$R^c{}_{[dab]} = 0. \tag{21}$$

The Bianchi/Jacobi identities are geometric/algebraic versions of the same property. The Ricci tensor is a matrix element:

$$R_{ab} = - \tfrac{s}{2L^2} f_{(b)}{}^{(cd)}{}_{(e)} f^{(e)}{}_{(cd)(a)} = - \tfrac{s}{L^2}(J^2)_{ab} \tag{22}$$

with $J^2 = \tfrac{1}{2} J^{(ab)} J_{(ab)}$ a Casimir operator of the Lorentz group. The scalar curvature is consequently its trace:

$$R = - \tfrac{s}{L^2} \mathrm{tr} J^2. \tag{23}$$

As $[J^2]^a{}_b = -3\delta^a{}_b$ from Eq.(12), we recover indeed the "geometrical" expression of Eq.(9). On the other hand Einstein's equation for the case,

$$R^a{}_b - \tfrac{1}{2}\delta^a{}_b R = \Lambda\,\delta^a{}_b, \tag{24}$$

acquires an algebraic version:

$$-\tfrac{s}{L^2}\left[J^2 - \tfrac{1}{2}(\mathrm{tr}\,J^2)I\right] = \Lambda\,I. \tag{25}$$

We find from Eq.(24) that the cosmological constant is $\Lambda = -R/4$, or

$$\Lambda = \tfrac{s}{4L^2}\mathrm{tr}\,J^2 = \tfrac{s}{L^2} \times \text{the eigenvalue of } J^2 \text{ in the vector representation.} \tag{26}$$

Acknowledgements

The authors thank FAPESP and CNPq, brazilian agencies, for financial support.

References

[1] Weinberg S 1972 Gravitation and Cosmology (New York: J. Wiley)
[2] Hawking S W and Ellis G F R 1973 The Large Scale Structure of Space-Time (Cambridge: Cambridge University Press)
[3] Narlikar J V 1993 Introduction to Cosmology (Cambridge: Cambridge University Press)
[4] Perlmutter S et al 1998 Nature 391 51; Perlmutter S et al 1999 Ap. J. 517 565; Riess A G et al 1998 Astron. J. 116 1009 and 1999 Astron. J. 118 2668. And de Bernardis P et al 2000 Nature 404 955
[5] Aldrovandi R and Pereira J G 1995 An Introduction to Geometrical Physics (Singapore: World Scientific)
[6] Chandrasekhar S 1992 The Mathematical Theory of Black Holes (Oxford: Clarendon Press)
[7] Gürsey F 1962 in Group Theoretical Concepts and Methods in Elementary Particle Physics, Istanbul Summer School of Theoretical Physics, edited by F. Gürsey (New York: Gordon and Breach)
[8] Aldrovandi R and Pereira J G 1986 Phys. Rev. D33 2788; 1988 J. Math. Phys. 29 1472
[9] Aldrovandi R 1991 J. Math. Phys. 32 2503; 1991 Phys. Lett. A155 459; Aldrovandi R and Barbosa A L 2000 Int. J. Theor. Phys. 39 2779

Inst. Phys. Conf. Ser. No 173: Section 2
Paper presented at 24th Int. Coll. Group Theoretical Methods in Physics, Paris, France, 15–20 July 2002
©*2003 IOP Publishing Ltd*

Are we permitted to deform the Poincaré group?

Henri Bacry

CPT, Université de la Méditéranée, Avenue de Luminy, 13900 Marseille , France.

Two preliminary conditions are involved to answer the title question :

1. Is it necessary to look for a deformation ?
2. Any acceptable deformation must necessarily preserve all the successes of the Poincaré group.

The answer to the first interrogation is yes. This is motivated by the fact that the Poincaré group and the Minkowski space suffer defects which have been denounced by all great physicists, namely Bohr, Born, Brillouin, De Broglie, Connes, Dirac, Dresden, Ehrenfest, Einstein, Heisenberg, Infeld, Landau, Lorentz, Pais, Peierls, Penrose, Poincaré, Rosenfeld, Sachs, Sakurai, Schrödinger, Schwinger, Slater, Wheeler, Wightman, Wigner. It would be too long to quote all these persons. We choose to retain only two of them :

> *The real existence of space-time points and the possibility of determining their coordinates is an assumption both in general relativity theory and in quantum mechanics -particularly in the field theories of the latter - but is very questionable in both.* [1].

> *Special relativity is only valid if we ignore the quantum nature of the electron, in particular its spin, and it would be worthwhile to modify the special relativity in taking account of this fact. In other words Maxwell theory would have to be reexamined in order to include a quantum electron. After having made the remark that Minkowski space time is not made of points, are we permitted to introduce point particles ?* (Einstein).

Although history is not our principal concern, we present here a historical table (figure 1) where we only want to underline some facts. First, Einstein has built the Minkowski spacetime using light rays, that is electromagnetic signals of short wavelengths as if he was ignoring the Maxwell theory of light. We note also that he always used the expression of light quanta, instead of electromagnetic quanta. Second, we underline the fact that the conservation laws associated with momenta denote as well the additivity laws which are in reality consequences of the Hopf algebra canonically related with the Poincaré group. Third, we want to remark that the boost transformations have no name, a fact which is intimately related with the difficulties of the Poincaré group, as it can be shown from the quotations of the great physicists we spoke about. We note that all successes of the Poincaré group only concerns angular momenta, energy and linear momenta, not the boosts.

All those facts lead to a search for some d eformation of the Poincaré group which preserve the physical momenta. An interesting deformation is the kappa-Poincaré group of Lukierski, Nowicki and Ruegg [2]. A way of describing it consists in defining the so-called Maślanka transformation

$$\mathbf{J}' = \mathbf{J} \tag{1}$$

$$\mathbf{P}' = \mathbf{P} \tag{2}$$

234

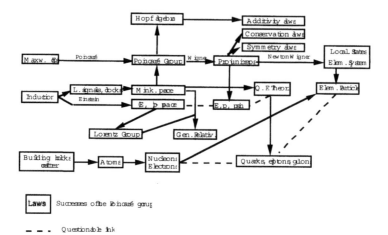

Figure 1. The Poincaré Group and History.

$$f(\mathbf{P}_0') = \frac{1}{\sqrt{g}}\operatorname{argcosh}(1 + g\mathbf{P_0}^2/2) \tag{3}$$

$$\mathbf{K}' = \frac{1}{2}\left[\sqrt{1 + g\mathbf{P_0}^2/4}, \mathbf{K}\right]_+ + \frac{\sqrt{1 + g\mathbf{P_0}^2/4} - \sqrt{1 + g\mathbf{M}^2/4}}{\mathbf{P_0}^2 - \mathbf{M}^2} \tag{4}$$

where $\mathbf{M}^2$ is defined by

$$\mathbf{M}^2 = \mathbf{P_0}^2 - \mathbf{P}^2. \tag{5}$$

Here f denotes an arbitrary function such that $f(x) \sim x$ for small x and g is a constant. The transformation defined by Eqs (1) to (5) is our generalization of Maślanka's map. It reduces to Maślanka's case in setting $f(x) = x$ and $g(x) = 1$.

In the Hopf algebra associated with the kappa-Poincaré group, the new momenta are the prime ones. One of the main inconvenients is the non commutativity of the coproduct. It means that for a system of two nucleons, we must know what is the first nucleon and what is the second one. The only deformed Poincaré group which does not have this ugly property is the Poincaré group itself. That is why we adopt the Maslanka formulae but in keeping the additivity of the old momenta. In such a case, the energy $\mathbf{P}_0'$ is no longer additive and we must imagine a deformation of the theory of gravitation (general relativity) in this direction. Two main consequences are the following

1. All unitary representations of the new interpretation of the Poincaré group become physical. In particular, there is no room for tachyons.

2. One can chose the value of the kappa constant in order to solve the problem of missing matter. In the same way physicists fell to find a missing planet to explain the motion of Mercury, a problem solved by general relativity, the dark matter does not exist in a theory which would replace Einstein's general relativity. Let us show how.

According to Eq. (3), a galaxy of a perfect gas of N stars of mass m would have a mass M given by the formula

$$\sinh(\mathbf{M}/2\kappa) = N\sinh(m/2\kappa). \tag{6}$$

We suppose that the galaxy is rotating and that the angular velocity is $\omega(r)$ at a distance r from the center. Let $n(r)$ be the density of the number of stars. Consider a star of mass m located at a distance R. It is submitted to the gravitational force and the centrifugal force. The first force is given by

$$G\frac{E(R)m}{R^2}, \tag{7}$$

where the energy $E(R)$ is given by the formula

$$\sinh(E(R)/2\kappa) = n(R)\sinh(m/2\kappa). \tag{8}$$

Now, the centrifugal force is given by

$$N(R) = \int_0^R n(r)2\pi r dr. \tag{9}$$

If we suppose that the constant κ is much larger than the mass of a star, we obtain:

$$N(R) = \sinh(m/2\kappa) = \sinh\left(\frac{\omega(r)^2 R^3}{2\kappa G}\right). \tag{10}$$

With some approximation, the "missing mass" $\mu(R)$ is given by the equation

$$\frac{\mu(R)}{N(R)m} \sim \frac{N(R)^2 m^2}{24\kappa^2} \tag{11}$$

which corresponds, for a galaxy of 10^{11} stars of mass $m = 4.10^{30}kg$, to

$$\frac{\mu(R)}{N(R)m} = \frac{2.10^{82}}{3\kappa^2}. \tag{12}$$

The missing mass is known to be large; it follows that the constant κ must be of the order of $10^{41}kg$, that is about 10^{11} the sun mass. A simple calculation shows that the missing mass for the Earth is equal to 10^{-34} times the mass of our planet, a value which is completely negligible.

Conclusion

We must invite the reader to examine these sentences of Dirac :

> One should keep the need for a sound mathematical basis dominating one's search for a new theory. Any physical or philosophical idea that one has must be adjusted to the mathematics; not the other way around.

References

[1] Wigner E.P., Int. Jour. Theor. Phys., **25**, 467 (1986).
[2] Lukierski J., Nowicki A. and Ruegg H., Physics Letters **B 293**, 344 (1992).

Inst. Phys. Conf. Ser. No 173: Section 2
Paper presented at 24th Int. Coll. Group Theoretical Methods in Physics, Paris, France, 15–20 July 2002
©*2003 IOP Publishing Ltd*

O(d+1,d+n+1)–invariant formulation of stationary heterotic string theory

N Barbosa–Cendejas

Escuela de Ciencias Físico–Matemáticas, UMSNH, Morelia, Michoacán, México

A Herrera–Aguilar

Instituto de Física y Matemáticas, UMSNH, Morelia, Michoacán, México

Abstract. We present an $O(d + 1, d + n + 1)$–invariant formulation of the stationary effective action of heterotic string theory which arises after the toroidal compactification of d dimensions. Such a formulation is written in terms of a matrix vector **W** endowed with an $O(d+1, d+n+1)$–invariant scalar product which linearizes the action of the $O(d+1, d+n+1)$ symmetry group on the coset space $O(d + 1, d + n + 1)/[O(d + 1) \times O(d + n + 1)]$. This fact opens a simple solution–generating technique which can be exploted in order to obtain exact solutions of the theory.

1. Introduction

In [1] it was shown that the effective field theory of heterotic string compactified down to three dimensions on a d–torus can be expressed in terms of a pair of matrix potentials in a form which is quite similar to the stationary Einstein–Maxwell (EM) theory in the language of the Ernst potentials [2]. For this reason these matrices were called *Matrix Ernst Potentials* (MEP). The isometries of the corresponding matrix target space were classified and the transformations that linearize the action of the charging symmetry group were pointed out in [3]. Later on, an even more compact formulation of this effective string theory was given in terms of a single rectangular matrix potential which linearizes as well the action of the charging symmetry group [4]. In this report we introduce a third matrix potential in order to define a matrix vector **W**, endowed with an $O(d + 1, d + n + 1)$–invariant scalar product, which linearizes the action of the whole symmetry group $O(d+1, d+n+1)$ on the coset space $O(d + 1, d + n + 1)/[O(d + 1) \times O(d + n + 1)]$ (see [5] for details). Thus, the application of a linear symmetry transformation on simple known solutions allows one to obtain new solutions in the framework of the heterotic string theory.

2. Effective action of heterotic string and matrix Ernst potentials

In the low–energy limit, heterotic string theory leads to an effective field theory which describes gravity coupled non–trivially to a set of matter fields. This effective action in D dimensions reads:

$$S^{(D)} = \int d^{(D)}x \, |G^{(D)}|^{\frac{1}{2}} \, e^{-\phi^{(D)}} (R^{(D)} + \phi^{(D)}_{;M} \phi^{(D);M} -$$

$$\frac{1}{12} H^{(D)}_{MNP} H^{(D)MNP} - \frac{1}{4} F^{(D)I}_{MN} F^{(D)IMN}) \tag{1}$$

where $F_{MN}^{(D)I}=\partial_M A_N^{(D)I}-\partial_N A_M^{(D)I}$ and $H_{MNP}^{(D)}=\partial_M B_{NP}^{(D)}-\frac{1}{2}A_M^{(D)I}F_{NP}^{(D)I}+$cycl. perms. of M, N, P; $G_{MN}^{(D)}$ is the metric, $B_{MN}^{(D)}$ is the anti–symmetric Kalb-Ramond field, $\phi^{(D)}$ is the dilaton and $A_M^{(D)I}$ is a set of $U(1)$ Abelian vector fields ($I=1,2,...,n$). For the critical case one must set $D=10$ and $n=16$, however, for the sake of generality and since our formalism does not depend on the dimensionality of the theory, we will leave the value of these parameters arbitrary. In [6]–[9] it was shown that after the compactification down to three dimensions on a d–torus we get the following fields

a) scalar fields ($p,q=1,2,...,d$)

$$G=\left(G_{pq}=G_{p+3,q+3}^{(D)}\right),\qquad B=\left(B_{pq}=B_{p+3,q+3}^{(D)}\right),$$

$$A=\left(A_p^I=A_{p+3}^{(D)I}\right),\qquad \phi=\phi^{(D)}-\frac{1}{2}\ln|\det G|,\qquad (2)$$

b) tensor fields

$$g_{\mu\nu}=e^{-2\phi}\left(G_{\mu\nu}^{(D)}-G_{p+3,\mu}^{(D)}G_{q+3,\nu}^{(D)}G^{pq}\right),$$

$$B_{\mu\nu}=B_{\mu\nu}^{(D)}-4B_{pq}A_\mu^pA_\nu^q-2\left(A_\mu^pA_\nu^{p+d}-A_\nu^pA_\mu^{p+d}\right),\qquad (3)$$

(we consider $B_{\mu\nu}=0$ in order to remove the effective cosmological constant).

c) vector fields $A_\mu^{(a)}=\left((A_1)_\mu^p,(A_2)_\mu^{p+d},(A_3)_\mu^{2d+I}\right)$ ($a=1,2,...,2d+n$)

$$(A_1)_\mu^p=\frac{1}{2}G^{pq}G_{q+3,\mu}^{(D)},\qquad (A_3)_\mu^{2d+I}=-\frac{1}{2}A_\mu^{(D)I}+A_q^IA_\mu^q,$$

$$(A_2)_\mu^{p+d}=\frac{1}{2}B_{p+3,\mu}^{(D)}-B_{pq}A_\mu^q+\frac{1}{2}A_p^IA_\mu^{2d+I}.\qquad (4)$$

In three dimensions all vector fields can be dualized on–shell:

$$\nabla\times\vec{A_1}=\frac{1}{2}e^{2\phi}G^{-1}\left(\nabla u+(B+\frac{1}{2}AA^T)\nabla v+A\nabla s\right),$$

$$\nabla\times\vec{A_3}=\frac{1}{2}e^{2\phi}(\nabla s+A^T\nabla v)+A^T\nabla\times\vec{A_1},$$

$$\nabla\times\vec{A_2}=\frac{1}{2}e^{2\phi}G\nabla v-(B+\frac{1}{2}AA^T)\nabla\times\vec{A_1}+A\nabla\times\vec{A_3}.\qquad (5)$$

Thus, the resulting three–dimensional theory describes the scalars G, B, A and ϕ and pseudoscalars u, v and s coupled to the metric $g_{\mu\nu}$.

We define two matrix Ernst potentials as follows [3]

$$\mathcal{X}=\begin{pmatrix}-e^{-2\phi}+v^TXv+v^TAs+\frac{1}{2}s^Ts & v^TX-u^T\\ Xv+u+As & X\end{pmatrix},\qquad \mathcal{A}=\begin{pmatrix}s^T+v^TA\\ A\end{pmatrix}\quad (6)$$

where $X=G+B+\frac{1}{2}AA^T$. They are $(d+1)\times(d+1)$ and $(d+1)\times n$ real matrices, respectively, and contain all the information about the matter content of the effective field theory. The physical meaning of these variables is as follows: the relevant information concerning the gravitatoinal field is encoded in X, whereas its rotational character is hidden in u; v is related to multidimensional components of the Kalb–Ramond field, and, finally, A and s stand for electric and magnetic potentials, respectively. In terms of MEP the effective three–dimensional Lagrangian of the theory takes the form

$$^3\mathcal{L}=-^3R+\text{Tr}[\frac{1}{4}\left(\nabla\mathcal{X}-\nabla\mathcal{A}\mathcal{A}^T\right)\mathcal{G}^{-1}\left(\nabla\mathcal{X}^T-\mathcal{A}\nabla\mathcal{A}^T\right)\mathcal{G}^{-1}+\frac{1}{2}\nabla\mathcal{A}^T\mathcal{G}^{-1}\nabla\mathcal{A}],\quad (7)$$

where $\mathcal{G} = \frac{1}{2}\left(\mathcal{X} + \mathcal{X}^T - \mathcal{A}\mathcal{A}^T\right)$.

The form of the Lagrangian (7) is very similar to that of the stationary EM theory. In fact, one can establish the following relationship between both theories

$$\mathcal{X} \longleftrightarrow -\mathcal{E}, \qquad \mathcal{A} \longleftrightarrow \Phi,$$

$$matrix\ transposition \longleftrightarrow complex\ conjugation. \tag{8}$$

where $\mathcal{E}$ and Φ are the complex Ernst potentials of the Einstein–Maxwell theory [2]. The relationship (8) allows one to generalize the results obtained in the EM theory to the heterotic string realm by making use of the MEP formalism. It is worth noticing that in the right hand side we have complex functions, whereas in the left hand side we have real matrices (hence the transposition instead of the complex conjugation) that obey the usual rules of matrix algebra.

3. O(d+1,d+n+1)–invariant formulation of the theory

In this Section we reformulate the matter sector of the three–dimensional effective field theory of the heterotic string (7) by introducing the new matrix variables Z_1 and Z_2 through the relations $\mathcal{X} = 2(Z_1 + \Sigma)^{-1} - \Sigma$ and $\mathcal{A} = \sqrt{2}(Z_1 + \Sigma)^{-1}Z_2$, where Z_1 and Σ are matrices of dimension $d + 1$, Z_2 is a $(d + 1) \times n$–matrix and $\Sigma = diag(-1, -1, 1, 1, ...1)$. Thus, the Lagrangian (7) takes the form

$$^3\mathcal{L} = -^3R + Tr\left(\Theta^{-1}\nabla Z_k Y_{kl} \nabla Z_l^T\right), \tag{9}$$

where the symmetric block–matrix reads

$$Y_{kl} = \begin{pmatrix} \Sigma + \Sigma Z_1^T \Theta^{-1} Z_1 \Sigma & \Sigma Z_1^T \Theta^{-1} Z_2 \\ Z_2^T \Theta^{-1} Z_1 \Sigma & I_n + Z_2^T \Theta^{-1} Z_2 \end{pmatrix},$$

$\Theta = \Sigma - Z_1 \Sigma Z_1^T - Z_2 Z_2^T$, I_n is the unit matrix of dimension n and $k, l = 1, 2$.

One can study the effective action under investigation in terms of other dynamical variables in which the group $O(d + 1, d + n + 1)$ acts linearly on the coset space $O(d + 1, d + n + 1)/O[(d + 1) \times (d + n + 1)]$. In order to achieve this aim, let us introduce the $O(d + 1, d + n + 1)$–matrix vector $\mathbf{W} = (W_1, W_2, W_3) \neq 0$ with components defined by the relations

$$Z_1 \equiv (W_2)^{-1}W_1, \quad Z_2 \equiv (W_2)^{-1}W_3, \tag{10}$$

where W_1 and W_2 are $(d + 1) \times (d + 1)$–matrices and the dimension of W_3 is $(d + 1) \times n$. Let us define as well the $O(d + 1, d + n + 1)$–invariant scalar product in the space of vectors $\mathbf{W}$

$$\begin{aligned}\left(\mathbf{W}, \mathbf{W}^T\right) &\equiv (W_1, W_2, W_3)\,\tilde{\mathcal{L}}\,(W_1, W_2, W_3)^T \\ &= -W_1 \Sigma W_1^T + W_2 \Sigma W_2^T - W_3 W_3^T,\end{aligned} \tag{11}$$

where the matrix $\tilde{\mathcal{L}}$ determines the indefinite signature $\tilde{\mathcal{L}} = diag(-\Sigma, \Sigma, -I_n)$ of the vector space.

In terms of the introduced vector our Lagrangian adopts the form

$$\begin{aligned}^3\mathcal{L} = -^3R - Tr&\left\{\left(\mathbf{W}, \mathbf{W}^T\right)^{-1}\left[\left(\nabla\mathbf{W}, \nabla\mathbf{W}^T\right)\right.\right. \\ &\left.\left. - \left(\nabla\mathbf{W}, \mathbf{W}^T\right)\left(\mathbf{W}, \mathbf{W}^T\right)^{-1}\left(\mathbf{W}, \nabla\mathbf{W}^T\right)\right]\right\},\end{aligned} \tag{12}$$

the corresponding equation of motion is

$$\nabla^2\mathbf{W} - 2\left(\mathbf{W}, \nabla\mathbf{W}^T\right)\left(\mathbf{W}, \mathbf{W}^T\right)^{-1}\nabla\mathbf{W} = 0, \tag{13}$$

which is nothing else that a matrix vector generalization of the Ernst equation for $\mathbf{W}$. Since we have introduced one more dynamical variable (we have now W_1, W_2, W_3 instead of Z_1 and Z_2), our field system is overdefined and we must ensure that the representation in terms of $\mathbf{W}$ is consistent with that in terms of Z_1 and Z_2. A consistency check consists of deriving the equations of motion in the Z_1, Z_2 language from the equation of motion for the matrix vector $\mathbf{W}$ making use of the definitions (10). It is straightforward to prove that this is indeed the case.

This formulation of the theory and its equation of motion is explicitly $O(d+1, d+n+1)$–invariant and is a direct generalization of the representation given in [10] and [11] in the framework of the stationary EM theory. The realization of the linear action of the $O(d+1, d+n+1)$ symmetry group on the coset space $O(d+1, d+n+1)/O[(d+1) \times (d+n+1)]$ is reached by means of the matrix transformation

$$\mathbf{W}' = \mathbf{W}U \tag{14}$$

where the matrix U satisfies the following condition

$$U\tilde{\mathcal{L}}U^T = \tilde{\mathcal{L}}, \tag{15}$$

i.e., U belongs to the $O(d+1, d+n+1)$ symmetry group.

4. Conclusion and discussion

We have presented a formulation of the toroidally compactified stationary heterotic string theory in terms of a matrix vector $\mathbf{W}$ which linearizes the action of the $O(d+1, d+n+1)$ symmetry group on the coset space $O(d+1, d+n+1)/[O(d+1) \times O(d+n+1)]$ and can be exploted for generating new solutions on the basis of known ones; namely, by applying the linear transformation (14) on a simple known solution one gets more complicated field configurations. Another interesting issue is the investigation of the full symmetry group of the theory expressed in terms of the matrix vector $\mathbf{W}$ since it introduces one more matrix dynamical variable in the formalism.

5. Acknowledgments

AHA is really grateful to S. Kousidou for encouraging him during the performance of this work. Both authors were supported by grants CONACYT-J34245-E and CIC-UMSNH-4.18.

References

[1] Herrera-Aguilar A and Kechkin O 1998 Int. J. Mod. Phys. A13 393–402.
[2] Ernst F J 1968 Phys. Rev. 168 1415–1417.
[3] Herrera-Aguilar A and Kechkin O 1999 Phys. Rev. D59 124006.
[4] Kechkin O 2002 Phys. Rev. D65 066006.
[5] Barbosa-Cendejas N and Herrera-Aguilar A 2003 hep–th/0202006; to appear in Gen. Rel. Grav. 35.
[6] Marcus N and Schwarz J H 1983 Nucl. Phys. B228 145–168.
[7] Hassan S F and Sen A 1992 Nucl. Phys. B375 103–118.
[8] Maharana J and Schwarz J H 1993 Nucl. Phys. B390 3–32.
[9] Sen A 1995 Nucl. Phys. B434 179–209.
[10] Kinnersley W 1977 J. Math. Phys. 18 1529–1537.
[11] Mazur P O 1983 Acta Phys. Pol. 14 219–234.

Inst. Phys. Conf. Ser. No 173: Section 2
Paper presented at 24th Int. Coll. Group Theoretical Methods in Physics, Paris, France, 15–20 July 2002
©*2003 IOP Publishing Ltd*

Metric-polynomial structures and gravitational Lagrangians

A Borowiec‡

Institute of Theoretical Physics, Wrocław University, Poland

Abstract. The Euler-Lagrange equations of motion for the Ricci type of gravitational Lagrangians are calculated by means of Palatini variational principle. It is shown that polynomial structures with compatible Einstein metrics are extremals of this problem.

1. Introduction

A *polynomial structure* on an n-dimensional differentiable manifold M is given by type $(1,1)$ tensor field $K \equiv K_\nu^\mu$, which satisfies polynomial equation $\pi(K) = 0$ for some polynomial $\pi(t) \doteq \sum_{i=0}^m a_i t^i$ of (constant) real coefficients $(a_i)_{i=0}^m$, $m \leq n$. Almost-complex $(\pi(t) = t^2 + 1)$ and almost-product $(\pi(t) = t^2 - 1)$ structures belong to the best known and the most fundamental examples of this kind. It has been recently shown that both of them appeared in a natural way from the first-order (Palatini) variational principle applied to a general class of non-linear Lagrangians depending on the Ricci squared invariant constructed out of a metric and a symmetric connection [1]. Moreover, Einstein equations of motion and Komar energy-momentum complex are *universal* for this class of Lagrangians [2].

In the present note, we are going to extend above results showing that more general Ricci type Lagrangians lead to more general polynomial structures and that the universality property remains still valid; both for the equations as for the energy-momentum. An alternative approach without polynomial relations has been recently proposed in [11].

1.1. Preliminaries and Notation

Let (M, g, Γ) be a n-dimensional (pseudo-) Riemannian manifold (M, g) equipped with a symmetric (i.e. torsion free) connection Γ. In the sequel we shall use lower case letters $r_{\beta\mu\nu}^\alpha$ and $r_{\beta\nu} = r_{\beta\alpha\nu}^\alpha$ to denote the Riemann and Ricci tensors of an arbitrary connection Γ

$$r_{\beta\mu\nu}^\alpha = r_{\beta\mu\nu}^\alpha(\Gamma) = \partial_\mu \Gamma_{\beta\nu}^\alpha - \partial_\nu \Gamma_{\beta\mu}^\alpha + \Gamma_{\sigma\mu}^\alpha \Gamma_{\beta\nu}^\sigma - \Gamma_{\sigma\nu}^\alpha \Gamma_{\beta\mu}^\sigma$$
$$r_{\mu\nu} = r_{\mu\nu}(\Gamma) = r_{\mu\alpha\nu}^\alpha \tag{1}$$

i.e. without assuming that Γ is the Levi-Civita connection of g.

Define a $(1, 1)$ tensor valued concomitant

$$S_\nu^\mu \equiv S_\nu^\mu(g, \Gamma) = g^{\mu\lambda} r_{(\lambda\nu)}(\Gamma) \tag{2}$$

of a metric g and a linear (torsionless) connection Γ. Here, $r_{(\lambda\nu)}$ stands for the symmetric part of $r_{\lambda\nu}$. One can further define a family of scalar concomitants of the Ricci type

$$s_k = \mathrm{tr} S^k \tag{3}$$

for $k = 1, \ldots n = \dim M$.

‡ This work was supported by Polish KBN # 2 P03B 144 19

Our goal in the present note is to apply a Palatini variational principle to the most general family of non-linear gravitational Lagrangians of the Ricci type

$$L_F(g, \Gamma) \doteq \sqrt{g}\, F(s_1, \ldots, s_n) \tag{4}$$

parameterized by a smooth real-valued function F of n-variables. Some special cases have been previously investigated in [1, 2, 3, 9].

2. Equation of Motion

In order to apply the so called "Palatini variational principle" (cf. [8, 6, 3]), we have to choose a metric g and a symmetric connection Γ as independent dynamical variables. Variation of L_F gives

$$\delta L_F = \sqrt{g}\left((\delta_g F)_{\alpha\beta} - \frac{1}{2} F g_{\alpha\beta}\right) \delta g^{\alpha\beta} + \sqrt{g}\, \delta_\Gamma F \tag{5}$$

where obviously $\delta F = \sum_{k=1}^{n} F'_k\, \delta s_k$, and $F'_k = \frac{\partial F}{\partial s_k}$. We see at once that

$$\delta_g s_k = k \operatorname{tr}(S^{k-1} \delta_g S) = k\, (S^{k-1})^\sigma_\alpha\, r_{(\beta\sigma)}\, \delta g^{\alpha\beta}$$

which is clear from $\delta s_k = k \operatorname{tr}(S^{k-1} \delta S)$. Accordingly

$$\delta_g F = \mathcal{F}^\sigma_\alpha\, r_{(\beta\sigma)} \delta g^{\alpha\beta}$$

where for simplicity we have introduced a $(1,1)$ tensor field concomitant

$$\mathcal{F} \doteq \sum_{k=1}^{n} k\, F'_k\, S^{k-1} \tag{6}$$

In a similar manner one calculates

$$\delta_\Gamma F = \mathcal{F}^\alpha_\sigma\, g^{\sigma\beta}\, \delta r_{(\alpha\beta)} \equiv \mathcal{F}^{\alpha\beta}\, \delta r_{(\alpha\beta)} \tag{7}$$

where the inverse metric g^{-1} has been used for rising the lower index in $\mathcal{F}$. Substituting all necessary terms into (5) gives then

$$\delta L_F = \sqrt{g}\left(\mathcal{F}^\sigma_\alpha\, r_{(\beta\sigma)} - \frac{1}{2} F g_{\alpha\beta}\right) \delta g^{\alpha\beta} - \sqrt{g}\, \mathcal{F}^{\alpha\beta}\, \delta r_{(\alpha\beta)} \tag{8}$$

Now, taking into account that $\delta r_{(\alpha\beta)} = \nabla_\mu \delta\Gamma^\mu_{\alpha\beta} - \nabla_{(\alpha} \delta\Gamma^\sigma_{\beta)\sigma}$, where ∇_α denotes the covariant derivative with respect to Γ and performing the "covariant" Leibniz rule one gets the variational decomposition formula under the form

$$\delta L_F = \sqrt{g}\left(\mathcal{F}^\sigma_\alpha\, r_{(\beta\sigma)} - \frac{1}{2} F g_{\alpha\beta}\right) \delta g^{\alpha\beta} - \nabla_\nu \left[\sqrt{g}\left(\mathcal{F}^{\alpha\beta} \delta^\nu_\lambda \right.\right.$$
$$\left.\left. - \mathcal{F}^{\nu\alpha} \delta^\beta_\lambda\right)\right] \delta\Gamma^\lambda_{\alpha\beta} + \partial_\mu \left[\sqrt{g}\, \mathcal{F}^{\alpha\beta}\, (\delta\Gamma^\mu_{\alpha\beta} - \delta^\mu_{(\beta} \delta\Gamma^\sigma_{\alpha)\sigma})\right] \tag{9}$$

This splits δL_F into the Euler-Lagrange part and the boundary term which scan be used for a construction of conserved currents (see e.g. [2, 10]). Therefore, the Euler-Lagrange field equations read as follows

$$\mathcal{F}^\sigma_{(\alpha}\, r_{(\beta)\sigma)} - \frac{1}{2} F\, g_{\alpha\beta} = 0 \tag{10}$$

$$\nabla_\nu \left[\sqrt{g}\, (\mathcal{F}^{(\alpha\beta)} \delta^\nu_\lambda - \mathcal{F}^{\nu(\alpha} \delta^{\beta)}_\lambda)\right] = 0 \tag{11}$$

(Compare for purely metric formalism presented in [4].)

Before proceeding further, it is convenient to introduce a $(0, 2)$ symmetric tensor field

$$h_{\alpha\beta} = r_{(\alpha\beta)}(\Gamma) \tag{12}$$

(a metric) which will be extremely useful for studying the symmetry properties of $\mathcal{F}$. For this purpose we shall employ a matrix notation. One has: $S = g^{-1} h$ with both g and h being symmetric matrices (c.f. equation (2)). It easily implies that $h S^k = g S^{k+1}$ and $S^k g^{-1} = S^{k+1} h^{-1}$ (provided that h^{-1} exists) are also symmetric matrices for arbitrary $k = 0, 1, \ldots$. Indeed since e.g. $h S^k = h g^{-1} \ldots g^{-1} h$ then it is self-transpose. In particular, $h \mathcal{F}$ in (10) and $\mathcal{F} g^{-1}$ in (11) (c.f. (8) and (15) below) are symmetric. In other words e.g., the matrix concomitant

$$\mathcal{F}^{\alpha\beta} \equiv \mathcal{F}^\alpha_\sigma g^{\sigma\beta} \tag{13}$$

is symmetric. These properties allow us to transform the Euler-Lagrange equations (10-11) into the form

$$S\mathcal{F} = \frac{1}{2} F I \tag{14}$$

$$\nabla_\nu \left(\sqrt{g}\, \mathcal{F}^{\alpha\beta} \right) = 0 \tag{15}$$

where I is a $n \times n$ identity matrix.

Equations (14) must be considered together with a condition obtained by taking the trace of (14). This gives

$$\sum_{k=1}^{n} k F'_k s_k = \frac{n}{2} F \tag{16}$$

The last equation (except the case it is identically satisfied) becomes a single (non-algebraic in general) equation on possible values of the Ricci scalars (remember that F and F'_k are given functions of the variables $(s_1, \ldots, s_n)$). Now, by using the characteristic equation techniques (see [11]), one is allowed to introduce a complementary system of $(n-1)$-equations that additionally relate values of the Ricci scalars. Thus, instead of the single equation (16) we may have at our disposal a system of n-equations with n-unknowns that provides us, in a regular case, in a set of numerical (i.e. constant) solutions $(s_1 = c_1, \ldots, s_n = c_n)$. (But this rather technical point will be consider in more details elsewhere [5]). Substituting back these constant roots into equation (16) we obtain a polynomial equation for the matrix S. In fact, with any set $c_1, \ldots, c_n$ of the (numerical) solutions of (16), one can associate a polynomial

$$\pi_{c_1, \ldots, c_n}(t) = \sum_{k=1}^{n} a_k t^{k-1} \tag{17}$$

with constant coefficients $a_k \doteq k \frac{\partial F}{\partial s_k}(c_1, \ldots, c_n)$, $k = 1 \ldots n$. As a consequence, (14) takes the form of polynomial equation for S

$$S\, \pi_{c_1, \ldots, c_n}(S) - \frac{n}{2} F(c_1, \ldots, c_n) I = 0 \tag{18}$$

which becomes now a substitute of (10). In this way S turns into a polynomial structure on M. Moreover, since $h = gS$ is already symmetric (c.f. (12)), this yields a compatibility condition between the initial metric g and the polynomial structure S under the form

$$g(SX, Y) = g(X, SY) \tag{19}$$

for any vector fields X, Y on M. In other words a pair (S, g) defines a *metric-compatible* polynomial structure with a nonstandard compatibility condition (19) (see e.g. [7] for typical compatibility condition).

244

3. Einstein metrics with compatible polynomial structures

From now on we assume that $S \doteq g^{-1}h$ is an invertible matrix (non-degenerate case) satisfying the polynomial equation (18) determined by some numerical solution $(c_1, \ldots, c_n)$ of (16). In particular, (18) implies that the determinant of S is a constant (as being functionally dependent of $c_1, \ldots, c_n$). Consequently, the determinant of g is up to a constant factor proportional to that of h, i.e. $\det g \sim \det h$. Invertibility of S is ensured by the condition $F(c_1, \ldots, c_n) \neq 0$. One has, of course, $S^{-1} \sim \pi_{c_1, \ldots, c_n}(S)$. Now, replacing $\det g$ in (15) by $\det h$ and making use of the ansätz (12) with $h^{-1} \sim \pi_{c_1, \ldots, c_n}(S) g^{-1}$ (c.f. (13)), gives

$$\nabla_\lambda (\sqrt{h} h^{\alpha\beta}) = 0$$

with $h^{\alpha\beta}$ being the inverse of $h_{\alpha\beta}$. This, in turn, in any dimension $n > 2$ (see [9, 2] for $n = 2$ case), forces Γ to be the Levi-Civita connection of h. Replacing back into (12) we find

$$h_{\mu\nu} = r_{(\mu\nu)}(\Gamma_{LC}(h)) \doteq R_{\mu\nu}(h) \tag{20}$$

the Einstein equations for the metric $h\S$. Finally, one can conclude, that a solution of our Palatini type variational problem, for Lagrangian (4), consists of two metrics (g, h) such that: the metric h is an Einstein metric, the $(1, 1)$ tensor field $S = g^{-1}h$ satisfies the metric compatibility condition (19) together with the polynomial equation (18) determined by some numerical solution of the *master equation* (16).

In particular, for f being a function of one variable the Lagrangian $L_f = \sqrt{g} f(s_1)$ reconstructs the Einstein theory [9, 3]. For $L_f = \sqrt{g} f(s_2)$, besides the Einstein equation, one gets a pseudo-Riemannian almost product structure ($S^2 = I$) and/or an almost-complex anti-Hermitian structure ($S^2 = -I$) [2, 1]. Some other examples will be considered in [5].

This shows that the use of Palatini formalism leads to results essentially different from the metric formulation when one deals with non-linear Ricci type Lagrangians: with the exception of special ("non-generic") cases we always obtain the Einstein equations as gravitational field equations. In this sense non-linear theories are equivalent to General Relativity at the classical level but their quantum contents and divergences could be slightly improved.

References

[1] Borowiec A Ferraris M Francaviglia M and Volovich I 1999 J.Math. Phys. 40 3446-3464
[2] Borowiec A Ferraris M Francaviglia M and Volovich I 1998 Class. Quantum Grav. 15 43-55
[3] Borowiec A and Francaviglia M 1998 Alternative Lagrangians for Einstein metrics Current Topics in Mathematical Cosmology Proc. Int. Sem. Math. Cosmol. Potsdam 1998 M. Rainer and H.-J. Schmidt, eds. (Singapore: WSPC) 361-368
[4] Borowiec A Francaviglia M and Smirichinski V 2000 Fourth-order Ricci gravity http://xxx.lanl.gov/gr-qc/0011103 ICGTM Group23 Dubna 2000 in print
[5] Borowiec A 2002 Einstein metrics with polynomial structures – in preparation
[6] Buchdahl H A 1979 J. Phys. A: Math. Gen. 12 1229
[7] Bureš J and Vanžura 1976 Kodai. Math. Sem. Rep. 27 345-352
[8] Ferraris M Francaviglia M and Reina C 1982 Gen. Rel. Grav. 14 243
[9] Ferraris M Francaviglia M and Volovich I 1997 Int. J. Mod. Phys. A12 5067
[10] Jakubiec A and Kijowski J 1989 J. Math. Phys. 30 1073
[11] Tapia V and Ujevic M 1998 Class. Quantum Grav. 15 3719

$\S$ Here a value of the cosmological constant is 1 due to the "unphysical" normalization made in (12) [5].

Inst. Phys. Conf. Ser. No 173: Section 2
Paper presented at 24th Int. Coll. Group Theoretical Methods in Physics, Paris, France, 15–20 July 2002
©2003 IOP Publishing Ltd

Octonions and M-theory

Luis J. Boya§

Departamento de Física Teórica. Facultad de Ciencias.
Universidad de Zaragoza, E-50009, Zaragoza, Spain.
email: luisjo@posta.unizar.es

Abstract. We explain how structures related to octonions are obiquitous in M-theory. All the exceptional Lie groups, and the projective Cayley line and plane appear in M-theory. Exceptional G_2-holonomy manifolds show up as compactifiyng spaces, and are related to the $M2$ Brane and 3-form. We review this evidence, which comes from the initial 11-dim structures. Relations between these objects are stressed, when extant and understood. We argue for the necessity of a better understanding of the role of the octonions themselves (in particular non-associativity) in M-theory.

PACS numbers: 02.40.Ky, 03.65.Fd

1. Introduction

If the current M-theory is a *unique theory*, one should expect it to make use of singular, non-generic mathematical structures. Now it is known that many of the special objects in mathematics are related to octonions [1], and therefore it is not surprising that this putative *theory-of-everything* should display geometric and algebraic structures derived from this unique non-associative division algebra.

Special algebraic objects related to octonions are the five exceptional simple Lie groups G_2, F_4, E_6, E_7, and E_8. Some sporadic (finite simple) groups seem also related to octonions (see e.g. Thomson's $E_2(3)$ [2]).The use of them in future directions of the theory is not to be discarded. The Cayley plane and its predecessor the octonionic projective line on one hand, and the geometries associated to the Magic Square [3] on the other, stand as fundamental octonionic geometries.

Very recently, the newly discovered (1996) compact manifolds of G_2 holonomy (a case of Joyce manifolds [4]) might play a fundamental role also in M-theory.

Also, the four coincidences in the list of simple Lie groups

$$A_1 = B_1 = C_1 \qquad B_2 = C_2 \qquad A_3 = D_3 \qquad D_2 = A_1 \times A_1$$
$$SU(2) = \mathrm{Spin}(3) = Sp(1) \quad \mathrm{Spin}(5) = Sp(2) \quad SU(4) = \mathrm{Spin}(6) \quad \mathrm{Spin}(4) = SU(2) \times SU$$

are related to some irreps. of the exceptional groups (Adams [5]); they also appear in physics and in compactifications in M-theory disguished in various forms [6].

What we lack at the moment is an understanding of the role of the octonions themselves in M-theory; for a previous discussion of the history of the role of octonions in physics see [7]

§ Presented at the the 24th International Colloquium on Group-Theoretical Methods in Physics. Paris, July 15-22, 2.002

2. Projective lines and planes

The first apparition of octonions in M-theory is as the fourth ladder in the "Brane Scan" of Townsend (1987) [8]. The four lists of classical supersymmetric p-Branes including instantons, embedded in D-dimensional space(-time) correspond precisely to the four division algebras $\mathbb{R}, \mathbb{C}, \mathbb{H}, \mathbb{O}$.

Real ladder: From $D = 1\ p = -1$ instanton (kink) to $D = 4\ p = 2$ membrane (domain wall; codimension one)

Complex ladder: From $D = 2\ p = -1$ instanton (vortex) to $D = 6\ p = 3$ "universe" (codimension two)

$\mathbb{H}$: Quaternionic ladder: From $D = 4\ p = -1$ instanton to $D = 10\ p = 5$ brane (codimension four)

Octonionic ladder: From $D = 8\ p = -1$, Fubini-Nicolai instanton [9] to $D = 11\ p = 2$ $M2$ membrane

(codimension eight)

Table I.- The Brane Scan (Townsend [8])

These four ladders can be thought of as "oxidation" of the corresponding instanton, in its turn associated to the fundamental line bundle for the projective spaces $\mathbb{KP}^1$, $\mathbb{K} = \mathbb{R}, \mathbb{C}, \mathbb{H}, \mathbb{O}$ [10]. The four are supersymmetrizable and are linked as the four Hopf bundles

Group

$\alpha:$ $S^0 \longrightarrow S^1 \longrightarrow S^1 = \mathbb{RP}^1$ $O(1)$

$\beta:$ $S^1 \longrightarrow S^3 \longrightarrow S^2 = \mathbb{CP}^1$ $U(1)$

$\gamma:$ $S^3 \longrightarrow S^7 \longrightarrow S^4 = \mathbb{HP}^1$ $Sp(1)$

$\delta:$ $S^7 \longrightarrow S^{15} \longrightarrow S^8 = \mathbb{OP}^1$ $\mathrm{Spin}(8)$

Table II.- Elementary Solitons and Projective Lines [10]

The relation of the series with the numbers $\mathbb{R} \ldots, \mathbb{O}$ can be made more precise [11], although the octonion case is more oscure [12]. The elementary objects (e.g. strings for $p = 1$) are "thin" (i.e., strictly p space dimensions) but the soliton has a width; as emphasized by Townsend ([8], [13]) the elementary "thin" membrane has a continuous excitation spectrum, but the membrane really grows a "core" due to gravitation.

Notice in Table I the $\mathbb{O}$-series appears as a "second coming" of the Real ladder: both end up in a membrane. Quantization seems to select only the $\mathbb{O}$ ladder.

The relation of the Cayley plane $\mathbb{OP}^2$ (R. Moufang, 1933) to the 11 D Sugra "corner" of M-theory is more recent: it comes from a nice paper of Ramond [14]. The Cayley plane is the 16 D rank one symmetric space (compact form)

$$F_4/B_4 : 1 \to \mathrm{Spin}(9) \to F_{4(-52)} \to \mathbb{OP}^2 \to 1 \qquad 52 = 36 + 26$$

Now the ratio ρ of the orders of the Weyl's groups is 3, as $\#$ Weyl(F_4)=1152 and Weyl(B_4)= $\mathbb{Z}_2{}^4 \odot S_4$, of order 384. But when a pair H, $H \subset G$ of semisimple Lie groups have the same rank, an important construction of Konstant *et al.* [15] generates, for each irrep of G, ρ irreps of H, where $\rho = [\text{Weyl}(G) : \text{Weyl}(H)]$ ($= 3$ in our case). In particular the Id(-entity) irrep of F_4 generates

$$\text{Id}(_F 4) \to +44(\text{graviton}) - 128(\text{gravitino}) + 84(3 - \text{form}) \quad \text{of Spin}(9)$$

i.e. precisely the Spin(9) content of 11 D Sugra, where Spin(9) is the little group in the light cone!

3. The E-series

The evidence for the $E_1 - E_{10}$ series in the descent of Sugra 11 D down to $D = 1$ is well-known, and first stated by Julia in 1982 [16]. Compactifying 11 D Sugra from the original 11 D to 3 D, all of the E series of *split* forms appear succesively; the moduli spaces of scalar fields are the homogeneous spaces. We recall only the non-compact/compact scalars:

		scalars
5 D	$E(6, +6)/Sp(4)$	42
4 D	$E(7, +7)/SU(8)$	35+35
3 D	$E(8, +8)/SO(16)$	128

On the other hand, the Heterotic Exceptional string, of course, makes an important and direct use of $E_8 \times E_8$, and its descent from 11 D M-theory has been clarified in the fundamental work of Witten and Horawa [17], through compactification in a segment. Finally, we recall that E_6, which appears naturally in some H-E string compactications, is a strong candidate for Grannd Unified Theories.

4. Manifolds of G_2 holonomy

In the "old" superstring compactification 10 $D \to 4$ D, Ricci-flat Calabi-Yau 3-folds were the objects of course. In M-theory with 11 D, their place is taken by manifolds with exceptional holonomy (of the Berger 1955 list [18]). In particular, G_2 compact holonomy manifolds are still Ricci flat and conserve 32/8=4 supercharges, that is, $N = 1$ Susy in 4 D as we want. These beasts are fairly new even for the mathematicians (Joyce manifolds, 1996 [4]). From the reduction of the tangent bundle of a compatifying space K_7, M-theory(11)$\to K_7 +$ Minkowski

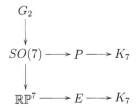

we see that G_2 holonomy involves some torsion properties in K_7, a hot topic today. The

248

structure is given in two diagrams

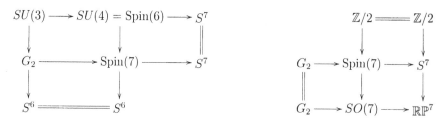

They come from the curious fact that G_2 is subgroup of Spin(7) (Adams [5]) and of $SO(7)$ (from the 7 irrep). For a review of the involved physics see [19]. The reason for the descent $SO(7)$ to G_2 is because the later conserves a 3-form, related to the octonionic product [20], and very likely to the 3-form extant in M-theory in the 11 D Sugra limit (see before). This is the only rationale, so far, for the presence of the octonions *themselves* in M-theory; we hope these matters will be clarified soon.

For alternative and/or complementary views of the items exposed here see [21], [22].

References

[1] J. Stillwell, Amer. Math. Monthly **105** (1998) 850–858

[2] D. Gorestein, *Finite Groups*, Harper & Row, 1968

[3] B. A. Rosenthal, *Geometry of Lie Groups*, Kluwer, 1997

[4] D. Joyce, *Manifolds with Exceptional Holonomy*, Oxford, 2000

[5] J. Adams, *Lectures on Exceptional Groups*, U. of Chicago, 1996

[6] D. Freed, *5 Lectures on Supersymmetry*, AMS, 1999

[7] F. Gürsey, *Symmetries in Physics*, World. Sci., 1987, p. 557

[8] P. Townsend *et al.*, Phys. Lett. **B 198** (1987) 441

[9] S. Fubini and H. Nicolai, Phys. Lett. **B 200** (1988) 301

[10] L.J. Boya, Lect. Not. Phys. **135** (1980) 265-273

[11] G. Sierra, Class. Q. Grav. **4** (1987) 227-236

[12] J.D. Stasheff *et al.*, Com. Math. Phys. **96** (1984) 431-437

[13] P.K.Townsend *et al.*, Phys. Lett. **B 189** (1987) 75

[14] T. Pengpan and P. Ramond, Phys. Rep. **315** (1999) 13; P. Ramond, Algebraic Dreams, hep-th 01-12-261.

[15] B. Gross *et al.*, Proc. Nat. Acad. Scien, 8441 (1998)

[16] E. Cremmer and B. Julia, Nucl. Phys. **B 139** (1979) 141

[17] P. Horawa, E. Witten, Nucl. Phys. **B 460** (1996) 506-524

[18] See e.g. A.L. Bessis, *Einstein Manifolds*, Springer 1985

[19] M. Duff, *M-theory on manifolds of G_2 holonomy*, hepth 02-01-062. See also the reprint volume by him *The World in 11 Dimensions*, I.O.P. Bristol, 1999

[20] N. Hitchin, *Stable forms and special metrics*, math.DG/01-07-101 (13-VII-01)

[21] J. Lukierski and F. Toppan, *Generalized Space-time Supersymmetries, Division Algebras and Octonionic M-theory*, hep-th/0203149

[22] P. Ramond, *Exceptional Structures in Physics*, These Proceedings

Inst. Phys. Conf. Ser. No 173: Section 2
Paper presented at 24th Int. Coll. Group Theoretical Methods in Physics, Paris, France, 15–20 July 2002
©2003 IOP Publishing Ltd

Symmetry and invariance in Leibniz's approach of natural phenomena. Application to dynamics

Naoum Daher

LPMO-CNRS 32, avenue de l'Observatoire F 25044 BESANCON Cedex

Abstract. In his critics of the local characters of Cartesian and Newtonian physics, Leibniz proposed some global symmetry and invariance principles in view of constructing what he called a "universal characteristic". When applied to mechanics, the Leibnizian formulation turns out to include Newtonian and Einsteinian dynamics. This formulation, in which energy is related to a geometrical ratio, is expressed through tree-like structures where Newtonian dynamics occupies the trunk while Einsteinian one occupies some of the branches. A second formulation is proposed where Einsteinian and Newtonian dynamics may be deduced from a unique second order differential equation. Both methods are complementary but they suggest very different interpretations and may be applied to other contexts.

1. Introduction

The two basic principles postulated in this work follow a line of thought developed by Leibniz against Cartesian and Newtonian visions of motion. These are called : "Leibnizian relativity principle" and "Finiteness principle".

Leibnizian relativity principle

Contrary to energy and impulse which are conservative quantities, nothing prevents the notion of velocity from being parametrized in different manners each of which constitute one point of view since it is not constrained by any conservation requirement. In addition, noting the parabolic character of Newtonian dynamics, when energy is expressed in terms of impulse (a possible world), one realizes that any regular even function leads to a parabolic one at first order. These mathematical properties extensively studied by Leibniz constitute the cornerstone of Leibniz's philosophy of multiplicity of points of view on possible worlds. Here, one shows that there exists an infinite number of possible worlds that yield locally the Newtonian world. In order to satisfy Leibniz's requirement, one should replace the unique Newtonian velocity concept v by the doubly multiple one v_k^K where k and K denote the two multiplicities associated with points of view and possible worlds. Contrary to the points of view that co-exist in each world, the different worlds exclude each other so that one should finally retain only the one that Leibniz calls "The best of all possible worlds". All these considerations are mathematically developed in the first section. In particular, the passage from v to v_k^K for each possible world (K : fixed) leads to an under determined mathematical problem. Geometrically, v_k^K corresponds to a family of infinite curves which locally coincide with v following a unique tangent (each world corresponds to a tree-like structure where the Newtonian velocity v occupies the trunk). In order to get a well-determined problem, one should postulate a link between the different branches of the Leibnizian tree. This is given in what we call "Leibnizian relativity principle" based on symmetry and invariance properties. In brief, one may say that through his philosophy of substance and motion and his development of differential calculus, Leibniz realized that Newtonian mechanics carries the seeds of its own destruction and generalization. In addition, because of the infinite multiplicity of points of view on motion, it turns out that a Leibnizian formulation does not only generalize Newtonian mechanics but it extends Einsteinian dynamics as well.

Finiteness principle

The second method that we call "Finiteness principle" concerns the infinite character of the spatio-temporal Newtonian velocity criticized by Leibniz. Its aim is to replace "action at a distance" by "contact action". This second approach violently contrasts with the first one in so far as the conceptual framework is concerned. Indeed, the existence of a tree-like structure of an infinite number of branches is conceptually very remote from our usual conception of dynamics inherited from Newtonian mechanics. As to the replacement of an "action at a distance" theory by a "contact action" one, it turns out that both theories remain very close to each other since they emerge from a unique differential equation where the difference manifests itself only at the level of the limit conditions. Last but not least, let us underline the fact that our final aim is not only to place in evidence new insights concerning Newtonian and Einsteinian dynamics through Leibnizian propositions but also to show the existence of a certain universality to which Leibniz was very attached. The latter becomes visible when the two above-mentioned principles are applied to other contexts.

Remark

Since different indices are required to account for the multiplicity of points of view and possible worlds, we shall not add to this the multidimensionality of space. Thus, a one dimensional model is considered.

2. On a Leibnizian interpretation and generalization of the classical laws of dynamics

Newtonian dynamics may be expressed through the following relations :

$$v = \int \frac{dP}{M} = \int \frac{dE_c}{P}, \quad (E_c,\ P,\ v) \rightarrow (0,\ 0,\ 0) \tag{1}$$

with

$$M = m > 0, \quad m : C^{te} \quad P = \pm\sqrt{2mE_c} \tag{2}$$

where m, P and E_c denote respectively mass, impulse and kinetic energy. The writing of Newtonian dynamics in the form of Eqs. (1) shows that the velocity is expressed in terms of the impulse and of kinetic energy and that it has the dimension of impulse over mass or energy over impulse. More essentially, it underlines the fact that impulse derives from kinetic energy and mass derives from impulse since Eqs. (1) are equivalent to $P \equiv \frac{dE_c}{dv}$, $M \equiv \frac{dP}{dv}$. This last property is the one that Leibniz favoured in order to obtain a certain unity of dynamics where only energy seems to be primary while the other two dynamical quantities (P,M) are derivative ones. (One may refer to the comments made by Poincaré at the end of Ref [1]). This property will be kept in the process of generalization and will constitute one cornerstone of the new construction to which one adds a new basic principle : "The Leibnizian relativity principle". The latter specifies the link between the different points of view on motion through the passage from v to v_k^K. Thus, the Leibnizian generalization takes on the following tree like forms :

$$v_k^K = \int \frac{dP}{M_k^K} = \int \frac{dE_c}{P_k^K}, \quad (E_c, \, P, \, v_k^K) \to (0, \, 0, \, 0) \tag{3}$$

with

$$M_k^K = m\left(1 + \frac{1+K}{2}\frac{P^2}{mE_d}\right)^{\frac{k}{1+K}} \tag{4}$$

$$P_k^K = \pm \left(1 + \frac{E_c}{E_d}\right)^{k-K} \sqrt{\frac{2mE_d}{1+K}\left[\left(1+\frac{E_c}{E_d}\right)^{1+K}-1\right]}.$$

The indices (k, K) denote respectively the multiplicities of points of view and possible worlds. These Leibnizian equations have been derived after having performed the following identifications :

$$^K P_K \equiv P, \quad ^K M_0 \equiv m \quad \forall K \tag{5}$$

and specified the "Leibnizian relativity principle". The latter defines kinetic energy in relation with the relative difference between two adjacent branches associated with the following invariant measure :

$$\frac{E_c}{E_d} = \left[\frac{^K P_{n+1} - ^K P_n}{^K P_n}\right] = \left[\frac{^K M_{n+1} - ^K M_n}{^K M_n}\right] \quad \forall K, \, n. \tag{6}$$

More precisely, the second equality shows the existence of an invariant measure associated with the elements $^K P_n$ and $^K M_n$ deduced from the properties of differential calculus. As to the first one, it expresses the association of the invariant measure with kinetic energy up to a multiplicative dimensional constant Ed. In order to select the best world among all possible worlds one shows that on substituting Eqs (4) into (3) with $k = K$ then the Leibnizian velocities may be explicitly integrated and inverted only for $K = 0$ and 1. The world of order one $(K = 1)$ is richer than that of order zero $(K = 0)$ since it includes it as a particular case. Notice that Eqs (4) show that the Newtonian world may be obtained either when $E_d \to \infty$ for any $k = K$ or when $k = K = 0$ for any E_d. As to the best world $(K = 1)$, the latter possesses the same structure as that of special relativity theory. However, let us note that in so far as the multiplicity of points of view is concerned the Leibnizian methodology is more general than that of Einstein since it leads to tree-like structures where Einstein's physics occupies only some of the branches. More precisely, proper and improper velocities correspond to the following elements $^1 v_0$ and $^1 v_3$ respectively where $k \neq K = 1$. As to the central element $^K v_K = ^1 v_1$, it corresponds to the so-called rapidity parameter [2, 3, 4]. Let us finally notice that if the attention is focused on total energy $E \equiv E_c + E_d$ rather than kinetic one, then one gets :

$$\frac{^1 M_{n+1}}{^1 M_n} = \frac{^1 P_{n+1}}{P_n^1} = r, \, r = \frac{E}{E_d} = \sqrt{1 + \frac{P^2}{mE_d}}. \tag{7}$$

Thus, total energy turns out to be associated with the geometrical ratio up to a multiplicative constant for homogeneity. These unusual properties can be obtained only if one adopts a Leibnizian methodology starting with the idea of an infinite multiplicity of subjective points of view on motion before generating objective worlds that link the conservative elements with each other through the ratio of subjective measures.

3. Leibnizian Finiteness principle

In the previous section, the development of Leibnizian structures that turned out to be compatible with special relativity tend to show that special relativity is fundamentally different from Newtonian mechanics at different levels. In the present section, we shall look at dynamics from a different angle leading to an opposite appreciation and conclusion in so far as the link between special relativity theory and Newtonian dynamics is concerned. Instead of adopting a multiplicity of points of view on motion, we eliminate the subjective element v from Eqs (1) and focus the attention on the objective relation that provides a direct link between energy and impulse.

$$\dot{P} - \frac{m}{P} = 0 \quad \dot{P} \equiv \frac{dP}{dE} \tag{8}$$

from which one may deduce the following second order differential equation :

$$\ddot{P} + \frac{m^2}{P^3} = 0 \tag{9}$$

At this point, let us forget about the Newtonian origin of Eq.(9) and consider it as a fundamental Eq. of dynamics. Thus, on integrating this equation and accounting for the two different limit conditions

$$P \to \infty, \quad \dot{P} \to 0 \tag{10}$$

$$P \to \infty, \quad \dot{P} \to \dot{P}_\infty \equiv \frac{1}{v_\infty}, \tag{11}$$

one obtains respectively a parabolic solution and a hyperbolic one. The first of these may be identified to Newton's theory while the second may be associated with Einsteinian dynamics. Let us recall that Leibniz criticized Newton's action at a distance which corresponds precisely to the limit condition given through (10), but the mechanicians of that period did not take these critics seriously. It is remarkable to note that the passage from Newtonian to Einsteinian dynamics is possible simply by replacing action at a distance (10) by contact-action (11). In spite of the numerous works that concern the links between classical and relativistic physics, it does not seem that these two Leibnizian procedures associated with dynamical relativity and finiteness have been considered previously.

4. Possible extension to other domains

The general Leibnizian conceptual framework has been applied to other contexts such as quantum mechanics and potential theory [4]. Our investigations in this direction led us to discover that different nonlinear approaches based on independent postulates turn out to occupy different branches of the same tree like structure. This leads to a certain unity that one can discover only by use of a Leibnizian procedure where a new type of principles is at work. The latter operates inside the tree so that the different branches harmonize with each other according to some symmetrical and invariance properties as shown through the Leibnizian relativity principle. By way of conclusion, let us notice that if Leibniz ideas about the construction of a Universal characteristic is, strictly speaking, unattainable, it remains an ideal that complements the numerous particular domains and opens new perspectives at the frontier of different disciplines.

5. Acknowledgments

I acknowledge the stimulating atmosphere of the "Epiphymaths" group of the university of Besançon where these ideas have been discussed. I am also grateful to Jean Merker, Claude-Alain Risset, François Bastien, Rachid Laydi, Bernard Cretin and Michel Planat for conversations and discussions concerning this subject.

References

[1] Leibniz, "La monadologie", Librairie Delagrave (1970).
[2] J.M. Lévy-Leblond "speed(s)". Am. J. Phys. **48** (5), (1980)
[3] N. Daher "L'espace, le temps et la matière dans la philosophie naturelle de Leibniz et leur rapport à la physique moderne", VII Internationaler Leibniz-Congress "NIHIL SINE RATIONE", Berlin, 10-14 sept. 2001.
[4] N. Daher "Démarche leibnizienne d'auto-organisation multiple et passage de lois indépendantes à des lois dépendantes d'échelle" (Annales françaises des Microtechniques et de Chronométrie Tome 50 sortie prévue fin 2002)

Inst. Phys. Conf. Ser. No 173: Section 2
Paper presented at 24th Int. Coll. Group Theoretical Methods in Physics, Paris, France, 15–20 July 2002

Gravitating lumps – 2002

D. V. Gal'tsov

Department of Theoretical Physics,
Moscow State University, 119899, Moscow, Russia
galtsov@grg.phys.msu.su

Abstract. In this talk we review some recent progress in the investigation of solitons and black holes in non-Abelian gauge theories coupled to gravity. Two new directions are discussed: attempts to find the rotating solitons and black holes with non-Abelian hair, and construction of static solitons and black holes in the non-Abelian Born-Infeld theory.

The soliton sector in the flat space gauge theories is quite well understood, the most notable example being the t'Hooft-Polyakov magnetic monopole. Unification approaches including gravitation stimulate an investigation of classical lumps in various field-theoretical models with full non-perturbative account of gravity. The whole new world of physically interesting classical solutions arises in gauge theories when gravity in switched on. ¿From one side, these solutions inherit topological features of classical lumps in the flat space theories (monopoles, sphalerons etc.), on the other hand, they are linked to typically gravitational "lumps" such as black holes. Some new solutions exist essentially due to gravity, while the other have the flat space ancestors, but non-perturbatively treated gravity substantially modifies them. It was found that certain statements of vacuum and electovacuum gravity, including the no-hair and uniqueness theorems for black holes, non-existence of globally regular solutions, the Birkhoff theorem and so on, do not hold for non-Abelian Yang-Mills (YM) fields. Thus, the investigation of gravitating gauge theories revealed many novel features rather surprising from the traditional viewpoint of general relativity (for a review see [1]). The aim of the present talk is to describe several new results related to the issue of rotation in soliton physics and construction of new classical particle-like solutions and black holes in the non-Abelian Born-Infeld theory emerging in the theory of superstrings.

First, recall, that the basic new soliton appearing in the gauge theories coupled to gravity is the Bartnik-McKinnon particle -like solution [2]. This is a stationary (saddle) point of the static truncation of the SU(2) EYM action

$$S_{EYM} = -\frac{1}{16\pi} \int \left(\frac{R}{G} + \frac{1}{g^2} F_{a\mu\nu} F_a^{\mu\nu} \right) \sqrt{-g} \, d^4 x, \tag{1}$$

with G, g denoting the Newton and the gauge coupling constants. The metric is spherically symmetric and is parameterized by two functions N, σ of the radial variable r:

$$ds^2 = N \sigma^2 dt^2 - \frac{dr^2}{N} - r^2 d\Omega, \tag{2}$$

while the YM field is purely magnetic:

$$A = (1 - w) \left(T_\theta \sin\theta d\varphi - T_\varphi \, d\theta \right), \tag{3}$$

where w is a real function of r, and T_θ, T_φ are spherical projections of the SU(2) generators. The particle-like solutions are distinguished by regularity of the metric at the origin implying that the function w takes the vacuum value (for definiteness, $w(0) = 1$). From the equations of motion one finds the series expansion

$$w = 1 - br^2 + O(r^4), \quad N = 1 - 4b^2r^2 + O(r^4), \tag{4}$$

with a free parameter b. The second condition is the finiteness of the ADM mass M, defined as the limit $m(r) \to M$, where the local mass-function is introduced via $N = 1 - 2m(r)/r$; the second metric function σ asymptotically must behave as $\sigma \to 1$. (To simplify formulas we set the Newton constant $G = 1$ everywhere if its explicit appearance is not needed for clarity.) To ensure this, the YM field asymptotically has to stay again in one of the vacua, the next term containing another free parameter p: $w = \pm(1 - p/r) + O(1/r^2)$ The free parameters b, M, p of the asymptotic solutions can be determined via the numerical matching, one finds a discrete increasing sequence b_n on the interval $[b_1 = 0.4537, b_\infty = 0.706)$, with masses converging to unity in the units of the rescaled Planck's length m_{Pl}/g. The function w_n has n zeroes and asymptotically tends to $(-1)^n$, solutions exist for any finite n.

For the lowest BK solution, $n = 1$, w is a decreasing function interpolating between $w = 1$ and $w = -1$ in exactly the same way as it does in the electroweak sphaleron. Odd-n solutions all have the sphaleron nature, while even-n ones are topologically trivial. A vacuum to vacuum path was constructed explicitly [3], showing that the Chern-Simons number is equal to $1/2$ for odd-n solutions and to zero for even n.

Soon after the discovery of the BK solutions, it was shown that there exist also black hole solutions with similar structure of fields outside the event horizon. For them, instead of the regularity conditions at the origin (4), the regularity of the event horizon is required. The (non-degenerate) event horizon is the largest linear zero of the metric function N : $N(r_h) = 0$, $N'(r_h) > 0$, where the YM function w starts with a finite value $w_h = w(r_h)$. This latter quantity is a parameter labeling the black hole solutions, another parameter being the horizon radius r_h. One finds the black hole solutions for any r_h and a discrete sequence of w_h within the interval $0 < w_h < 1$ [4]. The EYM black holes violate a naive no-hair conjecture, since the magnetic YM field outside the horizon is not associated with the conserved charges.

Recently some progress was achieved in constructing rotating counterparts to these solutions. The issue of rotation in pure gravity and Einstein-Maxwell theory is nicely solved in a very general way. In both cases, stationary spacetimes (admitting a Killing vector field which is timelike in an essential region) are described by three-dimensional gravity coupled sigma-models. Already this fact is sufficient to establish existence and uniqueness of rotating black holes [5, 6]. In establishing these results, one helpful fact is the circularity property, allowing for a simple Papapetrou form of the metric. Recall that circular spacetimes are those, admitting two commuting Killing vectors K_μ, $\tilde{K}_\mu$ which satisfy the Frobenius conditions

$$K_{[\mu}\tilde{K}_\nu \tilde{K}_{\lambda];\tau} = 0, \quad \tilde{K}_{[\mu}K_\nu K_{\lambda];\tau} = 0, \tag{5}$$

which mean that they are *hypersurface orthogonal*.

The *circularity theorem* asserts that the necessary and sufficient conditions for Eqs. (5) to hold are provided by the *Ricci-circularity* conditions (for a more recent discussion see [6]) $K_{[\mu}\tilde{K}_\nu R^\tau_{\lambda]} \tilde{K}_\tau = 0$, $\tilde{K}_{[\mu}K_\nu R^\tau_{\lambda]} K_\tau = 0$. Since the Ricci tensor R^τ_λ is involved here, the validity of (5) depends strongly on the matter system. In particular, these conditions are satisfied in the Einstein–Maxwell theory provided the Maxwell field is also stationary and axially symmetric.

For the EYM system this construction fails at both steps. First, no sigma-model representations can be found because of non-linearity of the equations for vector fields. Secondly, the circularity theorem no longer holds [7]. In order to fulfill the Ricci conditions, the YM field strength should have vanishing $F_{t\varphi}$, $\tilde{F}_{t\varphi}$ components, but this is not true already for the spherically symmetric configurations. Therefore, a general stationary axisymmetric solution of the EYM equations is not be describable by the Lewis-Papapetrou ansatz for the metric.

Dimensional reduction of the stationary axially symmetric *non-circular* 4D EYM system [8] gives rise to an Euclidean 2D dilaton gravity model

$$S_2 = \int e^\phi \left(R_2 + L_m \right) \sqrt{g} \, d^2 x, \tag{6}$$

where R_2 is the two-dimensional Ricci scalar, and the matter lagrangian L_m accounts both for the YM and metric variables including the 2D Yang-Mills field a_b, two 2D (effective) Higgs fields Φ, Ψ (matrix-valued), as well as two scalar moduli ψ, ω, the 'dilaton' ϕ, and two 2D Kaluza-Klein two-forms ω_{ab}, κ_{ab}. The 4D metric in terms of these variables reads:

$$ds^2 = e^\psi \left[dt + \Omega d\varphi + (\nu_a + \Omega k_a) dx^a \right]^2 - e^{2\phi - \psi} (d\varphi + k_a dx^a)^2 - e^{-\psi} g_{ab} dx^a dx^b, \tag{7}$$

and the Kaluza-Klein two-forms are defined as $\kappa_{ab} = \partial_a k_b - \partial_b k_a$, $\omega_{ab} = \partial_a \nu_b - \partial_b \nu_a + \Omega \kappa_{ab}$, which are actually scalars, e.g. $\kappa_{ab} = \kappa \epsilon_{ab}$.

The decomposition of the full 4D YM matrix-valued connection reads

$$A = \Phi \left[dt + \Omega d\varphi + (\nu_a + \Omega k_a) dx^a \right] + \Psi (d\varphi + k_a dx^a) + a_b dx^b. \tag{8}$$

It is essential that the variables in (8) can not depend only on x^a as in the Abelian case, such an assumption would exclude topologically non-trivial configurations. In fact, the symmetry of the gauge field under a spacetime isometry means that the action of the isometry can be compensated by a suitable gauge transformation, what on the infinitesimal level reads

$$\mathcal{L}_K A_\mu = D_\mu W, \tag{9}$$

with W being a Lie-algebra valued gauge function. In the case of the unique timelike Killing vector one can always choose $W = 0$, and hence the one-form A to be t-independent. But for the axial symmetry one has to take $W \neq 0$, that is $\partial_\varphi A \neq 0$, this is clear already from the flat space static axially symmetric ansatz given by Manton [9] for the winding number one and Rebbi and Rossi [10] for any winding number:

$$A_\mu = A_\mu^\rho T_\rho + A_\mu^\varphi T_\varphi + A_\mu^z T_z \tag{10}$$

with A_μ^i depending only on ρ, z and the rotated gauge generators are

$$T_\rho = \cos \nu\varphi T_x + \sin \nu\varphi T_y, \qquad T_\varphi = \nu^{-1} \partial_\varphi T_\rho. \tag{11}$$

This MRR ansatz involves twelve independent functions of two variables. Its generalization to the general non-circular stationary axisymmetric spacetime (7) leads to a complicated system of equations which was not fully explored yet.

Meanwhile, in the linearized theory, the number of the odd-parity zero modes on the static spherically symmetric background, which are potentially involved in description of the rotating EYM and EYMH configurations, is considerably less. Expanding the metric and the gauge field perturbations in the spin-weighted spherical harmonics, Volkov and Straumann [11] have found that only one metric mode and two gauge field modes contribute directly to the angular momentum. More recent analysis of perturbations of the EYMH system was performed in [12, 13, 14], showing, in particular, stability of the Schwarzschild and Reissner-Nordström black holes with respect to odd-parity EYM perturbations. It was confirmed also that only $l = 1$ zero modes, involving three real-valued functions, are responsible for the (infinitesimal) rotation of both EYM and EYMH solitons and black holes. At this level the circularity *does* hold. The set includes the metric perturbation $h_{t\varphi}$ and two isotopic variations of A_t. They satisfy a coupled system of linear ordinary differential equations of the second order, so the overall order of the system is *six*. A necessary condition for the globally defined continuously varying zero modes to exist is a non-trivial intersection of the moduli spaces of local solutions defined near the origin (for solitons) or the event horizon (for black holes) and at spatial infinity. Local analysis reveals that the moduli space at infinity is *four*-dimensional for the EYM system (without Higgs) and *three*-dimensional for the EYMH system [15]. In both cases the dimension of the moduli space near the origin (solitons) is *three*, while that at the black hole horizon is *four*. Thus, for solitons involving scalar fields, the intersection of two three-dimensional solution spaces inside the full six-dimensional space is trivial, that is, at best, only a discrete number of zero modes is possible. For black holes inside magnetic monopoles, there is a one-parameter intersection space, so the event horizons may be rotating. For the EYM (BK) particle-like solutions the intersection space is also one-dimensional, so the authors of [16] concluded that the rotating BK solutions should exist, and their electric charge Q and angular momentum J have to be proportional. Another important feature is that the asymptotic value of the time component of the gauge potential in the linearized solution must tends at infinity to some finite and possibly non-zero value V (contrary to the Abelian case this constant value is not pure gauge). From these three parameters only one is free for the regular solutions, while in the case of black holes two parameters from three are independent. Note that the usual staticity theorem does not hold in the YM case, and the stationary black hole is static (and the electric field is zero) provided the angular velocity of the horizon ω_h and these three parameters are related as follows [17]:

$$\omega_h J = QV.$$

According to this analysis, the black holes can be characterized by any two of the three quantities J, Q, V. So the linearized theory predicts the possibility of rotating EYM black holes of three kinds: first, there are solutions with $V = 0$, in this case Q, J and proportional. Secondly, there are uncharged stationary solutions (with $V \neq 0$), third there are non-static solutions with $J = 0$, but $\omega_h \neq 0$ (also with $V \neq 0$).

More recently the non-perturbative analysis has become available, with the results apparently not fully compatible with expectations based on the linearized theory. In the numerical work the metric was *assumed* to be in the Papapetrou form setting $\nu_a = \kappa_a = 0$ in (7) by hand. The gauge field was taken in the truncated MRR form with only six (from twelve) non-zero functions H_i, $i = 1, ..., 6$ of two variables r, θ:

$$\begin{aligned}
a_b dx^b &= \left[H_1 dr/r + (1 - H_2) d\theta \right] T_\varphi, \\
\Phi &= (H_5 T_r + H_6 T_\theta), \\
\Psi &= -\Phi\Omega - \nu \sin\theta \left[H_3 T_r + (1 - H_4) T_\theta \right],
\end{aligned} \tag{12}$$

subject to one differential condition $r \partial_r H_1 = \partial_\theta H_2$, imposed to fix the residual gauge invariance. Here the spherical basis is used in the isotopic space $T_r = T_\rho \sin\theta + T_z \cos\theta$, $T_\theta = T_\rho \cos\theta - T_z \sin\theta$. This ansatz is obtained within the flat space theory by imposing on the general MRR ansatz (10) the discrete symmetry under the product of the reflection through the xz plane and the charge conjugation, and it is compatible with the circularity assumption. To the best of our knowledge, it is the only ansatz which has been discussed in the context of the static and stationary axially symmetric EYM and EYMH configurations so far. In the first series of results by Kleihaus and Kunz [18] (confirmed in the later more detailed version [19]), the rotating EYM black holes were found with $V = 0$ and linearly related Q, J, but no other type of black holes with $V \neq 0$ and no rotating regular solutions either. Then van der Bij and Radu [20] confirmed the existence of black holes with $V = 0$ and suggested an explanation why the $V \neq 0$ solutions, predicted on the linearized level, are not encountered in the numerical analysis of the full non-linear problem. It turns out that one of the linearized YM modes satisfies a differential equation which asymptotically contains a term quadratic in

V, which can not be detected at the linearized theory level, but which is not small and which gives rise to oscillations of the relevant mode instead of the assumed decay. Perhaps the same reason is responsible for failure of attempts to find numerically the rotating globally regular solutions. In addition, in this paper the non-perturbative argument was given concerning the absence of the rotational excitations of Julia-Zee dyons, but, on the contrary, the possibility of the rotating EYMH configurations with the vanishing magnetic charge (such as monopole-antimonopole pairs). All this led the authors to conjecture that i)"For any regular solution in a gauge theory coupled to gravity, a non-vanishing angular momentum is incompatible with a net magnetic charge", and ii)"Any dyon solution with non-zero angular momentum necessarily contains an event horizon".

One aspect, related to superstrings, consists in investigation of solitons and black holes in the non-Abelian U(N) Dirac-Born-Infeld theory, arising as an effective theory on N D3-branes. D-branes are essentially non-perturbative objects of the string theory, which are defined as hyperplanes where open strings obey the Dirichlet boundary conditions. The ends of open strings move within D-branes generating the supersymmetric Yang-Mills theory governed by the Born-Infeld type lagrangian. The transversal coordinates (brane deformations) play the role of the Higgs scalars. When branes are coincident, one deals with the purely massless theory, while for separated branes there is a Higgs potential with the vacuum expectation value equal to the brane separation. The simplest case is the U(1) theory on a single D3-brane with the Born-Infeld (BI) action (in the flat spacetime)

$$S = \frac{\beta}{4\pi} \int \left\{ 1 - \sqrt{-\det(\eta_{\mu\nu} + \beta^{-1}F_{\mu\nu})} \right\} d^4x. \tag{13}$$

In the non-Abelian case the field strength $F_{\mu\nu}$ is matrix-valued, and one can a priori take the trace in different ways (see, e.g., [21] for details). Two definitions are the most reliable: the symmetrized trace suggested by Tseytlin, which well covers the lower orders of the perturbative string effective action, and the "ordinary trace-square root" form, which is a direct non-Abelian generalization of the four-dimensional U(1) BI action obtained by evaluating explicitly the determinant under the square root in (13). For the static spherically symmetric SU(2) magnetic ansatz (3) these definitions lead to somewhat different one-dimensional lagrangians

$$L_{tr} = \beta^2 r^2 (1 - \mathcal{R}), \quad \mathcal{R} = (1 + 2V^2 + 4K^2)^{1/2}, \tag{14}$$

where $V^2 = (1 - w^2)^2/(2\beta^2 r^4)$, $K^2 = w'^2/(2\beta^2 r^2)$, and [21]

$$l_{str} = \beta^2 r^2 \left[1 - (1 + V^2)^{1/2} + K^2 \mathcal{A}(1 + V^2)^{-1/2} \right] \tag{15}$$

with

$$\mathcal{A} = W^{-1} \arctan W, \quad W = (1 + V^2)^{1/2}(V^2 - K^2)^{-1/2}.$$

The BI lagrangians break the scale invariance of the ordinary Yang-Mills (or Maxwell) theory, introducing a new new scale parameter — the critical field β, which might be regarded as a manifestation of non-locality of the underlying dynamical object (string). Therefore, the no-go theorem for classical glueballs in the flat space YM theory is overruled, and one has to reconsider the possibility of such solutions in the non-Abelian Born-Infeld (NBI) theory. The proof of existence and the numerical glueball solutions in the flat space NBI theory were presented in [22] for the ordinary trace (14) and in [21] for the symmetrized trace (15) theories, they have qualitatively similar properties. Solutions form a sequence of the BK type, though the values of the parameter b_n in the expansion (4) are rather large (in natural units) and rapidly growing with the node number n, for example, $b_1 = 12.7$, $b_2 = 887$ in the (14); these values are even larger for the model (15) [21]. Solutions have sphaleron features and are expected to be unstable. When gravity is taken into account [23, 24], one finds that b_n, as a function of the dimensionless parameter $G\beta$, continuously interpolates between the values of [22] at small $G\beta$, and the BK values for large $G\beta$ [24].

Black hole solutions to the Einstein-NBI equations were also constructed [23, 24], outside the event horizon they are qualitatively similar to those in the EYM case. However, an internal structure of the ENBI black holes is drastically different. In fact, one could expect that violent oscillations inside the EYM black holes should be modified in the quantum theory: already in the first several inflation cycles the mass function attains values exceeding the Planck value [25]. Although the ENBI theory is still classical, it incorporates non-perturbatively the string quantum corrections in α', so one could expect that the problem of the over-Planckian masses in the EYM black holes will be resolved. This is indeed the case [24]: the internal behavior of the mass function now is perfectly smooth, though the singularity still bears some non-analytic features. Namely, the function w attains smoothly a finite value w_0 in the singularity, and there exist a two-parameter family of solutions for which the mass function is analytic at the Schwarzschild-type singularity. This family, however, is not generic like in the EYM case. But now a generic solution (three-parametric) can be obtained in the vicinity of the singularity in the series form, which is an expansion in terms of $r^{1/2}$. Mass function again attains a finite value, and one finds that the singularity is much weaker than in the Schwarzschild case. The scalar curvature diverges only as $R \sim r^{-3/2}$. Note that the non-analytic behavior of the mass function in the singularity is also observed in the U(1) Einstein-Born-Infeld theory coupled to dilaton [26].

Adding the (triplet) Higgs field to the NBI lagrangian, one finds the flat space monopole solutions [27, 28] exhibiting features, similar to those of the gravitating monopoles. In view of the existence of the flat-space NBI sphalerons, this is not surprising. Monopole (and dyon) solutions exist for the values of BI critical field β varying from infinity to a finite value β_{cr}, which was shown [29] to be a bifurcation point giving rise to the branch of excited monopoles with one-node w. Similarly to gravitating monopoles, there are further bifurcations where higher node sphaleron excitations pinch off. More detailed study [30] reveals that near the critical regime a large negative pressure appears in the monopole core which causes monopoles to shrink. For $\beta < \beta_{cr}$ only the pointlike magnetic monopoles exist as static solutions (these have the finite energy in the Born-Infeld theory). This behavior is reminiscent of the gravitational collapse, another aspect of similarity between the flat space NBI and the EYM theories.

The author wishes to thank the Organizers of GROUP-24 for a stimulating atmosphere of the Conference and financial support. This work was also supported in part by the Russian Foundation for Basic Research.

[1] Mikhail S. Volkov and Dmitri V. Gal'tsov. Gravitating non-Abelian solitons and black holes with Yang-Mills fields. *Phys. Rept.*, **319**:1–83, 1999, hep-th/9810070.

[2] R. Bartnik and J. McKinnon. Particlelike solutions of the Einstein – Yang – Mills equations. *Phys. Rev. Lett.*, **61**, 141–144, 1988.

[3] D.V. Gal'tsov and M.S. Volkov. Sphalerons in Einstein – Yang – Mills theory. *Phys. Lett.*, **B 273**, 255–259, 1991.

[4] M.S. Volkov and D.V.Gal'tsov. Non-Abelian Einstein – Yang – Mills black holes. *JETP Lett.*, **50**, 346–350, 1989; Black holes in Einstein – Yang – Mills theory. *Sov. J. Nucl. Phys.*, **51**, 747–753, 1990.
    H.P. Künzle and A.K.M. Masood ul Alam. Spherically symmetric static $SU(2)$ Einstein – Yang – Mills fields. *Journ.Math.Phys.*, **31**, 928–935, 1990.
    P. Bizon. Colored black holes. *Phys. Rev. Lett.*, **64**, 2844–2847, 1990.

[5] P. O. Mazur. Black hole uniqueness theorems. 1986, hep-th/0112038.

[6] M. Heusler. Black Hole Uniqueness Theorems . Cambridge University Press, 1996.

[7] M. Heusler and N. Straumann, The first law of black hole physics for a class of nonlinear matter models, *Class. Quant. Grav.*, **10**, 21–26, 1993.

[8] D. V. Gal'tsov. Einstein-Yang-Mills solitons: Towards new degrees of freedom. Proc. of the Int. Seminar *Mathematical Cosmology*, Potsdam, April 1998. Editors: M. Rainer, H.-J. Schmidt, World Scientific Publ., Singapore, 1998, 92-103, gr-qc/9808002.

[9] N.S. Manton. *Nucl. Phys.*, **B135**, 1978, 319.

[10] C. Rebbi and P. Rossi, Multimonopole solutions in the Prasad-Sommerfield limit, *Phys. Rev.*, **D22**, 2010–2017, 1980.

[11] M.S. Volkov and N. Straumann. Slowly rotating non-Abelian black holes. *Phys. Rev. Lett.*, **79**, 1428–1431, 1997.

[12] Othmar Brodbeck, Markus Heusler, and Olivier Sarbach. The generalization of the Regge-Wheeler equation for self- gravitating matter fields. *Phys. Rev. Lett.*, **84**, 3033, 2000, gr-qc/9906090.

[13] Olivier Sarbach, Markus Heusler, and Othmar Brodbeck. Perturbation theory for self-gravitating gauge fields. i: The odd-parity sector. *Phys. Rev.*, **D 62**, 084001, 2000, gr-qc/9910059.

[14] O. Sarbach, M. Heusler, and O. Brodbeck. Self-adjoint wave equations for dynamical perturbations of self-gravitating fields. *Phys. Rev.*, **D 63**, 104015, 2001, gr-qc/0010067.

[15] O. Brodbeck and M. Heusler. Stationary perturbations and infinitesimal rotations of static Einstein – Yang – Mills configurations with bosonic matter. *Phys. Rev.*, **D 56**, 6278–6283, 1997.

[16] O. Brodbeck, M. Heusler, N. Straumann, and M.S. Volkov. Rotating solitons and non-rotating, non-static black holes. *Phys. Rev. Lett.*, **79**, 4310–4313, 1997.

[17] D. Sudarsky and R.M. Wald. Extrema of mass, stationarity, and staticity, and solutions to the Einstein – Yang – Mills equations. *Phys. Rev.*, **D 46**, 1453–1447, 1992.

[18] B. Kleihaus and J. Kunz. Rotating hairy black holes. *Phys. Rev. Lett.*, **86**, 3704–3707, 2001, gr-qc/0012081.

[19] B. Kleihaus, J. Kunz, and F. Navarro-Lerida. Rotating Einstein-Yang-Mills black holes. *Phys. Rev.*, **D 66**, 104001, 2002, gr-qc/0207042.

[20] J. J. van der Bij and Eugen Radu. On rotating regular nonabelian solutions. *Int. J. Mod. Phys.*, **A17**, 1477-1490, 2002, gr-qc/0111046.

[21] V. V. Dyadichev and D. V. Gal'tsov. Sphaleron glueballs in NBI theory with symmetrized trace. *Nucl. Phys.*, **B590**, 504–518, 2000, hep-th/0006242.

[22] Dmitri Gal'tsov and Richard Kerner. Classical glueballs in non-Abelian Born-Infeld theory. *Phys. Rev. Lett.*, **84**, 5955–5958, 2000, hep-th/9910171.

[23] Marion Wirschins, Abha Sood, and Jutta Kunz. Non-Abelian Einstein-Born-Infeld black holes. *Phys. Rev.*, **D63**, 084002, 2001, hep-th/0004130.

[24] V. V. Dyadichev and D. V. Gal'tsov. Solitons and black holes in non-Abelian Einstein-Born- Infeld theory. *Phys. Lett.*, **B486**, 431–442, 2000, hep-th/0005099.

[25] D. V. Gal'tsov. Quantization near violent singularities in Einstein-Yang- Mills black holes. 2001, gr-qc/0101100.

[26] Gerard Clement and Dmitri Gal'tsov. Solitons and black holes in Einstein-Born-Infeld-dilaton theory. *Phys. Rev.*, **D62**, 124013, 2000, hep-th/0007228.

[27] N. Grandi, R. L. Pakman, F. A. Schaposnik, and Guillermo A. Silva. Monopoles, dyons and theta term in Dirac-Born-Infeld theory. *Phys. Rev.*, **D60**, 125002, 1999, hep-th/9906244.

[28] N. Grandi, E. F. Moreno, and F. A. Schaposnik. Monopoles in non-Abelian Dirac-Born-Infeld theory. *Phys. Rev.*, **D59**, 125014, 1999, hep-th/9901073.

[29] Dmitri Galtsov and Vladimir Dyadichev. D-branes and vacuum periodicity. Proceedings of the NATO Advanced Research Workshop "Non-commutative structures in mathematics and physics", Kiev, Ukraine, September 24-28, 2000; S. Duplij and J. Wess (Eds), Kluwer publ. 2001, p. 61-78, hep-th/0012059.

[30] Vladimir Dyadichev and Dmitri Galtsov. Monopoles in NBI theory and Born-Infeld collapse. *Phys. Rev.*, **D65**, 124026 , 2002, hep-th/0202177.

Inst. Phys. Conf. Ser. No 173: Section 2
Paper presented at 24th Int. Coll. Group Theoretical Methods in Physics, Paris, France, 15–20 July 2002
©2003 IOP Publishing Ltd

Group theoretical approach to QFT in de Sitter space

T. Garidi[*+], **E. Huguet**[*†], **J. Renaud**[*+]

* : Fédération de Recherches Astroparticules et Cosmologie.
\+ : LPTMC, Université Paris 7 Denis Diderot, boite 7020 ,F-75251 Paris Cedex 05, France;
garidi@ccr.jussieu.fr, renaud@ccr.jussieu.fr.
† : GEPI, Observatoire de Paris, 5.pl. J. Janssen, 92195 Meudon Cedex;
eric.huguet@obspm.fr.

Abstract. Quantum field theory in de Sitter space presents various well known problems (absence of global time like Killing vector field, zero mode problem in the minimally coupled scalar field and massless spin 2 field). After reviewing the group theoretical approach we will present some results concerning mainly the massless fields.

1. The de Sitter space

The metric g is determined by Einstein's equation : $G(g) + \Lambda g = T$. The choice $\Lambda = 0$ and $T = 0$ yields Special relativity on Minkowski space, whereas $\Lambda \neq 0$ and $T = 0$ yields Special relativity on de Sitter space. The de Sitter metric and the isometry group $SO_0(1,4)$ are respectively one parameter deformations of the Minkowski metric and of the proper Poincaré group. Despite their similarities, an important difference is the absence of a global time-like killing vector field thus, no "time" or "energy" can be globally defined.

Geometry, ambient space notations

The de Sitter space is conveniently seen as a hyperboloid embedded in a five-dimensional Minkowski space

$$X_H = \{x \in \mathbb{R}^5; x^2 = \eta_{\alpha\beta}x^\alpha x^\beta = -H^{-2}\},$$

where $\eta_{\alpha\beta} = \text{diag}(1, -1, -1, -1, -1)$, and with the metric $ds^2 = \eta_{\alpha\beta}dx^\alpha dx^\beta$. This embedding clarifies the relationship with unitary irreducible representations of the dS group, in particular because the Casimir operators are easy to identify.

A tensor field $\mathcal{K}(x)$ of rank l has to be viewed as a homogeneous function of the $\mathbb{R}^5$-variables x^α with some arbitrarily chosen degree λ of homogeneity. It can be chosen to be symmetric, traceless, divergenceless, and transverse ($x.\mathcal{K}(x) = 0$). The "intrinsic" field $h_{\mu\nu}(X)$ is locally determined by the tensor field $\mathcal{K}_{\alpha\beta}(x)$

$$h_{\mu\nu}(X) = \frac{\partial x^\alpha}{\partial X^\mu} \frac{\partial x^\beta}{\partial X^\nu} \mathcal{K}_{\alpha\beta}(x). \tag{1}$$

In the de Sitter arena we must know how to interpret familiar concepts such as time, energy, mass, and spin. This will partly be done using the group contraction procedure on the elementary systems (in Wigner's sense).

2. Classification of the unitary irreducible representations of the de Sitter group $SO_0(1,4)$

We now intend to clarify what we mean by "massive" or spin-s de Sitter fields. In as much as mass and spin are well-defined Poincaré concepts, we will consider exclusively the de Sitter elementary systems in the Wigner sense associated to a UIR of $SO_0(1,4)$ that admit a non-ambiguous massive, massless or spin-s UIR of the Poincaré group at the $H = 0$ contraction limit. Unitary irreducible representations (UIR) of $SO_0(1,4)$ are characterized by the eigenvalues of the two Casimir operators $Q^{(1)}$ and $Q^{(2)}$

$$Q^{(1)} = -\frac{1}{2}L_{\alpha\beta}L^{\alpha\beta}, \qquad Q^{(2)} = -W_\alpha W^\alpha, \tag{2}$$

with $W_\alpha = -\frac{1}{8}\epsilon_{\alpha\beta\gamma\delta\eta}L^{\beta\gamma}L^{\delta\eta}$, and where $L_{\alpha\beta} = M_{\alpha\beta} + S_{\alpha\beta}$ are the infinitesimal generators. The UIR's may be labelled by a pair of parameters $\Delta = (p,q)$ with $2p \in \mathbb{N}$ and $q \in \mathbb{C}$, in terms of which the eigenvalues of $Q^{(1)}$ and $Q^{(2)}$ are expressed as follows

$$Q^{(1)} = [-p(p+1) - (q+1)(q-2)]\mathrm{Id}, \quad Q^{(2)} = [-p(p+1)q(q-1)]\mathrm{Id}. \tag{3}$$

According to the possible values for p and q, three series of inequivalent representations may be distinguished: the principal, complementary and discrete series.

(i) Principal series representations‡ $U_{s,\nu}$: $\Delta = \left(s, \frac{1}{2} + i\nu\right)$ with

$$s = 0, 1, 2, \ldots \quad \text{and} \quad \nu \geq 0 \quad \text{or}$$
$$s = \frac{1}{2}, \frac{3}{2}, \ldots \quad \text{and} \quad \nu > 0.$$

The principal series contracts toward the direct sum of two massive Poincaré UIR's $\mathcal{P}(\pm m)$ with positive and negative energies [1].

(ii) Complementary series representations $V_{s,\nu}$: $\Delta = (s, \frac{1}{2} + \nu)$ with

$$s = 0 \quad \text{and} \quad \nu \in \mathrm{R}, \ 0 < |\nu| < \frac{3}{2} \quad \text{or}$$
$$s = 1, 2, 3, \ldots \quad \text{and} \quad \nu \in \mathrm{R}, \ 0 < |\nu| < \frac{1}{2}.$$

Here, the only physical representation in the sense of Poincaré limit is the scalar case corresponding to $\Delta = (0,1)$ also called conformally coupled massless case because it can be extended to an UIR of the conformal group.

(iii) Discrete series $\Pi_{p,0}$ and $\Pi_{p,q}^\pm$: $\Delta = (p,q)$ with

$$p = 1, 2, 3, \ldots \quad \text{and} \quad q = 0 \quad \text{or}$$
$$p = \frac{1}{2}, 1, \frac{3}{2}, 2, \ldots \quad \text{and} \quad q = p, p-1, \ldots, 1 \text{ or } \frac{1}{2}.$$

In this case, the only physical representations in the sense of Poincaré limit are those with $p = q = s$. They are called the massless representations of the dS group.

As a result, the principal series contracts towards the massive Poincaré UIR (for all s). The complementary series with $\Delta = (0,1)$ and discrete series with $\Delta = (s,s)$ lead to the massless scalar Poincaré UIR and to the massless Poincaré UIR (for all $s \neq 0$) respectively.

‡ Also called "massive" representations.

The ambient space approach allows to combine the geometry and the group theoretical aspects in a clear operational way. In particular, the field equation is given by $(Q_s - \langle Q_s \rangle) \mathcal{K}(x) = 0$, and can for a symmetric, traceless and divergenceless be rewritten as $(Q_0 - l(l+1) - \langle Q_s \rangle) \mathcal{K}(x) = 0$. With help of equation (1) the intrinsic field equation for a tensor field of rank l reads

$$\left(\Box_H + H^2 l(l+2) + H^2 \langle Q_s \rangle \right) h_{\mu\nu} = 0 . \tag{4}$$

This expression translates the clear group theoretical content into the commonly used intrinsic approach, and enables us to clearly identify the involved field.

3. Free quantum field

The field φ is an operator valued distribution satisfying a field equation, the microcausality requirement $[\varphi(x), \varphi(x')] = 0$, if x and x' are not causally connected, and the covariance condition $\underline{U}_g^{-1} \varphi(x) \underline{U}_g = \varphi(g \cdot x), \quad \forall x \in M, \ \forall g \in G$.

Canonical quantization

Let us first remind the usual quantization of a free scalar field. Let ϕ_k be a family of orthonormal§ solutions of the field equation, verifying $\sum_k \phi_k(x) \phi_k^*(x') - \phi_k^*(x) \phi_k(x') = -i \tilde{G}(x, x')$. The set of ϕ_k's span the Hilbert space $\mathcal{H}_p$. The quantum field $\varphi(x)$ is defined by

$$\varphi(x) = \sum_k \phi_k(x) A_k + \phi_k^*(x) A_k^\dagger, \tag{5}$$

together with the usual ccr. In the usual canonical quantization, these ccr are represented over the Fock space $\mathcal{H}_p$ by

$$A_k = a_k := a(\phi_k) \text{ and } A_k^\dagger = a_k^\dagger := a^\dagger(\phi_k). \tag{6}$$

Now, the quantum field as an operator valued distribution can be cast under the form $\varphi(f) = a\left(p(f)\right) + a^\dagger\left(p(f)\right)$, where f is a real test function and$\|$ $p(f) = \sum_k (\phi_k, f) \phi_k \in \mathcal{H}_p$. Actually, $p(f)$ is the unique element of $\mathcal{H}_p$ for which

$$\langle p(f), \psi \rangle = (f, \psi), \quad \forall \psi \in \mathcal{H}_p. \tag{7}$$

The above formulas make apparent that the definition of the quantum field φ is independent of the basis given by the modes ϕ_k. In particular the covariance of the field follows from that of p under the only hypothesis that $\mathcal{H}_p$ is invariant under the group action.

Thus the usual quantization and its covariance relies on the existence of a space $\mathcal{H}_p$ of solutions closed under the action of the group and with $\mathcal{H}_p \oplus \mathcal{H}_p^*$ containing every acceptable solution of the field equation. In the case of $SO_0(1, 4)$ the covariant quantization is possible in the massive case for all spins, since one can find a Hilbert space $\mathcal{H}_p$ closed under $SO_0(1, 4)$. On the contrary, such a closed Hilbert space does not exist for the so called minimally coupled scalar field, and the linear quantum gravity (zero mode problem).

§ $\langle \phi_1, \phi_2 \rangle = \int_\Sigma \phi_1^* \overset{\leftrightarrow}{\partial_\mu} \phi_2 \, d\sigma^\mu$. Note that $\langle \phi_k^*, \phi_{k'}^* \rangle = -\delta_{kk'}$.
$\|$ $(f, g) = \int_X f^*(x) g(x) d\mu(x)$

4. Quantization over a Krein Space

In situations where the Hilbert space $\mathcal{H}_p$ is not closed under the group action the above quantization scheme breaks down. Nevertheless, one can still obtain a covariant quantization by considering the whole space of solution : the Krein space $\mathcal{H} = \mathcal{H}_p \oplus \mathcal{H}_p^*$ which *is* closed under the group action. The usual scheme is modified through the ccr which are now represented over the Fock space $\mathcal{H}$ by

$$A_k = \frac{1}{\sqrt{2}}(a_k - b_k^\dagger) \text{ and } A_k^\dagger = \frac{1}{\sqrt{2}}(a_k^\dagger - b_k). \tag{8}$$

where $b_k = a(\phi_k^*)$. The resulting field is of Gupta Bleuler type: it contains a distinguished subspace of physical states. This method has been applied in order to covariantly quantize the minimally coupled scalar [2] and the massless spin-2 fields [3].

4.1. Minimally coupled scalar field

This field corresponds to the representation Π_{10} (scalar discrete series), that is $(Q_0 - \langle Q_0 \rangle)\phi = 0$ with $\langle Q_0 \rangle = 0$. The status of the zero mode problem has in this case been clarified by Allen's no-go theorem [4].

Nevertheless a covariant quantization can be performed over a Krein space. Such a procedure leads in addition to the result that vacuum expectation value of the stress tensor vanishes

$$\langle 0|T_{\mu\nu}|0 \rangle = 0. \tag{9}$$

It is worth noting that we consider the minimally coupled scalar field as a purely de Sitterian field in the sense that its UIR doesn't contract towards a Poincaré UIR.

4.2. Linear quantum gravity

The de Sitter metric is expanded in terms of a small perturbation $\mathcal{K}$. The linear approximation in Einstein's equation gives

$$(Q_2 + 6)\mathcal{K} + D_2 \partial_2 \cdot \mathcal{K} = 0, \tag{10}$$

where the operator D_2 is the generalized gradient symmetrizer operator, and ∂_2 is the generalized divergence. Now let us remind that the Casimir equation reads

$$(Q_2 - \langle Q_2 \rangle)\mathcal{K} = 0. \tag{11}$$

One can see that in the case of a symmetric, traceless and divergenceless tensor field, equation (10) becomes the Casimir equation (11) with $\langle Q_2 \rangle = -6$. This corresponds precisely to the representation $\Pi_{2,2}^+ \oplus \Pi_{2,2}^-$ which, according to the flat criterium, identifies this field as a massless spin 2 field. On the contrary to $\Pi_{1,0}$ this representation contracts towards a Poincaré UIR.

The general solution of the field equation (10) can be found in terms of a projection tensor of a minimally coupled scalar field

$$\mathcal{K}_{\alpha\beta} = D_{\alpha\beta}\phi. \tag{12}$$

Non surprisingly this field suffers from the zero mode problem. Again the Krein space quantization will allow us to quantize covariantly this type of field.

[1] Mickelsson J., Niederle J., Comm. Math. Phys. **27** (1972), 167.
[2] Gazeau J. P., Renaud J., Takook M.V., Class. Quantum Grav. **17** (2000), 1415 , gr-qc/9904023.
[3] Garidi T., Gazeau J. P., Renaud J., Rouhani S. and Takook M.V., *Linear covariant quantum gravity in de Sitter space*, in preparation.
[4] Allen B., *Vacuum states in de Sitter Space*, Phys. Rev. D **32** 12 (1985), 3136.

Inst. Phys. Conf. Ser. No 173: Section 2
Paper presented at 24th Int. Coll. Group Theoretical Methods in Physics, Paris, France, 15–20 July 2002
©2003 IOP Publishing Ltd

Time in Quantum Geometrodynamics

Nathan D. George‡, Adrian P. Gentle, Arkady Kheyfets§, and Warner A. Miller

Theoretical Division (T-6, MS B288)
Los Alamos National Laboratory, Los Alamos, NM 87545, USA

Abstract.

  Standard techniques of canonical gravity quantization on the superspace of 3–metrics are known to cause insurmountable difficulties in the description of time evolution. We forward a new quantization procedure on the superspace of true dynamic variables – geometrodynamic quantization. This procedure takes into account the states that are "off-shell" with respect to the constraints and thus circumvents the notorious problems of time. In this approach quantum geometrodynamics, general covariance, and the interpretation of time emerge together as parts of the solution to the total problem of geometrodynamic evolution.

  The standard approach to canonical quantum gravity [1, 2] is based on the classical dynamic picture of the evolving 3–geometry of a slicing of a spacetime manifold described by the lapse function N and the shift functions N^i. The canonical variables are the 3–metric components g_{ik} on a spatial slice Σ of the foliation induced by the spacetime 4–metric, and their canonical conjugate momenta π^{ik}. The customary variational procedure applied to the Hilbert action expressed in terms of these canonical variables yields Hamilton dynamics that, after applying the canonical quantization procedure on the superspace of 3–metrics (in both Dirac's and ADM square root Hamiltonian approaches), produces a quantum theory that appears to be incapable of providing a consistent description of time evolution for quantum gravitational systems. The source of the difficulties can be traced to mixing dynamical considerations with the requirements of general covariance and to restricting quantum states to the shell determined by constraints.

  The situation changes dramatically if York's analysis of gravitational degrees of freedom [4] is taken into account and actively utilized. According to York, the set of six parameters describing the slice 3–metric should be split into two subsets, $\{\beta_1, \beta_2\}$ (two functions) and $\{\alpha_1, \alpha_2, \alpha_3, \Omega\}$. The first of these is treated as the set of true gravitational degrees of freedom (the initial values for them can be given freely), while the second is considered to be the set of embedding variables. The α parameters are often referred to as coordinatization parameters, while Ω is called, depending on the context, the slicing parameter, the scale factor, or the many–fingered time parameter. Information relevant to dynamics is carried by β parameters, while α and Ω essentially describe time. The true dynamic variables form what we call a dynamic superspace while the embedding variables are treated as functional parameters.

  The idea is to develop geometrodynamics from the very beginning on the dynamic superspace instead of the superspace of 3–metrics or 3–geometries. The variational principle on the dynamic superspace or its phase space (formed by true dynamic variables $\{\beta_1, \beta_2\}$ and their conjugate momenta $\{\pi_{\beta_1}, \pi_{\beta_2}\}$) yields the dynamic equations describing the evolution of the true dynamic variables. All of these equations depend on lapse and shift and contain

‡ E-mail: ndg@lanl.gov
§ Permanent Address: Department of Mathematics, North Carolina State University, Raleigh, NC 27695-8205, USA.

embedding variables as functional parameters. These are treated as an external field and are determined by additional equations that do not follow from the variational principle on dynamic superspace. The quantization procedure is performed on the dynamic superspace (only β-s are quantized, i. e. generate commutation relations, while the embedding variables form a classical field). The Schrödinger equation is obtained by a quantization procedure from the Hamilton–Jacobi equation on the dynamic superspace and describes the time evolution of the state functional on the true dynamic superspace coupled with the external classical field determined by the embedding variables. Such a coupling can be achieved via a procedure similar to that of Hartree–Fock.

In a more detailed and precise description that follows, we omit indices on variables β and α for the sake of notational simplicity. They can be recovered easily whenever necessary.

We start from the standard Lagrangian $\mathcal{L}$ (written in terms of the 3-metric, shift and lapse) and the associated action (with appropriate boundary terms, as needed, to remove the second time derivatives terms) and we introduce the momenta conjugate to the true dynamic variables

$$\pi_\beta = \frac{\partial \mathcal{L}}{\partial \dot{\beta}}. \tag{1}$$

We then use these π_β's to form the geometrodynamic Hamiltonian $\mathcal{H}_{dyn}$,

$$\mathcal{H}_{dyn} = \pi_\beta \dot{\beta} - \mathcal{L}. \tag{2}$$

The arguments of the Hamiltonian $\mathcal{H}_{dyn}$ are described by the expression

$$\mathcal{H}_{dyn} = \mathcal{H}_{dyn}(\beta, \pi_\beta; \Omega, \alpha). \tag{3}$$

The variables following the semicolon are treated as describing an external field, while the ones preceding the semicolon are the coordinates and momenta of the true gravitational degrees of freedom, i.e. of the true geometrodynamics. The variation of β and π_β produce Hamilton equations on the dynamic superspace, while variation of the ends leads to the Hamilton–Jacobi equation

$$\frac{\delta S}{\delta t} = -\mathcal{H}_{DYN}\left(\beta, \frac{\delta S}{\delta \beta}; \Omega, \alpha\right). \tag{4}$$

Here S is a functional of β and, in addition, a function of t,

$$S = S[\beta; t]. \tag{5}$$

and $\frac{\delta}{\delta t}$ is defined by

$$\frac{\partial S}{\partial t} = \int \frac{\delta S}{\delta t} d^3 x. \tag{6}$$

The Hamilton–Jacobi equation (4) is incapable of providing any predictions as its solutions depend on the functional parameters Ω and α which are not yet known. One can complete the picture by adding the standard constraint equations of general relativity, obtained by variations of shift and lapse. These constraints should be satisfied once the solution for the geometrodynamic variables β, π_β (with appropriate initial data) is obtained and substituted. Using the symbols $[\beta]_s$, $[\pi_\beta]_s$ for such a solution, we have

$$\mathcal{H}^i\left([\beta]_s, [\pi_\beta]_s, \Omega, \alpha\right) = 0, \tag{7}$$

$$\mathcal{H}\left([\beta]_s, [\pi_\beta]_s, \Omega, \alpha\right) = 0. \tag{8}$$

These constraint equations should be treated as additional symmetries, or the equations for an external field. They do follow from the shift and lapse invariance of the action but their derivation in this new setting depends on the structure of the whole action integral. As a result, they cannot replace the full set of equations for geometrodynamic evolution (which is usually done on the superspace of 3–metrics). However, the resulting complete system of equations is equivalent to that of the standard geometrodynamics on the superspace of 3–geometries [6].

For the purpose of quantization, we make a transition to the corresponding Schrödinger equation based entirely on dynamics and ignoring the system symmetries

$$i\hbar \frac{\delta \Psi}{\delta t} = \widehat{\mathcal{H}}_{dyn}\left(\beta, \widehat{\pi}_\beta; \Omega, \alpha\right) \Psi \qquad \text{where} \qquad \widehat{\pi}_\beta = \frac{\hbar}{i}\frac{\delta}{\delta \beta}. \tag{9}$$

The Schrödinger equation (9) implies that commutation relations are imposed only on the true dynamic variables and treats the embedding variables as external classical fields. The state functional Ψ in this equation is a functional of β and a function of t,

$$\Psi = \Psi\left[\beta, t\right]. \tag{10}$$

This Schrödinger equation (with specific initial data) can be solved (cf., for instance the example of the Bianchi 1A cosmological model [5, 6]). The resulting solution Ψ_s of this Schrödinger equation is not capable of providing any definite predictions as it depends on four functional parameters Ω, α which remain at this stage undetermined. All expectations, such as the expectation values of β

$$< \beta >_s = \langle \Psi_s | \beta | \Psi_s \rangle = \int \Psi_s^* \beta \Psi_s \, \mathcal{D}\beta \tag{11}$$

or of $\widehat{\pi}_\beta$

$$< \pi_\beta >_s = \langle \Psi_s | \widehat{\pi}_\beta | \Psi_s \rangle = \int \Psi_s^* \widehat{\pi}_\beta \Psi_s \, \mathcal{D}\beta \tag{12}$$

also depend on these functional parameters. To specify these functions we resort to the constraint equations. The treatment of the constraints has nothing to do with the quantization of geometrodynamics. It merely introduces the coupling between the already quantized geometrodynamics and the classical field determined by the embedding variables. In other words, the constraints take care of the symmetries which are classical in nature to the extent that they are capable of doing so.

As in case of classical geometrodynamics, we impose the constraints on the solution of the dynamic equations (Schrödinger equation) with appropriate initial data and in this way, determine the unique values of Ω and α. It is possible that there are several ways to couple the constraints to the quantization of the true dynamic variables, β. Here we impose the four constraints only on the expectation values of the conformal dynamics

$$\mathcal{H}^i\left(< \beta >_s, < \pi_\beta >_s, \Omega, \alpha\right) = 0$$

$$\mathcal{H}\left(< \beta >_s, < \pi_\beta >_s, \Omega, \alpha\right) = 0. \tag{13}$$

Lapse and shift are assumed to be given either explicitly or by additional conditions.

Evolution can be described as follows. Initial data at $t = t_0$ consist of the initial state functional $\Psi = \Psi_0$ and the initial values (functions) of embedding variables. In addition, lapse and shift are supposed to be given either explicitly or by additional conditions. Equations (11), (12) yield the expectation values (functions) of the true dynamic variables and their conjugate momenta. The results are substituted into the constraints (13). After this, the constraints are solved with respect to the time derivatives of embedding variables. A step forward in time (say, with the increment Δt) is performed by integration of the constraints to evolve the embedding variables and by integration of the Schrödinger equation (9) to evolve the state functional. This concludes one step forward in time. The next step is performed by repeating the same operations in the same order.

One can be referred to [5, 6] for two particular examples illustrating such geometrodynamic evolution for the Bianchi 1A and Taub cosmologies respectively. The first one can and has been solved analytically, while the latter one has been solved numerically.

The resulting canonical gravity quantization procedure circumvents all the standard problems of time and removes all the obstacles for describing the time evolution of quantum gravitational systems. This has been achieved by including "off–shell" quantum states and imposing the constraints only on the expectation values of the dynamic variables.

It should be stressed that all three components of the evolution procedure described for quantum geometrodynamic systems — quantum dynamics itself, constraints enforcing the symmetries (general covariance), and the interpretation of time — emerge together as the solution to the total problem of geometrodynamic evolution.

References

[1] J. Butterfield and C. J. Isham, "Spacetime and the Philosophical Challenge of Quantum Gravity", in *Physics meets Philosophy at the Planck Scale*, eds. C. Callender and N. Huggett, Cambridge University Press (2000). See also gr–qc/9903072.

[2] K. V. Kuchař, "Canonical Quantum Gravity", in R. J. Gleiser, C. N. Kosameh and O. M. Moreschi, editors, *General Relativity and Gravitation* (IOP Publishing, Bristol, 1993).

[3] K. V. Kuchař, "Time and Interpretations of Quantum Gravity" in *Proc. 4th Canadian Conference on General Relativity and Relativistic Astrophysics* eds. G. Kunstatter, D. E. Vincent and J. G. Williams (World Scientific; Singapore, 1992).

[4] J. W. York, "Role of Conformal Three-Geometry in the Dynamics of Gravitation," *Phys. Rev. Lett.* **28**, 1082-1085 (1972).

[5] A. Kheyfets and W. A. Miller, "Quantum Geometrodynamics: Quantum-Driven Many-Fingered Time," *Phys. Rev.* **D51**, n. 2, 493-501

[6] A. Kheyfets, D. E. Holz and W. A. Miller, "The Issue of Time Evolution in Quantum Gravity," *Int. J. Mod. Phys.* **11**, n. 16, 2977-3002, (1996).

Inst. Phys. Conf. Ser. No 173: Section 2
Paper presented at 24th Int. Coll. Group Theoretical Methods in Physics, Paris, France, 15–20 July 2002
©2003 IOP Publishing Ltd

Conformal Lie algebras of spacetimes

Francisco J Herranz[1] and Mariano Santander[2]

[1] Departamento de Física, Universidad de Burgos, 09006 Burgos, Spain
[2] Departamento de Física Teórica, Universidad de Valladolid, 47011 Valladolid, Spain

Abstract. The vector fields of the conformal Lie algebras of the (1+1)-dimensional spacetimes with constant curvature are presented in a unified framework. The results cover relativistic and non-relativistic cases as well as the three classical two-dimensional Riemannian spaces. An apparently new conformal duality for spaces with non-zero curvature is obtained; this provides a new insight for the usual specific conformal transformations in flat spaces.

1. Introduction

The $(1 + 1)$D spacetimes [1] can be studied in a unified setting within a Cayley–Klein (CK) framework that covers the nine 2D real spaces with constant curvature [2]. This approach makes use of two *real* coefficients κ_1 and κ_2, in such a manner that the corresponding Lie groups of motions can collectively be considered as a parametrized family denoted $SO_{\kappa_1,\kappa_2}(3)$. The Lie brackets of the CK algebra $so_{\kappa_1,\kappa_2}(3)$ in the basis $\{P_1, P_2, J_{12}\}$ read

$$[J_{12}, P_1] = P_2 \qquad [J_{12}, P_2] = -\kappa_2 P_1 \qquad [P_1, P_2] = \kappa_1 J_{12}. \tag{1}$$

The CK space corresponds to the 2D symmetric homogeneous space

$$S^2_{[\kappa_1],\kappa_2} = SO_{\kappa_1,\kappa_2}(3)/SO_{\kappa_2}(2) \qquad SO_{\kappa_2}(2) = \langle J_{12} \rangle \tag{2}$$

so that J_{12} leaves an origin point O invariant, acting as the rotation around O, while P_1, P_2 generate translations along two basic directions. The metric of $S^2_{[\kappa_1],\kappa_2}$ has constant curvature κ_1 with signature $\mathrm{diag}(1, \kappa_2)$. In particular, we find nine different spaces:

Elliptic: $\mathbf{S}^2$	Euclidean: $\mathbf{E}^2$	Hyperbolic: $\mathbf{H}^2$
$S^2_{[+],+} = SO(3)/SO(?)$	$S^2_{[0],+} = ISO(2)/SO(2)$	$S^2_{[-],+} = SO(2,1)/SO(2)$
Oscillating NH: $\mathbf{NH}^{1+1}_+$	Galilean: $\mathbf{G}^{1+1}$	Expanding NH: $\mathbf{NH}^{1+1}_-$
$S^2_{[+],0} = ISO(2)/ISO(1)$	$S^2_{[0],0} = IISO(1)/ISO(1)$	$S^2_{[-],0} = ISO(1,1)/ISO(1)$
Anti-de Sitter: $\mathbf{AdS}^{1+1}$	Minkowskian: $\mathbf{M}^{1+1}$	De Sitter: $\mathbf{dS}^{1+1}$
$S^2_{[+],-} = SO(2,1)/SO(1,1)$	$S^2_{[0],-} = ISO(1,1)/SO(1,1)$	$S^2_{[-],-} = SO(2,1)/SO(1,1)$

Therefore we find the three classical Riemannian spaces for $\kappa_2 > 0$ with a metric of signature $\mathrm{diag}(+, +)$, three non-relativistic spacetimes for $\kappa_2 = 0$ with degenerate metric $\mathrm{diag}(+, 0)$ (NH means Newton–Hooke), and three relativistic spacetimes for $\kappa_2 < 0$ with a Lorentzian metric of signature $\mathrm{diag}(+, -)$. For the six $(1 + 1)$D spacetimes with $\kappa_2 \leq 0$, P_1, P_2, J_{12} are, in this order, time translation, space translation and boost generators, provided that $\kappa_1 = \pm 1/\tau^2$ and $\kappa_2 = -1/c^2$, where τ is the universe time radius and c the speed of light (κ_1 is proportional to the cosmological constant). Hence the contractions $\kappa_1 \to 0$ and $\kappa_2 \to 0$ correspond to the flat limit $\tau \to \infty$ and to the non-relativistic limit $c \to \infty$, respectively.

The aim of this contribution is to present the conformal Lie algebras for the nine real 2D CK spaces within this parametrized framework in terms of the curvature and signature. These results are obtained by searching for transformations that preserve cycles (geodesics, circles,...), instead of solving explicitly the conformal Killing equations for the metrics [3, 4, 5]; note that the latter approach exhibits difficulties for spaces with degenerate metrics such as the Newtonian spaces.

2. Vector model of the (1+1)-dimensional spacetimes

A 3D real matrix representation of $so_{\kappa_1,\kappa_2}(3)$ is given by:

$$P_1 = -\kappa_1 e_{01} + e_{10} \qquad P_2 = -\kappa_1\kappa_2 e_{02} + e_{20} \qquad J_{12} = -\kappa_2 e_{12} + e_{21} \qquad (3)$$

where e_{ij} is a 3D matrix with a non-zero entry 1 at row i and column j. The exponential of (3) leads to one-parameter subgroups of $SO_{\kappa_1,\kappa_2}(3)$ so that this acts as a group of linear transformations in an *ambient space* $\mathbb{R}^3$ with coordinates $\mathbf{x} = (x^0, x^1, x^2)$ called *Weierstrass coordinates*. Such an action is transitive on the *'sphere'* Σ obtained as the orbit of the origin $O = (1, 0, 0)$ which is identified with the CK space, that is,

$$\Sigma \equiv (x^0)^2 + \kappa_1(x^1)^2 + \kappa_1\kappa_2(x^2)^2 = 1 \longleftrightarrow S^2_{[\kappa_1],\kappa_2}. \qquad (4)$$

A point $\mathbf{x}$ can be parametrized in terms of two *geodesic parallel coordinates*, time t and space y, through the action of one-parameter subgroups as $\mathbf{x} = \exp(tP_1)\exp(yP_2)O$; this gives

$$x^0 = C_{\kappa_1}(t)C_{\kappa_1\kappa_2}(y) \qquad x^1 = S_{\kappa_1}(t)C_{\kappa_1\kappa_2}(y) \qquad x^2 = S_{\kappa_1\kappa_2}(y) \qquad (5)$$

where the curvature-dependent cosine $C_\kappa(x)$ and sine $S_\kappa(x)$ functions are defined by [2]:

$$C_\kappa(x) = \begin{cases} \cos\sqrt{\kappa}\,x & \kappa > 0 \\ 1 & \kappa = 0 \\ \cosh\sqrt{-\kappa}\,x & \kappa < 0 \end{cases} \qquad S_\kappa(x) = \begin{cases} \frac{1}{\sqrt{\kappa}}\sin\sqrt{\kappa}\,x & \kappa > 0 \\ x & \kappa = 0 \\ \frac{1}{\sqrt{-\kappa}}\sinh\sqrt{-\kappa}\,x & \kappa < 0 \end{cases}$$

The κ-tangent is $T_\kappa(x) = S_\kappa(x)/C_\kappa(x)$. Note that if $\kappa_2 > 0$, t is another spatial coordinate.

Within the vector model, *cycles*, that is, lines with constant geodesic curvature k_g, can be obtained as the intersection of the 'sphere' (4) with a plane in $\mathbb{R}^3$ (α_i are constants) [6]:

$$\alpha_0 x^0 + \alpha_1 x^1 + \alpha_2 x^2 = \alpha. \qquad (6)$$

If we substitute (5) in (6) we obtain the cycles expressed in parallel coordinates (t, y):

$$\alpha_0 C_{\kappa_1}(t) + \alpha_1 S_{\kappa_1}(t) = \frac{\alpha}{C_{\kappa_1\kappa_2}(y)} - \alpha_2 T_{\kappa_1\kappa_2}(y). \qquad (7)$$

Particular types of cycles are geodesics ($k_g = 0$), circles, equidistants and horocycles. For instance, *geodesics* are recovered from (7) by setting $\alpha = 0$, $\alpha_2 = -1$:

$$T_{\kappa_1\kappa_2}(y) = \alpha_0 C_{\kappa_1}(t) + \alpha_1 S_{\kappa_1}(t). \qquad (8)$$

In the flat limit $\kappa_1 = 0$, this equation reduces to $y = \alpha_0 + \alpha_1 t$.

3. Conformal Lie algebras

Let us consider the equations of geodesics (8). If we look for one-parameter geodesic-preserving transformations in the CK space $S^2_{[\kappa_1],\kappa_2}$, then we recover the generators of isometries that close the CK algebra $so_{\kappa_1,\kappa_2}(3)$ (1), the vector fields of which are given by

$$P_1 = -\partial_t \qquad P_2 = -\kappa_1\kappa_2 S_{\kappa_1}(t)T_{\kappa_1\kappa_2}(y)\,\partial_t - C_{\kappa_1}(t)\,\partial_y$$

$$J_{12} = \kappa_2 C_{\kappa_1}(t)T_{\kappa_1\kappa_2}(y)\,\partial_t - S_{\kappa_1}(t)\,\partial_y. \qquad (9)$$

If we now consider the cycle equation (7) in the curved case with $\kappa_1 \neq 0$ and search for one-parameter cycle-preserving transformations, then we find three additional generators beyond the motions: a dilation D and two generators of the so called Λ-translations L_1, L_2 [6] (these behave as usual translations only around the origin). The corresponding vector fields read

$$D = -\frac{S_{\kappa_1}(t)}{C_{\kappa_1\kappa_2}(y)}\,\partial_t - C_{\kappa_1}(t)S_{\kappa_1\kappa_2}(y)\,\partial_y$$

$$L_1 = -\frac{C_{\kappa_1}(t)}{C_{\kappa_1\kappa_2}(y)}\,\partial_t + \kappa_1 S_{\kappa_1}(t)S_{\kappa_1\kappa_2}(y)\,\partial_y \qquad L_2 = -C_{\kappa_1\kappa_2}(y)\,\partial_y. \qquad (10)$$

3.1. First conformal basis: $\{P_1, P_2, J_{12}, L_1, L_2, D\}$

The vector fields (9)–(10) span the *conformal Lie algebra*, $\mathrm{conf}_{\kappa_1,\kappa_2}$, of the CK space $S^2_{[\kappa_1],\kappa_2}$ with Lie brackets given by:

$$
\begin{array}{lll}
[J_{12}, P_1] = P_2 & [J_{12}, P_2] = -\kappa_2 P_1 & [P_1, P_2] = \kappa_1 J_{12} \\
[J_{12}, L_1] = L_2 & [J_{12}, L_2] = -\kappa_2 L_1 & [L_1, L_2] = -\kappa_1 J_{12} \\
[D, P_i] = L_i & [D, L_i] = P_i & [D, J_{12}] = 0 \qquad\qquad (11) \\
[P_1, L_1] = \kappa_1 D & [P_2, L_2] = \kappa_1\kappa_2 D & \\
[P_1, L_2] = 0 & [P_2, L_1] = 0. &
\end{array}
$$

These show a *conformal duality* between the generators of translations and Λ-translations:

$$P_i \leftrightarrow L_i \quad i = 1,2 \qquad J_{12} \leftrightarrow J_{12} \qquad D \leftrightarrow D. \qquad (12)$$

This map interchanges the set of conformal algebras as $\mathrm{conf}_{\kappa_1,\kappa_2} \leftrightarrow \mathrm{conf}_{-\kappa_1,\kappa_2}$, thus relating the conformal algebras of spaces with opposite curvatures and the same signature.

As an example of this unified approach, we display the conformal vector fields for the anti-de Sitter spacetime, $\mathbf{AdS}^{1+1} \equiv S^2_{[+],-}$, by setting $\kappa_1 = +1/\tau^2$ and $\kappa_2 = -1/c^2$:

$$P_1 = -\partial_t \qquad L_1 = -\frac{\cos(t/\tau)}{\cosh(y/c\tau)}\,\partial_t - c\sin(t/\tau)\sinh(y/c\tau)\,\partial_y$$

$$P_2 = \tfrac{1}{c}\sin(t/\tau)\tanh(y/c\tau)\,\partial_t - \cos(t/\tau)\,\partial_y \qquad L_2 = -\cosh(y/c\tau)\,\partial_y$$

$$J_{12} = -\tfrac{\tau}{c}\cos(t/\tau)\tanh(y/c\tau)\,\partial_t - \tau\sin(t/\tau)\,\partial_y \qquad\qquad (13)$$

$$D = -\frac{\tau\sin(t/\tau)}{\cosh(y/c\tau)}\,\partial_t - c\tau\cos(t/\tau)\sinh(y/c\tau)\,\partial_y.$$

3.2. Second conformal basis: $\{P_1, P_2, J_{12}, G_1, G_2, D\}$

Under the flat limit $\kappa_1 \to 0$, both L_i coincide with P_i: $L_1 = P_1 = -\partial_t$, $L_2 = P_2 = -\partial_y$. However as far as $\kappa_1 \neq 0$, we may take other two generators called *specific conformal transformations*: $G_i = (L_i - P_i)/\kappa_1$; these are

$$G_1 = \frac{1}{\kappa_1 C_{\kappa_1\kappa_2}(y)}\left(C_{\kappa_1\kappa_2}(y) - C_{\kappa_1}(t)\right)\partial_t + S_{\kappa_1}(t)S_{\kappa_1\kappa_2}(y)\,\partial_y \qquad\qquad (14)$$

$$G_2 = \kappa_2 S_{\kappa_1}(t)T_{\kappa_1\kappa_2}(y)\,\partial_t - \tfrac{1}{\kappa_1}\left(C_{\kappa_1\kappa_2}(y) - C_{\kappa_1}(t)\right)\partial_y.$$

In this second basis the commutation rules of $\mathrm{conf}_{\kappa_1,\kappa_2}$ read

$$
\begin{array}{lll}
[J_{12}, P_1] = P_2 & [J_{12}, P_2] = -\kappa_2 P_1 & [P_1, P_2] = \kappa_1 J_{12} \\
[J_{12}, G_1] = G_2 & [J_{12}, G_2] = -\kappa_2 G_1 & [G_1, G_2] = 0 \\
[D, P_i] = P_i + \kappa_1 G_i & [D, G_i] = -G_i & [D, J_{12}] = 0 \qquad\qquad (15) \\
[P_1, G_1] = D & [P_2, G_2] = \kappa_2 D & \\
[P_1, G_2] = -J_{12} & [P_2, G_1] = J_{12}. &
\end{array}
$$

Now the six conformal vector fields are independent for any value of κ_1, and under the contraction $\kappa_1 = 0$ we recover the known expressions for flat spaces [3]. For instance, for Minkowskian spacetime $\mathbf{M}^{1+1} \equiv S^2_{[0],-}$, with $\kappa_1 = 0$ ($\tau \to \infty$) and $\kappa_2 = -1/c^2$, they are

$$
\begin{aligned}
P_1 &= -\partial_t & G_1 &= \tfrac{1}{2}\left(t^2 + \tfrac{1}{c^2}y^2\right)\partial_t + t\,y\,\partial_y \\
P_2 &= -\partial_y & G_2 &= -\tfrac{1}{c^2}t\,y\,\partial_t - \tfrac{1}{2}\left(t^2 + \tfrac{1}{c^2}y^2\right)\partial_y \\
J_{12} &= -\tfrac{1}{c^2}y\,\partial_t - t\,\partial_y & D &= -t\,\partial_t - y\,\partial_y.
\end{aligned}
\tag{16}
$$

For the most contracted case, the Galilean spacetime $\mathbf{G}^{1+1} \equiv S^2_{[0],0}$ ($\tau, c \to \infty$), we obtain

$$
\begin{aligned}
P_1 &= -\partial_t & P_2 &= -\partial_y & J_{12} &= -t\,\partial_y & D &= -t\,\partial_t - y\,\partial_y \\
G_1 &= \tfrac{1}{2}t^2\partial_t + t\,y\,\partial_y & G_2 &= -\tfrac{1}{2}t^2\,\partial_y
\end{aligned}
\tag{17}
$$

We remark that the conformal duality $P_i \leftrightarrow L_i$ (12) is *invisible* in the conformal algebra of flat spaces with $\kappa_1 = 0$, for which $P_i \equiv L_i$, leaving the specific conformal transformations G_i as a kind of vestigial residue of the difference $L_i - P_i$. In this sense, the symmetry $P_i \leftrightarrow G_i$ in flat spaces can be seen as a residue of the deeper conformal duality (12).

3.3. Third conformal basis: $\{R_1, R_2, J_{12}, G_1, G_2, D\}$

If we define two new generators $R_i = P_i + \tfrac{1}{2}\kappa_1 G_i = \tfrac{1}{2}(P_i + L_i)$ with vector fields given by

$$
\begin{aligned}
R_1 &= -\frac{1}{2}\left(1 + \frac{C_{\kappa_1}(t)}{C_{\kappa_1\kappa_2}(y)}\right)\partial_t + \frac{1}{2}\kappa_1 S_{\kappa_1}(t)S_{\kappa_1\kappa_2}(y)\,\partial_y \\
R_2 &= -\frac{1}{2}\kappa_1\kappa_2 S_{\kappa_1}(t)T_{\kappa_1\kappa_2}(y)\,\partial_t - \frac{1}{2}\left(C_{\kappa_1}(t) + C_{\kappa_1\kappa_2}(y)\right)\partial_y,
\end{aligned}
\tag{18}
$$

the commutation relations of $\mathrm{conf}_{\kappa_1,\kappa_2}$ turn out to be

$$
\begin{array}{lll}
[J_{12}, R_1] = R_2 & [J_{12}, R_2] = -\kappa_2 R_1 & [R_1, R_2] = 0 \\
[J_{12}, G_1] = G_2 & [J_{12}, G_2] = -\kappa_2 G_1 & [G_1, G_2] = 0 \\
[D, R_i] = R_i & [D, G_i] = -G_i & [D, J_{12}] = 0 \\
[R_1, G_1] = D & [R_2, G_2] = \kappa_2 D & \\
[R_1, G_2] = -J_{12} & [R_2, G_1] = J_{12}.
\end{array}
\tag{19}
$$

Therefore, the curvature κ_1 disappears from the commutators so that all spaces in the family $S^2_{[\kappa_1],\kappa_2}$ with the same signature (i.e., same κ_2) have isomorphic conformal algebras. These are: $\mathrm{conf}_{\kappa_1,+} \equiv so(3,1)$ for the Riemannian spaces ($\kappa_2 > 0$), $\mathrm{conf}_{\kappa_1,0} \equiv iso(2,1)$ for the non-relativistic spacetimes ($\kappa_2 = 0$), and $\mathrm{conf}_{\kappa_1,-} \equiv so(2,2)$ for the relativistic ones ($\kappa_2 < 0$).

In relation with the usual approach to conformal groups, it is worthy to mentioning that all the vectors fields presented here, satisfy the conformal Killing equations for the metric of the CK space $S^2_{[\kappa_1],\kappa_2}$ [6]. We also stress that these results enable to study the *conformal completion* or *compactification* of spacetimes in a unified setting, which can be found in [7].

References

[1] Bacry H and Lévy-Leblond J M 1968 J. Math. Phys. **9** 1605
[2] Herranz F J, Ortega R and Santander M 2000 J. Phys. A: Math. Gen. **33** 4525
[3] Doubrovine B, Novikov S and Fomenko A 1982 Géométrie Contemporaine, Méthodes et Applications (Moscow: MIR)
[4] Berger M 1987 Geometry I (Berlin: Springer)
[5] Yaglom I M, Rozenfel'd B A and Yasinskaya E U 1966 Sov. Math. Surveys **19** 49
[6] Herranz F J and Santander M 2002 J. Phys. A: Math. Gen. **35** 6601
[7] Herranz F J and Santander M 2002 J. Phys. A: Math. Gen. **35** 6619

Inst. Phys. Conf. Ser. No 173: Section 2
Paper presented at 24th Int. Coll. Group Theoretical Methods in Physics, Paris, France, 15–20 July 2002
©2003 IOP Publishing Ltd

Orthogonal symmetry and matrix potentials in heterotic string theory

A Herrera–Aguilar

Instituto de Física y Matemáticas, UMSNH, Morelia, Michoacán, México

Abstract. The $3D$ stationary effective Lagrangian of the toroidally compactified heterotic string theory is presented as nonlinear σ–model parametrized by two matrix potentials coupled to gravity with the $SO(8, 24)$ group of symmetries. This form of the Lagrangian allows one to establish a relationship between the theory under consideration and the stationary Einstein–Maxwell theory written in terms of two complex Ernst potentials coupled to gravity with the $SU(2, 1)$ symmetry group. A clasification of the symmetries of the theory in terms of these potential languages is performed. A solution–generating technique for constructing charged solutions representing both single and pairs of axisymmetric interacting black holes by applying a Lie–Bäcklund transformation on a simple seed solution is outlined.

1. Introduction

It turns out that the stationary Einstein–Maxwell (EM) theory can be expressed as a nonlinear σ–model possessing the eight–parameter $SU(2, 1)$ symmetry group in terms of the so–called Ernst potentials [1]. Moreover, the classification of the symmetries of this theory in such a language adopts a simple form (see, for instance, [2]). A similar situation takes place for the toroidally compactified down to three dimensions effective field theory of the heterotic string in terms of the *Matrix Ernst Potentials* (MEP), but with the symmetry group $SO(8, 24)$. Due to this fact, one can think of certain relationship between the mathematical structure of both theories where, from one side, we have complex functions and, from the other, real matrices (see [3] and references therein). Such a relationship helps us to generalize the results obtained in the EM theory to the heterotic string realm. For example, the construction of single and double charged black hole solutions can be carried out in the framework of string theory by making use of the generalized Lie–Bäcklund matrix transformation of Harrison type.

2. Stationary Einstein–Maxwell theory and Ernst potentials

The general form of the stationary metric reads

$$ds^2 = -f \left(dt + \omega_\mu dx^\mu \right)^2 + f^{-1} \gamma_{\mu\nu} dx^\mu dx^\nu \tag{1}$$

where the functions f, ω and γ are independent of the temporal coordinate $x^4 = t$.

It is well known that the stationary formulation of the four–dimensional EM theory can be expressed in terms of the gravitational $\mathcal{E} = f + i\chi - \frac{1}{2} |\Phi|^2$ and electromagnetic Φ Ernst potentials [1], where the scalar potential χ is the related to vector ω through

$$\nabla \times \vec{\omega} = -f^{-2} \left[\nabla \chi + Im \left(\bar{\Phi} \nabla \Phi \right) \right].$$

The stationary Lagrangian of the EM theory has the form

$$^3\mathcal{L}_{EM} = -^3R + \frac{1}{2f^2} \left| \nabla \mathcal{E} + \bar{\Phi} \nabla \Phi \right|^2 - f^{-1} |\nabla \Phi|^2, \tag{2}$$

where $f = \frac{1}{2}\left(\mathcal{E} + \bar{\mathcal{E}} + \Phi\bar{\Phi}\right)$. The corresponding pair of field equations for the potentials is

$$\nabla^2\mathcal{E} - f^{-1}(\nabla\mathcal{E} + \bar{\Phi}\nabla\Phi) \cdot \nabla\mathcal{E} = 0,$$
$$\nabla^2\Phi - f^{-1}(\nabla\mathcal{E} + \bar{\Phi}\nabla\Phi) \cdot \nabla\Phi = 0. \tag{3}$$

Thus, in terms of these variables the Lagrangian (2) defines a σ–model on the symmetric coset space $SU(2,1)/S(U(2) \times U(1))$ as follows [4]

$$ds^2 = k_{\alpha\bar{\beta}}\nabla z^\alpha \cdot \nabla z^{\bar{\beta}}, \quad \text{where} \quad k_{\alpha\bar{\beta}} = \frac{1}{2f^2}\begin{pmatrix} 1 & \Phi \\ \bar{\Phi} & -(\mathcal{E} + \bar{\mathcal{E}}) \end{pmatrix}, \tag{4}$$

$\alpha, \bar{\beta} = 1, 2$; and the role of the local coordinates is played by the Ernst potentials $z^1 = \mathcal{E}$ and $z^2 = \Phi$. The isometries of this target space are classified as follows

$$
\begin{array}{llll}
\text{a) Scaling} & \mathcal{E}' = \alpha\mathcal{E}, & \Phi' = \sqrt{\alpha}\Phi; \\
\text{b) Electromagnetic rotation} & \mathcal{E}' = \mathcal{E}, & \Phi' = e^{i\beta}\Phi; \\
\text{c) Shift of the potential } \mathcal{E} & \mathcal{E}' = \mathcal{E} + i\gamma, & \Phi' = \Phi; & (5) \\
\text{d) Shift of the potential } \Phi & \mathcal{E}' = \mathcal{E} - \bar{\delta}\Phi - \frac{1}{2}|\delta|^2, & \Phi' = \Phi + \delta; \\
\text{e) Ehlers transformation} & \mathcal{E}' = \dfrac{\mathcal{E}}{1 + i\kappa\mathcal{E}}, & \Phi' = \dfrac{\Phi}{1 + i\kappa\mathcal{E}}; \\
\text{f) Harrison transformation [5]} & \mathcal{E}' = \dfrac{\mathcal{E}}{1 - \bar{\lambda}\Phi - \frac{1}{2}|\lambda|^2\mathcal{E}}, & \Phi' = \dfrac{\Phi + \lambda\mathcal{E}}{1 - \bar{\lambda}\Phi - \frac{1}{2}|\lambda|^2\mathcal{E}};
\end{array}
$$

where the constant parameters $\alpha, \beta, \gamma, \kappa$ are real, whereas δ, λ are complex.

3. Matrix Ernst potentials for heterotic string theory

Let us now consider the effective action of heterotic string theory in 10 dimensions:

$$S^{(10)} = \int d^{(10)}x \, |G^{(10)}|^{\frac{1}{2}} e^{-\phi^{(10)}}(R^{(10)} + \phi_{;M}^{(10)}\phi^{(10);M} -$$

$$\frac{1}{12}H_{MNP}^{(10)}H^{(10)MNP} - \frac{1}{4}F_{MN}^{(10)I}F^{(10)IMN}) \tag{6}$$

where $F_{MN}^{(10)I} = \partial_M A_N^{(10)I} - \partial_N A_M^{(10)I}$ and $H_{MNP}^{(10)} = \partial_M B_{NP}^{(10)} - \frac{1}{2}A_M^{(10)I}F_{NP}^{(10)I}$+cycl. perms. of M, N, P; $G_{MN}^{(10)}$ is the metric, $B_{MN}^{(10)}$ is the anti–symmetric Kalb-Ramond field, $\phi^{(10)}$ is the dilaton and $A_M^{(10)I}$ is a set of sixteen $U(1)$ Abelian vector fields ($I = 1, 2, ..., 16$). After the compactification down to three dimensions on a 7–torus we get the following fields [6]–[7]
 a) scalar fields

$$G = \left(G_{pq} = G_{p+3,q+3}^{(10)}\right), \qquad B = \left(B_{pq} = B_{p+3,q+3}^{(10)}\right),$$

$$A = \left(A_p^I = A_{p+3}^{(D)I}\right), \qquad \phi = \phi^{(10)} - \frac{1}{2}\ln|\det G|, \qquad \text{where} \quad \text{p,q=1,2,...,7. (7)}$$

 b)tensor fields

$$g_{\mu\nu} = e^{-2\phi}\left(G_{\mu\nu}^{(10)} - G_{p+3,\mu}^{(10)}G_{q+3,\nu}^{(10)}G^{pq}\right),$$

$$B_{\mu\nu} = B_{\mu\nu}^{(10)} - 4B_{pq}A_\mu^p A_\nu^q - 2\left(A_\mu^p A_\nu^{p+7} - A_\nu^p A_\mu^{p+7}\right), \tag{8}$$

(we set $B_{\mu\nu} = 0$ to remove the effective cosmological constant from our consideration).

c)vector fields $A_\mu^{(a)} = \left((A_1)_\mu^p, (A_2)_\mu^{p+7}, (A_3)_\mu^{14+I}\right)$ $(a = 1, 2, ..., 30)$

$$(A_1)_\mu^p = \frac{1}{2}G^{pq}G_{q+3,\mu}^{(10)}, \qquad (A_3)_\mu^{14+I} = -\frac{1}{2}A_\mu^{(10)I} + A_q^I A_\mu^q,$$

$$(A_2)_\mu^{p+7} = \frac{1}{2}B_{p+3,\mu}^{(10)} - B_{pq}A_\mu^q + \frac{1}{2}A_p^I A_\mu^{14+I}. \tag{9}$$

In three dimensions all vector fields can be dualized on–shell:

$$\nabla \times \vec{A_1} = \frac{1}{2}e^{2\phi}G^{-1}\left(\nabla u + (B + \frac{1}{2}AA^T)\nabla v + A\nabla s\right),$$

$$\nabla \times \vec{A_3} = \frac{1}{2}e^{2\phi}(\nabla s + A^T\nabla v) + A^T\nabla \times \vec{A_1},$$

$$\nabla \times \vec{A_2} = \frac{1}{2}e^{2\phi}G\nabla v - (B + \frac{1}{2}AA^T)\nabla \times \vec{A_1} + A\nabla \times \vec{A_3}. \tag{10}$$

Thus, the resulting three–dimensional theory describes the scalars G, B, A and ϕ and pseudoscalars u, v and s coupled to the metric $g_{\mu\nu}$.

We define two matrix Ernst potentials as follows [8]

$$\mathcal{X} = \begin{pmatrix} -e^{-2\phi} + v^T Xv + v^T As + \frac{1}{2}s^T s & v^T X - u^T \\ Xv + u + As & X \end{pmatrix}, \quad \mathcal{A} = \begin{pmatrix} s^T + v^T A \\ A \end{pmatrix} \tag{11}$$

where $X = G + B + \frac{1}{2}AA^T$. They are (8×8) and (8×16) real matrices, respectively, and contain all the information about the matter content of the effective field theory. The physical meaning of these variables is as follows: the relevant information concerning the gravitatoinal field is encoded in X, whereas its rotational character is hidden in u; v is related to multidimensional components of the Kalb–Ramond field, and, finally, A and s stand for electric and magnetic potentials, respectively. In terms of MEP the effective three–dimensional Lagrangian of the theory takes the form

$$^3\mathcal{L} = -{}^3R + \text{Tr}[\frac{1}{4}\left(\nabla\mathcal{X} - \nabla\mathcal{A}\mathcal{A}^T\right)\mathcal{G}^{-1}\left(\nabla\mathcal{X}^T - \mathcal{A}\nabla\mathcal{A}^T\right)\mathcal{G}^{-1} + \frac{1}{2}\nabla\mathcal{A}^T\mathcal{G}^{-1}\nabla\mathcal{A}] \tag{12}$$

where $\mathcal{G} = \frac{1}{2}\left(\mathcal{X} + \mathcal{X}^T - \mathcal{A}\mathcal{A}^T\right)$. The field equations for the matter part of this theory read

$$\nabla^2\mathcal{X} - 2(\nabla\mathcal{X} - \nabla\mathcal{A}\mathcal{A}^T)(\mathcal{X} + \mathcal{X}^T - \mathcal{A}\mathcal{A}^T)^{-1}\nabla\mathcal{X} = 0,$$

$$\nabla^2\mathcal{A} - 2(\nabla\mathcal{X} - \nabla\mathcal{A}\mathcal{A}^T)(\mathcal{X} + \mathcal{X}^T - \mathcal{A}\mathcal{A}^T)^{-1}\nabla\mathcal{A} = 0. \tag{13}$$

The Lagrangian (12) defines as well a nonlinear matrix σ–model on the symmetric coset space $SO(8, 24)/S[O(8) \times O(24)]$ in the following way

$$ds^2 = \frac{1}{4}Tr\left(\mathcal{G}^{-1}\nabla Z_k K_{kl}\nabla Z_l^T\right), \quad \text{where } K_{kl} = \begin{pmatrix} \mathcal{G}^{-1} & -\mathcal{G}^{-1}\mathcal{A} \\ \mathcal{A}^T\mathcal{G}^{-1} & 2I + \mathcal{A}^T\mathcal{G}^{-1}\mathcal{A} \end{pmatrix} \tag{14}$$

$k, l = 1, 2$; I is the (16×16) unit matrix and the role of the local coordinates is now played by the matrix Ernst potentials $Z_1 = \mathcal{X}$ and $Z_2 = \mathcal{A}$. By analogy, the isometries read

a) Scaling $\qquad\qquad\qquad\qquad\qquad \mathcal{X}' = \mathcal{S}^T\mathcal{X}\mathcal{S}, \qquad\qquad\qquad \mathcal{A}' = \mathcal{S}^T\mathcal{A};$

b) Electromagnetic rotation $\quad\quad \mathcal{X}' = \mathcal{X}, \qquad\qquad\qquad\qquad \mathcal{A}' = \mathcal{H}\mathcal{A};$

c) Shift of the potential $\mathcal{X} \quad\quad \mathcal{X}' = \mathcal{X} + \Lambda, \qquad\qquad\qquad \mathcal{A}' = \mathcal{A};$ $\qquad$ (15)

d) Shift of the potential $\mathcal{A} \quad\quad \mathcal{X}' = \mathcal{X} + \mathcal{A}\Gamma^T + \frac{1}{2}\Gamma\Gamma^T, \quad \mathcal{A}' = \mathcal{A} + \Gamma;$

e) Ehlers transformation $\quad\quad \mathcal{X}' = (1_8 + \mathcal{X}\Theta)^{-1}\mathcal{X}, \quad \mathcal{A}' = (1_8 + \mathcal{X}\Theta)^{-1}\mathcal{A};$

f) Harrison transformation $\quad\quad \mathcal{X}' = (1_8 - \mathcal{A}\Omega^T + \frac{1}{2}\mathcal{X}\Omega\Omega^T)^{-1}\mathcal{X},$

$$\mathcal{A}' = (1_8 - \mathcal{A}\Omega^T + \frac{1}{2}\mathcal{X}\Omega\Omega^T)^{-1}(\mathcal{A} - \mathcal{X}\Omega);$$

where 1_8 is the (8×8) unit matrix and the constant matrices obey the following restrictions $\det \mathcal{S} \neq 0$, $\mathcal{H}\mathcal{H}^T = I$, $\Lambda^T = -\Lambda$ and $\Theta^T = -\Theta$.

The form of the Lagrangian (12) is very similar to that of the stationary EM theory (2). In fact, one can establish the following relationship between both theories

$$\mathcal{X} \longleftrightarrow -\mathcal{E}, \quad \mathcal{A} \longleftrightarrow \Phi, \quad matrix\ transposition \longleftrightarrow complex\ conjugation; \quad (16)$$

thus, with the aid of the replacement (16), from the theory (2) one can obtain the theory (12).

4. Conclusion and Discussion

The relationship (16) establish some sort of matrix generalization of the stationary EM theory and allows us to extrapolate all the results obtained in the EM theory to the heterotic string realm. As a direct application one can construct the still missing full non–extremal charged rotating black hole solution in the framework of string theory [9] by applying the so–called normalized Harrison transformation:

$$\mathcal{X}' = (1_8 + \frac{1}{2}\Sigma\Omega\Omega^T)(1_8 - \mathcal{A}\Omega^T + \frac{1}{2}\mathcal{X}\Omega\Omega^T)^{-1}\left[\mathcal{X} + (\mathcal{A} - \frac{1}{2}\mathcal{X}\Omega)\Omega^T\Sigma\right] + \frac{1}{2}\Sigma\Omega\Omega^T\Sigma,$$

$$\mathcal{A}' = (1_8 + \frac{1}{2}\Sigma\Omega\Omega^T)(1_8 - \mathcal{A}\Omega^T + \frac{1}{2}\mathcal{X}\Omega\Omega^T)^{-1}(\mathcal{A} - \mathcal{X}\Omega) + \Sigma\Omega,$$

which is a matrix Lie–Bäcklund symmetry that preserves the asymptotics of the seed field configurations, on a neutral rotating black hole solution, i.e. with $\mathcal{A} = 0$; the construction of charged interacting rotating black hole solutions in $4D$ string theory can also be reached by applying this transformation on the seed potentials

$$\mathcal{X} = \frac{1}{Re\epsilon_2}\begin{pmatrix} Re\epsilon_1 & Re\epsilon_1 Im\epsilon_2 - Re\epsilon_2 Im\epsilon_1 \\ Re\epsilon_1 Im\epsilon_2 + Re\epsilon_2 Im\epsilon_1 & Re\epsilon_1|\epsilon_2|^2 \end{pmatrix}, \qquad \mathcal{A} = 0,$$

where $\epsilon_{1,2}$ are the gravitational Ernst potentials for rotating sources (see [10] for details).

5. Acknowledgments

The author is grateful to S. Kousidou for encouraging him during the performance of this work. This work was supported by grants CONACYT–J34245–E and CIC–UMSNH–4.18.

References

[1] Ernst F J 1968 Phys. Rev. 168 1415–1417 (1968).
[2] Kinnersley W 1977 J. Math. Phys. 18 1529–1537.
[3] Herrera-Aguilar A and Kechkin O 1998 Int. J. Mod. Phys. A13 393–402.
[4] Mazur P O 1983 Acta Phys. Pol. 14 219–234.
[5] Harrison B K 1968 J. Math. Phys. 9 1744–1752.
[6] Maharana J and Schwarz J H 1993 Nucl. Phys. B390 332.
[7] Sen A 1995 Nucl. Phys. B434 179–209.
[8] Herrera-Aguilar A and Kechkin O 1999 Phys. Rev. D59 124006.
[9] Youm D 1999 Phys. Rept. 316 1–232.
[10] Herrera-Aguilar A 2002 "Charging Interacting Rotating Black Holes in Heterotic String Theory", hep–th/0201126.

Inst. Phys. Conf. Ser. No 173: Section 2
Paper presented at 24th Int. Coll. Group Theoretical Methods in Physics, Paris, France, 15–20 July 2002
©*2003 IOP Publishing Ltd*

Orthogonal polynomials of several discrete variables and the 3nj-Wigner symbols: applications to spin networks

M Lorente

Departamento de Física, Universidad de Oviedo, 33007 Oviedo, Spain

Abstract. The use of orthogonal polynomials for integral models on the lattice is applied to the 3nj-symbols that appear in the coupling of several angular momenta. These symbols are connected to the Ponzano-Regge method to solve the Einstein equations on a discrete Riemannian manifold.

1. Classical orthogonal polynomials of one discrete variable

These polynomials satisfy a difference equation of hypergeometric type such that the difference derivatives of the some polynomial satisfy a similar equation. The discrete variable can be consider of two types.

a) *On homogeneous lattice*: $x = 0, 1, 2, \cdots$
The corresponding polynomials satisfy a difference equation

$$\sigma(x)\Delta\nabla p_n(x) + \tau(x)\Delta p_n(x) + \lambda_n p_n(x) = 0$$

with $\Delta f(x) = f(x+1) - f(x)$, $\nabla f(x) = f(x) - f(x-1)$, $\sigma(x)$ and $\tau(x)$ are functions of second and first order respectively;
an orthogonality relation

$$\sum_{x=a}^{b-1} p_n(x)\, p_m(x)\, \rho(x) = d_n^2\, \delta_{mn}$$

with $\rho(x)$ a weight function and d_n a normalization constant. To these polynomials correspond the Meixner, Kravchuk, Charlier and Hahn polynomials [1]

b) *On non homogeneous lattice*: $x = x(s)$, $s = 0, 1, 2, \cdots$ the polynomials satisfy a difference equation

$$\sigma\left[x(s)\right] \frac{\Delta}{\Delta x\left(s - \frac{1}{2}\right)} \frac{\nabla y_n(x)}{\nabla x(s)} + \frac{1}{2}\tau\left[x(s)\right]\left\{\frac{\Delta y_n(x)}{\Delta x(s)} + \frac{\nabla y_n(x)}{\nabla x(s)}\right\} + \lambda_n y_n(x) = 0;$$

an orthogonalit relation

$$\sum_{s=a}^{b-1} y_n(s)\, y_m(s)\, \rho_n(x)\, \Delta x\left(s - \frac{1}{2}\right) = d_n^2\, \delta_{nm}.$$

The corresponding polynomials are classified according to the latice function:
For $x(s) = s(s+1)$, we have the Racach and dual Hahn polynomials
For $x(s) = q^s$ or $\frac{q^s - q^{-s}}{2}$, we have the q-Kravchuk, q-Meixner, q-Charlier and q-Hahn polynomials.
For $x(s) = \frac{q^s + q^{-s}}{2}$ or $\frac{q^{is} + q^{-is}}{2}$, we have the q-Racach and q-dual Hahn polynomials [2].

2. Generalized Clebsch-Gordon coeficients and generalized $3nj$-Wigner symbols

If two angular momentum operators are coupled to give a total angular momentum $J = J_1 + J_2$ the new basis can be expressed in terms of the old ones

$$|j_1 j_2 jm\rangle = \sum_{m_1+m_2=m} \langle j_1 j_2 m_1 m_2 \mid j_1 j_2 jm \rangle |j_1 j_2 m_1 m_2\rangle .$$

The symmetry properties of the Clebsch-Gordon coefficients in this expansion are more patent if one sbstitutes them by the Wigner symbols

$$\langle j_1 j_2 jm \mid j_1 j_2 m_1 m_2 \rangle = (-1)^{j_1-j_2+j-m}\sqrt{2j+1} \begin{pmatrix} j_1 & j_2 & j \\ m_1 & m_2 & -m \end{pmatrix}$$

Similarly if we couple three angular momentum operator we obtain a new basis in terms of the old ones:

$$|j_1 j_2 j_3 j_{12} jm\rangle = \sum \langle j_1 j_2 j_3 m_1 m_2 m_3 \mid j_1 j_2 j_3 j_{12} jm \rangle |j_1 j_2 j_3 m_1 m_2 m_3\rangle$$

for the coupling $(J_1 + J_2) + J_3 = J$,

$$|j_1 j_2 j_3 j_{23} jm\rangle = \sum \langle j_1 j_2 j_3 m_1 m_2 m_3 \mid j_1 j_2 j_3 j_{23} jm \rangle |j_1 j_2 j_3 m_1 m_2 m_3\rangle$$

for the coupling $J_1 + (J_2 + J_3) = J$.
Both bases are related by some matrix $U(j_{12}, j_{23})$ that can be written in terms of generalized $6j$-Wigner symbol

$$U(j_{12}, j_{23}) = (-1)^{j_1+j_2+j_3+j}\sqrt{(2j_{12}+1)(2j_{23}+1)} \begin{Bmatrix} j_1 & j_2 & j_{12} \\ j_3 & j & j_{23} \end{Bmatrix}$$

In similar fashion can be written the generalized Clebsch-Gordon coefficients and generalized $3nj$-Wigner symbols [3]. The algebraic properties of these symbols can be represented by geometrical graphs [3].

3. $3nj$-symbols as orthogonal polynomials of several discrete variable

The 6j-symbols are proportional to the Racah polynomials through the following relation [2]

$$(-1)^{j_1+j_2+j_{23}}\sqrt{(2j_{12}+1)(2j_{23}+1)}\left\{\begin{pmatrix} j_1 & j_2 & j_{12} \\ j_3 & j & j_{23} \end{pmatrix}\right\} = \frac{\sqrt{\rho(x)}}{d_n}u_n^{(\alpha,\beta)}(x,a,b)$$

with
$x(s) = s(s+1) , \quad s = j_{23}$
$a = j_3 - j_2 , \quad b = j + j_3 + 1 , \quad n = j_{12} - j_1 + j_2 - j$
$\alpha = j_1 - j_2 - j_3 + j , \quad \beta = j_1 - j_2 + j_3 - j$
Using the assymptotic limit of the Racah polynomials and the connections between the Jacobi polynomials and the Wigner little functions one can prove the following approximation of the 6j-sumbols when $j_1 \sim j_2 \sim j_3 \sim j \gg j_{12}$

$$\begin{Bmatrix} j_1 & j_2 & j_{12} \\ j_3 & j & j_{23} \end{Bmatrix} \simeq \frac{(-1)^{j_2+j_3+j_{23}}}{\sqrt{j_1+j_2+1}\sqrt{j_3+j+1}}d_{j_1-j_2,j_3-j}^{j_{12}}(\vartheta) \tag{1}$$

with

$$\cos \theta = \frac{(2j_{23} + 1)^2 - (j_1 + j_2 + 1)^2 - (j_3 + j + 1)^2}{2 (j_1 + j_2 + 1) (j_3 + j + 1)}$$

The $3nj$-symbols of the first and second kind can be written in terms of $6j$-symbols, and therefore in terms of product of Racah polynomials, giving rise to orthogonal polynomials of several discrete variables. To illustrate this take, f.i., the $12j$-symbol of the second kind as a combination of $6j$-symbols.

$$\begin{Bmatrix} j_1 & j_2 & j_3 & j_4 \\ l_1 & l_2 & l_3 & l_4 \\ k_1 & k_2 & k_3 & k_4 \end{Bmatrix} = \sum_x (2x + 1)(-1)^{R_n + 4x} \begin{Bmatrix} j_1 & k_1 & x \\ k_2 & j_2 & l_1 \end{Bmatrix}$$

$$\begin{Bmatrix} j_2 & k_2 & x \\ k_3 & j_3 & l_2 \end{Bmatrix} \begin{Bmatrix} j_3 & k_3 & x \\ k_4 & j_4 & l_3 \end{Bmatrix} \begin{Bmatrix} j_4 & k_4 & x \\ k_1 & j_1 & l_4 \end{Bmatrix}$$

Here $R_n = \sum_{i=1}^{4} (j_i + l_i + k_i)$. Substituting each 6j-symbol for the corresponding Racah polynomial we obtain:

$$\begin{Bmatrix} j_1 & j_2 & j_3 & j_4 \\ l_1 & l_2 & l_3 & l_4 \\ k_1 & k_2 & k_3 & k_4 \end{Bmatrix} = \sum_x \frac{1}{2x + 1} \prod_{i=1}^{4} \frac{\sqrt{\rho(l_i)}}{d_{n_i}} u_{n_i}^{(\alpha_i, \beta_i)}(l_i) \equiv p_n (l_1\ l_2\ l_3\ l_4)$$

which is a polynomial of four discrete variables.
For the assymptotic limit we find

$$\begin{Bmatrix} j_1 & j_2 & j_3 & j_4 \\ l_1 & l_2 & l_3 & l_4 \\ k_1 & k_2 & k_3 & k_4 \end{Bmatrix} \approx \sum_x (2x + 1) \prod_{i=1}^{4} \frac{1}{j_i + k_i + 1} d_{j_i - k_i, j_{i+1} - k_{i+1(\bmod 4)}}^{x} (\vartheta_i)$$

These formulas can be easily generalized to any $3nj$-symbols of first and second kind.

4. Application to spin networks and to Ponzano-Regge integral action

Penrose has proposed a model for the space and time in which the underlying structure is given by a set of interactions between elementary units that satisfy the coupling of angular momentum operators, called spin networks [4]. One particular case of these networks can be described by the graphs of $3nj$-symbols. From different point of view Regge has proposed a method to calculate Einstein action by the approximation of curved riemannian manifold by a polyedron built up of triangles. Later Ponzano and Regge applied the properties of $6j$-symbols to calculate the sum action over this triangulation [5]
Let M be a riemannian manifold that is approximated by a polyedron with boundury D and it is decomposed into p tetrahedra T_k represented by $6j$-symbols.
The polyedron give rise to triangular faces f, represented by $3j$-symbols, and to q internal edges x_i, as well as to external ones l_i with respect to the boundary D.
Ponzano and Regge define the sum

$$S = \sum_{x_i} \prod_{k=1}^{p} T_k (-1)^{\varphi} \prod_{i=1}^{q} (2x_i + 1) \qquad (2)$$

When $l_i \to \infty$, $\hbar \to 0$, $\hbar l_i \to$ finite we recovered the continuous manifold. In order to compute the $6j$-symbols in te classical limit, we uses the assymptotic formula [2]

$$d^{j}_{mm'}(\theta) \approx (-1)^{m-m'} \sqrt{\frac{2}{\pi(j-m)}} \left(\frac{2j+m-m'+1}{2j-m+m'+1}\right)^{\frac{m+m'}{2}}$$

$$\frac{\cos\left[\left(j+\frac{1}{2}\right)\theta - \left(m-m'+\frac{1}{2}\right)\frac{\pi}{2}\right]}{\sqrt{\sin\theta}}$$

at $m \sim m' \sim 1$, $j \gg 1$. Substituting this expression in (1) with $m = j_1 - j_2$, $m' = j_4 - j_5$, $j = j_6$ and taking the edges of the tetrahedra $j_1 + \frac{1}{2}, \ldots, j_6 + \frac{1}{2}$, very large except j_6, we have

$$\begin{Bmatrix} j_1 & j_2 & j_3 \\ j_4 & j_5 & j_6 \end{Bmatrix} \approx \frac{1}{\sqrt{12\pi V}} \cos\left\{\left(j_6 + \frac{1}{2}\right)\theta - \left(j_1 + \frac{1}{2}\right)\frac{\pi}{2} + \right.$$

$$\left. + \left(j_2 + \frac{1}{2}\right)\frac{\pi}{2} - \left(j_4 + \frac{1}{2}\right)\frac{\pi}{2} + \left(j_5 + \frac{1}{2}\right)\frac{\pi}{2} + \frac{\pi}{4}\right\} =$$

$$= \frac{1}{\sqrt{12\pi V}} \cos\left\{\sum_{i=1}^{6}\left(j_i + \frac{1}{2}\right)\theta_i + \frac{\pi}{4}\right\} \tag{3}$$

where θ_i is the dihedral angle for the edge j_i and $V = \frac{1}{6}\left(j_1 + \frac{1}{2}\right)\left(j_4 + \frac{1}{2}\right)\left(j_6 + \frac{1}{2}\right)$ sen θ. Note the formula (3) has been proved rigorously by Roberts [5]. Introducing formula (3) in formula (2), Ponzano and Regge proved that it leads in the continuous limit to the integral action of the general relativity.

This work has been partially supported by M.I.C. (grant FM2000-0357) Spain

References

[1] M. Lorente, "Raising and lowering operators, factorization and difference operators of hypergeometric type", *J.Phys.A.Math.Gen 34* (2001) 569-588.
[2] Nikiforov, A.F. et al., "Classical orthogonal polynomials of a discrete variable", Springer 1991.
[3] Yutsis, A.P. et al., "Mathematical apparatus of the theory of angular momentum", Israel program for scientific translations, Jerusalem 1962.
[4] Penrose, R, "Angular momentum: an approach to combinatorial space-time" in Quantum theory and Beyond" (T. Bastin ed.) Cambridge 1970.
[5] Regge T., Williams R., "Discrete structures in gravity", *J. Math Phys. 41* (2000) 3964-3984.

e-mail: mlp@pinon.ccu.uniovi.es

Inst. Phys. Conf. Ser. No 173: Section 2
Paper presented at 24th Int. Coll. Group Theoretical Methods in Physics, Paris, France, 15–20 July 2002
©*2003 IOP Publishing Ltd*

Holomorphic realization of non-commutative space-time and gauge invariance

R M Mir-Kasimov

Joint Institute for Nuclear Research, Dubna, Russia
and Izmir Institute of High Technology, Izmir, Turkey

Abstract. The realization of the Poincare Lie algebra in terms of noncommutative differential calculus over the commutative algebra of functions is considered. The algebra of functions is defined on the spectrum of the unitary irreducible representations of the De Sitter group. Corresponding space-time carries the noncommutative geometry. Gauge invariance principle consistent with this noncommutative space is considered.

The momentum space of classical relativistic mechanics is the Lobachevsky space - the mass shell of the particle. In the Quantum Field Theory (QFT) the necessity to extend the S-matrix off the mass shell to describe the nontrivial interaction leads to the concept of Minkowski momentum space in which energy and momentum are the independent coordinates. Actually the concept of Minkowski momentum space is an additional axiom of QFT which is accepted silently as an obvious fact. But does the requirement of the Minkowski geometry uniquely follow from the basic physical principles or there exists the freedom to choose other geometries of the momentum space? It was shown (see [1]-[6] and further references there) that there are no restrictions on the geometry of the momentum space except the requirement that mass shell can be embedded into this space as a surface. As the quantum mechanical position operators are the generators of shifts of the momentum space, then fixing the geometry of the momentum space different from the pseudo-Euclidean one we unavoidable change the geometry of space-time, which becomes in general non-commutative [9]-[13]. In this contribution the QFT with the extension off the mass shell into the De Sitter momentum space is considered. We describe shortly the main features of the electromagnetic theory for this case. We consider the two-dimensional momentum space of constant curvature

$$p_L p^L = p_\mu p^\mu - p_2^2 = -M^2 c^2 \quad L = 0, 1, 2, \quad \mu = 0, 1 \tag{1}$$

M is "fundamental mass" [1]-[6]. (In the realistic case $L = 0, ..., 4$ and correspondingly $\mu = 0, ..., 3$) The correspondence with usual QFT, the "flat limit", is fulfilled, i.e. all relations convert into the usual ones, when the momentum is "small"

$$|p_\mu| \ll Mc, \qquad |p_2| \approx Mc \tag{2}$$

The equation of the mass shell

$$p_\mu^2 - m^2 = 0 \tag{3}$$

can be rewritten as

$$(m_2 - p_2)(m_2 + p_2) = 0, \quad m_2 = \sqrt{1 + m^2} = \cosh \mu, \ m = \sinh \mu \tag{4}$$

To each vector p^μ there corresponds two values of p_2. In the consequence each of the two brackets can vanish. We shall suppose that this happens for positive values of p_2

$$2(m_2 - p_2) \varphi(p, p_2) = 0 \tag{5}$$

This is generalized Klein-Gordon equation. In the "flat limit" (2) $p_2 = \sqrt{1 + p_\mu^2}$ $\approx 1 + \frac{p_\mu^2}{2}$ and (5) goes over into the standard Klein-Gordon equation. It can be shown [8] that in presence of the electromagnetic field (5) takes the form

$$\left[(p - eA)_L (p - eA)^{\dagger L} - 4 \sinh^2 \frac{\mu}{2} \right] \varphi(p, p_2) = 0 \tag{6}$$

The De Sitter vector-potential of the electromagnetic field A_L obeys the condition (1)

$$A_L A^L = -1 \tag{7}$$

which excludes the extra component A_4. The free case (5) corresponds to $A_L = V_L$ where $V = \{0_\mu, 1\}$ is the "momentum of the vacuum". Our problem here is to find the gauge transformations under which (6) is invariant. For this we need the configurational space conjugate to the De Sitter momentum space (1). The "plane waves" in this case, i.e., the kernels of the Fourier transform connecting the p-space (1) and the new quantum configurational space and at the same time the state vectors describing the free motion of the particle are the matrix elements $\langle \xi \mid p \rangle$ of the unitary irreducible representations of the isometries group of the space (1). To describe these matrix elements, we use the hyper-spherical coordinates

$$p^0 = \sinh \zeta, \; p^1 = \cosh \zeta \sin \omega, \; p^2 = \cosh \zeta \cos \omega, \tag{8}$$

$-\infty < \zeta < \infty$, $-\pi < \omega < \pi$. There are two principal series of the irreps

$$\sigma = i\Lambda - \frac{1}{2}, \; 0 \leq \Lambda < \infty, \text{ or } \sigma = k = 0, 1, 2, \ldots; \; n = 0, \pm 1, \pm 2, \ldots \tag{9}$$

We call the set $\xi = (\sigma, n)$ a point of quantum space. The plane waves obey the orthogonality and completeness conditions (see [3]). We shall limit ourselves with the continuous part of the spectrum.

$$\langle \xi \mid p \rangle = \langle z, \bar{z} \mid \zeta, \omega \rangle$$
$$= (\cosh \zeta)^{-\frac{z - \bar{z} + 1}{2}} e^{i\omega \frac{z + \bar{z}}{2}} F\left(\frac{z + \frac{1}{2}}{2}, \frac{-\bar{z} + \frac{1}{2}}{2}; \frac{1}{2}; \tanh^2 \zeta \right) \tag{10}$$
$$-2\xi^0 \tanh \zeta \cdot \kappa F\left(\frac{z + \frac{3}{2}}{2}, \frac{-\bar{z} + \frac{3}{2}}{2}; \frac{3}{2}; \tanh^2 \zeta \right)$$

where

$$\kappa = \frac{\Gamma\left(\frac{z + \frac{3}{2}}{2} \right) \Gamma\left(\frac{-\bar{z} + \frac{3}{2}}{2} \right)}{\Gamma\left(\frac{z + \frac{1}{2}}{2} \right) \Gamma\left(\frac{-\bar{z} + \frac{1}{2}}{2} \right)}, \quad z = i\Lambda + n = \sigma + n + \frac{1}{2} \tag{11}$$

In the flat limit $\langle \xi \mid p \rangle \to e^{ip_\mu x^\mu}$ As the hypergeometric function is the entire function of its parameters it can be easily seen that the plane wave $\langle z, \bar{z} \mid \zeta, \omega \rangle$ is an analytic function of its first argument z and anti-analytic function of its second argument $\bar{z}$. This function obeys the symmetry condition

$$\overline{\langle z, \bar{z} \mid \zeta, \omega \rangle} = \langle \bar{z}, z \mid \zeta, \omega \rangle \tag{12}$$

We shall call (10) the holomorphic form of the plane wave and the corresponding ξ-space the holomorphic realization of the quantum configurational space.

At the same time it is well known fact that hypergeometric functions don't obey any differential relation in its parameters, but obey the recurrence relations. This makes them the subjects to the non-commutative differential calculus. The holomorphic representation leads to the simplest form of such a calculus. We start with basic relations

$$[z, dz] = dz, \quad [\bar{z}, d\bar{z}] = d\bar{z}, \quad [\bar{z}, dz] = 0, \quad [z, d\bar{z}] = 0, \quad [z, \bar{z}] = 0 \quad (13)$$

in which in contrast with the standard (commutative) calculus the coordinates z, and $\bar{z}$ commute between themselves but don't commute with corresponding differentials. It can be shown that the comprehensive differential calculus based on the relations (13) exists [1]. We can introduce the generalized interior derivatives right $\overrightarrow{\partial}$ and left $\overleftarrow{\partial}$ of the function $f(z, \bar{z})$ as

$$d_z f = [dz, f] = \overrightarrow{\partial}_z f \, dz = dz \, \overleftarrow{\partial}_z f \quad (14)$$

and similar formulae for the non-commutative differentiation in $\bar{z}$. The Leibnitz rule is fulfilled for the exterior differentiations (14)

$$d_z(fg) = (d_z f)g + f(d_z g) \quad (15)$$

Referring the reader to [1, 3] for details we write down here the expressions for the momentum operators We deliver here the only information necessary for introducing the physical operators and refer the reader for the detailed theory of non-commutative differential forms on the commutative algebra of functions to [1, 12].

$$p^+ = 1 + \frac{2}{(z - \bar{z})} \left\{ \left(z + \frac{1}{2}\right) \overrightarrow{\partial}_{2z} - \left(\bar{z} + \frac{1}{2}\right) \overrightarrow{\partial}_{2\bar{z}} \right\}$$

$$p^- = 1 - \frac{2}{(z - \bar{z})} \left\{ \left(z + \frac{1}{2}\right) \overleftarrow{\partial}_{2z} - \left(\bar{z} + \frac{1}{2}\right) \overleftarrow{\partial}_{2\bar{z}} \right\}$$

$$p^0 = -\frac{2\,\xi^0\,\kappa}{(z - \bar{z})} \left\{ -\overrightarrow{\partial}_z \overrightarrow{\partial}_{\bar{z}} + \overleftarrow{\partial}_z \overrightarrow{\partial}_{\bar{z}} + \overrightarrow{\partial}_z + \overleftarrow{\partial}_z - \overrightarrow{\partial}_{\bar{z}} - \overleftarrow{\partial}_{\bar{z}} \right\} \quad (16)$$

$$p^+ = p^2 + ip^1 \qquad p^- = p^2 - ip^1$$

Operators p^L mutually commute and obey the De Sitter condition (1). Their common eigen-functions are the plane waves (10) with eigen-values (8).

Let us consider the gauge transformation localized in quantum ξ-space-time

$$\varphi'(\xi) = \Omega(\xi)\,\varphi(\xi) \qquad \Omega(\xi)^\dagger = \Omega(\xi)^{-1} \quad (17)$$

Unlike the usual theory the gauge transformation entangles the components of momenta:

$$\Omega^{-1}(\xi)\,P^L\Omega(\xi) = C^L{}_K(\xi)\,P^K, \quad \Omega^{-1} P_L \Omega = P_K C_L^{\dagger K}, \quad C^\dagger = C^{-1} \quad (18)$$

We write down explicitly the $C^L{}_K(z, \bar{z})$-matrix for the case when it depends on only on the variable $\frac{z - \bar{z}}{2}$ referring the reader for details to [4]

$$C^L{}_K\left(\frac{z - \bar{z}}{2}\right) = \left[\left(\partial^{(c)}_{\frac{z - \bar{z}}{2}} \Omega\left(\frac{z - \bar{z}}{2}\right) \right) - \frac{2i}{z - \bar{z}} \left(\partial^{(s)}_{\frac{z - \bar{z}}{2}} \Omega\left(\frac{z - \bar{z}}{2}\right) \right) \hat{\Sigma} \right]^L{}_K \quad (19)$$

where

$$\hat{\Sigma} = \begin{pmatrix} \frac{i}{2} & -M^{10} & -M^{20} \\ -\hat{M}^{10} & \frac{i}{2} & M^{12} \\ -M^{20} & -\hat{M}^{12} & \frac{i}{2} \end{pmatrix} \quad (20)$$

It is easily seen that in a consequence of (18) the De Sitter condition (1) is invariant in respect to the gauge transformations (17). It can be easily shown [8] that to make the theory gauge invariant we must introduce the complex De Sitter vector $\hat{A}^L$ of electromagnetic field and require that it transforms similarly to (18):

$$\Omega^{-1}\hat{A}^L\Omega = C^L{}_K\hat{A}^K, \quad \Omega^{-1}\hat{A}^\dagger_L\Omega = \hat{A}^\dagger_K C_L^{\dagger K} \quad (21)$$

It follows from (21) that the components of electromagnetic field do not commute with the gauge function and in a consequence do not commute between themselves. Introducing the covariant derivatives

$$\hat{D}^L = -i\left(p^L - \hat{A}^L\right) \tag{22}$$

we obtain the De Sitter invariant equations for the matter fields. The tensor of electromagnetic field is given as

$$\hat{F}^{KL} = \left[\hat{D}^L, \hat{D}^K\right] = \left[\hat{p}^L, \hat{A}^K\right] - \left[\hat{p}^K, \hat{A}^L\right] - \left[\hat{A}^L, \hat{A}^K\right] \tag{23}$$

The action of the electromagnetic field is

$$S = Tr\int \hat{F}^{KL}\hat{F}_{KL}\,d\Omega_\xi \tag{24}$$

and the non-commutative analog of the D'Alembert equation takes the form

$$\left[\hat{D}_K, \hat{F}^{KL}\right] = 0 \tag{25}$$

First pair of the Maxwell equations

$$-\left[D^K, \left(F^{ML}\right)\right] - \left[D^M, \left(F^{LK}\right)\right] - \left[D^L, \left(F^{KM}\right)\right]$$
$$= \left[D^K, \left[D^L, D^M\right]\right]\left[D^M, \left[D^K, D^L\right]\right]\left[D^L, \left[D^M, D^K\right]\right] = 0 \tag{26}$$

At the first sight these equations contain the cubic terms A_L^3. But thanks to the independent Jacoby identities the terms nonlinear in A cancel each other.

$$\left[p^K, \left[p^L, A^M\right]\right] - \left[p^K, \left[p^M, A^L\right]\right] + (\text{cyclic } K, L, M) = 0 \tag{27}$$

Second pair of Maxwell equations

$$\left[D_K, F^{KL}\right] = 4\pi J^L \tag{28}$$

Correspondence with the standard Maxwell theory is restored in the flat limit. To exclude the "extra" component of the 5-vector-potential of the electromagnetic field A_2 we use (7). The right limit for the Maxwell equations is realized for the weak electromagnetic fields:

$$\left|\frac{e}{c}A_\mu\right| \ll Mc, \quad \left|\frac{e}{c}A_2\right| \approx Mc, \quad \left[p_L, A^L\right] \to [p_\mu, A^\mu] = 0 \tag{29}$$

References

[1] Mir – Kasimov R M 2000 Physics of Particles and Nuclei 31 44-64
[2] Mir – Kasimov R M 2002 Foundations of Physics, 32 607-626
[3] Güven Z, Can Z, Mir-Kasimov R, and Oğuz 2001 Physics of Atomic Nuclei 64 2143-2155
[4] Mir – Kasimov R M 2002 Proc. Int. Conf. "Quantum Theories and Symmetries" (Singapore: World Scientific) p.487
[5] 1988 Kadyshevsky V G Nuclear Physics B 141 477-489
[6] Kadyshevsky V G and Fursaev D V 1990 Theor. and Math. Phys, 83 197
[7] 2002 Kadyshevsky V G and Sorin A S Theor. Math. Phys. 132 1079
[8] Mir – Kasimov R M 1991 Phys. Lett. B, 259 79 1997 Int.Journ.Mod.Phys. 12 24 1998 Physics of Atomic Nuclei 61 1951
[9] 1989 Woronowich D V Commun.Math.Phys. 122 125
[10] 1990 Dubois-Violette M, Kerner R and Madore J Journ.Math.Phys 31 323
[11] 1994 Connes A - Non-commutative Geometry Academic Press
[12] 1998 Dimakis A and Müller-Hoissen Journ.Math.Phys 40 1518
[13] 1986 E.Witten, Nucl.Phys. B268, 253 1999 N.Seiberg and E.Witten, JHEP 9909, 032

Inst. Phys. Conf. Ser. No 173: Section 2
Paper presented at 24th Int. Coll. Group Theoretical Methods in Physics, Paris, France, 15–20 July 2002
©2003 IOP Publishing Ltd

Four classes of modified relativistic symmetry transformations

Jerzy Lukierski *

Institute for Theoretical Physics, University of Wrocław, pl. Maxa Borna 9, 50-204 Wrocław, Poland

Anatol Nowicki *‡

Institute of Physics, University of Zielona Góra, ul. Podgórna 50, 65-246 Zielona Góra, Poland

Abstract. We discuss the nonlinear transformations of standard Poincaré symmetry in the context of recently introduced Doubly Special Relativity (DSR) theories. We introduce four classes of modified relativistic theories with three of them describing various DSR frameworks. We consider four examples of modified relativistic symmetries, which illustrate each of the considered class.

1. Introduction

The classical relativistic symmetries as described by classical Poincaré-Hopf algebra can be deformed due to the following two reasons (see also [1,2]):

i) One can introduce new nonlinear basis in enveloping algebra of classical Poincaré Lie algebra,

ii) One can deform the classical coalgebraic structure, which leads to quantum Poincaré algebras and by considering dual Hopf algebra structure also provides the quantum Poincaré groups.

In this talk we shall consider the modification of relativistic symmetry transformations due to the nonlinear change of basis in the algebraic sector. Recently Amelino-Camelia [3] introduced two categories of modified relativistic symmetries with two parameters c and κ – light velocity c and fundamental mass κ which can be identified with Planck mass – invariant under the modified Lorentz transformations:

i) Doubly special relativistic theories of first type, denoted by DSR1, with energy E unbounded and momentum $\vec{P}$ bounded by mass-like parameter κ: $|\vec{P}| \leq \kappa c$

ii) DSR2 theories, with both energy and moment bounded by κ: $|\vec{P}| \leq \kappa c$ and $E \leq \kappa c^2$. The first examples of DSR2 theories were provided by Magueijo and Smolin [4,5].

In order to complete the classification we shall introduce further two categories:

iii) DSR3 theories, with momentum unbounded and energy bounded: $E \leq \kappa c^2$.

iv) The theories with momentum as well as energy unbounded. In such a framework two parameters c and κ do not have the meaning of invariant parameters. We shall call these theories Smoothly Modified Special Relativity (SMSR) theories.

‡ Talk given by A. Nowicki
*Supported by the Polish State Committee for Scientific Research (KBN) grant No 5P03B05620

2. Nonlinear realizations of relativistic symmetries

We use the following notation in the description of classical D=4 Poincaré algebra:
– Lorentz algebra ($g_{\mu\nu} = (-1, 1, 1, 1)$)

$$[M_{\mu\nu}, M_{\rho\tau}] = i(g_{\mu\rho}M_{\nu\tau} + g_{\nu\tau}M_{\mu\rho} - g_{\mu\tau}M_{\nu\rho} - g_{\nu\rho}M_{\mu\tau}). \tag{1}$$

– covariance relations ($M_i = \frac{1}{2}\epsilon_{ijk}M_{jk}$, $N_i = M_{i0}$; c is the light velocity)

$$[M_i, \mathcal{P}_j] = i\epsilon_{ijk}\mathcal{P}_k, \quad [M_i, \mathcal{E}] = 0, \quad [N_i, \mathcal{P}_j] = \frac{i}{c}\delta_{ij}\mathcal{E}, \quad [N_i, \mathcal{E}] = ic\mathcal{P}_i. \tag{2}$$

– commuting four momenta $[P_\mu, P_\nu] = 0$
The momentum variables fulfill standard mass shell condition, described by mass Casimir

$$\mathcal{E}^2 - c^2\mathcal{P}^2 = \mu^2 c^4. \tag{3}$$

Mass shall (3) is covariant under the relativistic boost transformations

$$\mathcal{E}(\alpha) = \mathcal{E}\cosh(\alpha) - c(\vec{n} \cdot \vec{\mathcal{P}})\sinh(\alpha). \tag{4}$$

$$\vec{\mathcal{P}}(\alpha) = \vec{\mathcal{P}} + \left((\cosh\alpha - 1)\vec{n} \cdot \vec{\mathcal{P}} - \frac{\mathcal{E}}{c}\sinh\alpha\right)\vec{n}. \tag{5}$$

with the rapidity–velocity relation $\vec{v} = c\vec{n}\tanh\alpha$, $\vec{\alpha} = \alpha\vec{n}$.
We shall consider invertible nonlinear transformations of the momentum space in the form

$$\vec{\mathcal{P}} = \vec{\mathcal{P}}(E, \vec{P}) = \vec{P}\,g\left(\frac{E}{\kappa c^2}, \frac{\vec{P}^2}{\kappa^2 c^2}\right), \quad \mathcal{E} = \mathcal{E}(E, \vec{P}) = \kappa c^2 f\left(\frac{E}{\kappa c^2}, \frac{\vec{P}^2}{\kappa^2 c^2}\right). \tag{6}$$

with the dependence on dimensionfull mass-like parameter κ satisfying the conditions

$$\lim_{\kappa\to\infty} g = 1, \qquad \lim_{\kappa\to\infty}(\kappa c^2 f) = E. \tag{7}$$

This form of transformations imply only changes in covariance relations (2) i.e. $[N, P]$ depend on functions f, g but the classical Lorentz algebra (1) is not changed.
In transformed variables E, $\vec{P}$ the dispersion relation (3) is given by

$$\kappa^2 c^4 f^2\left(\frac{E}{\kappa c^2}, \frac{\vec{P}^2}{\kappa^2 c^2}\right) - c^2\vec{P}^2 g^2\left(\frac{E}{\kappa c^2}, \frac{\vec{P}^2}{\kappa^2 c^2}\right) = inv. = \mu^2 c^4. \tag{8}$$

and it is invariant under nonlinear transformations of boosts.
For the special choices of functions f and g we obtain three types of DSR and fourth class of SMSR theories, mentioned in the Introduction.

3. Doubly special relativity theories

3.1. DSR1 as nonlinear realization of Poincaré algebra - an example

We define nonlinear transformations of the momentum subalgebra as follows

$$\vec{\mathcal{P}} = \vec{P}e^{\frac{E}{\kappa c^2}}, \quad \mathcal{E} = \kappa c^2\left(\sinh\frac{E}{\kappa c^2} + \frac{\vec{P}^2}{2\kappa^2 c^2}e^{\frac{E}{\kappa c^2}}\right). \tag{9}$$

In this new basis $(E, \vec{P})$ the covariance relations (2) takes the form

$$[M_i, P_j] = i\epsilon_{ijk}P_k, \qquad [M_i, E] = 0,$$

$$[N_i, P_j] = i\kappa c\delta_{ij}\left[\sinh\left(\frac{E}{\kappa c^2}\right)e^{-\frac{P_0}{\kappa c}} + \frac{1}{2\kappa^2 c^2}(\vec{P})^2\right] - \frac{i}{\kappa c}P_iP_j,$$

$$[N_i, E] = icP_i. \tag{10}$$

DSR1 energy-momentum dispersion relation (κ-deformed mass Casimir) is given by

$$C_2 = \left(2\kappa\sinh\frac{E}{2\kappa c^2}\right)^2 - \frac{1}{c^2}\vec{P}^2 e^{\frac{E}{\kappa c^2}} = M^2. \tag{11}$$

Using the formulae (4-5) we get nonlinearly modified boosts transformations§

$$E(\alpha) = E + \kappa c^2 \ln W(\alpha, \vec{n}\vec{P}, E). \tag{12}$$

$$\vec{P}(\alpha) = W^{-1}(...)\left[\vec{P} + \left((\vec{n}\vec{P})(\cosh\alpha - 1) - \kappa cB(m, E)\sinh\alpha\right)\vec{n}\right]. \tag{13}$$

where

$$W(\alpha, \vec{n}\vec{P}, E) = 1 - \left(\frac{1}{\kappa c}(\vec{n}\cdot\vec{P})\sinh\alpha + B(m, E)(1 - \cosh\alpha)\right),$$

$$B(m, E) = 1 - \cosh\left(\frac{m}{\kappa}\right)e^{-\frac{E}{\kappa c^2}} = \frac{1}{2}\left(1 - e^{-\frac{2E}{\kappa c^2}} + \frac{\vec{P}^2}{\kappa^2 c^2}\right). \tag{14}$$

3.2. DSR2 as nonlinear realization of Poincaré algebra - an example

We assume that the classical Lorentz algebra is given by the formulae (1). We define nonlinear transformations of the momentum subalgebra as follows

$$\vec{\mathcal{P}} = \vec{P}\left(1 - \frac{E}{\kappa c^2}\right)^{-1}, \quad \mathcal{E} = E\left(1 - \frac{E}{\kappa c^2}\right)^{-1}. \tag{15}$$

In this new basis $(E, \vec{P})$ the covariance relations (2) take the form

$$[M_i, P_j] = i\epsilon_{ijk}P_k, \qquad [N_i, P_j] = \frac{i}{c}\left(\delta_{ij}E - \frac{P_iP_j}{\kappa}\right),$$

$$[M_i, E] = 0, \qquad [N_i, E] = 2ic\left(1 - \frac{E}{\kappa c^2}\right)P_i. \tag{16}$$

DSR2 energy-momentum dispersion relation is given by

$$C_2 = \frac{E^2 - c^2\vec{P}^2}{\left(1 - \frac{E}{\kappa c^2}\right)^2} = M^2 c^4. \tag{17}$$

Using the formulae (5-6) we get nonlinearly modified boost transformations

$$E(\alpha) = \left(E\cosh\alpha - c(\vec{n}\vec{P})\sinh\alpha\right)\mathcal{W}^{-1}(\alpha, \vec{n}\vec{P}, E). \tag{18}$$

$$\vec{P}(\alpha) = \left(\vec{P} + \vec{n}\left((\cosh\alpha - 1)\vec{n}\cdot\vec{P} - \frac{E}{c}\sinh\alpha\right)\right)\mathcal{W}^{-1}(\alpha, \vec{n}\vec{P}, E). \tag{19}$$

where

$$\mathcal{W}(\alpha, \vec{n}\vec{P}, E) = 1 + \frac{E}{\kappa c^2}(\cosh\alpha - 1) - \frac{(\vec{n}\vec{P})}{\kappa c}\sinh\alpha. \tag{20}$$

§ The transformation (13) has been firstly described for special choice of boost parameter $\vec{\alpha} = (0, 0, \alpha)$ in [6]; the general formula (13) was obtained firstly in [2] and further discussed in [7].

3.3. DSR3 as nonlinear realization of Poincaré algebra - an example

We assume that the classical Lorentz algebra is given by the formulae (1). We define nonlinear transformations of the four momentum subalgebra as follows

$$\vec{\mathcal{P}} = \vec{P}, \quad \mathcal{E} = E \left(1 + \frac{\vec{P}^2}{\kappa^2 c^2} \right)^{1/2} . \tag{21}$$

In this new basis $(E, \vec{P})$ the covariance relations (2) takes the form

$$[M_i, P_j] = i\epsilon_{ijk} P_k, \qquad [N_i, P_j] = \frac{i}{c} \delta_{ij} E \left(1 + \frac{\vec{P}^2}{\kappa^2 c^2} \right)^{1/2},$$

$$[M_i, E] = [P_\mu, P_\nu] = 0, \quad [N_i, E] = icP_i \left(1 + \frac{\vec{P}^2}{\kappa^2 c^2} \right)^{-1/2} \left(1 - \frac{E^2}{\kappa^2 c^4} \right) . \tag{22}$$

The energy-momentum dispersion relation is given by

$$C_2 = E^2 \left(1 + \frac{\vec{P}^2}{\kappa^2 c^2} \right) - c^2 \vec{P}^2 = M^2 c^4. \tag{23}$$

Then using the formulae (4-5) we get the nonlinear boost transformations

$$E(\alpha) = \left(\frac{1 + \vec{P}^2(\alpha)}{1 + \vec{P}^2} \right)^{-\frac{1}{2}} \left[E \cosh \alpha - c(\vec{n}\vec{P}) \left(1 + \frac{\vec{P}^2}{\kappa^2 c^2} \right)^{-1/2} \sinh \alpha \right] \tag{24}$$

$$\vec{P}(\alpha) = \vec{P} + \left((\cosh \alpha - 1)\vec{n} \cdot \vec{P} - \frac{E}{c} \left(1 + \frac{\vec{P}^2}{\kappa^2 c^2} \right)^{1/2} \sinh \alpha \right) \vec{n}. \tag{25}$$

4. SMSR as nonlinear realization of Poincaré algebra - an example

We assume that the classical Lorentz algebra is given by the formulae (2). We define nonlinear transformations of the momentum subalgebra as follows

$$\vec{\mathcal{P}} = \vec{P}, \quad \mathcal{E} = 2\kappa c^2 \sinh \left(\frac{E}{2\kappa c^2} \right). \tag{26}$$

In this new basis $(E, \vec{P})$ the covariance relations (2) take the form

$$[M_i, P_j] = i\epsilon_{ijk} P_k, \qquad [N_i, E] = icP_i \cosh^{-1} \left(\frac{E}{2\kappa c^2} \right),$$

$$[M_i, E] = [P_\mu, P_\nu] = 0, \quad [N_i, P_j] = 2i\kappa c \delta_{ij} \sinh \left(\frac{E}{2\kappa c^2} \right). \tag{27}$$

The corresponding energy-momentum dispersion relation is given by

$$C_2 = \left(2\kappa c^2 \sinh \frac{E}{2\kappa c^2} \right)^2 - c^2 \vec{P}^2 = M^2 c^4. \tag{28}$$

Using the formulae (4-5) we get nonlinear boost transformations

$$E(\alpha) = 2\kappa c^2 \text{arcsinh} \left[\sinh \left(\frac{E}{2\kappa c^2} \right) \cosh \alpha - \frac{(\vec{n}\vec{P})}{2\kappa c} \sinh \alpha \right]. \tag{29}$$

$$\vec{P}(\alpha) = \vec{P} + \left((\cosh \alpha - 1)\vec{n} \cdot \vec{P} - 2\kappa c \sinh \left(\frac{E}{2\kappa c^2} \right) \sinh \alpha \right) \vec{n}. \tag{30}$$

One can see easily that the values of $E(\alpha)$ and $P(\alpha)$ are not bounded.

5. Final Remarks

In this talk we mainly classified different nonlinear bases for classical Poincaré algebra. If we observe that the classical four momenta generators are endoved with primitive coproducts $(\mathcal{P}_\mu = (\mathcal{E}/c, \mathcal{P}_i))$

$$\Delta^{(0)} \mathcal{P}_\mu = \mathcal{P}_\mu \otimes 1 + 1 \otimes \mathcal{P}_\mu. \tag{31}$$

then by considering the formulas inverse to (6)

$$E = \kappa \, c^2 F \left(\frac{\mathcal{E}}{\kappa c^2}, \frac{\vec{\mathcal{P}}^2}{\kappa^2 c^2} \right), \qquad \vec{P} = \vec{\mathcal{P}} \, G \left(\frac{\mathcal{E}}{\kappa c^2}, \frac{\vec{\mathcal{P}}^2}{\kappa^2 c^2} \right). \tag{32}$$

one obtains the symmetric nonlinear coproducts for the nonlinear momentum $\vec{P}$ and nonlinear energy E. Such coproducts were considered in [2] as describing the deformed addition law for fourmomenta and further used in [8] in the notation without the notion of coproduct. We get the coproduct formulae:

$$\Delta E = \kappa \, c^2 F \left(\frac{\Delta \mathcal{E}}{\kappa c^2}, \frac{(\Delta \vec{\mathcal{P}})^2}{\kappa^2 c^2} \right), \qquad \Delta \vec{P} = \Delta(\vec{\mathcal{P}}) G \left(\frac{\Delta \mathcal{E}}{\kappa c^2}, \frac{(\Delta \vec{\mathcal{P}})^2}{\kappa^2 c^2}, \right). \tag{33}$$

where

$$\Delta \mathcal{E} = \mathcal{E}(E, \vec{P}) \otimes 1 + 1 \otimes \mathcal{E}(E, \vec{P}),$$
$$\Delta \vec{\mathcal{P}} = \vec{\mathcal{P}}(E, \vec{P}) \otimes 1 + 1 \otimes \vec{\mathcal{P}}(E, \vec{P}). \tag{34}$$

The quantum nonsymmetric coproduct is obtained if we modify primitive coproduct (31) by introducing Drinfeld twist T [1]

$$\Delta \mathcal{P}_\mu = T^{-1} \circ \Delta^{(0)}(\mathcal{P}_\mu) \circ T = \Delta^{(0)}(\mathcal{P}_\mu) + [r, \mathcal{P}_\mu] + \dots. \tag{35}$$

where r is the classical Poincaré r-matrix [9]. In such a way we obtain the quantum Poincaré algebras with nonlinear fourmomentum basis given by (32) and nonprimitive coproduct (33) with inserted formulae (35).

References

[1] J. Lukierski and A. Nowicki, hep-th/0209017.
[2] J. Lukierski and A. Nowicki, hep-th/0203065.
[3] G. Amelino-Camelia, D. Benetti and F. D'Andrea, hep-th/0201245.
[4] J. Magueijo, L.Smolin, Phys. Rev. Lett. **88**, 190403 (2002).
[5] J. Magueijo, L.Smolin, hep-th/0207085.
[6] N.R. Bruno, G. Amelino-Camelia, and J. Kowalski-Glikman, Phys. Lett. **B522**, 133 (2001)
[7] N.R. Bruno, gr-qc/0207076.
[8] S. Judes and M. Visser, gr-qc/0207085.
[9] S. Zakrzewski, Comm. Math. Phys. **185**, 285 (1997).

Inst. Phys. Conf. Ser. No 173: Section 2
Paper presented at 24th Int. Coll. Group Theoretical Methods in Physics, Paris, France, 15–20 July 2002
©2003 IOP Publishing Ltd

Cosmic Evolution as Inertial Motion in the Field Space of GR

V. Pervushin† and D. Proskurin†

† Bogoliubov Laboratory for Theoretical Physics,
Joint Institute for Nuclear Research, 141980 Dubna, Russia

E-mail: pervush@thsun1.jinr.ru

Abstract. The latest Supernova data on the redshift - luminosity-distance relation and the primordial elements abundance are treated as evidences for the conformal cosmology in the form of an *inertial motion* of the universe along geodesic lines in the field space of metric in General Relativity. The geometry of this space is determined by the method of the Cartan forms and the theory of nonlinear representation of an affine group $A(4)$ over the Lorentz one L. This "inertial motion" in the coset $A(4)/L$ is considered as a simplest model of quantum cosmology to study creation of the universe and creation of matter from the vacuum with the CMB temperature $2.7K$ as a constant of the "motion" and to discuss problems of cosmic singularity, cosmic initial data, the status of time in quantum theory, and the positive arrow of time.

Our idea is to consider the cosmic evolution of the universe in the Einstein General Relativity as a collective motion of all fields (metric and matter). The method of the Cartan forms and the theory of nonlinear representation of an affine group help us to determine a geometry of field space and to extend the Copernicus and Einstein principles of relativity (including the concepts of an inertial motion, geodesic line, relative and absolute "coordinates") to this field space.

We have shown that the averaging exact classical equations over the spatial volume reproduces the equations of the conformal version of the standard FRW cosmology. This means that *homogeneous approximation is not strictly necessary*. We can observe the cosmic evolution in the region where the energy density coincides with its averaging value.

We have shown that the cosmic evolution in the form of collective inertial motion of metric and matter along geodesic lines in the coset of an affine group $A(4)$ over the Lorentz one L corresponds to the choice of the rigid equation of state where the density is equal to pressure.

There are two standards of measurement: the absolute F that corresponds to the standard cosmology, and the relative $\bar{F}$ corresponding to the conformal cosmology.

Both the standards can be used for description of observational data. But the absolute standard cosmology requires three eras with different regimes (primordial inflation, radiation, and the present-day inflation); whereas the relative cosmology requires for description of observational data only one era that corresponds to the inertial motion of the Universe along a geodesic line in the field space. We have shown that "inertial" motions of the Universe (defined as ones with constant momenta) along a geodesic line of the factor-space $A(4)/L$ do not contradict data of observational cosmology including the primordial element abundance, and the latest Supernova data on the redshift - luminosity-distance relation.

However, in contrast to the standard FRW cosmology, the observable Universe in the relative cosmology has a constant volume and temperature but varying masses (including the Planck mass) like the Universe in the Hoyle-Narlikar conformal cosmology.

The next difference is the *Cosmic Relativity* that means that the time, energy, and cosmic initial data of the relativistic Universe are defined as similar quantities in Special Relativity, where for description of pure relativistic effects one uses two frames of references: the rest and comoving. A collective motion of the relativistic Universe can be covered by two Newton-like mechanical systems (with two wave functions in the initial *world field "space"* and *world geometrical "space")* and their relation in the form of the Levi-Civita canonical transformation.

The inertial motion in the coset A(4)/L gives us possible explanations of the creation of the universe in the initial *world field "space"*, the creation of time with its positive arrow in *world geometrical "space"*, the SN evolution of the cosmic scale factor as pure quantum relativistic effects.

The consistent and complete description of the creation of the quantum universe and its evolution allow us to consider the creation of matter in the quantum universe in the Standard Model of the electroweak and strong interactions. This Model points out on the creation of W, Z - vector bosons from the geometrical vacuum due to their mass singularity $\bar{m}_W(a \to 0) \to 0$. This mass singularity is the physical origin of the temperature of the matter. The inertial cosmic motion in the coset $A(4)/L$ leads to the definite temperature of the matter $T_I \simeq (m_W^2 H)^{1/3} K$ depending on the boson mass m_W and the Hubble parameter H. This temperature is an integral of the inertial motion and this integral coincides with the primordial value of the Hubble parameter $T_I = H_I$. These primordial bosons decay with the baryon number violation. The CMB radiation as the product of decay of the primordial bosons keeps in the rigid regime the primordial value of the temperature $T_{CMB} \simeq (m_{W0}^2 H_0)^{1/3} = 2.7K$, where m_{W0}, H_0 are the present - day values of boson mass and the Hubble parameter.

Thus, the new Cold Universe Scenario not only predicts CMB but also gives an opportunity to create a microscopic theory of CMB temperature.

Acknowledgments

We are grateful to B.M. Barbashov, D.B. Blaschke, A.A. Gusev, and S.I. Vinitsky for fruitful discussions.

References

[1] Pervushin V N 1975 *Teor. Mat. Fiz.* **22** 201
[2] Pervushin V N 1976 *Teor. Mat. Fiz.* **27** 16
[3] Kazakov D I, Pervushin V N and Pushkin S.V. 1977 *Teor. Mat. Fiz.* **31** 169
[4] Isaev G V, Pervushin V N and Pushkin S V 1979 *J. Phys. A: Math. Gen.* **12** 1499
[5] Blaschke D et al, Cosmological creation of vector bosons and CMB
    (Blaschke D et al Preprint 2001 gr-qc/0103114)
    Pervushin V N and Proskurin D V 2001 *Talk at the V International Conference on Cosmoparticle Physics (Cosmion-2001) dedicated to 80-th Anniversary of Andrei D. Sakharov* (21-30 May 2001, Moscow-St.Peterburg, Russia)
    (Preprint gr-qc/0106006)
[6] Behnke D. et al 2002 *Phys. Lett. B* **530** 20
    (Behnke D. et al 2001 gr-qc/0102039)
[7] Gyngazov L N, Pawlowski M, Pervushin V N, and Smirichinski V I 1998 *Gen. Rel. and Grav.* **30** 1749.
[8] Pawlowski M, Papoyan V V, Pervushin V N and Smirichinski V I 1998 *Phys. Lett. B* **444** 293
[9] Pervushin V N, Smirichinski V I 1999 *J. Phys. A: Math. Gen.* **32** 6191
[10] Pawlowski M and Pervushin V N 2001 *Int. J. Mod. Phys.* **16** 1715
    (Pawlowski M and Pervushin V N 2001 Preprint hep-th/0006116)
    Pervushin V N and Proskurin D V 2001 *Gravitation and Cosmology* **7** 89
[11] Barbashov B M and Pervushin V N 2001 *Theor. Math. Phys.* **127** 483
    (Barbashov B M and Pervushin V N 2000 Preprint hep-th/0005140)

[12] Barbashov B M, Pervushin V N and Pawlowski M 2001 *Phys. Particles and Nuclei* **32** 546
[13] Barbashov B M, Pervushin V N and Proskurin D 2002 *Theor. Math. Phys.* **132**
[14] Blaschke D et al *Yad. Fiz.* (submitted).
[15] Pavel H P and Pervushin V N 1999 *Int. J. Mod. Phys. A* **14** 2285

Inst. Phys. Conf. Ser. No 173: Section 2
Paper presented at 24th Int. Coll. Group Theoretical Methods in Physics, Paris, France, 15–20 July 2002
©2003 IOP Publishing Ltd

New Einstein-Hilbert Type Action and Unity of Nature

Kazunari Shima

Laboratory of Physics, Saitama Institute of Technology, Okabe-machi, Saitama 369-0293, Japan

Abstract. A new Einstein-Hilbert type (SGM) action describing gravitational interaction of Nambu-Goldstone(N-G) fermion of nonlinear supersymmetry(NL SUSY) is obtained by performing the Einstein gravity analogue geomtrical arguments in high symmetric four dimensioinal (SGM) spacetime. Some phenomenological implications and the linearlization of the high nonlinear SGM action are discussed.

1. Introduction

It is well known that in the monopole *phase*, (i.e. at the very short distances of spacetime), the degrees of freedom of spacetime are *fused* with the dimensions of (the linear representation of) the local symmetry, which allows to define a unified (composite) field strength of the monopole configuation through the symmetry breaking $SU(2) \times SO(3,1) \rightarrow U(1) \times SO(3,1)$. These phenomena suggest that spacetime itself would reveal unfamiliar features at the short distance by the identification(*fusion*) of the symmetries of spacetime with those of matter and that these ultimate spacetime may be described by a certain unified (composite) field strength(curvature), where the no-go theorem may become irrelevant and the fundamental SUSY Lagrangian with $N > 9$ may be written down. Also, we think that from the viewpoint of simplicity and beauty of nature it is interesting to attempt the accommodation of all observed particles in *a single* irreducible representation of a certain algebra(group), especially for spacetime with the boundary. The fundamental theory should be given by only the geometrical arguments of high symmetrical spacetime and its spontaneous breakdown, which is encoded in the geometrical argument of spacetime by itself. In this talk we would like to present a model along this scenario.

2. Superon-Graviton Model(SGM)

Among single irreducible representations of all SO(N) extended super-Poincaré(SP) symmetries, the massless irreducible representations of SO(10) SP algebra(SPA) is the only one that accommodates minimally all observed particles including the graviton[ks1][ks2]. By considering that (i)for the massless case the algebra of the supercharges of SO(10) SPA in the light-cone frame can be recast as those of the creation and annihilation operators of fermions and (ii)10 generators $Q^N (N = 1, 2, .., 10)$ of SO(10) SPA are decomposed $\underline{10} = \underline{5} + \underline{5}^*$ with respect to SU(5) following $SO(10) \supset SU(5)$ and span $2 \cdot 2^{10}$ dimensional massless irreducible representation of SO(10) SPA with the helicity up to 3, we can regard 10 generators $\underline{10} = \underline{5} + \underline{5}^*$ as the fundamental massless objects; *a superon-quintet* and *an antisuperon-quintet* with *spin* $\frac{1}{2}$ and that all the helicity states are the massless (gravitational) *eigenstates* of spacetime and matter with SO(10) SP symmetric structure, which are composed of *superon*. To survey the physical implications of *superon-graviton model(SGM)* for spacetime and *matter* we assign the following SM quantum numbers to superons and adopt the following symbols(and the conjugates for anti-superons).

$$\underline{5} = \left[Q_a(a = 1,2,3), Q_m(m = 4,5) \right] = [(\underline{3},\underline{1}; -\frac{1}{3}, -\frac{1}{3}, -\frac{1}{3}), (\underline{1},\underline{2}; 1,0)], \quad (1)$$

where we have specified $(SU(3), SU(2)$; electric charges $)$. Superon-quintet satisfy the Gell-Mann–Nishijima relation; $Q_e = I_z + \frac{1}{2}(B - L)$. Accordingly all $2 \cdot 2^{10}$ helicity states are specified uniquely with respect to $(SU(3), SU(2)$; electric charges $)$. Here we suppose boldly an ideal superHiggs-like mechanism, i.e. all unnecessary (for SM) higher helicity states become massive by absorbing the lower helicity states in $SU(3) \times SU(2) \times U(1)$ invariant way via [SO(10) SPA upon the Clifford vacuum] $\rightarrow [SU(3) \times SU(2) \times U(1)] \rightarrow [SU(3) \times U(1)]$. We have carried out the recombinations of the helicity states and found surprisingly that all the massless states necessary for the SM with three generations of quarks and leotons appear in the surviving massless states specified by the superon contents[2][3]. For three generations of leptons $[(v_e, e), (v_\mu, \mu), (v_\tau, \tau)]$, we take

$$\left[(Q_m \varepsilon_{ln} Q_l^* Q_n^*), (Q_m \varepsilon_{ln} Q_l^* Q_n^* Q_a Q_a^*), (Q_a Q_a^* Q_b Q_b^* Q_m^*) \right] \quad (2)$$

and for three generations of quarks $[(u,d), (c,s), (t,b)]$, we have *uniquely*

$$\left[(\varepsilon_{abc} Q_b^* Q_c^* Q_m^*), (\varepsilon_{abc} Q_b^* Q_c^* Q_l \varepsilon_{mn} Q_m^* Q_n^*), (\varepsilon_{abc} Q_a^* Q_b^* Q_c^* Q_d Q_m^*) \right] \quad (3)$$

and their conjugates respectively. For $SU(2) \times U(1)$ gauge bosons [W^+, Z, γ, W^-], $SU(3)$ color-octet gluons [$G^a(a = 1,2,..,8)$], [$SU(2)$ Higgs Boson], [(X,Y)] leptoquark bosons in GUTs, and [a color- and SU(2)-singlet neutral gauge boson from $\underline{3} \times \underline{3}^*$ (called S boson)] we have [$Q_4 Q_5^*, \frac{1}{\sqrt{2}}(Q_4 Q_4^* \pm Q_5 Q_5^*), Q_5 Q_4^*$], [$Q_1 Q_3^*, Q_2 Q_3^*, ..$] [$\varepsilon_{abc} Q_a Q_b Q_c Q_m$], [$Q_a^* Q_m$] and $Q_a Q_a^*$, (and conjugates) respectively. Among predicted new particles one lepton-type electroweak-doublet (v_Γ, Γ^-) with spin $\frac{3}{2}$ with the mass of the electroweak scale ($\leq Tev$) and doublly charged leptons ($> Tev$) are color singlets and can be observed directly. More to see the potential of SGM as a composite model of matter we interpret(reproduce) the Feynman diagrams of SM(GUT) in terms of the superon pictures, i.e. a single line of a propagating particle is replaced by multiple lines representing superons in the particle under two assumptions at the vertex; (i) the analogue of the OZI-rule of the quark model and (ii) the superon number consevation. Many remarkable new insights are obtained, e.g. in SM; naturalness of the mixing of K^0-$\overline{K^0}$, D^0-$\overline{D^0}$ and B^0-$\overline{B^0}$, no CKM-like mixings among the lepton generations, $v_e \leftrightarrow v_\mu \leftrightarrow v_\tau$ transitions beyond SM, strong CP-violation, small Yukawa couplings and no $\mu \longrightarrow e + \gamma$ despite compositeness, etc. and in (SUSY)GUT; proton is stable without R-parity by hand(absence of dangerous diagrams), etc.[ks1][ks2]. SGM may be the most economic model.

3. Fundamental Theory of Superon-Graviton Model(SGM)

By noting the supercharges Q of Volkov-Akulov(V-A) model[V-A] of the NL SUSY given by the supercurrents $J^\mu(x) = \frac{1}{i}\sigma^\mu \psi(x) - \kappa\{$the higher order terms of κ, $\psi(x)$ $\}$ satisfy the SP algebra, we find that the fundamental theory of SGM for spacetime and matter at(above) the Planck scale is SO(10) NL SUSY in the curved spacetime. We have written down the SGM action by performing the similar arguments to Einstein general relativity theory(EGRT) in high symmetric four dimensional (curved) *SGM spacetime*, where NL SUSY N-G fermion degrees of freedom $\psi(x)$ (i.e. the coset space coordinates of superGL(R)/GL(4R) representing N-G fermions) are embedded at every curved spacetime point[ks2]:

$$L_{SGM} = -\frac{c^3}{16\pi G}|w|(\Omega + \Lambda), \tag{4}$$

$$|w| = det w^a{}_\mu = det(e^a{}_\mu + t^a{}_\mu), \quad t^a{}_\mu = \frac{\kappa}{2i}\sum_{j=1}^{10}(\bar\psi^j\gamma^a\partial_\mu\psi^j - \partial_\mu\bar\psi^j\gamma^a\psi^j), \tag{5}$$

where $i = 1,2,..,10$, $\kappa^{-1} = \frac{c^3\Lambda}{16\pi G}$ is a fundamental volume of four dimensional spacetime of V-A, and Λ is a *small* cosmological constant related to the superon-vacuum coupling constant. Ω is a new scalar curvature analogous to the Ricci scalar curvature R of EGRT, whose explicit expression is obtained by just replacing $e^a{}_\mu(x)$ by $w^a{}_\mu(x)$ in Ricci scalar R. These results can be understood intuitively by observing that $w^a{}_\mu(x) = e^a{}_\mu(x) + t^a{}_\mu(x)$ defined by $\omega^a = w^a{}_\mu dx^\mu$, where ω^a is the NL SUSY invariant differential forms of V-A[V-A] and $w^a{}_\mu(x)$, invertible and $s^{\mu\nu}(x) \equiv w_a{}^\mu(x)w^{a\nu}(x)$ are a unified vierbein and a unified metric tensor in SGM spacetime. The SGM action (4) is invariant at least under global SO(10), ordinary GL(4R), the following new NL SUSY transformation;

$$\delta\psi^i(x) = \zeta^i + i\kappa(\bar\zeta^j\gamma^\rho\psi^j(x))\partial_\rho\psi^i(x), \quad \delta e^a{}_\mu(x) = i\kappa(\bar\zeta^j\gamma^\rho\psi^j(x))\partial_{[\rho}e^a{}_{\mu]}(x),\tag{6}$$

where $\zeta^i, (i = 1,..10)$ is a constant spinor and $\partial_{[\rho}e^a{}_{\mu]}(x) = \partial_\rho e^a{}_\mu - \partial_\mu e^a{}_\rho$, the following GL(4R) transformations due to (6);

$$\delta_\zeta w^a{}_\mu = \xi^\nu\partial_\nu w^a{}_\mu + \partial_\mu\xi^\nu w^a{}_\nu, \quad \delta_\zeta s_{\mu\nu} = \xi^\kappa\partial_\kappa s_{\mu\nu} + \partial_\mu\xi^\kappa s_{\kappa\nu} + \partial_\nu\xi^\kappa s_{\mu\kappa},\tag{7}$$

where $\xi^\rho = i\kappa(\bar\zeta^j\gamma^\rho\psi^j(x))$, and the following local Lorentz transformation on $w^a{}_\mu$;

$$\delta_L w^a{}_\mu = \varepsilon^a{}_b w^b{}_\mu \tag{8}$$

with the local parameter $\varepsilon_{ab} = (1/2)\varepsilon_{[ab]}(x)$ or accordingly on ψ and $e^a{}_\mu$

$$\delta_L\psi(x) = -\frac{i}{2}\varepsilon_{ab}\sigma^{ab}\psi, \quad \delta_L e^a{}_\mu(x) = \varepsilon^a{}_b e^b{}_\mu + \frac{\kappa}{4}\varepsilon^{abcd}\bar\psi\gamma_5\gamma_d\psi(\partial_\mu\varepsilon_{bc}). \tag{9}$$

The commutators of two new NL SUSY transformations (6) on $\psi(x)$ and $e^a{}_\mu(x)$ are GL(4R), i.e. new NL SUSY is the square-root of GL(4R);

$$[\delta_{\zeta_1}, \delta_{\zeta_2}]\psi = \Xi^\mu\partial_\mu\psi, \quad [\delta_{\zeta_1}, \delta_{\zeta_2}]e^a{}_\mu = \Xi^\rho\partial_\rho e^a{}_\mu + e^a{}_\rho\partial_\mu\Xi^\rho, \tag{10}$$

where $\Xi^\mu = 2i\kappa(\bar\zeta_2\gamma^\mu\zeta_1) - \xi_1^\rho\xi_2^\sigma e_a{}^\mu(\partial_{[\rho}e^a{}_{\sigma]})$. They show the closure of the algebra. SGM action (4) is invariant at least under[st1

$$[\text{global NL SUSY}] \otimes [\text{local GL}(4,\text{R})] \otimes [\text{local Lorentz}] \otimes [\text{global SO(N)}], \tag{11}$$

which is isomorphic to SO(10)SP of the linear representation of SGM.

4. Toward Low Energy Theory of SGM

The linearlization of such a high nonlinear theory is inevitable to obtain a renormalizable field theory which is equivalent to the model. As a flat space limit of SGM, we have shown that N=2 V-A model is equivalent to the spontaneously broken N=2 linear SUSY *vector* $J^P = 1^-$ gauge supermultiplet model with spontaneously broken SU(2) structure[stt].

We conjecture that any global linear SUSY (unified) model is equivalent to a nonlinear SUSY model. These results are favorable to the SGM scenario based upon the composite (eigenstates) nature of all elementary particles except graviton. SGM case is now in progress and will appear in the near future.

300

References

[1] Shima K 1999 European. Phys. J. C7 341-348.
[2] Shima K 2001 Phys. Lett. B501 237-244.
[3] Volkov D V and Akulov V P 1973 Phys. Lett. B46 109-110.
[4] Shima K and Tsuda M 2001 Phys. Lett. B507 260-263.
[5] Shima K Tanii Y and Tsuda M 2002 hep-th/0110102 Phys. Lett.B in press.

Inst. Phys. Conf. Ser. No 173: Section 2
Paper presented at 24th Int. Coll. Group Theoretical Methods in Physics, Paris, France, 15–20 July 2002
©2003 IOP Publishing Ltd

Fuzzy D-branes on group manifolds

J. Pawelczyk[†], H. Steinacker[*]

[*] Institut für theoretische Physik, LMU München, Theresienstr. 37, D-80333 München
[†] Institute of Theoretical Physics,Warsaw University, Hoża 69, PL-00-681 Warsaw, Poland

Abstract. We propose an algebraic description of (untwisted) D-branes on compact group manifolds G using quantum algebras related to $U_q(\mathfrak{g})$. It reproduces the known characteristics of D-branes in the WZW models, in particular their configurations in G, energies as well as the set of harmonics.

1. Introduction

This report is a brief review of the quantum algebraic description of D-branes on group manifolds as proposed in [1].

The structure of D-branes in a B field background has attracted much attention recently. The case of flat branes in a constant B background has been studied extensively and leads to quantum spaces with a Moyal-Weyl star product. A rather different situation is given by D-branes on compact Lie groups G, which carry a B field which is not closed. It is known from their CFT (conformal field theory) descriptions [2] that stable branes are given by certain conjugacy classes in the group manifold. On the other hand, it is expected that these branes are formed as bound states of $D0$-branes. Attempting to unify these various approaches, we proposed in [5] a matrix description of D-branes on $SU(2)$. This was generalized in [1], giving a simple and compact description of all (untwisted) D-branes on group manifolds G in terms of quantum algebras related to $U_q(\mathfrak{g})$. The main result is that a model based on the reflection equation algebra (RE) leads to precisely the same branes as the WZW model. It not only reproduces their configurations in G, i.e. the positions of the corresponding conjugacy classes, but also gives the same (noncommutative) algebra of functions on the branes.

2. CFT and the classical description of untwisted D-branes

The CFT description is given in terms of a WZW model, which is specified by a compact group group G and an integer level k [6]. We concentrate on the case $G = SU(N)$, but all constructions work for $SO(N)$ and $USp(N)$ as well.

The WZW branes can be described by certain boundary states of the Hilbert space of closed strings. We consider here only "symmetry-preserving branes" (untwisted branes), given by the Cardy (boundary) states. They are labeled [2, 7] by a finite set of integral weights

$$\lambda \in P_k^+ = \{\lambda \in P^+; \ \lambda \cdot \theta \le k\} \tag{1}$$

(here θ is the highest root of $\mathfrak{g}$), corresponding to integrable irreps of $\widehat{\mathfrak{g}}$. Hence untwisted branes are in one-to-one correspondence with $\lambda \in P_k^+$. The energy of the brane λ is given by

$$E_\lambda = \prod_{\alpha > 0} \frac{\sin\left(\pi \frac{\alpha \cdot (\lambda + \rho)}{k + g^\vee}\right)}{\sin\left(\pi \frac{\alpha \cdot \rho}{k + g^\vee}\right)} \tag{2}$$

The CFT also contains the description of branes as quantum manifolds, in terms of boundary primary fields. Their number is finite for any compact WZW model. In the $k \to \infty$ limit, these boundary primaries generate the (noncommutative) algebra of functions on the branes [8, 7], see also Section 4. For finite k, the corresponding algebra as given in [8] is not associative. It becomes associative after "twisting", so that it can be considered as algebra of functions of a quantum manifold. Then the primaries become modules of the quantum group $U_q(\mathfrak{g})$.

On a semi-classical level, the D-branes are simply conjugacy classes of the group manifold, of the form

$$\mathcal{C}(t) = \{gtg^{-1}; \quad g \in G\}. \tag{3}$$

One can assume that t belongs to a maximal torus T of G. These conjugacy classes are invariant under the adjoint action of the vector subgroup $G_V \hookrightarrow G_L \times G_R$ of the group of motions on G. This reflects the breaking $\hat{\mathfrak{g}}_L \times \hat{\mathfrak{g}}_R \to \hat{\mathfrak{g}}_V$.

A lot of information about the spaces $\mathcal{C}(t)$ can be obtained from the harmonic analysis, i.e. by decomposing scalar fields on $\mathcal{C}(t)$ into harmonics under the action of the (vector) symmetry G_V. One finds

$$\mathcal{F}(\mathcal{C}(t)) \cong \bigoplus_{\lambda \in P^+} mult_{\lambda^+}^{(K_t)} V_\lambda. \tag{4}$$

Here λ runs over all dominant integral weights P^+, V_λ is the corresponding highest-weight G-module, and $mult_{\lambda^+}^{(K_t)}$ is the dimension of the subspace of $V_{\lambda^+} = V_\lambda^*$ which is invariant under the stabilizer K_t of t.

As discussed above, there is only a finite set of stable D–branes on G (up to global motions) in the CFT description, one for each integral weight $\lambda \in P_k^+$. They correspond to $\mathcal{C}(t_\lambda)$ for

$$t_\lambda = q^{2(H_\lambda + H_\rho)}, \qquad q = e^{\frac{i\pi}{k+g^\vee}} \tag{5}$$

Here $g^\vee$ is the dual Coxeter number. The location of these branes in G is encoded in

$$s_n = \text{tr}(g^n) = \text{tr}(t^n), \quad g \in \mathcal{C}(t), \ n \in \mathbb{N} \tag{6}$$

which are invariant under the adjoint action. For the classes $\mathcal{C}(t_\lambda)$, they can be easily calculated:

$$s_n = \text{tr}_{V_N} \left(q^{2n(H_\rho + H_\lambda)}\right) = \sum_{\nu \in V_N} q^{2n(\rho+\lambda)\cdot\nu} \tag{7}$$

where V_N is the defining representation. The s_n completely characterize $\mathcal{C}(t_\lambda)$, and their quantum analogs (11) can be calculated exactly. An equivalent characterization of these conjugacy classes is provided by a characteristic equation of the form $P_\lambda(M) = 0$ [1].

3. Quantum algebras and symmetries for branes

We now define the quantum space describing G and its branes in terms of a non-commutative algebra $\mathcal{M}$, which transforms under a quantum symmetry. The quantized algebra $\mathcal{M}$ of functions on G is generated by elements M_j^i with indices i, j in the defining representation of G, subject to some commutation relations and constraints. The relations are given by the so-called reflection equation (RE) [9], which in a short notation reads

$$R_{21} M_1 R_{12} M_2 = M_2 R_{21} M_1 R_{12}. \tag{8}$$

Here R is the $\mathcal{R}$ matrix of $U_q(\mathfrak{g})$ in the defining representation. For $q = 1$, this reduces to $[M_j^i, M_l^k] = 0$. Because $\mathcal{M}$ should be a quantization of G, there must be further constraints.

In the case $G = SU(N)$, these are $\det_q(M) = 1$ where $\det_q$ is the so-called quantum determinant (12), and suitable reality conditions imposed on the generators M_j^i. The RE appeared more than 10 years ago in the context of the boundary integrable models [9].

$\mathcal{M}$ is covariant under the the transformation

$$M_j^i \rightarrow (s^{-1}Mt)_j^i \tag{9}$$

where s_j^i and t_j^i generate algebras $\mathcal{G}_L$ and $\mathcal{G}_R$ respectively, which both coincide with the well–known quantum groups $Fun_q(G)$ as defined in [10]; i.e. $s_2 s_1 R = R s_1 s_2$, $t_2 t_1 R = R t_1 t_2$ etc. This (co)action is consistent with RE if we impose that (the matrix elements of) s and t commute with M, and in addition satisfy $s_2 t_1 R = R t_1 s_2$. Formally, $\mathcal{M}$ is then a right $\mathcal{G}_L \otimes^R \mathcal{G}_R$ - comodule algebra; see [1] for further details. Notice that (9) is a quantum analog of the action of the classical isometry group $G_L \times G_R$ on classical group element g. Furthermore, $\mathcal{G}_L \otimes^R \mathcal{G}_R$ can be mapped to a vector Hopf algebra $\mathcal{G}_V$ with generators r, by $s_j^i \otimes 1 \rightarrow r_j^i$ and $1 \otimes t_j^i \rightarrow r_j^i$. The (co)action of $\mathcal{G}_V$ on M is then

$$M_j^i \rightarrow (r^{-1}Mr)_j^i. \tag{10}$$

The (generic) central elements of the algebra (8) are given by

$$c_n = \mathrm{tr}_q(M^n) \equiv \mathrm{tr}_{V_N}(M^n\, v) \in \mathcal{M}, \tag{11}$$

where $v = \pi(q^{-2H_\rho})$ is a numerical matrix which satisfies $S^2(r) = v^{-1}rv$ for the generator r of $\mathcal{G}_V$. These elements c_n are independent for $n = 1, 2, ..., rank(G)$. One can also show that the c_n are invariant under $\mathcal{G}_V$. As we shall see, the c_n for $n = 1, ...rank(G)$ fix the position of the brane configuration on the group manifold, i.e. they are quantum analogs of the s_n (7).

There is another central term $\det_q(M)$, the quantum determinant, which is invariant under the full chiral quantum algebra $\mathcal{G}_L \otimes^R \mathcal{G}_R$. Hence we can impose the constraint

$$\det_q(M) = 1. \tag{12}$$

Furthermore, there is a realization (algebra homomorphism) of the RE algebra (8) in terms of the algebra $U_q(\mathfrak{g})$, given by $M = (\pi \otimes id)(\mathcal{R}_{21}\mathcal{R}_{12})$ where π is the defining representation. (8) is then a consequence of the Yang-Baxter equation for $\mathcal{R}$ [10] .

4. Representations of $\mathcal{M}$ and quantum D–branes

We claim that the quantized orbits corresponding to the D-branes of interest here are described by irreps π_λ of $\mathcal{M}$. On any irrep, the Casimirs c_n (11) take distinct values, i.e. they become constraints. In view of their explicit form, an irrep of $\mathcal{M}$ should be considered as quantization of a conjugacy class $\mathcal{C}(t)$, whose position depends on c_n. To confirm this interpretation, one can calculate the position of the branes on the group manifold, and study their geometry by performing the harmonic analysis on the branes.

The "good" irreps of the algebra $\mathcal{M}$ coincide with the highest weight representations V_λ of $U_q(\mathfrak{g})$ for $\lambda \in P_k^+$: they are unitary [11], have positive quantum-dimension, and are in one-to-one correspondence with the integrable modules of the affine Lie algebra $\hat{\mathfrak{g}}$. As shown in [1], **these irreps π_λ of $\mathcal{M}$ for $\lambda \in P_k^+$ describe precisely the stable D-branes $\mathcal{C}(t_\lambda)$,** denoted by D_λ. It is an algebra of maps from V_λ to V_λ which transforms under the quantum adjoint action of $U_q(\mathfrak{g})$. For "small" weights, this algebra coincides with $Mat(V_\lambda)$. There is clearly a one–to–one correspondence between the (untwisted) branes in string theory and these D_λ, since both are labeled by $\lambda \in P_k^+$.

Position of D_λ The values of the Casimirs c_n on D_λ are as follows [1]:

$$c_1(\lambda) = \text{tr}_{V_N} \left(q^{2(H_\rho + H_\lambda)} \right), \tag{13}$$

$$c_n(\lambda) = \sum_{\nu \in V_N;\, \lambda + \nu \in P_k^+} q^{2n((\lambda + \rho)\cdot\nu - \lambda_N \cdot \rho)} \frac{\dim_q(V_{\lambda+\nu})}{\dim_q(V_\lambda)}, \quad n \geq 1. \tag{14}$$

Here λ_N is the highest weight of the defining representation V_N, and the sum in (14) goes over all $\nu \in V_N$ such that $\lambda + \nu$ lies in P_k^+.

The value of $c_1(\lambda)$ agrees precisely with the corresponding value (7) of s_1 on $\mathcal{C}(t_\lambda)$. For $n \geq 2$, the agreement of $c_n(\lambda)$ with s_n on $\mathcal{C}(t_\lambda)$ is only approximate, becoming exact for large λ. The discrepancy can be blamed to operator–ordering ambiguities. M also satisfies a characteristic equation [1] similar to the classical one. Therefore the position and "size" of the branes essentially agrees with the results from CFT. Furthermore, the energy of the D-brane is given by the quantum dimension of the representation space V_λ,

$$\dim_q(V_\lambda) = \text{tr}_{V_\lambda}(q^{2H_\rho}) = \prod_{\alpha > 0} \frac{\sin(\pi \frac{\alpha \cdot (\lambda + \rho)}{k + g^\vee})}{\sin(\pi \frac{\alpha \cdot \rho}{k + g^\vee})} = E_\lambda. \tag{15}$$

The space of harmonics on D_λ For simplicity, assume that λ is not too large. Then

$$D_\lambda \cong Mat(V_\lambda) = V_\lambda \otimes V_\lambda^* \cong \oplus_\mu N_{\lambda\lambda+}^\mu V_\mu, \tag{16}$$

since the tensor product is completely reducible. Here $N_{\lambda\lambda+}^\mu$ are the usual fusion rules of $\mathfrak{g}$. This has a simple geometrical meaning if μ is small enough: comparing with (4), one can show [1] that

$$D_\lambda \cong \mathcal{F}(\mathcal{C}(t_\lambda')) \tag{17}$$

up to some cutoff in μ, where $t_\lambda' = \exp(2\pi i \frac{H_\lambda}{k + g^\vee})$. This differs slightly from (5), by a shift $\lambda \to \lambda + \rho$. The structure of harmonics on D_λ is however in complete agreement with the CFT results. Moreover, it is known [7] that the structure constants of the corresponding boundary operators are essentially given by the $6j$ symbols of $U_q(\mathfrak{g})$, which in turn are precisely the structure constants of the algebra of functions on D_λ. Therefore our quantum algebraic description not only reproduces the correct set of boundary fields, but also essentially captures their algebra in (B)CFT.

Particularly interesting examples of degenerate conjugacy classes are the complex projective spaces $\mathbb{C}P^{N-1}$, which in the simplest case of $G = SU(2)$ become (q-deformed) fuzzy spheres [12] $S_{q,N}^2$.

References

[1] Pawelczyk J and Steinacker H, Nucl.Phys. B638 (2002) 433-458
[2] Alekseev A Yu and Schomerus V, Phys. Rev. D60 (1999) 061901
[3] Myers R C, JHEP 9912 (1999) 022
[4] Alekseev A Yu, Recknagel A and Schomerus V, JHEP 0005(2000) 010
[5] Pawelczyk J and Steinacker H, JHEP 0112 (2001) 018
[6] Fuchs J, "Affine Lie Algebras and Quantum Groups", Cambridge University Press 1992.
[7] Felder G, Fröhlich J, Fuchs J, Schweigert C, J.Geom.Phys. 34 (2000) 162
[8] Alekseev A Yu, Recknagel A and Schomerus V, JHEP 9909 (1999) 023
[9] Sklyanin E, J.Phys. A 21 (1988) 2375
[10] Faddeev L D, Reshetikhin N Yu, Takhtajan L A, Algebra Anal. 1 178 (1989)
[11] Steinacker H, Rev. Math. Phys. 13, No. 8 (2001) 1035-1054
[12] Grosse H, Madore J, Steinacker H, J.Geom.Phys. 38 (2001) 308-342

Inst. Phys. Conf. Ser. No 173: Section 2
Paper presented at 24th Int. Coll. Group Theoretical Methods in Physics, Paris, France, 15–20 July 2002
©*2003 IOP Publishing Ltd*

Fractals in Cosmology

Francesco Sylos Labini

Laboratoire de Physique Theorique, Université Paris XI, Bâtiment 211, F-91405 Orsay, France

Andrea Gabrielli

INFM Sezione Roma1, Dip. di Fisica, Universitá "La Sapienza", P.le A. Moro, 2, I-00185 Roma, Italy.

Abstract. Cosmology is based more on observational and testable grounds than ever in the past. As in many other scientific areas, a statistical-mathematical filter is necessary to process data and obtain the correct information to be compared with theories and hypothesis. The new approach we have proposed here can been see as filter to interpret the new many data which are appearing in Cosmology. The methods of modern Statistical Physics, which have been successfully applied in many different fields, represent a new general framework for the study of cosmic structures, both from a phenomenological and theoretical points of view. This new approach includes as a particular case the old analytical liquid-like approach (homogeneous mass density fields with small fluctuations), but it is able to shed light in more complex cases as matter distribution in the universe seems to be.

1. Introduction

Cosmic structures represent a very interesting playground for the methods of Statistical Physics and Self-Organization of complex structures. There are two main areas. The first deals with the very irregular spatial structures developed by the clustering of galaxies. The second broad area is provided by the Cosmic Microwave Background Radiation (hereafter CMBR) which is extremely smooth a part very small-amplitude fluctuations. A general theory should link these two areas which appear quite different. Many new data for both areas are now available and much more are expected in the near future, creating big expectations, interest and animating challenges. Indeed, on the observational side, the data in Cosmology have been growing exponentially in the last ten years, and in the coming decade we will have a huge amount of new data, in particular for the three-dimensional observations (via redshift) of matter distribution and CMBR anisotropies. Cosmology therefore is based more on observational and testable grounds than ever in the past. On the other hand, modern Statistical Physics can be able to provide the concepts which are crucial to study and understand these data. Let us firstly briefly discuss the new problems posed by intrinsically irregular structures (fractals, etc.); we will then comment about the application of these ideas to cosmology.

A mathematical fractal (mass) distribution is scale-invariant and self-similar [12]: this means that at any scale it shows the same degree of irregularities and hence it is characterized by structures and voids of all sizes. Such a system represents the most extreme case of correlated mass distributions: Its average density is zero in the infinite volume limit and the *conditional* average density seen by an occupied point at a distance r from it decays to zero as a power law $r^{-\gamma}$ with the distance from the occupied origin with $0 < \gamma \le d$ (the fractal dimension of the mass distribution being defined as $D = d - \gamma$). This means that the density field itself coincides with the fluctuation field itself and is intrinsically scale invariant. For these distributions one-point statistical properties are not well-defined in a finite sample. For this reason a fractal is non-homogeneous at any scale and the concept of average density in a finite sample centered on an occupied point has not an intrinsic meaning because it depends on the sample size. Example of intrinsical irregular systems are very well-known in physics since thirty years [11]. For instance, there is a link between fractals and critical systems: if we take a liquid-gas coexisting phase at the critical point and define a new density field as the subset of over-densities we obtain a fractal mass distribution.

A characterizing feature of fractal distributions is that the asymptotic average density of a fractal distribution is zero. Clearly in practical situations one has only finite samples, and hence a possible scale-invariant behavior can be defined only between a lower and an upper cut-off. The lower cut-off is defined by the elementary brick of the self-similar structure and the upper cut-off can be defined as the length at which the distribution changes its statistical nature and does not show self-similarity anymore at larger scales, or it can be the size of the available samples. In any finite sample one finds a positive value of the (unconditional) average density by estimating it, for example, by $\langle n \rangle = N/V$ (where N is the number of points contained in the volume V). By changing the sample size one will obtain different value of the estimated average density. This implies that the concept of average density is not well-defined. From a physical point of view this implies that one cannot treat fluctuations within usual perturbation theories of some dynamics, as fluctuations never become small with respect to the average density.

For this reason the statistical methods developed to study systems with small fluctuations around a well-defined value of the average density (as for example the case of liquids [5]) are often not appropriate to study intrinsical irregular systems. Therefore estimating the average density (or the amplitude of fluctuations with respect to the average density) in a finite sample for a fractal, without testing whether such a quantity is independent on the sample size, one finds a series of mathematical inconsistencies, which in cosmology have been addressed as physical phenomena. This occurs because one considers fluctuations around the finite size estimate of the average density: such a situation requires and assumes that the average density is a well-defined (sample independent) quantity. From the point of view of stochastic processes, the statistical properties of fluctuating processes with zero average density (in the ensemble definition) present a series of peculiar features which must be investigated within an appropriate mathematical framework and which are related to the subtle properties of conditional probabilities (e.g. [4]).

In the past years we have worked extensively on the application of the methods previously discussed to galaxy and galaxy cluster distributions [2, 15]. Our studies have generated a large debate in the field. However, even if the discussion about the quality of the present data is clearly an important issue, our main point is a *methodological one*. Indeed in Cosmology, for more than twenty years, the basic standard approach has been to describe these observations in terms of a theory of small amplitude fluctuations and smooth, analytical structures. On the contrary, available galaxy surveys are characterized by large scale structures and voids up to the size of the available samples [15]. The analysis of these data from the new perspective of scaling and complex systems has given rise to an entirely new picture both from conceptual and practical point of view [14, 15]. The restriction to the standard analysis has led instead to an apparent discrepancy between the different database and to a number of paradoxical situations which also brought to the introduction of a variety of ad-hoc hypotheses to explain them. For example, one of this puzzling problems was that galaxy and clusters correlation appear to be quite different even if clusters are made of galaxies [2]. The main results of this new approach are the following: (i) Eliminating the a-priori assumption of homogeneity *within a given sample*, the various experimental data are in good agreement among them and permit the identification of well-defined statistical properties typical of irregular structures

[15]. (ii) Galaxy and clusters samples are characterized by a well defined power law density-density correlations and by the lack of a well defined average density up to available sample size [15, 7, 8, 9]. Such a result contradicts the standard small amplitude approach for the description of galaxy clustering at any scale. (iii) Some apparent paradoxes like the galaxy-cluster mismatch, the luminosity segregation and similar ones are automatically eliminated by the new approach. (iv) The so-called "correlation length" derived within the standard cosmological approach (in this context it is defined differently from Statistical Physics where it was introduced [13, 11]) is shown to scale linearly with the sample depth, resulting therefore in several spurious concepts, treated as physical problems. This novel picture is leading to a new framework for what concerns the properties of the observed cosmic structures and it also implies new fascinating theoretical challenges. The price to pay is the elimination of some of the traditional a-priori assumptions of the field which are small amplitude, smooth structures and homogeneity at small scales. Therefore, this is one of the reasons why our approach has led to a large debate in the field (e.g [1, 16, 7, 8, 9]).

The results touch directly a more fundamental theoretical point. In cosmology, the basic idea of the traditional interpretation is that the matter density field (made by both visible - baryonic for $\leq 4\%$ of the whole matter content in the universe - and dark and non-baryonic matter - $\geq 96\%$ -) is very smooth, and the extraordinary isotropy of the CMBR supports this scenario. This picture can be possibly appropriate for the early time matter distribution, but it is inappropriate to describe galaxy spatial distribution which is characterized by large clusters and large voids with fractal-like properties up to sample scales. The usual replies to this criticism are that (i) galaxies are homogeneously distributed at scales larger than $\sim 5 Mpc$ - as it results from the standard correlation analysis which assumes homogeneity a-priori, and that (ii) the global matter field is mostly made of dark matter (non-baryonic), which is actually very smooth with small Gaussian fluctuations even today on the scale of few Megaparsec. Therefore, the underlying idea is that only the peaks of this smooth continuous field are the regions leading to galaxies and this should resolve the apparent puzzle introduced before [6]. In our opinion even in this interpretation, one cannot achieve a satisfactory picture neither theoretical nor in relation to the analysis of observations [3]. The problem is that available three dimensional samples of the galaxy distribution are strongly characterized by their discreteness. Before applying directly the standard statistical techniques used in cosmology which were developed for continuous mass distributions with small and weakly correlated fluctuations (i.e. liquids), one should test whether the approximation of the discrete distribution with a continuous one is permitted. We have proposed to apply a more general statistical analysis able to distinguish the case in which the "continuum" approximation is acceptable (e.g. liquids) to the case in which the discreteness of the distribution is too "strong" up to some large scales, to be reduced to a continuous field. Moreover in this second case, while the standard approach fails, this new one permits to characterize the system through the functions and the parameters typical of fractal analysis. On the other side the new methods allow one to verify the smooth case, if the average mass density calculated in the sample does not depend on the sample size, and if the density fluctuation correlations are weak enough. Only at this point one can apply the standard statistical analysis. In other words within this new statistical framework one can test the hypothesis of homogeneity of the distribution in the sample, while within the old one homogeneity is assumed a-priori. By applying this new approach, we have verified that we are in the case of a strongly irregular discrete distribution of galaxies at least up to the present sample scales. This is confirmed by the fact that galaxy distribution shows voids and clusters with scale comparable to the available galaxy samples.

The existence of a cross-over to homogeneity at larger scales is still an open problem. However a point must be stressed: the possible identification of this *homogeneity scales* at scales larger than the present ones will not eliminate our criticism to the standard statistical approach. This is true for two reasons: (i) with our new method we showed that galaxy distribution is highly irregular up to the present scales $(30 - 50\ Mpc/h)$ which poses fundamental problems to the theoretical interpretation, (ii) the identification of the homogeneity scale can be done only through this new method.

For some reasons the appearance of fractal features in the galaxy spatial distribution leads to a radically new perspective and this is hard to accept. But it is based on the best data and analysis available. Up to the present available scales, it is neither a conjecture nor a model, it is a fact. The theoretical problem is that there is no dynamical theory to explain how such a distribution of matter could have arisen from the pretty smooth initial state (see e.g. [10]). However, the fact that something can be hard to explain theoretically has nothing to do with whether it is true or not. Facing a hard problem is far more interesting than hiding it under the rug by an inconsistent procedure.

References

[1] Chown, M., New Scientist, **2200**, 22, (1999)

[2] Coleman, P.H. & Pietronero, L., Phys.Rep., **231**, 311, (1992)

[3] Gabrielli, A., Sylos Labini, F., and Durrer, R., Astrophys.J. Letters, **531**, L1, (2000)

[4] Gabrielli, A. & Sylos Labini, F., Europhys.Lett., **54**, 1, (2001)

[5] Hansen, J.P. and McDonald, I.R., *"Theory of simple fluids"*, (Academic Press, London, 1976)

[6] Kaiser, N., Astrophys. J. Lett., **284**, L9, (1984)

[7] Joyce, M., Montuori, M., Sylos Labini, F., Astrophys. Journal, **514**, L5, (1999)

[8] Joyce, M., Montuori, M., Sylos Labini, F., Pietronero, L., Astron.Astrophys, **344**, 387, (1999)

[9] Joyce, M. and Sylos Labini, F., Astrophys.J.Lett. **554**, L1, (2001)

[10] Joyce, M., Anderson, P. W., Montuori, M., Pietronero, L. and Sylos Labini, F., Europhys.Letters **50**, 416, (2000)

[11] Ma, S.K., *"The Modern Theory of Critical Phenomena"* , (Benjamin Reading, 1976)

[12] Mandelbrot, B.B., *"The Fractal Geometry of Nature"*, (Freeman, New York, 1983)

[13] Peebles, P.J.E., *"Large Scale Structure of the Universe"*, (Princeton University Press, 1980)

[14] Pietronero, L., Montuori, M. and Sylos Labini, F., in the Proceedings of the Conference *"Critical Dialogues in Cosmology"*, Ed. Turok, N., p.24, (World Scientific, Singapore, 1997)

[15] Sylos Labini, F., Montuori, M., Pietronero, L., Physics Reports, **293**, 66, (1998)

[16] Wu, K.K., Lahav, O. and Rees, M., Nature, **225**, 230, (1999)

Inst. Phys. Conf. Ser. No 173: Section 2
Paper presented at 24th Int. Coll. Group Theoretical Methods in Physics, Paris, France, 15–20 July 2002
©2003 IOP Publishing Ltd

Quantization of scalar fields in curved background, deformed Hopf algebra and entanglement

A Iorio, G Lambiase and G Vitiello

Dipartimento di Fisica "E.R.Caianiello", Università di Salerno, 84100 Salerno, Italy,
INFN, Gruppo Collegato di Salerno and INFM, Sezione di Salerno

Abstract. A suitable deformation of the Hopf algebra of the creation and annihilation operators for a complex scalar field, initially quantized in Minkowski space–time, induces the canonical quantization of the same field in a generic gravitational background. The deformation parameter q turns out to be related to the gravitational field. The entanglement of the quantum vacuum appears to be robust against interaction with the environment.

1. Introduction

We shortly report on two main results of some recent works [1, 2] on the quantization of a scalar field in curved background: i) a suitable deformation of the Hopf algebra for a complex scalar operator field, initially quantized in Minkowski space–time, induces the canonical quantization of the same field in a generic gravitational background. The deformation parameter q thus turns out to be related to the gravitational field. ii) The entanglement of the quantum vacuum appears to be robust against interaction with the environment.

Thermal properties of quantum field theory (QFT) in curved space–time can be derived in this deformed algebra setting. On the other hand, it is well known the intimate relationship between space–times with an event horizon and thermal properties [3, 4]. In particular, it has been shown [4] that global thermal equilibrium over the whole space–time implies the presence of horizons in this space–time. We find that the doubling of the degrees of freedom implied by the coproduct map of the deformed Hopf algebra turns out to be most appropriate for the description of the modes on both sides of the horizon. The entanglement between inner and outer particles with respect to the event horizon appears to be rooted in the background curvature and it is therefore robust against interaction with the environment.

2. Quantization and deformed Hopf algebra

We consider a complex scalar operator field $\phi(x)$, initially quantized in Minkowski space–time. To study the quantization procedure in curved space–time, we treat the gravitational field as a classical background. We start with few notions on the deformation of the Hopf algebra [5, 6]. We shall focus on the case of bosons for simplicity.

The coproduct is a homomorphism which duplicates the algebra, $\Delta : \mathcal{A} \to \mathcal{A} \otimes \mathcal{A}$. The operational meaning of the coproduct is that it provides the prescription for operating on two modes. Associated to that, there is the *doubling* of the degrees of freedom of the system. Our finding is that in the presence of a single event horizon such a doubling perfectly describes the modes on the two sides of the horizon [1, 2] (see also [7]).

The bosonic Hopf algebra for a single mode (the case of modes labelled by the momentum is straightforward), also called $h(1)$, is generated by the set of operators $\{a, a^\dagger, H, N\}$ with commutation relations:

$$[a, a^\dagger] = 2H, \quad [N, a] = -a, \quad [N, a^\dagger] = a^\dagger, \quad [H, \bullet] = 0, \tag{1}$$

where H is a central operator, constant in each representation. The Casimir operator is given by $C = 2NH - a^\dagger a$. In $h(1)$ the coproduct is defined by $\Delta\mathcal{O} = \mathcal{O}\otimes 1 + 1\otimes\mathcal{O} \equiv \mathcal{O}^{(+)} + \mathcal{O}^{(-)}$, where $\mathcal{O}$ stands for a, $a^\dagger$, H and N. The q-deformation of $h(1)$ is the Hopf algebra $h_q(1)$:

$$[a_q, a_q^\dagger] = [2H]_q, \quad [N, a_q] = -a_q, \quad [N, a_q^\dagger] = a_q^\dagger, \quad [H, \bullet] = 0, \tag{2}$$

where $N_q \equiv N$, $H_q \equiv H$ and $[x]_q = \dfrac{q^x - q^{-x}}{q - q^{-1}}$. The Casimir operator is given by $C_q = N[2H]_q - a_q^\dagger a_q$. The coproduct stays the same for H and N, while for a_q and $a_q^\dagger$ now it changes. In the fundamental representation, obtained by setting $H = 1/2, C = 0$, it is written as

$$\Delta a_q = a_q \otimes q^{1/2} + q^{-1/2} \otimes a_q = a^{(+)}q^{1/2} + q^{-1/2}a^{(-)},$$
$$\Delta a_q^\dagger = a_q^\dagger \otimes q^{1/2} + q^{-1/2} \otimes a_q^\dagger = a^{(+)\dagger}q^{1/2} + q^{-1/2}a^{(-)\dagger}, \tag{3}$$

where self-adjointness requires that q can only be real or of modulus one. In this representation $h(1)$ and $h_q(1)$ coincide. The differences appear in the coproduct. Note that $[a^{(\sigma)}, a^{(\sigma')\dagger}] = [a^{(\sigma)}, a^{(\sigma')}] = 0$, $\sigma \neq \sigma'$ with $\sigma \equiv \pm$. Now the key point is that, by setting $q = q(\epsilon) \equiv e^{2\epsilon(p)}$, suitable linear combinations of the deformed copodruct operation (3) (where the momentum label is introduced) give [1]:

$$d_p^{(\sigma)}(\epsilon) = d_p^{(\sigma)} \cosh\epsilon(p) + \bar{d}_{\tilde{p}}^{(-\sigma)\dagger} \sinh\epsilon(p),$$
$$\bar{d}_{\tilde{p}}^{(-\sigma)\dagger}(\epsilon) = d_p^{(\sigma)} \sinh\epsilon(p) + \bar{d}_{\tilde{p}}^{(-\sigma)\dagger} \cosh\epsilon(p), \tag{4}$$

where $d_p^{(\sigma)} \equiv \sum_k F(k, p) a_k^{(\sigma)}$, $\bar{d}_p^{(\sigma)} \equiv \sum_k F(k, p) \bar{a}_k^{(\sigma)}$ and $\{F(k,p)\}$ is a complete orthonormal set of functions [8], $p \in \mathbf{Z}^{n-1}$, as for $k = (k_1, \mathbf{k})$, and $p = (\Omega, \mathbf{p})$, $\tilde{p} = (\Omega, -\mathbf{p})$. We use $q(p) = q(\tilde{p})$. In general $k \neq p$. $a_k^{(\sigma)}$ and $\bar{a}_k^{(\sigma)}$ are the two (annihilation) operator modes of the complex scalar field $\phi(x)$ (for each of the sides $\pm$ of the horizon). Eqs. (4) are recognized to be the Bogolubov transformations obtained in the quantization procedure in the gravitational background in the semiclassical approximation [8]. We thus see that use of the deformed coproducts is equivalent to such a quantization procedure.

The generators of (4) is $g(\epsilon) = \sum_p \sum_\sigma \epsilon(p)[d_p^{(\sigma)}\bar{d}_{\tilde{p}}^{(-\sigma)} - d_p^{(\sigma)\dagger}\bar{d}_{\tilde{p}}^{(-\sigma)\dagger}]$ and $G(\epsilon) \equiv \exp g(\epsilon)$ is a unitary operator at finite volume. The Hilbert–Fock space $\mathcal{H}$ associated to the Minkowski space is built by repeated action of $(d_p^{(\sigma)\dagger}, \bar{d}_{\tilde{p}}^{(-\sigma)\dagger})$ on the vacuum state $|0_M\rangle$. The generator $G(\epsilon)$ maps vectors of $\mathcal{H}$ to vectors of another Hilbert space $\mathcal{H}_\epsilon$: $\mathcal{H} \to \mathcal{H}_\epsilon$. In particular,

$$|0(\epsilon)\rangle = G(\epsilon)|0_M\rangle, \tag{5}$$

where $|0(\epsilon)\rangle$ is the vacuum state of the Hilbert space $\mathcal{H}_\epsilon$ annihilated by the new operators $(d_p^{(\sigma)}(\epsilon)$, $\bar{d}_{\tilde{p}}^{(-\sigma)}(\epsilon))$. We use the short-hand notation for the Hilbert spaces ($\mathcal{H}$ stands for $\mathcal{H} \otimes \mathcal{H}$), as well as for the states (for instance $|0_M\rangle$ stands for $|0_M\rangle \otimes |0_M\rangle$). The group underlying this construction is $SU(1,1)$. By inverting Eq. (5), $|0_M\rangle$ can be expressed as a $SU(1,1)$ generalized coherent state [9] of Cooper-like pairs

$$|0_M\rangle = \frac{1}{Z} \exp\left[\sum_\sigma \sum_p \tanh\epsilon(p) d_p^{(\sigma)\dagger}(\epsilon)\bar{d}_{\tilde{p}}^{(-\sigma)\dagger}(\epsilon)\right] |0(\epsilon)\rangle, \tag{6}$$

where $Z = \prod_p \cosh^2 \epsilon(p)$. Moreover, $\langle 0(\epsilon)|0(\epsilon)\rangle = 1, \forall \epsilon$, and $\langle 0(\epsilon)|0_M\rangle \to 0$ and $\langle 0(\epsilon)|0(\epsilon')\rangle \to 0$ as $V \to \infty$, $\forall \epsilon, \epsilon', \epsilon \neq \epsilon'$, i.e. $\mathcal{H}$ and $\mathcal{H}_\epsilon$ become unitarily inequivalent in the infinite-volume limit. In this limit ϵ labels the set $\{H_\epsilon, \forall \epsilon\}$ of the infinitely many unitarily inequivalent representations of the canonical commutation relations [6, 10, 11].

The physical meaning of having two distinct momenta k and p for states in the Hilbert spaces $\mathcal{H}$ and $\mathcal{H}_\epsilon$, respectively, is the occurrence of two *different* reference frames: the M-frame (Minkowski) and the M_ϵ-frame. To explore the physics in the M_ϵ-frame, one has to construct a diagonal operator H_ϵ which plays the role of the Hamiltonian in the M_ϵ-frame. In order to do that one has to use the generator of the boosts. Thus one finds [1]

$$H_\epsilon = G(\epsilon)\mathcal{M}_{10}G^{-1}(\epsilon) = \sum_\sigma \sum_p \sigma\Omega[d_p^{(\sigma)\dagger}(\epsilon)d_p^{(\sigma)}(\epsilon) + \bar{d}_{\bar{p}}^{(\sigma)}(\epsilon)\bar{d}_{\bar{p}}^{(\sigma)\dagger}(\epsilon)]$$

$$= H^{(+)}(\epsilon) - H^{(-)}(\epsilon). \tag{7}$$

Here $\mathcal{M}_{10}$ denotes the deformed generator of the boosts. Eq. (7) gives the wanted Hamiltonian in the M_ϵ-frame, as also suggested by the customary results of QFT in curved space-time [8].

3. Entropy and entanglement

The condensate structure of the vacuum (6) suggests to consider the thermal properties of the system. The entropy operator is $S^{(\sigma)}(\epsilon) = \mathcal{S}^{(\sigma)}(\epsilon) + \bar{\mathcal{S}}^{(\sigma)}(\epsilon)$ with $\mathcal{S}^{(\sigma)}(\epsilon)$ given by $(\sigma \equiv \pm)$

$$\mathcal{S}^{(\sigma)}(\epsilon) = -\sum_p [d_p^{(\sigma)\dagger}(\epsilon)d_p^{(\sigma)}(\epsilon) \ln \sinh^2 \epsilon(p) - d_p^{(\sigma)}(\epsilon)d_p^{(\sigma)\dagger}(\epsilon) \ln \cosh^2 \epsilon(p)]. \tag{8}$$

$\bar{\mathcal{S}}^{(\sigma)}(\epsilon)$ has a similar form (with $d_p \to \bar{d}_p$). The total entropy operator is $S_\epsilon = S^{(+)}(\epsilon) - S^{(-)}(\epsilon)$ and it is invariant under the Bogoliubov transformations. Similarly one may introduce the free energy as [13, 12]

$$\mathcal{F}^{(+)}(\epsilon) \equiv \langle 0_M|H^{(+)}(\epsilon) - \frac{1}{\beta}S^{(+)}(\epsilon)|0_M\rangle. \tag{9}$$

with $\beta \equiv T^{-1}$. Stationarity of $\mathcal{F}^{(+)}(\epsilon)$ gives

$$\mathcal{N}_{d(\epsilon)}^{(+)} = \sinh^2 \epsilon(p) = \frac{1}{e^{\beta\Omega} - 1}, \tag{10}$$

and similarly for $\mathcal{N}_{\bar{d}(\epsilon)}^{(+)}$. Eq. (10) shows that for vanishing T the deformation parameter ϵ vanishes too. In that limit thermal properties as well as the event horizon are lost, and M_ϵ-frame $\to M$-frame. Moreover, i) β is related to the event horizons, and being β constant in time the M_ϵ space–time is static and stationary; ii) the gravitational field itself vanishes as $\epsilon \to 0$. The vanishing of the gravitational field occurs either if the M-frame is far from the gravitational source where space-time is flat, or if there exists a reference frame locally flat, i.e. the M-frame is a free–falling reference frame. This clearly is a realization of the equivalence principle, which manifests itself when "ϵ-effects" are shielded.

We now consider the entanglement. The expansion of $|0_M\rangle$ in (6) contains terms such as

$$\sum_p \tanh \epsilon(p) \left(|1_p^{(+)}, \bar{0}\rangle \otimes |0, \bar{1}_p^{(-)}\rangle + |0, \bar{1}_p^{(+)}\rangle \otimes |1_p^{(-)}, \bar{0}\rangle\right) + \dots, \tag{11}$$

where, we denote by $|n_p^{(\sigma)}, \bar{m}_p^{(\sigma)}\rangle$ a state of n particles and m "antiparticles" in whichever sector (σ). For the generic n^{th} term, it is $|n_p^{(\sigma)}, \bar{0}\rangle \equiv |1_{p_1}^{(\sigma)}, \dots, 1_{p_n}^{(\sigma)}, \bar{0}\rangle$, and similarly for antiparticles. By introducing a well known notation, $\uparrow$ for a particle, and $\downarrow$ for an antiparticle, the two-particle state in (11) can be written as

$$|\uparrow^{(+)}\rangle \otimes |\downarrow^{(-)}\rangle + |\downarrow^{(+)}\rangle \otimes |\uparrow^{(-)}\rangle, \tag{12}$$

which is an entangled state of particle and antiparticle living in the two sectors $(\pm)$. The generic n^{th} term in (11) shares exactly the same property as the two-particle state, but this time the $\uparrow$ describes a *set* of n particles, and $\downarrow$ a *set* of n antiparticles. The mechanism of the entanglement, induced by the q-deformation, takes place at all orders in the expansion, always by grouping particles and antiparticles into two sets. Thus the whole vacuum $|0_M\rangle$ is an infinite superposition of entangled states (a similar structure also arises in the temperature-dependent vacuum of Thermo-Field Dynamics [13] (see also [14])):

$$|0_M\rangle = \sum_{n=0}^{+\infty} \sqrt{W_n}|\text{Entangled}\rangle_n \,, \quad W_n = \prod_p \frac{\sinh^{2n_p} \epsilon(p)}{\cosh^{2(n_p+1)} \epsilon(p)} \,, \tag{13}$$

with $0 < W_n < 1$ and $\sum_{n=0}^{+\infty} W_n = 1$. The probability of having entanglement of two sets of n particles and n antiparticles is W_n. At finite volume, being W_n a decreasing monotonic function of n, the entanglement is suppressed for large n. It appears then that only a finite number of entangled terms in the expansion (13) is relevant. Nonetheless this is only true at finite volume (the quantum mechanics limit), while the interesting case occurs in the infinite volume limit, which one has to perform in a QFT setting.

The entanglement is generated by $G(\epsilon)$, where the field modes in one sector (σ) are coupled to the modes in the other sector $(-\sigma)$ via the deformation parameter $q(\epsilon)$. Since the deformation parameter describes the background gravitational field (environment), it appears that the origin of the entanglement *is* the environment, in contrast with the usual quantum mechanics view, which attributes to the environment the loss of the entanglement. In the present treatment such an origin for the entanglement makes it quite robust. One further reason for the robustness is that this entanglement is realized in the limit to the infinite volume *once and for all* since then there is no unitary evolution to disentangle the vacuum: at infinite volume one cannot "unknot the knots". Such a non-unitarity is only realized when *all* the terms in the series (13) are summed up, which indeed happens in the $V \to \infty$ limit [2].

References

[1] Iorio A, Lambiase G and Vitiello G 2001 Annals of Phys. 294 234-250
[2] Iorio A, Lambiase G and Vitiello G 2002 arXiv:hep-th/0204034
[3] Israel W 1976 Phys. Lett. A 57 107-110
[4] Sanchez N and Whiting B F 1986 Phys. Rev. D 34 1056-1071
[5] Celeghini E, Palev T D and Tarlini M 1991 Mod. Phys. Lett. B 5 187-194
    Kulish P P and Reshetikhin N Y 1989 Lett. Math. Phys. 18 143-149
[6] Celeghini E, De Martino S, De Siena S, Iorio A, Rasetti M and Vitiello G 1998 Phys. Lett. A 244 455-461
[7] Martellini M, Sodano P and Vitiello G 1978 Nuovo Cim. A 48 341-358
[8] Birrel N D and Davies P C W 1982 Quantum Fields in Curved Space (Cambridge: University Press)
    Takagi S 1986 Progress of Theor. Phys. Suppl. 88
[9] Perelomov A 1986 Generalized Coherent States and Their Applications (Berlin: Springer)
[10] Iorio A and Vitiello G 1994 Mod. Phys. Lett. B 8 269-276
    Iorio A and Vitiello G 1995 Annals Phys. 241 496-506
[11] Celeghini E, De Martino S, De Siena S, Rasetti M and Vitiello G 1995 Annals Phys. 241 50-67
    Celeghini E, Rasetti M and Vitiello G 1991 Phys. Rev. Lett. 66 2056-2059
[12] Celeghini E, Rasetti M and Vitiello G 1992 Annals Phys. 215 156-170
[13] Takahashi Y and Umezawa H 1996 Int. J. Mod. Phys. B 10 1755-1805 (and 1975 Collect. Phenomen. 2 55)
    Umezawa H 1993 Advanced field theory: micro, macro and thermal concepts (New York: AIP)
[14] Mi D, Song H S and An Y 2001 Mod. Phys. Lett. A 16 655-662

Conformal Theories and Integrable Systems

Inst. Phys. Conf. Ser. No 173: Section 3
Paper presented at 24th Int. Coll. Group Theoretical Methods in Physics, Paris, France, 15–20 July 2002
©*2003 IOP Publishing Ltd*

Generalized Non-Abelian Toda Models of Dyonic type

J.F. Gomes, G.M. Sotkov and A.H. Zimerman

Instituto de Física Teórica - IFT/UNESP, Rua Pamplona 145, 01405-900, São Paulo - SP, Brazil

Abstract. The construction of a class of non-abelian Toda models admitting dyonic type soliton solutions is reviewed.

The particle-like (soliton) solutions of a large class of 2-D integrable models are known to be an important tool in the description of non-perturbative properties of 4-D Yang-Mills theories, superstrings, matrix strings, etc. Together with the simplest integrable models like sine-Gordon and the A_n-abelian Toda models exhibiting topological charges only, one has also to consider more general non-abelian Toda models admitting solitons carrying both, Noether and topological charges (dyons). The aim of this talk is to discuss the systematic construction of integrable Toda models in terms of gauged Wess-Zumino-Witten (WZW) model. In fact, we are interested in a class of relativistic invariant integrable models related to non abelian embeddings $\mathcal{G}_0 \subset \mathcal{G}$. These are classified according to a grading operator Q which decomposes the lie algebra $\mathcal{G}$ into integer graded subspaces, i.e. $\mathcal{G} = \oplus_a \mathcal{G}_a$, $[Q, \mathcal{G}_a] = a \, \mathcal{G}_a, a \in Z$. As it is well known, the WZW action,

$$
\begin{aligned}
S_{WZW} = &-\frac{k}{4\pi} \int d^2 x Tr(g^{-1}\partial g g^{-1}\bar\partial g) \\
&+ \frac{k}{24\pi} \int_D \epsilon^{ijk} Tr(g^{-1}\partial_i g g^{-1}\partial_j g g^{-1}\partial_k g) d^3 x,
\end{aligned}
\tag{1}
$$

describes the dynamics of a matrix field $g(z, \bar z) \in G$ lying in a group manifold G of a finite dimensional Lie algebra $\mathcal{G}$. The equations of motion are given by

$$
\partial \bar J = \bar \partial J = 0, \quad \text{where} \quad J = g^{-1}\partial g, \quad \bar J = \bar\partial g g^{-1},
\tag{2}
$$

The Q-decomposition of $\mathcal{G}$ allows to write the group element g in the Gauss form, i.e.

$$
g = NBM
\tag{3}
$$

where $N = e^{\mathcal{G}_<}$, $B = e^{\mathcal{G}_0}$, $M = e^{\mathcal{G}_>}$ and $\mathcal{G}_<, \mathcal{G}_>$ denote the negative and positive grade subalgebras respectively. If we now seek the action for the field B in the zero grade subgroup G_0, we need to introduce a set of constraints in order to eliminate the degrees of freedom associated to the positive and negative subgroups. A consistent set of constraints can be encoded within the specification of the two constant grade ± 1 operators $\epsilon_\pm$ [1]. The reduced action corresponding to the conformal G-Toda models has the form

$$
S = S_{WZW}(B) + \frac{k}{2\pi} \int Tr\left(\epsilon_+ B \epsilon_- B^{-1}\right) d^2 x.
\tag{4}
$$

The equations of motion are then given by the Leznov-Saveliev equations [2],

$$
\bar\partial \left(B^{-1}\partial B\right) + [\epsilon_-, B^{-1}\epsilon_+ B] = 0, \qquad \partial \left(\bar\partial B B^{-1}\right) - [\epsilon_+, B\epsilon_- B^{-1}] = 0 \tag{5}
$$

The systematic approach in deriving the action (4) from (1) consists in introducing a gauged WZW action with auxiliary gauge fields $a \in \mathcal{G}_<, \bar a \in \mathcal{G}_>$ playing the role of Lagrange multipliers and subsequently integrating over these auxiliary fields (see for instance [1]). For $\mathcal{G}$ an infinite dimensional affine Kac-Moody algebra, the same arguments can be generalized replacing the finite dimensional WZW model by the two-loop WZW model constructed in [3]. For instance, considering the affine Kac-Moody algebra $\hat{\mathcal{G}}$ and the grading operator $Q = \tilde{h}\hat{d} + \sum_{i=1}^{rank\mathcal{G}} \frac{2\lambda_i \cdot H}{\alpha_i^2}$, where $\tilde{h}$ is the dual Coxeter number of $\mathcal{G}$ and λ_i are the fundamental weights, we find that the zero grade subalgebra corresponds to the Cartan subalgebra of $\mathcal{G}$ with two new generators added, i.e., $\mathcal{G}_0 = \{h_i, i = 1, \cdots, rank\ \mathcal{G}, \hat{d}, \hat{c}\}$, where $\hat{d}$ and $\hat{c}$ are the derivation and central charge generators respectively. The zero grade subgroup is then parametrized by $B' = Be^{\eta\hat{d}+\nu\hat{c}} = e^{\sum_{i=1}^{rank\mathcal{G}} \phi_i h_i} e^{\eta\hat{d}+\nu\hat{c}}$ and for $\epsilon_\pm = \sum_{i=1}^{rank\mathcal{G}} E_{\pm\alpha_i}^{(0)} + E_{\mp\psi}^{(\mp1)}$, where ψ denotes the highest root of $\mathcal{G}$, we obtain the conformal affine Toda models (CAT) proposed in [3], [4]. The fields η and ν have been naturaly introduced in order to ensure conformal invariance. If we consider the centerless Kac-Moody algebra (loop algebra), the η and ν fields decouple from the Toda fields and the conformal invariance of the theory is broken. For such case, we find from action (4), the Lagrangian density,

$$\mathcal{L} = \frac{1}{2}\eta_{ij}\partial\phi_i\bar{\partial}\phi_j - \sum_{i=1}^{rank\mathcal{G}} \frac{2}{\alpha_i^2}e^{k_{ij}\phi_j} - e^{-k_{\psi j}\phi_j} \tag{6}$$

where $\eta_{ij} = tr(h_i h_j)$, $k_{\psi j} = \frac{2\psi\cdot\alpha_j}{\alpha_j^2}$, and k_{ij} is the Cartan matrix.

Non-abelian structure in $\mathcal{G}_0$ can be introduced by supressing a fundamental weight in the definition of the grading operator. Consider for instance the centerless affine algebra $\mathcal{G} = A_n^{(1)} = \hat{SL}(n+1)$ with grading defined by $Q = n\hat{d} + \sum_{i=1}^{n-1} \frac{2\lambda_i \cdot H}{\alpha_i^2}$ and $\epsilon_\pm = \sum_{i=1}^{n-1} E_{\pm\alpha_i}^{(0)} + E_{\mp(\alpha_1+\cdots+\alpha_{n-1})}^{(\mp1)}$. Because of the absence of $\lambda_n \cdot H$ in the definition of Q, the zero grade subalgebra acquires a non-abelian character, $\mathcal{G}_0 = SL(2) \otimes U(1)^{n-1}$ and the zero grade subgroup is parametrized by $B = e^{\tilde{\chi}E_{-\alpha_n}}e^{\lambda_n \cdot HR + \sum_{i=1}^{n-1} \phi_i h_i}e^{\tilde{\psi}E_{\alpha_n}}$. Notice from the equations of motion (5) that

$$\partial Tr(X\bar{J}) = \bar{\partial}Tr(XJ) = 0, \quad X \in \mathcal{G}_0^0 = \{X, [X, \epsilon_\pm] = 0\} \tag{7}$$

which is consistent with the subsidiary constraint $Tr(X\bar{J}) = Tr(XJ) = 0$ where $X \in \mathcal{G}_0^0$ and the Toda fields parametrize the factor group $g_0^f \in G_0/G_0^0$. The factor group element can be realized as axial or vector gauging, i.e. $B = \alpha_0(g_0^f)\alpha_0'$ for $\alpha_0' = \alpha_0$ or $\alpha_0' = \alpha_0^{-1}$, $\alpha_0 \in \mathcal{G}_0^0$ respectively. This fact leads to a pair of actions related to each other by canonical transformation (T-duality) [5]. For the simplest non-abelian case, $\mathcal{G}_0 = SL(2) \otimes U(1)^{n-1}$, $\mathcal{G}_0^0 = U(1)$, $g_0^f \in \frac{SL(2)\otimes U(1)^{n-1}}{U(1)}$, i.e. $g_0^f = e^{\chi E_{-\alpha_n}}e^{\sum_{i=1}^{n-1} \phi_i h_i}e^{\psi E_{\alpha_n}}$. In order to implement systematicaly the subsidiary constraint and derive the effective action we consider again the gauged WZW action, now with auxiliary gauged fields $a_0, \bar{a}_0 \in \mathcal{G}_0^0$ and integrate over them. The general construction for such case is presented in [6]. Its lagrangian density for axial gauging of the $U(1)$ symmetry is obtained in the form,

$$\mathcal{L} = \frac{1}{4}\sum_{i,j=1}^{n-1} k_{ij}\partial^\mu\phi_i\partial_\mu\phi_j + \frac{1}{2}\frac{(\partial^\mu\psi\partial_\mu\chi + \epsilon^{\mu\nu}\partial_\mu\psi\partial_\nu\chi)}{1 + \frac{n+1}{2n}\psi\chi e^{\phi_{n-1}}}e^{-\phi_{n-1}} - V$$

$$V = \sum_{i=1}^{n-1} e^{k_{ij}\phi_j} + e^{(\phi_1+\phi_{n-1})}(1 + \psi\chi e^{-\phi_{n-1}}) \tag{8}$$

The one- and two-soliton solutions were constructed in [6] by using vertex operators within the dressing formalism. The Backlund transformation relating the vacuum and one-soliton solutions was constructed for both axial and vector models in [7]. The model (8) presents a global $U(1)$ symmetry

$$\psi' = e^{\alpha}\psi, \quad \chi' = e^{-\alpha}\chi, \quad \phi_i' = \phi_i. \tag{9}$$

For the case of imaginary coupling $\beta^2 = \frac{2\pi}{k}$, (and field rescaling $\phi_i \to i\beta\phi_i$, $\psi_i \to i\beta\psi_i$, $\chi_i \to i\beta\chi_i$) the model (8) manifests the following discrete symmetry

$$\psi'' = e^{i\pi(\frac{N}{n}+s_1)}\psi, \quad \chi'' = e^{i\pi(\frac{N}{n}+s_2)}\chi, \quad \phi_i'' = \phi_i + \frac{2\pi i}{\beta}\frac{N}{n}, \quad N \in Z \tag{10}$$

where s_2, s_2 are both even (odd) integers. These symmetries are responsible for solitonic solutions carrying both, nontrivial electric and topological charges. The specific structure of the solitons of the ungauged models with local $U(1)$ symmetry was studied in [8].

The next example consists in enlarging the non-abelian structure by considering $\mathcal{G}_0 = SL(2) \otimes SL(2) \otimes U(1)^{n-2}$ and $\mathcal{G}_0^0 = U(1) \otimes U(1)$. This is accomplished with the grading operator $Q = (n-1)\hat{d} + \sum_{i=2}^{n-1}\frac{2\lambda_i \cdot H}{\alpha_i^2}$ and $\epsilon_{\pm} = \sum_{i=2}^{n-1} E_{\pm\alpha_i}^{(0)} + E_{\mp(\alpha_2+\cdots+\alpha_{n-1})}^{(\mp 1)}$. The subsidiary condition (7) is implemented with $X = \{\lambda_1 \cdot H, \lambda_n \cdot H\}$ in the gauged WZW action. The simplest example of such class of models is provided with $\mathcal{G} = A_3^{(1)}$, i.e., $n = 3$. The lagrangian density is given by,

$$\mathcal{L} = \partial\phi\bar{\partial}\phi + \frac{1}{\Delta}\left((1 + \frac{3}{4}\psi_3\chi_3 e^{-\phi})\bar{\partial}\psi_1\partial\chi_1 e^{-\phi} + (1 + \frac{3}{4}\psi_1\chi_1 e^{-\phi})\bar{\partial}\psi_3\partial\chi_3 e^{-\phi}\right.$$

$$\left. + \frac{1}{4}(\chi_1\psi_3\bar{\partial}\psi_1\partial\chi_3 + \chi_3\psi_1\bar{\partial}\psi_3\partial\chi_1)e^{-2\phi}\right) - V \tag{11}$$

with $V = e^{-2\phi} + e^{2\phi}(1 + \psi_3\chi_3 e^{-\phi})(1 + \psi_1\chi_1 e^{-\phi})$ and $\Delta = 1 + \frac{3}{4}\psi_1\chi_1 e^{-\phi} + \frac{3}{4}\psi_3\chi_3 e^{-\phi} + \frac{1}{2}\psi_1\chi_1\psi_3\chi_3 e^{-2\phi}$. The model (11) presents a global invariance under $U(1) \otimes U(1)$. The multicharged topological soliton solutions are constructed in [9], where the soliton spectra is also derived.

Following the same line of reasoning, we next introduce models with non-abelian global symmetry. Consider $\mathcal{G}_0 = SL(3) \otimes U(1)$ and $\mathcal{G}_0^0 = SL(2) \otimes U(1)$. Such algebraic structure can be realized within $\mathcal{G} = A_2^{(1)}$ with the homogeneous gradation, $Q = \hat{d}$ and $\epsilon_{\pm} = \lambda_2 \cdot H^{(\pm 1)}$. The additional zero grade constraints are now implemented with $X = \{E_{\pm\alpha_1}, h_1, h_2\}$. The effective action for the factor group $g_0^f \in \frac{SL(3)\otimes U(1)}{SL(2)\otimes U(1)}$ parametrized as $g_0^f = e^{\chi_1 E_{-\alpha_2}}e^{\chi_2 E_{-\alpha_1-\alpha_2}}e^{\psi_1 E_{\alpha_2}}e^{\psi_2 E_{\alpha_1+\alpha_2}}$, is given by,

$$\mathcal{L} = \frac{1}{\Delta}\left(\bar{\partial}\psi_1\partial\chi_1(1 + \psi_1\chi_1 + \psi_2\chi_2) + \bar{\partial}\psi_2\partial\chi_2(1 + \psi_1\chi_1)\right.$$

$$\left. -\frac{1}{2}(\psi_1\chi_2\bar{\partial}\psi_2\partial\chi_1 + \chi_1\psi_2\bar{\partial}\psi_1\partial\chi_2)\right) - V \tag{12}$$

where $V = \frac{2}{3} + \psi_1\chi_1 + \psi_2\chi_2$ and $\Delta = (1 + \psi_1\chi_1)^2 + \psi_2\chi_2(1 + \frac{3}{4}\psi_1\chi_1)$. Notice that this is the simplest example in which $\mathcal{G}_0^0$ is non-abelian. In fact, the model (12) describe electrically charged solitons with isospin.

The examples of the affine non-abelian Toda models we have discussed are single out by the requirement to admit soliton solutions with both, topological and Noether charges. They are based on the intermediate (dyonic) gradation, which interpolate between the principal and the homogeneous gradations. The principal one, giving rise to the abelian Toda models with solitons of topological charges only. The homogeneous one leads to the so called homogeneous sine-Gordon models characterized by its nontopological solitons carrying Noether charges only [10].

Acknowledgements: Work partially supported by CNPq

318

References

[1] Balog J, Feher L, O'Raifeartaigh L, Forgacs P, Wipf A 1990, Ann. of Phys 76
[2] Leznov A N, Saveliev M V 1992 Group Theoretical Methods for Integration of Nonlinear Dynamical Systems, Progress in Physics, Vol. 15 , (Birkhauser Verlag, Berlin)
[3] Aratyn H, Ferreira L A, Gomes J F and Zimerman A H 1991, Phys. Lett. B254, 372
[4] Babelon O. and Bonora L 1990, Phys. Lett. B244, 220
[5] Gomes J F, Gueuvoghlanian E P , Sotkov G M and Zimerman A H 2001 Ann. of Phys 289 232, hepth/0007116
[6] Gomes J F, Gueuvoghlanian E P , Sotkov G M and Zimerman A H 2001 Nucl. Phys. B606 441, hepth/0007169
[7] Gomes J F, Gueuvoghlanian E P , Sotkov G M and Zimerman A H 2001 Nucl. Phys. B598615, hepth/0011187
[8] Gomes J F, Gueuvoghlanian E P , Sotkov G M and Zimerman A H 2002 JHEP 0207 001 hepth/0205228
[9] Cabrera-Carnero I, Gomes J F, Sotkov G M and Zimerman A H 2002, Nucl. Phys. B634433, hepth/0201047
[10] Fernandez-Pousa C R, Gallas M V, Hollowood T J and Miramontes J L 1997 Nucl. Phys. B484609; Nucl. Phys. B499 673; Fernandez-Pousa C R and J.L. Miramontes J L 1998 Nucl. Phys. B518 745

Inst. Phys. Conf. Ser. No 173: Section 3
Paper presented at 24th Int. Coll. Group Theoretical Methods in Physics, Paris, France, 15–20 July 2002
©2003 IOP Publishing Ltd

Local scale invariance, conformal invariance and dynamical scaling

Malte Henkel

Laboratoire de Physique des Matériaux CNRS UMR 7556, Université Henri Poincaré
Nancy I, B.P. 239, F - 54506 Vandœuvre lès Nancy Cedex, France

Abstract. Building on an analogy with conformal invariance, local scale transformations consistent with dynamical scaling are constructed. Two types of local scale invariance are found which act as dynamical space-time symmetries of certain non-local free field theories. Physical applications include uniaxial Lifshitz points and ageing in simple ferromagnets.

Scale invariance is a central notion of modern theories of critical and collective phenomena. We are interested in systems with strongly anisotropic or dynamical criticality. In these systems, two-point functions satisfy the scaling form

$$G(t, r) = b^{2x} G(b^\theta t, b r) = t^{-2x/\theta} \Phi \left(r t^{-1/\theta} \right) = r^{-2x} \Omega \left(t r^{-\theta} \right) \tag{1}$$

where t stands for 'temporal' and r for 'spatial' coordinates, x is a scaling dimension, θ the anisotropy exponent (when t corresponds to physical time, $\theta = z$ is called the dynamical exponent) and Φ, Ω are scaling functions. Physical realizations of this are numerous, see [1] and references therein. For isotropic critical systems, $\theta = 1$ and the 'temporal' variable t becomes just another coordinate. It is well-known that in this case, scale invariance (1) with a constant rescaling factor b can be replaced by the larger group of conformal transformations $b = b(t, r)$ such that angles are preserved. It turns out that in the case of one space and one time dimensions, conformal invariance becomes an important dynamical symmetry from which many physically relevant conclusions can be drawn [2].

Given the remarkable success of conformal invariance descriptions of equilibrium phase transitions, one may wonder whether similar extensions of scale invariance also exist when $\theta \neq 1$. Indeed, for $\theta = 2$ the analogue of the conformal group is known to be the Schrödinger group [3, 4] (and apparently already known to Lie). While applications of the Schrödinger group as dynamical space-time symmetry are known [5], we are interested here in the more general case when $\theta \neq 1, 2$. We shall first describe the construction of these *local scale transformations*, show that they act as a dynamical symmetry, then derive the functions Φ, Ω and finally comment upon some physical applications. For details we refer the reader to [6].

The defining axioms of our notion of *local scale invariance* from which our results will be derived, are as follows (for simplicity, in $d = 1$ space dimensions).

(i) We seek space-time transformations with infinitesimal generators X_n, such that time undergoes a Möbius transformation

$$t \to t' = \frac{\alpha t + \beta}{\gamma t + \delta} \quad ; \quad \alpha \delta - \beta \gamma = 1 \tag{2}$$

and we require that even after the action on the space coordinates is included, the commutation relations

$$[X_n, X_m] = (n - m) X_{n+m} \tag{3}$$

remain valid. This is motivated from the fact that this condition is satisfied for both conformal and Schrödinger invariance.

(ii) The generator X_0 of scale transformations is

$$X_0 = -t\partial_t - \frac{1}{\theta}r\partial_r - \frac{x}{\theta} \tag{4}$$

with a scaling dimension x. Similarly, the generator of time translations is $X_{-1} = -\partial_t$.

(iii) Spatial translation invariance is required.

(iv) Since the Schrödinger group acts on wave functions through a projective representation, generalizations thereof should be expected to occur in the general case. Such extra terms will be called *mass terms*. Similarly, extra terms coming from the scaling dimensions should be present.

(v) The generators when applied to a two-point function should yield a finite number of independent conditions, i.e. of the form $X_n G = 0$.

Proposition 1: *Consider the generators*

$$X_n = -t^{n+1}\partial_t - \sum_{k=0}^{n}\binom{n+1}{k+1}A_{k0}r^{\theta k+1}t^{n-k}\partial_r - \sum_{k=0}^{n}\binom{n+1}{k+1}B_{k0}r^{\theta k}t^{n-k} \tag{5}$$

where the coefficients A_{k0} and B_{k0} are given by the recurrences $A_{n+1,0} = \theta A_{n0}A_{10}$, $B_{n+1,0} = \frac{\theta}{n-1}(nB_{n0}A_{10} - A_{n0}B_{10})$ for $n \geq 2$ where $A_{00} = 1/\theta$, $B_{00} = x/\theta$ and in addition one of the following conditions holds: (a) $A_{20} = \theta A_{10}^2$ (b) $A_{10} = A_{20} = 0$ (c) $A_{20} = B_{20} = 0$ (d) $A_{10} = B_{10} = 0$. These are the most general linear first-order operators in ∂_t and ∂_r consistent with the above axioms (i) and (ii) and which satisfy the commutation relations $[X_n, X_{n'}] = (n - n')X_{n+n'}$ for all $n, n' \in \mathbb{Z}$.

Closed but lengthy expressions of the X_n for all $n \in \mathbb{Z}$ are known [6]. In order to include space translations, we set $\theta = 2/N$ and use the short-hand $X_n = -t^{n+1}\partial_t - a_n\partial_r - b_n$. We then define

$$Y_m = Y_{k-N/2} = -\frac{2}{N(k+1)}\left(\frac{\partial a_k(t,r)}{\partial r}\partial_r + \frac{\partial b_k(t,r)}{\partial r}\right) \tag{6}$$

where $m = -\frac{N}{2} + k$ and k is an integer. Clearly, $Y_{-N/2} = -\partial_r$ generates space translations.

Proposition 2: *The generators X_n and Y_m defined in eqs. (5,6) satisfy the commutation relations*

$$[X_n, X_{n'}] = (n - n')X_{n+n'} \quad , \quad [X_n, Y_m] = \left(n\frac{N}{2} - m\right)Y_{n+m} \tag{7}$$

in one of the following three cases: (i) B_{10} arbitrary, $A_{10} = A_{20} = B_{20} = 0$ and N arbitrary. (ii) B_{10} and B_{20} arbitrary, $A_{10} = A_{20} = 0$ and $N = 1$. (iii) A_{10} and B_{10} arbitrary, $A_{20} = A_{10}^2$, $B_{20} = \frac{3}{2}A_{10}B_{10}$ and $N = 2$.

In each case, the generators depend on two free parameters. The physical interpretation of the free constants $A_{10}, A_{20}, B_{10}, B_{20}$ is still open. In the cases (ii) and (iii), the generators close into a Lie algebra, see [6] for details. For case (i), a closed Lie algebra exists if $B_{10} = 0$.

Turning to the mass terms, we now restrict to the projective transformations in time, because we shall only need those in the applications later. It is enough to give merely the 'special' generator X_1 which reads for $B_{10} = 0$ as follows [6]

$$X_1 = -t^2\partial_t - Ntr\partial_r - Nxt - \alpha r^2\partial_t^{N-1} - \beta r^2\partial_r^{2(N-1)/N} - \gamma\partial_r^{2(N-1)/N}r^2 \tag{8}$$

where α, β, γ are free parameters (the cases (ii,iii) of Prop. 2 do not give anything new). Furthermore, it turns out that the relation $[X_1, Y_{N/2}] = 0$ for N integer is only satisfied in one of the two cases (I) $\beta = \gamma = 0$ which we call *Type I* and (II) $\alpha = 0$ which we call *Type II*. In both cases, all generators can be obtained by repeated commutators of $X_{-1} = -\partial_t$,

$Y_{-N/2} = -\partial_r$ and X_1, using (7). Commutators between two generators Y_m are non-trivial and in general only close on certain 'physical' states. One might call such a structure a *weak Lie algebra*. These results depend on the construction [6] of *commuting* fractional derivatives satisfying the rules $\partial_r^{a+b} = \partial_r^a \partial_r^b$ and $[\partial_r^a, r] = a\partial_r^{a-1}$ (the standard Riemann-Liouville fractional derivative is not commutative, see e.g. [7]).

For $N = 1$, the generators of both Type I and Type II reduce to those of the Schrödinger group. For $N = 2$, Type I reproduces the well-known generators of $2D$ conformal invariance (without central charge) and Type II gives another infinite-dimensional group whose Lie algebra is isomorphic to the one of $2D$ conformal invariance [6].

Dynamical symmetries can now be discussed as follows, by calculating the commutator of the 'Schrödinger-operator' S with X_1. We take $d = 1$ and $B_{10} = 0$ for simplicity.

Proposition 3: *The realization of Type I sends any solution $\psi(t,r)$ with scaling dimension $x = 1/2 - (N-1)/N$ of the differential equation*

$$S\psi(t,\boldsymbol{r}) = \left(-\alpha\partial_t^N + \left(\frac{N}{2}\right)^2 \partial_r^2\right)\psi(t,r) = 0 \tag{9}$$

into another solution of the same equation.

Proposition 4: *The realization of Type II sends any solution $\psi(t,r)$ with scaling dimension $x = (\theta-1)/2 + (2-\theta)\gamma/(\beta+\gamma)$ of the differential equation*

$$S\psi(t,r) = \left(-(\beta+\gamma)\partial_t + \frac{1}{\theta^2}\partial_r^\theta\right)\psi(t,r) = 0 \tag{10}$$

into another solution of the same equation.

In both cases, S is a Casimir operator of the 'Galilei'-subalgebra generated from $X_{-1}, Y_{-N/2}$ and the generalized Galilei-transformation $Y_{-N/2+1}$. The equations (9,10) can be seen as equations of motion of certain free field theories, where x is the scaling dimension of that free field ψ. These free field theories are non-local, unless N or θ are integers, respectively.

From a physical point of view, these wave equations suggest that the applications of Types I and II are very different. Indeed, eq. (9) is typical for equilibrium systems with a scaling anisotropy introduced through competing uniaxial interactions. Paradigmatic cases of this are so-called Lifshitz points which occur for example in magnetic systems when an ordered ferromagnetic, a disordered paramagnetic and an incommensurate phase meet (see [8] for a recent review). On the other hand, eq. (10) is reminiscent of a Langevin equation which may describe the temporal evolution of a physical system. In any case, causality requirements can only be met by an evolution equation of first order in ∂_t.

Next, we find the scaling functions Φ, Ω in eq. (1) from the assumption that G transforms covariantly under local scale transformations.

Proposition 5: *Local scale invariance implies that for Type I, the function $\Omega(v)$ must satisfy*

$$\left(\alpha\partial_v^{N-1} - v^2\partial_v - Nx\right)\Omega(v) = 0 \tag{11}$$

together with the boundary conditions $\Omega(0) = \Omega_0$ and $\Omega(v) \sim \Omega_\infty v^{-Nx}$ for $v \to \infty$. For Type II, we have

$$\left(\partial_u + \theta(\beta+\gamma)u\partial_u^{2-\theta} + 2\theta(2-\theta)\gamma\partial_u^{1-\theta}\right)\Phi(u) = 0 \tag{12}$$

with the boundary conditions $\Phi(0) = \Phi_0$ and $\Phi(u) \sim \Phi_\infty u^{-2x}$ for $u \to \infty$.

Here $\Omega_{0,\infty}$ and $\Phi_{0,\infty}$ are constants. The ratio β/γ turns out to be universal and related to x. From the linear differential equations (11,12) the scaling functions $\Omega(v)$ and $\Phi(u)$ can be found explicitly using standard methods [6].

Given these explicit results, the idea of local scale invariance can be tested in specific models. Indeed, the predictions for $\Omega(v)$ coming from Type I with $N = 4$ nicely agree with cluster Monte Carlo data for the spin-spin and energy-energy correlators of the $3D$ ANNNI model at its Lifshitz point [6, 9]. On the other hand, the predictions of Type II have been tested extensively in the context of ageing ferromagnetic spin systems to which we turn now.

Consider a ferromagnetic spin system (e.g. an Ising model) prepared in a high-temperature initial state and then quenched to some temperature T at or below the critical temperature T_c. Then the system is left to evolve freely (for recent reviews, see [10, 11]). It turns out that clusters of a typical time-dependent size $L(t) \sim t^{1/z}$ form and grow, where z is the dynamical exponent. Furthermore, two-time observables such as the response function $R(t, s; \boldsymbol{r} - \boldsymbol{r}') = \delta \langle \sigma_{\boldsymbol{r}}(t) \rangle / \delta h_{\boldsymbol{r}'}(s)$ depend on *both* t and s, where $\sigma_{\boldsymbol{r}}$ is a spin variable and $h_{\boldsymbol{r}'}$ the conjugate magnetic field. This breaking of time-translation invariance is called *ageing*. We are mainly interested in the autoresponse function $R(t, s) = R(t, s; \boldsymbol{0})$. One finds a dynamic scaling behaviour $R(t, s) \sim s^{-1-a} f_R(t/s)$ with $f_R(x) \sim x^{-\lambda_R/z}$ for $x \gg 1$ and where λ_R and a are exponents to be determined.

In order to apply local scale invariance to this problem, we must take into account that time translation invariance does *not* hold. The simplest way to do this is to remark that the Type II-subalgebra spanned by X_0, X_1 and the Y_m leaves the initial line $t = 0$ invariant, see (8). Therefore the autoresponse function $R(t, s)$ is fixed by the two covariance conditions $X_0 R = X_1 R = 0$. Solving these differential equations and comparing with the above scaling forms leads to [6]

$$R(t, s) = r_0 (t/s)^{1+a-\lambda_R/z} (t - s)^{-1-a} \quad , \quad t > s \tag{13}$$

where r_0 is a normalization constant. Therefore the functional form of R is completely fixed once the exponents a and λ_R/z are known. Similarly, the spatio-temporal response $R(t, s; \boldsymbol{r}) = R(t, s) \Phi \left(r(t - s)^{-1/z} \right)$, with the scaling function $\Phi(u)$ determined by (12).

The prediction (13) has been confirmed recently in several physically distinct systems undergoing ageing, see [12, 13, 14, 15] and references therein. These confirmations (which go beyond free field theory) suggest that (13) should hold independently of (i) the value of the dynamical exponent z (ii) the spatial dimensionality $d > 1$ (iii) the numbers of components of the order parameter and the global symmetry group (iv) the spatial range of the interactions (v) the presence of spatially long-range initial correlations (vi) the value of the temperature T (vii) the presence of weak disorder. Evidently, additional model studies are called for to test this conjecture further.

Summarizing, we have shown that local scale transformations exist for any θ, act as dynamical symmetries of certain non-local free field theories and appear to be realized as space-time symmetries in some strongly anisotropic critical systems of physical interest.

References

[1] Cardy J L 1996 *Scaling and Renormalization in Statistical Mechanics* (Cambridge University Press)
[2] Belavin A A, Polyakov A M and Zamolodchikov A B 1984 Nucl. Phys. **B241** 333
[3] Niederer U 1972 Helv. Phys. Acta **45** 802
[4] Hagen C R 1972 Phys. Rev. **D5** 377
[5] Henkel M 1994 J. Stat. Phys. **75** 1023
[6] Henkel M 2002 Nucl. Phys. **B** in press; (hep-th/0205256)
[7] Hilfer R (ed) 2000 *Applications of Fractional Calculus in Physics* (World Scientific: Singapore)
[8] Diehl H W 2002 Acta physica slovaka **52**, 271
[9] Pleimling M and Henkel M 2001 Phys. Rev. Lett. **87**, 125702
[10] Cates M E and Evans M R (eds) 2000 *Soft and Fragile Matter* (Bristol: IOP).
[11] Godrèche C and Luck J-M 2002 J. Phys. Cond. Mat. **14**, 1589
[12] Henkel M, Pleimling M, Godrèche C and Luck J-M 2001 Phys. Rev. Lett. **87**, 265701
[13] Cannas S A, Stariolo D A and Tamarit F A 2001 Physica **A294**, 362
[14] Calabrese P and Gambassi A 2002 Phys. Rev. **E65**, 066120 and cond-mat/0207487
[15] Picone A and Henkel M 2002 J. Phys. **A35**, 5575

Inst. Phys. Conf. Ser. No 173: Section 3
Paper presented at 24th Int. Coll. Group Theoretical Methods in Physics, Paris, France, 15–20 July 2002
©*2003 IOP Publishing Ltd*

Loop models from Coulomb gases and supersymmetry: Goldstone phases in 2D polymers

Jesper Lykke Jacobsen

LPTMS, Université Paris-Sud, Bâtiment 100, 91405 Orsay, France

Abstract. We review the different approaches to the various phases of two-dimensional loop models possessing an O(n) symmetry. These approaches are based, respectively, on a Liouville theory for an associated interface model (Coulomb gas method) and on non-linear sigma models endowed with supersymmetry. On the level of the corresponding lattice models, the difference is that the supersymmetric formulation allows for loop self-intersections. In both approaches, the choice between the various phases (dilute/dense/compact) is linked to the geometry of the target space. It is argued that the inclusion of self-intersections is a relevant perturbation in the dense phase, inducing a flow to a symmetry-broken Goldstone phase. The corresponding predictions for the central charge and various scaling dimensions are checked numerically. Finally, we discuss the coupling of self-intersecting loop models to quantum gravity. Such models are related to the problem of enumerating alternating knots.

1. Introduction

An old idea going back to de Gennes [1] is to relate various polymer problems to standard ϕ^4 theory for an N-component scalar field ϕ possessing an O(N) symmetry. The Euclidean continuum actions reads

$$S = \int d^d x \left\{ \frac{1}{2} \partial \phi \cdot \partial \phi + \frac{r}{2} \phi \cdot \phi + \frac{\lambda}{8} (\phi \cdot \phi)^2 \right\} \tag{1}$$

The diagrammatic expansion of S will lead to a gas of closed loops, each weighted by a factor of N, and taking $N \to 0$ gives us the polymer limit.

This model exhibits a second-order phase transition transition in dimension $d > 2$. In $d = 2$, our main concern in this note, there is only a transition for $N \leq 2$: its is of second order for $-2 \leq N \leq 2$ and of first order for $N < -2$.

The $(\phi \cdot \phi)^2$ term in the action allows for loop crossings. Such crossings can be disfavoured by choosing the sign of the initial coupling, $\lambda > 0$, and field theory then predicts a crossover to the infinitely repulsive limit. In this way the polymers become self-avoiding.

In this note, based on recent work done in collaboration with N. Read and H. Saleur [2], we shall examine the possibility of having a distinct Goldstone phase in $d = 2$. From the solution of the O(N) model it is known that a distinct low-temperature phase exists for $N \leq 2$. In particular, we examine the role of strict self-avoidance in this phase. From Landau-theory one expects the symmetry to be spontaneously broken to O($N - 1$), the resulting behaviour being described by a nonlinear sigma-model with target space O(N)/O($N - 1$) $\cong S^{N-1}$. In $d = 2$, such behaviour is only possible for $N < 2$, by the Mermin-Wagner theorem.

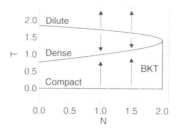

Figure 1. Phase diagram of the truncated O(N) model on the honeycomb lattice.

2. Lattice realisations in two dimensions

The O(N) model at temperature T can be realised on a lattice from the partition function

$$Z = \sum_{\{\vec{S}_i\}} \prod_{\langle ij \rangle \in \mathcal{E}} \exp\left(T^{-1} \vec{S}_i \cdot \vec{S}_j\right),$$ (2)

where $G = (\mathcal{V}, \mathcal{E})$ is some lattice and $\vec{S}_i$ = designates an N-component spin, normalised according to $\vec{S}_i \cdot \vec{S}_i = N$. Note that the diagrammatic (high-temperature) expansion of this model allows for loop crossings and for multiply occupied edges.

A simpler model was introduced by Nienhuis [3] by truncating the high-T expansion:

$$Z = \sum_{\{\vec{S}_i\}} \prod_{\langle ij \rangle \in \mathcal{E}} \left(1 + T^{-1} \vec{S}_i \cdot \vec{S}_j\right) = \sum_{\text{loops}} N^L T^{-E}.$$ (3)

In this rewriting, a configuration consists of L loops occupying E lattice edges. Loop crossings are suppressed by construction by taking G to be the honeycomb lattice.

The phase diagram of this model is shown on Fig. 2, with critical lines and renormalization group flows indicated. Apart from the critical-point (dilute) and low-temperature (dense) phases, it possesses distinct fully-packed (compact) [4, 5] and Berezinskii-Kosterlitz-Thouless phases [6, 7]. However, whether this truncated model describes the generic behaviour of the O(N) model at low T is questionnable.

3. Field theoretic approaches

There exists two distinct field-theoretical approaches for describing the long-distance behaviour (critical fluctuations) of loop models. In both of these, the phase one wishes to describe must be selected by a careful choice of target space.

3.1. Coulomb gas construction

This is based on mapping the loop model to an interface model, whose rough (critical) phase is equivalent to that of a gas of electromagnetic particles, by Coulomb's law in $d = 2$.

The interface model is defined by letting oriented loops act as contour lines for the height. The target space of heights is one-dimensional for the dilute and dense phases, but two-dimensional for the compact phase on the honeycomb lattice [5]. Note that loops can be independently oriented exactly because they are non-intersecting. In fact, the orientation is tantamount to augmenting the original O(N) symmetry to U(N).

Exact results for the Coulomb gas are available from conformal field theory, by means of a Liouville field theory construction [8]. For example, in the case of a one-dimensional target the central charge c and conformal weights h_k of operators inserting k-valent vertices are [9]

$$c = 1 - \frac{6(g-1)^2}{g}, \tag{4}$$

$$h_k = \frac{g}{2}k^2 - \frac{(g-1)^2}{4g}, \tag{5}$$

with $N = -2\cos(\pi g)$, and $0 < g \le 1$ (dense phase) or $1 \le g < 2$ (dilute phase).

3.2. Supersymmetric approach

Alternatively, ϕ^4 theory can be linked to supersymmetry through a non-linear sigma model [10]. In this approach, $N = \operatorname{Str} 1$ in the defining representation of the underlying superalgebra. Target spaces are graded supersymmetric spaces. This approach is more rigorous than the Coulomb gas, but it requires N to be integer. Note in particular that loop crossings are now allowed by construction.

Again, the considered phase is linked to the choice of target space [11]. E.g., dense loops can be described by taking $\frac{\mathrm{U}(N+n|n)}{\mathrm{U}(1)\times\mathrm{U}(N+n-1|n)} \cong \mathbf{CP}^{N+n-1|n}$, the supersymmetric generalisation of a complex projective space, and dilute loops require the target $\frac{\mathrm{OSp}(N+2n|2n)}{\mathrm{OSp}(N+2n-1|2n)} \cong S^{N+2n-1|2n}$, a supersphere.

In any case, there is a cancellation of "unused" indices, leading to a liberty (expressed by the parameter n) in the choice of the number of bosonic and fermionic degrees of freedom. Accordingly, the RG flow of the sigma model coupling constant g_σ is independent of n.

4. Model of crossing loops

To understand the effects of relaxing strict self-avoidance we have studied [2] a particular model on the square lattice. At each vertex, two loop segments avoid one another with unit weight, or cross with a weight w. This model is equivalent to perturbing the self-avoiding loop gas by a 4-leg operator. Conformal dimensions h_k of the k-leg operators in the $w = 0$ theory are given by Eq. (5). To appreciate the special significance of the $k = 4$ operator, note that: odd-k operators act like a magnetic field; $k = 2$ is redundant (it just marks a loop); $k \ge 6$ operators are always irrelevant. On the other hand, the 4-leg operator is irrelevant in the dilute phase and relevant in the dense one. This means that Nienhuis' description should be valid for the dilute phase, while the supersymmetric approach yields the generic Goldstone behaviour in the dense phase.

In our model, loops are fully packed, and so Goldstone behaviour should set in for $w > 0$. Among the predictions of supersymmetry [11, 2], let us cite: $c = N - 1$ with periodic boundary conditions; $c = N - 1 + 3n$ when twisted loop carry weight $N' = N + 4n$; all h_k are zero. These predictions are verified by numerical transfer matrix calculations [2].

5. Coupling to gravity: Knot diagrams

Another interesting issue is to consider crossing loop models on random lattices, i.e., to couple the models discussed above to two-dimensional quantum gravity. It was shown in Ref. [12] how counting the resulting planar diagrams with tetravalent crossing and tangent vertices (see Fig. 5) allows one to enumerate alternating knots modulo topological equivalences.

326

Figure 2. Knot diagram with tetravalent crossing and tangent vertices.

The $w = 0$ case is the standard gravitational $O(N)$ model, whose exact solution is known [13]. Its central charge equals that of the regular-lattice dense loops phase, Eq. (4), and its scaling exponents are obtained by applying the KPZ relation [14] to Eq. (5). Zinn-Justin originally conjecture that this result should also apply to the knot diagrams [15]. However, in the light of the regular-lattice results it would seem that the $w > 0$ flow should rather be to the Goldstone phase with $c = N - 1$.

This cannot be tested on the level of c for the exactly solvable cases $N = 1$ [16] and $N = 2$ [17], since in both cases the two conjectures for the central charge coincide. Further evidence can be gained by computerised enumeration of the knot diagrams [12]. Indeed, for $N \to 0$ the number of knots with p crossings is expected to scale like $\mu^p p^{-\alpha}$, where the exponent α is linked to c [14]. Assuming the supersymmetric scenario ($c = -1$), this leads to $\alpha = \frac{13+\sqrt{13}}{6} \sim 2.76759$. The precision obtained in Ref. [12] is not sufficient to convincingly distinguish this from $c = -2$ [15]. However, recent numerical work [18] has permitted to evaluate the derivative $\alpha'(N)|_{N=1} = 0.301 \pm 0.001$, which is in excellent agreement with the supersymmetric result of $3/10$.

References

[1] de Gennes P G 1972, Phys. Lett. A **38**, 339
[2] Jacobsen J L, Read N and Saleur H 2002, cond-mat/0205033
[3] Nienhuis B 1982, Phys. Rev. Lett. **49**, 1062
[4] Batchelor M T, Suzuki J and Yung C M 1994, Phys. Rev. Lett. **73**, 2646
[5] Kondev J, de Gier J and Nienhuis B 1996, J. Phys. A **29**, 6489
[6] Berezinskii V L 1971, Sov. Phys. JETP **34**, 610
[7] Kosterlitz J M and Thouless D J 1973, J. Phys. C **6**, 1181
[8] Jacobsen J L and Kondev J 1998, Nucl. Phys. B **515**, 701
[9] Duplantier B and Saleur H 1987, Nucl. Phys. B **290**, 291
[10] Parisi G and Sourlas N 1980, J. de Physique Lettres **41**, L403
[11] Read N and Saleur H 2001, Nucl. Phys. B **613**, 409
[12] Jacobsen J L and Zinn-Justin P 2001, J. Knot Th. Ramif. **10**, 1233
[13] Kostov I K 1989, Mod. Phys. Lett. A **4**, 217
[14] Knizhnik V G, Polyakov A M and Zamolodchikov A B 1988, Mod. Phys. Lett. A **3**, 819
[15] Zinn-Justin P 2001, in *Random matrices and their applications*, MSRI Publications 40
[16] Sundberg C and Thistlethwaite M 1998, Pac. J. Math. **182** 329
[17] Zinn-Justin P and Zuber J-B 2000, J. Knot Th. Ramif. **9** 1127
[18] Schaeffer G and Zinn-Justin P 2002, unpublished

Inst. Phys. Conf. Ser. No 173: Section 3
Paper presented at 24th Int. Coll. Group Theoretical Methods in Physics, Paris, France, 15–20 July 2002
©2003 IOP Publishing Ltd

Magnetic properties of quasi-one-dimensional strongly correlated systems

Pierre Pujol

Laboratoire de Physique, ENS-Lyon
46 Allée d'Italie, 69364 Lyon Cedex 07
and
Physics Department, Boston University
590 Commonwealth Avenue, Boston, MA, 02215.

Abstract.
    The present article contains a brief review of theoretical work on magnetization properties of strongly correlated systems in 1D. We will focus on the study of the magnetization curves of quantum spin chains and ladders, and generalizations such as the presence of charge carriers or impurities. We show that plateaux appear in these curves and give the conditions under which such plateaux are expected to appear.

The study of quasi-one dimensional strongly correlated electron systems (SCES's) has received a lot of attention in the past few years. This interest was mainly triggered by the synthesis of materials which in a wide range of temperatures can be well modeled by a 3D system of (almost) decoupled spin chains, spin ladders or more generally, Hubbard chains and ladders [1]. Oshikawa and collaborators [2] have undertaken the first systematic study of such systems in a magnetic field, by extending the Lieb-Schultz-Mattis theorem. They provided a necessary condition for the appearance of magnetization plateaux in 1D systems. When the magnetization $\langle M \rangle$ is normalized to saturation values ± 1, this condition for the appearance of a plateau with magnetization $\langle M \rangle$ can be cast in the form

$$SV \left(1 - \langle M \rangle\right) \in Z. \tag{1}$$

Here S is the size of the local spin and V the number of spins in the unit cell for the translation operator acting on the magnetization $\langle M \rangle$ ground-state. It should be noted that translational invariance can be spontaneously broken in the ground-state and then V would be larger than the unit cell of the Hamiltonian.

Spin ladders in a magnetic field constitute a class of systems where the full phase diagram was explored and where it was checked when the necessary condition (1) becomes also sufficient (see for example [3, 4, 5] and references therein). A similar analysis was also done for periodically modulated $S = 1/2$ Heisenberg spin chains with period q (so-called q-merized chains) [6].

The field theory description of spin chains is a useful technique for treating problems of weakly coupled spin ladders, weak dimerization in chains, etc. as we will illustrate below for the case of $S = 1/2$. For higher spin chains, non-Abelian bosonization has proven to be better suited. For spin 1/2 chains, the Abelian bosonization technique has proven to be very efficient. It describes the low energy, large scale behavior of the system, and can be extended to the case of an easy axis anisotropy. More specifically, the continuum limit of the Hamiltonian

$$H_{XXZ} = J \sum_{x=1}^{L} \left\{ \Delta S_x^z S_{x+1}^z + \frac{1}{2} \left(S_x^+ S_{x+1}^- + S_x^- S_{x+1}^+ \right) \right\} - h \sum_{x=1}^{L} S_x^z \tag{2}$$

is given by the Tomonaga-Luttinger Hamiltonian

$$H = \frac{1}{2} \int dx \left(vK(\partial_x \tilde{\phi})^2 + \frac{v}{K}(\partial_x \phi)^2 \right).$$ (3)

The bosonic field ϕ^i and its dual $\tilde{\phi}^i$ are given by the sum and difference of the light-cone components, respectively. The constant $K = K(\langle M \rangle, \Delta)$ governs the conformal dimensions of the bosonic vertex operators and can be obtained exactly from the Bethe Ansatz solution of the XXZ chain (see e.g. [4] for a detailed summary). One has $K = 1$ for the $SU(2)$ symmetric case ($\Delta = 1$) and it is related to the radius R of [4] by $K^{-1} = 2\pi R^2$.

In terms of these fields, the spin operators read

$$S_x^z = \frac{1}{\sqrt{2\pi}} \partial_x \phi + a : \cos(2k_F x + \sqrt{2\pi}\phi) : + \frac{\langle M \rangle}{2},$$ (4)

$$S_x^{\pm} = (-1)^x : e^{\pm i \sqrt{2\pi}\tilde{\phi}} \left(b \cos(2k_F x + \sqrt{2\pi}\phi) + c \right) :,$$ (5)

where the colons denote normal ordering with respect to the ground-state with magnetization $\langle M \rangle$. The Fermi momentum k_F is related to the magnetization of the chain as $k_F = (1 - \langle M \rangle)\pi/2$. An XXZ anisotropy and/or the external magnetic field modify the scaling dimensions of the physical fields through K and the commensurability properties of the spin operators, as can be seen from (4), (5).

To build a ladder, one couple N identical chains with a transversal coupling J'

$$H = \sum_{a=1}^{N} H_{XXZ}^a + J' \sum_{x,a=1}^{a=N} \vec{S}_x^a \cdot \vec{S}_x^{a+1}.$$

For simplicity we used here periodic boundary conditions (PBC's) along the transverse direction. One obtains in the continuum a collection of identical Hamiltonians like (3), with perturbation terms which couple the fields of the different chains. After a careful renormalization group (RG) analysis, one can show that at most one degree of freedom, given by the combination of fields $\phi_D = \sum_a \phi_a$, remains massless. The large scale effective action for the ladder systems is then given again by a Hamiltonian (3) for ϕ_D and the perturbation term

$$H_{pert} = \lambda \int dx \, \cos(2Nk_F x + \sqrt{2\pi}\phi_D),$$ (6)

where $k_F = (1 - \langle M \rangle)\pi/2$ is related to the total magnetization $\langle M \rangle$.

The key point is to identify the values of the magnetization for which the perturbation operator (6) can play an important rôle. In fact, this operator is commensurate at values of the magnetization given by (1) with $S = 1/2$ and $V = N$. If this operator turns out to be also relevant in the RG sense (this depends on the parameters of the effective Hamiltonian (3), the model will have a finite gap, implying a plateau in the magnetization curve. The field-theoretical treatment is valid when the transversal coupling J' can be treated as a perturbation, $J' \ll J$. For the opposite limit, a strong coupling analysis is more appropriate. The details of these computations can be found in [4]. We just mention here that the values of the magnetizations for which the plateau appears in this limit coincide with the ones obtained by bosonization. Figure (1) shows the magnetization curve of a spin $1/2$ three leg ladder with anti-ferromagnetic transversal couplings. As expected from (1) for this case, a plateau at $M = 1/3$ is clearly visible.

Another perturbation relevant for experimental situations is the presence of periodically modulated couplings between the spins [6]: $J_x = J$ if $x \neq nq$ and $J_x = (1 - \delta)J \equiv J'$ if

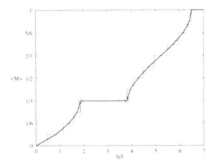

Figure 1. Magnetization curves of a spin $1/2$ three leg ladder with anti-ferromagnetic transverse couplings and chains with different lengh: $L = 8$, $L = 6$ and $L = 4$. The thick full line indicates the expected form in the thermodynamic limit.

$x = nq$, this amounts to a similar effective field theory for the description of the system. In this case the result for the appearance of plateaux is:

$$\frac{p}{2}(1 - \langle M \rangle) \in Z \tag{7}$$

This analysis can be generalized to the presence of charge carriers (see [7] for a single q-merized chain and [8] for a two leg ladder). As an example, let us consider the Hubbard model, representing interacting electrons with spin $1/2$ in a q-merized chain. The Hamiltonian is given by

$$H = -\sum_{x,\alpha} t(x)\,(c^\dagger_{x+1,\alpha}c_{x,\alpha} + H.c.) + U\sum_x c^\dagger_{x,\uparrow}c_{x,\uparrow}c^\dagger_{x,\downarrow}c_{x,\downarrow}$$

$$+\sum_{x,\alpha} \mu(x)\,c^\dagger_{x,\alpha}c_{x,\alpha} - \frac{h}{2}\sum_x (c^\dagger_{x,\uparrow}c_{x,\uparrow} - c^\dagger_{x,\downarrow}c_{x,\downarrow})\,, \tag{8}$$

where $t(x)$ and $\mu(x)$ are taken as periodic in the variable x with period q. The bosonized version of this Hamiltonian can be written as a Gaussian part, consisting basically in two copies of (3) where the fields represent in the case of zero magnetic field the charge and spin degrees of freedom. There are also several perturbation terms arising from interactions among fermions and the q-merization. We just mention that using the same arguments of commensurability and relevance as above, we can show that for $qn \in Z$, we have a charge gap. If the condition is further constrained to $qn/2 \in Z$, for zero magnetic field we have also a spin gap, implying a plateau with $\langle M \rangle = 0$. In the case of non-zero magnetic field, if one of the conditions

$$\frac{q}{2}(n \pm \langle M \rangle) \in Z\,, \tag{9}$$

is satisfied, and the doping is kept fixed (as is natural from the point of view of experimental realizations of doped systems), the system has a magnetization plateau, but still exhibits massless behavior as well, like in the specific heat which vanishes linearly as the temperature goes to zero. If both conditions are simultaneously satisfied, the system is gaped in the charge and spin sectors; this situation is in fact the generalization to arbitrary doping of the results for p-merized Heisenberg chains discussed above.

Finally, the effect of disorder on plateau systems, such as the q-merized XXZ chains, has also been studied [9, 10]. A direct field theory treatment is in general very complicated due to the non-local effective theory arising because of the impurities [11]. In this case a lattice real space decimation procedure proposed by Dasgupta and Ma [12] is more appropriate. It

330

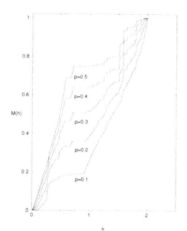

Figure 2. Magnetization curves of dimerized XX chains with different concentrations p of impurities.

was found that, for a disorder represented by a binary distribution of the bonds, plateaux do not disappear but are rather shifted in a precise amount which can be predicted by means of a simple argument to be [9] $\langle M \rangle = 1 + \frac{2}{q}(p-1)$. Generalizations to other kind of impurities with discrete probability distributions give rise to other families of plateaux, which can be classified (see [10] for the details). For example, a plateau is expected to appear at $\langle M \rangle = 2p - p^2$ for a disorder simulating site centered impurities (see figure (2)). Conversely, continuous distributions of disorder erase completely the plateau structure.

Acknowledgements: The author is grateful to D. C. Cabra, M. D. Grynberg, A. Honecker, A. De Martino, S. Peysson, J. Rech and P. Simon for their collaboration on the different aspects of this subject.

References

[1] E. Dagotto, T.M. Rice, Science **271**, 618 (1996); T.M. Rice, Z. Phys. **B103**, 165 (1997); E. Dagotto, Rep. Prog. Phys. **62**, 1525 (1999).
[2] M. Oshikawa, M. Yamanaka, I. Affleck, Phys. Rev. Lett. **78**, 1984 (1997).
[3] D.C. Cabra, A. Honecker, P. Pujol, Phys. Rev. Lett. **79**, 5126 (1997).
[4] D.C. Cabra, A. Honecker, P. Pujol, Phys. Rev. **B58**, 6241 (1998).
[5] D.C. Cabra, M. D. Grynberg, A. Honecker, P. Pujol, cond-mat/0010376.
[6] D.C. Cabra, M.D. Grynberg, Phys. Rev. **B59**, 119 (1999);
    A. Honecker, Phys. Rev. **B59**, 6790 (1999);
    R.M. Wießner, A. Fledderjohann, K.-H. Mütter, Eur. Phys. J. **B15**, 475 (2000).
[7] D.C. Cabra, A. De Martino, A. Honecker, P. Pujol, P. Simon, Phys. Lett. **A268**, 418 (2000).
[8] D.C. Cabra, A. De Martino, P. Pujol, P. Simon, Europhys. Lett. **57** (3), 402 (2002).
[9] D.C. Cabra, A. De Martino, M.D. Grynberg, S. Peysson, P. Pujol, Phys. Rev. Lett. **85**, 4791 (2000).
[10] P. Pujol, J. Rech, Phys. Rev. **B66**, 104401 (2002).
[11] T. Giamarchi, H. J. Schulz Phys. Rev. **B37**, 325 (1988).
[12] C. Dasgupta, S. K. Ma Phys. Rev. **B22**, 1305 (1980).

Inst. Phys. Conf. Ser. No 173: Section 3
Paper presented at 24th Int. Coll. Group Theoretical Methods in Physics, Paris, France, 15–20 July 2002
©*2003 IOP Publishing Ltd*

Joint Description of Periodic SL(2,R) WZNW Model and Its Coset Theories

George Jorjadze[a] and **Gerhard Weigt**[b]

[a]Razmadze Mathematical Institute, Georgia
[b]DESY Zeuthen, Germany

Abstract. Liouville, $SL(2,R)/U(1)$ and $SL(2,R)/R_+$ coset structures are completely described by gauge invariant Hamiltonian reduction of the $SL(2,R)$ WZNW theory.

1. Introduction

Wess-Zumino-Novikov-Witten (WZNW) models are fascinating two-dimensional conformal field theories with reach symmetry and dynamical structures. Their cosets form important classes of integrable theories. An outstanding example is the $SL(2,R)$ WZNW model which is related to the Liouville theory and other cosets with interesting ('black hole') space-time properties. Using Hamiltonian reduction one can show what happens under this reduction with the $SL(2,R)$ WZNW fields, the symplectic structure, the Kac-Moody currents, the Sugawara energy-momentum tensor, and most importantly how the general $SL(2,R)$ solution reduces to those of the gauged theories. Free-field parametrisation allows canonical quantisation, and we present as a typical quantum result a causal quantum group commutator.
  The talk is based on a series of papers [1]-[5].

2. SL(2,R) WZNW theory

WZNW models are invariant under left (chiral) and right (anti-chiral) multiplications of the WZNW field $g(z, \bar{z})$. These Kac-Moody symmetries provide integrability of the theory and give its general solution as a product of chiral and anti-chiral fields, $g(z)$ respectively $\bar{g}(\bar{z})$, which for periodic boundary conditions in σ have monodromies $g(z + 2\pi) = g(z)M$ and $\bar{g}(\bar{z} - 2\pi) = M^{-1}\bar{g}(\bar{z})$ with $M \in SL(2,R)$. $z = \tau + \sigma$, $\bar{z} = \tau - \sigma$ are light cone coordinates. The conformal symmetry is generated by the traceless Sugawara energy-momentum tensor.
  The basic (anti-)chiral Poisson brackets follow by inverting the corresponding symplectic form of the $SL(2,R)$ theory. But piecing together the chiral and anti-chiral results surprisingly simple causal non-equal time Poisson brackets follow for the $SL(2,R)$ WZNW fields [5]

$$\{ g_{ab}(z, \bar{z}), g_{cd}(y, \bar{y}) \} = \frac{\gamma^2}{4} \Theta \left[2g_{ad}(z, \bar{y}) \, g_{cb}(y, \bar{z}) - g_{ab}(z, \bar{z}) \, g_{cd}(y, \bar{y}) \right], \quad (1)$$

where $\Theta = \epsilon(z - y) + \epsilon(\bar{z} - \bar{y})$ is the causal factor. The stair-step function $\epsilon(z) = 2n + 1$ for $2\pi n < z < 2\pi(n+1)$ ensures that causality. This Poisson bracket was derived for hyperbolic monodromy in the 'fundamental' interval $z - y \in (-2\pi, 2\pi)$ where $\epsilon(z - y) = sign\,(z - y)$, and it holds therefore also on the line. Eq.(1) can be generalised even outside this domain [2].
  From (1) one can derive, e.g., the canonical Poisson brackets, the Kac-Moody algebra and Poisson bracket relations of the energy-momentum tensor with itself or any other field.
  The causal Poisson brackets encode the full WZNW dynamics.

3. Coset theories

Gauging the SL(2,R) WZNW theory with respect to the three different types of one-dimensional subgroups $h = e^{\alpha t} \in SL(2, R)$

$$e^{\alpha t_0} = \begin{pmatrix} \cos\alpha & -\sin\alpha \\ \sin\alpha & \cos\alpha \end{pmatrix}, \quad e^{\alpha t_2} = \begin{pmatrix} e^{\alpha} & 0 \\ 0 & e^{-\alpha} \end{pmatrix}, \quad e^{\alpha t_+} = \begin{pmatrix} 1 & 0 \\ \alpha & 1 \end{pmatrix}, \quad (2)$$

and considering axial respectively vector gauging $g \mapsto hgh$ and $g \mapsto hgh^{-1}$ one finds six integrable [3] coset theories. The subgroups are called compact ($t = t_0$), non-compact ($t = t_2$) and nilpotent ($t = t_+$), where the t_n are elements of the $sl(2, R)$ algebra given by the Pauli matrices $t_0 = -i\sigma_2$, $t_1 = \sigma_1$, $t_2 = \sigma_3$, and $t_+ = t_0 + t_1$ with $t_+^2 = 0$ is a nilpotent element. The cosets will be described in terms of gauge invariant components of the WZNW field. Taking into account the condition $\det g = 1$ and parametrising the WZNW field by

$$g = cI + v^n t_n = \begin{pmatrix} c - v_2 & -v_1 - v_0 \\ -v_1 + v_0 & c + v_2 \end{pmatrix}, \quad \text{with} \quad c^2 + v^n v_n = 1, \quad (3)$$

for each of the coset models only a couple of these field components will be gauge invariant.

It is worth to consider both Lagrangean and Hamiltonian reduction.

3.1. Lagrangean reduction by gauging

The considered $SL(2, R)$ WZNW theory on the cylinder can be gauged in the standard manner and it yields for the compact axial respectively vector cases the gauge invariant Lagrangeans with the euclidean target space geometries of a cigar (the 'euclidean black hole') and a trumpet

$$\mathcal{L}_G^{(1)}\Big| = \frac{1}{\gamma^2} \frac{\partial_z v_1 \partial_{\bar{z}} v_1 + \partial_z v_2 \partial_{\bar{z}} v_2}{1 + v_1^2 + v_2^2}, \quad \mathcal{L}_G^{(2)}\Big| = \frac{1}{\gamma^2} \frac{\partial_z c \partial_{\bar{z}} c + \partial_z v_0 \partial_{\bar{z}} v_0}{c^2 + v_0^2 - 1}. \quad (4)$$

Whereas the target space of $\mathcal{L}_G^{(1)}\big|$ is R^2, for $\mathcal{L}_G^{(2)}\big|$ the unit disk $c^2 + v_0^2 < 1$ is missing and this Lagrangean is singular at the disk boundary, but both coset theories are mutually related [5].

For the non-compact cases we obtain two equivalent minkowskian ('black hole') actions

$$\mathcal{L}_G^{(3)}\Big| = \frac{1}{\gamma^2} \frac{\partial_z v_1 \partial_{\bar{z}} v_1 - \partial_z v_0 \partial_{\bar{z}} v_0}{1 + v_1^2 - v_0^2}, \quad \mathcal{L}_G^{(4)}\Big| = \frac{1}{\gamma^2} \frac{\partial_z v_2 \partial_{\bar{z}} v_2 - \partial_z c \partial_{\bar{z}} c}{1 + v_2^2 - c^2}, \quad (5)$$

which are analytically related to (4). The target space is R^2 and $\mathcal{L}_G^{(3)}\big|$, e.g., has two singularity lines $v_0 = \pm\sqrt{1 + v_1^2}$. There are so three different regular domains in the target space

$$v_0 > \sqrt{1 + v_1^2}, \quad -\sqrt{1 + v_1^2} < v_0 < \sqrt{1 + v_1^2}, \quad v_0 < -\sqrt{1 + v_1^2}, \quad (6)$$

and this coset theory has to be investigated in each of them separately.

Finally, for the nilpotent gaugings only two identical gauged Lagrangeans arise for the field $V = g_{12}(z, \bar{z})$ whereas the other gauge invariant components v_2 or c simply disappear

$$\mathcal{L}_G^{(5)}\Big| = \mathcal{L}_G^{(6)}\Big| = \frac{1}{\gamma^2} \frac{\partial_z V \partial_{\bar{z}} V}{V^2}. \quad (7)$$

$V = 0$ is a singularity of the Lagrangian, but for the regular parametrisation $V = \pm e^{\gamma\phi}$ we get as a result free-field theories only

$$\mathcal{L}_G^{(5)}\Big| = \mathcal{L}_G^{(6)}\Big| = \partial_z \phi \partial_{\bar{z}} \phi. \quad (8)$$

Note that the Liouville theory does not arise by this standard gauging.

3.2. Hamiltonian reduction by constraints

Hamiltonian reduction is an alternative but more flexible method to construct and investigate coset theories. Here the constrained Kac-Moody currents $J_0 = 0 = \bar{J}_0$, $J_2 = 0 = \bar{J}_2$ and $J_+ = 0 = \bar{J}_+$ provide the cosets (4), (5) and (7) respectively. But both, the axial and the vector gauged Lagrangeans arise by one and the same constraints [4, 5]. Although these systems are described by different components of the WZNW field they live on the same constrained surface, and are therefore mutually related with each other. It is important to mention that the two first current constraints are of second class and the nilpotent gauging is of first class.

   Imposing to the $SL(2, R)$ WZNW theory the alternative nilpotent constraints $J_+ = \rho$, $\bar{J}_+ = \bar{\rho}$ with non-vanishing constants ρ and $\bar{\rho}$, and write the only gauge invariant field component as $g_{12}(z, \bar{z}) = \psi(z)\bar{\psi}(\bar{z}) + \chi(z)\bar{\chi}(\bar{z})$, then $\psi(z) = g_{11}(z)$, $\chi(z) = g_{12}(z)$ etc., satisfy constant Wronskians $\psi(z)\chi'(z) - \psi'(z)\chi(z) = \rho\gamma^2$ etc.. The following identification

$$e^{-\gamma\varphi(z,\bar{z})} = \psi(z)\bar{\psi}(\bar{z}) + \chi(z)\bar{\chi}(\bar{z}) \tag{9}$$

leads us to the Liouville equation with the 'cosmological' constant given by $\mu = -\rho\bar{\rho}\gamma^3$

$$\partial_{z\bar{z}}\varphi + \mu e^{2\gamma\varphi} = 0. \tag{10}$$

The equation (9) obviously also provides the general solution of the Liouville equation.
   Hamiltonian reduction is in fact a method for integrating coset theories [4].

4. Reduction of Poisson brackets

Since the nilpotent constraints are of first class, for the gauge invariant Liouville exponential $e^{-\gamma\varphi} = g_{12}(z, \bar{z})$ the reduced non-equal time Poisson bracket can be read off directly from the relation (1) without any further calculations. For the (anti-)chiral fields $\psi(z) = g_{11}(z)$, $\bar{\psi}(\bar{z}) = \bar{g}_{11}(\bar{z})$, $\chi(z) = g_{12}(z)$, $\bar{\chi}(\bar{z}) = \bar{g}_{12}(\bar{z})$ the classical form of the exchange algebra results, which quantum mechanically become the celebrated Gervais-Neveu quantum group relations.

   It might be worth to note that the gauge invariant nilpotent reduction of the Sugawara energy-momentum tensor immediately generates the Liouville form with the standard classical 'improvement' term included. But in this case there do not exist coset currents.

   The situation is different if we reduce the $SL(2, R)$ WZNW theory by the second class constraints. Using the isomorphism between $SL(2, R)$ and $SU(1, 1)$ there is a natural complex structure given, e.g. for the euclidean case (4), by the complex coordinates $u = v_1 + iv_2$ and $x = c + iv_0$ which are related by $|x|^2 - |u|^2 = 1$. $u(z, \bar{z})$ and $x(z, \bar{z})$ are the physical fields of the axial respectively vector gauged cosets. These gauge invariant fields can be expressed similarly as in the Liouville case (9) by $u(z, \bar{z}) = \psi(z)\bar{\psi}(\bar{z}) + \chi(z)\bar{\chi}(\bar{z})$, but now in terms of the complex fields $\psi(z) = g_{11}(z) + ig_{12}(z)$, $\chi(z) = g_{21}(z) + ig_{22}(z)$ etc., and with the non-constant Wronskians $\psi(z)\chi'(z) - \psi'(z)\chi(z) = 2W(z)$ and the anti-chiral one. Here $W(z) = J_1(z) + iJ_2(z)$ is the parafermionic coset current [3, 4]. The algebra of the coset fields is given by Dirac brackets [5], and there are causal relations for each coset

$$\{ u(z, \bar{z}), u(y, \bar{y}) \}_D = \gamma^2 \, \Theta \, [u(z, \bar{y}) \, u(y, \bar{z}) - u(z, \bar{z}) \, u(y, \bar{y})],$$
$$\{ u(z, \bar{z}), u^*(y, \bar{y}) \}_D = \gamma^2 \, \Theta \, x(z, \bar{y}) \, x^*(y, \bar{z}),$$
$$\{ x(z, \bar{z}), x(y, \bar{y}) \}_D = \gamma^2 \, \Theta \, [x(z, \bar{y}) \, x(y, \bar{z}) - x(z, \bar{z}) \, x(y, \bar{y})],$$
$$\{ x(z, \bar{z}), x^*(y, \bar{y}) \}_D = \gamma^2 \, \Theta \, u(z, \bar{y}) \, u^*(y, \bar{z}), \tag{11}$$

and non-causal connections which with the notation $2E = \epsilon(z - y)$, $2\bar{E} = \epsilon(\bar{z} - \bar{y})$ are

$$\{ u(z, \bar{z}), x(y, \bar{y}) \}_D = \gamma^2 \, \Theta \, x(z, \bar{y}) \, u(y, \bar{z}) - \gamma^2 \, E \, u(z, \bar{z}) \, x(y, \bar{y}),$$
$$\{ u(z, \bar{z}), x^*(y, \bar{y}) \}_D = \gamma^2 \, \Theta \, u(z, \bar{y}) \, x^*(y, \bar{z}) - \gamma^2 \, \bar{E} \, u(z, \bar{z}) \, x^*(y, \bar{y}). \tag{12}$$

   As expected the axial and vector gauged theories form a coupled algebra.

5. Canonical quantisation

The canonical quantisation of the cosets can be performed in the same way as it has been done for the Liouville theory [1]. Here one uses the general solution of the coset as a canonical transformation between the non-linear coset fields and free fields. The quantisation will be defined by replacing the Poisson brackets of the canonical free fields by commutators. Non-linear expressions in the free fields will be normal ordered. But calculations with normal ordered operators usually yield anomalous contributions. Such anomalies can be avoided by quantum mechanically deforming the composite operators of the cosets. The deformations are determined by requiring the classical symmetry transformations, and locality, to be valid as commutator relations. As a result, we show the non-equal time commutator for the Liouville exponential $u(z, \bar{z}) = e^{-\gamma \varphi(z, \bar{z})}$ which is written here for convenience as a Moyal bracket [2]

$$\{\check{u}(z, \bar{z}), \, \check{u}(y, \bar{y})\}_* = \frac{1}{\hbar} \, \sin(\hbar \gamma^2/4) \left[\epsilon(z - y) + \epsilon(\bar{z} - \bar{y}) \right] \times \tag{13}$$

$$\left[\check{u}(z, \bar{y}) * \check{u}(y, \bar{z}) + \check{u}(y, \bar{z}) * \check{u}(z, \bar{y}) - \frac{\check{u}(z, \bar{z}) * \check{u}(y, \bar{y}) + \check{u}(y, \bar{y}) * \check{u}(z, \bar{z})}{2 \cos(\hbar \gamma^2/4)} \right].$$

Its expansion in $\hbar$ reproduces the Poisson bracket. We can define by (13) other operators and their commutators. The operator of $e^{-2\gamma \varphi(z, \bar{z})}$ simply follows by differentiation for equal time.

Further results will be discussed in the lecture notes [6].

6. Final remarks

There is a complete classical understanding of the whole set of $SL(2, R)$ theories. Quantum mechanically the Liouville theory is best worked out, but still incomplete. The zero mode structure and the Hilbert space require further intensive study. Coset currents only exist for the non-nilpotent gauged $SL(2, R)$ theories as parafermions. They generate quantum mechanically a dilaton which might render the classically non-dynamical metric dynamical.

Besides being interesting in its own right the $SL(2, R)$ WZNW model and its cosets also appear in many applications, in particular in string calculations. An exact and complete quantum mechanically treatment of the $SL(2, R)$ family would be helpful to understand, e.g., the AdS_3/CFT correspondence, which is an intensively discussed contemporary problem.

Minor knowledge exists for the quantum mechanical $SL(2, R)$ WZNW model.

References

[1] H.J. Otto, G. Weigt, Construction of exponential Liouville field operators for closed string models, *Z. Phys.* **C31**, (1986) 219; G. Weigt, Canonical quantisation of the Liouville theory, quantum group structures, and correlation functions, Pathways to Fundamental Theories, World Scientific Publishing Co., (1993) 227, hep-th/9208075.

[2] G. Jorjadze, G. Weigt, Poisson structure and Moyal quantisation of the Liouville theory, *Nucl.Phys.* **B 619**, (2001) 232, hep-th/0105306; G. Jorjadze, Hamiltonian description of singular solutions of the Liouville equation, *Teor. Mat. Fiz.* **65** n.3 (1985) 323.

[3] U. Müller, G. Weigt, Analytical solution of the SL(2,R)/U(1) WZNW black hole model, *Phys. Lett.* **B 400**, (1997) 21, hep-th/9702095; The complete solution of the classical SL(2,R)/U(1) gauged WZNW field theory, *Comm. Math. Phys.* **205**, (1999) 421, hep-th/9805215; Integration of the SL(2,R)/U(1) gauged WZNW model with periodic boundary conditions, *Nucl.Phys.* **B 586**, (2000) 457, hep-th/9909057.

[4] C. Ford, G. Jorjadze, G. Weigt, Integration of the SL(2,R)/U(1) gauged WZNW theory by reduction and quantum parafermions, *Theor.Math.Phys.* **128**, (2001) 1046, hep-th/0003246.

[5] C. Ford, G. Jorjadze, G. Weigt, Causal Poisson brackets of the SL(2,R) WZNW model and its coset theories, *Phys. Lett.* **B 514**, (2001) 413, hep-th/0106060.

[6] G. Jorjadze, G. Weigt, Quantisation of gauged SL(2,R) WZNW theories, talk given by G. J. at the International Symposium Ahrenshoop, August 26-30, 2002, Wernsdorf, Germany, to be published.

Section 4

Particle and Nuclear Physics

Inst. Phys. Conf. Ser. No 173: Section 4
Paper presented at 24th Int. Coll. Group Theoretical Methods in Physics, Paris, France, 15–20 July 2002
©2003 IOP Publishing Ltd

Representations and interactions from field equations on an extended spin space

J Besprosvany

Instituto de Física, Universidad Nacional Autónoma de México, Apartado Postal 20-364, México 01000, D. F., México

Abstract. We consider a Dirac equation set on an extended spin space that contains fermion and boson solutions. The underlying Clifford algebra of the space, at given dimension, determines and classifies the Poincaré and scalar symmetries, and the resulting representations. The standard field equations can be equivalently written in terms of such degrees of freedom, and are similarly constrained. At 9+1 dimension, one finds restrictions on the interactions and representations that relate to the standard model. Information on the coupling constants is also provided.

1. Introduction

The current theory of elementary particles, the standard model (SM) is successful in describing their behavior, but it is phenomenological. The origin of the interaction groups, the particles' spectrum and representations, and parameters has remained largely unexplained. Still, the generalization of features of the model into larger structures with a unifying principle has suggested connections among the observables. Thus, additional dimensions in Kaluza-Klein theories are associated with gauge interactions, and larger groups in grand-unified theories[1] put some restrictions on them. Spin is a physical manifestation of the fundamental representation of the Lorentz group and it is more so in relation to space, which uses the vector representation. By setting Poincaré-invariant field equations on an extended spin space[2], described by a Clifford algebra, this contribution finds restrictions on and classifies the symmetries and the representation solutions, at given dimension. They contain spin-1/2 and boson solutions with a fixed scalar representation. A field theory can be formulated in terms of such degrees of freedom. By constraining only the dimension and Poincaré algebra of the spin space on which the fields are formulated, one finds the interactions and representations of the electroweak-lepton SM sector at 5+1 d, and of the SM at 9+1 d, the minimal space that contains them.

2. Extended Dirac equation

The Dirac equation

$$\gamma_0(i\partial_\mu\gamma^\mu - M)\Psi = 0, \tag{1}$$

uses an extended spin space when Ψ represents a matrix instead of, as traditionally, a four-entries (column) spinor. Eq. 1 contains four conditions over four spinors in a 4×4 matrix. There are, then, additional possible transformations and symmetry operations that further classify Ψ. The Dirac-operator transformation $(i\partial_\mu\gamma^\mu - M) \to U(i\partial_\mu\gamma^\mu - M)U^{-1}$ induces the left-hand side of the transformation

$$\Psi \to U\Psi U^\dagger, \tag{2}$$

and Ψ is postulated to transform as indicated on the right-hand side.

U and Ψ can be classified in terms of Clifford algebras. In four dimensions (4-d) U is conventionally a 4×4 matrix containing symmetry operators as the Poincaré generators, but it can contain others, although, e. g., in the chiral massless case it can only carry an additional $U(2)$ scalar symmetry[2]. More symmetry operators appear if Eq. 1, $\mu = 0, ..., 3$, is assumed within the larger Clifford algebra C_N, $\{\gamma_\eta, \gamma_\sigma\} = 2g_{\eta\sigma}$, $\eta, \sigma = 0, ..., N - 1$, where N is the (assumed even) dimension, whose structure is helpful in classifying the available symmetries, and which is represented by $2^{N/2} \times 2^{N/2}$ matrices. The usual 4-d Lorentz symmetry, generated in terms of $\sigma_{\mu\nu} = \frac{i}{2}[\gamma_\mu, \gamma_\nu]$, $\mu, \nu = 0, ..., 3$, is maintained and U contains also γ_a, $a = 4, ..., N - 1$, and their products as possible symmetry generators. Indeed, these elements are scalars for they commute with the Poincaré generators, which contain $\sigma_{\mu\nu}$, and they are also symmetry operators of the massless Eq. 1, bilinear in the γ_μ matrices, which is not necessarily the case for mass terms (containing γ_0). In addition, their products with $\gamma_5 = -i\gamma_0\gamma_1\gamma_2\gamma_3$ are Lorentz pseudoscalars. As $[\gamma_5, \gamma_a] = 0$, we can classify the (unitary) symmetry algebra as $\mathcal{S}_{N-4} = \mathcal{S}_{(N-4)R} \times \mathcal{S}_{(N-4)L}$, consisting of the projected right-handed $\mathcal{S}_{(N-4)R} = \frac{1}{2}(1 + \gamma_5)U(2^{(N-4)/2})$ and left-handed $\mathcal{S}_{(N-4)L} = \frac{1}{2}(1 - \gamma_5)U(2^{(N-4)/2})$ components.

The solutions of Eq. 1 do not span all the matrix complex space, but this is achieved by considering also solutions of

$$\Psi\gamma_0(-i\overleftarrow{\partial_\mu}\,\gamma^\mu - M) = 0, \tag{3}$$

consistent with the transformation in Eq. 2, (the Dirac operator transforming accordingly).

It is not possible to find always solutions that simultaneously satisfy equations of the type 1 and 3 (except trivially), which means they are not simultaneously on-shell, but they satisfy at least one and therefore the Klein-Gordon equation. Indeed, the solutions of eqs. 1 and 3, can be generally characterized as bosonic since Ψ can be understood to be formed of states as $\sum_{i,j} a_{ij}|w_i\rangle\langle w_j|$, where the column $|w_i\rangle$, and row $\langle w_j|$ states can have a spinor interpretation.

Generalized operators acting on this tensor-product space ($|w_i\rangle \times \langle w_j| \times$ configuration or momentum space) further characterize the solutions. Positive-energy solutions, according to Eq. 1 are interpreted as negative-energy solutions from the right-hand side. This problem is overcome if we assume the hole interpretation for the $\langle w_j|$ components, which amounts to the requirement that operators generally acting from the right-hand side acquire a minus, and that the commutator be used for operator evaluation. Thus, the 4-d plane-wave solution combination $\frac{1}{4}[(1 - \gamma_5)\gamma_0(\gamma_1 - i\gamma_2)]e^{-ikx}$, with $k^\mu = (k, 0, 0, k)$, is a massless vector$-$axial ($V - A$) state propagating along $\hat{z}$ with left-handed circular polarization, normalized covariantly according to $\langle\Psi_A|\Psi_B\rangle = tr\Psi_A^\dagger\Psi_B$, the generalized inner product for the solution space. In fact, combinations of solutions of Eqs. 1 and 3 can be formed with a well-defined Lorentz index: vector $\gamma_0\gamma_\mu$, pseudo-vector $\gamma_5\gamma_0\gamma_\mu$, scalar γ_0, pseudoscalar $\gamma_0\gamma_5$, and antisymmetric tensor $\gamma_0[\gamma_\mu, \gamma_\nu]$. For example, $A_\mu^C(x) = \frac{i}{2}\gamma_0\gamma_\mu e^{-ikx}$ is a combination that transforms under parity into $A^{C\mu}(\tilde{x})$, $\tilde{x}_\mu = x^\mu$, that is, as a vector. We may also view $\frac{1}{2}\gamma_0\gamma_\mu$ as an orthonormal polarization basis, $A_\mu = tr\frac{1}{2}\gamma_\mu A^\nu\frac{1}{2}\gamma_\nu\ddagger$; just as n_μ in $A_\mu = g_{\mu\nu}A^\nu = n_\mu \cdot A^\nu n_\nu$. In fact, the sum of Eqs. of 1 and 3 implies [3] for a Ψ containing $\gamma_0 A\!\!\!/ = A^\mu\gamma_0\gamma_\mu$ that A^μ satisfies the free Maxwell's equations.

Solutions contain also products of γ_a matrices that define their scalar-group representation. For given N, there are variations of the symmetry algebra depending on the chosen Poincaré generators and Dirac equation, respectively, through the projection operators $\mathcal{P}_P$, $\mathcal{P}_D \in \mathcal{S}_{N-4}$, $[\mathcal{P}_P, \mathcal{P}_D] = 0$. $\mathcal{P}_P$ acts as in, e.g., $\mathcal{P}_P\sigma_{\mu\nu}$, and $\mathcal{P}_D$ modifies Eqs. 1 and 3

‡ As for $\bar{\psi} = \psi^\dagger\gamma_0$, a unitary transformation can be applied to the fields and operators to convert them to a covariant form.

through $\mathcal{P}_D\gamma_0(i\partial_\mu\gamma^\mu - M)$. Together, they characterize the Lorentz and scalar-group solution representations. We require $\text{rank}\mathcal{P}_D \leq \text{rank}\mathcal{P}_P$, for otherwise pieces of the solution space exist that do not transform properly. For $\mathcal{P}_D \neq 1$ Lorentz operators act trivially on one side of the solutions containing $1 - \mathcal{P}_P$, since $(1 - \mathcal{P}_P)\mathcal{P}_P = 0$, so we also get fermions. Fig. 1(a) depicts the distribution of Lorentz-representation solutions according to the matrix space they occupy in $\mathcal{S}_6$, when $\mathcal{P}_P = \mathcal{P}_D \neq 1$. Although it refers to $N = 10$ case, it is general for any N.

	F		F	
F	V		S,A	
		(b)		(c)
F	S,A		V	
				(d)

Table 1. Arrangement of $\mathcal{S}_6$ scalar components of $N = 9 + 1$ solutions. (a) $\mathcal{S}_6$ is divided into four 6-d 8×8 matrix blocks, with fermion (F), vector (and axial-) (V), and scalar (and pseudo-) and antisymmetric (S,A) terms.

3. Interactive field equations

An interactive field theory can be constructed in terms of the above degrees of freedom. We consider a vector and fermion non-abelian gauge-invariant theory. The expression for the kinetic component of the Lagrangian density $\mathcal{L}_V = -\frac{1}{4}F^a_{\mu\lambda}g^{\lambda\eta}\delta_{ab}F^{b\mu}_\eta = -\frac{1}{4N_o}tr\mathcal{P}_D F^a_{\mu\lambda}\gamma_0\gamma^\lambda G_a F^{b\mu}_\eta\gamma_0\gamma_\eta G_b$ shows $\mathcal{L}_V$ is equivalent to a trace over combinations over normalized components $\frac{1}{\sqrt{N_o}}\gamma_0\gamma_\mu G_a$ with coefficients $F^a_{\mu\nu} = \partial_\mu A^a_\nu - \partial^a_\nu A^a_\mu + gA^b_\mu A^c_\nu C^a_{bc}$, g the coupling constant, $\gamma_\mu \in \mathcal{C}_N$, $G_a \in \mathcal{S}_{N-4}$ the group generators, C^a_{bc} the structure constants, and $N_o = trG_aG_a$, where for non-abelian irreducible representations we use $trG_iG_j = 2\delta_{ij}$.

Similarly, the interactive part of the fermion gauge-invariant Lagrangian $\mathcal{L}_f = \frac{1}{2}\psi^{\alpha\dagger}\gamma_0(i\overleftrightarrow{\partial}_\mu - gA^a_\mu G_a)\gamma^\mu\psi^\alpha$, with ψ^α a massless spinor with flavor α, can be written $\mathcal{L}_{int} = -g\frac{1}{2N_o}tr\mathcal{P}_D A^a_\mu\gamma_0\gamma^\mu G_a j^{ba}_\lambda\gamma_0\gamma_\lambda G_b$, with $j^{a\alpha}_\mu = tr\Psi^{\alpha\dagger}\gamma_0\gamma_\mu G_a\Psi^\alpha$ containing $\Psi^\alpha = \psi^\alpha\langle\alpha|$, and $\langle\alpha|$ is a row state accounting for the flavor. $\mathcal{L}_{int}$ is written in terms of $\gamma_0\cancel{A}$, and $\gamma_0 j^{a\alpha}$, that is, the vector field and the current occupy the same spin space. This connection and the quantum field theory (QFT) understanding of this vertex as the transition operator between fermion states, exerted by a vector particle, with the coupling constant as a measure of the transition probability, justifies the interpretation for it $\frac{1}{2}gA^{a\mu}j^{a\alpha}_\mu = A^{a\mu}\frac{1}{\sqrt{N_o}}tr\Psi^{\alpha\dagger}\gamma_0\gamma_\mu G_a\Psi^\alpha$, leading to the identification $g \to 2\sqrt{\frac{K}{N_o}}$, K correcting for over-counted reducible representations, which will not be further needed. The theoretical assignment of g complements QFT, in which the coupling constant is set experimentally. It should be also understood as tree-level information, while the values are modified by the presence of a virtual cloud of fields, at given energy. Although in QFT the coupling constant is obtained perturbatively in terms of powers of the bare, which takes infinite values absorbed through renormalization, we may take the view that renormalization is a calculational device and that its physical value is a manifestation of the bare one; this is feasible for small coupling constants, which can give small corrections. Energy corrections are also necessary for a more detailed calculation. Applications of the above rule are obtained in Refs. [2] and [4].

It is apparent that the minimal algebra that includes the SM groups requires $N = 9 + 1$, with $\mathcal{S}_6$, on which we will concentrate. There are limited ways in which we may represent the SM interactions in such a matrix space and only one giving the correct fermion and boson quantum numbers. In order to have fermions we need $\mathcal{P}_P \neq 1$. Account of the quark quantum numbers requires that their left-handed $SU(3) \times SU(2)_L$, and right-handed $SU(3) \times U(1)_Y$ symmetry generators be direct-product reducible representations occupying, respectively, 6×6 matrix pieces of $\mathcal{S}_{6L}$ and $\mathcal{S}_{6R}$. The remaining 2×2 matrix into which $\mathcal{S}_{6L}$ is broken is associated to the $SU(2)_L$ acting on the $SU(3)$-singlet leptons, and that of $\mathcal{S}_{6R}$ to a $U(1)$ describing the right-handed leptons hypercharge, and an inert $U(1)$ that gives rise to two fermion generations (all this applies also to antiparticles). There are additional $U(1)$ symmetries which can be assigned in correspondence to SM symmetries. More details are found in Ref. [4]

In summary, the formulation of field equations on an extended spin of a given dimension provides restrictions on the interactions and representations. There are particular configurations that produce information on the standard model.

References

[1] Georgi H and Glashow S L 1974 Phys. Rev. Lett. 32 438-441

[2] Besprosvany J 2000 Int. J. Theor. Phys. 39 2797-2836 hep-th/0203114; Besprosvany J 2001 Nuc. Phys. B (Proc. Suppl.) 101 323-329

[3] Bargmann V and Wigner E P 1948 Proc. Nat. Acad. Sci. (USA) 34 211

[4] Besprosvany J hep-th/0203122

Inst. Phys. Conf. Ser. No 173: Section 4
Paper presented at 24th Int. Coll. Group Theoretical Methods in Physics, Paris, France, 15–20 July 2002
©2003 IOP Publishing Ltd

Mass Relations for the Quark-Diquark Model

S. Catto[†], J. Huntley, N. Moh and D. Tepper
University Center and The Graduate School, the City University of New York
365 Fifth Avenue, New York, NY 10016
[†] Also at the Center for Theoretical Physics, The Rockefeller University
New York, New York 10021-6399

Abstract. Quark model with potentials derived from QCD, including the quark-diquark model for excited hadrons gives mass formulae in very good agreement with experiment and goes a long way in explaining the approximate symmetries and supersymmetries of the hadronic spectrum, including the symmetry breaking mechanism.

1. Introduction

There is a good phenomenological evidence that in a rotationally excited baryon a quark-diquark $(q - D)$ structure is favored over a three quark (qqq) structure. Regge trajectories for mesons and baryons are closely parallel; both have a slope of about $0.9 GeV^{-2}$. At large spin two of the quarks form a diquark $(D = qq)$, a bilocal object at one end of a bag, the remaining quark being at the other. For the light quarks we had shown ([1],[2],[3]) that the underlying quark-diquark symmetry leads to supersymmetric $SU(6/21)$ symmetry between mesons and baryons. This new scheme uses split octonion algebra that produces the algebraic description of color degrees of freedom, supresses color-symmetric and space-symmetric quark configurations, and leads to existence of exotic meson (diquark-antidiquark) states for which there is now some experimental evidence. In this note we shall deal with derivation of hadronic mass formulae only. A new extended relativistic mass formulae based on octonionic supersymmetry of hadrons will be discussed in another publication.

2. Mass Formulae

The ground state energy eigenvalue E of the Hamiltonian can be estimated by using the Heisenberg uncertainty principle. This leads to the replacement of r by Δr and p_r by

$$\Delta p_r = \frac{1}{2}(\Delta r)^{-1}, \quad (h = 1). \tag{1}$$

Then E as a function of Δr is minimized for the value of r_0 of Δr. The r_0 corresponds to the Bohr radius for the bound state. The confining energy associated with this Bohr radius is obtained from the linear confining potential $S(r) = br$, so that the effective masses of the constituents become

$$M_1 = m_1 + \frac{1}{2}S_0, \quad M_2 = m_2 + \frac{1}{2}S_0, \quad (S_0 = br_0) \tag{2}$$

For a meson m_1 and m_2 are the current quark masses while M_1 and M_2 can be interpreted as the constituent quark masses. Note that even in the case of vanishing quark masses associated with perfect chiral symmetry, confinement results in non zero constituent masses that spontaneously break the $SU(2) \times SU(2)$ symmetry of the u, d quarks.

Let us illustrate this method on the simplified spin free Hamiltonian involving only the scalar potential. In the center of mass system, $p^{(1)} + p^{(2)} = 0$, or $p^{(1)} = -p^{(2)} = p$. The semi-relativistic hamiltonian of the system is then given by

$$E_{12} \ \Phi = \sum_{i=1}^{2} \sqrt{(m_i + \frac{1}{2}br)^2 + p^2} \ \Phi. \tag{3}$$

Taking $m_1 = m_2 = m$ for the quark-antiquark system, we have

$$E_{12} \ \Phi = 2\sqrt{(m + \frac{1}{2}br)^2 + p_r^2 + \frac{\ell(\ell+1)}{r^2}} \ \Phi, \tag{4}$$

where we have written the momentum part in spherical coordinates.

Putting

$$b = \mu^2, \qquad \rho = \mu \ r, \tag{5}$$

for the $q - \bar{q}$ system we find E_{12} by minimizing the function

$$E_{q\bar{q}} = 2\sqrt{(m + \frac{1}{2}\mu\rho)^2 + \frac{\mu^2}{\rho^2}(\ell + \frac{1}{2})^2}. \tag{6}$$

For u and d quarks, m is small and can be neglected so that

$$E^2 = \mu^2[\rho^2 + \rho^{-2}(2\ell + 1)^2] \tag{7}$$

which has a minimum for

$$\rho^2 = \rho_0^2 = 2\ell + 1, \tag{8}$$

giving

$$E_{min}^2 = E^2(\rho_0) = 4\mu^2(\ell + \frac{1}{2}). \tag{9}$$

Thus, we obtain a linear Regge trajectory with

$$\alpha' = \frac{1}{4}\mu^{-2} = \frac{b}{4}. \tag{10}$$

Also $J = \ell + S$, where S arises from the quark spins. Experimentally

$$\alpha' = 0.88(GeV)^{-2} \tag{11}$$

for mesons giving the value 0.54 GeV for μ. A more accurate calculation (see [4]) gives

$$\alpha' = (2\pi\mu^2)^{-1}, \qquad \mu \sim 0.43GeV. \tag{12}$$

The constituent quark mass can be defined in two ways

$$M_c(\ell) = \frac{1}{2}E_{min} = \mu\sqrt{\ell + \frac{1}{2}}, \tag{13}$$

or

$$m'_c(\ell) = S_0 = \frac{1}{2}\mu\rho_0 = \frac{\mu}{\sqrt{2}}\sqrt{\ell + \frac{1}{2}}. \tag{14}$$

The first definition gives for $\ell = 0$,

$$M_c = 0.31GeV \quad for \quad \mu = 0.43 \tag{15}$$

in the case of u and d quarks.

When the Coulomb like terms are introduced in the simplified Hamiltonian (4) with negligible quark masses one obtains

$$E = \frac{\mu}{\rho}[-\bar{\alpha} + \sqrt{\rho^4 + (2\ell + 1)^2}] \tag{16}$$

with

$$\bar{\alpha} = \frac{4}{3}\alpha_s \ \ for(q\bar{q}), \ \ \bar{\alpha} = \frac{2}{3}\alpha_s \ \ for(qq). \tag{17}$$

In the energy range around 1 GeV, α_s is of order of unity. Estimates range from 0.3 to 3. Minimization of E gives

$$E_0 = \mu u_0^{\frac{-1}{4}}(-\bar{\alpha} + \sqrt{u_0 + (2\ell + 1)^2}) \tag{18}$$

where

$$u_0(\epsilon) = \rho_0^4 = (2\ell + 1)^2(1 + \frac{1}{2}\beta^2 + \epsilon\sqrt{2}\beta\sqrt{\ell + \frac{1}{8}\beta^2}), \tag{19}$$

$$\epsilon = \pm 1, \ \ \beta = \frac{\bar{\alpha}}{(2\ell + 1)}. \tag{20}$$

The minimum E_0 is obtained for $\epsilon = -1$, giving to second order in β:

$$E_0 = \mu\sqrt{2(2\ell + 1)}(1 - \frac{\beta}{\sqrt{2}} - 3\frac{\beta^2}{8}). \tag{21}$$

Linear Regge trajectories are obtained if β^2 is negligible. Then for mesons

$$E_0^2 = 4\mu^2\ell + 2\mu^2(1 - \sqrt{2}\bar{\alpha}). \tag{22}$$

The β^2 is negligible for small ℓ only if we take the lowest estimate for α_s, giving 0.4 for $\bar{\alpha}$ in the $q\bar{q}$ case. For mesons with u, d constituents, incorporating their spins through the Breit term we obtain approximately

$$m_\rho \simeq m_\omega = E_0 + \frac{c}{4}, \ \ m_\pi = E_0 - \frac{3c}{4}, \ \ c = K\frac{\Delta V}{M_q^2} \tag{23}$$

where M_q is the constituent quark mass. This gives

$$E_0 = \frac{(3m_\rho + m_\pi)}{4} = 0.61 GeV. \tag{24}$$

The Regge slope being of the order of $1 GeV$ an average meson mass of the same order is obtained from Eq.(21) in the linear trajectory approximation. To this approximation $\bar{\alpha}$ should be treated like a parameter rather than be placed by its value derived from QCD under varying assumptions. Using Eq.(12) for μ one gets a better fit to the meson masses by taking $\alpha_s \sim 0.2$.

Turning now to baryon masses, we must first estimate the diquark mass. We have for the qq system

$$M_D = \mu(\sqrt{2} - \frac{2}{3}\alpha_s), \tag{25}$$

that is slightly higher than the average meson mass

$$\tilde{m} = \mu(\sqrt{2} - \frac{4}{3}\alpha_s). \tag{26}$$

Here we note that E is not very sensitive to the precise value of the QCD running coupling constant in the GeV range. Taking $\alpha_s \sim 0.3$ changes E^{qq} from 0.55 to $0.56 GeV$.

Note that Eq.(25) gives $m_D = 0.55 GeV$. For excited $q - \bar{q}$ and $q - D$ systems if the rotational excitation energy is large compared with μ, then both the m_D and the Coulomb term $-\frac{4}{3}\frac{\alpha_s}{r}$ (same for $q - D$ and $q - \bar{q}$ systems) can be neglected. Thus, for both $(q - D)$ [excited baryon] and $q - \bar{q}$ [excited meson] systems we have Eq.(9), namely

$$(E^{q-D})^2 \sim (E^{q-\bar{q}})^2 \sim 4\mu^2\ell + 2\mu^2 \tag{27}$$

giving again Eq.(10), i.e.

$$(\alpha')_{q-D} = (\alpha')_{q-\bar{q}} \cong \frac{1}{4\mu^2} \quad or \quad (\frac{1}{2\pi\mu^2}) \tag{28}$$

as an explanation of hadronic supersymmetry in the nucleon and meson Regge spectra. We also have, extrapolating to small ℓ:

$$\Delta(M^2)^{q-D} = \Delta(m^2)^{q-\bar{q}} = 4\mu^2\Delta\ell = \frac{1}{\alpha'}\Delta\ell. \tag{29}$$

For $\Delta\ell = 1$ we find

$$m_\Delta^2 - m_N^2 = m_\rho^2 - m_\pi^2. \tag{30}$$

This relationship is same as the one proved in our earlier paper ([1]) through the assumption that $U(6/21)$ symmetry is broken by an operator that behaves like $s = 0$, $I = 0$ member of 35×35 representations of $SU(6)$, which is true to 5%. It corresponds to a confined quark approximation with $\alpha_s = 0$.

The potential model gives a more accurate symmetry breaking ($\alpha_s \sim 0.2$):

$$\frac{9}{8}(m_\rho^2 - m_\pi^2) = m_\Delta^2 - m_N^2 \tag{31}$$

with an accuracy of 1%.

This mass squared formula arises from the second order iteration of the $q - D$, $q - \bar{q}$ Dirac equation. The factor $\frac{9}{8}$ comes from

$$\frac{1}{2}(\frac{4}{3}\alpha_s)^2 = \frac{8}{9}\alpha_s^2. \tag{32}$$

At this point it is more instructive to derive a first order mass formula. Since the constituent quark mass M_q is given by Eq.(13) ($\ell = 0$), we have

$$M_q = \frac{\mu}{\sqrt{2}}, \tag{33}$$

so that

$$\bar{m} = 2M_q(1 - \frac{\sqrt{2}}{3}\alpha_s) \simeq 1.9M_q. \tag{34}$$

When the baryon is regarded as a $q - D$ system, each constituent gains an effective mass $\frac{1}{2}\mu\rho_0$ which was approximately the effective mass of the quark in the meson. Hence, the effective masses of q and D in the baryon are

$$m'_q \simeq M_q, \quad m'_D = M_D + M_q \simeq 3M_q. \tag{35}$$

The spin splittings for the nucleon N and the Δ are given by the Breit term

$$\Delta M = K\Delta V \frac{\mathbf{S}_q \cdot \mathbf{S}_D}{m'_q m'_D}. \tag{36}$$

For the nucleon with spin $\frac{1}{2}$ the term $\mathbf{S}_q \cdot \mathbf{S}_D$ gives -1 while it has the value $\frac{1}{2}$ for Δ with spin $\frac{3}{2}$. Using the same K for mesons and baryons which are both considered to be a bound state of a color triplet with a color antitriplet we can relate the baryon splitting ΔM to the meson splitting Δm for which $\mathbf{S}_q \cdot \mathbf{S}_{\bar{q}}$ takes the values $\frac{1}{4}$ and $\frac{-3}{4}$. Hence we find

$$\Delta M = M_\Delta - M_N = \frac{3}{2} \cdot \frac{K\Delta V}{m'_q m'_D} = \frac{1}{2} \cdot \frac{K\Delta V}{M_q^2}, \quad \text{and} \quad \Delta m = \frac{K\Delta V}{M_q^2} \quad (37)$$

which leads to a linear mass formula

$$\Delta M = \frac{1}{2}\Delta m \quad (38)$$

which is well satisfied, and has been verified before using the three quark constituents for the baryon([3]).

The formation of diquarks which behave like antiquarks as far as QCD is concerned is crucial to hadronic supersymmetry and to quark dynamics for excited hadrons. The splittings in the mass spectrum are well understood on the basis of spin-dependent terms derived from QCD. This approach to hadronic physics has led to many in depth investigations recently. For extensive references we refer to recent papers by Lichtenberg and collaborators ([5]) and by Klempt ([6]). To see the symmetry breaking effect, note that the mass of a hadron will take the approximate form

$$m_{12} = m_1 + m_2 + K\frac{\mathbf{S}_1 \cdot \mathbf{S}_2}{m_1 m_2} \quad (39)$$

where m_i and $\mathbf{S}_i$ ($i = 1, 2$) are respectively the constituent mass and the spin of a quark or a diquark. The spin-dependent Breit term will split the masses of hadrons of different spin values. If we assume $m_q = m_{\bar{q}} = m$, where m is the constituent mass of u or d quarks, and denote the mass of a diquark as m_D, then this approximation gives

$$m_\pi = (m_{q\bar{q}})_{s=0} = 2m - K\frac{3}{4m^2}, \quad m_\rho = (m_{q\bar{q}})_{s=1} = 2m + K\frac{1}{4m^2} \quad (40)$$

$$m_\Delta = (m_{qD})_{s=3/2} = m + m_D + K\frac{1}{2mm_D} \quad (41)$$

$$m_N = (m_{qD})_{s=1/2} = m + m_D - K\frac{1}{mm_D}, \quad (42)$$

Eliminating m, m_D and K, we obtain a mass relation

$$\frac{8}{3} \cdot \frac{2m_\Delta + m_N}{3m_\rho + m_\pi} = 1 + \frac{3}{2} \cdot \frac{m_\rho - m_\pi}{m_\Delta - m_N} \quad (43)$$

which agrees with experiment to 13%.

† Work Supported by DOE grants DE-AC 0276 ER3074 and 3075, and PSC-CUNY Research Awards.

References

[1] Catto S and Gürsey F 1985 Nuovo Cim. 86 A, 201-218
[2] Catto S and Gürsey F 1988 Nuovo Cim. 99A, 685-699
[3] Catto S 1994 Colored Supersymmetry of Mesons and Baryons Based on Octonionic Algebras. Symmetries in Science VI, 129-148. Ed. B. Gruber (Plenum Publishers)
[4] Eguchi T 1975 Phys. Lett. 59, 457; Johnson K and Thorn C.B 1976 Phys. Rev. 13, 1934
[5] Lichtenberg D.B., et.al. 1982 Phys. Rev. Lett. 48, 1653; Lichtenberg D.B. and Namgung W 1984 Lett. Nuovo Cimento 41, 597; Lichtenberg D.B., et.al., 1993 Rev. Mod. Phys. 65, 1199.
[6] Klempt E 2002 Baryon Resonances and Strong QCD. arXiv:nucl-ex/ 0203002

Inst. Phys. Conf. Ser. No 173: Section 4
Paper presented at 24th Int. Coll. Group Theoretical Methods in Physics, Paris, France, 15–20 July 2002
©2003 IOP Publishing Ltd

On the role of quantum deformation in isovector pairing interactions in nuclei

K. D. Sviratcheva † , C. Bahri † , J. P. Draayer † and A. I. Georgieva ‡†

† Louisiana State University, Department of Physics and Astronomy, Baton Rouge, Louisiana, 70808-4001 USA
‡ Institute of Nuclear Research and Nuclear Energy, Bulgarian Academy of Sciences, Sofia 1784, Bulgaria

In this article we consider the physical importance of q deformation in nuclear physics [1] by considering a q deformed generalization of a classical nuclear theory. The role of the deformation parameter is explored by comparing the non-deformed and deformed results with experimental data. This is accomplished by relating physical quantities, other than the Hamiltonian itself, with generators of the Cartan subalgebra of the principal symmetry group. This is important because the latter do not change as a function of the deformation while the other generators do. In this way the effects of the deformation on the underlying physics are isolated from other changes.

The specific example considered is the $Sp(4)$ model which is known to be a reasonable theory for studying pairing correlations in nuclei. The model allows for a systematic classification of nuclei within a given shell [2]. And since the strength of the two-body nucleon-nucleon interaction is normally assumed to be constant for all the nuclei within a major shell, changes from one nucleus to another are driven solely by the number of nucleons. In addition, the q-deformation of the $sp(4)$ algebra can be used to explore higher-order correlations within this theory without changing the nature of the operators that measure physical quantities and with or without changing the overall strength of the interaction. A feature that distinguishes this approach from others is that all changes enter into the theory in a very prescriptive, algebraic way.

The $sp_q(4)$ algebra is constructed in terms of q-deformed creation and annihilation operators $\alpha^\dagger_{m,\sigma}$ and $\alpha_{m,\sigma}$, $(\alpha^\dagger_{m,\sigma})^* = \alpha_{m,\sigma}$, which create and annihilate a particle of type σ in a state of total angular momentum $j = \frac{2k+1}{2}$, $k = 0, 1, 2, ...$, with projection m along the z axis $(-j \leq m \leq j)$. The deformed single-particle operators are defined through their anticommutation relation for every σ and m [3]:

$$\{\alpha_{m,\sigma}, \alpha^\dagger_{m',\sigma'}\}_{q^{\pm 1}} = q^{\pm \frac{N_\sigma}{2\Omega_j}} \delta_{m,m'}\delta_{\sigma,\sigma'} \qquad \{\alpha^{(\dagger)}_{m,\sigma}, \alpha^{(\dagger)}_{m',\sigma'}\} = 0 \qquad (1)$$

where by definition the q anticommutator is given as $\{A, B\}_k = AB + q^k BA$. In the limit $q \to 1$, the operators $\alpha^\dagger_{m,\sigma}$ $(\alpha_{m,\sigma})$ revert to their classical counterpart $c^\dagger_{m,\sigma}$ $(c_{m,\sigma})$, $\sigma = \pm 1$. For a given σ, the dimension of the fermion space is $2\Omega_j = 2j + 1$.

The generators of $sp_q(4)$ are the deformed analogues of the "classical" operators [4, 5] that create (annihilate) a pair of fermions coupled to total angular momentum and parity $J^\pi = 0^+$. Consequently, they constitute boson-like objects, with the properties of the components of a tensor of first rank $F_{0,\pm 1}$ $(G_{0,\pm 1})$, where $F_{\frac{\sigma+\sigma'}{2}} \equiv F_{\sigma,\sigma'}$ $(G_{\frac{\sigma+\sigma'}{2}} \equiv G_{\sigma,\sigma'})$, $\sigma, \sigma' = \pm 1$:

$$F_{\sigma,\sigma'} = \frac{1}{\sqrt{2\Omega_j}\sqrt{(1+\delta_{\sigma,\sigma'})}} \sum_{m=-j}^{j} (-1)^{j-m}\alpha^\dagger_{m,\sigma}\alpha^\dagger_{-m,\sigma'} = F_{\sigma',\sigma} = (G_{\sigma,\sigma'})^\dagger \quad (2)$$

The raising and lowering number preserving generators of $sp_q(4)$ are:

$$E_{\pm 1, \mp 1} = \frac{1}{\sqrt{2\Omega_j}} \sum_{m=-j}^{j} \alpha^{\dagger}_{m,\pm 1} \alpha_{m,\mp 1}. \tag{3}$$

In addition to the generators (2) and (3), the operators that count the number of fermions of each kind, $N_{\pm 1} = \sum_{m=-j}^{j} c^{\dagger}_{m,\pm 1} c_{m,\pm 1}$, remain non-deformed and form the Cartan subalgebra of $sp_q(4)$. These ten operators close on the symplectic $sp_q(4)$ algebra [3]. We use commutation relations that are symmetric with respect to the exchange of the deformation parameter $q \leftrightarrow q^{-1}$.

In analogy with the microscopic "classical" approach [6], the most general Hamiltonian of a system with $Sp_q(4)$ dynamical symmetry, which preserves the total number of particles, can be expressed through the group generators as

$$H_q = -(\epsilon_j^q - (\frac{1}{2} - 2\Omega)C_q - \frac{D_q}{4})N - -2\Omega C_q[K_0]^{*2} - \Omega D_q[T_0]^{*2} - O_q$$

$$- G_q F_0 G_0 - F_q(F_{+1}G_{+1} + F_{-1}G_{-1}) - \frac{1}{2}E_q(\{T_+, T_-\} - \left[\frac{N}{2\Omega}\right]), \tag{4}$$

where by definition, $[X]_k = \frac{q^{kX}-q^{-kX}}{q^k-q^{-k}}$, $([X] \equiv [X]_1)$, $[X]^{*2} = \Omega[X]([X+1] + [X-1])$. In (4), $\epsilon_j^q > 0$ is the Fermi level of the nuclear system, G_q, F_q, E_q, C_q and D_q are constant interaction strength parameters that may be different then the corresponding non-deformed phenomenological parameters. The constant $O_q = -2\Omega C_q[\Omega]^{*2}$ sets the energy of zero particles to be zero. An important feature of the phenomenological Hamiltonian (4) is that it not only breaks the isospin symmetry ($D_q \neq 0$), but it also mixes states with definite isospin values ($F_q \neq G_q$). Hence the model describes the behavior of the valence particles in a nucleus in addition to the mean-field of the doubly-magic core. In the q-deformed case, the Hamiltonian of the system can be expressed in terms of the q-deformed first and second order Casimir operators of four different two-dimensional unitary q subalgebras, $u_q^\mu(2) \supset u^\mu(1) \oplus su_q^\mu(2)$ ($\mu = \{T, 0, \pm\}$), presented in **Table 1.** The difference between the pp and nn interactions is set by the deformed algebra, which introduces the coefficients $\rho_\pm = (q^{\pm 1} + q^{\pm\frac{1}{2\Omega}})/2$, $(\rho_\pm \overset{\varkappa \to -\varkappa}{\to} \rho_\mp)$.

With the help of an expansion of the eigenvalues of the deformed pairing Hamiltonian in orders of $\varkappa$ ($q = e^\varkappa$) [2], we see that the deformed terms of (4) include all higher orders in the non-deformed pairing energies, which are their zeroth order approximation.

Table 1. Realizations of the unitary subalgebras of $sp_{(q)}(4)$, $\mu = \{T, 0, \pm\}$

$U^\mu(1)$	$SU_q^\mu(2)$	$C_2(SU_q^\mu(2))$
$N = N_{+1} + N_{-1}$	$T_\pm \equiv E_{\pm 1, \mp 1}$ $T_0 \equiv \tau_0 = \frac{N_1 - N_{-1}}{2}$	$2\Omega_j(\{T_-, T_+\} + [T_0]^{*2}_{2\Omega})$
τ_0	F_0, G_0 $K_0 \equiv \frac{N}{2} - \Omega_j$	$2\Omega_j(\{G_0, F_0\} + [K_0]^{*2}_{2\Omega})$
$N_{\mp 1}$	$F_{\pm 1}, G_{\pm 1}$ $K_{\pm 1} = \frac{N_{\pm 1} - \Omega_j}{2}$	$\Omega_j(\{G_{\pm 1}, F_{\pm 1}\} + \rho_\pm[K_{\pm 1}]^{*2}_{1\Omega})$

In this way, the deformation parameter q governs the character of the Hamiltonian, yielding the original two-body form in the $q \to 1$ limit. Relative to the "classical" model, the q-deformation introduces higher-order many-body interactions into the theory. These interaction, as will be seen below, turn out to be non-negligible and important for a description of nuclear properties.

To investigate the role of the q deformation within the framework of the $Sp(4)$ model, we fit eigenvalues of the deformed Hamiltonian to the experimental energies of the lowest 0^+ isovector ($T = 1$) states. A fit with all possible parameters $P^{\{i\}} = \{G_q,\ F_q,\ E_q,\ C_q$ and $D_q\}$ yields the same values (within uncertainties) for the non-deformed counterparts ($q \to 1$). This reveals a decoupling of the q parameter from the pairing strengths, which suggests that the origin of the deformation is not in the strength of the interaction but rather it is a true many-body character which accounts for neglected residual interactions. Once the general interaction strength parameters were fixed, the deviation in the predicted energies for each nucleus from the experimental values was reduced with respect to q using a minimization procedure, $|\langle H(q, P^{\{i\}}) \rangle - E_{\exp}|^2$ (4). The non-linear pairing interaction turns out to be

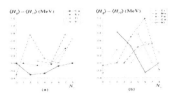

Figure 1. The energy difference between q-deformed and non-deformed total energies for isotopes with a ^{40}Ca core

very important around closed shells, where the q parameter has significant values and the experimental energy is exactly reproduced. This is further confirmed by the significant value of the q parameter for the $N = Z$ nuclei and the ones around the closed shells, where strong pairing correlations are expected. Around the middle of the shell ($N = 2\Omega$) the deformation theory does not restore the experimental values and does not differ much from the "classical" case. This means that for these nuclei the pairing interaction alone is insufficient to describe the energy spectra. We need to point out that nuclei with a closed proton shell or a closed neutron shell, for which the non-deformed model provides a good description, call for a negligible (consistent with $q = 1$) contribution for the non-linear terms. This is an expected result since the core is regarded as a passive background and the model does not take into account its influence.

The results for values of the parameter q together with the experimental energies of the first excited 2^+ states of nuclei as classified in the diamond of the $sp(4)$ irrep space for a given nuclear shell are plotted on Figure 2. They show that for the nuclei considered, the deformation follow these energies, which estimate the the magnitude of the pairing gap associated with the breaking of a pair. In this way, the nature of the q deformation is seen in the pairing correlations. Based on the $Sp(4)$ classification scheme, a smooth function can

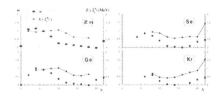

Figure 2. Deformation parameter $\varkappa$ and excitation energies of 2^+_1 (solid line) states in MeV as a function of neutron numbers ($= N_-$) for various isotopes taking ^{56}Ni as a core.

be found to describe optimum values for the q parameter. The interpretation of this effect comes from the fact that the q deformation of the $Sp_q(4)$ model includes nonlinear (higher order) np, pp and nn pairing interactions. In this way it is clearly decoupled from the strength parameters, which have global values for the single or multiple j shells. The specific values of the deformation parameter q for each nucleus with valence nucleons classified in these shells point out its local meaning as a characteristic of the residual pairing interaction. To conclude, we note that we are able to make a reliable prediction for the lowest isovector 0^+ state energies in nuclei with unknown energy spectrum. The energies inherit not only the global properties of the nuclear interaction but also the specific local features through the q deformation.

This work was partially supported by the US National Science Foundation through a regular grant (9970769) and a cooperative agreement (9720652) that includes matching from the Louisiana Board of Regents Support Fund.

[1] D. Bonatsos, et al, Phys. Rev. A **62**, 013203 (2000)
[2] K. D. Sviratcheva, J. P. Draayer, and A. I. Georgieva *An Algebraic Pairing Model with $Sp(4)$ Symmetry and its Deformation,* ArXivV/nucl-th/0204070.
[3] K. D. Sviratcheva, A. I. Georgieva, V. G. Gueorguiev, J. P. Draayer, and M. I. Ivanov, *J. Phys. A: Math. Gen.* **34**, 8365 (2001)
[4] K. T. Hecht, *Nucl. Phys.* **63**, 177 (1965), *Phys. Rev.* **139**, B794 (1965), *Nucl. Phys.* **A102**, 11 (1967)
[5] B. H. Flowers, *Proc. Roy. Soc.* (London) **A212**, 248 (1952)
[6] A. Klein and E. Marshalek, *Rev. Mod. Phys.* **63**, 375 (1991)

Inst. Phys. Conf. Ser. No 173: Section 4
Paper presented at 24th Int. Coll. Group Theoretical Methods in Physics, Paris, France, 15–20 July 2002
©2003 IOP Publishing Ltd

351

Mixed-Symmetry Shell-Model Calculations

V. G. Gueorguiev and J. P. Draayer

Department of Physics and Astronomy, Louisiana State University,
Baton Rouge, Louisiana 70803, USA

Abstract. The one-dimensional harmonic oscillator in a box problem is used to introduce the concept of an oblique-basis shell-model theory. The method is applied to nuclei by combining traditional spherical shell-model states with SU(3) collective configurations. An application to ^{24}Mg, using the realistic two-body interaction of Wildenthal, is used to explore the validity of this oblique-basis, mixed-symmetry shell-model concept. The applicability of the theory to the lower pf-shell nuclei $^{44-48}$Ti and ^{48}Cr using the Kuo-Brown-3 interaction is also discussed. While these nuclei show strong SU(3) symmetry breaking due mainly to the single-particle spin-orbit splitting, they continue to yield enhanced B(E2) values not unlike those expected if the symmetry were not broken. Other alternative basis sets are considered for future oblique-basis shell-model calculations. The results suggest that an oblique-basis, mixed-symmetry shell-model theory may prove to be useful in situations where competing degrees of freedom dominate the dynamics.

Two dominate but often competing modes characterize the structure of atomic nuclei. One is the single-particle shell structure underpinned by the validity of the mean-field concept; the other is the many-particle collective behavior manifested through nuclear deformation. The spherical shell model is the theory of choice when single-particle behavior dominates [1]. When deformation dominates, the Elliott SU(3) model can be used successfully [2]. This manifests itself in two dominant elements in the nuclear Hamiltonian: the single-particle term, $H_0 = \sum_i \varepsilon_i n_i$, and a collective quadrupole-quadrupole interaction, $H_{QQ} = Q \cdot Q$. It follows that a simplified Hamiltonian $H = \sum_i \varepsilon_i n_i - \chi Q \cdot Q$ has two solvable limits associated with these modes.

To probe the nature of such a system, we consider a simpler problem: the one-dimensional harmonic oscillator in a box of size $2L$ [3]. As for real nuclei, this system

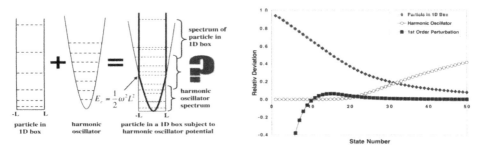

Figure 1. Left graph shows the structure of the interaction potential of a particle in an one-dimensional box subject to a harmonic oscillator restoring force toward the center of the box. Right graph shows the relative deviations from the exact energy eigenvalues for $\omega = 16$, $L = \pi/2$, $\hbar = m = 1$. The open circles represent deviation of the exact energy eigenvalue from the corresponding harmonic-oscillator eigenvalue $(1 - E_{ho}/E_{exact})$, the solid diamonds are the corresponding relative deviation from the energy spectrum of a particle in a 1D box, and the solid squares are the first-order perturbation theory results.

352

has a finite volume and a restoring force whose potential is of a harmonic oscillator type, $\omega^2 x^2/2$. For this model, shown in fig.1, there is a well-defined energy scale which measures the strength of the potential at the boundary of the box, $E_c = \omega^2 L^2/2$. The value of E_c determines the type of low-energy excitations of the system. Specifically, depending the value of E_c there are three spectral types:

(1) For $\omega \to 0$ the energy spectrum is simply that of a particle in a box.
(2) At some value of ω, the energy spectrum begins with E_c followed by the spectrum of a particle in a box perturbed by the harmonic oscillator potential.
(3) For sufficiently large ω there is a harmonic oscillator spectrum below E_c followed by the perturbed spectrum of a particle in a box.

The last scenario (3) is the most interesting one since it provides an example of a two-mode system. For this case the use of two sets of basis vectors, one representing each of the two limits, has physical appeal, especially at energies near E_c. One basis set consists of the harmonic oscillator states; the other set consists of basis states of a particle in a box. We call this combination a mixed-mode / oblique-basis approach. In general, the oblique-basis vectors form a nonorthogonal and overcomplete set. Even thought a mixed spectrum is expected around E_c, our numerical study, that includes up to 50 harmonic oscillator states below E_c, shows that the first order perturbation theory in energy using particle in a box wave functions as the zero order approximation to the exact functions works quite well after

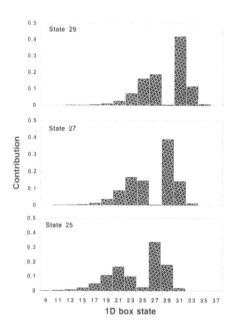
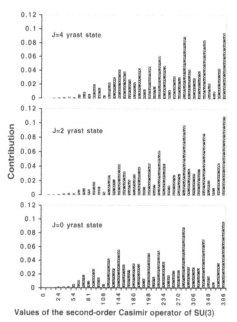

Figure 2. Coherent structure with respect to the non-zero components of the 25th, 27th and 29th exact eigenvector in the basis of a free particle in an one-dimensional box. Parameters of the toy Hamiltonian are $\omega = 16$, $L = \pi/2$, $\hbar = m = 1$. Right graph shows the coherent structure of the first three yrast states in ^{48}Cr calculated using realistic single-particle energies with Kuo-Brown-3 two body interaction ($KB3$). On the horizontal axis is C_2 of $SU(3)$ with contribution of each $SU(3)$ state to the corresponding yrast state on the vertical axis.

the breakdown of the harmonic oscillator spectrum. This observation is demonstrated in the right graph of fig.1 which shows the relative deviations from the exact energy spectrum for a particle in a box.

Although the spectrum seems to be well described using first order perturbation theory based on particle in a box wave functions, the exact wave functions near E_c have an interesting structure. For example, the zero order approximation to the wave function used to calculate the energy may not be present at all in the structure of the exact wave function as is the case shown in the left graph of fig.2. Another feature also seen in fig.2 is the common shape of the distribution of the non-zero components along the particle in a box basis. The right graph of fig.2 shows this same effect in nuclei which is usually attributed to coherent mixing [4, 5].

An application of the theory to ^{24}Mg [6], using the realistic two-body interaction of Wildenthal [7], demonstrates the validity of the mixed-mode shell-model scheme. In this case the oblique-basis consists of the traditional spherical states, which yield a diagonal representation of the single-particle interaction, together with collective SU(3) configurations, which yield a diagonal quadrupole-quadrupole interaction. The results shown in fig.3 were obtained in a space that spans less than 10% of the full-space. They reproduce, within 2% of the full-space result, the correct binding energy as well as the low-energy spectrum and the dominate structure of the states that have greater than 90% overlap with the full-space results. In contrast, for a m-scheme spherical shell-model calculation one needs about 60% of the full space to obtain results comparable with the oblique basis results.

Studies of the lower pf-shell nuclei $^{44-48}Ti$ and ^{48}Cr [4], using the realistic Kuo-Brown-3 (KB3) interaction [8], show strong SU(3) symmetry breaking due mainly to the single-particle spin-orbit splitting. Thus the KB3 Hamiltonian could also be considered a two-mode system. This is further supported by the behavior of the yrast band B(E2) values that seems to be insensitive to fragmentation of the SU(3) symmetry. Specifically, the quadrupole collectivity as measured by the B(E2) strengths remains high even though the SU(3) symmetry is rather badly broken. This has been attributed to a quasi-SU(3) symmetry [5] where the observables behave like a pure SU(3) symmetry while the true eigenvectors exhibit a strong

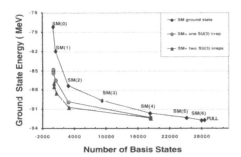
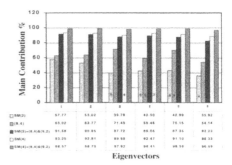

Figure 3. Left graph shows the calculated ground-state energy for ^{24}Mg as a function of various model spaces. SM(n) denotes spherical shell model calculation with up to n particles outside of the $d_{5/2}$ sub-shell. Note the dramatic increase in binding (3.3 MeV) in going from SM(2) to SM(2)+(8,4)&(9,2) (a 0.5% increase in the dimensionality of the model space). Enlarging the space from SM(2) to SM(4) (a 54% increase in the dimensionality of the model space) adds 4.2 MeV in the binding energy. The right graph shows representative overlaps of pure SM(n), pure SU(3), and oblique-basis results with the exact full sd shell eigenstates. A number within a bar denotes the state with the overlap shown by the bar if it is different from the number for the exact full-space calculation shown on the abscissa. For example, for SM(2) the third eigenvector has the largest overlap with the fourth exact eigenstate, not the third, while the fifth SM(2) eigenvector has greatest overlap with the third exact eigenstate.

coherent structure with respect to each of the two bases. This provides strong justification for further study of the implications of two-mode shell-model studies.

Future research may provide justification for an extension of the theory to multi-mode oblique shell-model calculations. An immediate extension of the current scheme might use the eigenvectors of the pairing interaction [9] within the Sp(4) algebraic approach to the nuclear structure [10], together with the collective SU(3) states and spherical shell model states. Hamiltonian driven basis sets can also be considered. In particular, the method may use eigenstates of the very-near closed shell nuclei obtained from a full shell-model calculation to form Hamiltonian driven J-pair states for mid-shell nuclei [11]. This type of extension would mimic the Interacting Boson Model (IBM) [12] and the so-called broken-pair theory [11]. In particular, the three exact limits of the IBM [13] can be considered to comprise a three-mode system. Nonetheless, the real benefit of this approach is expected when the system is far away of any exactly solvable limit of the Hamiltonian and the spaces encountered are too large to allow for exact calculations.

Acknowledgments

We acknowledge support from the U.S. National Science Foundation under Grant No. PHY-9970769 and Cooperative Agreement No. EPS-9720652 that includes matching from the Louisiana Board of Regents Support Fund. V. G. Gueorguiev is grateful to the Louisiana State University Graduate School for awarding him a dissertation fellowship and a travel grant to attend the XXIV International Colloquium on Group Theoretical Methods in Physics held June 15-20, 2002 in Paris, France.

References

[1] R. R. Whitehead, Nucl. Phys. **A182**, 290 (1972); R. R. Whitehead, A. Watt, B. J. Cole, and I. Morrision, Advances in Nuclear Physics **9**, ed. M. Baranger, and E. Vogt (Plenum Press, New York, 1977)

[2] J. P. Elliott, Proc. Roy. Soc. London Ser. **A 245**, 128 (1958); **A 245**, 562 (1958); J. P. Elliott and H. Harvey, Proc. Roy. Soc. London Ser. **A 272**, 557 (1963); J. P. Elliott and C. E. Wilsdon Proc. Roy. Soc. London Ser. **A 302**, 509 (1968)

[3] G. B. Armen and A. R. P. Rau, "The limitations of fixed-basis calculations in quantum mechanics", unpublished

[4] V. G. Gueorguiev, J. P. Draayer, and C. W. Johnson, Phys. Rev. C **63**, 14318 (2001)

[5] P. Rochford and D. J. Rowe, Phys. Lett. **B210**, 5 (1988); A. P. Zuker, J. Retamosa, A. Poves, and E. Caurier, Phys. Rev. C **52**, R1741 (1995); G. Martinez-Pinedo, A. P. Zuker, A. Poves, and E. Caurier, Phys. Rev. C **55**, 187 (1997); D. J. Rowe, C. Bahri, and W. Wijesundera, Phys. Rev. Lett. **80**, 4394 (1998); A. Poves, J. Phys. G **25**, 589 (1999); D. J. Rowe, S. Bartlett, and C. Bahri, Phys. Lett. **B 472**, 227 (2000)

[6] V. G. Gueorguiev, W. E. Ormand, C. W. Johnson, and J. P. Draayer, Phys. Rev. C **65**, 024314 (2002)

[7] B. H. Wildenthal, Prog. Part. Nucl. Phys. **11**, 5 (1984); *wpn* interaction file from B. A. Brown (www.nscl.msu.edu/~brown/database.htm)

[8] T. Kuo and G. E. Brown, Nucl. Phys. **A114**, 241 (1968); A. Poves and A. P. Zuker, Phys. Rep. **70**, 235 (1981)

[9] J. Dukelsky, C. Esebbag, and P. Schuck, Phys. Rev. Lett. **87**, 066403 (2001) (cond-mat/0107477)

[10] K. D. Sviratcheva, A. I. Georgieva, V. G. Gueorguiev, J. P. Draayer, and M. I. Ivanov, J. Phys. A **34**, 8365 (2001) (nucl-th/0104051)

[11] K. L. G. Heyde, "The Nuclear Shell Model," ed. J. M. Irvine (Springer-Verlag, Berlin Heidelberg, 1990)

[12] F. Iachello, "The interacting boson model," (Cambridgeshire University Press, New York, 1987)

[13] M. Moshinsky and Y. F. Smirnov, "The Harmonic Oscillator in Modern Physics", Contemporary Concepts in Physics Volume **9**, ed. H. Feshbach (Harwood Academic Publishing, Amsterdam, 1996)

Inst. Phys. Conf. Ser. No 173: Section 4
Paper presented at 24th Int. Coll. Group Theoretical Methods in Physics, Paris, France, 15–20 July 2002
©2003 IOP Publishing Ltd

The Boson/Fermion Statistic for $SU(3)$ Colour requires Quark Confinement

William P. Joyce

Department of Physics & Astronomy, University of Canterbury, Private Bag 4800, Christchurch, New Zealand.

Abstract. We extend the boson/fermion statistic of particle spins over $SU(2)$ to $SU(3)$ colour. This leads unavoidably to the Pauli exclusion principle and quark state confinement. We are forced to conclude that quark confinement is as fundamental as Pauli's exclusion principle. Moreover, we do not require a "confining force" to explain quark state confinement.

PACS numbers: 03.50.Dc, 03.30.+p

We begin with the boson/fermion particle statistic for $SU(2)$ spin and show how to extend it to $SU(3)$ colour. This statistic is important since it underlies the Pauli exclusion principle. As we shall see the extension to $SU(3)$ not only leads to this fundamental principle but to quark state confinement. Historically $SU(3)$ colour was introduced to avoid the symmetric states of certain fermionic resonances, see Kaku [1], p375. This result is independent of the QCD Lagrangian and temperature scale.

We begin by considering a collection of free particles under an exact symmetry corresponding to $SU(2)$ spin or $SU(3)$ colour. A state space for a single particle is given by an irreducible representation of the symmetry group. We label these state spaces $a_1,...,a_n$. The combined state space of an ensemble of n free particles $a_1, ..., a_n$ is given by $a_1 \otimes \cdots \otimes a_n$. For $SU(2)$ a permutation of the particles introduces a phase factor of ± 1. This is generated by adjacent transpositions introducing a phase factor $\gamma_{a,b}$ for the interchange of two particles a and b given by

$$\gamma_{a,b} = \begin{cases} 1 & : a \text{ or } b \text{ integer spin} \\ -1 & : \text{otherwise} \end{cases} \tag{1}$$

In particular n identical bosons (integer spin) have completely symmetric states, while n identical fermions (half integer spin) have completely anti–symmetric states. The anti–symmetry of the fermion state is equivalent to the Pauli exclusion principle.

For $SU(3)$ colour the phases are not so simple. Moreover, one has to take into account the order particles are coupled into an ensemble. We represent this by a rooted binary tree where each branch point has its own level, and the leaves are labeled by particles. For example the ensemble $a \otimes b \otimes c \otimes d \otimes e$ has the three coupling schema given in figure 1.

Figure 1. The three coupling schema corresponding to $(a \otimes b) \otimes (c \otimes (d \otimes e))$.

The distinction between these orders of coupling cannot be distinguished by bracketing. In fact they all correspond to $(a \otimes b) \otimes (c \otimes (d \otimes e))$. We introduce families of phases for recoupling as given in figure 2.

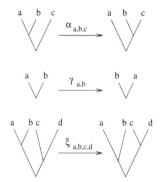

Figure 2. The three generative families of recoupling phases corresponding to associativity (α), commutativity (γ) and deformativity (ξ).

For $SU(2)$ spin the associativity and deformativity phase factors are $+1$. However, because of the triality or $\mathbb{Z}_3$ grading of the representations of $SU(3)$ neither of these two families can be trivial. The phases are class functions of triality. For example $\gamma_{a,b} = \gamma_{c,d}$ whenever a and b, and c and d have the same triality. Furthermore, we require a choice that is coherent. That is any two alternative sequences of recouplings between the same source and target give the same phase. This is guaranteed if the symmetry condition $\gamma_{a,b}\gamma_{b,a} = 1$, the square conditions $\xi_{a,b,c,d}\xi_{c,d,a,b} = 1$ and $\xi_{a,b,c,d} = \xi_{a,b,d,c}$, and the hexagon and deformed pentagon conditions of figures 3 and 4 all hold for any a, b, c, d. The deformed pentagon and

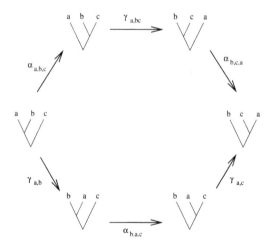

Figure 3. The hexagon condition

hexagon conditions are respectively given by

$$\alpha_{a \otimes b,c,d}\xi_{a,b,c,d}\alpha_{a,b,c \otimes d} = \alpha_{a,b,c}\alpha_{a,b \otimes c,d}\alpha_{b,c,d} \tag{2}$$

$$\alpha_{a,b,c}\gamma_{a,b \otimes c}\alpha_{b,c,a} \qquad = \gamma_{a,b}\alpha_{b,a,c}\gamma_{a,c} \tag{3}$$

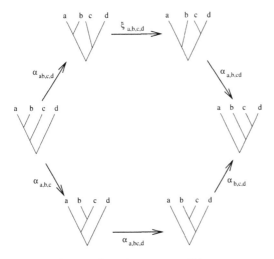

Figure 4. The deformed pentagon condition

Underlying these conditions is the structure of a symmetric premonoidal category. The definition and coherence proof may be found in Joyce [2]. More generally this mathematical structure forms the basis of the Racah–Wigner calculus [3, 4], and ultimately many–body quantum theory. A complete and comprehensive treatment is given in Joyce [5, 6, 7, 8, 9].

We next extend the boson/fermion particle statistic from $SU(2)$ to $SU(3)$. A full account for a general symmetry group is given in Joyce [10]. The commutativity phases must be given by

$$\gamma_{a,b} = \begin{cases} 1 & a \text{ or } b \text{ has triality } 0 \\ -1 & : \text{otherwise} \end{cases} \tag{4}$$

One can show that with this choice the hexagon condition is satisfied for

$$\alpha_{a,b,c} = \begin{cases} 1 & a, b, c \text{ or } a \otimes b \text{ has triality } 0 \\ -1 & : \text{otherwise} \end{cases} \tag{5}$$

It now easily to show that the pentagon condition gives

$$\xi_{a,b,c,d} = \begin{cases} 1 & a, b, c, d, a \otimes b \text{ or } c \otimes d \text{ has tirality } 0 \\ -1 & : \text{otherwise} \end{cases} \tag{6}$$

If there are four free quarks in the universe then we can form the ensemble $([3] \otimes [3]) \otimes ([3] \otimes [3])$. A state of this space must satisfy $\psi = \xi_{[3],[3],[3],[3]}\psi$. But the right hand side is $-\psi$ and the only possible state is $\psi = 0$. Closer inspection reveals that the only possible states are those of triality 0, and what is more they must be viewed as inextricably bound, otherwise only the trivial state is possible. Typical allowed bound ensembles are the vacuum, free gluon, free hadron and free meson. Moreover, gluons can pass freely across boundaries providing a mechanism for mediating the strong interaction. This property is a confinement to ensembles of triality 0. The fact that the hadron and meson ensembles are also confined to a small spatial region at low temperature is a property of QCD. Even at high temperature one cannot resolve free quarks even though they may not be spatially bound. If one does observe free quarks then a gross violation of the Pauli Exclusion principle at high temperatures will be required.

358

Acknowledgements

This research was supported by the New Zealand Foundation for Research, Science and Technology. Contract number UOCX0102

References

[1] M. Kaku *Quantum Field Theory: a Modern Introduction* Oxford University Press N. Y. (1993)
[2] W. P. Joyce 2002 *Natural Associativity without the Pentagon Condition* submitted to TAC, available at http://www.arXiv.org/abs/math.CT/0109088
[3] W. P. Joyce 2001 *Diagram Projection Rules for Recoupling Diagrams in the Racah–Wigner Category*, J. Math. Phys. **42**, 1346–1363
[4] W. P. Joyce, P. H. Butler and H. J. Ross 2002 *The Racah–Wigner category* Can. J. Phys. **80**(6), 613–632
[5] W. P. Joyce 2002 *Vertex Calculus in Premonoidal Categories: I. The Racah–Wigner calculus and Natural Statistics*, submitted to J. Math. Phys.
[6] W. P. Joyce 2002 *Vertex Calculus in Premonoidal Categories: II. Coupling and the Clebsch–Gordon Vertex*, submitted to J. Math. Phys.
[7] W. P. Joyce 2002 *Vertex Calculus in Premonoidal Categories: III. Recoupling and Recoupling Coefficients*, submitted to J. Math. Phys.
[8] W. P. Joyce 2002 *Vertex Calculus in Premonoidal Categories: IV. The Wigner Symbols*, submitted to J. Math. Phys.
[9] W. P. Joyce 2002 *Vertex Calculus in Premonoidal Categories: V. Many–particle Quantum Theory and Scattering*, submitted to J. Math. Phys.
[10] W. P. Joyce 2002 *Quark Confinement as a Consequence of the Extension of the Boson/Fermion Statistic to $SU(3)$ Colour* submitted to J. Math. Phys.

Inst. Phys. Conf. Ser. No 173: Section 4
Paper presented at 24th Int. Coll. Group Theoretical Methods in Physics, Paris, France, 15–20 July 2002
©2003 IOP Publishing Ltd

How to define colour charge in Quantum Chromodynamics

J Kijowski
Center for Theoretical Physics, Polish Academy of Sciences
al. Lotników 32/46, 02-668 Warsaw, Poland

G Rudolph
Institut für Theoretische Physik, Universität Leipzig
Augustusplatz 10/11, 04109 Leipzig, Germany

Abstract. This paper contains a definition of the global, additive, colour charge operator in lattice QCD. The charge takes values in the center of the gauge group $SU(3)$.

1. Introduction

One of the basic challenges in quantum chromodynamics (QCD) consists in constructing an effective microscopic theory of interacting hadrons out of this gauge theory. For this purpose, nonperturbative methods for describing the low energy regime should be developed. In particular, the observable algebra and the superselection structure of this theory should be investigated. The present paper is a step in this direction.

In QED, superselection sectors are labelled by different values of the *global* electric charge, defined by "summing up" *local* Gauss laws, which are linear and built from gauge invariant operators. This way one obtains the following gauge invariant conservation law: The global electric charge is equal to the electric flux through a 2-sphere at infinity.

On the contrary, in QCD the local Gauss law is neither built from gauge invariant operators nor is it linear. There are two strategies which one may follow: either we sum up gauge-dependent quantities and then try to extract a suitable gauge-invariant quantity from this sum, or we first extract gauge-invariant information from the local Gauss laws, which we sum up subsequently. The first strategy cannot lead to a reasonable result, because using an appropriate gauge one may always annihilate the sum (integral – in the continuum case) of gauge-dependent, $su(3)$-valued objects. Within the second option one has to find an invariant, which is additive.

The main point of the present paper is to show that it is possible to define such a quantity. Contrary to electrodynamics, the invariant we construct here leads to operators whose eigenvalues belong to $\mathbb{Z}_3$ (which is canonically identified with the dual of the center of $SU(3)$). Calculating the value of this invariant on the both sides of the local Gauss equation of QCD we obtain, after summation, the global law: the global $\mathbb{Z}_3$-valued colour flux at the boundary must be equal to the global $\mathbb{Z}_3$-valued colour charge.

Our main result is universal in the following sense: Suppose somebody had constructed a nonperturbative version of QCD rigorously. Then our construction of the global charge would apply. Since, however, such a theory is not at our disposal, for a rigorous discussion we have to restrict ourselves to the case of QCD approximated on a finite lattice. For basic notions concerning lattice gauge theories (including fermions) we refer to [1] and references therein. Thus, we consider QCD on a finite, regular cubic lattice in the Hamiltonian approach. Our starting point is the notion of the algebra of field operators. By imposing the local Gauss law and gauge invariance we obtain the observable algebra $\mathcal{O}(\Lambda)$ for QCD on the lattice. For an analogous investigation of the superselection structure of QED on the lattice we refer to [5] and for a (rather heuristic) discussion of continuum QCD within the functional integral approach (using only gauge invariant quantities) we refer to [6]. For QED, the mathematical structure of $\mathcal{O}(\Lambda)$ has been completely clarified, see [7].

We stress that standard methods from algebraic quantum field theory for models which do not contain massless particles, see [2], do not apply here. Some progress towards an implementation of similar ideas for theories with massless particles has been made, for the case of quantum electrodynamics (QED) see [3] and [4] and further references therein.

2. Field Algebra in the Lattice QCD

Gluonic potentials of lattice QCD are described by commuting fields $U^A{}_B(x,y)$ assigned to every oriented lattice link (x,y) ($A, B = 1, 2, 3$ are colour indices), fulfilling relations

$$(U^A{}_B(x,y))^* U_A{}^C(x,y) = \delta^C{}_B \, \mathbf{1} \, , \tag{1}$$

$$\epsilon_{ABC} \, U^A{}_D(x,y) U^B{}_E(x,y) U^C{}_F(x,y) = \epsilon_{DEF} \, \mathbf{1} \, , \tag{2}$$

together with the transformation law under the change of the link orientation:

$$U^A{}_B(y,x) = (U_B{}^A(x,y))^* \, .$$

Moreover, there are colour electric fields $E^A{}_B(y,x)$ assigned to oriented lattice links. They are traceless (i. e. $E^A{}_A(y,x)) \equiv 0$), selfadjoint,

$$(E^A{}_B(x,y))^* = E_B{}^A(x,y) \, , \tag{3}$$

and fulfill $su(3)$-commutation relations:

$$[E^A{}_B(x,y), E^C{}_D(u,z)] = \delta_{xu}\delta_{yz} \left(\delta^C{}_B E^A{}_D(x,y) - \delta^A{}_D E^C{}_B(x,y) \right) \, . \tag{4}$$

Under the change of the link orientation they transform as follows:

$$E^A{}_B(y,x) = -U^A{}_D(y,x) U^C{}_B(x,y) E^D{}_C(x,y) \, . \tag{5}$$

Finally, E and U satisfy canonical commutation relations between left invariant vector fields and functions on $SU(3)$. The local Gauss law at x has the form

$$\sum_y E^A{}_B(x,y) = \rho^A{}_B(x) \, , \tag{6}$$

with the sum taken over all points y adjacent to x, where the matter density

$$\rho^A{}_B(x) = \sum_a \left(\psi^{*aA}(x)\psi^a{}_B(x) - \frac{1}{3}\delta^A_B \psi^{*aC}(x)\psi^a{}_C(x) \right) \tag{7}$$

is built from the quark and anti-quark fields ψ and ψ^*, fulfilling canonical anticommutation relations. Consequently, also $\rho^A{}_B(x)$ fulfils the $su(3)$-commutation relations.

3. The $\mathbb{Z}_3$-valued local colour charge

Consider a collection of operators $F^A{}_B$ in a Hilbert space $\mathcal{H}$, fulfilling $F^A{}_A = 0$ and $(F^A{}_B)^* = F_B{}^A$, realizing the canonical commutation relations for the Lie algebra $su(3)$:

$$[F^A{}_B, F^C{}_D] = \delta^C{}_B F^A{}_D - \delta^A{}_D F^C{}_B \, . \tag{8}$$

We have two basic types of such operators in lattice QCD: the electric field $E^A{}_B(x,y)$ on each lattice link, see (4), and the charge operator $\rho^A{}_B(x)$ at each lattice site. We assume integrability of the Lie algebra representations under consideration. This means that for each F there exists a unitary representation $\bar{F}(g) \in B(\mathcal{H})$ of the group $SU(3)$, associated with F.

It is easy to check that if F and G are two commuting representations of $su(3)$ then so is $F + G$. Indeed, if $\bar{F}(g)$ and $\bar{G}(g)$ are representations of $SU(3)$ corresponding to F and G, then $F + G$ may be obtained by differentiating the representation $SU(3) \ni g \to \bar{F}(g)\bar{G}(g) \in B(\mathcal{H})$, where $B(\mathcal{H})$ denotes the C^*-algebra of bounded operators on $\mathcal{H}$. Moreover, $-F^*$ is also a representation of $su(3)$, corresponding to the following representation of $SU(3)$: $SU(3) \ni g \to (\bar{F}(g^{-1}))^* \in B(\mathcal{H})$.

Such a collection of operators is an *operator domain* in the sense of Woronowicz (see [10]). We are going to construct an operator function on this domain, i. e. a mapping $F \to \varphi(F)$ which satisfies $\varphi(UFU^{-1}) = U\varphi(F)U^{-1}$ for an arbitrary isometry U. For this purpose, consider the two gauge-invariant, self-adjoint and commuting (Casimir) operators K_2 and K_3 of F:

$$K_2 = F^A{}_B F^B{}_A \,, \tag{9}$$

$$K_3 = \frac{1}{2}\left(F^A{}_B F^B{}_C F^C{}_A + F^A{}_B F^C{}_A F^B{}_C\right) \,. \tag{10}$$

The Hilbert space $\mathcal{H}$ splits into the direct sum of subspaces $\mathcal{H}_\alpha$ on which F acts irreducibly. Each of these subspaces is a common eigenspace of K_2 and K_3. Denoting the highest weight characterizing a given irreducible representation by (m,n), with m and n being nonnegative integers, the eigenvalues k_2 and k_3 of K_2 and K_3 are given by:

$$k_2 = \frac{2}{3}(m^2 + mn + n^2 + 3m + 3n) \,, \tag{11}$$

$$k_3 = \frac{1}{9}(m-n)(3 + 2m + n)(3 + m + 2n) \,. \tag{12}$$

It is easy to check that the above formulae may be uniquely solved with respect to m and n yielding:

$$m = M(k_2, k_3) := \sqrt{\tfrac{2}{3}(k_2 + 2)}\left(\cos\left(\frac{1}{3}\arccos\frac{\sqrt{6}k_3}{\sqrt{(k_2+2)^3}} + \frac{2}{3}\pi\right) + \right.$$

$$\left. 2\cos\left(\frac{1}{3}\arccos\frac{\sqrt{6}k_3}{\sqrt{(k_2+2)^3}}\right)\right) - 1 \,, \tag{13}$$

$$n = N(k_2, k_3) := -\sqrt{\tfrac{2}{3}(k_2 + 2)}\left(2\cos\left(\frac{1}{3}\arccos\frac{\sqrt{6}k_3}{\sqrt{(k_2+2)^3}} + \frac{2}{3}\pi\right) + \right.$$

$$\left. \cos\left(\frac{1}{3}\arccos\frac{\sqrt{6}k_3}{\sqrt{(k_2+2)^3}}\right)\right) - 1 \,. \tag{14}$$

Using these functions we define a function with values in $\mathbb{Z}_3$, which will be identified with the dual to the center $\mathcal{C}$ of the gauge group $SU(3)$:

$$f(k_2, k_3) := (M(k_2, k_3) - N(k_2, k_3)) \bmod 3 \,. \tag{15}$$

Since K_2 and K_3 are commuting and self-adjoint, there exists an operator-valued function:

$$\varphi(F) = f(K_2(F), K_3(F)) \,. \tag{16}$$

This means that $\varphi(F)$ takes eigenvalues $-1, 0, 1$ and that every irreducible subspace $\mathcal{H}_\alpha$ is an eigenspace of $\varphi(F)$ with eigenvalue $m - n \bmod 3$.

362

Theorem: The operator function φ is additive for commuting F and G:

$$\varphi(F + G) = \varphi(F) + \varphi(G) \,. \tag{17}$$

4. The Global Charge and the Flux Law

Applying φ to the local Gauss law (6) and using the above additivity we obtain:

$$\sum_y \varphi(E(x,y)) = \varphi(\rho(x)) \,. \tag{18}$$

The quantity on the right hand side is the (gauge invariant) local colour charge density carried by the quark field. Using the transformation law (5), one can show that

$$K_2(E(x,y)) = K_2(E(y,x)) \,, \qquad K_3(E(x,y)) = -K_3(E(y,x)) \,, \tag{19}$$

which implies on every lattice bond (x, y):

$$\varphi(E(x,y)) + \varphi(E(y,x)) = 0 \,. \tag{20}$$

Taking the sum of equations (18) over all lattice sites $x \in \Lambda$, all terms on the left hand side cancel, except for contributions coming from the boundary. This way we obtain the total flux through the boundary $\partial\Lambda$ of Λ:

$$\Phi_{\partial\Lambda} := \sum_{x \in \partial\Lambda} \varphi(E(x,\infty)) \,, \tag{21}$$

where by $E(x, \infty)$ we denote the colour electric charge along the external link, connecting the point x on the boundary of Λ with the "rest of the world". On the right hand side we get the (gauge invariant) global colour charge, carried by the matter field

$$Q_\Lambda = \sum_{x \in \Lambda} \varphi(\rho(x)) \,. \tag{22}$$

Both quantities appearing in the global Gauss law

$$\Phi_{\partial\Lambda} = Q_\Lambda \,, \tag{23}$$

take values in the center $\mathbb{Z}_3$ of $SU(3)$. The "sum modulo three" is the composition law in $\mathbb{Z}_3$.

References

[1] E. Seiler, *Gauge Theories as a Problem of Constructive Quantum Field Theory and Statistical Mechanics*, Lecture Notes in Phys., vol. 159, Springer (1982); E. Seiler, "Constructive Quantum Field Theory: Fermions", in *Gauge Theories: Fundamental Interactions and Rigorous Results*, eds. P. Dita, V. Georgescu, R. Purice

[2] S. Doplicher, R. Haag and J. Roberts, Commun. Math. Phys. 23 (1971) 199

[3] J. Fröhlich, Commun. Math. Phys. 66 (1979), 223

[4] D. Buchholz, Commun. Math. Phys. 85 (1982) 49; D. Buchholz, Phys. Lett. B174 (1986) 331

[5] J. Kijowski, G. Rudolph and A. Thielman, Commun. Math. Phys. 188 (1997) 535

[6] J. Kijowski, G. Rudolph and M. Rudolph, Ann. Inst. H. Poincaré Vol. 68, n. 3, (1998) 285

[7] J. Kijowski, G. Rudolph and C. Sliwa, Lett. Math. Phys. 43 (1998) 299

[8] J. Kijowski, G. Rudolph, Jour. Math. Phys. **43** (2002) p. 1796 – 1808

[9] A. Klimyk, K. Schmüdgen, *Quantum Groups and Their Representations*, Springer 1997

[10] S. L. Woronowicz, *Duality in the C*-algebra theory*, Proc. Intern. Congr. Mathematicians Warsaw 1983, pp. 1347-1356, PWN Warszawa.
     S. L. Woronowicz, Rev. Math. Phys. 7 (1995) 481

Inst. Phys. Conf. Ser. No 173: Section 4
Paper presented at 24th Int. Coll. Group Theoretical Methods in Physics, Paris, France, 15–20 July 2002
©*2003 IOP Publishing Ltd*

Geometrisation of electromagnetic and weak interactions and geometrical mechanism of P and T violation

O A Olkhov

Semenov Institute of Chemical Physics, Moscow, Russia

Abstract. Early we showed that interacting electron-positron and electromagnetic fields can be considered as some microscopic distortion of Euclidean properties of the Minkovsky 4-space-time. The known Dirac and Maxwell equations happens to be the group-theoretic relations describing this distortion. Here we apply above geometrical approach to obtain equations for neutrino interacting with its weak field. These equations demonstrate geometrical mechanism of P and T violation. Equations are also proposed for gravitational field and its microscopic quantum sources.

1. Introduction

Early we suggested topological interpretation of the Dirac and Maxwell equations for interacting electron-positron and electromagnetic fields [1]

$$i\gamma_1\left(\frac{\partial}{\partial x_1} - ieA_1\right)\psi - \sum_{\alpha=2}^{4}i\gamma_\alpha\left(\frac{\partial}{\partial x_\alpha} - ieA_\alpha\right)\psi = m\psi, \tag{1}$$

$$F_{kl} = \frac{\partial A_k}{\partial x_l} - \frac{\partial A_l}{\partial x_k}, \tag{2}$$

$$\sum_{i=1}^{4}\frac{\partial F_{ik}}{\partial x_i} = j_k, \quad j_k = e\psi * \gamma_1\gamma_k\psi. \tag{3}$$

Here $\hbar = c = 1$, $x_1 = t$, $x_2 = x$, $x_3 = y$, $x_4 = z$, F_{kl} is the electromagnetic field tensor, A_k is the 4-potential, γ_k are the well-known Dirac matrices, ψ is the Dirac bispinor, m and e are electron mass and charge.

It was shown that Eqs.(1-3) can be interpreted as group-theoretic relations that accounts for the topological and metric properties of some unified microscopic nonmetrized closed space-time 4-manifold [2, 3, 4]. Eqs.(1-3) represent above properties as symmetry properties of the manifold covering space and this space happens to be the so called "space with semimetrical parallel translation" [5]. Dirac's spinors ψ_l serve in (1) as basis vectors of the manifold fundamental group representation, while the components of the electromagnetic field tensor F_{ik} prove to be components of a curvature tensor of the manifold covering space. Energy, momentum components, mass, charge, spin and particle-antiparticle states appear to be geometrical characteristics of the above manifold.

2. Geometrisation of weak interaction

Does above topological interpretation reflect any physical reality? It is impossible to give an answer being within considered classical relativistic electrodynamics only. New approach did not here lead to predictions of new effects and it did not explain anything that was not explained before (as, for example, deviations from Newtonian gravitational law predicted by the theory of general relativity. In this section we apply topological approach to the geometrisation of weak interaction and we show that it leads to some new results.

Let us now show that, in a one-particle approximation adopted in this work (low energies), weak interaction can be represented as a manifestation of the torsion in the covering space of a 4-manifold representing weak field and its sources. In due time, Einstein attempted at including electromagnetic field into a unified geometrical description of physical fields by "adding" torsion to the Riemannian space–time curvature, which reflects the presence of a gravitational field in general relativity [6]. Since the curvature of covering space corresponds within our approach to the electromagnetic field, we shall attempt to include weak interaction into the topological approach by including torsion in this space.

It is known that weak interaction breaks the mirror space-time symmetry. On the other hand above symmetry can be violated in the space with torsion (left screw looks like a right one in the mirror). So it is natural to assume that within our topological approach torsion may be connected with weak interaction. Let us first consider the case where the electromagnetic field is absent, i.e., the curvature of covering space is zero. A space with torsion but without curvature is called the space with absolute parallelism [6]. Thus, the challenge is to determine how does free-particle Eq. (1) change if the interparticle interaction is due only to the torsion, which transforms the covering space into a space with absolute parallelism.

Let us denote the torsion tensor by S_{lm}^k; then the problem can be formulated as follows. It is necessary to "insert" the tensor S_{lm}^k or some of its components into Eq. (1) so that the resulting equation remains invariant about the Lorentz transformations and adequately describes the experimental data (e.g., violation of spatial and time symmetry by weak interaction). Among spaces with torsion, there are so-called "spaces with semisymmetric parallel translation" [5, 7]. The torsion tensor S_{lm}^k for such spaces is defined by the antisymmetric part of connection and can be represented in the form

$$S_{lm}^k = K_l A_m^k - K_m A_l^k. \tag{4}$$

Here, K_l is a certain vector and A_l^k is the identity tensor. The vector K_l has the property that the infinitesimal parallelogram remains closed upon parallel translation in the hyperplanes perpendicular to this vector.

One may thus assume that in the presence of vector K_l the spatial isotropy breaks in such a way that the isotropy is retained only in the indicated hyperplanes. On the other hand unlike the case of electromagnetic interaction we can now fix the invariant orientation in every point of the cavering space. It follows from the fact that in addition to vector K_l we may introduce into consideration one another vector B_l

$$B_l = S_{lp}^p = K_l A_p^p - K_p A_l^p. \tag{5}$$

Having two nonparallel vectors K_l and B_l we can define the antisymmetric second rank tensor (bivector) T_{lm}

$$T_{lm} = \frac{1}{2}(K_l B_m - B_l K_m). \tag{6}$$

And it is known that every bivector defines an orientation in the space [7].

After taking into account above symmetry properties we obtained the following analogy of the Dirac Eq.(1) for neutrino interacting with its weak field

$$i\sigma_1 \bigtriangledown_1 \varphi - \sum_{\alpha=2}^{4} i\sigma_\alpha \bigtriangledown_\alpha \varphi + \frac{g^2}{2} T_{lm}\sigma_m \bigtriangledown_l \varphi = m\varphi, \tag{7}$$

where σ_1 is a two row unite matrix, $\sigma_{2,3,4}$ are the Pauli matices ($\sigma_{x,y,z}$), φ is a two component spinor and $\bigtriangledown_l$ are covariant derivatives

$$\bigtriangledown_l = \frac{\partial}{\partial x_l} - igD_l. \tag{8}$$

Here D_l is a connection in considered covering space. Note that we introduce the interaction constant g by inserting $B \rightarrow igB$, $K \rightarrow igK$, $D \rightarrow igD$. We see that the last term at the left side of Eq.(7) is of the second order of the small interaction constant g but this term plays the main role in the space-time symmetry violation.

Above term corresponds within Lagrangian formalism to the new additional pseudoscalar term L_{ad} (in standard Lagrangian)

$$L_{ad} = g^2 T_{lm}\varphi^+ \sigma_m \bigtriangledown_l \varphi. \tag{9}$$

The fields D_l in Eqs.(7,8) can be obtained using the symmetry properties of the covering space with semisymmetric transfer. This symmetry have to be the same after rotations within the 3-hyperplane perpendicular to the vector K_l. For two component spinors above rotations are represented by matrices of $SU(2)$-representation. So this symmetry happens to be just the $SU(2)$-gauge invariance used in standard model of weak interaction for calculating of weak fields. It means that fields D_l in Egs.(7,8) can be expressed in terms of twelve Yang-Mills fields D_l^α ($\alpha = x, y, z$) [8]

$$D_l = \frac{1}{2}(D_l^x \sigma_x + D_l^y \sigma_y + D_l^z \sigma_z). \tag{10}$$

Then the weak field strength has the standard form

$$G_{lm} = \frac{\partial D_l}{\partial x_m} - \frac{\partial D_m}{\partial x_l} - ig(D_l D_m - D_m D_l). \tag{11}$$

3. Gravitational interaction

Gravitational field is geometrised in general relativity by considering classical trajectories of macroscopic bodies as geodesics into the curved Riemannian space-time. We suggest here a hypothesis for geometrisation of gravitational field generated by microscopic gravitational field sources (elementary gravitational charges).

The main idea looks as follows. Electrons are microscopic elementary sources of elecgtromagnetic field and they both (field and sources) are represented within suggested topological approach by special 4-manifold. Topological properties of this manifold are described by the Dirac-Maxwell equations (1-3). Neutrinos are microscopic elementary sources of weak field and they both (field and sources) are represented by another special 4-manifold and topological properties of this manifold are described by equations (7-11). It would appear reasonable that elementary sources of gravitational field are some microscopic particles and that they both (field and these sources) can be considered as some another special

366

4-manifold and can be described by group-theoretical relations which are similar to Eqs.(1-3) and Egs.(7-11).

We obtained above relations for gravitational interaction assuming that the covering space for corresponding manifold is the Riemann space (the same one as in general relativity). Instead of spinor fields describing electromagnetic and weak interactions we use here vector field ϕ_l. Then the analogy of equation for free Dirac field will here be the known Lorentz invariant Proca equation for free ϕ_l-field describing free particles with mass m and spin 1

$$\frac{\partial \phi_{lm}}{\partial x_m} = m^2 \phi_l, \quad \phi_{lm} = \frac{\partial \phi_l}{\partial x_m} - \frac{\partial \phi_m}{\partial x_l}. \tag{12}$$

We take the gravitational interaction into account replacing in Eq.(12) usual derivatives by covariant ones where connections correspond to the Riemann covering space. Resulting equations have the form

$$\frac{\partial \phi_{lm}}{\partial x_m} - \Gamma^p_{ml} \phi_{pm} - \Gamma^p_{mm} \phi_{lp} = m^2 \phi_l, \tag{13}$$

$$\phi_{lm} = \frac{\partial \phi_l}{\partial x_m} - \frac{\partial \phi_m}{\partial x_l}. \tag{14}$$

$$R_{ik} = \frac{\partial \Gamma^l_{il}}{\partial x_k} - \frac{\partial \Gamma^l_{ik}}{\partial x_l} - \Gamma^l_{ik} \Gamma^p_{lp} + \Gamma^l_{ip} \Gamma^p_{kl}. \tag{15}$$

$$\frac{\partial R_{ik}}{\partial x_i} = -im \left(\phi^p * \frac{\partial \phi_p}{\partial x_k} - \phi^p \frac{\partial \phi_p *}{\partial x_k} \right). \tag{16}$$

Here R_{ik} is the Ricci tensor and the expression on the right side of Eq.(16) is a gravitational current of considered vector field.

Proposed elementary gravitational charge m (mass m is not here yet defined) represents may be one else candidate for rather numerious family of so called WINP-particles (Weakly Interacting Neutral Particle [9]). The question about possibilities of experimental confirmations will be considered in details elsewere. Note in conclusion that in distinction to general relativity the Riemann space appears here not as a real curved space-time but as effective covering space used for formal discription of more compicated geometrical object– microscopic nonmetrised topological space-time closed 4-manifold.

References

[1] Bjorken J D and Drell S D Relativistic quantum mechanics, McGraw-Hill Book Company, 1964

[2] Olkhov O A Proceedings of the 7th International Symposium on Particles, Strings and Cosmology, Lake Tahoe, California, 10-16 December 1999. Singapure-New Jersey-Hong-Kong, World Scientific, 2000, p.160, quant-ph/0101137

[3] Olkhov O A Chemical Physics (russian), 21,N1(2002)49; e-print hep-th/0201020

[4] Olkhov O A Proceedings of the XXIV Workshop on High Energy Physics and Field Theory, Protvino, 27-29 June 2001, Protvino 2001, p.327, e-print hep-th/0205120

[5] Schouten J A and Struik D J Einführung in die neueren Methoden der Differentialgeometrie, I. Groningen–Batavia.: Noordhoff, 1935

[6] Einstein A Riemann Geometrie mit Aufrechterhaltung des Fernparallelismus. Sitzungsber: preuss. Akad. Wiss., phys-math. K1., 1928, 217-221

[7] Schouten J A Tensor analysis for physicists. Oxford, 1952

[8] Yang C N and Mills R L 1954 Phys. Rev 96 191

[9] Klapdor-Kleingrothaus H V and Zuber K Teilchenastrophysik, Stuttgart, 1997

Inst. Phys. Conf. Ser. No 173: Section 4
Paper presented at 24th Int. Coll. Group Theoretical Methods in Physics, Paris, France, 15–20 July 2002
©2003 IOP Publishing Ltd

Weak Chiral Symmetry in the Standard Model
– Deformed Cyclic Symmetry for Chiral Family Mixings –

Ikuo S Sogami and Kouzou Koizumi

Department of Physics, Kyoto Sangyo University, Kyoto 603-8555, Japan

Abstract. In a field theory with generalized covariant derivatives, a weak chiral symmetry is proved to exist in the standard model and a deformed representation of cyclic symmetry is proposed to describe coupling constants of the Yukawa interaction.

1. Introduction

While the vector (gauge) interactions work to preserve symmetries in the standard model, the scalar (Higgs) interaction has a role to break down the symmetries. Connes proposed a new gauge theory to treat both types of interactions in a unified way [1, 2]. In his theory, all bosonic fields are accommodated together in the Dirac operator. Stimulated by Connes' theory, one of the present authors has developed a unified theory of gauge and Higgs fields by using a generalized concept of covariant derivatives [3, 4, 5].

In this note we report results of our recent study on possible symmetries hidden in the quark sector of the standard model which is reformulated in our theory with the generalized covariant derivative (GCD) [5].

2. Total quark field and GCD

We combine all chiral quark fields into a single multi-spinor field, called a total quark field, defied by the expression

$$\Psi(x) = \sum_{i=1}^{3} \left\{ \psi_{ui}^{L}(x)|u[i]\rangle_L + \psi_{di}^{L}(x)|d[i]\rangle_L + \psi_{ui}^{R}(x)|u[i]\rangle_R + \psi_{di}^{R}(x)|d[i]\rangle_R \right\} \tag{1}$$

where ψ_{fi}^{h} ($f = u, d$, $h = L, R$ and $i = 1, 2, 3$) represents generically an h-handed component field of an f quark belonging to the i-th family. The bra-ket symbols which are subjected to the orthonormal conditions $_h\langle f[i]|f'[j]\rangle_{h'} = \delta_{ff'}\delta_{hh'}\delta_{ij}$ are used to specify entry-positions of the component fields in the total quark field $\Psi(x)$. The total quark field $\Psi(x)$ satisfy the relations

$$\bar{\Psi}(x)\Psi(x) = 0, \quad \bar{\Psi}(x)\gamma_5\Psi(x) = 0. \tag{2}$$

In our formalism [5], the concept of covariant derivative is generalized so as to include the Higgs field as well as the gauge fields. As a basic postulate it is assumed that the GCD of the total quark field Ψ takes the following form

$$\mathcal{D}_\mu = \partial_\mu - \{\text{gauge fields}\}_\mu - \frac{i}{4}\gamma_\mu\{\text{Higgs field}\}. \tag{3}$$

To determine the Higgs part of the $\mathcal{D}_\mu$ explicitly, it is necessary to find appropriate algebras of chiral transformations and family mixings acting on the total quark field.

3. Deformed representation of permutation group

From phenomenological point of view, one of the most reasonable pattern of the Yukawa couplings is now considered to have a quasi-democratic structure in which all component fermions interact with nearly the same strength. In this report we construct such a quasi-democratic structure by means of a deformed representaion $\text{Rep}(S_3)$ of the permutation group S_3. For the transpositions (23), (31) and (12) which generate the S_3 group, let us introduce the following unitary matrix representations

$$
p_1(\theta_1) = \begin{pmatrix} 1 & 0 & 0 \\ 0 & 0 & e^{-i\theta_1} \\ 0 & e^{i\theta_1} & 0 \end{pmatrix}, \quad
p_2(\theta_2) = \begin{pmatrix} 0 & 0 & e^{i\theta_2} \\ 0 & 1 & 0 \\ e^{-i\theta_2} & 0 & 0 \end{pmatrix}, \quad
p_3(\theta_3) = \begin{pmatrix} 0 & e^{-i\theta_3} & 0 \\ e^{i\theta_3} & 0 & 0 \\ 0 & 0 & 1 \end{pmatrix} \quad (4)
$$

where $\theta_a = \theta_{a+3}$ $(a = 1, 2, 3 : \sum_{a=1}^{3} \theta_a = 0)$ are angles of deformation. The unitary representations of the cyclic permutations (123) and $(321) = (123)^2$ are given, respectively, by

$$
C_0(\theta_1, \theta_3, \theta_3) = p_1 p_2 = p_2 p_3 = p_3 p_1 = \begin{pmatrix} 0 & 0 & e^{i\theta_2} \\ e^{i\theta_3} & 0 & 0 \\ 0 & e^{i\theta_1} & 0 \end{pmatrix} \quad (5)
$$

and

$$
C_0^2(\theta_1, \theta_3, \theta_3) = p_2 p_1 = p_3 p_2 = p_1 p_3 = C_0^{-1}(\theta_1, \theta_3, \theta_3) = C_0^{\dagger}(\theta_1, \theta_3, \theta_3). \quad (6)
$$

Note that $C_0^{b-a} = p_a p_b$. The representation $\text{Rep}(C_3) = \{C_0, C_0^2, C_0^3 = I\}$ of the cyclic group C_3 forms a subgroup of the representation group $\text{Rep}(S_3) = \{I, p_1, p_2, p_3, C_0, C_0^2\}$.

4. Algebra of chiral family mixings

It is necessary to introduce operators for chiral transformations and family mixings which work on the total quark field. With the matrix of family mixings C_0 and the chiral projection Dirac matrices $L = \frac{1}{2}(I - \gamma_5)$ and $R = \frac{1}{2}(I + \gamma_5)$, let us define

$$
\varpi_a^{(f)} = \sum_{i,j=1}^{3} [C_0^{2a}(\theta_1^{(f)}, \theta_2^{(f)}, \theta_3^{(f)})]_{ij} \{L|f[i]\rangle_{LR}\langle f[j]| + R|f[i]\rangle_{RL}\langle f[j]|\} \quad (7)
$$

and

$$
\varpi_a^{(f)\dagger} = \sum_{i,j=1}^{3} [C_0^{a}(\theta_1^{(f)}, \theta_2^{(f)}, \theta_3^{(f)})]_{ij} \{L|f[i]\rangle_{RL}\langle f[j]| + R|f[i]\rangle_{LR}\langle f[j]|\} \quad (8)
$$

for the up $(f = u)$ and down $(f = d)$ quark sectors. These operators are nilpotent, $i.e.$, $\varpi_a^{(f)}\varpi_b^{(f)} = 0$ and $\varpi_a^{(f)\dagger}\varpi_b^{(f)\dagger} = 0$. The idempotency $C_0^3 = I$ results in the periodic condition $\varpi_{a+3} = \varpi_a$. The products of these operators $\varpi_a^{(f)}$ and $\varpi_a^{(f)\dagger}$ given by

$$
\mathcal{P}_{b-a}^{(f)} = \varpi_a^{(f)}\varpi_b^{(f)\dagger}, \quad \overline{\mathcal{P}}_{a-b}^{(f)} = \varpi_a^{(f)\dagger}\varpi_b^{(f)} \quad (9)
$$

are proved to satisfy the following relations:

$$
\mathcal{P}_a^{(f)}\overline{\mathcal{P}}_b^{(f)} = \overline{\mathcal{P}}_a^{(f)}\mathcal{P}_b^{(f)} = 0, \quad \mathcal{P}_a^{(f)}\mathcal{P}_b^{(f)} = \mathcal{P}_{a+b}^{(f)}, \quad \overline{\mathcal{P}}_a^{(f)}\overline{\mathcal{P}}_b^{(f)} = \overline{\mathcal{P}}_{a+b}^{(f)}. \quad (10)
$$

Here we define new operators

$$
\mathcal{P} = \sum_{f=u,d} \mathcal{P}_3^{(f)}, \quad \overline{\mathcal{P}} = \sum_{f=u,d} \overline{\mathcal{P}}_3^{(f)} \quad (11)
$$

which satisfy the relations $\mathcal{P}^2 = 0$, $\overline{\mathcal{P}}^2 = 0$ and $\mathcal{P} + \overline{\mathcal{P}} = I$. These are projection operators acting on the total quark field as

$$\mathcal{P}\Psi(x) = \Psi(x), \quad \overline{\mathcal{P}}\Psi(x) = 0. \tag{12}$$

With the projection operator $\mathcal{P}$ and the element $\mathfrak{g}_i$ of the Lie algebra of the gauge group G=SU(3)×SU(2)×U(1)=$\{e^{i\mathfrak{g}_i\theta_i(x)}\}$ of the standard model, it is possible to construct a group

$$G_{\text{wch}} = \{e^{i\mathfrak{g}_i\theta_i(x)\mathcal{P}}\}. \tag{13}$$

We call this group G_{wch} a weak chiral group associated with the original group G. Owing to the relation (12), the groups G and G_{wch} have the same action on the total quark field. Therefore, the fermionic part of the standard model Lagrangian $\mathcal{L}_f$ is automatically invariant under this new group G_{wch}.

5. GCD, Unified field strength and bosonic Lagrangian density

To make the bosonic part of the standard model Lagrangian $\mathcal{L}_b$ invariant under G and G_{wch}, we postulate that the GCD for the total quark field $\mathcal{D}_\mu$ satisfies the following criterions:

$$e^{-i\mathfrak{g}_i\theta_i(x)}\mathcal{D}_\mu e^{i\mathfrak{g}_i\theta_i(x)} = \mathcal{D}_\mu, \quad e^{-i\mathfrak{g}_i\theta_i(x)\mathcal{P}}\mathcal{D}_\mu e^{i\mathfrak{g}_i\theta_i(x)\mathcal{P}} = \mathcal{D}_\mu. \tag{14}$$

The most general form of the GCD satisfying these conditions is found to be

$$\mathcal{D}_\mu = \partial_\mu - ig_3\mathcal{A}_\mu^{(3)}(x) - ig_2\mathcal{A}_\mu^{(2)}(x) - ig_1\mathcal{A}_\mu^{(1)}(x)$$

$$- \frac{i}{4}\gamma_\mu \sum_{a=1}^{3} \left\{ \xi_a^{(u)}\tilde{\Phi}(x)\varpi_a^{(u)\dagger} + \xi_a^{(d)}\Phi(x)\varpi_a^{(d)\dagger} + \xi_{2a}^{(u)*}\varpi_a^{(u)\dagger}\tilde{\Phi}^\dagger(x) + \xi_{2a}^{(d)*}\varpi_a^{(d)\dagger}\Phi^\dagger(x) \right.$$

$$\left. + \eta_a^{(u)*}\varpi_a^{(u)}\tilde{\Phi}^\dagger(x) + \eta_a^{(d)*}\varpi_a^{(d)}\Phi^\dagger(x) + \eta_{2a}^{(u)}\tilde{\Phi}(x)\varpi_a^{(u)} + \eta_{2a}^{(d)}\Phi(x)\varpi_a^{(d)} \right\}. \tag{15}$$

In this expression $\mathcal{A}_\mu^{(3)}(x)$, $\mathcal{A}_\mu^{(2)}(x)$ and $\mathcal{A}_\mu^{(1)}(x)$ are gauge fields of the groups G and G_{wch}, and $\Phi(x)$ and $\tilde{\Phi}(x)$ are scalar fields defined by

$$\Phi(x) = \sum_{f=u,d}\sum_{i=1}^{3} [\phi(x)_f|f[i]\rangle_L({}_L\langle u[i]| + {}_L\langle d[i]|)], \quad \tilde{\Phi}(x) = i\mathcal{P}_{\text{conj}}\,\Phi(x)^* \tag{16}$$

where $\phi(x)$ is the origninal Higgs field and

$$\mathcal{P}_{\text{conj}} = \sum_{i=1}^{3} (-i|u[i]\rangle_{LL}\langle d[i]| + i|d[i]\rangle_{LL}\langle u[i]|). \tag{17}$$

The GCD thus constructed enables us to define the unified field strength for all bosonic fields $\mathcal{F}_{\mu\nu}$ by the commutator

$$[\mathcal{D}_\mu, \mathcal{D}_\mu] = -i\rho\mathcal{F}_{\mu\nu} \tag{18}$$

which leads to the bosonic Lagrange $\mathcal{L}_b$ as follows:

$$\mathcal{L}_b = -\frac{1}{4!}\frac{1}{4}\text{Tr}(\mathcal{F}_{\mu\nu}\gamma_5\mathcal{F}_{\mu\nu}\gamma_5) + c\text{Tr}(\sigma_{\mu\nu}\mathcal{F}_{\mu\nu}) \tag{19}$$

where c is an arbitrary constant [5].

6. Mass matrices and Cabbibo-Kobayashi-Maskawa matrix

In the low energy region where the spontaneous breakdown of the G and G_{wch} symmetries takes place, the GCD in equation (15) results in the mass matrices

$$M^{(f)} = \left(\xi_1^{(f)} C_0^{(f)} + \xi_2^{(f)} C_0^{(f)2} + \xi_3^{(f)} I \right) \langle \phi^0 \rangle \tag{20}$$

for the $f = u$ and d sectors, where $\langle \phi^0 \rangle$ is the vacuum expectation value of the Higgs field ϕ. The $M^{(f)}$ are normal matrices satisfying $M^{(f)\dagger} M^{(f)} = M^{(f)} M^{(f)\dagger}$ which are diagonalized by the unitary tranformations as

$$U^{(f)\dagger} M^{(f)\dagger} M^{(f)} U^{(f)} = U^{(f)\dagger} M^{(f)} M^{(f)\dagger} U^{(f)} = \mathrm{Diag} \left(\lambda_1^{(f)}, \lambda_2^{(f)}, \lambda_3^{(f)} \right) \tag{21}$$

where the eigenvalues are derived to be

$$\lambda_{1,2}^{(f)} = y^{(f)} + 2|z^{(f)}| \cos \left(\theta^{(f)} \pm \frac{2\pi}{3} \right), \quad \lambda_3^{(f)} = y^{(f)} + 2|z^{(f)}| \cos \theta^{(f)} \tag{22}$$

in which $y^{(f)}$ and $z^{(f)}$ are given by

$$y^{(f)} = \left(|\xi_1^{(f)}|^2 + |\xi_2^{(f)}|^2 + |\xi_3^{(f)}|^2 \right) |\langle \phi^0 \rangle|^2,$$

$$z^{(f)} = |z^{(f)}| e^{i\theta^{(f)}} = \left(\xi_1^{(f)} \xi_2^{(f)*} + \xi_2^{(f)} \xi_3^{(f)*} + \xi_3^{(f)} \xi_1^{(f)*} \right) |\langle \phi^0 \rangle|^2. \tag{23}$$

The unitary matrices for diagonalization $U^{(f)}$ are obtained by

$$U^{(f)} = \frac{1}{\sqrt{3}} \left[p_1 \left(\theta_1 - \frac{2\pi}{3} \right) + p_2 \left(\theta_2 + \frac{2\pi}{3} \right) + e^{i\frac{2\pi}{3}} p_3 (\theta_3) \right] \tag{24}$$

which lead readily to the Cabbibo-Kobayashi-Maskawa matrix $V_{\mathrm{CKM}} = U^{(u)\dagger} U^{(d)}$.

7. Discussion

In this way we have reformulated the standard model of elementary particles in the theory with the GCD. This new scheme provide a flexible framework which is convenient to study relations between the fermionic and bosonic sectors and to investigate the structure of the Yukawa interaction. The eigenvalues in equation (22) can explain the detailed mass spectra of the up and down quark sectors. However, the Cabbibo-Kobayashi-Maskawa matrix derived from the unitary matrices in equation (24) can not describe well the experimental results.

To obtain a reasonable scheme which explains all the observed data, it seems necessary to formulate a new deformed representation of the cyclic group. We investigate a scheme with a new generator C of the cyclic group defined by the replacement

$$C_0 \rightarrow C = U^\dagger C_0 U \tag{25}$$

where $U \in \mathrm{Rep}(S_3)$ is a unitary matrix generated by p_α in equation (4).

References

[1] Connes A 1985 Pub. Math. IHES **62** 41; *Noncommutative Geometry* (Academic Press, 1994) and references therein.
[2] Connes A and Lott J 1990 Nucl. Phys. Proc. Supp. **B18** 295
[3] Sogami S I 1995 Prog.Theor.Phys. **94** 117; 1996 *ibidem* **95** 637
[4] Sogami S I, Tanaka H and Shirafuji T 1999 Prog.Theor.Phys. **101** 903
[5] Sogami S I 2001 Prog.Theor.Phys. **105** 483

Inst. Phys. Conf. Ser. No 173: Section 4
Paper presented at 24th Int. Coll. Group Theoretical Methods in Physics, Paris, France, 15–20 July 2002
©2003 IOP Publishing Ltd

Stability of Monopole Condensation in SU(2) QCD

Y. M. Cho and M. L. Walker

Department of Physics, College of Natural Sciences, Seoul National University, Seoul
151-747, Korea

Abstract. We resolve the controversy on the stability of the monopole condensation in the
one-loop effective action of $SU(2)$ QCD by calculating the imaginary part of the effective
action with two different methods at one-loop order. Our result confirms that the effective
action for the magnetic background has no imaginary part but the one for the electric
background has a negative imaginary part. This assures that the monopole condensation is
indeed stable, but the electric background becomes unstable due to the pair-annihilation of
gluons.

PACS numbers: 11.15.Bt, 14.80.Hv, 12.38.Aw

1. Introduction

It has long been argued that monopole condensation can explain the confinement of color
through the dual Meissner effect [1, 2, 3]. However previous attempts to demonstrate
monopole condensation in $SU(2)$ QCD with the one-loop effective action using the
background field method [4, 5] have failed, primarily because their magnetic condensations
were unstable due to an imaginary part in the effective action [4, 5] that destabilized them
through the pair creation of gluons.

A recent calculation by Cho, Lee, and Pak [6, 7] with a gauge independent separation of
the classical background from the quantum field found no imaginary part in the presence
of the non-Abelian monopole background, but a negative imaginary part in the presence
of the pure color electric background. This means that in QCD the non-Abelian monopole
background produces a stable monopole condensation, but the color electric background
becomes unstable by generating a pair annihilation of the valence gluon at one-loop level.
The stability of monopole condensation is controversial, so it is imperative that we verify it
with an independent method.

In the massless limit of gauge theories the imaginary part of the effective action is
proportional to g^2, where g is the gauge coupling constant. This allows us to calculate the
imaginary part by perturbative methods [10], which is the purpose of this paper. Our result
confirms that the effective action indeed has no imaginary part in the presence of the monopole
background but has a negative imaginary part in the presence of the color electric background.

2. Abelian Decomposition of QCD

We start from the Cho decomposition [12, 13] of the gauge potential into the restricted potential $\hat{A}_\mu$ and the valence potential $\vec{X}_\mu$. Let $\hat{n}$ be the unit isovector which selects the color charge direction everywhere in space-time, and let [2, 3]

$$\vec{A}_\mu = A_\mu \hat{n} - \frac{1}{g}\hat{n} \times \partial_\mu \hat{n} + \vec{X}_\mu = \hat{A}_\mu + \vec{X}_\mu,$$

$$(A_\mu = \hat{n} \cdot \vec{A}_\mu, \; \hat{n}^2 = 1, \; \hat{n} \cdot \vec{X}_\mu = 0), \tag{1}$$

where A_μ is the "electric" potential. Notice that the restricted potential $\hat{A}_\mu$ is precisely the connection which leaves $\hat{n}$ invariant under parallel transport,

$$\hat{D}_\mu \hat{n} = \partial_\mu \hat{n} + g\hat{A}_\mu \times \hat{n} = 0. \tag{2}$$

The gauge transformation is given by

$$\delta \hat{n} = -\vec{\alpha} \times \hat{n}, \;\; \delta \vec{A}_\mu = \frac{1}{g} D_\mu \vec{\alpha},$$

$$\delta A_\mu = \frac{1}{g}\hat{n} \cdot \partial_\mu \vec{\alpha}, \;\; \delta \hat{A}_\mu = \frac{1}{g}\hat{D}_\mu \vec{\alpha} \qquad \delta \vec{X}_\mu = -\vec{\alpha} \times \vec{X}_\mu. \tag{3}$$

This tells that $\hat{A}_\mu$ by itself describes an $SU(2)$ connection enjoying the full $SU(2)$ gauge degrees of freedom. Furthermore, the valence potential $\vec{X}_\mu$ forms a gauge covariant vector field under the gauge transformation. Note that the decomposition is gauge independent.

$\hat{A}_\mu$ has a dual structure,

$$\hat{F}_{\mu\nu} = \partial_\mu \hat{A}_\nu - \partial_\nu \hat{A}_\mu + g\hat{A}_\mu \times \hat{A}_\nu = (F_{\mu\nu} + H_{\mu\nu})\hat{n},$$
$$F_{\mu\nu} = \partial_\mu A_\nu - \partial_\nu A_\mu,$$
$$H_{\mu\nu} = -\frac{1}{g}\hat{n} \cdot (\partial_\mu \hat{n} \times \partial_\nu \hat{n}) = \partial_\mu \tilde{C}_\nu - \partial_\nu \tilde{C}_\mu, \tag{4}$$

where $\tilde{C}_\mu$ is the "magnetic" potential of the monopoles [2, 3]. Thus, one can identify the non-Abelian monopole potential by

$$\vec{C}_\mu = -\frac{1}{g}\hat{n} \times \partial_\mu \hat{n}, \tag{5}$$

in terms of which the magnetic field is expressed by

$$\vec{H}_{\mu\nu} = \partial_\mu \vec{C}_\nu - \partial_\nu \vec{C}_\mu + g\vec{C}_\mu \times \vec{C}_\nu = H_{\mu\nu}\hat{n}. \tag{6}$$

This provides the gauge independent separation of the monopole field $H_{\mu\nu}$, from the color electromagnetic field $F_{\mu\nu}$.

With the decomposition (1), one has

$$\vec{F}_{\mu\nu} = \hat{F}_{\mu\nu} + \hat{D}_\mu \vec{X}_\nu - \hat{D}_\nu \vec{X}_\mu + g\vec{X}_\mu \times \vec{X}_\nu, \tag{7}$$

3. Diagrammatic Calculation of Imaginary Part of QCD Effective Action

For an arbitrary background B_μ there are four Feynman diagrams that contribute to the order g^2. Two are shown schematically in Fig. 1, where the curly line and the dotted line represent

Figure 1. The Feynman diagrams that contribute to the imaginary part of the effective action at g^2 order

the valence gluon and the ghost, respectively. The other two are tadpole diagrams with a quadratic divergence, so they don't appear in the final result, and are not shown.

Dimensional regularisation gives

$$\Delta \mathcal{L}_{eff} = -\frac{1}{8\pi^2} \int_0^1 dx \Big[B_\mu (4(p^2 g_{\mu\nu} - p_\mu p_\nu) + p_\mu p_\nu (1-2x)^2$$
$$- \frac{1}{2} g_{\mu\nu} p^2 (4x - 4x^2)) B_\nu \times \ln(\frac{p^2}{\mu^2}(4x - 4x^2)) \Big] \tag{8}$$

for the sum of these diagrams where only the log terms have been kept since only they contribute to the imaginary part. They will only do so however, when p is timelike, since positive μ^2 corresponds to spacelike momenta. Furthermore, since this is an Abelianized theory, a purely magnetic background corresponds to purely spacelike momenta. It follows that a purely magnetic background has no imaginary part and is stable. The same argument does not apply to pure electric backgrounds so we integrate by parts, obtaining

$$\Delta L_{eff} = \frac{1}{8\pi^2} g^2 \int_0^1 dx \Big(B_\mu (4x(p^2 g_{\mu\nu} - p_\mu p_\nu)$$
$$+ (x - 2x^2 + \frac{4}{3}x^3) p_\mu p_\nu - \frac{1}{2} g_{\mu\nu} p^2 (2x^2 - \frac{4}{3}x^2)) B_\nu \frac{1-2x}{x(1-x) - i\epsilon} \Big), \tag{9}$$

where we have discarded the wave function renormalisation. Introducing the standard imaginary infinitesimal to the denominator to find the imaginary part, we observe that the singularity occurs at the end of the integral, so the contour describes a quarter arc and not a semicircle and the correct prescription for obtaining the imaginary part is $\frac{1}{2} i\pi$ and not $i\pi$.

We duplicate the result of Cho, Lee, and Pak [6, 7], *ie.* that there is no imaginary part for the magnetic background while a purely electric background has an imaginary part of $-\frac{11b^2}{96\pi}$, where the magnetic and electric field strengths are given by

$$a = \frac{g}{2} \sqrt{\sqrt{G^4 + (G\tilde{G})^2} + G^2},$$

$$b = \frac{g}{2} \sqrt{\sqrt{G^4 + (G\tilde{G})^2} - G^2}, \tag{10}$$

respectively.

4. Schwinger's Method

With the Abelian formalism of QCD, we can follow Schwinger step by step to obtain the imaginary part of the QCD effective action. For the pure magnetic case (i.e., for $b = 0$), the effective action found by Cho, Lee and Pak [6, 7] gives us

$$\Delta S = i\mathrm{Tr}\left[\ln(-\tilde{D}^2 + 2a) + \ln(-\tilde{D}^2 - 2a)\right],$$

$$= -i\mathrm{Tr}\int_0^\infty \frac{ds}{s}\left[\exp(-i(\tilde{D}^2 - 2a)s) + \exp(-i(\tilde{D}^2 + 2a)s)\right]. \tag{11}$$

The perturbative expansion of

$$\mathrm{Tr}\,U(s)_\pm \equiv \mathrm{Tr}\exp(-i(\tilde{D}^2 \mp 2a)s), \tag{12}$$

to second order in $g\tilde{C}_\mu$ and a (remember that a contains a factor of g) followed the steps taken in Schwinger's seminal paper, after some tedium, yields

$$\Delta S = -\frac{1}{2\pi^2}\int dk \int_0^1 dv \frac{v^2(1 - v^2/12)}{1 - v^2}a^2. \tag{13}$$

The corresponding calculation for the pure electric background (i.e., for $a = 0$), gives us

$$\Delta S = \frac{1}{2\pi^2}\int dk \int_0^1 dv \frac{v^2(1 - v^2/12)}{1 - v^2}b^2. \tag{14}$$

Since this is an Abelianized theory, Schwinger's argument that only the electric field has the timelike fourier components needed to generate particle/antiparticle pairs (gluons in this work, electrons in his), applies. The imaginary component of the magnetic field is therefore zero while that of the electric background is found by repeating the corresponding steps at the end of the last section. We again confirm the result of [6, 7].

5. Discussion

We have checked the stability of the monopole condensation with two independent perturbative methods. Both produce a result in agreement with [6, 7], confirming the stability of the monopole condensation. and confirming that magnetic confinement is indeed the correct confinement mechanism. That the magnetic condensate is indeed the true vacuum of QCD, one must calculate the effective action with an arbitrary background in the presence of the quarks is demonstrated for $SU(2)$ QCD, at least at one loop level, in [6, 16].

Acknowledgements

One of the authors (YMC) thanks S. Adler and F. Dyson for the fruitful discussions, and Professor C. N. Yang for the continuous encouragements. The work is supported in part by Korea Research Foundation (Grant KRF-2001 -015-BP0085) and by the BK21 project of Ministry of Education.

References

[1] Y. Nambu, Phys. Rev. **D10**, 4262 (1974); S. Mandelstam, Phys. Rep. **23C**, 245 (1976); A. Polyakov, Nucl. Phys. **B120**, 429 (1977); G. 't Hooft, Nucl. Phys. **B190**, 455 (1981).

[2] Y. M. Cho, Phys. Rev. **D21**, 1080 (1980); J. Korean Phys. Soc. **17**, 266 (1984); Phys. Rev. **D62**, 074009 (2000).

[3] Y. M. Cho, Phys. Rev. Lett. **46**, 302 (1981); Phys. Rev. **D23**, 2415 (1981); W. S. Bae, Y. M. Cho, and S. W. Kimm, Phys. Rev. **D65**, 025005 (2002).

[4] G. K. Savvidy, Phys. Lett. **B71**, 133 (1977); N. Nielsen and P. Olesen, Nucl. Phys. **B144**, 485 (1978); C. Rajiadakos, Phys. Lett. **B100**, 471 (1981).

[5] A. Yildiz and P. Cox, Phys. Rev. **D21**, 1095 (1980); M. Claudson, A. Yilditz, and P. Cox, Phys. Rev. **D22**, 2022 (1980); S. Adler, Phys. Rev. **D23**, 2905 (1981); W. Dittrich and M. Reuter, Phys. Lett. **B128**, 321, (1983); C. Flory, Phys. Rev. **D28**, 1425 (1983); S. K. Blau, M. Visser, and A. Wipf, Int. J. Mod. Phys. **A6**, 5409 (1991); M. Reuter, M. G. Schmidt, and C. Schubert, Ann. Phys. **259**, 313 (1997).

[6] Y. M. Cho, H. W. Lee, and D. G. Pak, Phys. Lett. **B 525**, 347 (2002); Y. M. Cho and D. G. Pak, Phys. Rev. **D65**, 074027 (2002).

[7] Y. M. Cho and D. G. Pak, J. Korean Phys. Soc. **38**, 151 (2001); in *Proceedings of TMU-Yale Symposium on Dynamics of Gauge Fields*, edited by T. Appelquist and H. Minakata (Universal Academy Press, Tokyo) (1999).

[8] Y. M. Cho and D. G. Pak, Phys. Rev. Lett. **86**, 1947 (2001); W. S. Bae, Y. M. Cho, and D. G. Pak, Phys. Rev. **D64**, 017303 (2001).

[9] Y. M. Cho and D. G. Pak, hep-th/0010073, submitted to Phys. Rev. **D**, Rapid Communication.

[10] V. Schanbacher, Phys. Rev. **D26**, 489 (1982); L. Freyhult, hep-th/0106239.

[11] J. Schwinger, Phys. Rev. **82**, 664 (1951).

[12] L. Faddeev and A. Niemi, Phys. Rev. Lett. **82**, 1624 (1999); Phys. Lett. **B449**, 214 (1999).

[13] S. Shabanov, Phys. Lett. **B458**, 322 (1999); **B463**, 263 (1999); H. Gies, hep-th/0102026.

[14] See for example, C. Itzikson and J. Zuber, *Quantum Field Theory* (McGraw-Hill) 1985; M. Peskin and D. Schroeder, *An Introduction to Quantum Field Theory* (Addison-Wesley) 1995; S. Weinberg, *Quantum Theory of Fields* (Cambridge Univ. Press) 1996.

[15] A similar calculation was first carried out by Honerkamp. See, J. Honerkamp, Nucl. Phys. **B48**, 269 (1972).

[16] Y. M. Cho and D. G. Pak, hep-th/0006051, submitted to Phys. Rev. **D**.

Atomic and Molecular Physics, Theoretical Chemistry, Condensed Matter

Inst. Phys. Conf. Ser. No 173: Section 5
Paper presented at 24th Int. Coll. Group Theoretical Methods in Physics, Paris, France, 15–20 July 2002
©2003 IOP Publishing Ltd

Algebraic Models and Quantum Deformations in Molecular Thermodynamics

Maia N Angelova

School of Informatics, Northumbria University, Newcastle upon Tyne, UK

Abstract. Lie-algebaic and quantum-algebraic techniques are used in the analysis of thermodynamic properties of molecules at high temperatures. In the framework of the algebraic models, the local anharmonic effects are described by a Morse-like potential, associated with the $SU(2)$ algebra. The vibrational high-temperature properties, such as mean energy and specific heat, are studied in terms of the parameters of the algebraic models. The concept of a critical temperature associated with the anharmonic vibrations is introduced. q-bosons are applied to molecular thermodynamics. A quantum deformation, associated with the model is discussed and its effect on the thermodynamic properties is studied.

1. Introduction

The development of algebraic approaches in molecular physics has brought new insights into the interactions in complex molecules and powerful methods for computing spectroscopic properties [1, 2]. The algebraic models exploit the isomorphism between the $SU(2)$ algebra and the one-dimensional Morse oscillator. The $SU(2)$ anharmonic models [3]-[5] combine Lie algebraic techniques, describing the interatomic interactions, with discrete symmetry techniques associated with the local symmetry of the molecules. The algebraic models are developed to analyze molecular vibrational spectra [1]-[5]. They provide a systematic procedure for studying vibrational excitations in a simple form by describing the stretching and bending modes in a unified scheme based on $SU(2)$ algebras. The first steps in applying the algebraic approach to molecular thermodynamics are made in [7, 8] where the partition function and the basic thermodynamic functions of diatomic molecules are derived in terms of the parameters of the model.

For one-dimensional Morse oscillator, the anharmonic effects are described by boson operators [1] written in terms of the generators of $SU(2)$,

$$\hat{b} = \frac{\hat{J}_+}{\sqrt{N}}, \quad \hat{b}^\dagger = \frac{\hat{J}_-}{\sqrt{N}}, \quad \hat{v} = \frac{\hat{N}}{2} - \hat{J}_z \tag{1}$$

where $\hat{N}$ is the number operator and N is the total number of bosons fixed by the potential shape. The value of N is dependent on the depth D, the width d of the Morse potential well and the reduced mass μ of the oscillator [1, 4],

$$N + 1 = \left(\frac{8\mu D d^2}{\hbar^2} \right)^{\frac{1}{2}}. \tag{2}$$

The operators $\hat{b}$ and $\hat{b}^\dagger$ satisfy the commutation relations,

$$\left[\hat{b}, \hat{v}\right] = \hat{b}, \quad \left[\hat{b}^\dagger, \hat{v}\right] = -\hat{b}^\dagger, \quad \left[\hat{b}, \hat{b}^\dagger\right] = 1 - \frac{2\hat{v}}{N} \tag{3}$$

The harmonic limit is obtained when $N \to \infty$, in which case $[\hat{b}, \hat{b}^\dagger] \to 1$ giving the usual boson commutation relations. The one-dimensional Morse Hamiltonian can be written in terms of the anharmonic boson operators $\hat{b}$ and $\hat{b}^\dagger$,

$$H_M \sim \frac{1}{2}\left(\hat{b}\hat{b}^\dagger + \hat{b}^\dagger\hat{b}\right) \tag{4}$$

with vibrational energies

$$\varepsilon_v = \hbar\omega_0\left(v + \frac{1}{2} - \frac{v^2}{N}\right), \ v = 1, 2, \ldots, \left[\frac{N}{2}\right] \tag{5}$$

where ω_0 is the harmonic oscillator frequency. The Morse phonon operator $\hat{v}$ has an eigenvalue v which gives the number of quanta in the oscillator.

The spectrum of the Morse potential leads to a deformation of the harmonic oscillator algebra. This deformation is derived using a quantum analogue of the anharmonic oscillator [9], where the anharmonic vibrations are described as anharmonic q-bosons. In this paper we discuss further application of the algebraic approach to the vibrational high-temperature thermodynamics and the impact of the quantum deformations onto basic thermodynamic properties. We define a critical temperature of the model and consider a possible physical interpretation of a quantum deformation.

2. Applications of the algebraic approach to vibrational thermodynamics

The algebraic approach is applied to molecular thermodynamics by deriving the appropriate partition functions. The anharmonic effects are accounted for by using the vibrational energies (5) and the fixed number of anharmonic bosons N [7, 8]. Here, we summarize the results for diatomic molecules.

At high temperatures the vibrational partition function in the Morse-like spectrum is,

$$Z_N = \frac{1}{2}\sqrt{\frac{N_0\pi}{\alpha}}e^{-\alpha(N_0+1)}\mathrm{erf}\,i\left(\sqrt{\alpha N_0}\right). \tag{6}$$

where $\alpha = \frac{\hbar\omega_0}{2k_BT}$ and $N_0 = \left[\frac{N}{2}\right]$ are the parameters of the algebraic model and $\mathrm{erf}\,i\left(\sqrt{\alpha N_0}\right)$ is the error function. The dependence on the temperature T is given by the parameter α. The anharmonic contributions are essential for $\alpha < 0.5$, where $\alpha = 0.5$ corresponds to the characteristic vibrational temperature of the molecule. When $N_0 \to \infty$, the harmonic limit of the model is obtained, $Z_\infty \sim e^{-\alpha}/2\alpha$, which is precisely the harmonic vibrational partition function of a diatomic molecule at high temperatures. The expression for the partition function (6) can be generalized to polyatomic molecules, by combining the present results with the use of a local-mode model where each interatomic potential is of the Morse form [3].

In the framework of the algebraic model, the mean vibrational energy at high temperatures is,

$$U_N = \frac{\hbar\omega_0}{2}\left(1 + N_0 + \frac{1}{2\alpha} - \sqrt{\frac{N_0}{\alpha\pi}}\frac{e^{\alpha N_0}}{\mathrm{erf}\,i\left(\sqrt{\alpha N_0}\right)}\right). \tag{7}$$

where the partition function (6) has been used in derivation of the result. When $N_0 \to \infty$, the classical mean energy of a diatomic molecule at high temperatures is obtained.

The vibrational part of the specific heat is $C_N = \frac{\partial U_N}{\partial T}$, which gives the following dependence of C_N on the parameters α and N_0 at high-temperatures,

$$C_N = \frac{k_B}{2} + k_B\sqrt{\frac{\alpha N_0}{\pi}}\frac{e^{\alpha N_0}}{\mathrm{erf}\,i\left(\sqrt{\alpha N_0}\right)}\left(\alpha N_0 - \frac{1}{2} - \sqrt{\frac{\alpha N_0}{\pi}}\frac{e^{\alpha N_0}}{\mathrm{erf}\,i\left(\sqrt{\alpha N_0}\right)}\right) \tag{8}$$

When $N_0 \to \infty$, the harmonic limit of the model gives $C_\infty \sim k_B$.

Figure 1 represents the vibrational specific heat, C_N/k_B, as a function of the combined parameter αN_0. The graph shows an anomaly: the vibrational specific heat has a maximum

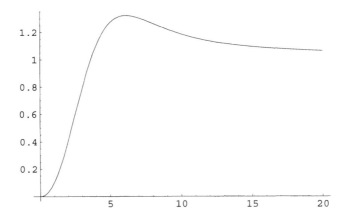

Figure 1. Vibrational specific heat C_N/k_B as a function of αN_0.

for a value of the parameter $\alpha_C N_0$. As N_0 is fixed and constant for the molecule, the maximum is reached for $\alpha = \alpha_C$ which corresponds to a temperature T_C. We call this temperature a critical temperature of the model. The anomaly of the specific heat is a result of the finite number N of bosons in the system. When all bosons in the oscillator are excited to the state with the maximal energy ε_{N_0}, the specific heat reaches its maximum. The numerical solution gives $\alpha_C N_0 = 6.133$. For the molecule $^1H^{35}Cl$, the total number of anharmonic bosons is $N = 56$, the total number of quanta in the oscillator is $N_0 = 28$, $\alpha_C = 0.219$ which gives $T_C = 9815K$, the characteristic vibrational temperature is $4300K$. The critical value α_C decreases as the number of anharmonic bosons increases. The anomaly of the specific heat disappears as $N_0 \to \infty$, $\alpha_C \to 0$.

The comparison of $\Delta\varepsilon = \varepsilon_{N_0} - \varepsilon_0$ with the dissociation energy of the molecule shows that $\Delta\varepsilon > DE$. At T_C when all bosons in the system can be excited, some of the molecules may start dissociating whilst others may still be in stable molecular states. Our model, in its present form, does not account for the effects of the dissociation. In addition, this simple version of the model does not yet include the contributions of the translational and rotational degrees of freedom which at temperatures close to T_C may be substantial. The critical temperature T_C can be considered as a temperature above which the model is no longer valid in its current form and other effects take place.

3. Quantum deformations

In [9], we have shown that the anharmonic bosons $b, b^\dagger$ from equation (9) can be obtained as an approximation of the q-bosons (see *e.g.* [10]). The q-bosons are defined by the following commutation relations:

$$[a, a^\dagger] = q^{\hat{n}}, \quad [\hat{n}, a] = -a, \quad [\hat{n}, a^\dagger] = a^\dagger \tag{9}$$

where the deformation parameter q is in general a complex number. The anharmonic commutation relations (3) can be recovered for real values of the deformation q close to 1, $q < 1$, and an expansion of q in terms of the first-order of a parameter p, $p \equiv 1/(1-q)$,

$$q^{\hat{n}} = 1 - \frac{\hat{n}}{p} \,. \tag{10}$$

We will refer to the expansion (10) as a linear deformation. Substituting the approximation (10) in the commutation relations (9) and identifying the parameter p with $N/2$, $\hat{n}$ with $\hat{v}$ and the creation and annihilation operators $a, a^{\dagger}$, with $b, b^{\dagger}$ respectively, the $SU(2)$ anharmonic commutation relations (3) are recovered.

Thus, the $SU(2)$ commutation relations (3) can be considered as a deformation of the harmonic oscillator commutation relations, with a deformation parameter $p = N/2$. This gives a possible physical realization for the linear deformation (10): the quantum deformation parameter p is the fixed number N_0 of the anharmonic bosons in the oscillator. Using the relation between the fixed number of anharmonic bosons N and the characteristics of the Morse potential (2), we conclude that the quantum deformation is also determined by the depth, the width and in general the shape of the Morse potential well. For the molecule $^1H^{35}Cl$, $p = 28$ which gives $q = 27/28$.

Using the linear deformation, we can write the Morse Hamiltonian (4) and the vibrational energies (5) in terms of the quantum deformation parameter p. Further, substituting $N_0 = p$ in the expressions (6), (7) and (8), we obtain the basic thermodynamic properties of the diatomic molecules as functions of the deformation parameter p. The critical temperature T_C is proportional to the quantum deformation parameter. For large values of p, $(q \rightarrow 1)$, the classic harmonic case is restored. Work is in progress on expansions of q in terms of higher-orders of p and the effect of the higher-order parameters on the Hamiltonian, vibrational energies and the related thermodynamic properties.

4. Conclusion

We have obtained the thermodynamic properties of diatomic molecules in terms of the parameters of the algebraic model. We can also interpret these properties in terms of a linear quantum deformation, which is associated with the fixed total number of anharmonic bosons and the shape of the Morse potential. We can write the algebraic model and the thermodynamic properties of diatomic molecules in terms of the quantum deformation parameter.

References

[1] Frank A and van Isacker P 1994 Algebraic Methods in Molecular and Nuclear Structure Physics (New York: John Wiley) .
[2] Iachello F and Levine R 1995 Algebraic Theory of Molecules (Oxford: OUP).
[3] Lemus R and Frank A 1994 Chem. Phys. **101** 8321.
[4] Frank A et al 1996 Annals of Physics **252** 211-238.
[5] Carvajal M et al 2000 Chem. Phys. **260** 420.
[6] Angelova M, Dobrev V and Frank A, *J Phys A* Math. Gen **34** (2001) L503-L509.
[7] Angelova M and Frank A 2002 Proc. Int. Symp. Quantum Theory and Symmetries New Jersey: World Scientific 243-248.
[8] Angelova M and Frank A 2002 preprint cond-mat/0203413.
[9] Angelova M, Dobrev V and Frank A, *J Phys A* Math. Gen **34** (2001) L503-L509.
[10] Biedenharn L C 1989 J. Phys. A: Math. Gen. **22**, L873-L878.

Inst. Phys. Conf. Ser. No 173: Section 5
Paper presented at 24th Int. Coll. Group Theoretical Methods in Physics, Paris, France, 15–20 July 2002
©2003 IOP Publishing Ltd

The Bilbao Crystallographic Server

E. Kroumova[1], C. Capillas[1], M. I. Aroyo[1], J. M. Perez-Mato[1],
S. Ivantchev[1], G. Madariaga[1], A. Kirov[2], H. Wondratschek[3],
H. T. Stokes[4] and D. M. Hatch[4]

[1] Depto Física de la Materia Condensada, Universidad del País Vasco, Bilbao, Spain
[2] Condensed Matter Physics Dept, Sofia University, Sofia, Bulgaria
[3] Institut für Kristallographie der Universität Karlsruhe, D-76128 Karlsruhe, Germany
[4] Dept. of Physics and Astronomy, Brigham Young University, Provo, Utah 84602, USA

Abstract. The *Bilbao Crystallographic Server* is a web site with crystallographic databases and programs freely available on-line. Currently, the server gives access to the databases containing the data from the International Tables for Crystallography, Vol. A, *Space Group Symmetry* and Vol. A1, *Symmetry Relations between Space Groups*. The accompanying software is divided into several shells according to different topics. The software is useful in studies of crystal-structure symmetry, phase transitions and solid state problems in general.

1. Databases and programs

1.1. Databases

The databases form the core of the *Bilbao Crystallographic Server* (http://www.cryst.ehu.es) [1]. The information can be accessed using simple retrieval tools. The available databases are:

- The data on the 230 space groups as listed in the *International Tables for Crystallography* (IT), Vol. A [2]. Access tools are the programs **GENPOS** for the generators and general positions and **WYCKPOS** for the Wyckoff positions.
- The maximal subgroups of indices 2, 3 and 4 of the space groups contained in the forthcoming IT Vol.A1 [3]. Its access tool is the program **MAXSUB**.
- **k**-vector database containing figures of the Brillouin zones and classification tables of the **k**-vectors for the space groups [4]. The program **KVEC** is the access tool.
- A database **ICSDB** of incommensurate structures containing information about modulated structures and composites.

1.2. Crystallographic computing programs

The crystallographic computing programs form the second shell of the server. They are aimed at obtaining crystallographic information not directly contained in the databases.

The program **SUBGROUPGRAPH** [5] provides the group-subgroup relations between two space groups $\mathcal{G} > \mathcal{H}$ of index i. Input of the program are the IT space-group numbers of $\mathcal{G}$ and $\mathcal{H}$ and their index i. The output comprises the chains of maximal subgroups between $\mathcal{G}$ and $\mathcal{H}$, the determination of all different subgroups $\mathcal{H}_j$ of the type of $\mathcal{H}$ and index i and their distribution into conjugacy classes relative to $\mathcal{G}$.

The program **SUPERGROUPS** [6] calculates the different supergroups $\mathcal{G} > \mathcal{H}$ of a space group $\mathcal{H}$ of a given index i. The input of the program is the IT numbers of the groups

$\mathcal{G}$ and $\mathcal{H}$ and their index. The output of the program contains all supergroups $\mathcal{G}_j > \mathcal{H}$ of the type of $\mathcal{G}$ and index i with the transformations that relate the bases of $\mathcal{G}_j$ and $\mathcal{H}$ as well as the coset decomposition of $\mathcal{G}_j$ relative to $\mathcal{H}$.

The program **WYCKSPLIT** [7] obtains the relations between the atomic orbits of a pair $\mathcal{G} > \mathcal{H}$. The input consists of the space-group numbers of the group-subgroup pair and the transformation that relates them. The output contains the splitting of the Wyckoff positions of $\mathcal{G}$ into the corresponding Wyckoff positions of $\mathcal{H}$.

1.3. Representations

The third shell includes programs for the application of representation theory. The program **REPRES** constructs little-group and full-group irreducible representations (irreps) for a given space group and **k**-vector, **CORREL** deals with the correlations between the irreps of group-subgroup related space groups for a given **k**-vector. The program **POINT** lists character tables of crystallographic point groups, Kronecker multiplication tables of their irreps, and further useful symmetry information.

1.4. Solid-state applications

The next shell includes software packages for specific solid-state applications two of which are of considerable interest for phase-transition problems.

The program **PSEUDO** [8] performs a systematic search of pseudosymmetry. For a crystal structure S_o specified by its space group $\mathcal{H}$, its cell parameters and its coordinates of the atoms in the asymmetric unit, the program searches for a pseudosymmetry with respect to the minimal supergroups $\mathcal{G} > \mathcal{H}$, supplied by the program SUPERGROUPS. The pseudosymmetry search checks directly the compatibility of the additional symmetry operations of the supergroup $\mathcal{G}$ with the initial crystal structure S_o. The structure S_o is considered pseudosymmetric with respect to a supergroup $\mathcal{G} > \mathcal{H}$ if all atoms deviate from the ideal positions in the high-symmetry structure S with the space group $\mathcal{G}$ less than some previously determined tolerance value. The output of the program includes the pseudosymmetry supergroup $\mathcal{G}$, represented by its coset representatives relative to $\mathcal{H}$, the ideal atomic positions of S and the atomic displacements necessary to obtain S from the initial crystal structure S_o.

The software package **SYMMODES** (Symmetry Modes) performs a group-theoretical analysis of a structural phase transition characterized by a symmetry change $\mathcal{G} \rightarrow \mathcal{H}$, with $\mathcal{G} > \mathcal{H}$. The package combines modules of the programs **SUBGROUPGRAPH** and **WYCKSPLIT** with the program **SOPD** (Subgroup Order-Parameter Displacements) which is a dedicated segment of the package ISOTROPY‡. The main steps of the symmetry analysis provided by the program are:

(i) Given the space-groups of $\mathcal{G}$ and $\mathcal{H}$ and the index i, the program constructs the lattice of maximal subgroups relating $\mathcal{G}$ and $\mathcal{H}$. All possible subgroups $\mathcal{H}_j$ of the type of $\mathcal{H}$ and index i are listed and distributed into classes of conjugate subgroups with respect to $\mathcal{G}$. The particular group-subgroup graph of maximal subgroups for any group $\mathcal{H}_j < \mathcal{G}$ and the corresponding transformation matrices are obtained by specifying the subgroup $\mathcal{H}_j$ from the list of all possible subgroups of the type of $\mathcal{H}$.

(ii) For a given symmetry break $\mathcal{G} \rightarrow \mathcal{H}_j$ and a crystal structure specified by the Wyckoff positions of the occupied atomic orbits, the program calculates: (1) the number and the polarization vectors of the primary and secondary modes characterized by the irreps of $\mathcal{G}$

---

‡ The program ISOTROPY is available at www.physics.byu.edu/~stokesh/isotropy.html

and the corresponding isotropy subgroups from the $\mathcal{G} > \mathcal{H}_j$ graph; (2) the splitting of the Wyckoff positions during the symmetry break $\mathcal{G} \to \mathcal{H}_j$.

There are two more programs in the shell on solid-state applications: (i) the program **SAM** which calculates symmetry adapted modes on the Γ point and classifies them by their IR and Raman activity, and (ii) the program **NEUTRON** for computing phonon selection rules applicable in inelastic neutron scattering results.

2. Example: Phase-transition studies of Aurivillius compounds

The aim of the example is to show how the combination of different programs available on the server can be used for a symmetry study of displacive phase transitions. Due to space reasons, in the following, we have included an illustrative selection of our results on Aurivillius ferroelectric transitions (for more details the reader is referred to [9]).

There exists a large number of displacive ferroelectrics among the compounds of the Aurivillius family $Bi_2A_{n-1}B_nO_{3n+3}$, where A is a combination of cations such as Na^+, K^+, Ca^{2+}, Bi^{3+}, etc., and B of the type Fe^{3+}, Cr^{3+}, Ti^{4+}, Nb^{5+}, etc., with n=1,...,4. In general, the room-temperature phase of the compounds can be described as a slight distortion of an ideal $I4/mmm$ structure. It consists of $A_{n-1}B_nO_{3n+1}$ blocks of perovskite type regularly interchanged with Bi_2O_2 layers. The distortion that reduces the tetragonal symmetry to the polar orthorhombic symmetry can be decomposed into two parts: (i) a cell deformation to an orthorhombic $Fmmm$ symmetry, and (ii) a distortion of displacive type which reduces the centro-symmetric orthorhombic symmetry to a polar one.

Our study of the Aurivillius ferroelectric phase transitions includes the following main points: (i) determination of the prototype structures and the corresponding global distortions that relate the high-symmetry phase to the experimentally known polar phases; (ii) calculation of the primary and secondary symmetry modes for all occupied atomic orbits of the prototype phase; (iii) symmetry-mode analysis of the global distortion which results in the determination of the amplitudes (contributions) of the symmetry modes in the distortion characterizing the phase transition.

Step1: Prototype structure For the determination of the prototype structure we have applied the pseudosymmetry approach PSEUDO by assuming that the known low-symmetry structure is a slight distortion of a high-symmetry (prototype) one. Following the chains of minimal supergroups of the space groups of the known polar phases of the Aurivillius compounds, all compounds have been flagged as pseudosymmetric for Fmmm with relative displacements less than 1 Å. For example, the graph of minimal supergroups which relate the group $Pca2_1$ (symmetry group of the polar phase of Aurivillius compounds for n=1) to $Fmmm$ (the prototype phase) is given in Figure 1.

Step2: Symmetry modes Once the exact group-subgroup relation between the space groups of the polar and non-polar phases is known it is possible to determine the primary and secondary modes relevant for the studied phase transition by the program SYMMODES. In the case of Aurivillius compounds all symmetry modes for the occupied Wyckoff positions in the prototype structure have as isotropy groups some of the maximal subgroups of $Fmmm$, *i.e.* no mode has the space group of the polar phase as an isotropy group. In other words, there is no symmetry mode that can reduce the symmetry from $Fmmm$ to the space groups describing the polar phases. The results of SYMMODES include also the polarization vectors of the modes. For example, for each of the occupied orbits there exists a ferroelectric mode with isotropy group $Fmm2$ which is characterized by displacements along the polar axis.

Step3: Symmetry-mode analysis The total distortion characterizing the displacive phase transition is calculated from the comparison of the known polar structure and the calculated prototype. Using the orthogonality properties of the symmetry modes, determined in Step

2, it is straightforward to determine the contributions of each of the relevant modes in the total distortion. The example of the symmetry-mode analysis for Aurivillius compounds with n=1 (Figure 2) shows some typical features of the studied ferroelectric phase transition: (i) strong contribution of a ferroelectric mode specified by displacements of the A and B atoms of the perovskite group and of the atoms of the Bi-O layers. It is accompanied by octahedra rotations of the perovskite oxygen atoms along the polar and the large b axis. It is the combination of these modes (and not a single primary mode) which reduces the symmetry from the prototype phase to the experimentally observed polar phases. This situation is rather unusual in a continuous or quasi-continuous transition and points out to an atypical transition mechanism which requires further investigation (see [10] for other such cases).

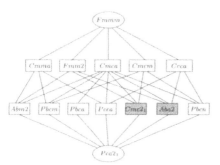

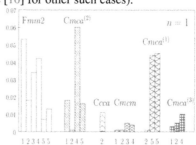

Figure 1. Group-subgroup graph relating the prototype group $Fmmm$ to $Pca2_1$, the symmetry group of the polar phase of Aurivillius compounds with n=1.

Figure 2.   Symmetry-mode amplitudes in the total distortion of the symmetry break $Fmmm > Pca2_1$ for Aurivillius compounds with n=1. The modes are specified by the isotropy groups and numbers which distinguish the different atoms: 1-Bi; 2-O (Bi_2O_2 layer); 3-B; 4 and 5-O (octahedra).

Acknowledgements: This work has been supported by MCYT-Spain (HA2000-020), and DAAD-Germany(314/AI-e-dr) within the program 'Acciones Integradas Hispano-Alemanas'; DGESIC (Project No PB98-0244) and UPV (Project No 063.310-G19/98).

References

[1] Kroumova, E., M.I. Aroyo, J.M. Perez-Mato, A. Kirov, C. Capillas, S. Ivantchev & H. Wondratschek (2002). "*Bilbao Crystallographic Server*: useful databases and tools for phase-transition studies." *Phase Transitions*, (in print).

[2] *International Tables for Crystallography (2002). Space Group Symmetry*, Vol.A. Ed. T. Hahn, 5th ed. Dordrecht, Kluwer Academic Publishers.

[3] *International Tables for Crystallography (to appear). Symmetry Realtions between Space Groups*, Vol. A1. Eds. H. Wondratschek & U. Müller, Dordrecht, Kluwer Academic Publishers.

[4] Aroyo, M. I. & H. Wondratschek (1995). Crystallographic viewpoints in the classification of space-group representations, *Z. Kristallogr.* **210**, 243.

[5] Ivantchev, S., E. Kroumova, G. Madariaga, J. M. Perez-Mato & M. I. Aroyo (2000). SUBGROUPGRAPH: a computer program for analysis of group-subgroup relations between space groups, *J. Appl. Cryst.* **33**, 1190.

[6] Ivantchev, S., E. Kroumova, M. I. Aroyo, J. Igartua & J. M. Perez-Mato (2002). SUPERGROUPS: a computer programs for determination of supergroups of space groups, *J. Appl. Cryst.* **35**, 511.

[7] Kroumova, E., M. I. Aroyo & J. M. Perez-Mato (1998). WYCKSPLIT: a computer program for determination of the relations of Wyckoff positions for a group-subgroup pair, *J. Appl. Cryst.* **31**, 646.

[8] Kroumova, E., M. I. Aroyo, J. M. Perez-Mato. S. Ivantchev, J. M. Igartua & H. Wondratschek (2001). PSEUDO – a program for a pseudosymmetry search, *J. Appl. Cryst.*, **34**, 783.

[9] Kroumova, E. (2002). Desarrollo de métodos y algoritmos para el análisis cristalográfico y la predicción de transiciones de fase estructurales. PhD Thesis, Universidad del Pais Vasco, Bilbao.

[10] Hatch, D. M. & H. T. Stokes (2002). Complete listing of order parameters for a crystalline phase transition: A solution to the generalized inverse Landau problem, *Phys. Rev. B*, **65**, 0141131.

Inst. Phys. Conf. Ser. No 173: Section 5
Paper presented at 24th Int. Coll. Group Theoretical Methods in Physics, Paris, France, 15–20 July 2002
©2003 IOP Publishing Ltd

Quantum algebraic symmetries in atomic nuclei and metal clusters

Dennis Bonatsos[†], B. A. Kotsos[#], P. P. Raychev[*], P. A. Terziev[*]

[†] Institute of Nuclear Physics, N.C.S.R. Demokritos, GR-15310 Aghia Paraskevi, Attiki, Greece
[#] Department of Electronics, Technological Education Institute, GR-35100 Lamia, Greece
[*] Institute for Nuclear Research and Nuclear Energy, Bulgarian Academy of Sciences, 72 Tzarigrad Road, BG-1784 Sofia, Bulgaria

Abstract. Using irreducible tensor operators under $su_q(2)$ a rotationally invariant Hamiltonian which provides a good description of nuclear rotational spectra is constructed and its relation to existing nuclear models is considered. Using the same techniques a 3-dimensional q-deformed harmonic oscillator with $u_q(3) \supset so_q(3)$ symmetry is constructed, compared to the modified oscillator of Nilsson, and used for the successful description of magic numbers and supershells in metal clusters.

1. Atomic nuclei

An irreducible tensor operator (ITO) of rank k is the set of $2k + 1$ operators $T_{k,\kappa}^{(q)}$ ($\kappa = k$, $k - 1$, $k - 2$, ..., $-k$), which satisfy with the generators of the $su_q(2)$ algebra (L_0, L_+, L_-) the commutation relations [1]

$$[L_0, T_{k,\kappa}^{(q)}] = \kappa T_{k,\kappa}^{(q)}, \qquad [L_\pm, T_{k,\kappa}^{(q)}]_{q^\kappa} = \sqrt{[k \mp \kappa][k \pm \kappa + 1]} T_{k,\kappa\pm1}^{(q)} q^{-L_0}, \quad (1)$$

where q-numbers and q-operators are defined as $[x] = (q^x - q^{-x})/(q - q^{-1})$, while q-commutators are defined by $[A, B]_{q^\alpha} = AB - q^\alpha BA$. We can construct an irreducible tensor operator of rank 1 using as building blocks the generators of $su_q(2)$. It turns out that this ITO will consist of the operators

$$J_{\pm 1} = \mp \frac{1}{\sqrt{[2]}} q^{-L_0} L_\pm, \qquad J_0 = \frac{1}{[2]}(qL_+L_- - q^{-1}L_-L_+). \quad (2)$$

One can now try to build out of these operators the scalar square of the angular momentum operator. For this purpose one needs the definition of the tensor product of two irreducible tensor operators, which has the form [1]-[4]

$$[A_{j_1}^{(q)} \otimes B_{j_2}^{(q)}]_{j,m}^{(1/q)} = \sum_{m_1,m_2} \langle j_1 m_1 j_2 m_2 | jm \rangle_{1/q} A_{j_1,m_1}^{(q)} B_{j_2,m_2}^{(q)}, \quad (3)$$

in which deformed Clebsch–Gordan coefficients [1] appear. It turns out that the relevant tensor products take the form [5]

$$[J \otimes J]_{1,m}^{(1/q)} = -\sqrt{\frac{[2]}{[4]}} \left\{ q^{-2L_0} + (q - q^{-1})J_0 \right\} J_m = -\sqrt{\frac{[2]}{[4]}} Z J_m, \quad (4)$$

where $m = 0, \pm 1$ and

$$Z = q^{-2L_0} + (q - q^{-1})J_0 = 1 + \frac{(q - q^{-1})^2}{[2]} C_2^{(q)}, \quad (5)$$

with $C_2^{(q)}$ being the second order Casimir operator of $su_q(2)$. Therefore the operators

$$J'_m = \frac{J_m}{Z}, \qquad m = +1, 0, -1 \tag{6}$$

also form an ITO, while Eq. (4) can be written in the form

$$\left[\frac{J}{Z} \otimes \frac{J'}{Z}\right]^{(1/q)}_{1,m} = -\sqrt{\frac{[2]}{[4]}} \frac{J_m}{Z} \Rightarrow [J' \otimes J']^{(1/q)}_{1,m} = -\sqrt{\frac{[2]}{[4]}} J'_m. \tag{7}$$

The scalar product of two irreducible tensor operators is defined as [1, 3]

$$(A_j^{(q)} \cdot B_j^{(q)})^{(1/q)} = (-1)^{-j}\sqrt{[2j+1]}[A_j^{(q)} \times B_j^{(q)}]^{(1/q)}_{0,0}. \tag{8}$$

Substituting the irreducible tensor operators J'_m in this definition we obtain [5]

$$(J' \cdot J')^{(1/q)} = \frac{1 - Z^{-2}}{(q - q^{-1})^2}. \tag{9}$$

We have therefore determined the scalar square of the angular momentum operator. We can assume at this point that this quantity can be used (up to an overall constant) as the Hamiltonian for the description of rotational spectra, defining

$$H = A\frac{1 - Z^{-2}}{(q - q^{-1})^2}, \tag{10}$$

where A is a constant.

The eigenvalues of the Hamiltonian in the $su_q(2)$ basis $|\ell, m\rangle$ are

$$E = \langle H \rangle = A\frac{1}{4\sinh^2 \tau}\left(1 - \frac{\cosh^2 \tau}{\cosh^2((2\ell+1)\tau)}\right), \tag{11}$$

where $q = e^\tau$ is a real number, as in [1]. One can prove that this Hamiltonian is rotationally invariant, i.e. that it commutes with the generators of the usual su(2). The detailed proof shall be given elsewhere [5].

With the help of Taylor expansions and the Bernoulli numbers B_n, and using the functions

$$f_n(\tau) = (-1)^{n+1}(2\tau)^n(n+1)!$$
$$\sum_{k=0}^{\infty} \frac{2^{2n+2k+4}(2^{2n+2k+4}-1)B_{2n+2k+4}}{(2n+2k+2)!(2n+2k+4)}\binom{n+k+1}{n+1}\tau^{2k}, \tag{12}$$

$$f_0(\tau) = -\sum_{k=0}^{\infty}\frac{2^{2k+4}(2^{2k+4}-1)B_{2k+4}}{(2k+2)!(2k+4)}(k+1)\tau^{2k} = \frac{\sinh\tau}{\tau\cosh^3\tau}, \tag{13}$$

the spectrum of Eq. (11) can be put into the exact form [5]

$$\frac{E}{A} = \left(\frac{\tau^2 \cosh^2 \tau}{\sinh^2 \tau}\right)\sum_{n=0}^{\infty}\frac{(-1)^n(2\tau)^n}{(n+1)!}f_n(\tau)(\ell(\ell+1))^{n+1}. \tag{14}$$

It is clear that Eq. (14) is an expansion in terms of $\ell(\ell+1)$, as expected because of the rotational invariance of the Hamiltonian.

In the limit of $|\tau| \ll 1$ one is entitled to keep in Eq. (12) only the term with $k = 0$. Using Taylor expansions once more one can put the spectrum into the approximate compact form [5]

$$E \approx \frac{A}{(2\tau)^2}\tanh^2(2\tau\sqrt{\ell(\ell+1)}). \tag{15}$$

The extended form of the Taylor expansion of E is then

$$E \approx A \left(\ell(\ell+1) - \frac{2}{3}(2\tau)^2(\ell(\ell+1))^2 + \frac{17}{45}(2\tau)^4(\ell(\ell+1))^3 - \cdots \right). \quad (16)$$

Eq. (15) will be referred to as the "hyperbolic tangent formula".

It should be remembered at this point that the second order Casimir operator of $su_q(2)$ has also been used for the description of rotational nuclear spectra [6], with $q = e^{i\tau}$, $\tau \in \mathbf{R}$ (and $q^n \neq 1$, $n \in \mathbf{N}$). In this case the spectrum has the form

$$E_\ell^{(\tau)} = A \frac{\sin(\ell\tau)\sin((\ell+1)\tau)}{\sin^2(\tau)}, \qquad q = e^{i\tau}. \quad (17)$$

Using the spherical Bessel functions of the first kind $j_n(x)$ one can put this spectrum into the exact form [5]

$$E_\ell^{(\tau)} = \frac{A}{j_0^2(\tau)} \sum_{n=0}^{\infty} \frac{(-1)^n(2\tau)^n}{(n+1)!} j_n(\tau) \{\ell(\ell+1)\}^{n+1}, \quad (18)$$

which is an expansion in terms of $\ell(\ell+1)$, as expected from the rotational invariance of the Hamiltonian, the proof of which is given elsewhere [5]. In the limit $\tau \ll 1$ the following approximate compact form of the spectrum can be obtained [5]

$$E_\ell^{(\tau)} \approx A \frac{\sin^2(\tau\sqrt{\ell(\ell+1)})}{\tau^2}, \quad (19)$$

which shall be referred to as the "sinus formula". This result is similar to the expression proposed for the unified description of nuclear rotational spectra by G. Amal'sky [7]

$$E_\ell = \varepsilon_0 \sin^2 \left(\frac{\pi}{N} \sqrt{\ell(\ell+1)} \right), \quad (20)$$

where ε_0 is a phenomenological constant ($\varepsilon_0 \approx 6.664$ MeV) which remains the same for all nuclei, while N is a free parameter varying from one nucleus to the other. The equivelence of both the "hyperbolic tangent formula" and the "sinus formula" to the Harris formalism [8], which is known [9] to be equivalent to the Variable Moment of Inertia (VMI) model [10], can also be proved [5].

Numerical tests of the above formulae against the experimental spectra of the Th isotopes lead to the following conclusions [5]:

1) The "sinus formula" is a very accurate approximation of Eq. (17) (the "original $su_q(2)$ formula"). Furthermore, the "hyperbolic tangent formula" is a very accurate approximation of Eq. (11) (the "$su_q(2)$ ITO formula").

2) The "hyperbolic tangent formula" provides better fits of rotational spectra than the "sinus formula". However, both expressions are "too rigid" for describing transitional or vibrational spectra.

2. Metal clusters

Using similar techniques [2, 3] one can construct a 3-dimensional q-deformed harmonic oscillator (3-dim q-HO), the space of which consists of the completely symmetric irreducible representations of the quantum algebra $u_q(3)$. In this space a deformed angular momentum algebra, $so_q(3)$, can be defined [3]. The Hamiltonian of the 3-dim q-HO is defined so that it satisfies the following requirements:

a) It is an $so_q(3)$ scalar, i.e. the energy is simultaneously measurable with the q-deformed angular momentum related to the algebra $so_q(3)$ and its z-projection.

b) It conserves the number of bosons, in terms of which the quantum algebras $u_q(3)$ and $so_q(3)$ are realized.

c) In the limit $q \to 1$ it is in agreement with the Hamiltonian of the usual 3-dimensional harmonic oscillator.

It has been proved [3] that the Hamiltonian of the 3-dim q-HO satisfying the above requirements takes the form

$$H_q = \hbar\omega_0 \left\{ [\mathcal{N}]q^{\mathcal{N}+1} - \frac{q(q - q^{-1})}{[2]} C_q^{(2)} \right\},$$ (21)

where $\mathcal{N}$ is the number operator and $C_q^{(2)}$ is the second order Casimir operator of $so_q(3)$.

The energy eigenvalues of the 3-dim q-HO are then [3]

$$E_q(n, l) = \hbar\omega_0 \left\{ [n]q^{n+1} - \frac{q(q - q^{-1})}{[2]} [l][l + 1] \right\},$$ (22)

where n is the number of vibrational quanta and l is the eigenvalue of the angular momentum, obtaining the values $l = n, n - 2, \ldots, 0$ or 1.

For small values of the deformation parameter τ (where $q = e^\tau$) one can expand Eq. (22) in powers of τ obtaining [3]

$$E_q(n, l) = \hbar\omega_0 n - \hbar\omega_0 \tau \left(l(l + 1) - n(n + 1) \right)$$

$$-\hbar\omega_0 \tau^2 \left(l(l + 1) - \frac{1}{3} n(n + 1)(2n + 1) \right) + \mathcal{O}(\tau^3).$$ (23)

The first two terms resemble the eigenvalues of Nilsson's modified oscillator

$$E_{nl} = \hbar\omega \left(n + \frac{3}{2} \right) - \hbar\omega\mu' \left(l(l + 1) - \frac{1}{2} n(n + 3) \right).$$ (24)

It has been found that the 3-dim q-HO provides a very good description of the magic numbers appearing in alkali metal clusters [11], as well as of the supershells [12] appearing in these systems. The advantages of the 3-dim q-HO over Nilsson's modified oscillator in the description of supershells have also been considered [12].

References

[1] Smirnov Yu F, Tolstoy V N and Kharitonov Yu I 1991 Yad. Fiz. 53 959; 1993 Yad. Fiz. 56 223

[2] Raychev P P, Roussev R P, Terziev P A, Bonatsos D and Lo Iudice N 1996 J. Phys. A: Math. Gen. 29 6939

[3] Raychev P P, Roussev R P, Lo Iudice N and Terziev P A 1998 J. Phys. G: Nucl. Part. 24 1931

[4] Smirnov Yu F, Tolstoy V N and Kharitonov Yu I 1991 Yad. Fiz. 53 1746; 1992 Yad. Fiz. 55 2863

[5] Bonatsos D, Kotsos B A, Raychev P P and Terziev P A 2002 to be published

[6] Bonatsos D and Daskaloyannis C 1999 Prog. Part. Nucl. Phys. 43 537

[7] Amal'sky G M 1993 Yad. Fiz. 56 70

[8] Harris S M 1965 Phys. Rev. 138 B509

[9] Klein A, Dreizler R M and Das T K 1970 Phys. Lett. B 31 333

[10] Mariscotti M A J, Scharff–Goldhaber G and Buck B 1969 Phys. Rev. 178 1864

[11] Bonatsos D, Karoussos N, Lenis D, Raychev P P, Roussev R P and Terziev P A 2000 Phys. Rev. A 62 013203

[12] Bonatsos D, Lenis D, Raychev P P and Terziev P A 2002 Phys. Rev. A 65 033203

Inst. Phys. Conf. Ser. No 173: Section 5
Paper presented at 24th Int. Coll. Group Theoretical Methods in Physics, Paris, France, 15–20 July 2002
©2003 IOP Publishing Ltd

Broken symmetry and symmetry classification of magnetic ordering in doped crystals

B R Gadjiev

Frank Laboratory of Neutron Physics, JINR, 141980 Dubna, Moscow reg., Russia
Institute of Physics, Academy Science of Azerbaijan, 37000 Baku, Azerbaijan Republic
e-mail: gadjiev@nf.jinr.ru

Abstract. The problem of the spontaneous breaking of the symmetry in the crystals with variable composition is discussed.

1. Introduction

The paper is devoted to the problem of the spontaneous broken symmetry in doped compounds.

Neutron diffraction research of crystals with variable composition has revealed the basic feature of the phase diagram of the doped compounds [1]. Thus with variation of substitutional atom concentration and external fields in doped structures the sets of modification of the structure and magnetic ordering are discovered [1,2,3,4,5].

In the doped structures with an incommensurate phase with variations of substitutional atom concentration, the space group (SG) of symmetry of the structure being invariant, the variation of the wave vector of modulation of the structure is observed. As a rule, the wave vector of magnetic structure varies along a certain direction in a Brillouin zone corresponding to the SG symmetry of the structure. In this context we shall further discuss the problems of magnetic ordering of the $U(Pd_xFe_{1-x})_2Ge_2$ type compounds [2].

We shall consider manganites of the $(Ln_yLn^*_{1-y})_{1-x}A_xMnO_3$ type (where Ln is a rare - earth and A is a alkaline - earth or a alkali metal), with a structure perovskite as an example of the structure without incommensurate phase. In these compounds depending on the concentration of substitutional atom the paramagnetic phase has either D^6_{3d}, or D^{16}_{2h} SG group of symmetry [3,4]. At low temperatures in perovskite like compounds one can observe ferromagnetic phase that undergoes a concentration phase transition to an antiferromagnetic phase with the increase of substitutional atom concentration.

2. The problem

Knowledge of symmetry of the order parameter is necessary for the solution of the problem of spontaneous breaking of symmetry. Knowledge of SG symmetry of paramagnetic phase and wave vector of modulation of superstructure allows to define representation of the group of symmetry of the structure, according to which the order parameter will be transformed [5].

The thermodynamic potential functional (TPF) is constructed by the invariants of representation D of the SG of high symmetry phase G_0, which determines the lowering of symmetry $G_0 \Rightarrow G_1$, where G_1 is a subgroup of group G_0. For definition of the order parameter symmetry in doped structures it is necessary to notice, that, strictly speaking, in contrast to the ideal crystals distributions of atoms have no SG of symmetry. Namely, the substitution of the atoms in the structure breaks the periodicity of the ideal crystals.

Let us discuss the influence of doping on the magnetic orderings in periodic structures, assuming that the doping induces variation with accuracy up to that the atomic structure does not change the SG symmetry of the structure.

While discussing theoretically the spontaneous symmetry breaking in doped crystals we should have in mind that the approximation preservation of the SG of the structure requires at least an account of the interaction of major order parameter η with a secondary one, namely with deformation u. Besides, the account of the energy, contributed by the substitution of atoms of the structure, is necessary.

3. Field effects on inhomogeneous superstructures

Let us consider the influence of a field on inhomogeneous superstructures. We could expect, that under the influence of a field the wave vector of superstructure would deviate from the initial symmetric direction.

Let us define the character of the wave vector of the superstructure variation under the influence of a weak field in the case of phase transitions when the Lifshitz invariant is taken in account in the Landau thermodynamic potential expansion :

$$F = \frac{\alpha}{2}\eta^2 + \frac{\beta}{4}\eta^4 + \sigma k\eta^2 + \frac{\kappa}{2}k^2\eta^2 + \frac{c}{2}u_{zz}^2 + \gamma k\eta^2 u_{zz}. \tag{1}$$

Here $\alpha = \alpha_0 + ma$, $\beta = \beta_0 + mb$, $\sigma = \sigma_0 + mc$, $\kappa = \kappa_0 + me$ [5]. From the equilibrium condition we obtain the values of the wave vector k as a linear function of the spontaneous deformation and substitutional atom concentration m. Besides, it is possible to show that the critical temperature T_c linearly depends on the components of the spontaneous deformation $T_c = (\sigma^2/\kappa + \delta(2 - \gamma/\kappa)u_{zz})/\alpha_0$. Therefore it is possible to assert that the wave vector of the superstructure in the doped structures is a linear function of a spontaneous deformation and of a substitutional atom concentration.

In the case where the Lifshitz invariant is absent the density of TPF of the system is represented by the expression.

$$f = \frac{\alpha}{2}\eta^2 + \frac{\beta}{4}\eta^4 + \frac{\delta}{2}\left(\frac{d\eta}{dx}\right)^2 + \frac{\lambda}{2}\left(\frac{d^2\eta}{dx^2}\right)^2 + \frac{\chi}{2}\eta^2\left(\frac{d\eta}{dx}\right)^2. \tag{2}$$

Here $\delta = \delta_0 + md$, $\lambda = \lambda_0 + ml$, $\chi = \chi_0 + ms$ [5]. The account of the influence of the deformation results in variation of the thermodynamic potential parameters $\beta_0 = \beta - 2f_0^2/c$, $\delta_0 = \delta + 2e\kappa/c$ and $\chi_0 = \chi - \delta_0 f_0/c$, and also in the shift of the critical temperature $T_c^* = T_c - am/\alpha_0 - 2ef_0/(\alpha_0 c)$ [5].

The variation of the thermodynamic potential parameters could break the conditions, limiting realization of phases characterized by the solutions $\eta = \rho\operatorname{sn}(px)$, $\eta = \rho\operatorname{dn}(px)$ and $\eta = \rho\operatorname{cn}(px)$ [5]. Therefore there exist critical value deformations or concentration of doped atoms that could result in threshold variation of a wave vector modulation of superstructure.

4. Magnetic orderings in compounds of the $U(Pd_{1-x}Fe_x)_2Ge_2$ type

The phase of symmetry D_{2h}^{17} corresponding to real tetragonal structures could arises as a result of the phase transition from the latent phase with symmetry SG O_h^9. The transition from the latent phase to the corresponding superstructure is described by an order parameter that is transformed according to the irreducible representation (IR) of the group O_h^9, belonging to the point $\vec{k}_8 = (0, 0, \frac{2\pi}{na})$ [5,6].

The phase of the symmetry D_{2h}^{17} arise for odd values $n \geq 3$, when only two rays of $\vec{k}_8$ are active. The six-dimensional IR representation $\tau_1(\vec{k}_8)$ is passive and consequently the phase transition $O_h^9 \Rightarrow D_{2h}^{17}$ in the general case is carried out through an incommensurate phase.

Let us consider the structure with a SG D_{2h}^{17} in the paramagnetic phase. Let us consider the two-dimensional IR $D^{*\vec{k}_{10}}$ of the SG D_{2h}^{17}, corresponding to the vector $\vec{k}_{10} = (0, 0, \frac{2\mu\pi}{\tau_z})$ [6].

In accordance with the condition of invariance the TPF is expressed as:

$$f(x) = \frac{\alpha}{2}\rho^2 + \frac{\beta}{4}\rho^4 + \gamma\rho^n \cos n\varphi - \delta\rho^2(\frac{d\varphi}{dx}) + \frac{k}{2}\rho^2(\frac{d\varphi}{dx})^2, \tag{3}$$

where the parameters of the thermodynamic potential are linear functions of concentration of doped atoms. With the increase of a doping concentration the temperature of the phase transition from the paramagnetic phase to the modulated one is lowered $T_i(m) = T_i(m = 0) - am$.

Spatial dependence of the phase of the order parameter is defined by the Jacobi elliptic function $\varphi(x) = \frac{2}{n}\text{am}(px, \kappa)$, where $\kappa \in [0, 1]$. The temperature interval in which the incommensurate phase exists depends on the doping concentration according to the equation

$$T_i(m) - T_c = (\frac{k}{\gamma})^{\frac{2}{n-2}}(\frac{\beta}{\alpha_0})^{1/2}(\frac{\pi\delta}{4k})^{\frac{4}{n-2}} - am. \tag{4}$$

It is clear that as the doping concentration increases, the temperature region in which the incommensurate phase exists becomes narrower.

In the commensurate phase the expression for the magnetic moment in the n-th unit cell of the crystal in terms of the magnetic moment in the first cell is $\vec{M}_{\vec{n}} = \vec{M}_1 e^{ik\vec{n}} + \vec{M}_1 e^{-ik\vec{n}}$, where $\vec{n}$ is the vector of the lattice, $\vec{k}$ is the wave vector of the superstructure. It follows from this expression that in the case $\vec{M}_1 = \frac{M}{2}(\vec{m}_1 + ip\vec{m}_2)$, $\vec{m}_1\vec{m}_2 = 0$ and $\vec{m}_1^2 = \vec{m}_2^2$ with the ellipticity parameter $p \neq 0$ the magnetic moment $\vec{M}_{\vec{n}} = M[\vec{m}_1 \cos k\vec{n} - p\vec{m}_2 \sin k\vec{n}]$, describes the elliptical helix in the space. If we set $p = 0$ we have a structure of the spin wave type $\vec{M}_{\vec{n}} = M\vec{m}_1 \cos k\vec{n}$. For $\vec{m}_1 \parallel \vec{k}$ the last expression describes the LSW structure and for $\vec{m}_1 \perp \vec{k}$ it describes the TSW structure.

5. Magnetic orderings in compounds of the $(La_{0.25}Pr_{0.75})_{0.7}Ca_{0.3}MnO_3$ type

The phase of symmetry O_h^1 corresponding to ABO_3 type crystals could arise as a result of phase transitions from a latent phase with the symmetry SG O_h^9. The phase transition $O_h^9 \Rightarrow O_h^1$ is described by the order parameter that is transformed according to the IR of the group O_n^9 belonging to the point $\vec{k}_{12}$ [6]. In the framework of the Landau theory the observed structures of perovskites ABO_3 could be obtained in the result of the phase transition described by the reducible representation (RR) $\tau_5(\vec{k}_{11}) \oplus \tau_8(\vec{k}_{13})$ of the SG O_h^1 [6]. Thus the symmetry analysis shows that the phase with the group of symmetry D_{2h}^{16} corresponds to the solution of the type $(\eta_1, 0, 0, 0, \zeta_2, \zeta_3)$, and the phase of symmetry D_{3d}^6 corresponds to the solution of the type $(0, 0, 0, \zeta_1, \zeta_2, \zeta_3)$, where η and ζ are order parameters, which is transformed according to the RR $\tau_5(\vec{k}_{11}) \oplus \tau_8(\vec{k}_{13})$ of the group O_h^1.

The structures of perovskites $Pr_{0.7}Sr_{0.3-x}{}_xMnO_3$ and $Pr_{0.7-x}{}_xSr_{0.3}MnO_3$ under observation are the function of the substitutional atom concentration and are either D_{3d}^6, or D_{2h}^{16}. At low temperatures these compounds have a ferromagnetic ordering phase that with the increase of temperature is transformed into a paramagnetic phase. The integer basis of the invariants of the RR $A_{1g} \oplus A_{2g}$ of D_{3d}^6 and $A_g \oplus B_{1g}$ of D_{2h}^{16} coincide. Therefore, in both cases the symmetry classification of the magnetic ordering in the doped structures $Pr_{0.7}Sr_{0.3-x}{}_xMnO_3$ and $Pr_{0.7-x}{}_xSrMnO_3$ leads to a TPF of the form

$$g(x) = \frac{\alpha_2}{2}\eta^2 + \frac{\beta_2}{4}\eta^4 + \frac{a}{2}u^2 + \frac{b}{4}u^4 + \gamma\eta^2 u. \tag{5}$$

The equation of state corresponding to (5) has a solution (I) $u = 0, \eta = 0$, (II) $u^2 = -a/b, \eta = 0$, (III) $u \neq 0, \eta \neq 0$. In phases (I) and (II) the average values of magnetization are equal to zero. In phase (III) it is not equal to zero and consequently the phase is ferromagnetic.

The structures in more complicated doped compounds $(La_{0.25}Pr_{0.75})_{0.7}Ca_{0.3}MnO_3$ with on isotope O^{18} and isotope O^{16} have a SG symmetry $Pnma$. For the symmetry classification of the magnetic ordering in the doped structure $(La_{0.25}Pr_{0.75})_{0.7}Ca_{0.3}MnO_3$ it is necessary to consider the RR $D^{\vec{k}_{20}} \oplus D^{\vec{k}_{22}} \oplus D^{\vec{k}_{19}} \oplus A_g$ [6], which according to symmetry arguments leads to the model of the TPF as

$$f(x) = g(x) + \frac{\alpha}{2}(\varphi_1^2 + \varphi_2^2 + \psi^2) + \frac{\beta}{4}(\varphi_1^2 + \varphi_2^2)^2 + \frac{\upsilon}{2}\varphi_1^2\varphi_2^2 +$$

$$+ \frac{g_3}{2}\psi^2(\varphi_1^2 + \varphi_2^2) + \frac{\beta_1}{4}\psi^4 + \gamma\eta^2 u + \zeta^2(\varphi_1^2 + \varphi_2^2)u + \xi\psi^2 u. \tag{6}$$

The equation of state corresponding to (6) has 19 solutions. Thus for the initial paramagnetic phase we obtain (I) $\varphi_1 = \varphi_2 = \psi = \eta = u = 0$. The phase corresponding to the solution, (II) $\varphi_1 = \varphi_2 = \psi = \eta = 0, u \neq 0$ is also paramagnetic.

The solution of the equations of state corresponding to the antiferromagnetically-ordering phase with the wave vectors of modulation $\vec{k}_{20}$ and $\vec{k}_{22}$ [6] have the form (III) $\varphi_1 \neq 0, \varphi_2 \neq 0, \psi \neq 0, u \neq 0, \eta = 0$. The solutions of equations of state corresponding to the ferromagnetically ordering phase have the form (IV) $\varphi_1 = \varphi_2 = \psi = 0, u \neq 0, \eta \neq 0$. It is easy to show that the values of displacement in phases (III) and (IV) are not equal. In $(La_{0.25}Pr_{0.75})_{0.7}Ca_{0.3}MnO_3$ with an isotope O^{16} with the lowering of temperature the change of phases is observed (I)$\Rightarrow$(III), whereas in the same compounds with complete substitution $O^{16} \Rightarrow O^{18}$ with lowering of temperature the change of phase (I)$\Rightarrow$(III)$\Rightarrow$(IV) is observed.

References

[1] Coey J M D, Viret M, von Molner S 1999 Advances in Physics 48 12 167
[2] Duh H M, Lyubutin I S, Jiang I M, Hwang G H, Lain K D J. 1996 Magn. and Magn. Mater. 153 86
[3] Boujelben W, Cheikh-Rouhou A, Joubert J C 2001 Eur.Phys. J. B 24 419-423
[4] Balagurov A M, Pomyakushin V Yu, Sheptyakov D, Aksenov V L 1999 Phys. Rev. 60 1 383-387
[5] Gadjiev B R 2000 Low Temperature Physics 26 12 874-879
[6] Kovalev D V 1986 Irreducible and Induced Representation and Corepresentation of Fedorov Groups [in Russian] (Moscow, Nauka)

Inst. Phys. Conf. Ser. No 173: Section 5
Paper presented at 24th Int. Coll. Group Theoretical Methods in Physics, Paris, France, 15–20 July 2002
©2003 IOP Publishing Ltd

Duality and conformal twisted boundaries in the Ising model

Uwe Grimm

Applied Mathematics Department, Faculty of Mathematics and Computing,
The Open University, Walton Hall, Milton Keynes MK7 6AA, UK

E-mail: u.g.grimm@open.ac.uk

Abstract. There has been recent interest in conformal twisted boundary conditions and their realisations in solvable lattice models. For the Ising and Potts quantum chains, these amount to boundary terms that are related to duality, which is a proper symmetry of the model at criticality. Thus, at criticality, the duality-twisted Ising model is translationally invariant, similar to the more familiar cases of periodic and antiperiodic boundary conditions. The complete finite-size spectrum of the Ising quantum chain with this peculiar boundary condition is derived.

Dedicated to the memory of Sonia Stanciu

1. Introduction

Quantum spin chains are one-dimensional models of interacting quantum systems. They have been used to model magnetic properties of materials, in particular when strongly anisotropic behaviour suggests a one-dimensional modelling. They also arise as limits of two-dimensional lattice models of classical statistical mechanics, in an anisotropic limit where the lattice spacing in one space direction vanishes. In particular, the Ising quantum chain discussed below is related to the classical Ising model in this way; and the parameter in the Hamiltonian corresponds to the temperature variable of the two-dimensional classical model.

At criticality, quantum spin chains as well as two-dimensional lattice models possess scaling limits that correspond to $(1+1)$-dimensional conformal field theories. The critical exponents describing the non-analytic behaviour of thermodynamic quantities are given by conformal dimensions of certain conformal operators. For translationally invariant quantum chains, the scaling limit corresponds to a conformal field theory on the torus, and the partition function is a quadratic expression in terms of Virasoro characters, distinguishing left and right moving excitations. For free or fixed boundary conditions, we have a conformal field theory on the half plane, with a partition function which is linear in Virasoro characters.

If the spin chain possesses global symmetries, we can define toroidal boundary conditions, i.e., specific twists at the boundary that do not destroy translational invariance. However, this does not seem to yield all possible boundary conditions one might expect from the conformal field theory. Recently, such "conformal twisted boundary conditions" have attracted growing attention [1, 2], and were realised in solvable lattice models [3]. Here, we consider the simplest possible example of such an exotic boundary condition. At least in this case, it is once more related to a symmetry of the model, which turns out to be duality [4]. Furthermore, the complete spectrum of the Hamiltonian is obtained exactly [5], even for finite chains of length N. The scaling limit partition function is computed, verifying the result expected from a mapping to the XXZ Heisenberg quantum spin chain which originally led to the discovery of these duality twisted boundary conditions in the Ising model [4].

Before we come to the particular example of the Ising quantum chain, we first discuss the construction of toroidal boundary conditions in more generality.

2. Quantum Chains and Translational Invariance

For simplicity, we consider quantum spin chains with nearest neighbour couplings only. In this case, the Hamiltonian for a system of N spins has the form

$$H = \sum_{j=1}^{N} H_{j,j+1}, \qquad H_{j,j+1} = \sum_{a,b} \varepsilon_{a,b} \sigma_j^a \sigma_{j+1}^b + \sum_a \delta_a \sigma_j^a, \tag{1}$$

where $\varepsilon_{a,b}$ and δ_a are arbitrary constants which are independent of j. The Hamiltonian is expressed in terms of local spin operators σ_j^a acting on a tensor product space $\mathscr{V}$,

$$\sigma_j^a = \mathbf{1}_V^{\otimes(j-1)} \otimes \sigma^a \otimes \mathbf{1}_V^{\otimes(N-j)}, \qquad \mathscr{V} = V^{\otimes N} = \underbrace{V \otimes V \otimes \ldots \otimes V}_{N \text{ factors}}, \tag{2}$$

where $\mathbf{1}_V$ denotes the identity operator on the vector space V. Let us assume periodic boundary conditions for the moment, i.e., $H_{N,N+1} \equiv H_{N,1}$, so the last spin couples to the first in the same way as the neighbouring spins along the chain. We define a unitary translation operator T by its action on the local spin operators

$$T \sigma_j^a T^{-1} = \sigma_{j+1}^a, \ 1 \le j \le N-1, \qquad T \sigma_N^a T^{-1} = \sigma_1^a, \tag{3}$$

which obviously commutes with the Hamiltonian H, so $THT^{-1} = H$. As $T^N = \mathbf{1}$, the identity operator on $\mathscr{V}$, the eigenvalues of T are of the form $\exp(2\pi i k/N)$, with $k = 0, 1, \ldots, N-1$, and define the lattice momenta of the one-dimensional chain.

Consider now the case where the Hamiltonian has a global symmetry, i.e., it commutes with an operator $Q = g \otimes g \otimes \ldots \otimes g = g_1 g_2 \ldots g_N$, where g belongs to a representation of some group. We can then define a modified Hamilton operator $\tilde{H}$ by setting

$$\tilde{H}_{j,j+1} = H_{j,j+1}, \ 1 \le j \le N-1, \qquad \tilde{H}_{N,N+1} \equiv \tilde{H}_{N,1} = g_1 H_{N,1} g_1^{-1}, \tag{4}$$

so $\tilde{H}$ differs from H only in the coupling term at the boundary, which is twisted by the local transformation g. On first view, this may appear to be no longer translationally invariant, due to the different coupling between the first and the last spin. However, the transformation $Q = g_1 g_2 \ldots g_N$ is a symmetry of the model, which means that the nearest neighbour coupling $H_{j,j+1}$ commutes with the product $g_j g_{j+1}$. We thus can define a modified translation operator $\tilde{T} = g_1 T$ which commutes with $\tilde{H}$,

$$\begin{aligned}
\tilde{T} \tilde{H}_{j,j+1} \tilde{T}^{-1} &= g_1 T H_{j,j+1} T^{-1} g_1^{-1} = H_{j+1,j+2} = \tilde{H}_{j+1,j+2}, \quad 1 \le j \le N-1, \\
\tilde{T} \tilde{H}_{N-1,N} \tilde{T}^{-1} &= g_1 T H_{N-1,N} T^{-1} g_1^{-1} = g_1 H_{N,1} g_1^{-1} = \tilde{H}_{N,1}, \\
\tilde{T} \tilde{H}_{N,1} \tilde{T}^{-1} &= g_1 T g_1 H_{N,1} g_1^{-1} T^{-1} g_1^{-1} = g_1 g_2 T H_{N,1} T^{-1} g_2^{-1} g_1^{-1} \\
&= g_1 g_2 H_{1,2} (g_1 g_2)^{-1} = H_{1,2} = \tilde{H}_{1,2},
\end{aligned} \tag{5}$$

the corresponding boundary conditions are known as toroidal boundary conditions.

The Hamiltonian of the Ising quantum chain is given by

$$H_{j,j+1} = -\frac{1}{4} \left(\sigma_j^z + \sigma_{j+1}^z + 2\lambda \sigma_j^x \sigma_{j+1}^x \right), \qquad H = -\frac{1}{2} \sum_{j=1}^{N} \sigma_j^z + \lambda \sigma_j^x \sigma_{j+1}^x, \tag{6}$$

where σ^x and σ^z are Pauli matrices, so $V \cong \mathbb{C}^2$. It has global spin reversal symmetry, i.e., H commutes with the operator $Q = \sigma^z \otimes \sigma^z \otimes \ldots \otimes \sigma^z = \prod_{j=1}^{N} \sigma_j^z$. Corresponding to this C_2 symmetry we have periodic (H^{P} with $g = \mathbf{1}_V$) and antiperiodic (H^{A} with $g = g^{-1} = \sigma^z$) boundary conditions, the latter yielding a change in sign of the $\sigma_N^x \sigma_1^x$ coupling term, because $\sigma_1^z \sigma_1^x \sigma_1^z = -\sigma_1^x$. It turns out that it is useful to consider the mixed-sector Hamiltonians [7] $H^+ = H^{\mathrm{P}} P_+ + H^{\mathrm{A}} P_-$ and $H^- = H^{\mathrm{A}} P_+ + H^{\mathrm{P}} P_-$ instead, where $P_{\pm} = (\mathbf{1} \pm Q)/2$ are projectors.

3. Duality Twist in the Ising Quantum Chain

Duality is a symmetry that relates the ordered and disordered phases of the classical Ising model. It provides an equality between the partition functions at two different temperatures, the critical temperature being mapped onto itself. In the quantum chain language, duality relates the Hamiltonians H with parameters λ and $1/\lambda$; the critical point corresponds to $\lambda = 1$.

In order to understand the duality transformation, it is advantageous to rewrite the mixed-sector Hamiltonians $H^{\pm}$ as follows

$$H^{\pm}(\lambda) = -\sum_{j=1}^{2N-1} [(e_{2j-1} - \tfrac{1}{2}) + \lambda(e_{2j} - \tfrac{1}{2})] - [(e_{2N-1} - \tfrac{1}{2}) - \lambda(e_{2N}^{\pm} - \tfrac{1}{2})], \quad (7)$$

where the Temperley-Lieb operators e_j are given by

$$e_{2j-1} = \frac{1}{2}(1 + \sigma_j^z), \quad e_{2j} = \frac{1}{2}(1 + \sigma_j^x \sigma_{j+1}^x), \quad e_{2N}^{\pm} = \frac{1}{2}(1 \pm Q\sigma_N^x \sigma_1^x). \quad (8)$$

Defining invertible operators $g_j = (1+i)e_j - 1$, with $g_j^{-1} = g_j^*$ and $i^2 = -1$, the appropriate duality transformations are $D^+ = g_1 g_2 \ldots g_{2N-1}$ and $D^- = D^+ \sigma_N^x$ [4]. The corresponding duality maps are $D^{\pm}H^{\pm}(\lambda) = \lambda H^{\pm}(1/\lambda)D^{\pm}$. Evidently, $D^{\pm}$ act on the operators e_j like a translation, i.e., $D^{\pm}e_j = e_{j+1}D^{\pm}$ for $1 \le j \le 2N-2$, and at the boundary $D^{\pm}e_{2N-1} = e_{2N}^{\pm}D^{\pm}$ and $D^{\pm}e_{2N}^{\pm} = e_1 D^{\pm}$. Thus the squares of the duality transformations $D^{\pm}$ commute with the corresponding Hamiltonians $H^{\pm}$ and are nothing but the appropriate translation operators $T^{\pm} = (D^{\pm})^2$ of the mixed-sector Hamiltonians [4].

At criticality, when $\lambda = 1$, duality itself becomes a symmetry, as $D^{\pm}H^{\pm}(1) = H^{\pm}(1)D^{\pm}$. Thus we can define corresponding twisted boundary conditions. This works in a slightly different way as for the periodic and antiperiodic boundary conditions discussed above, as we have to consider an odd number of generators e_j. The corresponding mixed-sector Hamiltonians are given by [4]

$$\tilde{H}^{\pm} = -\sum_{j=1}^{2N-2} (e_j - \tfrac{1}{2}) - (e_{2N-1}^{\pm} - \tfrac{1}{2}), \quad (9)$$

where the operators e_j, for $1 \le j \le 2N-2$, are defined as in equation (8) above, and where $e_{2N-1}^{\pm} = (1 \pm Q\sigma_N^y \sigma_1^x)/2$. So the duality-twisted Ising Hamiltonian contains coupling terms of the type $\pm \sigma_N^y \sigma_1^x$ at the boundary, and, in particular, does *not* contain a term σ_N^z.

The Hamiltonians $\tilde{H}^{\pm}$ are translationally invariant [4], the corresponding translation operators $\tilde{T}^+ - (\tilde{D}^{\pm})^2$ can be constructed as above as the squares of the appropriate duality transformations $\tilde{D}^+ = g_1 g_2 \ldots g_{2N-2}$ and $\tilde{D}^- = \tilde{D}^+ \sigma_N^z$, which commute with the critical Hamiltonians $\tilde{H}^+$ and $\tilde{H}^-$, respectively, of equation (9).

4. Spectrum and Partition Function

The spectrum of the duality-twisted Ising quantum chain can be calculated by a modified version [5] of the standard approach. Essentially, the Hamiltonians $\tilde{H}^{\pm}$ (9) are rewritten in terms of fermionic operators by means of a Jordan-Wigner transformation, and the resulting bilinear expressions in fermionic operators are subsequently diagonalised by a Bogoliubov-Valatin transformation, see [5] for details. The diagonal form of the Hamiltonian is

$$\tilde{H}^{\pm} = \sum_{k=0}^{N-1} \Lambda_k \eta_k^{\dagger} \eta_k + E_0 \mathbf{1} \quad (10)$$

where $\eta_k^\dagger$ and η_k are fermionic creation and annihilation operators, respectively. The energies of the elementary fermionic excitations are given by

$$\Lambda_k = 2|\sin(\tfrac{p_k}{2})|, \qquad p_k = \frac{4k\pi}{2N-1}, \qquad k = 0,1,2,\ldots,N-1. \tag{11}$$

The ground-state energy E_0 is

$$-E_0 = \sum_{k=0}^{N-1} \sin\left(\tfrac{k\pi}{\tilde{N}}\right) = \frac{1+\cos(\tfrac{\pi}{2\tilde{N}})}{2\sin(\tfrac{\pi}{2\tilde{N}})} = \frac{2\tilde{N}}{\pi} - \frac{\pi}{24(\tilde{N})} + O[(\tilde{N})^{-3}] \tag{12}$$

which shows the expected finite-size corrections of a translational invariant critical quantum chain with an effective number of sites of $\tilde{N} = N - 1/2$, reminiscent of the fact that it is related to the XXZ Heisenberg quantum chain with an odd number $2\tilde{N} = 2N-1$ of sites [4]. With the appropriate finite-size scaling, the linearised low-energy spectrum in the infinite system is

$$\frac{\tilde{N}}{2\pi}(\tilde{H}^\pm - \frac{\tilde{N}}{N}E_0^P \mathbf{1}) \xrightarrow{N \to \infty} \sum_{r=0}^{\infty}[ra_r^\dagger a_r + (r+\tfrac{1}{2})b_r^\dagger b_r] + \frac{1}{16} \tag{13}$$

where the fermionic operators a_k and b_k follow from the η_k by suitable renumbering, and where $E_0^P = 1/\sin(\tfrac{\pi}{2N})$ denotes the ground-state energy of the N-site Ising quantum chain with periodic boundary conditions. The conformal partition functions are given by the combinations $(\chi_0 + \chi_{1/2})\bar{\chi}_{1/16}$ and $\chi_{1/16}(\bar{\chi}_0 + \bar{\chi}_{1/2})$, respectively, of characters χ_Δ of irreducible representations with highest weight Δ of the $c = 1/2$ Virasoro algebra, corresponding to operators with conformal spin $1/16$ and $7/16$.

5. Concluding Remarks

For the Ising and Potts quantum chains [4], "exotic" conformal twisted boundary conditions can be realised by means of twists related to duality, which is a symmetry of the model at criticality. It would be interesting to know whether this is a more general feature. If this is the case, this observation might help to identify non-trivial symmetries in quantum chains or two-dimensional solvable lattice models of statistical mechanics.

References

[1] Petkova V B and Zuber J-B 2001 Generalised twisted partition functions *Phys. Lett.* B **504** 157–64
[2] Coquereaux R and Schieber G 2001 Twisted partition functions for ADE boundary conformal field theories and Ocneanu algebras of quantum symmetries *J. Geom. Phys.* **42** (2002) 216–258
[3] Chui C H O, Mercat C, Orrick W P and Pearce P A 2001 Integrable lattice realizations of conformal twisted boundary conditions *Phys. Lett.* B **517** 429–35
[4] Grimm U and Schütz G 1993 The spin-1/2 XXZ Heisenberg chain, the quantum algebra $U_q[sl(2)]$, and duality transformations for minimal models *J. Stat. Phys.* **71** 921–64
[5] Grimm U 2002 Spectrum of a duality-twisted Ising quantum chain *J. Phys. A: Math. Gen.* **35** (2002) L25–30
[6] Grimm U 1990 The quantum Ising chain with a generalized defect *Nucl. Phys.* B **340** 633–58
[7] Baake M, Chaselon P and Schlottmann M 1989 The Ising quantum chain with defects. II. The so(2n) Kac-Moody spectra *Nucl. Phys.* B **314** 625–45

Inst. Phys. Conf. Ser. No 173: Section 5
Paper presented at 24th Int. Coll. Group Theoretical Methods in Physics, Paris, France, 15–20 July 2002
©2003 IOP Publishing Ltd

Quantum Frenkel-Kontorova Model

Choon-Lin Ho[1] and Chung-I Chou[2]
[1]Department of Physics, Tamkang University, Tamsui 25137, Taiwan
[2]Institute of Physics, Academia Sinica, Taipei 11529, Taiwan

Abstract. This paper presents a simple variational approach to the quantum Frenkel-Kontorova model.

1. Introduction

The Frenkel-Kontorova (FK) model [1] is a simple one-dimensional model used to study incommensurate structures appearing in many condensed-matter systems, such as charge-density waves, magnetic spirals, and adsorbed monolayers. These modulated structures arise as a result of the competition between two or more length scales. The FK model describes a chain of atoms connected by harmonic springs subjected to an external sinusoidal potential. In an important development in the study of the classical FK model, Aubry [2] first made use of the connection between the FK model, the so-called "standard map", and the Kolmogorov-Arnold-Moser (KAM) theorem to reveal many interesting features of the FK model. Particularly, he showed that when the mean distance (also called the winding number) between two successive atoms is rational, the system is always pinned. But when the winding number is irrational, there exists a critical external field strength below (above) which the system is unpinned (pinned). This transition is called by Aubry a "transition by breaking of analyticity", and is closely connected with the breakup of a KAM torus. It is very analogous to a phase transition, and various critical exponents and questions of universality have been extensively studied in the past.

Needless to say, quantum effects are very important in the FK model. However, unlike the classical case, study of quantum FK models is rather scanty. It was first considered in a quantum Monte Carlo (QMC) analysis in [3]. Their main observation is that the map appropriate to describe the quantum case is no longer the standard map, but rather a map with a sawtooth shape.

Previous theoretical attempts at obtaining the sawtooth map require one to go beyond the independent-particle approximation. In [4], however, we showed that all the essential features observed in the QMC studies can indeed be obtained from an independent-particle picture of the many-body ground state. Our strategy is to derive an effective Hamiltonian for the quantum FK model by adopting Dirac's time-dependent variational principle together with the Jackiw-Kerman (JK) function [5] as the single particle state. The JK wavefunction can be viewed as the Q-representation of the squeezed state.

2. Effective Hamiltonian

The Hamiltonian of the quantum FK model is given by

$$\mathcal{H} = \sum_i \left[\frac{\hat{p}_i^2}{2m} + \frac{\gamma}{2} \left(\hat{q}_{i+1} - \hat{q}_i \right)^2 - V \cos(l_0 \hat{q}_i) \right]. \tag{1}$$

Here $\hat{q}_i$ and $\hat{p}_i$ are the position and momentum operators, respectively, of the ith atom, γ the elastic constant of the spring, V and $2\pi/l_0$ are the strength and the period of the external potential. It is convenient to use the dimensionless variables $\hat{Q}_i = l_0\hat{q}_i$, $\hat{P}_i = l_0\hat{p}_i/\sqrt{m\gamma}$, and $K = Vl_0^2/\gamma$. With these new variables, we obtain the following dimensionless Hamiltonian H

$$H = \sum_i \left[\frac{\hat{P}_i^2}{2} + \frac{1}{2}\left(\hat{Q}_{i+1} - \hat{Q}_i\right)^2 - K\cos(\hat{Q}_i) \right]. \tag{2}$$

We have $\mathcal{H} = \gamma H/l_0^2$. The effective Planck constant is $\tilde{\hbar} = \hbar l_0^2/\sqrt{m\gamma}$. For the classical FK model, the Aubry transition occurs at the critical value $K_c = 0.971635\cdots$.

To study the ground state properties of the quantum FK model in (2), we adopt here the time-dependent variational principle pioneered by Dirac. In this approach, one first constructs the effective action $\Gamma = \int dt\,\langle\Psi, t|i\hbar\partial_t - H|\Psi, t\rangle$ for a given system described by H and $|\Psi, t\rangle$. Variation of Γ is then the quantum analogue of the Hamilton's principle. The time-dependent Hartree-Fock approximation emerges when a specific ansatz is made for the state $|\Psi, t\rangle$. We now assume the trial wavefunction of the ground state of our quantum FK system to have the Hartree form $|\Psi, t\rangle = \prod_i |\psi_i, t\rangle$, where the normalized single-particle state $|\psi_i, t\rangle$ is taken to be the JK wavefunction [5]:

$$\langle Q_i|\psi_i, t\rangle = \frac{1}{(2\pi\tilde{\hbar}G_i)^{1/4}}$$

$$\times \exp\left\{ -\frac{1}{2\tilde{\hbar}}\left(Q_i - x_i\right)^2 \left[\frac{1}{2}G_i^{-1}\right.\right.$$

$$\left.\left. - 2i\Pi_i \right] + \frac{i}{\tilde{\hbar}}p_i\left(Q_i - x_i\right) \right\}. \tag{3}$$

The real quantities $x_i(t)$, $p_i(t)$, $G_i(t)$ and $\Pi_i(t)$ are variational parameters the variations of which at $t = \pm\infty$ are assumed to vanish. Squeezed state function in the form of the JK wavefunction has the advantage that the physical meanings of the variational parameters contained in the JK wavefunction are most transparent, as we shall show below. Furthermore, the JK form is in the general Gaussian form so that integrations are most easily performed.

It is not hard to check that x_i and p_i are the expectation values of the operators $\hat{Q}_i$ and $\hat{P}_i$: $x_i = \langle\Psi|\hat{Q}_i|\Psi\rangle$, $p_i = \langle\Psi|\hat{P}_i|\Psi\rangle$. Also, one has $\langle\Psi|(\hat{Q}_i - x_i)^2|\Psi\rangle = \tilde{\hbar}G_i$, and $\langle\Psi|i\hbar\partial_t|\Psi\rangle = \sum_i(p_i\dot{x}_i - \tilde{\hbar}G_i\dot{\Pi}_i)$, where the dot represents derivative with respect to time t. It is now clear that $\tilde{\hbar}G_i$ is the mean fluctuation of the position of the i-th atom, and that $G_i > 0$. From the form of the effective action one sees that p_i and Π_i are the canonical conjugates of x_i and G_i, respectively. The Dirac variational principle leads to the following effective Hamiltonian

$$H_{eff} = \langle\Psi|H|\Psi\rangle$$

$$= \sum_i \frac{1}{2}\left[p_i^2 + \tilde{\hbar}\left(\frac{1}{4}G_i^{-1} + 4\Pi_i^2 G_i\right) \right]$$

$$+ \sum_i \frac{1}{2}\left(x_{i+1} - x_i\right)^2$$

$$+ \sum_i \frac{\tilde{\hbar}}{2}\left(G_{i+1} + G_i\right)$$

$$- \sum_i K\exp\left(-\frac{\tilde{\hbar}}{2}G_i\right)\cos x_i. \tag{4}$$

We can obtain the equations for the equilibrium states in the Hartree-Fock approximation by directly varying the effective Hamiltonian H_{eff} with respect to the variables p_i, Π_i, x_i and G_i, which give, respectively,

$$p_i = 0 \quad , \quad 4\Pi_i G_i = 0 , \tag{5}$$

$$x_{i+1} - 2x_i + x_{i-1} = K \exp\left(-\frac{\tilde{\hbar}}{2} G_i\right) \sin x_i , \tag{6}$$

$$\frac{1}{4} G_i^{-2} - K \exp\left(-\frac{\tilde{\hbar}}{2} G_i\right) \cos x_i - 2 = 4\Pi_i^2 . \tag{7}$$

The second equation in (5) implies $\Pi_i = 0$ as $G_i > 0$. This in turn means that the right hand side of eq.(7) is equal to zero:

$$\frac{1}{4} G_i^{-2} - K \exp\left(-\frac{\tilde{\hbar}}{2} G_i\right) \cos(x_i) - 2 = 0 . \tag{8}$$

In the limit $\tilde{\hbar} = 0$, eq.(6) is equivalent to the standard map.

3. Numerical results

We numerically solve for the set of variables x_i and G_i which characterize the ground state using the Newton method. In all our numerical computations the winding number $P/Q = 610/987$, which is an approximation of the golden mean winding number $(\sqrt{5}-1)/2$, is used with the periodic boundary condition $x_{i+Q} = x_i + 2\pi P$. This winding number is much more accurate than those used in previous works to approximate the golden mean number, thus giving us better accuracy in the computations of physical quantities related to the ground state.

Having obtained the values of x_i which give the mean positions of the quantum atoms in the chain, we can compare the results with the classical configuration by plotting the so-called g-function, defined by

$$g_i \equiv K^{-1} (x_{i+1} - 2x_i + x_{i-1}) \tag{9}$$

versus the actual atomic positions x_i. From (6), we also have

$$g_i = \exp\left(-\frac{\tilde{\hbar}}{2} G_i\right) \sin x_i . \tag{10}$$

Here G_i is related to x_i by eq.(8). We see from this equation that quantum fluctuations G_i will modify the shape of the classical $sine$-map.

In Fig. 1 we show the graphs of the g-function for the case $K = 5$. The curve defined by (10) with G_i satisfying (8) are shown here as dashed curves for different $\tilde{\hbar}$. In the classical limit ($\tilde{\hbar} = 0$) this curve is simply the standard map ($sine$-curve). As $\tilde{\hbar}$ increases, the amplitude of the curve decreases. For sufficiently large $\tilde{\hbar}$, the curve resembles more closely a "sawtooth" shape. This is first noted in QMC study in [3]. Here we see that it comes out very naturally from the equation of motion (8) and (10). We have therefore demonstrated that the sawtooth map could be recovered in the independent-particle approximation. In the supercritical case ($K = 5$), when $\tilde{\hbar} < \tilde{\hbar}_c \approx 6.58$, the positions x_i of the atoms cover only a subset of the g-curves. This is in accord with the fact that the atoms are in the pinning phase.

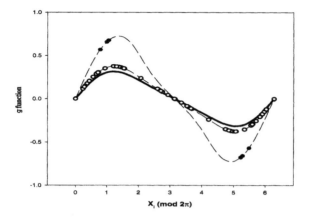

Figure 1. *g-function plotted against actual atomic positions for* $K = 5$ *and winding number* $P/Q = 610/987$ *at* $\hbar = 2$ *(black dots),* 6 *(white dots) and* 7 *(black curve) (the dashed curves represent eq.(10) with* G_i *satisfying (8).*

[1] Frenkel Y I and Kontorova T 1938 Sov. Phys.-JETP 13 1.
[2] Aubry S 1983 J. Phys. (Paris) 44 147; Physica 7D 240; *ibid* 8D 381.
[3] Borgonovi F Guarneri I and Shepelyansky D 1989 Phys. Rev. Lett. 63 2010; 1990 Z. Phys. B79 133.
[4] Ho C L and Chou C I 2001 Phys. Rev. E63 016203.
[5] Jackiw R and Kerman A 1979 Phys. Lett. A71 158.

Inst. Phys. Conf. Ser. No 173: Section 5
Paper presented at 24th Int. Coll. Group Theoretical Methods in Physics, Paris, France, 15–20 July 2002
©2003 IOP Publishing Ltd

Geometric mechanics for applications in diatomic and triatomic molecular dynamics

Florence J. Lin

Department of Mathematics, University of Southern California, 1042 Downey Way, Los Angeles, CA 90089-1113 USA

Abstract. Geometric mechanics provides a framework for treating classical molecular dynamics as an N-body system. The Poisson formulation used here has two advantages: It generalizes the previous symplectic formulation, and it is more generally accessible. The following topics are discussed: (1) For diatomic systems, relevant phase spaces describing the dynamics for Hamiltonian systems with symmetry are reduced, and the total energy is the sum of the translational, rotational, and internal energy terms. (2) For diatomic systems, the topology of the energy-momentum map involved in reduction provides a unified treatment including both bound and scattering trajectories. (3) For noncolinear triatomic molecules, the velocity of the angle conjugate to the total angular momentum includes contributions from the momenta conjugate to "internal" vibrations. For the case of zero total angular momentum, the net angle of overall rotation is a geometric phase and is also coupled with vibration.

1. Introduction

Classical molecular dynamics of two- and three-body systems have been discussed in the symplectic formulation [1], [2]. In the more general Poisson formulation, the momentum phase space is described by a Poisson manifold [3], i.e., a surface with a Poisson bracket. For the two-body system, the result of reduction [4] by Lie group symmetries is merely stated and leads to an expression for the total energy as a sum of potential and kinetic energies and leads to the coupling of the dynamics in the sense described below. For the three-body system, reduction leads to two descriptions of the coupling of the rotational and internal motions. One description uses Hamilton's equations to obtain an explicit expression [2] for the contributions to the velocity of the angle conjugate to the total angular momentum due to "internal" motions in equation (22). Another description involves a geometric phase [5] analogous to that observed for the falling cat with zero total angular momentum [6].

2. Diatomic molecular dynamics

2.1. Dynamics of a two-body system on a Poisson manifold

The Hamiltonian $H\colon \mathbb{R}^8 \to \mathbb{R}$ for a two-body molecular system in two-dimensional space is written as the sum of a kinetic energy plus a potential energy V [1], i.e.,

$$H(\mathbf{r}_1, \mathbf{r}_2, \mathbf{p}_1, \mathbf{p}_2) = \frac{1}{2m_1}\|\mathbf{p}_1\|^2 + \frac{1}{2m_2}\|\mathbf{p}_2\|^2 + V(\|\mathbf{r}_2 - \mathbf{r}_1\|) \ , \tag{1}$$

where $\mathbf{r}_i$ and $\mathbf{p}_i$ are the coordinate and momentum, respectively, of the i-th atom with mass m_i for $i = 1, 2$, and $V(r)$ is the potential energy of interaction for the two atoms at a separation r.

The dynamics are described by Hamilton's equations

$$\dot{\mathbf{z}} = \mathbf{X}_H(\mathbf{z}) \quad \text{for } \mathbf{z} = (\mathbf{r}_1, \mathbf{r}_2, \mathbf{p}_1, \mathbf{p}_2) \in \mathbb{R}^8 \tag{2}$$

where $\mathbf{X}_H$ is the Hamiltonian vector field defined by

$$\mathbf{X}_H(\mathbf{z}) = \{\mathbf{z}, H\} \quad \text{for all } \mathbf{z} \in \mathbb{R}^8 \ . \tag{3}$$

Here and in each case below, the Hamiltonian vector field is determined by an appropriate Hamiltonian and Poisson bracket. Here, the Poisson bracket $\{ \, , \, \}$ is

$$\{F, G\} = \sum_{j=1}^{2} \left(\frac{\partial F}{\partial r_1^j} \frac{\partial G}{\partial (p_1)_j} - \frac{\partial F}{\partial (p_1)_j} \frac{\partial G}{\partial r_1^j} + \frac{\partial F}{\partial r_2^j} \frac{\partial G}{\partial (p_2)_j} - \frac{\partial F}{\partial (p_2)_j} \frac{\partial G}{\partial r_2^j} \right) \ , \tag{4}$$

where F and G are functions of $\mathbf{r}_1, \mathbf{r}_2, \mathbf{p}_1$, and $\mathbf{p}_2$.

2.2. Poisson reduction to the translating, center-of-mass frame

Make a change of variables [1], [2] from the Cartesian laboratory coordinate system $(\mathbf{r}_1, \mathbf{r}_2, \mathbf{p}_1, \mathbf{p}_2) \in \mathbb{R}^8$ to $(\mathbf{r}, \mathbf{r}_{cm}, \mathbf{p}, \mathbf{p}_{lin}) \in \mathbb{R}^8$, where $\mathbf{r} \in \mathbb{R}^2$ is the relative coordinate defined by $\mathbf{r} = \mathbf{r}_2 - \mathbf{r}_1$, and $\mathbf{r}_{cm} \in \mathbb{R}^3$ is the coordinate of the center of mass $\mathbf{r}_{cm} = \frac{m_1}{m_1+m_2}\mathbf{r}_1 + \frac{m_2}{m_1+m_2}\mathbf{r}_2$. Define the conjugate momenta in terms of the Lagrangian $\mathcal{L}$, i.e., $\mathbf{p} = \partial\mathcal{L}/\partial\dot{\mathbf{r}} = -\frac{m_2}{m_1+m_2}\mathbf{p}_1 + \frac{m_1}{m_1+m_2}\mathbf{p}_2$, $\mathbf{p}_{lin} = \partial\mathcal{L}/\partial\dot{\mathbf{r}}_{cm} = \mathbf{p}_1 + \mathbf{p}_2$.

Since $\mathbf{r}_{cm}$ is a cyclic coordinate of the Hamiltonian, the total linear momentum $\mathbf{p}_{lin}$ is constant. Define the momentum map $\mathbf{J}_{\mathbf{p}_{lin}}: \mathbb{R}^8 \to \mathbb{R}^2$ corresponding to the conservation of $\mathbf{p}_{lin}$. The Hamiltonian $H(\mathbf{r}, \mathbf{r}_{cm}, \mathbf{p}, \mathbf{p}_{lin})$ is expressible as a reduced Hamiltonian $H_{\mathbf{p}_{lin}}(\mathbf{r}, \mathbf{p})$: $P_{\mathbf{p}_{lin}} \to \mathbb{R}$

$$H_{\mathbf{p}_{lin}}(\mathbf{r}, \mathbf{p}) = \frac{1}{2\frac{m_1 m_2}{m_1+m_2}}\|\mathbf{p}\|^2 + V_{\mathbf{p}_{lin}}(\|\mathbf{r}\|) \tag{5}$$

with amended potential energy $V_{\mathbf{p}_{lin}}(\|\mathbf{r}\|)$

$$V_{\mathbf{p}_{lin}}(\|\mathbf{r}\|) = V(\|\mathbf{r}\|) + \frac{1}{2(m_1+m_2)}\|\mathbf{p}_{lin}\|^2 \ . \tag{6}$$

The dynamics drop to

$$\dot{\mathbf{z}}_{\mathbf{p}_{lin}} = \mathbf{X}_{H_{\mathbf{p}_{lin}}}(\mathbf{z}_{\mathbf{p}_{lin}}) \quad \text{for } \mathbf{z}_{\mathbf{p}_{lin}} \in \mathbb{R}^4 \tag{7}$$

on $P_{\mathbf{p}_{lin}} = \mathbf{J}_{\mathbf{p}_{lin}}^{-1}(\mathbf{p}_{lin})/\mathbb{R}^2 \cong \mathbb{R}^4$, i.e., the quotient of the inverse image of the momentum map by the symmetry group, with $\mathbf{z}_{\mathbf{p}_{lin}} = (\mathbf{r}, \mathbf{p})$. The Hamiltonian vector field $\mathbf{X}_{H_{\mathbf{p}_{lin}}}$ is defined by $\mathbf{X}_{H_{\mathbf{p}_{lin}}}(\mathbf{z}_{\mathbf{p}_{lin}}) = \{\mathbf{z}_{\mathbf{p}_{lin}}, H_{\mathbf{p}_{lin}}\}_{\mathbf{p}_{lin}}$. The Poisson bracket is defined

$$\{F, G\}_{\mathbf{p}_{lin}} = \sum_{j=1}^{2} \left(\frac{\partial F}{\partial r^j} \frac{\partial G}{\partial p_j} - \frac{\partial F}{\partial p_j} \frac{\partial G}{\partial r^j} \right) \ , \tag{8}$$

where F and G are functions of $\mathbf{r}$ and $\mathbf{p}$.

2.3. Poisson reduction to the rotating, internal frame: energy terms, coupling of dynamics, and unified treatment of scattering and bound trajectories

Make a change of variables from the Cartesian relative coordinate system $(\mathbf{r}, \mathbf{p}) \in \mathbb{R}^2 \times \mathbb{R}^2$, where $\mathbf{p} = (p_x, p_y)$ and $\mathbf{r} = (x, y)$, to the polar relative coordinate system, where $\mathbf{r} = (r, \theta) \in (\mathbb{R}^+, S^1)$ with its conjugate momentum $\mathbf{p}$.

Since θ is a cyclic coordinate of $H_{\mathbf{p}_{\text{lin}}}$, the angular momentum p_θ is constant. Define the momentum map $\mathbf{J}_{\mathbf{p}_{\text{lin}},p_\theta}: \mathbb{R}^+ \times S^1 \times \mathbb{R}^2 \to \mathbb{R}^1$ corresponding to the conservation of p_θ. The Hamiltonian $H_{\mathbf{p}_{\text{lin}},p_\theta}(r,p_r)$ is

$$H_{\mathbf{p}_{\text{lin}},p_\theta}(r,p_r) = \frac{1}{2\frac{m_1 m_2}{m_1+m_2}}p_r^2 + V_{\mathbf{p}_{\text{lin}},p_\theta}(r) \tag{9}$$

with $V_{\mathbf{p}_{\text{lin}},p_\theta}(r)$

$$V_{\mathbf{p}_{\text{lin}},p_\theta}(r) = V_{\mathbf{p}_{\text{lin}}}(r) + \frac{1}{2\frac{m_1 m_2}{m_1+m_2}r^2}p_\theta^2 . \tag{10}$$

Thus, the total energy $H_{\mathbf{p}_{\text{lin}},p_\theta}(r,p_r)$ is the sum of the potential energy plus the kinetic energy terms corresponding to vibration, rotation, and translation.

The dynamics drop to

$$\dot{\mathbf{z}}_{\mathbf{p}_{\text{lin}},p_\theta} = \mathbf{X}_{H_{\mathbf{p}_{\text{lin}},p_\theta}}(\mathbf{z}_{\mathbf{p}_{\text{lin}},p_\theta}) \quad \text{for } \mathbf{z}_{\mathbf{p}_{\text{lin}},p_\theta} \in \mathbb{R}^+ \times \mathbb{R} \tag{11}$$

on $P_{\mathbf{p}_{\text{lin}},p_\theta} = \mathbf{J}_{\mathbf{p}_{\text{lin}},p_\theta}^{-1}(\mathbf{p}_{\text{lin}}, p_\theta)/S^1 \cong \mathbb{R}^+ \times \mathbb{R}$, i.e., the quotient of the inverse image of the momentum map by the symmetry group, with $\mathbf{z}_{\mathbf{p}_{\text{lin}},p_\theta} = (r,p_r)$. The Hamiltonian vector field $\mathbf{X}_{H_{\mathbf{p}_{\text{lin}},p_\theta}}$ is defined by $\mathbf{X}_{H_{\mathbf{p}_{\text{lin}},p_\theta}}(\mathbf{z}_{\mathbf{p}_{\text{lin}},p_\theta}) = \{\mathbf{z}_{\mathbf{p}_{\text{lin}},p_\theta}, H_{\mathbf{p}_{\text{lin}},p_\theta}\}_{\mathbf{p}_{\text{lin}},p_\theta}$, and the reduced Poisson bracket is defined by

$$\{F,G\}_{\mathbf{p}_{\text{lin}},p_\theta} = \frac{\partial F}{\partial r}\frac{\partial G}{\partial p_r} - \frac{\partial F}{\partial p_r}\frac{\partial G}{\partial r} , \tag{12}$$

where F and G are functions of r and p_r. As the time derivative of r is independent of θ, the differential equation for r is decoupled from θ. However, this time derivative does depend on p_θ, which appears in the amended potential as a constant in the energy of the centrifugal barrier, so the dynamics in r are coupled with the dynamics in θ in this sense (cf. [7, pp. 46-49]).

Further, for the reduced dynamics, examination of the bifurcation set of the energy-momentum map [1, fig. 1] shows that it extends the previous bifurcation set of the energy-momentum map for non-negative energies corresponding to scattering dynamics [8, fig. 13] to include negative energies corresponding to bound trajectories as well.

3. Triatomic molecular dynamics

3.1. Dynamics of a three-body system on a Poisson manifold

The Hamiltonian for three-body molecular dynamics on $\mathbb{R}^{12}$ is [2]

$$H(\mathbf{r}_1,\mathbf{r}_2,\mathbf{r}_3,\mathbf{p}_1,\mathbf{p}_2,\mathbf{p}_3) = \frac{1}{2m_1}\|\mathbf{p}_1\|^2 + \frac{1}{2m_2}\|\mathbf{p}_2\|^2 + \frac{1}{2m_3}\|\mathbf{p}_3\|^2$$
$$+ V(\|\mathbf{r}_2 - \mathbf{r}_1\|, \|\mathbf{r}_3 - \mathbf{r}_2\|, \|\mathbf{r}_1 - \mathbf{r}_3\|) , \tag{13}$$

where m_i, $\mathbf{r}_i$, and $\mathbf{p}_i$ for $i = 1, 2, 3$ are the masses, coordinates, and momenta, respectively, of the three bodies and V is the interatomic potential energy function. The Poisson bracket on the phase space is

$$\{F,G\} = \sum_{j=1}^{2}\sum_{i=1}^{3}\left(\frac{\partial F}{\partial r_i^j}\frac{\partial G}{\partial (p_i)_j} - \frac{\partial F}{\partial (p_i)_j}\frac{\partial G}{\partial r_i^j}\right) , \tag{14}$$

where F and G are functions of $\mathbf{r}_1, \mathbf{r}_2, \mathbf{r}_3, \mathbf{p}_1, \mathbf{p}_2,$ and $\mathbf{p}_3$. Hamilton's equations are written

$$\dot{\mathbf{z}} = \mathbf{X}_H(\mathbf{z}) \quad \text{for } \mathbf{z} = (\mathbf{r}_1,\mathbf{r}_2,\mathbf{r}_3,\mathbf{p}_1,\mathbf{p}_2,\mathbf{p}_3) \in \mathbb{R}^{12} \tag{15}$$

with the Hamiltonian vector field $\mathbf{X}_H$ defined by $\mathbf{X}_H(\mathbf{z}) = \{\mathbf{z}, H\}$ for $\mathbf{z} \in \mathbb{R}^{12}$.

3.2. Poisson reduction to the translating, center-of-mass frame

Make a change of variables to the center-of-mass frame [2] with reduced Hamiltonian $H_{\mathrm{Plin}}(\mathbf{r}, \mathbf{s}, \mathbf{p_r}, \mathbf{p_s})$, amended potential V_{Plin}, and reduced Poisson bracket $\{\,,\,\}_{\mathrm{Plin}}$

$$\{F, G\}_{\mathrm{Plin}} = \sum_{j=1}^{2}\left(\frac{\partial F}{\partial r^j}\frac{\partial G}{\partial (p_r)_j} - \frac{\partial F}{\partial (p_r)_j}\frac{\partial G}{\partial r^j} + \frac{\partial F}{\partial s^j}\frac{\partial G}{\partial (p_s)_j} - \frac{\partial F}{\partial (p_s)_j}\frac{\partial G}{\partial s^j}\right). \quad (16)$$

3.3. Poisson reduction to a rotating, internal frame

Make a change of variables from $(\mathbf{r},\mathbf{s},\mathbf{p_r},\mathbf{p_s})$ to $(\rho_1,\rho_2,\rho_3,\phi,P_{\rho_1},P_{\rho_2},P_{\rho_3},P_\phi)$ and define θ_i to be the oriented exterior angle at the i-th vertex [2], [9], [10] to obtain the reduced Hamiltonian $H_{\mathrm{Plin},P_\phi}(\rho_1,\rho_2,\rho_3,P_{\rho_1},P_{\rho_2},P_{\rho_3})$, amended potential $V_{\mathrm{Plin},P_\phi}(\rho_3,\rho_1,\rho_2)$, and reduced Poisson bracket $\{\,,\,\}_{\mathrm{Plin},P_\phi}$

$$\{F, G\}_{\mathrm{Plin},P_\phi} = \sum_{i=1}^{3}\left(\frac{\partial F}{\partial \rho_i}\frac{\partial G}{\partial P_{\rho_i}} - \frac{\partial F}{\partial P_{\rho_i}}\frac{\partial G}{\partial \rho_i}\right) . \quad (17)$$

In the Eckart frame [11], the coupling of the rotational and internal motions in the kinetic energy vanishes, i.e.,

$$\left[\frac{1}{3}\left(\frac{\sin\theta_3}{m_3\rho_2} - \frac{\sin\theta_2}{m_2\rho_3}\right)P_{\rho_1} + \frac{1}{3}\left(\frac{\sin\theta_1}{m_1\rho_3} - \frac{\sin\theta_3}{m_3\rho_1}\right)P_{\rho_2} + \frac{1}{3}\left(\frac{\sin\theta_2}{m_2\rho_1} - \frac{\sin\theta_1}{m_1\rho_2}\right)P_{\rho_3}\right]P_\phi = 0 . \quad (18)$$

3.4. Poisson internal dynamics for two bond lengths and one bond angle

Make a change of variables from interatomic distances (ρ_1,ρ_2,ρ_3) to internal coordinates (S_1,S_2,S_3), where S_1 is ρ_2, the bond length between m_1 and m_3; S_2 is ρ_1, the bond length between m_2 and m_3; and S_3 is θ, the internal bond angle at m_3 [2], [11]. Let P_1, P_2, and P_3 be the corresponding conjugate momenta. The Poisson bracket is

$$\{F, G\}_{\mathrm{Plin},P_\phi,\mathrm{Eck}} = \left[\sum_{i=1}^{2}\left(\frac{\partial F}{\partial \rho_i}\frac{\partial G}{\partial P_{\rho_i}} - \frac{\partial F}{\partial P_{\rho_i}}\frac{\partial G}{\partial \rho_i}\right)\right] + \frac{\partial F}{\partial \theta}\frac{\partial G}{\partial P_\theta} - \frac{\partial F}{\partial P_\theta}\frac{\partial G}{\partial \theta} . \quad (19)$$

3.5. Coupling of rotation and vibration shown using Hamilton's equations

Hamilton's equations show that rotation and vibration are coupled. An explicit expression for the angular velocity $\dot{\phi}$ for the angle conjugate to the total angular momentum rotation is given dynamically by Hamilton's equations in the center-of-mass frame [2]

$$\dot{\phi} = \frac{\partial H_{\mathrm{Plin}}}{\partial P_\phi} = \dot{\phi}_{\mathrm{conj}} + \dot{\phi}_{\mathrm{inter}} \quad (20)$$

with

$$\dot{\phi}_{\mathrm{conj}} = \left[\frac{2}{9}\sum m_i^{-1}\left(\frac{1}{\rho_j^2} + \frac{1}{\rho_k^2} - \frac{2\cos\theta_i}{\rho_j\rho_k}\right)\right]P_\phi \quad (21)$$

$$\dot{\phi}_{\mathrm{inter}} = \frac{1}{3}\left(\frac{\sin\theta_3}{m_3\rho_2} - \frac{\sin\theta_2}{m_2\rho_3}\right)P_{\rho_1} + \frac{1}{3}\left(\frac{\sin\theta_1}{m_1\rho_3} - \frac{\sin\theta_3}{m_3\rho_1}\right)P_{\rho_2}$$
$$+ \frac{1}{3}\left(\frac{\sin\theta_2}{m_2\rho_1} - \frac{\sin\theta_1}{m_1\rho_2}\right)P_{\rho_3} . \quad (22)$$

[The summation $\sum a_{ijk}$ means $a_{123} + a_{231} + a_{312}$.] The terms $\dot{\phi}_{\mathrm{conj}}$ and $\dot{\phi}_{\mathrm{inter}}$ are due to the momentum conjugate to ϕ and due to Coriolis interactions [12] with the other "internal" vibrational momenta within the plane of the atoms, respectively. Even when P_ϕ vanishes, "internal" vibrations contribute to the rotation of the molecule due to the terms in equation (22). This provides a dynamical explanation of Guichardet's [13] observation that vibrational motions can not, in general, be separated from rotational motions.

3.6. Coupling of rotation and vibration shown using conservation of total angular momentum

Conservation of the total angular momentum also shows that rotation and vibration are coupled. Following Herzberg [12], write the total angular momentum as $\mathbf{P}_\phi = \ell + \mathbf{p}$, where ℓ is the overall angular momentum and $\mathbf{p}$ is the vibrational angular momentum. Then, following Marsden *et al.* [5], write the magnitude of the total angular momentum in terms of angular velocities with ℓ and $\mathbf{p}$ playing the roles of the two momenta. For the case of zero total angular momentum, the net angle of overall rotation is a geometric phase

$$\Delta\phi_\ell = -\int \frac{I_{\theta_\mathbf{p}}}{I_{\theta_\ell} + I_{\theta_\mathbf{p}}} d\theta_{\mathbf{p}\ell} \quad . \tag{23}$$

The geometric phase is the holonomy of the mechanical $\mathcal{A}_{\text{mech}}$ connection, where

$$\mathcal{A}_{\text{mech}} = d\phi_\ell + \frac{I_{\theta_\mathbf{p}}}{I_{\theta_\ell} + I_{\theta_\mathbf{p}}} d\theta_{\mathbf{p}\ell} \quad , \tag{24}$$

ϕ_ℓ is the angle of overall rotation with respect to a fixed direction, $\theta_\mathbf{p}$ is the angle corresponding to the vibrational angular momentum with respect to the fixed direction, $\theta_{\mathbf{p}\ell}$ is the difference $\theta_\mathbf{p} - \theta_\ell$ between the angles, and $I_{\theta_\mathbf{p}}$ and I_{θ_ℓ} are the moments of inertia corresponding to the angles $\theta_\mathbf{p}$ and θ_ℓ, respectively.

Acknowledgment

Support for travel from the Women In Science and Engineering Program, Office of the Provost, University of Southern California is gratefully acknowledged.

References

[1] Lin F J and Marsden J E 1992 J. Math. Phys. 33 1281-1294
[2] Lin F J 1997 Phys. Lett. A 234 291-300
[3] Marsden J E and Ratiu T S 1994 Introduction to Mechanics and Symmetry (New York: Springer-Verlag)
[4] Marsden J and Ratiu T 1986 Lett. Math. Phys. 11 161-170
[5] Marsden J, Montgomery R, and Ratiu T 1990 Mem. Amer. Math. Soc. 88, No. 436 (Providence, RI: American Mathematical Society)
[6] Frohlich C 1980 Scientific American 242(3) 154-164
[7] Weston R E Jr. and Schwarz H A 1972 Chemical Kinetics (Englewood Cliffs, NJ: Prentice-Hall)
[8] Ford K W and Wheeler J A 1959 Ann. Phys. (NY) 7 287-322
[9] Murnaghan F D 1936 Am. J. Math. 58 829-832
[10] van Kampen E R and Wintner A 1937 Am. J. Math. 59 153-166
[11] Wilson E B Jr., Decius J C, and Cross P C 1980 Molecular Vibrations: The Theory of Infrared and Raman Vibrational Spectra (New York: Dover), republication of McGraw-Hill edition of 1955

[12] Herzberg G 1945 Molecular Spectra and Molecular Structure: II. Infrared and Raman Spectra of Polyatomic Molecules (New York: Van Nostrand Reinhold)
[13] Guichardet A 1984 Ann. Inst. H. Poincaré, Phys. Théor. 40 329-342

Inst. Phys. Conf. Ser. No 173: Section 5
Paper presented at 24th Int. Coll. Group Theoretical Methods in Physics, Paris, France, 15–20 July 2002
©2003 IOP Publishing Ltd

Regular fractal structure of the energy matrix in 1D Heisenberg open spin-$\frac{1}{2}$ chains

F Pan[†,‡], J P Draayer[‡], L -R Dai[†], and D Zhang[†]

[†]Department of Physics, Liaoning Normal University, Dalian 116029, P. R. China
[‡]Department of Physics and Astronomy, Louisiana State University, Baton Rouge, LA 70803, USA

Abstract. A permutation group approach to a one-dimensional Heisenberg open spin-$\frac{1}{2}$ chain with nearest neighbor interactions is studied. Regular fractal structures appear in matrices of the Hamiltonian constructed under the outer-product basis of the permutation group, which enables one to solve the eigenvalue problem in a systematic manner.

The one-dimensional Heisenberg quantum spin-$\frac{1}{2}$ chain with nearest neighbor interactions is a basic model of physics. Solutions of this simple but important model have been studied extensively since the pioneering work of Bethe [1]. Besides semiclassical treatments [2]-[3], a lot of work has been focused on the Bethe ansatz [4]-[7]. The Bethe ansatz is certainly very useful in understanding the large N limit, where N is the number of sites, and has been developed into a powerful self-consistent method in dealing with a large number of quantum many-body problems. However, the Bethe ansatz solution is only a formal expression, one that is neither easy nor straightforward to use because its use requires the solution of a set of highly non-linear algebraic equations. Numerical algorithms for systematically solving such non-linear equations are generally not available. Since this limitation (numerical and computational) exists, a long-standing argument as to whether Bethe ansatz solutions are complete or not remains unresolved [6, 8].

The Hamiltonian can be written as

$$\hat{H} = 2J \sum_{i=1}^{N-1} \left(\hat{s}_i \cdot \hat{s}_{i+1} + \frac{1}{4} \right), \tag{1}$$

where the extra constant $\frac{1}{4}$ simplifies expressions that enter in a permutation group approach to the solution. It is well-known that (1) can be rewritten as

$$\hat{H} = J \sum_{i=1}^{N-1} g_i, \tag{2}$$

where $g_i = 2 \left(\hat{s}_i \cdot \hat{s}_{i+1} + \frac{1}{4} \right)$, with $i = 1, 2, \cdots, N-1$, are generators of the spin-component permutation group S_N. Let $c_{i\sigma}^\dagger$ be the creation operator of an electron at the i-th site with third component of spin σ, and $|0\rangle$ be the vacuum state. Then, $c_{1\sigma_1}^\dagger c_{2\sigma_2}^\dagger \cdots c_{k\sigma_k}^\dagger |0\rangle$ are basis vectors of the k-electron tensor product space $V^{(1)} \otimes V^{(2)} \otimes \cdots \otimes V^{(k)}$, where each space $V^{(i)}$ is two dimensional. The action of g_i ($i \leq k$) on the k-electron basis vector is given by

$$g_i \left(c_{1\sigma_1}^\dagger \cdots c_{i\sigma_i}^\dagger c_{i+1\sigma_{i+1}}^\dagger \cdots c_{k\sigma_k}^\dagger \right) |0\rangle = c_{1\sigma_1}^\dagger \cdots c_{i\sigma_{i+1}}^\dagger c_{i+1\sigma_i}^\dagger \cdots c_{k\sigma_k}^\dagger |0\rangle. \tag{3}$$

The generators g_i ($i = 1, 2, \cdots, N-1$) satisfy the following well-known relations

$$g_i g_{i+1} g_i = g_{i+1} g_i g_{i+1}, \quad g_i g_j = g_j g_i \text{ for } |i - j| \geq 2, \quad g_i^2 = 1. \tag{4}$$

Let $Y_m^{[\lambda]}$ be a standard Young tableau, and $|Y_m^{[\lambda]}\rangle$ the corresponding orthogonal basis vector, $\langle Y_m^{[\lambda]} | Y_{m'}^{[\lambda]}\rangle = \delta_{mm'}$. Here $[\lambda] \equiv [\lambda_1, \lambda_2, \cdots, \lambda_N]$, with $\lambda_1 \geq \lambda_2 \geq \cdots \geq \lambda_N$ and $\sum_{i=1}^{N} \lambda_i = N$, stands for a standard Young diagram with N boxes [9, 10], and m denotes a component of an irreducible representation (irrep) of S_N in the standard basis. Also, by construction, the $|Y_m^{[\lambda]}\rangle$ form a Yamanouchi basis relative to operations on the indices $(1, \cdots, N)$, where m can be understood either as a Yamanouchi symbol or the indice of the basis components in so-called decreasing page order of a Yamanouchi symbol [10]. Let $g_i Y_m^{[\lambda]}$ be the Young tableau obtained by interchanging the numbers i and $i+1$ in $Y_m^{[\lambda]}$. It is understood that if the resultant tableau is not a standard one the corresponding basis vector $|g_i Y_m^{[\lambda]}\rangle$ is set to zero. The irreducible representation of S_N in the standard basis, i.e. a basis adapted to the group chain $S_N \supset S_{N-1} \supset \cdots \supset S_2$, is given by

$$g_i |Y_m^{[\lambda]}\rangle = \frac{1}{d_i} |Y_m^{[\lambda]}\rangle + \left(\frac{(d_i + 1)(d_i - 1)}{d_i^2} \right)^{\frac{1}{2}} |g_i Y_m^{[\lambda]}\rangle, \tag{5}$$

where d_i is the axial distance from the box i to the box $i + 1$ in the Young tableau $Y_m^{[\lambda]}$ with movement upward and to the right being counted as positive.

It should be clear that Hamiltonian (2) is diagonalizable within a given irrep $[\lambda]$ of S_N. Furthermore, it can easily be proven that Hamiltonian (2) is invariant under $U(2)$ transformations that are generated by the total spin operators $\hat{S}_\mu$ ($\mu = 0, +, -$) and the total electron number operator $\hat{N}$ with

$$\hat{S}_+ = \sum_{i=1}^{N} c_{i\uparrow}^\dagger c_{i\downarrow}, \quad \hat{S}_- = \sum_{i=1}^{N} c_{i\downarrow}^\dagger c_{i\uparrow}, \quad \hat{S}_0 = \frac{1}{2} \sum_{i=1}^{N} \left(c_{i\uparrow}^\dagger c_{i\uparrow} - c_{i\downarrow}^\dagger c_{i\downarrow} \right), \quad \hat{N} = \sum_{i=1}^{N} \sum_{\sigma} c_{i\sigma}^\dagger c_{i\sigma}. \tag{6}$$

Therefore, Hamiltonian (2) should also be diagonalizable under an irrep $[\lambda]$ of $U(2)$. Due to the Schur-Weyl duality relation between the permutation group S_N and $U(2)$ in this case, any irrep $[\lambda]$ of $U(2)$ with exactly N boxes is simultaneously the same irrep of S_N. According to the Schur-Weyl duality relation, Hamiltonian (2) for N electrons on an N-site lattice with total spin $S = \frac{1}{2}(N - 2n)$ can be diagonalized in the irrep $[N - n, n]$ of S_N with basis vectors $|Y_m^{[N-n,n]}; S = \frac{1}{2}(N - 2n), S_0 = S\rangle$ which are simultaneously basis vectors of $U(2)$. Nevertheless, there are two drawbacks in this direct diagonalization process. Firstly, there is no regularity in the structure to the energy matrix for $n \geq 2$ due to the nature of the matrix elements that enter in (5). Therefore, one can only construct the energy matrix on a case-by-case basis for $n \geq 2$, which with increasing values of N and n very quickly becomes an intractable task. Secondly, the basis vectors $|Y_m^{[N-n,n]}; S = \frac{1}{2}(N - 2n), S_0 = S\rangle$ should be expanded in terms of products of single electron states, of which the expansion coefficients should be evaluated separately according to representation theory of the permutation group. Evaluation of these expansion coefficients is of the same level of complexity as diagonalization of Hamiltonian (2) within basis vectors of the $[N-n, n]$ irrep of S_N.

In the following, we will deal with this problem in an alternative way to avoid these drawbacks. It can be proven that eigenstates of (2) with $S_0 = \frac{1}{2}(N - 2n)$ can be written as

$$| \zeta, [N - \nu, \nu]; S = \frac{1}{2}(N - 2\nu), S_0 = \frac{1}{2}(N - 2n)\rangle = \sum_\omega \alpha_\omega^{(\zeta)} Q_\omega |[n] \downarrow; (\omega_1^0)\rangle |[N - n] \uparrow; (\omega_2^0)\rangle, \tag{7}$$

where $\nu = 0, 1, 2, \cdots, n$, $(\omega_1^0) = (12 \cdots n)$, $(\omega_2^0) = (n+1\ n+2 \cdots N)$, ζ is an additional quantum number needed to distinguish different eigenstates with the same total spin and its third component, $\alpha_\omega^{(\zeta)}$ expansion coefficients to be determined by the corresponding eigenequation, (ω) is the so-called normal-ordering sequences, $(\omega) = (\omega_1, \omega_2)$, $(\omega_1) = (a_1, a_2, \cdots, a_n)$, and $(\omega_2) = (a_{n+1}, a_{n+2}, \cdots, a_N)$ with $a_1 < a_2 < \cdots < a_n$, $a_{n+1} < a_{n+2} < \cdots < a_N$. General wavefunction with any allowed S_0 value can be obtained by applying the lowering $\hat{S}_-$ or raising $\hat{S}_+$ operators onto (7). Equation (7) is constructed according to the Littlewood rules of the outer-product $[n] \times [N - n] \downarrow [N - \nu, \nu]$ of the permutation group $S_n \times S_{N-n}$, $[n] \otimes [N - n] = \sum_{\nu=0}^n \oplus [N - \nu, \nu]$ as long as $N \geq 2n$. In our construction, we always choose n to satisfy condition $N \geq 2n$. Q_ω with $Q_\omega(\omega^0) = (\omega)$ in (7) is the left coset representative in the decomposition $S_n \times S_{N-n} \downarrow S_N$, $S_N = \sum_\omega Q_\omega (S_n \times S_{N-n})$. It should be noted that the number $N(\omega)$ of the left coset representatives Q_ω is equal to the sum of the dimensions of the irreps of S_N occurring in the outer-product, namely $N(\omega) = \sum_{\nu=0}^n \dim([N - \nu, \nu])$ with $\dim([N - \nu, \nu]) = N!(N - 2n + 1)/n!(N - n + 1)!$. The uncoupled basis vectors on the right-hand-side of (7) can be expressed in terms of the electron creation operators as

$$|[n] \downarrow; (a_1, a_2, \cdots, a_n)\rangle = c_{a_1\downarrow}^\dagger c_{a_2\downarrow}^\dagger \cdots c_{a_n\downarrow}^\dagger |0\rangle,$$

$$|[N - n] \uparrow; (a_{n+1}, a_{n+2}, \cdots, a_N)\rangle = c_{a_{n+1}\uparrow}^\dagger c_{a_{n+2}\uparrow}^\dagger \cdots c_{a_N\uparrow}^\dagger |0\rangle. \tag{8}$$

One also needs to first arrange the ordering of the basis vectors $\{Q_\omega|[n] \downarrow; (\omega_1^0)\rangle|[N-n] \uparrow; (\omega_2^0)\rangle \equiv |(\omega_1), (\omega_2)\rangle\}$. The ordering of the sequences (ω_1, ω_2) is specified in the following way. The (ω_1) part of (ω_1, ω_2) is regarded as a vector of dimension n. If the last nonzero component of the vector $(\omega_1) - (\omega_1')$ is less than zero, we arrange that (ω_1, ω_2) precedes (ω_1', ω_2'). For example, if $n = 2$ and $N = 4$, the ordering of the basis vectors is $\{|(12), (34)\rangle;$ $|(13), (24)\rangle, |(23), (14)\rangle; |(14), (23)\rangle, |(24), (13)\rangle, |(34), (12)\rangle\}$. After diagonalizing the energy matrix $h(N, n)$, with matrix elements given by

$$\langle(\omega_1), (\omega_2)|\hat{H}/J|(\omega_1'), (\omega_2')\rangle, \tag{9}$$

one simultaneously obtains eigenenergies and the corresponding expansion coefficients $\alpha_\omega^{(\zeta)}$ for all irreps $[N - \nu, \nu]$ of S_N with $\nu = 0, 1, 2, \cdots, n$. Direct calculation shows that the energy matrix $h(N, n)$ has the following simple structure:

$$h(N,n) = \begin{pmatrix} M_0^N(n) & \tilde{I}_{M_0^N(n)} & & & & \\ I_{M_0^N(n)} & M_1^N(n) & \tilde{I}_{M_1^N(n)} & & & \\ & I_{M_1^N(n)} & M_2^N(n) & \tilde{I}_{M_2^N(n)} & & \\ & & \ddots & \ddots & \ddots & \\ & & & I_{M_{N-n-2}^N(n)} & M_{N-n-1}^N(n) & \tilde{I}_{M_{N-n-1}^N(n)} \\ & & & & I_{M_{N-n-1}^N(n)} & M_{N-n}^N(n) \end{pmatrix},$$

$$\tag{10}$$

where $M_\mu^N(n)$ $(\mu = 0, 1, \cdots, N - n)$ is the energy sub-matrix formed under the subspace spanned by $\{|(\omega)\rangle \equiv Q_\omega|[n] \downarrow; (\omega_1^0)\rangle |[N-n] \uparrow; (\omega_2^0)\rangle\}$ with $(\omega) = (a_1, a_2, \cdots, a_{n-1}, n+\mu)$, $a_1 < a_2 < \cdots < a_{n-1}$, are different indices less than $n + \mu$, $I_{M_\mu^N(n)}$ is a $\dim(M_{\mu-1}^N(n)) \times \dim(M_\mu^N(n))$ matrix that is formed by adding $\dim(M_\mu^N(n)) - \dim(M_{\mu-1}^N(n))$ rows with 0 entries to the bottom of the $\dim(M_{\mu-1}) \times \dim(M_{\mu-1})$ identity matrix

$$I_{M_{\mu-1}} = \begin{pmatrix} 1 & 0 & & & & & \\ 0 & 1 & 0 & & & & \\ & \ddots & \ddots & \ddots & & & \\ & & 0 & 1 & 0 & \\ & & & 0 & 1 & \\ 0 & \cdots & \cdots & \cdots & 0 & \\ \vdots & & & & \vdots & \\ 0 & \cdots & \cdots & \cdots & 0 \end{pmatrix}. \tag{11}$$

Other entries not written explicitly in (10) are all zero. $\tilde{I}_{M_\mu}$ in (10) is the transposition of I_{M_μ}. Similarly to the tridiagonal case, the energy matrix (10) is called quasi-tridiagonal. The dimension of $M_\mu^N(n)$ is $(n + \mu - 1)!/\mu!(n - 1)!$. It is important to note that the energy sub-matrix $M_\mu^N(n)$ has the following recurring structure:

$$M_\mu^N(n) = \begin{pmatrix} M_0^N(n-1) - 2 + \delta_{N\,\mu+n} & \tilde{I}_{M_0^N(n-1)} & & & & \\ I_{M_0^N(n-1)} & M_1^N(n-1) - 2 + \delta_{N\,\mu+n} & \tilde{I}_{M_1^N(n-1)} & & & \\ & I_{M_1^N(n-1)} & M_2^N(n-1) - 2 + \delta_{N\,\mu+n} & \tilde{I}_{M_2^N(n-1)} & & \\ & & \ddots & \ddots & \ddots & \\ & & I_{M_{\mu-2}^N(n-1)} & M_{\mu-1}^N(n-1) - 2 + \delta_{N\,\mu+n} & \tilde{I}_{M_{\mu-1}^N(n-1)} \\ & & & I_{M_{\mu-1}^N(n)} & M_\mu^N(n-1) + \delta_{N\,\mu+n} \end{pmatrix} \tag{12}$$

for $\mu = 0, 1, 2, \cdots, N - n$, starting with

$$M_0^N(1) = N - 2 + \delta_{N1}, \quad M_1^N(1) = N - 3 + \delta_{N1} \quad \cdots,$$
$$M_{N-2}^N(1) = N - 3 + \delta_{N1}, \quad M_{N-1}^N(1) = N - 2 + \delta_{N1}. \tag{13}$$

Clearly, the energy matrix has self-similarity for different n values. Such a regular fractal structure of the energy matrix can be used to construct the energy matrix for any N and n values easily.

This work was supported by the U.S. National Science Foundation through a regular grant (9970769) and a Cooperative Agreement (9720652) that includes matching from the Louisiana Board of Regents Support Fund, and by the Natural Science Foundation of China (Grant No. 10175031) as well as by the Natural Science Foundation of Liaoning Province (Grant No. 2001101053).

References

[1] Bethe H 1931 Z. Phys. 71 205
[2] Anderson P W 1952 Phys. Rev. 86 694
[3] Haldane F D 1983 Phys. Rev. Lett. 50 1153
[4] Faddeev L D and Takhtajan L A 1979 Russ. Math. Survey 34 11
[5] Kulish P P and Sklyanin E K 1979 Phys. Lett. A70 461
[6] Essler F H L, Korepin V E and Schoutens K 1992 J. Phys. A25 4115
[7] Baxter R J 1982 Exactly Solved Models in Statistical Mechanics (New York: Academic)
[8] Siddharthan R 1998 Cond-mat/9804210
[9] Chen J Q 1989 Group Representation Theory for Physicists (Singapore: World Scientific)
[10] Pan F and Chen J Q 1993 J. Math. Phys. 34 4305

Inst. Phys. Conf. Ser. No 173: Section 5
Paper presented at 24th Int. Coll. Group Theoretical Methods in Physics, Paris, France, 15–20 July 2002
©2003 IOP Publishing Ltd

New applications of the group theory in the kinetic theory of gases

V. L. Saveliev

Institute of Ionosphere, Almaty 480020, Kazakhstan

Abstract. On the basis of a recently discovered collision group, a new renormalized form of the Boltzmann equation is obtained. A new class of discrete velocity models for gas mixtures is constructed.

1. Parameterization of collisions in the Boltzmann equation by a rotation matrix

Applications of the group theory remain underutilized in kinetic theory although they deserve to be [1-3]. Recently [2], it has been shown that if we parameterize the two-particle collision by a matrix $\hat{R}$ belonging to the group of rotations O_3^+ then the transformation of particle velocities due to collision becomes a linear one:

$$\xi' = \hat{S}\xi, \quad \xi = \begin{pmatrix} \mathbf{v} \\ \mathbf{u} \end{pmatrix}, \quad \hat{S} = \begin{pmatrix} \dfrac{m + \hat{R}}{1 + m} & \dfrac{1 - \hat{R}}{1 + m} \\ \dfrac{m(1 - \hat{R})}{1 + m} & \dfrac{1 + m\hat{R}}{1 + m} \end{pmatrix}, \quad m = \frac{m_1}{m_2} \tag{1}$$

and the scattering matrices $\hat{S}\left(\hat{R}\right)$, $\hat{R} \in O_3^+$ constitute a group that is isomorphic to the group of rotations O_3^+ :

$$\hat{S}(\hat{R}_1) \cdot \hat{S}(\hat{R}_2) = \hat{S}(\hat{R}_1 \cdot \hat{R}_2), \qquad \hat{S}^{-1}(\hat{R}) = \hat{S}(\hat{R}^{-1}) \tag{2}$$

To rewrite the collision integral from the conventional form [4] to the case when a collision is parameterized by rotation matrix [2], integration over directions of the relative velocity vector $\mathbf{v}'_r$ should be replaced by integration on the invariant measure [5,6] over the group O_3^+ ($d\Omega/4\pi \rightarrow d\hat{R}/8\pi^2$):

$$d\hat{R} = d\hat{R}_0\hat{R} = d\hat{R}\hat{R}_0 = d\hat{R}^{-1}, \qquad \int d\hat{R} = 8\pi^2. \tag{3}$$

The collision integral now takes the following form:

$$I(f,\psi) = I(F) = \int b\left(\mathrm{v_r},\mu\right) \left[f(\mathbf{v}')\psi(\mathbf{u}') - f(\mathbf{v})\psi(\mathbf{u})\right] \frac{d\hat{R}}{2\pi} d\mathbf{u}$$

$$= \int b\left(\mathrm{v_r},\mu\right) \left[F\left(\hat{S}(\hat{R})\xi\right) - F(\xi)\right] \frac{d\hat{R}}{2\pi} d\mathbf{u}, \tag{4}$$

where $b\left(\mathrm{v_r},\mu\right) = \mathrm{v_r}\sigma_\theta\left(\mathrm{v_r},\mu\right)$, $\mathbf{v}_r = \mathbf{v} - \mathbf{u}$ is a relative velocity, $\mu\left(\hat{R}, \mathbf{v}_r\right) = \mathbf{v}_r \cdot \hat{R}\mathbf{v}_r/\mathrm{v}_r{}^2$, $F(\xi) = f \circ \psi\,(\xi) = f(\mathbf{v})\psi(\mathbf{u})$ is the two-particle velocity distribution function.

2. Construction of new class of discrete velocity models for gas mixtures

The form of collision operator (4) obtained above is very useful for an application to the discrete velocity models [2]. A collision integral for these models can be constructed by replacing in eq.(4) an averaging over rotations from group O_3^+ by the averaging over K elements from some discrete subgroup $O_3(K) \subset O_3$:

$$I_d(f, \Psi) = \frac{4\pi}{K} \int d\mathbf{u} \sum_{k=1}^{K} b\left(\mathbf{v}_r, \hat{R}_k\right) \left[F\left(\hat{S}(\hat{R}_k)\xi\right) - F(\xi) \right] . \tag{5}$$

Presenting the distribution function $f_\alpha(\mathbf{v})$ by an expansion in terms of Dirac delta-functions concentrated on the velocities from the discrete velocity sets L_α ,

$$f_\alpha(\mathbf{v}) = \sum_{\mathbf{v}_0 \in L_\alpha} \bar{f}_\alpha(\mathbf{v}_0)\delta(\mathbf{v} - \mathbf{v}_0) , \tag{6}$$

and accordingly, the two-particle distribution function of species α and β by an expansion concentrated on bivectors from ensembles $L_{\alpha\beta}$ which are invariant under scattering transformations corresponding to subgroup $O_3(K)$,

$$F_{\alpha\beta}(\xi) = \sum_{\xi_0 \in L_{\alpha\beta}} \bar{f}_\alpha(\mathbf{v}_0)\bar{f}_\beta(\mathbf{u}_0)\,\delta(\mathbf{v} - \mathbf{v}_0)\,\delta(\mathbf{u} - \mathbf{u}_0) , \tag{7}$$

we obtain an analog of the Boltzmann equation for discrete models of gas mixtures with M species in the following explicit form:

$$\frac{\partial \bar{f}_\alpha(\mathbf{v}_0)}{\partial t} + \mathbf{v}_0 \cdot \frac{\partial \bar{f}_\alpha(\mathbf{v}_0)}{\partial \mathbf{r}} = \sum_\beta \sum_{\mathbf{u}_0 \in L_\beta(\mathbf{v}_0)} \frac{4\pi}{K} \sum_{k=1}^{K} b_{\alpha\beta}\left(\mathbf{v}_r, \hat{R}_k\right) \left[\bar{f}'_\alpha \bar{f}'_\beta - \bar{f}_\alpha \bar{f}_\beta \right] ,$$

$$\mathbf{v}_0 \in L_\alpha; \quad \alpha, \beta = 1, ... M.$$

Here L_α are sets of discrete velocities $\mathbf{v}_{0\alpha}$ for each mixture species (depend only on the mass m_α); $L_{\alpha\beta}$ are sets of colliding pairs $(\mathbf{v}_{0\alpha}, \mathbf{u}_{0\beta})^T \in L_{\alpha\beta}$; M is number of mixture components; a set $L_\beta(\mathbf{v}_0) \equiv \{\mathbf{u}_0 | (\mathbf{v}_0, \mathbf{u}_0) \in L_{\alpha\beta}\}$ means a set of all $\mathbf{u}_0$ for which the pair $(\mathbf{v}_0, \mathbf{u}_0)$ with a given $\mathbf{v}_0$ is contained in the invariant ensemble $L_{\alpha\beta}$.

2.1. Invariant ensembles of pairs of discrete velocities

To construct the invariant ensembles $L_{\alpha\beta}$ we represent the center of mass velocity and the momentum in the center of mass reference frame of the first particle in the pair by linear combinations of the basis vectors $\mathbf{w}_1, \mathbf{w}_2, \mathbf{w}_3$ and $m_0\mathbf{w}_1, m_0\mathbf{w}_2, m_0\mathbf{w}_3$, respectively (here m_0 is some unit mass).

$$\begin{aligned} \mathbf{w}(k_1, k_2, k_3) &= k_1\mathbf{w}_1 + k_2\mathbf{w}_2 + k_3\mathbf{w}_3 \\ \mathbf{p}(l_1, l_2, l_3) &= l_1 m_0\mathbf{w}_1 + l_2 m_0\mathbf{w}_2 + l_3 m_0\mathbf{w}_3. \end{aligned} \tag{9}$$

A set of six numbers $l_1, l_2, l_3, k_1, k_2, k_3$ in accordance with (9) will define discrete velocity pairs before and after collision as follows:

$$\mathbf{v}_\alpha = \left(\frac{m_0}{m_\alpha}l_1 + k_1\right)\mathbf{w}_1 + \left(\frac{m_0}{m_\alpha}l_2 + k_2\right)\mathbf{w}_2 + \left(\frac{m_0}{m_\alpha}l_3 + k_3\right)\mathbf{w}_3 \in L_\alpha;$$

$$\mathbf{u}_\beta = \mathbf{v}_\beta(-l_1, -l_2, -l_3, k_1, k_2, k_3) \in L_\beta; l_1, l_2, l_3, k_1, k_2, k_3 = 0, \pm 1, \pm 2, ...;$$

$$\mathbf{v}'_\alpha = \left(\frac{m_0}{m_\alpha}l_1\hat{R} + k_1\right)\mathbf{w}_1 + \left(\frac{m_0}{m_\alpha}l_2\hat{R} + k_2\right)\mathbf{w}_2 + \left(\frac{m_0}{m_\alpha}l_3\hat{R} + k_3\right)\mathbf{w}_3$$

$$\mathbf{u}'_{0\beta} = \mathbf{v}'_\beta(-l_1, -l_2, -l_3, k_1, k_2, k_3) ; \tag{10}$$

To provide an invariance of the ensemble of velocity pairs (10), the vectors $\mathbf{w}_1, \mathbf{w}_2, \mathbf{w}_3$ should be the basic vectors of Bravais lattice invariant with respect to the discrete subgroup $O_3(K)$,

$$\hat{R}\mathbf{w}_i = \sum_k \mathbf{w}_k T_{ki}\left(\hat{R}\right), \tag{11}$$

where $T_{ki}\left(\hat{R}\right)$ are unimodular integer matrices. It is known in the theory of crystals [5] that there are 14 types of Bravais lattices and 7 point symmetry groups of these lattices. For example, the simple cubic lattice Γ_c is invariant under transformations from the group of cube symmetry $O_h(48)$.

3. Renormalization of the Boltzmann collision integral

In accordance with general properties [5, 6] of Lie groups we can express scattering matrices and representations of scattering matrices in the Hilbert space of functions on bivector ξ in the exponential form:

$$\hat{S} = e^{\phi\hat{c}}, \quad \hat{c} = \frac{1}{1+m}\begin{pmatrix} \hat{n} & -\hat{n} \\ -m\hat{n} & m\hat{n} \end{pmatrix}, \quad \hat{n}\mathbf{v}_r \overset{df}{=} \mathbf{n} \times \mathbf{v}_r \tag{12}$$

$$F\left(\hat{S}^{-1}\xi\right) = e^{\phi\hat{\sigma}}F(\xi), \quad \hat{\sigma} = -\frac{1}{1+m}\mathbf{n} \cdot \mathbf{v}_r \times \left(\frac{\partial}{\partial\mathbf{v}} - m\frac{\partial}{\partial\mathbf{u}}\right) \tag{13}$$

The invariant measure for the group of proper rotations is as follows when rotations are parameterized by the angle of rotation ϕ and the direction $\mathbf{n}$ of the rotation axis (or ϕ, φ and θ, the latter being the angle between vectors $\mathbf{v}_r$ and $\mathbf{n}$, φ is azimuthal angle) [5,6]:

$$d\hat{R} = 2(1 - \cos\phi)d\phi d\Omega_{\mathbf{n}}, \quad d\Omega_{\mathbf{n}} = \sin\theta d\theta d\varphi,$$
$$(0 \leq \phi, \theta \leq \pi; \ 0 \leq \varphi \leq 2\pi) \tag{14}$$

Making use of (12) and (13), we obtain the following form for the Boltzmann collision integral:

$$I(f, \psi) = \int b\left(\mathbf{v}, \mu\right)\left[f(\mathbf{v}')\psi(\mathbf{u}') - f(\mathbf{v})\psi(\mathbf{u})\right]\frac{d\hat{R}}{2\pi}d\mathbf{u}$$

$$= \int d\mathbf{u}\frac{d\hat{R}}{2\pi}b\left(\mathbf{v}, \mu\right)\left[e^{-\phi\hat{\sigma}} - 1\right]f(\mathbf{v})\psi(\mathbf{u})$$

$$= \int d\mathbf{u}\frac{d\hat{R}}{2\pi}\left[e^{-\phi\hat{\sigma}} - 1\right]b\left(v, \mu\right)f(\mathbf{v})\psi(\mathbf{u}). \tag{15}$$

To separate the singular part (in the case of infinite cross sections) of the collision integral from the regular one, we can use for scattering operator $e^{-\phi\hat{\sigma}}$ the Taylor series with a residual term:

$$e^{-\phi\hat{\sigma}} = 1 - \phi\hat{\sigma} + \ldots + \frac{(-1)^{n-1}}{(n-1)!}\phi^{n-1}\hat{\sigma}^{n-1} + \frac{(-1)^n}{n!}\phi^n\hat{\sigma}^n\int_0^1 d\alpha q_n\left(\alpha\right)e^{-\alpha\phi\hat{\sigma}}, \tag{16}$$

where $n = 1, 2, \cdots$ and $q_n(\alpha) = n\left(1 - \alpha\right)^{n-1}$, $\int_0^1 d\alpha q_n(\alpha) = 1$. Finally, the collision integral can be rewritten (exactly for any $n > 0$) in the following divergence form [1]:

$$I(f, \psi) = -\frac{\partial}{\partial\mathbf{v}} \cdot \mathbf{J}(f, \psi), \tag{17}$$

where the flow in the velocity space is given by

$$
\mathbf{J} = \frac{-1}{1+m} \int d\mathbf{u}\, \mathbf{v}_r \times \left\langle \mathbf{n} \left\{ \sum_{k=1}^{2k<n} \frac{\phi^{2k}\hat{\sigma}^{2k-1}}{(2k)!} \right. \right.
$$
$$
\left. \left. + \frac{\phi^{n}\hat{\sigma}^{n-1}}{2 \cdot n!} \left(e^{\alpha\phi\hat{\sigma}} + (-1)^{n} e^{-\alpha\phi\hat{\sigma}} \right) \right\} \right\rangle f\psi, \quad (18)
$$

$$
\langle ... \rangle = \int d\alpha \frac{d\phi}{\pi} d\Omega_{\mathbf{n}} q_n (\alpha) (1 - \cos\phi) b(v, \mu) [...] .
$$

As a result, the Boltzmann equation can be rewritten in the form of the Liouville equation:

$$
\frac{\partial f}{\partial t} + \mathbf{v} \cdot \frac{\partial f}{\partial \mathbf{r}} + \frac{\partial}{\partial \mathbf{v}} \cdot (\frac{e_1}{m_1}\mathbf{E} + \frac{e_1}{m_1 c}\mathbf{v} \times \mathbf{B} + \mathbf{g} + \frac{1}{m_1}\mathbf{F}_{\text{coll}})f = 0, \quad (19)
$$

where, in addition to the usual electromagnetic and gravity forces, we have non-local friction force $\mathbf{F}_{coll} = m_1 \mathbf{J}/f(\mathbf{v})$, which depends on the distribution functions $f(\mathbf{v})$ and $\psi(\mathbf{v})$. In the case $n = 1$ this force has the following expression:

$$
\mathbf{F}_{coll} = \frac{m_1 m_2}{m_1 + m_2} \int d\mathbf{u} d\alpha \frac{d\hat{R}}{2\pi} b(\mathbf{v}_r, \mu) \frac{\phi}{2} [\mathbf{v}_r \times \mathbf{n}]
$$
$$
\frac{[f(\mathbf{v}'_\alpha)\psi(\mathbf{u}'_\alpha) - f(\mathbf{v}'_{-\alpha})\psi(\mathbf{u}'_{-\alpha})]}{f(\mathbf{v})}, \quad (20)
$$

where the new "post-collisional" velocities can be expressed as

$$
\mathbf{v}'_\alpha = \mathbf{v} + \frac{[(1 - \cos\alpha\phi)\hat{n}^2 + \sin(\alpha\phi)\hat{n}](\mathbf{v} - \mathbf{u})}{1 + m},
$$
$$
\mathbf{u}'_\alpha = \mathbf{u} - \frac{m[(1 - \cos\alpha\phi)\hat{n}^2 + \sin(\alpha\phi)\hat{n}](\mathbf{v} - \mathbf{u})}{1 + m}. \quad (21)
$$

Equation (19) allows us to consider the distribution function $f(\mathbf{v}, \mathbf{r}, t)$ as a density in the phase space of the points which are moving along smooth trajectories under the influence of non local force. The points do not jump any more as it was in the case of the classical Boltzmann equation. This equation provides new opportunities for numerical simulation of gas flows.

3.1. References

[1] V.L. Saveliev and K. Nanbu, "Collision group and renormalization of the Boltzmann collision integral", Phys. Rev. E 65, 051205, p.1-9, (2002)
[2] V.L. Saveliev, "A Parameterization of Collisions in the Boltzmann Equation by a Rotation Matrix and Boltzmann Collision Integral in Discrete Models of Gas Mixtures" Rarefied Gas Dynamics: 22nd Int. Symp., ed. by T. J. Bartel and M. A. Gallis, AIP Conf. Proc.No. 585 (AIP. Melville. NY. 2001), p.101
[3] V.L. Saveliev, "Temperature and mass dependence of the Boltzmann linear collision operator from the group theory point of view," J Math Phys 37, 6139-6151 (1996).
[4] J. H. Ferziger and H.G. Kaper, Mathematical Theory of Transport Processes in Gases, (North-Holland, Amsterdam, 1972).
[5] M. Hamermesh, Group Theory and its Application to Physical Problems. (Addison - Wesley. Reading. MA. 1964).
[6] A.O. Barut and R. Raczka, Theory of Group Representations and Applications, 2nd rev. ed. (World Scientific. Singapore. 1986).

Inst. Phys. Conf. Ser. No 173: Section 5
Paper presented at 24th Int. Coll. Group Theoretical Methods in Physics, Paris, France, 15–20 July 2002
©*2003 IOP Publishing Ltd*

Ladder Operators for Integrable One-Dimensional Lattice Models

M C Takizawa ‡ and J R Links

Department of Mathematics, University of Queensland, Brisbane, QLD 4072, Australia

Abstract. A generalised ladder operator is used to construct the conserved operators for any one-dimensional lattice model derived from the Yang-Baxter equation. As an example, the low order conserved operators for the XYh model are calculated explicitly.

1. Introduction

The method for constructing integrable one-dimensional lattice models from solutions of the Yang-Baxter equation is well known (eg. see [1]). In principle, the conserved operators can be obtained by series expansion of the family of commuting transfer matrices. A more practical approach is to use the ladder operator which permits a recursive method through repeated commutators to obtain the conserved operators.

For models where the solution of the Yang-Baxter equation has the difference property, it has been established [2, 3, 4] that the ladder operator is a lattice analogue of the boost operator for Lorentz invariant systems. Recently it has been shown that for the Hubbard model, which is not Lorentz invariant in the continuum limit as a consequence of spin-charge separation, and is reflected by the fact that the solution of the Yang-Baxter equation does not have the difference property, the ladder operator still exists [5].

The present work extends [5] to develop a general theory for the construction of the ladder operator for any integrable system obtained through the Yang-Baxter equation. The theory will be applied to analyse the conservation laws for the XY model in a transverse magnetic field.

2. Integrable Lattice Models using the Quantum Inverse Scattering Method

We begin with a vector-dependent solution of the Yang-Baxter equation

$$R_{12}(\vec{u}, \vec{v})R_{13}(\vec{u}, \vec{w})R_{23}(\vec{v}, \vec{w}) = R_{23}(\vec{v}, \vec{w})R_{13}(\vec{u}, \vec{w})R_{12}(\vec{u}, \vec{v})$$

where $\vec{u}$, $\vec{v}$ and $\vec{w}$ are m-component vectors. Throughout, we assume the regularity property $R(\vec{u}, \vec{u}) = P$. Define a set of m local Hamiltonians

$$h_l\{i\} = P \cdot \left. \frac{\partial R_{l(l+1)}(\vec{u}, \vec{v})}{\partial u_i} \right|_{\vec{u} = \vec{v}}, \quad i = 1, .., m$$

with the corresponding global Hamiltonians acting on a one-dimensional lattice of length L given by

$$H\{i\} = \sum_{l=0}^{L-1} h_l\{i\}.$$

‡ mct@maths.uq.edu.au

Throughout, periodic boundary conditions are assumed on all summations which are evaluated over the length of the lattice. Note it is implicit that all the operators $h\{i\}$ are in fact functions of $\vec{v}$.

The transfer matrix is constructed through

$$T(\vec{u}, \vec{v}) = \mathrm{tr}_a \left(R_{a(L-1)}(\vec{u}, \vec{v})...R_{a1}(\vec{u}, \vec{v}) R_{a0}(\vec{u}, \vec{v}) \right)$$

where a refers to the auxiliary space, which by the standard argument gives rise to a commutative family in the first variable; i.e.

$$[T(\vec{u}, \vec{v}), T(\vec{w}, \vec{v})] = 0, \quad \forall \vec{u}, \vec{w}. \tag{1}$$

It can also be easily verified that

$$[H\{i\}, T(\vec{u}, \vec{v})] = 0, \quad \forall \vec{u}. \tag{2}$$

It is convenient, however, to define the conserved operators as

$$t\{\vec{n}\} = \left[\frac{\partial^{n_1 + ... + n_m}}{\partial u_1^{n_1}...\partial u_m^{n_m}} \ln T(\vec{u}, \vec{v}) \right]$$

where they appear in the series expansion

$$\ln T(\vec{u}, \vec{v}) = \sum_{\vec{n}} \frac{(u_1 - v_1)^{n_1}...(u_m - v_m)^{n_m}}{n_1!...n_m!} t\{\vec{n}\}. \tag{3}$$

Thus it follows from (1) that

$$[t\{\vec{n}\}, t\{\vec{k}\}] = 0, \quad \forall \vec{n}, \vec{k}$$

and moreover from (2)

$$[H\{i\}, t\{\vec{n}\}] = 0, \quad \forall i, \vec{n}.$$

Note that $\vec{n}$ is an m-component vector with non-negative integer entries. Introducing the notation $\{\vec{\varepsilon}_i\}_{i=1}^m$ for the basis of the m-dimensional vector space, we can write

$$\vec{n} = \sum_{i=1}^m n_i \vec{\varepsilon}_i.$$

3. Recursion Formula for Calculating the Conserved Operators

For each of the index labels i we define a ladder operator

$$B\{i\} = \sum_{l=0}^{L-1} l h_l\{i\}$$

with the coefficients l taken from the set of integers modulo L. For any function ϕ admitting a Taylor's series expansion we have

$$[B\{i\}, \phi(\mathscr{T})] = \mathscr{T}.H\{i\}.\phi'(\mathscr{T})$$

where $\mathscr{T} = T(\vec{u}, \vec{u})$ and ϕ' denotes the derivative of ϕ. Choosing ϕ to be the logarithm now gives

$$[B\{i\}, \ln \mathscr{T}] = H\{i\}.$$

It can be shown that

$$[B\{i\}, T(\vec{u}, \vec{v})] = -\frac{\partial T(\vec{u}, \vec{v})}{\partial v_i}. \tag{4}$$

As a result we obtain the following recursion formula from (4) and the expansion (3)

$$t\{\vec{n} + \vec{\varepsilon}_i\} = [B\{i\}, t\{\vec{n}\}] + \frac{\partial t\{\vec{n}\}}{\partial v_i} \tag{5}$$

The first few terms in (3) can be identified immediately

$$t\{\vec{0}\} = \ln \mathscr{T}, \qquad t\{\vec{\varepsilon}_i\} = H\{i\}. \tag{6}$$

In principle, through repeated use of (4) expressions for all the operators $t\{\vec{n}\}$ may be obtained.

Applying the recursion (5), the second order conserved currents can be obtained by the following formula:

$$t\{\vec{\varepsilon}_i + \vec{\varepsilon}_j\} = \frac{1}{2}\sum_l [h_l\{j\}, h_{l-1}\{i\}] + \frac{1}{2}\sum_l [h_l\{i\}, h_{l-1}\{j\}] + \frac{1}{2}\frac{\partial H\{j\}}{\partial v_i} + \frac{1}{2}\frac{\partial H\{i\}}{\partial v_j}. \tag{7}$$

4. The XYh Model

The XY model in a transverse magnetic field has the following Hamiltonian:

$$H = \sum_{i=1}^{N}(J_x\sigma_i^x\sigma_{i+1}^x + J_y\sigma_i^y\sigma_{i+1}^y + h\sigma_i^z) \qquad J_x, J_y, h \text{ const.}$$

This model is known to be integrable [6]. Barouch and Fuchssteiner [7], Araki [8], and Grabowski and Mathieu [9] have explicitly calculated the low order conserved operators. These results have been reproduced using the generalised ladder operator method.

4.1. R Matrix of the XYh Model

Bazhanov and Stroganov [10] constructed an elliptic parametrization for the Boltzmann vertex weights of the XYh model. In this parametrization, the weights are meromorphic functions of 3 complex variables, $\vec{u} = (u_1, u_2)$ and $\vec{v} = (v_1, v_2)$, where only the first vector entry contains the difference property.

The R matrix is

$$R(\vec{u}, \vec{v}) = \begin{pmatrix} R_{11}^{11} & 0 & 0 & R_{22}^{11} \\ 0 & R_{12}^{12} & R_{21}^{12} & 0 \\ 0 & R_{12}^{21} & R_{21}^{21} & 0 \\ R_{11}^{22} & 0 & 0 & R_{22}^{22} \end{pmatrix}$$

with

$$R_{22}^{22} = \rho(1 - e(u_1 - v_1)e(u_2)e(v_2)) \qquad R_{11}^{11} = \rho(e(u_1 - v_1) - e(u_2)e(v_2))$$

$$R_{21}^{21} = \rho(e(u_2) - e(u_1 - v_1)e(v_2)) \qquad R_{12}^{12} = \rho(e(v_2) - e(u_1 - v_1)e(u_2))$$

$$R_{21}^{12} = R_{12}^{21} = \frac{\rho\sqrt{e(u_2)s(u_2)}\sqrt{e(v_2)s(v_2)}(1 - e(u_1 - v_1))}{s\left(\frac{u_1 - v_1}{2}\right)}$$

$$R_{22}^{11} = R_{11}^{22} = -ik\rho\sqrt{e(u_2)s(u_2)}\sqrt{e(v_2)s(v_2)}(1 + e(u_1 - v_1))s\left(\frac{u_1 - v_1}{2}\right)$$

where ρ is an arbitrary constant and s and e are the respective elliptic functions sn and $(cn + isn)$. By imposing the condition $R(\vec{u}, \vec{u}) = P$, we obtain the value of $\rho = \frac{1}{1 - e^2(v_2)}$.

4.2. Local Hamiltonians

The local Hamiltonians $h_l\{1\}$ and $h_l\{2\}$ are given as follows:

$$h_l\{1\} = \rho[A(\sigma^x \otimes \sigma^x) + B(\sigma^y \otimes \sigma^y) + C(I \otimes I) + D(I \otimes \sigma^z + \sigma^z \otimes I)]$$
$$h_l\{2\} = \rho[E(I \otimes I) + F(\sigma^x \otimes \sigma^y - \sigma^y \otimes \sigma^x)]$$

where $\sigma^x, \sigma^y, \sigma^z$ are the Pauli sigma matrices and

$$A = -\frac{1}{2}ie(v_2)(1 + ks(v_2)), \quad B = -\frac{1}{2}ie(v_2)(1 - ks(v_2)), \quad k \text{ const.}$$

$$C = s(v_2)e(v_2), \qquad\qquad D = \frac{1}{2}ic(v_2)e(v_2),$$

$$E = -id(v_2)e(v_2)^2, \qquad F = \frac{1}{2}d(v_2)e(v_2).$$

where d and c are the elliptic functions dn and cn respectively.

4.3. Second Order Conserved Currents

The second order conserved currents can be obtained by the formula (7):

$$t\{2\vec{e}_1\} = t\{2\vec{e}_2\} = \sum_l \{\alpha(\sigma^x \otimes \sigma^z \otimes \sigma^y - \sigma^y \otimes \sigma^z \otimes \sigma^x)$$
$$+ \beta(\sigma^x \otimes \sigma^y \otimes I - \sigma^y \otimes \sigma^x \otimes I)\} \quad \alpha, \beta \text{ const.}$$

and

$$t\{\vec{e}_1 + \vec{e}_2\} = \sum_l \{\gamma(\sigma^x \otimes \sigma^z \otimes \sigma^x) + \zeta(\sigma^y \otimes \sigma^z \otimes \sigma^y)$$
$$+ \eta(\sigma^x \otimes \sigma^x \otimes I + \sigma^y \otimes \sigma^y \otimes I)$$
$$- (\gamma + \zeta)(\sigma^z \otimes I \otimes I)\} \qquad \gamma, \zeta, \eta \text{ const.}$$

in agreement with [7]-[9].

References

[1] Faddeev L D 1995 *Int. J. Mod. Phys. A* **10** 1845
[2] Thacker H B 1986 *Physica (Amsterdam) D* **18** 348
[3] Sogo K and Wadati M 1983 *Prog Theor. Phys.* **69** 431
[4] Tetel'man M G 1982 *Sov. Phys. JETP* **55** 306
[5] Links J R, Zhou H Q, McKenzie R H and Gould M D 2001 *Phys. Rev. Lett.* **86** 5096
[6] Krinsky S 1972 *Phys. Lett. A* **39** 169
[7] Barouch E and Fuchssteiner B 1985 *Stud. Appl. Math.* **73** 221
[8] Araki H 1990 *Commun. Math. Phys.* **132** 155
[9] Grabowski M P and Mathieu P 1996 *J.Phys. A* **29** 7635
[10] Bazhanov V V and Stroganov Yu G 1985 *Teor. Mat. Fiz.* **62** 377

Inst. Phys. Conf. Ser. No 173: Section 5
Paper presented at 24th Int. Coll. Group Theoretical Methods in Physics, Paris, France, 15–20 July 2002
©*2003 IOP Publishing Ltd*

Macroscopic properties of A-statistics and A-superstatistics

T D Palev‡ and J Van der Jeugt

Ghent University, Krijgslaan 281-S9, B-9000 Gent, Belgium.

Abstract. In the context of Lie (super)algebras of type A, generalized quantum statistics is considered. For $sl(n + 1)$ this is A-statistics, and for $sl(1|n)$ this is A-superstatistics. The creation and annihilation operators, their triple relations, and their Fock spaces are given. We consider a 'free' Hamiltonian and some macroscopic properties are studied : the grand partition function, average number of particles, etc. Symmetric functions play a role in expressions for these thermodynamical quantities.

Many years after Green [1] introduced para-Bose and para-Fermi statistics, as generalizations of Bose-Einstein and Fermi-Dirac statistics, the algebraic structure of these statistics became clear. It was shown [2, 3] that n pairs of parafermions generate the Lie algebra $so(2n+1) = B_n$, and that n pairs of parabosons generate [4] the Lie superalgebra $osp(1|2n) = B(0, n)$. So para-Fermi and para-Bose statistics fall into the class of B-statistics. In this paper, we shall consider the generalized statistics related to the Lie algebra $A_n = sl(n + 1)$ (A-statistics) and to the Lie superalgebra $A(0, n) = sl(1|n)$ (A-superstatistics) and some of their thermodynamical quantities. We first summarize the properties of A-statistics, given in [5], in order to compare them with those of A-superstatistics (the details of which will be published elsewhere).

The CAO's (creation and annihilation operators) of A-statistics are defined [6] as Jacobson generators $a_1^\pm, a_2^\pm, \ldots, a_n^\pm$ of $sl(n + 1)$ satisfying

$$[[a_i^+, a_j^-], a_k^+] = \delta_{kj} a_i^+ + \delta_{ij} a_k^+,$$
$$[[a_i^+, a_j^-], a_k^-] = -\delta_{ki} a_j^- - \delta_{ij} a_k^-, \qquad (1)$$
$$[a_i^+, a_j^+] = [a_i^-, a_j^-] = 0.$$

The Lie algebra generated by the $2n$ elements $a_i^\pm$ ($i = 1, \ldots, n$) subject to the relations (1) is $sl(n + 1)$ [6]. In terms of the usual Weyl generators e_{ij} ($i, j = 0, 1, \ldots, n$) of $sl(n + 1)$, one can take $a_i^+ = e_{i0}$ and $a_i^- = e_{0i}$. In [5], the Fock spaces $W(p, n)$ of A-statistics are constructed. These are labelled by $p \in \mathbb{N}$, and can be defined by a vacuum vector $|0\rangle$ and the relations

$$a_k^- |0\rangle = 0, \qquad a_i^- a_j^+ |0\rangle = \delta_{ij} p |0\rangle. \qquad (2)$$

$W(p, n)$ is a finite-dimensional irreducible $sl(n + 1)$-module with a basis consisting of all vectors $(a_1^+)^{l_1} (a_2^+)^{l_2} \ldots (a_n^+)^{l_n} |0\rangle$ subject to the restriction

$$|l| = l_1 + l_2 + \cdots + l_n \leq p. \qquad (3)$$

The usual Hermitian form on $W(p, n)$ gives an inner product, and the vectors

$$|p; l_1, \ldots, l_n\rangle = \sqrt{\frac{(p - |l|)!}{p!}} \frac{(a_1^+)^{l_1} \cdots (a_n^+)^{l_n}}{\sqrt{l_1! l_2! \cdots l_n!}} |0\rangle, \qquad (4)$$

‡ Permanent address : INRNE, Tsarigradsko Chaussee 72, 1784 Sofia, Bulgaria.

$(|l| = l_1 + l_2 + \cdots + l_n \le p)$ constitute an orthonormal basis. The action of the CAO's of A_n on this basis is given by :

$$a_i^+|p; l_1, \ldots, l_n\rangle = \sqrt{(l_i + 1)(p - |l|)}\,|p; l_i + 1\rangle, \tag{5}$$

$$a_i^-|p; l_1, \ldots, l_n\rangle = \sqrt{l_i(p - |l| + 1)}\,|p; l_i - 1\rangle, \tag{6}$$

where in the rhs we have written only the label that changes. The Fock space $W(p, n)$ is also an irreducible $gl(n + 1)$ module, and denoting the Weyl generators e_{ii} by N_i, one has

$$N_i|p; l_1, \ldots, l_n\rangle = l_i|p; l_1, \ldots, l_n\rangle, \quad i = 1, \ldots, n. \tag{7}$$

The Hamiltonian for which macroscopic properties will be studied is the "free" Hamiltonian $H = \sum_{i=1}^{n} \epsilon_i N_i$. This Hamiltonian satisfies $[H, a_i^{\pm}] = \pm\epsilon_i a_i^{\pm}$, so a_i^+ (a_i^-) can be interpreted as an operator creating (annihilating) a particle (quasiparticle, excitation) on orbital i with energy ϵ_i. Since

$$H|p; l_1, l_2, \ldots, l_n\rangle = (\epsilon_1 l_1 + \epsilon_2 l_2 + \cdots + \epsilon_n l_n)|p; l_1, l_2, \ldots, l_n\rangle, \tag{8}$$

the state $|p; l_1, \ldots, l_n\rangle$ can be interpreted as a state with l_1 particles on the first orbital, l_2 particles on the second, etc. In our picture, H has the form of a free Hamiltonian, and the interaction is inherently introduced via the statistical properties of A-statistics. These statistical properties follow from the structure of the Fock spaces; in particular, (3) leads to the Pauli principle of A-statistics : if the system is in the representation $W(p, n)$, it cannot accommodate more than p particles.

In order to study macroscopic properties of A-statistics, denote by τ the (fundamental) temperature of the system and let μ_i be the chemical potential for particles on orbital i. The probability $\mathcal{P}(p, n; r)$ for the system to be in a (quantum) state $r = (l_1, \ldots, l_n)$ (with $N_r = l_1 + \cdots + l_n$ and energy $E_r = l_1\epsilon_1 + \cdots + l_n\epsilon_n$) is given by :

$$\mathcal{P}(p, n; r) = \exp\left(\sum_{i=1}^{n}(\frac{\mu_i - \epsilon_i}{\tau})l_i\right)/Z(p, n), \tag{9}$$

where $Z(p, n)$ is the grand partition function (GPF) :

$$Z(p, n) = \sum_{0 \le l_1 + \cdots + l_n \le p} \left(\exp(\frac{\mu_1 - \epsilon_1}{\tau})\right)^{l_1} \cdots \left(\exp(\frac{\mu_n - \epsilon_n}{\tau})\right)^{l_n}. \tag{10}$$

Using the notation $x_i = \exp(\frac{\mu_i - \epsilon_i}{\tau})$, $i = 1, \ldots, n$, one can show that this is equal to

$$\sum_{0 \le l_1 + \cdots + l_n \le p} x_1^{l_1} x_2^{l_2} \cdots x_n^{l_n} = \sum_{k=0}^{p} h_k(x_1, \ldots, x_n) = h_p(x_1, \ldots, x_n, 1), \tag{11}$$

where h_k are the *complete symmetric functions* [7]. Once this has been realized, also various other thermodynamical quantities (distribution function, average energy, ...) can be expressed in terms of these symmetric functions [5]. In that same paper, some special cases are considered. One of these is the degenerate case, where all orbitals i have the same energy and chemical potential; then the various thermodynamical functions simplify and they can easily be compared with the case of Bose-Einstein statistics. Another special case is that of equidistant energy levels, also leading to interesting specializations of the general thermodynamical functions.

Let us now describe the situation of A-superstatistics. Here, the CAO's are defined as (odd) Jacobson generators $f_i^{\pm}$ $(i = 1, \ldots, n)$ of the Lie superalgebra $sl(1|n)$ satisfying

$$\{f_i^+, f_j^+\} = \{f_i^-, f_j^-\} = 0,$$

$$[\{f_i^+, f_j^-\}, f_k^+] = \delta_{jk}f_i^+ - \delta_{ij}f_k^+, \tag{12}$$

$$[\{f_i^+, f_j^-\}, f_k^-] = -\delta_{ik}f_j^- + \delta_{ij}f_k^-.$$

The Lie superalgebra generated by the $2n$ odd elements $f_i^{\pm}$ ($i = 1, \ldots, n$) subject to the relations (12) is $A(0, n) = sl(1|n)$ [8]. In terms of the usual Weyl generators e_{ij} ($i, j = 0, 1, \ldots, n$) of $sl(1|n)$, one can again take $a_i^+ = e_{i0}$ and $a_i^- = e_{0i}$. The Fock spaces $V(p, n)$ of A-superstatistics are also labelled by $p \in \mathbb{N}$, and defined by

$$f_i^-|0\rangle = 0, \qquad f_i^- f_j^+|0\rangle = p\,\delta_{ij}|0\rangle. \tag{13}$$

These Fock spaces are finite-dimensional unitary irreducible $sl(1|n)$-modules, with a basis consisting of vectors $(f_1^+)^{\theta_1}(f_2^+)^{\theta_2}\cdots(f_n^+)^{\theta_n}|0\rangle$, where

$$\theta_i \in \{0, 1\}, \qquad |\theta| \equiv \sum_{i=1}^{n} \theta_i \leq p. \tag{14}$$

For $p < n$, these are so-called atypical representations of $sl(1|n)$; for $p \geq n$, these representations are typical and in that case $\dim V(p, n) = 2^n$. The usual Hermitian form on $V(p, n)$ gives again an inner product, and the vectors

$$|p; \theta\rangle \equiv |p; \theta_1, \ldots, \theta_n\rangle = \sqrt{\frac{(p - |\theta|)!}{p!}}(f_1^+)^{\theta_1}(f_2^+)^{\theta_2}\cdots(f_n^+)^{\theta_n}|0\rangle, \tag{15}$$

subject to the condition (14), constitute an orthonormal basis of $V(p, n)$. Under the action of the CAO's, this basis transforms as follows :

$$f_i^-|p; \theta\rangle = \theta_i(-1)^{\theta_1 + \cdots + \theta_{i-1}}\sqrt{p - |\theta| + 1}\,|p; \theta_i - 1\rangle,$$
$$f_i^+|p; \theta\rangle = (1 - \theta_i)(-1)^{\theta_1 + \cdots + \theta_{i-1}}\sqrt{p - |\theta|}\,|p; \theta_i + 1\rangle. \tag{16}$$

If we denote again the Weyl generator e_{ii} by N_i, and introduce the Hamiltonian $H = \sum_{i=1}^{n} \epsilon_i N_i$, then $[H, f_i^{\pm}] = \pm\epsilon_i f_i^{\pm}$, so f_i^+ (f_i^-) can be interpreted as an operator creating (annihilating) a particle on orbital i with energy ϵ_i. Here, the statistics is closer related to Fermi-Dirac statistics in the sense that on each orbital there can be at most one particle. On the other hand, A-superstatistics differs from Fermi-Dirac statistics in the sense that also the order of statistics p plays a role, and no more than p particles can be accommodated if the system is in the representation $V(p, n)$.

Let us now investigate some macroscopic properties of A-superstatistics. With a notation similar to that of A-statistics, the grand partition function becomes :

$$Z(p, n) = \sum_{\substack{0 \leq \theta_1 + \cdots + \theta_n \leq p \\ \theta_i \in \{0,1\}}} \left(\exp(\frac{\mu_1 - \epsilon_1}{\tau})\right)^{\theta_1} \cdots \left(\exp(\frac{\mu_n - \epsilon_n}{\tau})\right)^{\theta_n}. \tag{17}$$

Putting $x_i = \exp(\frac{\mu_i - \epsilon_i}{\tau})$ ($i = 1, \ldots, n$), this becomes

$$Z(p, n) = \sum_{\substack{0 \leq \theta_1 + \cdots + \theta_n \leq p \\ \theta_i \in \{0,1\}}} x_1^{\theta_1} x_2^{\theta_2} \cdots x_n^{\theta_n} = \sum_{k=0}^{p} e_k(x_1, \ldots, x_n). \tag{18}$$

Herein, e_k are the *elementary symmetric functions* [7]. So where the complete symmetric functions play a role in A-statistics, the elementary symmetric functions take over this role in A-superstatistics. For the typical case with $p \geq n$, one finds $Z(p \geq n, n) = (1 + x_1)\cdots(1 + x_n)$.

Next, let us also construct the distribution function, i.e. the average number of particles of A-superstatistics of order p. From the definition, one has

$$\bar{N}(p, n) = \sum_{\substack{0 \leq \theta_1 + \cdots + \theta_n \leq p \\ \theta_i \in \{0,1\}}} |\theta|\frac{x_1^{\theta_1} x_2^{\theta_2} \cdots x_n^{\theta_n}}{Z(p, n)}. \tag{19}$$

Using the symmetric functions, this can be written as

$$\bar{N}(p,n) = p - \frac{\sum_{k=0}^{p-1}(p-k)e_k(x_1,\ldots,x_n)}{\sum_{k=0}^{p}e_k(x_1,\ldots,x_n)}, \tag{20}$$

from which it is also clear that the average number of particles is less than p. In the typical case with $p \geq n$, one finds $\bar{N}(p \geq n, n) = \sum_{i=1}^{n} x_i/(1+x_i)$.

As a special example of A-superstatistics, let us consider the degenerate case where all orbitals have the same energy and chemical potential, so where all x_i are the same and denoted by x. Then the earlier computed functions simplify. For example, $Z(p,n) = \sum_{k=0}^{p}\binom{n}{k}x^k$. This reduces further to $(1+x)^n$ when $p \geq n$, i.e. it coincides in that case with the GPF for a Fermi system with n distinct orbitals having the same energy. Let us also consider the specialization of the distribution function :

$$\bar{N}(p,n) = \frac{\sum_{k=0}^{p}k\binom{n}{k}x^k}{\sum_{k=0}^{p}\binom{n}{k}x^k} = \frac{nx}{1+x} - n\binom{n-1}{p}\frac{x^{p+1}}{(1+x)Z(p,n)}. \tag{21}$$

Clearly, for $p \geq n$, this becomes $\bar{N}(p \geq n, n) = nx/(1+x)$, coinciding with the distribution function of a Fermi system with n distinct orbitals having the same energy. For $p < n$, the last term in (21) is responsible for the deviation from such a Fermi system. To have an idea of such deviations, it is instructive to consider an example. Let us fix $p = 2$ (so at most two particles present), and consider the graph of $\bar{N}(p,n)$ for some n-values, say $n = 2, 3, 4, 5, 6$. This gives us the average number of particles of A-superstatistics of order $p = 2$ when n orbitals of equal energy (but distinct "internal labels") are available. These graphs are shown in Figure 1. As energy variable we have taken $y = (\epsilon - \mu)/\tau$, where $x = e^{-y}$. For $n = p = 2$, the graph coincides with the distribution function of a Fermi system with 2 distinct orbitals; for $n > p$, the function is similar but the half-filling is shifted to the right.

Figure 1. Graphs of $\bar{N}(p,n)$, for fixed $p = 2$, and $n = 2, 3, 4, 5, 6$. The graph of $\bar{N}(p,2)$ is the closest to the (horizontal) y-axis, then $\bar{N}(p,3)$, etc.

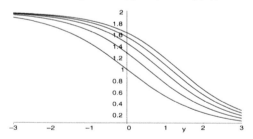

References

[1] Green H S 1953 *Phys. Rev.* **90** 270
[2] Kamefuchi S and Takahashi Y 1962 *Nucl. Phys.* **36** 177
[3] Ryan C and Sudarshan E C G 1963 *Nucl. Phys.* **47** 207
[4] Ganchev A Ch and Palev T D 1980 *J. Math. Phys.* **21** 797
[5] Jellal A, Palev T D and Van der Jeugt J 2002 *J. Phys. A: Math. Gen.* **34** 10179
[6] Palev T D and Van der Jeugt J 2002 *J. Math. Phys.* **43** 3871
[7] Macdonald I G 1995 *Symmetric Functions and Hall Polynomials* (Oxford: Clarendon)
[8] Palev T D 1980 *J. Math. Phys.* **21** 1293

Inst. Phys. Conf. Ser. No 173: Section 5
Paper presented at 24th Int. Coll. Group Theoretical Methods in Physics, Paris, France, 15–20 July 2002
©*2003 IOP Publishing Ltd*

Admissible partitions and the square of the Vandermonde determinant

Brian G Wybourne

Instytut Fizyki,Uniwersytet Mikołaja Kopernika, 87-100 Toruń, POLAND

Abstract. The expansion of the second power of the Vandermonde determinant as a finite sum of Schur functions is considered.

1. Introduction

Laughlin[1] has described the fractional quantum Hall effect in terms of a wavefunction

$$\Psi_{Laughlin}^m(z_1,\ldots,z_N) = \prod_{i<j}^N (z_i - z_j)^{2m+1} \exp\left(-\frac{1}{2}\sum_{i=1}^N |z_i|^2\right) \tag{1}$$

The Vandermonde alternating function in N variables is defined as

$$V(z_1,\ldots,z_N) = \prod_{i<j}^N (z_i - z_j) \tag{2}$$

$$\frac{\Psi_{Laughlin}}{V} = V^{2m} = \sum_{\lambda \vdash n} c^\lambda s_\lambda \tag{3}$$

where $n = mN(N-1)$ and the s_λ are Schur functions. The coefficients c_λ are signed integers.

Dunne[2] and Di Francesco *et al*[3] have discussed properties of the expansions while Scharf *et al*[4] have given specific algorithms for computing the expansions for $m = 1$ with N from 2 to 9. The author has extended these results to $N = 10$ leading to a number of new conjectures.

1.1. Expansion of the Laughlin wavefunction

Henceforth we consider the case where $m = 1$. The partitions, (λ), indexing the Schur functions are of weight $N(N-1)$. For a given N the partitions are bounded by a highest partition $(2N - 2, 2N - 4, \ldots, 0)$ and a lowest partition $((N - 1)^{N-1})$ with the partitions being of length N and $N - 1$.

Let

$$n_k = \sum_{i=0}^k \lambda_{N-i} - k(k+1)k = 0, 1, \ldots, N - 1 \tag{4}$$

Di Francesco *et al*[3] define *admissible partitions* as satisfying Eq(4) with *all* $n_k \geq 0$. They computed the number of admissible partitions A_N for $N \leq 29$ and conjectured that A_N was the number of distinct partitions arising in the expansion, Eq(3), *provided none of the coefficients vanished.*

The conjecture has been shown[4] to fail for $N \geq 8$. We find the number of admissible partitions associated with vanishing coefficients as

$$(N = 8) \quad 8, \quad (N = 9) \quad 66, \quad (N = 10) \quad 389$$

The coefficients of s_λ and s_{λ_r} are equal if[2]

$$(\lambda_r) = (2(N-1) - \lambda_N, \ldots, 2(N-1) - \lambda_1) \tag{5}$$

We list the 8 partitions for $N = 8$ as reverse pairs

$$
\begin{array}{lll}
\{13\ 11\ 985^2 41\} & \{13\ 10\ 9^2 6531\} & (Q1) \\
\{13\ 11\ 9854^2 2\} & \{13\ 10\ 987531\} & (Q2) \\
\{13\ 11\ 976541\} & \{12\ 10^2 96531\} & (Q3) \\
\{12\ 11\ 97^2 4^2 1\} & \{12\ 10^2 7^2 532\} & (Q4)
\end{array}
$$

1.2. The q−discriminant

Let $q\mathbf{x} = (qx_1, qx_2, \ldots, qx_N)$ and the q−discriminant of $\mathbf{x}$ be

$$D_N(q; \mathbf{x}) = \prod_{1 \leq i \neq j \leq N} (x_i - qx_j) \tag{6}$$

and

$$R_N(q; \mathbf{x}) = \prod_{1 \leq i \neq j \leq N} (x_i - qx_j)(qx_i - x_j) \qquad = \sum_\lambda c^\lambda(q) s_\lambda(\mathbf{x}) \tag{7}$$

So that

$$V_N^2(\mathbf{x}) = \prod_{1 \leq i < j \leq N} (x_i - x_j)^2 = R_N(1; \mathbf{x}) \tag{8}$$

Introduce q−polynomials such that

$$R_N(q; \mathbf{x}) = \sum_\lambda c^\lambda(q) s_\lambda(\mathbf{x}) \tag{9}$$

$$R_N(q; \mathbf{x}) = \frac{(-1)^{N(N-1)/2}}{(1-q)^N} \sum_{\nu \subseteq (N-1)^N} ((-q)^{|\nu|} + (-q)^{N^2 - |\nu|})$$
$$\times s_{(N-1)^N / \nu}(\mathbf{x}) s_{\nu'}(\mathbf{x})$$

Such expansions have been evaluated as polynomials in q for all admissible partitions for $N = 2 \ldots 6$ with many examples for $N = 7, 8, 9$.

N=2	[1]	q	$\{2\}$
	[-3]	$-(q^2 + q + 1)$	$\{1^2\}$
N=3	[1]	q^3	$\{42\}$
	[-3]	$-q^2(q^2 + q + 1)$	$\{41^2\} + \{3^2\}$
	[6]	$+q(q^2 + q + 1)(q^2 + 1)$	$\{321\}$
	[-15]	$-(q^2 + q + 1)(q^4 + q^2 + q + 1)$	$\{2^3\}$
N=4	[1]	q^6	$\{642\}$
	[-3]	$-q^5(q^2 + q + 1)$	$\{641^2\} + \{63^2\} + \{5^2 2\}$
	[6]	$+q^4(q^2 + q + 1)(q^2 + 1)$	$\{6321\} + \{543\}$
	..	.....	....

The q−polynomials for the four pairs of partitions designated earlier as $Q(1) \ldots Q(4)$ are

$$Q(1) - q^{17}(q^2 - q + 1)^2(q^2 + 1)^2(q^2 + q + 1)^5(1 - q)^4$$
$$Q(2) + q^{16}(q^2 - q + 1)^2(q^2 + 1)(q^2 + q + 1)^6(1 - q)^4$$
$$Q(3) + q^{16}(q^2 - q + 1)^2(q^2 + 1)^3(q^2 + q + 1)^5(1 - q)^4$$
$$Q(4) + q^{14}(q^2 - q + 1)^2(q^2 + q + 1)^5(1 - q)^4$$
$$\times (q^{10} + q^9 + 3q^8 + 4q^6 + q^5 + 4q^4 + 3q^2 + q + 1)$$

Note the factor $(q - 1)^4$ which vanishes for $q = 1$.

1.3. A conjecture

The following conjecture has been verified to hold for $N \le 10$
   If a q–polynomial is of the form $(-1)^\phi q^p Q(q)$ then under $N \to N + 1$

$$\phi \to \phi, \ p \to p + N, \ Q(q) \to Q(q), \ \{\lambda\} \to \{2N - 2, \lambda\}$$

Define

$$QS(N) = \sum_\lambda c_\lambda(q)$$

then

$$QS(N) = \prod_{x=0}^{[N/2]} (-3x + 1) \prod_{x=0}^{[(N-1)/2]} (6x + 1)$$

   Di Francesco *etal*[3] establish the remarkable result that the sum of the squares of the coefficients of the second power of the Vandermonde with $q = 1$ is

$$\frac{(3N)!}{N!(3!)^N}$$

What is the corresponding result for the q–polynomials? For $N = 4$ one finds
$$q^{24} + 6q^{23} + 22q^{22} + 58q^{21} + 128q^{20} + 242q^{19}$$
$$+ 418q^{18} + 646q^{17} + 929q^{16} + 1210q^{15} + 1490q^{14}$$
$$+ 1670q^{13} + 1760q^{12} + 1670q^{11} + 1490q^{10} + 1210q^9$$
$$+ 646q^8 + 418q^6 + 242q^5 + 128q^4 + 58q^3 + 22q^2 + 6q + 1$$

Note the polynomial is symmetrical and unimodal! Can the general result be found?

Acknowledgments

This work has benefited from interaction with R C King and J-Y Thibon and is supported in part by the Polish KBN Grant 5P03B 5721.

References

[1] Laughlin R B 1983 *Phys. Rev. Lett.* 50 1395
[2] Dunne G V 1993 Int. J. Mod. Phys. B7 4783
[3] Di Francesco P, Gaudin M, Itzykson C and Lesage F 1994 Int. J. Mod. Phys. A9 4257
[4] Scharf T, Thibon J-Y and Wybourne B G 1994 *J. Phys. A: Math. Gen.* 27 4211

Section 6

Noncommutative Geometry

Inst. Phys. Conf. Ser. No 173: Section 6
Paper presented at 24th Int. Coll. Group Theoretical Methods in Physics, Paris, France, 15–20 July 2002
©2003 IOP Publishing Ltd

Yang-Mills and Born-Infeld actions on finite group spaces

P Aschieri[1], **L Castellani**[2] **and A P Isaev** [3]

[1] Sektion Physik der Ludwig-Maximilians-Universität, Theresienstr. 37, D-80333 München, Germany
[2] Dipartimento di Scienze, Università del Piemonte Orientale, Dipartimento di Fisica Teorica and I.N.F.N. Via P. Giuria 1, 10125 Torino, Italy.
[3] Bogoliubov Laboratory of Theoretical Physics, JINR, 980 Dubna, Moscow Region, Russia

Abstract. Discretized nonabelian gauge theories living on finite group spaces G are defined by means of a geometric action $\int Tr \, F \wedge *F$. This technique is extended to obtain a discrete version of the Born-Infeld action.‡

1. Introduction

The base space in a gauge theory is usually a manifold; we consider here a finite group G as base space, and the Lie group $\mathcal{G} = U(N)$ as fiber. One can associate to G a "manifold" structure by constructing on G a differential calculus. This can be done following the general procedure first studied by Woronowicz [1] in the noncommutative context of quantum groups. The differential calculus on G not only determines the "manifold" G, but has been proven to be a very useful tool in the formulation of gauge and gravity theories with base space G [2, 3, 4, 5, 6]. Following [5], we here show that given a differential calculus on G, a Yang-Mills action can be naturally constructed using just geometric objects: differential forms, invariant metric, *-Hodge operator, Haar measure. We similarly construct a discretized version of Born-Infeld theory. These actions can be generalized [5] to the case where the base space is $M^D \times G$, with M^D a continuous D-dimensional manifold. Then one can study gauge theories on M^D that are "Kaluza-Klein" reduced from those on $M^D \times G$: one has just to reinterpret the $M^D \times G$ gauge action as a new M^D one. For example, approximating the circle S_1 with the cyclic group Z_n we obtain on M^D a gauge action plus extra scalar fields from a pure gauge action on $M^D \times Z_n$. This provides an alternative and different procedure from the usual Kaluza-Klein massive modes truncation in $M^D \times S_1 \to M^D$. In some cases one can obtain a Higgs field and potential; a similar mechanism is present in the Connes-Lott standard Model [7].

Noncommutative structures in string/brane theory have emerged in the last years and are the object of intense research. Our study is in this general context. The noncommutativity we discuss is mild, in the sense that fields commute between themselves (in the classical theory), and only the commutations between fields and differentials, and of differentials between themselves, are nontrivial. This noncommutativity is in a certain sense complementary to the one of $\star$ (Moyal)-deformed theories, where (for constant θ) only functions do not commute. It is tempting to interpret the product space $M^D \times G$ as a bundle of n (D-1)-dimensional branes evolving in time, n being the dimension of the finite group G.

The paper is organized as follows. In Section 2 we recall some basic facts about the differential geometry of finite groups, and construct those geometric tools needed in Section

‡ Talk presented by P Aschieri. We thank Chiara Pagani for useful discussions. Work partially supported by a Marie Curie Fellowship of the EC programme IHP, contract number MCFI-2000-01982.

3 for the study of Yang-Mills theories. In Section 4 an action for the Born-Infeld theory is presented.

2. Differential geometry on finite groups

Let G be a finite group of order n with generic element g and unit e. Consider $Fun(G)$, the set of complex functions on G. The left and right actions of the group G on $Fun(G)$ read $\mathcal{L}_g f|_{g'} = f(g\,g')$, $\mathcal{R}_g f|_{g'} = f(g'\,g)$ $\forall f \in Fun(G)$. The vectorfields t_g on G are defined by their action on $Fun(G)$: $t_g f|_{g'} = f(g'g) - f(g') = (\mathcal{R}_g f - f)|_{g'}$, with $g \neq e$. They are nonlocal and left-invariant: $\mathcal{L}_{g'}(t_g f) = t_g(\mathcal{L}_{g'} f)$. We also have $\mathcal{R}_{g'}(t_g f) = t_{g'gg'^{-1}}(\mathcal{R}_{g'} f)$. The differential of an arbitrary function $f \in Fun(G)$ is then given by

$$df = \sum_{g \neq e} (t_g f)\, \theta^g \ , \tag{1}$$

where θ^g are one-forms; by definition they span the linear space dual to that of the vectorfields t_g: $\langle t_g, \theta^{g'} \rangle = \delta_g^{g'}$. From the Leibniz rule for the differential $d(fh) = (df)h + f(dh)$, $\forall f, h \in Fun(G)$, we have that functions do not commute with one-forms:

$$\theta^g f = (\mathcal{R}_g f)\theta^g \qquad (g \neq e) \ . \tag{2}$$

Similarly, left and right invariance of the differential $\mathcal{L}_g\, df = d(\mathcal{L}_g f)$, $\mathcal{R}_g\, df = d(\mathcal{R}_g f)$ implies that θ^g are left-invariant one-forms $\mathcal{L}_{g'}(\theta^g) = \theta^g$ and that

$$\mathcal{R}_{g'}(\theta^g) = \theta^{g'g\,g'^{-1}} \ . \tag{3}$$

A generic one form ρ can always be written in the θ^g basis as $\rho = \sum_{g \neq e} f_g \theta^g$, with $f_g \in Fun(G)$. We have here described the so-called universal differential calculus on G. Smaller calculi can be obtained setting $\theta^g = 0$ for some $g \in G$. Because of (3), the new differential is still left and right invariant iff, given $\theta^g = 0$, also $\theta^{g'g g'^{-1}} = 0$. In other words, (bicovariant) differential calculi are in 1-1 correspondence with unions of conjugacy classes of G.

The algebra $Fun(G)$ has a natural *-conjugation: $f^*(g) \equiv \overline{f(g)}$, where $\bar{}$ is complex conjugation. This *-conjugation can be extended to the differential algebra, the reality condition $(df)^* = d(f^*)$ then implying $(\theta^g)^* = -\theta^{g^{-1}}$. If we set $\theta^g = 0$ then the one-forms in both the $\{\theta^g\}$ and the $\{\theta^{g^{-1}}\}$ conjugacy classes should be set to zero.

An exterior product, compatible with the left and right actions of G, can be defined as (here $\otimes$ by definition satisfies $\rho f \otimes \rho' = \rho \otimes f\rho'$ with ρ, ρ' generic one-forms)

$$\theta^g \wedge \theta^{g'} = \theta^g \otimes \theta^{g'} - \theta^{gg'g^{-1}} \otimes \theta^g = \theta^g \otimes \theta^{g'} - (\mathcal{R}_g \theta^{g'}) \otimes \theta^g \ , \quad (g, g' \neq e) . \tag{4}$$

The metric $< , >$ maps couples of 1-forms ρ, σ into $Fun(G)$, and is required $< , >$ to satisfy the properties $< f\rho, \sigma h >= f < \rho, \sigma > h$, $< \rho f, \sigma >=< \rho, f\sigma >$, and to be symmetric on left-invariant one-forms. Up to a normalization the above conditions imply

$$g^{rs} \equiv\, < \theta^r, \theta^s >\, \equiv \delta_{s^{-1}}^r \ . \tag{5}$$

We can generalize $< , >$ to tensor products of left-invariant one-forms as follows:

$$< \theta^{i_1} \otimes \cdots \otimes \theta^{i_k}, \theta^{j_1} \otimes \cdots \otimes \theta^{j_k} >\, \equiv\, < \theta^{i_1}, \theta^{j_1'} >< \theta^{i_2}, \theta^{j_k'} > \ldots < \theta^{i_k}, \theta^{j_k} > \tag{6}$$

where $j_p' = (i_{p+1}i_{p+2}...i_k)j_p(i_{p+1}i_{p+2}...i_k)^{-1}$, i.e. $\theta^{j_p'} = R_{i_{p+1}i_{p+2}...i_k}\theta^{j_p}$. The pairing (6) is extended to all tensor products by $< f\rho, \sigma h >= f < \rho, \sigma > h$ where now ρ and σ are generic tensor products of same order. Then, using (2) and (6), one can prove that $< \rho f, \sigma >=< \rho, f\sigma >$ for any function f. Moreover $< , >$ is left and right invariant.

Finally, if there exists a form vol of maximal degree, then it can be chosen left-invariant, right-invariant and real. The Hodge dual is then defined by $\rho \wedge *\sigma =< \rho, \sigma > vol$. It is left linear; if vol is central it is also right linear: $*(f\rho h) = f(*\rho)h$.

3. Gauge theories on finite groups spaces

The gauge field of a Yang-Mills theory on a finite group G is a matrix-valued one-form $A(g) = A_h(g)\theta^h$. The components A_h are matrices whose entries are functions on G, $A_h = (A_h)^\alpha_\beta$, $\alpha, \beta = 1, ...N$. As in the usual case, $\mathcal{G}$ gauge transformations are given by

$$A' = -(dT)T^{-1} + TAT^{-1} \; , \tag{7}$$

where $T(g) = T(g)^\alpha_\beta$ is an $N \times N$ representation of a $\mathcal{G}$ group element; its matrix entries belong to $Fun(G)$. The 2-form field strength F is given by the familiar expression

$$F = dA + A \wedge A \; ,$$

and satisfies the Bianchi identity: $dF + A \wedge F - F \wedge A = 0$. Note that $A \wedge A \neq 0$ even if the gauge group $\mathcal{G}$ is abelian. Thus $U(1)$ gauge theory on a finite group space looks like a nonabelian theory, a situation occurring also in gauge theories with $\star$-product noncommutativity. Under the gauge transformations (7), F varies homogeneously: $F' = TFT^{-1}$. We also have

$$F = U \wedge U \; , \tag{8}$$

where $U_h = 1 + A_h$, $U = \sum_{h \neq e} U_h \theta^h = \sum_{h \neq e} \theta^h + A$. From (7) we see that also U transforms covariantly: $U' = TUT^{-1}$. Defining the components $F_{h,g}$ as:

$$F \equiv F_{h,k} \, \theta^h \otimes \theta^k \tag{9}$$

eq. (8) yields:

$$F_{h,k} = U_h \left(\mathcal{R}_h \, U_k \right) - U_k (\mathcal{R}_k \, U_{k^{-1}hk}) \; . \tag{10}$$

We now have all the geometric tools needed to construct a Yang-Mills action. The Yang-Mills action is the geometrical action quadratic in F given by

$$A_{YM} = -\int Tr(F \wedge *F) = -\int Tr < F, F > vol = -\sum_{s \in G} Tr < F, F > \tag{11}$$

the right-hand side being just the Haar measure of the function $Tr < F, F >$. Recalling the properties of the pairing $< , >$, the proof of gauge invariance of $Tr < F, F >$ is immediate: $Tr < F', F' >= Tr < TFT^{-1}, TFT^{-1} >= TrT < F, F > T^{-1} = Tr < F, F >$.

The metric (5) is an euclidean metric (as is seen using a real basis of one-forms) and as usual we require (11) to be real and positive definite. This restricts the gauge group $\mathcal{G}$ and imposes reality conditions on the gauge potential A. Positivity of (11) requires $-(Tr < F, F >) \geq 0$. Explicitly $< F, F >= F_{r,s} < \theta^r \otimes \theta^s, \theta^m \otimes \theta^n > \mathcal{R}_{n^{-1}m^{-1}} F_{m,n} = F_{r,s} \mathcal{R}_{rs} F_{s^{-1}r^{-1}s,s^{-1}}$, and therefore

$$-Tr < F, F >= -(F_{r,s})^\alpha_\beta (\mathcal{R}_{rs} F_{s^{-1}r^{-1}s,s^{-1}})^\beta_\alpha \; ; \tag{12}$$

we see that (12) is positive if $(\mathcal{R}_{rs} F_{s^{-1}r^{-1}s,s^{-1}})^\beta_\alpha = -(F^*_{r,s})^\alpha_\beta$. This holds if (use (10))

$$U = -U^\dagger \quad \text{i.e} \quad A^\dagger = -A \; ; \tag{13}$$

here hermitian conjugation on matrix valued one forms A (or U) is defined as:

$$A^\dagger = (A_h \theta^h)^\dagger = (\theta^h)^* A_h^\dagger = -\theta^{h^{-1}} A_h^\dagger = -\theta^h A_{h^{-1}}^\dagger = -(\mathcal{R}_h A_{h^{-1}}^\dagger) \, \theta^h \; . \tag{14}$$

Finally gauge trasformations must preserve antihermiticity of A, and this is the case if the representation T of $\mathcal{G}$ is unitary. We thus obtain that the action (11) has maximal gauge group $\mathcal{G} = U(N)$. Notice that writing $A = (A_h)^\alpha_\beta \theta^h$, where $\alpha, \beta = 1, ...N$, the reality condition

$A^\dagger = -A$ is equivalent to $\overline{A_h(g)}^\beta_\alpha = A_{h^{-1}}(hg)^\alpha_\beta$ and is thus not local (not fiberwise). It follows that A has values in $M_{N \times N}(\mathbf{C})$ and not in the Lie algebra of $U(N)$. Nevertheless $A^\dagger = -A$ is a good reality condition because it cuts by half the total number of components of A. This can be seen counting the real components of the antihermitian field A: they are $N^2 \times n \times d$, where n is the number of points of G, and the "dimension" d counts the number of independent left-invariant one-forms. In conclusion, when $\mathcal{G} = U(N)$, we have a bona fide pure gauge action, where the number of components of A is consistent with the dimension of the gauge group.

The action (11) can be expressed in terms of the link fields U_h: substituting (9), (10) into (11) leads to: $A_{YM} = 2\sum_G Tr[U_k U_k^\dagger U_h U_h^\dagger - U_k^\dagger U_h (\mathcal{R}_h U_k)(\mathcal{R}_k U_{k^{-1}hk}^\dagger)]$. We could as well start with a Yang-Mills action on $M^D \times G$ and then obtain ($\mu, \nu = 1, ...D$)

$$A_{YM} = -\int_{M^D \times G} Tr\, F \wedge *F = -\int_{M^D} d^D x \sum_G Tr\, [F_{\mu\nu} F_{\mu\nu}$$

$$+ \frac{1}{2} D_\mu U_k (D_\mu U_k)^\dagger + 2 U_k U_k^\dagger U_h U_h^\dagger - 2 U_k^\dagger U_h (\mathcal{R}_h U_k)(\mathcal{R}_k U_{k^{-1}hk}^\dagger)]$$

This action describes a Yang-Mills theory on M^D minimally coupled to the scalar fields U_g, with a nontrivial quartic potential.

4. Born-Infeld theory on finite group spaces

Due to space limitations we only consider here a Born-Infeld action on abelian finite group space G. In the commutative case we have $A_{BI} = \int_{M^D} d^D x \sqrt{\det(\delta_{\mu\nu} + F_{\mu\nu})}$. The analogue of $\delta_{\mu\nu} + F_{\mu\nu}$ becomes simply $E_{g,h} \equiv \delta_{g,h^{-1}} + i F_{g,h}$ (the i is because a constant $F_{g,h}$ is an antihermitian matrix if $g = g^{-1}$ and $h = h^{-1}$). The matrix $E_{g,h}$ transforms under $U(N)$ gauge variations in the same way as $F_{g,h}$: $E'_{g,h} = T E_{g,h} R_{gh} T^{-1}$. We need now a gauge covariant definition of determinant for a matrix transforming as $E_{g,h}$. We define

$$\det_G E_{g,h} = \epsilon^{g_1,...,g_p} E_{g_1,h_1} (\mathcal{R}_{g_1 h_1} E_{g_2,h_2}) (\mathcal{R}_{g_1 h_1 g_2 h_2} E_{g_3,h_3}) \cdots$$

$$\cdots (\mathcal{R}_{g_1 h_1 g_2 h_2 ... g_{p-1} h_{p-1}} E_{g_p,h_p}) \, \epsilon^{h_1,...,h_p}$$

$$\equiv \epsilon^{g_1,...,g_p} E_{g_1,h_1,g_2,h_2,...,g_p,h_p} \, \epsilon^{h_1,...,h_p}, \tag{15}$$

where $\epsilon^{h_1,...,h_p}$ is the usual antisymmetric epsilon tensor (this definition can be generalized to the case of nonabelian G [5]). Then one can prove that $\det_G E'_{g,h} = T \det_G E_{g,h} T^{-1}$. This determinant is also real and the discrete Born-Infeld action reads

$$A_{BI}^G = \int_G Tr\, \sqrt{\det_G (\delta_{g,h^{-1}} + i F_{g,h})} \, vol = \sum_G Tr\, \sqrt{\det_G (\delta_{g,h^{-1}} + i F_{g,h})}.$$

References

[1] S.L. Woronowicz, Commun. Math. Phys. **122**, (1989) 125.
[2] K. Bresser, F. Müller-Hoissen, A. Dimakis and A. Sitarz, J.Phys.**A29** (1996) 2705, [q-alg/9509004];
[3] L. Castellani, Commun. Math. Phys. **218** (2001) 609, [gr-qc/9909028]; Class. Quantum Grav. **17** (2000) 3377, [hep-th/0005210]; L. Castellani and C. Pagani, Annals Phys. **297** (2002) 295, [hep-th/0109163].
[4] S. Majid, Commun. Math. Phys. **225** (2002) 131 [math.qa/0006150]; S. Majid and E. Raineri, *Electromagnetism and Gauge Theory on the Permutation Group* S_3, [hep-th/0012123]; F. Ngakeu, S. Majid and D. Lambert, J. Geom. Phys. **42** (2002) 259 [math.qa/0107216].
[5] P. Aschieri, L. Castellani and A. P. Isaev, *Discretized Yang-Mills and Born-Infeld actions on finite group geometries*, [hep-th/0201223].
[6] A. Dimakis and F. Muller-Hoissen, *Differential geometry of group lattices*, [math-ph/0207014].
[7] A. Connes and J. Lott, Nucl. Phys. Proc. Suppl. **18B**, 29 (1991).

Inst. Phys. Conf. Ser. No 173: Section 6
Paper presented at 24th Int. Coll. Group Theoretical Methods in Physics, Paris, France, 15–20 July 2002
©2003 IOP Publishing Ltd

Noncommutative Geometry on the Algebra $U(su_2)$ and Quantization of Coadjoint Orbits

Eliezer Batista‡

Dep. de Matemática, Universidade Federal de Santa Catarina CEP: 88.040-900, Florianopolis-SC, Brazil

Shahn Majid §

School of Mathematical Sciences, Queen Mary, University of London, Mile End Rd, London E1 4NS, UK

Abstract. We study the standard angular momentum algebra $[x_i, x_j] = \imath\lambda\epsilon_{ijk}x_k$ as a noncommutative manifold $\mathbb{R}_\lambda^3$. There is a natural 4D differential calculus and it is possible to obtain its cohomology and Hodge * operator. We solve the spin 0 wave equation and some aspects of the Maxwell or electromagnetic theory. The space $\mathbb{R}_\lambda^3$ has a natural quantum isometry group given by the quantum double $D(U(su(2))) = \mathbb{C}(SU(2)) \rtimes U(su_2)$ which is a singular limit of the q-Lorentz group.

1. Introduction

There has been an increasing interest in recent years in the noncommutativity of the classical space or spacetime itself as a source of new physically interesting effects. In this work, we explore the algebra $\mathbb{R}_\lambda^3 \cong U(su(2))$ which can be seen as a formal deformation quantisation of the coadjoint orbit method [3]. Our goal is to use modern quantum group methods to take this further by developing the noncommutative differential geometry of this quantum space. We will see that $\mathbb{R}_\lambda^3$, has an isometry quantum group given by the Drinfeld quantum double $D(U(su(2)))$. This theory enables one to show that exactly this noncommutative space is needed for a geometrical picture underlying the approach to 3d quantum gravity of with Euclidean signature and vanishing cosmological constant [7].

Then we introduce a natural bicovariant differential calculus on $\mathbb{R}_\lambda^3$. The cotangent directions or basic forms are given literally by Pauli matrices plus an additional generator θ. There are non-commutation relations between functions and 1-forms and every derivative can be performed as a graded commutation with θ. The origin of the θ direction can be explained as the remnant of the time direction dt of a standard 4-dimensional noncommutative spacetime $\mathbb{R}_q^{1,3}$ [5] in a certain scaling limit as $q \to 1$. In this limit, the q-Lorentz group action degenerates into an action of the above quantum double isometry group, see the ref. [1].

We briefly discuss the Hodge * operator and solve the resulting wave equations for spin 0 and spin 1 (the Maxwell equations). Among the solutions of interest are the plane waves. The spheres S_λ^2 inside $\mathbb{R}_\lambda^3$ inherit a 3-dimensional differential geometry. This case could be viewed as a slightly different approach to fuzzy spheres[4] that is more adapted to their classical limit $\lambda \to 0$. For the relationship between $\mathbb{R}_\lambda^3$ and fuzzy spheres, see the reference [1].

‡ Supported by CAPES, proc. BEX0259/01-2.
§ Royal Society University Research Fellow

2. The Noncommutative Space $\mathbb{R}^3_\lambda$

The noncommutative space $\mathbb{R}^3_\lambda$, comes from the quantization of coadjoint orbits of the group $SU(2)$ in $\mathbb{R}^3$. The coadjoint orbit method [3] can be made on by considering the isomorphism $\mathfrak{su}(2)^* \cong \mathbb{R}^3$, the algebra of functions $C^\infty(\mathbb{R}^3)$ has a natural Poisson structure, which given in terms of the coordinate functions x_a closes a Lie algebra isomorphic to $\mathfrak{su}(2)$.

In the quantization process, the commutative algebra of functios is replaced by an algebra of operators depending on a lenght parameter λ, which we will denote by $\mathbb{R}^3_\lambda$ such that the Poisson bracket is associated to the commutator. We will call the generators of $\mathbb{R}^3_\lambda$ by x_a as well, they satisfy the commutation relations

$$[x_a, x_b] = i\lambda \epsilon_{abc} x_c. \tag{1}$$

We can see the isomorphism $\mathbb{R}^3_\lambda \cong U(\mathfrak{su}(2))$ by the rescaling, $J_a = \frac{1}{\lambda} x_a$.

The usual universal enveloping algebra algebra $U(\mathfrak{su}(2))$ has a structure of cocommutative Hopf algebra generated by 1 and J_a, $a = 1, 2, 3$ with relations [5]

$$[J_a, J_b] = i\epsilon_{abc} J_c, \quad \Delta(J_a) = J_a \otimes 1 + 1 \otimes J_a, \quad \epsilon(J_a) = 0, \quad S(J_a) = -J_a. \tag{2}$$

The algebra $U(\mathfrak{su}(2))$ is dually paired with the commutative Hopf algebra $\mathbb{C}(SU(2))$ generated by coordinate functions $t^i{}_j$, for $i, j = 1, 2$ on $SU(2)$ satisfying the determinant relation $t^1{}_1 t^2{}_2 - t^1{}_2 t^2{}_1 = 1$ and with:

$$\Delta(t^i{}_j) = \sum_{k=1}^{2} t^i{}_k \otimes t^k{}_j, \quad \epsilon(t^i{}_j) = \delta^i{}_j, \quad St^i{}_j = t^i{}_j{}^{-1}.$$

The pairing between these two algebras is defined on the generators by $\langle J_a, t^i{}_j \rangle = \frac{1}{2}\sigma_a{}^i{}_j$, where $\sigma_a{}^i{}_j$ are the i, j entries of the Pauli matrices. We omit here a discussion of unitarity, see [5] for further details.

There is a canonical action of the quantum double $D(U(\mathfrak{su}(2)))$ on $\mathbb{R}^3_\lambda$, providing an exact quantum symmetry group of that. Because $U(\mathfrak{su}(2))$ is cocommutative, its quantum double $D(U(\mathfrak{su}(2)))$ is a usual the crossed product [5]

$$D(U(\mathfrak{su}(2))) = \mathbb{C}(SU(2))_{\mathrm{Ad}_L^*} \rtimes U(\mathfrak{su}(2))$$

where the action is the coadjoint action on $\mathbb{C}(SU(2))$. This crossed product is isomorphic as a vector space with $\mathbb{C}(SU(2)) \otimes U(\mathfrak{su}(2))$. Otherwise we can view its algebra structure as generated by $U(\mathfrak{su}(2))$ and $\mathbb{C}(SU(2))$ with cross relations

$$[J_a, t^i{}_j] = \frac{1}{2}\left(t^i{}_k \sigma^k_{aj} - \sigma_{ak}^i t^k{}_j\right). \tag{3}$$

Meanwhile the coproducts are the same as those of $U(\mathfrak{su}(2))$ and $\mathbb{C}(SU(2))$.

The action of $U(\mathfrak{su}(2))$ on $\mathbb{R}^3_\lambda$ is given by the left adjoint action. Then J_a simply act by

$$J_a \triangleright f(x) = \lambda^{-1}[x_a, f(x)], \quad \forall f(x) \in \mathbb{R}^3_\lambda \tag{4}$$

in the generators (4) reads, $J_a \triangleright x_a = \imath \epsilon_{abc} x_c$. And of $\mathbb{C}(SU(2))$ acts by the co-regular action [5] which can be written as

$$t^i{}_j \triangleright f(x) = \langle t^i{}_j, f(x)_{(1)} \rangle f(x)_{(2)}, \quad \text{e.g.} \quad t^i{}_j \triangleright x_a = \frac{\lambda}{2}\sigma_a{}^i{}_j 1 + \delta^i{}_j x_a.$$

Let us analyse the classical limit of this action. When $\lambda \to 0$ the algebra $\mathbb{R}^3_\lambda$ becomes the polynomial algebra of coordinate functions on $\mathbb{R}^3$. The same parameter λ can be introduced into the quantum double by means of a redefinition of the generators of $\mathbb{C}(SU(2))$, defining, $t^i{}_j = \delta^i{}_j + \lambda M^i{}_j$ and the momentum generators $P_a = \sigma_a{}^i{}_j M^j{}_i$ ($a = 1, 2, 3$), and $P_0 = \mathrm{Tr}(M)$.

The condition on the determinant, $t^1_1t^2_2 - t^1_2t^2_1 = 1$, implies that $\text{Tr}(M) = -\lambda\det(M)$. In terms of the momentum generators, the co-product writes

$$\Delta P_a = P_a \otimes 1 + 1 \otimes P_a + \mathcal{O}(\lambda)$$

so that we have the usual additive coproduct in the $\lambda \to 0$ limit. Note that P_0 depends on the other P_a because the trace is proportional to the determinant. Meanwhile, left coadjoint action and resulting cross relations in the double become

$$\text{Ad}^*_{L J_a}(P_b) = \imath\epsilon_{abc}P_c, \quad [J_a, P_b] = \imath\epsilon_{abc}P_c.$$

The action of these scaled generators can also be written as derivative operators, compatible with the calculus of the next section. In the limit $\lambda \to 0$, the action of J_a and of P_a become respectively the usual rotations in three dimensional Euclidean space and translation operators, so we recover the classical action of $U(iso(3))$ on $\mathbb{R}^3$.

3. The 4-Dimensional Calculus on $\mathbb{R}^3_\lambda$

The purpose of this section is to construct a bi-covariant calculus on the algebra $\mathbb{R}^3_\lambda$. That is to specify $(\Omega^1(\mathbb{R}^3_\lambda), \text{d})$, where Ω^1 is an $U(\mathfrak{su}(2))$-bimodule. Here we write $\mathbb{R}^3_\lambda$ as generated by x_a, $a = 1, 2, 3$, and with the Hopf algebra structure given explicitly in terms of the generators as (1) and the additive coproduct as before (remember that $\mathbb{R}^3_\lambda \cong U(\mathfrak{su}(2))$). We have a derivation $\text{d} : U \to \Omega^1$ which obeys the Leibniz rule, $\text{d}(hg) = (\text{d}h)g + h(\text{d}g)$, and Ω^1 is spanned by elements of the form $(\text{d}h)g$. By bicovariant [8] we mean that there are left and right coactions of $U(\mathfrak{su}(2))$ in Ω^1 which are bimodule homomorphisms, and compatible with d. Given a bicovariant calculus one can find invariant forms $\omega(h) = \sum(\text{d}h_{(1)})Sh_{(2)}$, $\forall h \in U(\mathfrak{su}(2))$. The particular form of the co-product and the relations shows that $\text{d}\xi = \omega(\xi)$ for all $\xi \in \mathfrak{su}(2)$. The calculi can be classified by left ideals in $\ker \epsilon \subset U(\mathfrak{su}(2))$ that are stable by the left adjoint co-action [8]. Because of the co-commutativity, all ideals in $\mathbb{R}^3_\lambda$ are ad-invariant.

In order to construct an ideal of $\ker \epsilon$, consider a two dimensional representation $\rho :$ $\mathbb{R}^3_\lambda \to M_2(\mathbb{C})$, given by $\rho(x_a) = \frac{1}{2}\sigma_a$. Where these matrices are the Pauli matrices, σ_a. The kernel of $\rho \mid_{\ker \epsilon}$ is a 2-sided ideal in $\ker \epsilon$. Then we have $M_2(\mathbb{C}) \equiv \ker \epsilon / \ker \rho$. This isomorphism allows us to identify the basic one forms with 2×2 matrices. Then the first order differential calculus is $\Omega^1(\mathbb{R}^3_\lambda) = M_2(\mathbb{C}) \otimes \mathbb{R}^3_\lambda$. The exterior derivative operator is

$$\text{d}f(x) = \lambda^{-1}\sum\rho(f(x)_{(1)} - \epsilon(f(x)_{(1)})1)f(x)_{(2)}.$$

In particular, $\text{d}\xi = \lambda^{-1}\rho(\xi)$, $\forall\xi \in \mathfrak{su}(2)$, and for a monomial $\xi_1 \ldots \xi_n$ we have

$$\text{d}(\xi_1 \ldots \xi_n) = \lambda^{-1}\sum_{k=1}^{n}\sum_{\sigma\in S_{(n,k)}}\rho(\xi_{\sigma(1)} \cdots \xi_{\sigma(k)})\xi_{\sigma(k+1)} \cdots \xi_{\sigma(n)},$$

where σ is a (n, k)-shuffle [6].

Our basis of invariant one forms is $\text{d}x_a = \frac{1}{2}\sigma_a$, and $\theta = \text{id}$. The compatibility conditions of this definition of the derivative with the Leibniz rule result in some commutation relations between the generators of the algebra and the basic one forms as listed below:

$$x_a\text{d}x_b = (\text{d}x_b)x_a + \frac{\imath\lambda}{2}\epsilon_{abc}\text{d}x_c + \frac{\lambda}{4}\delta_{ab}\theta, \quad x_a\theta = \theta x_a + \lambda\text{d}x_a. \tag{5}$$

This calculus is four dimensional, in the sense that one has four basic 1-forms, but in the classical limit $\lambda \to 0$, the calculus turns out to be commutative and the extra one dimensional subspace generated by the 1-form θ, decouples totally. This extra dimension can be seen as a reminiscent of the time coordinate in the q-Minkowski space $\mathbb{R}^{1,3}_q$ [5] in the limit $q \to 1$.

We can also construct the full exterior algebra $\Omega^{\cdot}(\mathbb{R}^3_\lambda) = \bigoplus_{n=0}^{\infty} \Omega^n(\mathbb{R}^3_\lambda)$. In our case the braiding used in [8] is the trivial flip homomorphism because the right invariant basic 1-forms are also left invariant. Hence our basic 1-forms are totally anticommutative and their usual antisymmetric wedge product generates the usual exterior algebra on the vector space $M_2(\mathbb{C})$. The full $\Omega^{\cdot}(\mathbb{R}^3_\lambda)$ is generated by these and elements of $\mathbb{R}^3_\lambda$ with the commutation relations (5) and with the exterior differentiation given by the graded commutator with θ, that is

$$\mathrm{d}\omega = \omega \wedge \theta - (-1)^{\deg\omega}\theta \wedge \omega.$$

In particular, the basic 1-forms $M_2(\mathbb{C})$ are all closed, among which θ is not exact. The cohomologies are $H^0 = \mathbb{C}.1$, $H^1 = \mathbb{C}.\theta$, and $H^2 = H^3 = H^4 = \{0\}$ [1].

4. Hodge $*$-Operator and Electromagnetic Theory

The above geometry also admits a metric structure, that is a nondegenerate and invariant bilinear form $\eta \in \Lambda^1 \otimes \Lambda^1$ [6]. For the case of $\mathbb{R}^3_\lambda$ we can define the metric

$$\eta = \mathrm{d}x_1 \otimes \mathrm{d}x_1 + \mathrm{d}x_2 \otimes \mathrm{d}x_2 + \mathrm{d}x_3 \otimes \mathrm{d}x_3 - \theta \otimes \theta. \tag{6}$$

With this metric structure, it is possible to define Hodge $*$-operator, $* : \Omega^k \to \Omega^{4-k}$ given by

$$*(\omega_{i_1} \ldots \omega_{i_k}) = \frac{1}{(4-k)!}\epsilon_{i_1\ldots i_k i_{k+1}\ldots i_4}\eta^{i_{k+1}j_1} \ldots \eta^{i_4 j_{4-k}}\omega_{j_1} \ldots \omega_{j_{4-k}},$$

with $\omega_a = \mathrm{d}x_a$, $a = 1, 2, 3$, $\omega_4 = \theta$. The explicit expressions are in ref. [1].

With the definition of the Hodge $*$-operator, one can write the wave operator $\Box = *\mathrm{d}*\mathrm{d}$. An electromagnetic theory is the analysis of solutions $A \in \Omega^1(\mathbb{R}^3_\lambda)$ of the Maxwell equation $\Box A = \mathrm{J}$, where J is a 1-form which can be interpreted as a "physical" source.

First let us take a look on modes on $\Omega^0(\mathbb{R}^3_\lambda) = \mathbb{R}^3_\lambda$, the general eigenfunctions of $\Box$ in degree 0 are the plane waves:

$$\Box e^{\imath k \cdot x} = -\frac{1}{\lambda^2}\left\{ 4\sin^2\left(\frac{\lambda \mid k \mid}{2}\right) + \left(\cos\left(\frac{\lambda \mid k \mid}{2}\right) - 1\right)^2 \right\} e^{\imath k \cdot x}.$$

On $\Omega^1(\mathbb{R}^3_\lambda)$, if we write $A = (\mathrm{d}x_a)A^a + \theta A^0$ for functions A_μ, then the curvature $F = \mathrm{d}A$, can be splitted into "electric" and "magnetic" parts, as for usual spacetime. Three classes of solutions of the Maxwell equation were found: The first group was the self propagating modes, for J $= 0$. The second is the electrostatic solution, for J $= \theta$. In this case the solution has the dependence on the coordinates given a field strength increasing with the radius, this is a kind of solution exhibiting a confinement behaviour. There's an intrinsic technical difficulty to find realistic solutions with radial decay. Because it is necessary to complete our calculus in order to get some consistent rules to deal with inverses. The third solution is a magnetic, for J $= k \cdot \mathrm{d}x = \sum_a k^a \mathrm{d}x_a$, which solution has the same aspect as the classical magnetic field around an electric current in the direction of the vector k.

References

[1] E. Batista and S. Majid: preprint hep-th/0205128 (2002).
[2] X.Gomes and S. Majid: Lett. Math. Phys. (to appear) and math.QA/0110323.
[3] A.A. Kirillov: "Elements of the theory of representations", Springer Verlag (1976).
[4] J. Madore: J. Class. Quant. Grav. 9 (1) (1992) 69-87.
[5] S. Majid: "Foundations of Quantum Group Theory", Cambridge University Press (1997).
[6] S. Majid: Commun. Math. Phys. 225 (2002) 131-170.
[7] B.J. Schroers: preprint math.QA/0006228.
[8] S.L. Woronowicz: Commun. Math. Phys. 122 (1989) 125-170.

Inst. Phys. Conf. Ser. No 173: Section 6
Paper presented at 24th Int. Coll. Group Theoretical Methods in Physics, Paris, France, 15–20 July 2002

Quantum de Rham complex with $d^3 = 0$ differential on $GL_{p,q}(2)$

N Bazunova

Tallinn Technical University, Tallinn, Estonia

Abstract. In this work, we construct the de Rham complex with differential operator d satisfying the Q-Leibniz rule, where Q is a complex number, and the condition $d^3 = 0$, $d^2 \neq 0$ on the quantum group $GL_{p,q}(2)$ assuming that that de Rham complex with that differential on the "representation space" of this group, i.e. on the quantum plane $xy = qyx$, is already constructed.

1. Introduction

The series of papers [1]-[5] is devoted to the $\mathbb{Z}_3$-graded generalization of differential calculus. The main idea of such generalization is to replace the representation of the group $\mathbb{Z}_2$ by the representation of the group $\mathbb{Z}_3$. This approach makes it possible to obtain the $\mathbb{Z}_3$-graded Grassmann, Lie and Clifford algebras [2]. Also in [2] the external differential satisfying the property $d^3 = 0$, $d^2 \neq 0$ and q-Leibniz rule

$$d(\omega \theta) = (d\omega)\theta + q^{|\omega|}\omega \theta,$$

where ω, θ are the differential forms, $|\omega|$ is degree of ω, q is a primitive root of unity of the order three, was defined on a manifold.

In this paper, we develop the considered in the paper [6] idea of differentiation with differential d satisfying the condition $d^3 = 0$, $d^2 \neq 0$ and the Q-Leibniz rule

$$d(\omega \theta) = (d\omega)\theta + Q^{|\omega|}\omega \theta,$$

where ω, θ are the differential forms, $|\omega|$ is degree of ω, $Q \in \mathbb{C}$, on the quantum plane (or the quadratic algebra) and, assuming that the quantum plane $xy = qyx$ is "a representation space" of the quantum group $GL_{p,q}(2)$, we construct the de Rham complex with the differential d ($d^3 = 0$, $d^2 \neq 0$) on quantum group $GL_{p,q}(2)$.

In the paper [6], it was shown, that in the case of dimension two, the second order differentials d^2x, d^2y which appear as result of condition $d^2 \neq 0$ can determine the q- or p-deformed quantum plane in depending on value of parameter Q.

Following to the the paper [7], we consider both quantum planes (determined by x and y, and d^2x and d^2y) as the parts of the de Rham complex on the quantum plane $xy = qyx$. All the de Rham complex on the quantum plane $xy = qyx$ is defined by the commutation relations between x, y, between d^2x, d^2y, and the cross-commutation relations between coordinates and their first and second order differentials.

Two quantum planes and all cross-commutation relations are compatible with the action of the group $GL_{p,q}(2)$. Following to the paper [8], we assume that differential don't vanish on the generators of group $GL_{p,q}(2)$. Requiring that all relations are preserved under action of $GL_{p,q}(2)$, we can find all commutation relations which connect the generators of group $GL_{p,q}(2)$, and their first and second order differentials. Here the second order differentials also appear as result of condition $d^2 \neq 0$.

2. General formalism

Let $\mathscr{A}_q(2)$ be two-dimensional quantum plane determined by the relation

$$xy = qyx. \tag{1}$$

Introduce the following notation

$$X_1 = \begin{pmatrix} x \\ x \\ y \\ y \end{pmatrix}, \quad X_2 = \begin{pmatrix} x \\ y \\ x \\ y \end{pmatrix}, \quad P = \begin{pmatrix} 1 & 0 & 0 & 0 \\ 0 & 0 & 1 & 0 \\ 0 & 1 & 0 & 0 \\ 0 & 0 & 0 & 1 \end{pmatrix}, \tag{2}$$

(P is the matrix of the permutation). Here we assume that the matrices dX_1, dX_2, d^2X_1, d^2X_2 are written in the same form as X_1 and X_2 with first and second order differentials of coordinates x and y, and shall omit the sign of tensor multiplication between all these matrices.

Now we can rewrite the formula (1) in the matrix form:

$$X_2X_1 = \frac{1}{q}R_{12}X_1X_2, \tag{3}$$

where R_{12} is the Hecke-R-matrix:

$$R_{12} = \begin{pmatrix} q & 0 & 0 & 0 \\ 0 & 1 & qp^{-1} & 0 \\ 0 & q-p^{-1} & 0 & 0 \\ 0 & 0 & 0 & q \end{pmatrix}. \tag{4}$$

The de Rham complex $(\Omega(\mathscr{A}_q), d)$ on quantum plane (1) with differential d satisfying the condition $d^3 = 0$ and the Q-Leibniz rule

$$d(\omega\theta) = d\omega\,\theta + Q^{|\omega|}\,d\theta, \tag{5}$$

where $Q \in \mathbb{C}$, $|\omega|$ is the grade of the form ω, and ω, $\theta \in \Omega(\mathscr{A}_q)$, consists from a first order differential calculus and its prolongations. We assume that the first order differential calculus on (1) coincides with the Wess-Zumino type differential calculus [7], that is the commutation relations between coordinates x, y and its first differentials are determined by

$$X_2\,dX_1 = pR_{12}dX_1 X_2 \tag{6}$$

The matrix R_{12} satisfies two Wess-Zumino conditions:

$$(E_{12} - \frac{1}{q}R_{12})(E_{12} + pR_{12}) = 0,$$
$$R_{12}R_{23}R_{12} = R_{23}R_{12}R_{23}.$$

The prolongation of first order differential calculus is obtained by the differentiation of (6) three times and consists the relations between coordinates x, y and their first and second order differentials dx, dy, d^2x, d^2y:

$$X_2\,d^2X_1 = pR_{12}\,d^2X_1 X_2 + (QpR_{12} - P)dX_1\,dX_2, \tag{7}$$
$$([2]_Q P - Q^2 pR_{12})dX_2\,d^2X_1 = ([2]_Q QpR_{12} - P)d^2X_1\,dX_2, \tag{8}$$
$$[3]_Q d^2X_2\,d^2X_1 = [3]_Q Q^2 pR_{12}\,d^2X_1\,d^2X_2. \tag{9}$$

Now we assume, that Q in not a cubic root of unity, i.e. the second order differentials are connected by the commutation relations:

$$d^2X_2 d^2X_1 = Q^2 pR_{12} d^2X_1 d^2X_2. \tag{10}$$

So the de Rham complex on quantum plane (1) is determined by the relations (6), (7), (8), and (10).

Let $GL_{p,q}(2)$ be a quantum automorphism group [8] of the quantum plane (1) and the elements a, b, c, d are generators of $GL_{p,q}(2)$. Then this group preserves the relations (1) (or (3)) under the transformation

$$\begin{pmatrix} x' \\ y' \end{pmatrix} = \begin{pmatrix} a & b \\ c & d \end{pmatrix} \begin{pmatrix} x \\ y \end{pmatrix} \quad \text{or in the matrix form} \quad X' = TX \tag{11}$$

Now we assume that differential d don't vanish on the group $GL_{p,q}(2)$. Denote by $\dot{T}$ and $\ddot{T}$ the matrices composed from the first and second order differentials of generators a, b, c, d correspondingly. Then, differentiating of (11) two times, we get

$$dX' = \dot{T}X + TdX \quad \text{and} \quad d^2X' = \ddot{T}X + \dot{T}dX + Td^2X, \tag{12}$$

and require that all relations (1), (6), (7), (8), and (10) are preserved under transformations (11), (12). We assume that the coordinates x and y, their first and second order differentials commute with the generators a, b, c, d of the group $GL_{p,q}(2)$ and their first and second order differentials. Also we assume that the grade of the coordinates and generators is equal to zero, the grade of the first and second order differentials is equal to one and two correspondingly.

The transformation (11) give us the well-known "RTT"-relations [9]:

$$R_{12}T_1T_2 = T_2T_1R_{12},$$

where $T_1 = T \otimes E$ and $T_2 = E \otimes T$.

The commutation relations between the generators a, b, c, d, and their first order differentials

$$T_2\dot{T}_1R_{12} = pqR_{12}\dot{T}_1T_2$$

follow from deformation of (6).

From the transformation of relations (7), we get two sorts of commutation relations which connect the generators and their first order and second order differentials

$$T_2\ddot{T}_1R_{12} = pqR_{12}\ddot{T}_1T_2 + (QpqR_{12} - qP)\dot{T}_1\dot{T}_2,$$
$$T_2\dot{T}_1R_{12} = R_{12}\dot{T}_2T_1 + ([2]_Q)^{-1}(QR_{12} - p^{-1}P)(T_1\dot{T}_2 + \dot{T}_1T_2pPR_{12}),$$

where $[2]_Q = 1 + Q$.

We have to agree the obtained relations with the following from transformed relations (8)

$$\Psi\dot{T}_2\ddot{T}_1R_{12} = q\Phi\ddot{T}_1\dot{T}_2,$$
$$\Psi([2]_Q\dot{T}_2T_1pR_{12} + T_2\ddot{T}P) = \Phi(\ddot{T}_1T_2pR_{12} + [2]_Q\dot{T}_1\dot{T}_2),$$
$$\Psi(\dot{T}_2T_1(QpR_{12} - P) + [2]_QT_2\dot{T}P) = \Phi[2]_Q\ddot{T}_1T_2,$$
$$\Psi\dot{T}_2T_1pR_{12} = \Phi T_1\dot{T}_2,$$
$$\Psi T_2T_1\Psi^{-1}\Phi = \Phi T_1T_2,$$

where $\Psi = [2]_QP - Q^2 pR_{12}$, $\Phi = [2]_QQpR_{12} - P$, and with the following from transformed relations (9)

$$\ddot{T}_2 T_1 R_{12} = Q^2 pq R_{12}(\ddot{T}_1 T_2 p P R_{12} + T_1 \ddot{T}_2),$$
$$\ddot{T}_2 T_1 p R_{12} + \dot{T}_2 \dot{T}_1 P = Q^2 p R_{12}(\ddot{T}_1 \dot{T}_2 p P R_{12} + \dot{T}_1 \ddot{T}_2),$$
$$\ddot{T}_2 T_1 p R_{12} + T_2 \ddot{T}_1 P = Q^2 p R_{12}(\ddot{T}_1 T_2 p P R_{12} + T_1 \ddot{T}_2),$$
$$\ddot{T}_2 T_1 (Q p R_{12} - P) + ([2]_Q)^2 \dot{T}_2 \dot{T}_1 P = Q^2 p R_{12}(\ddot{T}_1 T_2 (Q p P R_{12} - E) + ([2]_Q)^2 \dot{T}_1 \dot{T}_2$$
$$\dot{T}_2 T_1 \Psi^{-1} \Phi + T_2 \dot{T}_1 P = Q^2 p R_{12}(\dot{T}_1 T_2 P \Psi^{-1} \Phi + [2]_Q T_1 \dot{T}_2),$$
$$T_2 T_1 P = Q^2 p R_{12} T_1 T_2.$$

The author is very grateful to the organizers of the XXIV International Colloquium on Group Theoretical Methods in Physics for financial support and hospitality.

References

[1] Kerner R 1996 Rend. Semin. Mat.: Torino 54, No.4 319-336

[2] Abramov V, Kerner R, and Le Roy B 1997 J. Math. Phys. 38 No.3 1650-1669

[3] Kerner R, Niemeyer B 1998 Lett. Math. Phys. 45, No.2, 161-176

[4] Kerner R 1998 Czech. J. Phys. 48 No.11 1387-1394

[5] Kerner R, Abramov V 1999 Rep. Math. Phys. 43 No.1-2 179-194

[6] Bazunova N, Borowiec A, and Kerner R 2001 Czech. J. Phys. 51 No12 1266-1271

[7] Wess J and Zumino B 1990 Nucl. Phys. B: Proc. Suppl. 18B 302-312

[8] Manin Yu 1992 Theor. Math. Phys. 92 No3 997-1019

[9] Faddeev L D, Reshetikhin N Yu, and Takhtadzhan L.A. 1989 Algebraic analysis: Pap. Dedicated to Prof. Mikio Sato on the Occas. of his Sixtieth Birthday Vol.1 129-139

Inst. Phys. Conf. Ser. No 173: Section 6
Paper presented at 24th Int. Coll. Group Theoretical Methods in Physics, Paris, France, 15–20 July 2002
©*2003 IOP Publishing Ltd*

Realization of q-deformed spacetime as star product by a Drinfeld twist

Christian Blohmann

Ludwig-Maximilians-Universität München, Sektion Physik
Lehrstuhl Prof. Wess, Theresienstr. 37, D-80333 München

Max-Planck-Institut für Physik, Föhringer Ring 6, D-80805 München

Abstract. Covariance ties the noncommutative deformation of a space into a quantum space closely to the deformation of the symmetry into a quantum symmetry. Quantum deformations of enveloping algebras are governed by Drinfeld twists, inner automorphisms which relate the deformed to the undeformed coproduct. While Drinfeld twists naturally define a covariant star product on the space algebra, this product is in general not associative and does not yield a quantum space. It is reported that, nevertheless, there are certain Drinfeld twists which realize the quantum plane, quantum Euclidean 4-space, and quantum Minkowski space.

1. Introduction

From the beginnings of quantum field theory it had been argued that the pathological ultraviolet divergences should be remedied by limiting the precision of position measurements. This is one of the main motivations to study noncommutative geometries, which imply a space uncertainty in a natural and fundamental way. From experience we know that, if spacetime is noncommutative, the noncommutativity can only be small. This suggests to describe noncommutative spacetime as perturbative deformation of ordinary, commutative Minkowski space. The algebraic aspects of a deformation can be separated from the analytic questions of continuity and convergence by considering formal power series. In such a framework a noncommutative geometry is a formal deformation in the sense of Gerstenhaber [1] of the function algebra on the space manifold. Such formal deformations have appeared naturally in the context of gauge theories on noncommutative spaces [2, 3].

Algebraically, physical spacetime is characterized by the Minkowski algebra of spacetime functions and covariance with respect to the Lorentz symmetry. The symmetry distinguishes Minkowski space from other 4-dimensional flat spaces, such as Euclidean 4-space. Covariance ties the deformation of the symmetry closely to the deformation of the space. Quantum deformations of the enveloping algebra which describes this symmetry are known to be governed by Drinfeld twists, inner automorphisms which relate the deformed to the undeformed coproduct [4, 5]. Therefore, one ought to be able to use these twists in order to deform the space algebra into the according quantum space as it was suggested in [6]. It will be shown that for quantum Minkowski space this is indeed possible.

2. The problem

2.1. Covariant quantum spaces

Let $\mathbf{g}$ be the Lie algebra of the symmetry group of a space and $\mathscr{X}$ be the function algebra of this space. The elements $g \in \mathbf{g}$ of the Lie algebra act on $\mathscr{X}$ as derivations, $g \triangleright xy =$

$(g \triangleright x)y + x(g \triangleright y)$ for $x, y \in \mathscr{X}$. A generalized way of writing this is

$$g \triangleright xy = (g_{(1)} \triangleright x)(g_{(2)} \triangleright y) \tag{1}$$

for all $g \in \mathscr{U}(\mathbf{g})$, where the coproduct is defined as $g_{(1)} \otimes g_{(2)} \equiv \Delta(g) := g \otimes 1 + 1 \otimes g$ on the generators $g \in \mathbf{g}$ and extended to a homomorphism on the enveloping algebra. This covariance condition makes sense for the action of a general Hopf algebra on an associative algebra $\mathscr{X}$. It ties a deformation $\mu \to \mu_\hbar$ of the multiplication map $\mu(x \otimes y) := xy$ of the space algebra to the deformation $\Delta \to \Delta_\hbar$ of the coproduct of the symmetry Hopf algebra if covariance is to be preserved,

$$g \triangleright xy = (g_{(1)} \triangleright x)(g_{(2)} \triangleright y) \quad \overset{\hbar}{\longrightarrow} \quad g \triangleright (x \star y) = (g_{(1_\hbar)} \triangleright x) \star (g_{(2_\hbar)} \triangleright y), \tag{2}$$

where

$$x \star y := \mu_\hbar(x \otimes y) \quad \text{and} \quad g_{(1_\hbar)} \otimes g_{(2_\hbar)} := \Delta_\hbar(g). \tag{3}$$

A large class of deformations which are covariant in this sense are quantum spaces.

2.2. Star products by Drinfeld twists

In the case of quantum spaces, the deformed coproduct belongs to the Drinfeld-Jimbo deformation $\mathscr{U}_\hbar(\mathbf{g})$ of the enveloping Hopf algebra [7, 8]. Drinfeld has observed [4, 5] that as $\hbar$-adic algebras $\mathscr{U}_\hbar(\mathbf{g})$ and $\mathscr{U}(\mathbf{g})[[\hbar]]$ are isomorphic and that the deformed coproduct $\Delta_\hbar$ of the Hopf algebra $\mathscr{U}_\hbar(\mathbf{g}) \cong (\mathscr{U}(\mathbf{g})[[\hbar]], \Delta_\hbar, \varepsilon_\hbar, S_\hbar)$ is related to the undeformed coproduct Δ by an inner automorphism. That is, there is an invertible element $\mathscr{F} \in (\mathscr{U}(\mathbf{g}) \otimes \mathscr{U}(\mathbf{g}))[[\hbar]]$ with $\mathscr{F} = 1 \otimes 1 + \mathcal{O}(\hbar)$, called Drinfeld twist, such that

$$\Delta_\hbar(g) = \mathscr{F} \Delta(g) \mathscr{F}^{-1}. \tag{4}$$

Comparing the covariance condition (2) of the deformed multiplication,

$$g \triangleright \mu_\hbar(x \otimes y) = \mu_\hbar(\Delta_\hbar(g) \triangleright [x \otimes y]) = \mu_\hbar(\mathscr{F} \Delta(g) \mathscr{F}^{-1} \triangleright [x \otimes y]) \tag{5}$$

with the covariance property (1) of the undeformed product, we see that Eq. (5) is naturally satisfied if we define the deformed product by

$$\mu_\hbar(x \otimes y) := \mu(\mathscr{F}^{-1} \triangleright [x \otimes y]) \quad \Leftrightarrow \quad x \star y := (\mathscr{F}^{-1}_{[1]} \triangleright x)(\mathscr{F}^{-1}_{[2]} \triangleright y), \tag{6}$$

as it was observed in [6] (suppressing in a Sweedler like notation the summation of $\mathscr{F} = \sum_i \mathscr{F}_{1i} \otimes \mathscr{F}_{2i} \equiv \mathscr{F}_{[1]} \otimes \mathscr{F}_{[2]}$). Since the elements of the Lie algebra $\mathbf{g}$ act on the undeformed space algebra $\mathscr{X}$ as derivations, $\mathscr{F}^{-1}$ acts as $\hbar$-adic differential operator on $\mathscr{X} \otimes \mathscr{X}$. Hence, writing out the $\hbar$-adic sum of $\mathscr{F}^{-1} = 1 \otimes 1 + \sum_k \hbar^k \mathscr{F}_k^{-1}$ we can define the bidifferential operators

$$B_k(x, y) := \mu(\mathscr{F}_k^{-1} \triangleright [x \otimes y]) = (\mathscr{F}_{k[1]}^{-1} \triangleright x)(\mathscr{F}_{k[2]}^{-1} \triangleright y), \tag{7}$$

such that the star product (6) can be written in the more familiar form [9]

$$x \star y := xy + \hbar B_1(x, y) + \hbar^2 B_2(x, y) + \dots \tag{8}$$

2.3. The problem of associativity

Even though the twist $\mathscr{F}$ yields by Eq. (4) a coassociative coproduct, Eq. (6) will in general not define an associative product. The associativity condition $(x \star y) \star z = x \star (y \star z)$ for $\mu_\hbar$ can be expressed with the Drinfeld coassociator

$$\Phi := (\Delta \otimes \mathrm{id})(\mathscr{F}^{-1})(\mathscr{F}^{-1} \otimes 1)(1 \otimes \mathscr{F})(\mathrm{id} \otimes \Delta)(\mathscr{F}), \tag{9}$$

as

$$(\Phi_{[1]} \triangleright x)(\Phi_{[2]} \triangleright y)(\Phi_{[3]} \triangleright z) = xyz. \tag{10}$$

for all $x, y, z \in \mathscr{X}$.

For a given $\mathscr{U}_\hbar(\mathbf{g})$-covariant quantum space, is there a Drinfeld twist $\mathscr{F}$ which yields by Eq. (6) the associative product of the quantum space? We will answer this question positively for three important cases: the quantum plane, quantum Euclidean 4-space, and quantum Minkowski space.

3. Constructing covariant star products

3.1. The general approach

To our knowledge, no Drinfeld twist for the Drinfeld-Jimbo quantum enveloping algebra of a semisimple Lie algebra has ever been computed. This indicates that it will be rather difficult to answer this question on an algebraic level. The representations of Drinfeld twists, however, can be expressed by Clebsch-Gordan coefficients [10, 11]. Therefore, we propose the following approach, which tackles the problem on a representation theoretic level:

Consider a $\mathscr{U}_\hbar(\mathbf{g})$-covariant quantum space algebra $\mathscr{X}_\hbar$ and its undeformed limit, the $\mathscr{U}(\mathbf{g})$-covariant space algebra $\mathscr{X}$.

  (i) Determine the irreducible highest weight representations of all possible Drinfeld twists from Δ to $\Delta_\hbar$.
 (ii) Determine the basis $\{T_m^j\}$ of the quantum space $\mathscr{X}_\hbar$ which completely reduces $\mathscr{X}_\hbar$ into (possibly degenerate) irreducible highest weight-j representations of $\mathscr{U}_\hbar(\mathbf{g})$.
(iii) Calculate the multiplication map $\mu_\hbar$ of $\mathscr{X}_\hbar$ with respect to this basis. The undeformed limit $\mu = \lim_{\hbar \to 0} \mu_\hbar$ yields the commutative multiplication map with respect to this basis.
 (iv) Check on the level of representations if one of the twists realizes the deformed multiplication by Eq. (6) as linear map with respect to this basis.

Since this procedure reduces the algebraic problem to a representation theoretic one, it works well for cases where the representation theory is well understood, such as for the quantum spaces of $\mathscr{U}_\hbar(\mathrm{su}_2)$, $\mathscr{U}_\hbar(\mathrm{so}_4)$, and $\mathscr{U}_\hbar(\mathrm{sl}_2(\mathbf{C}))$.

3.2. Example: the quantum plane

The $\hbar$-adic quantum plane generated by x and y with commutation relations $xy = qyx$, $q := e^\hbar$, is a $\mathscr{U}_\hbar(\mathrm{su}_2)$-covariant space. Let us denote by ρ^j the structure map of the spin-j representation of $\mathscr{U}_\hbar(\mathrm{su}_2)$. The results of the proposed approach are:

  (i) The irreducible representations of the Drinfeld twists can be expressed by the q-deformed and undeformed Clebsch-Gordan coefficients [12] as

$$(\rho^{j_1} \otimes \rho^{j_2})(\mathscr{F})^{m_1 m_2}{}_{m_1' m_2'} = \sum_{j,m} \eta(j_1, j_2, j) \begin{pmatrix} j_1 & j_2 & j \\ m_1 & m_2 & m \end{pmatrix}_q \begin{pmatrix} j_1 & j_2 & j \\ m_1' & m_2' & m \end{pmatrix}, \tag{11}$$

  where $\eta(j_1, j_2, j) \in \mathbf{C}[[\hbar]]$ is some complex formal power series [10, 11].
 (ii) A basis of the irreducible spin-j $\mathscr{U}_\hbar(\mathrm{su}_2)$-subrepresentation of the quantum plane is

$$T_m^j = \begin{bmatrix} 2j \\ j+m \end{bmatrix}_{q^{-2}}^{\frac{1}{2}} x^{j-m} y^{j+m}, \text{ where } \begin{bmatrix} j \\ k \end{bmatrix}_q \text{ is the } q\text{-binomial coefficient.} \tag{12}$$

(iii) The multiplication map with respect to this basis is

$$\mu_\hbar(T^{j_1}_{m_1} \otimes T^{j_1}_{m_1}) = \begin{pmatrix} j_1 & j_2 & j_1+j_2 \\ m_1 & m_2 & m_1+m_2 \end{pmatrix}_q T^{j_1+j_2}_{m_1+m_2}. \tag{13}$$

For the undeformed limit $\mu_\hbar \to \mu$ the q-Clebsch-Gordan coefficient has to be replaced by the undeformed Clebsch-Gordan coefficient.

(iv) The twist $\mathscr{F}$ which yields $\mu_\hbar$ by (6) can now be read off using the orthogonality of the Clebsch-Gordan coefficients to be the one with $\eta(j_1, j_2, j) = 1$ in Eq. (11).

3.3. Quantum Minkowski space

It can be shown [11] that there is also a twist $\mathscr{F}_{so_4}$ of $\mathscr{U}_\hbar(so_4)$ which realizes quantum Euclidean 4-space and a twist $\mathscr{F}_{sl_2(C)}$ of the quantum Lorentz algebra $\mathscr{U}_\hbar(sl_2(C))$ which realizes quantum Minkowski space by (6). These twists are composed out of the twist $\mathscr{F}$ which realizes the quantum plane and the universal $\mathscr{R}$-matrix of $\mathscr{U}_\hbar(su_2)$ as

$$\mathscr{F}_{so_4} = \mathscr{F}_{13}\mathscr{F}_{24}, \qquad \mathscr{F}_{sl_2(C)} = \mathscr{R}_{23}^{-1}\mathscr{F}_{13}\mathscr{F}_{24}, \tag{14}$$

where we use tensor leg notation, $\mathscr{F}_{13} = \mathscr{F}_{[1]} \otimes 1 \otimes \mathscr{F}_{[2]} \otimes 1$, etc. These expressions are plausible, considering the fact that $\mathscr{U}_\hbar(so_4)$, is the product of two copies of $\mathscr{U}_\hbar(su_2)$ and that $\mathscr{U}_\hbar(sl_2(C))$ is $\mathscr{U}_\hbar(so_4)$ twisted by the $\mathscr{R}$-matrix.

4. Conclusion

By definition, Drinfeld twists yield the deformation of an enveloping algebra into a quantum enveloping algebra. It was shown that out of all twists of the Drinfeld-Jimbo algebras $\mathscr{U}_\hbar(su_2)$ and $\mathscr{U}_\hbar(so_4)$, and the quantum Lorentz algebra $\mathscr{U}_\hbar(sl_2(C))$ there are certain twists which realize their fundamental covariant quantum spaces, the quantum plane, quantum Euclidean 4-space, and quantum Minkowski space, respectively, as covariant star products on the undeformed, commutative space algebras. In other words, these particular twists describe the deformation of space and symmetry completely.

Therefore, it can be expected that these twists also describe constructions which are solely based on this deformation, such as the realization of the quantum Minkowski space algebra within the undeformed Poincaré algebra or the formal equivalence of deformed and undeformed gauge theory which was conjectured by Seiberg and Witten [2].

References

[1] M. Gerstenhaber, Ann. Math. **79**, 59 (1964).
[2] N. Seiberg and E. Witten, JHEP **09**, 032 (1999), hep-th/9908142.
[3] J. Madore, S. Schraml, P. Schupp, and J. Wess, Eur. Phys. J. **C16**, 161 (2000), hep-th/0001203.
[4] V. G. Drinfeld, Leningrad Math. J. **1**, 321 (1990).
[5] V. G. Drinfeld, Leningrad Math. J. **1**, 1419 (1990).
[6] H. Grosse, J. Madore, and H. Steinacker, hep-th/0103164.
[7] V. G. Drinfeld, Soviet Math. Dokl. **32**, 254 (1985).
[8] M. Jimbo, Lett. Math. Phys. **10**, 63 (1985).
[9] F. Bayen, M. Flato, C. Fronsdal, A. Lichnerowicz, and D. Sternheimer, Ann. Phys. (NY) **111**, 61 (1978).
[10] T. L. Curtright, G. I. Ghandour, and C. K. Zachos, J. Math. Phys. **32**, 676 (1991).
[11] C. Blohmann, math.QA/0209180.
[12] A. Klymik and K. Schmüdgen, *Quantum Groups and Their Representations* (Springer, 1997).

Inst. Phys. Conf. Ser. No 173: Section 6
Paper presented at 24th Int. Coll. Group Theoretical Methods in Physics, Paris, France, 15–20 July 2002
©*2003 IOP Publishing Ltd*

Equivariant noncommutative index on braided orbits

D. Gurevich[+], P. Saponov[++]

[+] ISTV, Université de Valenciennes, 59304 Valenciennes, France
[++] Theory Department of Institute for High Energy Physics, 142284 Protvino, Russia

1. Introduction.

The notion of noncommutative (NC) index was introduced by A. Connes. In its simplest form it is defined as a pairing

$$\text{Ind} : K_0(A) \times K^0(A) \to \mathbb{Z}. \tag{1}$$

Here A is a given associative algebra, $K^0(A)$ is the Grothendieck group of the monoid of classes of its (finite dimensional) representations, and $K_0(A)$ is the group of the monoid of classes of projective finitely generated A-modules. The pairing (1) consists in the following.

Let us fix a representation $\pi_U : A \to \text{End}(U)$ and an idempotent[‡] $e \in A \otimes \text{End}(W)$ (by fixing a base in W we can identify $\text{End}(W)$ with the matrix space and treat elements of $A \otimes \text{End}(W)$ as matrices with coefficients in A). Then the pairing (1) is defined by

$$\text{Ind}(e, \pi_U) = \text{tr}(\pi_U^{(1)}(e)) = \text{tr}(\pi_U(\text{Tr}(e))) \tag{2}$$

where $\text{Tr} = \text{id} \otimes \text{tr} : A \otimes \text{End}(W) \to A$ is the map associated to the usual matrix trace $\text{tr} : \text{End}(W) \to \mathbb{K}$ and $\pi_U^{(1)} = \pi_U \otimes \text{id}$ is a natural extension of π_U to $A \otimes \text{End}(W)$

$$\pi_U^{(1)} : A \otimes \text{End}(W) \to \text{End}(U) \otimes \text{End}(W).$$

Hereafter, $\mathbb{K}$ is the basic field (always $\mathbb{R}$ or $\mathbb{C}$).

Note that this definition is perfectly adapted to the algebras looking like $U(su(n))$ whose representation category is not braided[§]. However, if such a category is braided the usual trace in the formula (2) should be replaced by its categorical analogue. This is motivated by the fact that in a braided category the usual trace is an irrelevant operator (similarly to a super-category where only the appropriate super-trace is a categorical morphism and should be of use).

In this note we suggest a braided version of the NC index (we call it *equivariant* w.r.t. to action of the quantum group $U_q(sl(n))$) for braided NC analogues of the coadjoint orbits in $sl(n)^*$. In the particular case of the so-called quantum sphere (it corresponds to $n = 2$) we compute the index in question.

2. Braided NC orbits.

First, let us describe the algebra we deal with. Let V, $\dim(V) = n$ be the basic (vector) module of the quantum group (QG) $U_q(sl(n))$ and $R = R_q : V^{\otimes 2} \to V^{\otimes 2}$ be the corresponding braiding (i.e., the image of the universal quantum R-matrix). Consider the system

$$R L_1 R L_1 - L_1 R L_1 R - \hbar(R L_1 - L_1 R) = 0, \quad L_1 = L \otimes \text{id}, \ \hbar \in \mathbb{K}. \tag{3}$$

[‡] In the sequel we identify a projective module with the corresponding idempotent.
[§] We use the term *braided* for a monoidal (quasi)tensor category $\mathscr{C}$ of vector spaces with a braiding $R : U \otimes V \to V \otimes U$, $U, V \in \text{Ob}(\mathscr{C})$ which is different from the usual flip. A typical example is that of $U_q(sl(n))$-modules.

Here $L = (l_i^j)$, $1 \leq i, j \leq n$ is the matrix with entries l_i^j and $L_1 = L \otimes \mathrm{id}$. We call this system modified reflection equation (mRE). Remark that for $\hbar = 0$ we have the usual (nonmodified) RE. The algebra generated by the entries l_i^j subject to the mRE is called mRE algebra. It will be denoted $\mathscr{L}_{\hbar,q}$.

Note that in some sense this algebra is a braided analogue of $U(gl(n)_\hbar)$ (given a Lie algebra g with the bracket $[\,,\,]$ we denote $g_\hbar$ the Lie algebra whose bracket is $\hbar[\,,\,]$). Remark that, in contrast with the QG $U_q(sl(n))$ the algebra $\mathscr{L}_{\hbar,q}$ is one-sided $U_q(sl(n))$-module and it is more adapted to describing braided analogue of coadjoint orbits and projective modules over them.

The algebra $\mathscr{L}_{\hbar,q}$ possesses a central element $\mathbf{l} = C_j^i l_i^j$ belonging to the space span (l_i^j) (note that in the classical limit the matrix $C = (C_j^i)$ becomes identical). Let us consider the quotient $\mathscr{SL}_{\hbar,q} = \mathscr{L}_{\hbar,q}/\{\mathbf{l}\}$ ($\{I\}$ stands for the ideal generated by a given set I). This quotient is a braided analogue of the algebra $U(sl(n)_\hbar)$.

Similarly to $U(sl(n)_\hbar)$ the algebra $\mathscr{SL}_{\hbar,q}$ possesses a nontrivial center $Z(\mathscr{SL}_{\hbar,q})$. Let us consider a character

$$\chi : Z(\mathscr{L}_{\hbar,q}) \to \mathbb{K}$$

and the quotient $\mathscr{SL}_{\hbar,q}(\chi) = \mathscr{SL}_{\hbar,q}/\{I^\chi\}$ where I^χ is the ideal generated by the elements

$$z - \chi(z), \quad z \in Z(\mathscr{L}_{\hbar,q}).$$

This quotient is a braided NC analogue of of generic orbits in $sl(n)^*$ for generic χ and $\hbar \neq 0$. (For $\hbar = 0$ it becomes "braided commutative" analogue.)

3. Quantum sphere.

Let us consider a particular case of this construction for $n = 2$. The braiding R takes in some base the following form

$$R = \begin{pmatrix} q & 0 & 0 & 0 \\ 0 & \lambda & 1 & 0 \\ 0 & 1 & 0 & 0 \\ 0 & 0 & 0 & q \end{pmatrix} \qquad \lambda = q - q^{-1}.$$

Then by putting

$$a = l_1^1, b = l_1^2, c = l_2^1, d = l_2^2$$

we get the following mRE system

$$
\begin{array}{ll}
qab - q^{-1}ba = \hbar b & q(bc - cb) = (\lambda a - \hbar)(d - a) \\
qca - q^{-1}ac = \hbar c & q(cd - dc) = c(\lambda a - \hbar) \\
ad - da = 0 & q(db - bd) = (\lambda a - \hbar)b.
\end{array}
$$

By eliminating the central element $\mathbf{l}$ as above we get the system

$$
\begin{aligned}
q^2 gb - bg &= \hbar(q + q^{-1})b \\
gc - q^2 cg &= -\hbar(q + q^{-1})c \\
(q^2 + 1)(bc - cb) + (q^2 - 1)g^2 &= \hbar(q + q^{-1})g.
\end{aligned}
$$

where $g = a - d$.

Note that being equipped with an involution this algebra becomes Podles' quantum sphere (standard as $\hbar = 0$ and nonstandard unless). In what follows we mainly deal with the quantum sphere (we keep this term although we do not use any involution). However, we want to emphasize that our approach is valid in a more general setting (cf. [GLS2]).

4. Category of $\mathscr{S}\mathscr{L}_{\hbar,q}$-modules.

It can be shown (for example by using [LS]) that any $U_q(sl(2))$-module is $\mathscr{S}\mathscr{L}_{\hbar,q}$-one and therefore all $\mathscr{S}\mathscr{L}_{\hbar,q}$-modules form a braided category. We will label irreducible objects of this category V_k, $k = 2\times$spin, $k = 0,1,2,....$ Let us consider the objects $\text{End}(V_k)$ (called internal morphisms). They are completely identical to similar objects in the category of $SL(2)$-modules.

The only (but very important) difference is that the categorical traces in different categories are different. Let us precise that a map $\text{End}(V) \to \mathbb{K}$ is called *categorical trace* if it is a properly normalized morphism of the category (the normalization should be chosen in such a way that the corresponding dimension $\dim = \text{tr id}$ would become an additive and multiplicative functional on Grothendieck ring of the category, cf. [GLS1] for details).

It is well known that the categorical trace (denoted tr_q) in the category in question has the following form

$$\text{tr}_q X = \text{tr} B_k X, \; X \in \text{End}(V_k)$$

where tr is the usual trace and $B_k \in \text{End}\,V_k$ whose form is indicated in [GLS2]. Note that the categorical dimension of the space V_k is equal to $[k]_q = \frac{q^k - q^{-k}}{q - q^{-1}}$.

In the sequel π_k stands for a representation of $\mathscr{S}\mathscr{L}_{\hbar,q}$ in $\text{End}(V_k)$, i.e. we assume that the image of $\mathbf{l}$ vanishes.

5. Projective modules.

Now, let us describe a way to explicitly construct some projective modules over the algebra $\mathscr{S}\mathscr{L}_{\hbar,q}(\chi)$. Their construction is based on the property of the matrix L to satisfy some sort of the Cayley-Hamilton (CH) identity. Namely, if R is the braiding corresponding to QG $U_q(sl(n))$ in the basic space V then the matrix L from (3) satisfies the following CH identity

$$L^n + \sum_{i=0}^{n-1} \sigma_{n-i}(L) L^i = 0,$$

where the coefficients $\sigma_i(L)$ belong to the center $Z(\mathscr{L}_{\hbar,q})$ of the algebra $\mathscr{L}_{\hbar,q}$ (by imposing the condition $\mathbf{l} = 0$ we can treat $\sigma_i(L)$ as elements of $Z(\mathscr{S}\mathscr{L}_{\hbar,q})$). Upon passing to the quotient $\mathscr{S}\mathscr{L}_{\hbar,q}(\chi)$ we get a polynomial identity with numerical coefficients.

For $n = 2$ this identity takes the form

$$L^2 - q^{-1}\hbar L + \sigma\,\text{id} = 0.$$

where

$$\sigma = -[2]_q^{-1}([2]_q^{-1}g^2 + q^{-1}bc + qcb) \in Z(\mathscr{S}\mathscr{L}_{\hbar,q})$$

is the unique (up to a factor) "braided quadratic Casimir element". In this case the character is defined by the image of $\sigma : \chi(\sigma) = \alpha$. The corresponding algebra will be denoted $\mathscr{S}\mathscr{L}_{\hbar,q}(\alpha)$. Thus, being reduced to this algebra the CH identity becomes

$$L^2 - q^{-1}\hbar L + \alpha\,\text{id} = 0. \tag{4}$$

We would like to point out that the element σ becomes scalar (i.e. a multiple of the identity operator) as the algebra $\mathscr{S}\mathscr{L}_{\hbar,q}$ is represented in the space $\text{End}(V_k)$. Let α_k be the corresponding factor. Thus, the algebra $\mathscr{S}\mathscr{L}_{\hbar,q}(\alpha)$ has a (unique up to an isomorphism) finite dimensional representation for a distinguished value of α namely, $\alpha = \alpha_k$.

Let us denote $\mu_i(1)$, $i = 0, 1$ the roots of the polynomial equation corresponding to (4) and assume them to be distinct. Then it is not difficult to see that the matrices

$$e_i = \prod_{j \neq i} \frac{(L - \mu_j)}{(\mu_i - \mu_j)} \in \mathscr{SL}_{\hbar,q}(\alpha) \otimes \text{End}(V), \quad 0 \leq i \leq 1 \tag{5}$$

are idempotents.

In [GS2], [GLS2] a way is discussed of constructing some matrices $L_{(m)}$, $m = 2, 3, \ldots$ belonging to $\mathscr{SL}_{\hbar,q}(\alpha) \otimes \text{End}(V_m)$ such that each of them is subject to some nontrivial polynomial identity ("higher CH identities"). Assuming the roots $\mu_i(m)$, $0 \leq i \leq m$ of polynomials corresponding to any such identity to be distinct we can introduce idempotents $e_i(m)$ making use of (5).

6. Index on quantum sphere.

Thus, we have introduced some elements of $K_0(\mathscr{SL}_{\hbar,q}(\alpha))$ and $K_0(\mathscr{SL}_{\hbar,q}(\alpha_k))$ for the quantum sphere. Now, let us present the result of their pairing by means of the categorical trace as explained above (note that this pairing is possible iff $\alpha = \alpha_k$).

Proposition 1 *Let $k \geq m$. Then the eigenvalues $\mu_i(m)$ being evaluated at the representation π_k are distinct and for any m we can order the idempotents $e_i(m)$ is such a way that*

$$\text{Ind}\,(e_i(m), \pi_k) = [m + k - 2i + 1]_q, \ 0 \leq i \leq m.$$

Note that index on the standard quantum sphere was considered in [Ha] (also cf. [HM]). We want to point out the main innovations of our approach. First, we use the categorical trace instead of the usual one. Second, we consider the representation theory of the algebra $\mathscr{SL}_{\hbar,q}$ similar to that of $U(sl(2))$. Thus, all our constructions have a specialization at the classical case $q = 1$ (this particular case was previously considered in [GS2]). By the contrary, in [Ha] the author uses the representation theory of the (standard) quantum sphere constructed in [P]. This representation theory has no specialization at the classical case.

Third, we introduce projective modules over quantum sphere (standard or not) explicitly via the CH identity whereas the method previously used in this area is based on the so-called Hopf-Galois extension. This method is not explicit and can be hardly useful for computations of numerical characteristics of higher dimensional braided orbits.

References

[GLS1]   D.Gurevich, R.Leclercq, P.Saponov *Traces in braided categories*, JGP, to be published.
[GLS2]   D.Gurevich, R.Leclercq, P.Saponov *Equivariant noncommutative index on braided sphere*, K-theory, submitted.
[GS1]   D.Gurevich, P.Saponov *Quantum line bundles via Cayley-Hamilton identity*, J. Phys. A: Math. Gen. 34 (2001), pp. 4553 – 4569.
[GS2]   D.Gurevich, P.Saponov *Quantum line bundles on noncommutative sphere*, J. Phys A, submitted.
[Ha]   P.Hajac *Bundles over the quantum sphere and noncommutative index theorem*, K-theory 21 (2001), pp.141–150.
[HM]   P.Hajac, S.Majid *Projective module description of the Q-monopole*, CMP 206 (1999), pp. 247–264.
[LS]   V.Lyubashenko, A.Sudbery *Generalized Lie algebras of A_n type*, J. Math. Phys. 39 (1998), pp. 3487–3504.
[P]   P.Podles *Quantum spheres*, LMP 14 (1987), pp. 193-202.

Inst. Phys. Conf. Ser. No 173: Section 6
Paper presented at 24th Int. Coll. Group Theoretical Methods in Physics, Paris, France, 15–20 July 2002
©2003 IOP Publishing Ltd

Noncommutative Planar Particles: Higher Order Versus First Order Formalism and Supersymmetrization

J. Lukierski‡

Institute for Theoretical Physics, University of Wrocław,
pl. Maxa Borna 9, 50-204 Wrocław, Poland

P. Stichel

An der Krebskuhle 21, D-33619 Bielefeld, Germany

W.J. Zakrzewski

Department of Mathematical Sciences, Science Laboratories, University of Durham, South
Road, Durham DH1 3LE, UK

Abstract.
We describe the supersymmetrization of two formulations of free noncommutative planar
particles – in coordinate space with higher order Lagrangian [1] and in the framework of
Faddeev and Jackiw [2,3], with first order action. In nonsupersymmetric case the first
formulation after imposing subsidiary condition eliminating internal degrees of freedom
provides the second formulation. In supersymmetric case one can also introduce the split
into "external" and "internal" degrees of freedom both describing supersymmetric models.

1. Introduction

In [1] the present authors introduced the following nonrelativistic higher order action for D=2
(planar) particle:

$$L_1^{(0)} = \frac{m\ddot{x}_1^2}{2} - k\epsilon_{ij}\dot{x}_i\ddot{x}_j \,. \tag{1}$$

The canonical quantization of (1) implies the consideration of $x_i, \dot{x}_i$ as independent degrees
of freedom (see e.g. [4]) with the following canonically conjugated two momenta [1,4]

$$p_i = \frac{\partial L^{(0)}}{\partial \dot{x}_i} - \frac{d}{dt}\frac{\partial L^{(0)}}{\partial \ddot{x}_i} = m\dot{x}_i - 2k\epsilon_{ij}\ddot{x}_j \,, \tag{2}$$

$$\hat{p}_i = k\epsilon_{ij}\dot{x}_j \,. \tag{3}$$

The relation (3) introduces a second class constraint, i.e. after the introduction of Dirac
brackets the Lagrangian system (1) is described by six degrees of freedom $Y_A = (x_i, p_i, v_i = \dot{x}_i)$. One gets the following set of Dirac brackets [1]:

$$\{Y_A, Y_B\} = \begin{pmatrix} 0 & 1_2 & 0 \\ -1_2 & 0 & 0 \\ 0 & 0 & -\frac{1}{2k}\epsilon \end{pmatrix} \tag{4}$$

‡ Talk given by J. Lukierski

In order to get the first order formulation of the action (1) one can use the technique proposed by Faddeev and Jackiw [2,3]. The action (1) is equivalent to the following one§

$$L^{(0)} = \frac{m v_i^2}{2} - k \epsilon_{ij} v_i \dot{v}_j + p_i (\dot{x}_i - v_i) , \qquad (5)$$

with six canonical variables (x_i, v_i, p_i). The canonical quantization of (5) using Dirac brackets leads again to the relations (4).

Next we introduce the variables [5,6]

$$Q_i = -2k(v_i - p_i) , \qquad P_i = p_i$$
$$X_i = x_i + \epsilon_{ij} Q_j , \qquad (6)$$

one gets the following set of canonical Poisson brackets (PB)

$$\{X_i, X_j\} = -2k\epsilon_{ij} , \qquad \{P_i, P_j\} = 0 ,$$
$$\{X_i, P_j\} = \delta_{ij} , \qquad (7)$$

and

$$\{Q_i, Q_j\} = 2k \, \epsilon_{ij} . \qquad (8)$$

The action (5) takes the form

$$L^{(0)} = L_{\text{ext}}^{(0)} + L_{\text{int}}^{(0)} , \qquad (9)$$

where

$$L_{\text{ext}}^{(0)} = P_i \dot{X}_i - k\epsilon_{ij} P_i \dot{P}_j - \frac{1}{2} \vec{P}^2 , \qquad (10)$$

$$L_{\text{int}}^{(0)} = -\frac{1}{4k} \epsilon_{ij} Q_i \dot{Q}_j + \frac{1}{8k^2} \vec{Q}^2 . \qquad (11)$$

We note that the external and internal degrees of freedom are dynamically independent, and following Duval and Horvathy [7] we can consider the part (10) of the action as the first order action describing noncommutative particles. We observe that our model permits easily the consistent introduction of a scalar potential [5], electromagnetic interactions [5–7] and general Lagrangian framework [8].

We see that the action (10) describes an invariant sector of the model (1) in "external" phase space X_i, P_i. The internal degrees of freedom (11) can be related with nonvanishing anyonic spin [9]. The aim of this note is to supersymmetrize both actions (1) and (10) and discuss the relation between such supersymmetric models. We will show that the split (9) into dynamically independent supersymmetric parts with external and internal degrees of freedom can be performed again.

2. Supersymmetrization of Higher Order Action and its First Order Form

Let us consider for simplicity N=1 supersymmetric quantum mechanics. We introduce the real field $X_i(t, \theta)$ with one Grassmann variable θ

$$x_i(t) \longrightarrow X_i(t, \theta) = x_i(t) + i\theta\psi(t) , \qquad (12)$$

where $\theta^2 = \theta\psi_j + \psi_j\theta = \psi_i\psi_j + \psi_j\psi_i = 0$. Introducing the supersymmetric covariant derivative

$$D = \frac{\partial}{\partial\theta} - i\theta\frac{\partial}{\partial t} \Rightarrow D^2 = -i\frac{\partial}{\partial t} = -H , \qquad (13)$$

§ The equivalence of (5) and (1) can be seen in a clear way if we consider the generating functionals based on both actions (5) and (1) - the last term in (5) shall introduce the functional Dirac delta function replacing v_i by $\dot{x}_l$.

we get the following supersymmetric extension of (1)

$$
\begin{aligned}
L_{\text{SUSY}}^{(0)} &= i \int d\theta \left(\frac{m}{2} \dot{X}_i DX_i - k\epsilon_{ij} \ddot{X}_i DX_j \right) \\
&= \frac{m}{2} \left(\dot{x}_i^2 + i\psi_i \dot{\psi}_i \right) - k\epsilon_{ij} \left(\dot{x}_i \ddot{x}_j - i\dot{\psi}_i \dot{\psi}_j \right).
\end{aligned}
\tag{14}
$$

If $k = 0$ we obtain the standard case of N=1 nonrelativistic spinning particle, with fermionic second class constraints. If $k \neq 0$ the fermionic momenta become independent from fermionic coordinates ψ_i.

Using the Faddeev-Jackiw method we extend supersymmetrically the action (5) as follows:

$$
\begin{aligned}
L_{\text{SUSY}}^{(0)} &= \frac{mv_i^2}{2} - k\epsilon_{ij} v_i \dot{v}_j + \frac{im}{2} \psi_i \rho_i \\
&\quad + i k\epsilon_{ij} \rho_i \rho_j + p_i(\dot{x}_i - v_i) + \chi_i(\dot{\psi}_i - \rho_i).
\end{aligned}
\tag{15}
$$

The field equation for ρ_i is purely algebraic where ρ_i and χ_i are fermionic. Substituting

$$
\chi_i = \frac{im}{2} \psi_i - 2ik\,\epsilon_{ij} \rho_j,
\tag{16}
$$

we find that (we put for simplicity further $m = 1$)

$$
\begin{aligned}
L_{\text{SUSY}}^{(0)} &= \frac{v_i^2}{2} - k\epsilon_{ij} v_i \dot{v}_j + \frac{i}{2} \psi_i \dot{\psi}_i \\
&\quad + 2ik\epsilon_{ij} \dot{\psi}_i \rho_j - ik\epsilon_{ij} \rho_i \dot{\rho}_j + p_i(\dot{x}_i - v_i).
\end{aligned}
\tag{17}
$$

For the fermionic sector of the action (17) one obtains with the use of Dirac brackets the following PB algebra

$$
\{\psi_i, \psi_j\} = 0, \qquad \{\psi_i, \rho_j\} = \frac{i}{2k} \epsilon_{ij}, \qquad \{\rho_i, \rho_j\} = i\frac{1}{4k^2} \delta_{ij}
\tag{18}
$$

In order to split (15) into external and internal sector we introduce besides the variables (6) also new fermionic variables

$$
\psi_i \longrightarrow \tilde{\psi}_i = \psi_i - 2k\epsilon_{ij} \rho_j.
\tag{19}
$$

We get

$$
L_{\text{SUSY}}^{(0)} = L_{\text{SUSY;ext}}^{(0)} + L_{\text{SUSY;int}}^{(0)},
\tag{20}
$$

where

$$
L_{\text{SUSY;ext}}^{(0)} - P_i \dot{X}_i - k\epsilon_{ij} P_i \dot{P}_j - \frac{1}{2} P_i^2 + \frac{i}{2} \tilde{\psi}_i \dot{\tilde{\psi}}_i,
\tag{21}
$$

$$
L_{\text{SUSY;int}}^{(0)} = -\frac{1}{4k} \epsilon_{ij} Q_i \dot{Q}_j + \frac{1}{8k^2} \vec{Q}^2 - 2ik^2 \rho_k \dot{\rho}_k - ik\epsilon_{ij} \rho_i \rho_k,
\tag{22}
$$

The new fermionic coordinates satisfy the following PB algebra

$$
\{\tilde{\psi}_i, \tilde{\psi}_j\} = i\delta_{ij}, \qquad \{\tilde{\psi}_i, \rho_j\} = 0.
\tag{23}
$$

We see therefore that again the supersymmetric action (15) or (20) can be split into dynamically independent external and internal sectors (see (21)–(22)).

3. Supersymmetry in External and Internal Sectors

The actions (21) and (22) describe the supersymmetric extensions respectively of external and internal actions (10) and (11). These actions are invariant under the following set of supersymmetry transformations:
  i) in external sector (see (21))

$$\delta X_i = i \epsilon \tilde{\psi}_i ,$$
$$\delta \tilde{\psi}_i = - \epsilon P_i ,$$
$$\delta P_i = 0 \tag{24}$$

  ii) in internal sector (see (22))

$$\delta Q_i = 2i k \epsilon \epsilon_{ij} \rho_j ,$$
$$\delta \rho_i = \frac{1}{4k^2} \epsilon Q_i , \tag{25}$$

where ϵ is a constant Grassmann number.

The supercharge corresponding to (22) is given by the formula

$$Q_{\text{ext}} = i \tilde{\psi}_i P_i \tag{26}$$

and we get consistently the external Hamiltonian (see (21))

$$-\frac{i}{2}\{Q_{\text{ext}}, Q_{\text{ext}}\} = \frac{1}{2} P_i^2 = H^{(0)}_{\text{SUSY;ext}} . \tag{27}$$

Similarly from (7) and (22) we obtain the tranformation (25) if

$$Q_{\text{int}} = i Q_i \rho_i \tag{28}$$

and our internal Hamiltonian (see (22)) is given by

$$-\frac{i}{2}\{Q_{\text{int}}, Q_{\text{int}}\} = -\frac{1}{8k^2} \vec{Q}^2 + ik \epsilon_{ij} \rho_i \rho_j = H^{(0)}_{\text{SUSY;int}} . \tag{29}$$

We would like to add here that one could consider only the external part, described by the action (20), as describing supersymmetric planar particles. The quantum mechanical states describing the internal sector can be eliminated by subsidiary conditions.

4. Final Remarks

We would like to mention that
  i) We have considered here the N=1 world line supersymmetry. It is quite straightforward to extend the above considerations to N=2 by employing the N=2 D=1 superfields.
  ii) We have discussed here, for simplicity, only the free case. The supersymmetrization of the models with gauge intersections considered in [6] is under active consideration.

5. References

[1] J. Lukierski, P. Stichel and W.J. Zakrzewski, Ann. Phys. **260**, 224 (1997).
[2] L. Faddeev and R. Jackiw, Phys. Rev. Lett. **60**, 1968 (1988).
[3] R. Jackiw, in "Constraints Theory and Quantization Methods", ed. F. Colomo et. al., World Scientific, Singapore, 1994, p. 163.
[4] A. Barut and G.H. Mullen, Ann. Phys. **20**, 203 (1964).
[5] P.A. Horvathy and M.S. Plynskey, JHEP 06, 033 (2002).
[6] J. Lukierski, P. Stichel and W.J. Zakrzewski, hep-th/0207149.
[7] C. Duval, P.A. Horvathy, Phys. Lett. **B479**, 284 (2000).
[8] A.A. Deriglazov, hep-th/0208072; hep-th/0208200.
[9] C. Duval, P.A. Horvathy, hep-th/0209166.

Inst. Phys. Conf. Ser. No 173: Section 6
Paper presented at 24th Int. Coll. Group Theoretical Methods in Physics, Paris, France, 15–20 July 2002
©*2003 IOP Publishing Ltd*

A locally trivial quantum Hopf bundle

R. Matthes

Fachbereich Physik der TU Clausthal, Leibnizstr. 10, D-38678 Clausthal-Zellerfeld, Germany

Abstract.
We describe a locally trivial quantum principal $U(1)$-bundle over the quantum space S^2_{pq} which is a noncommutative analogue of the usual Hopf bundle. We also provide results concerning the structure of its total space algebra (irreducible $*$-representations and topological K-groups) and its Galois aspects (Galois property, existence of a strong connection, non-cleftness).

1. Introduction

In this note, we describe an example of a principal bundle in the setting of noncommutative geometry, which meets two possible (still provisional) definitions: It is a locally trivial quantum principal bundle in the sense of [BK96] as well as a Hopf-Galois extension [M-S93]. Besides giving the definition of these notions and a description of the bundle [CM00], [CM02], we provide a list of results obtained in [HMS] concerning the structure of the total space algebra and the Galois aspects of the bundle.

2. Quantum principal bundles

2.1. Hopf-Galois extensions

Dualizing the corresponding classical structure "à la Gelfand-Neumark", one arrives at the following items which show up in the definition of quantum principal bundles:

- There is some algebra P replacing the total space of a principal bundle.
- There is some Hopf algebra H replacing the structure group, coacting on P on the right, i.e., there is an algebra homomorphism $\Delta_R : P \to P \otimes H$ with $(\Delta_R \otimes \mathrm{id}) \circ \Delta_R = (\mathrm{id} \otimes \Delta) \circ \Delta_R$ and $(\mathrm{id} \otimes \varepsilon) \circ \Delta_R = \mathrm{id}$.
- There is another algebra replacing the base space, which coincides with the subalgebra of coinvariants of the coaction of H on P, $B = P^{coH} := \{p \in P \mid \Delta_R(p) = p \otimes 1\}$. The bundle projection is the embedding $B \subset P$, denoted by $\iota : B \to P$.

$B \subset P$ is called H-extension in the above context [M-S93]. For a classical principal bundle with base space M, total space P and structure group G, the right action is assumed to be free. This assumption can be restated as bijectivity of the map $X \times G \to X \times_M X$, $(x, g) \mapsto (x, xg)$. At the level of algebras, this means bijectivity of the map

$$can : P \otimes_B P \longrightarrow P \otimes H, \ p \otimes p' \mapsto pp'_{(0)} \otimes p'_{(1)}.$$

Here we use Sweedler notation, $\Delta_R(p) = p_{(0)} \otimes p_{(1)}$. An H-extension is called Hopf-Galois if can is bijective. This is essentially the notion of an algebraic quantum principal bundle (see, e.g., [BM93]).

2.2. Locally trivial quantum principal bundles

There is another approach to quantum principal bundles emphasizing the idea of gluing which is behind the definition of classical fibre bundles [BK96], [CM02]. In order to state this definition, we need an algebraic notion of covering:

A covering of an algebra B is a family $(J_i)_{i \in I}$ of ideals with zero intersection. Let $\pi_i : B \to B_i := B/J_i$, $\pi_j^i : B_i \to B_{ij} := B/(J_i + J_j)$ be the quotient maps. A covering $(J_i)_{i \in I}$ is called complete if the homomorphism $B \ni b \mapsto (\pi_i(b))_{i \in I} \in \{(b_i)_{i \in I} \in \prod_{i \in I} B_i \mid \pi_j^i(b_i) = \pi_i^j(b_j)\}$ is surjective (it is always injective). Finite coverings by closed ideals in C*-algebras and two-element coverings are always complete. A locally trivial H-extension is an H-extension $B \subset P$ supplied with the following local data:

(i) B has a complete finite covering $(J_i)_{i \in I}$.

(ii) There are given surjective homomorphisms $\chi_i : P \to B_i \otimes H$ (local trivializations) such that

(a) $\chi_i \circ \iota = \pi_i \otimes 1$ ($\iota : B \to P$),
(b) $(\chi_i \otimes \mathrm{id}) \circ \Delta_R = (\mathrm{id} \otimes \Delta) \circ \chi_i$ (right colinearity),
(c) $(\ker \chi_i)_{i \in I}$ is a complete covering of P.

As in the classical situation, locally trivial bundles can be reconstructed from transition functions related to the covering of the base algebra. More precisely, every locally trivial principal fibre bundle with fixed base algebra B and Hopf algebra H is determined by the following data:

- a complete finite covering $(J_i)_{i \in I}$
- a family of transition functions, i.e., of homomorphisms $\tau_{ij} : H \to Z(B_{ij})$ (center) fulfilling $\tau_{ii} = 1\varepsilon$, $\tau_{ji} \circ S = \tau_{ij}$ (S the antipode of H), and the cocycle condition $\pi_k^{ij} \circ \tau_{ij} = m_{B_{ijk}} \circ ((\pi_j^{ik} \circ \tau_{ik}) \otimes (\pi_i^{jk} \circ \tau_{kj})) \circ \Delta$.

The total space algebra is then given as the gluing

$$P = \{(f_i)_{i \in I} \in \oplus_{i \in I} B_i \otimes H \mid (\pi_j^i \otimes \mathrm{id})(f_i) = \varphi_{ij} \circ (\pi_i^j \otimes \mathrm{id})(f_j)\},$$

where $\varphi_{ij}(b \otimes h) = b\tau_{ji}(h_{(1)}) \otimes h_{(2)}$. The remaining data of the corresponding locally trivial H-extension are as follows:

$$\Delta_R((f_i)_{i \in I}) = ((\mathrm{id} \otimes \Delta)(f_i))_{i \in I}, \quad \chi_i((f_i)_{i \in I}) = f_i, \quad \iota(b) = (\pi_i(b) \otimes 1)_{i \in I}.$$

3. Description of the locally trivial $U(1)$-bundle $S_{pq}^3 \to S_{pq}^2$

3.1. Quantum discs

We use the following subfamily of a two-parameter family of quantum discs defined in [KL93] whose *-algebra is $\mathcal{O}(D_q) := \mathbb{C}\langle x, x^* \rangle / (x^*x - qxx^* - (1-q))$, $0 < q < 1$. The irreducible *-representations of $\mathcal{O}(D_q)$ are an S^1-family of one-dimensional representations, given by $\pi_\theta(x) = e^{i\theta}$ (classical points), and an infinite-dimensional representation π_q in a separable Hilbert space representing the generator x as a one-sided weighted shift. The classical points define an embedding of S^1 into D_q, i.e., $\mathcal{O}(D_q) \ni x \xrightarrow{\phi_q} u \in \mathcal{O}(S^1) := \mathbb{C}\langle u, u^* \rangle / (u^*u - 1, u\, u^* - 1)$. Since $\|\pi(x)\| = 1$ for any *-representation of π in some $B(\mathcal{H})$, the C*-closure $C(D_q)$ of $\mathcal{O}(D_q)$ is well-defined (using bounded *-representations). One knows that $C(D_q) \simeq \mathcal{T}$ (Toeplitz or shift algebra). Using the above-mentioned irreducible *-representations, one may heuristically interpret D_q as a diffuse membrane spanned by a classical S^1.

3.2. Quantum two-spheres (quantum cones)

They are defined as a gluing of two quantum discs along the classical "boundary" S^1:
$\mathcal{O}(S^2_{pq}) := \mathcal{O}(D_p) \oplus_\phi \mathcal{O}(D_q) = \{(f,g) \in \mathcal{O}(D_p) \oplus \mathcal{O}(D_q) \mid \phi_p(f) = \phi_q(g)\}, \ 0 < p, q < 1.$
The $*$-algebra $\mathcal{O}(S^2_{pq})$ can be identified with the quotient of the free algebra generated by f_1, f_1^*, f_0 by the ideal J defined by the relations $f_0^* = f_0, \ f_1^* f_1 - q f_1 f_1^* = (p-q) f_0 + (1-p)\mathbf{1}, \ (1-f_0)(f_1 f_1^* - f_0) = 0$. There are an S^1-family of one-dimensional and two nonequivalent infinite dimensional $*$-representations in a separable Hilbert space. The latter represent f_0 as a diagonal operator and f_1 as a one-sided weighted shift. Again $\|\rho(f_0)\| = \|\rho(f_1)\| = 1$ for any bounded $*$-representation. The C^*-closure $C(S^2_{pq})$ is defined using such representations. One knows $C(S^2_{pq}) \simeq C(D_p) \oplus_\phi C(D_q), \simeq C(S^2_{\mu c}), \ |\mu| < 1, c > 0$ (Podleś spheres [P-P87]). Thus, the glued two-spheres are homeomorphic to the so-called equilateral Podleś spheres. Using, as for the disc, the irreducible representations, one may visualize S^2_{pq} as a top of a diffuse cone, with edge S^1.

3.3. The $\mathcal{O}(U(1))$-extension $\mathcal{O}(S^2_{pq}) \subset \mathcal{O}(S^3_{pq})$

Note that $\mathcal{O}(S^2_{pq}) = \mathcal{O}(D_p) \oplus_\phi \mathcal{O}(D_q)$ has a canonical covering consisting of the kernels of the first and second projections, $J_1 = \ker pr_1$, $J_2 = \ker pr_2$. One has canonical identifications $\mathcal{O}(S^2_{pq})/J_1 = \mathcal{O}(D_p)$, $\mathcal{O}(S^2_{pq})/J_2 = \mathcal{O}(D_q)$, $\mathcal{O}(S^2_{pq})/(J_1 + J_2) = \mathcal{O}(S^1)$. These are the data of the base algebra. The desired extension results from gluing $\mathcal{O}(D_p) \otimes \mathcal{O}(U(1))$ and $\mathcal{O}(D_q) \otimes \mathcal{O}(U(1))$ by means of one transition function $\tau : \mathcal{O}(U(1)) \longrightarrow \mathcal{O}(S^1)$, $u \mapsto \underline{u}$, following the general method of Subsection 2.2. The corresponding gluing $\mathcal{O}(S^3_{pq})$ of two quantum solid tori along their set $\mathbb{T}^2$ of classical points is fully analogous to the geometrical picture in the case of the usual $U(1)$-Hopf bundle (Heegard splitting of S^3). It turns out that $\mathcal{O}(S^3_{pq})$ is isomorphic to the quotient of the free $*$-algebra generated by a, b by the ideal generated by the relations

$$ab = ba, \ ab^* = b^*a, \ a^*b^* = b^*a^*, \ a^*b = ba^*,$$

$$a^*a - qaa^* = 1 - q, \quad b^*b - pbb^* = 1 - p,$$

$$(1 - aa^*)(1 - bb^*) = 0.$$

The structural $*$-homomorphisms of the locally trivial $U(1)$-extension in terms of the generators a, b are:

$$\Delta_R(a) = a \otimes u, \quad \Delta_R(b) = b \otimes u^*,$$

$$\chi_p(a) = 1 \otimes u, \quad \chi_p(b) = x \otimes u^*, \quad \chi_q(a) = y \otimes u, \quad \chi_q(b) = 1 \otimes u^*,$$

$$\iota(f_1) = ba, \quad \iota(f_0) = bb^*.$$

4. Further results

4.1. Structure of S^3_{pq}

- The classes of irreducible $*$-representations of $\mathcal{O}(S^3_{pq})$ in bounded operators are classified: There is a $\mathbb{T}^2$-family of one dimensional representations and two S^1-families of infinite-dimensional representations in a separable Hilbert space. In the first of these two families, a is a multiple of the unit operator, and b is a one-sided weighted shift. In the second family a and b exchange their roles. Since again the norms of a and b are 1 in any bounded representation, one can define the C^*-algebra $C(S^3_{pq})$ using such representations.

- A vector space basis of $\mathcal{O}(S_{pq}^3)$ can be exhibited.

- $C(S_{pq}^3)$ is a 2-graph C^*-algebra.

- The K-groups of $C(S_{pq}^3)$ coincide with the K-groups of the classical S^3, i.e., $K_0(C(S_{pq}^3)) = K_1(C(S_{pq}^3)) = \mathbb{Z}$.

4.2. Hopf-Galois (bundle) aspects

- The $\mathcal{O}(U(1))$-extension $\mathcal{O}(S_{pq}^2) \subset \mathcal{O}(S_{pq}^3)$ has the Galois property. (Idea of proof: Find a lift l of the translation map and use a general argument of Schneider.)

- The lift l of the translation map is a strong connection in the sense of [H-PM96]. Consequently, the $\mathcal{O}(U(1))$-extension $\mathcal{O}(S_{pq}^2) \subset \mathcal{O}(S_{pq}^3)$ is relatively projective [BH].

- As a further consequence of the existence of a strong connection, all associated modules (vector bundles) are finitely generated projective. In particular, using the strong connection one can for any winding number give explicitly a projector matrix corresponding to the associated line bundle.

- The $\mathcal{O}(U(1))$-extension $\mathcal{O}(S_{pq}^2) \subset \mathcal{O}(S_{pq}^3)$ is non-cleft (not a crossed product). This is proved using a trace on $\mathcal{O}(S_{pq}^2)$, which is defined as the operator trace composed with the difference of the two irreducible infinite dimensional representations. The Chern-Connes pairing of this trace with the K_0-class of the projector defining the associated line bundle with winding number -1 just gives this number, which proves the above claim (cf. [HM99]).

Acknowledgements: This work was supported by the Deutsche Forschungsgemeinschaft and the Mathematisches Forschungsinstitut Oberwolfach, where this note was completed during a stay under the Research in Pairs programme. Also, it is a pleasure to thank D. Calow, P.M. Hajac and W. Szymanski for many hours of discussion and joint work.

References

[BH]   Brzeziński T., Hajac P.M.: Relatively projective extensions of Galois type. Preprint of the University of Wales, Swansea, MRR-01-15

[BM93]   Brzeziński, T., Majid, S.: Quantum group gauge theory on quantum spaces, Commun. Math. Phys. **157**, 591–638 (1993); Erratum **167**, 235 (1995), hep-th/9208007

[BK96]   Budzyński R.J., Kondracki W.: Quantum principal fibre bundles: Topological aspects, Rep. Math. Phys. **37**, 365–385 (1996), hep-th/9401019

[CM00]   Calow D., Matthes R.: Covering and gluing of algebras and differential algebras, J. Geom. Phys. **32**, 364–396 (2000)

[CM02]   Calow D., Matthes R.: Connections on locally trivial quantum principal fibre bundles, J. Geom. Phys. **41**, 114–165 (2002), math.QA/0002228

[H-PM96]   Hajac P.M.: Strong connections on quantum principal bundles, Commun. Math. Phys. **182**, 579–617 (1996)

[HM99]   Hajac P.M., Majid S.: Projective module description of the q-monopole, Commun. Math. Phys. **206**, 247–264 (1999)

[HMS]   Hajac P.M., Matthes R., Szymański W.: Locally trivial quantum Hopf fibration, in preparation

[KL93]   Klimek S., Lesniewski A.: A two-parameter quantum deformation of the unit disc, J. Funct. Anal. **115**, 1–23 (1993)

[M-S93]   Montgomery, S.: *Hopf Algebras and Their Actions on Rings.* Regional Conference Series in Mathematics no. 82, AMS, 1993

[P-P87]   Podleś, P.: Quantum spheres, Lett. Math. Phys. **14**, 193–202 (1987)

[S-HJ90]   Schneider H.-J.: Principal homogeneous spaces for arbitrary Hopf algebras, Israel J. Math. **72**, 167–195 (1990)

Inst. Phys. Conf. Ser. No 173: Section 6
Paper presented at 24th Int. Coll. Group Theoretical Methods in Physics, Paris, France, 15–20 July 2002
©2003 IOP Publishing Ltd

$\mathbb{Z}_n$-Graded Topological Generalizations of Supersymmetry and Orthofermion Algebra

A Mostafazadeh

Department of Mathematics, Koç University, Rumelifeneri Yolu, 80910 Sariyer, Istanbul, Turkey

Abstract. We review various generalizations of supersymmetry and discuss their relationship. In particular, we show how supersymmetry, parasupersymmetry, fractional supersymmetry, orthosupersymmetry, and the $\mathbb{Z}_n$-graded topological symmetries are related.

1. Introduction

The advent of supersymmetric quantum mechanics (SQM) in the 1980s [19] and its remarkable applications [7] have since motivated many researchers to seek for generalizations of SQM. Most of these generalizations are algebraic in nature in the sense that they are defined in terms of an operator algebra involving a central element called the Hamiltonian H and a number of noncentral operators $\mathcal{Q}_a$ and $\mathcal{Q}_a^\dagger$ called the symmetry generators such that this operator algebra generalizes the algebra of SQM, namely [17, 19]

$$\mathcal{Q}_a^2 = \mathcal{Q}_a^{\dagger 2} = 0, \quad [\mathcal{Q}_a, H]_- = 0, \quad [\mathcal{Q}_a, \mathcal{Q}_b^\dagger]_+ = 2\delta_{ab}H, \tag{1}$$

where $[A, B]_\pm := AB \pm BA$ and $a, b = 1, 2, \cdots, \mathcal{N}$ for some $\mathcal{N} \in \mathbb{Z}^+$. This is known as the algebra of type $N = 2\mathcal{N}$ supersymmetry (SUSY). In what follows we shall only consider the case $\mathcal{N} = 1$ and drop the label $a = 1$. The general case ($\mathcal{N} > 1$) may be treated similarly.

An algebraic generalization of the $N = 2$ SUSY corresponds to an associate operator algebra involving H, $\mathcal{Q}$, $\mathcal{Q}^\dagger$, and a set of more general defining relations. Typical examples are Parasupersymmetry (PSUSY), fractional supersymmetry (FSUSY), and orthosupersymmetry (OSUSY). Note that in ordinary unitary quantum mechanics, one demands H to be Hermitian. Therefore, one considers unitary (*-representations) of the underlying operator algebra in a Hilbert space.

An alternative approach to generalize SQM is to adopt the grading and the topological properties of SQM as the guiding principle [14, 1]. This leads to a set of generalizations of SQM, called the $\mathbb{Z}_n$-graded topological symmetries (TS), for which one can define certain integer-valued topological invariants that generalize the Witten index of SUSY [19].

The aim of this article is to provide a brief review of the origins of the above mentioned generalizations of SQM and their relationships.

2. SUSY Algebra and Its Statistical Generalizations

Consider the Hamiltonian of the Bose-Fermi oscillator [6]

$$H := N_+ + N_-, \tag{2}$$

where $N_\pm := a_\pm^\dagger a_\pm$ is the number operator for a bosonic or fermionic degree of freedom depending on whether its subscript is $+$ or $-$ respectively, and $a_\pm$ is the corresponding annihilation operator. Then one can check that H together with $Q := \sqrt{2}a_+^\dagger a_-$ satisfies the $N = 2$ SUSY algebra (1). Here one uses the algebraic identities of Bose and Fermi statistics. If one identifies the subscript $-$ with a parafermion [8] of order p and assumes relative bosonic statistics [9], then H and $Q := a_+^\dagger a_-$ satisfy the algebra of PSUSY of order p, [18],

$$Q^{p+1} = 0, \quad [Q, H]_- = 0, \quad \sum_{k=0}^{p} Q^{p-k} Q^\dagger Q^k = 2p Q^{p-1} H. \tag{3}$$

Similarly if a_α, with $\alpha = 1, 2, \cdots, p$, denote the annihilation operators associated with an orthofermion of order p, H and $Q_\alpha := a_+^\dagger a_\alpha$ satisfy the algebra of OSUSY of order p, [10],

$$Q_\alpha Q_\beta = 0, \quad [Q_\alpha, H]_- = 0, \quad Q_\alpha Q_\beta^\dagger + \delta_{\alpha\beta} \sum_{\gamma=1}^{p} Q_\gamma^\dagger Q_\gamma = 2\delta_{\alpha\beta} H. \tag{4}$$

Note that both (3) and (4) reduce to the $N = 2$ SUSY algebra for $p = 1$. Therefore, PSUSY and OSUSY are generalizations of SUSY. Next consider the operator algebra for the FSUSY of order $F = 2, 3, \cdots$, namely $Q^F = H$, [3]. Again if we set $F = 2$ we obtain the algebra of $N = 1$ SUSY for which $Q = Q^\dagger$. Therefore FSUSY is also an algebraic generalization of SUSY.

3. Topological Properties of SUSY and PSUSY of Order 2

Given a supersymmetric quantum system, one can show that the difference of the number of zero-energy bosonic and fermionic states, which is called the Witten index, remains invariant under arbitrary SUSY-preserving continuous deformations of the system [19]. This observation serves as the basis of the supersymmetric proofs of the celebrated Atiyah-Singer index theorem, [4]. The fact that SUSY has a rich topological content raises the natural question whether its generalizations share similar properties.

In order to define the Witten index for a general supersymmetric system, one uses the double ($\mathbb{Z}_2$) grading of its Hilbert space. The Hilbert space $\mathcal{H}$ always admits a grading operator τ fulfilling $\tau = \tau^\dagger = \tau^{-1}$ and $[\tau, Q]_+ = 0$. This operator splits $\mathcal{H}$ into the direct sum of its two eigenspaces $\mathcal{H}_\pm$. The elements of $\mathcal{H}_+$ and $\mathcal{H}_-$ are respectively called 'bosonic' and 'fermionic' state vectors. We shall instead use the term: 'vectors with definite grade $+$ or $-$'. Another important ingredient that one uses to establish topological invariance of the Witten index is the particular spectral degeneracy structure (SDS) of supersymmetric systems: the energy spectrum is nonnegative, and positive-energy eigenstates come in degenerate pairs with opposite grade. The proof of the fact that every supersymmetric system has this particular SDS is equivalent to the problem of finding unitary irreducible representation (irrep) of the SUSY algebra (1). It turns out that there are only two types of irreps namely the trivial 1-dim. irrep in which $H = Q = 0$ and the 2-dim. irreps labeled by $E \in \mathbb{R}^+$ in which $H = EI$, I is the 2×2 identity matrix, $\tau = \sigma_3$, $Q = \sqrt{E/2}(\sigma_1 - i\sigma_2)$, and σ_i are Pauli matrices.

In [5] the authors explored the topological invariants of the extended and generalized SUSY. These are essentially SUSYs with more than one grading operator. The first study of the topological properties of a genuine generalization of SUSY is [12] where the question of defining an analog of the Witten index for PSUSY of order 2 was addressed. Again this question may be reduced to finding unitary irreps of the algebra of PSUSY of order 2. It turns out that there are three types of irreps: the trivial 1-dim. irreps labeled by $E \in \mathbb{R}$ in which $Q = 0$ and $H = E$, the 2-dim. irreps that coincide with those of the SUSY algebra, and the 3-dim. irreps labeled by $E \in \mathbb{R}^+$ and $t \in [-1, 1]$ in which $H = EI$, I is the 3×3 identity matrix, and

$$\tau = \sqrt{E} \begin{pmatrix} 1 & 0 & 0 \\ 0 & 1 & 0 \\ 0 & 0 & -1 \end{pmatrix}, \qquad \mathcal{Q} = \sqrt{E} \begin{pmatrix} 0 & 0 & i(1+t) \\ 0 & 0 & \sqrt{1-t^2} \\ i(1-t) & \sqrt{1-t^2} & 0 \end{pmatrix}. \quad (5)$$

One can use the above results on the representation theory of the PSUSY algebra of order 2 to infer that only for a special type of PSUSY of order 2 one can define a topological invariant. These correspond to the systems with nonnegative spectrum whose positive energy eigenvalues are all triply degenerate. These systems have been studied and classified in [13]. The extension of the results of [12, 13] to PSUSY of order $p > 2$ has not been possible mainly due to difficulties associated with the representation theory of the algebra (3) for $p > 2$.

4. Topological Generalizations of SUSY

The study of the topological properties of SUSY and PSUSY of order 2 reveals the fact that the basic ingredients responsible for these properties are their grading and degeneracy structures. This leads to the point of view that one should define a set of generalizations of SUSY by requiring that they possess appropriate grading and degeneracy structures. By definition these symmetries involve a set of integer-valued topological invariants and include SUSY as a special case. They are consequently called topological symmetries (TS). TSs were initially defined for systems with a $\mathbb{Z}_2$-graded Hilbert space in [14]. Their $\mathbb{Z}_n$-graded generalization was subsequently considered in [1].

$\mathbb{Z}_n$-graded TSs are described by n positive integers $m_1, \cdots, m_n$ as follows: 1. The Hilbert space $\mathcal{H}$ is $\mathbb{Z}_n$-graded, i.e., there are (nonzero) subspaces $\mathcal{H}_\ell$ such that $\mathcal{H} = \mathcal{H}_1 \oplus \cdots \oplus \mathcal{H}_n$. The elements of $\mathcal{H}_\ell$ is said to have (definite) grade ℓ; 2. The Hamiltonian maps $\mathcal{H}_\ell$ to $\mathcal{H}_\ell$; 3. The energy spectrum is nonnegative; 4. For every positive energy eigenvalue E, there is a $\lambda_E \in \mathbb{Z}^+$ such that the eigenspace of E is spanned by $\lambda_E m_1$ vectors of grade 1, $\lambda_E m_2$ vectors of grade 2, $\cdots$, and $\lambda_E m_n$ vectors of grade n. Given this definition one can easily show that the integers $\Delta_{ij} := m_i n_j^{(0)} - m_j n_i^{(0)}$, with $n_\ell^{(0)}$ denoting the number of zero-energy states of grade ℓ, are topological invariants.

It is quite remarkable that the rather general definition of TS is indeed sufficient to determine the underlying operator algebra. It turns out that the algebras of SUSY, PSUSY of order 2, and FSUSY of arbitrary order are among the algebras supporting TSs. In particular, for $n = 2$ and $m_1 = m_2 = 1$ one derives a unique operator algebra that is identical with the SUSY algebra (1) with $N = 2$. Similarly, for $n = m_1 = 2$ and $m_2 = 1$ one obtains the PSUSY algebra (3) with $p = 2$. Finally, for arbitrary n and $m_1 = m_2 = \cdots = m_\ell = 1$ one obtains an algebra defined by the FSUSY relation $H = \mathcal{Q}^{n+1}$ together with a couple of additional relations, [1].

5. Orthofermion Algebra and Topological Symmetries

In [15], it has been shown that the orthofermion algebra of order p, [11]

$$a_\alpha a_\beta = 0, \qquad a_\alpha a_\beta^\dagger + \delta_{\alpha\beta} \sum_{\gamma=1}^{p} a_\gamma^\dagger a_\gamma = \delta_{\alpha\beta}, \quad (6)$$

has a unique nontrivial unitary irrep which has dimension $p + 1$, that in this representation a_α is represented by a matrix with entries $[a_\alpha]_{ij} = \delta_{i,1}\delta_{j,\alpha+1}$, $i, j = 1, \cdots, p + 1$, and that every representation of (6) is completely reducible to copies of the above irrep and the trivial irrep. Furthermore, one can check that the operators $L := a_1 + \sum_{\alpha=2}^{p} a_{\alpha-1}^\dagger a_\alpha$ and $J := L + a_p^\dagger$

respectively satisfy $L^{p+1} = 0$, $\sum_{k=0}^{p} L^{p-k} L^{\dagger} L^k = p L^{p-1}$, and $J^{p+1} = 1$. This in turn implies that every system with an OSUSY of order p has a $\mathbb{Z}_n$-graded TS, a PSUSY of order p, and a FSUSY of order $p + 1$, [15, 2].

For the case $p = 3$, one can actually find a realization of this type of TS using Fredholm (possibly differential) operators acting in an inner product (Hilbert) space. This realization has been studied and the relation between the topological invariants Δ_{ij} and the analytic indices of the associated operators has been discussed in [16].

References

[1] K Aghababaei Samani and A Mostafazadeh, Nucl. Phys. B **595**, 467 (2001).

[2] K Aghababaei Samani and A Mostafazadeh, Mod. Phys. Lett. A **17**, 131 (2002).

[3] C Ahn, D Bernard, and A Leclair, Nucl. Phys. B **346**, 409 (1990); L Baulieu and E G Floratos, Phys. Lett. B **258**, 171 (1991); R Kerner, J. Math. Phys. **33**, 403 (1992); S Durand, Phys. Lett. B **312**, 115 (1993) and Mod. Phys. Lett. A **8**, 1795 (1993); ibid 2323 (1993); A T Filippov, A P Isaev, and R D Kurdikov, Mod. Phys. Lett. A **7**, 2129 (1993); N Mohammedi, Mod. Phys. Lett. A **10**, 1287 (1995); N Fleury and M Rausch de Traubenberg, Mod. Phys. Lett. A **11**, 899 (1996); J A de Azćarraga and A. Macfarlane, J. Math. Phys. **37**, 1115 (1996); R S Dunne, A Macfarlane, J A de Azćarraga, and J C Pérez Bueno, Int. J. Mod. Phys. Lett. A **12**, 3275 (1997). H Ahmedov H and O F Dayi, J. Phys. A: Math. Gen. **32**, 6247 (1999).

[4] L Alvarez-Gaume, Commun. Math. Phys. **90**, 161 (1983) and J. Phys. A: Math. Gen. **16**, 4177 (1983); P Windey, Acta. Phys. Pol. B **15**, 453 (1984); A Mostafazadeh, J. Math. Phys. **35**, 1095 (1994).

[5] N V Borisov, K N Ilinski, and V M Uzdin, Phys. Lett. A **169**, 422 (1992); A D Dolgallo and K N Ilinski, Ann. Phys. **236**, 219 (1994).

[6] B. DeWitt, *Supermanifolds*, 2nd Ed. (Cambridge Uni. Press, Cambridge, 1992).

[7] L E Gendenshtein and I V Krive, Sov. Phys. Usp. **28**, 645-666 (1985); F Cooper, A Khare, and U Sukhatme, Phys. Rep. **251**, 267-385 (1995); G Junker, *Supersymmetric Methods in Quantum and Statistical Physics* (Springer-Verlag, Berlin, 1996).

[8] H. S. Green, Phys. Rev. **90**, 270 (1953).

[9] O W Greenberger and A M Messiah, Phys. Rev. **138**, 1155 (1965).

[10] A Khare, A K Mishra and G Rajasekaran, Int. J. Mod. Phys. A **8**, 1245 (1993).

[11] A K Mishra and G Rajasekaran, Pramana J. Phys. **36**, 537 (1991); **38**, L411 (1992); and **45**, 91 (1995); ibid Mod. Phys. Lett. A **7**, 3425 (1992); See also A K Mishra, Phys. Rev. B **63**, 132405 (2001).

[12] A. Mostafazadeh, Int. J. Mod. Phys. Lett. A **11**, 1057 (1996).

[13] A. Mostafazadeh, Int. J. Mod. Phys. Lett. A **12**, 2725 (1997).

[14] A Mostafazadeh and K Aghababaei Samani, Mod. Phys. Lett. A **15**, 175 (2000).

[15] A Mostafazadeh, J. Phys. A: Math. Gen. **34**, 8601 (2001).

[16] A Mostafazadeh, Nucl. Phys. B **624**, 500 (2002).

[17] H Nicolai, J. Phys. A: Math. Gen. **9**, 1497 (1976)

[18] M Tomiya, J. Phys. A: Math. Gen. **25**, 4699 (1992); A Khare, J. Phys. A: Math. Gen. **25**, L749 (1992) and J. Math. Phys. **34**, 1277 (1993); See also V A Rubakov and V P Spiridonov, Mod. Phys. Lett. A **3**, 1337 (1988).

[19] E. Witten, Nucl. Phys. B **202**, 253 (1982).

Inst. Phys. Conf. Ser. No 173: Section 6
Paper presented at 24th Int. Coll. Group Theoretical Methods in Physics, Paris, France, 15–20 July 2002
©2003 IOP Publishing Ltd

Homogeneous algebras, parastatistics and combinatorics

Todor Popov‡

Institute for Nuclear Research and Nuclear Energy,
Tsarigradsko Chaussée 72, BG-1784, Sofia, Bulgaria

Abstract. As shown in [3] the concepts developed for quadratic algebras such as dual algebra, Koszul complexes and Koszul algebra [15, 13] have counterparts for homogeneous algebras of any order $N(N \geq 2)$. Here we apply these generalized notions on two particular types of cubic algebras. The first one is the parafermionic(parabosonic) algebra *generated (only) by the creation operators* of a system with D degrees of freedom quantized according to the fermionic (bosonic) parastatistics. The second algebra is the algebra of the associative monoid of the Young tableaux with entries in $\{1, \ldots, D\}$. This monoid called the plactic monoid [12] is a useful combinatorial tool, one of its applications is the proof of the Littlewood-Richardson rule. It turns out that these two algebras share many features. Koszul algebras are class of regular algebras which contains the algebras of polynomials. The notion of Koszul algebra for N-homogeneous algebras was first introduced in [2] inspired by the classification of Artin-Schelter [1] of a class of very regular algebras. We will see where the parafermionic(parabosonic) and the plactic algebras are placed in this classification. This talk is a short review of [9].

1. Homogeneous algebras

A *homogeneous algebra of degree N or N-homogeneous algebra* is an algebra of the form [3]

$$\mathcal{A} = A(E, R) = T(E)/(R) \tag{1}$$

where E is a finite-dimensional vector space over (the ground field) $\mathbb{K}$, $T(E)$ is the tensor algebra of E and (R) is the two-sided ideal of $T(E)$ generated by a vector subspace R of $E^{\otimes N}$. The homogeneity of (R) implies that $\mathcal{A}$ is a graded algebra $\mathcal{A} = \oplus_{n \in \mathbb{N}} \mathcal{A}_n$ with $\mathcal{A}_n = E^{\otimes n}$ for $n < N$ and

$$\mathcal{A}_n = E^{\otimes n} / \sum_{r+s=n-N} E^{\otimes r} \otimes R \otimes E^{\otimes s} \text{ for } n \geq N \tag{2}$$

where we have set $E^{\otimes^0} = \mathbb{K}$ as usual. Thus $\mathcal{A}$ is a graded algebra $\mathcal{A} = \oplus_{n \in \mathbb{N}} \mathcal{A}_n$ which is connected ($\mathcal{A}_0 = \mathbb{K}$) generated in degree 1 and such that the $\mathcal{A}_n$ are finite-dimensional vector spaces. So the *Poincaré series* $P_{\mathcal{A}}(t) = \sum_n \dim(\mathcal{A}_n) t^n$ of $\mathcal{A}$ is well defined.

Given a N-homogeneous algebra $\mathcal{A} = A(E, R)$, *its dual* $\mathcal{A}^!$ is defined to be [3] the N-homogeneous algebra $\mathcal{A}^! = A(E^*, R^\perp)$ where E^* is the dual vector space of E and where $R^\perp \subset E^{*\otimes^N}$ is the annihilator of R, $R^\perp = \{\omega \in (E^{\otimes^N})^* \mid \omega(x) = 0, \ \forall x \in R\}$, with the canonical identification $E^{*\otimes^N} = (E^{\otimes^N})^*$. One has $(\mathcal{A}^!)^! = \mathcal{A}$.

‡ tpopov@inrne.bas.bg

2. The parafermionic algebra $\mathcal{B}$

For a review of the physics of the paraquantization see [10] and also the classical book [14]. All we need here is (a part) of the parastatistical commutation relations. The parafermionic creation operators themselves generate an algebra which we will refer to as (with some abuse) parafermionic algebra $\mathcal{B}$

$$[a_k^*, [a_\ell^*, a_m^*]] = 0 \tag{3}$$

where $k, \ell, m \in \{1, \ldots, D\}$ i.e. we are dealing with D degrees of freedom. $\mathcal{B}$ can be interpreted as spanning the Fock space for parafermionic statistics with D degrees of freedom and of arbitrary parafermionic order. The modes a_k^* span a D dimensional vector space over $\mathbb{C}$. The parabosonic algebra $\tilde{\mathcal{B}}$ is very similar, everything we are going to do for $\mathcal{B}$ can be done for $\tilde{\mathcal{B}}$ replacing commutators with supercommutators in (3). The relations (3) are homogeneous of degree $N = 3$ and according to the general construction $\mathcal{B} = A(\mathbb{C}^D, R_\mathcal{B})$ with the subspace $R_\mathcal{B} \subset (\mathbb{C}^D)^{\otimes^3}$ spanned by

$$\{[[x, y]_\otimes, z]_\otimes \mid x, y, z \in \mathbb{C}^D\} \tag{4}$$

where $[x, y]_\otimes = x \otimes y - y \otimes x$. The linear group $GL(D)$ acting on $\mathbb{C}^D$ leaves $R_\mathcal{B}$ invariant therefore its action passes to the quotient $\mathcal{B}$ respecting the grading. The $\mathcal{B}$ decomposes into irreducible representations $\mathcal{B}^\lambda$ of $GL(D)$ labelled by the Young diagrams and each $\mathcal{B}^\lambda$ appears with multiplicity 1, $\mathcal{B} = \oplus_\lambda \mathcal{B}^\lambda$ [14]. In every $\mathcal{B}^\lambda$ there is a linear basis labelled by the Young tableaux hence the whole $\mathcal{B}$ admits a homogeneous basis labelled by the Young tableaux. Using this the Poincaré series of $\mathcal{B}$ has been obtained in [4]

$$P_\mathcal{B}(t) = \left(\frac{1}{1-t}\right)^D \left(\frac{1}{1-t^2}\right)^{\frac{D(D-1)}{2}} \tag{5}$$

From Formula (5) one deduces that $\mathcal{B}$ has polynomial growth.

For $\mathcal{B}^! = A(\mathbb{C}^{D*}, R_\mathcal{B}^\perp)$, $R_\mathcal{B}^\perp \subset (\mathbb{C}^{D*})^{\otimes^3}$ is the linear span of the set

$$\{\alpha \otimes \beta \otimes \gamma - \gamma \otimes \beta \otimes \alpha, \ \theta^{\otimes^3} \mid \alpha, \beta, \gamma, \theta \in \mathbb{C}^{D*}\} \tag{6}$$

The relations of $\mathcal{B}^!$ read from (6) imply that the symmetrized product and the antisymmerized product of 3 elements of $\mathcal{B}_1^! = \mathbb{C}^{D*}$ vanish and that the product of 5 elements of $\mathcal{B}_1^! = \mathbb{C}^{D*}$ also vanishes. Therefore one has $\mathcal{B}_n^! = 0$ for $n \geq 5$. The $GL(D)$ action leaves $R_\mathcal{B}^\perp$ invariant and therefore acts on $\mathcal{B}^!$ by automorphisms which preserve the degree. The labels of irreducible subspaces of this action reads as follows

$$(\bullet)_0 \oplus (\square)_1 \oplus \left(\boxplus \oplus \square\square\right)_2 \oplus \left(\boxminus\right)_3 \oplus \left(\boxplus\right)_4 \tag{7}$$

where the parenthesis corresponds to the homogeneous component and where $\bullet$ is the empty Young diagram corresponding to the trivial 1-dimensional representation. From (7) one computes the Poincaré series of $\mathcal{B}^!$

$$P_{\mathcal{B}^!}(t) = 1 + Dt + D^2 t^2 + \frac{1}{3} D(D^2 - 1)t^3 + \frac{1}{12} D^2(D^2 - 1)t^4 \tag{8}$$

3. The plactic algebra $\mathcal{P}$

The set of Young tableaux can be endowed with the structure of associative monoid, the so called *plactic monoid* [12],[11]. The algebra of the plactic monoid refered to as *the plactic algebra* is generated by an ordered set of elements $\{e_1, \ldots, e_D\}$ (for tableaux with entries in $\{1, \ldots, D\}$) subjects of the relations

$$
\left.
\begin{aligned}
e_\ell e_m e_k &= e_\ell e_k e_m \quad \text{if } k < \ell \leq m \\[2mm]
e_k e_m e_\ell &= e_m e_k e_\ell \quad \text{if } k \leq \ell < m
\end{aligned}
\right\}
\tag{9}
$$

for $k, \ell, m \in \{1, \ldots, D\}$. These relations are the *Knuth relations* [11].

The plactic algebra $\mathcal{P}$ is a cubic algebra: if we take as $\{e_1, \ldots, e_D\}$ the canonical basis of $\mathbb{C}^D$ we can write $\mathcal{P} = A(\mathbb{C}^{\otimes D}, R_\mathcal{P})$ with subspace $R_\mathcal{P} \subset (\mathbb{C}^D)^{\otimes^3}$ associated to the Knuth relations (9). In contrast to $R_\mathcal{B}$, $R_\mathcal{P}$ depends on the basis (e_k) and even on the ordered set $\{1, \ldots, D\}$. Thus there is no natural action of $GL(D)$ on $\mathcal{P}$. Nevertheless $\mathcal{P}$ admits a homogeneous linear basis labelled by the Young tableaux [11]. Thus the basis of $\mathcal{P}$ is indexed in the same manner as one of the $\mathcal{B}$ which implies that $\mathcal{P}$ has the same Poincaré series as $\mathcal{B}$, i.e. $P_\mathcal{P}(t) = P_\mathcal{B}(t)$.

Let (θ^k), $k \in \{1, \ldots, D\}$ be the basis of $\mathbb{C}^{D*}$ dual to the basis (e_ℓ) of $\mathbb{C}^D$, i.e. such that $\langle \theta^k, e_\ell \rangle = \delta_{k\ell}$. One has $\mathcal{P}^! = A\left(\mathbb{C}^{D*}, R_\mathcal{P}^\perp\right)$ and $R_\mathcal{P}^\perp \subset (\mathbb{C}^{D*})^{\otimes^3}$ is the subspace associated with the relations

$$
\left.
\begin{aligned}
\theta^j \theta^k \theta^i + \theta^j \theta^i \theta^k \quad &\text{with } i < j \leq k \\
\theta^i \theta^k \theta^j + \theta^k \theta^i \theta^j \quad &\text{with } i \leq j < k \\
\theta^i \theta^j \theta^k \quad &\text{with } i \leq j \leq k \\
\theta^k \theta^j \theta^i \quad &\text{with } i < j < k
\end{aligned}
\right\}
\tag{10}
$$

Using (10) one sees that $\mathcal{P}_n^! = 0$ for $n \geq 5$ and that moreover, in an obvious sense, $\mathcal{P}^!$ has the same content (7) in Young diagrams as $\mathcal{B}^!$ (i.e. homogeneous linear basis labelled by the corresponding Young tableaux). So one also has $P_{\mathcal{P}^!}(t) = P_{\mathcal{B}^!}(t)$.

It turns out that the parafermionic algebra $\mathcal{B}$, the parabosonic algebra $\tilde{\mathcal{B}}$ and the plactic algebra $\mathcal{P}$ are special points of one continuous family of algebras.

We now work out in details the simpler case $D = 2$. The relations of $\mathcal{B}$, $\tilde{\mathcal{B}}$ and $\mathcal{P}$ for $D = 2$ are obtained from the relations

$$
\left.
\begin{aligned}
e_2 e_1^2 + q e_1^2 e_2 - (q+1) e_1 e_2 e_1 &= 0 \\[2mm]
e_2^2 e_1 + q e_1 e_2^2 - (q+1) e_2 e_1 e_2 &= 0
\end{aligned}
\right\}
\tag{11}
$$

for the values $q = 1$, $q = -1$ and $q = 0$ respectively. Thus we get the cubic algebra $\mathcal{A}_q = A((\mathbb{C}^2), R_q)$ with R_q determined by (12). The $GL(2)$ symmetry of $\mathcal{B}$ generalizes to $GL_q(2)$ symmetry of $\mathcal{A}_q$. The plactic algebra $\mathcal{P} = \mathcal{A}_0$ is the singular point where the $GL_q(2)$ symmetry degenerates. The family $\mathcal{A}_q$ is a subfamily of a family $\mathcal{A}_{q,r}$ with relations

$$
\left.
\begin{aligned}
e_2 e_1^2 + qr e_1^2 e_2 - (q+r) e_1 e_2 e_1 &= 0 \\[2mm]
e_2^2 e_1 + qr e_1 e_2^2 - (q+r) e_2 e_1 e_2 &= 0
\end{aligned}
\right\}
\tag{12}
$$

When $qr \neq 0$ the $GL_q(2)$ symmetry extends to $GL_{p,q}(2)$ symmetry with $p = q/r^2$ (for $GL_{p,q}(2)$ see e.g. [6]). In fact both the $GL_{q/r^2,q}(2)$ and $GL_{r/q^2,r}(2)$ take place due to $\mathcal{A}_{q,r} = \mathcal{A}_{r,q}$. Relations (12) characterize the class of cubic regular Artin-Schelter algebras of type S_1 whenever $qr \neq 0$, [1].

4. Homological properties, Koszul algebras

As shown in [3], a N-homogeneous algebra $\mathcal{A} = A(E, R)$ is canonically associated with a N-complex $K(\mathcal{A})$ generalizing the Koszul complex of a quadratic algebra which is defined as follows, (for N-complexes see [7], [8])

$$\cdots \xrightarrow{d} \mathcal{A} \otimes (\mathcal{A}_n^!)^* \xrightarrow{d} \mathcal{A} \otimes (\mathcal{A}_{n-1}^!)^* \xrightarrow{d} \cdots \xrightarrow{d} \mathcal{A} \otimes (\mathcal{A}_1^!)^* \xrightarrow{d} \mathcal{A} \longrightarrow 0 \qquad (13)$$

$$(\mathcal{A}_n^!)^* = E^{\otimes n} \quad \text{for} \quad n < N, (\mathcal{A}_n^!)^* = \cap_{r+s=n-N} E^{\otimes^r} \otimes R \otimes E^{\otimes^s} \quad \text{for } n \geq N \qquad (14)$$

the differential d on $K(\mathcal{A})$ being the restriction of the mapping $\mathcal{A} \otimes E^{\otimes^{n+1}} \to \mathcal{A} \otimes E^{\otimes^n}$

$$a \otimes (e_0 \otimes e_1 \otimes \cdots \otimes e_n) \mapsto (ae_0) \otimes (e_1 \otimes \cdots \otimes e_n). \qquad (15)$$

The inclusions $(\mathcal{A}_n^!)^* \subset R \otimes E^{\otimes^{n-N}}$ for $n \geq N$ imply $d^N = 0$ i.e. $K(\mathcal{A})$ is a N-complex. $K(\mathcal{A})$ splits into N-subcomplexes homogeneous for the total degree

$$K^{(n)}(\mathcal{A}) = \oplus_m \mathcal{A}_{n-m} \otimes (\mathcal{A}_m^!)^*. \qquad (16)$$

For a quadratic algebra $\mathcal{A}$, i.e. $N = 2$, $K(\mathcal{A})$ is an ordinary complex called the *Koszul complex* of $\mathcal{A}$. When $K(\mathcal{A})$ has trivial homology the quadratic algebra $\mathcal{A}$ is called *Koszul algebra*. By trivial homology we mean acyclicity in positive degrees $H(K^{(n)}(\mathcal{A})) = 0$ for $n > 0$ and $H(K^{(0)}(\mathcal{A})) = \mathbb{C}$.

How the notion of Koszul algebra generalizes for other N-homogeneous algebras with $N > 2$?

The natural guess to require complete acyclicity in positive degrees of the N-complex $K(\mathcal{A})$ was proven to be too strong for $N \geq 3$ [3], it leaves no room for nontrivial algebras. A meaningful generalization of the notion of Koszul algebra was defined in [2] requiring triviality of the homology of an ordinary complex. This (ordinary) complex was shown in [3] to be a contraction of the N-complex $K(\mathcal{A})$ and moreover was shown there to be the only contraction which can have trivial homology and leave room for nontrivial algebras. The complex in question, denoted $C_{N-1,0}(K(\mathcal{A}))$, is obtained from $K(\mathcal{A})$ by alternation of d and d^{N-1}

$$\cdots \xrightarrow{d^{N-1}} \mathcal{A} \otimes (\mathcal{A}_{N+1}^!)^* \xrightarrow{d} \mathcal{A} \otimes (\mathcal{A}_N^!)^* \xrightarrow{d^{N-1}} \mathcal{A} \otimes (\mathcal{A}_1^!)^* \xrightarrow{d} \mathcal{A} \longrightarrow 0 \qquad (17)$$

Accordingly a N-homogeneous algebra $\mathcal{A}$ will be said to be a *Koszul algebra* whenever $C_{N-1,0}(K(\mathcal{A}))$ has trivial homology. The following result was shown in [9].

Proposition: *Let $\mathcal{A}$ be a N-homogeneous algebra which is Koszul and*

$$Q_{\mathcal{A}}(t) = \sum_n (\dim(\mathcal{A}_{nN}^!)t^{nN} - \dim(\mathcal{A}_{nN+1}^!)t^{nN+1}) \quad \text{then one has} \quad P_{\mathcal{A}}(t)Q_{\mathcal{A}}(t) = 1.$$

This proposition is very useful to compute the Poincaré series $P_{\mathcal{A}}(t)$ of a Koszul algebra $\mathcal{A}$ when $\mathcal{A}^!$ is small, see e.g. in [5]. In the case $N = 2$, one has $Q_{\mathcal{A}}(t) = P_{\mathcal{A}^!}(-t)$ so one sees that this theorem generalizes the well-known result $P_{\mathcal{A}}(t)P_{\mathcal{A}^!}(-t) = 1$ for quadratic Koszul algebras [15],[13]. When $N \geq 3$ the series $Q_{\mathcal{A}}(t)$ contains only a part of the terms of $P_{\mathcal{A}^!}(t)$.

In the case of $\mathcal{B}$ and $\mathcal{P}$ one knows $P_{\mathcal{A}}(t)$ (5) and $P_{\mathcal{A}^!}(t)$ (8) , consequently one knows $Q_{\mathcal{A}}(t)$ so it is straightforward to verify that $P_{\mathcal{A}}(t)Q_{\mathcal{A}}(t) = 1$ holds when $D = 2$ and doesn't hold for $D \geq 3$. One concludes that $\mathcal{B}$ and $\mathcal{P}$ are not Koszul algebras for $D \geq 3$. For the case $D = 2$ it is easy to verify that $\mathcal{B}$, $\tilde{\mathcal{B}}$ and $\mathcal{P}$ are Koszul algebras. As mentioned above $\mathcal{B}$,

$\tilde{\mathcal{B}}$ in this case (D=2) are particular cases of cubic regular Artin-Schelter algebras [1] so they are also Gorenstein while $\mathcal{P}$ is only a limiting case of such algebras where the Gorenstein property is lost, (see in [9]).

I wish to extend my heartfelt gratitude to Michel Dubois-Violette for his inspiring guidance and also to Patricia Flad for her technical support.

References

[1] M. Artin, W. F. Schelter. Graded algebras of global dimension 3. *Adv. Math.* **66** (1987) 171-216.
[2] R. Berger. Koszulity for nonquadratic algebras. *J. Algebra* **239** (2001) 705-734.
[3] R. Berger, M. Dubois-Violette, M. Wambst. *J. Algebra* **261**(2003) 172-185.
[4] S. Chaturvedi. any order. hep-th/9509150.
[5] A. Connes, M. Dubois-Violette. Yang-Mills algebra. *Lett. Math. Phys.* **61**(2002) 149-158.
[6] V. K. Dobrev. Duality for the matrix group $GL_{p,q}(2, \mathbb{C})$ *J.Math.Phys.* **33** (1992) 3419-1430.
[7] M. Dubois-Violette. $d^N = 0$: Generalized homology. *K-Theory* **14** (1998) 371-404.
[8] M. Dubois-Violette. math.QA/0005256.
[9] M. Dubois-Violette, T. Popov. *Lett. Math. Phys.* **61**(2002) 159-170.
[10] M. Flato, C. Fronsdal. *J. Geom. Phys.* **6** (1989) 293-309.
[11] W. Fulton. Young tableaux. Cambridge University Press 1997.
[12] A. Lascoux, M.P. Schützenberger. Le monoïde plaxique. *Quaderni de "La ricerca scientifica"* **109**, Roma, CNR (1981) 129-156.
[13] Yu.I. Manin. Quantum groups and non-commutative geometry. CRM Univ. de Montréal 1988.
[14] Y. Ohnuki, S. Kamefuchi. Quantum field theory and parastatistics. Springer-Verlag 1982.
[15] S.B. Priddy. Koszul resolutions. *Trans. Amer. Math. Soc.* **152** (1970) 39-60.

Inst. Phys. Conf. Ser. No 173: Section 6
Paper presented at 24th Int. Coll. Group Theoretical Methods in Physics, Paris, France, 15–20 July 2002
©2003 IOP Publishing Ltd

Lie algebras of order F and extensions of the Poincaré algebra

M. Rausch de Traubenberg

Laboratoire de Physique Théorique, CNRS UMR 7085, Université Louis Pasteur, F-67084 Strasbourg cedex, France

Abstract. F−Lie algebras are natural generalisations of Lie algebras ($F = 1$) and Lie superalgebras ($F = 2$). We give finite dimensional examples of F−Lie algebras obtained by an inductive process from Lie algebras and Lie superalgebras. Matrix realizations of the F−Lie algebras constructed in this way from $osp(2|m)$ are given. We obtain a non-trivial extension of the Poincaré algebra by an Inönü-Wigner contraction of a certain F−Lie algebras with $F > 2$.

1. Introduction

Describing the laws of physics in terms of underlying symmetries has always been a powerful tool. Lie algebras and Lie superalgebras are central in particle physics, and the space-time symmetries can be obtained by an Inönü-Wigner contraction of certain Lie (super)algebras. F−Lie algebras [1, 2, 3], a possible extension of Lie (super)algebras, have been considered some times ago as the natural structure underlying fractional supersymmetry (FSUSY) [1, 4, 5, 6] (one possible extension of supersymmetry). In this contribution we show how one can construct many examples of finite dimensional F−Lie algebras from Lie (super)algebras and finite-dimensional FSUSY extensions of the Poincaré algebra are obtained by Inönü-Wigner contraction of certain F−Lie algebras.

2. F−Lie algebras

The natural mathematical structure, generalizing the concept of Lie superalgebras and relevant for the algebraic description of fractional supersymmetry was introduced in [1] and called an F−Lie algebra. We do not want to go into the detailed definition of this structure here and will only recall the basic points, useful for our purpose. More details can be found in [1].

Let F be a positive integer and $q = e^{2i\frac{\pi}{F}}$. We consider now a complex vector space S which has an automorphism ε satisfying $\varepsilon^F = 1$. We set $A_k = S_{q^k}$, $1 \le k \le F - 1$ and $B = S_1$ (S_{q^k} is the eigenspace corresponding to the eigenvalue q^k of ε). Hence,

$$S = B \oplus A_1 \oplus \cdots \oplus A_{F-1}.$$

We say that S is an F−Lie algebra if:

(i) B, the zero graded part of S, is a Lie algebra.
(ii) A_i ($i = 1, \ldots, F - 1$), the i graded part of S, is a representation of B.
(iii) There are symmetric multilinear B−equivariant maps

$$\{ , \ldots, \} : \mathscr{S}^F(A_k) \to B,$$

where $\mathscr{S}^F(D)$ denotes the F−fold symmetric product of D. In other words, we assume that some of the elements of the Lie algebra B can be expressed as F−th order symmetric products of "more fundamental generators".

(iv) The generators of S are assumed to satisfy Jacobi identities ($b_i \in B$, $a_i \in A_k$, $1 \le k \le F-1$):

$$[[b_1,b_2],b_3] + [[b_2,b_3],b_1] + [[b_3,b_1],b_2] = 0,$$
$$[[b_1,b_2],a_3] + [[b_2,a_3],b_1] + [[a_3,b_1],b_2] = 0,$$
$$[b,\{a_1,\ldots,a_F\}] = \{[b,a_1],\ldots,a_F\} + \cdots + \{a_1,\ldots,[b,a_F]\},$$
$$\sum_{i=1}^{F+1} [a_i,\{a_1,\ldots,a_{i-1},a_{i+1},\ldots,a_{F+1}\}] = 0. \tag{1}$$

The first three identities are consequences of the previously defined properties but the fourth is an extra constraint.

More details (unitarity, representations, *etc.*) can be found in [1, 3]. Let us first note that no relation between different graded sectors is postulated. Secondly, the sub-space $B \oplus A_k \subset S$ ($k = 1,\ldots,F-1$) is itself an F-Lie algebra. From now on, F-Lie algebras of the types $B \oplus A_k$ will be considered.

Most of the examples of F-Lie algebras are infinite dimensional (see *e.g.* [1, 5]). However in [3] an inductive theorem to construct finite-dimensional F-Lie algebras was proven:

Theorem 1 *Let g_0 be a Lie algebra and g_1 a representation of g_0 such that*
(i) $S_1 = g_0 \oplus g_1$ is an F-Lie algebra of order $F_1 \ge 1$ ‡;
(ii) g_1 admits a g_0-equivariant symmetric form μ_2 of order $F_2 \ge 1$.
Then $S = g_0 \oplus g_1$ admits an F-Lie algebra structure of order $F_1 + F_2$, which we call the F-Lie algebra induced from S_1 and μ_2.

By hypothesis, there exist g_0-equivariant maps $\mu_1 : \mathscr{S}^{F_1}(g_1) \longrightarrow g_0$ and $\mu_2 : \mathscr{S}^{F_2}(g_1) \longrightarrow \mathbf{C}$. Now, consider $\mu : \mathscr{S}^{F_1+F_2}(g_1) \longrightarrow g_0 \otimes \mathbf{C} \cong g_0$ defined by

$$\mu(f_1,\cdots,f_{F_1+F_2}) = \tag{2}$$
$$\frac{1}{F_1!}\frac{1}{F_2!} \sum_{\sigma \in S_{F_1+F_2}} \mu_1(f_{\sigma(1)},\cdots,f_{\sigma(f_{F_1})}) \otimes \mu_2(f_{\sigma(f_{F_1+1})},\cdots,f_{\sigma(f_{F_1+F_2})}),$$

where $f_1,\cdots,f_{F_1+F_2} \in g_1$ and $S_{F_1+F_2}$ is the group of permutations on $F_1 + F_2$ elements. By construction, this is a g_0-equivariant map from $\mathscr{S}^{F_1+F_2}(g_1) \longrightarrow g_0$, thus the three first Jacobi identities are satisfied. The last Jacobi identity, is more difficult to check and is a consequence of the corresponding identity for the F-Lie algebra S_1 and a factorisation property (see [3] for more details).

3. Finite dimensional F-Lie algebras

An interesting consequence of the theorem of the previous section is that it enables us to construct an F-Lie algebras associated to *any* Lie (super)algebras.

‡ Strictly speaking this theorem is not valid for $F_1 = 1$. In this case the notion of graded 1-Lie algebra has to be introduced [3]. $S = g_0 \oplus g_1$, is a graded 1-Lie algebra if (i) g_0 a Lie algebra and g_1 is a representation of g_0 isomorphic to the adjoint representation, (ii) there is a g_0-equivariant map $\mu : g_1 \to g_0$ such that $[f_1,\mu(f_2)] + [f_2,\mu(f_1)] = 0, f_1,f_2 \in g_1$.

3.1. Finite dimensional F−Lie algebras associated to Lie algebras

Consider the graded 1−Lie algebra $S = g_0 \oplus g_1$ where g_0 is a Lie algebra, g_1 is the adjoint representation of g_0 and $\mu : g_1 \to g_0$ is the identity. Let $J_1, \cdots, J_{\dim g_0}$ be a basis of g_0, and $A_1, \cdots, A_{\dim g_0}$ the corresponding basis of g_1. The graded 1−Lie algebra structure on S is then:

$$[J_a, J_b] = f_{ab}{}^c J_c, \qquad [J_a, A_b] = f_{ab}{}^c A_c, \qquad \mu(A_a) = J_a, \tag{3}$$

where $f_{ab}{}^c$ are the structure constants of g_0, The second ingredient to construct an F−Lie algebra is to define a symmetric invariant form on g_1. But on g_1, the adjoint representation of g_0, the invariant symmetric forms are well known and correspond to the Casimir operators [7]. Then, considering a Casimir operator of order m of $g_1 \cong g_0$, we can induce the structure of an F−Lie algebra of order $m+1$ on $S_{m+1} = g_0 \oplus g_1$. One can give explicit formulae for the bracket of these F−Lie algebras as follows. Let $h_{a_1 \cdots a_m}$ be a Casimir operator of order m (for $m = 2$, the Killing form $g_{ab} = \mathrm{Tr}(A_a A_b)$ is a primitive Casimir of order two). Then, the F−bracket of the F−Lie algebra is

$$\{A_{a_1}, A_{a_2}, \cdots, A_{a_{m+1}}\} = \sum_{\ell=1}^{m+1} h_{a_1 \cdots a_{\ell-1} a_{\ell+1} \cdots a_{m+1}} J_{a_\ell} \tag{4}$$

For the Killing form this gives

$$\{A_a, A_b, A_c\} = g_{ab} J_c + g_{ac} J_b + g_{bc} J_a. \tag{5}$$

If $g_0 = sl(2)$, the F−Lie algebra of order three induced from the Killing form is the F−Lie algebra of [8].

3.2. Finite dimensional F−Lie algebras associated to Lie superalgebras

The construction of F−Lie algebras associated to Lie superalgebras is more involved. We just give here a simple example (for more details see [3]): the F−Lie algebra of order 4 $S = g_0 \oplus g_1$ induced from the (i) Lie superalgebra $osp(2|2m) = (so(2) \oplus sp(2m)) \oplus \mathbf{C}^2 \otimes \mathbf{C}^{2m}$, and (ii) the quadratic form $\varepsilon \otimes \Omega$, where ε is the invariant symplectic form on $\mathbf{C}^2$ and Ω the invariant symplectic form on $\mathbf{C}^{2m}$. Let $\{S_{\alpha\beta} = S_{\beta\alpha}\}_{\substack{1 \le \alpha \le 2m \\ 1 \le \beta \le 2m}}$ be a basis of $sp(2m)$ and $\{h\}$ be a basis of $so(2)$. Let $\{F_{q\alpha}\}_{\substack{q = \pm 1 \\ 1 \le \alpha \le 2m}}$ be a basis of $\mathbf{C}^2 \otimes \mathbf{C}^{2m}$. Then the four brackets of S take the following form

$$\{F_{q_1\alpha_1}, F_{q_2\alpha_2}, F_{q_3\alpha_3}, F_{q_4\alpha_4}\} = \varepsilon_{q_1 q_3} \Omega_{\alpha_1 \alpha_3} \left(\delta_{q_2+q_4} S_{\alpha_2 \alpha_4} + \varepsilon_{q_2 q_4} \Omega_{\alpha_2 \alpha_4} h\right)$$
$$+ \text{ perm.} \tag{6}$$

It is interesting to notice that this F−Lie algebra admits a simple matrix representation [3]: $g_0 = \left\{ \begin{pmatrix} q & 0 & 0 \\ 0 & -q & 0 \\ 0 & 0 & S \end{pmatrix}, q \in \mathbf{C}, S \in sp(2n) \right\} \cong so(2) \oplus sp(2n)$ and $g_1 =$

$$\left\{ \begin{pmatrix} 0 & 0 & F_+ \\ 0 & 0 & F_- \\ -\Omega F_-^t & -i\Omega F_+^t & 0 \end{pmatrix}, F_\pm \in \mathcal{M}_{1,2n}(\mathbf{C}) \right\}.$$

4. Finite-dimensional FSUSY extensions of the Poincaré algebra

It is well known that supersymmetric extensions of the Poincaré algebra can be obtained by Inönü-Wigner contraction of certain Lie superalgebras. In fact, one can also obtain some FSUSY extensions of the Poincaré algebra by Inönü-Wigner contraction of certain $F-$Lie algebras as we now show with one example [3]. Let $S_3 = sp(4) \oplus \text{ad} \, sp(4)$ be the real $F-$lie algebra of order three induced from the real graded $1-$Lie algebra $S_1 = sp(4) \oplus \text{ad} \, sp(4)$ and the Killing form on ad $sp(4)$ (see eq. 5). Using vector indices of $so(1,3)$ coming from the inclusion $so(1,3) \subset so(2,3) \cong sp(4)$, the bosonic part of S_3 is generated by $M_{\mu\nu}, M_{\mu 4}$, with $\mu, \nu = 0, 1, 2, 3$ and the graded part by $J_{\mu\nu}, J_{4\mu}$. Letting $\lambda \to 0$ after the Inönü-Wigner contraction,

$$
\begin{aligned}
M_{\mu\nu} &\to L_{\mu\nu}, & M_{\mu 4} &\to \tfrac{1}{\lambda} P_\mu \\
J_{\mu\nu} &\to \tfrac{1}{\sqrt[3]{\lambda}} Q_{\mu\nu}, & J_{4\mu} &\to \tfrac{1}{\sqrt[3]{\lambda}} Q_\mu,
\end{aligned}
\tag{7}
$$

one sees that $L_{\mu\nu}$ and P_μ generate the $(1+3)D$ Poincaré algebra and that $Q_{\mu\nu}, Q_\mu$ are the fractional supercharges in respectively the adjoint and vector representations of $so(1,3)$. This $F-$Lie algebra of order three is therefore a non-trivial extension of the Poincaré algebra where translations are cubes of more fundamental generators. The subspace generated by $L_{\mu\nu}, P_\mu, Q_\mu$ is also an $F-$Lie algebra of order three extending the Poincaré algebra in which the trilinear symmetric brackets have the simple form·

$$
\{Q_\mu, Q_\nu, Q_\rho\} = \eta_{\mu\nu} P_\rho + \eta_{\mu\rho} P_\nu + \eta_{\rho\nu} P_\mu,
\tag{8}
$$

where $\eta_{\mu\nu}$ is the Minkowski metric.

5. Conclusion

In this paper a sketch of the construction of $F-$Lie algebras associated to Lie (super)algebras were given. More complete results, such as a criteria for simplicity, representation theory, matrix realizations *etc.*, was given in [3].

References

[1] Rausch de Traubenberg M and Slupinski M. J. 2000 J. Math. Phys 41 4556-4571 [hep-th/9904126].
[2] Rausch de Traubenberg M and Slupinski M. J. 2002 *Proceedings of Institute of Mathematics of NAS of Ukraine*, p 548-554, Vol. 43, Editors A.G. Nikitin, V.M. Boyko and R.O. Popovych, Kyiv, Institute of Mathematics [arXiv:hep-th/0110020].
[3] Rausch de Traubenberg M and Slupinski M. J. 2002 *Finite-dimensional Lie algebras of order F*, arXiv:hep-th/0205113, to appear in J. Math. Phys.
[4] Durand S 1993 Mod. Phys. Lett A 8 2323–2334 [hep-th/9305130].
[5] Rausch de Traubenberg M and Slupinski M. J. 1997 Mod. Phys. Lett. A 12 3051-3066 [hep-th/9609203].
[6] Rausch de Traubenberg M 1998 hep-th/9802141 (Habilitation Thesis, in French).
[7] Chevalley C and Eilenberg S 1948 Trans. Amer. Math. Soc. 63 85-124.
[8] Ahmedov H, Yildiz A and Ucan Y 2001 J. Phys. A 34 6413-6424 [math.rt/0012058].

Inst. Phys. Conf. Ser. No 173: Section 6
Paper presented at 24th Int. Coll. Group Theoretical Methods in Physics, Paris, France, 15–20 July 2002
©*2003 IOP Publishing Ltd*

Noncommutative differential geometric method to fractal geometry

A fractal method for infinite dimensional Clifford algebras

Julian Lawrynowicz [+], **Osamu Suzuki**[++]

[+]Institute of Physics, University of Lodz, ul. Pomorska, Lodz, Poland
[++]Department of Computer and System Analysis,College of Humanities and Sciences,Nihon University,156 Setagaya,Tokyo, Japan

Abstract. Infinite dimensional Clifford algebras are discussed by use of fractal geometry and the following results are obtained which are stated as Theorems I,II:
(1)Cuntz algebras are introduced and infinite dimensional Clifford algebras are defined as the inductive limit of finite dimensional Clifford algebras in the Cuntz algebra.
(2)Self similar fractal sets are introduced and representarions of infinite dimensional Clifford algebras on self similar fractal sets are constructed and criterions of their unitary equivalences are given.

1. Introduction

It is well known that fractal methods are quite useful to describe phenomenas of complex systems and describe the amorphas physics and the growth of cities, and baccterias([11],[14]). From a mathematical point of view, we must say that we have still not established the fractal geometry. In fact, we have concentrated ourselves only to calculating the Hausdorff dimensions of fractal sets or studying the measure theory with respect to the Hausdorff measures([5],[7]).
In this paper we shall show that a fractal geometry will contribute to an analysis of infinite dimensional Clifford algebras. At first we describe our idea of this paper. When we look at the construction of the algebra, we can observe a self similarity in its construction. In fact, we have a recursive construction from the Clifford algebra of lower order to that of higher order:For the generators $A_j (j = 1, 2, .., 2p - 1)$ of $Cl_{2p-1}(\mathbf{C})$, putting

$$\begin{pmatrix} A_j & 0 \\ 0 & -A_j \end{pmatrix} \begin{pmatrix} 0 & I_2 \\ I_2 & 0 \end{pmatrix} \begin{pmatrix} 0 & iI_2 \\ -iI_2 & 0 \end{pmatrix} (j = 1, 2, .., 2p - 1), \tag{1}$$

we have the generators of $Cl_{2p+1}(\mathbf{C})$. The construction given in the above manner is called "basic construction" of finite dimensional Clifford algebras in this paper([6],[10]). This construction tempts us to apply the fractal method to an analysis on infinite dimensional Clifford algebras.

2. Cuntz algebras, fractal sets and infinite dimensional Clifford algebras

In this section we give preliminaries for the statement of our results. By the basic construction several authors have chosen the definition of the infinite Clifford algebra $Cl(\infty, \mathbf{C})$ by the inductive limit of finitely dimensional Clifford algebras([9]):

$$Cl(\infty; \mathbf{C}) = \underline{\lim} Cl_{2p-1}(\mathbf{C}). \tag{2}$$

An element ϕ of (2) determines just an ordered sequence:$\phi = \lim_{n \to \infty} \phi_n$ and we do not know the behavior when n tends to infinity. Here we take the Cuntz algebra and make the inductive limit in this algebra. A C^*-algbera $\mathcal{O}(N)$ is called the Cuntz algebra $\mathcal{O}(N)$ of order N, when it is generated by $\{S_j\}(j = 1, 2, .., N)$ satisfying the following commutation relations:

$$(1)S^*_j S_j = 1(j = 1, 2, .., N), \quad (2)\sum_{j=1}^{N} S_j S^*_j = 1. \tag{3}$$

In this paper we choose N as $N = 2^q$ with a positive integer q. In this case we see that

$$Cl_{2p-1}(\mathbf{C}) \subset \mathcal{O}(N) \tag{4}$$

for arbitrary $p(\geq q)$. Hence we can get the following definition of the infinite dimensional Clifford algebra:

Definition 1 (Infinite dimensional Clifford algebra)
Taking the inductive limit (2) in the Cuntz algebra $\mathcal{O}(N)$, we have the definition of the infinite dimensional Clifford algebra:

$$Cl_N(\infty; \mathbf{C}) = \underline{\lim} Cl_{2p-1}(\mathbf{C}) \cap \mathcal{O}(N). \tag{5}$$

Next we proceed to the realization of the infinite dimensional Clifford algebra on self similar fractal sets. This idea can be realized by use of the representations of the Cuntz algebras on self similar fractal sets. Here we define self similar fractal sets in the following manner:

Definition 2 (A self similar fractal set)
For a system of self similar contractible mappings $\tau_j : K_0 \mapsto K_0 (j = 1, 2, .., N)$ with the separation condition (i.e.,$\tau_i(K_0^\circ) \cap \tau_j(K_0^\circ) = \emptyset (i \neq j)$), we define a self similar fractal set K:

$$K = \cap_{n=1}^{\infty} K_n, \text{ where } K_n = \cup_{j=1}^{N} \tau_j(K_{n-1}). \tag{6}$$

Here we denote the open kernel of E by E°.

3. Main results

In this section we state our main theorems. The proofs will be given in the third part of our paper with the same title. At first we notice that the commutation relations (3) give an algebraic description of the division of the total space into N-parts. By this fact we can give representations of a Cuntz algebra $\mathcal{O}(N)$ on self similar fractal sets([12],[13]). Here a homorphism $\pi : Cl_N(\infty.\mathbf{C}) \mapsto B(L^2(K, d\mu))$ is called a representation, where $L^2(K, d\mu)$ implies the L^2-space with respect to the Hausdorff measure on K and $B(H)$ implies the algebra of bounded operators on the Hilbert space H. Taking (4) into an accont and restricting the representations on $Cl_N(\infty; \mathbf{C})$, we have the following theorem:

Theorem I (Existence Theorem)([12],[13])
There exists a representation of the $Cl_N(\infty; \mathbf{C})$ on self similar fractal sets which is called *Hausdorff representation*.

By use of this theorem we can "confine " every finite dimensional Clifford algebra on fractal sets. The fractal sets corresponding to the construction (1) will be given at the end of this paper.

Next we proceed to the unitary equivalence of the representations. By use of the Kakutani's dichotomy theorem on fractal sets, we can prove the following theorem:

Theorem II (Equivalence theorem)([12],[13])
Two representations of the $Cl_N(\infty; \mathbf{C})$ on self similar fractal sets are unitary equivalenct if and only if

$$\lambda_i{}^D = \lambda_i'{}^{D'}, \tag{7}$$

where λ_i and $\lambda_i'(i = 1, 2, .., N)$ are contraction ratios of the defining operators of the fractal sets and D and D' are the Hausdorff dimensions.

Remarks
(1)We notice that the Hausdorff dimension itself does not give the invariance of the representations.
(2)We notice that the basic construction is not unique. We have another example of the basic construction:

$$\begin{pmatrix} 0 & A_j \\ A_j & 0 \end{pmatrix} \begin{pmatrix} 0 & I_2 \\ I_2 & 0 \end{pmatrix} \begin{pmatrix} I_2 & 0 \\ 0 & -I_2 \end{pmatrix} (j = 1, 2, .., 2p - 1). \tag{8}$$

Then we have another infinite dimensional Clifford algebra $Cl'(\infty; \mathbf{C})$. We see that although they are unitary equivalnet each other on finite dimensional subalgebras, they are not unitary equivalent on the total algebras. This will be discussed in the forthcoming paper.

Recently M.Abe and K.Kawamura have discovered many interesting recurrence formulas for Fermion algebras([1],[2],[3]). Their key ideas is the use of the embeding of Fermion algebras in the Cuntz algebra. The basic constructions might be understood as one of their recurrence formulas. They derive many non equivalent Fermion algebras and we can see the merit of the use of the Cuntz algebras. Their results can be transported on fractal sets by use of the Hausdorff representations and their equivalences can be discussed by Theorem I,II.

4. Examples

Example 1 (Geometric representations on fractal sets of Serpinski's carpet type)
We choose four self similar contractible mappings $\tau_{i,j}(i, j = 1, 2)$ between the rectangle K_0 on the plane $\mathbf{R}^2$ satisfying the separation conditions and we make the fractal set of Serpinski's carpet type. Here we take the representation of $Cl_4(\infty, \mathbf{C})$ on the Serpinski's carpet. Then the basic construction (1) can be realized on the fractal set and the inclusion mappings

$$Cl_5(\mathbf{C}) \hookrightarrow Cl_7(\mathbf{C}) \hookrightarrow \dots. \tag{9}$$

can be realized by the construction. By this reason we call the representation *geometric representation*. We consider the space $\Gamma_c(K_0, \mathbf{C})$ of locally constant complex valued functions on the open kernel of K_0. Then we see that

$$\Gamma_c(K_n, \mathbf{C}) \cong M(2^n, \mathbf{C}). \tag{10}$$

By use of the natural inclusion induced by the basic construction$\Gamma_c(K_n, \mathbf{C}) \hookrightarrow \Gamma_c(K_{n+1}, \mathbf{C})$, we have the inductive limit

$$E = \lim \Gamma_c(K_n, \mathbf{C}), \tag{11}$$

which we call the *wavelet line bundle* of the fractal set K. We notice that a section ϕ of E can be expressed as $\phi = \lim \phi_n$, where $\phi_n \in \Gamma(K_n, \mathbf{C})$. In order to control the behavior tending n to infinity, we have to put the condition on sections, which can be done by the representation of the Cuntz algebra. In this case we see that we can choose

$$L^2(K, d\mu_D) = \lim \Gamma_c(K_n, \mathbf{C}). \tag{12}$$

This can be discussed in terms of the *Haar expansion* on the fractal set. This will be discussed in the forthcoming paper.

Example 2 (Representations on the Cantor set)

Next we treat the representation of $Cl_2(\infty, \mathbf{C})$. Then we have representations on the Cantor sets and the Koch curves for examples. In this case we have not clear geometric meanings of the representations. Here we notice that the representation of the $\mathcal{O}(2)$ derives that of $\mathcal{O}(4)$. Let K be a fractal set defined by $\sigma_i : K_0 \mapsto K_0(i = 1, 2)$. Then, putting $\tilde{\sigma}_{i,j} = (\sigma_i, \sigma_j) : \tilde{\sigma}_{i,j} : K_0 \times K_0 \mapsto K_0 \times K_0(i, j = 1, 2)$, we have a fractal set of Serpinski carpet type and we have a geometric representation on $K \times K$. The correspondence can be stated in the following manner:

$$(\pi(S_1), \pi(S_2)) \longrightarrow (\pi(S_1 S_1)), \pi(S_1 S_2)), \pi(S_2 S_1), \pi(S_2 S_2)). \tag{13}$$

References

[1] M.Abe and K.Kawamura: Recursive Fermion systems in Cuntz algebra. I(Embeddings of Fermion algberas into Cuntz algebra), to appear in Comm. Math. Phys..

[2] M.Abe and K.Kawamura: Nonlinear transformation group of CAR Fermion algbera, Preprints in RIMS-1334(2001)

[3] M.Abe and K.Kawamura: Recursive Fermion systems in Cuntz algebra. II in preparation(2002)

[4] J. Cuntz: Simple C^*-algebras generated by isometries, Comm. Math. Phys. **57** (1977), 173-185.

[5] K.Falconer: Fratal Geometry, Mathematical foundations and Applications,John Wiley and Sons, U.S.A.(1990)

[6] I.Furuoya, S,Kanemaki, J.Lawrynowicz and O.Suzuki: Hermitian Hurwitz pairs, Deformations of Mathematical tructures II. Hurwitz Type Structures and Applications to Surface Physics. Kluwer Academic, Dordrecht(1994)

[7] S and S. Ishimura:Fractal Mathematics, Tokyo-Tosho, Japan(in japanese) (1990)

[8] S.Kakutani: On equivalence of infinite product measures Ann. Math., 47 (1948), 214-224.

[9] Kostant,B. Sternberg,S.: Symplectic reduction, BRS cohomology, and infinite-dimensional Clifford algebras, Ann. Phys., 176(1987), 49-113

[10] J.Lawrynowicz, K.Nono, O.Suzuki, and N.Fujimoto:A basic bonstruction of real pre-Hurwtiz algebras, Rech. sur les deform. de la Societe des Science et des Lettres de Lodz vol. XXXIV (2001), 77-89.

[11] B.B.Mandelbrot: The fractal geometry of nature, Freeman, San Fransisco(1982)

[12] M.Mori, O.Suzuki and Y.Watatani: Representations of Cuntz algebras on fractal sets, in preparation.

[13] M.Mori, O.Suzuki and Y.Watatani: Noncommutative differential geometric method to fractal geometry(I)(Representations of Cuntz algebras of Hausdorff type on fractal sets) to appear from the Proc. of 3th. ISSAC Int. Nat. Conf. Berlin 2001

[14] K.Murayama and T.Ninomiya:Solid State Commun. 53,125(1985)

Inst. Phys. Conf. Ser. No 173: Section 6
Paper presented at 24th Int. Coll. Group Theoretical Methods in Physics, Paris, France, 15–20 July 2002
©*2003 IOP Publishing Ltd*

Non-Commutative Standard Model

Xavier Calmet, Branislav Jurčo, Peter Schupp, Julius Wess, Michael Wohlgenannt

Sektion Physik. Universität München
Theresienstraße 37, D-80333 München

Abstract. We consider the Standard Model on a non-commutative space and expand its action in the non-noncommutativity parameter θ. No new particles are introduced, the structure group is $SU(3) \times SU(2) \times U(1)$. We derive the action to leading order in θ. At zeroth order the action coincides with the ordinary Standard Model on commutative space-time. The most striking features are couplings between quarks, gluons and electroweak bosons and many new vertices in the charged and neutral currents. The Higgs mechanism can be applied. We consider a "minimally" deformed Standard Model, where QED is not deformed. There is some ambiguity in chosing kinetic terms for the gauge fields, which still are to be resolved.

1. Introduction

This is a short outline of the generalisation of the Standard Model of particle physics to non-commutative space-time given in [1]. Earlier attempts have been plagued with some difficulties, such as charge quantisation in QED, dealing with different gauge groups as $U(n)$ and tensor product of gauge groups, see references in [1]. We are going to address each of these problems. Non-commutative coordinates are introduced by non-trivial commutation relations

$$[\hat{x}^\mu, \hat{x}^\nu] = i\theta^{\mu\nu}(\hat{x}). \tag{1}$$

Most commonly, $\theta(x)$ is chosen to be a constant, linear or quadratic in $\hat{x}$. In his contribution to this conference, Todor Popov discusses the case of $\theta(\hat{x})$ depending on some other power of $\hat{x}$. However, we will restrict ourselves to the canonical case,

$$\theta^{\mu\nu}(\hat{x}) = \theta^{\mu\nu} \in \mathbb{R}. \tag{2}$$

The coordinates $\hat{x}^\mu$ generate the non-commutative algebra of functions

$$\hat{\mathcal{A}} = \mathbb{C}[[\hat{x}^\mu]]/\mathcal{I}, \tag{3}$$

where $\mathcal{I}$ is the ideal generated by relations (1). We want to work in the $*$-formalism, therefore we translate the non-commutativity to a non-commutative multiplication $*$ in the ordinary algebra of functions

$$\mathcal{A} = \mathbb{C}[[\hat{x}^\mu]]\Big/\{[\hat{x}^\mu, \hat{x}^\nu] = 0\} \tag{4}$$

via the algebra isomorphism $W : \mathcal{A} \to \hat{\mathcal{A}}$. This isomorphism is accomplished by choosing a basis in $\hat{\mathcal{A}}$ and identifying basis elements. The $*$-multiplication is defined by

$$W(f * g) := W(f)W(g), \quad f, g \in \mathcal{A}. \tag{5}$$

In the canonical case, this is just the Weyl-Moyal product of functions

$$(f * g)(x) = \mu \circ \exp(\frac{i}{2}\theta^{ij}\partial_i \otimes \partial_j)f(x) \otimes g(x). \tag{6}$$

We have the ordinary differential calculus, i.e., $\partial_i * f = \partial_i f$. The ordinary integrals have the following crucial properties:
$\int d^n x(f * g)(x) = \int d^n x(g * f)(x) = \int d^n x f(x)g(x)$. The former is called trace property.

2. Non-Commutative Gauge Theories

Non-commutative gauge theories are based on only a few principles.

2.1. Covariant Coordinates

Let us assume that a field Ψ transforms under a gauge transformation in the following way,

$$\delta\Psi = i\Lambda * \Psi, \tag{7}$$

where Λ is the non-commutative gauge parameter. It is clear that the product $x^\mu * \Psi$ does not transform covariantly. Therefore we have to introduce covariant coordinates X^μ and a vector potential A^μ, in the very same way as one introduces covariant derivatives in the commutative theory; $X^\mu = x^\mu + A^\mu$, s.t. $\delta(X^\mu * \Psi) = i\Lambda * (X^\mu * \Psi)$.

2.2. Classical Limit and Locality

In the following we will denote non-commutative (nc) quantities with hats. We want to expand them in terms of the non-commutativity parameter θ. In each order, they may be local functions of the corresponding commutative quantity and the commutative gauge field,

$$\hat{\Psi} = \hat{\Psi}[\Psi, A], \ \hat{\Lambda} = \hat{\Lambda}[A], \ \hat{A} = \hat{A}[A]. \tag{8}$$

These are the so called Seiberg-Witten maps [2]. A very important requirement is that we regain the classical theory and the classical fields in the limit $\theta \to 0$.

2.3. Gauge equivalence conditions

The Seiberg-Witten maps are solutions of gauge equivalence conditions, e.g., for fields: $\delta_\lambda\hat{\Psi}[\Psi, A] = i\hat{\Lambda} * \hat{\Psi}[\Psi, A]$, where $\delta_\lambda\Psi = i\Lambda\Psi$ is the commutative gauge transformation. I.e., you may first apply the Seiberg-Witten transformation and then a (commutative) gauge transformation or first a (nc) gauge transformation and then the Seiberg-Witten map, the outcome will be the same. Gauge transformations and Seiberg-Witten maps commute.

As a consequence, nc gauge fields and parameters will be enveloping algebra valued. The infinite number of parameters can be reduced to the classical parameters. Therefore arbitrary symmetry groups can be tackled [3].

3. Non-Commutative Standard Model

We study the symmetry group $SU(3)_C \times SU(2)_L \times U(1)_Y$ of the Standard Model, without introducing new particles or parameters. The straight forward method is to write down Standard Model Lagrangian and to replace $\cdot$ by $*$ and fields Ψ, A by $\hat{\Psi}[\Psi, A]$, $\hat{A}[A]$. But it is not quite that easy. We have to deal with the following problems: Tensor product of gauge groups, charge quantisation in nc QED, gauge invariance of the Yukawa couplings, ambiguities in the choice of kinetic terms for the gauge fields.

3.1. Tensor Product of Gauge Groups

We choose the most natural and symmetric tensor product. We take the classical tensor product and plug it into the Seiberg-Witten map,

$$V_\mu = g' A_\mu(x) Y + g \sum_{a=1}^{3} B_{\mu a} T_L^a + g_S \sum_{a=1}^{8} G_{\mu a} T_S^a. \tag{9}$$

The non-commutative gauge field is given by $\hat{V}_\mu = \hat{V}_\mu[V]$ [1].

3.2. Charge Quantisation in nc QED

In nc QED only charges $\pm q$, 0 can be accounted for, once q is fixed [4,5]. Other charges $q^{(n)}$ cannot be absorbed into the respective field $\hat{a}_\mu^{(n)}$, because of the commutator in

$$\hat{f}_{\mu\nu}^{(n)} = \partial_\mu \hat{a}_\nu^{(n)} - \partial_\nu \hat{a}_\mu^{(n)} - i e q^{(n)} [\hat{a}_\mu^{(n)} \, \overset{\star}{,} \, \hat{a}_\nu^{(n)}],$$

$$D_\mu \hat{\psi}^{(n)} = \partial_\mu \hat{\psi}^{(n)} - i e q^{(n)} \hat{a}_\mu^{(n)} \hat{\psi}^{(n)}.$$

Therefore, we need a different gauge field $\hat{a}_\mu^{(n)}$ for each charge $q^{(n)}$. There are not too many degrees of freedom, since

$$\hat{a}_\mu^{(n)} = a_\mu + \frac{e q^{(n)}}{4} \theta^{\sigma\tau} \{\partial_\sigma a_\mu, a_\tau\}$$

$$+ \frac{e q^{(n)}}{4} \theta^{\sigma\tau} \{f_{\sigma\mu}, a_\tau\} + \mathcal{O}(h^2)$$

can be expanded in terms of the commutative field a^μ.

3.3. Yukawa couplings

Following the above method of quantisation, the Yukawa couplings are given by

$$S_{Yukawa} = \int d^4x \Bigg(- \sum_{i,j=1}^{3} W^{ij} \Big((\bar{\hat{L}}_L^{(i)} \star \rho_L(\hat{\Phi})) \star \hat{e}_R^{(j)} + \bar{\hat{e}}_R^{(i)} \star (\rho_L(\hat{\Phi})^\dagger \star \hat{L}_L^{(j)}) \Big)$$

$$- \sum_{i,j=1}^{3} G_u^{ij} \Big((\bar{\hat{Q}}_L^{(i)} \star \rho_{\bar{Q}}(\hat{\bar{\Phi}})) \star \hat{u}_R^{(j)} + \bar{\hat{u}}_R^{(i)} \star (\rho_{\bar{Q}}(\hat{\bar{\Phi}})^\dagger \star \hat{Q}_L^{(j)}) \Big) \tag{10}$$

$$\sum_{i,j=1}^{3} G_d^{ij} \Big((\bar{\hat{Q}}_L^{(i)} \star \rho_Q(\hat{\Phi})) \star \hat{d}_R^{(j)} + \bar{\hat{d}}_R^{(i)} \star (\rho_Q(\hat{\Phi})^\dagger \star \hat{Q}_L^{(j)}) \Big) \Bigg),$$

where $\hat{L}_L$, $\hat{Q}_L$, $\hat{e}_R$, $\hat{u}_R$, $\hat{d}_R$ are the nc fields for the left handed leptons and quarks and for the right handed ones respectively. The Higgs field Φ commutes with generators of $U(1)$ and $SU(3)$, only in the case of commutative space-time. Therefore the Higgs needs to transform from both sides in order to "cancel the charges" from fields on either side (e.g., considering the first coupling in (10), $\hat{L}_L$ and $\hat{e}_R$) and to render the couplings gauge invariant. The expansion of Φ transforming on the left and on the right under arbitrary gauge groups is called hybrid SW-map,

$$\hat{\Phi}[\Phi, V, V'] = \Phi + \frac{1}{2} \theta^{\mu\nu} V_\nu \Big(\partial_\mu \Phi - \frac{i}{2} (V_\mu \Phi + \Phi V'_\mu) \Big),$$

$$+ \frac{1}{2} \theta^{\mu\nu} \Big(\partial_\mu \Phi - \frac{i}{2} (V_\mu \Phi + \Phi V'_\mu) \Big) V'_\nu + \mathcal{O}(\theta^2),$$

with $\delta \hat{\Phi} = i\hat{\Lambda} * \hat{\Phi} - i\hat{\Phi} * \hat{\Lambda}$. In the above example, we have: $V_\mu = -\frac{1}{2}g'A_\mu + gB_\mu^a T_L^a$, $V_\mu' = g'A_\mu$. The covariant derivative for $\hat{\Phi}$ is given by $\hat{D}_\mu \hat{\Phi} = \partial_\mu \hat{\Phi} - i\hat{V}_\mu * \hat{\Phi} - i\hat{\Phi} * \hat{V}_\mu'$. In the classical limit, the Higgs fields transforms under a gauge transformation in the usual way.

3.4. Kinetic Terms for the Gauge Bosons

The representations for the gauge bosons in the trace of the kinetic terms are not uniquely determined by gauge invariance. There is no physical criterium at hand - such as renormalizability in commutative case - so why not choose the simplest option, which we call minimal NCSM. The simplest choice is the usual $SU(2)$ and $SU(3)$ traces and a $U(1)$ trace using $Y = \frac{1}{2} \begin{pmatrix} 1 & 0 \\ 0 & -1 \end{pmatrix}$ as charge generator.

Considering a standard model originating from a SO(10) theory, these terms have a unique noncommutative generalisation [7].

4. Conclusions

- Higgs mechanism and Yukawa sector can be implemented in NCSM.
- To 0th order of the expansion in θ we get the commutative Standard Model.
- The different interactions cannot be considered seperately in the NCSM, since we had to introduce the overall gauge field V_μ; they mix due to the SW-maps.
- No coupling of the Higgs to the electromagnetic Photon occurs, in the minimal version.
- There is also no self interaction of the $U(1)_Y$ boson. Vertices with five and six gauge bosons for $SU(3)_C$ and $SU(2)_L$ occur.
- We encounter no UV/IR mixing due to θ expansion.
- The important task is to search for processes which may be sensitive to non-commutativity: e.g., $Z \to \gamma\gamma$ in nonminimal NCSM [6].

References

[1] X. Calmet, B. Jurčo, P. Schupp, J. Wess, M. Wohlgenannt, "Non-Commutative Standard Model", Eur. Phys. J. C **23** (2002) 363 [hep-th/0111115].

[2] N. Seiberg, E. Witten, "String Theory and Noncommutative Geometry", JHEP **9909** (1999) 032 [hep-th/9908142].

[3] B. Jurčo, L. Möller, S. Schraml, P. Schupp, J. Wess, "Construction of non-Abelian gauge theories on noncommutative spaces", Eur. Phys. J. C **21** (2001) 383 [hep-th/0104153].

[4] M. Hayakawa, "Perturbative analysis on infrared aspects on noncommutative QED on R^4", Phys. Lett. B **478** (2000) 394 [hep-th/9912094].

[5] M. Chaichian, P. Prešnajder, M.M. Sheikh-Jabbari, A. Tureanu, "Non-commutative gauge field theories: A no-go theorem", Phys. Lett. B **526** (2002) 132 [hep-th/0107037].

[6] W. Behr, N.G. Deshpande, G. Duplančić, P. Schupp, J. Trampetić, J. Wess, "The $Z \to \gamma\gamma, gg$ Decays in the Noncommutative Standard Model", hep-ph/0202121.

[7] P. Aschieri, B. Jurčo, P. Schupp, J. Wess, "Non-Commutative GUTs, Standard Model and C,P,T", hep-th/0205214.

Methods and Applications of Quantum Theory

65th Birthday Tribute to Allan Solomon

As chairman of this afternoon's session I was asked by the organisers of this conference to make a small gesture towards the first speaker of this session, **Professor Allan Solomon**, of the Open University UK, who is the head of their *Quantum Processes Group*, on the occasion of his 65th birthday.

Since we all know Allan and we appreciate and like his work, I immediately and willingly agreed. Therefore I have decided to excise a couple of minutes from Allan's first contribution this afternoon (I believe he would agree, but I have not asked him, this is the sole liberty of a chairman that I take) to tell you something more about Allan's achievements, tastes and work methods as well as a few impressions from a close collaboration with him that I have had the privilege to enjoy over the last 5 years or so.

In my judgement Allan is an exceptionally gifted, versatile and profound theoretical physicist, as well as an excellent lecturer, and a wonderful colleague and friend. His official title is Professor of Mathematical Physics, and my feeling is that in this description the *Physics* comes first. He has graduated from, and been a research student in, the best schools including Trinity College Dublin, Cambridge University and the University of Paris.

His initial interest during his post-graduate work centred on the application of group theoretical methods to particle physics, where he applied developments of the SU(3) theory of Gell-Mann and Ne'eman to the electromagnetic spectrum of baryons.

Then comes the period of Allan's activity in industrial research, when he worked at the Republic Aviation Corporation of Farmingdale, Long Island, on cosmic radiation and its effect on the then-proposed U.S. supersonic transport plane (SST).

He returned to academic life in 1966 as an Assistant Professor at the Polytechnic Institute of Brooklyn. His interests in Group Theoretical Methods were enhanced through a collaboration with Yuval Ne'eman on spectrum generating algebras and dynamical groups in the late 60's.

Although initially these concepts were applied to particle physics, Allan was the first to extend them to the treatment of condensed matter problems. In his seminal paper of 1970 he applied the non-compact group SU(1,1) to the theory of Superfluid Helium Four, a boson system. Subsequently, in a long collaboration with Joseph Birman, Allan opened up a new avenue with applications of the group-theoretical method to fermionic systems with competing interactions, such as coexisting antiferromagnetism and superconductivity.

Eventually in 1987 Allan and Joe produced a group theoretical unifying scheme based on SU(8) which contains most of the coexisting phases in such systems, which in many cases have been observed experimentally.

Following periods in the USA, the Institute for Advanced Studies in Ireland and the Department of Physics in Tel Aviv University, Allan began his career in 1971 at the Open University, where he now works. He extended his interests to quantum optics and the theory of squeezed light, spurred on by a collaboration with Jacob Katriel (Haifa Technion) and Mario Rasetti (Politecnico di Torino) in the late Eighties, involving ingenious applications of group concepts, and even more elaborate structures called Quantum Groups and q-deformations, on which latter topic he has written pioneering papers.

In the last couple of years he has concentrated on the problem of Quantum Control, where once again the use of group theory proves extremely valuable. (He will himself talk about some of this work in a minute).

I have had the privilege and pleasure of working with Allan for several years now. I must say this was and continues to be a real source of inspiration for me, although it is not always easy to be able to adhere to his highest standards as a true leader of the field. I am talking here about scientific and ethical standards.

On a more human side, all who meet Allan know what a charming story teller he is, probably the best available - and he is a fine musician too! Those who have heard him playing trumpet and his constructive approach to improving on jokes keep asking themselves what is he doing with Lie algebras!

Dear Allan: We can now look back at what you have achieved and tell you that you are on the right track. Just continue on this path. Since your birthday has arrived may we thank you, Allan, for all you have done for physics and the physics community, and wish you and your family every happiness and continued success in the future.

I therefore declare the *Allan Solomon Session* of Group 24 open and I have the pleasure to invite him to be the first speaker.

Karol Penson, Paris , July 2002.

Inst. Phys. Conf. Ser. No 173: Section 7
Paper presented at 24th Int. Coll. Group Theoretical Methods in Physics, Paris, France, 15–20 July 2002
©2003 *IOP Publishing Ltd*

Dissipative 'Groups' and the Bloch Ball

Allan I. Solomon‡ and Sonia G. Schirmer§

Quantum Processes Group, The Open University, Milton Keynes MK7 6AA

Abstract. We show that a quantum control procedure on a two-level system including dissipation gives rise to the semi-group corresponding to the Lie algebra $gl(3, R) \oplus R^3$. The physical evolution may be modelled by the action of this semi-group on a 3-vector as it moves inside the Bloch sphere, in the Bloch ball.

1. Introduction

Recent developments in quantum computing have emphasized the need for a realistic analysis of dissipation in systems which have the potential for use as qubits. In this note we discuss the effects of control and dissipation on a two-level system. For a single qubit *pure* state, it is well known that the unitary evolution may be visualised as the movement of a vector, the Bloch vector, on the surface of a 2-sphere, the Bloch sphere. In this note we extend the idea to a two-level *mixed* state. For this system, unitary evolution is on a spherical shell within the Bloch Sphere. Dissipation causes more general motion within the Bloch ball. This motion corresponds to the action of a certain *semi-group*.

We also show that the effects of dissipation may not be compensated by the interaction with the external control. However, taking into account the effects of measurement, which may be modelled by certain projection operators, the dissipative effects may indeed be modified, allowing more effective control of the system.

2. Dynamics of dissipative quantum control systems

In pure-state quantum mechanics the state of the system is usually represented by a wavefunction $|\Psi\rangle$, which is an element of a Hilbert space $\mathcal{H}$. For dissipative quantum systems, however, a quantum statistical mechanics formulation is necessary since dissipative effects can and do convert pure states into statistical ensembles and vice versa. In this case, the state of the system is represented by a density operator $\hat{\rho}$, whose diagonal elements determine the populations of the energy eigenstates, while the off-diagonal elements determine the coherences between energy eigenstates, which distinguish coherent superposition states $|\Psi\rangle = \sum_{n=1}^{N} c_n |n\rangle$ from statistical ensembles of energy eigenstates (i.e., mixed states) $\hat{\rho} = \sum_{n=1}^{N} w_n |n\rangle\langle n|$. For a non-dissipative system the time evolution of the density matrix $\hat{\rho}(t)$ with $\hat{\rho}(t_0) = \hat{\rho}_0$ is governed by

$$\hat{\rho}(t) = \hat{U}(t)\hat{\rho}_0\hat{U}(t)^\dagger, \tag{1}$$

where $\hat{U}(t)$ is the time-evolution operator satisfying the Schrodinger equation

$$i\hbar\frac{d}{dt}\hat{U}(t) = \hat{H}(\mathbf{f})\hat{U}(t), \qquad \hat{U}(0) = \hat{I}, \tag{2}$$

‡ a.i.solomon@open.ac.uk
§ sgs29@cam.ac.uk

where $\hat{I}$ is the identity operator. $\hat{\rho}(t)$ also satisfies the quantum Liouville equation

$$i\hbar \frac{d}{dt}\hat{\rho}(t) = [\hat{H}(\mathbf{f}), \hat{\rho}(t)] = \hat{H}(\mathbf{f})\hat{\rho}(t) - \hat{\rho}(t)\hat{H}(\mathbf{f}). \tag{3}$$

$\hat{H}(\mathbf{f})$ is the total Hamiltonian of the system, which depends on a set of control fields f_m:

$$\hat{H}(\mathbf{f}) = \hat{H}_0 + \sum_{m=1}^{M} f_m(t)\hat{H}_m, \tag{4}$$

where $\hat{H}_0$ is the internal Hamiltonian and $\hat{H}_m$ is the interaction Hamiltonian for the field f_m for $1 \leq m \leq M$.

The advantage of the Liouville equation (3) over the unitary evolution equation (1) is that it can easily be adapted to dissipative systems by adding a dissipation (super-)operator $\mathcal{L}_D[\hat{\rho}(t)]$:

$$i\hbar\dot{\rho}(t) = [\hat{H}_0, \hat{\rho}(t)] + \sum_{m=1}^{M} f_m(t)[\hat{H}_m, \hat{\rho}(t)] + i\hbar\mathcal{L}_D[\hat{\rho}(t)]. \tag{5}$$

In general, uncontrollable interactions of the system with its environment lead to two types of dissipation: phase decoherence (dephasing) and population relaxation (decay). The former occurs when the interaction with the enviroment destroys the phase correlations between states, which leads to a decay of the off-diagonal elements of the density matrix:

$$\dot{\rho}_{kn}(t) = -\frac{i}{\hbar}([\hat{H}(\mathbf{f}), \hat{\rho}(t)])_{kn} - \Gamma_{kn}\rho_{kn}(t) \tag{6}$$

where Γ_{kn} (for $k \neq n$) is the dephasing rate between $|k\rangle$ and $|n\rangle$. The latter happens, for instance, when a quantum particle in state $|n\rangle$ spontaneously emits a photon and decays to another quantum state $|k\rangle$, which changes the populations according to

$$\dot{\rho}_{nn}(t) = -\frac{i}{\hbar}([\hat{H}(\mathbf{f}), \hat{\rho}(t)])_{nn} + \sum_{k \neq n} [\gamma_{nk}\rho_{kk}(t) - \gamma_{kn}\rho_{nn}(t)] \tag{7}$$

where $\gamma_{kn}\rho_{nn}$ is the population loss for level $|n\rangle$ due to transitions $|n\rangle \rightarrow |k\rangle$, and $\gamma_{nk}\rho_{kk}$ is the population gain caused by transitions $|k\rangle \rightarrow |n\rangle$. The population relaxation rate γ_{kn} is determined by the lifetime of the state $|n\rangle$, and for multiple decay pathways, the relative probability for the transition $|n\rangle \rightarrow |k\rangle$. Phase decoherence and population relaxation lead to a dissipation superoperator (represented by an $N^2 \times N^2$ matrix) whose non-zero elements are

$$\begin{aligned} (\mathcal{L}_D)_{kn,kn} &= -\Gamma_{kn} & k \neq n \\ (\mathcal{L}_D)_{nn,kk} &= +\gamma_{nk} & k \neq n \\ (\mathcal{L}_D)_{nn,nn} &= -\sum_{n \neq k} \gamma_{kn}. \end{aligned} \tag{8}$$

Population decay and dephasing allow us to overcome kinematical constraints such as unitary evolution to create statistical ensembles from pure states, and pure states from statistical ensembles, which is important for many applications such as optical pumping. However, there are instances when this is not desirable such as in quantum computing, where these effects destroy quantum information. Thus, there are situations when we would like to prevent decay and dephasing. A cursory glance at the quantum Liouville equation for coherently driven, dissipative systems (5) suggests that it might be possible to prevent population and phase relaxation by applying suitable control fields such that

$$\sum_{m=1}^{M} f_m(t)[\hat{H}_m, \hat{\rho}(t)] + i\hbar\mathcal{L}_D[\hat{\rho}(t)] = 0. \tag{9}$$

Unfortunately, however, a more careful analysis reveals that this is *not* possible, in general, as we shall now show explicitly for a two-level system, or qubit in quantum computing parlance.

3. Dynamics of a 2-level system subject to control, decay and dephasing

The Hamiltonian for a driven two-level system with energy levels $E_1 < E_2$ is

$$\hat{H}[\mathbf{f}(t)] = \hat{H}_0 + f_1(t)\hat{H}_1 + f_2(t)\hat{H}_2 \tag{10}$$

where $\hat{H}_0$ is the internal Hamiltonian and $\hat{H}_1$ and $\hat{H}_2$ represent interaction Hamiltonians with independent (real-valued) control fields $f_1(t)$ and $f_2(t)$,

$$\hat{H}_0 = \begin{bmatrix} E_1 & 0 \\ 0 & E_2 \end{bmatrix}, \quad \hat{H}_1 = d_1 \begin{bmatrix} 0 & 1 \\ 1 & 0 \end{bmatrix}, \quad \hat{H}_2 = d_2 \begin{bmatrix} 0 & -i \\ i & 0 \end{bmatrix}.$$

d_1, d_2 are the (real-valued) dipole moments for the transition and $\omega = (E_2 - E_1)/\hbar$ is the transition frequency.

We can re-write the Liouville equation in matrix form in a higher dimensional space, often referred to as Liouville space. Straightforward computation shows that

$$\frac{d}{dt}|\rho(t)\rangle\rangle = \mathcal{L}|\rho(t)\rangle\rangle = [(1/i\hbar)(\mathcal{L}_0 + f_1(t)\mathcal{L}_1 + f_2(t)\mathcal{L}_2) + \mathcal{L}_D]|\rho(t)\rangle\rangle \tag{11}$$

where $|\rho(t)\rangle\rangle = (\rho_{11}(t), \rho_{12}(t), \rho_{21}(t), \rho_{22}(t))^T$

$$\mathcal{L}_0 = \begin{pmatrix} 0 & 0 & 0 & 0 \\ 0 & -\hbar\omega & 0 & 0 \\ 0 & 0 & +\hbar\omega & 0 \\ 0 & 0 & 0 & 0 \end{pmatrix} \quad \mathcal{L}_1 = d_1 \begin{pmatrix} 0 & -1 & +1 & 0 \\ -1 & 0 & 0 & +1 \\ +1 & 0 & 0 & -1 \\ 0 & +1 & -1 & 0 \end{pmatrix}$$

$$\mathcal{L}_2 = d_2 \begin{pmatrix} 0 & -i & -i & 0 \\ +i & 0 & 0 & -i \\ +i & 0 & 0 & -i \\ 0 & +i & +i & 0 \end{pmatrix} \quad \mathcal{L}_D = \begin{pmatrix} -\gamma_{21} & 0 & 0 & \gamma_{12} \\ 0 & -\Gamma & 0 & 0 \\ 0 & 0 & -\Gamma & 0 \\ \gamma_{21} & 0 & 0 & -\gamma_{12} \end{pmatrix}$$

γ_{12} is the rate of population relaxation from $|2\rangle$ to $|1\rangle$, γ_{21} is the rate of population relaxation from $|1\rangle$ to $|2\rangle$ (usually zero), and Γ is the dephasing rate.

Notice that the matrix elements of the Liouville operators $\mathcal{L}_1$ and $\mathcal{L}_2$ are zero where the matrix elements of the dissipation operator $\mathcal{L}_D$ are non-zero, and vice versa. Thus, no matter how we choose the control fields, we cannot cancel the effect of the dissipative terms. The best we can do is to use coherent control to implement quantum error correction schemes to restore decayed/dephased quantum states to their correct values. An early contribution to the theory of continuous feedback for such systems is contained in [1], while a more recent scheme of continuous quantum error correction involving weak measurements and feedback has been proposed in [2].

4. Dynamical Semi-group and the Bloch Ball

The more usual real vector form for ρ is $\rho_B \equiv (\rho_{1,2} + \rho_{2,1}, i(\rho_{1,2} - \rho_{2,1}), \rho_{1,1} - \rho_{2,2}, \rho_{1,1} + \rho_{2,2})$. In the general case this is referred to as the *coherence vector*[3]. Since for the systems under consideration we shall take $\rho_{1,1} + \rho_{2,2}$ as constant (no population loss) only the first three components of ρ_B transform under the dynamics. For *pure states* the norm is constant - thus generating motion on the surface of a 2-sphere, the Bloch sphere. For our more general scenario, motion takes place in the interior of this sphere, the *Bloch ball*.

The family of control hamiltonians generate the Lie algebra $u(2)$, in this case a completely controllable system[4]. The Lie algebra generated by the matrices corresponding to Eq.(3) acting on ρ_B is the inhomogeneous algebra $gl(3, R) \oplus R^3$. The evolution of the system in time is determined by $\exp(\mathcal{L}t)$. Noting that the eigenvalues of $\mathcal{L}_D$ are (always) negative, the demand that our set of operators remain bounded gives in effect a *semi-group*, with only unlimited *positive* values of t permitted. These considerations may be generalized to any dimensions [5].

5. Conclusions

In this note we showed how the effects of control and dissipation can be treated geometrically, by the movement of the coherence vector inside the Bloch ball. We incidentally noted that dissipative effects could not be compensated by control dynamics alone. Without measurements and feedback, dissipation usually forces the system into an equilibrium state. One can see that without any control this state corresponds to a point on the z-axis of the Bloch sphere; with constant controls one can show that the state converges to a point on an ellipse inside the Bloch ball. The coordinates of these attractors can easily be expressed as a function of the dissipative terms.

Intuitively, error correction is not possible with control fields alone because the dissipative terms tend to pull us inside the Bloch ball and, without quantum measurement feedback, we can't get back to the surface of the ball because the control fields can only perform rotations.

The analysis can be treated by traditional Lie algebraic methods, by use of a dynamical Lie algebra. In the specific case of a qubit treated here, this algebra is the inhomogeneous semi-direct sum $gl(3, R) \oplus R^3$. The control fields generate a rotation algebra within this larger algebra. Imposing boundedness conditions on the evolution of the dynamics obtained by exponentiation of this algebra leads to a semi-group description of the evolution.

Although due to restrictions of space we have treated explicitly only the case of a two-level system, the methods can be generalised without difficulty to any finite level system. For an N-level system, the coherence vector is essentially an $N^2 - 1$ real component vector (for population-preserving dynamics) whose motion is restricted to the $N^2 - 2$-dimensional Bloch sphere for non-dissipative systems, and to the interior of the corresponding Bloch ball in general. The associated inhomogeneous real algebra is given by semi-direct sum $gl(N^2 - 1, R) \oplus R^{(N^2-1)}$, and this gives rise by means of exponentiation to the corresponding semi-group which determines the dynamics. Finally, for *quasi-spin* systems, defined by symmetric population decay rates, the inhomogeneous (translation-like) terms disappear.

References

[1] Wiseman H 1994 Phys. Rev. A 49 2133
[2] Ahn C, Doherty, A C and Landahl, A J 2002 Phys. Rev. A 65 042301
[3] Lendi, K 1987 N-Level systems and Applications to Spectroscopy Lecture Notes in Physics 286 (Berlin: Springer-Verlag)
[4] Schirmer S G, Fu H and Solomon, A I 2001 Phys. Rev. A 63 063410
[5] Solomon A I and Schirmer S G, 2002 To be published

Inst. Phys. Conf. Ser. No 173: Section 7
Paper presented at 24th Int. Coll. Group Theoretical Methods in Physics, Paris, France, 15–20 July 2002
©*2003 IOP Publishing Ltd*

Gauging Quantum-Mechanical Space-Time Symmetry

V Aldaya and J L Jaramillo

Instituto de Astrofísica de Andalucía, Apartado Postal 3004, 18080 Granada, Spain

J Guerrero

Depto. de Matemática Aplicada, Universidad de Murcia, 30100 Murcia, Spain

Abstract. We claim that the U(1) phase invariance in Quantum Theory constitutes an essential ingredient of the (quantum) space-time symmetry. If we identify this group with a 1-dimensional subgroup of some internal rigid symmetry, then turning the space-time subgroup of the Poincaré group into a local symmetry automatically promotes the original internal symmetry to the gauge level in a non-direct product way. This mechanism provides, in particular, a non-trivial mixing of the electromagnetic and gravitational forces.

1. Introduction

Let us start this talk by reminding you the fundamental character of the $U(1)$ or Phase Invariance in Quantum Mechanics. To this end we shall consider the behaviour of the Schrödinger equation corresponding to the free quantum particle $i\hbar\frac{\partial}{\partial t}\Psi = -\frac{\hbar^2}{2m}\nabla^2\Psi$ under the Galilei transformations: $x' = x + a + vt$, $t' = t + b$. The Schrödinger equation acquires an extra term,

$$i\hbar\frac{\partial}{\partial t'}\Psi + i\hbar v\frac{\partial\Psi}{\partial x'} = -\frac{\hbar^2}{2m}\nabla'^2\Psi \tag{1}$$

which can be compensated by also transforming the wave function. Allowing for a non-trivial phase factor in front of the transformed wave function of the form $\Psi' = e^{\frac{im}{\hbar}(vx+\frac{1}{2}v^2t)}\Psi$, the Schrödinger equation becomes strictly invariant, i.e. $i\hbar\frac{\partial}{\partial t'}\Psi' = -\frac{\hbar^2}{2m}\nabla'^2\Psi'$.

The need for a transformation in Ψ accompanying the space-time transformation to accomplish full invariance strongly suggests the adoption of a central extension $\tilde{G}$ of the Galilei group as the basic (quantum-mechanical) space-time symmetry for the free particle [1]. The constant $\hbar$ is required to keep the exponent in the Ψ transformation dimensionless.

The successive composition of two transformation in $\tilde{G}$ leads to the group law:

$$b'' = b' + b, \qquad a'' = a' + a + v'b, \qquad v'' = v' + v \tag{2}$$
$$e^{i\phi''} = e^{i\phi'}e^{i\phi}e^{\frac{im}{\hbar}[a'v+b(v'v+\frac{1}{2}v'^2)]}$$

The main feature of the central extension is that of making non-trivial the commutator between the generators associated with translations $X_a \equiv P$ and boosts $X_v \equiv K$: $[P, K] = -mX_\phi$, *mimicking* the Poisson bracket between p and x provided that we impose on the wave function the $U(1)$-function condition $X_\phi\Psi \equiv \frac{\partial\Psi}{\partial\phi} = i\Psi$ (i.e. homogeneous of degree one on $e^{i\phi}$).

The usual way of introducing an interaction in Physics is through the *gauge principle*, which requires the invariance of the Lagrangian of free matter under a *gauge* group obtained from an original symmetry group, the *rigid* group, by making the group parameter to depend on space-time variables.

In this talk, we shall approach the problem in the simplest and most economical way, in a Particle Mechanics (versus Field Theoretical) framework, leaving the more mathematically involved field formulation for the future. To be precise, we face the situation that arises when promoting to the "local" level the space-time translations of the centrally extended space-time symmetry (either Galilei or Poincaré group), rather than the space-time symmetry itself.

The way of associating a physical dynamics with a specific symmetry can be accomplished by means of the standard co-adjoint orbits method of Kirillov [2], where the Lagrangian is seen as the local potential of the corresponding symplectic form, or through a generalized group approach to quantization which is related to the co-homological structure of the symmetry and leads directly to the quantum theory (see [3] and references there in).

2. The electromagnetic interaction

To illustrate technically the present revisited Minimal Coupling Theory, let us consider the simpler case of the non-relativistic pure Lorentz force, keeping rigid the space-time translations. For this aim, we consider the Lie algebra $\tilde{\mathcal{G}}$ of the centrally extended Galilei group $\tilde{G}$ (only non-zero commutators):

$$
\begin{array}{lll}
[X_{v^i}, X_t] = X_{x^i} & [X_{v^i}, X_{x^j}] = m\delta_{ij}X_\phi & \\
[X_{J^i}, X_{J^j}] = \epsilon_{ij.}{}^k X_{J^k} & [X_{J^i}, X_{v^j}] = \epsilon_{ij.}{}^k X_{v^k} & [X_{J^i}, X_{x^j}] = \epsilon_{ij.}{}^k X_{x^k}
\end{array} \tag{3}
$$

which leaves strictly invariant the extended Poincaré-Cartan form $\Theta = p_i dx^i - \frac{p^2}{2m}dt + d\phi$, $L_{X_\alpha}\Theta = 0, \forall X_\alpha \in \tilde{\mathcal{G}}$. This 1-form is defined on the extended phase space parametrized by (x^i, p_j, ϕ), where $e^{i\phi} \in U(1)$ is the phase transforming non-trivially under the Galilei group. It generalizes the Lagrangian and constitutes a potential for the symplectic form ω on the solution manifold (on trajectories $s(t)$, $(p_i dx^i - \frac{p^2}{2m}dt)|_{s(t)} = (p_i\dot{x}^i - \frac{p^2}{2m})|_{s(t)}dt$).

Local $U(1)$ transformations generated by $f \otimes X_\phi$, f being a real function $f(\vec{x},t)$, are incorporated into the scheme by adding to (3) the extra commutators [4]:

$$
[X_a, f \otimes X_\phi] = (L_{X_\alpha}f) \otimes X_\phi \tag{4}
$$

Keeping the invariance of Θ under $f \otimes X_\phi$ requires adding to Θ the term $A = A_i dx^i + A_0 dt$ whose components transform under $U(1)(\vec{x},t)$ as the space-time gradient of the function f.

The algebra (3)+(4) is infinite-dimensional but the dynamical content of it is addressed by the (co-homological) structure of the finite-dimensional subalgebra generated by $\tilde{\mathcal{G}}$ and the generators $f \otimes X_\phi$ with only linear functions. Thus, a very economical trick for dealing with this sort of infinite-dimensional algebra consists in proceeding with the above mentioned 15-dimensional electromagnetic subgroup and then imposing the generic constraint $A^\mu = A^\mu(\vec{x},t)$ on the symplectic structure. Let us call this group $\tilde{G}_E$, and the generators associated with linear functions in $f \otimes X_\phi$, X_{A^μ}.

The non-zero commutators of $\tilde{\mathcal{G}}_E$ are (omitting rotations, they operate in the usual way):

$$
\begin{array}{lll}
[X_t, X_{v^i}] = -X_{x^i} & [X_t, X_{A^0}] = -qX_\phi & [X_{x^i}, X_{v^j}] = m\delta_{ij}X_\phi \\
[X_{v^i}, X_{A^i}] = \delta_{ij}X_{A^0} & [X_{x^i}, X_{A^j}] = q\delta_{ij}X_\phi &
\end{array} \tag{5}
$$

where we have performed a new central extension parametrized by what proves to be the electric charge q.

The co-adjoint orbits of the group $\tilde{G}_E$ with non-zero electric charge have dimension 4+4 as a consequence of the Lie algebra cocycle piece $\Sigma(X_t, X_{A^0}) = -q$, which lends dynamical (symplectic) content to the time variable. This is a property inherited from the (centrally extended) conformal group from which $\tilde{\mathcal{G}}_E$ is an Inönü-Wigner contraction. In the case of the conformal group [5] the symplectic character of time is broken by means of a dynamical constraint (or by choosing a Poincaré vacuum) and the dimension 3+3 of the phase space is restored. Here the constraint $A^\mu = A^\mu(\vec{x}, t)$ also accomplishes this task at the same time as it introduces the notion of electromagnetic potential.

On a general orbit with $q \neq 0$ the extended Poincaré-Cartan form acquires the expression:

$$\Theta = m\vec{v} \cdot d\vec{x} - \frac{1}{2}m\vec{v}^2 dt + q\vec{A} \cdot d\vec{x} - qA^0 dt + d\phi \tag{6}$$

After imposing the above-mentioned constraint on A^μ, we compute the kernel of the presymplectic form $d\Theta$, i.e. a vector field X such that $i_X d\Theta = 0$:

$$X = \frac{\partial}{\partial t} + \vec{v} \cdot \frac{\partial}{\partial \vec{x}} + \frac{q}{m}\left[\left(\frac{\partial A_j}{\partial x^i} - \frac{\partial A_i}{\partial x^j}\right)v^j - \frac{\partial A^0}{\partial x^i} - \frac{\partial A_i}{\partial t}\right]\frac{\partial}{\partial v_i} - [\frac{1}{2}m\vec{v}^2 + q(\vec{v} \cdot \vec{A} - A^0)]\frac{\partial}{\partial \phi}, \tag{7}$$

It defines the equations of motion of a charged particle moving in an electromagnetic field:

$$m\frac{d\vec{v}}{dt} = q\left[\vec{v} \wedge (\vec{\nabla} \wedge \vec{A}) - \vec{\nabla}A^0 - \frac{\partial \vec{A}}{\partial t}\right], \tag{8}$$

which are nothing more than the standard Lorenz force equation. The same can be repeated with the centrally extended Poincaré group $\tilde{P}$ (see [5] and references therein) by promoting to "local" the $U(1)$ transformations and considering the finite-dimensional subgroup $\tilde{P}_E$ analogous to $\tilde{G}_E$.

3. Electro-gravity mixing

Let us consider now the gravitational interaction. To this end, we start directly with the centrally extended Poincaré group $\tilde{P}$ and see how the fact that the translation generators produce the central term under commutation with some other generators (boosts) plays a singular role in the relationship between local space-time translations and local $U(1)$ transformations. Symbolically denoting the generators of translations by P, P_0, those of boosts by K and the central one by X_ϕ, we find:

$$[K, f \otimes P] \simeq (L_K f) \otimes P + f \otimes (P_0 + X_\phi), \tag{9}$$

which means that turning the translations into local symmetry entails also the local nature of the $U(1)$ phase. We expect, in this way, a non-trivial mixing of gravity and electromagnetism into an infinite-dimensional electro-gravitational group.

We shall follow identical steps as those given in the former example. The generators of local space-time translations associated with linear functions will be called $X_{h^{\mu\nu}}$, and the corresponding parameters $h^{\mu\nu}$ will also be constrained, in the form $h^{\mu\nu} = h^{\mu\nu}(\vec{x}, x^0)$, on the symplectic orbits. However, the co-homological structure of this finite-dimensional electro-gravitational subgroup, $\tilde{P}_{EG}$ is richer than that of $\tilde{P}_E$ and the exponentiation of the Lie algebra $\tilde{\mathcal{P}}_{EG}$ must be made order by order (the explicit calculations will be kept up to order 3 in the group law). This will be enough to recognize the standard part of the interaction, i.e. the ordinary Lorentz force and the geodesic equations, although the latter in a quasi-linear approximation in terms of the metric $g^{\mu\nu} \equiv \eta^{\mu\nu} + h^{\mu\nu}$. But in addition, and associated with a new Lie algebra co-homology constant, κ, different from m and q and mixing both interactions, a new term appears in the Lorentz force made of the gravitational potential $h^{\mu\nu}$.

We shall not dwell on explicit calculations in this talk and simply give the resulting equations of motion at the "non-relativistic" limit stated by the Inönü-Wigner contraction with respect to the subgroup generated by (x_0, L_{ij}, A_k). The contracted algebra reads:

$$
\begin{aligned}
[x_0, L_{0i}] &= x_i, & [x_0, h_{0i}] &= x_i, & [x_0, h_{00}] &= -2m\phi, & [x_0, A_0] &= q\phi \\
[x_i, L_{0j}] &= -(m + \kappa q)\delta_{ij}\phi & [x_i, L_{jk}] &= -\delta_{ij}x_k + \delta_{ik}x_j \\
[x_i, h_{0j}] &= m\delta_{ij}\phi & [x_i, A_j] &= -q\delta_{ij}\phi \\
[L_{0i}, L_{jk}] &= -\delta_{ij}L_{0k} + \delta_{ki}L_{0j} & [L_{0i}, h_{0j}] &= -\delta_{ij}h_{00} + \kappa\delta_{ij}A_0 \\
[L_{0i}, A_j] &= -\delta_{ij}A_0 & [L_{ij}, L_{kl}] &= -\delta_{kj}L_{il} + \delta_{lj}L_{ik} + \delta_{ik}L_{jl} - \delta_{il}L_{jk} \\
[L_{ij}, h_{0k}] &= -\delta_{kj}h_{0i} + \delta_{ik}h_{0j} & [L_{ij}, h_{kl}] &= -\delta_{kj}h_{il} - \delta_{lj}L_{ik} + \delta_{ik}L_{jl} + \delta_{il}L_{jk} \\
[L_{ij}, A_k] &= -\delta_{kj}A_i + \delta_{ki}A_j & [h_{0i}, A_j] &= -\delta_{ij}A_0
\end{aligned}
\tag{10}
$$

Writing $\vec{h}$ for (h^{0i}), we finally derive from this algebra the following Lagrangian and equations of motion, which at this contraction limit are indeed exact:

$$
L = \frac{1}{2}(m + \kappa q)\,\dot{\vec{x}}^2 - q(A^0 - \frac{\kappa}{8}\vec{h}^2) + q(\vec{A} - \frac{\kappa}{2}\vec{h})\cdot\dot{\vec{x}} + m(h^{00} + \frac{1}{4}\vec{h}^2) - m\vec{h}\cdot\dot{\vec{x}} \tag{11}
$$

$$
(m + \kappa q)\frac{d\vec{v}}{dt} = q\left[\vec{v} \wedge (\vec{\nabla} \wedge \vec{A}) - \vec{\nabla}A^0 - \frac{\partial \vec{A}}{\partial t}\right] - m\left[\vec{v} \wedge (\vec{\nabla} \wedge \vec{h}) - \vec{\nabla}h_{00} - \frac{\partial \vec{h}}{\partial t}\right]
$$

$$
+ \frac{m}{4}\vec{\nabla}(\vec{h}\cdot\vec{h}) - \frac{\kappa q}{2}\left[\vec{v} \wedge (\vec{\nabla} \wedge \vec{h}) - \frac{1}{4}\vec{\nabla}(\vec{h}\cdot\vec{h}) - \frac{\partial \vec{h}}{\partial t}\right] \tag{12}
$$

The fourth term in the r.h.s. is quite new and represents another Lorentz-like force (proportional to q) generated by the gravitational potential and which must not be confused with the previous one. It is worth mentioning that the constant m in front of the term $\vec{\nabla}h_{00}$ in (12), naturally interpreted as a gravitational coupling, could acquire a different constant value, let us say g, allowed by the Lie algebra co-homology. Nevertheless, it must be made equal to m to recover the standard physics when switching the constant κ off. In this way, the equivalence principle between inertial and gravitational mass, in this co-homological setting, follows from the natural requirement of absence of a pathological mixing between electromagnetism and gravity when $\kappa = 0$.

As far as the magnitude of the new Lie algebra co-homology constant κ, it is limited by experimental clearance for the difference between particle and anti-particle mass, which for the electron is about $10^{-8}m_e$. Even though this is a small value, extremely dense rotating bodies could be able to produce measurable forces. In the other way around, a mixing of electromagnetism and gravity predicts a mass difference between charged particles and anti-particles, which could be experimentally tested.

Since the present theory has been formulated on symmetry grounds, it can be quantized on the basis of the group approach to quantization referred in [3]. Also, a natural yet highly non-elementary extension of the present theory to Quantum Field Theory is in course.

Finally, as commented above, seeing the group $U(1)$ as a Cartan subgroup of a larger internal symmetry group, like $SU(2) \otimes U(1)$, would result in additional phenomenology. Then, in a QFT version, the production of Z_0 particles out of gravity might be permitted.

References

[1] V. Bargmann, Ann. Math. **59**, 1 (1954)

[2] A.A. Kirillov, *Elements of the Theory of Representations*, Springer-Verlag (1976).

[3] V. Aldaya, J. Navarro-Salas and A. Ramírez, Commun. Math. Phys., **121**, 541 (1989).

[4] In general, $[f \otimes X_\alpha, g \otimes X_\beta] = (fL_{X_\alpha}g) \otimes X_\beta - (gL_{X_\beta}f) \otimes X_\alpha + (fg) \otimes [X_\alpha, X_\beta]$.

[5] V. Aldaya, M. Calixto and J.M. Cerveró, Commun. Math. Phys. **200**, 325 (1999).

[6] R.M. Wald, *General Relativity*, The University of Chicago Press (1984).

Inst. Phys. Conf. Ser. No 173: Section 7
Paper presented at 24th Int. Coll. Group Theoretical Methods in Physics, Paris, France, 15–20 July 2002
©*2003 IOP Publishing Ltd*

Quantum XOR operation using macroscopic polarization states of light in Mach-Zehnder interferometer

A.Yu.Leksin, A.P.Alodjants, A.V.Prokhorov, S.M.Arakelian

Department of Physics and Applied Mathematics, Vladimir State University
600000,Vladimir, Russia

Abstract. In the paper we discuss the quantum optical element XOR (XOR-gate) on the basis of switching of macroscopic polarization states in Mach-Zehnder interferometer for single-photon pulses. It is shown that the state at the output of the optical system has the Schrödinger cat macroscopic superposition properties in polarization characteristics.

PACS numbers: 42.50.Dv, 03.67.Lx

1. Introduction

Among the current fundamental problems of quantum physics which has a principal practical meaning, the problem of developing of the element basis and logic gates for quantum computing presents a special interest [1].

In quantum optics it has already been suggested that nonlocality and interference of single-photon and entangled states occurring in interferometers of different geometry are absolutely necessary– e.g. [2, 3].

In this paper the theory of formation of nonclassical states of light in the Mach-Zehnder interferometer, having Kerr nonlinearity in one of its arms, has been developed. Previously, the optical system under consideration has repeatedly been studied in aspect of generation of quadrature squeezed and SU(2) squeezed light [4, 5] as well as in precise measurement of small phase shifts in optics [5, 6]. A conceptual physical aspect being in the focus of our study is the generation of entangled macroscopic polarization states at the optical system output having nonclassical properties of the "Schrödinger' cat" states superposition [7].

2. Photon switching and quantum XOR-gate

Let us consider successively the state transformation for a two-mode (b and c) light field in different parts of the Mach-Zehnder interferometer with Kerr-like nonlinear medium in one of its arms (Fig.1) in the Schrödinger formalism.

For definiteness sake let us consider the case where only one photon is carried to the interferometer input b, that is the initial state of the whole system

$$|\Psi\rangle = |1, 0\rangle |\alpha_+\rangle |\alpha_-\rangle, \tag{1}$$

where $\alpha_\pm$ are complex parameters of the polarization coherent states $|\alpha\rangle_\pm$ that looks like:

$$\alpha_\pm = r_{1,2} e^{\mp i\theta}\alpha, \qquad r_{1,2} = \frac{\cos\eta \pm \sin\eta}{\sqrt{2}} \tag{2}$$

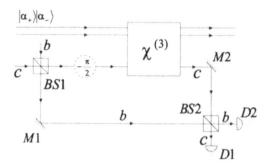

Figure 1. Mach-Zehnder interferometer with a Kerr-like medium of nonlinearity $\chi^{(3)}$; $BS1,2$ are the *polarization* beam splitters, $-\frac{\pi}{2}$ is the phase plate, $M1,2$ are the mirrors, $D1,2$ are the photon detectors

Here η and θ are the angles characterizing orientation and ellipticity respectively for polarization ellipse(their changes range: $-\pi/4 \leq \eta \leq \pi/4$, $-\pi/2 \leq \theta < \pi/2$); $|\alpha|^2 = |\alpha_+|^2 + |\alpha_-|^2$ is total average number of photons.

As a result we shall get the following state at the nonlinear medium output:

$$|\Psi\rangle_{NL} = \cos\psi|1,0\rangle|\alpha_+\rangle|\alpha_-\rangle + i\sin\psi|0,1\rangle|\alpha_+ e^{-i\varphi_+}\rangle|\alpha_- e^{-i\varphi_-}\rangle, \qquad (3)$$

where ψ is the phase which is created by the beam-splitter $BS1$; $\varphi_\pm$ is the phase shift of a single photon per each of the polarization components.

This transformation is useful for realizing of quantum logic operation XOR if it is assumed that the control qubit state is governed by the mode in the states b and c, in contrast with the target qubit state being controlled – by the state of the orthogonally polarized components pair a_+ and a_-. For the control qubit let us represent the logic states "0" and "1" in relation to the availability of a photon in one arm or the other, that is:

$$|0\rangle_C \equiv |1,0\rangle, \qquad |1\rangle_C \equiv |0,1\rangle. \qquad (4)$$

Thus, the beam-splitter $BS1$ forms control qubit as the Boolean states superposition $|0\rangle_C$ and $|1\rangle_C$:

$$|\Psi\rangle_C = \cos\psi|0\rangle_C + i\sin\psi|1\rangle_C. \qquad (5)$$

For the target qubit let us represent the logic states by two linearly polarized components of coherent radiation respectively, when $\eta = 0$, $\theta = 0$, $\varphi_+ = \varphi_- = \pi/2$:

$$|0\rangle_T \equiv \left|\frac{\alpha}{\sqrt{2}}\right\rangle_+ \left|\frac{\alpha}{\sqrt{2}}\right\rangle_-, \qquad |1\rangle_T \equiv \left|\frac{i\alpha}{\sqrt{2}}\right\rangle_+ \left|\frac{i\alpha}{\sqrt{2}}\right\rangle_-. \qquad (6)$$

If the coherent radiation has a large number of photons, that is when $|\alpha|^2 \gg 1$ (semiclassical limit), they can be considered orthogonal.

Thus, the control qubit "affects" the target qubit through the nonlinear medium; in its turn the target qubit can be as a linear superposition of two states $|0\rangle_T$ $|1\rangle_T$:

$$(\cos\psi|0\rangle_C + i\sin\psi|1\rangle_C)(C_1|0\rangle_T + C_2|1\rangle_T) \xrightarrow{U_{NL}}$$

$$\xrightarrow{U_{NL}} C_1\cos\psi|0\rangle_C|0\rangle_T + C_2\cos\psi|0\rangle_C|1\rangle_T +$$

$$+iC_1 \sin \psi |1\rangle_C |1\rangle_T + iC_2 \sin \psi |1\rangle_C |0\rangle_T, \tag{7}$$

$|C_1|^2 + |C_2|^2 = 1$. In expression (7) the imaginary unity can be eliminated if the element effecting the phase-shift by the angle $-\pi/2$ is placed in the upper arm of the interferometer.

Therefore, at the output of the nonlinear system the system state will be described by the right part of expression (11), which corresponds to the result of the quantum logic XOR operation.

Now let us show that the state $|\Psi\rangle_{NL}$ (3) allows us to develop a macroscopic superposition of the Schrödinger cat states at the output of the optical system – see Fig.1. After the beam-splitter $BS2$ we have for the state vector $|\Psi\rangle$:

$$|\Psi\rangle_{out} = |1,0\rangle \left(\cos \psi \cos \xi |\alpha_+\rangle |\alpha_-\rangle - \sin \psi \sin \xi |\alpha_+ e^{-i\varphi+}\rangle |\alpha_- e^{-i\varphi-}\rangle \right) +$$

$$+i|0,1\rangle \left(\cos \psi \sin \xi |\alpha_+\rangle |\alpha_-\rangle + \sin \psi \cos \xi |\alpha_+ e^{-i\varphi+}\rangle |\alpha_- e^{-i\varphi-}\rangle \right), \tag{8}$$

where ξ is the phase shift created by the beam-splitter $BS2$.

If a photon detector is available at each of the interferometer output ports (Fig.1), then after the registration of photon by one of these detectors it will be possible to speak about the selective measurement which is carried out with the system or, in other words, about the partial reduction of the wave function (8). In this case the state $|\Psi\rangle_{out}$ proceeds over to one of the following (unnormalized) states:

$$|\Psi\rangle_{D1} = \cos \psi \sin \xi |\alpha_+\rangle |\alpha_-\rangle + \sin \psi \cos \xi |\alpha_+ e^{-i\varphi+}\rangle |\alpha_- e^{-i\varphi-}\rangle, \tag{9a}$$
$$|\Psi\rangle_{D2} = \cos \psi \cos \xi |\alpha_+\rangle |\alpha_-\rangle - \sin \psi \sin \xi |\alpha_+ e^{-i\varphi+}\rangle |\alpha_- e^{-i\varphi-}\rangle, \tag{9b}$$

which are the macroscopic superposition of coherent states and determine the properties of polarization components $\alpha_\pm$ at the output of the nonlinear medium. These states are entangled, obviously. In fact, it is provided by correlation in the interferometer arms of the two polarization components α_+ and α_- of the coherent radiation.

Fig.2 shows the dependence of the photon number distribution function $W_{N,D1} \equiv |\langle 0,1|\langle N_+|\langle N_-|\Psi\rangle_{out}|^2$ via values N_+ for the different control parameters. It can be seen that when nonlinear medium properties (nonlinear phases φ_+ and φ_-) vary, photon statistics may be Poissonian (curve 1), and/or may have the view of "hump crests" which is the characteristics of superpositional Schrödinger cat states (curve 3), and/or as well may demonstrate strong oscillations (curve 2), being the characteristics of interference of two quantum states.

Thus, when detectors $D1$ and $D2$ are available, selective measurements carried out with considered optical system is characterized by the state vector $|\Psi\rangle_{out}$ in (8) enable to find nonclassical properties of the superposition of macroscopic polarization states (9) which are controlled by nonlinear interaction parameters.

3. Conclusion

Let's discuss shortly the possibilities of the experimental observation of the given states.

In the problem under discussion there is one important question connected with the media selection with a large (of the order of 1.5 rad) values of the nonlinear optical phase shift $\varphi_\pm$ per a photon. A great progress has been made in this area lately [8]. First, it's necessary to mention the effect of electromagnetically induced transparency which enables to get giant nonlinear cross-modulation effect for interacting waves. In the recent experiment with the cooling of sodium atoms in the gaseous state to the temperature of the order of few nK, the authors of the paper [9] managed to get the light group velocity of 17 m/s, the value of Shtark

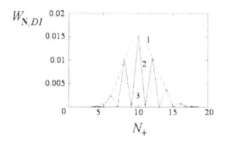

Figure 2. Dependence of the photon number distribution function $W_{\mathbf{N},D1}$ on the photon number N_+. The values of the control parameters of the problem are: $\eta = 0$, $n_0 = 21$; 1) $\psi = \xi = \pi/4$, $\varphi_+ = \varphi_- = \pi$, $N_- = N_+$; 2) $\psi = \xi = \pi/4$, $\varphi_+ = \varphi_- = \pi/2$, $N_- = N_+$; 3) $\psi = \pi/4$, $\xi = -\pi/4$, $\varphi_+ = \pi/8$, $\varphi_- = -\pi/8$, $N_- = n_0 - N_+$

shift is $1.3 \cdot 10^6$ rad/s, corresponding to value of the nonlinear refractive index of the order $0.18 \, \mathrm{cm}^2/\mathrm{W}$ (for the laser radiation intensity of $40 \, \mathrm{mW/cm}^2$). From the physical point of view it is spoken about the use of coherent Bose-Einstein condensate in Fig.1 as a nonlinear medium. This condensate interacts with the single -photon light pulses.

Acknowledgments

This work has been supported by the Russian Foundation for Basic Researches (grant No. 01-02-17478), within the framework of scientific programmes of the Ministry of Science, Industry and Technology, and also of the Ministry of Education of Russian Federation

References

[1] Cirac J I and Zoller P 2000 Nature 404 579; Briegel H.-J, Calarco T, Jaksch D, Cirac J I and Zoller P 2000 J.Mod.Opt. 47 415
[2] Turchette Q A, Hood C J, Lange W, Mabuchi H and Kimble H J 1995 Phys.Rev.Letts. 75, 4710
[3] Chuang I L and Yamamoto Y 1995 Phys.Rev.A 52 3489; Brandt H E 1998 Progress in Quantum Electronics 22 257
[4] Yurke B, McCall S L and Klauder J R 1986 Phys.Rev.A 33 4033
[5] Caves C M 1981 Phys.Rev.D 23 1693
[6] Alodjants A P and Arakelian S M 1999 Gravitation and Cosmology No.4 p.253
[7] Schleich W, Pernigo M and Fam Le Kien 1991 Phys.Rev.A 44 2172
[8] Schmidt H and Iamoglu A 1996 Opt.Lett. 21 1936; ibid 1998 Opt.Lett. 21 1007
[9] Hau L V, Harris S E, Dutton Z and Behroozi C H 1999 Nature 397 594

Inst. Phys. Conf. Ser. No 173: Section 7
Paper presented at 24th Int. Coll. Group Theoretical Methods in Physics, Paris, France, 15–20 July 2002
©*2003 IOP Publishing Ltd*

Semigroup techniques for the efficient classical simulation of optical quantum information

S D Bartlett

Department of Physics and Centre for Advanced Computing – Algorithms and Cryptography,
Macquarie University, Sydney, NSW 2109, Australia

Abstract. A framework to describe a broad class of physical operations (including unitary transformations, dissipation, noise, and measurement) in a quantum optics experiment is given. This framework provides a powerful tool for assessing the capabilities and limitations of performing quantum information processing tasks using current experimental techniques. The Gottesman-Knill theorem is generalized to the infinite-dimensional representations of the group stabilizer formalism and further generalized to include non-invertible semigroup transformations, providing a theorem for the efficient classical simulation of operations within this framework. As a result, we place powerful constraints on obtaining computational speedups using current techniques in quantum optics.

1. Introduction

Information processing using the rules of quantum mechanics may allow tasks that cannot be performed using classical laws [1]. The efficient factorization algorithm of Shor [2] and secure quantum cryptography [3] are two examples. Of the many possible realizations of quantum information processes, optical realizations have the advantange of negligible decoherence: light does not interact with itself, and thus a quantum state of light can be protected from becoming entangled with the environment. Several proposed optical schemes [4, 5, 6, 7] offer significant potential for quantum information processing.

In order to prove theorems regarding the possibilities and limitations of optical quantum computation, one must construct a framework for describing all types of physical processes (unitary transformations, projective measurements, interaction with a reservoir, etc.) that can be used by an experimentalist to perform quantum information processing. Most frameworks currently employed (e.g., [7]) are restricted to describing only unitary transformations. However, such transformations are a subset of all possible physical processes. Non-unitary transformations such as dissipation, noise, and measurement must also be described within a complete framework. The new results of Knill et al [5] show that photon counting measurements allow for operations that are "difficult" with unitary transformations alone; thus, non-unitary processes may be a powerful resource in quantum information processing and must be considered in any framework that attempts to address the capabilities of quantum computation with optics.

In this paper, we show that unitary transformations, measurements and any other physical process can be described in the unified formalism of completely positive (CP) maps. Also, a broad class of these maps which includes linear optics and squeezing transformations, noise processes, amplifiers, and measurements with feedforward that are typical to quantum optics experiments can be described within the framework of a *Gaussian semigroup*. This framework allows us to place limitations on the potential power of certain quantum information processing tasks.

One important goal is to identify classes of processes that can be efficiently simulated on a classical computer; such processes cannot possibly be used to provide any form of "quantum speedup". The Gottesman-Knill (GK) theorem [8, 1] for qubits and the CV classical simulatability theorems of Bartlett et al [9, 10] provide valuable tools for assessing the classical complexity of a quantum optical process. It is shown here that semigroup techniques provide a powerful formalism with which one can address issues of classical simulatability. In particular, a classical simulatability result is presented for a general class of quantum optical operations, and thus a no-go theorem for quantum computation with optics is proven using semigroup techniques.

2. Semigroup Description of Gaussian operations

Consider an optical quantum information process involving n coupled electromagnetic field modes, with each mode described as a quantum harmonic oscillator. The two observables for the (complex) amplitudes of a single field mode serve as canonical operators for this oscillator. A system of n coupled oscillators, then, carries an irreducible representation of the Heisenberg-Weyl algebra hw(n), spanned by the $2n$ canonical operators $\{q_i, p_i, i = 1, \ldots, n\}$ along with the identity operator I. These operators satisfy the commutation relations $[q_i, p_j] = i\hbar\delta_{ij}I$. We express the $2n$ canonical operators in the form of a phase space vector z with components $z_i = q_i$ and $z_{n+i} = p_i$ for $i = 1, \ldots, n$. These operators satisfy $[z_i, z_j] = i\hbar\Sigma_{ij}$, with Σ the skew-symmetric $2n \times 2n$ matrix

$$\Sigma = \begin{pmatrix} 0 & I_n \\ -I_n & 0 \end{pmatrix} \tag{1}$$

and I_n the $n \times n$ identity matrix. For a state ρ represented as a density matrix, the *means* of the canonical operators is a vector defined as the expectation values $\xi = \langle z \rangle_\rho$, and the *covariance matrix* is defined as

$$\Gamma = \langle (z - \xi)(z - \xi)^\dagger \rangle_\rho - i\Sigma. \tag{2}$$

A *Gaussian state* (a state whose Wigner function is Gaussian and thus possesses a quasiclassical description) is completely characterized by its means and covariance matrix [11]. Coherent states, squeezed states, and position- and momentum-eigenstates are all examples of Gaussian states.

We define $\mathcal{C}_n$ to be the group of linear transformations of the canonical operators $\{z_i\}$ [9]; this group corresponds to the infinite-dimensional (oscillator) representation of the "Clifford group" employed by Gottesman [8]. For a system of n oscillators, it is the unitary representation of the group ISp($2n, \mathbb{R}$) (the inhomogeneous linear symplectic group in $2n$ phase space coordinates) [12] which is the semi-direct product of phase-space translations (the Heisenberg-Weyl group HW(n)) plus one- and two-mode squeezing (the linear symplectic group Sp($2n, \mathbb{R}$)). Phase space displacements are generated by Hamiltonians that are linear in the canonical operators; a displacement operator $X(\alpha) \in$ HW(n) is defined by a real $2n$-vector α. A symplectic transformation $M(A) \in$ Sp($2n, \mathbb{R}$), with A a real matrix satisfying $A^\dagger \Sigma A = \Sigma$, is generated by a Hamiltonian that is a homogeneous quadratic polynomial in the canonical operators. A general element $C \in \mathcal{C}_n$ can be expressed as a product $C(\alpha, A) = X(\alpha)M(A)$, and transforms the canonical operators as

$$C(\alpha, A) : z \to z' = zA + \alpha. \tag{3}$$

The group $\mathcal{C}_n$ consists of unitary transformations that map Gaussian states to Gaussian states; however, unitary transformations do not describe all physical processes. In the following, we include other (non-unitary) CP maps that correspond to processes such as dissipation or measurement. We define the *Gaussian semigroup*, denoted $\mathcal{K}_n$, to be the set of Gaussian CP maps [11] on n modes: a Gaussian CP map takes any Gaussian state to a Gaussian state. Because Gaussian CP maps are closed under composition but are not necessarily invertible, they form a semigroup. A general element $T \in \mathcal{K}_n$ is defined by its action on the canonical operators as

$$T(\boldsymbol{\alpha}, A, G) : \boldsymbol{z} \to \boldsymbol{z}' = \boldsymbol{z}A + \boldsymbol{\alpha} + \boldsymbol{\eta}, \tag{4}$$

where $\boldsymbol{\alpha}$ is a real $2n$-vector, A and G are $2n \times 2n$ real matrices, and A is no longer required to be symplectic. Eq. (4) includes the transformations (3) plus additive noise processes [13] described by quantum stochastic noise operators (the vector $\boldsymbol{\eta}$) with expectation values equal to zero and covariance matrix

$$\langle \boldsymbol{\eta}\boldsymbol{\eta}^\dagger \rangle_{\rho_R} - i\Sigma = G - iA^\dagger \Sigma A. \tag{5}$$

Here, ρ_R is a Gaussian 'reservoir' state which, in order to define a CP map, must be chosen such that the noise operators satisfy the quantum uncertainty relations. This condition is satisfied if the noise operators define a positive definite density matrix, which leads to the condition

$$G + i\Sigma - iA^\dagger \Sigma A \geq 0. \tag{6}$$

The group $\mathcal{C}_n$ is recovered for $G = 0$.

The action of the Gaussian semigroup on the means and covariance matrix is straightforward and given by

$$T(\alpha, A, G) : \begin{cases} \boldsymbol{\xi} \to \boldsymbol{\xi}' = \boldsymbol{\xi}A + \boldsymbol{\alpha} \\ \Gamma \to \Gamma' = A^\dagger \Gamma A + G. \end{cases} \tag{7}$$

Because the means and covariance matrix completely define a Gaussian state, the resulting action of the Gaussian semigroup on Gaussian states can be easily calculated via this action.

The Gaussian semigroup $\mathcal{K}_n$ represents a broad framework to describe several important types of processes in a quantum optical circuit. The group $\mathcal{C}_n \subset \mathcal{K}_n$ comprises the unitary transformations describing phase–space displacements and squeezing (both one– and two–mode). Introduction of noise to the circuit (e.g., via linear amplification) is also in $\mathcal{K}_n$. Furthermore, the Gaussian semigroup describes certain measurements in the quantum circuit. These include measurements where the outcome is discarded (thus evolving the system to a mixed state) or retained (where the system follows a specific quantum trajectory defined by the measurement record [14]). Finally, the Gaussian semigroup includes Gaussian CP maps conditioned on the outcome of such measurements. For details and examples of all of these types of Gaussian semigroup transformations, see [10].

3. Classical Simulation of Gaussian Semigroup Processes

Using the framework of the Gaussian semigroup, it is straightforward to prove the classical simulatability result of Bartlett and Sanders [10].

Theorem: Any quantum information process that initiates in a Gaussian state and that performs only Gaussian semigroup maps can be *efficiently* simulated using a classical computer.

Proof: Recall that any Gaussian state is completely characterized by its means and covariance matrix. For any quantum information process that initiates in a Gaussian state and involves only Gaussian semigroup maps, one can follow the evolution of the means and the covariance matrix rather than the quantum state itself. For a system of n coupled oscillators, there are

$2n$ independent means and $2n^2 + n$ elements in the (symmetric) covariance matrix; thus, following the evolution of these values requires resources that are polynomial in the number of coupled systems.   *QED*

Because most current experimental techniques in quantum optics are describable by Gaussian semigroup maps, this theorem places a powerful constraint on the capability of achieving quantum computational speedups (tasks that are not efficient on any classical machine) using quantum optics.

4. Conclusions

Semigroup techniques provide a powerful tool for constructing and assessing new quantum information protocols using quantum optics. These techniques have been used to show that algorithms or circuits consisting of only Gaussian semigroup maps can be efficiently simulated on a classical computer, and thus do not provide the ability to perform quantum information processing tasks efficiently that cannot be performed efficiently on a classical machine. Eisert et al [15] use related techniques to show that local Gaussian semigroup transformations are insufficient for distilling entanglement: an important process for quantum communication and distributed quantum computing. Most current quantum optics experiments consist only of Gaussian semigroup transformations; thus, the challenge is to exploit this semigroup to prove new theorems, limitations and possibilities for quantum information processing using optics.

Acknowledgments

This project has been supported by Macquarie University and the Australian Research Council. The author thanks B. C. Sanders for helpful discussions.

References

[1] Nielsen M A and Chuang I L 2000 Quantum Computation and Quantum Information (Cambridge: Cambridge University Press)
[2] Shor P W 1994 Proceedings, 35th Annual Symposium on Foundations of Computer Science (Los Alamitos: IEEE Press)
[3] Bennett C H and Brassard G 1984 Proceedings of IEEE International Conference on Computers, Systems and Signal Processing (New York: IEEE Press) p 175
[4] Chuang I L and Yamamoto Y 1995 Phys. Rev. A 52, 3489
[5] Knill E et al 2001 Nature (London) 409, 46
[6] Gottesman D et al 2001 Phys. Rev. A 64 012310
[7] Lloyd S and Braunstein S L 1999 Phys. Rev. Lett. 82, 1784
[8] Gottesman D 1999 Proceedings of the XXII International Colloquium on Group Theoretical Methods in Physics, eds. Corney S P et al (Cambridge: International Press) p 32
[9] Bartlett S D et al 2002 Phys. Rev. Lett. 88 097904
[10] Bartlett S D and Sanders B C 2002 Efficient classical simulation of optical quantum circuits *Preprint* quant-ph/0204065
[11] Lindblad G 2000 J. Phys. A: Math. Gen. 33, 5059
[12] Wünsche A 2002 J. Opt. B: Quantum Semiclass. Opt. 4, 1
[13] Gardiner C W and Zoller P 2000 Quantum Noise (2nd Ed.) (Berlin: Springer-Verlag)
[14] Carmichael H J 1993 An Open Systems Approach to Quantum Optics (Berlin: Springer)
[15] Eisert J et al 2002 Phys. Rev. Lett. 89

Inst. Phys. Conf. Ser. No 173: Section 7
Paper presented at 24th Int. Coll. Group Theoretical Methods in Physics, Paris, France, 15–20 July 2002
©2003 IOP Publishing Ltd

Groups and Semigroups in Scattering Theory – Poincaré Transformations and Decaying States

Arno Böhm[a] with H. Kaldass and S. Wickramasekara

[a] Department of Physics, University of Texas at Austin, Austin, Texas 78712

1. Introduction

In this talk we discuss how to combine aspects of the theory of stable relativistic particles and of the theory of non-relativistic decaying particles into a relativistic theory of unstable particles and resonances. For the theory of stable relativistic particles we use Wigner's description by unitary irreducible representations (UIR) of the Poincaré group [1] which has been the foundation for relativistic quantum mechanics and field theory [2]. For the non-relativistic decaying particles we use a time asymmetric quantum mechanics in which the Hilbert space axiom, that allows only symmetric (unitary group) time evolution, is replaced by time asymmetric boundary conditions which lead to an irreversible, semigroup time evolution [3]. In the limit in which certain contributions to the energy continuum are neglected, this new quantum mechanics leads to the Weisskopf-Wigner approximate methods [4] for decay, and to the Breit-Wigner energy distribution [5] for the resonance lineshape. We thus generalize and combine two fundamental contributions of Wigner into a description of relativistic resonances and decaying states. The central concept of this description is the relativistic Gamow vector, which is a generalized eigenvector (ket) of the *self-adjoint* invariant square mass operator $M^2 = P_\mu P^\mu$ ($P^\mu = P_1^\mu + P_2^\mu$ in resonance scattering $1+2 \rightarrow R \rightarrow 3+4$) with a *complex* eigenvalue, a relativistic Breit-Wigner energy distribution, and an exponential time evolution governed by a semigroup of continuous operators in the space of kets.

The relativistic Gamow state vector enables us to give a unique, unambiguous definition for the mass and width of a relativistic resonance. This problem is best illustrated in the case of the Z-boson:

The Review of Particle Properties [6] gives two definitions of the mass and width of the Z-boson and lists two different values which are obtained from fitting two different formulas for the lineshape to the same experimental data. The value M_Z obtained from the fit to the "relativistic Breit-Wigner with energy dependent width" of the on-shell renormalization scheme is

$$M_Z = (91.1871 \pm 0.0021)\,\text{GeV}; \quad \Gamma_Z = (2.4945 \pm 0.0024)\,\text{GeV}.$$

Another value $\bar{M}_Z$ is obtained from the "relativistic Breit-Wigner of the S-matrix pole". M_Z and $\bar{M}_Z$ differ from each other by about 10 times the experimental error [6].

$$\bar{M}_Z = M_Z - 34\text{MeV} = M_Z - 10 \times \Delta M_{exp}.$$

An obvious question is: What is the right definition of the Z-boson mass and width and what are their numerical values? The semigroup representations of the Poincaré transformations furnished by the relativistic Gamow vectors, described in Section 3 below, suggest a third definition of the real mass M_R and the width Γ_R which gives

$$M_R = (91.1626 \pm 0.0031)\,\text{GeV} = M_Z - 0.026\,\text{GeV}; \quad \Gamma_R = (2.4934 \pm 0.0024)\,\text{GeV}.$$

2. Non-Relativistic Gamow Vectors

In the Weisskopf-Wigner approximate method of decaying states [4], an approximate exponential decay law is obtained for the decay rate of an unstable state R into a decay channel η, $R \to \eta$:

$$\dot{\mathscr{P}}_\eta(t) \approx \Gamma_\eta e^{-\Gamma t}, \tag{1}$$

where $\dot{\mathscr{P}}_\eta(t)$ denotes the decay rate and $\mathscr{P}_\eta(t)$, the decay probability of $R \to \eta$. The above equation defines the lifetime τ of the unstable state as $\tau = \hbar/\Gamma$. The well known deviation from the exact exponential decay law (1) is a mathematical consequence of Hilbert space quantum theory [7] and need not be a feature of other mathematical theories to which the Weisskopf-Wigner scheme is an approximation, especially if, as we do, one admits generalized vectors like the Dirac kets.

Another manifestation of a quasistable state besides the exponential decay law (1) is the phenomenon of resonances. The signature of a resonance is a Breit-Wigner energy distribution. The j-th partial wave amplitude for non relativistic resonance scattering $\eta \to R \to \eta$ takes the form

$$a_j(E) = \frac{\frac{1}{2}\Gamma_\eta}{E - (E_R - i\Gamma_R/2)} + B(E) \tag{2}$$

where the first term is the Breit-Wigner energy distribution, and $B(E)$ is a slowly varying background. Equation (2) defines the resonance energy E_R and the width Γ_R.

With decaying states being described by an exponential decay law (1), and resonances by a Breit-Wigner amplitude (2), it is natural to ask whether resonances and decaying states are two different phenomena or two different manifestations of the same entity. Related to this is the question: Do Γ in (1) and Γ_R in (2) have anything to do with each other, e.g., does the equality

$$\Gamma_R = \Gamma(= \hbar/\tau = \text{inverse lifetime}) \tag{3}$$

hold exactly, approximately or not at all? This is a particularly relevant question for the relativistic case where one either measures τ (for $\frac{\Gamma}{M} \sim 10^{-10}$) or Γ_R (for $\frac{\Gamma_R}{M_R} \sim 10^{-1}$) and does not come close in accuracy to an experimental test of (3).

To answer this question, a state vector for the decaying state ψ^G needs to be defined and used to calculate the decay probabilities $\mathscr{P}_\eta(t)$ and rates $\dot{\mathscr{P}}_\eta(t)$ from the fundamental principles of quantum theory. This means that (1) should be obtained by calculating the Born probabilities for the decay products (observables) in the state ψ^G:

$$\mathscr{P}_\eta(t) = \text{Tr}(\Lambda_\eta |\psi^G\rangle\langle\psi^G|) = \left(|\langle\psi_\eta^-|\psi^G\rangle|^2 \quad \text{if} \quad \Lambda_\eta = |\psi_\eta^-\rangle\langle\psi_\eta^-|\right) \tag{4}$$

Here

$$\Lambda_\eta = |\psi_\eta^-\rangle\langle\psi_\eta^-| \tag{5}$$

is the projection operator on the space of decay products η (decay channel η), and the decay rate should be obtained from the Born probability by differentiation:

$$\dot{\mathscr{P}}_\eta(t) = \frac{d\mathscr{P}_\eta(t)}{dt}. \tag{6}$$

In the non-relativistic theory one uses for ψ^G the non-relativistic Gamow vectors [3].

The non-relativistic Gamow vectors were obtained [3] from resonance poles of the S-matrix. They are kets with complex energy $E_R - i\Gamma_R/2$, and have an integral representation in terms of the Lippmann-Schwinger kets $|E,b^-\rangle$ (out-plane wave states) given by:

$$\psi^G = |E_R - i\Gamma_R/2, b^-\rangle \equiv \frac{i}{2\pi} \int_{-\infty_{II}}^{+\infty} dE \frac{|E,b^-\rangle}{E - (E_R - i\Gamma_R/2)} \tag{7}$$

where $-\infty_{II}$ indicates that the integration along the negative semi-axis is in the second Riemann sheet of the complex energy surface.

The Gamow vectors can be given a rigorous mathematical meaning within a Rigged Hilbert Space [8] theory of scattering [3]. This theory, unlike the conventional theory in Hilbert space, distinguishes meticulously between the in-states defined by the preparation apparatus (detector), $\phi^+ \in \Phi_-$, and the out-observables (usually called out-states) defined by the registration apparatus, $\psi^- \in \Phi_+$. The deviation from conventional Hilbert space theory is that the space of observables Φ_+ and the space of states Φ_- are *distinct* dense subspaces of the same Hilbert space $\mathcal{H}$. This is in contrast to the standard theory where $\Phi_+ = \Phi_- (= \mathcal{H})$. We mention here while omitting the mathematical details that $\Phi_\pm$ are postulated to be Hardy spaces [11] of the upper and lower complex half-plane of the second sheet of the S-matrix. Then, the space of in-states ϕ^+ is spanned by the Lippmann-Schwinger kets $|E,b^+\rangle$ and the space of observables ψ^- is spanned by the Lippmann-Schwinger kets $|E,b^-\rangle$ which are usually given as, respectively, the in and out solutions of the Lippmann-Schwinger equations [9]. The precise mathematical characterization of the Lippmann-Schwinger kets in our theory is as antilinear functionals over $\Phi_\pm$: $|E,b^\mp\rangle \in \Phi_\pm^\times$. Therefore, the mathematical arena for this new quantum theory is not the Hilbert space but a pair of Rigged Hilbert Spaces [10]:

$$\Phi_+ \subset \mathcal{H} \subset \Phi_+^\times \quad \text{for observables or "out} - \text{states" defined by the detector} \tag{8a}$$

$$\Phi_- \subset \mathcal{H} \subset \Phi_-^\times \quad \text{for prepared in} - \text{states defined by the preparation apparatus} \tag{8b}$$

where $\Phi_+(\Phi_-)$ is the Hardy space of the upper (lower) complex plane. This means that the energy wave functions $\langle^+E|\phi^+\rangle (\langle^-E|\psi^-\rangle)$ constitute the space of well behaved Hardy functions [11], i.e., smooth, rapidly decreasing functions with analytic extensions into the open upper (lower) complex half plane.

Like the Lippmann-Schwinger kets $|E,b^-\rangle$, the Gamow vectors (7) are antilinear functionals over the space of observables Φ_+: $\psi^G \in \Phi_+^\times$. Moreover, they are generalized eigenvectors of the Hamiltonian with the complex eigenvalue $E_R - i\Gamma_R/2$, i.e.,

$$\langle H\psi_\eta^- | E_R - i\Gamma_R/2^- \rangle = \langle \psi_\eta^- | H^\times | E_R - i\Gamma_R/2^- \rangle$$
$$= (E_R - i\Gamma_R/2)\langle \psi_\eta^- | E_R - i\Gamma_R/2^- \rangle \tag{9}$$
$$\text{for all } \psi_\eta^- \in \Phi_+$$

In (9), $H^\times$ is the conjugate operator of the self-adjoint and semibounded Hamiltonian H, $H^\times \supset H^\dagger$. The adequacy of the Rigged Hilbert Space theory of resonances is further demonstrated by the fact that these state vectors with complex energy are obtained without dropping the self-adjointness of the Hamiltonian. The initial motivation for the definition of Gamow vectors (7) and the construction of the RHS of Hardy class (8) was to obtain vectors associated with the S-matrix pole which have the property (9) and from which it was hoped to obtain the exponential decay law [3].

A remarkable, unexpected property of the Gamow vectors is that not only do they lead to an exponential time evolution but also this evolution is given by a semigroup. Precisely, one can calculate that the Gamow vectors are generalized eigenvectors of the time evolution operator $e^{-iH^\times t}$ *only* for $t \geq t_0(=0)$, where $t_0 = 0$ is interpreted as the time at which state containing the resonance has been prepared and the registration (counting) of the decay products can begin.

$$\langle e^{iHt}\psi_\eta^- | E_R - i\Gamma_R/2^- \rangle = \langle \psi_\eta^- | e^{-iH^\times t} | E_R - i\Gamma_R/2^- \rangle$$
$$= e^{-iE_R t} e^{-\Gamma_R t/2} \langle \psi_\eta^- | E_R - i\Gamma_R/2^- \rangle \tag{10}$$
$$\text{for } t \geq 0 \text{ only}.$$

Omitting the arbitrary ψ_η^- in (10), we write this as the functional equation for kets:

$$e^{-iH^\times t} | E_R - i\Gamma_R/2^- \rangle = e^{-iE_R t} e^{-\Gamma_R t/2} | E_R - i\Gamma_R/2^- \rangle \tag{11}$$
$$\text{for } t \geq 0 \text{ only}.$$

The semigroup time evolution of Gamow vectors representing exponentially decaying states is a manifestation of a fundamental time asymmetry in quantum physics. It is mathematically described by the boundary conditions (8) for the state vectors, $\phi^+ \in \Phi_-$, and the observable vectors, $\psi^- \in \Phi_+$.

From (11), it follows that the width Γ_R of the Breit-Wigner amplitude (2) is the same as the exponential decay constant Γ of (1)

$$\Gamma_R = \Gamma \tag{12a}$$

and hence for the Gamow state vector ψ^G we calculate the exact relation between lifetime τ and width Γ_R:

$$\tau = \frac{1}{\Gamma_R} \tag{12b}$$

Thus the Gamow vector describes a state with a Breit-Wigner energy distribution and an exponential time evolution which is observed as a resonance in the scattering amplitude (2) and as an exponentially decaying particle in the decay rate (1).

3. Relativistic Gamow Vectors

To extend the Gamow vector into the relativistic domain we start with Wigner's irreducible unitary representations (URI) of the Poincaré group [1]. The UIR of the Poincaré group provide the definition for elementary relativistic stable states. These UIR are characterized by two quantum numbers: m^2 and j which are interpreted as the quantum numbers the mass square and the spin of the relativistic stable particle. For the representation spaces $[j, m^2]$ one uses the basis vectors of $|[j, m^2], b\rangle$ where b denote the additional quantum numbers, for which one has various choices depending upon the complete set of commuting observables that one takes. For Wigner's canonical basis system the choice for b is the momentum and a component of the spin $\{\vec{p}, j_3\}$ [2]. One could as well choose $\{\vec{\hat{p}}, j_3\}$ [12], with $\vec{\hat{p}} = \vec{p}/m$ being the spatial components of the 4-velocity $\hat{p} = p/m$. The 4-velocity is the preferred choice for the derivation of the relativistic Gamow vectors for reasons specified below. A relativistic stable particle state characterized by m^2 and j, $f_{[j, m^2]}$, is the continuous superposition of the basis vectors $|[j, m^2], b\rangle$ with some measure μ:

$$f_{[j, m^2]} = \int d\mu(b) |[j, m^2], b\rangle f(b) \tag{12}$$

In (12), $f(b)$ is a well behaved (Schwartz) function of b, $f(b) \in \mathcal{S}(R^3)$, and the Lorentz invariant measure, $d\mu(b) = d^3\hat{p}/2\hat{p}^0$, is chosen. The integration in (12) is understood to comprise summation over all discrete degeneracy quantum numbers (e.g., j_3) encapsulated in b. The same applies to (13) and (18) below.

In a scattering process of two incoming particles, 1 and 2, and two outgoing particles, 3 and 4:

$$1+2 \rightarrow 3+4,$$

the system of out-observable vectors $\psi_3 \times \psi_4 \equiv \psi^-$, can be expressed in terms of "out-states" basis vectors, which span a two-particle irreducible representation of the Poincaré group. From the direct product basis of the two particle space one obtains new basis vectors using the Clebsch-Gordan coefficients of the Poincaré group [12]. These basis vectors are labelled by the total invariant mass square $s = (p_3 + p_4)^2$ and the total angular momentum j of the 3,4 system and again by the labels b and some other degeneracy labels which we do not want to discuss here (and which are irrelevant for two spinless in- and out- particles). Thus, these basis vectors are denoted by $|[j,s],b^-\rangle$ with b denoting the spatial component of the 4-velocity of the two particle system, $\vec{\hat{p}} = (\vec{p}_3 + \vec{p}_4)/\sqrt{s}$, the third component of the total angular momentum, j_3 (and possibly other discrete degeneracy quantum numbers). An out-observable with a fixed value for s and j, $\psi_{[j,s]}^-$, can thus be expanded in analogy to (12) in terms of $|[j,s],b^-\rangle$ as

$$\psi_{[j,s]}^- = \int d\mu(b)|[j,s],b^-\rangle \psi(b), \tag{13}$$

The general out-observable vector $\psi^- = \psi_3 \times \psi_4$ is the linear combination of the $\psi_{[j,s]}^-$:

$$\psi^- = \sum_j \int ds \psi_{[j,s]}^- \psi_j^-(s) = \sum_j \int ds \int d\mu(b)|[j,s],b^-\rangle \psi^-(s,b) \tag{14}$$

Here $d\mu(b) = d^3\hat{p}/2\hat{p}^0$, $\psi(b)$ are Schwartz functions of $b = \vec{\hat{p}}$, and $\psi_j^-(s)$ are very well behaved Hardy functions in the lower half plane [13, 14]. In analogy to the non-relativistic $|E,b^-\rangle$, the $|[j,s],b^-\rangle$ are antilinear functionals on the space of observables Φ_+, i.e., $|[j,s],b^-\rangle \in \Phi_+^\times$. These are the kets which one usually assumes to fulfill the Lippmann-Schwinger equation. Because resonances appear in one partial wave with a fixed value of j, we consider in (14) only the term with that value of j.

To obtain the relativistic Gamow vector [14], we start from the relativistic Breit-Wigner of the S-matrix pole

$$a_j^{BW}(s) = \frac{R}{s - s_R} = \frac{R}{s - \bar{M}_Z^2 + i\bar{M}_Z\bar{\Gamma}_Z} = \frac{R}{s - (M_R - i\frac{\Gamma_R}{2})^2}, \tag{15}$$

and define decaying Gamow kets in terms of Lippmann-Schwinger kets $|[j,s],b^-\rangle \in \Phi_+^\times$. According to the new hypothesis (8a), for every $\psi^- \in \Phi_+$ the complex conjugate of the wave function $\overline{\psi^-(s)} = \overline{\langle -b,[j,s]|\psi^-\rangle} = \langle \psi^-|[j,s],b^-\rangle$ is a very well behaved Hardy function in the lower half energy plane of the second sheet of the S-matrix. Therefore, one can take convolutions of these functions with a Cauchy kernel along a path in the lower half plane and define:

$$\langle \psi^-|[j,s_R],b^-\rangle \equiv -\frac{i}{2\pi} \oint ds \langle \psi^-|[j,s],b^-\rangle \frac{1}{s - s_R}$$

$$= \frac{i}{2\pi} \int_{-\infty_{II}}^{+\infty} ds \langle \psi^-|[j,s],b^-\rangle \frac{1}{s - s_R}. \tag{16}$$

The first equality is the well known Cauchy's theorem and the second is Titchmarsh's theorem for Hardy functions. Since the above equation is valid for all $\psi^- \in \Phi_+$ representing out-observables η (decay products of $R \to \eta$), we can omit ψ^- and obtain the integral representation of the Gamow vectors:

$$|[j,\mathsf{s}_R],b^-\rangle = \frac{i}{2\pi} \int_{-\infty_{II}}^{\infty} ds \frac{|[j,\mathsf{s}],b^-\rangle}{\mathsf{s}-\mathsf{s}_R}. \tag{17}$$

In (15), (16), and (17), s_R is the resonance pole position of the analytically continued relativistic S-matrix $S_j(\mathsf{s})$ on the second sheet, and $(\bar{M}_Z, \bar{\Gamma}_Z)$ and (M_R, Γ_R) in (15) are different parameterizations of s_R. The 4-velocity $\hat{p}$ in (16) and (17) is required to be real also for complex values of s. While we analytically extend the Lippmann-Schwinger kets $|[j,\mathsf{s}],\vec{p}\,\rangle$ in the variable s from its physical values $(m_1+m_2)^2 \leq \mathsf{s} < \infty$ into the complex lower half plane and in particular to the s_R, we want to keep the degeneracy quantum numbers b for the basis kets fixed and real. This we cannot do for $b = \vec{p}$ because for complex s, $\vec{p}$ will also have to be complex. However, we can this do for $b = \vec{\hat{p}}$, and therefore we choose as basis vectors $|[j,\mathsf{s}],\vec{\hat{p}}\rangle$ and obtain the representation of the Poincaré transformations with "minimally complex" momentum $\vec{p} = \sqrt{\mathsf{s}_R}\vec{\hat{p}}$.

The relativistic Gamow vectors describing the state associated to the pole at s_R are, in analogy to (13), the continuous superpositions of the basis vectors $|[j,\mathsf{s}_R],b^-\rangle$:

$$\psi_{[j,\mathsf{s}_R]}^G = \int d\mu(b)|[j,\mathsf{s}_R],b^-\rangle \psi(b) \tag{18}$$

where $\psi(b) \in \mathscr{S}(R^3)$, and $-j \leq j_3 \leq j$. The space of Gamow vectors $\{\psi_{[j,\mathsf{s}_R]}^G|$ for all $\psi(b) \in \mathscr{S}(R^3)\} = \Phi_+^\times([j,\mathsf{s}_R])$ is the vector space which corresponds to the exact (i.e., with $-\infty_{II} < \mathsf{s} < \infty$) Breit-Wigner amplitude:

$$a_j^{BW_i}(\mathsf{s}) = \frac{R_i}{\mathsf{s}-\mathsf{s}_R}, \quad -\infty_{II} < \mathsf{s} < \infty. \tag{19}$$

This space $\Phi_+^\times([j,\mathsf{s}_R])$ is a representation space of an irreducible representation $[j,\mathsf{s}_R]$ of the Poincaré *semigroup* (Λ,x) of proper orthochronous Lorentz transformations and space-time translations into the forward light cone [15]:

$$\mathscr{P}_+ = \{(\Lambda,x)|\Lambda \in \overline{SO(3,1)}, \det\Lambda = +1, \Lambda_0^0 \geq +1, x^2 \equiv t^2 - \vec{x}^2 \geq 0, t \geq 0\}. \tag{20}$$

These "causal" (forward) Poincaré Semigroup representations $[j,\mathsf{s}_R]$ are characterized by:

(i) Spin parity j given by the j-th partial wave amplitude $a_j(\mathsf{s}) = a_j^{BW}(\mathsf{s}) + B_j(\mathsf{s})$ (or the j-th partial S-matrix $S_j(\mathsf{s}) = 2ia_j(\mathsf{s}) + 1$).

(ii) Complex mass square s_R (with $\text{Im}\,\mathsf{s}_R < 0$) given by the resonance pole position on the 2nd sheet of $S_j(\mathsf{s})$ (and the requirement of minimal complexity $\hat{p} \equiv p/\sqrt{\mathsf{s}} = $ real).

The Gamow kets (17) are generalized eigenvectors of the full Hamiltonian $P^0 = H = H^{\text{free}} + V$:

$$H^\times|[j,\mathsf{s}_R],\vec{\hat{p}} = \vec{0}, j_3^-\rangle = \sqrt{\mathsf{s}_R}|[j,\mathsf{s}_R],\vec{\hat{p}} = \vec{0}, j_3^-\rangle \tag{21}$$

and of the momentum operators $P^i(i = 1,2,3)$:

$$P^i|[j,\mathsf{s}_R],\vec{\hat{p}}, j_3^-\rangle = \sqrt{\mathsf{s}_R}\hat{p}^i|[j,\mathsf{s}_R],\vec{\hat{p}}, j_3^-\rangle \tag{22}$$

Moreover, the space $\Phi_+^\times([j,\mathsf{s}_R])$ is an eigenspace of the total invariant mass square operator $P_\mu P^\mu$, $P^\mu = P_1^\mu + P_2^\mu$

$$(P_\mu P^\mu)^\times|\psi_{[j,\mathsf{s}_R]}^G\rangle = \mathsf{s}_R|\psi_{[j,\mathsf{s}_R]}^G\rangle \tag{23}$$

The transformation of $|[j,s_R],\vec{p},j_3^-\rangle$ under $(\Lambda,x)\in\mathscr{P}_+$ are found to be given by [15]:

$$U^\times(\Lambda,x)|[j,s_R],\vec{p},j_3^-\rangle =$$
$$e^{-i\gamma\sqrt{s_R}(t-\vec{x}.\vec{v})}\sum_{j_3'}D_{j_3'j_3}^j(W(\Lambda^{-1},\hat{p}))|[j,s_R],\Lambda^{-1}\vec{p},j_3'^-\rangle \tag{24}$$

only for $t\geq 0$

where $\vec{v}$ is the three-velocity, $\vec{p}=\gamma\vec{v}$, and $\gamma=1/\sqrt{1-\vec{v}^2}=\sqrt{1+\vec{p}^2}=\hat{p}^0$, and where $W(\Lambda,\hat{p})=L^{-1}(\Lambda\hat{p})\Lambda L(\hat{p})$ is the Wigner rotation. $L(\hat{p})$ and therewith $W(\Lambda,p)$ depends upon $\hat{p}$ not upon the momentum $p=\sqrt{s}\hat{p}$. It is this property that allows us to construct the representations $[j,s_R]$ by analytic continuation of the Lippmann-Schwinger kets to the Gamow kets

$$|[j,s],\vec{p},j_3^-\rangle \rightarrow |[j,s_R],\vec{p},j_3^-\rangle \tag{25}$$

in such a way that $\hat{p}$ remain unaffected and always real. We will call these representations $[j,s_R]$ the "minimally complex" representations. In these representations $[j,s_R]$ the homogeneous Lorentz transformations $(\Lambda,0)$ are represented in the usual way as for the UIR $[j,m^2]$.

We note that the time asymmetric transformation (24) has its origin in the fact that the Lippmann-Schwinger scattering states, $|[j,s],\vec{p},j_3^-\rangle\in\Phi_+^\times$ ("out-plane waves"), furnish a representation for the semigroup $\mathscr{P}_+$ [15]:

$$U^\times(\Lambda,x)|[j,s],\vec{p},j_3^-\rangle = e^{-ip.x}\sum_{j_3'}D_{j_3'j_3}^j(W(\Lambda^{-1},p))|[j,s],\Lambda^{-1}\vec{p},j_3'^-\rangle$$

only for $t\geq 0$ \tag{26}

in contrast to the one-particle free states $|[j,m^2],\vec{p},j_3\rangle\in\Phi^\times\supset\mathscr{H}(m^2,j)$ which furnish a representation for the whole Poincaré group:

$$U^\times(\Lambda,x)|[j,m^2],\vec{p},j_3\rangle = e^{-ip.x}\sum_{j_3'}D_{j_3'j_3}^j(W(\Lambda^{-1},\hat{p}))|[j,m^2],\Lambda^{-1}p,j_3'\rangle$$

for $-\infty<t<+\infty$ \tag{27}

This is the well known transformation of the Wigner basis vectors of the UIR space $\mathscr{H}(j,m^2)$. Here, the operator $U^\times(\Lambda,x)$ is the extension of $U^\dagger(\Lambda,x)=U(\Lambda^{-1},-\Lambda^{-1}x)$ to the space $\Phi^\times\supset\mathscr{H}(j,m^2)\supset\Phi$, and Φ is the completion of the space of differentiable vectors $\mathscr{D}\subset\mathscr{H}(j,m^2)$ with respect to the countable norm topology given by the Nelson operator [16]. This is in contrast to (26) where $U^\times(\Lambda,x)\supset U^\dagger(\Lambda,x)$ is the extension of the unitary $U^\dagger$ to the space $\Phi_+^\times$ of (8) [13].

For the special case of the time evolution of the Gamow vectors in the rest frame, (24) reduces to:

$$|\psi_{s_R}^G(t)^-\rangle \equiv e^{-iH^\times t}|[j,s_R],b_{rest}^-\rangle = e^{-i\sqrt{s_R}t}|[j,s_R],b_{rest}^-\rangle$$
$$= e^{-iM_Rt}e^{-\Gamma_Rt/2}|[j,s_R],b_{rest}^-\rangle$$

for $t\geq 0$ only \tag{28}

This is the analogue of the non-relativistic equation (11).

¿From the evolution equation (28) follows that the lifetime of the particle represented by the relativistic Gamow vector is given by $\tau=\frac{1}{\Gamma_R}$, where the width $\Gamma_R=-2\text{Im}\sqrt{s_R}$ and s_R of (15) is the pole position of the relativistic S-matrix. This means the lifetime of the relativistic

Gamow states is not $\frac{1}{\bar{\Gamma}_Z}$ or any other $\frac{1}{\Gamma_Z}$ defined by any other "relativistic Breit-Wigner" than (15), but it is $\frac{1}{\Gamma_R}$ given by the second parameterization in (15).

Returning to the problem of the definition of the mass and width of the Z-boson mentioned in the introduction, we conclude from (28) that if we want (12b) to hold also for the Z-boson, then the "width" of the relativistic particle should be defined by Γ_R of (15). The real resonance mass is then [17]

$$M_R = \mathrm{Re}\sqrt{s_R}\,(= 91.1626 \pm .0031\,\mathrm{GeV} \text{ for the } Z-\text{boson})$$

and not the peak position $\bar{M}_Z = M_R\sqrt{1 - 1/4(\Gamma_R/M_R)^2}$ of the relativistic Breit-Wigner or any other parameterization of the lineshape used and reported in the Review of Particle Properties [6].

4. Summary

We have reported some results of a mathematical theory that describes time symmetric as well as time asymmetric, non-relativistic as well as relativistic, quantum physics. This theory is an extension of the Rigged Hilbert Space formulation of quantum mechanics which about 1965 gave a mathematical justification to Dirac's kets and Dirac's continuous basis vector expansion [18]. At that time one chose one RHS $\Phi \subset \mathcal{H} \subset \Phi^\times$ where Φ was usually taken to be a Schwartz space. One did not assign different spaces to the states ϕ and observables ψ but assumed $\{\phi\} = \{\psi\} = \Phi$. Time asymmetry was not an issue because one was not concerned with decaying states, and for scattering one assumed asymptotic completeness, meaning that the Lippmann-Schwinger kets with $i\varepsilon$ and $-i\varepsilon$ span the same space Φ

In order to describe decay, Hardy functions were needed and the in- and out- wave functions had to have different analyticity properties. This led to two RHS's of Hardy class with complementary analyticity properties:

$$\Phi_\pm \subset \mathcal{H} \subset \Phi_\pm^\times \quad \text{for} \left\{ \begin{array}{c} \text{observables} \\ \text{states} \end{array} \right\} \tag{29}$$

and to the distinction between states and observables, and therewith to time asymmetry.

The dual spaces $\Phi_\pm^\times$ contain, besides the Dirac-Lippmann-Schwinger kets $|[j,s],b^\mp\rangle \in \Phi_\pm^\times \left(\left(\begin{array}{c} \text{out}- \\ \text{in}+ \end{array} \right) \text{plane waves} \right)$, also Gamow kets $|[j,s_R],b^-\rangle \in \Phi_+^\times$ which are associated to the resonance pole at $s_R = (\bar{M}_Z^2 - i\bar{M}_Z\bar{\Gamma}_Z) = (M_R - i\Gamma_R/2)^2$ of the S-matrix. They have the properties required of decaying states.

Relativistic Gamow kets are the basis vectors of a semigroup representation of causal Poincaré transformations characterized by $[j,s_R]$ representing spin j and complex mass $\sqrt{s_R} = M_R - i\frac{\Gamma_R}{2}$. ¿From the exponential time evolution *derived* for the relativistic Gamow vector it follows that the "width" of the relativistic Breit-Wigner Γ_R (not $\bar{\Gamma}_Z$, Γ_Z or any other Γ) is the inverse lifetime $\tau = \hbar/\Gamma_R$. The semigroup character of the representation $[j,s_R]$, expressed by (24) and for the time in the rest frame by (28), is a manifestation of time asymmetry in relativistic quantum theory.

Acknowledgements

This contribution is based on a lecture by one of us (AB) at the XXIV International Colloquium on Group Theoretical Methods in Physics, Paris, 2002. We would like to thank the organizers for their hospitality and the Welch foundation for its support.

[1] E. P. Wigner, Ann. Math. (2) **40**, 149 (1939); V. Bargmann and E. P. Wigner, Proc. Nat. Acad. Sci. USA **34**, 211 (1948).

[2] S. Weinberg, *The Quantum Theory of Fields*, Vol. 1, (Cambridge University Press, 1995).

[3] A. Bohm, Lett. Math. Phy. **3**, 455 (1978); A. Bohm, J. Math. Phys. **22** 2813 (1981). A. Bohm, S. Maxson, M. Loewe and M. Gadella, Physica A **236**, 485 (1997); A. Bohm, *Quantum Mechanics*, third edition, (Springer, 1993), Chapter XXI

[4] V. Weisskopf and E. P. Wigner, Z. f. Physik **63**, 54 (1930); **65**, 18 (1930); W. Heitler, *Quantum Theory of Radiation*, Oxford (1954); M. L. Goldberger, K. M. Watson, *Collision Theory*, Wiley, New York (1964).

[5] G. Breit and E. Wigner, Phys. Rev. **49**, 519 (1936).

[6] D. E. Groom et. al., Particle Data Group, The European Physical Journal C 15, 1 (2000).

[7] L. A. Khalfin, JETP **33**, 1371 (1957); JETP **6**, 1053 (1958); L. Fonda, G. C. Ghirardi and A. Rimini, Repts. on Prog. in Phys. **41**, 587 (1978), and references thereof.

[8] I. M. Gel'fand and N. Ya. Vilenkin, *Generalized Functions*, Vol. 4 (Academic Press, New York, 1964); K. Maurin, *Generalized Eigenfunction Expansions and Unitary Representations of Topological Groups* (Polish Scientific Publishers, Warzawa, 1968);

[9] B. A. Lippmann and J. Schwinger, Phys. Rev. **79** (1950) 469; M. Gell-Mann and H. L. Goldberger, Phys. Rev. **91** (1953) 398; W. Brenig and R. Haag, Fortschr. Physik **7** (1959) 183.

[10] A. Bohm and M. Gadella, *Dirac Kets, Gamow Vectors, and Gel'fand Triplets*, Lecture Notes in Physics, Vol. 348 (Springer, Berlin, 1989); M. Gadella, J. Math. Phys. **24**, 1462 (1983).

[11] P. L. Duren, $\mathcal{H}^p$ *Spaces* (Academic Press, New York, 1970); A. Bohm, S. Maxson, M. Loewe and M. Gadella, Physica A **236**, 485 (1997): Appendix.

[12] A. Bohm and H. Kaldass, Phys. Rev. A **60**, 4606 (1999).

[13] S. Wickramasekara, *Differentiable Representations of Finite Dimensional Lie Groups in Rigged Hilbert Spaces*, Dissertation, The University of Texas at Austin (1999); S. Wickramasekara and A. Bohm, *Relativistic Gamow Vectors II: The Rigged Hilbert Space*, preprint.

[14] A. Bohm, S. Wickramasekara and H. Kaldass, *Relativistic Gamow Vectors I: Derivation from Poles of the S-Matrix*, preprint.

[15] A. Bohm, S. Wickramasekara and H. Kaldass, *Relativistic Gamow Vectors III: Transformations Under the Causal Poincaré Semigroup*, preprint.

[16] A. Bohm, *Generalized Eigenvectors and Group Representations*, in *Studies in Mathematical Physics*, A. O. Barut, Editor, NATO Advanced Study Series, D. Reidel Pub, 197 (1973).

[17] Arno R. Bohm, N. L. Harshman. *On the Mass and Width of the Z-boson and Other Relativistic Quasistable Particles*, Nucl .Phys. B**581**, 91 (2000); LANL hep-ph/0001206.

[18] E. Roberts, J. Math. Phys. **7**, 1097 (1966); A. Bohm, *Boulder Lectures in Theoretical Physics 1966*, Vol. 9A, (Gordon and Breach, New York, 1967); J. P. Antoine, J. Math. Phys. **10**, 53 (1969); **10** 2276 (1969); O. Melsheimer, J. Math. Phys. **15**, 902; 917 (1974).

Inst. Phys. Conf. Ser. No 173: Section 7
Paper presented at 24th Int. Coll. Group Theoretical Methods in Physics, Paris, France, 15–20 July 2002
©2003 IOP Publishing Ltd

Quantum ground-mode computation with static gates

Giuseppe Castagnoli ‡ and David Ritz Finkelstein §

Abstract. We develop a computation model for solving Boolean networks by implementing wires through quantum ground-mode computation and gates through identities following from angular momentum algebra and statistics. Gates are represented by three-dimensional (triplet) symmetries due to particle indistinguishability and are identically satisfied throughout computation being constants of the motion. The relaxation of the wires yields the network solutions. Such gates cost no computation time, which is comparable with that of an easier Boolean network where all the gate constraints implemented as constants of the motion are removed. This model computation is robust with respect to decoherence and yields a generalized quantum speed-up for all NP problems.

‡ Information Technology Division and Quantum Labs, Elsag spa, 16154 Genova, Italy
§ School of Physics, Georgia Institute of Technology, Atlanta, GA 30332, USA

1. Introduction

The prevailing approach to quantum computation is an evolution of classical reversible-algorithmic computation (a sequence of elementary logically reversible transformations, represented by unitary transformations). As well known, entanglement, interference and measurement yield in principle dramatic "speed-ups" over the corresponding classical algorithms in solving some problems. In spite of this outstanding result, it is in general recognized that this form of computation faces two, possibly basic, difficulties. Its speed-ups rely on quantum interference, which requires computation reversibility. Decoherence may then limit computation size below practical interest. Only two speed-ups of practical interest have been found so far (factoring and database search), and none since 1996.

Reversible-algorithmic computation is not the most general form of quantum computation. Its limitations justify reconsidering quantum ground-mode computation (Castagnoli 1998, Farhi et al. 2001, Kadowaki 2002, among others), a formerly neglected approach, still believed to be mathematically intractable. It is a quantum version of classical ground-state computation (Kirkpatrick & Selman 1994, among others), which is a well-developed approach competitive with algorithmic computation for solving Boolean networks (a most general problem).

A Boolean network is a set of nodes (Boolean variables) variously connected by gates and wires that impose equations on the variables they connect (fig. 1). A Boolean assignment satisfying all gates and wires is a network solution.

In quantum ground-mode computation, one sets up a quantum network whose energy is minimum when all gates and wires are satisfied. Coupling the network with a heat-bath of suitably decreasing temperature, relaxes the network to its ground mode, a mixture of solutions (we assume there is at least one). Measuring the node variables (Hermitian operators with eigenvalues 0 and 1) yields a solution. It is believed that this form of computation yields a (still ill-defined) speed-up over its classical counterpart: quantum tunneling reduces network trapping in local energy minima (e.g. Kadowaki 2002). However, long simulation times seriously limit research on this approach.

Here we develop a hybrid mode of computation. Only wires are implemented by ground-mode computation. Gates are always identically satisfied throughout the computation being constants of the motion. Gates are implemented by quantum symmetries due to particle indistinguishability. We show that relaxation-computation time is comparable with that an easier (loosely constrained) logical network where all the gate constraints implemented by quantum symmetries are removed. We plausibly conjecture that for this computation mode all hard to solve (NP) networks become easy (P). Decoherence is not a problem in principle since the network mode can be a mixture.

This model computation still belongs to the realm of principles, like other literature on quantum ground-mode computation, while algorithmic-reversible computation is now almost a technology. Nevertheless, it is worth starting over with a new approach that might overcome fundamental limitations of algorithmic computation.

2. Computation model

We use a network normal form with just wires and triodes (fig.1). Each triode τ – properly a partial gate – connects three nodes labeled τ, x - τ, y - τ, z (replaced by collective indices in fig.1) imposing the equation $q_{\tau,x} + q_{\tau,y} + q_{\tau,z} = 2$, where q's are Boolean variables and + denotes arithmetical sum. The three solutions are the rows of table I. Each wire w connects two nodes i, j imposing $q_i = q_j$, table II. The network exemplified in fig. 1, with $Q = 6$ nodes, $W = 4$ wires, and $T = 2$ triodes, has just one solution: $q_3 = q_5 = 0$, $q_1 = q_2 = q_4 = q_6 = 1$.

$q_{\tau,x}$	$q_{\tau,y}$	$q_{\tau,z}$
0	1	1
1	0	1
1	1	0

Table I

q_i	q_j
0	0
1	1

Table II

$q_{\tau,x}$	$q_{\tau,y}$	$q_{\tau,z}$	sy
0	0	0	s
0	1	1	t
1	0	1	t
1	1	0	t

Table III

Fig. 1. A network

All network nodes belong to triodes. The node variables are physically represented by associating each triode τ with a proton pair with spin vectors $\frac{1}{2}\sigma_{\tau,1}, \frac{1}{2}\sigma_{\tau,2}$ in units $\hbar = 1$. Two independent spin vectors have total spin $\frac{1}{2}(\sigma_{\tau,1} + \sigma_{\tau,2})$ with $s_{\tau,z} = \pm 1, 0$, and define three commuting Hermitian operators with eigenvalues 0 and 1: $q_{\tau,x} = s_{\tau,x}^2$, $q_{\tau,y} = s_{\tau,y}^2$, $q_{\tau,z} = s_{\tau,z}^2$, each corresponding to a network node. For the composition of angular momentum, the eigenvalues of the three operators of each proton pair must satisfy the XOR gate equation (table III). Its four rows correspond to the singlet and the three triplet pair modes of τ, spanning the Hilbert space $\mathcal{H}_\tau^{(4)}$. We use $\mathcal{H}_N^{(4)} = \bigotimes_{\tau=1}^{T} \mathcal{H}_\tau^{(4)}$ as the network space.

To model the triodes physically we assume that the spatial wave function of each proton pair τ is frozen antisymmetric in a stable ground mode (like in orthohydrogen nuclei) throughout the computation. Angular momentum composition and triplet symmetry are extra-dynamical: the triode Hamiltonians are zero in $\mathcal{H}_N^{(4)}$. Let $\mathcal{H}_\tau^{(3)}$ be the space spanned by the three triplet modes of triode τ. $\mathcal{H}_N^{(3)} = \bigotimes_{\tau=1}^{T} \mathcal{H}_\tau^{(3)} \subset \mathcal{H}_N^{(4)}$ is the network subspace with all triodes satisfied.

The wire frustration Hamiltonian in $\mathcal{H}_N^{(4)}$ is $H_N^{(4)} = g \sum_{\{i,j\}} (q_i - q_j)^2$, where $\{i, j\}$ is the set of all wires and g a coefficient to provide the dimension of energy. $H_N^{(4)}$ is quadrilinear in the $s_{\tau,w}$. By using two auxiliary spin $1/2$ variables for each wire, the wire Hamiltonian can be made bilinear, as in the Ising model (Castagnoli & Finkelstein 2002, briefly "I"). $H_N^{(4)}$ is symmetric under all $X_{12,\tau}$, the exchange operators of the two protons of each triode τ, since the q's are. Triplet symmetry is thus a constant of motion of $H_N^{(4)}$.

The heat-bath is a photon filled cavity, with Hilbert space $\mathcal{H}_B$. $\mathcal{H}^{(4)} := \mathcal{H}_N^{(4)} \otimes \mathcal{H}_B$ is the "system" (=network+bath) space, $\mathcal{H}^{(3)} := \mathcal{H}_N^{(3)} \otimes \mathcal{H}_B$ its subspace with triplet symmetry (all triodes satisfied). We denote by $H_B^{(4)}(t)$ the heat-bath Hamiltonian and define the network-bath coupling in $\mathcal{H}^{(4)}$ by $H_I^{(4)}(t) = g \sum_\tau \left[\vec{B}_\tau(t) \cdot \vec{\sigma}_{\tau,1} + \vec{B}_\tau(t) \cdot \vec{\sigma}_{\tau,2} \right]$. Each proton spin is coupled to a small random Gaussian time-varying magnetic field $\vec{B}_\tau(t)$ of the photon at the site of the spin. Indistinguishability requires that the two protons of the same triode τ experience the same magnetic field (we assume no overlap between spatial wave functions of different proton pairs). Therefore triplet symmetry (triodes satisfaction) is a constant of motion of $H_I^{(4)}(t)$, and also of the system Hamiltonian $H^{(4)}(t) = H_N^{(4)} + H_B^{(4)}(t) + H_I^{(4)}(t)$ ($H_N^{(4)}$ is already symmetric). Let $|\psi, t\rangle$ be the mode of the system, developed by $H^{(4)}(t)$. The relaxation of the network mode is described by the statistical operator $\rho_N(t) := \mathrm{Tr}_B(|\psi, t\rangle \langle \psi, t|)$, where Tr_B means trace over the bath degrees of freedom. If $\rho_N(t)$ starts in $\mathcal{H}^{(3)}$, under $H^{(4)}(t)$ it remains in it.

With a suitable time-variation of $\vec{B}_\tau(t)$, $H^{(4)}(t)$ relaxes the network to its zero point. A direct estimate of this relaxation time is likely mathematically intractable, and simulation is very lengthy. We take a shortcut that also sheds light on the nature of this hybrid computation. We compare the network relaxation time with that of an easier network obtained by replacing all triodes (Table I) by XOR gates (Table III): as if proton indistinguishability were suspended – each proton pair replaced by a deuteron. Restriction to $\mathcal{H}^{(3)}$ vanishes: a network of XOR gate and wires is loosely constrained and easy to solve. In particular all $q_i = 0$ is always a solution. A XOR network is solvable in poly(Q) time in classical computation and, reasonably, also in the present form of computation (for more detail, see I). The asymmetric Hamiltonian of the comparison system in $\mathcal{H}^{(4)}$ is $H^\blacklozenge(t) = H_N^{(4)} + H_B^{(4)}(t) + H_I^\blacklozenge(t)$, where $H_I^\blacklozenge(t) = g \sum_\tau \left[\vec{B}_{\tau,1}(t) \cdot \vec{\sigma}_{\tau,1} + \vec{B}_{\tau,2}(t) \cdot \vec{\sigma}_{\tau,2} \right]$ is the asymmetric coupling. Now we have two independent random Gaussian time-varying magnetic fields at each proton site, such that $\vec{B}_\tau(t) = \left[\vec{B}_{\tau,1}(t) + \vec{B}_{\tau,2}(t) \right]/2$ is the actual heat-bath (the sum of two Gaussian distributions is Gaussian). Let $|\varphi, t\rangle$ be the mode of the comparison system, developed by $H^\blacklozenge(t)$. The comparison network relaxation is described by $\rho_N^\blacklozenge(t) := \mathrm{Tr}_B(|\varphi, t\rangle \langle \varphi, t|)$. Let $P := 2^{-T} \prod_{\tau=1}^T [1 + X_{12,\tau}]$ be the network symmetrization operator. It projects $\mathcal{H}^{(4)}$ on $\mathcal{H}^{(3)}$. Clearly $P H_I^\blacklozenge(t) P = H_I^{(4)}(t)$, thus $P H^\blacklozenge(t) P = H^{(4)}(t)$ ($H_N^{(4)}$ is already symmetric, P is the identity in $\mathcal{H}_B$).

We show that the development of the actual system (hard triode network and bath) is the continuous projection on $\mathcal{H}^{(3)}$ of the development of the comparison system (easy XOR network and bath) in $\mathcal{H}^{(4)}$. Let $|\psi, t\rangle \subset \mathcal{H}^{(3)}$ so that $P|\psi, t\rangle = |\psi, t\rangle$. Under $H^{(4)}(t)$, it develops into $|\psi, t + dt\rangle = \left(1 - iH^{(4)}(t)\,dt\right)|\psi, t\rangle$; under $H^\blacklozenge(t)$ into $|\varphi, t + dt\rangle := \left(1 - iH^\blacklozenge(t)\,dt\right)|\psi, t\rangle$, in general non-symmetric. We restore particle indistinguishability by projecting $|\varphi, t + dt\rangle$ on $\mathcal{H}^{(3)}$, symmetrizing it: $P|\varphi, t + dt\rangle = \left(P^2 - iP H^\blacklozenge(t) P\right)|\psi, t\rangle = \left(1 - iH^{(4)}(t)\,dt\right)|\psi, t\rangle = |\psi, t + dt\rangle$. Thus the continuous projection of the comparison development yields the actual development.

Computation time is by assumption poly(Q) for the comparison easy network. To estimate that of the actual hard network, we decompose a ΔT into $N = \Delta T / \Delta t$ consecutive time slices $\Delta t_i \equiv [t_i, t_{i+1}]$ of equal length Δt. Within each Δt_i, we consider the relaxation $\rho_N^\blacklozenge(t)$ of the comparison XOR network in $\mathcal{H}^{(4)}$. At the end of each Δt_i, we project $\rho_N^\blacklozenge(t)$ on

$\mathcal{H}^{(3)}$, then take the limit $\Delta t \to 0$. This yields the relaxation of the actual network $\rho_N(t)$.

Within each Δt_i we consider the decomposition $\rho_N^{\blacklozenge}(t) := \rho_0(t) + \rho_F(t) + \rho_V(t)$. $\rho_0(t)$ describes networks with satisfied triodes and wires, namely solutions of the actual network; its probability is $p_0(t) := \mathrm{Tr}\rho_0(t)$. $\rho_F(t)$ describes networks with satisfied triodes and at least one frustrated wire; $p_F(t) := \mathrm{Tr}\rho_F(t)$. $\rho_V(t)$ describes networks with at least one violated triode, wires are either satisfied or frustrated; $p_V(t) := \mathrm{Tr}\rho_V(t)$. All possible modes of the comparison network are considered, thus $p_0(t) + p_F(t) + p_V(t) = 1$. $p_V(t)$ goes to zero with Δt and is annihilated by each projection.

The actual network-bath interaction soon randomly generates a $\rho_0(t_h)$, a mixture of solutions of the actual network, with extremely small probability $p_0(t_h) = O\left(1/2^Q\right)$. For a given confidence level, t_h does not depend on Q. For $t > t_h$ we apply the projection method. $p_0(t_h) = O\left(1/2^Q\right)$ becomes the nucleus of condensation of the network solutions.

Within each and every Δt_i, we take a constant-average logarithmic rate of decrease k of the comparison network frustration energy: $E_N(t_{i+1}) = (1 - k\Delta t) E_N(t_i)$. There is no error in taking a constant rate – see later. The relaxation time constant $1/k$ is by assumption poly(Q). Since $E_N(t) = Tr\rho_F(t)H_N^{(4)}$ (no contribution from $\rho_0(t)$, second order infinitesimal from $\rho_V(t)$) $E_N(t)$ and $p_F(t)$ go to zero together (see I for more detail). Thus on average:

$$p_F(t_{i+1}) = (1 - k\Delta t)\, p_F(t_i). \tag{1}$$

$p_F(t)$ decrease implies an increase of $p_0(t) + p_V(t)$. It is reasonable and conservative to consider the increase of $p_V(t)$ dominant. The relaxation of the comparison network is quicker because triodes can be violated. Note that we compare relaxation rates, not directions: the comparison network can head toward $\mathcal{H}_N^{(4)} \sim \mathcal{H}_N^{(3)}$, the actual network remains in $\mathcal{H}_N^{(3)}$. Furthermore, $H_N^{(4)}$ does not couple $\rho_0(t)$ with $\rho_F(t)$ or $\rho_V(t)$ ($H_N^{(4)}\rho_0(t) = \rho_0(t)H_N^{(4)} = 0$). Therefore $p_0(t)$ neither decreases nor increases on average. Since $p_F(t)$ decreases and $p_0(t)$ does not, the ratio $p_F(t)/p_0(t)$ decreases. When we project on $\mathcal{H}_N^{(3)}$ at the end of Δt_i, we remain with a smaller $p_F(t)$ and a larger $p_0(t)$ (probability of solutions of the actual network). For (1), with $p_0(t)$ unaltered within each Δt_i and not too close to 1 (say $p_0(t) < 3/10$, so that $p_F(t)$ remains close to 1), and $\Delta t \to 0$, we have approximately for the actual network:

$$p_0(t_h + \Delta T) \approx p_0(t_h)e^{k\Delta T} \approx \frac{1}{2^Q}\, e^{k\Delta T}, \tag{2}$$

as readily checked (see also I). The probability of having solutions of the actual network becomes $O\left(1\right)$ in a $\Delta T \approx Q/k = Q$poly(Q)=poly(Q).

Using a different k_i for each Δt_i, with average k ($\sum_i k_i\Delta t_i = k\Delta T$), yields the same result: $e^{k\Delta T}$ in (2) should be replaced by $\prod_i e^{k_i\Delta t_i} = e^{k\Delta T}$.

In conclusion, particle indistinguishability yields a new form of quantum computation where the gates of a Boolean network are always satisfied as constants of the motion. This model computation appears to be promising for achieving: (i) robust quantum computation and (ii) a generalized speed-up in all NP problems.

Many ideas propounded in this work were developed through discussions with A. Ekert.

References

Castagnoli, G. 1998 *Physica* D **120**, 48.
Castagnoli, G. & Finkelstein, D. arXiv:quant-ph/0209084 v2 13 September 2002.
Farhi, E. Goldstone, J., Gutmann, S., Lapan, J., Lundgren, A. & Preda, D. 2001 *Science* **292**, 472
Kadowaki, T. arXiv:quant-ph/0205020 v1 5 May 2002.
Kirkpatrick, S. & Selman, D. 1994 *Science* **264**, 1297

Inst. Phys. Conf. Ser. No 173: Section 7
Paper presented at 24th Int. Coll. Group Theoretical Methods in Physics, Paris, France, 15–20 July 2002
©2003 IOP Publishing Ltd

Complex Coherent States and Non Equilibrium Quantum Thermodynamics

E Celeghini
University of Florence and INFN-FI, Italy

Abstract. A possible thermodynamic evolution to equilibrium of a quantum statistical system is discussed. The description is entirely algebraic and it is done by means of complex coherent states built in the Universal Enveloping Algebra of the complex algebra $\mathcal{A}_1$. Finite dimensional representations of $\mathcal{A}_1$ give fermion distributions while infinite dimensional ones describe bosons. The scheme is thermodynamic in the sense that fluctuations are not considered and this allows us to resort to an evolution operator that realizes, for $t \to \infty$, a projector on the state of equilibrium. The shrinking of the Hilbert space is related to non unitary evolution that change the normalization of the state vectors.

The theory is invariant under time reversal: vectors at $t \to \pm\infty$ differ only in phases and both correspond to the equilibrium probability distribution. Equilibrium and non-equilibrium thermodynamic quantities can be calculated at microscopic level in a framework that is in between thermodynamic and statistical mechanics.

Our approach is a physically founded example of non-linear quantum theory.

1. Introduction

Thermodynamics works quite well and it is not invariant under time reversal also if T invariance is always assumed in microscopic statistical mechanics. This paradox has been discussed for more than a century but the original Boltzmann idea - in spite to the fact that, perhaps, ruined his life - seems to be always the best possible solution [1]. To be schematic, he suggests that the apparent violation of time reversal is an effect of fluctuations and that sufficiently long observations (perhaps many times the life of universe) would show that the evolution of a system is, in any case, time reversal invariant [2]. The alternative idea - that T invariance has been loosed in some way in macroscopic systems - is less popular today, essentially because we are unable to find an appropriate mechanism for the breaking.

We propose here a thermodynamic description in Fock space $\mathcal{F}$ of a quantum statistical system out of equilibrium that is T invariant. Because we give up to the possibility of describe fluctuations - and this is the reason of the word *thermodynamics* in the title - we can give a description in terms of an evolution operator E(t).

T invariance - for its part - implies that E(t) is invertible, allowing us to adopt the mathematical apparatus of group theory, more tractable of that of semi-groups required in a T violating theory.

Our evolution operator will be indeed realized in the scheme of coherent states of the complex algebra $\mathcal{A}_1$ and it is such that any generic state at $t = t_0$ is projected, for $t \to \pm\infty$, in two vectors different between themselves in phases but corresponding to the same probability distribution: that of equilibrium quantum statistical mechanics.

Numerical results confirm the Boltzmann idea of many different scale of time, suggesting the presence of meta-stable states.

From an interdisciplinary point of view, it could be of interest that, in an application oriented paper, a new approach to non linear theories appears.

2. Complex Coherent States

Coherent states are usually defined for real Lie algebras $\mathcal{L} = \{V \equiv \bar{r} \cdot \bar{X}; \ r_i \in \mathcal{R}, \ [X_i, X_j] = f_{ij}{}^k X_k, \ X_i{}^\dagger = -X_i, (f_{ij}{}^k)^* = f_{ij}{}^k\}$. If D^α is a representations and $|\psi\rangle \in D^\alpha$ a vector belonging to it, they are[3]

$$|\bar{r}, \psi\rangle \equiv e^{\bar{r} \cdot \bar{X}} |\psi\rangle. \tag{1}$$

To be consistent with the Liouville theorem, reality of algebra is essential as it imposes that the operator $e^{\bar{r} \cdot \bar{X}}$ is unitary, saving the Hilbert space structure: real (we specify to distinguish with the ones defined below) coherent states are indeed complete (in reality, over-complete).

For the same reasons, to describe a dissipative evolution we need, on the contrary, non unitary operator. We thus generalize coherent states (1) to complex algebras:

$$|\bar{c}, \psi\rangle \equiv e^{\bar{c} \cdot \bar{X}} |\psi\rangle \tag{2}$$

where the only difference between eq.(1) and eq.(2) is that $c_i \in \mathcal{C}$.

Because equilibrium quantum statistical mechanics is related to su_2 and $su_{1,1}$[4], we restrict here ourselves to coherent states of their common complexification $\mathcal{A}_1$.

3. Evolution Operator and Algebraic Master Equations

Let us rephrase the fundamental formulas of Equilibrium Quantum Statistical Mechanics in a way suitable to a non-equilibrium extension. We describe, as usual[5], a many identical particles system at equilibrium by the Fock space $\mathcal{F}$ vector:

$$|\psi\rangle = \sum_{\{n_i\}} a_{\{n_i\}} |n_1, n_2, \ldots\rangle \tag{3}$$

where $|a_{\{n_i\}}|^2$ is $\prod_i \binom{g_i}{n_i} h_i{}^{n_i}$ for fermions and $\prod_i \binom{g_i + n_i - 1}{n_i} h_i{}^{n_i}$ for bosons (n_i is the number of particles in the i level, g_i is the real or apparent multiplicity and $h_i \equiv e^{-\alpha - \beta \epsilon_i}$).

On the other hand, to be able to give an operatorial description of convergence at equilibrium we adopt a weak version of random phases postulate including in the statement the word *macroscopic*. It sounds, for us, as:

Do not exist macroscopic (i.e. related to extensive properties of the system at equilibrium) *operators relating different states of Fock basis* $\{|n_1, n_2, \ldots \rangle\}$.

Absence of correlation between the states is thus, in our definition, no more a general property of the formalism but simply an *accidental* property of operators at equilibrium, not inevitably shared by all other operators as in the standard formulation.

We are thus allowed to look for an evolution operator $E(t)$ on $\mathcal{F}$ that brings all states to equilibrium; i.e. such that for every vector $|\chi(t_0)\rangle = \sum_{\{n_i\}} b_{\{n_i\}}(t_0) |n_1, n_2, \ldots\rangle \in \mathcal{F}$ we have

$$\lim_{t \to \pm \infty} E(t) |\chi(t_0)\rangle = |\psi_\pm\rangle,$$

where $|\psi_\pm\rangle$ are both consistent with eq.(3).

The essential result presented in this paper is that, for generic $\{c_i\}$ and generic $|\chi(t_0)\rangle$, $E(t) \equiv e^{\bar{c} \cdot \bar{X} t}$ is this operator, where X_i's are the $\mathcal{A}_1$ generators. More explicitly, quantum statistical distributions are complex coherent states of $\mathcal{A}_1$ and, for both $t \to \pm \infty$, they reduce to states of equilibrium. In particular, in agreement with [4], finite-dimensional representations, unitary for the real form su_2, give rise to quantum statistics of fermions, while the infinite-dimensional ones, unitary for the real form $su_{1,1}$, give rise to quantum statistics of bosons.

Note that the operator $e^{\bar{c}\cdot\bar{X}t}$, in spite of being a linear operator on $\mathcal{F}$ (thus allowing us to use all the known mathematical apparatus of linear operators) describes a non-linear evolution. The reason is that a physical state in quantum physics is described not by a vector of Hilbert space but by a ray: more formally the true relation is $|\chi\rangle \in \mathcal{F}/\mathcal{C}$, as all vectors different for a c-number factor describe the same system. This implies only a phase freedom (always not irrelevant as gauge theories are related to it) if evolution is unitary (as assumed usually in Quantum Theory) but, when we have dissipation, evolution operator introduces - by means of normalization - also a non-linear effect.

This evolution operator $E(t)$ is equivalent to a set of master equations, we call - as derived from the algebra - algebraic master equations. They are

$$\sum_{\{n_i\}} \left[b_{\{n_i\}}(t+t_0) - b_{\{n_i\}}(t)e^{\bar{c}\cdot\bar{X}t} \right] |n_1, n_2, \ldots\rangle = 0 \tag{4}$$

in integral form and

$$\sum_{\{n_i\}} \left[\frac{d}{dt} b_{\{n_i\}}(t) - b_{\{n_i\}}(t)\bar{c}\cdot\bar{X} \right] |n_1, n_2, \ldots\rangle = 0 \tag{5}$$

in differential form.

Note that, unlike variables in normal master equation and non-linear physics, $b_{\{n_i\}}(t) \in \mathcal{C}$ and, as the physical meaning in related only to their quotients, they often diverge. Still the connection with probabilities is non-linear as the amplitudes are not normalized and $P_{\{n_i\}}(t) = \frac{|b_{\{n_i\}}(t)|^2}{\sum |b_{\{n_i\}}(t)|^2}$. Note at last that the $c_i, i = 1, 2, 3$ are the unique parameters.

For all these reasons, the found non-linear differential equations do not have the freedom of classical ones, and we obtain - from $\mathcal{A}_1$ complex coherent states - in generic evolution stable and unstable fixed points, sometime torus and never chaos.

4. Examples

To clarify the procedure let us discuss in some details fermions in one degree of freedom. As $n = 0, 1, \ldots, g$ we have to consider the representation of dimension $g + 1$ i.e. the representation $D^{g/2}$ of su_2, where $g = 2j$ and $n = j + m$. The simplest case is the two dimensional representation of $\mathcal{A}_1$, $D^{1/2}$. If we specialize the evolution operator as e^{irJ_1t} (with $r \in \mathcal{R}^+$, remember that $J_1^\dagger = -J_1$), we have:

$$\lim_{t \to \pm\infty} e^{irJ_1t} = \lim_{t \to \pm\infty} \begin{pmatrix} ch(rt) & sh(rt) \\ sh(rt) & ch(rt) \end{pmatrix} \propto \begin{pmatrix} 1 & \pm 1 \\ +1 & 1 \end{pmatrix}$$

i.e. for $t \to \pm\infty$, this evolution operators realizes (up to an irrelevant numerical factor) two different projectors, projecting every vector $|\chi(t_0)\rangle \in \mathcal{F}$ on two different vectors $|\psi_\pm\rangle$, both corresponding to the quantum statistical distribution of fermions with $g = 1$ and temperature infinite.

This evolution is not peculiar of this case but, on the contrary, in all irreducible representations results are similar. For g arbitrary, always in one degree of freedom, fixed states in $\mathcal{F}/\mathcal{C}$ are obtained by the request

$$c_+\sqrt{(g-n+1)n}\, a_{n-1} + c_-\sqrt{(g-n)(n+1)}\, a_{n+1} + c_3\left(n - g/2\right) a_n \propto a_n$$

and it can be shown that the number of fixed points is of the order of g, but that only one is stable for $t \to +\infty$ and only one is stable for $t \to -\infty$, and they correspond to $a_n = \sqrt{\binom{g}{n}} h_\pm^{n/2}$,

where the temperature of the system is determined by $h_\pm^{1/2} = \frac{c_3 \pm \sqrt{c_3^2 + 4c_+ c_-}}{2c_-}$.

To have T invariance, we must thus have $|h_+| = |h_-|$ and we obtain $c_+c_-/(c_3)^2$ real, $c_+c_-/(c_3)^2 < -1/4$ and $|h_\pm| = |c_+/c_-|$. A detailed analysis of independent parameters and of they physical interpretation will be done elsewhere[6].

5. Conclusions

We have exhibited one degree of freedom only, but - as in many degrees of freedom $E(t)$ factorizes - the extension to them is simply related to the iterated coalgebra (see e.g.[7]).

The discussion of bosons is simply a little more elaborated then that of fermions as we have to deal with infinite dimensional representations of $\mathcal{A}_1$: multiplicity g is again related to the highest weight of the representation D_+^k of $su_{1,1}$: $g = 2k$.

$E(t)$ is, of course, related to convergence to equilibrium of fluctuations but it is also the operator driving a statistical system from one macroscopic status into another: in other words it is the operator that describes macroscopic transition of the system. As an example let us quote a system of two cans where fermions, for $t < 0$, are all in the first one; when, at $t = 0$, we turn on the tap it is $E(t)$ that controls the flux of fermions into the second can. Any thermodynamic evolution can be analogously described, allowing to evaluate specific heath and all other thermodynamic variables.

To conclude, let us remember that the presence of computers of increasing power is changing the same meaning of *computability*: while few years ago the previous discussion could look academical as the computation of a significant number of degrees of freedom where impossible, today the situation is quite different.

References

[1] L. Boltzmann Ann. Phys. (Leipzig) **57** (1896) 773; translated and reprinted in S.G.Brush, Kinetic Theory 2, Pergamon, Elmsford, N.Y. (1996).
[2] J.L.Lebowitz Physics Today **46** (1993) 32; Rev. Mod. Phys.**71** (1999) S346.
[3] A.Perelomov, Generalized Coherent States and their Applications, Springer, Berlin (1996).
[4] E.Celeghini and M.Rasetti, Phys. Rev. Lett. **80** (1998) 3424.
[5] K.Huang, Statistical Mechanics, Wiley, N.Y.(1987).
[6] E.Celeghini, in preparation.
[7] J.Fuchs, Affine Lie Algebras and Quantum Groups, Cambridge U. P., N.Y. (1995).

Inst. Phys. Conf. Ser. No 173: Section 7
Paper presented at 24th Int. Coll. Group Theoretical Methods in Physics, Paris, France, 15–20 July 2002
©2003 IOP Publishing Ltd

Deformation quantization of damped systems

D Chruściński

Institute of Physics, Nicholas Copernicus University, ul. Grudziądzka 5/7, PL-87-100 Toruń, Poland

Abstract. Both classical and quantum damped systems give rise to complex spectra and corresponding resonant states. We investigate how resonant states, which do not belong to the Hilbert space, fit the phase space formulation of quantum mechanics. It turns out that one may construct out of a pair of resonant states an analog of a stationary Wigner function.

Quantum mechanics teaches us that if $\widehat{A}$ is a s.a. operator acting on a Hilbert space $\mathcal{H}$ then the corresponding eigenvalue problem

$$\widehat{A}\psi = \lambda\psi \tag{1}$$

leads to real λ if $\psi \in \mathcal{H}$. However, λ needs not be real if $\psi \notin \mathcal{H}$. Let D be a dense, nuclear subset in $\mathcal{H}$. One defines the Gelfand triplet [1] (or the rigged Hilbert space)

$$D \subset \mathcal{H} \subset D^{*}. \tag{2}$$

Now, $\psi \in D^{*}$ defines a generalized eigenvector of $\widehat{A}$ corresponding to a generalized eigenvalue λ if

$$(\widehat{A}\varphi, \psi) = \lambda(\varphi, \psi) \quad \text{for any } \varphi \in D. \tag{3}$$

Generalized eigenvectors corresponding to complex λ are usually called resonant states. Such states play important role in quantum mechanics and it is widely believed that they are responsible for the irreversible dynamics of physical systems (see e.g. [2]). Recently, it was observed by Kossakowski [3] that quantized simple damped systems (e.g. damped harmonic oscillator) give rise to discrete complex spectra and hence the corresponding eigenvectors may be interpreted as resonant states. Moreover, it was shown [4] that the damping behavior in a classical system may be also interpreted as appearance of resonant states for the corresponding Koopman operator.

In the present paper we shall study the phase space formulation of quantum damped systems. Clearly, this formulation, called also the deformation quantization, is perfectly equivalent to the standard Hilbert space approach. However, as we already mentioned, resonant states lie outside the Hilbert space, and hence, it would be interesting to find how they fit phase space approach. Any vector $\psi \in \mathcal{H}$ gives rise to a Wigner function W_{ψ} on a classical phase space P. For the system with one degree of freedom one has:

$$W_{\psi}(x, p) := \frac{1}{2\pi} \int dy\, e^{-ipy}\overline{\psi}(x - \hbar y/2)\, \psi(x + \hbar y/2). \tag{4}$$

As was shown already by Wigner [5] this function is real and produces marginal probability densities $\int W_{\psi}(x, p)dx$ and $\int W_{\psi}(x, p)dp$. The classical limit of W_{ψ} reproduces a classical probability distribution on P. Moreover, if ψ is an eigenvalue of the Hamilton operator $\widehat{H}$, i.e. $\widehat{H}\psi = E\psi$, then the corresponding Wigner function W_{ψ} satisfies the following eigenvalue problem:

$$H \star W_{\psi} = W_{\psi} \star H = EW_{\psi}, \tag{5}$$

where H is a classical Hamiltonian on P and the $\star$-product $f \star g$ defines a "quantum deformation" of a usual commutative product of functions $f \cdot g$:

$$f \star g := f \exp\left[\frac{i\hbar}{2}\left(\overleftarrow{\partial}_x \overrightarrow{\partial}_p - \overleftarrow{\partial}_p \overrightarrow{\partial}_x\right)\right] g . \tag{6}$$

For more information see e.g. the review [6] (see also recent paper [7]).

As an example consider harmonic oscillator defined by $H_{\mathrm{ho}} = \frac{\omega}{2}(x^2 + p^2)$. Applying the above rules one finds that the eigen-equation (5) reproduces the standard oscillator spectrum $E_n = \hbar\omega(n + 1/2)$ and

$$W_n = \frac{(-1)^n}{\pi\hbar} e^{-\xi/2} L_n(\xi) , \tag{7}$$

where $\xi := 2(x^2 + p^2)/\hbar$ and L_n stand for the Laguerre polynomials [6]. It turns out that oscillator Wigner functions satisfy the following orthonormality condition

$$W_n \star W_m = \frac{1}{2\pi\hbar} \delta_{nm} W_n , \tag{8}$$

and hence one obtains the following resolution of identity on P:

$$\sum_n W_n = \frac{1}{2\pi\hbar} , \tag{9}$$

which is phase space analog of the Hilbert space formula $\sum_n P_n = \mathbb{1}$, where P_n is a 1-dimensional projector onto the eigenspace generated by ψ_n.

Now, we apply this scheme to the simple damped system described by the following equation:

$$\dot{x} = -\gamma x , \tag{10}$$

where $\gamma > 0$ is a damping constant. Clearly, this system is not Hamiltonian. However, following [8] we may lift an arbitrary non-Hamiltonian dynamics $\dot{x} = F(x)$ on $\mathbb{R}$ to the Hamiltonian one on $\mathbb{R}^2$. Let us define the corresponding Hamiltonian $H(x,p) = pF(x)$. The Hamilton equations read as follows:

$$\dot{x} = \{x, H\} = F(x) \quad \text{and} \quad \dot{p} = \{p, H\} = -p\partial_x F(x) . \tag{11}$$

Therefore, for the damped system (10) one obtains:

$$H_{\mathrm{d}}(x, p) = -\gamma x p . \tag{12}$$

This system was analyzed in [4] where the spectrum of the quantum Hamiltonian

$$\widehat{H}_{\mathrm{d}} = -\frac{\gamma}{2}\left(\widehat{xp} + \widehat{px}\right) , \tag{13}$$

was found:

$$\mathrm{Spec}(\widehat{H}_{\mathrm{d}}) = \left\{ i\hbar\gamma\left(n + \frac{1}{2}\right) \,\Big|\, n \in \mathbb{Z} \right\} . \tag{14}$$

Although $\widehat{H}_{\mathrm{d}}$ is s.a. on $L^2(\mathbb{R})$ its spectrum is purely imaginary and discrete. It means that the corresponding eigenvectors do not belong to $L^2(\mathbb{R})$. Indeed, it is easy to see [4] that for

$$\psi_n^+(x) := x^n \quad \text{and} \quad \psi_n^-(x) := (-i\hbar)^n \delta^{(n)}(x) , \quad n = 0, 1, 2, \dots , \tag{15}$$

one has:

$$\widehat{H}_{\mathrm{d}}\psi_n^\pm = \pm i\hbar\gamma\left(n + \frac{1}{2}\right)\psi_n^\pm . \tag{16}$$

Evidently, these states living outside the Hilbert space $L^2(\mathbb{R})$ cannot be used to construct stationary Wigner functions. Defining

$$W_n^\pm(x,p) \propto \int dy\, e^{-ipy} \overline{\psi_n^\pm}(x - \hbar y/2)\, \psi_n^\pm(x + \hbar y/2) , \tag{17}$$

one obtains

$$W_n^\pm(x,p,t) = e^{\pm(2n+1)\gamma t}\, W_n^\pm(x,p,0) , \tag{18}$$

which shows that $W_n^\pm$ are non-stationary. To find the analogs of stationary Wigner functions let us solve for the eigenvalue problem (5):

$$H_{\mathrm{d}} \star F = F \star H_{\mathrm{d}} = EF . \tag{19}$$

In analogy to the oscillator Wigner functions W_n one finds

$$F_n^+ = \frac{(-1)^n}{\pi\hbar}\, e^{-\eta/2}\, L_n(\eta) , \tag{20}$$

where $\eta := -4ixp/\hbar$ and the corresponding imaginary eigenvalues read

$$E_n = i\hbar\gamma \left(n + \frac{1}{2} \right) , \quad n = 0, 1, 2, \ldots . \tag{21}$$

However, there is another family of stationary functions F_n^-

$$F_n^- := \overline{F_n^+} , \tag{22}$$

which satisfies

$$H_{\mathrm{d}} \star F_n^- = F_n^- \star H_{\mathrm{d}} = \overline{E}_n F_n^- . \tag{23}$$

It follows immediately from the following property:

$$\overline{f \star g} = \overline{g} \star \overline{f} , \tag{24}$$

which may be easily proved using the definition of the $\star$-product (6).

Clearly, $F_n^\pm$ contrary to W_n are not real. However, they enjoy many properties of the usual Wigner functions. In particular they are normalized and give rise to the well defined marginal probability distributions. In particular one has

$$\int F_0^\pm(x,p)\, dx = \delta(p) \quad \text{and} \quad \int F_0^\pm(x,p)\, dp = \delta(x) . \tag{25}$$

The above property seems to violate the Heisenberg uncertainty principle – the particle is localized both in x and p variables. Clearly, we lose the probabilistic interpretation of $F_n^\pm$ since the corresponding eigenvectors $\psi_n^\pm$ (15) do not belong to the Hilbert space $L^2(\mathbb{R})$. Interestingly, $F_n^\pm$ satisfy the following condition:

$$F_n^\pm \star F_m^\pm = \frac{1}{2\pi\hbar}\, \delta_{nm}\, F_n^\pm , \tag{26}$$

in perfect analogy to (8). Therefore, one obtains the corresponding resolution of identity

$$\sum_n F_n^\pm = 2 \sum_n \mathrm{Re}\, F_n^+ = \frac{1}{2\pi\hbar} . \tag{27}$$

Finally, it would be interesting to find relation between the resonant states $\psi_n^\pm$ defined in (15) and $F_n^\pm$. Using some simple algebraic manipulations it is easy to show that

$$F_n^+(x,p) = \frac{(-1)^n}{\pi\hbar} \int dy\, e^{-ipy} \overline{\psi_n^+}(x - \hbar y/2)\, \psi_n^-(x + \hbar y/2) . \tag{28}$$

This shows that the analog of a stationary Wigner function $F_n^{\pm}$ is constructed out of a pair of resonant states (ψ_n^+, ψ_n^-). Clearly, if ψ is a proper eigenvectors corresponding to a real eigenvalue E, then using the above prescription for F one recovers the Wigner function corresponding to ψ.

Comparing the spectra of harmonic oscillator H_{ho} and damped system H_{d} one finds striking similarity, that is, they are related by the following relation:

$$\omega = \pm i\gamma \,. \tag{29}$$

It is not an accident. Note, that performing the following canonical transformation:

$$x = \frac{1}{\sqrt{2}}(X + P) \quad \text{and} \quad p = \frac{1}{\sqrt{2}}(X - P)\,, \tag{30}$$

one obtains

$$H_{\text{d}} = -\gamma x p = \frac{\gamma}{2}(P^2 - X^2)\,, \tag{31}$$

i.e. in the new variables (X, P), H_{d} corresponds formally to the harmonic oscillator with purely imaginary frequency $\omega = \pm i\gamma$.

Summarizing, we considered only one example of a damped system (10). However this toy model shows many general features of more complicated systems. For more detailed discussion see the forthcoming paper [9].

It is pleasure to thank professor Andrzej Kossakowski for very interesting discussions. This work was partially supported by the Polish State Committee for Scientific Research (KBN) Grant no 2P03B01619.

References

[1] Gelfand I M and Vilenkin N J 1964 Generalized functions Vol IV (Academic Press: New York)
[2] Bohm A Doebner H-D and Kielanowski P 1998 Irreversability and Causality, Semigroups and Rigged Hilbert Spaces Lecture Notes in Physics Vol 504 (Berlin: Springer)
[3] Kossakowski A 2002 Open. Sys. Information Dyn. 9 1-18
[4] Chruściński D 2002 Resonant states and classical damping LANL preprint math-ph/0206009
[5] Wigner E 1932 Phys. Rev. 40 749
[6] O'Conell R Scully M and Wigner E 1984 Phys. Rep. 106 121
[7] Zachos C 2002 Int. J. Mod. Phys. A 17 297-316
[8] Pontriagin L S Boltański V G Gamkrelidze R V and Miscenko E F 1962 The Mathematical Theory of Optimal Precesses (New York: Wiley)
[9] Chruściński D 2002 Wigner functions for damped systems (in preparation)

Inst. Phys. Conf. Ser. No 173: Section 7
Paper presented at 24th Int. Coll. Group Theoretical Methods in Physics, Paris, France, 15–20 July 2002
©*2003 IOP Publishing Ltd*

Coherent state realizations of su($n + 1$) in terms of subgroup functions

Hubert de Guise* and Marco Bertola[†]

* Department of Physics, Lakehead University, Thunder Bay, Ontario, P7B 5E1, Canada
[†] Centre de Recherche Mathématique, Université de Montréal, C.P. 6128 Succ. A, Montreal, Qc, H3C 3J7, Canada

Abstract. We present a new kind of coherent state realizations, where generators of the algebra $su(n + 1)$ (or, more properly, its complex extension) are expressed in terms of simple actions and subgroup functions. Phase operators and other potential applications to quantum optics are briefly discussed.

1. Physical context and general outlook

The Lie algebra $su(n + 1)$ is the natural tool to describe a collection of $(n + 1)$–level atoms. Transitions from level i to level j are expressed as

$$\hat{C}_{ij} = a_i^\dagger a_j , \tag{1}$$

where i and j can take values from 1 to $n + 1$, $a_i^\dagger$ creates an excitation in level i whereas a_j removes an excitation from level j. The $(n + 1)^2$ operators $\hat{C}_{ij} = a_i^\dagger a_j$ close under commutation to form the $u(n + 1)$ algebra:

$$[\hat{C}_{ij}, \hat{C}_{k\ell}] = \delta_{i\ell}\hat{C}_{jk} - \delta_{jk}\hat{C}_{\ell j} , \qquad i, j = 1, \ldots, n + 1 ,$$

The algebra $su(n + 1)$ is obtained by selecting from $u(n + 1)$ the subset of generators of the form $\hat{C}_{ij}$, $i \neq j$ together with n diagonal operators $\hat{h}_i = \hat{C}_{ii} - \hat{C}_{i+1,i+1}$.

Unitary irreducible representations (UIR) of su($n+1$) constructed from the $\hat{C}_{ij}$ operators have highest weights of the type $(\lambda, 0, \ldots)$, where λ is the total number of excitations in the system. Basis states $|\nu_1\nu_2\ldots\rangle$ are labeled by $n + 1$ non–negative integers ν_i, denoting the excitation populations of the levels. Thus, for two–level atoms, $|ab\rangle$ denotes a system with a quanta in the first level and b quanta in the second. Obviously, total number of quanta $\nu_1 + \nu_2 + \ldots = \lambda$ is conserved in the transitions, and provides a physical interpretation to the first Dynkin label of the UIR $(\lambda, 0, \ldots)$.

For UIRs of the type $(\lambda, 0, 0, \ldots)$, the eigenvalues $\nu_i - \nu_{i+1}$ of the Cartan elements $\hat{h}_i$, which represent population inversions, are sufficient to completely specify a state. Thus, one can intuitively imagine that it should be possible to express the action of every generator in terms of Cartan generators and their eigenstates, which are just functions on the n-torus. It is expressions of this type, along with some generalizations, which are discussed in this contribution.

2. Realizations on the n–torus

Note that a considerably more elaborate discussion of these realizations can be found in [1].

2.1. Some examples and features

The simplest example of the type of realization that we are interested in is the realization Γ_1 of $su(2)$, given by

$$\Gamma_1 : \hat{h}_1 \mapsto -i\frac{d}{d\varphi}, \quad \hat{C}_{12} \mapsto \frac{1}{2}e^{2i\varphi}\left(\lambda + i\frac{d}{d\varphi}\right), \quad \hat{C}_{21} \mapsto \frac{1}{2}e^{-2i\varphi}\left(\lambda - i\frac{d}{d\varphi}\right).$$

With the identification $\hat{h}_1 \Leftrightarrow \hat{L}_0, \hat{C}_{12} \Leftrightarrow \hat{e}_+, \hat{C}_{21} \Leftrightarrow \hat{e}_-$, one easily verifies that Γ_1 reproduces the commutation relations of the complex extension of $su(2)$. Γ_1 acts naturally on functions over the subgroup U(1)⊂SU(2); basis states in the UIR are mapped to toroidal states:

$$|\nu_1, \nu_2\rangle \mapsto \frac{e^{i(\nu_1 - \nu_2)\varphi}}{\sqrt{\pi}}, \quad \hat{X}|\nu_1, \nu_2\rangle \mapsto \Gamma_1(\hat{X})\frac{e^{i(\nu_1 - \nu_2)\varphi}}{\sqrt{\pi}} \quad \nu_1 + \nu_2 = \lambda. \quad (2)$$

Using the U(1) inner product $\langle \psi | \xi \rangle = \int_0^\pi d\varphi\, \psi^* \xi$, one verifies that Γ_1 is not a Hermitian realization. However, as $e^{i\lambda\varphi}$ is a highest integral weight vector, Γ_1 must be equivalent to a Hermitian realization γ_1 via a similarity transformation $\mathcal{K}$: $\gamma_1 = \mathcal{K}^{-1}\Gamma_1\mathcal{K}$. The intertwining operator $\mathcal{K}$ can be worked out explicitly, but it is much more interesting to note that γ_1 and Γ_1 are closely related to polar representations of an operator. For instance:

$$\gamma_1(\hat{C}_{12}) = \frac{1}{2}e^{2i\varphi}\sqrt{\Gamma_1(\hat{C}_{21})\Gamma_1(\hat{C}_{12})} \quad \Leftrightarrow \quad \gamma_1(\hat{e}_+) = \frac{1}{2}e^{2i\varphi}\sqrt{\Gamma_1(\hat{e}_-)\Gamma_1(\hat{e}_+)}.$$

Γ_1 already exhibits features found in realizations of $su(n+1)$ on the n–torus: the Cartan element $\hat{h}_1$ acts by simple differentiation; ladder operators can be written in two parts: a $U(1)^{\otimes n}$ function which is preceded by an operator acting diagonally on exponential functions. The $\frac{1}{2}$ factor, which is specific to $su(2)$, becomes $1/(n+1)$ in the general case.

In the example of $su(3)$, we find, for instance:

$$\Gamma_2(\hat{h}_1) = -i\frac{\partial}{\partial\varphi_1}, \quad \Gamma_2(\hat{h}_2) = -i\frac{\partial}{\partial\varphi_2}, \quad \Gamma_2(\hat{C}_{13}) = -\frac{\sqrt{2}}{3}e^{i(\varphi_1 + \varphi_2)}\left[\lambda + i\frac{\partial}{\partial\varphi_1} + 2i\frac{\partial}{\partial\varphi_2}\right].$$

The advertised similarities with su(2) are obvious: Cartan elements represented by constant coefficient differential operators, ladder operators, like $\hat{C}_{13}$, expressed as diagonal operators on U(1)×U(1), premultiplied by a U(1)×U(1) factor. Again, Γ_2 is not hermitian but equivalent to a Hermitian γ_2, related to a "polar decomposition" of γ_2, where, for instance, $\gamma_2(\hat{C}_{13}) = -\frac{\sqrt{2}}{3}e^{i(\varphi_1 + \varphi_2)}\sqrt{\Gamma_2(\hat{C}_{31}) \cdot \Gamma_2(\hat{C}_{13})}$.

2.2. Application: Polar decomposition and $su(3)$ phase states

It has already been observed that there is a natural relation between expressions such as $\gamma_1(\hat{C}_{12})$, $\Gamma_1(\hat{C}_{12})$ and the polar decomposition of an operator. Polar decompositions were introduced by Dirac in an attempt to quantize the phase of a field.

In a three–level system, one naturally defines two "fundamental" phase operators, each associated with the phase part of the polar decomposition of the simple generators $\hat{C}_{12}$ and $\hat{C}_{23}$. From the explicit expressions of $\hat{C}_{ij}$ found in [1], one finds

$$\hat{E}_{12} = e^{i(2\varphi_1 - \varphi_2)}, \qquad \hat{E}_{23} = e^{i(-\varphi_1 + 2\varphi_2)}.$$

An operator like $\hat{E}_{12}$ is reputed to be the exponential of some Hermitian phase operator $\hat{\varphi}_{12} = 2\hat{\varphi}_1 - \hat{\varphi}_2$, corresponding to a phase observable. It is easy to verify that

$$\left[\frac{1}{2}\Gamma(\hat{h}_1), \hat{E}_{12}\right] = \hat{E}_{12} \qquad \left[\frac{1}{2}\Gamma(\hat{h}_2), \hat{E}_{23}\right] = \hat{E}_{23}, \qquad \left[\hat{E}_{12}, \hat{E}_{23}\right] \sim 1/\lambda,$$

indicating first that $\hat{E}_{12}$ and $\hat{E}_{23}$ are conjugate to $\hat{h}_1$ and $\hat{h}_2$, respectively, and that

$$\hat{E}_{12}\hat{E}_{23} \neq \hat{E}_{23}\hat{E}_{12} \Rightarrow \mathrm{e}^{i\hat{\varphi}_{12}}\mathrm{e}^{i\hat{\varphi}_{23}} \neq \mathrm{e}^{i\hat{\varphi}_{23}}\mathrm{e}^{i\hat{\varphi}_{12}} \Rightarrow [\hat{\varphi}_{12},\hat{\varphi}_{23}] \neq 0, \tag{3}$$

i.e. phase operators do not commute in the quantum domain. This effect disappears in the limit of large number of excitations, as can be understood as follows. Write $\hat{E}_{12}$ as

$$\hat{E}_{12} = \sum_{\nu_3=0}^{\lambda} \sum_{p=0}^{\lambda-\nu_3} |\lambda - \nu_3 - p + 1, p - 1, \nu_3\rangle\langle\lambda - \nu_3 - p, p, \nu_3|. \tag{4}$$

Since all entries in $|\nu_1\nu_2\nu_3\rangle$ must be non–negative, $\hat{E}_{12}$ annihilates any state of the form $|\lambda - \nu_3, 0, \nu_3\rangle$, containing no excitation in mode 2; there are $\lambda + 1$ such states. Using a similar argument, one shows that $\hat{E}_{23}$ must annihilate states containing no excitation in mode 3, of which there are again $\lambda+1$. Thus, the number of entries where $(\hat{E}_{23}\hat{E}_{12})_{ij} \neq (\hat{E}_{12}\hat{E}_{23})_{ij}$ grows like λ. On the other hand, the total number of states in the UIR is $\frac{1}{2}(\lambda + 1)(\lambda + 2)$. Hence we expect the number of states for which the commutator does not vanish relative to the total number of states in the UIR to decay like $1/\lambda$.

3. su(n+1) states on the subgroup $\mathcal{H}$: the example of $su(3)$.

3.1. Basis state: the example of su(3) on SU(2)×U(1).

In order to obtain basis states in terms of subgroup functions, start with $|\chi_\lambda\rangle$ a highest weight, and "translate" $|\chi_\lambda\rangle$ by a sufficiently general $g^{-1} \in SU(3)$ to get the coherent state $g^{-1}|\chi_\lambda\rangle$. A "rotation" by $\omega^{-1} \in \mathcal{H} \subset SU(3)$ will analyse $g^{-1}|\chi_\lambda\rangle$ in its subgroup components. The component proportional to the basis state $|\psi\rangle$ is then projected to get $\langle\psi|\,\omega^{-1}\,g^{-1}|\chi_\lambda\rangle$. It is usually convenient to take the complex conjugate to obtain the final expression

$$|\psi\rangle \mapsto \psi_g(\omega) \equiv \langle\chi_\lambda|\,g\,\omega\,|\psi\rangle \tag{5}$$

As an example, consider an $su(3)$ states $|\nu; I\rangle$, where [2]

$$|\nu; I\rangle = \sum_{m_1 m_2 m_3} (\tfrac{1}{2}\nu_1, m_1; \tfrac{1}{2}\nu_2, m_2|I, m_2)(I, M; \tfrac{1}{2}\nu_3, m_3|\tfrac{1}{2}\lambda, m_3)$$
$$\times |\tfrac{1}{2}\nu_1, m_1\rangle|\tfrac{1}{2}\nu_2, m_2\rangle|\tfrac{1}{2}\nu_3, m_3\rangle, \tag{6}$$

with $(s_1, m_1; s_2, m_2|S, m_2)$ is an SU(2) Clebsch–Gordan coefficient, and where

$$|j_i\, m_i\rangle = \frac{(a_{i+}^\dagger)^{j_i+m_i}\,(a_{i-}^\dagger)^{j_i-m_i}}{\sqrt{(j_i+m_i)!(j_i-m_i)!}}, \qquad m_i = \tfrac{1}{2}(\nu_i - \nu_{i+1}), \qquad \nu_1 + \nu_2 + \ldots = \lambda + 2\mu.$$

The construction of $|\nu; I\rangle$ thus requires two types of bosons for each field, the extra indices $\pm$ describing internal degrees of freedom, identified with, say, any two orthogonal polarizations of the photon. The polarization–summed bilinears

$$\hat{C}_{ij} = a_{i+}^\dagger a_{j+} + a_{i-}^\dagger a_{j-}, \qquad i, j = 1, \ldots, n+1, \tag{7}$$

generate transformations leaving polarization unaffected. In addition to the SU(2) label I, basis states $|\nu; I\rangle$ also carry a U(1)×U(1) label via $\hat{h}_1|\nu; I\rangle = (\nu_1 - \nu_2)|\nu; I\rangle$, $\hat{h}_2|\nu; I\rangle = (\nu_1 + \nu_2 - 2\nu_3)|\nu; I\rangle$. Here, it is convenient to use $\hat{h}_2 = \hat{C}_{11} + \hat{C}_{22} - 2\hat{C}_{33}$, as it is invariant under the subgroup. To obtain states on SU(2)× U(1), apply directly Eq.(5) to obtain:

$$\begin{aligned}
|\nu; I\rangle \mapsto \langle\chi_\lambda|\,g\,\omega\,|(\nu_1\nu_2\nu_3); I\rangle &= \sum_{\nu_1'\nu_2'\nu_3'}\langle\chi_\lambda|\,g\,|(\nu_1'\nu_2'\nu_3'); I\rangle\,\langle(\nu_1'\nu_2'\nu_3'); I|\omega\,|(\nu_1\nu_2\nu_3); I\rangle, \\
&= \sum_k \langle\chi_\lambda|\,g\,|(\nu_1'\nu_2'\nu_3'); I\rangle\,\mathcal{D}_{k,m}^I(\alpha, \beta, \gamma)\,\mathrm{e}^{ip\varphi}, \\
&= \sum_k a_{pIk}\,\mathcal{D}_{k,m}^I(\alpha, \beta, \gamma)\,\mathrm{e}^{ip\varphi},
\end{aligned}$$

with a_{pIK} a coefficient depending on the group element g, $\omega = (\alpha, \beta, \gamma) \cdot e^{ip\varphi} \in SU(2) \times U(1)$, $\mathcal{D}^l_{k,m}$ an SU(2) Wigner function, $e^{ip\varphi}$ a U(1) function, $k = \frac{1}{2}(\nu'_1 - \nu'_2)$, $m = \frac{1}{2}(\nu_1 - \nu_2)$.

3.2. Obtaining Γ: the example of the 2–torus.

The SU(3) highest weight for the UIR $(\lambda, 0)$ is defined via the action of six elements of $su(3)$:

$$\hat{C}_{12}|\chi_\lambda\rangle = 0, \ \hat{C}_{23}|\chi_\lambda\rangle = 0, \ \hat{C}_{32}|\chi_\lambda\rangle = 0, \ \hat{C}_{13}|\chi_\lambda\rangle = 0, \ \hat{h}_1|\chi_\lambda\rangle = \lambda|\chi_\lambda\rangle, \ \hat{h}_2|\chi_\lambda\rangle = 0.$$

To the above six generators, one adds $g^{-1}\hat{h}_1 g$ and $g^{-1}\hat{h}_2 g$ to obtain a set of eight linearly independent traceless matrices. From $\psi_g(\varphi_1, \varphi_2)$, define

$$\Gamma_2(\hat{X})\psi_g(\varphi_1, \varphi_2) \equiv \langle\chi_\lambda| \, g\,\omega\, \hat{X} \, |\psi\rangle, \quad \omega = e^{i\varphi_1\hat{h}_1} \, e^{i\varphi_2\hat{h}_2} \in U(1) \times U(1).$$

If $\hat{X}$ is $\hat{h}_1$, we immediately find the expression of $\hat{h}_1$ as $\partial/\partial\varphi_1$. $\Gamma_2(\hat{h}_2)$ is found in a similar way. If $\hat{X}$ is a ladder operator, for instance $\hat{C}_{12}$, then conjugate $\hat{C}_{12}$ by $\omega = e^{i\varphi_1\hat{h}_1} \, e^{i\varphi_2\hat{h}_2}$ to get $\omega\,\hat{C}_{12} = e^{i(2\varphi_1 - \varphi_2)}\hat{C}_{12}\,\omega$, conjugate again $\hat{C}_{12}$ by g and expand $\langle\chi_\lambda|g\,\hat{C}_{12}\,g^{-1}$ using the aforementioned set of eight generators to obtain $\Gamma_2(\hat{C}_{12})$.

3.3. Obtaining $su(3)$ on SU(2)×U(1)

Representations on the torus being limited to those with highest weights $(\lambda, 0, \dots)$, it would be interesting from a mathematical point of view to construct UIRs with more general highest weights. The next simplest UIRs of $su(n + 1)$ are of the type $(\lambda, \mu, 0, \dots)$. As mentioned before, these have applications in the description of polarized photons.

It is possible to express $su(3)$ generators in terms of operators that have simple actions in terms on SU(2)×U(1) subgroup functions. The coefficients a_{pIk} in the basis states are determined by a recursion relation connecting coefficients with identical values of p, I but different values of k and by a normalization condition.

To obtain Γ_3 on SU(2)×U(1), we need the four subgroup generators to provide a right action, plus another copy of these four generators conjugated by the group element g to provide a left action. Using this set for expansion purposes, and following a prescription adapted from the torus case, one obtains, for instance,

$$
\begin{aligned}
\Gamma_3(\hat{C}_{12}) &= -ie^{i\gamma}\left(-i\frac{\partial}{\partial\beta} - \frac{1}{\tan\beta}\frac{\partial}{\partial\gamma}\right), \qquad \Gamma_3(\hat{h}_1) = -2i\frac{\partial}{\partial\gamma} \\
\Gamma_3(\hat{C}_{13}) &= \mathcal{D}^{1/2}_{1/2, 1/2}(\alpha, \beta, \gamma)\,e^{i\varphi}\left(-\frac{\sqrt{2}}{3}(2\lambda + \mu) + \hat{\mathbf{C}}_{12} - \frac{\sqrt{2}}{2}(\hat{\mathbf{h}}_1 + \frac{1}{3}\hat{\mathbf{h}}_2)\right) \\
&\quad + \mathcal{D}^{1/2}_{-1/2, 1/2}(\alpha, \beta, \gamma)\,e^{i\varphi}\left(\frac{1}{3\sqrt{2}}(\lambda + 2\mu) + \frac{1}{3}\hat{\mathbf{h}}_2\right)
\end{aligned}
$$

where $\hat{\mathbf{C}}_{12}$, $\hat{\mathbf{h}}_1$, $\hat{\mathbf{h}}_2$ act on $\mathcal{D}^l_{km}(\alpha, \beta, \gamma)e^{ip\varphi}$ by left action, v.g $\hat{\mathbf{C}}_{12}\mathcal{D}^l_{km}(\alpha, \beta, \gamma) = \sqrt{(I + k)(I - k + 1)}\mathcal{D}^l_{k-1,m}(\alpha, \beta, \gamma)$.

Elements in the $su(2) \times u(1)$ subalgebra act on basis states as right vector fields. Other generators contain a mixture of diagonal operators and operators acting on the left, premultiplied by subgroup functions.

Reference

[1] de Guise H and Bertola M 2002 J.Math.Phys. 43 3425–3444
[2] Rowe D J, Sanders B C and de Guise H 1999 3604–3615

Inst. Phys. Conf. Ser. No 173: Section 7
Paper presented at 24th Int. Coll. Group Theoretical Methods in Physics, Paris, France, 15–20 July 2002
©*2003 IOP Publishing Ltd*

Metric formulation of Galilean invariance in five dimensions

M de Montigny, F C Khanna and A E Santana

Theoretical Physics Institute, Univ. of Alberta, Edmonton, Canada T6G 2J1

Abstract. This is a summary of recent publications and works in progress. We describe a five-dimensional metric formulation of Galilean covariance and illustrate it with various examples in field theory. As a first illustration, we recover the two Galilean limits of electromagnetism first examined by Le Bellac and Lévy-Leblond. Then we describe the field theoretical formulation of some fluids and superfluids models : Navier-Stokes equation, Takahashi Lagrangian for irrotational fluids and Thellung-Ziman Lagrangian for Helium II. Our last examples treat non-relativistic Bhabha equations for spin 0 and 1 particles, and Dirac equation for spin 1/2.

1. Introduction

There exists a wealth of low-energy systems where any new methods or results concerning Galilean invariance are likely to be useful. Landau's theory of superfluid state of helium is just one example. The general program presented hereafter consists in investigating the physical applications of a metric formulation of Galilei invariance such that one can use tensor analysis, etc. Hereafter we summarize the references [1]-[5], which contain more details.

A *Galilean five-vector* is such that a boost acts on it as $x^{\mu'} = \Lambda^{\mu'}_{\nu} x^{\nu}$:

$$
\begin{pmatrix} x^{1'} \\ x^{2'} \\ x^{3'} \\ x^{4'} \\ x^{5'} \end{pmatrix} = \begin{pmatrix} 1 & 0 & 0 & -V_1 & 0 \\ 0 & 1 & 0 & -V_2 & 0 \\ 0 & 0 & 1 & -V_3 & 0 \\ 0 & 0 & 0 & 1 & 0 \\ -V_1 & -V_2 & -V_3 & \frac{1}{2}\mathbf{V}^2 & 1 \end{pmatrix} \begin{pmatrix} x^1 \\ x^2 \\ x^3 \\ x^4 \\ x^5 \end{pmatrix} \tag{1}
$$

with $\mathbf{V}$ the relative velocity. The scalar product $g_{\mu\nu} A^{\mu} B^{\nu}$ of two Galilei-vectors A and B is invariant under transformation (1) if we define the Galilean metric by

$$
g^{\mu\nu} = g_{\mu\nu} = \begin{pmatrix} 1 & 0 & 0 & 0 & 0 \\ 0 & 1 & 0 & 0 & 0 \\ 0 & 0 & 1 & 0 & 0 \\ 0 & 0 & 0 & 0 & -1 \\ 0 & 0 & 0 & -1 & 0 \end{pmatrix}. \tag{2}
$$

Galilean one-forms transform as

$$
x_{\mu'} = g_{\mu'\alpha'} x^{\alpha'} = \overbrace{g_{\mu'\alpha'} \Lambda^{\alpha'}_{\beta} g^{\beta\nu}}^{\Lambda^{\nu}_{\mu'}} x_{\nu}. \tag{3}
$$

Hereafter we often consider components with units of length :

$$
(x^1, \ldots, x^5) = \left(\mathbf{x}, vt, \frac{s}{v} \right). \tag{4}
$$

For a real field $\tilde{\phi}$, the projection onto the Newtonian space-time is defined as

$$\tilde{\phi}(x) \equiv \phi(\mathbf{x}, t) + a_0 s \tag{5}$$

with a_0 a constant. For a complex field $\tilde{\psi}$ we use the definition :

$$\tilde{\psi}(x) \equiv e^{i a_0 s} \psi(\mathbf{x}, t). \tag{6}$$

Thereafter we use mostly $a_0 = \pm 1$ or $\pm m$.

The new coordinate s can be seen as the dual of the mass. Therefore the mass does enter as a remnant of the fifth component of the particle's momentum, starting from an apparently massless theory in five dimensions!

2. Galilean electromagnetism

More details for this section can be found in [1]. Here we illustrate the formalism by recovering the two 'Galilean limits' of electromagnetism derived nearly thirty years ago by Le Bellac and Lévy-Leblond. Throughout this section only we define the embedding of the Newtonian space-time into the de Sitter space by : $(\mathbf{x}, t) \hookrightarrow x = (\mathbf{x}, t, 0)$, so that $\partial_k = \nabla_k, \partial_4 = \partial_t$ and $\partial_5 = 0$. The Galilean limits will be obtained by defining two embeddings of the five-potential $A_\mu = (\mathbf{A}, A_4, A_5)$ which transforms as

$$\begin{aligned}
\mathbf{A}' &= \mathbf{A} + \mathbf{V} A_5 \\
A_{4'} &= A_4 + \mathbf{V} \cdot \mathbf{A} + \tfrac{1}{2} \mathbf{V}^2 A_5 \\
A_{5'} &= A_5.
\end{aligned} \tag{7}$$

Consider the five-dimensional electromagnetic antisymmetric tensor :

$$F_{\mu\nu} \equiv \partial_\mu A_\nu - \partial_\nu A_\mu = \begin{pmatrix}
0 & b_3 & -b_2 & c_1 & d_1 \\
-b_3 & 0 & b_1 & c_2 & d_2 \\
b_2 & -b_1 & 0 & c_3 & d_3 \\
-c_1 & -c_2 & -c_3 & 0 & a \\
-d_1 & -d_2 & -d_3 & -a & 0
\end{pmatrix} \tag{8}$$

so that

$$\begin{aligned}
\mathbf{b} &= \nabla \times \mathbf{A} \\
\mathbf{c} &= \nabla A_4 - \partial_4 \mathbf{A} \\
\mathbf{d} &= \nabla A_5 - \partial_5 \mathbf{A} \\
a &= \partial_4 A_5 - \partial_5 A_4.
\end{aligned} \tag{9}$$

The Maxwell equations :

$$\partial_\mu F_{\alpha\beta} + \partial_\alpha F_{\beta\mu} + \partial_\beta F_{\mu\alpha} = 0 \tag{10}$$

$$\partial_\nu F^{\mu\nu} = j^\mu \tag{11}$$

become

$$\begin{aligned}
\nabla \cdot \mathbf{b} &= 0 \\
\nabla \times \mathbf{c} + \partial_4 \mathbf{b} &= \mathbf{0} \\
\nabla \times \mathbf{d} + \partial_5 \mathbf{b} &= \mathbf{0} \\
\nabla a - \partial_4 \mathbf{d} + \partial_5 \mathbf{c} &= \mathbf{0}
\end{aligned} \tag{12}$$

and

$$\begin{aligned}
\nabla \times \mathbf{b} - \partial_5 \mathbf{c} - \partial_4 \mathbf{d} &= \mathbf{j} \\
\nabla \cdot \mathbf{c} - \partial_4 a &= -j_4 \\
\nabla \cdot \mathbf{d} + \partial_5 a &= -j_5.
\end{aligned} \tag{13}$$

The *electric limit* corresponds to the embedding :

$$(\mathbf{A}_e, \phi_e) \hookrightarrow A_e = (\mathbf{A}_e, 0, -\mu_0\epsilon_0\phi_e). \tag{14}$$

Let us just mention that, for instance, equations (12) and (13) become

$$\begin{aligned}
\nabla \times \mathbf{E}_e &= \mathbf{0} \\
\nabla \cdot \mathbf{B}_e &= 0 \\
\nabla \times \mathbf{B}_e - \mu_0\epsilon_0\partial_t\mathbf{E}_e &= \mu_0\mathbf{j}_e \\
\nabla \cdot \mathbf{E}_e &= \tfrac{1}{\epsilon_0}\rho_e
\end{aligned} \tag{15}$$

as obtained by Le Bellac and Lévy-Leblond. The *magnetic limit* corresponds to the embedding

$$(\mathbf{A}_m, \phi_m) \hookrightarrow A_m = (\mathbf{A}_m, \phi_m, 0) \tag{16}$$

and, using equations (12) and (13), leads to

$$\begin{aligned}
\nabla \times \mathbf{E}_m &= -\partial_t\mathbf{B}_m \\
\nabla \cdot \mathbf{B}_m &= 0 \\
\nabla \times \mathbf{B}_m &= \mu_0\mathbf{j}_m \\
\nabla \cdot \mathbf{E}_m &= \tfrac{1}{\epsilon_0}\rho_m.
\end{aligned} \tag{17}$$

3. Fluids and superfluids

Hereafter we summarize references [2, 3]. Consider the functional Lagrangian :

$$\tilde{\mathcal{L}}[\tilde{\rho}, \tilde{\phi}] = -\frac{1}{2}\tilde{\rho}\partial_\mu\tilde{\phi}\partial^\mu\tilde{\phi} - V(\tilde{\rho}). \tag{18}$$

With $\tilde{\rho}(x) \equiv \rho(\mathbf{x}, t)$ and the embedding for ϕ mentioned previously, we obtain

$$\frac{1}{2}\nabla\phi \cdot \nabla\phi + \partial_t\phi = -V'. \tag{19}$$

The gradient of this expression, with $\mathbf{v} = \nabla\phi$ and $\nabla(V') = \frac{1}{\rho}\nabla p$ (p : pressure) gives the Navier-Stokes equation :

$$\partial_t\mathbf{v} + (\mathbf{v} \cdot \nabla)\mathbf{v} = -\frac{1}{\rho}\nabla p. \tag{20}$$

In [3] we have found the Takahashi model for compressible irrotational barotropic fluids with pressure proportional to the square of the mass density with

$$\tilde{\mathcal{L}} = \frac{\rho_0}{8v_0^2}\left(\partial^\mu\tilde{\phi}\partial_\mu\tilde{\phi} - 2v_0^2\right)^2. \tag{21}$$

On the other hand the similar, but gauged, Lagrangian :

$$\tilde{\mathcal{L}} \propto D_\mu\tilde{\psi}D^\mu\tilde{\psi}^* \equiv (\partial_\mu\tilde{\psi} + i\tilde{A}_\mu\tilde{\psi})(\partial^\mu\tilde{\psi}^* - i\tilde{A}^\mu\tilde{\psi}^*) \tag{22}$$

with $\tilde{A}_\mu = k\partial_\mu\tilde{\rho}$, provides another model suggested by Takahashi.

Now consider equation (18) with Clebsch-like transformation $\partial\tilde{\phi} \to \partial\tilde{\phi} + \tilde{\alpha}\partial\tilde{\beta}$:

$$\tilde{\mathcal{L}} = -\frac{\tilde{\rho}}{2v_0^2}(\partial_\mu\tilde{\phi} + \tilde{\alpha}\partial_\mu\tilde{\beta})(\partial^\mu\tilde{\phi} + \tilde{\alpha}\partial^\mu\tilde{\beta}) - V(\tilde{\rho}). \tag{23}$$

If we define $\tilde{\alpha}(x) = \alpha(\mathbf{x}, t)$, $\tilde{\beta}(x) = \beta(\mathbf{x}, t)$ and $\tilde{\rho}(x) = \rho(\mathbf{x}, t)$ then the Lagrangian of equation (23) after projection onto the Newtonian space-time becomes

$$\mathcal{L} = \frac{\rho}{v_0^2}\left(\partial_t\phi - \frac{1}{2}\nabla\phi \cdot \nabla\phi + \alpha(\partial_t\beta - \frac{1}{2}\alpha\nabla\beta \cdot \nabla\beta) - \alpha\nabla\phi \cdot \nabla\beta\right) - V(\rho). \tag{24}$$

530

This may be expressed as

$$\mathcal{L} = \frac{\rho}{v_0^2}\left(\partial_t\phi + \alpha\partial_t\beta - \frac{1}{2}\mathbf{v}^2\right) - V(\rho) \tag{25}$$

where $\mathbf{v} = -\nabla\phi - \alpha\nabla\beta$. This Lagrangian was employed by Thellung and Ziman.

4. Linear first-order wave equations : spin 0, 1 and 1/2

More details can be found in [4, 5]. The momentum-space version of the Duffin-Kemmer-Petiau (DKP) equation is

$$(\beta^\mu p_\mu - ik)\Psi = 0 \tag{26}$$

with matrices β satisfying the DKP algebra :

$$\beta^\mu\beta^\lambda\beta^\nu + \beta^\nu\beta^\lambda\beta^\mu = g^{\mu\lambda}\beta^\nu + g^{\nu\lambda}\beta^\mu \tag{27}$$

If we introduce a DKP spinor

$$\Psi \equiv \begin{pmatrix} \mathbf{A} \\ \theta \\ \varphi \\ \phi \end{pmatrix} \tag{28}$$

and consider the harmonic oscillator by performing the substitution $\mathbf{p} \to \mathbf{p} + i\omega\eta\mathbf{r}$, we find

$$E\phi = \left(\frac{\mathbf{p}^2}{2m} + \frac{1}{2}m\omega^2\mathbf{r}^2 - \frac{3}{2}\hbar\omega\right)\phi. \tag{29}$$

For spin one, we consider a 15-dimensional spinor and after a few steps, we obtain the Hamiltonian :

$$E\mathbf{A} = \left[\frac{\mathbf{p}^2}{2m} + \frac{1}{2}m\omega^2\mathbf{r}^2 - \frac{3}{2}\hbar\omega - \frac{\omega}{\hbar}\mathbf{L}\cdot\mathbf{S}\right]\mathbf{A}. \tag{30}$$

For the spin 1/2 Dirac field, we replace the β with γ in equation (26), with the Clifford algebra

$$\{\gamma^\mu, \gamma^\nu\} = \gamma^\mu\gamma^\nu + \gamma^\nu\gamma^\mu = 2g^{\mu\nu}. \tag{31}$$

The Hamiltonian obtained thereby for the non-relativistic oscillator is

$$E\varphi = \left(\frac{\mathbf{p}^2}{2m} + \frac{1}{2}m\omega^2\mathbf{r}^2 - \frac{3}{2}\hbar\omega - \frac{2}{\hbar}\omega\mathbf{L}\cdot\mathbf{S}\right)\varphi. \tag{32}$$

NSERC (Canada), CAPES and CNPq (Brazil) are acknowledged for financial support.

References

[1] de Montigny M, Khanna F C and Santana A E Non-relativistic wave equation with external gauge field (in preparation)
[2] de Montigny M, Khanna F C and Santana A E Lorentz-like covariant equations of non-relativistic fluids (in preparation)
[3] de Montigny M, Khanna F C and Santana A E 2001 J. Phys. A : Math. Gen. 34 10921-10937
[4] de Montigny M, Khanna F C, Santana A E and Santos E S 2001 J. Phys. A : Math. Gen. 34 8901-8917
[5] de Montigny M, Khanna F C, Santana A E, Santos E S and Vianna J D M 2000 J. Phys. A : Math. Gen. 33 L273-L278

Inst. Phys. Conf. Ser. No 173: Section 7
Paper presented at 24th Int. Coll. Group Theoretical Methods in Physics, Paris, France, 15–20 July 2002
©*2003 IOP Publishing Ltd*

Nonperturbative approach to potentials in impenetrable boxes based on quasi-exact solvability

A. de Souza Dutra, V. G. C. S. dos Santos and A. M. Stuchi
UNESP-Campus de Guaratinguetá-DFQ
Av. Dr. Ariberto Pereira Cunha, 333
12516-410 Guaratinguetá SP Brasil

Abstract. In this work we develop an approach to obtain analytical expressions for potentials in an impenetrable box. In this kind of system the expression has the advantage of being valid for arbitrary values of the box length, and respect the correct quantum limits. The similarity of this kind of problem with the quasi exactly solvable potentials is explored in order to accomplish our goals. Problems related to the break of symmetries and simultaneous eigenfunctions of commuting operators are discussed.

1. Introduction

Quantum systems under non trivial boundary conditions, corresponding to penetrable and nonpenetrable walls, do simulate the effect of atoms or molecules in the neighbour of a central particle. The dependence of the eigenvalues on the box size, allows one to define the effect of the pressure over the system. Some examples would include that of the proton-deuteron transformation as a source of energy in dense stars [1], the determination of the escape rate from galactic and globular clusters [2], and the understanding of the spectral line shift under pressure [3]. One good list of phenomena associated to this kind of physical system is presented in [4]. In the last years, as a consequence of the developing of systems like quantum wells and quantum dots, many of quantum systems which were supposed to be only of academic interest, became realistic possibilities to be tested. In fact this gives rise to a plenty of new physical possibilities. In particular spatially confined systems, which have been poorly investigated, now are in conditions to be confronted with experimental data [5].

First of all, we illustrate the method by treating the harmonic oscillator bounded in a "one-dimensional box" of length a. Some similar problems were treated previously in the literature [6], [7], but in general only solutions valid for a specific value of the potential parameter are taken and, every time one needs the energy spectra for another value of that parameter, the numerical calculation must be repeated. This implies that one hardly can visualize a qualitative behavior of the system, besides there is the obvious unpleasant need of repeated numerical calculations each time one needs to change the value of the potential parameter.

2. Harmonic Oscillator in a one dimensional box

We start by treating the harmonic oscillator in a box, similarly to the treatment given to the quasi-exactly solvable potentials [8]-[10], which did lead us to obtain analytically approximated solutions for the eigenvalues of anharmonic oscillators under usual boundary conditions [11].

Remembering that the general solution for the harmonic oscillator is given by something like

$$\Psi_n\left(x\right) = P_n\left(x\right)\exp\left(-\alpha x^2\right),\tag{1}$$

where $P_n(x)$, in principle, is a polynomial of infinite degree (associated to the parabolic cylindrical functions), which however is taken finite (Hermite polynomials) in the case of boundary conditions at the infinity, in order to get a convergent function. Here however, the exponential does not guarantees that the wavefunction vanishes at the walls of the interval, so that this condition determines the energy spectrum. Notwithstanding, when we do insist to using finite polynomials, which are nothing but the known Hermite ones, whose energy is well defined and given by $E_n = \hbar\omega\left(n + \frac{1}{2}\right)$, we get an equation for the oscillator frequency which, once solved, give us those frequencies which generate polynomials whose zeroes are at the walls of the box. At this point it is important to stress the similarity of this problem with that of the so called quasi-exactly solvable potentials, in the sense that in this case only for some specific frequencies the eigenfunction will be elementary or, in other words, it is represented by a finite polynomial, the Hermite ones, times an usual exponentially decaying factor. Once one gets those frequencies, one must to substitute them on the above expression for the energy, so obtaining the corresponding exact energy for the harmonic oscillator in a box.

In order to obtain the frequency and the energy for an oscillator in a box whose walls are at $-a/2$ and $a/2$, by applying the idea which was outlined above, we use the condition $H_n\left(\pm\frac{a}{2}\right) = 0$, so that $H_0(x) = 1$, does not give us any solution, the polynomial $H_2\left(\pm\frac{a}{2}\right)$ generates one solution, $H_4\left(\pm\frac{a}{2}\right)$ two solutions, so for and so on. To be more precise, let us exemplify with case of $H_2\left(\pm\frac{a}{2}\right)$. In this case the equation which implies that the wavefunction vanishes at the boundary is given by $H_2\left(\pm\frac{a}{2}\right) = a_0\left(-2 + 4\xi\right) = 0$, where $\xi = \sqrt{\frac{m\omega}{\hbar}}$. Remembering that in this case $E_2 = \frac{5}{2}\hbar\omega$, one gets finally: $\omega = \frac{2\hbar}{ma^2}$, $E = 5\frac{\hbar^2}{ma^2}$. As can be easily verified, the wavefunction of the harmonic oscillator for the above frequency and energy vanishes at the boundaries, not presenting any nodes between the walls. So it corresponds to the ground state of the oscillator in a box with impenetrable walls. Now, if one takes the following even case of $H_4\left(\pm\frac{a}{2}\right)$, it is straightforward to show that there are now two possible solutions of (2), respectively corresponding to the ground state and the second excited one, so for and so on. Analogously a similar structure appears in the odd cases.

In this way, by solving a sufficiently great number of polynomials, we can plot the energy as a function of the frequency for each one of the energy levels, and then try to get a functional relation between these several levels. So obtaining an approximate analytical expression for the dependence on the frequency, the length of the box and the quantum number.

At this point, in order to help our quest for a suitable analytical expression for the energy, we remember that in the limit of zero frequency one recalls the free particle in a box. On the other hand, when the length of the box goes to infinity, one should recover the well known harmonic oscillator spectra. Finally for great values of the principal quantum number, once again the energy becomes close to that of a free particle in a box. As a consequence of these physical constraints, we write the expression for the energy as

$$\mathcal{E} = \left(\frac{2ma^2}{\pi^2\hbar^2}\right)E = n^2 + \frac{8}{\pi^2}\left(n - \frac{1}{2}\right)\omega_a + g\left(\omega_a\right), \, n = 1, 2, 3, \ldots \tag{2}$$

where $\omega_a = \frac{ma^2}{\hbar}\frac{1}{4}\omega$ and the function $g\left(\omega_a\right)$ should have as appropriate limits: $g\left(\omega_a = 0\right) = 0$, $g\left(\omega_a \to \infty\right) = 0$. Furthermore, one should also have this function going to zero when n

increases or at most growing more slowly than n^2. Using this hypothesis as a guide we try the following form for $g_n(\omega_a)$:

$$g_n(\omega_a) = c_0(n)\,\omega_a\,e^{-\sum_{j=1}^{J} c_j(n)\,\omega_a^j}. \tag{3}$$

After performing the necessary calculations we verify that indeed, this function has the expected behavior supposed above. Besides, by obtaining the roots of Hermite polynomials up to two hundred degree, we find an approximated expression for the coefficients $c_j(n)$, taking $J = 3$, whose expressions are given below:

$$c_0(n) = \frac{1}{(0.405231 + 0.810579\,n)}, \tag{4}$$

$$c_1(n) = \frac{0.0104832}{n} - \frac{0.00588616}{n^2} - \frac{0.00187449}{n^3}, \tag{5}$$

$$c_2(n) = 10^{-6}\left[-1.24 + 1.35\cos h\,(0.1762\,n - 0.12)\right]^{-1},\ n > 1, \tag{6}$$

$$c_3(n) = 10^{-8}\left[-2.5 + 2.6\cos h\,(0.086\,n - 0.278)\right]^{-1},\ n > 3, \tag{7}$$

Note that for the last two coefficients, the first elements were separated in order to getting better fittings. In these cases one have $c_2(1) = 3.70973\,10^{-6}$, $c_3(1) = -1.54146\,10^{-6}$, $c_3(2) = -8.78283\,10^{-7}$, $c_3(3) = 4.56951\,10^{-8}$. In fact, the coefficients really approaches to zero for larger values of n as supposed. These calculations were performed for $n \leq 20$, so that for higher values of n, one must extrapolate it, but it should be expected good results due to the behavior of $g_n(\omega_a)$ when n becomes greater and greater.

The comparison of the energy coming from the above analytical approximation with pure numerical values, in the case of the range of frequencies considered, shows that the error was always less than 0.07% along the range of the parameters verified which due to technique reasons, depends on the energy level studied and grows for higher quantum numbers. Below we present a Table with the corresponding ranges and their respective maximum percentual errors for the first twenty levels.

n	$\omega_{a_{max}}$	$\delta\%_{max}$	n	$\omega_{a_{max}}$	$\delta\%_{max}$
1	2.0	9.0×10^{-5}	11	60.5	0.018
2	6.0	0.072	12	67.3	0.018
3	10.9	9.1×10^{-3}	13	74.1	0.017
4	16.3	0.02	14	81	0.017
5	22.1	0.016	15	87.9	0.016
6	28.1	9.3×10^{-5}	16	94.9	0.016
7	34.4	0.012	17	101.9	0.015
8	40.8	0.017	18	109	0.013
9	47.2	0.019	19	116.1	0.016
10	53.8	0.019	20	123.2	8.9×10^{-3}

3. Coulomb Potential in a sphere

In this section we work with the more mathematically involved case of the Coulomb potential in a sphere of radius a. Notwithstanding, the basic program to be followed is the same of the previous section. This was done for the first nine energy levels, using one hundred degree polynomials. For each energy level, we plotted the exact energies coming from the solutions described above as a function of the parameter g, and fitted it with a polynomial of the third degree given by

$$E_{n,l} = \sum_{m=0}^{3} C_m^{(n,l)} g^m \qquad (8)$$

where the coefficients $C_m^{(n,l)}$ are presented in the Table below.

n,l m	0	1	2	3
1,0	-1.00421	0.502846	0.0152158	0.00487229
2,0	-4.50591	0.918442	0.0402514	0.00162719
2,1	-2.05892	0.384483	0.00167988	0.000448289
3,0	-9.69984	0.938899	-0.0182804	0.000282908

The energy expression coming from the use of the above parameters is in good accordance with the exact numerical data and, as happened in the harmonic oscillator case, does have an increasing range of validity with the increasing of the principal quantum number, as can be seen in the Table below. In these ranges, the error is quite small, about $10^{-2}\%$.

n,l	g_{max}	g_{min}	$\delta\%_{max}$
1,0	2.0	1.8	4.7×10^{-5}
2,0	7.1	6.15	4.3×10^{-3}
2,1	6.0	5.0	6.2×10^{-4}
3,0	15.5	12.9	9.4×10^{-3}

By analyzing the data, one can verify that the range was limited by the maximum polynomial used (in this case we used polynomials of degree 100). So, in order to increase the precision and the range of validity, one should use higher degrees for the Laguerre polynomials.

One interesting feature observed was that, in this case the energy is no more degenerate in the angular momentum quantum number, in contrast with the usual boundary condition at the infinite. So indicating the breaking of some symmetry of the system. As the spherical symmetry was preserved by the boundary condition used, we must to look for another broken symmetry to be responsible for this behavior. In fact, the symmetry related to the conservation of the Runge-Lenz vector is that one. It is interesting to observe that technically the Runge-Lenz operator and the Hamiltonian of the system are still commuting but, now due to the boundary conditions, the eigenfunctions of the Hamiltonian of this complete set of commuting operators are not shared with those of the Runge-Lenz operator. It is important to stress that this is a generic property of this kind of system.

Acknowledgments: The authors are grateful to FAPESP and CNPq for partial financial support.

[1] F. C.Auluck, Proc. Natl. Inst. Sci. India. **7** (1941) 133.
[2] S. Chandrasekar, Astroph. J. **97** (1943) 263.
[3] S. R. De Groot e C. A. Ten Seldam, J. Math. Physics (Utrecht) **4** (1937) 981.
[4] P. O. Fröman, S. Yngve e N. Fröman, J.Math. Phys. **28** (1987) 1813.
[5] M. A. Reed, Sci. Am. **268** (1993) 118.
[6] V. C. Aguilera-Navarro, H. Iwamoto, E. Ley Koo and A. H. Zimerman, Il Nuovo Cimento **62** (1981) 91.
[7] J. L. Marin and S. A. Cruz, Am J. Phys. **56** (1988) 1134.
[8] M. A. Shiffman, Int. J. Mod. Phys. A **4** (1989) 2897 and **4** (1989) 3305.
[9] A. de Souza Dutra and H. Boschi Filho, Phys. Rev. A **44** (1991) 4721.
[10] A. G. Ushveridze, Quasi-Exactly Solvable Models in Quantum Mechanics, Institute of Physics, Bristol, 1994, and references therein.
[11] A. de Souza Dutra, A. S. de Castro and H. Boschi Filho, Phys. Rev. A **51** (1995) 3480.
[12] A. S. de Castro and A. de Souza Dutra, Phys. Lett. A **269** (2000) 281.

Inst. Phys. Conf. Ser. No 173: Section 7
Paper presented at 24th Int. Coll. Group Theoretical Methods in Physics, Paris, France, 15–20 July 2002
©2003 IOP Publishing Ltd

Critical temperature for entanglement transition in Heisenberg Models

Hongchen Fu[†], Allan I Solomon[†] and Xiaoguang Wang[‡]

[†] Quantum Processes Group, The Open University, Milton Keynes, MK7 6AA, U.K.
[‡] Department of Physics, Macquarie University, Sydney, New South Wales 2109, Australia

Abstract. We study thermal entanglement in some low-dimensional Heisenberg models. It is found that in each model there is a critical temperature above which thermal entanglement is absent.

1. Introduction

Entanglement[1] plays an important role in quantum computation and quantum information processing. With appropriate coding, a system of interacting spins, such as described by a Heisenberg hamiltonian, can be used to model a solid-state quantum computer. It is therefore of some significance to study thermal entanglement in Heisenberg models. We find that for each model there is a corresponding critical temperature for transition to the entanglement regime, and the entanglement only occurs below this critical temperature.

2. Measures of entanglement

A pure state described by the wave function $|\Psi\rangle$ is *non-entangled* if it can be factorized as $|\Psi\rangle = |\Psi_1\rangle \otimes |\Psi_2\rangle$. Otherwise, it is entangled. A typical example of an entangled state is the Bell state for a bipartite system of two qubits:

$$\frac{1}{\sqrt{2}}(|01\rangle - |10\rangle) \tag{1}$$

For such a bipartite system the most popular entanglement measure is the *entanglement of formation*. For a pure state the entanglement of formation is defined as the reduced entropy of either subsystem[2].

For the two-qubit system one can use *concurrence*[3] as a measure of the entanglement. Let ρ_{12} be the density matrix of the pair which may represent either a pure or a mixed state. The concurrence corresponding to the density matrix is defined as

$$C_{12} = \max \{\lambda_1 - \lambda_2 - \lambda_3 - \lambda_4, 0\}, \tag{2}$$

where the quantities λ_i are the square roots of the eigenvalues of the operator

$$\varrho_{12} = \rho_{12}(\sigma_1^y \otimes \sigma_2^y)\rho_{12}^*(\sigma_1^y \otimes \sigma_2^y) \tag{3}$$

in descending order. The eigenvalues of ϱ_{12} are real and non-negative even though ϱ_{12} is not necessarily Hermitian. The entanglement of formation is a monotonic function of the concurrence, whose values range from zero, for an non-entangled state, to one, for a maximally entangled state.

3. Heisenberg models

The general N-qubit Heisenberg XYZ model in a magnetic field B is described by the Hamiltonian

$$H = \frac{1}{2} \sum_{n=1}^{N} \left(\sigma_n^x \sigma_{n+1}^x + \sigma_n^y \sigma_{n+1}^y + \sigma_n^z \sigma_{n+1}^z \right) + \sum_{n=1}^{N} B_n \sigma_n^z \tag{4}$$

where we assume cyclic boundary conditions $N + 1 \equiv 1$. The Gibbs state of a system in thermodynamic equilibrium is represented by the density operator

$$\rho(T) = \exp(-H/kT)/Z, \tag{5}$$

where $Z = \text{tr}[\exp(-H/kT)]$ is the partition function, k is Boltzmann's constant which we henceforth take equal to 1, and T is the temperature.

As $\rho(T)$ represents a thermal state, the entanglement in the state is called *thermal entanglement*. At $T = 0$, $\rho(0)$ represents the ground state which is pure for the non-degenerate case and mixed for the degenerate case. The ground state may be entangled. At $T = \infty$, $\rho(\infty)$ is a completely random mixture and cannot be entangled.

4. Thermal entanglement in the 2-site Heisenberg model

The density matrix can be obtained[4] as

$$\rho(T) = A \begin{pmatrix} e^{-B/T} & & & \\ & \cosh(J/T) & -\sinh(J/T) & \\ & -\sinh(J/T) & \cosh(J/T) & \\ & & & e^{B/T} \end{pmatrix} \tag{6}$$

where $A = (2\cosh(J/T) + 2\cosh(B/T))^{-1}$, and the concurrence

$$C = \max \left\{ \frac{\sinh(J/T) - 1}{\cosh(J/T) + \cosh(B/T)}, 0 \right\}. \tag{7}$$

As the denominator is always positive, the entanglement condition is

$$\sinh(J/T) - 1 > 0 \quad \text{or} \quad T < 1.134J. \tag{8}$$

from which we conclude that

- There is a critical temperature $T_c \sim 1.134J$. The thermal state is entangled when $T < T_c$.
- The critical temperature is independent of the magnetic field B.
- Entanglement occurs only for the antiferromagnetic case ($J > 0$).

5. Thermal entanglement in the 3-site Heisenberg model

We now consider pairwise entanglement in the 3-site Heisenberg model with uniform magnetic field, and also in the presence of a magnetic impurity[5]. The reduced density matrix of two sites can be written as

$$\rho_{12} = \frac{2}{3Z} \begin{pmatrix} u & & & \\ & w & y & \\ & y & w & \\ & & & v \end{pmatrix} \tag{9}$$

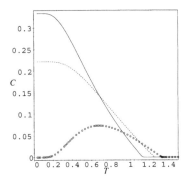

Figure 1. Concurrence as a function of T for different magnetic fields $B = 1$(solid line), 3/2(dashed line), and 2(circle point line).

The concurrence may be readily obtained as

$$C = \frac{4}{3Z} \max \left\{ |y| - \sqrt{uv}, 0 \right\},$$ (10)

In the case of a uniform magnetic field, the entanglement between any two sites is the same due to cyclic symmetry. We therefore need only consider the entanglement between sites 1 and 2. Then

$$
\begin{aligned}
u(B) = v(-B) &= \frac{3}{2} e^{3\beta B} + \frac{1}{2} e^{\beta B} (2z + z^{-2}) \\
w &= \cosh(\beta B)(2z + z^{-2}) \\
y &= \cosh(\beta B)(z^{-2} - z) \\
Z &= 2 \cosh(3\beta B) + 2 \cosh(\beta B)(2z + z^{-2}).
\end{aligned}
$$ (11)

where $(z = \exp(\beta J))$.

If $B = 0$, one can easily find that the sites 1 and 2 are entangled if and only if

$$2|z^{-2} - z| - 3 - 2z - z^{-2} > 0$$ (12)

from which we conclude that

- There is no entanglement when $J > 0$;
- Entanglement occurs when $J < 0$ and $T < T_c$, where the critical temperature is given by $-1.27J = 1.27|J|$.
- The maximal concurrence is 1/3, which occurs for $T \to 0$.

Fig.1 plots the concurrence against τ for different B. From these graphs we see that there exists a critical temperature above which the entanglement vanishes. It is also noteworthy that the critical temperature increases as the magnetic field B increases.

We now consider the case of a single impurity field on the third site; thus $B_1 = B_2 = 0$ and $B_3 = BJ > 0$. In this case the cyclic symmetry is violated and we have to consider the entanglement between sites 1 and 2, and between sites 1 and 3, separately.

Fig.2 plots the concurrence C_{12} and C_{13} against scaled temperature $\tau = kt/|J|$ for different magnetic fields B. From Fig.2(a) we see that when the magnetic field is located at the third site both the antiferromagnetic and ferromagnetic cases are entangled in the range $0 < \tau \leq \tau_c$, where the critical temperature τ_c depends on B. Fig.2(a) also suggests that

538

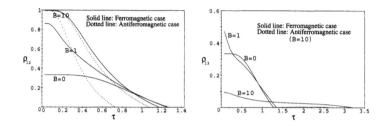

Figure 2. Concurrence C_{12} C_{13}against τ for different B. For antiferromagnetic case (dotted line), $B = 10$.

the concurrence C_{12} tends to 1, namely that the $(1, 2)$ entanglement becomes maximal, when $\tau \to 0$ for large enough B, in both the antiferromagnetic and ferromagnetic cases.

In contrast to the $(1, 2)$ case, the entanglement between sites 1 and 3 increases to a maximum with increasing B and then decreases. The lower the τ, the smaller the B at which the concurrence reaches its maximum value. For smaller B, entanglement occurs only in the ferromagnetic case ($J < 0$), while for large enough B (e.g. $B = 10$ in our units), weak entanglement occurs in both the antiferromagnetic and ferromagnetic cases.

[1] Schrödinger E 1935 Naturwissenschaften 23 807
[2] Werner R F 1989 Phys.Rev.A 40 4277
[3] Hill S and Wootters W K 1997 Phys.Rev. Lett. 78 5022; Wootters W K 1998 Phys.Rev. Lett. 80 2245; Coffman V, Kundu J and Wootters W K 2000 Phys.Rev. A 61 052306.
[4] Wang X 2001 Phys.Rev. A 64 012313 Wang X 2001 Phys. Lett. A 281 101
[5] Wang X, Fu H and Solomon A I 2001 J. Phys. A 34 11307.

Inst. Phys. Conf. Ser. No 173: Section 7
Paper presented at 24th Int. Coll. Group Theoretical Methods in Physics, Paris, France, 15–20 July 2002
©2003 IOP Publishing Ltd

Nonlinear Extension of Quantum Mechanics: Remarks and Recent Results

H-D Doebner

Department of Physics, Technical University of Clausthal, Germany
E-mail: asi@pt.tu-clausthal.de

G A Goldin

Departments of Mathematics and Physics, Rutgers University, New Brunswick, NJ, USA
E-mail: gagoldin@dimacs.rutgers.edu

Abstract. We discuss the problem of extending the framework of quantum mechanics so that observables are represented by nonlinear operators in a Hilbert space or manifold. Our motivation includes the derivation of certain nonlinear Schrödinger equations from diffeomorphism group representations. In this context we review some groups of nonlinear gauge transformations.

1. The problem of extending quantum mechanics

Quantum theory is based intimately on linear structures; for example: pure states belong to a (linear) Hilbert space $\mathcal{H}$, while mixed states are (linear) trace class operators; observables are described by (linear) self-adjoint operators in $\mathcal{H}$; representations of symmetry groups and algebras and the time-evolution equations (Schrödinger or von Neumann) are linear; inner products give us transition probability amplitudes, while the average outcomes of measurements occur in general as expectation values; the "collapse of the wave function" when a measurement occurs is described by linear projection onto a subspace of $\mathcal{H}$; and for N-particle systems the Hilbert space is the tensor product of one-particle spaces, with the corresponding N-particle observables obtained as operators extended linearly from product states to the full tensor product space. The obvious nonlinearities in classical mechanics and in classical field theory are mapped through quantization into a linear framework.

This reliance on linearity is indeed remarkable. Is such linearity correctly seen as an empirical property of nature, or is it an artifact of our formalism? Can there be a natural extension of quantum mechanics, as anticipated for example by Wigner [1], based on a Hilbert space but with *nonlinear* operators representing observables? A "superposition" principle might then hold only in the sense that a linear combination of states is again a possible state; but linear combinations of eigenfunctions in general fail to be eigenfunctions.

Introduction of nonlinear operators should if possible be motivated by some fundamental principle, with the usual linear formulation occurring as a limiting case. New physical effects are to be expected, with precision measurements establishing an empirical upper bound to the magnitude of any deviation from linear quantum mechanics [2].

We have been involved in formulating two sets of arguments that generate possible nonlinear time-evolutions in nonrelativistic quantum mechanics. The first method starts either with quantum kinematics described by generalized position and momentum operators on a smooth configuration space, or with the local current algebra of mass and momentum densities on the physical space [3, 4, 5, 6, 7]. Particular classes of nonlinear pure state time-evolution equations occur when we require compatibility with certain representations of these infinite-dimensional Lie algebras. Such an algebra is a canonical geometrical object: the semidirect sum of the scalar functions and the (tangent) vector fields on a manifold M (with compact support). It is the differential of the semidirect product of the additive group of scalar functions with the group of diffeomorphisms of M (with compact support). The irreducible representations are characterized in part by a real quantum number (in the sense of Wigner), that we call D. The continuity equation for density and current then leads to a family of nonlinear Schrödinger equations (NLSEs) [8, 9]. If the time-evolution of mixed states is well-defined, a von-Neumann equation results with a Lindblad term and nonlinearities [10, 11].

The second method begins with the usual linear (Schrödinger) time-evolution for pure states, and a characterization of quantum-mechanical measurements that does not presuppose linearity [12]. Transformations that preserve the outcomes of all measurements are considered to be gauge transformations; and suitable groups of nonlinear gauge transformations generate nonlinear time-evolutions [13, 14]. We review this approach briefly in Section 2. Both of these methods are building blocks in ordinary quantum mechanics, that are suggestive of nonlinear quantum mechanics. The families of nonlinear time-evolution equations that result are (unexpectedly) closely related.

Certainly, even if mathematically respectable NLSEs can be obtained from fundamental principles, the nonlinear extension of quantum mechanics is far from obvious. This is because the remainder of the usual linear framework (together with its standard interpretation) seems to disallow nonlinear time-evolutions for pure states. There are several obstructions: (1) Nonlinear evolution of pure states yields an evolution of mixed states that is not well-defined; e.g., see [11]. Some new notion of mixed states is required. (2) Nonlinear operators cannot be essentially self-adjoint; thus the usual probability interpretation based on the spectral decomposition of self-adjoint operators must be reformulated. (3) When nonlinear time-evolution is permitted, nonlocal effects appear quite generally in correlation measurements [15]; one possibility of avoiding them is based on the "open systems" approach proposed in [16]. (4) Consider a system of two particles, evolving nonlinearly according to the particular NLSEs obtained from diffeomorphism group representations, and prepared in a correlated pure state in the tensor product Hilbert space. Nonlocal effects (apparent interactions) then occur as a result of the nonlinearity in the single-particle Schrödinger equations; such nonlocalities may possibly be avoided using the construction of Polchinski for the two-particle Hamiltonian [17, 18]. We emphasize that this list of obstructions is not complete, and that the methods for possibly avoiding unphysical effects stemming from the nonlinearity may not be universally applicable. In our opinion, the question of the best mathematically and physically consistent nonlinear generalization of quantum mechanics remains open.

Apart from these difficulties, we have at present no known experimental result that suggests the need for a fundamental nonlinear term in the Schrödinger equation. However, we can use the NLSEs we have obtained to investigate an empirical upper bound for the coefficient D that occurs. For example, a NLSE of our type for the hydrogen atom is given by

$$i\hbar\partial_t\psi = [-\frac{\hbar^2}{2m}\nabla^2 + V(|\mathbf{x}|) + i\frac{\hbar}{2}D\frac{\nabla^2\rho}{\rho}]\psi, \tag{1}$$

where m is the reduced mass, V is the Coulomb potential, and $\rho = \bar{\psi}\psi$. Following [9] we find stationary states of (1) and eigenvalues from an affiliated linear Schrödinger equation, shifted from the usual eigenvalues. Interpreting these eigenvalues as energy levels, one compares for instance the relative frequency shifts for the hydrogen atom between the ground states and first excited states of the nonlinear equation (1) and the usual linear equation; i.e., one looks at

$$\Delta_{10} = \frac{\nu_{10}^{NL} - \nu_{10}^{L}}{\nu_{10}^{L}} \tag{2}$$

where ν_{10}^{NL} and ν_{10}^{L} are the respective frequencies. The accuracy of precision measurements that show no significant deviations from the linear result [19, 20] implies an upper bound for $|D|$ of approximately $10^{-7}\hbar/m$ [21]. For a possible logarithmic nonlinearity of the form $D(\ln\rho)\psi$ [22], one finds a related bound for $|D|$ from scattering experiments [23] [24].

2. Nonlinear Schrödinger equations from nonlinear gauge transformations

Groups of onlinear gauge transformations were introduced in [13], analyzed in [14], and generalized in [25] and [26]. Once can motivate them as follows. Consider a transformation $\mathcal{N}$, linear or nonlinear, defined and invertible on a suitable dense set in $\mathcal{H}$. If $\psi(\mathbf{x},t)$ satisfies a linear Schrödinger equation, then $\psi' = \mathcal{N}[\psi]$ satisfies (in general) a nonlinear equation that is, by construction, linearizable. Now we impose some physically-motivated conditions on $\mathcal{N}$.

First, take as a working conjecture the view that in quantum mechanics all measurements consist fundamentally of sequences of positional measurements performed at successive times. Then if we interpret $|\psi(\mathbf{x},t)|^2$ as the positional probability density for a single-particle state, we impose the condition that $|\psi'(\mathbf{x},t)|^2 = |\psi(\mathbf{x},t)|^2$ for all ψ as a necessary requirement for $\mathcal{N}$ to transform quantum states to physically equivalent states. Next assume the value of $\psi'(\mathbf{x},t)$ to depend strictly locally on ψ; that is, to be a function only of $\mathbf{x}, t$, and $\psi(\mathbf{x},t)$ independent of any values at other points in space or time, and independent of space or time derivatives of ψ. Obviously these properties hold for ordinary gauge transformations in quantum mechanics, and respect composition of tranformations. Finally consider the extension $\mathcal{N}^{(N)}$ of $\mathcal{N}$ to N-particle product states, and require that it be consistent with a separation condition,

$$\mathcal{N}^{(N)}[\psi_1 \otimes \ldots \otimes \psi_N] = \mathcal{N}[\psi_1] \otimes \ldots \otimes \mathcal{N}[\psi_N]. \tag{3}$$

¿From these conditions we derived a transformation group parameterized by smooth real-valued functions $\gamma(t)$, $\Lambda(t)$, and $\theta(\mathbf{x},t)$, with $\Lambda(t) \neq 0$. Writing $\ln\psi = T + iS$, we have

$$\begin{pmatrix} S' \\ T' \end{pmatrix} = \begin{pmatrix} \Lambda & \gamma \\ 0 & 1 \end{pmatrix} \begin{pmatrix} S \\ T \end{pmatrix} + \begin{pmatrix} \theta \\ 0 \end{pmatrix}. \tag{4}$$

When $\Lambda(t) \equiv 1$ and $\gamma(t) \equiv 0$, we have ordinary gauge transformations $\psi' = \psi\exp[i\theta(\mathbf{x},t)]$. Otherwise, if ψ obeys a linear Schrödinger equation, ψ' obeys a physically equivalent NLSE. This family of NLSEs, being linearizable, provides a kind of laboratory for interpreting and investigating nonlinear quantum mechanics. Furthermore, by analogy with electromagnetic fields in quantum mechanics, we can follow a procedure of "gauge generalization" to obtain a larger family of equations that are not physically equivalent to the linear equations [14]. It is surprising that the resulting NLSEs are the same as those obtained from diffeomorphism group representations, so that our two quite different sets of assumptions and mathematical procedures lead to the same result.

Let $\hat{\mathbf{j}} = (1/2i)[\overline{\psi}\nabla\psi - (\nabla\overline{\psi})\psi]$ and introduce the real nonlinear functionals

$$R_1 = \frac{\nabla \cdot \hat{\mathbf{j}}}{\rho}, \quad R_2 = \frac{\nabla^2\rho}{\rho}, \quad R_3 = \frac{\hat{\mathbf{j}}^2}{\rho 2}, \quad R_4 = \frac{\hat{\mathbf{j}} \cdot \nabla\rho}{\rho 2}, \quad R_5 = \frac{(\nabla\rho)2}{\rho 2}, \tag{5}$$

542

so that $\nabla^2 \psi = [iR_1 + (1/2)R_2 - R_3 - (1/4)R_5]\psi$. Then in the case $\theta \equiv 0$, with Λ and γ time-independent, our equations take the form

$$i\partial_t \psi = (i \sum_{j=1}^{2} \nu_j R_j + \sum_{j=1}^{5} \mu_j R_j + U)\psi, \tag{6}$$

where the ν_j and μ_j are real coefficients. The current $\hat{\mathbf{j}}$ and the coefficients ν_j and μ_j all change under the nonlinear gauge transformations (4), but one can write a gauge invariant current and gauge-invariant parameters in terms of the coefficients, from which the physical content is to be derived: for example, $\nu_1 \mu_2 - \mu_1 \nu_2 = \hbar 2/8m2$. In particular, existence of a gauge-invariant current means we can use it to describe measurement outcmes, reducing the need to rely exclusively on predictions from positional probability densities.

The class of gauge transformations and corresponding nonlinear equations can be further enlarged if we modify (4) in the natural way, letting

$$\begin{pmatrix} S' \\ T' \end{pmatrix} = \begin{pmatrix} \Lambda & \gamma \\ \lambda & \kappa \end{pmatrix} \begin{pmatrix} S \\ T \end{pmatrix} + \begin{pmatrix} \theta \\ \phi \end{pmatrix}, \tag{7}$$

where $\lambda(t)$, $\kappa(t)$, and $\phi(\mathbf{x}, t)$ are additional real-valued functions and the determinant of the matrix is nonvanishing. This procedure is justified by the existence of generalized expressions for probability density and current that are invariant under the enlarged group [25]. It results in a new symmetry between T and S, and in the case of time-independent coefficients with $\phi \equiv 0$, a simple modification of Eq. (6) so that the summations in the imaginary and real terms on the right both extend from $j = 1$ to 5.

Generalizations of nonlinear gauge transformations that permit dependence on spatial derivatives of ψ have also been found, and are discussed in [26].

3. Outlook

Based on the results of current investigations we speculate that a mathematically consistent theoretical framework for nonlinear quantum mechanics is possible. Future precision experiments or quantum information technologies might require a theory that allows physically small deviations from linearity, though at present there is no empirical indication of its necessity. At deeper levels of physics, nonlinear quantum mechanics may also prove to be a useful tool. For example in the realm of string-theoretic approaches, it has been shown that Galilean-invariant equations of the type of Eq. (6) can describe the quantum dynamics of D_0-branes [27].

References

[1] Wigner E P 1939 Ann. Math. 40 149
[2] Weinberg S 1989 Phys. Rev. Lett. 62 485; 1989 Ann. Phys. 194 336
[3] Goldin G A 1971 J. Math. Phys. 12 462-487
[4] Angermann B, Doebner H-D, and Tolar J 1983 in Lec. Notes in Mathematics 1037 171
    (Berlin: Springer)
[5] Goldin G A, Menikoff R, and Sharp D H 1983 Phys. Rev. Lett. 51 2246-2249
[6] Goldin G A 1992 Int. J. Mod. Phys. B 6 1905-1916
[7] Doebner H-D, Stovicek P, and Tolar J 2001 Rev. Math. Phys. 13 799-845
[8] Doebner H-D and Goldin G A 1992 Phys. Lett. A 162 397-401
[9] Doebner H-D and Goldin G A 1994 J. Phys. A: Math. Gen. 27 1771-1780

[10] Goldin G A and Svetlichny G 1994 in Quantization and Infinite-Dimensional Systems, ed. Antoine J-P et al. 245-255 (London: Plenum)

[11] Doebner H-D and Hennig D 1995 in Symmetries in Science VIII, ed. Gruber B 85-90 (London: Plenum)

[12] Mielnik B 1974 Commun. Math. Phys. 37 221

[13] Doebner H-D and Goldin G A 1996 Phys. Rev. A 54 3764-3771

[14] Doebner H-D, Goldin G A, and Nattermann P 1999 J. Math. Phys. 40 49-63

[15] Gisin N 1990 Phys. Lett. A 143 1

[16] Czachor M and Doebner H-D 2002 Phys. Lett. A 301 139-162

[17] Czachor M 1997 Phys. Lett. A 225 1

[18] Polchinski J Phys. Rev. Lett. 66 397

[19] Ghezali S, Laurent P, Lea S N, and Clairon A 1996 Europhys. Lett. 36 25

[20] Simon E, Laurent P, and Clairon A 1998 Phys. Rev. A 57 436

[21] Doebner H-D, Manko V I, and Scherer W 2000 Phys. Lett. A 268 17-24

[22] Bialynicki-Birula I and Mycielski J 1976 Ann. Phys. 100 62

[23] Shimony A 1979 Phys. Rev. A 20 394

[24] Shull C G, Atwood D K, Arthur J, and Horne M A 1980 Phys. Rev. Lett. 44 765

[25] Goldin G A 2000 in Quantum Theory and Symmetries, ed. Doebner H-D et al. 111-123 (Singapore: World Scientific)

[26] Goldin G A and Shtelen V M 2002 J. Math. Phys. 43 2180-2186

[27] Mavromatos N E and Szabo R J 2001 Int. J. Mod. Phys. A 16 209-250

Inst. Phys. Conf. Ser. No 173: Section 7
Paper presented at 24th Int. Coll. Group Theoretical Methods in Physics, Paris, France, 15–20 July 2002
©2003 IOP Publishing Ltd

Quantization of the Linearized Kepler Problem

Julio Guerrero and José Miguel Pérez

Depto. de Matemática Aplicada, Universidad de Murcia, 30100 Murcia, Spain

Abstract.
   The linearized Kepler problem is considered, as obtained from the Kustaanheimo-Stiefel (K-S) transformation, both for negative and positive energies. The symmetry group for the Kepler problem turns out to be $SU(2,2)$. For negative energies, the Hamiltonian of Kepler problem can be realized as the sum of the energies of four harmonic oscillator with the same frequency, with a certain constraint. For positive energies, it can be realized as the sum of the energies of four repulsive oscillator with the same (imaginary) frequency, with the same constraint. The quantization for the two cases, negative and positive energies is considered, using group theoretical techniques and constraints. The case of zero energy is also discussed.

1. KS Regularization of the Kepler problem.

In this work we affront the task of the quantization of the Kepler problem, given by the Hamiltonian defined on $\mathbb{R}_0^3 \times \mathbb{R}^3$, $\mathcal{H} = \frac{\vec{Y} \cdot \vec{Y}}{2m} - \frac{\gamma}{r}$, where $r = \sqrt{\mathbf{X}^2}$, $(\mathbf{X}, \mathbf{Y}) \in \mathbb{R}_0^3 \times \mathbb{R}^3$. For this purpose we shall use the linearization provided by the KS regularization, as introduced by P. Kustaanheimo and E. Stiefel, in the spinorial version due to Jost (see [1]). The KS transformation regularizes the Kepler problem and linearizes it, showing that the dynamical group of the Kepler problem is $SU(2,2)$. The linearization means that the Kepler problem, for the case of negative energy, can be seen as a system of **4 harmonic oscillators in resonance** subject to a constraint. For the case of positive energy, it turns to be a system of **4 repulsive harmonic oscillators in resonance** subject to constraints. Finally, as we shall see, the singular case of zero energy can be expressed as **4 free particles** subject to constraints.

   The key point in the KS transformation is the commutativity of the diagram (see [1]):

$$
\begin{array}{ccc}
(z, w) \in (I^{-1}(0))' & \xrightarrow{\mathfrak{C}} & (\eta, \zeta) \in I^{-1}(0) \subset \mathbb{C}^4 \\
\pi \downarrow & \circlearrowleft & \downarrow \widehat{\pi} \\
(\vec{x}, \vec{y}) \in \mathbb{R}_0^3 \times \mathbb{R}^3 & \xrightarrow{\nu^{-1}} & (q, p) \in T^+ S^3
\end{array}
\tag{1}
$$

   In this diagram ν is Moser transformation (see [1]), which allows us to see $T^+ S^3$ as an embedded manifold in $\mathbb{R}_0^3 \times \mathbb{R}^3$, where $T^+ S^3 = \{(q,p) \in \mathbb{R}^8, ||q|| = 1, < p, q >= 0, p \neq 0\}$, is named **Kepler manifold**.

   The KS transformation is the map π, which can be seen as a symplectic lift of the Hopf fibration, $\pi_0 : \mathbb{C}_0^2 \to \mathbb{R}_0^3$, $z = (z_1, z_2) \mapsto \pi_0(z) := < z, \vec{\sigma} z >$, ($\vec{\sigma}$ are Pauli matrices):

$$
\pi : T^* \mathbb{C}_0^2 \to T^* \mathbb{R}_0^3, \ (z, w) \mapsto (\vec{x} = \pi_0(z), \vec{y} = \text{Im} < w, \vec{\sigma} z > / < z, z >),
\tag{2}
$$

such that, $\pi^* \theta_{\mathbb{R}_0^3} = \theta_{\mathbb{C}_0^2}|_{(I^{-1}(0))'} = 2\text{Im} < w, dz > (= \theta_{\eta\zeta} = \text{Im}(< \eta, d\eta > - < \zeta, d\zeta >)$ up to a total differential) and $\theta_{\mathbb{R}_0^3}$ is the canonical potential form restricted to $\mathbb{R}_0^3$. The map $\mathfrak{C} = \frac{1}{\sqrt{2}} \begin{pmatrix} \sigma_0 & \sigma_0 \\ \sigma_0 & -\sigma_0 \end{pmatrix}$ provides the injection of collision states. The function $I = \frac{1}{2}(< \eta, \eta > - < \zeta, \zeta >)$ defines the regularized space $I^{-1}(0)$ ($(I^{-1}(0))'$ doesn't contain collision states), which is diffeomorphic to $\mathbb{C}_0^2 \times S^3$ while $I^{-1}(0)/U(1)$ is diffeomorphic to $\mathbb{R}_0^3 \times S^3$.

The transformation, $\vec{\mathbb{X}} = \frac{1}{\sqrt{mk}}\vec{x}$, $\vec{\mathbb{Y}} = k\sqrt{m}\vec{y}$, with $\rho = \sqrt{\vec{x}^2}$, relates the variables in the Kepler problem to the variables used in ν. The map $\hat{\pi}$ is a symplectomorphism between $I^{-1}(0)/U(1)$ and T^+S^3, with the symplectic structures restricted to the corresponding spaces. The Kepler Hamiltonian for negative energy is associated with $\mathcal{J} = \frac{1}{2}(<\eta,\eta> + <\zeta,\zeta>)$, which corresponds to a system of 4 harmonic oscillators in resonance 1-1-1-1.

We can proceed in the same manner for the case of positive energies changing the Kepler manifold by $T^+H^n = \{(q,p) :< q,q >= -1, < q,p >= 0, p_0 > ||\vec{p}||\}$. For this case, the Kepler Hamiltonian is associated with $-P_0$ (see below), with the additional constrain $-P_0 > 0$ (besides $I^{-1}(0)$). However, due to its singular character there is no possibility of considering the zero energy case in this (geometrical) way. We shall use, in this case, a group theoretical argument to study it.

The potential 1-form $\theta_{\eta\zeta}$ is left invariant by the Lie group $U(2,2)$, which also leaves invariant the constraint I when acting on $\mathbb{C}_0^4$. This is thus the dynamical group for the Kepler problem. A convenient basis for the Lie algebra $u(2,2)$ is given by the components of the momentum map associated with its action on $\mathbb{C}_0^4$ (here I is central):

$$I, \quad \mathcal{J}, \quad \vec{M} = -\frac{1}{2} <\eta,\vec{\sigma}\eta>, \quad \vec{N} = \frac{1}{2} <\zeta,\vec{\sigma}\zeta>,$$
$$Q = (-Im <\eta,\zeta>, Re <\eta,\vec{\sigma}\zeta>), \quad P = (Re <\eta,\zeta>, Im <\eta,\vec{\sigma}\zeta>). \tag{3}$$

Table I: KS regularization with physical constants
$\vec{\mathbb{X}} = \frac{1}{\sqrt{mk}}(\vec{Q} - \vec{R'}), \qquad \vec{\mathbb{Y}} = k\sqrt{m}\frac{\vec{P}}{\|P\|+P_0},$
$\mathcal{H} = \frac{\vec{\mathbb{Y}}^2}{2m} - \frac{\gamma}{\|\vec{\mathbb{X}}\|} = \frac{k}{2(\|P\|+P_0)}(k(\|P\| - P_0) - 2\gamma\sqrt{m}), \qquad A\vec{M} = \vec{L} = \vec{\mathbb{X}} \times \vec{\mathbb{Y}} = \vec{M} + \vec{N},$
$\vec{R'} = \vec{M} - \vec{N}, \qquad \mathcal{R}\vec{L} = \frac{\vec{\mathbb{Y}}\times\vec{L}}{m} - \gamma\frac{\mathbb{X}}{\|\mathbb{X}\|} = \frac{\vec{R'}(kP_0+\gamma\sqrt{m})+\vec{Q}(k\|P\|-\gamma\sqrt{m})}{\sqrt{m}(\|P\|+P_0)}.$

2. Quantization of Kepler problem: $E < 0$.

The KS transformation reveals that the Kepler problem for negative energies can be seen as the Hamiltonian system $(\mathbb{C}^4, \theta_{(\eta,\zeta)}, \mathcal{J})$ restricted to $I^{-1}(0)$. Defining $\mathbf{C} = (\mathbf{C}_1, \mathbf{C}_2) = (\eta, \zeta^+)$, $\mathbf{C}_i \in \mathbb{C}^2$, the Hamiltonian $\mathcal{J}$ adopts the form $\mathcal{H}_{har} = \omega \mathbf{C}\mathbf{C}^+$ which corresponds to four harmonic oscillators. The quantization of this system can be obtained from the group law of the corresponding symmetry group (a central extension of it by $U(1)$, rather, see [3]):

$$\lambda'' = \lambda' + \lambda, \qquad \mathbf{C}'' = \mathbf{C}' e^{-i\lambda} + \mathbf{C}, \qquad \mathbf{C}''^+ = \mathbf{C}'^+ e^{i\lambda} + \mathbf{C}^+,$$
$$\varsigma'' = \varsigma'\varsigma \exp[\tfrac{i}{2}(i\mathbf{C}'\mathbf{C}^+ e^{-i\lambda} - i\mathbf{C}'^+\mathbf{C} e^{i\lambda})], \tag{4}$$

where $\mathbf{C}, \mathbf{C}^+ \in \mathbb{C}^4$, $\varsigma \in U(1)$ and $\lambda = \omega t \in \mathbb{R}$. We can obtain the quantum version of this system using any geometrical (like Geometric Quantization, see [2]) or group-theoretical method, like Group Approach to Quantization (GAQ, see [3]), the one used here.

The resulting wave functions (defined on the group) are $\psi = \varsigma e^{-\frac{1}{2}\mathbf{C}\mathbf{C}^+}\phi(\mathbf{C}^+, \lambda)$, and Schrödinger equation for this system is $i\frac{\partial\phi}{\partial\lambda} = i\mathbf{C}^+\frac{\partial\phi}{\partial\mathbf{C}^+}$. In this formalism, quantum operators are constructed from the right-invariant vector fields on the group (4), and in this case **creation** and **annihilation** operators are given by $\widehat{\mathbf{C}}^+ = X_{\mathbf{C}}^R$ and $\widehat{\mathbf{C}} = X_{\mathbf{C}^+}^R$, respectively. Since the momentum map (3) is expressed as quadratic functions on $\mathbf{C}$ and $\mathbf{C}^+$, we can resort to Weyl prescription to obtain the quantization of these functions on the (right) enveloping algebra of the group (4). In this way we obtain a Lie algebra of quantum operators isomorphic to the one satisfied by the momentum map (3) with the Poisson bracket associated with $\theta_{\eta\zeta}$. The Hamiltonian operator and the quantum version of the constrain, when acting on wave functions are given by ($\mathcal{W} = \varsigma e^{-\frac{1}{2}\mathbf{C}\cdot\mathbf{C}^+}$): $\widehat{\mathcal{J}}\psi = -\frac{1}{2}\mathcal{W}(2 + \mathbf{C}^+\frac{\partial}{\partial\mathbf{C}^+})\phi$ and $\widehat{I}\psi = -\frac{1}{2}\mathcal{W}(\mathbf{C}_1^+\frac{\partial}{\partial\mathbf{C}_1^+} - \mathbf{C}_2^+\frac{\partial}{\partial\mathbf{C}_2^+})\phi$. To obtain the quantum version of the Kepler manifold (that is, the Hilbert space of states of the Hydrogen atom for $E < 0$), we must impose the constraint $\widehat{I}\psi = 0$. This means that the energy of the first two oscillators must equal the energy of the other two. It is easy to check that the operators in the (right) enveloping algebra of the group (4) preserving the constraint (see [5, 6] for a characterization of these operators) is the algebra $su(2,2)$ of the quantum version of the momentum map (3). These operators act irreducibly on the constrained Hilbert space, as can be checked computing the Casimirs of $su(2,2)$, which are constant. The quantum operators commuting with the Hamiltonian (and providing the degeneracy of the spectrum) are $\widehat{M}$ and $\widehat{N}$. They define two commuting $su(2)$ algebras in the same representation $((\widehat{M})^2 = (\widehat{N})^2 = \frac{1}{4}(\widehat{J})^2 - \frac{1}{4})$, and linear combinations of them provide us with the angular momentum and the Runge-Lenz vector (see Table I).

The relation between the Kepler Hamiltonian $\mathcal{H}$ and the Hamiltonian $\mathcal{J}$ is $\mathcal{H} = -\frac{m\gamma^2}{2\mathcal{J}^2}$. If we act on eigenstates of the number operator for each oscillator, $\psi_{n_1,n_2,n_3,n_4} \approx (C_{11}^+)^{n_1}(C_{12}^+)^{n_2}(C_{13}^+)^{n_3}(C_{14}^+)^{n_4}$, and taking into account that: $\widehat{\mathcal{E}}\psi_{n_1,n_2,n_3,n_4} = \widehat{\mathcal{J}}\psi_{n_1,n_2,n_3,n_4} = \frac{1}{2}(2 + \sum n_i)\psi_{n_1,n_2,n_3,n_4}$, we recover the **spectrum of the Hydrogen atom**, $E_n = -\frac{m\gamma^2}{2n^2}$, $n = 1 + n_1 + n_2$. The degeneracy is provided by the dimension of the representations of the algebra $su(2) \times su(2)$, which turn to be n^2 (if spin 1/2 is considered, the degeneracy is doubled).

3. Quantization of Kepler problem: $E > 0$.

The KS transformation, for the case of positive energies, maps the Kepler Hamiltonian to the function $-P_0$, with the constraints $I = 0$ and $-P_0 > 0$ and with the same potential 1-form $\theta_{(\eta,\varsigma)}$. Performing the change of variables:

$$q_i = \frac{1}{2}(\alpha_{i+1} + \nu_{i+1}), \ i = 0,1,2,3, \ p_i = \frac{1}{2}(\alpha_{i+2} - \nu_{i+2}), \ i = 0,2, \ p_i = \frac{1}{2}(\nu_i - \alpha_i), \ i = 1,3$$

$z_1 = q_0 + i\,q_1$, $z_2 = q_2 + i\,q_3$, $w_1 = p_0 + i\,p_1$ and $w_1 = p_2 + i\,p_3$, the new Hamiltonian $-P_0$ can be written as $\sim \alpha\nu$, which corresponds to four repulsive oscillators in resonance 1-1-1-1. The (extended) symmetry group for this system is given by (see [4]):

$$\lambda'' = \lambda' + \lambda, \qquad \alpha'' = \alpha' e^\lambda + \alpha, \qquad \nu'' = \nu' e^{-\lambda} + \nu,$$
$$\varsigma'' = \varsigma'\varsigma\,\exp[\tfrac{i}{2}(\alpha\,\nu' e^{-\lambda} - \alpha'\,\nu\,e^\lambda)], \tag{5}$$

where $\alpha,\nu \in \mathbb{R}^4$ and $\lambda := \omega t \in \mathbb{R}$. The Hamiltonian for this system is $\mathcal{H}_{rep} = -\omega\,\alpha\,\nu$. Applying GAQ we obtain that the wave functions are $\psi = \varsigma\,e^{\frac{-i\,\alpha\,\nu}{2}}\,\phi(\nu,\lambda)$, and the Schrödinger equation is $i\frac{\partial\phi}{\partial\lambda} = i\,\nu\,\frac{\partial\phi}{\partial\nu}$.

With the same procedure as in the case of negative energies, we construct a realization of the algebra (3) resorting to the enveloping algebra of the group (5). The quantum version of P_0 and the constraint are $(\mathcal{W} = \varsigma e^{-\frac{i}{2}\alpha\nu})$: The representation defined by the quantum version of (3) is irreducible, since the Casimirs are constant. The additional constraint $-\widehat{P}_0 > 0$ restricts further the algebra of physical operators, being generated by the **angular momentum** $\widehat{L}$ and the **Runge-Lenz** vector $\widehat{Q}$. These operators close the **Lorentz algebra**, $[\widehat{L}_i, \widehat{Q}_j] = -2i\,\epsilon_{ijk}\widehat{Q}_k$, $[\widehat{L}_i, \widehat{L}_j] = -2i\,\epsilon_{ijk}\widehat{L}_k$ y $[\widehat{Q}_i, \widehat{Q}_j] = +2i\,\epsilon_{ijk}\widehat{L}_k$, as expected.

Again, using the relation between the Kepler Hamiltonian and P_0, $\mathcal{H} = \frac{m\gamma^2}{2P_0^2}$, we obtain the spectrum of the Hydrogen atom for positive energies, $E_\tau = \frac{m\gamma^2}{2\tau^2}$; $\tau = 2 + \tau_1 + \tau_2$.

4. Quantization of Kepler problem: $E = 0$.

For the case of zero energy, it is not clear which is the Kepler manifold, and some of the expressions obtained for negative and positive energies have no limit when the energy (related to k) goes to zero. We propose a candidate for the linearized zero energy Kepler problem, using a group-theoretical argument. The idea is to look, in the $su(2,2)$ algebra, for an appropriate Runge-Lenz vector $\vec{\mathcal{RL}}$ (angular momentum doesn't change, but Runge-Lenz vector depends on the energy). From Table I, we observe that $\mathcal{H} = -\frac{k^2}{2} \mapsto \vec{\mathcal{RL}} = \frac{k}{\sqrt{m}}\vec{R}'$ and $\mathcal{H} = \frac{k^2}{2} \mapsto \vec{\mathcal{RL}} = \frac{k}{\sqrt{m}}\vec{Q}$. This suggests the choice, for zero energy (and for $k \neq 0$), $\vec{\mathcal{RL}} = \frac{k}{2\sqrt{m}}(\vec{R}' + \vec{Q})$, implying that the Hamiltonian is the sum of the two Hamiltonian (with equal opposite energies!), $\mathbf{E}_0 = \frac{1}{2}(\mathcal{J} - P_0) = \frac{1}{2}(||P|| - P_0)$. This can be achieved if we impose the constraint (zero energy) $||P|| - P_0 = \frac{2\gamma\sqrt{m}}{k}$, derived from our choice of Runge-Lenz vector. The constraint $I = 0$ is also satisfied in this case, by construction. Now we perform the following change of variables:

$$b_0 = q_0, \quad a_0 = q_1, \quad B_0 = p_1, \quad A_0 = p_0, \quad b_1 = q_2, \quad a_1 = q_3, \quad B_1 = p_3, \quad A_1 = p_2,$$

obtaining that the Hamiltonian is written $\mathbf{E}_0 = \frac{1}{2}\sum_i(A_i^2 + B_i^2)$, that is, a system of **four free particles**. It must be stressed that the variables a_i, A_i, $i = 0,1$ satisfy commutation relations with opposite sign to that of b_i, B_i, $i = 0,1$, as can be seen from the potential 1-form:

$$\theta = -B_0 dh_0 + h_0\, dB_0 - B_1 db_1 + b_1 dB_1 + A_0 da_0 - a_0 dA_0 + A_1 da_1 - a_1 dA_1. \tag{6}$$

The (extended) symmetry group for the system of four free particles is given by:

$$\lambda'' = \lambda' + \lambda, \quad \mathbf{a}'' = \mathbf{a} + \mathbf{a}' + \mathbf{A}'\lambda, \quad \mathbf{b}'' = \mathbf{b} + \mathbf{b}' + \mathbf{B}'\lambda, \quad \mathbf{A}'' = \mathbf{A} + \mathbf{A}', \quad \mathbf{B}'' = \mathbf{B} + \mathbf{B}'$$
$$\varsigma'' = \varsigma'\varsigma\, \exp\{i\,[(\mathbf{a}'\,\mathbf{A} + \lambda\,(\mathbf{A}'\,\mathbf{A} + \tfrac{1}{2}\mathbf{A}'^2)]\}\, \exp\{i\,[(\mathbf{b}\,\mathbf{B}' + \tfrac{1}{2}\mathbf{B}'^2\lambda]\},$$

where the relative sign in the canonical structure of the (a_i, A_i) and (b_i, B_i) variables has been taken into account in the 2-cocycle. Repeating the procedure of the non-zero energy cases, we obtain a new realization of the algebra $su(2,2)$ on this space, which is irreducible.

The energy and constraint operators are $(\mathcal{W} = \varsigma e^{-i\mathbf{b}\cdot\mathbf{B}} e^{\frac{i}{2}(\mathbf{B}^2 - \mathbf{A}^2)\lambda})$: $\widehat{\mathbf{E}}_0\psi = -\frac{1}{2}\mathcal{W}\sum_i(A_i^2 + B_i^2)\varphi$, and $\widehat{I}\psi = i\mathcal{W}\sum_i(A_i\frac{\partial}{\partial B_i} - B_i\frac{\partial}{\partial A_i})\varphi$. If we impose $E = 0$, we obtain the new constraint $\sum_i(A_i^2 + B_i^2) = \frac{2\gamma\sqrt{m}}{k}$. Therefore k defines a foliation in spheres. The operators preserving this new constraint are again the angular momentum $\widehat{\vec{L}}$ and the Runge-Lenz vector $\widehat{\vec{S}} = \widehat{\vec{R}'} + \widehat{\vec{Q}}$, closing **the Euclidean algebra** $e(3)$, as expected. This guarantees that we have made the correct choice of Hamiltonian and Runge-Lenz vector for $E = 0$.

An interesting application of these results is the fact that the linearization is preserved in some perturbed problems, such as the lunar problem or the Stark effect, see [7].

A similar study can be found in [8], where the quantization of the Kepler problem for $E \neq 0$ is considered in the Weyl-Wigner-Moyal formalism using the KS transformation.

References

[1] M. Kummer, Comm. Math. Phys. **84**, 133 (1982).
[2] N.M.J. WoodHouse, *Geometric Quantization (2nd ed.)*, Oxford University Press (1991).
[3] V. Aldaya and J.A. de Azcárraga, J. Math. Phys. **23**, 1297 (1982).
[4] V. Aldaya, A. de Azcárraga and K.B.Wolf, J. Math. Phys. **25**, 506 (1984).
[5] V. Aldaya, M. Calixto, J. Guerrero, Comm. Math. Phys. **178**, 399 (1996).
[6] J. Guerrero, M. Calixto and V. Aldaya, J. Math. Phys. **40** 3773 (1999).
[7] J. Guerrero and J.M. Pérez, in preparation.
[8] J.M. Gracia-Bonda, Phys. Rev. **A30**, 691 (1984)

Inst. Phys. Conf. Ser. No 173: Section 7
Paper presented at 24th Int. Coll. Group Theoretical Methods in Physics, Paris, France, 15–20 July 2002
©*2003 IOP Publishing Ltd*

Alternative Hamiltonians and Wigner Quantization

A Horzela

H.Niewodniczański Institute of Nuclear Physics,
ul. Eliasza-Radzikowskiego 152, 31 342 Krakow, Poland

Abstract. Alternative Hamiltonians of classical mechanics, *i.e.*, canonically inequivalent Hamiltonians leading to the same Newton equations of motion, generate different phase space descriptions of the same mechanical systems. Relation of such descriptions to different algebraic descriptions of quantum systems consistent with the same time evolution - so called Wigner quantization - is presented and illustrated on example of two particles system moving on a line.

1. Introduction

It was noticed more than 100 years ago [1] that Lagrangians and, in the consequence, Hamiltonians, which, through the minimal action principle, lead to the same Newton equations of motion can not be chosen uniquely even if formalism makes those which differ by canonical transformations equivalent. In the framework of the Lagrangian formalism of classical mechanics extensive studies of the problem performed during last 30 years led to the conclusion that ambiguity in the choice of a Lagrangian always exists for one dimensional models while for multidimensional systems it occurs in the case of their integrability [2]. As it was pointed out in [3] such an ambiguity of the formalism of classical mechanics mirrors in quantum physics as a nonuniqueness of the choice of quantization procedures, both within the Hamiltonian based approach, as well as when the path integral method is used. Alternative quantization schemes were discovered more than 50 years ago [4]. Within this approach the identification of coordinate and velocity operators as a canonical pair is rejected. Instead of that it is required that suitable commutators are consistent with given equations of motion governed by usual Hamiltonians. This condition is sufficient to find solutions challenging a canonical one. New quantum systems have properties uncommon for canonically quantized and they have been investigated for many physically interesting examples [5]. The search for alternative quantization schemes may be done also in the case when the Hamiltonian is supposed to be a generator of the time evolution which is not fixed as a quantum analogue of its standard classical form. In such a case the set of relations satisfied by quantum mechanical observables, including their equations of motion in the Heisenberg picture and relations reflecting the transformation properties with respect to spacetime symmetries - Galilean ones - can be investigated under the requirement that it closes to a well defined algebraic structure [6]. For simple oscillator-like examples the solutions have been found within the class of Lie algebras different from the linear envelope of the Heisenberg algebra. Systems quantized in such a way obey nonstandard properties: they generate new uncertainty relations and require a noncommutative geometry for description of the underlying spacetime [7]. In this note we are going to show how the problem of alternative, *i.e.*, giving the same equations of motion but canonically inequivalent [8] - [10], Hamiltonians links to the problem of noncanonical quantization schemes or how inequivalent phase-space canonical descriptions of the same system influence its quantum properties.

2. Alternative Hamiltonians for Newton equations

In order to investigate a classical system analogous to the Wigner quantum system studied in [7] we consider two mass points of masses m_1 and m_2 moving on a line. We assume that their mutual interaction is given by a distance dependent potential force and, moreover, both mass points are driven by a constant external force. The Newton equations of motion are

$$m_1\ddot{x}_1 = f_{\text{int}}(x_1 - x_2) + C = -\frac{\partial V_{\text{int}}(x_1 - x_2)}{\partial x_1} + \frac{d(Cx_1)}{dx_1},$$
$$m_2\ddot{x}_2 = -f_{\text{int}}(x_1 - x_2) + C = -\frac{\partial V_{\text{int}}(x_1 - x_2)}{\partial x_2} + \frac{d(Cx_2)}{dx_2}. \tag{1}$$

In the center of mass and the relative motion coordinates

$$R = (m_1 + m_2)^{-1}(m_1 x_1 + m_2 x_2), \qquad r = x_1 - x_2 \tag{2}$$

these equations become

$$\begin{aligned}
\ddot{R}(t) &= \frac{2C}{m_1 + m_2}, \\
\ddot{r}(t) &= \frac{m_1 + m_2}{m_1 m_2} f_{\text{int}}(r) + \frac{(m_2 - m_1)}{m_1 m_2}C.
\end{aligned} \tag{3}$$

The transformation (2) leads to a canonical transformation from variables x_1, p_1, x_2, p_2 to R, P, r, p where the center of mass momentum P and the relative momentum p are

$$P = (p_1 + p_2), \qquad p = \frac{m_2}{m_1 + m_2}p_1 - \frac{m_1}{m_1 + m_2}p_2. \tag{4}$$

The Hamilton equations written down in terms of R, P, r, p take the form

$$\begin{aligned}
\dot{R} &= \frac{\partial H(R, P, r, p)}{\partial P}, & \dot{P} &= -\frac{\partial H(R, P, r, p)}{\partial R}, \\
\dot{r} &= \frac{\partial H(R, P, r, p)}{\partial p}, & \dot{p} &= -\frac{\partial H(R, P, r, p)}{\partial r}
\end{aligned} \tag{5}$$

and their consistency conditions with the Newton equations (3) can be expressed in terms of the Poisson brackets

$$\frac{\partial^2 H}{\partial R \partial P}\frac{\partial H}{\partial P} - \frac{\partial^2 H}{\partial P^2}\frac{\partial H}{\partial R} + \frac{\partial^2 H}{\partial r \partial P}\frac{\partial H}{\partial p} - \frac{\partial^2 H}{\partial p \partial P}\frac{\partial H}{\partial r} =$$

$$\left\{\frac{\partial H}{\partial P}, H\right\} = \ddot{R} = \frac{2C}{m_1 + m_2}, \tag{6}$$

$$\frac{\partial^2 H}{\partial r \partial p}\frac{\partial H}{\partial p} - \frac{\partial^2 H}{\partial p^2}\frac{\partial H}{\partial r} + \frac{\partial^2 H}{\partial R \partial p}\frac{\partial H}{\partial P} - \frac{\partial^2 H}{\partial P \partial p}\frac{\partial H}{\partial R} =$$

$$\left\{\frac{\partial H}{\partial p}, H\right\} = \ddot{r} = \frac{m_1 + m_2}{m_1 m_2} f_{\text{int}}(r) + \frac{m_2 - m_1}{m_1 m_2}C.$$

Obviously the standard Hamiltonian of the system (1)

$$\begin{aligned}
H(R, P, r, p) &= \\
\frac{1}{2(m_1 + m_2)}P^2 &+ \frac{m_1 + m_2}{2m_1 m_2}p^2 + V_{\text{int}}(r) - C\left(2R + \frac{m_2 - m_1}{m_1 m_2}r\right)
\end{aligned} \tag{7}$$

is the solution of (6) which provides us usual relations between canonical momenta and velocities

$$P = (m_1 + m_2)\dot{R}, \qquad p = \frac{m_1 m_2}{m_1 + m_2}\dot{r}. \tag{8}$$

Here we would like to stress that (7) is a very particular solution of (6). It is singled out by a requirement that one searches for a solution being a sum of terms which depend on each of variables R, P, r and p separately [8]-[10]. Moreover, looking for solutions of (6) it is unproductive to consider a free case. Then (6) is solved by an arbitrary function of momenta and to avoid such an ambiguity we have included external constant force into description of the system. In order to find alternative solutions one can observe that differentiation of (6) with respect to P and p gives conditions expressible in terms of the Poisson brackets

$$\left\{\frac{\partial^2 H}{\partial P^2}, H\right\} = 0, \quad \left\{\frac{\partial^2 H}{\partial p^2}, H\right\} = 0, \quad \left\{\frac{\partial^2 H}{\partial P \partial p}, H\right\} = 0. \tag{9}$$

They are satisfied not only by (7) but also by

$$H(R, P, r, p) = H_{\text{CM}}(R, P) + H_{\text{int}}(r, p),$$

$$\frac{\partial^2 H_{\text{CM}}}{\partial P^2} = \lambda_{\text{CM}}^2 H_{\text{CM}}, \quad \frac{\partial^2 H_{\text{int}}}{\partial p^2} = \lambda_{\text{int}}^2 H_{\text{int}}. \tag{10}$$

Solutions of (10), if put into (6), lead to

$$H_{\text{CM}}(R, P) = \alpha_{\text{CM}}(R) \exp\left(\lambda_{\text{CM}} P\right) + \beta_{\text{CM}}(R) \exp\left(-\lambda_{\text{CM}} P\right),$$

$$\alpha_{\text{CM}}(R)\beta_{\text{CM}}(R) = \frac{1}{2\left(m_1 + m_2\right)\lambda_{\text{CM}}^2}\left(CR + C_{\text{CM}}\right), \tag{11}$$

and

$$H_{\text{int}}(r, p) = \alpha_{\text{int}}(r) \exp\left(\lambda_{\text{int}} p\right) + \beta_{\text{int}}(r) \exp\left(-\lambda_{\text{int}} p\right),$$

$$\alpha_{\text{int}}(r)\beta_{\text{int}}(r) = \frac{m_1 + m_2}{2 m_1 m_2 \lambda_{\text{int}}^2}\left(V_{\text{int}}(r) + \frac{m_2 - m_1}{m_1 m_2} Cr + C_{\text{int}}\right), \tag{12}$$

with C_{CM} and C_{int} denoting arbitrary constants.

3. Galilean covariance

Consider now the model of the previous section for vanishing external force, *i.e.*, for $C = 0$. The alternative total Hamiltonian (10), composed of the center of mass, (11), and the relative motion, (12), alternative Hamiltonians, generate the Poisson brackets

$$\begin{aligned}\{R, P\} &= 1, & \{R, r\} &= 0, & \{R, p\} &= 0, \\ \{r, p\} &= 1, & \{r, P\} &= 0, & \{p, P\} &= 0,\end{aligned} \tag{13}$$

$$\begin{aligned}\{R, H\} &= \dot{R}, & \{P, H\} &= 0, \\ \{r, H\} &= \dot{r}, & \{p, H\} &= -\frac{\partial H}{\partial r},\end{aligned} \tag{14}$$

$$\left\{\dot{R}, R\right\} = -\lambda_{\text{CM}}^2 H_{\text{CM}}(R, P), \quad \left\{\dot{R}, P\right\} = 0, \tag{15}$$

$$\begin{aligned}\{\dot{r}, r\} &= -\lambda_{\text{int}}^2 H_{int}(r, p), \\ \{\dot{r}, p\} &= \left(\frac{\partial H_{\text{int}}(r, p)}{\partial p}\right)^{-1} \frac{\partial}{\partial r}\left(\frac{\lambda_{\text{int}}^2}{2} H_{\text{int}}(r, p)) - \frac{m_1 + m_2}{m_1 m_2} V(r)\right).\end{aligned} \tag{16}$$

The generators of the (1+1) extended Galilei group: $\mathcal{H}, \Pi$ and K which denote the generators of the time and the space translations, and the generator of the Galilean boost, respectively, and $\mathcal{M}$ which is the parameter of the central extension, satisfy the Poisson brackets

$$\{K, \mathcal{H}\} = \Pi, \quad \{K, \Pi\} = \mathcal{M}, \quad \{\Pi, \mathcal{H}\} = 0, \tag{17}$$

$$\{R, \mathcal{H}\} = \dot{R}, \quad \{\dot{R}, \mathcal{H}\} = 0, \quad \{r, \mathcal{H}\} = \dot{r}, \quad \{\dot{r}, \mathcal{H}\} = -\frac{m_1 + m_2}{m_1 m_2} \frac{dV}{dr},$$
$$\{K, R\} = t, \quad \{K, \dot{R}\} = 1, \quad \{K, r\} = 0, \quad \{K, \dot{r}\} = 0, \qquad (18)$$
$$\{R, \Pi\} = 1, \quad \{\dot{R}, \Pi\} = 0, \quad \{r, \Pi\} = 0, \quad \{\dot{r}, \Pi\} = 0.$$

If we compare (13) - (15) with (17) and (18) we see that the variables which describe the center of mass motion, namely the Hamiltonian H_{CM}, the momentum P and the coordinate R, can not act as a representation of the Galilei group. But solving (6) we can take H_{int} as an alternative solution and leave H_{CM} as a canonical one. The first equation in (15) becomes

$$\{\dot{R}, R\} = -\frac{1}{m_1 + m_2}, \qquad (19)$$

while the rest of (13) - (16) remains unchanged. Now the identification

$$\mathcal{H} = H, \quad K = (m_1 + m_2)\left(R - t\dot{R}\right), \quad \Pi = (m_1 + m_2)\dot{R} \qquad (20)$$

is allowed exactly like it happens in the framework of the standard formalism. The Galilean generators may be represented in terms of the center of mass variables but we have to accept that the phase-space descriptions of the center of mass and the relative motion are governed by different schemes.

4. Conclusions

Quantum analogues of the Poisson brackets (13) - (16) (with a replacement (19) used) are commutators which we have found for a system of two particles interacting harmonically and quantized according to the Wigner approach [6], [7]. Investigating this example we have got a noncanonical Lie algebra of quantum mechanical observables paying the price that the center of mass variables are quantized canonically while alternative quantization rules are applied to the internal variables only. Such dichotomy in quantization scheme has enabled us to identify observables connected with the center of mass motion with the Galilean generators. Here we have shown that the same occurs in classical mechanics. The existence of alternative Hamiltonians justifies different phase-space descriptions of various degrees of freedom of the physical system. Moreover, the alternative descriptions satisfy all fundamental requirements and we do not see any argument to reject them as a challenger to the canonical quantization.

Acknowledgment: The author is deeply grateful to Professors E.Kapuścik and V. I. Man'ko for illuminating discussions on the subject.

References

[1] Helmholtz H 1887 Z.Reine Angew. Math. 100 137
[2] Santilli R M 1982 Foundations of Theoretical Mechanics (Springer); Henneaux M 1982 Ann.Phys.(NY) 140 45; Morandi G, Ferrario C, Lo Vecchio G, Marmo G and Rubano C 1990 Phys. Rep. 188 147
[3] Dodonov V V, Man'ko V I, Skarzhinsky V D 1981 Hadronic J. 4 1734
[4] Wigner E P 1950 Phys. Rev. 77 711
[5] Palev T D and Stoilova N I 1997 J. Math. Phys. 38 2506, and references therein
[6] Kapuścik E, Horzela A and Uzes Ch A 1997 Proc. of Quantum Groups Symposium at Group21, Goslar 1996, (Eds. H. -D. Doebner and V. K. Dobrev), Heron Press Science Series, p. 148
[7] Horzela A 2000 Czech J.of Physics 50 1245
[8] Degasperis A and Ruisenaars S N M 2001 Ann. Phys. (NY) 293 92
[9] Cisło J and Łopuszański J 2001 J. Math. Phys. 42 5163
[10] Horzela A 2002 Czech J.of Physics 52 1239

Inst. Phys. Conf. Ser. No 173: Section 7
Paper presented at 24th Int. Coll. Group Theoretical Methods in Physics, Paris, France, 15–20 July 2002
©2003 IOP Publishing Ltd

Spectral properties of Landau operator with δ cylinder interaction

M N Hounkonnou† ‡ § **and G Honnouvo**† ‡

† International Chair in Mathematical Physics and Applications (ICMPA) 01 B.P.: 2628
Porto-Novo, Benin
‡ Unité de Recherche en Physique Théorique (URPT), Institut de Mathématiques et de
Sciences Physiques (IMSP) 01 B.P.: 2628 Porto-Novo, Bénin

Abstract. Using the theory of self-adjoint extensions, we study the spectral properties of
the Landau operator with δ interactions on a cylinder of radius R for a charged spin particle

system, formally given by the Hamiltonian $H_B = (p - A)^2 + \xi V(r)$, $\left(V(r) = \delta(R - r) \right)$,

acting in $L^2(\mathbf{R}^2)$. ξ is a real constant. The potential vector has the form $A = (B/2)(-y, x)$
and $B > 0$.

In this paper, we provide the self-adjoint extensions, the resolvent and the spectral
analysis of the operator defined by $H_B + V(r)$, where H_B is the free Landau Hamiltonian
acting in the Hilbert space $\mathcal{H} = L^2(\mathbf{R}^2)$ and $V(r) = \xi \delta(r - R)$, with $\xi \in \mathbf{R}, R > 0$. A natural
starting point for the construction of the self-adjoint Hamiltonian in this case is to define this
operator in the domain of smooth functions vanishing for $r = R$ *i.e.* $C_0^\infty(\mathbf{R}^2 \setminus \{\partial \overline{\Gamma(O, R)}\})$.
$\overline{\Gamma(O, R)}$ is a closed circle of radius R centered at the origin of $\mathbf{R}^2$. Then, we look for the
possible s.a. extensions. Each of the extensions will be characterized by a specific behavior,
i.e. boundary conditions for the elements of the domain near $r = R$.

Consider the corresponding radial equation for δ - cylinder interaction, formally given
by the expression [1]:

$$\left[-\frac{d^2}{dr^2} + \left(\frac{m}{r} + \frac{B}{2}r \right)^2 - \frac{1}{4r^2} + \xi_m \delta(r - R) \right] f_m(k, r) = k f_m(k, r). \quad (1)$$

Then, we assume the function $f_m(k, r)$ continuous at $r = R$, i.e. $f_m(k, R_+) = f_m(k, R_-) \equiv$
$f_m(k, R)$. Integrating the equation (1) between $r = R - \epsilon$ and $r = R + \epsilon$ and taking
the limit when $\epsilon \longrightarrow 0$, we have: $f'_m(k, R_+) - f'_m(k, R_-) = \xi_m f(k, R)$. Let us consider
in $L^2(\mathbf{R}^2)$ the closed and non-negative operator $\dot{H}_B = \overline{H_B|_{C_0^\infty(\mathbf{R}^2 \setminus \{\partial \overline{\Gamma(O, R)}\})}}$, with the
domain $D(\dot{H}_B) = \{f \in L^2(\mathbf{R}^2) \cap H_{loc}^{2,2}(\mathbf{R}^2) / f(\partial \overline{\Gamma(O, R)}) = 0, \ H_B f \in L^2(\mathbf{R}^2)\}$, where
$H_{loc}^{m,n}(\Omega)$ is the local Sobolev space of indices (m,n). Let us now decompose the Hilbert space
$\mathcal{H} = L^2(\mathbf{R}^2)$ as follows: $L^2(\mathbf{R}^2) = L^2(\mathbf{R}^+) \bigotimes L^2(S^1)$, S^1 being the unit circle in $\mathbf{R}^2$. Let

$$U : \left\{ \begin{array}{c} L^2((0, \infty); r dr) \longrightarrow L^2((0, \infty); dr) \equiv L^2((0, \infty)) \\ f \longmapsto (Uf)(r) = \sqrt{r} f(r). \end{array} \right. \quad (2)$$

We obtain the following decomposition of $L^2(\mathbf{R}^2)$:

$L^2(\mathbf{R}^2) = \bigoplus_{m=-\infty}^{m=+\infty} U^{-1}(L^2(\mathbf{R}^+)) \bigotimes \left[\frac{e^{im\phi}}{\sqrt{2\pi}} \right]$, $m \in \mathbf{Z}$. Provided this decomposition,
$\dot{H}_B = \bigoplus_{m=-\infty}^{m=+\infty} U^{-1} \dot{h}_{B,m} U \bigotimes \mathbb{1}$, where the operator $\dot{h}_{B,m}$ in $L^2(]0, \infty[)$ is defined by

$$\dot{h}_{B,m} = -\frac{d^2}{dr^2} + \left(\frac{m}{r^2} + \frac{B}{2}r \right)^2 - \frac{1}{4r^2}, \quad (3)$$

§ To whom correspondence should be addressed (hounkonnou@yahoo.fr)

with the domain

$$\mathcal{D}(\dot{h}_{B,m}) = \left\{ f \in L^2(]0,\infty[, dr) \cap H^{2,2}_{loc}(]0,\infty[) \,;\; f(R_\pm) = 0 \,; \right.$$

$$\left. \left(-\frac{d^2}{dr^2} + \left(\frac{m}{r} + \frac{B}{2}r\right)^2 - \frac{1}{4r^2}\right) f \in L^2((0,\infty)) \right\}, \quad m \in \mathbf{Z}. \qquad (4)$$

The adjoint operator $\dot{h}^*_{B,m}$ of $\dot{h}_{B,m}$ is defined by

$$\dot{h}^*_{B,m} = -\frac{d^2}{dr^2} + \left(\frac{m}{r} + \frac{B}{2}r\right)^2 - \frac{1}{4r^2},$$

with the domain

$$D(\dot{h}^*_{B,m}) = \{ f \in L^2(]0,\infty[, dr) \cap H^{2,2}_{loc}(]0,\infty[-\{R\}); f(R_+) = f(R_-)$$

$$\equiv f(R); \left(-\frac{d^2}{dr^2} + \left(\frac{m}{r} + \frac{B}{2}r\right)^2 - \frac{1}{4r^2}\right) f \in L^2(]0,\infty[)\}, \quad m \in \mathbf{Z}. \,(5)$$

From the above space decomposition, we obtain $\dot{H}^*_B = \bigoplus_{m=-\infty}^{m=+\infty} U^{-1} \dot{h}^*_{B,m} U \otimes \mathbb{1}$. The indicial equation reads

$$h^*_{B,m} f_m(k,r) = k f_m(k,r), \qquad (6)$$

or equivalently

$$[-\frac{d^2}{dr^2} + \left(\frac{m}{r} + \frac{B}{2}r\right)^2 - \frac{1}{4r^2}] f_m(k,r) = k f_m(k,r). \qquad (7)$$

Next, selecting, in the two-dimensional space of solutions, the solution which satisfies the boundary conditions at $r = R$, we arrive at the function

$$f_{B,m}(k,r) = \begin{cases} N_{B,m} G^{(0)}_{B,m}(k,R) \times F^{(0)}_{B,m}(k,r) \,; & r \leq R \,, \\ N_{B,m} F^{(0)}_{B,m}(k,R) \times G^{(0)}_{B,m}(k,r) \,; & r \geq R \,, \end{cases} \qquad (8)$$

where

$$F^{(0)}_{B,m}(k,r) = r^{1/2+|m|} e^{-\frac{1}{4}Br^2} {}_1F_1\left(\tfrac{1}{2}(|m| + m + 1 - \tfrac{k}{B}), |m| + 1; \tfrac{B}{2}r^2\right),$$

$$G^{(0)}_{B,m}(k,r) = r^{1/2+|m|} e^{-\frac{1}{4}Br^2} U\left(\tfrac{1}{2}(|m| + m + 1 - \tfrac{k}{B}), |m| + 1; \tfrac{B}{2}r^2\right), \qquad (9)$$

$$N_{B,m} = \left(\|P_{B,m}(k)\|_{L^2(]0,\infty))}\right)^{-1} \qquad (10)$$

with

$$P_{B,m}(k,r) = \begin{cases} G^{(0)}_{B,m}(k,R) \times F^{(0)}_{B,m}(k,r) \,; & r \leq R \,, \\ F^{(0)}_{B,m}(k,R) \times G^{(0)}_{B,m}(k,r) \,; & r \geq R \,. \end{cases} \qquad (11)$$

${}_1F_1(a;b;z)$ and $U(a;b;z)$ denote the (ir)regular confluent hypergeometric functions, respectively [2]. Since the indicial equation admits one solution, $\dot{h}_{B,m}$ has deficiency indices $(1,1)$ and, consequently, all self-adjoint (s.a) extensions of $\dot{h}_{B,m}$ are given by a 1-parameter family of (s.a.) operators [3] which are defined by

$$h_{B,m,\xi_m} = -\frac{d^2}{dr^2} + \left(\frac{m}{r} + \frac{B}{2}r\right)^2 - \frac{1}{4r^2}, \qquad (12)$$

with the domain

$$D(h_{\alpha,m,\xi_m}) = \Big\{ f \in L^2(]0,\infty[,dr) \cap H^{2,2}_{loc}(]0,\infty[-\{R\});$$

$$f(R_+) = f(R_-) \equiv f(R);\ f'(R_+) - f'(R_-) = \xi_m f(R);$$

$$\left(-\frac{d^2}{dr^2} + \left(\frac{m}{r} + \frac{B}{2}r\right)^2 - \frac{1}{4r^2}\right) f \in L^2(]0,\infty[)\Big\},$$

$$m \in \mathbf{Z},\ -\infty < \xi_m \le +\infty. \tag{13}$$

The case $\xi_m = 0$ coincides with the free kinetic energy Hamiltonian $h_{\alpha,m,0}$ for fixed quantum number m. Let $\xi = \{\xi_m\}_{m\in\mathbf{Z}}$ and introduce in $L^2(\mathbf{R}^2)$ the operator

$$H_{B,\xi} = \bigoplus_{m=-\infty}^{m=+\infty} U^{-1} h_{B,m,\xi_m} U \bigotimes \mathbb{1}. \tag{14}$$

By definition, $H_{B,\xi}$ is the rigorous mathematical formulation of the formal expression $H_B + V(r)$. It provides a slight generalization of this operator, since ξ may depend on $m \in \mathbf{Z}$.

Theorem 0.1 *(i) The resolvent of h_{B,m,ξ_m} is given by*

$$(h_{B,m,\xi_m} - k)^{-1} = (h_{B,m,0} - k)^{-1} + \mu_m(k) \left(f_{B,m}(-\overline{k}),.\right) f_{B,m}(k),$$

$$k \in \rho(h_{B,m,\xi_m}) \cap \rho(h_{B,m,0}),\ m \in \mathbf{Z}, \tag{15}$$

where

$$\mu_m(k) = \xi_m \Big[N_{B,m} \Big(G'^{(0)}_{B,m}(k,R) F^{(0)}_{B,m}(k,R)$$

$$-G^{(0)}_{B,m}(k,R) F'^{(0)}_{B,m}(k,R) - \xi_m G^{(0)}_{B,m}(k,R) F^{(0)}_{B,m}(k,R)\Big)\Big]^{-1} \tag{16}$$

and the free resolvent with integral kernel

$$g_{m,k}(r,r') = \begin{cases} N_{B,m} G^{(0)}_{B,m}(k,r) \times F^{(0)}_{B,m}(k,r');\ & r' \le r, \\ N_{B,m} F^{(0)}_{B,m}(k,r) \times G^{(0)}_{B,m}(k,r');\ & r' \ge r. \end{cases} \tag{17}$$

We note that $g_{m,k}(R,r) = f_{B,m}(k,r)$.
(ii) The resolvent of $H_{B,\xi}$ is given by

$$(H_{B,\xi} - k)^{-1} = (H_{B,0} - k)^{-1}$$

$$+ \bigoplus_{m=-\infty}^{m=+\infty} \mu_m(k) \left(|.|^{-1} f_{B,m}(-\overline{k})\frac{e^{im\phi}}{\sqrt{2\pi}},.\right) |.|^{-1} f_{B,m}(k)\frac{e^{im\phi}}{\sqrt{2\pi}},$$

$$k \in \rho(H_{B,\xi}) \cap \rho(H_{B,0}). \tag{18}$$

Spectral properties of h_{B,m,ξ_m} are provided by the following theorem where $\sigma_{ess}(.)$, $\sigma_{sc}(.)$ and $\sigma_p(.)$ denote the essential spectrum, singularly continuous spectrum and point spectrum, respectively.

Theorem 0.2 : *For all $\xi_m \in (-\infty,\infty)$, we have: $\sigma_{ess}(h_{B,m,\xi_m}) = \emptyset$, $\sigma_{sc}(h_{B,m,\xi_m}) = \emptyset$ and $\sigma_p(h_{B,m,\xi_m}) = \Big\{ E \in \mathbf{R}/G'^{(0)}_{B,m}(E,R) F^{(0)}_{B,m}(E,R) - G^{(0)}_{B,m}(E,R) F'^{(0)}_{B,m}(E,R) - \xi_m G^{(0)}_{B,m}(E,R) F^{(0)}_{B,m}(E,R) = 0 \Big\}$. The negative eigenvalues of h_{B,m,ξ_m} are obtained from the equation $\mu_m(E)^{-1} = 0$; $E < 0$, which has at most one solution $E_0 < 0$.*

The results for the point interaction at the origin appear as a particular case of the cylinder interaction investigated here. Indeed, when $R \to 0$, we recover the boundary conditions corresponding to the nonrelativistic point interaction investigated in [4]. Finally, let us emphasize that the properties for point interaction placed at any point x could be found using the transformation relation $t_x H_\alpha t_{-x}$ given in [5], where t_x is a translation application of vector x, H_α being the Hamiltonian perturbed by point interaction at the origin $r = 0$.

References

[1] Chakraborty T and Pietiläinen P 1995 *The Quantum Hall Effects, Fractional and Integral*, (Berlin Heidelberg: Springer-Verlag)

[2] Abramowitz M and Stegun I A 1972 *Handbook of Mathematical Functions* (New York: Dover)

[3] Akhiezer N I and Glazman I M 1981 *Theory of Linear Operators in Hilbert Space* vol **2** (Pitman / Boston)

[4] Gesztesy F Holden H and Šeba P 1988 *"On point interactions in magnetic field systems" Proc. on the Schrödinger Operators, Standard and Non-standard (Dubna USSR)*, eds Exner P and Šeba P (World Scientific)

[5] Albeverio S Gesztesy F Hoegh-Krohn R and Holden H 1988, *Solvable Models in Quantum Mechanics* Texts and Monographs in Physics (Berlin: Springer Verlag)

Inst. Phys. Conf. Ser. No 173: Section 7
Paper presented at 24th Int. Coll. Group Theoretical Methods in Physics, Paris, France, 15–20 July 2002
©2003 IOP Publishing Ltd

The Jaynes-Cummings model and raising and lowering operators

V Hussin

Centre de Recherches Mathématiques, Université de Montréal, C. P. 6128, Succ. Centre-ville, Montréal, (QC) H3C 3J7, Canada

L M Nieto

Departamento de Física Teórica, Universidad de Valladolid, 47005 Valladolid, Spain

Abstract. Ladder operators acting on the energy eigenstates of the Jaynes-Cummings model in the rotating wave approximation are constructed using the fact that they satisfy some reasonable additionnal properties. In fact, we show that it is possible to write the Jaynes-Cummings Hamiltonian as a product of such ladder operators. It may be relevant for the study of new sets of coherent states for this model.

1. Introduction

The Jaynes-Cummings model is a non-linear model which is well-known and used in atomic physics and quantum optics [1]. It describes the interaction of a two levels atomic system with a quantized electromagnetic field. In the rotating wave approximation, it is exactly solvable and the energy spectrum and associated eigenstates can be computed in a closed form [2]. This model is the fundamental theoretical tool in the study of cavity QED for analysing ion traps [3]. Hence, it has become very valuable for researchers working in the quickly developing domain of quantum information processing [4].

In this paper, ladder operators associated with the energy eigenstates are constructed for this model. More constraints are then introduced like the factorization of the Jaynes-Cummings Hamiltonian.

2. The Jaynes-Cummings model and its energy spectrum

We begin with the Hamiltonian of the Jaynes-Cummings model in the rotating-wave approximation, which is [5, 6]

$$H_{JC} = \hbar\omega \left(a^+ a + \frac{1}{2} \right) \sigma_0 + \frac{1}{2} \hbar\omega_0 \sigma_3 + \hbar\kappa (a^+ \sigma_- + a\sigma_+). \tag{1}$$

In this expression, a^+ and a are the photon creation and annihilation operators (the usual ladder operator for the harmonic oscillator), σ_0 is the 2×2 identity matrix, $\{\sigma_1, \sigma_2, \sigma_3\}$ are the usual Pauli matrices, and $\sigma_\pm = \sigma_1 \pm i\sigma_2$.

The parameters that appear in equation (1) are ω the field mode frequency, ω_0 the atomic frequency and κ the coupling constant between the radiation and the atom. In a realistic experiment we can imagine that the atomic frequency ω_0 is fixed and that the field mode frequency ω varies. Hence, we introduce the "detuning" parameter ϵ, which can be positive or negative, and such that

$$\omega = \omega_0(1 + \epsilon), \quad \text{with} \quad |\epsilon| \approx 0. \tag{2}$$

So we will use the following form of the Hamiltonian (1):

$$H = \frac{H_{JC}}{\hbar\omega_0} = (1 + \epsilon)\left(a^+a + \frac{1}{2}\right)\sigma_0 + \frac{1}{2}\sigma_3 + \lambda(a^+\sigma_- + a\sigma_+), \tag{3}$$

where we have introduced the new parameter

$$\lambda = \frac{\kappa}{\omega_0} \geq 0. \tag{4}$$

The eigenstates can now be presented in a quite elegant form if we use the Fock space formalism:

$$\mathcal{F} = \mathcal{F}_b \otimes \mathcal{F}_f = \left\{ |n, -\rangle = \begin{pmatrix} 0 \\ |n\rangle \end{pmatrix}, \ |n, +\rangle = \begin{pmatrix} |n\rangle \\ 0 \end{pmatrix}, \ n \in \mathbb{N} \right\}, \tag{5}$$

with $\{|n\rangle\}$ the eigenstates of the photon number operator $N = a^+a$.

The matrix representation of the Hamiltonian H given in (3) is:

$$H = \begin{pmatrix} (1 + \epsilon)N + 1 + \frac{\epsilon}{2} & \lambda a \\ \lambda a^+ & (1 + \epsilon)N + \frac{\epsilon}{2} \end{pmatrix}. \tag{6}$$

The normalized energy eigenstates of H are

$$|\varepsilon_0^-\rangle = |0, -\rangle, \tag{7}$$

$$|\varepsilon_{n+1}^-\rangle = \sin\theta(n)\,|n, +\rangle + \cos\theta(n)\,|n+1, -\rangle, \tag{8}$$

$$|\varepsilon_n^+\rangle = \cos\theta(n)\,|n, +\rangle - \sin\theta(n)\,|n+1, -\rangle, \tag{9}$$

where

$$\sin\theta(n) = \frac{\lambda\sqrt{n+1}}{R(n+1)}, \tag{10}$$

$$\cos\theta(n) = \frac{\frac{\epsilon}{2} + \lambda r(n+1)}{R(n+1)}. \tag{11}$$

In the previous expressions, we have

$$r(n+1) = \sqrt{\delta + n + 1}, \tag{12}$$

$$R(n+1) = \sqrt{\left[\frac{\epsilon}{2} + \lambda r(n+1)\right]^2 + \lambda^2(n+1)}, \tag{13}$$

with

$$\delta = \left(\frac{\epsilon}{2\lambda}\right)^2 \geq 0. \tag{14}$$

The corresponding energy eigenvalues are:

$$\varepsilon_0^- = \frac{\epsilon}{2}, \tag{15}$$

$$\varepsilon_{n+1}^- = (1 + \epsilon)(n + 1) + \lambda r(n + 1). \tag{16}$$

$$\varepsilon_n^+ = (1 + \epsilon)(n + 1) - \lambda r(n + 1), \tag{17}$$

Let us mention that the set $\{\varepsilon_n^-\}$ contains positive and increasing values while it is not the case for $\{\varepsilon_n^+\}$. See [7] for more comments on this fact.

3. Raising and lowering operators

We are now looking for a lowering operator $\mathcal{M}^-$ that satisfy the following properties:

$$\mathcal{M}^-|\varepsilon_0^-\rangle = 0, \qquad \mathcal{M}^-|\varepsilon_{n+1}^-\rangle = k(n)\,|\varepsilon_n^-\rangle, \; n \in \mathbb{N}, \tag{18}$$

and

$$\mathcal{M}^-|\varepsilon_0^+\rangle = 0, \qquad \mathcal{M}^-|\varepsilon_n^+\rangle = s(n)\,|\varepsilon_{n-1}^+\rangle, \quad n \in \mathbb{N}/\{0\}. \tag{19}$$

The corresponding raising operator is given by $\mathcal{M}^+ = (\mathcal{M}^-)^\dagger$.

In the sequel the functions $k(n)$ and $s(n)$, which have to be determined, are assumed to be real and the operator $\mathcal{M}^-$ takes the form:

$$\mathcal{M}^- = \begin{pmatrix} m_1(N)\,a & m_2(N)\,a^2 \\ m_3(N) & m_4(N)\,a \end{pmatrix}. \tag{20}$$

By assuming (18) and (19), we get the following expressions for the four functions appearing in the definition of the lowering operator $\mathcal{M}^-$:

$$m_1(n) = \frac{\sin\theta(n)\sin\theta(n+1)\,k(n+1) + \cos\theta(n)\cos\theta(n+1)s(n+1)}{\sqrt{n+1}}, \tag{21}$$

$$m_2(n) = \frac{\sin\theta(n)\cos\theta(n+1)\,k(n+1) - \cos\theta(n)\sin\theta(n+1)\,s(n+1)}{\sqrt{n+1}\sqrt{n+2}}, \tag{22}$$

$$m_3(n) = \cos\theta(n-1)\sin\theta(n)\,k(n) - \sin\theta(n-1)\cos\theta(n)\,s(n), \tag{23}$$

$$m_4(n) = \frac{\cos\theta(n-1)\cos\theta(n)\,k(n) + \sin\theta(n-1)\sin\theta(n)\,s(n)}{\sqrt{n+1}}. \tag{24}$$

The functions $k(n)$ and $s(n)$ are still free and will be fixed in the following.

4. Factorization and the Jaynes-Cummings Hamiltonian

Let us also notice that the form of $\mathcal{M}^-$ and $\mathcal{M}^+$ leads to an expression of $\mathcal{M}^+\mathcal{M}^-$ similar to the Jaynes-Cummings Hamiltonian. So we get

$$H = \mathcal{M}^+\mathcal{M}^- + \varepsilon_0^+ \sigma_0, \tag{25}$$

when

$$k_1^2(n) = \varepsilon_{n+1}^- - \varepsilon_0^+ = (1+\epsilon)n + \lambda(r(1) + r(n+1)) \tag{26}$$

and

$$s_1^2(n) = \varepsilon_n^+ - \varepsilon_0^+ = (1+\epsilon)n + \lambda(r(1) - r(n+1)). \tag{27}$$

For example, in the very particular case of resonance between the electromagnetic field and the atom, that is when $\varepsilon = 0$, we have

$$m_1(N) = \frac{k(N+1) + s(N+1)}{2\sqrt{N+1}}, \tag{28}$$

$$m_2(N) = \frac{k(N+1) - s(N+1)}{2\sqrt{N+1}\sqrt{N+2}}, \tag{29}$$

$$m_3(N) = \frac{k(N) - s(N)}{2}, \tag{30}$$

$$m_4(N) = \frac{k(N) + s(N)}{2\sqrt{N+1}}, \tag{31}$$

and

$$k^2(N) = N + \lambda(1 + \sqrt{N+1}), \tag{32}$$

$$s^2(n) = N + \lambda(1 - \sqrt{N+1}). \tag{33}$$

The possible case where some of the differences $\varepsilon_n^+ - \varepsilon_0^+$ are negative is not acceptable with our assumption that the function $s(n)$ be real. If we set $\epsilon = 0$, it is easy to see that the quantity (27) will be positive (or zero) when $\lambda \leq 1 + \sqrt{2}$. At $\lambda = 1 + \sqrt{2}$, we see that $s_1^2(0) = s_1^2(1) = 0$ which means that $\varepsilon_1^+ = \varepsilon_0^+$ and the values of (27) are all positive for $n > 1$. For $\lambda > 1 + \sqrt{2}$, (27) becomes negative for some values of n. A detailed analysis of this phenomena is given in [8].

Let us finally mention that coherent states $|z, \alpha\rangle$ may then be defined as eigenstates of $\mathcal{M}^-$ and expressed as a linear combination of the energy eigenstates (7)–(9). These are such that the mean value of the energy may be computed directly and we get

$$\langle H \rangle = |\alpha|^2 + \varepsilon_0^+. \tag{34}$$

This work is partly supported by research grants from NSERC of Canada and FCAR du gouvernement du Québec, as well as by the Spanish DGES (PB98-0370), MCYT (BFM2002-03773), and Junta de Castilla y León (VA085/02).

References

[1] Jaynes E T and Cummings F 1963 Proc. IEEE **51** 89
[2] Louisell W H 1973 Quantum Statistical Properties of Radiation, Wiley Series in Pure and Applied Optics (New York: John Wiley & Sons)
[3] Cirac J I, Parkins A S, Blatt R, and Zoller P 1996 Nonclassical States of Motion in Ion Traps, in Advances in Atomic, Molecular, and Optical Physics, Vol 37, B Benderson and H Walther (Eds) (San Diego: Academic Press)
[4] Nielsen M A and Chuang I L 2002 Quantum Computation and Quantum Information (Cambridge: Cambridge University Press)
[5] Gea-Banacloche J 1992 Optics Comm. **88** 531
[6] Narozhny N B, Sanchez-Mondragon J J and Eberly J H 1981 Phys. Rev. A **23** 236
[7] Daoud M and Hussin V 2002 J. Phys. A: Math. Gen. (to be published)
[8] Hussin V and Nieto L M 2002 CRM preprint

Differential form of the star product for SU(2) Stratonovich-Weyl symbols

A. B. Klimov†and P. Espinoza‡

† Departamento de Física, Universidad de Guadalajara, Revolución 1500, 44410, Guadalajara, Jal., México.
‡ Departamento de Ciencias Básicas, Universidad de Guadalajara, Enrique Díaz de León 1, 47460, Lagos de Moreno, Jal., México.

Abstract. An exact differential form for the star product is derived and its asymptotic form in the limit of large spin is analyzed. As an immediate application we obtain evolution equations for a class of s-parametrized SU(2) quasi-distribution functions.

1. A differential form for the star product

The phase-space methods allows us to treat quantum mechanics as a statistical theory in a classical phase-space [1]-[4], i.e., both states and observables are considered as functions on a given phase space, in such a way that average values are computed by integrating over the phase space of some quasiprobability distribution function with the Weyl symbol of a corresponding operator. For spin-like systems (possessing the $SU(2)$ dynamical symmetry group) we associate each operator $\hat{f}$ with its (s-parametrized) symbol $W_f^{(s)}(\theta, \phi)$, a c-number function defined on the sphere $(\theta, \phi) \in S_2$. The function $W_f^{(s)}$ is defined by the invertible map (the Stratonovich-Weyl correspondence [5]-[8])

$$W_f^{(s)}(\theta, \phi) = \text{Tr}\left(\hat{f}\hat{w}_s(\theta, \phi),\right) \tag{1}$$

where the operator $\hat{w}_s(\theta, \phi)$ has the form

$$\hat{w}_s(\theta, \phi) = \frac{2\sqrt{\pi}}{\sqrt{2S+1}} \sum_{L=0}^{2S} \sum_{M=-L}^{L} \left(C_{SS,L0}^{SS}\right)^{-s} Y_{LM}^*(\theta, \phi)\, \hat{T}_{LM}^{(S)}, \tag{2}$$

where $Y_{LM}(\theta, \phi)$ are the spherical harmonics and $\hat{T}_{LM}^{(S)}$ are the irreducible tensor operators [9]. In the case when the operator $\hat{f}$ is the density matrix ρ, the Weyl symbol $W_\rho^{(s)}(\theta, \phi)$ gives us a quasi-distribution function. The value $s = 0$ corresponds to the Stratonovich-Wigner function W, meanwhile $s = \pm 1$ leads to contravariant P-symbol and covariant Q-symbol respectively [10].

The cornerstone of the phase-space approach is the so-called star (or twisted) product. This star product allows us to replace the standard manipulations with operators in the Hilbert space by a differential or integral operator acting on the product of Weyl's symbols. Although the integral form for the star product is easy to obtain from the Stratonovich-Weyl correspondence [7]-[8], it is not very suitable for calculations. For practical purposes the differential form is more convenient. The differential form for the star product of two Stratonovich-Weyl symbols is determined by the condition

$$W_{fg}^{(s)} = W_f^{(s_1)} *_s W_g^{(s_2)} = \hat{L}_{fg}^{(s)}(\theta, \phi) \left[W_f^{(s_1)} W_g^{(s_2)} \right] \tag{3}$$

for any two operators $\hat{f}, \hat{g}$, where $\hat{L}_{fg}^{(s)}(\theta, \phi)$ is a differential operator. This allows us to write down a differential evolution equation for quasi-distribution functions,

$$i\partial_t W_\rho^{(s)} = \{W_H^{(s_1)}, W_\rho^{(s_2)}\}_M = W_H^{(s_1)} *_s W_\rho^{(s_2)} - W_\rho^{(s_2)} *_s W_H^{(s_1)}, \tag{4}$$

where H is the Hamiltonian of the system and $\{,\}_M$ are the so-called Moyal brackets. The operator $\hat{L}_{fg}^{(s)}(\theta, \phi)$ can be represented in the following form [11]:

$$\hat{L}_{fg}^{(s)}(\theta, \phi) = N_S \sum_j a_j \tilde{F}^{s-1}(\mathcal{L}^2) \left[\left(S^{+(j)} \tilde{F}^{1-s_1}(\mathcal{L}^2) \right)_f \otimes \left(S^{-(j)} \tilde{F}^{1-s_2}(\mathcal{L}^2) \right)_g \right] \tag{5}$$

where

$$N_S = \sqrt{2S+1} \left[(2S+1)!(2S)!\right]^{(s_1+s_2-s)/2}, \qquad a_j = \frac{(-1)^j}{j!\,(2S+j+1)!}, \tag{6}$$

$$S^{\pm(j)} = \Pi_{k=0}^{j-1} \left(k \cot\theta - \partial_\theta \mp \frac{i}{\sin\theta}\partial_\phi \right) \tag{7}$$

and $\tilde{F}(\mathcal{L}^2)$ is a function of the Casimir operator on the sphere,

$$\mathcal{L}^2 = - \left[\partial_\theta^2 + \cot\theta\partial_\theta + \sin^{-2}\theta\partial_\phi^2 \right], \tag{8}$$

such that $\tilde{F}(\mathcal{L}^2)Y_{LM}(\theta, \phi) = \sqrt{(2S+L+1)!\,(2S-L)!}Y_{LM}(\theta, \phi)$. The operators with subscript "f" act only on the $W_f^{(s_1)}(\theta, \phi)$, the operators with subscript "g" act only on the $W_g^{(s_2)}(\theta, \phi)$, whereas the external operator $\tilde{F}^{s-1}(\mathcal{L}^2)$ acts on the whole product. Another representation for $\hat{L}_{fg}^{(s)}$ is obtained by performing a formal summation in (5),

$$\hat{L}_{fg}^{(s)}(\theta, \phi) = N_S \int \frac{d\psi}{2\pi} \tilde{F}^{s-1}(\mathcal{J}^2) \sigma \left(S_f^+ \otimes S_g^- \right) \left(\tilde{F}^{1-s_1}(\mathcal{J}^2) \right)_f \otimes \left(\tilde{F}^{1-s_2}(\mathcal{J}^2) \right)_g, \tag{9}$$

where $S^{\pm,0}$ are the angular momentum operators in the rotating frame [9]:

$$S^\pm = ie^{\mp i\psi} \left(\pm\cot\theta\partial_\psi + i\partial_\theta \mp \frac{1}{\sin\theta}\partial_\phi \right), \qquad S^0 = -i\partial_\psi, \tag{10}$$

and $\mathcal{J}^2 = S^0(S^0+1) - S^+S^-$ is the corresponding Casimir operator, $[\mathcal{J}^2, S^{\pm,0}] = 0$. The function $\sigma(z)$ is defined as

$$\sigma(z) = \sum_j a_j z^j = z^{-S-1/2} J_{2S+1}(2\sqrt{z}), \tag{11}$$

where $J_\nu(x)$ is the Bessel function.

In particular, for $P_f = W_f^{(s=1)}$, $Q_f = W_f^{(s=-1)}$ and $W_f = W_f^{(s=0)}$ symbols we obtain

$$P_{fg}(\theta, \phi) = (2S+1)! \sum_j a_j \left(S^{+(j)} P_f(\theta, \phi) \right) \left(S^{-(j)} P_g(\theta, \phi) \right), \tag{12}$$

$$Q_{fg}(\theta, \phi) = \frac{1}{(2S)!} \sum_j a_j \tilde{F}^{-2}(\mathcal{L}^2) \left(S^{+(j)} \tilde{F}^2(\mathcal{L}^2) Q_f(\theta, \phi) \right) \left(S^{-(j)} \tilde{F}^2(\mathcal{L}^2) Q_g(\theta, \phi) \right), \tag{13}$$

$$W_{fg}(\theta, \phi) = \sqrt{2S+1} \sum_j a_j \tilde{F}^{-1}(\mathcal{L}^2) \left(S^{+(j)} \tilde{F}(\mathcal{L}^2) W_f(\theta, \phi) \right) \left(S^{-(j)} \tilde{F}(\mathcal{L}^2) W_g(\theta, \phi) \right), \tag{14}$$

where a_j is defined in (6).

2. Evolution equations for linear and quadratic Hamiltonians

Using the explicit form of the star product operator (5) one can obtain evolution equation for the (s-parametrized) quasi-distribution function $W_\rho^{(s)}(\theta, \phi)$ [12]. As an example let us consider quadratic Hamiltonians, which we separate into linear H_1 and quadratic H_2 parts, $H = H_1 + H_2$, such that the corresponding (symmetrized, $s = 0$) symbol is represented as

$$W_H^{(0)}(\theta, \phi) = W_{H_1}(\theta, \phi) + W_{H_2}(\theta, \phi). \tag{15}$$

The evolution equation (4) takes the form

$$\partial_t W_\rho^{(s)}(\theta, \phi) = \frac{1}{\sqrt{S(S+1)}} \{W_\rho^{(s)}(\theta, \phi), W_{H_1}(\theta, \phi)\}_P +$$

$$\chi \tilde{F}^{s-1}(\mathcal{L}^2) \left[\left(\frac{1}{2}\mathcal{L}^2 + 2S + 4\right) \left\{ \tilde{F}^{1-s}(\mathcal{L}^2) W_\rho^{(s)}(\theta, \phi), W_{H_2}(\theta, \phi) \right\}_P - \right.$$

$$\left. \frac{1}{2} \left\{ \mathcal{L}^2 \tilde{F}^{1-s}(\mathcal{L}^2) W_\rho^{(s)}(\theta, \phi), W_{H_2}(\theta, \phi) \right\}_P - \frac{1}{2} \left\{ \tilde{F}^{1-s}(\mathcal{L}^2) W_\rho^{(s)}(\theta, \phi), \mathcal{L}^2 W_{H_2}(\theta, \phi) \right\}_P \right], \tag{16}$$

where $\chi = [(2S+3)(2S-1)S(S+1)]^{-1/2}$, $\mathcal{L}^2$ is the Casimir operator (8) and $\{, \}_P$ means the Poisson brackets on the sphere,

$$\{, \}_P = \frac{1}{\sin \theta} (\partial_\phi \otimes \partial_\theta - \partial_\theta \otimes \partial_\phi). \tag{17}$$

In the case of linear Hamiltonians, just the first term appears in the right hand side of the evolution equation (16), leading to the classical Liouville equation [7].

For example, for the simplest nonlinear (quadratic) Hamiltonian $H = S_z^2$ we obtain the following evolution equations for the $P(\theta, \phi)$, $W(\theta, \phi)$ and $Q(\theta, \phi)$ symbols

$$\partial_t P = -[2(S+1)\cos\theta + \sin\theta\partial_\theta] \partial_\phi P, \tag{18}$$

$$\partial_t W = -\left[(S + \frac{1}{2})\Phi(\mathcal{L}^2)\cos\theta + \varepsilon\Phi^{-1}(\mathcal{L}^2)\left(\frac{3}{2}\cos\theta + \sin\theta\partial_\theta\right)\right]\partial_\phi W, \tag{19}$$

$$\partial_t Q = -[2S\cos\theta - \sin\theta\partial_\theta] \partial_\phi Q, \tag{20}$$

where the function $\Phi(\mathcal{L}^2)$ is defined as

$$\Phi(\mathcal{L}^2) = \left[2 - \varepsilon^2 (2\mathcal{L}^2 + 1) + 2\sqrt{1 - \varepsilon^2 (2\mathcal{L}^2 + 1) + \varepsilon^4\mathcal{L}^4}\right]^{1/2}, \tag{21}$$

and $\varepsilon = (2S + 1)^{-1}$.

3. Large dimensions of representation

In the limit of large spin, $S \gg 1$, ($s = s_1 = s_2$) the operator (9) is approximated as

$$\hat{L}_{fg}^{(s)}(\theta, \phi) \approx \int \frac{d\psi}{2\pi} \exp\left[-\frac{s\varepsilon}{2} \left(S_f^- \otimes S_g^+ + S_f^+ \otimes S_g^-\right) + \frac{\varepsilon}{2} \left(S_f^- \otimes S_g^+ - S_f^+ \otimes S_g^-\right)\right], \tag{22}$$

where $\varepsilon = (2S + 1)^{-1} \ll 1$, leading to a form of the star product similar to that for the Heisenberg-Weyl group [4], [13]. In particular one obtains

$$P_{fg}(\theta, \phi) = \int \frac{d\psi}{2\pi} \exp\left[-\varepsilon S_f^- \otimes S_g^+\right] P_f(\theta, \phi) P_g(\theta, \phi),$$

$$Q_{fg}(\theta, \phi) = \int \frac{d\psi}{2\pi} \exp\left[\varepsilon \left(S_f^+ \otimes S_g^-\right)\right] Q_f(\theta, \phi) Q_g(\theta, \phi),$$

$$W_{fg}(\theta, \phi) = \int \frac{d\psi}{2\pi} \exp\left[\frac{\varepsilon}{2} \left(S_f^+ \otimes S_g^- - S_f^- \otimes S_g^+\right)\right] W_f(\theta, \phi) W_g(\theta, \phi).$$

From the other hand, expanding (5) (or (9)) as a series on powers of ε we obtain

$$\hat{L}_{fg}^{(s)}(\theta, \phi) = I_f \otimes I_g + \frac{\varepsilon}{2}\left[\left((1-s)S_f^{-(1)} \otimes S_g^{+(1)} - (1+s)S_f^{+(1)} \otimes S_g^{-(1)}\right)\right] + O(\varepsilon^2). \quad (23)$$

Taking into account that

$$S_f^{+(1)} \otimes S_g^{-(1)} - S_f^{-(1)} \otimes S_g^{+(1)} = 2i\{f, g\}_P, \quad (24)$$

we obtain that in the limit $S \gg 1$ the evolution equation (4) for the quasi-distribution function takes the form

$$\partial_t W_\rho^{(s)} \approx 2\varepsilon\{W_\rho^{(s)}, W_H^{(s)}\}_P + O(\varepsilon^2), \quad (25)$$

which is similar to the Heisenberg-Weyl case when the Moyal brackets are reduced to the classical Poisson brackets (in the flat space) in the limit $\hbar \to 0$. It is important to note that quantum corrections to the classical evolution equation (25) (which can be obtained by further expansion of the star product operator (5) in series of ε) essentially depend on the type of s-ordering of the quasi-distribution function. In particular, it can be proved [12] that the first order (quantum) correction vanishes *only* for the symmetrically ordered ($s = 0$) Stratonovich-Wigner quasidistribution function.

[1] Wigner E P 1932 Phys. Rev. **40** 749 .
[2] Hillery M, O'Connel R F, Scully M O and Wigner E P 1984 Phys. Rep. **106** 121.
[3] Lee H-W 1995 Phys. Rep. **259** 147.
[4] Moyal J E 1949 Proc. Cambridge Philos. Soc. **45** 99.
[5] Stratonovich R L 1956 Sov. Phys. JETP **31** 1012.
[6] Agarwal G S 1981 Phys. Rev. A **24** 2889.
[7] Várilly J C and Gracia-Bondía J M 1989 Annals of Physics **190** 107.
[8] Brif C and Mann A 1999 Phys. Rev. A **59** 971.
[9] Varshalovich D A, Moskalev A N and Khersonskiĭ V K 1988 *Quantum Theory of Angular Momentum*, (World Scientific, Singapore).
[10] Beresin P A 1975 Comm. Math. Phys. **40** 153.
[11] Klimov A B and Espinoza P 2002 J. Phys. A **35** to be published.
[12] Klimov A B 2002 J. Math. Phys. **43** 2202.
[13] Cohen L J 1966 J. Math. Phys. **7** 781; *ibid* 1976 **17** 1863.

Inst. Phys. Conf. Ser. No 173: Section 7
Paper presented at 24th Int. Coll. Group Theoretical Methods in Physics, Paris, France, 15–20 July 2002
©*2003 IOP Publishing Ltd*

On the combined effect of different symmetries on quantum mechanical potentials

G. Lévai†
Institute of Nuclear Research of the Hungarian Academy of Sciences,
PO Box 51, H–4001 Debrecen, Hungary

Abstract. We study the interrelation of $\mathcal{PT}$ symmetry, supersymmetry and Lie symmetries in the description of one-dimensional quantum mechanical potentials. Compared to their Hermitian counterparts, the complex $\mathcal{PT}$ potentials usually possess a second set of normalizable solutions, and the two sets can be distinguished by the quasi-parity quantum number $q = \pm 1$. We demonstrate that this doubling mechanism leads to larger potential algebras and also to the doubling of the SUSY partners of these potentials.

With the introduction of $\mathcal{PT}$ symmetric quantum mechanics [1] the investigation of non-Hermitian quantum mechanical problems has received much attention in the past couple of years. In $\mathcal{PT}$ symmetric quantum mechanics the potentials defined in one dimension are invariant under the simultaneous action of the space and time reflection operations $\mathcal{P}$ and $\mathcal{T}$, and have the property $[V(-x)]^* = V(x)$. The first notable finding was that despite being complex, these potentials often have real bound-state energy spectrum, and this was interpreted as the consequence of $\mathcal{PT}$ symmetry. However, it was soon noticed that $\mathcal{PT}$ symmetry is neither a necessary, nor a sufficient condition for having real energy spectrum in a complex potential. It is not a sufficient condition, because the energy eigenvalues may also appear in complex conjugated pairs, in which case the eigenfunctions cease to be eigenfunctions of the $\mathcal{PT}$ operator, and this scenario has been interpreted as the spontaneous breakdown of PT symmetry [1]. Neither is $\mathcal{PT}$ symmetry a necessary condition, because there are complex non-$\mathcal{PT}$ symmetric potentials with real energy eigenvalues [2, 3, 4].

More recently $\mathcal{PT}$ symmetric quantum mechanics was put into a more general context as the special case of pseudo-Hermiticity [5]. A Hamiltonian is said to be η-pseudo-Hermitian if $H^\dagger = \eta H \eta^{-1}$ holds, where † denotes the adjoint operation. Based on these general arguments it was demonstrated [5] that a Hamiltonian is pseudo-Hermitian if and only if its eigenvalues are real or come in complex conjugate pairs, as was the observation for PT symmetric potentials. In this context $\mathcal{PT}$ symmetric Hamiltonians can be interpreted as $\mathcal{P}$-pseudo-Hermitian. The modified inner product [6, 7] and the the pseudo-norm [6] also received a natural foundation winthis framework [5]. It has to be noted that the term psudo-Hermiticity has already been introduced long before the formulation of $\mathcal{PT}$ symmetric quantum mechanics [8].

After the first numerical examples [1], a number of exactly solvable $\mathcal{PT}$ symmetric potentials have been derived, mainly as the $\mathcal{PT}$ symmetric versions of conventional solvable potentials (see e.g. [9] and references therein). The analysis of these problems showed that due to the generally less strict boundary conditions, $\mathcal{PT}$ symmetric potentials have two sets of normalizable solutions, which can be distinguished with the introduction of the quasi-parity quantum number $q = \pm 1$ [10]. Besides the relatively trivial shape-invariant potentials [9], this doubling mechanism was also found in the spectrum of more complicated solvable $\mathcal{PT}$ symmetric potentials [11].

† E-mail: levai@atomki.hu

These unusual features naturally raise the question how other symmetries of the same potentials are affected by $\mathcal{PT}$ symmetry. We are particularly interested in constructions based on supersymmetric quantum mechanics (SUSYQM) [12] and Lie algebras. The doubling of the basis states due to $q = \pm 1$ implies that the superpotential also has to carry the quasi-parity quantum number, and also that some algebras associated with the basis states of conventional potentials have to be enlarged.

Let us modify the standard SUSYQM formalism by adding the q quasi-parity quantum number to the SUSYQM shift operators

$$A^{(q)} = \frac{\mathrm{d}}{\mathrm{d}x} + W^{(q)}(x) \qquad A^{\dagger(q)} = -\frac{\mathrm{d}}{\mathrm{d}x} + W^{(q)}(x) \tag{1}$$

through the superpotential $W^{(q)}(x) = -\frac{\mathrm{d}}{\mathrm{d}x} \ln \psi_{0,-}^{(q)}(x)$, where $\psi_{n,-}^{(q)}(x)$ is the n'th normalizable wavefunction with quasi-parity q. Substituting directly $A^{(q)}$ and $A^{\dagger(q)}$ in the factorized form of the Hamiltonian, the two sets of solutions would belong to two potentials shifted with respect to each other with an energy constant, because the ground-state energies $E_{0,-}^{(\pm q)}$ would be zero for $q = 1$ and -1 alike, by construction [12]. In order to avoid this, let us write the "bosonic" Hamiltonian in the factorized form $\mathbf{H}_- = A^{\dagger(q)} A^{(q)} + \varepsilon^{(q)} = A^{\dagger(-q)} A^{(-q)} + \varepsilon^{(-q)}$, containing the q-dependent factorization energies $\varepsilon^{(\pm q)} = E_{0,-}^{(\pm q)}$. Then $\mathbf{H}_-$ becomes independent of q, and its eigenvalue equation takes the form

$$\mathbf{H}_- \psi_{n,-}^{(q)} = [A^{\dagger(\pm q)} A^{(\pm q)} + \varepsilon^{(\pm q)}] \psi_{n,-}^{(q)} = E_{n,-}^{(q)} \psi_{n,-}^{(q)} . \tag{2}$$

The "fermionic" partner Hamiltonians $\mathbf{H}_+^{(\pm q)}$, however, will depend on q:

$$\mathbf{H}_+^{(\pm q)} \psi_{n,+}^{(q)} = [A^{(\pm q)} A^{\dagger(\pm q)} + \varepsilon^{(\pm q)}] \psi_{n,+}^{(q)} = E_{n,+}^{(q)} \psi_{n,+}^{(q)} . \tag{3}$$

With Eqs. (2) and (3) one can easily prove the $A^{(\pm q)} \psi_{n,-}^{(q)}$ functions are eigenfunctions of the $\mathbf{H}_+^{(\pm q)}$ "fermionic" Hamiltonians, and the corresponding energy eigenvalues are the same as those of the q-independent "bosonic" Hamiltonian:

$$\mathbf{H}_+^{(q)} A^{(q)} \psi_{n,-}^{(q)} = E_{n,-}^{(q)} A^{(q)} \psi_{n,-}^{(q)} , \tag{4}$$

$$\mathbf{H}_+^{(-q)} A^{(-q)} \psi_{n,-}^{(q)} = E_{n,-}^{(q)} A^{(-q)} \psi_{n,-}^{(q)} . \tag{5}$$

However, there is a difference between (4) and (5) that in the former case $A^{(q)} \psi_{n,-}^{(q)} = 0$ holds by construction, so the partner of the ground-state "bosonic" level is missing from the spectrum of $\mathbf{H}_+^{(q)}$ [12], while the situation is different for (5), so there the number of levels is the same in the "bosonic" and "fermionic" Hamiltonians.

We illustrate this procedure with the example of the Scarf II potential [13]. There the superpotential is $W^{(q)}(x) = -\frac{1}{2}(q\alpha + \beta + 1)\tanh x - \frac{1}{2}(\beta - q\alpha)\mathrm{sech}x$, which generates the "bosonic" potential

$$V_-(x) = -\frac{1}{\cosh^2 x}\left[\left(\frac{\alpha+\beta}{2}\right)^2 + \left(\frac{\alpha-\beta}{2}\right)^2 - \frac{1}{4}\right] + \frac{2\mathrm{i}\sinh x}{\cosh^2 x}\left(\frac{\beta+\alpha}{2}\right)\left(\frac{\beta-\alpha}{2}\right) \tag{6}$$

if the factorization energies $\varepsilon^{(q)} = -\frac{1}{4}(q\alpha + \beta + 1)^2$ are used. The "fermionic" partner potentials then take the form [13]

$$V_+^{(q)}(x) = -\frac{1}{\cosh^2 x}\left[\left(\frac{q\alpha+\beta+2}{2}\right)^2 + \left(\frac{q\alpha-\beta}{2}\right)^2 - \frac{1}{4}\right]$$
$$+ \frac{2\mathrm{i}\sinh x}{\cosh^2 x}\left(\frac{\beta+q\alpha+2}{2}\right)\left(\frac{\beta-q\alpha}{2}\right) . \tag{7}$$

The results obtained for the Scarf II potential have significantly different implications for unbroken and broken $\mathcal{PT}$ symmetry, corresponding to real and imaginary values of α [13]. In the former case the "fermionic" partner potentials (7) are $\mathcal{PT}$ symmetric, and the energy eigenvalues remain real. In the latter case, however, the coupling parameters of both the even and odd component of the potential become complex due to the imaginary value of α, therefore the "fermionic" potentials cease to be $\mathcal{PT}$ symmetric.

We note that a similar system of partner potentials has been obtained [14] from two essentially different supersymmetric constructions; i.e. the parasupersymmetric scheme (where a three- rather than two-dimensional matrix representation is used) and second-order supersymmetry (where A and $A^\dagger$ in (1) are second-order differential operators).

Isospectral potentials can also be discussed in terms of potential algebras [15], the ladder operators of which connect degenerate states of potentials with *different* depth, but of the same type. The practical equivalence of the SUSYQM construction and one based on an su(1,1) (or su(2)) potential algebra has been demonstrated [16] for B and A class shape-invariant potentials, which contain the Morse potential and various Scarf and Pöschl–Teller potentials. Complexified potentials have been studied in terms of the sl(2,C) [17, 18] and su(1,1)≃so(2,1) [19] algebras, and it was established that these algebras act on one set of the wavefunctions, similarly to the case of Hermitian problems. In order to handle both sets of solutions with $q = \pm 1$, a larger algebra is needed.

In a systematic study [20] the so(2,2)∼so(2,1)⊕so(2,1) algebra has been proposed to describe all the states of the $\mathcal{PT}$ symmetric type A potentials, including the Scarf II and the generalized Pöschl–Teller potentials. This algebra is defined as

$$[J_z, J_\pm] = \pm J_\pm \qquad [J_+, J_-] = -2J_z\,, \qquad [J_i, K_j] = 0 \tag{8}$$

$$[K_z, K_\pm] = \pm K_\pm \qquad [K_+, K_-] = -2K_z\,, \qquad i, j = +, -, z\,. \tag{9}$$

The differential realization of this algebra for the Scarf II potentials is [20]

$$J_\pm = e^{\pm i\phi}\left(\pm\frac{\partial}{\partial x} - \tanh x(J_z \pm \frac{1}{2}) + \frac{i}{\cosh x}K_z\right)\,, \qquad J_z = -i\frac{\partial}{\partial\phi}\,, \tag{10}$$

$$K_\pm = e^{\pm i\chi}\left(\pm\frac{\partial}{\partial x} - \tanh x(K_z \pm \frac{1}{2}) + \frac{i}{\cosh x}J_z\right)\,, \qquad K_z = -i\frac{\partial}{\partial\chi}\,. \tag{11}$$

The basis states are written in terms of the $\psi(x)$ physical wavefunctions as $\Psi(x, \phi, \chi) = e^{i(m\phi + m'\chi)}\psi(x)$. As it can be seen from (10) and (11), the structure of the so(2,2) generators is essentially the same as the that of the supersymmetric shift operators A and $A^\dagger$, and the α and β parameters can be related to the m and m' eigenvalues of the J_z and K_z generators. In fact, direct calculation also shows that these operators have the same effect on the wavefunctions in the two symmetry-based schemes [20, 13].

It has to be noted that the so(2,2) algebra (or its compact version so(4)) plays the role of a potential algebra only for a limited number of potentials, i.e. the members of the A factorization type, while the supersymmetric construction presented here for $\mathcal{PT}$ symmetric potentials might have wider applicability. This is somewhat different for the Morse potential belonging to type B factorization: it is possible to define an sl(2,C) potential algebra associated with it [17, 18], but it is not $\mathcal{PT}$ symmetric. It can be made $\mathcal{PT}$ symmetric by defining it along a bent contour of the complex x plane [21, 9].

Acknowledgments

This work was supported by the OTKA grant No. T031945.

568

References

[1] Bender C M and Boettcher S 1998 *Phys. Rev. Lett.* **24** 5243
[2] Cannata F, Junker G and Trost J 1998 *Phys. Lett. A* **246** 219
[3] Khare A and Mandal B P 2000 *Phys. Lett. A* **272** 53
[4] Bagchi B and Roychoudhury R 2000 *J. Phys. A: Math. Gen.* **33** L1
[5] Mostafazadeh A 2002 *J. Math. Phys.* **43** 205, 2814, 3944
[6] Znojil M 2001 *Preprint* quant-ph/0103054 v3
[7] Japaridze G S 2002 *J. Phys. A: Math. Gen.* **35** 1709
[8] Lee T D and Wick G C 1969 *Nucl. Phys. B* **9** 209
[9] Lévai G and Znojil M 2000 *J. Phys. A: Math. Gen.* **33** 7165.
[10] Bagchi B, Quesne C and Znojil M 2001 *Mod. Phys. Lett. A* **16** 2047
[11] Znojil M, Lévai G, Roychoudhury R and Roy P 2001 *Phys. Lett. A* **290** 249
[12] Cooper F, Khare A and Sukhatme U 1995 *Phys. Rep.* **251** 267
[13] Lévai G and Znojil M 2002 *J. Phys. A: Math. Gen.*, in press (*Preprint* quant-ph/0206032)
[14] Bagchi B, Mallik S and Quesne C 2002 *Int. J. Mod. Phys. A* **17** 51
[15] Alhassid Y, Gürsey F and Iachello F 1986 *Ann. Phys. (N.Y.)* **167** 181
[16] Lévai G 1994 *J. Phys. A:Math. Gen.* **27** 3809
[17] Bagchi B and Quesne C 2000 *Phys. Lett. A* **273** 285
[18] Bagchi B and Quesne C 2002 *Phys. Lett. A* **300** 18
[19] Lévai G, Cannata F and Ventura A 2001 *J. Phys. A: Math. Gen.* **34** 839
[20] Lévai G, Cannata F and Ventura A 2002 *J. Phys. A: Math. Gen.* **35** 5041
[21] Znojil M 1999 *Phys. Lett. A* **264** 108

Inst. Phys. Conf. Ser. No 173: Section 7
Paper presented at 24th Int. Coll. Group Theoretical Methods in Physics, Paris, France, 15–20 July 2002
©2003 IOP Publishing Ltd

The application of Lie algebras to the separation of degrees of freedom in problems with many degrees of freedom

Harry J. Lipkin

Department of Particle Physics Weizmann Institute of Science, Rehovot 76100, Israel
School of Physics and Astronomy, Raymond and Beverly Sackler Faculty of Exact Sciences, Tel Aviv University, Tel Aviv, Israel
High Energy Physics Division, Argonne National Laboratory, Argonne, IL 60439-4815, USA

Abstract. Lie algebras can furnish a simple description of elementary excitations involving only a small number of the degrees of freedom of a complex system. Examples are active and spectator excitations and collective and single particle excitations.

1. Introduction

I am very pleased to be honored at this meeting by the award of the Wigner Medal. I learned many things from my former teacher Eugene Wigner, and always remember his remark: "I believe that this theory is wrong. But you know, the old quantum theory of Bohr and Sommerfeld was wrong, too. And it is hard to see how we could ever have reached the correct quantum theory without first going through this stage." I have been following this advice and throughout my career have pursued approaches believed to be wrong by conventional wisdom.

At a time when most physicists referred derisively to the "Gruppenpest", many young theorists who later became famous did not attend Giulio Racah's lectures on "Group theory and spectroscopy" at the Institute for Advanced Study at Princeton. They were sure that group theory was completely useless. I learned the connection between group theory and physics from Racah and later found myself in the position of educating "pedestrians" on what I had learned from Racah[1].

There are many examples where Lie Algebras can be used to separate elementary excitations involving different kinds of degrees of freedom

(i) Active and spectator degrees of freedom
  - Cooper Pairs and Quasiparticles[2].
  - Seniority in the nuclear and atomic shell models
  - Nuclear Excitons and Phonons in a Mössbauer Crystal[3].

(ii) Independent particle and collective degrees of freedom
  - Translation Invariance and Center of Mass Motion in Many Fermion Systems
    3N-3 relative co-ordinates vs, Pauli permutations of 3N co-ordinates
  - Collective rotations and "intrinsic" particle co-ordinates
  - Collective vibrations, quadrupole and monople (breathing)

A few simple examples are discussed in this paper.

2. Active and spectator degrees of freedom - Cooper Pairs and Quasiparticles

A simple example of a state with quasiparticle and Cooper pair degrees of freedom was the "seniority" classification introduced by Racah in atomic and nuclear physics[2]. Racah's seniority anticipated Cooper pairs and BCS.

Racah's treatment for a fixed number n of particles was generalized by the introduction of the algebra of second-quantized operators into many-body physics. Consider an n-particle state $|n, v\rangle$ which has v quasiparticles and $(n - v)/2$ Cooper pairs. The Anderson-Kerman quasispin algebra[4] can be used to define creation and destruction operators for a simplified version of Cooper pairs[2]..

$$S_+ = \sum_{k=1}^{N} c_{k\uparrow}^\dagger c_{-k\downarrow}; \quad S_- = \sum_{k=1}^{N} c_{k\downarrow}^\dagger c_{-k\uparrow}; \quad [S_+, S_-] = 2S_z \tag{1}$$

$$|n, v\rangle = (S_+)^{(n-v)/2} |v, v\rangle ; \quad S_- |v, v\rangle = 0 \tag{2}$$

The state $|v, v\rangle$ has v quasiparticles and no Cooper pairs. The number of Cooper pairs $(n - v)/2$ is defined by the eigenvalue of the quasispin Casimir operator $S(S + 1)$

For n particles in a j-j coupling shell, all states are in an irreducible representation of $SU(2j + 1)$. Racah noted that the subgroup $Sp(2j + 1)$ defines the transformations that leave invariant pairs coupled to J=0 (anticipating Cooper pairs). The seniority number v defined by this $Sp(2j + 1)$ algebra is the number of unpaired particles (quasiparticle excitations).

Racah anticipated BCS in showing that $J = 0$ fermion pairs are not bosons. The standard BCS energy gap in an n-particle state is seen by simple algebra to be produced by the Pauli effect in overlapping fermion pairs. For any simple two-body interaction that binds the $J = 0$ two-body state more than any other two-body state the energy E(n,o) of a state with n/2 pairs and no quasiparticles is raised by breaking a pair by an amount which is a motonically increasing function of n. For large values of n.

$$E(n, 2) - E(n, 0) \gg E(2, 2) - E(2, 0). \tag{3}$$

3. Noncompact Groups and Angular Momentum Algebras with a Wrong Sign

The spectrum-generating algebra of the harmonic oscillator applied to a many-particle system shows how collective boson excitations are described with a Lie algebra.

The spectrum generating algebra for a single harmonic oscillator is[5]

$$t_1 = (a^\dagger a^\dagger + aa)/4; \quad t_2 = (a^\dagger a^\dagger - aa)/4i; \quad t_3 = (a^\dagger a + aa^\dagger)/4 \tag{4}$$

For a system of many harmonic oscillators, define

$$\vec{T} = \sum_k \vec{t}_k; \quad [T_1, T_2] = -iT_3; \quad [T_2, T_3] = iT_1; \quad [T_3, T_1] = iT_2 \tag{5}$$

These operators satisfy angular momentum commutation rules with one wrong sign. To fix the sign at the price of making two operators antihermitean, we define

$$J_1 = iT_1; \quad J_2 = iT_2; \quad J_3 = T_3 \tag{6}$$

$$[J_1, J_2] = iJ_3; \quad [J_2, J_3] = iJ_1; \quad [J_3, J_1] = iJ_2 \tag{7}$$

The Casimir operator commutes with the generators,

$$J^2 = J_3^2 + J_1^2 + J_2^2 = T_3^2 - T_1^2 - T_2^2; \quad [J^2, J_3] = [J^2, J_1] = [J^2, J_2] = 0 \tag{8}$$

The eigenvalues of the casimir operator are $J(J-1)$; with the eigenvalues of $J_3 = J, J+1, J+2, \ldots\ldots\ldots$

For a single harmonic oscillator the algebra generates the complete set of states

$$|2n\rangle = (t_1 + it_2)^n |0\rangle; \quad |2n+1\rangle = (t_1 + it_2)^n \cdot a^\dagger |0\rangle \tag{9}$$

For a system of many harmonic oscillators the simultaneous eigenstates of J^2 and J_3 are

$$|n, v\rangle = (T_1 + iT_2)^{(n-v)/2} |v, v\rangle; \quad (T_1 - iT_2)|v, v\rangle = 0 \tag{10}$$

The operator $(T_1 + iT_2)$ creates a boson pair in a "singlet" state; $|v, v\rangle$ is a state of v bosons containing no singlet pairs. The "active" (v) bosons and spectator boson pair degrees of freedom are directly analogous to the v quasiparticles and $(n-v)/2$ Cooper pairs in the fermion system.

For a system of particles in a harmonic oscillator potential the operator $(T_1 + iT_2)$ creates a "breathing mode" collective excitation.

4. Simple Soluble Models of Many-particle Systems

The Lipkin-Meshkov-Glick Model[6] is an example of the use of a Lie algebra to create soluble models. The model has two shells with the same degeneracy and an unperturbed ground state with the lower shell filled and the upper shell empty, as in a closed shell nucleus. A "monopole-monopole" interaction scatters pairs of particles between corresponding states in the two levels. A quasispin algebra for monopole excitations is used to construct a Hamiltonian with a two-body interaction which is a function of the generators of the algebra, and therefore commutes with the algebra's Casimir operator.

$$H = \epsilon S_z + V[(S_+)^2 + (S_-)^2]; \quad [H, S^2] = 0 \tag{11}$$

The Hamiltonian matrix therefore breaks down into submatrices with the same eigenvalue of S^2. This simplifies obtaining an exact solution which can be used to test approximation methods; e.g. perturbation theory and RPA.

Other more complicated models have been created using other groups; e.g. an O(5) model combining monopole and pairing quasispins.

5. Clashing of Translation Invariance and Pauli Principle in Center -of-Mass Motion of identical particles

In many-body systems with both individual particle and collective degrees of freedom, the individual particle coordinates exhaust the total number of degrees of freedom of the system, and adding collective degrees of freedom gives too many. For a simple example of the problems involved and the clashing of different symmetries, consider a system of N identical particles with interactions having both translational and Galilean invariance described by the Hamiltonian H, total momentum $\vec{P}$ and center-of-mass co-ordinate $\vec{X}$

$$H = \sum_{\mu=1}^{N} \frac{\vec{p}_\mu^2}{2M_\mu} + \sum_{\mu,\nu=1}^{N} V_{\mu\nu}(\vec{x}_\mu - \vec{x}_\nu); \quad \vec{P} = \sum_{\mu=1}^{N} \vec{p}_\mu; \quad \vec{X} = \sum_{\mu=1}^{N} \vec{x}_\mu \tag{12}$$

The translational and Galilean invariance are described by the symmetry and spectrum-generating algebras

$$[H, \vec{P}] = 0; \quad [H, \vec{X}] = \vec{P}/M; \quad M = \sum_{\mu=1}^{N} M_\mu \tag{13}$$

The spectrum generating feature is seen by noting that for a state ψ_o with zero total momentum $\vec{P}$ and energy E_i we can define a continuous set of momentum eigenstates ψ_K such that

$$H\psi_o = E_i\psi_o; \quad \vec{P}\psi_o = 0; \quad \psi_K \equiv e^{i\vec{K}\cdot\vec{X}}\psi_o; \quad H\psi_K = \left\{E_i + \frac{(\hbar K)^2}{2M}\right\}\psi_K \quad (14)$$

But the exact problem does not simply describe the localized nature of a bound wave function and the permutation symmetry required by the Pauli principle. The relative coordinates $\vec{x}_\mu - \vec{X}$ describe the localization and are easily transformed into one another by the permutation symmetry. But there are too many and they are not all independent. And the continuous spectrum with wave functions having equal magnitudes over all space are difficult to treat in perturbation theory.

The "oscillator trick" eliminates these difficulties by replacing the exact symmetry and the spectrum-generating algebra of Galilean invariance by a spurious spectrum-generating algebra which leaves the relative degrees of freedom unchanged. Define a modified Hamiltonian H' and a new spectrum-generating algebra

$$H' = H + (1/2) \cdot M\omega^2|\vec{X}|^2; \quad [H', \vec{P}] = M\omega^2\vec{X} \neq 0 \quad (15)$$

The new spectrum-generating feature is seen by noting that for any eigenstate ψ_o of the original Hamiltonian H with zero total momentum $\vec{P}$ and energy E_i one can define a discrete set of eigenstates ψ_n of the modified Hamiltonian H' such that

$$\psi_n \equiv \phi_n(\vec{X})\psi_o; \quad H'\psi_n = [E_i + (n + 3/2)\hbar\omega]\psi_n \quad (16)$$

where $\phi_n(\vec{X})$ denotes an eigenfunction of the harmonic oscillator Hamiltonian with the eigenvalue $(n + 3/2)\hbar\omega$

This new problem has the correct relative motion and relative energy and a spurious oscillator spectrum for the center-of-mass motion. It provides a basis for approximate treatment of the problem with localized wave functions satisfying the Pauli principle.

References

[1] Harry J. Lipkin, Lie Groups for Pedestrians, Second Edition. North-Holland Publishing Co. Amsterdam (1966), unabridged republication, Dover Publications, New York (2002)

[2] Harry J. Lipkin, Quantum Mechanics, North-Holland Publishing Co. Amsterdam (1973) Chapter 10, pp.281-294

[3] J. P. Hannon and G. T. Trammell, Phys. Rev. Lett. 61, 653 (1988)

[4] A. K. Kerman, Ann. Phys. 12, 300 (1961); P. W. Anderson, Phys. Rev. 112, 164 (1958).

[5] S. Goshen and H. J. Lipkin, Ann. Phys 6, 301 (1959)

[6] Harry J. Lipkin, N. Meshkov and A.J. Glick, Nucl. Phys. 62, 188, 199 and 211 (1965)

Inst. Phys. Conf. Ser. No 173: Section 7
Paper presented at 24th Int. Coll. Group Theoretical Methods in Physics, Paris, France, 15–20 July 2002
©*2003 IOP Publishing Ltd*

Multivariable Hermite polynomials, nonlinear coherent states and photon distributions

Olga V Man'ko

P.N. Lebedev Physical Institute, Moscow, Russia

Abstract. Nonlinear coherent states and photon distributions expessed in terms of multivariable Hermite polynomials are discussed. Tomograms of the squeezed photon and trapped ion states (including nonlinear coherent states) are constructed.

1. Introduction

In the process of parametric excitation of the electromagnetic field oscillators in a cavity with moving boundaries (nonstationary Casimir effect [1]), photons can be created in squeezed states [2]. The photon statistics of multimode squeezed states differs essentially from Poissonian statistics of coherent light [3] and can be described by multivariable Hermite polynomials. The other important parametric process is time evolution of a trapped ion, which is also modelled by oscillator with time-dependent frequency [4–6]. This oscillator can be either in squeezed state or in nonlinear coherent state [7, 8]. Recently the tomographic-probability representation of quantum states was applied to formulate the standard quantum mechanics in which the states are described by the probability (tomogram) instead of wave function [9]. Our aim is to obtain tomograms of multimode squeezed states and number states explicitly and establish some integral relations for multivariable Hermite polynomials describing the photon statistics of these states in terms of the tomograms. Another our goal is to study tomograms of nonlinear coherent states of trapped ions.

2. Photon distributions

For a cavity with moving boundaries, the electromagnetic field can be created in multimode squeezed state described by Gaussian density operator. A most general mixed squeezed state of the N-mode light with a Gaussian density operator $\hat{\rho}$ is described by the Wigner function $W(\mathbf{p}, \mathbf{q})$ of generic Gaussian form (see, for example, [3])

$$W(\mathbf{p}, \mathbf{q}) = \{\det \mathcal{M}\}^{-1/2} \exp\left[-\frac{1}{2}\big(\mathbf{Q} - \langle\mathbf{Q}\rangle\big)\mathbf{M}^{-1}\big(\mathbf{Q} - \langle\mathbf{Q}\rangle\big)\right], \tag{1}$$

with $2N$-dimensional vector $\mathbf{Q} = (\mathbf{p}, \mathbf{q})$ which consists of N components $p_1, \ldots, p_N$ and N components $q_1, \ldots, q_N$, where $2N$ parameters $\langle p_i\rangle$ and $\langle q_i\rangle$ $(i = 1, 2, \ldots, N)$, being combined into vector $\langle\mathbf{Q}\rangle$, are mean values of the quadratures $\langle\mathbf{p}\rangle = \text{Tr}\,\hat{\rho}\hat{\mathbf{p}}$ and $\langle\mathbf{q}\rangle = \text{Tr}\,\hat{\rho}\hat{\mathbf{q}}$. A $2N \times 2N$ real symmetric dispersion matrix

$$\mathcal{M} = \begin{pmatrix} M^{(1)} & M^{(2)} \\ M^{(3)} & M^{(4)} \end{pmatrix}$$

consists of $2N^2 + N$ variances and covariances of quadratures and contains four $N \times N$ blocks $M^{(s)}$, $\mathcal{M}_{\alpha\beta} = \frac{1}{2}\langle\hat{Q}_\alpha\hat{Q}_\beta + \hat{Q}_\beta\hat{Q}_\alpha\rangle - \langle\hat{Q}_\alpha\rangle\langle\hat{Q}_\beta\rangle$ with $\alpha, \beta = 1, 2, \ldots, 2N$.

The photon distribution function of this state was calculated in [3, 10] and it reads

$$\mathcal{P}_{\mathbf{n}} = \mathcal{P}_0 \frac{H_{\mathbf{nn}}^{\{R\}}(\mathbf{y})}{\mathbf{n}!}, \qquad \mathbf{n} = (n_1, n_2, \ldots, n_N), \tag{2}$$

where the function $H_{\mathbf{nn}}^{\{R\}}(\mathbf{y})$ is multidimensional Hermite polynomial.

The probability to have no photons is

$$\mathcal{P}_0 = \left[\det\left(\mathbf{M} + \frac{1}{2}\mathbf{I}_{2N}\right)\right]^{-1/2} \exp\left[-\langle\mathbf{Q}\rangle(2\mathbf{M} + \mathbf{I}_{2N})^{-1}\langle\mathbf{Q}\rangle\right], \tag{3}$$

where we introduced the matrix $\mathbf{R} = 2\mathbf{U}^\dagger(1 + 2\mathbf{M})^{-1}\mathbf{U}^* - \sigma_{Nx}$ along with the matrix

$$\sigma_{Nx} = \left(\begin{array}{cc} 0 & \mathbf{I}_N \\ \mathbf{I}_N & 0 \end{array}\right).$$

The argument of Hermite polynomial reads

$$\mathbf{y} = 2\mathbf{U}^t(\mathbf{I}_{2N} - 2\mathbf{M})^{-1}\langle\mathbf{Q}\rangle, \quad \text{with} \quad \mathbf{U} = \frac{1}{\sqrt{2}}\left(\begin{array}{cc} -i\mathbf{I}_N & i\mathbf{I}_N \\ \mathbf{I}_N & \mathbf{I}_N \end{array}\right)$$

being the $2N$-dimensional unitary matrix in which $\mathbf{I}_N$ is the $N \times N$-identity matrix; also the notation $\mathbf{n}! = n_1! n_2! \cdots n_N!$ was introduced.

Using the explicit Gaussian form of Wigner function (1), one can calculate the symplectic tomogram of the quantum state $w(\mathbf{X}, \mu, \nu)$

$$w(\mathbf{X}, \mu, \nu) = \left\langle\prod_{i=1}^{N}\delta\left(X_i - \mu_i\hat{q}_i - \nu_i\hat{p}_i\right)\right\rangle, \tag{4}$$

with $\mathbf{X} = (X_1, \ldots, X_N)$, $\mu = (\mu_1, \ldots, \mu_N)$, and $\nu = (\nu_1, \ldots, \nu_N)$.

It has the standard form of N-dimensional Gaussian distribution

$$w_G(\mathbf{X}, \mu, \nu) = (2\pi)^{-N/2}(\det\sigma)^{-1/2}\exp\left\{-\frac{1}{2}\left(\mathbf{X} - \langle\mathbf{X}\rangle\right)\sigma^{-1}\left(\mathbf{X} - \langle\mathbf{X}\rangle\right)\right\}, \tag{5}$$

with means $\langle\mathbf{X}\rangle = (\langle X_1\rangle, \langle X_2\rangle, \ldots, \langle X_n\rangle)$, $\langle X_k\rangle = \mu_k\langle q_k\rangle + \nu_k\langle p_k\rangle$, and $N \times N$-dispersion matrix $\quad \sigma_{ik} = \mu_i\mu_k M_{ik}^{(4)} + \nu_i\nu_k M_{ik}^{(1)} + \mu_i\nu_k M_{ik}^{(2)} + \nu_i\mu_k M_{ik}^{(3)}$.

In formulas (4), (5), there is no summation over repeated indices. ·

One can show that the state $|\mathbf{n}\rangle$ with n_k photons in kth mode has the tomogram of the form

$$w_{\mathbf{n}}(\mathbf{X}, \mu, \nu) = \prod_{k=1}^{N}\left[\pi\left(\mu_k^2 + \nu_k^2\right)\right]^{-1/2}\exp\left[-\frac{X_k^2}{\mu_k^2 + \nu_k^2}\right]\frac{1}{2^{n_k}n_k!}H_{n_k}^2\left(\frac{X_k}{\sqrt{\mu_k^2 + \nu_k^2}}\right). \tag{6}$$

The photon distribution (2) can be expressed in terms of tomograms (5) and (6):

$$\mathcal{P}_{\mathbf{n}} = (2\pi)^{-1}\int w_G(\mathbf{X}, -\mu, -\nu)\, w_{\mathbf{n}}(\mathbf{Y}, \mu, \nu)\prod_{k=1}^{N}e^{i(X_k + Y_k)}dX_k\, dY_k\, d\mu_k\, d\nu_k. \tag{7}$$

In view of (2), formula (7) provides the possibility to calculate a new overlap integral with Hermite polynomial squared. An analogous overlap integral, which is the Frank–Condon factor for a polyatomic molecule, has been calculated in [11]. The corresponding result uses the formula ($\mathbf{n} = n_1, n_2, \ldots, n_N$, $\mathbf{m} = m_1, m_2, \ldots m_N$ and $m_i, n_i = 0, 1, \ldots$)

$$\int H_{\mathbf{m}}^{\{R\}}(\mathbf{x})H_{\mathbf{n}}^{\{r\}}(\Lambda\mathbf{x} + \mathbf{d})\exp(-\mathbf{x}m\mathbf{x} + \mathbf{c}\mathbf{x})d\mathbf{x} = \frac{\pi^{N/2}}{\sqrt{\det m}}\exp\left(\frac{1}{4}\mathbf{c}m^{-1}\mathbf{c}\right)H_{\mathbf{nm}}^{\{\rho\}}(\mathbf{y}), \tag{8}$$

where the symmetric $2N \times 2N$-matrix

$$\rho = \begin{pmatrix} R_1 & R_{12} \\ \tilde{R}_{12} & R_2 \end{pmatrix},$$

with $N \times N$-blocks R_1, R_2, R_{12}, is expressed in terms of symmetric $N \times N$-matrices R, r, m and $N \times N$-matrix Λ in the form

$$R_1 = R - \left(Rm^{-1}R\right)/2, \qquad R_2 = r - \left(r\Lambda m^{-1}\tilde{\Lambda}r\right)/2, \qquad \tilde{R}_{12} = -\left(r\Lambda m^{-1}R\right)/2.$$

Here the matrix $\tilde{\Lambda}$ is transposed matrix Λ, and $\tilde{R}_{12}$ is transposed matrix R_{12}.

The $2N$-vector $\mathbf{y}$ is expressed in terms of N-vectors $\mathbf{c}$ and $\mathbf{d}$ in the form

$$\mathbf{y} = \rho^{-1} \begin{pmatrix} \mathbf{y}_1 \\ \mathbf{y}_2 \end{pmatrix}, \quad \text{with} \quad \mathbf{y}_1 = \tfrac{1}{4}\left(Rm^{-1} + m^{-1}R\right)\mathbf{c}, \quad \mathbf{y}_2 = \tfrac{1}{4}\left(r\Lambda m^{-1} + m^{-1}\tilde{\Lambda}r\right)\mathbf{c} + r\mathbf{d}.$$

This overlap integral provides several new relations for special functions [3].

3. Trapped ion

A trapped ion is an important nonstationary quantum system. Since an ion in a Paul trap is described by the model of a parametric oscillator [12, 4–6] we review its properties. For a parametric oscillator with an arbitrary time dependence of the frequency and the Hamiltonian $H = -\left(\partial^2/2\partial x^2\right) + \left(\omega^2(t)x^2/2\right)$, where we put $\hbar = m = \omega(0) = 1$ and use expressions for the position and momentum operators in the coordinate representation, there is the time-dependent integral of motion found in [13], namely, $A = \left(i/\sqrt{2}\right)\left[\varepsilon(t)\hat{p} - \dot{\varepsilon}(t)\hat{q}\right]$, where $\ddot{\varepsilon}(t) + \omega^2(t)\varepsilon(t) = 0$, $\varepsilon(0) = 1$, $\dot{\varepsilon}(0) = i$, which satisfies the commutation relation: $[A, A^\dagger] = 1$.

For a trapped ion, the frequency time dependence is taken to be periodic [12], i.e., $\omega^2(t) = 1 + \kappa^2 \sin^2 \Omega t$. It is easy to show that Gaussian packet solutions to the Schrödinger equation may be introduced and interpreted as coherent states (since they are eigenstates of the invariant A) of the form [13]

$$\Psi_\alpha (x, t) = \Psi_0 (x, t) \exp \left\{ -\frac{|\alpha|^2}{2} - \frac{\alpha^2 \varepsilon^*(t)}{2\varepsilon(t)} + \frac{\sqrt{2}\alpha x}{\varepsilon} \right\}, \tag{9}$$

where $\Psi_0 (x, t) = \pi^{-1/4} [\varepsilon(t)]^{-1/2} \exp \left[i\dot{\varepsilon}(t)x^2/2\varepsilon(t)\right]$ is an analog of the oscillator ground state and α is a complex number.

Analogs of number states, which are excited states of an ion in a Paul trap, are obtained by expansion of (9) into a power series in α, namely,

$$\Psi_m (x, t) = \left(\frac{\varepsilon^*(t)}{2\varepsilon(t)}\right)^{m/2} \frac{1}{\sqrt{m!}} \Psi_0 (x, t) H_m \left(\frac{x}{|\varepsilon(t)|}\right), \tag{10}$$

and these squeezed and correlated number states are eigenstates of the invariant $A^\dagger A$.

There exist the integrals of motion $B = A f\left(A^\dagger A\right)$ and $B^\dagger = f\left(A^\dagger A\right) A^\dagger$, which are determined by a function f of the invariants A and $A^\dagger$.

By generalizing the notion of coherent states to the operator case (nonlinearly transformed annihilation operator), we introduce the eigenfunctions of the invariant B, $B\Psi_\beta (x, t) = \beta \Psi_\beta (x, t)$, which are nonlinear coherent states.

One can show that the tomogram of a trapped ion in nonlinear coherent state has the form

$$w_\beta (X, \mu, \nu, t) = \left(\sum_{n=0}^{\infty} \frac{|\beta|^{2n}}{[n]!([f(n)]!)^2} \right)^{-1} \sum_{m=0}^{\infty} \sum_{n=0}^{\infty} \frac{\beta^n \beta^{*m}}{\sqrt{n!}\sqrt{m!}\,[f(n)]!\,[f(m)]!} w_{nm} (X, \mu, \nu, t),$$

with
$$w_{nm}(X,\mu,\nu,t) = \left(\pi\left[\mu^2(t) + \nu^2(t)\right]\right)^{-1/2} 2^{-(n/2+m/2)} (n!\,m!)^{-1/2}$$
$$\times \frac{[\nu(t) + i\mu(t)]^n\,[\nu(t) - i\mu(t)]^m}{[\mu^2(t) + \nu^2(t)]^{(n/2+m/2)}} \exp\left(-\frac{X^2}{\mu^2(t) + \nu^2(t)}\right)$$
$$\times H_n\left(\frac{X}{\sqrt{\mu^2(t) + \nu^2(t)}}\right) H_m\left(\frac{X}{\sqrt{\mu^2(t) + \nu^2(t)}}\right),$$

where $\nu(t) = (\dot\varepsilon - \dot\varepsilon^*/2i)\,\nu + (\varepsilon - \varepsilon^*/2i)\,\mu$ and $\mu(t) = (\dot\varepsilon + \dot\varepsilon^*/2)\,\nu + (\varepsilon + \varepsilon^*/2)\,\mu$.

The photon distribution function of nonlinear coherent states can be easily calculated using overlap integral of the type (7). The distribution depends on the character of nonlinearity described by the function $f(n)$.

4. Conclusion

To conclude, the main result of the paper is the explicit expression for the tomogram of photon multimode squeezed state and integral relations for multivariable Hermite polynomials. We have also shown that there exist deformed integrals of motion for a trapped ion. These integrals of motion are time-dependent functions which depend nonlinearly on quadrature components. The eigenfunctions of deformed annihilation operator are nonlinear coherent states. For electromagnetic field in a cavity, photon statistics of nonlinear coherent states is determined by nonlinear function $f(n)$. If this function differs essentially from unity, the photon statistics is deformed Poissonian one.

Acknowledgments

This study was supported by the Russian Foundation for Basic Research under Projects Nos. 00-02-16516 and 03-02-16408.

References

[1] Dodonov V V 2001 Nonstationary Casimir effect and analytic solutions for quantum fields in cavities with moving boundaries *Contemporary Optics and Electrodynamics* Pt 1 ed M. Evans *Advances in Chemical Physics* (New York: Willey) Vol 114 Pt 1 p 309

[2] Dodonov V V, Klimov A B and Man'ko V I 1990 *Phys. Lett.* A **49** 255

[3] Dodonov V V, Man'ko Olga and Man'ko V I 1994 *Phys. Rev.* A **49** 2993; **50** 813

[4] Schrade G, Man'ko V I, Schleich W P and Glauber R J 1995 *Quantum Semiclass. Opt.* **7** 307

[5] Man'ko Olga 1997 *Phys. Lett.* A **228** 29

[6] Man'ko O V 1996 Symplectic tomography of nonclassical states of a trapped ion *Preprint* IC/96/39 (Trieste: Abdus Salam International Centre for Theoretical Physics); 1996 *J. Russ. Laser Res.* **17** 439

[7] Man'ko V I, Marmo G, Sudarshan E C G and Zaccaria F 1996 f-Oscillators in *Proceedings of the Fourth Wigner Symposium* (Guadalajara, Mexico, July 1995) eds N. M. Atakishiyev, T. H. Seligman and K.-B. Wolf (Singapore: World Scientific) p 421; 1997 *Phys. Scr.* **55** 528

[8] De Matos Filho R L and Vogel W 1996 *Phys. Rev.* A **54** 4560

[9] Mancini S, Man'ko V I and Tombesi P 1996 *Phys. Lett.* A **213** 1; 1997 *Found. Phys.* **27** 801

[10] Dodonov V V, Man'ko V I and Semjonov V V 1984 *Nuovo Cim.* B **83** 145

[11] Malkin I A, Man'ko V I and Trifonov D A 1973 *J. Math. Phys.* **14** 576

[12] Glauber R J 1993 in *Recent Developments in Quantum Optics, Proceedings of the International Conference on Quantum Optics* (Hyderabad, India, January 1991) ed R. Inguva (New York: Plenum Press) p 1

[13] Malkin I A and Man'ko V I 1970 *Phys. Lett.* A **32** 243

Inst. Phys. Conf. Ser. No 173: Section 7
Paper presented at 24th Int. Coll. Group Theoretical Methods in Physics, Paris, France, 15–20 July 2002
©2003 IOP Publishing Ltd

Hidden SL(2,R) Symmetry for Time-Dependent Schrödinger Equations

Susumu Okubo

Department of Physics and Astronomy, University of Rochester Rochester, NY 14627, USA

Abstract. It has been found that the one-dimensional time-dependent Schrödinger equation for linear potential possesses a hidden semi-direct product group of $SL(2, R)$ with a 2-dimensional translation group where the time variable transforms as $t \to t' = (ct+d)/(at+b)$ satisfying $bc - ad = 1$ together with appropriate transformation for the space coordinate x.

Many solutions of Schrödinger equations are known [1-4] to possess dynamical (or hidden) symmetries which are not apparent at first glance. In this note, we report that some Schrödinger equations will exhibit hidden global $SL(2R)$ symmetry in which the time variable t changes accordingly to

$$t \to t' = \frac{ct + d}{at + b} \quad , \quad bc - ad = 1 \tag{1}$$

together with corresponding transformation on the spatial coordinate x.

Let us consider an equation of the form

$$\frac{\partial}{\partial t} \psi(t, x) = k \left\{ \frac{\partial^2}{\partial x^2} - V(x) \right\} \psi(t, x) \tag{2}$$

where the case of $k = -2mi/\hbar$ being purely imaginary gives the Schrödinger equation while the real value for k realizes a diffusion-like equation. Consider first a simple case of the inverse quadratic potential, i.e.,

$$V(x) = \lambda/x^2 \quad . \tag{3}$$

We can then easily prove the following Proposition.

Proposition 1

Let $\psi(t, x)$ satisfy Eq. (2) with Eq. (3). A new wave function given by

$$\psi'(t, x) = \left(\frac{1}{at + b} \right)^{\frac{1}{2}} \exp \left\{ -\frac{a}{4k(at + b)} x^2 \right\} \psi(t', x') \tag{4}$$

will then also satisfy the same equation, i.e.,

$$\frac{\partial}{\partial t} \psi'(t, x) = k \left\{ \frac{\partial^2}{\partial x^2} - \frac{\lambda}{x^2} \right\} \psi'(t, x) \tag{5}$$

where t' and x' are given by

$$t' = \frac{ct + d}{at + b} \quad , \quad x' = \frac{x}{at + b} \quad . \tag{6}$$

Here a, b, c, and d are arbitrary real constants satisfying $bc - ad = 1$. Moreover, if we introduce a linear operator $U(M)$ in the solution space of Eq. (2) by

$$U^{-1}(M)\psi(t,x) = \psi'(t,x) \tag{7}$$

for 2×2 matrix M of the form

$$M = \begin{pmatrix} c & d \\ a & b \end{pmatrix} \quad , \quad \det M = 1 \quad , \tag{8}$$

then we have

$$U(M)U(M') = U(MM') \tag{9}$$

for two $SL(2,R)$ matrices M and M'. ∎

We note that Eq. (9) offers a infinite dimensional realization of the $SL(2,R)$ group. However, a more interesting (but also more complicated) case is obtained for the case of one-dimensional linear potential:

$$\frac{\partial}{\partial t}\psi(t,x) = k\left\{\frac{\partial^2}{\partial x^2} - (\alpha + \beta x)\right\}\psi(t,x) \quad , \tag{10}$$

where α and β are some constants. In this case, the underlying symmetry group G turns out to be the semi-direct product

$$G = SL(2,R) ⓢ T_2 \tag{11}$$

for a 2-dimensional translation group T_2. Let $\Lambda \in G$ be a generic element G of the form

$$\Lambda = \left\{M, \begin{pmatrix} \mu \\ \sigma \end{pmatrix}\right\} \tag{12}$$

where M is given by Eq. (8) and $\begin{pmatrix} \mu \\ \sigma \end{pmatrix}$ is a real 2-dimensional vector on which M acts. It is often more convenient to set

$$\xi = \{t, x\} \tag{13}$$

collectively, on which $\Lambda \in G$ is assumed to act as

$$\Lambda\xi = \{t', x'\}$$

where

$$t' = \frac{ct + d}{at + b} \quad , \tag{14}$$

$$x' = \frac{x}{at+b} + \mu - \sigma\frac{ct+d}{at+b} + k^2\beta\left\{\left(\frac{ct+d}{at+b}\right)^2 - \frac{t^2}{at+b}\right\} \quad . \tag{15}$$

We next introduce a function $K(\xi|\Lambda) \equiv K(t,x|\Lambda)$ by

$$K(\xi|\Lambda) = \frac{1}{\sqrt{at+b}}\exp\left\{A(t|\Lambda) + B(t|\Lambda)x + C(t|\Lambda)x^2\right\} \tag{16}$$

where

$$C(t|\Lambda) = -\frac{1}{4k}\frac{a}{at+b} \quad , \tag{17}$$

$$B(t|\Lambda) = -\frac{\sigma}{2k}\frac{1}{at+b} + \frac{k\beta}{2}\left\{\frac{2(ct+d)}{(at+b)^2} - t - \frac{bt}{at+b}\right\} \quad , \tag{18}$$

$$A(t|\Lambda) = -\frac{1}{4k}\sigma\mu + \alpha k\left[\frac{ct+d}{at+b} - t\right] + \frac{\sigma^2}{4k}\frac{ct+d}{at+b}$$

$$+ k\beta\left\{\mu\frac{ct+d}{at+b} - \sigma\left[\left(\frac{ct+d}{at+b}\right)^2 - \frac{1}{2}\frac{t^2}{at+b}\right]\right\}$$

$$+ k^3\beta^2\left\{\frac{2}{3}\left(\frac{ct+d}{at+b}\right)^3 + \frac{1}{12}t^3 + \frac{b}{4}\frac{t^3}{at+b} - \frac{t^2(ct+d)}{(at+b)^2}\right\} \tag{19}$$

We can then prove readily the following Proposition.

Proposition 2

For any $\psi(t,x)$ satisfying Eq. (10), the second wave function defined by

$$\psi'(t,x) = K(t,x|\Lambda)\psi(t',x') \tag{20}$$

will also satisfy Eq. (10), i.e.,

$$\frac{\partial}{\partial t}\psi'(t,x) = k\left\{\frac{\partial^2}{\partial x^2} - (\alpha + \beta x)\right\}\psi'(t,x) \quad . \quad \blacksquare \tag{21}$$

In order to show that the underlying symmetry group is indeed $G = SL(2,R) \circledS T_2$, we consider the 2nd transformation by $\Lambda' \in G$ of the form

$$\Lambda' = \left\{M', \begin{pmatrix} \mu' \\ \sigma' \end{pmatrix}\right\} \quad , \quad \det M' = 1 \tag{22}$$

and define the group product $\Lambda \circ \Lambda'$ in G by

$$\Lambda \circ \Lambda' - \left\{MM', \begin{pmatrix} \mu \\ \sigma \end{pmatrix} + M\begin{pmatrix} \mu' \\ \sigma' \end{pmatrix}\right\} \quad . \tag{23}$$

After some long calculations, we then find

Proposition 3

We have

(i) $\quad \Lambda\{\Lambda'\xi\} = (\Lambda \circ \Lambda')\xi \tag{24}$

(ii) $\quad K(\xi|\Lambda')K(\Lambda'\xi|\Lambda) = \exp\{\omega(\Lambda,\Lambda')\}K(\xi|\Lambda \circ \Lambda') \tag{25}$

where $\omega(\Lambda,\Lambda')$ is a constant given by

$$\omega(\Lambda,\Lambda') = \frac{1}{4k}\{(\mu a - \sigma c)\mu' + (\mu b - \sigma d)\sigma'\} \tag{26}$$

which satisfies the co-cycle condition

$$\omega\left(\Lambda_1, \Lambda_2\right) + \omega\left(\Lambda_1 \circ \Lambda_2, \Lambda_3\right) = \omega\left(\Lambda_2, \Lambda_3\right) + \omega\left(\Lambda_1, \Lambda_2 \circ \Lambda_3\right) \tag{27}$$

as well as

$$\omega\left(\Lambda_2^{-1}, \Lambda_1^{-1}\right) = -\omega\left(\Lambda_1, \Lambda_2\right) \quad . \quad \blacksquare \tag{28}$$

Introducing a linear operator $U(\Lambda)$ in the solution space of Eq. (10) by

$$U^{-1}(\Lambda)\psi(\xi) = \psi'(\xi) \quad , \tag{29}$$

then Eqs. (22) and (24) imply the validity of

$$U(\Lambda)U\left(\Lambda'\right) = \exp\left\{-\omega\left(\Lambda, \Lambda'\right)\right\} U\left(\Lambda \circ \Lambda'\right) \tag{30}$$

so that $U(\Lambda)$ offers a projective representation of the group $G = SL(2, R) \; \text{ⓢ} \; T_2$.

In this connection, it may be of interest to note that a set of all $\Lambda \in G$ of form

$$\Lambda = \left\{ \begin{pmatrix} 1 & \lambda \\ 0 & 1 \end{pmatrix} , \begin{pmatrix} \mu \\ \sigma \end{pmatrix} \right\}$$

gives the Galilean sub-group of G. Also, some time-dependent solutions of Eq. (10) are intimately related to the bound-state problem as follows. For example, the function given by

$$\xi(t, x) = \exp\left\{ -k(\alpha + \beta x)t + \frac{1}{3} k^3 \beta^2 t^3 \right\} \tag{31}$$

is a solution of Eq. (10). Choosing $k = -i$, and integrating on the time variable t just above on the entire real axis as

$$u(x) \equiv \int_{-\infty+i0}^{\infty+i0} dt\, \xi(t, x) = 2 \int_0^{\infty} dt \cos\left\{ (\alpha + \beta x)t + \frac{1}{3} \beta^2 t^3 \right\} \quad , \tag{32}$$

it satisfies

$$\left\{ \frac{d^2}{dx^2} - \beta x \right\} u(x) = \alpha u(x) \quad . \tag{33}$$

If we impose the boundary condition $u(0) = 0$ at $x = 0$, then the eigenvalue α for Eq. (29) is determined by

$$\int_0^{\infty} dt \cos\left\{ \alpha t + \frac{1}{3} \beta^2 t^3 \right\} = 0 \quad , \tag{34}$$

since $u(x) \in L^2(0, \infty)$ for $\beta > 0$.

This paper is supported in part by the U.S. Department of Energy contract no. DE-FG02-91ER40685.

References

[1] Wu J and Alhassid Y 1990 J. Math. Phys. 31 557-562
[2] Sezgin M, Verdiyev A Y and Verdiyer Y A 1998 J. Math. Phys. 39 1910-1918
[3] Nieto M M and Truax D R 2000 J. Math. Phys. 41 2753-2787
[4] Doebner H D and Goldin G 2002 Some Remarks on Non-linear Quantum Mechanics (talk presented at this Conference)

Inst. Phys. Conf. Ser. No 173: Section 7
Paper presented at 24th Int. Coll. Group Theoretical Methods in Physics, Paris, France, 15–20 July 2002
©2003 IOP Publishing Ltd

The Particle-Operator and Particle States in Quantum Field Theory

M Omote[1] **and S Kamefuchi**[2]

[1]Department of Physics, Keio University, Hiyoshi, Yokohama, Japan
[2]Institute of Quantum Science, Nihon University, Tokyo, Japan

Abstract. By using the method of particle-operators, proposed previously, we examine the problem of particle states (spatially localized) with emphasis upon the relativistic case. A physical interpretation is also given of the results which seem to contradict locality, covariance and causality.

1. Introduction

In a previous paper [1] we proposed the method of wave- and particle- operators, which enable us to define waves and particles, respectively,in a purely quantum-mechanical manner. In what follows we study the application mainly of particle- operators to relativistic fields.

As in the nonrelativistic case we can discuss various properties of relativistic particles by use of particle-operators. Although there exist a host of papers written so far on this subject (cf. for example Odaka [2]) our method differs from those in that it provides a simple and concise, quantum-field-theoretical reformulation of the problem, and that the explicit expressions for the particle-operators are given of the general case of mass $m \neq 0$ and arbitrary spin S. Furthermore, we state at the end our basic attitude towards some well-known mathematical results which appear to contradict basic physical requirements such as locality, covariance and causality.

2. Nonrelativistic cases

To illustrate our method of particle-operators, let us begin by considering a free de Broglie field $\psi(\vec{x}, t)$ such that

$$[\psi(\vec{x}, t), \psi^\dagger(\vec{x}', t)]_\mp = \delta(\vec{x} - \vec{x}'),$$
$$[\psi(\vec{x}, t), \psi(\vec{x}', t)]_\mp = [\psi^\dagger(\vec{x}, t), \psi^\dagger(\vec{x}', t)]_\mp = 0; \tag{1}$$

kinematical quantities A such as energy, momentum, $\cdots$ are of the form

$$A = \int d^3x \psi^\dagger(\vec{x}, t) \mathcal{A}(\vec{\nabla}, \vec{x}) \psi(\vec{x}, t), \tag{2}$$

and in particular the matter density ρ is given as

$$\rho(\vec{x}, t) = \psi^\dagger(\vec{x}, t)\psi(\vec{x}, t). \tag{3}$$

In this case, the amount of matter contained in a spatial domain v is given by $N_v \equiv \int_v d^3x \rho(\vec{x}, t)$. Then, for the state $|\vec{x}\rangle \equiv \psi^\dagger(\vec{x}, 0)|0\rangle$, where $|0\rangle$ denotes the vacuum state, we have $N_v|\vec{x}\rangle = |\vec{x}\rangle \; (= 0)$ when $\vec{x} \in v \; (\vec{x} \notin v)$. Note furthermore that $|\vec{x}\rangle$, at the same time, is the eigenstate of the number operator $N \equiv N_{v \to \infty}$ with eigenvalue 1. These results, thus, enable us to identify $|\vec{x}\rangle$ with the state of a particle at $\vec{x}$. Similarly, for n-particle states

$|\vec{x}_1, \vec{x}_2, \cdots, \vec{x}_n\rangle$. Since the particle states can be constructed by appling $\psi^\dagger$'s to $|0\rangle$, let us say that the particle-operator Φ for the present case is $\Phi = \psi(\vec{x}, t)$. The position operator $\vec{X}$ is naturally to be defined by

$$\vec{X} \equiv \int d^3x \; \Phi^\dagger(\vec{x}, t) \vec{x} \; \Phi(\vec{x}, t). \tag{4}$$

It should be noted here that the argument $\vec{x}$ of $\Phi(\vec{x}, t)$ is directly related to the position of a particle.

As seen above, (1), (2) and (3) are the basic relations that the particle-operator Φ has to satisfy.

3. Relativistic cases

Given a relativistic field ψ, the problem lies in how to constuct the Φ that satisfy the relations corresponding to (1), (2) and (3). For the case of a free neutral scalar field $\phi(x)$ such that

$$\phi(x) = \frac{1}{\sqrt{2(2\pi)^3}} \int dw_{\vec{k}} \; \{a_{\vec{k}} e^{ikx} + a_{\vec{k}}^\dagger e^{-ikx}\} \tag{5}$$

with $kx = \vec{k}\vec{x} - k_0 t, k_0 = \sqrt{\vec{k}^2 + m^2}, dw_{\vec{k}} = d^3k/k_0$ and $[a_{\vec{k}}, a_{\vec{k}'}^\dagger] = k_0 \delta(\vec{k} - \vec{k}')$, etc., the corresponding Φ is

$$\Phi(x) = \frac{1}{\sqrt{(2\pi)^3}} \int dw_{\vec{k}} \; k_0^{1/2} \, a_{\vec{k}} e^{ikx}. \tag{6}$$

It is easy to see that $\vec{X}$ agrees basically with Newton-Wigner's results [3], and the velocity operator $\vec{V} \equiv [\vec{X}, H]/i$ has the right property as expected from relativity. In this sense our method corresponds to a field-theoretical version of Newton-Wigner's.

For the case of general spin $S \geq 1/2$ we adopt a symmetric multi-spinor with $n(= 2S)$ spinor indices, that is, a Bargmann-Wigner field $\psi_{\rho_1 \rho_2 \cdots \rho_n}(x)$:

$$\psi_{\rho_1 \rho_2 \cdots \rho_n}(x) = \sqrt{\frac{m}{(2\pi)^3}} \int dw_{\vec{p}} \sum_{n_1=0}^{n} \Big\{ U_{n_1, \rho_1 \rho_2 \cdots \rho_n}(\vec{p}) \, a_{n_1}(\vec{p}) e^{ipx}$$
$$+ V_{n_1, \rho_1 \rho_2 \cdots \rho_n}(\vec{p}) \, b_{n_1}^\dagger(\vec{p}) e^{-ipx} \Big\}. \tag{7}$$

Here

$$U_{n_1, \rho_1 \rho_2 \cdots \rho_n}(\vec{p}) \equiv \mathcal{N}_{n_1} \sum_{(12)} u_{1\rho_1}^{(1)}(\vec{p}) u_{1\rho_2}^{(2)}(\vec{p}) \cdots u_{1\rho_{n_1}}^{(n_1)}(\vec{p}) \cdot u_{2\rho_{n_1+1}}^{(n_1+1)}(\vec{p}) \cdots u_{2\rho_n}^{(n)}(\vec{p}),$$

$$a_{n_1}(\vec{p}) \equiv \mathcal{N}_{n_1} \sum_{(12)} a_{(11 \cdots 12 \cdots 2)}, \quad \mathcal{N}_{n_1} \equiv \left(n_1!(n - n_1)!/n! \right)^{1/2}, \tag{8}$$

$u_r^{(k)}(\vec{p})$ $(r = 1, 2)$ is the solution of the Dirac equation with $E_{\vec{p}} = \sqrt{\vec{p}^2 + m^2}$, and with the superscript (k) being for the spinors to which the kth set of Dirac matrices $\vec{\alpha}^{(k)}, \beta^{(k)}$ apply $(k = 1, 2, \cdots, n)$, $\sum_{(12)}$ is the summation over all different sequences of n_1 of 1's and $(n - n_1)$ of 2's, and the operators $a_{(\cdots)}$ and $b_{(\cdots)}$ satisfy (anti-) commutation relations for $n =$ even (odd) such that $[a_{n_1}(\vec{p}), a_{n_1'}^\dagger(\vec{p}')]_\mp = E_{\vec{p}} \delta_{n_1 n_1'} \delta(\vec{p} - \vec{p}')$, etc..

In this case the particle-operators Φ_{n_1} are given by

$$\Phi_{n_1}(x) = \frac{1}{\sqrt{(2\pi)^3}} \int dw_{\vec{p}} \sqrt{E_p} \, a_{n_1}(\vec{p}) \, e^{ipx} \quad (n_1 = 0, 1, \cdots, n), \tag{9}$$

and the antiparticle-operators $\bar{\Phi}_{n_1}$ are obtained from (9) by the replacement $a_{n_1}(\vec{p}) \to b_{n_1}(\vec{p})$.

The case $n = 1$ corresponds to that of a Dirac field $\psi_\rho(x)$. As clear from (7) and (9), $\Phi_\rho \neq \psi_\rho$. This means, as is well-known, that the argument $\vec{x}$ of $\psi_\rho(\vec{x}, t)$ does not represent the position of a particle. In other words, $\psi_\rho(x)$ cannot be the Schrödinger amplitude as originally thought by Dirac. Our result for this case can also be related to what Foldy and Wouthuysen called the mean position of a Dirac particle [4]. In fact, their mean position is a sum of contributions from the particle and antiparticle positions in our sense.

4. Discussion and Conclusion

For simplicity, let us consider the case of $\phi(x)$ given by (5). When expressed in terms of local quantities $\phi(x)$ and $\pi(x) = \dot{\phi}(x)$, $\Phi(x)$ given by (6) is of a nonlocal form: $\Phi(x) \equiv (1/\sqrt{2})\{(\sqrt{-\Delta + m^2})^{1/2}\phi(x) + i(\sqrt{-\Delta + m^2})^{-1/2}\pi(x)\}$. Thus, the nonlocal deviation of $\Phi(x)$ from $\phi(x)$ is effectively restricted to the region with the size of the Compton wavelength.

With the Lorentz-boost operator $\vec{K}$, Φ has the commutation relation $[\Phi(x), \vec{K}] = i(\vec{x}\partial_t + t\vec{\nabla})\Phi(x) - 1/(2\sqrt{-\Delta + m^2}) \cdot \vec{\nabla}\Phi(x)$, so that the second term violates relativistic covariance. Accordingly, for the state $|\vec{0}\rangle$ of a particle situated at $\vec{x} = 0, t = 0$, its Lorentz-transformed state $|\vec{0}\rangle'$ becomes $|\vec{0}\rangle - (im/4\pi^2) \int d^3x \, K_1(m|\vec{x}|)/|\vec{x}| \cdot (\vec{\tau}\vec{\nabla})|\vec{x}\rangle$, where K_1 is a modified Bessel function and $\vec{\tau}$ a constant vector. In fact, the transformed object spreads over a spatial region with the size of the Compton wavelength.

Futhermore, $\Phi(x)$ obeys the field equation $i\partial_t\Phi(x) = \sqrt{-\Delta + m^2}\,\Phi(x)$, and the solution can be written as $\Phi(\vec{x}, t) = \int d^3x' G(\vec{x} - \vec{x}', t)\,\Phi(\vec{x}', 0)$, where $G(\vec{x}, t)$ for $t^2 - r^2 > 0(< 0)$ is proportional to the Hankel function $H_2^{(2)}(m\sqrt{t^2 - r^2})$ $(H_2^{(1)}(im\sqrt{r^2 - t^2}))$. As a consequence, the Φ-wave propagates also into the space-like region.

The above results appear to contradict the basic physical requirements: locality, covariance and causality. The difficulty arises clearly from our use of quantities such as $\Phi(x), |\vec{x}\rangle, \cdots$. On the other hand, it is commonly understood that no such difficulties show up, so far as we stick only to the local quantities $\phi(x)$ and $\dot{\pi}(x)$ in describing the theory.

We should say therefore that the above results suggest the following, concerning the very nature of physical theories. A theory, in general, consists of those theoretical elements which are not necessarily consistent with each other and/or with basic physical requirements. Hence, in order to have a 'good theory' it is necessary to choose, from among such elements, a suitable set of those which are consistent with each other and with basic physical requirements. For example, the renormalization theory remains a good theory, so far as only the renormalized quantities are employed in describing the theory.

If such a view is to be accepted, we then have to give up using Φ's and $|\vec{x}\rangle$'s, and to resume using the covariant $\phi(x)$ and $\pi(x)$. That means that we give up talking about the Schrödinger amplitudes, that is, those for finding particles at certain spatial points. In other words, the spatial position of a particle, being determinable only within the order of the Compton wavelength, cannot remain a good concept in the relativistic regime. As the history shows, here also mathematical formality must be respected more than physical meaning.

References

[1] Kamefuchi S and Omote M, in press
[2] Odaka K 1998 Proc. of the 5th Wigner Symposium (Vienna, 25-29 August 1997), eds.Kasperkovitz P and Gren D World Scientific in Singapore 487-489

584

[3] Newton T D and Wigner E P 1945 Rev. Mod. Phys. 21 400- 406
[4] Foldy L L and Wouthuysen S A 1950 Phys. Rev. 78 29-36 ; Tani S 1951 Prog. Theor. Phys. 6 267-282

Inst. Phys. Conf. Ser. No 173: Section 7
Paper presented at 24th Int. Coll. Group Theoretical Methods in Physics, Paris, France, 15–20 July 2002
©2003 IOP Publishing Ltd

Combinatorics of Boson Normal Ordering: the Dobiński Formula Revisited

Karol A. Penson‡ and Allan I. Solomon§

Laboratoire de Physique Thorique des Liquides,
Universit Paris VI, 75252 Paris Cedex 05, France

Abstract. We derive explicit formulas for the normal ordering of powers of arbitrary monomials of boson operators. These formulas lead to generalisations of conventional Bell and Stirling numbers and to appropriate generalisations of the Dobiński relations. These new combinatorial numbers are shown to be coherent state matrix elements of powers of the monomials in question. It is further demonstrated that such Bell-type numbers, when considered as power moments, give rise to positive measures on the positive half-axis, which in many cases can be written in terms of known functions.

The standard boson commutation relation $[a, a^\dagger] = 1$ can be realised by identifying *formally* $a = \frac{d}{dx}$ and $a^\dagger = x$, since $[\frac{d}{dx}, x] = 1$. In the present note we shall use both forms. Integer sequences arise naturally when considering the action of $(x\, d/dx)^n$ on $f(x)$ which in general can be written as

$$(x\frac{d}{dx})^n f(x) = \sum_{k=1}^{n} S(n,k) x^k (d/dx)^k f(x) \tag{1}$$

or, alternatively [1]

$$(a^\dagger a)^n = \sum_{k=1}^{n} S(n,k)(a^\dagger)^k a^k. \tag{2}$$

The *Stirling numbers of the second kind* $S(n,k)$ appearing in Eqs.(1) and (2) have been known for over 250 years [2]. Eq.(2) exemplifies the *normal ordering problem*, that is, finding the form of $(a^\dagger a)^n$ with the powers of a on the right. Although explicit expressions for $S(n,k)$ are known [3], of particular interest here are the *Bell numbers* $B(n)$ given by the sums

$$B(n) = \sum_{k=1}^{n} S(n,k), \qquad n = 1, 2, \ldots \tag{3}$$

with $B(0) = 1$ by convention ($S(n,0) = \delta_{n,0}$). A closed-form expression for $B(n)$ can be found by considering the action of $(xd/dx)^n$ on $f(x) = e^x$ giving:

$$(1/e^x) \sum_{k=0}^{\infty} \frac{k^n}{k!} x^k = \sum_{k=1}^{n} S(n,k) x^k, \tag{4}$$

which for $x = 1$ reduces to

$$(1/e) \sum_{k=0}^{\infty} \frac{k^n}{k!} = \sum_{k=1}^{n} S(n,k) = B(n). \tag{5}$$

Equations (4) and (5) are the celebrated Dobiński formulas ([2],[3],[4],[5]) which have been the subject of much combinatorial interest. Eq.(5) represents the integer $B(n)$ as an infinite

‡ penson@lptl.jussieu.fr
§ a.i.solomon@open.ac.uk

series, which is however *not* a power series in n. An immediate consequence of Eqs.(4) and (5) is that $B(n)$ is the n-th moment of a (singular) probability distribution, consisting of weighted Dirac delta functions located at the positive integers (the so-called Dirac comb) :

$$B(n) = \int_0^\infty x^n W(x)dx, \quad n = 0, 1, \dots \tag{6}$$

where

$$W(x) = (1/e) \sum_{k=1}^\infty \frac{\delta(x-k)}{k!}. \tag{7}$$

The discrete measure $W(x)$ serves as a weight function for a family of orthogonal polynomials $C_n^{(1)}(x)$, the Charlier polynomials [6]. The exponential generating function (EGF) of the sequence $B(n)$ can be obtained from Eq.(5) as

$$e^{(e^\lambda - 1)} = \sum_{k=0}^\infty B(n)\frac{\lambda^n}{n!}. \tag{8}$$

This equation is related via Eq.(2) to a formula giving the normal ordered form of $e^{\lambda a^\dagger a}$ [7],[8]

$$e^{\lambda a^\dagger a} = \mathcal{N}(e^{\lambda a^\dagger a}) =: e^{a^\dagger a(e^\lambda - 1)} : \tag{9}$$

The symbol $\mathcal{N}$ denotes normal ordering, while :: refers to normal-ordering with neglect of non-commutativity. We stress that in the derivation of Eq.(9) in [7] and [8] no use has been made of the Stirling and Bell numbers. It may readily be seen that Eq.(8) is the expectation value of Eq.(9) in the coherent state $|z\rangle$ defined by $a|z\rangle = z|z\rangle$ at the value $|z| = 1$, using Eqs.(2) and (3). This circumstance has been used recently to re-establish the link between the matrix element $\langle z|e^{\lambda a^\dagger a}|z\rangle$ and the properties of Stirling and Bell numbers [9].

The purpose of this note is to show that the above results on functions of $a^\dagger a$ may be extended to functions of $(a^\dagger)^r a^s$, $(r, s = 1, 2, \dots)$ with $r \geq s$; thus extending the Dobiński relations.

With this in mind we pose the following questions:

(i) What extensions of the conventional Stirling and Bell numbers occur in the normal ordering of $[(a^\dagger)^r a^s]^n$?

(ii) Can the generalised Bell numbers $B_{r,s}(n)$ so defined be represented by an infinite series of the type of Eq.(6) - that is, do they satisfy a generalised Dobiński formula ?

(iii) May one consider the $B_{r,s}(n)$ as the n-th moments of a positive weight function $W_{r,s}(x)$ on the positive half-axis, and may this latter be explicitly obtained ?

In this note we indicate affirmative answers to these questions.

To this end we generalize Eq.(2) by defining for $r \geq s$:

$$[(a^\dagger)^r a^s]^n = (a^\dagger)^{n(r-s)} \sum_{k=s}^{ns} S_{r,s}(n, k)(a^\dagger)^k a^k \tag{10}$$

or, alternatively,

$$[x^r(d/dx)^s]^n = x^{n(r-s)} \sum_{k=s}^{ns} S_{r,s}(n, k)x^k (d/dx)^k. \tag{11}$$

Eqs.(10) and (11) introduce generalized Stirling numbers $S_{r,s}(n, k)$ which imply an extended definition of generalized Bell numbers:

$$B_{r,s}(n) \equiv \sum_{k=s}^{ns} S_{r,s}(n, k). \tag{12}$$

Note that the Stirling numbers $S_{r,1}(n,k)$ were studied in [10]. Also $B_{1,1}(n) = B(n)$ of Eq.(3).

We have found a representation of the numbers $B_{r,s}(n)$ as an infinite series, which is a generalization of the Dobiński formula Eq.(5). For $r = s$ one obtains:

$$B_{r,r}(n) = (1/e) \sum_{k=0}^{\infty} \frac{1}{k!} \left[\frac{(k+r)!}{k!} \right]^{n-1}, \quad n = 1, 2 \ldots \tag{13}$$

with $B_{r,r}(0) = 1$ by convention. For $r > s$ the corresponding formula is:

$$B_{r,s}(n) = [(r-s)^{s(n-1)}/e] \sum_{k=0}^{\infty} \left[\prod_{j=1}^{s} \frac{\Gamma(n + \frac{k+j}{r-s})}{\Gamma(1 + \frac{k+j}{r-s})} \right], \quad B_{r,s}(0) = 1. \tag{14}$$

The formula Eq.(5) and its extensions Eqs.(13) and (14) share a common feature, namely, the fact that they give rise to a series of integers is by no means evident!

A general family of sequences arising from Eq.(14) has the form $(p, r = 1, 2, \ldots)$:

$$B_{pr+p,pr}(n) = (1/e) \left[\prod_{j=1}^{r} \frac{(p(n-1)+j)!}{(pj)!} \right] \times$$

$$\times \, {}_rF_r(pn+1, \ldots, pn+1+p(r-1); 1+p, \ldots, 1+p+p(r-1); 1), \tag{15}$$

where ${}_rF_r$ is the hypergeometric function. Knowledge of the generalized Stirling numbers in Eq.(10) solves the normal ordering problem for $[(a^\dagger)^r a^s]^n$. We are able to give the appropriate generating functions for the sequences $B_{r,s}(n)$. It then follows that, at least formally, we can furnish the generating functions for $S_{r,s}(n,k)$ as well [11]. Additionally, it turns out that in certain circumstances one may obtain explicit expressions for them. We quote two such cases:

$$S_{r,r}(n,k) = \sum_{p=0}^{k-r} \frac{(-1)^p [\frac{(k-p)!}{(k-p-r)!}]^n}{(k-p)! p!} \quad (r \leq k \leq rn) \tag{16}$$

and

$$S_{2,1}(n,k) = \frac{n!}{k!} \binom{n-1}{k-1} \quad (1 \leq k \leq n) \tag{17}$$

which are the so-called unsigned Lah numbers [3, 10].

For those pairs (r, s) for which we have an explicit expression for $S_{r,s}(n)$ we may generalize Eq.(9) to obtain the normal ordered form of $e^{\lambda(a^\dagger)^r a^s}$. For example, the matrix element $\langle z | e^{\lambda(a^\dagger)^r a} | z \rangle$ leads to the following normally ordered expression:

$$e^{\lambda(a^\dagger)^r a} = \mathcal{N}(e^{\lambda(a^\dagger)^r a}) =: \exp\{[(1 - \lambda(u^\dagger)^{r-1}(r-1))^{\frac{1}{r-1}} - 1]a^\dagger a\} : \tag{18}$$

We apply the method to Eq.(18) which gave the EGF of Eq.(8) and take the expectation of Eq.(9) in the coherent state $|z\rangle$ to get:

$$\langle z | e^{\lambda(a^\dagger)^r a} | z \rangle =: \exp\{[(1 - \lambda(z^*)^{r-1}(r-1))^{\frac{1}{r-1}} - 1]|z|^2\} : \tag{19}$$

which evaluates at $z = 1$ giving

$$\langle z | e^{\lambda(a^\dagger)^r a} | z \rangle_{z=1} =: \exp\{(1 - \lambda(r-1))^{\frac{1}{r-1}} - 1\} : \tag{20}$$

which is precisely the EGF for the numbers $B_{r,1}(n)$ [10].

For general $r > s$ the corresponding $B_{r,s}(n)$ grow much more rapidly than $n!$ and thus may not be obtained via the usual form of EGF. One instead defines the EGF in terms of $B_{r,s}(n)/(n!)^t$ where t is an integer chosen to ensure that $\sum_{n=0}^{\infty} B_{r,s}(n)/(n!)^{t+1}$ has a finite radius of convergence. As a result one obtains variants of Eq.(19) involving different hypergeometric functions [11].

The analogue of Eq.(6) in the general case

$$B_{r,s}(n) = \int_0^\infty x^n W_{r,s}(x)dx \tag{21}$$

leads to weight functions $W_{r,s}(x)$ which may be shown to be positive through use of properties of the Mellin transform, and for some of which analytic expressions may be obtained. For $r = s$ we obtain discrete measures, giving a "rarefied" form of the Dirac comb, while for $r > s$ we obtain continuous measures. We conclude this note with one example of the latter kind:

$$W_{2r,r}(x) = \frac{1}{e\,r} x^{\frac{2-3r}{2r}} e^{-x^{\frac{1}{r}}} I_r(2x^{\frac{1}{2r}}) \tag{22}$$

where $I_1(y)$ is the modified Bessel function of the first kind.

References

[1] Katriel J and Duchamp G 1995 J. Phys. A **28** 7209
[2] Yablonsky S V 1989 Introduction to Discrete Mathematics (Moscow: Mir Publishers)
[3] Comtet L 1974 Advanced Combinatorics (Dordrecht: Reidel)
[4] Constantine G M and Savits T H 1994 SIAM J. Discrete Math. **7** 194.
[5] Wilf H 1994 Generatingfunctionology (New York: Academic)
[6] Koekoek R and Swarttouv R F 1998 The Askey scheme of hypergeometric polynomials and its q-analogue Dept. of Technical Mathematics and Informatics, Report No. 98-17 Delft University of Technology
[7] Klauder J R and Sudarshan E C G 1968 Fundamentals of Quantum Optics (New York: Benjamin)
[8] Louisell W H 1977 Radiation and Noise in Quantum Electronics (Florida: Krieger)
[9] Katriel J 2000 Phys. Lett. **A273** 159
[10] Lang W 2000 J. Int. Seqs. **12** Article 00.2.4 (www.research.att.com/~njas/sequences/)
[11] Penson K A and Solomon A I 2002 to be published.

Inst. Phys. Conf. Ser. No 173: Section 7
Paper presented at 24th Int. Coll. Group Theoretical Methods in Physics, Paris, France, 15–20 July 2002
©*2003 IOP Publishing Ltd*

Non-Hermitian Hamiltonians with real and complex eigenvalues: An sl(2,C) approach

B Bagchi[a] **and C. Quesne**[b]

[a] Department of Applied Mathematics, University of Calcutta, India
[b] Physique Nucléaire Théorique et Physique Mathématique, Université Libre de Bruxelles, Belgium

Abstract. Potential algebras are extended from Hermitian to non-Hermitian Hamiltonians and shown to provide an elegant method for studying the transition from real to complex eigenvalues for a class of non-Hermitian Hamiltonians associated with the complex Lie algebra A_1.

1. Introduction

In recent years there has been much interest in non-Hermitian Hamiltonians with real, bound-state eigenvalues. In particular, PT-symmetric Hamiltonians (such that $(PT)H(PT)^{-1} = H$, where P is the parity and T the time reversal) have been conjectured to have a real bound-state spectrum except when the symmetry is spontaneously broken, in which case their complex eigenvalues should come in conjugate pairs [1]. It is also known that PT symmetry is not a necessary condition for the occurrence of real or complex-conjugate pairs of eigenvalues. A more general condition, namely pseudo-Hermiticity of the Hamiltonian (i.e., the existence of a Hermitian linear automorphism η such that $\eta H \eta^{\dagger} = H^{\dagger}$) has been identified as an explanation for the existence of this phenomenon for some non-PT-symmetric Hamiltonians [2, 3]. It should be noted that all such non-Hermitian Hamiltonians require a generalization of the normalization condition corresponding to an indefinite scalar product (see [4] and references quoted therein).

Very recently there has been a growing interest in determining the critical strengths of the interaction, if any, at which PT symmetry (or some generalization thereof) becomes spontaneously broken. Among the various techniques that have been employed to contruct and study non-Hermitian Hamiltonians with real or complex spectra, algebraic methods provide powerful approaches. In the present communication, we show how potential algebras can be used for such a purpose.

2. Potential algebras for Hermitian and non-Hermitian Hamiltonians

Potential algebras refer to Lie algebras whose generators connect eigenfunctions $\psi_n^{(m)}(x)$ corresponding to the same eigenvalue (i.e., $E_n^{(m)}$ is constant), but to different potentials $V_m(x)$ of a given family. Here m is some parameter (generally related to the potential strength), which may change by one unit under the action of the generators.

Potential algebras were introduced for Hermitian Hamiltonians as real Lie algebras, the simplest example being that of sl(2,R) $\simeq$ so(2, 1) [5, 6]. The latter is generated by J_0, J_+, J_-, satisfying the commutation relations

$$[J_0, J_{\pm}] = \pm J_{\pm}, \qquad [J_+, J_-] = -2J_0, \tag{1}$$

and the Hermiticity properties $J_0^\dagger = J_0$, $J_\pm^\dagger = J_\mp$. Such operators are realized as differential operators

$$J_0 = -i\frac{\partial}{\partial\phi}, \qquad J_\pm = e^{\pm i\phi}\left[\pm\frac{\partial}{\partial x} + \left(i\frac{\partial}{\partial\phi} \mp \frac{1}{2}\right)F(x) + G(x)\right], \tag{2}$$

depending upon a real variable x and an auxiliary variable $\phi \in [0, 2\pi)$, provided the two real-valued functions $F(x)$ and $G(x)$ in (2) satisfy coupled differential equations

$$F' = 1 - F^2, \qquad G' = -FG. \tag{3}$$

The sl(2,R) Casimir operator, $J^2 = J_0^2 \mp J_0 - J_\pm J_\mp$, then becomes a second-order differential operator.

For bound states, to which we restrict ourselves here, one considers unitary irreducible representations of sl(2,R) of the type D_k^+, spanned by states $|km\rangle$, $k \in \mathbf{R}^+$, $m = k+n$, $n \in \mathbf{N}$, such that $J_0|km\rangle = m|km\rangle$ and $J^2|km\rangle = k(k-1)|km\rangle$. In the realization (2), these states are given by $|km\rangle = \Psi_{km}(x, \phi) = \psi_{km}(x)e^{im\phi}/\sqrt{2\pi}$, where $\psi_{km}(x) = \psi_n^{(m)}(x)$ satisfies the Schrödinger equation

$$-\psi_n^{(m)\prime\prime} + V_m\psi_n^{(m)} = E_n^{(m)}\psi_n^{(m)}. \tag{4}$$

In (4), the potential $V_m(x)$ is defined in terms of $F(x)$ and $G(x)$ by

$$V_m = \left(\tfrac{1}{4} - m^2\right)F' + 2mG' + G^2, \tag{5}$$

and the energy eigenvalues are given by $E_n^{(m)} = -\left(m - n - \tfrac{1}{2}\right)^2$. The eigenfunction $\psi_0^{(m)}(x)$ can be easily constructed by solving the first-order differential equation $J_-\psi_0^{(m)}(x) = 0$, while the remaining eigenfunctions $\psi_n^{(m)}(x)$ can be obtained from the action of J_+ on $\psi_{n-1}^{(m-1)}(x)$. Imposing the regularity condition for bound states $\psi_n^{(m)}(\pm\infty) \to 0$ restricts the allowed values of n to $n = 0, 1, \ldots, n_{\max} < m - \tfrac{1}{2}$.

Detailed inspection of the system of differential equations (3) shows that it admits three classes of solutions corresponding to the (nonsingular) Scarf II, the (singular) generalized Pöschl-Teller, and the Morse potentials, respectively [7, 8].

The transition from Hermitian to non-Hermitian Hamiltonians can now be performed by replacing real Lie algebras by complex ones [9, 10] (see also [11, 12] for a related approach). In the case of sl(2,R), we find the algebra known as A_1 in Cartan's classification of complex Lie algebras (which we shall refer to as sl(2,C), considered here as a complex algebra, not a real one as is usely the case). Its generators still satisfy the commutation relations (1), but their Hermiticity properties remain undefined. This means that the realization (2) is still applicable with $F(x)$ and $G(x)$ now some complex-valued functions.

For bound states, we restrict ourselves to irreducible representations spanned by states $|km\rangle$, for which both $k = k_R + ik_I$ and $m = m_R + im_I$ may be complex with $m_R = k_R + n \in \mathbf{R}$, $m_I = k_I \in \mathbf{R}$, and $n \in \mathbf{N}$. Equations (4) and (5) remain valid and we get some complexified forms of the Scarf II, generalized Pöschl-Teller, and Morse potentials, given by

$$\text{I}: \qquad V_m = \left(b^2 - m^2 + \tfrac{1}{4}\right)\mathrm{sech}^2\,\tau - 2mb\,\mathrm{sech}\,\tau\tanh\tau, \tag{6}$$

$$\text{II}: \qquad V_m = \left(b^2 + m^2 - \tfrac{1}{4}\right)\mathrm{cosech}^2\,\tau - 2mb\,\mathrm{cosech}\,\tau\coth\tau, \tag{7}$$

$$\text{III}: \qquad V_m = b^2 e^{\mp 2x} \mp 2mb e^{\mp x}, \tag{8}$$

where $b = b_R + ib_I$ and $\tau = x - c - i\gamma$ with $b_R, b_I, c \in \mathbf{R}$, $-\tfrac{\pi}{4} \leq \gamma < \tfrac{\pi}{4}$. For generic values of the parameters, such potentials are neither PT-symmetric nor pseudo-Hermitian. The corresponding energy eigenvalues become $E_n^{(m)} = -\left(m_R + im_I - n - \tfrac{1}{2}\right)^2$ and are therefore complex if $m_I \neq 0$. From the explicit form of $\psi_0^{(m)}(x)$, it can be shown that the potentials (6) – (8) have at least one regular eigenfunction (namely that corresponding to $n = 0$) provided $m_R > 1/2$ and $b_R > 0$, where the second condition applies only to class III.

It is worth noting that the same potentials (6) – (8) can alternatively be derived in the framework of supersymmetric quantum mechanics by considering a complex superpotential $W(x) = \left(m - \frac{1}{2}\right) F(x) - G(x)$ [13]. They turn out to be shape-invariant as their real counterparts.

3. Some examples

As a first example, let us consider a special case of class I potentials,

$$V(x) = -V_1 \operatorname{sech}^2 x - iV_2 \operatorname{sech} x \tanh x, \qquad V_1 > 0, \qquad V_2 \neq 0, \tag{9}$$

which is both PT-symmetric and P-pseudo-Hermitian. It corresponds to $c = \gamma = 0$ in (6), while the other four parameters m_R, m_I, b_R, b_I are related to V_1 and V_2 through four quadratic equations,

$$b_R^2 - b_I^2 - m_R^2 + m_I^2 + \tfrac{1}{4} = -V_1, \tag{10}$$
$$b_R b_I - m_R m_I = 0, \tag{11}$$
$$m_R b_R - m_I b_I = 0, \tag{12}$$
$$2(m_R b_I + m_I b_R) = V_2. \tag{13}$$

On solving the latter to express m_R, m_I, b_R, b_I in terms of V_1, V_2, taking the regularity condition $m_R > 1/2$ into account, and inserting the m_R and m_I values into $E_n^{(m)}$, we find one critical strength (for a given sign of V_2) corresponding to $|V_2| = V_1 + \frac{1}{4}$. For $|V_2| < V_1 + \frac{1}{4}$, there are (in general) two series of real eigenvalues (instead of one for the real Scarf II potential), given by

$$E_{n,\pm} = -\left[\tfrac{1}{2}\left(\sqrt{V_1 + \tfrac{1}{4} + |V_2|} \pm \sqrt{V_1 + \tfrac{1}{4} - |V_2|}\right) - n - \tfrac{1}{2}\right]^2, \tag{14}$$

where $n = 0, 1, 2, \ldots < \frac{1}{2}\left(\sqrt{V_1 + \frac{1}{4} + |V_2|} \pm \sqrt{V_1 + \frac{1}{4} - |V_2|} - 1\right)$. When $|V_2|$ reaches the value $V_1 + \frac{1}{4}$, the two series of real energy eigenvalues merge and for higher $|V_2|$ values they move into the complex plane so that we get a series of complex-conjugate pairs of eigenvalues,

$$E_{n,\pm} = -\left[\tfrac{1}{2}\left(\sqrt{|V_2| + V_1 + \tfrac{1}{4}} \pm i\sqrt{|V_2| - V_1 - \tfrac{1}{4}}\right) - n - \tfrac{1}{2}\right]^2, \tag{15}$$

where $n = 0, 1, 2, \ldots < \frac{1}{2}\left(\sqrt{|V_2| + V_1 + \frac{1}{4}} - 1\right)$. This agrees with results obtained elsewhere by another method [14].

Another example of occurrence of just one critical strength is provided by a PT-symmetric and P-pseudo-Hermitian special case of class II potentials,

$$V(x) = V_1 \operatorname{cosech}^2 \tau - V_2 \operatorname{cosech} \tau \coth \tau, \qquad V_1 > -\tfrac{1}{4}, \qquad V_2 \neq 0, \tag{16}$$

which is defined on the entire real line (contrary to its real counterpart).

A rather different situation is depicted by the potential

$$V(x) = (V_{1R} + iV_{1I})e^{-2x} - (V_{2R} + iV_{2I})e^{-x}, \qquad V_{1R}, V_{1I}, V_{2R}, V_{2I} \in R, \tag{17}$$

which is the most general class III potential (for the upper sign choice in (8)) and has no special property for generic values of the parameters.

By proceeding as in the first example, we find that the regularity conditions $m_R > 1/2$ and $b_R > 0$ impose that V_{1I} be non-vanishing and

$$(V_{1R} + \Delta)^{1/2} V_{2R} + \nu(-V_{1R} + \Delta)^{1/2} V_{2I} > \sqrt{2}\Delta, \tag{18}$$

where ν denotes the sign of V_{1I} and $\Delta \equiv \sqrt{V_{1R}^2 + V_{1I}^2}$.

The results for the bound-state energy eigenvalues strongly contrast with those obtained hereabove. Indeed real eigenvalues belonging to a single series,

$$E_n = -\left[\frac{V_{2R}}{\sqrt{2}|V_{1I}|}(-V_{1R} + \Delta)^{1/2} - n - \frac{1}{2}\right]^2, \tag{19}$$

where $n = 0, 1, 2, \ldots < (V_{2R}/\sqrt{2}|V_{1I}|)(-V_{1R} + \Delta)^{1/2} - \frac{1}{2}$, only occur for a special value of V_{2I}, namely $V_{2I} = \nu(-V_{1R} + \Delta)^{1/2}(V_{1R} + \Delta)^{-1/2} V_{2R}$, while complex eigenvalues,

$$E_n = -\left\{\frac{1}{2\sqrt{2}\Delta}\left[(V_{1R} + \Delta)^{1/2} - i\nu(-V_{1R} + \Delta)^{1/2}\right](V_{2R} + iV_{2I})\right.$$
$$\left. - n - \frac{1}{2}\right\}^2, \tag{20}$$

where $n = 0, 1, 2, \ldots < \frac{1}{2\sqrt{2}\Delta}\left[(V_{1R} + \Delta)^{1/2} V_{2R} + \nu(-V_{1R} + \Delta)^{1/2} V_{2I}\right] - \frac{1}{2}$ and which do not form complex-conjugate pairs, occur for all the remaining values of V_{2I}.

Such results can be interpreted by choosing the parametrization $V_{1R} = A^2 - B^2$, $V_{1I} = 2AB$, $V_{2R} = \gamma A$, $V_{2I} = \delta B$, where A, B, γ, δ are real, $A > 0$, and $B \neq 0$. The complexified Morse potential (17) then becomes $V(x) = (A + iB)^2 e^{-2x} - (2C + 1)(A + iB)e^{-x}$, where $C = [(\gamma - 1)A + i(\delta - 1)B]/[2(A + iB)]$. Its (real or complex) eigenvalues can be written in a unified way as $E_n = -(C - n)^2$, while the regularity condition (18) amounts to $(\gamma - 1)A^2 + (\delta - 1)B^2 > 0$.

For $\delta = \gamma > 1$, and therefore $C = \frac{1}{2}(\gamma - 1) \in \mathbf{R}^+$, $V(x)$ is pseudo-Hermitian under imaginary shift of the coordinate [3]. It has only real eigenvalues corresponding to $n = 0, 1, 2, \ldots < C$, thus exhibiting no symmetry breaking over the whole parameter range. For the values of δ different from γ, the potential indeed fails to be pseudo-Hermitian. In such a case, C is complex as well as the eigenvalues. Nevertheless, the eigenfunctions associated with $n = 0, 1, 2, \ldots < \operatorname{Re} C$ remain regular. Note that the existence of regular eigenfunctions with complex energies for general complex potentials is a well-known phenomenon (see e.g. [15]).

References

[1] Bender C M and Boettcher S. 1998 Phys. Rev. Lett. 80 (1998) 5243–5246
[2] Mostafazadeh A 2002 J. Math. Phys. 43 205–214
[3] Ahmed Z 2001 Phys. Lett. A 290 19–22
[4] Bagchi B, Quesne C and Znojil M 2001 Mod. Phys. Lett. A 16 2047–2057
[5] Alhassid Y, Gürsey F and Iachello F 1983 Ann. Phys., NY 148 346–380
[6] Frank A and Wolf K B 1984 Phys. Rev. Lett. 52 1737–1739
[7] Wu J and Alhassid Y 1990 J. Math. Phys. 31 557–562
[8] Englefield M J and Quesne C 1991 J. Phys. A: Math. Gen. 24 3557–3574
[9] Bagchi B and Quesne C 2000 Phys. Lett. A 273 285–292
[10] Bagchi B and Quesne C 2002 Phys. Lett. A 300 18–26
[11] Lévai G and Znojil M 2000 J. Phys. A: Math. Gen. 33 7165–7180
[12] Lévai G and Znojil M 2001 Mod. Phys. Lett. A 16 1973–1981
[13] Bagchi B, Mallik S and Quesne C 2001 Int. J. Mod. Phys. A 16 2859–2872
[14] Ahmed Z 2001 Phys. Lett. A 282 343–348, 287 295–296
[15] Baye D, Lévai G and Sparenberg J-M 1996 Nucl. Phys. A 599 435–456

Inst. Phys. Conf. Ser. No 173: Section 7
Paper presented at 24th Int. Coll. Group Theoretical Methods in Physics, Paris, France, 15–20 July 2002
©2003 IOP Publishing Ltd

The Stationary Maxwell-Dirac Equations

Chris Radford

School of Mathematics, Statistics and Computer Science
University of New England, Armidale 2351
Australia
chris@turing.une.edu.au

Abstract. The Maxwell-Dirac equations are the equations for electronic matter, the "classical" theory underlying QED. The system combines the Dirac equations with the Maxwell equations sourced by the Dirac current.

In this talk we will be interested in properties of isolated systems (such as a single electron). Foremost among such properties must be "localisation". Does the Maxwell-Dirac system admit particle like solutions?

For stationary, isolated Maxwell-Dirac systems there is a strong localisation result, which will be the main focus for this talk.

1. Introduction

The Maxwell-Dirac system consists of the Dirac equation

$$\gamma^\alpha(\partial_\alpha - i\,e\,A_\alpha)\psi + im\psi = 0, \tag{1}$$

with electromagnetic interaction given by the potential A_α; and the Maxwell equations (sourced by the Dirac current, j^α),

$$F_{\alpha\beta} = \partial_\alpha A_\beta - \partial_\beta A_\alpha,$$
$$\partial^\alpha F_{\alpha\beta} = -4\pi e j_\beta = -4\pi e\bar{\psi}\gamma_\beta\psi. \tag{2}$$

Most studies of the Dirac equation treat the electromagnetic field as given and ignore the Dirac current as a source for the Maxwell equations, i.e. these treatments ignore the electron "self-field". A comprehensive survey of these results can be found in the book by Thaller[1]. This is not surprising, inclusion of the electron self-field via the Dirac current leads to a very difficult, highly non-linear set of partial differential equations. So difficult in fact that the existence theory and solution of the Cauchy problem was not completed until 1997 – seventy years after Dirac first wrote down his equation! In a stunning piece of non-linear analysis , worked out over an almost twenty year period, Flato, Simon and Taflin ([2]) solved the Cauchy problem for small initial data. Other contributors to this work on the existence of solutions would include Gross [3], Chadam [4], Georgiev [5], Esteban *et al* [6], and Bournaveas [7].

There are no known non-trivial, exact solutions to the Maxwell-Dirac equations in $1 + 3$ dimensions – all known solutions involve some numerical work. These solutions do, however, exhibit interesting non-linear behaviour which would not have been apparent through perturbation expansions. The particular solutions found in [8] and [9] exhibit just this sort of behaviour – localisation and charge screening. See also Das [10] and the more recent work of Finster, Smoller and Yau [11].

In [12] a strong localisation result is established in the form of the following theorem.

Main Theorem *A stationary, isolated Maxwell-Dirac system has no embedded eigenvalues, i.e.* $-m \leq E \leq m$.

If $|E| < m$ then the Dirac field, decays exponentially as $|x| \to \infty$.
*If $|E| = m$ then the system is "asymptotically static" and, if the system has non-vanishing
total charge, decays exponentially as $|x| \to \infty$.*

We will first describe, precisely, the terms used in the theorem and then give an overview
of the sequence of theorems which lead to this "main theorem".

2. The Maxwell-Dirac Equations

In this section we give a very brief account of the 2-spinor formulation of the Maxwell-Dirac
equations, details may be found in [8]. That paper also demonstrates how one may use the
2-spinor formalism to "solve" the Dirac equations for the electromagnetic potential and a set
of "reality conditions".

For 2-spinors u_A and v^B (see [13] for an exposition of the 2-spinor formalism) we have

$$\psi = \begin{pmatrix} u_A \\ \overline{v}^{\dot{B}} \end{pmatrix}, \text{ with}$$

$$u_C v^C \neq 0 \text{ (non-degeneracy),}$$

where $A, B = 0, 1$, $\dot{A}, \dot{B} = \dot{0}, \dot{1}$ are two-spinor indices. The Dirac equations are

$$(\partial^{A\dot{A}} - i\,e\,A^{A\dot{A}})u_A + \frac{im}{\sqrt{2}}\overline{v}^{\dot{A}} - 0, \tag{3}$$

$$(\partial^{A\dot{A}} + i\,e\,A^{A\dot{A}})v_A + \frac{im}{\sqrt{2}}\overline{u}^{\dot{A}} = 0,$$

where $\partial^{A\dot{A}} \equiv \sigma^{\alpha A\dot{A}}\partial_\alpha$, and $A^{A\dot{A}} = \sigma^{\alpha A\dot{A}}A_\alpha$; here $\sigma^{\alpha A\dot{A}}$ are the Infeld-van der Waerden
symbols.

The Maxwell equations are,

$$\partial^\alpha F_{\alpha\beta} = -4\pi e\, j_\beta = -4\pi e\, \sqrt{2}\sigma_\beta^{A\dot{A}}(u_A \overline{u}_{\dot{A}} + v_A \overline{v}_{\dot{A}}). \tag{4}$$

3. Isolated, Stationary Maxwell-Dirac Systems

We recall the definitions of [14] for stationary and isolated systems,

Definition *A Maxwell-Dirac system is said to be* stationary *if there is a gauge in which
$\psi = e^{i\omega t}\phi$, with the bi-spinor ϕ independent of t. Such a gauge will be referred to as a
stationary gauge.*

In most physical processes that we would wish to model using the Maxwell-Dirac system
we would be interested in isolated systems – systems where the fields and sources are largely
confined to a compact region of $\mathbf{R}^3$. This requires that the fields decay sufficiently quickly as
$|x| \to \infty$.

The best language for the discussion of such decay conditions and other regularity issues
is the language of weighted function spaces; specifically weighted classical and Sobolev
spaces. In [14] the weighted Sobolev spaces, $W_\delta^{k,p}$, were used following the definitions of
[15]. These definitions have the advantage that the decay rate is explicit: under appropriate
circumstances a function in $W_\delta^{k,p}$ behaves as $|x|^\delta$ for large $|x|$. An element, f, of $W_\delta^{k,p}$ has
$\sigma^{-\delta+|\alpha|-\frac{3}{p}}\partial^{|\alpha|}f$ in L^p for each multi-index α for which $0 \leq |\alpha| \leq k$; here $\sigma = \sqrt{1 + |x|^2}$ and
we are working on $\mathbf{R}^3$ (or some appropriate subset thereof) – see [15] or [16] and [17] (the
later papers use a different indexing of the Sobolev spaces).

Suppose we have a stationary system and we are in a stationary gauge for which $A^\alpha \to 0$ as $|x| \to \infty$. Write, $u_A = e^{-iEt}U_A$ and $\bar{v}^{\dot{A}} = e^{-iEt}\bar{V}^{\dot{A}}$ with U_A, V_A and A^α all independent of time, t. Note that $u_C v^C = U_C V^C$ is a gauge and Lorentz invariant complex scalar function, this means we can introduce a (unique up to sign) "spinor dyad" $\{o_A, \iota_B\}$ with $\iota^A o_A = 1$. The dyad is defined as follows, let $U_C V^C = Re^{ix}$ – where R and χ are real functions – then write,

$$U_A = \sqrt{R}e^{i\frac{\chi}{2}}o_A \text{ and } V_A = \sqrt{R}e^{i\frac{\chi}{2}}\iota_A.$$

Note that we must have $R > 0$ (almost everywhere) because of our non-degeneracy condition ($j^\alpha j_\alpha \neq 0$).

Definition *A stationary Maxwell-Dirac system will be said to be isolated if, in some stationary gauge, we have*

$$\psi = e^{-iEt}\sqrt{R}\begin{pmatrix} e^{\frac{ix}{2}}o_A \\ e^{-\frac{ix}{2}}\bar{\iota}^{\dot{A}} \end{pmatrix},$$

with E constant and $\sqrt{R} \in W^{3,2}_{-\tau}(E_\rho)$; $e^{\frac{i}{2}\chi}o_A$, $e^{\frac{i}{2}\chi}\iota_A \in W^{3,2}_\epsilon(E_\rho)$ and $A^\alpha \in W^{2,2}_{-1+\epsilon}(E_\rho)$, for some $\tau > \frac{3}{2}$, $\rho > 0$ and any $\epsilon > 0$.

This definition ensures, after use of the Sobolev inequality and the multiplication lemma, that $\psi = o(r^{-\tau+\epsilon})$ and $A^\alpha = o(r^{-1+\epsilon})$. Notice our conditions place regularity restrictions on the fields in the region E_ρ only; In the interior of B_ρ there are no regularity assumptions. A minimal condition that one may impose on the Dirac field is that it have finite total charge in the region E_ρ, this amounts to

$$\int_{E_\rho} j^0 \, dx = \int_{E_\rho} \left(|U_0|^2 + |U_1|^2 + |V^0|^2 + |V^1|^2\right) \, dx$$

$$= \int_{E_\rho} R\left(|o_0|^2 + |o_1|^2 + |\iota^0|^2 + |\iota^1|^2\right) \, dx < \infty.$$

This, of course, simply means that U_A and V^A are in $L^2(E_\rho)$. So U_A and V^A would have L^2 decay at infinity; roughly, they would decay faster than $|x|^{-\frac{3}{2}}$, i.e. we require at least $\tau > \frac{3}{2}$.

The spherically symmetric solution of [8] provides an excellent example of an *isolated*, *stationary* and *static* Maxwell-Dirac system.

Next using standard theorems on the inversion of the Laplacian one then establishes that

$$A^0 - \frac{q_0}{|x|} \in W^{5,2}_{-\eta}(E_\rho), \quad q_0 \text{ a constant and} \tag{5}$$

$$A^j \in W^{5,2}_{-\eta}(E_\rho), \quad j = 1, 2, 3; \text{ and } \eta = 2(\tau - 1) > 1. \tag{6}$$

The constant q_0 is the total electric charge of the system (i.e. the electric charge of the Dirac field plus the charge due to any external sources in B_ρ); this is easily seen by taking a Gauss integral over the sphere at infinity of the electrostatic field (given by the gradient of A^0).

4. No Embedded Eigenvalues and Exponential Decay

In [12] we established that the Dirac Hamiltonian operator H of a stationary and isolated Maxwell-Dirac system has the same essential spectrum as the free operator, and that there are no embedded ("energy") eigenvalues E, i.e. $-m \leq E \leq m$.

Exploiting the properties of weighted Sobolev spaces and making use of the maximum principle for the Klein-Gordon equations derived from our Dirac equations one easily proves that, at least in the case $|E| < m$, that the Dirac field decays exponentially. In fact,

$$R \leq C_0 \frac{e^{-\sqrt{2}k|x|}}{|x|}.$$

596

The case $|E| = m$ is a little harder, but in a sense more interesting. One finds that the current vector in this case must be asymptotic (at spatial infinity) to a static vector, ie $j^\alpha \to j^0 \delta_0^\alpha$. In this case we also find exponential decay, provided the total charge of the system, q_0, does not vanish. In fact in the $|E| = m$ case we have the strong result,

$$C_1 \frac{e^{-4\sqrt{2}m\lambda\sqrt{|x|}}}{|x|^{\frac{3}{2}}} < R < C_2 \frac{e^{-4\sqrt{2}m\lambda\sqrt{|x|}}}{|x|^{\frac{3}{2}}},$$

where $\lambda > 0$, $\lambda^2 = -\varepsilon e \frac{q_0}{m}$ is necessarily positive and $\frac{E}{m} = \varepsilon = \pm 1$.

This establishes the essential ingredients of the Main Theorem, the interested reader is refered to [12] for detailed proofs.

References

[1] B. Thaller, *The Dirac Equation*, Springer-Verlag Texts and Monographs in Physics, 1992.
[2] M. Flato, J. C. H. Simon, E. Taflin, *Asymptotic Completeness, Global Existence and the Infrared Problem for the Maxwell-Dirac Equations*, Memoirs of the AMS **127**(606) (1997).
[3] L. Gross, *The Cauchy Problem for the Coupled Maxwell-Dirac Equations*, Comm. Pure Appl. Math. **19** (1966), pp. 1-5.
[4] J. Chadam, *Global Solutions of the Cauchy Problem for the (Classical) Coupled Maxwell-Dirac System in One Space Dimension*, J. Funct. Anal. **13** (1973), pp. 495-507.
[5] V. Georgiev, *Small Amplitude Solutions of the Maxwell-Dirac Equations*, Indiana Univ. Math. J. **40**(3) (1991), pp. 845-883.
[6] M. Esteban, V. Georgiev, E. Séré, *Stationary Solutions of the Maxwell-Dirac and Klein-Gordon-Dirac Equations*, Calc. Var. **4** (1996), pp. 265-281.
[7] N. Bournaveas, *Local Existence for the Maxwell-Dirac Equations in Three Space Dimensions*, Comm. Part. Diff. Eq. **21**(5& 6) (1996), pp. 693-720.
[8] C. J. Radford, *Localised Solutions of the Dirac-Maxwell Equations*, J. Math. Phys. **37**(9) (1996), pp. 4418-4433.
[9] H. S. Booth and C. J. Radford, *The Dirac-Maxwell Equations with Cylindrical Symmetry*, J. Math. Phys. **38**(3) (1997), pp. 1257-1268.
[10] A. Das, *General Solutions of the Maxwell-Dirac Equations in 1+1 Dimensional Spacetime and a Spatially Confined Solution*, J. Math. Phys. **34**(10)
[11] F. Finster, J. Smoller, and S-T, Yau, *Particle-Like Solutions of the Einstein-Dirac-Maxwell Equations*, preprint gr-qc 9802012 (1998).
[12] C. J. Radford, *The Stationary Maxwell-Dirac Equations*, submitted Annales Henri Pioncaré (2002).
[13] R. Penrose and W. Rindler, *Spinors and Space-Time* Vol. 1 and 2, Cambridge Monographs in Mathematical Physics (Cambridge U. P., Cambridge) (1992).
[14] C. J. Radford and H. S. Booth, *Magnetic Monopoles, Electric Neutrality and the Static Maxwell-Dirac Equations* J. Phys. A, Math. and Gen. **32**, (1999), pp. 5807-5822.
[15] R. Bartnik, *The Mass of an Asymptotically Flat Manifold*, Comm. Pure and Appl. Math. **94**, (1986), pp. 661-693.
[16] Y. Choquet-Bruhat and D. Christodoulou, *Elliptic Systems in $H_{s,\delta}$ Spaces on Manifolds which are Euclidean at Infinity*, Acta Math. **146**, (1981), pp. 126-150.
[17] C. Amrouche, V. Girault and J. Giroire, *Weighted Sobolev Spaces for Laplace's Equation in R^n*, J. Math. Pures Appl., **73**, (1994), pp. 579-606.

Inst. Phys. Conf. Ser. No 173: Section 7
Paper presented at 24th Int. Coll. Group Theoretical Methods in Physics, Paris, France, 15–20 July 2002
©2003 *IOP Publishing Ltd*

Confined systems through Variational Supersymmetric Quantum Mechanics

E Drigo Filho

Instituto de Biociências, Letras e Ciências Exatas, IBILCE-UNESP, Brazil

R M Ricotta

Faculdade de Tecnologia de São Paulo, FATEC/SP-CEETPS-UNESP, Brazil

Abstract. The energy states of the confined harmonic oscillator and the Hulthén potentials are evaluated using the Variational Method associated to Supersymmetric Quantum Mechanics.

1. Introduction

Recently a new methodology based on the association of the variational method with Supersymmetric Quantum Mechanics, (SQM), formalism has been introduced. Its starting point is the association of an *Ansatz* for the superpotential. Through the superalgebra the wave function is evaluated, the so-called trial wave function containing the variational parameters, which will be varied until the energy expectation value reaches its minimum. This new methodology has been successfully applied to get answers of atomic systems using the Hulthén, the Morse, the Coulomb potentials and lately the 3-dimensional confined hydrogen atom, [1]-[2]. Here we show results obtained for the $1s$ and $2p$ states for the confined Hulthén and the Harmonic Oscillator potentials. They are very good when compared to recent results obtained from other approximative methods as well as numerical exact results, [3]-[5].

2. The variational method associated to Supersymmetric Quantum Mechanics

Consider a system described by a given potential V_1. The associated Hamiltonian H_1 can be factorized in terms of bosonic operators, in $\hbar = c = 1$ units,

$$H_1 = -\frac{1}{2}\frac{d^2}{dr^2} + V_1(r) = A_1^+ A_1^- + E_0^{(1)} \tag{1}$$

where $E_0^{(1)}$ is the lowest eigenvalue. Notice that the function $V_1(r)$ includes the barrier potential term. The bosonic operators are defined in terms of the so called superpotential $W_1(r)$,

$$A_1^\pm = \frac{1}{\sqrt{2}}\left(\mp\frac{d}{dr} + W_1(r)\right). \tag{2}$$

As a consequence of the factorization of the Hamiltonian H_1, the Riccati equation must be satisfied,

$$W_1^2 - W_1' = 2\left(V_1(r) - E_0^{(1)}\right). \tag{3}$$

Through the superalgebra, the eigenfunction for the lowest state is related to the superpotential W_1 by

$$\Psi_0^{(1)}(r) = N exp(- \int_0^r W_1(\bar{r})d\bar{r}). \tag{4}$$

However, if the potential is non-exactly solvable, the Hamiltonian is not exactly factorizable, i.e., there is no superpotential that satisfies the Riccati equation. On the other hand, the Hamiltonian can be factorized in terms of an approximated superpotential giving rise to an effective potential. Based on physical arguments, an *Ansatz* for the superpotential is proposed and, through the superalgebra, the trial wave function is evaluated, (equation (4)). By doing this we naturally introduce free parameters, $\{\mu\}$. Therefore this is trial wave function of the variational method, Ψ_μ which depends on a set of parameters $\{\mu\}$. The approach consists in varying these parameters in the expression for the expectation value of the energy until it reaches its minimum value, which is an upper limit of the energy level.

We remark that, in fact, by making an *Ansatz* in the superpotential corresponds to be dealing with an effective potential V_{eff} that has a similar form as the original pontential, i.e.,

$$V_{eff}(r) = \frac{1}{2} \left(\bar{W}_1^2 - \bar{W}_1' \right) + E(\bar{\mu}) \tag{5}$$

where $\bar{W}_1 = W_1(\bar{\mu})$ is the superpotential for $\mu = \bar{\mu}$, the parameter that minimises the energy expectation value.

3. The Confined Harmonic Oscillator

The radial Hamiltonian equation for the Harmonic Oscillator potential, written in atomic units is given by

$$H = -\frac{1}{2}\frac{d^2}{dr^2} + \frac{l(l+1)}{2r^2} + \frac{r^2}{2} \tag{6}$$

As the Harmonic Oscillator potential is symmetric, the confinement is introduced by an infinite potential barrier at radius $r = R$. Thus we make the following *Ansatz* for the superpotential

$$W(r) = -\frac{\mu}{r} + \frac{\mu_1}{R-r} + \mu_2 r \tag{7}$$

which depends on R, the radius of confinement, and three variational parameters, μ, μ_1 and μ_2. The first and the last terms are already known from the non-confined case. The second term deals with the confinement.

Our trial wavefunction for the variational method is obtained from the superalgebra through equation (4), using the superpotential given by the *Ansatz* made in equation (7). It is given by

$$\Psi(\mu, \mu_1, \mu_2, r) \propto r^\mu (R-r)^{\mu_1} e^{-\mu_2 r^2/2}. \tag{8}$$

It depends of three free parameters, μ, μ_1 and μ_2 and vanishes at $r = R$.

The energy E_{VSQM} is obtained by minimisation of the energy expectation value with respect to the three parameters, μ, μ_1 and μ_2. The results are given in the two tables below, together with the exact numerical, E_{EXACT}, those obtained from perturbative calculations, E_{pert}, [3], the WKB energies, E_{WKB} and the variational energies, E_{VAR}, [4].

Table 1. Energy eigenvalues (in Rydbergs) for different values of R and for $l = 0$, (1s state).

R	E_{EXACT} [3]	E_{VSQM}	E_{pert} [3]	E_{WKB} [4]	E_{VAR} [4]
1.0	5.0755	5.0865	-	5.0627	5.1313
2.0	1.7648	1.7664	1.7588	1.9882	1.7739
3.0	1.5061	1.5069	1.1532	1.5061	1.5105
4.0	1.5000	1.5002	- 4.3417	1.5000	1.5033
5.0	1.5000	1.5001	- 48.0763	1.5000	1.5025
10.0	1.5000	1.5000	-	1.5000	-

Table 2. Energy eigenvalues (in Rydbergs) for different values of R and for $l = 1$, (2p state).

R	E_{EXACT} [3]	E_{VSQM}	E_{pert} [3]	E_{WKB} [5]	E_{VAR} [5]	$E_{centri0}$ [5]
1.0	10.2822	10.2847	-	10.2643	10.3188	10.2876
1.5	4.9036	4.9046	4.9034	4.9084	4.9169	4.9068
2.0	3.2469	3.2471	3.2434	3.2490	3.2469	3.3081
2.5	2.6881	2.6891	-	2.7079	2.6901	2.6835
3.0	2.5313	2.5322	2.3104	2.5310	2.5337	2.5313
4.0	2.5001	2.5003	-1.5656	2.5001	2.5015	2.5001
5.0	2.5000	2.5001	-34.0359	2.5000	2.5012	2.5000
10.0	2.5000	2.5000	-	2.5000	-	

4. The Confined Hulthén Potential

The radial Hamiltonian equation for the Hulthén potential, written in atomic units is given by

$$H = -\frac{1}{2}\frac{d^2}{dr^2} + \frac{l(l+1)}{2r^2} - \frac{\delta e^{-\delta r}}{1 - e^{-\delta r}} \tag{9}$$

The trial wavefunction for this case will be given by the wave function of the unconfined case, as given in [1], added by the confining term

$$\Psi(\mu, \mu_1, r) \propto \left(1 - e^{-\mu r}\right)^{(l+1)} (R - r)^{\mu_1} e^{-br}. \tag{10}$$

with $b = -\frac{\mu}{2} + \frac{1}{l+1}$. It depends of two free parameters, μ and μ_1 and vanishes at $r = R$.

As before, the energy E_{VSQM} is obtained by minimisation of the energy expectation value with respect to the two parameters, μ and μ_1. The results are given in the table below, which also presents the exact values, E_{EXACT}, the results from $1/N$ calculations, $E_{1/N}$, [4] and those obtained from a new WKB calculation, E_{CENTRI}, [5].

Table 3. Energy eigenvalues (in Rydbergs) for different values of R, $n = 0$ and $l = 0, 1$, (states $1s$ and $2p$) and $\delta = 0.1$.

R	l	E_{EXACT} [3]	E_{VSQM}	$E_{1/N}$[4]	E_{CENTRI} [5]
7.0	0	-0.45111	-0.45043	-0.45181	
	1	-0.04069	-0.04037	-0.00324	-0.03976
8.0	0	-0.45122	-0.45076	-0.45193	
	1	-0.05783	-0.05762	-0.05293	-0.05510
9.0	0	-0.45125	-0.45090	-0.45188	
	1	-0.06728	-0.06712	-0.06389	-0.06612
10.0	0	-0.45125	-0.45098	-0.45179	
	1	-0.07257	-0.07243	-0.07008	-0.07257
25.0	0	-0.45125	-0.45125	-0.45131	
	1	-0.07918	-0.07915	-0.07920	-0.07921
50.0	0	-0.45125	-0.45125	-0.45126	
	1	-0.07918	-0.07918	-0.07920	-0.07920

5. Conclusions

The variational method associated with SQM was used to get the energy states of quantum confined systems. This was achieved through an *Ansatz* for the superpotential. Using the superalgebra the trial wave function was evaluated which depends on free parameters introduced through the *Ansatz*. The parameters were varied until the energy expectation value reached its minimun value. The results are better for increasing values of the radius R when compared to results obtained from other methods and exact numerical results. We stress that the great advantage of our method is the achievement of the trial wave function through the *Ansatz* for the superpotential. This allowed us a previous comparative analysis between the original potential and the effective potential which approaches the infinite at the neighbourhood of the barrier. Following this line of reasoning, the wave function, evaluated through the superalgebra vanishes at the border because it finds a potential barrier that increases until becoming impenetrable at $r = R$. These border effects become more perceptive for smaller values of R. For increasing values of R, smaller will be the border effects so that the variational results get better. In other words, for large values of R the variational SQM *Ansatz* provides fast converging results and the results for the non-confined system are recovered.

In conclusion, we remark that the association of the superalgebra of SQM with the variational method provides an appropriate approach to analyse confined systems.

EDF would like to acknowledge the financial support from FAPESP and CNPq.

6. References

[1] Drigo Filho E and Ricotta R M 1995 Mod. Phys. Lett. A10 1613-
[2] Drigo Filho E and Ricotta R M 2000 Phys. Lett.A269 269-276 ; 2000 Mod. Phys. Lett. A15 1253-1259 ; 2002 Phys. Lett. A299 137-143
[3] Aguilera-Navarro V C, Gomes J F, Zimerman A H and Ley Koo E 1983 J. Phys. A: Math. Gen. 16 2943-2952
[4] Sinha A, Roychoudhury R and Varshni Y P 2001 Can. J. Phys. 79 939-945 and ibid. 2000 Can. J. Phys. 78 141-152
[5] Sinha A, quant-ph/0205122

Inst. Phys. Conf. Ser. No 173: Section 7
Paper presented at 24th Int. Coll. Group Theoretical Methods in Physics, Paris, France, 15–20 July 2002

Quantum version of Hamilton-Jacobi theory for Time-dependent Systems

Seiji Sakoda[1], Minoru Omote[2], and Susumu Kamefuchi[3]

[1] Department of Applied Physics, National Defense Academy, Yokosuka, Japan
[2] Department of Physics, Keio University, Hiyoshi, Yokohama, Japan
[3] Atomic Energy Research Institute, Nihon University, Kanda-Surugadai, Tokyo, Japan

Abstract. By constructing a quantum conserved quantity even for a system with time-dependent Hamiltonian, we develop the quantum version of Hamilton-Jacobi theory, that is applicable for non-conserved systems as well as conserved ones. A novel viewpoint, Hamilton-Jacobi picture, of the quantum mechanics will be introduced.

1. Introduction

Quantum mechanics, as well as the classical one, of a non-conserved system described by a time-dependent Hamiltonian has been studied intensively in the literature from various interest. A typical example of several interesting features of such a system will be the Berry phase[1] in quantum mechanics and the corresponding Hannay angle[2] in classical mechanics. To find a Berry phase we usually rely on the adiabatic approximation in solving the time-dependent Schrödinger equation. If we wish to learn the quantum mechanics of such a system entirely, we cannot be satisfied with any approximation. A method of obtaining the exact solution of the Schrödinger equation for time-dependent systems was developed by M. Omote and S. Kamefuchi[3]. In their method it is quite significant to find a difference of two kinds of phases, the geometrical phase and the dynamical one, to yield a formal solution of the Schrödinger equation in terms of eigenvectors of some conserved quantity. As is well known[4], the phase of a state vector in quantum mechanics has its classical counterpart in the differential equation of Hamilton-Jacobi(HJ). Therefore it will be interesting to view the quantum mechanics of non-conserved systems from the quantum version of the Hamilton-Jacobi theory. We develop our method by referring the *generalized harmonic oscillator* which is described by a Hamiltonian $H(q, p; t) = X(t)p^2 + 2Y(t)pq + Z(t)q^2$.

2. The partial differential equation of Hamilton-Jacobi for time-dependent systems

While for a conserved system, there is a systematic way[4, 5] of finding the Hamilton's principal function S by separation of variables to reduce the problem to find the corresponding characteristic function W, there seems no such a systematic method of solving the differential equation to find S for non-conserved systems. Therefore we begin with

$$H\left(q, \frac{\partial S}{\partial q}; t\right) + \frac{\partial S}{\partial t} = 0, \quad S = S(q, P; t), \tag{1}$$

where P is the new canonical momentum after the canonical transformation. To solve this equation we first introduce a map between $(q(t),p(t))$ and $(Q,P) = (q(0),p(0))$ ($i.e.$ initial values of canonical variables) by

$$\begin{pmatrix} p(t) \\ q(t) \end{pmatrix} = \begin{pmatrix} A(t) & B(t) \\ C(t) & D(t) \end{pmatrix} \begin{pmatrix} P \\ Q \end{pmatrix}, \quad A(t)D(t) - B(t)C(t) = 1. \tag{2}$$

We may regard (2) as a canonical transformation : $(p,q) \longleftrightarrow (P,Q)$. Then the generating function S_1 for this canonical transformation should satisfy

$$p = \frac{1}{D}(P+Bq) = \frac{\partial S_1}{\partial q}, \quad Q = \frac{1}{D}(q-CP) = \frac{\partial S_1}{\partial P} \tag{3}$$

and can be resolved by

$$S_1(q,P,t) = \frac{1}{2D(t)}\left(B(t)q^2 + 2Pq - C(t)P^2\right). \tag{4}$$

If we make a Legendre transform, we obtain another kind of the generating function

$$\tilde{S}_1(q,Q,t) = S_1(q,P,t) - PQ = \frac{1}{2C(t)}\left(A(t)q^2 - 2Qq + D(t)Q^2\right). \tag{5}$$

We can also consider a transformation to other variables which may not be expressed as a linear transformation. For example, there is a conserved quantity expressed as a bilinear form of the original variables :

$$N \equiv \alpha(t)p^2 + 2\beta(t)pq + \gamma(t)q^2, \quad \frac{dN}{dt} = \{N,H\} + \frac{\partial N}{\partial t} = 0. \tag{6}$$

Conditions for N to be conserved are given by

$$\begin{cases} \dot{\alpha}(t) &=& 4Y(t)\alpha(t) - 4X(t)\beta(t) \\ \dot{\beta}(t) &=& 2Z(t)\alpha(t) - 2X(t)\gamma(t) \\ \dot{\gamma}(t) &=& 4Z(t)\beta(t) - 4Y(t)\gamma(t) \end{cases} \quad (\alpha(t)\gamma(t) - \beta^2(t) = \kappa = \text{const.}), \tag{7}$$

where κ is some positive constant. After eliminating $\beta(t)$ and $\gamma(t)$, we obtain a nonlinear differential equation for $\alpha(t)$[6] :

$$\ddot{\alpha} - \frac{\dot{\alpha}^2}{2\alpha} - \frac{\dot{X}}{X}\dot{\alpha} + 4\left\{2(XZ-Y^2) + \frac{\dot{X}Y - X\dot{Y}}{X}\right\}\alpha - \frac{8\kappa X^2}{\alpha} = 0. \tag{8}$$

Though complicated, this equation and above linear ones can be solved in terms of the matrix element of the linear transformation (2) as

$$\begin{cases} \alpha(t) &=& D^2(t)\alpha - 2D(t)C(t)\beta + C^2(t)\gamma \\ \beta(t) &=& -B(t)D(t)\alpha + (A(t)D(t) + B(t)C(t))\beta - A(t)C(t)\gamma, \\ \gamma(t) &=& B^2(t)\alpha - 2A(t)B(t)\beta + A^2(t)\gamma \end{cases} \tag{9}$$

where α, β, and γ being the initial value of $\alpha(t)$, $\beta(t)$, and $\gamma(t)$, respectively. A candidate of the canonical conjugate to N will be

$$\theta^*(t) = \frac{1}{2\sqrt{\kappa}}\sin^{-1}\left(\sqrt{\frac{\kappa}{N\alpha(t)}}q\right) = \frac{1}{4i\sqrt{\kappa}}\log\frac{\alpha(t)p + (\beta(t) + i\sqrt{\kappa})q}{\alpha(t)p + (\beta(t) - i\sqrt{\kappa})q}, \tag{10}$$

but it is not a constant of motion(Note that $\theta^*(t)$ reduces to a linear function of t when the parameters in the Hamiltonian become constant. Therefore it should be regarded as a generalization of the angle variable to non-conserved systems.). To obtain a conserved canonical conjugate variable θ, we need an additional function of t. Then we define

$$\theta = \theta^*(t) - \int_0^t d\tau \frac{X(\tau)}{\alpha(\tau)} \quad \text{to find} \quad \{\theta, N\} = 1, \quad \frac{dN}{dt} = \frac{d\theta}{dt} = 0. \tag{11}$$

If we use a relation

$$\frac{X(t)}{\alpha(t)} = \frac{1}{2\sqrt{\kappa}} \frac{d}{dt} \tan^{-1} \left(\frac{\sqrt{\kappa}C(t)}{D(t)\alpha - C(t)\beta} \right), \tag{12}$$

we can rewrite θ as

$$\theta = \frac{1}{2\sqrt{\kappa}} \sin^{-1} \left(\sqrt{\frac{\kappa}{N\alpha}} Q(q, N, t) \right) \tag{13}$$

in which $Q(q, N, t) = q(0)$ must be expressed in terms of q, N, and t. Then the difference between θ and $\theta^*(t)$ is given by $\theta - \theta^*(t) = -(1/2\sqrt{\kappa}) \tan^{-1}(\sqrt{\kappa}C(t)/(D(t)\alpha - C(t)\beta))$. Although the difference is just a function of t and has no effect in classical mechanics, it will play a role in quantum mechanics through the definition of the phase of a state vector.

3. Quantum version of the Hamilton-Jacobi theory

If we work within the Heisenberg picture, it is easy to formulate a quantum theory for non-conserved systems from the classical one developed in the previous section. However we will begin with the Schrödinger picture to introduce a novel viewpoint of quantum mechanics. Assuming that the Schrödinger and Heisenberg pictures coincide to each other at $t = 0$, we can find the time dependence of a Heisenberg operator $\mathcal{O}_H(t)$ as

$$\mathcal{O}_H(t) = U^\dagger(t, 0) \mathcal{O}_S(t) U(t, 0), \quad U(t, 0) \equiv \text{T} \exp \left(-i \int_0^t H_S(\tau) d\tau \right). \tag{14}$$

Equation of motion for a conserved quantity is given by

$$i \frac{d}{dt} \mathcal{Q}_H(t) = [\mathcal{Q}_H(t), H_H(t)] + i \frac{\partial}{\partial t} \mathcal{Q}_H(t) = 0 \tag{15}$$

which can also be expressed in terms of the Schrödinger operator

$$i \frac{\mathcal{D}}{\mathcal{D}t} \mathcal{Q}_S(t) \equiv [\mathcal{Q}_S(t), H_S(t)] + i \frac{\partial}{\partial t} \mathcal{Q}_S(t) = 0. \tag{16}$$

Formal solutions of Schrödinger equation[3] can be expressed in terms of eigenvectors of a self-adjoint conserved quantity even for non-conserved systems. Let us suppose $\mathcal{D}\Lambda_S(t)/\mathcal{D}t = 0$ and $\Lambda_S(t)$ to be self-adjoint, whose eigenvalues and eigenvectors being given by

$$\Lambda_S(t)|n;t\rangle = \lambda_n|n;t\rangle, \quad \sum_n |n;t\rangle\langle n;t| = 1, \quad \langle m;t|n;t\rangle = \delta_{mn}. \tag{17}$$

Then a solution of the Schrödinger equation can be expressed as

$$|\psi(t)\rangle = U(t, 0)|\psi(0)\rangle = \sum_n \exp \left(i \int_0^t \theta_n(\tau) d\tau \right) |n;t\rangle\langle n;0|\psi(0)\rangle \tag{18}$$

where the phase $\theta_n(t)$ being given by the difference between the geometrical and the dynamical phases[1, 7]

$$\theta_n(t) \equiv \langle n;t| \left(i \frac{\partial}{\partial t} - H_S(t) \right) |n;t\rangle. \tag{19}$$

If we introduce a viewpoint of quantum mechanics that diagonalizes a Schrödinger operator corresponding to a conserved quantity introduced in the classical theory, we can easily solve the Schrödinger equation. For example, we may consider an eigenvector of $Q_S(t) = Q(q_S, p_S, t)$, corresponding to the initial value of $q(t)$ in classical mechanics, to obtain

$$Q_S(t)|Q;t\rangle = Q|Q;t\rangle, \quad |Q;t\rangle = U(t,0)|Q\rangle \tag{20}$$

excepting an arbitrary phase factor of the eigenvector. Then it resolves the Schrödinger equation automatically. Indeed, we see that

$$\langle q|Q;t\rangle = \frac{1}{\sqrt{2\pi}} \exp\left(i\tilde{S}_1(q,Q,t) - \frac{1}{2}\log iC(t)\right) \tag{21}$$

yields a Feynman kernel

$$\langle q_F|U(t,0)|q_I\rangle = \int dQ \langle q_F|U(t,0)|Q\rangle \langle Q|q_I\rangle = \int dQ \langle q_F|Q;t\rangle \delta(Q - q_I)$$

$$= \frac{1}{\sqrt{2\pi i C(t)}} \exp\left(\frac{i}{2C(t)}\left(A(t)q_F^2 - 2q_F q_i + D(t)q_I^2\right)\right). \tag{22}$$

We may call this novel viewpoint of quantum mechanics as Hamilton-Jacobi picture[8]. It is also possible to perform the same procedure for the conserved quadratic quantity N(Though its conjugate has no operator counterpart in quantum mechanics). We may regard it as a description of quantum mechanics in terms of action-angle variables. As was already mentioned previously, the definition of the phase of a state vector must be treated carefully otherwise we may fail to obtain the exact solution of the Schrödinger equation. For example, if we define

$$a(t) \equiv \frac{1}{\sqrt{2\sqrt{\kappa}\alpha(t)}} \left(\left(\sqrt{\kappa} + i\beta(t)\right)q + i\alpha(t)p\right) \tag{23}$$

as the annihilation operator, the phase in (18) for the n-th excited state should be

$$\int_0^t \theta_n(\tau)d\tau = -(n+1/2)\int_0^t \frac{X(\tau)}{\alpha(\tau)}d\tau = -(n+1/2)(\theta^*(t) - \theta) \tag{24}$$

while no phase is required if we define $a(t)e^{i\theta^*(t)}$ to be the annihilation operator. Note that if we write wave functions in Hamilton-Jacobi picture in the form $\psi = e^{iS}$, the phase S satisfies the quantum version of the partial differential equation of Hamilton-Jacobi

$$H\left(q, \frac{\partial S}{\partial q}, t\right) + \frac{\partial S}{\partial t} = iX\frac{\partial^2 S}{\partial q^2} + iY. \tag{25}$$

4. Summary

For quantum mechanics of time-dependent systems, we have considered the quantum version of the partial differential equation of Hamilton-Jacobi, as an equivalent method, instead of solving the Schrödinger equation. The technique developed here is in some sense specific to the quadratic system. However, the novel viewpoint of quantum mechanics, the Hamilton-Jacobi picture, will be useful for analyses of other time-dependent systems.

[1] M. V. Berry, Proc. R. Soc. A 392(1984) 45,
[2] J. H. Hannay, J. Phys. A: Math. 18(1985)221,
[3] M. Omote and S. Kamefuchi, Phys. Lett. A206(1995)273,
[4] H. Goldstein, Classical Mechanics (Addison-Wesley Publishing Company),
[5] V. I. Arnold, Mathematical Methods of Classical Mechanics (New York:Springer),
[6] H. Lewis Jr. and W. B. Riesenfield, J.Math. phys. 10(1964)1458,
[7] Y. Aharonov and J. Anderson, Phys. Rev. Lett. 58(1987)1593,
[8] M. Omote, S. Sakoda, and S. Kamefuchi, *Classical and Quantum Behavior of Generalized Oscillators - action variable, angle variable and quantum phase* - (hep-th/0106115)

Inst. Phys. Conf. Ser. No 173: Section 7
Paper presented at 24th Int. Coll. Group Theoretical Methods in Physics, Paris, France, 15–20 July 2002
©2003 IOP Publishing Ltd

Heun equation and Inozemtsev models

Kouichi Takemura

Department of Mathematical Sciences, Yokohama City University, 22-2 Seto, Kanazawa-ku,
Yokohama 236-0027, JAPAN

Abstract. The BC_N elliptic Inozemtsev model is a quantum integrable systems with N-particles whose potential is given by elliptic functions. Eigenstates and eigenvalues of this model are investigated.

1. Introduction

The BC_N Inozemtsev model [2] is a system of quantum mechanics with N-particles whose Hamiltonian is given by

$$H = -\sum_{j=1}^{N} \frac{\partial^2}{\partial x_j^2} + 2l(l+1) \sum_{1 \leq j < k \leq N} (\wp(x_j - x_k) + \wp(x_j + x_k)) \tag{1}$$

$$+ \sum_{j=1}^{N} \sum_{i=0}^{3} l_i(l_i + 1)\wp(x_j + \omega_i),$$

where $\wp(x)$ is the Weierstrass $\wp$-function with periods $(1, \tau)$, $\omega_0 = 0$, $\omega_1 = \frac{1}{2}$, $\omega_2 = -\frac{\tau+1}{2}$, $\omega_3 = \frac{\tau}{2}$ are half periods, and l and l_i ($i = 0, 1, 2, 3$) are coupling constants.

It is known that the BC_N Inozemtsev model is quantum completely integrable. More precisely, there exist operators of the form $H_k = \sum_{j=1}^{N} \left(\frac{\partial}{\partial x_j}\right)^{2k} + $ (lower terms) ($k = 2, \ldots, N$) such that $[H, H_k] = 0$ and $[H_{k_1}, H_{k_2}] = 0$ ($k, k_1, k_2 = 2, \ldots, N$). Note that the BC_N Inozemtsev model is a universal completely integrable model of quantum mechanics with B_N symmetry, which follows from the classification due to Ochiai, Oshima and Sekiguchi [5]. For the case $N = 1$, finding eigenstates of the Hamiltonian is equivalent to solving the Heun equation [5, 7]. In this sense, the BC_N Inozemtsev model is a generalization of the Heun equation.

In this report, we are going to investigate eigenvalues and eigenstates of the BC_N Inozemtsev model.

A possible approach is to use the quasi-exact solvability. If the coupling constants l, l_0, l_1, l_2, l_3 satisfy some equation, the Hamiltonian H (see (1)) and the commuting operators of conserved quantities preserve some finite dimensional space of doubly periodic functions [1, 10]. On the finite dimensional space, eigenvalues are calculated by solving the characteristic equation, and eigenfunctions are obtained by solving linear equations. In this sense, a part of eigenvalues and eigenfunctions is obtained exactly and this is the reason to use the phrase "quasi-exact solvability".

Another approach is to use a method of perturbation. By the trigonometric limit $p = \exp(\pi\sqrt{-1}\tau) \rightarrow 0$, the Hamiltonian H of the BC_N elliptic Inozemtsev model tends to the Hamiltonian of the BC_N Calogero-Moser-Sutherland model H_T (see (2)), and eigenvalues and eigenstates of the BC_N Calogero-Moser-Sutherland model are known (see Proposition 2.1).

Based on the eigenstates for the case $p = 0$, we can obtain eigenvalues and eigenstates of the BC_N elliptic Inozemtsev model $(p \neq 0)$ as formal power series in p. This procedure is sometimes called the algorithm of perturbation. Generally speaking, convergence of the formal power series obtained by perturbation is not guaranteed a priori, but for the case of the BC_N elliptic Inozemtsev model, the convergence radius of the formal power series in p is shown to be non-zero (see Corollary 3.4), and it is seen that this perturbation is holomorphic. As a result, real-holomorphy of the eigenvalues in p and the completeness of the eigenfunctions are shown.

In section 2, some results on the BC_N trigonometric Calogero-Moser-Sutherland model are reviewed. In section 3, we present some propositions about the perturbation of the BC_N elliptic Inozemtsev model from the trigonometric model. In section 4, we give some comments for future problems.

2. Trigonometric limit

In this section, we will consider the trigonometric limit $(\tau \to \sqrt{-1}\infty)$. and review some results on the trigonometric model. Assume $l, l_0, l_1 \geq 0$ and set $p = \exp(\pi\sqrt{-1}\tau)$. Then $p \to 0$ as $\tau \to \sqrt{-1}\infty$.

If $p \to 0$, then $H \to H_T + C_T$, where

$$H_T = -\sum_{j=1}^{N} \frac{\partial^2}{\partial x_j^2} + \sum_{1 \leq j < k \leq N} \left(\frac{2\pi^2 l(l+1)}{\sin^2 \pi(x_j - x_k)} + \frac{2\pi^2 l(l+1)}{\sin^2 \pi(x_j + x_k)} \right) \tag{2}$$
$$+ \sum_{j=1}^{N} \left(\frac{\pi^2 l_0(l_0+1)}{\sin^2 \pi x_j} + \frac{\pi^2 l_1(l_1+1)}{\cos^2 \pi x_j} \right),$$

and $C_T = -\frac{N(N-1)\pi^2}{3} l(l+1) - \frac{N\pi^2}{3} \sum_{i=0}^{3} l_i(l_i+1)$. The operator H_T is nothing but the Hamiltonian of the BC_N trigonometric Calogero-Moser-Sutherland model.

Now we solve the spectral problem for H_T by using hypergeometric functions. Set

$$\Phi_T(x) = \prod_{j=1}^{N} (\sin \pi x_j)^{l_0+1} (\cos \pi x_j)^{l_1+1} \tag{3}$$
$$\prod_{1 \leq j_1 < j_2 \leq N} (\sin \pi(x_{j_1} - x_{j_2}) \sin \pi(x_{j_1} + x_{j_2}))^{l+1}.$$

Let $D = \{(x_1, \ldots, x_N) \in \mathbf{R}^N | 0 < x_1 \leq \ldots \leq x_N < 1\}$. A function $f(x_1, \ldots, x_N)$ is $W(B_N)$-invariant iff $f(x_1, \ldots, x_i, \ldots, x_j, \ldots x_N) = f(x_1, \ldots, x_j, \ldots, x_i, \ldots x_N)$ for all $i < j$ and $f(x_1, \ldots, x_i, \ldots x_N) = f(x_1, \ldots, -x_i, \ldots x_N)$ for all i. The Hilbert space $\mathbf{H}$ is defined by

$$\mathbf{H} = \left\{ f : \mathbf{R}^N \to \mathbf{C} \text{ measurable} \,\middle|\, \begin{array}{l} \int_D |f(x)|^2 dx < +\infty, \\ \frac{f(x)}{\Phi_T(x)} \text{ is } W(B_N)\text{-invariant a.e. } x, \\ \frac{f(x+n)}{\Phi_T(x+n)} = \frac{f(x)}{\Phi_T(x)}, \forall n \in \mathbf{Z}^N \text{ a.e. } x \end{array} \right\}. \tag{4}$$

Proposition 2.1 *Let* $\mathcal{M}_N = \{(\lambda_1, \ldots, \lambda_N) \in \mathbf{Z}^N | \lambda_1 \geq \ldots \geq \lambda_N \geq 0\}$. *There exists a complete orthonormal system* $\{v_\lambda\}_{\lambda=(\lambda_1,\ldots,\lambda_N)\in\mathcal{M}_N}$ *on the Hilbert space* $\mathbf{H}$ *such that*

$$H_T v_\lambda = E_\lambda v_\lambda,$$

where $E_\lambda = \pi^2 \sum_{j=1}^{N} (2\lambda_j + l_0 + l_1 + 2 + 2(l+1)(N-j))^2$.

Note that the function v_λ is expressed as the product of the ground state $\Phi_T(x)$ and the BC_N Jacobi polynomial $\psi_\lambda(z_1,\ldots,z_N)$ $(z_j = \exp(2\pi\sqrt{-1}x_j),\ j = 1,\ldots,N)$. Essential selfadjointness of the operator H_T on the Hilbert space $\mathbf{H}$ is obtained by applying Proposition 2.1.

3. Perturbation from the trigonometric model

We apply a method of perturbation and have an algorithm for obtaining eigenvalues and eigenfunctions as formal power series of p. The proofs of propositions in this report are obtained by imitating the proofs of the corresponding propositions in [4, 8]

Set $p = \exp(\pi\sqrt{-1}\tau)$. For the Hamiltonian of the BC_N Inozemtsev model, we adopt a notation $H(p)$ instead of H. The operator $H(p)$ admits the following expansion:

$$H(p)(= H) = H_T + C_T + \sum_{k=1}^{\infty} V_k(x)p^k, \tag{5}$$

where H_T is the Hamiltonian of trigonometric model defined in (2), $V_k(x)$ is some function in variables $x_1,\ldots,x_N$, and C_T is a constant defined in section 2.

Based on the eigenvalues E_λ $(\lambda \in \mathcal{M}_N)$ and the eigenfunctions v_λ of the operator H_T, we determine eigenvalues $E_\lambda(p) = E_\lambda + C_T + \sum_{k=1}^{\infty} E_\lambda^{\{k\}}p^k$ and normalized eigenfunctions $v_\lambda(p) = v_\lambda + \sum_{k=1}^{\infty}\sum_{\mu\in\mathcal{M}_N} c_{\lambda,\mu}^{\{k\}}v_\mu p^k$ of the operator $H(p)$ as formal power series in p. In other words, we will find $E_\lambda(p)$ and $v_\lambda(p)$ that satisfy equations

$$H(p)v_\lambda(p) = (H_T + C_T + \sum_{k=1}^{\infty} V_k(x)p^k)v_\lambda(p) = E_\lambda(p)v_\lambda(p), \tag{6}$$

$$\langle v_\lambda(p), v_\lambda(p)\rangle = 1,$$

as formal power series of p.

First we calculate coefficients $\sum_{\mu\in\mathcal{M}_N} d_{\lambda,\mu}^{\{k\}}v_\mu = V_k(x)v_\lambda$ $(k \in \mathbf{Z}_{>0},\ \lambda \in \mathcal{M}_N)$. Next we compute $E_\lambda^{\{k\}}$ and $c_{\lambda,\mu}^{\{k\}}$ for $k \geq 1$ and $\lambda, \mu \in \mathcal{M}_N$. By comparing coefficients of $v_\mu p^k$ in (6), we obtain recursive relations for $E_\lambda^{\{k\}}$ and $c_{\lambda,\mu}^{\{k\}}$.

Now we present results which are obtained by applying the Kato-Rellich theory. We use definitions written in Kato's book [3] freely. It is shown that the operator $H(p)$ $(-1 < p < 1)$ is essentially selfadjoint on the Hilbert space $\mathbf{H}$, because the operator $\sum_{k=1}^{\infty} V_k(x)p^k$ is bounded. Let $\tilde{H}(p)$ $(-1 < p < 1)$ be the unique extension of $H(p)$ to the selfadjoint operator.

Proposition 3.1 *The operators $\tilde{H}(p)$ form a holomorphic family of type (A) for $-1 < p < 1$.*

Proposition 3.2 *The spectrum $\sigma(\tilde{H}(p))$ contains only point spectra and it is discrete. The multiplicity of each eigenvalue is finite.*

Combining Theorem 3.9 in [3, VII-§3.5], propositions in this report and the selfadjointness of $\tilde{H}(p)$, the following theorem is proved:

Theorem 3.3 *All eigenvalues of $\tilde{H}(p)$ $(-1 < p < 1)$ can be represented as $E_\lambda(p)$ $(\lambda \in \mathcal{M}_N)$, which is real-holomorphic in $p \in (-1, 1)$ and $E_\lambda(0) = E_\lambda + C_T$. The eigenfunction $v_\lambda(p)$ of the eigenvalue $E_\lambda(p)$ is holomorphic in $p \in (-1, 1)$ as an element in L^2-space, and the eigenvectors $v_\lambda(p)$ $(\lambda \in \mathcal{M}_N)$ form a complete orthonormal family on $\mathbf{H}$.*

As an application of the theorem, the convergence of formal power series of eigenvalues in the variable p obtained by the algorithm of perturbation is shown.

Corollary 3.4 *Let* $E_\lambda(p)$ $(\lambda \in \mathcal{M}_N)$ *(resp.* $v_\lambda(p)$*) be the formal eigenvalue (resp. eigenfunction) of the Hamiltonian* $H(p)$ *defined by (6). If* $|p|$ *is sufficiently small then the power series* $E_\lambda(p)$ *converges and as an element in* L^2 *space the power series* $v_\lambda(p)$ *converges.*

4. Comments

In this report, holomorphy of perturbation for the Hamiltonian of the BC_N Inozemtsev model from the trigonometric one is established. Relationship between the perturbation and the complete integrability should be clarified. More precisely, holomorphy of perturbation for commuting operators of conserved quantities should be shown, although it is not succeeded as of this writing. For the elliptic Calogero-Moser-Sutherland model of type A_N, holomorphy of perturbation for commuting operators is shown in [4, 6].

In [7, 9], Bethe Ansatz method for the BC_1 Inozemtsev model is proposed and some results on the finite gap integration for the BC_1 Inozemtsev model are established. We hope that some progress on the Bethe Ansatz method or the finite gap integration for the BC_N Inozemtsev model will be made.

References

[1] Finkel F, Gomez-Ullate D, Gonzalez-Lopez A, Rodriguez M A, and Zhdanov R 2001 New spin Calogero-Sutherland models related to B_N-type Dunkl operators *Nuclear Phys. B* **613** 472–496

[2] Inozemtsev V I 1989 Lax representation with spectral parameter on a torus for integrable particle systems *Lett. Math. Phys.* **17** 11–17

[3] Kato T 1980 Perturbation theory for linear operators, corrected printing of the second ed. (Berlin: Springer-Verlag).

[4] Komori Y and Takemura K 2002 The perturbation of the quantum Calogero-Moser-Sutherland system and related results *Commun. Math. Phys.* **227** 93–118

[5] Ochiai H, Oshima T, and Sekiguchi H 1994 Commuting families of symmetric differential operators *Proc. Japan. Acad.* **70** 62–66.

[6] Takemura K 2000 On the eigenstates of the elliptic Calogero-Moser model *Lett. Math. Phys.* **53** 181-194

[7] Takemura K 2001 The Heun equation and the Calogero-Moser-Sutherland system I: the Bethe Ansatz method *Preprint* math.CA/0103077

[8] Takemura K 2001 The Heun equation and the Calogero-Moser-Sutherland system II: the perturbation and the algebraic solution *Preprint* math.CA/0112179

[9] Takemura K 2002 The Heun equation and the Calogero-Moser-Sutherland system III: the finite gap property and the monodromy *Preprint* math.CA/0201208

[10] Takemura K 2002 Quasi-exact solvability of Inozemtsev models *Preprint* math.QA/0205274

Inst. Phys. Conf. Ser. No 173: Section 7
Paper presented at 24th Int. Coll. Group Theoretical Methods in Physics, Paris, France, 15–20 July 2002
©2003 IOP Publishing Ltd

Group-theoretical and algebraic aspects of Bose-Einstein condensation of interacting particles in traps

O K Vorov[1], M S Hussein[2], and P Van Isacker[1]
[1]Grand Accélérateur National d'Ions Lourds, BP 5027, F-14076, Caen, Cedex 5, France
[2]Instituto de Fisica, Universidade de Sao Paulo CP 66318, 05315-970, SP, Brasil

Abstract. Trapped bosonic atoms with manipulated interaction strength near the sign flip of the scattering length are considered. Analytic solutions for the rotating ground states are obtained within algebraic approach, using permutation symmetry and group-theoretical properties of Bose operators. It is shown that the form of the ground states at any angular momentum is universal, and it does not depend on the details of the interaction. The ground states are either "collective rotations" or "condensed vortex states", depending on the sign of the "modified Born scattering length".

Newly developed techniques [1] for manipulating the strength of the effective interaction between trapped atoms [2] open the possibility to experimentally realize a Bose gas with weak interatomic interactions that are either attractive or repulsive. The limit $\langle V \rangle / \hbar \omega \to \pm 0$, where $\langle V \rangle$ is a typical expectation value of the interaction and $\hbar \omega$ is the quantum energy of the confining potential, is now reached experimentally via the Feshbach resonance [1]. Of special interest are the *ground states* of a rotating system[3, 4, 5, 6] at given angular momenta L, the so-called *yrast states* [5, 6, 7, 8, 9, 10]. So far, the problem was usually studied in dimension $k=2$ for the contact interaction $V \sim a^{sc} \delta(\vec{r})$, either attractive ($a^{sc} < 0$) or repulsive ($a^{sc} > 0$) where a^{sc} is the scattering length. The case $a^{sc} < 0$ was solved analytically in [5], while the case $a^{sc} > 0$ was studied numerically in [8]. The conjecture for the ground state wave function[8] was confirmed analytically in [9], and generalized to a universality class of predominantly repulsive interactions[9]. A universality class of predominantly attractive interactions has been constructed in[10].

Here, we discuss the group-theoretical aspects of the solution of the problem for the case of arbitrary interactions[11]. In the functional space $\{V\}$ of all possible interactions $V(r)$, we may restrict our attention to those of physical interest, $\{V_{phys}\}$, of which we require that the force $-\frac{dV}{dr}$ changes sign only once from repulsive at short to attractive at long distance:

$$\frac{dV}{dr} < 0, \quad r < R; \qquad \frac{dV}{dr} > 0, \quad r \geq R; \qquad R < 1. \tag{1}$$

Since the crossover at R occurs for atomic reasons, we may assume it to be smaller than the trapping size, $R < \omega^{-1/2}$ (or $R < 1$ in units $\omega = 1$). In addition, we assume that the force does not grow at $r \to \infty$ and the quantum eigenvalue problem is free of divergencies. In practice, this implies that $V(r)$ does not grow as (or faster than) r^{-k} at $r \to 0$. It will be shown that the entire functional space $\{V_{phys}\}$ is divided into two distinct classes $\{V_{phys}^-\}$ and $\{V_{phys}^+\}$ of (effectively) attractive or repulsive interactions. Within each class the energies of the yrast states depend in a simple way on the interaction while their wave functions remain the same. The ground states in the two classes differ qualitatively. When the interaction is changed from effectively attractive to repulsive, the transition between the two classes can be visualized as a quantum phase transition, with the relative interparticle angular momentum as a discrete order parameter.

The Hamiltonian of N spinless bosons of mass m in a k-dimensional symmetric harmonic trap $(k=2,3,...)$ reads

$$H = \sum_i^N \left(\frac{\vec{p}_i^2}{2m} + \frac{m\omega^2 \vec{r}_i^2}{2} \right) + \sum_{i>j}^N V(r_{ij}) \equiv H_0 + V, \tag{2}$$

where H_0 describes noninteracting particles, $\vec{r}_i(\vec{p}_i)$ is the k-dimensional position (momentum) vector of i-th boson, and V is the two-body interaction with $r_{ij}\equiv|\vec{r}_i-\vec{r}_j|$. The Hamiltonian commutes with $k(k-1)/2$ generators of $O(k)$ algebra, $L^{\alpha\beta}=(r^\alpha p^\beta - r^\beta p^\alpha)$, where $\alpha,\beta=1,..,k$ denote the spatial components. The Casimir operator $\vec{L}^2$ of the $SO(k)$ algebra and $[k/2]$ generators of its Cartan subalgebra

$$\mathcal{C} \equiv \left\{ H, \quad \{L^{12}, L^{34}, ...\}, \quad \vec{L}^2 \equiv \sum_{\alpha>\beta}^k L^{\alpha\beta}L^{\alpha\beta} \right\}, \qquad [\mathcal{C}_a, \mathcal{C}_b] = 0, \tag{3}$$

are in involution, forming a mutually commuting set. Hereafter, $|0_L\rangle$ denotes the ground state at fixed L and with maximum value of the conserved component, L^{xy}. That is, $L^{xy}|0_L\rangle \equiv L^{12}|0_L\rangle = L|0_L\rangle$ and $\vec{L}^2|0_L\rangle = L(L+k-2)|0_L\rangle$ with $\vec{L}^2 \equiv \sum_{\alpha>\beta}^k (r^\alpha p^\beta - r^\beta p^\alpha)^2$ where $\alpha,\beta=1,..,k$ denote the spatial components. Other rotationally degenerate wave functions can be obtained from $|0_L\rangle$ by applying the standard angular momentum ladder operators (e.g., $L_- = L_x - iL_y$, in $k=3$). In the limit $\hbar\omega \gg V$, the determination of $|0_L\rangle$ requires the diagonalization of the interaction within the Hilbert space of the symmetrized states $S|\mu\rangle$

$$S|\mu\rangle, \quad |\mu\rangle \equiv z_1^{l_1} z_2^{l_2} ... z_N^{l_N} |0\rangle, \quad \sum_{i=0}^N l_i = L, \tag{4}$$

where $z_i = x_i + iy_i$, S is the symmetrization operator, and $|0\rangle \equiv e^{-\frac{1}{2}\sum \vec{r}_k^2}$, in the convention $\hbar = m = \omega = 1$. The states involving $z_i^* = x_i - iy_i$ are separated from (4) by an energy $\delta E = n\hbar\omega$ and their admixtures can be neglected for $V \ll \hbar\omega$. The dimensionality of the basis (4) grows exponentially with L[7]. Within the subspace (4) the Hamiltonian (2) is

$$H = L + (Nk)/2 + W, \quad W = S \sum_{i>j} w(\hat{l}_{ij})S, \tag{5}$$

where W is the interaction V, projected [9] onto the subspace (4) with

$$w(l) \equiv \int_0^\infty V(\sqrt{2t}) e^{-t} \frac{t^{l+k/2-1}}{\Gamma(l+k/2)} dt, \quad \hat{l}_{ij} = \frac{1}{2}(a_i^+ - a_j^+)(a_i - a_j). \tag{6}$$

$\hat{l}_{ij}$ the relative angular momentum between atoms i and j written in terms of the ladder operators $a_i^+ = \frac{z_i}{2} - \frac{\partial}{\partial z_i^*}$ and $a_i = \frac{z_i^*}{2} + \frac{\partial}{\partial z_i}$. The total *internal angular momentum* $J = \sum_{i>j} \hat{l}_{ij}$ is an exactly conserved quantity with eigenvalues $J = \frac{Nj}{2}$, $j=0,2,3,..,L$[9], commuting with the total angular momentum $L \equiv \sum_i a_i^+ a_i$.

We split [9] the projected interaction W into $W = W_0 + W_S$ such that the first term is simple enough to find its lowest eigenvalue E_0 and its associated eigenstate $|0\rangle$,

$$W_0|0\rangle = E_0|0\rangle, \quad W_S \equiv S \sum_{i>j} v_S(ij)S \geq 0, \quad W_S|0\rangle = 0. \tag{7}$$

The state $|0\rangle$ will also be the ground state of the total interaction $W = W_0 + W_S$ if (i) W_S is *non-negative definite*, and (ii) $|0\rangle$ is annihilated by W_S[9]. In general, the operator W_0 can be written as a power series $W_0 = S \sum_{i>j} v_0(\hat{l}_{ij})S$, $v_0(l) = \sum_{m=0} c_m l^m$, where c_m are hitherto unknown coefficients that need to be fixed such that conditions (7) are satisfied. The following solution for the c_m satisfies (7)

$$c_m = \theta(2 - m)[\theta(1 - m)w(0) - \Delta_2\theta(\Delta_2)/2], \quad \theta(x) = \begin{cases} 0, x \le 0, \\ 1, x > 0 \end{cases} \quad (8)$$

here $\Delta_{2n}\equiv w(0)-w(2n)$. The eigenvalues $\lambda_n^\pm$ (for $\Delta_2 \gtrless 0$) of the operator $v_S(ij)$ are

$$\lambda_n = n\theta(\Delta_2)\Delta_2 - \Delta_{2n} \ge 0. \quad (9)$$

As will be shown below, the inequalities are satisfied for all interactions V_{phys} defined in (1). The ground state and its energy as a function of L ($L\le N$) are

$$|0_L\rangle = S \prod_{j=1}^{L} \sum_{i=1}^{N} (a_i^+ - \theta(\Delta_2)a_j^+)|0\rangle, \quad (10)$$

$$E_0(L) = L + \frac{Nk}{2} + \frac{N(N-1)w_0}{2} - \frac{NL\Delta_2\theta(L-1)\theta(\Delta_2)}{4}.$$

The ground state depends on the interaction $V(r)$ only via the *sign* of the control parameter

$$\Delta_2 = \frac{2^{1-k/2}}{\Gamma(k/2)} \int_0^\infty V(r)r^{k-1}e^{-\frac{r^2}{2}}\left(1 - \frac{r^4}{k(k+2)}\right) dr. \quad (11)$$

The two options are distinguished by the value of the *order parameter* $J = \sum_{i>j} \hat{l}_{ij}$:

$$\Delta_2 \le 0: \quad |0_L^-\rangle = (\mathcal{A}^+)^L|0\rangle, \quad J|0_L^-\rangle = 0, \quad (12)$$

$$\Delta_2 \ge 0: \quad |0_L^+\rangle = \hat{S}\tilde{a}_1^+\tilde{a}_2^+...\tilde{a}_L^+|0\rangle, \quad J|0_L^+\rangle = \frac{NL}{2}|0_L^+\rangle. \quad (13)$$

where we used the collective boson operators $\tilde{a}_i^+ = a_i^+ - \mathcal{A}$, and $\mathcal{A} = \frac{1}{N}\sum_{j=1}^{N} a_j^+$. The wave functions $|0_L^-\rangle$ (collective rotations) and $|0_L^+\rangle$ (condensed vortex states) are nondegenerate ground states for $\Delta_2 \ne 0$. For the potentials defined by the equation

$$\Delta_2 = 0 \quad (14)$$

the states $|0_L^\mp\rangle$ become degenerate. We call the manifold of such potentials *separatrix*. The *separatrix* divides the interactions $V_{phys}(r)$ into the two classes $\{V_{phys}^-\}$ and $\{V_{phys}^+\}$, differing by the form of the ground state. For $k=2$ $|0_L^\mp\rangle$ accord with the results obtained for the attractive[5] and repulsive[8],[9] contact interactions, respectively.

Let $\{V\}$ be the complete manifold of all potentials [not necessarily the physical potentials (1)] as indicated by the big disc in Fig.1. Within this extended manifold, we can still define the subclasses that have $\lambda_n^-\ge 0$ and $\lambda_n^+\ge 0$ with the ground states (12) and (13), respectively. Their boundaries Λ^- and Λ^+ are in general distinct, leaving room marked by "?" when the ground state is not (10). Within physical class $\{V_{phys}\}$ (1) [small disc in Fig.1.], the two merge and coincide with the separatrix $\Delta_2=0$ (14). We show that $\lambda_n^-\ge 0$ for $\Delta_2\le 0$ and $\lambda_n^+\ge 0$ for $\Delta_2\ge 0$, for $V(r)\epsilon V_{phys}$. We write $\Delta_{2n}\equiv \int_0^\infty \phi_{2n}F dt$ where $F\equiv\frac{-dV(\sqrt{2t})}{dt}$. By (1) the quantity Δ_{2n} is the sum of the positive ($t<\tau\equiv\frac{R^2}{2}$) and negative ($t<\tau$)areas

$$\Delta_{2n} = \int_0^\tau h_n\phi_2 F dt + \int_\tau^\infty h_n\phi_2 F dt, \quad \phi_{2n} \equiv \sum_{m=1}^{2n} \frac{e^{-t}t^{k/2+m-1}}{\Gamma(k/2+m)}. \quad (15)$$

Here $h_n\equiv\frac{\phi_{2n}}{\phi_2}$ are positive-valued functions. If $\Delta_2\le 0$, the second (negative) term prevails for $n=1$. It will then prevail for any $n>1$, because the functions $h_{n>1}$ increase monotonically.

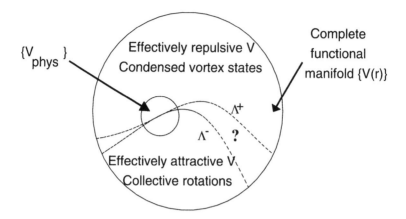

Figure 1. Global phase diagram in the functional space $\{V(r)\}$. Within $\{V_{phys}\}$, the boundaries Λ^- and Λ^+ merge with the *separatrix* $\Delta_2=0$ (14).

That is, if $\Delta_2\leq0$, then all $\Delta_{2n}\leq0$ and $\lambda_n^-\geq0$. If $\Delta_2\geq0$, the first (positive) term in (15) prevails for $n=1$. We introduce the monotonically decreasing functions $\tilde{h}\equiv n-h_n$. They are positive for $t<t_n$ and negative for $t>t_n$ with all $t_n>1$. By (1) we have $\tau<t_n$ and we write

$$\lambda_n^+ = \int_0^\tau \tilde{h}_n\phi_2 F dt + \int_\tau^{t_n} \tilde{h}_n\phi_2 F dt + \int_{t_n}^\infty \tilde{h}_n\phi_2 F dt.$$

The second term is the only negative contribution. The corresponding area is however smaller than that of the first term, because $\tilde{h}_n$ decreases monotonically and $\tau<t_n$. Thus $\lambda_n^+\geq0$ for $\Delta_2\geq0$. To sum up, the inequalities (9) hold throughout, and the solution (10) covers the universality class $\{V_{phys}\}$. By similar arguments, one can append the class $\{V_{phys}(r)\}$ (1) by potentials with constant sign of F and $\frac{d^2V(\sqrt{2}t)}{dt^2}$, like $\delta(\vec{r})$, $\frac{1}{r}$, $log(r)$, $e^{-r/a}$, $\frac{e^{-r/a}}{r}$, e^{-r^2/a^2} *etc.*

In summary, we have solved the problem of rotating ground states of weakly interacting trapped Bose atoms. The resulting phase diagram gives a complete classification of these ground states. The conclusion is that the ground state is either a collective rotation or a condensed vortex state. The work was supported by CEA (France) and FAPESP (Brazil).

References

[1] Roberts J L *et al* 2001, Phys. Rev. Lett. **86**, 4211.
[2] Anderson M H *et al* 1997, Science **269**, 198.
[3] Madison K W *et al* 2001, Phys. Rev. Lett. **86**, 4443.
[4] Butts D A and Rokhsar D S 1999, Nature **397**, 327.
[5] Wilkin N K, Gunn J M, and Smith R A 1998, Phys. Rev. Lett. **80**, 2265.
[6] Dalfovo F *et al* 1999, Rev. Mod. Phys. **71**, 463.
[7] Mottelson B 1999, Phys. Rev. Lett. **83**, 2695.
[8] Bertsch G F and Papenbrock T 1999, Phys. Rev. Lett. **83**, 5412.
[9] Hussein M S and Vorov O K 2002, Phys. Rev. A**65**, 035603; 2002 Physica B**312**, 550; 2002 Ann. Phys. (N.Y.) **298**, 248.
[10] Hussein M S and Vorov O K 2002, Phys. Rev. A **65**, 053608.
[11] Vorov O K, Hussein M S, Van Isacker P, subm. to Phys. Rev. Lett., cond-mat/0207339.

Inst. Phys. Conf. Ser. No 173: Section 7
Paper presented at 24th Int. Coll. Group Theoretical Methods in Physics, Paris, France, 15–20 July 2002
©2003 IOP Publishing Ltd

Quantum coding with systems with finite Hilbert space

A. Vourdas

Department of Computing, University of Bradford, Bradford BD7 1DP, United Kingdom

Abstract. Angle states and operators are defined in a $(2j + 1)$-dimensional angular momentum Hilbert space H through Fourier transform. Displacement operators in the corresponding quantum phase space which in this case is a toroidal lattice, are generators of $SU(2j + 1)$ transformations in H. In this context, a concatenated code is studied. In the first step the code is the space H_A spanned by the direct products of N angular momentum states with the same m. In the second step the code is the space H_B spanned by the direct products of M angle states of the space H_A, with the same m.

1. Introduction

Coding adds redundant information to a message so that in spite of partial corruption of the encoded message by noise, it is possible to recover the original message. A lot of the work on quantum coding has been based on qubits associated with quantum systems with a two-dimensional angular momentum ($j = 1/2$) Hilbert space. Most recently the use of d-dimensional Hilbert spaces (qudits) has been considered [1]. In ref.[2] we have generalized Shor's coding method [3] for the case of qudits. In this paper we review briefly and extend further this work.

Our code uses $(2j + 1)$-dimensional angular momentum Hilbert spaces and it is concatenated, in two steps. In the first step it introduces 'J-redundancy' and in the second step 'θ-redundancy'. The combined effect is general redundancy. The code protects quantum information against 'small' noise that can act only on a few qudits.

2. Finite systems

Finite quantum systems have been studied in various contexts by various authors [4]. In refs [5] we have studied the angle-angular momentum quantum phase space. In this section we introduce the notation and review briefly some of these ideas in the context of qudits.

We denote as $|J, jm>$ the usual angular momentum states. m belongs to $\mathcal{Z}(2j + 1)$ (integers modulo $2j + 1$). The states $|J, jm>$ span the $2j + 1$-dimensional Hilbert space H. The finite Fourier transform is defined as:

$$F = (2j+1)^{-1/2} \sum_{m,n} \omega(mn)|J; j\ m\rangle\langle J; j\ n|; \quad \omega(\alpha) = \exp\left[i\frac{2\pi\alpha}{2j + 1}\right]; \quad (1)$$

In ref.[5] we have introduced the θ-basis of angle states $|\theta; j\ m\rangle = F|J; j\ m\rangle$. We have also introduced the angle operators $\theta_z = FJ_zF^\dagger$, $\theta_+ = FJ_+F^\dagger$, $\theta_- = FJ_-F^\dagger$, which obey the $SU(2)$ algebra. The θ-operators act on the θ-states in an analogous way to the J-operators acting on the J-states.

The corresponding quantum phase space J_z-θ_z is a toroidal lattice $(\mathcal{Z}(2j+1)\times\mathcal{Z}(2j+1))$. Displacement operators in this phase space are defined as:

$$X = \exp\left[-i\frac{2\pi}{2j+1}\theta_z\right]; \qquad Z = \exp\left[i\frac{2\pi}{2j+1}J_z\right] \tag{2}$$

$$X^{2j+1} = Z^{2j+1} = 1; \qquad X^\beta Z^\alpha = Z^\alpha X^\beta \omega(-\alpha\beta) \tag{3}$$

where α,β are integers in $\mathcal{Z}(2j+1)$. They perform displacements along the J_z and θ_z axes, as follows:

$$X^\beta|J;j\,m\rangle = |J;j\,m+\beta\rangle; \qquad X^\beta|\theta;j\,m\rangle = \omega(-\beta m)|\theta;j\,m\rangle \tag{4}$$

$$Z^\alpha|J;j\,m\rangle = \omega(m\alpha)|J;j\,m\rangle; \qquad Z^\alpha|\theta;j\,m\rangle = |\theta;j\,m+\alpha\rangle \tag{5}$$

It has been explained in Ref.[6] that the displacement operators $Z^\alpha X^\beta$, are generators of $SU(2j+1)$ transformations in the Hilbert space H. Therefore infinitesimal $SU(2j+1)$ transformations can be written as $g = 1 + \sum_{\alpha,\beta}\lambda_{\alpha\beta}X^\alpha Z^\beta$ where $\lambda_{\alpha\beta}$ are infinitesimal coefficients.

3. J-redundancy

We consider states in the direct product $H^N \equiv H \otimes ... \otimes H$. In this $(2j+1)^N$-dimensional space we consider the $(2j+1)$-dimensional subspace spanned by the vectors

$$H_A = \{|J_A;j\,m\rangle \equiv |J;j\,m\rangle \otimes ... \otimes |J;j\,m\rangle, m = -j, ..., j\} \tag{6}$$

The Hilbert space H_A is isomorphic to the Hilbert space H through the mapping $|J_A;j\,m\rangle \leftrightarrow |J;j\,m\rangle$. Given some states and operators in H, we use the same notation for their counterparts in H_A with an additional index A. We call Π_A the projection operators in H_A and we use the notation $W_i \equiv 1 \otimes ... \otimes W \otimes ... \otimes 1$ $(i = 1, ..., N)$ for operators acting on H^N, with the operator W acting on the i Hilbert space H. We can show that $Z_A = Z_i\Pi_A$.

The Fourier operator F_A acting on the states of the Hilbert space H_A, is given by:

$$F_A = (2j+1)^{-1/2}\sum_{m,n}\omega(mn)|J;j\,m\rangle\langle J;j\,n| \otimes ... \otimes |J;j\,m\rangle\langle J;j\,n| \tag{7}$$

The states $|\theta;j\,m\rangle$ in H correspond to the states $|\theta_A;j\,m\rangle \equiv F_A|J_A;j\,m\rangle$ which form an orthonormal basis in H_A. It is easily seen that

$$\Pi_A|\theta;j\,m\rangle \otimes ... \otimes |\theta;j\,m\rangle = (2j+1)^{(1-N)/2}|\theta_A;j\,Nm\rangle \tag{8}$$

where Nm is defined modulo $(2j+1)$. We can also prove

$$[\Pi_A, X_1...X_N] = 0; \qquad X_A = X_1...X_N\Pi_A \tag{9}$$

The states $|J_A;j\,m\rangle$ are far from each other in the sense that the operator X_A that performs shifts among them, is equal to the product of all X_i (Eq. (9)). Therefore, it is unlikely that noise will perform such transformations causing errors. In contrast the operator $Z_A = Z_i\Pi_A$ that performs shifts between the various $|\theta_A, j\,m\rangle$ states requires the action of only one Z_i.

We next consider $SU(2j+1)$ transformations on the states in the Hilbert space H_A. Infinitesimal action of these transformations can be written as:

$$g_A = 1 + \sum_{\alpha,\beta}\lambda_{\alpha\beta}X_A^\alpha Z_A^\beta = \Pi_A[1 + \sum_{\alpha,\beta}\lambda_{\alpha\beta}(X_1^\alpha...X_N^\alpha)Z_i^\beta] \tag{10}$$

It is seen here that transformations that contain X_A^α are performed with the $X_1^\alpha...X_N^\alpha$ while transformations that contain Z_A^β are performed with the Z_i^β. So we have redundancy in the J-direction only.

The above are unitary transformations. A rather general class of noise transformations describing the interaction of a quantum system with density matrix ρ with the environment, can be written as

$$\rho' = \sum_i E_\ell \rho E_\ell^\dagger; \qquad E_\ell = \sum \lambda_{\ell;\alpha_1\beta_1...\alpha_N\beta_N}(X_1^{\alpha_1} Z_1^{\beta_1})...(X_N^{\alpha_N} Z_N^{\beta_N}) \qquad (11)$$

Our coding so far provides protection against noise in the J-direction only. In the next section we introduce redundancy in the θ-direction and show that the combined effect is redundancy in any direction.

4. General redundancy

We consider the space $(H_A)^M \equiv H_A \otimes ... \otimes H_A$ which is clearly a $(2j+1)^M$-dimensional subspace of the space H^{NM}. The operator $P_A = \Pi_A \otimes ... \otimes \Pi_A$ projects the space H^{NM} to the space $(H_A)^M$. We use the notation $W_{A\mu} \equiv \Pi_A \otimes ... \otimes W_A \otimes ... \otimes \Pi_A$ ($\mu = 1, ..., M$) for operators acting on $(H_A)^M$, with the operator W_A acting on the μ Hilbert space H_A. Clearly $\Pi_{A\mu} = P_A$ for any μ. For a product of two operators $W_A V_A$ it is easily seen that $(W_A V_A)_\mu = W_{A\mu} V_{A\mu}$. Using this notation we can write $Z_{A\mu} = (Z_i \Pi_A)_\mu = Z_{i\mu} P_A$. In $Z_{i\mu}$ the indices i and μ refer to the positions in the words considered at the first and second step of the concatenated code, correspondingly.

We consider the $(2j+1)$-dimensional subspace spanned by the vectors

$$H_B = \{|\theta_B; j\ m\rangle \equiv |\theta_A; j\ m\rangle \otimes ... \otimes |\theta_A; j\ m\rangle, m = -j, ..., j\} \qquad (12)$$

The Hilbert space H_B, is isomorphic to the Hilbert space H_A and also to the Hilbert space H, through the mapping: $|\theta_B; j\ m\rangle \leftrightarrow |\theta_A; j\ m\rangle \leftrightarrow |\theta; j\ m\rangle$. We call P_B the projection operator in H_B. We can show that

$$X_B = X_{A\mu} P_B = X_{1\mu}...X_{N\mu} P_B \qquad (13)$$

The Fourier operator F_B acting on the states of the Hilbert space H_B, given by:

$$F_B = (2j+1)^{-1/2} \sum_{m,n} \omega(mn)|\theta_A; j\ m\rangle\langle\theta_A; j\ n| \otimes ... \otimes |\theta_A; j\ m\rangle\langle\theta_A; j\ n| \qquad (14)$$

The states $|J, j\ m\rangle$ in H correspond to the states $|J_B, j\ m\rangle \equiv F_B|\theta_B; j\ m\rangle$ which form an orthonormal basis in H_B. It is easily seen that

$$P_B|J_A, j\ m\rangle \otimes ... \otimes |J_A, j\ m\rangle = (2j+1)^{(1-M)/2}|J_B, j\ Mm\rangle \qquad (15)$$

For the product $Z_{A1}...Z_{AM}$ we show that

$$[P_B, Z_{A1}...Z_{AM}] = 0; \qquad Z_B = Z_{i1}...Z_{iM} P_B \qquad (16)$$

Here i can take any value from 1 to N.

In the previous section we have introduced redundancy in the J-direction. In this section we have introduced further redundancy in the θ-direction. The operator X_B that performs shifts among the states $|J_B; j\ m\rangle$ is equal to the product of all $X_{i\mu}$ and the operator Z_B that performs shifts among the states $|\theta_B; j\ m\rangle$, is equal to the product of all $Z_{i\mu}$. Therefore, it is unlikely that small noise will perform such transformations causing errors.

We next consider $SU(2j+1)$ transformations on the states in the Hilbert space H_B. Infinitesimal action of these transformations can be written as:

$$g_B = 1 + \sum_{\alpha,\beta} \lambda_{\alpha\beta} X_B^\alpha Z_B^\beta = P_B[1 + \sum_{\alpha,\beta} \lambda_{\alpha\beta}(X_{1\mu}^\alpha...X_{N\mu}^\alpha)(Z_{i1}^\beta...Z_{iM}^\beta)] \qquad (17)$$

It is seen that we have redundancy in all directions.

Noise transformations can be written as

$$\rho' = \sum_i E_\ell \rho E_\ell^\dagger; \qquad E_\ell = \sum \lambda_{\ell;\alpha_{11}\beta_{11}...\alpha_{NM}\beta_{NM}} (X_{11}^{\alpha_{11}} Z_{11}^{\beta_{11}})...(X_{NM}^{\alpha_{NM}} Z_{NM}^{\beta_{NM}}) \quad (18)$$

Our coding provides protection against noise in any direction provided that we have 'small noise' which acts only on some qudits.

5. Discussion

We have considered a $(2j + 1)$-dimensional angular momentum Hilbert space H and through Fourier transform we have introduced angle states and operators . The corresponding quantum phase space J_z-θ_z is a toroidal lattice. Displacement operators in this phase space, are generators of $SU(2j + 1)$ transformations in H.

In this context we have studied a concatenated code that involves two steps. In the first one the space H^N is considered and the code is the subspace H_A spanned by the direct products of N angular momentum states with the same m. In the second step the space H_A^M is considered and the code is the subspace H_B spanned by the direct products of M angle states with the same m. We have shown that this introduces general redundancy in any direction. The proposed scheme will protect quantum information against 'small' errors which occur on some the components.

The limit $j \to \infty$ is an interesting one, and further work in this direction is in progress.

References

[1] D. Gottesman, A. Kitaev, J. Preskill, Phys. Rev. A64, 012310 (2001)
   D. Gottesman, Lecture Notes Computer Science 1509, 302 (1999)
   S.D. Bartlett, H. de Guise, B.C. Sanders, Phys. Rev. A65, 052316 (2002)
[2] A. Vourdas, Phys. Rev. A65, 042321 (2002)
[3] P. Shor, Phys. Rev. A52, 2493 (1995)
[4] H. Weyl, Theory of Groups and Quantum Mechanics (Dover, New York, 1950);
   J. Schwinger, Proc. Nat. Acad. Sci. U.S.A. 46, 570 (1960); Quantum Kinematics and Dynamics
   (Benjamin, New York, 1970)
   L. Auslander, R. Tolimieri Bull. Am. Math.Soc. 1, 847 9(1979)
   R. Balian and C. Itzykson, C.R. Acad. Sci. 303, 773 (1986)
   W.K. Wootters and B.D. Fields, Ann. Phys (N.Y) 191, 363 (1989)
   M.L. Mehta, J.Math. Phys. 28, 781 (1987)
   V.S. Varadarajan, Lett. Math. Phys. 34, 319 (1995)
[5] A. Vourdas, Phys. Rev. A41, 1653 (1990); A43, 1564 (1991)
   A. Vourdas, C. Bendjaballah, Phys.Rev. A47, 3523 (1993)
   A. Vourdas, J.Phys.A29, 4275 (1996)
[6] D.B. Fairlie, P. Fletcher, C.K. Zachos, J. Math. Phys. 31, 1088 (1990)

Inst. Phys. Conf. Ser. No 173: Section 7
Paper presented at 24th Int. Coll. Group Theoretical Methods in Physics, Paris, France, 15–20 July 2002
©2003 IOP Publishing Ltd

A Note on the Optimal Teleportation

Wen-Li Yang §

Institute of Modern Physics, Northwest University, Xian 710069, China
Physikalisches Institut der Universität Bonn, 53115 Bonn, Germany

Abstract. We study optimal teleportation based on the Bell measurements. The optimal transmission fidelity is calculated and shown to be related to the fully entangled fraction of the quantum resource, rather than the singlet fraction as in the standard teleportation protocol.

PACS: 03.67.Hk; 89.70.+c; 03.65.-w   AMS classification: 81P68; 68P30.

By using a quantum resource (a nonlocal entangled state), the teleportation protocols give ways to transmit an unknown quantum state ρ from a sender (Alice) to a receiver (Bob) who are spatially separated. These teleportation processes can be viewed as quantum channels. When the maximally entangled pure state $|\Phi>$ is used as the quantum resource, the standard teleportation protocol T_0 proposed in [1] provides an ideal noiseless quantum channel $\Lambda_{T_0}^{(|\Phi><\Phi|)}(\rho) = \rho$. However due to decoherence Alice and Bob usually share a mixed entangled state χ. Recently, an explicit expression for the quantum channel associated with the standard teleportation protocol T_0 with an arbitrary mixed state χ resource has been given in [2, 3].

In this talk, we consider the following problem. Alice and Bob previously only share a pair of particles in an arbitrary mixed entangled state χ. In order to teleport an unknown state to Bob, Alice first performs a joint Bell measurement on her particles (particle 1 and particle 2) and tell her result to Bob by the classical communication channel. Then Bob, instead of the *Pauli* rotation like in the standard teleportation protocol [1], tries his best to choose a particular unitary transformation which depends on the quantum resource χ, so as to get the maximal transmission fidelity. We call our teleportation protocol the optimal teleportation based on the Bell measurement. We derive an explicit expression for the quantum channel associated with the optimal teleportation with an arbitrary mixed state resource. The transmission fidelity of the corresponding quantum channel is given in term of the *fully entangled fraction* of the quantum resource.

Let $\{|i>, i = 0, ..., n-1\}$, $n < \infty$, be an orthogonal normalized basis of an n-dimensional Hilbert space $\mathcal{H}$. We consider the following three-tensor Hilbert space: $\mathcal{H} \otimes \mathcal{H} \otimes \mathcal{H}$ where Alice has the first and the second Hilbert space, and the third one belongs to Bob. Let h and g be $n \times n$ matrices such that $h|j> = |(j+1) \mod n>$, $g|j> = \omega^j|j>$, with $\omega = exp\{\frac{-2i\pi}{n}\}$. We introduce n^2 linear-independent $n \times n$-matrices $U_{st} = h^t g^s$, which satisfy

$$U_{st}U_{s't'} = \omega^{st'-ts'}U_{s't'}U_{st}, \quad tr(U_{st}) = n\delta_{s0}\delta_{t0}. \tag{1}$$

$$\begin{cases} tr\left(U_{st}U_{s't'}^+\right) = n\delta_{tt'}\delta_{ss'}, \\ U_{st}\,U_{st}^+ = I_{n\times n}, \end{cases} \tag{2}$$

§ wlyang@th.physik.uni-bonn.de

$\{U_{st}\}$ form a complete basis of $n \times n$-matrices, namely, for any $n \times n$ matrix W, W can be expressed as

$$W = \frac{1}{n}\sum_{s,t} tr(U_{st}^+ W)U_{st}. \tag{3}$$

From $\{U_{st}\}$, we can introduce the generalized Bell-states,

$$|\Phi_{st}> = (1 \otimes U_{st}^*)|\Phi> = \frac{1}{\sqrt{n}}\sum_{i,j}(U_{st})_{ij}^*|ij>, \quad \text{and} \quad |\Phi_{00}> = |\Phi>, \tag{4}$$

$|\Phi_{st}>$ are all maximally entangled states and form a complete orthogonal normalized basis of $\mathcal{H} \otimes \mathcal{H}$ shared by Alice and Bob.

For any state χ shared by Alice and Bob, let us introduce the *singlet fraction* [4]: $F = <\Phi|\chi|\Phi>$, and the *fully entangled fraction* [4] of a state χ by

$$\mathcal{F}(\chi) = max\left\{<\Phi|(1 \otimes U^+)\chi(1 \otimes U)|\Phi>\right\}, \quad \text{for all } UU^+ = U^+U = I_{n\times n}. \tag{5}$$

Since the group of unitary transformations in n-dimensions is compact, there exists an unitary matrix W_χ such that

$$\mathcal{F}(\chi) = <\Phi|(1 \otimes W_\chi^+)\chi(1 \otimes W_\chi)|\Phi>. \tag{6}$$

Suppose now Alice and Bob previously share a pair of particles in an arbitrary mixed entangled state χ. To transform an unknown state to Bob, Alice first performs a joint Bell measurement based on the generalized Bell-states Eq.(4) on her parties. According to the measurement results of Alice, Bob chooses particular unitary transformations $\{T_{st}\}$ to act on his particle. The transmission fidelity of the quantum channel

$$f(\chi) = \overline{<\phi_{in}|\Lambda^{(\chi)}(\{T\})(|\phi_{in}><\phi_{in}|)|\phi_{in}>}, \tag{7}$$

quantifies how well the channel is.
Then we have [5]

Theorem 1 *The teleportation protocol defined by $\{T_{st}\}$, when used with an arbitrary mixed state with density matrix χ as a resource, acts as a quantum channel*

$$\Lambda^{(\chi)}(\{T\})(\rho) = \frac{1}{n^2}\sum_{s,t}\sum_{s',t'}<\Phi_{st}|\chi|\Phi_{s't'}>$$

$$\times \left\{\sum_{\gamma\beta}T_{\gamma\beta}^+U_{st}U_{\gamma\beta}\,\rho\,U_{\gamma\beta}^+U_{s't'}^+T_{\gamma\beta}\right\}, \tag{8}$$

and the transmission fidelity is

$$f(\chi) = \frac{1}{n(n+1)}\sum_{\gamma\beta}<\Phi|\left(1 \otimes (T_{\gamma\beta}U_{\gamma\beta}^+)^+\right)\chi\left(1 \otimes T_{\gamma\beta}U_{\gamma\beta}^+\right)|\Phi>$$

$$+\frac{1}{n+1}. \tag{9}$$

For the standard teleportation protocol [2, 3]: *(if one choose $T_{st} = U_{st}$)*

$$\Lambda_s^{(\chi)}(\rho) = \frac{1}{n^2}\sum_{s,t}<\Phi_{st}|\chi|\Phi_{st}>U_{st}\,\rho\,U_{st}^+,$$

$$f_s = \frac{nF(\chi)}{n+1}+\frac{1}{n+1}.$$

Obviously when the term $< \Phi| \left(1 \otimes (T_{\gamma\beta}U_{\gamma\beta}^+)^+\right) \chi \left(1 \otimes T_{\gamma\beta}U_{\gamma\beta}^+\right) |\Phi >$ is maximized, i.e., $T_{\gamma\beta}U_{\gamma\beta}^+ = W_\chi$, one gets the maximal fidelity. Recalling the definition of the *fully entangled fraction* Eq.(5) and Eq.(6), we arrive at our main result:

Theorem 2 *The optimal teleportation based on the Bell measurements, when used with an arbitrary mixed state with density matrix χ as a resource, acts as a general trace-preserving quantum channel*

$$\Lambda_O^{(\chi)}(\rho) = \frac{1}{n^2} \sum_{s,t} \sum_{s',t'} < \Phi_{st} |\chi| \Phi_{s't'} >$$
$$\times \left\{ \sum_{\gamma\beta} U_{\gamma\beta}^+ W_\chi^+ U_{st} U_{\gamma\beta} \, \rho \, U_{\gamma\beta}^+ U_{s't'}^+ W_\chi U_{\gamma\beta} \right\}. \tag{10}$$

The corresponding transmission fidelity is given by

$$f_{max}(\chi) = \frac{n\mathcal{F}(\chi)}{n+1} + \frac{1}{n+1}, \tag{11}$$

where $\mathcal{F}(\chi)$ is the fully entangled fraction Eq.(5) and W_χ is the unitary matrix which fulfills such a fully entangled fraction Eq.(6).

Our results show that the maximally transmission fidelity of the teleportation based on the Bell measurement depends on the *fully entangled fraction* only, whereas that of a standard teleportation depends on the singlet fraction [3]. What "entanglement" for the mixed state we learn from teleportation is just *fully entangled fraction*. Hence not all the *"entanglement"* of a mixed state is useful for teleportation due to the fact that there exists some entangled state, as teleportation source, which have less transmission fidelity than that of no entangled state [4]. Our result also agrees with the fidelity formula of the general optimal teleportation given by the Horodecki family [6].

Summarizing, we obtain the explicit expression of the output state of the optimal teleportation, with arbitrary mixed entangled state as resource, in terms of some noisy quantum channel. This allow us to calculate the transmission fidelity of the quantum channel. It is shown that the transmission fidelity depends only on the *fully entangled fraction* of the quantum resource shared by the sender and the receiver. The fidelity in our optimal teleportation protocol is in general greater than the one in standard teleportation protocol [1, 2, 3].

Acknowledgments: This work has been done collaborating with S. Albeverio and S.-M. Fei. The support from the Alexander-von-Humboldt Foundation is acknowledged. We would like to thank Prof. von Gehlen for his continuous encouragements.

References

[1] Bennett C H, Brassard G, Crepeau C, Jozsa R, Peres A and Wooters W K 1993 *Phys. Rev. Lett* **70** 1895-1899.
[2] Bowen G and Bose S 2001 *Phys. Rev. Lett.* **87** 267901-267904.
[3] Albeverio S, Fei S M and Yang W L 2002 *Teleportation with an arbitrary mixed resource as a trace-preserving quantum channel.*
[4] Horodecki M, Horodecki P, Horodecki R 2001 *Mixed-state entanglement and quantum communication* quant-ph/0109124.
[5] Albeverio S, Fei S M and Yang W L 2002 *Phys. Rev.* **A66** 012301-012304.
[6] Horodecki M, Horodecki P and Horodecki R 1999 *Phys. Rev.* **A60** 1888-1898.

Inst. Phys. Conf. Ser. No 173: Section 7
Paper presented at 24th Int. Coll. Group Theoretical Methods in Physics, Paris, France, 15–20 July 2002
©2003 IOP Publishing Ltd

Generalized Relativistic Dynamics as a Dynamics of Third Order

R M Yamaleev

Universidad Nacional Autonoma de México, México

Abstract.
Within the framework of Nambu's $n + 1$-order phase space formalism we propose an hierarchy of the dynamic systems ordered by n. In this hierarchy the first level, $n = 1$, occupies Newtonian mechanics, the second level, $n = 2$, belongs to the relativistic mechanics. The third level with $n = 3$ is the subject of investigation of the present paper. We suggest two extensions of the Lorentz-force equations. In the first one, the dynamic equations are given by Weierstrass equations for elliptic functions, in the second, by Jacobi equations for elliptic functions.

1. Introduction

In 1972 Y.Nambu proposed the phase space formalism of $(n + 1)$-order [1]. As it has been shown in Refs.[2], [3] the Nambu's formalism is related with the metric forms of order higher than quadratic forms. In Ref.[4] has been suggested a new formulation of the relativistic dynamics related with the Nambu's three-order phase space formalism. Within the framework of this formulation the energy and momentum of the relativistic particle are composed of two types of energies according to the Vieta's formulae of the quadratic polynomial. This polynomial is a characteristic polynomial of the relativistic dynamics. The general formulation of the mechanics deals with the hierarchy ordered by number n, where n is the order of a characteristic polynomial of the dynamics. The first level of this hierarchy, $n = 1$, occupies Newtonian mechanics, the second level, $n = 2$, belongs to the relativistic mechanics. Now, naturally, of special interest for investigations is the third level, $n = 3$, of this hierarchy.

We start from the Nambu's four-order phase space formalism. Within the framework of this formulation the energy and momentum of the particle are composed of three types of energies according to the Vieta's formulae of the cubic polynomial. This polynomial is a characteristic polynomial of the generalized relativistic dynamics. We investigate two possible extensions of the relativistic mechanics. Both are obtained from the unique *root equations* for the momenta of the subsystems. In terms of the momenta of the composed system we obtain two different sets of equations, one is given by Weierstrass equation, and the other, by Jacobi equations for elliptic functions.

2. Characteristic equation of the relativistic dynamics

Consider a motion of the particle with charge e in the external e.m. fields $\vec{E}$ and $\vec{B}$. The relativistic equations of motion with respect to the proper time τ are given by the Lorentz-force equations

$$\frac{d\vec{p}}{d\tau} = \frac{e}{mc}(\vec{E}\,p_0 + [\vec{p}\times\vec{B}]), \quad \frac{dp_0}{d\tau} = \frac{e}{mc}(\vec{E}\cdot\vec{p}), \quad \frac{d\vec{r}}{d\tau} = \frac{\vec{p}}{m}, \quad \frac{dt}{d\tau} = \frac{p_0}{mc}. \tag{1}$$

In the direction of the momentum Eqs.(1) are reduced into equations for *hyperbolic oscillator*:

$$\frac{dp}{dw} = p_0, \quad \frac{dp_0}{dw} = p, \quad \text{with } dw = \frac{1}{mc}(\vec{U}\cdot\vec{n})\,d\tau, \tag{2}$$

where $\vec{U} = e\vec{E}$ for the electromagnetic field and $\vec{U} = -\vec{\nabla}V(r)$ for the potential field.
  Eqs.(1) imply two first integrals of motion, for the mass and the energy

$$p_0^2 - p^2 = (Mc)^2, \quad cp_0 + V(r) = E_0. \tag{3}$$

Define two kinds of energies by

$$\mathcal{E}_h = \frac{p_h^2}{2m_h} + V, \quad \mathcal{E}_e = \frac{p_e^2}{2m_e} + V.$$

Then, as it has been shown in Ref.[4],

$$\mathcal{E}_h = E_0 - Mc^2, \quad \mathcal{E}_e = E_0 + Mc^2.$$

In this way we arrive to the model of the relativistic particle the energy-momentum and mass of which are composed of the energies and the masses of two subparticles. In the sequel let us use the scale

$$p_h = \frac{p_h}{\sqrt{2m_h}}, \quad p_e = \frac{p_e}{\sqrt{2m_e}}, \quad c = 1.$$

Proposition
  Relativistic Lorentz-force equations can be cast into Nambu's three-order phase space formalism within the framework of the following mapping:
the direct mapping is defined by the Vieta's formulae

$$p = p_h p_e, \quad M = \frac{1}{2}(p_e^2 - p_h^2), \quad p_0 = \frac{1}{2}(p_h^2 + p_e^2); \tag{4}$$

the inverse mapping is defined by solutions of the quadratic equation

$$X^2 - 2p_0 X + p^2 = 0. \tag{5}$$

This quadratic equation is *a characteristic equation of the relativistic dynamics*.
The equations for the triplet $\{p_h, p_e, r\}$ are given by Nambu's equations with two Hamilton-Nambu functions $H_1 = \mathcal{E}_h$, $H_2 = \mathcal{E}_e$:

$$\frac{dp_h}{d\tau} = -\left(\frac{dV(r)}{d\vec{r}}\cdot\vec{n}\right)\frac{p_e}{m_e}, \quad \frac{dp_e}{d\tau} = -\left(\frac{dV(r)}{d\vec{r}}\cdot\vec{n}\right)\frac{p_h}{m_h}, \quad \frac{dr}{d\tau} = \frac{p_h\,p_e}{m_h\,m_e}. \tag{6}$$

For the *proof* see Ref.[4].

3. Dynamics with cubic characteristic equation

The scheme presented above can be naturally generalized to the dynamics with cubic characteristic equation. This dynamics describes a motion of the particle the energy and momentum of which are composed of the sub-momenta $\{p_h, p_e, p_g\}$, the masses $\{m_h, m_e, m_g\}$ and the energies:

$$\mathcal{E}_i = \frac{p_i^2}{2m_i} + V(r), \ i = h, e, g.$$

For the sake of convenience we use the scale:

$$p_h = \frac{p_h}{\sqrt{2m_h}}, \ p_e = \frac{p_e}{\sqrt{2m_e}}, \ p_g = \frac{p_g}{\sqrt{2m_g}} \text{ and } c = 1.$$

Relations between the momenta and the sub-momenta are given according to the following mapping:
the direct mapping is done by Vieta's formulae

$$P_1 = \frac{1}{3}(p_h^2 + p_e^2 + p_g^2), \ P_2 = \frac{1}{2}(p_h^2 p_e^2 + p_e^2 p_g^2 + p_g^2 p_h^2), \ P = p_h p_e p_g. \quad (7)$$

The inverse mapping is defined as the solutions of the cubic equation

$$X^3 - 3P_1 X^2 + 2P_2 X - P^2 = 0. \quad (8)$$

The Nambu's equations with Hamilton-Nambu functions $H_1 = \frac{1}{2}(p_e^2 - p_h^2), H_2 = \frac{1}{2}(p_g^2 - p_h^2)$ are written as

$$\frac{dp_h}{du} = p_g p_e \frac{1}{\mu}, \ \frac{dp_e}{du} = p_g p_h \frac{1}{\mu}, \ \frac{dp_g}{du} = p_e p_h \frac{1}{\mu}, \ du = \frac{1}{2m}(\vec{U} \cdot \vec{n}) \, d\tau, \quad (9)$$

where $dim[\mu] = dim[p_h] = dim[p_e] = dim[p_g]$. These equations we consider as *a root equations* of $n = 3$ order of dynamics.

From Eqs.(9) by using the mapping (7), we derive the following equations for P, P_2, P_1 with respect to the evolution parameter $w = 2u$:

$$\frac{dP}{dw} = P_2 \frac{1}{\mu}, \ \frac{dP_2}{dw} = 3PP_1 \frac{1}{\mu}, \ \frac{dP_1}{dw} = P \frac{1}{\mu}. \quad (10)$$

From these equations we find two invariants

$$R_1 = -2P_2 + 3P_1^2, \ R_0 = P_1^3 - R_1 P_1 - P^2. \quad (11)$$

From (10) by using (11), we get

$$\mu^2 (\frac{dP_1}{dw})^2 = P_1^3 - R_1 P_1 - R_0.$$

The solution of this equation is given by the Weierstrass elliptic function
$P_1(w) = \mu \, \rho(u; 4R_1/\mu^4, 4R_0/\mu^6)$.
The following cubic relationship between P_1 and P^2 holds

$$P_1^3 - R_1 P_1 - R_0 = P^2.$$

This cubic polynomial can be regarded as an analogue of the relativistic quadratic energy-momentum-mass formula.

The motion in the electromagnetic fields are given by the generalized Lorentz-force equations [5]:

$$\frac{d\vec{P}}{d\tau} = \frac{e}{m\mu}P_2\vec{E} + \frac{e}{m}[\vec{P}\times\vec{B}], \quad \frac{dP_2}{d\tau} = \frac{3e}{m\mu}(\vec{E}\cdot\vec{P})P_1, \quad \frac{dP_1}{d\tau} = \frac{e}{m\mu}(\vec{E}\cdot\vec{P}).(12)$$

Notice that the solutions of root equations (9) are given by by Jacoby elliptic functions of imaginary argument

$$p_h = \sqrt{2m}\,sc(u,k'), \quad p_e = \sqrt{2m}\,nc(u,k'), \quad p_g = \mu\,dc(u,k'), \quad k' = \frac{2m}{\mu^2}, \quad u = \frac{1}{2}w.(13)$$

In the next, we shall construct a mapping between the triplets $\{\mathcal{P}, \mathcal{P}_1, \mathcal{P}_2\}$ and $\{p_h, p_e, p_g\}$ as relationships between two kinds of Jacobi functions sc, nc, dc with the arguments $w = 2u$ and u, correspondingly.

The components of the momentum of the corporeal particle will obey the generalized Lorentz-force equations

$$\frac{d\vec{\mathcal{P}}}{d\tau} = \frac{e}{m\lambda}\vec{E}\,\mathcal{P}_1\mathcal{P}_2 + \frac{e}{m}[\vec{\mathcal{P}}\times\vec{B}], \quad \frac{d\mathcal{P}_1}{d\tau} = \frac{e}{m\lambda}(\vec{E}\cdot\vec{\mathcal{P}})\mathcal{P}_2, \quad \frac{d\mathcal{P}_2}{d\tau} = \frac{e}{m\lambda}(\vec{E}\cdot\vec{\mathcal{P}})\mathcal{P}_1,(14)$$

with $\lambda^2 = \frac{1}{2}\mu^2 m$.

In the direction of the vector of momentum the solutions of Eqs.(14) are given by the set of Jacobi elliptic functions of imaginary argument:

$$\mathcal{P} = m\,sc(w,k'), \quad \mathcal{P}_1 = m\,nc(w,k'), \quad \mathcal{P}_2 = \lambda\,dc(w,k').$$

Then the mapping $\{\mathcal{P}, \mathcal{P}_1, \mathcal{P}_2\} \rightarrow \{p_h, p_e, p_g\}$ is obtained by using relationships between Jacoby functions of integral and one-half arguments[6]. The direct mapping is defined by

$$\frac{\mathcal{P}}{m} = \frac{2}{\mu}\frac{p_h p_e p_g}{p_e^2 - p_h^2 p_g^2 \frac{1}{\mu^2}}, \quad \frac{\mathcal{P}_1}{m} = \frac{p_e^2 + p_h^2 p_g^2 \frac{1}{\mu}}{p_e^2 - p_h^2 p_g^2 \frac{1}{\mu^2}}, \quad \mathcal{P}_2 = \sqrt{\frac{m}{2}}\mu\frac{2mp_g^2 + p_h^2 p_e^2}{\mu p_e^2 - p_h^2 p_g^2}.(15)$$

The inverse mapping is given by

$$p_h^2 = \lambda\frac{\mathcal{P}_1 - m}{\mathcal{P}_2 + \lambda}, \quad p_e^2 = \lambda\frac{\mathcal{P}_1 + \frac{m}{\lambda}\mathcal{P}_2}{\mathcal{P}_2 + \lambda}, \quad p_g^2 = \mu^2\frac{\frac{m}{\lambda}\mathcal{P}_2 + \mathcal{P}_1}{\mathcal{P}_1 + m}. \tag{16}$$

When μ runs to ∞, the parameter k' tends to 0. At this limit $\mathcal{P}_1 \rightarrow p_0$, $\mathcal{P}_2 \rightarrow \lambda$, and formulae (15) are reduced to the formulae of projective momenta on Lobachevsky space[7]:

$$\frac{p}{m} = \frac{2p_h p_e}{p_e^2 - p_h^2}, \quad \frac{p_0}{m} = \frac{p_e^2 + p_h^2}{p_e^2 - p_h^2}.$$

[1] Y.Nambu Y 1973 Phys.Rev. D7 2405
[2] Vainerman L, Kerner R 1996 J.Math.Phys. 37 2553
    Abramov V, Kerner R 2000 J.Math.Phys. 41 5598
[3] Yamaleev R M 2000 Advances Applied Clifford Algebras 12 2 1
[4] Yamaleev R M 1999 Ann. of Phys.(N.Y.) 277 1 1
    Yamaleev R M 2000 Ann. of Phys. (N.Y.) 285 2 1
[5] Yamaleev R M 2001 Ann. of Phys. (N.Y.) 292 2 157
[6] Akhiezer N I 1970 Elements of theory of elliptic functions (Moscow: Nauka)
[7] Mir-Kasimov R M 2000 Physics of Particles and Nuclei(Dubna, Russia) 31 1 44

Inst. Phys. Conf. Ser. No 173: Section 7
Paper presented at 24th Int. Coll. Group Theoretical Methods in Physics, Paris, France, 15–20 July 2002
©*2003 IOP Publishing Ltd*

Reorganization of energy bands in quantum finite particle systems

B I Zhilinskii

Université du Littoral, UMR de CNRS 8101, 145 av. M. Schumann, Dunkerque 59140 France

Abstract. The qualitative phenomenon of the redistribution of energy levels between bands in the energy spectrum of finite particle quantum systems (atoms and molecules) is studied from the point of view of quantum, semi-quantum and purely classical approaches. Relation with topological quantum numbers, classical and quantum monodromy, defects of regular lattices is shown.

1. Introduction

Presence of energy bands in the spectra of small atomic and molecular quantum systems is a well known characteristic feature related to existence of several typical energetic scales. Vibrational structure of electronic states, rotational structure of vibrational states, fine spin and hyperfine nuclear spin structures are simplest examples of band organization which reflect in classical terms the existence of slow and fast motions and their approximate separation. Within families of effective quantum Hamiltonians the modification of the band structure is typically observed as a result of relatively strong coupling. The purpose of the present note is to look at the phenomenon of the reorganization of energy bands under the variation of some strict or approximate integral of motion for molecular systems from several different points of view. The main idea is to understand the qualitative structure and its possible generic modifications taking into account the topological structure of the underlying classical dynamics and molecular symmetry.

Several recent review articles [1, 2, 3, 4] discuss qualitative approach to intramolecular dynamics and in particular the redistribution of energy levels between bands. We only breifly remind here the relation between redistribution, topological quantum numbers, and quantum monodromy whereas interpretation in terms of lattice defects is mainly concerned.

2. Topological quantum numbers

The relation between topological Chern numbers and the redistribution of energy levels between rotational energy bands associated with several close vibrational molecular quantum states was first conjectured in [5]. Many concrete examples of redistributions observed in real molecules were described in [6, 7], where the relation between redistribution and symmetry was emphasized, namely special symmetry selection rules for redistribution were formulated. Rigourous mathematical statement which associates to each rotational band its topological quantum number, Chern class of the complex line fibre bundle, was formulated in [8, 9]. Number of energy levels in each band was expressed in terms of the Chern class of the complex fibre bundle. This bundle characterizes the rotational structure of vibrational states in the semi-quantum model which treats rotational (slow) motion as classical and vibrational (fast) motion as quantum.

626

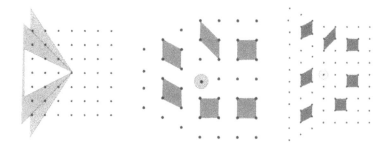

Figure 1. Construction of lattice defect by removing or adding one of the solid angles shown on the left picture. Reconnection of the lattice is done by vertical parallel shift of all vertices. The number of removed (added) vertices varies linearly. Three different angles correspond to removing (adding) respectively one, two and three vertices during the first step in the horizontal direction. On the reconstructed lattice the singularity point is shown by shadowed disk. The circular path of the elementary cell shows that removing and adding the same solid angle result in monodromy with different sign but with the same absolute value of the nondiagonal element.

Generalization of semi-quantum model to the case of effective Hamiltonians with several slow (classical) dynamical degrees of freedom leads to a new very interesting phenomenon of the formation of topologically coupled energy bands [10, 4]. Concrete model introduced in [10] deals with CP_2 (complex projective space) as the classical phase space for slow variables. The coupling with three quantum states is needed to see the formation of topologically coupled bands [4]. Number of states in the band can again be expressed through topological invariants using Atiyah-Singer index formula within, for example, Fedosov version of deformation quantization [11].

3. Quantum monodromy

Completely classical treatment [12] of the model problem of coupled angular momenta introduced in [5] to study the redistribution phenomenon shows the appearance of classical Hamiltonian monodromy [13, 14, 15] for this model problem. Manifestation of classical monodromy in corresponding quantum problem [16, 17, 12, 18] can be most easily detected if we interpret the quantum monodromy as a defect of a regular lattice of quantum states. For two degree of freedom system the evolution of elementary cell $(n_1, n_2; n_1 + 1, n_2; n_1 + 1, n_2 + 1; n_1, n_2 + 1)$ formed by quantum states with consequent quantum numbers around the singularity characterizes the monodromy and does not depend on the choice of the basis of the lattice [12, 18]. The relation between classical (quantum) monodromy and Chern indices can be seen [18] through the generalization of the Duistermaat-Heckman theorem [19]. At the same time the analogy with the description of defects in regular lattices [20, 21, 22] seems to be quite important for generalization of the quantum monodromy and corresponding phenomenon of the reorganization of energy bands.

4. Lattice defects

Lattice of quantum states with monodromy can be formally constructed in a way similar to construction of dislocations and disclinations of crystals. The presence of defect for integrable system means that global action-angle variables are not defined whereas local action-angle

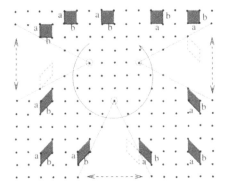

Figure 2. Regular lattice with three singular points corresponding each to elementary monodromy. For two singular points the removed solid angle is filled by vertical shift of columns of points whereas for the third singular point the shift is horizontal. The circular path around three singular points results in the rotation of the elementary cell on $\pi/2$.

Figure 3. Reconstruction of lattice after 1:2 rational cut. The identification of two rays after cutting is done through parallel transport of vertices. Left and right figures show respectively the transfer over cut of elementary cell of dimension 1×1 and 1×2.

variables exist [23, 13]. Figure 1 illustrates the construction of lattice defect by adding or removing the same solid angle. To ensure the presence of isolated point defect for a two dimensional lattice special matching rules should be introduced to identify the two boundary of the cut. Namely, the deformation of the lattice after cut is done by parallel vertical displacement of lattice points up to identification of points on the boundary. The formation of an isolated point defect after such reconstruction is possible only for very special cuts. Three possible choices of cuts are shown in Figure 1, left. Two other subfigures show reconstructed lattices after removing and adding the smallest possible solid angle. The so obtained defects in these two cases are characterised by elementary matrices $\left(\begin{smallmatrix} 1 & 1 \\ 0 & 1 \end{smallmatrix} \right)$ or $\left(\begin{smallmatrix} 1 & -1 \\ 0 & 1 \end{smallmatrix} \right)$ which are defined up to conjugation with arbitrary $SL(2, Z)$ matrix due to an arbitrary choice of the lattice basis. The author of the present note has conjectured recently that only one kind of defects (associated with removing solid angle) is relevant for Hamiltonian systems. The proof of this conjecture is given in [24]. In non-Hamiltonian systems both cases of signs of elementary monodromy matrices are possible as demonstrated in [25].

It seems reasonable to conjecture further that the elementary monodromy matrix with inversed sign should appear as a generic obstruction to the existence of global quantum numbers for integrable PT-invariant non-Hermitian Hamiltonian systems [26].

In the case of multiple elementary defects the global monodromy matrix associated with the close path going around several defects can be arbitrary [27]. Figure 2 shows, for example, that three elementary $\left(\begin{smallmatrix} 1 & 1 \\ 0 & 1 \end{smallmatrix}\right)$ defects (remind that the monodromy matrix of the defect is defined up to $SL(2,Z)$ conjugation) result in the fourth order monodromy matrix $\left(\begin{smallmatrix} 0 & 1 \\ -1 & 0 \end{smallmatrix}\right)$. From the point of view of crystal defects such global monodromy matrix corresponds to a $\pi/2$-disclination. In a similar way standard $\pi/3$ disclination for the triangular lattice can be constructed as a result of two defects with elementary monodromy matrices. In [27] it is proved that inversed sign monodromy matrix can be formed as a global monodromy matrix for eleven (this is in fact a minimal number of elementary defects needed for such realization) properly chosen elementary defects.

Using analogy with the "partial disclination" it is possible to form more general defects of lattice of quantum states by making a "rational cut" of the lattice. Figure 3 shows an example of such 1 : 2 cut. The reconstruction of the lattice after such cut is done again by parallel transport of vertices in vertical direction. It is important to note now that the matching rules imposed for the 1 : 2 cut result in splitting all vertices lying on the boundary of the cut in two subsets: vertices which are present on both sides of the cut and should be identified and vertices presented on one side with no partners to be identified on the another side. This leads to the line defects on the reconstructed lattice with the point defect at the top of the solid angle. The transfer of the elementary cell through the line defect can not be unambigously defined. At the same time for the 1×2 cell the monodromy is well defined but the new feature now is the appearence of half integer (rational in general) entries for the monodromy matrix [28].

References

[1] Zhilinskii B 1996 Spectrochim. Acta A 52 881-900
[2] Michel L and Zhilinskii B 2001 Phys. Rep. 341 11-84
[3] Zhilinskii B 2001 Phys. Rep. 341 85-171
[4] Faure F and Zhilinskii B 2002 Acta Appl. Math. 70 265-282
[5] Pavlov-Verevkin V B Sadovskii D and Zhilinskii B 1987 Europhys. Lett. 6 573-578
[6] Zhilinskii B and Brodersen S 1994 J. Mol. Spectrosc. 163 326-338
[7] Brodersen S and Zhilinskii B 1995 J. Mol. Spectrosc. 169 1-17
[8] Faure F and Zhilinskii B 2000 Phys. Rev. Lett. 85 960-963
[9] Faure F and Zhilinskii B 2001 Lett. Math. Phys. 55 219-238
[10] Faure F and Zhilinskii B 2002 ArXive: quant-ph/0204100 Phys. Lett. A (in press)
[11] Fedosov B 2000 Comm. Math. Phys. 209 29-49
[12] Sadovskii D and Zhilinskii B 1999 Phys. Lett. A 256 235-244
[13] Duistermaat J J 1980 Comm.Pure Appl. Math. 33 687-706
[14] Cushman R H and Bates L M 1997 Global aspects of classical integrable systems (Basel: Birkhäuser)
[15] Matveev V 1996 Sb. Math. 187 495-524
[16] Cushman R H and Duistermaa J J 1988 Bull. Am. Math. Soc. 19 475-479
[17] San Vu Ngoc 1999 Comm. Math. Phys. 203 465-479
[18] Grondin L Sadovskii D and Zhilinskii B 2002 Phys. Rev. A 65 012105 1-15
[19] Duistermaa J J and Heckman G J 1982 Invent. Math. 69 259-269
[20] Mermin N D 1979 Rev. Mod. Phys. 51 591-648
[21] Michel L 1980 Rev. Mod. Phys. 52 617-651
[22] Kleman M 1983 Points Lines and Walls (Chicester: Wiley).
[23] Nekhoroshev N N 1972 Trans.Moscow Math. Soc. 26 180-198
[24] Cushman R H and San Vu Ngoc 2002 Annales Henri Poincaré 3 in press
[25] Cushman R H and Duistermaa J J 2001 J. Diff. Eq. 172 42-58
[26] Bender C M and Boettcher S 1998 Phys. Rev. Lett. 80 5243-5246
[27] Cushman R H and Zhilinskii B 2002 J. Phys. A: Math. Gen. 35 L415-L419
[28] Nekhoroshev N N Sadovskii D and Zhilinskii B 2002 submitted for publication

Inst. Phys. Conf. Ser. No 173: Section 7
Paper presented at 24th Int. Coll. Group Theoretical Methods in Physics, Paris, France, 15–20 July 2002
©2003 IOP Publishing Ltd

$\mathcal{PT}$ symmetry and supersymmetry

M Znojil

Nuclear Physics Institute, 250 68 Řež, Czech Republic

Abstract. Pseudo-Hermitian (so called PT symmetric) Hamiltonians are featured within a re-formulated Witten's supersymmetric quantum mechanics. An unusual form of the supersymmetric partnership between the spiked harmonic oscillators is described.

1. Supersymmetry for pedestrians, or bosons and fermions in Witten's picture

In the review [1] of the Witten's (or so called supersymmetric) quantum mechanics one finds the specific Fock vacuum

$$\langle q|0\rangle = \left[\begin{array}{c} \exp(-q^2/2)/\sqrt{\pi} \\ 0 \end{array} \right], \qquad q \in (-\infty, \infty). \tag{1}$$

In the upper line we recognize the ground state of the one-dimensional harmonic oscillator $H^{(HO)} = p^2 + q^2$ and see that "bosons" may be created/annihilated by the action of the respective operators $-\partial_r + q$ and $\partial_r + q$ (to be denoted as $B^{(-1/2)}$ and $A^{(-1/2)}$ here). The solvable harmonic-oscillator character of this elementary model enables us to define the three two-by-two matrices

$$\mathcal{F}^\dagger = \left[\begin{array}{cc} 0 & 0 \\ 1 & 0 \end{array} \right], \qquad \mathcal{F} = \left[\begin{array}{cc} 0 & 1 \\ 0 & 0 \end{array} \right], \qquad \mathcal{N}_{\mathcal{F}} = \left[\begin{array}{cc} 0 & 0 \\ 0 & 1 \end{array} \right] \tag{2}$$

which create and annihilate a "fermion" and determine the fermionic number, respectively. Formally, this enables us to introduce the two partner Hamiltonians $H_{(L)} = H^{(HO)} - 1$, $H_{(R)} = H^{(HO)} + 1$ and identify the underlying symmetry with the superalgebra $sl(1/1)$ generated by the following three operator matrices

$$\mathcal{H} = \left[\begin{array}{cc} H_{(L)} & 0 \\ 0 & H_{(R)} \end{array} \right], \qquad \mathcal{Q} = \left[\begin{array}{cc} 0 & 0 \\ A & 0 \end{array} \right], \qquad \tilde{\mathcal{Q}} = \left[\begin{array}{cc} 0 & B \\ 0 & 0 \end{array} \right]. \tag{3}$$

One easily verifies that $\{\mathcal{Q}, \tilde{\mathcal{Q}}\} = \mathcal{H}$ while $\{\mathcal{Q}, \mathcal{Q}\} = \{\tilde{\mathcal{Q}}, \tilde{\mathcal{Q}}\} = 0$ and $[\mathcal{H}, \mathcal{Q}] = [\mathcal{H}, \tilde{\mathcal{Q}}] = 0$.

Our forthcoming considerations may by summarized as a generalization of the above harmonic-oscillator-based model to D dimensions. Beyond its obvious phenomenological and methodical appeal, the mathematical motivation for such a construction stems from the well known requirements of absence of the centrifugal-type singularities in the general Witten's formalism [1]. Indeed, within the standard Hermitian quantum mechanics, all ways of suppression of this difficulty seem to remain unclear up to these days [2]. In contrast, the weakening of the Hermiticity (to the so called $\mathcal{PT}$ symmetry – see below) appears to be amazingly efficient in this context [3, 4].

2. $\mathcal{PT}$ symmetry for pedestrians, interpreted as a regularization of a spike in the force

Our key idea dates back to Buslaev and Grecchi [5] who proposed a specific regularization of the non-vanishing centrifugal term in $D \neq 1$ dimensions (in their case, for some specific anharmonic oscillator examples) via a constant complex shift of the coordinate $r = |\vec{q}|$, better understood as a transition to the (increasingly popular [6]) non-Hermitian formalism of the so called $\mathcal{PT}$ symmetric quantum mechanics [7]. In its present implementation, this merely means an extension of the real-line symmetry $\mathcal{P}r = -r$ in (2.1) to the complex plane of $r \in \mathbb{C}$. Thus, we require the invariance of our Hamiltonian with respect to the parity $\mathcal{P}$ *multiplied by* the time reversal mimicked by the complex conjugation, $\mathcal{T}i = -i$.

2.1. Illustration: $\mathcal{PT}$ symmetric version of the D−dimensional harmonic oscillator

For the sake of brevity we shall only assign here our supersymmetry generators to the generalized class of the spiked harmonic oscillator Hamiltonians

$$H^{(\alpha)} = -\frac{d^2}{dr^2} + \frac{\alpha^2 - 1/4}{r^2} + r^2, \qquad \alpha > 0.$$

Their Buslaev's and Grecchi's regularization will use $r = x - i\varepsilon$ with real x and has thoroughly been studied in ref. [8]. Its normalizable wave functions

$$\psi(r) \equiv \mathcal{L}_N^{(\varrho)} = \langle r|N, \varrho \rangle = \frac{N!}{\Gamma(N + \varrho + 1)} \cdot r^{\varrho+1/2} \exp(-r^2/2) \cdot L_N^{(\varrho)}(r^2) \quad (4)$$

and energies

$$E = E_N^{(\varrho)} = 4N + 2\varrho + 2, \qquad \varrho = -Q \cdot \alpha$$

are both labeled by an additional quantum number $Q = \pm 1$ of the so called quasi-parity. In the context of fields, this concept proves closely related to the well known charge-conjugation symmetry $\mathcal{C}$ [9].

3. $\mathcal{PT}$ symmetric supersymmetry and the pedestrian's spiked harmonic oscillator

We noticed in ref. [4] that the complexification of r regularizes the Witten's spiked harmonic oscillator (SHO) superpotential

$$W^{(\gamma)}(r) = -\frac{\partial_r \langle r|0, \gamma \rangle}{\langle r|0, \gamma \rangle} = r - \frac{\gamma + 1/2}{r} \qquad r = r(x) = x - i\varepsilon, \quad x \in \mathbb{R}$$

as well as all the related operator matrix elements in the $sl(1/1)$ generators (3),

$$A^{(\gamma)} = \partial_r + W^{(\gamma)}, \qquad B^{(\gamma)} = -\partial_r + W^{(\gamma)}, \qquad \gamma \neq 0, \pm 1, \ldots, \quad (5)$$

$$H_{(L)} = B \cdot A = \hat{p}^2 + W^2 - W', \qquad H_{(R)} = A \cdot B = \hat{p}^2 + W^2 + W'.$$

This returns us to the $D = 1$ oscillator of section 1 at the "exceptional" value of $\gamma = -1/2$ while at all the complex γ we have the generalized SUSY partners,

$$H_{(L)}^{(\gamma)} = H^{(\alpha)} - 2\gamma - 2, \quad H_{(R)}^{(\gamma)} = H^{(\beta)} - 2\gamma, \quad \alpha = |\gamma|, \qquad \beta = |\gamma + 1|.$$

Whenever $\gamma \in (-\infty, \infty)$ is real (while $\alpha = |\gamma|$ and $\beta = |\gamma + 1|$ are defined as positive), we have to distinguish between the following three different SUSY regimes characterized by the unbroken $\mathcal{PT}$ symmetry,

$$\begin{cases} I. & \text{large negative } \gamma = -\alpha < -1, \quad \text{dominant } \alpha = \beta + 1 \\ II. & \text{small negative } \gamma = -\alpha > -1, \quad \text{both small, } \alpha + \beta = 1 \\ III. & \text{positive } \gamma = \alpha > 0, \quad \text{dominant } \beta = \alpha + 1 \end{cases}.$$

The energies (arranged in the descending order) form a (once degenerate) *completely real* quadruplet at each N,

SUSY partner energies	*I.*	*II.*	*III.*
$E_{(L)}^{(\beta)}$	$4N + 4\alpha$	$4N + 4$	$4N + 4$
$E_{(L)}^{(\alpha)}$	$4N + 4\alpha$	$4N + 4\alpha$	$4N$
$E_{(L)}^{(-\beta)}$	$4N + 4$	$4N + 4\alpha$	$4N - 4\alpha$
$E_{(L)}^{(-\alpha)}$	$4N$	$4N$	$4N - 4\alpha$

It is amusing to notice that up to the regular case (with $\alpha = 1/2$) there always exist two alternative $\gamma = \pm\alpha$ to a given $\alpha > 0$. Thus, each α also has the *two different* partners β such that $\beta_1 = |\alpha - 1| < \alpha < \beta_2 = \alpha + 1$. Finally, in the domain II, all our SUSY construction remains perfectly valid even in the Hermitian limit $\varepsilon \to 0$ [10].

4. Non-standard $\mathcal{PT}$ symmetric supersymmetries

4.1. Working at a fixed parameter α

The respective annihilation and creation of ref. [4] was mediated by the *second-order* differential operators

$$A^{(-\gamma-1)} \cdot A^{(\gamma)} = A^{(\gamma-1)} \cdot A^{(-\gamma)} = \mathbf{A}(\alpha)$$

$$B^{(-\gamma)} \cdot B^{(\gamma-1)} = B^{(\gamma)} \cdot B^{(-\gamma-1)} = \mathbf{B}(\alpha)$$

with the "norm" $c_5(N, \gamma) = -4\sqrt{(N + 1)(N + \gamma + 1)}$ and property

$$\mathbf{A}(\alpha) \cdot \mathcal{L}_{N+1}^{(\gamma)} = c_5(N, \gamma) \, \mathcal{L}_N^{(\gamma)}, \qquad \mathbf{B}(\alpha) \cdot \mathcal{L}_N^{(\gamma)} = c_5(N, \gamma) \, \mathcal{L}_{N+1}^{(\gamma)}.$$

Hamiltonian $H^{(\alpha)} = [\mathbf{A}(\alpha)\,\mathbf{B}(\alpha) - \mathbf{B}(\alpha)\,\mathbf{A}(\alpha)]/8$ satisfies commutation relations

$$\mathbf{A}(\alpha)\,H^{(\alpha)} - H^{(\alpha)}\,\mathbf{A}(\alpha) \equiv 4\,\mathbf{A}(\alpha), \quad H^{(\alpha)}\,\mathbf{B}(\alpha) - \mathbf{B}(\alpha)\,H^{(\alpha)} \equiv 4\,\mathbf{B}(\alpha)$$

of the Lie algebra $sl(2, \mathbb{R})$ with the normalized generators $\mathbf{A}(\alpha)/\sqrt{32}$, $\mathbf{B}(\alpha)/\sqrt{32}$ and $H^{(\alpha)}/4$. As a consequence, the new, $\mathcal{PT}$ SUSY results from Eq. (3), with A, B and $H_{(L/R)}$ replaced by $\mathbf{A}(\alpha)$, $\mathbf{B}(\alpha)$ and $\mathbf{G}_{(\mathbf{L/R})} = (H^{(\alpha)} \mp 2)^2 - 4\alpha^2$, respectively. The SHO eigenvectors themselves may then be obtained as solutions of the differential equations of the fourth order (cf. ref. [11]) which, in our case, read

$$\mathbf{G}_{(L)} \left| N^{(\gamma)} \right\rangle = \Omega_N^{(\gamma)} \left| N^{(\gamma)} \right\rangle, \qquad \mathbf{G}_{(R)} \left| N^{(\gamma)} \right\rangle = \Omega_{N+1}^{(\gamma)} \left| N^{(\gamma)} \right\rangle.$$

where $\Omega_N^{(\gamma)} = 16\,N\,(N + \gamma)$. This is our present main result.

4.2. SUSY constructions at the complex γ

Marginally, let us note that even the complex choice of γ (when the $\mathcal{PT}$ symmetry itself is broken) may lead to the partially real SUSY spectrum of energies. In order to show that, one has to derive a few identities for the Laguerre polynomials in (4) showing that the operators (5) change merely the subscripts or superscripts [4]. In the regime with the spontaneously broken $\mathcal{PT}$ symmetry we may distinguish between the two options,

$$\begin{cases} \delta > 0 \text{ in } \gamma = i\,\delta, & \alpha = i\,\delta, & \beta = 1 + \alpha \\ \eta > 0 \text{ in } \gamma = -i\,\eta, & \alpha = i\,\eta, & \beta = 1 - \alpha \end{cases}$$

and get the partially real energy multiplets

$$\begin{cases} E_{(L)}^{(+\alpha)} = 4n, & E_{(L)}^{(-\alpha)} = E_{(R)}^{(-\beta)} = 4n - 4\alpha, & E_{(R)}^{(+\beta)} = 4n + 4 \\ E_{(L)}^{(-\alpha)} = 4n, & E_{(L)}^{(+\alpha)} = E_{(R)}^{(-\beta)} = 4n + 4\alpha, & E_{(R)}^{(+\beta)} = 4n + 4 \end{cases}$$

Similarly, at $\gamma = N + i\,q\,\delta$ with $q = \pm 1$ and $\delta > 0$, i.e., with no $\mathcal{PT}$ symmetry at all, we get

$$E_{(L)} = \begin{cases} 4n - 4\gamma \\ 4n \end{cases} , \qquad E_{(R)} = \begin{cases} 4n - 4\gamma \\ 4n + 4 \end{cases}$$

for the indices $\beta_1 = N - 1 + i\,\delta$, $\alpha = N + i\,\delta$ and $\beta_2 = N + 1 + i\,\delta$, still giving the partially real energy spectra.

Acknowledgements

Work partially supported by the grant of GA AS CR Nr. A 1048004.

References

[1] Cooper F, Khare A and Sukhatme U 1995 Phys. Rep. 251 267

[2] Jevicki A and Rodriguez J 1984 Phys. Lett. B 146 55;
    Das A and Pernice S 1999 Nucl. Phys. B 561 357;
    Gangopadhyaya A and Mallow J W 2002 "Supersymmetry in the Half-Oscillator - Revisited", LANL arXiv: hep-th/0206133;
    Das A and Pernice S 2002 "Comment on 'Supersymmetry in the half-oscillator revisited'", LANL arXiv: hep-th/0207112

[3] Znojil M 2000 Annihilation and creation operators in non-Hermitian supersymmetric quantum mechanics, arXiv hep-th/0012002

[4] Znojil M 2002 J. Phys. A: Math. Gen. 35 2341

[5] Buslaev V and Grecchi V 1993 J. Phys. A: Math. Gen. 26 5541

[6] Bender C M or Lévai G or Mostafazadeh A or Quesne C 2002 Group 24 conference proceedings

[7] Bender C M, Boettcher S and Meisinger P N 1999 J. Math. Phys. 40 2201

[8] Znojil M 1999 Phys. Lett. A 259 220

[9] Streater R F and Wightman A S 1964 PCT, spin and statistics and all that (New York: Benjamin);
    Bender C M, Brody D C and Jones H F 2002 Complex extension of quantum mechanics, arXiv: quant-ph/0208076

[10] Znojil M 2002 Re-establishing supersymmetry between harmonic oscillators in $D \neq 1$ dimensions, arXiv: hep-th/0203252

[11] Bagchi B, Mallik S and Quesne C 2002 Int. J. Mod. Phys. A 17 51

Algebras, Groups and their Representations

Inst. Phys. Conf. Ser. No 173: Section 8
Paper presented at 24th Int. Coll. Group Theoretical Methods in Physics, Paris, France, 15–20 July 2002
©2003 IOP Publishing Ltd

New infinite-dimensional Lie algebras of $U(N_+, N_-)$-tensor operators and applications

M Calixto

Department of Applied Mathematics and Statistics, Polytechnic University of Cartagena,
Paseo Alfonso XIII 52, 30203 Cartagena, Spain

Abstract. The structure constants for Moyal brackets of an infinite basis of functions on the algebraic manifolds M of pseudo-unitary groups $U(N_+, N_-)$ are provided. They generalize the Virasoro and $\mathcal{W}_\infty$ symmetries to higher dimensions. These infinite-dimensional Lie-algebras provide also the arena for non-linear integrable field theories in higher dimensions, residual gauge symmetries of higher-extended objects in the light-cone gauge and C^*-algebras for tractable non-commutative versions of symmetric curved spaces.

The general study of infinite-dimensional algebras and groups, their quantum deformations (in particular, central extensions) and representation theory has not progressed very far, except for some important achievements in one- and two-dimensional systems, and there can be no doubt that a breakthrough in the subject would provide new insights into the two central problems of modern physics: unification of all interactions and exact solvability in QFT and statistics.

The aforementioned achievements refer mainly to Virasoro and Kac-Moody symmetries, which have played a fundamental role in the analysis and formulation of conformally-invariant (quantum and statistical) field theories in one and two dimensions, and systems in higher dimensions which in some essential respects are one- or two-dimensional (e.g. String Theory). Generalizations of the Virasoro symmetry, as the algebra $\mathrm{diff}(S^1)$ of reparametrisations of the circle, lead to the infinite-dimensional Lie algebras of area-preserving diffeomorphisms $\mathrm{sdiff}(\Sigma)$ of two-dimensional surfaces Σ. These algebras naturally appear as a residual gauge symmetry in the theory of relativistic membranes [1], which exhibits an intriguing connection with the quantum mechanics of space constant (e.g. vacuum configurations) $SU(N)$ Yang-Mills potentials in the limit $N \to \infty$ [2]; the argument that the internal symmetry space of the $U(\infty)$ pure Yang-Mills theory must be a functional space, actually the space of configurations of a string, was pointed out in Ref. [3]. Moreover, the $\mathcal{W}_\infty$ and $\mathcal{W}_{1+\infty}$ algebras of area-preserving diffeomorphisms of the cylinder [4] generalize the underlying Virasoro gauged symmetry of the light-cone two-dimensional induced gravity discovered by Polyakov [5] by including all positive conformal-spin currents [6], and induced actions for these $\mathcal{W}$-gravity theories have been proposed [7, 8]. Additionally, the $\mathcal{W}_{1+\infty}$ (dynamical) symmetry has been identified by [9] as the set of canonical transformations that leave invariant the Hamiltonian of a two-dimensional electron gas in a perpendicular magnetic field, and appears to be relevant in the classification of all the universality classes of *incompressible quantum fluids* and the identification of the quantum numbers of the excitations in the Quantum Hall Effect. Higher-spin symmetry algebras where introduced in [10] and could provide a guiding principle towards the still unknown "M-theory".

It is remarkable that area-preserving diffeomorphisms, higher-spin and $\mathcal{W}$ algebras can be seen as distinct members of a one-parameter family $\mathcal{L}_\mu(su(2))$ —or the non-compact version $\mathcal{L}_\mu(su(1,1))$— of non-isomorphic [11] infinite-dimensional Lie-algebras of $SU(2)$ —and $SU(1,1)$— tensor operators, more precisely, the factor algebra $\mathcal{L}_\mu(su(2)) =$

$\mathcal{U}(su(2))/\mathcal{I}_\mu$ of the universal enveloping algebra $\mathcal{U}(su(2))$ by the ideal $\mathcal{I}_\mu = (\hat{C} - \hbar^2\mu)\mathcal{U}(su(2))$ generated by the Casimir operator $\hat{C}$ of $su(2)$ (μ denotes an arbitrary complex number). The structure constants for $\mathcal{L}_\mu(su(2))$ and $\mathcal{L}_\mu(su(1,1))$ are well known for the Racah-Wigner basis of tensor operators [12], and they can be written in terms of Clebsch-Gordan and (generalized) $6j$-symbols [1, 6, 13]. Another interesting feature of $\mathcal{L}_\mu(su(2))$ is that, when μ coincides with the eigenvalue of $\hat{C}$ in an irrep D_j of $SU(2)$, that is $\mu = j(j+1)$, there exists and ideal χ in $\mathcal{L}_\mu(su(2))$ such that the quotient $\mathcal{L}_\mu(su(2))/\chi \simeq sl(2j+1, C)$ or $su(2j+1)$, by taking a compact real form of the complex Lie algebra. That is, for $\mu = j(j+1)$ the infinite-dimensional algebra $\mathcal{L}_\mu(su(2))$ collapses to a finite-dimensional one. This fact was used in [1] to approximate $\lim_{\substack{\mu\to\infty \\ \hbar\to 0}} \mathcal{L}_\mu(su(2)) \simeq \text{sdiff}(S^2)$ by $su(N)|_{N\to\infty}$ ("large number of colours").

The generalization of these constructions to general unitary groups proves to be quite unwieldy, and a canonical classification of $U(N)$-tensor operators has, so far, been proven to exist only for $U(2)$ and $U(3)$ (see [12] and references therein). Tensor labelling is provided in these cases by the Gel'fand-Weyl pattern for vectors in the carrier space of the irreps of $U(N)$.

In [14], a quite appropriate basis of operators for $\mathcal{L}_{\vec{\mu}}(u(N_+, N_-))$, $\vec{\mu} = (\mu_1, \ldots, \mu_N)$, $N \equiv N_+ + N_-$ was provided. The structure constants of this infinite-dimensional Lie algebra were calculated for the particular case of the boson realization of the $U(N_+, N_-)$ Lie algebra generators in terms of N oscillator variables. In this case, Moyal bracket captures the essence of more general deformations.

The particular set of operators in $\mathcal{U}(u(N_+, N_-))$ is the following:

$$\hat{L}^I_{|m|} \equiv \prod_\alpha (\hat{G}_{\alpha\alpha})^{I_\alpha - (\sum_{\beta>\alpha}|m_{\alpha\beta}| + \sum_{\beta<\alpha}|m_{\beta\alpha}|)/2} \prod_{\alpha<\beta} (\hat{G}_{\alpha\beta})^{|m_{\alpha\beta}|}$$

$$\hat{L}^I_{-|m|} \equiv \prod_\alpha (\hat{G}_{\alpha\alpha})^{I_\alpha - (\sum_{\beta>\alpha}|m_{\alpha\beta}| + \sum_{\beta<\alpha}|m_{\beta\alpha}|)/2} \prod_{\alpha<\beta} (\hat{G}_{\beta\alpha})^{|m_{\alpha\beta}|} \tag{1}$$

where $\hat{G}_{\alpha\beta}$, $\alpha, \beta = 1, \ldots, N$, are the $U(N_+, N_-)$ Lie-algebra (step) generators with commutation relations:

$$\left[\hat{G}_{\alpha_1\beta_1}, \hat{G}_{\alpha_2\beta_2}\right] = \hbar(\eta_{\alpha_1\beta_2}\hat{G}_{\alpha_2\beta_1} - \eta_{\alpha_2\beta_1}\hat{G}_{\alpha_1\beta_2}), \tag{2}$$

and the indefinite metric $\eta = \text{diag}(1, \overset{N_+}{\ldots}, 1, -1, \overset{N_-}{\ldots}, -1)$ is used to raise and lower indices; the upper (generalized spin) index $I \equiv (I_1, \ldots, I_N)$ of $\hat{L}$ in (1) represents a N-dimensional vector, which is taken to lie on an half-integral lattice; the lower index m symbolizes a integral upper-triangular $N \times N$ matrix, and $|m|$ means absolute value of all its entries. Thus, the operators $\hat{L}^I_m$ are labelled by $N+N(N-1)/2 = N(N+1)/2$ indices, in the same way as wave functions ψ^I_m in the carrier space of irreps of $U(N)$. The $U(N_+, N_-)$ Casimir operators are polynomials of degree $1, 2, \ldots, N$ of step operators $\hat{G}_\alpha$ as follows: $\hat{C}_1 = \hat{G}^\alpha_\alpha, \hat{C}_2 = \hat{G}^\beta_\alpha\hat{G}^\alpha_\beta, \ldots$

The manifest expression of the structure constants f for the commutators

$$\left[\hat{L}^I_m, \hat{L}^J_n\right] = \hat{L}^I_m\hat{L}^J_n - \hat{L}^J_n\hat{L}^I_m = f^{IJl}_{mnK}[\vec{\mu}]\hat{L}^K_l \tag{3}$$

of a pair of operators (1) of $\mathcal{L}_{\vec{\mu}}(u(N_+, N_-))$ entails an unpleasant and difficult computation, because of inherent ordering problems. However, the essence of the full quantum algebra $\mathcal{L}_{\vec{\mu}}(u(N_+, N_-))$ can be still captured in a classical construction by extending the Poisson-Lie bracket

$$\left\{L^I_m, L^J_n\right\}_{\text{PL}} = (\eta_{\alpha_1\beta_2}G_{\alpha_2\beta_1} - \eta_{\alpha_2\beta_1}G_{\alpha_1\beta_2})\frac{\partial L^I_m}{\partial G_{\alpha_1\beta_1}}\frac{\partial L^J_n}{\partial G_{\alpha_2\beta_2}} \tag{4}$$

of a pair of functions L_m^I, L_n^J on the commuting coordinates $G_{\alpha\beta}$ to its deformed version, in the sense of Ref. [15]. To perform calculations with (4) is still rather complicated because of non-canonical brackets for the generating elements $G_{\alpha\beta}$. Nevertheless, there is a standard boson operator realization $G_{\alpha\beta} \equiv a_\alpha \bar{a}_\beta$ of the generators $G_{\alpha\beta}$ of $u(N_+, N_-)$ in terms of N oscillator variables $(a_\alpha, \bar{a}_\beta)$, for which things simplify greatly. Indeed, we shall understand that the quotient by the ideal generated by polynomials $G_{\alpha_1\beta_1}G_{\alpha_2\beta_2} - G_{\alpha_1\beta_2}G_{\alpha_2\beta_1}$ is taken, so that the Poisson-Lie bracket (4) coincides with the standard Poisson bracket

$$\left\{L_m^I, L_n^J\right\}_{\mathrm{P}} = \eta_{\alpha\beta}\left(\frac{\partial L_m^I}{\partial a_\alpha}\frac{\partial L_n^J}{\partial \bar{a}_\beta} - \frac{\partial L_m^I}{\partial \bar{a}_\beta}\frac{\partial L_n^J}{\partial a_\alpha}\right) \tag{5}$$

for the Heisenberg-Weyl algebra. There is basically only one possible deformation of the bracket (5) —corresponding to a full symmetrization— that fulfils the Jacobi identities [15], which is the Moyal bracket [16]:

$$\left\{L_m^I, L_n^J\right\}_{\mathrm{M}} = L_m^I * L_n^J - L_n^J * L_m^I = \sum_{r=0}^{\infty} 2\frac{(\hbar/2)^{2r+1}}{(2r+1)!}P^{2r+1}(L_m^I, L_n^J), \tag{6}$$

where $L * L' \equiv \exp(\frac{\hbar}{2}P)(L, L')$ is an invariant associative $*$-product and

$$P^r(L, L') \equiv \Upsilon_{\iota_1 \jmath_1}\ldots\Upsilon_{\iota_r \jmath_r}\frac{\partial^r L}{\partial x_{\iota_1}\ldots\partial x_{\iota_r}}\frac{\partial^r L'}{\partial x_{\jmath_1}\ldots\partial x_{\jmath_r}}, \tag{7}$$

with $x \equiv (a, \bar{a})$ and $\Upsilon_{2N\times 2N} \equiv \begin{pmatrix} 0 & \eta \\ -\eta & 0 \end{pmatrix}$. We set $P^0(L, L') \equiv LL'$; see also that $P^1(L, L') = \{L, L'\}_{\mathrm{P}}$. It is worthwhile mentioning that the Moyal bracket (6) was identified as the primary quantum deformation $\mathcal{W}_\infty$ of the classical algebra w_∞ of area-preserving diffeomorphisms of the cylinder (see Ref. [17]).

With these simplifications, the manifest expression of the structure constants f for the Moyal bracket (6) is the following:

$$\left\{L_m^I, L_n^J\right\}_{\mathrm{M}} = \sum_{r=0}^{\infty} 2\frac{(\hbar/2)^{2r+1}}{(2r+1)!}\eta^{\alpha_0\alpha_0}\ldots\eta^{\alpha_{2r}\alpha_{2r}}f_{mn}^{IJ}(\alpha_0,\ldots,\alpha_{2r})L_{m+n}^{I+J-\sum_{j=0}^{2r}\delta_{\alpha_j}},$$

$$f_{mn}^{IJ}(\alpha_0,\ldots,\alpha_{2r}) = \sum_{\wp\in\Pi_2^{(2r+1)}}(-1)^{\ell_\wp+1}\prod_{s=0}^{2r}f_\wp(I_{\alpha_{\wp(s)}}^{(s)}, m)f_\wp(J_{\alpha_{\wp(s)}}^{(s)}, -n),$$

$$f_\wp(I_{\alpha_{\wp(s)}}^{(s)}, m) = I_{\alpha_{\wp(s)}}^{(s)} + (-1)^{\theta(s-\ell_\wp)}\Big(\sum_{\beta>\alpha_{\wp(s)}}m_{\alpha_{\wp(s)}\beta} - \sum_{\beta<\alpha_{\wp(s)}}m_{\beta\alpha_{\wp(s)}}\Big)/2,$$

$$I_{\alpha_{\wp(s)}}^{(s)} = I_{\alpha_{\wp(s)}} - \sum_{t=(\ell_\wp+1)\theta(s-\ell_\wp)}^{s-1}\delta_{\alpha_{\wp(t)},\alpha_{\wp(s)}}, \quad I^{(0)} = I^{(\ell_\wp+1)} \equiv I,$$

$$\theta(s-\ell_\wp) = \begin{cases} 0, & s\leq \ell_\wp \\ 1, & s>\ell_\wp \end{cases}, \quad \delta_{\alpha_j} = (\delta_{1,\alpha_j},\ldots,\delta_{N,\alpha_j}), \tag{8}$$

where $\Pi_2^{(2r+1)}$ denotes the set of all possible partitions $\wp$ of a string $(\alpha_0,\ldots,\alpha_{2r})$ of length $2r+1$ into two substrings

$$\overbrace{(\alpha_{\wp(0)},\ldots,\alpha_{\wp(\ell)})}^{\ell_\wp}\overbrace{(\alpha_{\wp(\ell+1)},\ldots,\alpha_{\wp(2r)})}^{2r+1-\ell_\wp} \tag{9}$$

of length $\ell_\wp$ and $2r+1-\ell_\wp$, respectively. The number of elements $\wp$ in $\Pi_2^{(2r+1)}$ is clearly $\dim(\Pi_2^{(2r+1)}) = \sum_{\ell=0}^{2r+1}\frac{(2r+1)!}{(2r+1-\ell)!\ell!} = 2^{2r+1}$.

For $r = 0$, there are just 2 partitions: $(\alpha)(\cdot)$, $(\cdot)(\alpha)$, and the leading (classical, $\hbar \to 0$) structure constants are, for example:

$$f_{mn}^{IJ}(\alpha) = J_\alpha \left(\sum_{\beta > \alpha} m_{\alpha\beta} - \sum_{\beta < \alpha} m_{\beta\alpha} \right) - I_\alpha \left(\sum_{\beta > \alpha} n_{\alpha\beta} - \sum_{\beta < \alpha} n_{\beta\alpha} \right). \tag{10}$$

They reproduce in this limit the Virasoro commutation relations for the particular generators $V_k^{(\alpha\beta)} \equiv L_{ke_{\alpha\beta}}^{\delta_\alpha}$, where $k \in Z$ and $(e_{\alpha\beta})_{\mu\nu} = \delta_{\alpha\mu}\delta_{\beta\nu}$. Indeed, there are $N(N-1)$ *non-commuting* Virasoro sectors in (8), corresponding to each positive root in $SU(N_+, N_-)$, with classical commutation relations:

$$\left\{ V_k^{(\alpha\beta)}, V_l^{(\alpha\beta)} \right\}_{\text{P}} = \eta^{\alpha\alpha} \text{sign}(\beta - \alpha)\,(k - l) V_{k+l}^{(\alpha\beta)}. \tag{11}$$

Note the close resemblance between the algebra (8) —and the leading structure constants (10)— and the quantum deformation $\mathcal{W}_\infty \simeq \mathcal{L}_0(su(1,1))$ of the algebra of area-preserving diffeomorphisms of the cylinder [6, 17], although we recognize that the case discussed in this letter is far richer.

A thorough study of the Lie-algebra cohomology (central extensions) of $\mathcal{L}_{\vec{\mu}}(u(N_+, N_-))$ and its irreps still remains to be accomplished; it requires a separate attention and shall be left for future works [18]. Two-cocycles provide the essential ingredient to construct invariant geometric action functionals on coadjoint orbits of $\mathcal{L}_{\vec{\mu}}(u(N_+, N_-))$ —see e.g. [8] for the derivation of the Wess-Zumino-Witten action of $D = 2$ matter fields coupled to chiral $\mathcal{W}_\infty$ gravity background from $\mathcal{W}_\infty \simeq \mathcal{L}_0(su(1,1)))$.

References

[1] Hoppe J 1982 MIT Ph.D. Thesis; 1989 Int. J. Mod. Phys. A 4 5235
[2] Floratos E G, Iliopoulos J and Tiktopoulos G 1989 Phys. Lett. B 217 285
[3] Gervais J L and Neveu A 1981 Nucl. Phys. B 192 463
[4] Bakas I 1989 Phys. Lett. B 228 57
[5] Polyakov A M 1981 Phys. Lett. B 103 207
[6] Pope C N, Shen X and Romans L J 1990 Nucl. Phys. B 339 191
[7] Bergshoeff E, Pope C N, Romans L J, Sezgin E, Shen X and Stelle K S 1990 Phys. Lett. B 243 350
[8] Nissimov E, Pacheva S and Vaysburd I 1992 Phys. Lett. B 288 254
[9] Cappelli A and Zemba G R 1997 Nucl. Phys. B 490 595
[10] Fradkin E S and Vasiliev M A 1987 Ann. Phys. (NY) 77 63
[11] Bordemann M, Hoppe J and Schaller P 1989, Phys. Lett. B 232 199
[12] Biedenharn L C and Louck J D 1981 The Racah-Wigner algebra in quantum theory (Addison-Wesley, New York, MA)
    Biedenharn L C and Lohe M A 1995 Quantum group symmetry and q-tensor algebras (World Scientific, Singapore)
[13] Fradkin E S and Linetsky V Y 1991 J. Math. Phys. 32 1218
[14] M. Calixto 2000 J. Phys. A: Math. Gen. 33 L69-L75; 2000 Mod. Phys. Lett. A 15 939-944; 2001 Class. Quantum Grav. 18 3857-3884
[15] Bayen F, Flato M, Fronsdal C, Lichnerowicz A and Sternheimer D 1978 Ann. Phys. (NY) 111 61
[16] Moyal J E 1949 Proc. Cambridge Philos. Soc. 45 99
[17] Fairlie D B and Nuyts J 1990 Commun. Math. Phys. 134 413
[18] M. Calixto, "Group, Tensor Operator, Poisson and Diffeomorphism Algebras of $U(N_+, N_-)$", in progress.

Inst. Phys. Conf. Ser. No 173: Section 8
Paper presented at 24th Int. Coll. Group Theoretical Methods in Physics, Paris, France, 15–20 July 2002
©*2003 IOP Publishing Ltd*

Invariants of Lie algebras having a frobeniusian structure

R Campoamor-Stursberg

Depto. Geometría y Topología
Fac. CC. Matemáticas U.C.M.
E-28040 Madrid (Spain)

E-mail: `rutwig@nfssrv.mat.ucm.es`

Abstract. We prove that if $\mathfrak{g}'$ is a contraction of a Lie algebra $\mathfrak{g}$ then the number of functionally independent invariants of $\mathfrak{g}'$ is at least that of $\mathfrak{g}$. Applying this fact to frobeniusian Lie algebras we deduce the existence of Lie algebras with non-zero Levi factor and having only trivial invariants.

1. Contractions of Lie algebras. Generalized Casimir invariants.

The study of the orbits of Lie algebras by the action of the general linear group leads naturally to interesting questions concerning the elements in a neighborhood of an algebra, their closure and limiting processes. Contractions of Lie algebras appeared when comparing similar but non-isomorphic symmetry groups related to physical systems [1], and have shown their effectiveness in the formalization of physical theories. During the last years, the study of the existing relations of the invariants of a Lie algebra with those of its contractions has become a relevant problem. As important examples we can cite the kinematical Lie algebras [2, 3], deformation theory and the classification of rigid structures [4]. Traditionally contractions of Lie algebras are presented as limits [5], although other authors have approached the contraction problem from the point of view of group actions [6]. A Lie algebra $\mathfrak{g} = (\mathbb{K}^n, \mu)$ may be considered as an element μ of the manifold $Hom\left(\bigwedge^2 \mathbb{K}^n, \mathbb{K}^n\right)$ via the skew-symmetric bilinear map $\mu : \mathfrak{g} \otimes \mathfrak{g} \to \mathfrak{g}$ defining the Lie bracket on $\mathfrak{g}$. Thus we can identify the Lie algebra $\mathfrak{g}$ with its law μ. The set $\mathcal{L}^n$ of Lie algebras is then a subset of the manifold $Hom\left(\bigwedge^2 \mathbb{K}^n, \mathbb{K}^n\right)$ on which the general linear group $GL(n, \mathbb{K})$ acts by :

$$(g \circ \mu)(x, y) = g^{-1}\left(\mu\left(gx, gy\right)\right), \quad g \in GL(n, \mathbb{K}) \, ; x, y \in \mathbb{K}^n \tag{1}$$

Clearly the orbit $\mathcal{O}(\mathfrak{g})$ under this action are the Lie algebras isomorphic to $\mathfrak{g}$. A Lie algebra μ_∞ is called a contraction of a Lie algebra μ_0 if $\mu_\infty \in \overline{\mathcal{O}(\mu_0)}$, the Zariski closure of the orbit (nontrivial if μ_∞ lies in the boundary of the orbit). This geometrical definition is nothing more than a topological reformulation of the classical concept of Inönü-Wigner contractions [1]. As known, these contractions can be viewed as singular changes of basis, starting from a fixed basis $\{X_1, .., X_n\}$ of a Lie algebra $\mathfrak{g}$. That is, considering a sequence of endomorphisms $\{f_p(\epsilon_1, .., \epsilon_r)\}_{p \in \mathbb{N} \cup \{0\}}$ (where f_0 can be taken as the identity and ϵ_i designates the parameters), for any p we have:

$$\mu_p := f_p^{-1} \circ \mu_0\left(f_p, f_p\right). \tag{2}$$

If the limit exists, it also represents a Lie algebra, and the law of the contraction is given by $\mu_\infty = \lim_{p\to\infty}\mu_p$. Therefore, if $\{C_{ij}^k\}$ are the structure constants of $\mathfrak{g}_0 = (\mathbb{K}^n, \mu_0)$ over a basis $\{X_1, .., X_n\}$ and $\{C_{ij}^k(p)\}$ the constants of $\mathfrak{g}_p = (\mathbb{K}^n, \mu_p)$, the law of μ_∞ is given by

$$\widetilde{C_{ij}^k} = \lim_{p\to\infty} C_{ij}^k(p). \tag{3}$$

Let G be a Lie group and $\mathfrak{g}$ the corresponding Lie algebra. The coadjoint representation of G is given by the mapping:

$$ad^* : G \to GL(\mathfrak{g}^*) : \left(ad_g^*F\right)(x) = F(ad_{g^{-1}}x), \; g \in G, F \in \mathfrak{g}^*, x \in \mathfrak{g} \tag{4}$$

We say that a function $F \in C^\infty(\mathfrak{g}^*)$ is an invariant for the coadjoint representation if

$$F(ad_{g^{-1}}x) = F(x) \tag{5}$$

for any $g \in G$. The usual method to compute the invariants of a Lie algebra is making use of the theory of linear partial differential equations [7]. Let $\{X_1, .., X_n\}$ be a basis of $\mathfrak{g}$ and let $\{C_{ij}^k\}$ be its structure constants over this basis. We can represent $\mathfrak{g}$ in the space $C^\infty(\mathfrak{g}^*)$ by the differential operators

$$\widehat{X}_i = -C_{ik}^k.x_k\frac{\partial}{\partial x_j} \tag{6}$$

where $[X_i, X_j] = C_{ij}^k X_k \; (1 \le i < j \le n)$. Then a function $F \in C^\infty(\mathfrak{g}^*)$ is an invariant if and only if it is a solution of the following system:

$$\left\{\widehat{X}_iF = 0, \; 1 \le i \le n\right\} \tag{7}$$

This reduces the determination of the invariants to a system of linear first-order partial differential equations. Solutions will be called generalized Casimir invariants (polynomial solutions naturally correspond to classical Casimir operators). For any given Lie algebra $\mathfrak{g} = (\mathbb{K}^n, \mu)$, the number of functionally independent invariants of the coadjoint representation can be computed from the brackets [8]. More specifically, let $A_{ijk}(\mu) := \left(C_{ij}^k x_k\right)$ be the matrix which represents the commutator table over the basis $\{X_1, .., X_n\}$, the $\{C_{ij}^k\}$ being the structure constants (for the chosen representative $\mu \in \mathcal{O}(\mathfrak{g})$). The matrix is clearly skew-symmetric, which implies that the rank is necessarily even. The cardinal $\mathcal{N}$ of a fundamental set of invariants of $\mathfrak{g}$ is given by:

$$\mathcal{N}(\mathfrak{g}) = \dim \mathfrak{g} - \sup\{rankA(\mu) \mid \mu \in \mathcal{O}(\mathfrak{g})\}. \tag{8}$$

2. The contraction formula

The contractions of the simple Lie algebra $\mathfrak{sl}(3, \mathbb{C})$ have been analyzed in detail, and their invariants have been obtained as limits of the invariants of $\mathfrak{sl}(3, \mathbb{C})$, at least in the case of Inönü-Wigner contractions [9]. In particular, from this analysis we get that contractions are expected to have more invariants than the algebra they come from. We will now prove that this important observation generalizes indeed to arbitrary continuous contractions of Lie algebras.

Proposition 1 *If $\mathfrak{g}_1 = (\mathbb{K}, \mu_1)$ is a contraction of $\mathfrak{g}_0 = (\mathbb{K}, \mu_0)$, then $\mathcal{N}(\mathfrak{g}_1) \ge \mathcal{N}(\mathfrak{g}_0)$.*

Let $\{X_1, .., X_n\}$ be a basis of $\mathfrak{g}_0$ and $f_p(\epsilon_1, .., \epsilon_r)$ be the sequence of endomorphisms such that $\mu_1 = \lim_{p\to\infty}\mu_p$. It is clear that on the transformed basis $\{f_pX_1, .., f_pX_n\}$ we obtain the matrix $A_{ijk}^p = \left(C_{ij}^k(p)\right)$. By application of elementary techniques of linear algebra we obtain that:

$$rank(A_{ijk}(\mu_0)) \ge rank\left(\lim_{p\to\infty}\left(A_{ijk}^p(\mu_p)\right)\right) \tag{9}$$

This inequality holds for any representative μ_0 of the Lie algebra $\mathfrak{g}_0$ and any family $f_p \in GL(n, \mathbb{K})$ realizing the contraction. Since any contraction can be realized as a deformation [6] (this is the non-trivial step, since it is based on a characterization of contractions in terms of inverse limits), the maximal rank of commutation matrices $A_{ijk}(\mu_1)$ of representatives μ_1 of $\mathfrak{g}_1$ is lower or equal to the rank of some commutator matrix of a representative of $\mathfrak{g}_0$. Therefore we obtain that:

$$sup\{rank(A_{ijk}(\mu_0))\} \geq sup\left\{rank\left(\lim_{p \to \infty}(A^p_{ijk}(\mu_1))\right)\right\} \tag{10}$$

from which the assertion follows.

This result coincides with the intuition that contractions have "less brackets" than the Lie algebra they come from. In particular, if $\mathfrak{g}$ is a contraction of a simple Lie algebra of rank p, then $\mathfrak{g}$ has at least p functionally independent invariants.

Corollary 1 *If $\mathfrak{g}$ has a contraction without any non-trivial invariants, then $\mathfrak{g}$ itself has only trivial invariants.*

Although this result cannot be formulated in terms of deformations (there exist deformations which are not related to a contraction [6]), for certain kinds of deformations it is possible. Recall that a jump deformation μ_t of a Lie algebra $(\mathbb{K}^n, \mu_0)$ is a formal deformation $\mu_t = \mu_0 + t\phi_1 + t^2\phi_2 + \ldots (\phi_i \in Hom(\wedge^2\mathbb{K}^n, \mathbb{K}^n))$ which remains constant for generic $t \neq 0$. That is, if u is an additional variable and coefficient are extended to $\mathbb{K}((u))[[u]]$, we have an isomorphism $\mu_t \simeq \mu_{(1+u)t}$.

Proposition 2 *Let $\mathfrak{g}$ be a Lie algebra satisfying $\mathcal{N}(\mathfrak{g}) = 0$. Then any jump deformation $\mathfrak{g}'$ also satisfies $\mathcal{N}(\mathfrak{g}') = 0$.*

This result is for interest for the expansion problem, and suggests to analyze the invariants of a Lie algebra in relation with the invariants of its contractions. A recent example of this is the study of the invariants of the (2+1) kinematical algebras [10]

3. Frobeniusian Lie algebras

The contraction formula enables us to obtain some criteria on the number of invariants of a Lie algebra. This will be of special interest for those classes of Lie algebras which, having an additional structure, can be classified up to contraction. Let $\mathfrak{g} = (\mathbb{K}^{2n}, \mu)$ be a Lie algebra. We say that $\mathfrak{g}$ is frobeniusian (or that it admits a frobeniusian structure) if there exists a linear form $\omega_\mu \in \mathfrak{g}^*$ such that

$$\bigwedge^n d\omega_\mu \neq 0. \tag{11}$$

Frobeniusian Lie algebras have been classified up to contraction in [11].

Theorem 1 *Let $\mathfrak{g} = (\mathbb{R}^{2n}, \mu)$ be a frobeniusian Lie algebra. Then $\mathfrak{g}$ contracts to some element of the following family $\mathfrak{g}(\alpha_1, .., \alpha_s, \beta_1, .., \beta_{n-1-s})$:*

$$\left.\begin{array}{l} [X_1, X_2] = X_1 \\ [X_{2r+1}, X_{2r+2}] = X_1, \ 1 \leq r \leq n - 1 \\ [X_2, X_{4k-1}] = \alpha_k X_{4k-1} + \beta_k X_{4k+1}, \ k \leq s \\ [X_2, X_{4k}] = (-1 - \alpha_k) X_{4k} - \beta_k X_{4k+2}, \ k \leq s \\ [X_2, X_{4k+1}] = -\beta_k X_{4k-1} + \alpha_k X_{4k+1}, \ k \leq s \\ [X_2, X_{4k+2}] = \beta_k X_{4k} + (-1 - \alpha_k) X_{4k+2}, \ k \leq s \\ [X_2, X_{4s+2k-1}] = -\frac{1}{2}X_{4s+2k-1} + \beta_{k+s-1}X_{4s+2k}, \ 2 \leq k \leq n - 2s \\ [X_2, X_{4s+2k}] = -\beta_{k+s-1}X_{4k+2s-1} - \frac{1}{2}X_{4s+2k}, \ 2 \leq k \leq n - 2s \end{array}\right\}, \tag{12}$$

where $0 \leq s \leq \left[\frac{n-1}{2}\right]$ and $(\alpha_1, .., \alpha_s, \beta_1, .., \beta_{n-1-s}) \in \mathbb{R}^{n-1}$. The algebras $\mathfrak{g}(\alpha_1, .., \alpha_s, \beta_1, .., \beta_{n-1-s})$ are called frobeniusian model Lie algebras (over $\mathbb{C}$ we obtain the models by complexification of the algebras above).

Proposition 3 Let $\mathfrak{g}$ be a frobeniusian Lie algebra over $\mathbb{K} = \mathbb{R}, \mathbb{C}$. Then $\mathfrak{g}$ has no non-trivial generalized Casimir invariants.

The proof follows easily from system (7). Now one can ask whether any frobeniusian Lie algebra must be solvable, as happens for dimension four. If such an algebra is not solvable, then it necessarily has a non-zero Levi factor. In dimension six there are only two possibilities, either $\mathfrak{sl}(2, \mathbb{R})$ or $\mathfrak{so}(3)$. Consider the non-semisimple and non-solvable Lie algebra $\mathfrak{sl}(2, \mathbb{R}) \overrightarrow{\oplus}_R A_{3,1}$, where $R = D_{\frac{1}{2}} \oplus D_0$ is the representation of $\mathfrak{sl}(2, \mathbb{R}) = \left\{ X_1, X_2, X_3 \mid [X_2, X_3] = X_1, [X_1, X_i] = 2(-1)^i X_i, i = 2, 3 \right\}$ describing the semidirect sum and $A_{3,1}$ is the solvable Lie algebra defined by $\{ X_4, X_5, X_6 \mid [X_i, X_6] = X_i, i = 4, 5 \}$. Let $\{\omega_1, .., \omega_6\}$ be a basis of $\left(\mathfrak{sl}(2, \mathbb{R}) \overrightarrow{\oplus}_R A_{3,1}\right)^*$ and consider the linear form $\omega = \omega_1 + \omega_4 + \omega_5 \in \left(\mathfrak{sl}(2, \mathbb{R}) \overrightarrow{\oplus}_R A_{3,1}\right)^*$. It satisfies

$$d\omega \wedge d\omega \wedge d\omega = 12\omega_1 \wedge \omega_2 \wedge \omega_3 \wedge \omega_4 \wedge \omega_5 \wedge \omega_6 \neq 0 \tag{13}$$

and therefore $\mathfrak{sl}(2, \mathbb{R}) \overrightarrow{\oplus}_R A_{3,1}$ is a frobeniusian Lie algebra, which contracts to some element $\mathfrak{g}(\alpha_1, .., \alpha_s, \beta_1, .., \beta_{n-1-s})$. Indeed this is the only Lie algebra in dimension six having no invariants and non-trivial Levi subalgebra. From this example we also deduce the existence of frobeniusian Lie algebras whose Levi decomposition is non-trivial in any dimension $n \geq 6$. In fact, take the Lie algebras $\mathfrak{sl}(2, \mathbb{R}) \overrightarrow{\oplus}_R (A_{3,1} \oplus k\mathfrak{r}_2)$ with $R = D_{\frac{1}{2}} \oplus (2k+1) D_0$ and $k \geq 1$, where $\mathfrak{r}_2$ is the two dimensional nonabelian Lie algebra. It is straightforward to verify that all these algebras admit a linear form ω whose differential $d\omega$ is symplectic. The interest of the algebras $\mathfrak{sl}(2, \mathbb{R}) \overrightarrow{\oplus}_R (A_{3,1} \oplus k\mathfrak{r}_2)$ is that they constitute the simplest example of Lie algebras $\mathfrak{g}$ having nonzero Levi subalgebra and such that $\mathcal{N}(\mathfrak{g}) = 0$. Lie algebras with Levi factor $\mathfrak{so}(3)$ are of importance for multidimensional cosmologies [12], and the nonexistence of invariants for such an extended model should lead to interesting properties of the corresponding cosmology. Another important task that should be analyzed is whether for discrete contractions [13] the conclusion of proposition 1 also holds. For groups of low order the answer is positive, but since graded contractions depend both on the grading group and the matrix describing the graduation, it will not be possible to find a procedure that works for all the groups and grading matrices.

References

[1] Inönü E and Wigner E P 1953 Proc. Natl. Acad. Sci. U. S. 39 510-525
[2] Bacry H and Lévy-Leblond J M 1968 J. Math. Phys. 9 1605-1614
[3] Cariñena J F, del Olmo M and Santander M 1981 J. Phys. A: Math. Gen. 14 1
[4] Campoamor-Stursberg R 2002 J. Phys. A: Math. Gen. 35 6293-6306
[5] Weimar-Woods E 1995 J. Math. Phys. 36 4519-4548
[6] Fialowski A and O'Halloran J 1990 Comm. Algebra 18 4121-4140
[7] Pecina-Cruz J N 1994 J. Math. Phys. 35 3146-3162
[8] Beltrametti E G and Blasi A 1966 Phys. Lett. 20 62-64
[9] Ait Abdelmalek, Leng X, Patera J and Winternitz P 1996 J. Phys. A: Math. Gen. 29 7519-7543
[10] Herranz F J and Santander M 1999 J. Phys. A: Math. Gen. 32 3743-3754
[11] Goze M 1981 C. R. A. S. Paris Sér. A 293 425-428
[12] Demianski M, Golda Z, Sokolowski, L M Szydlowski M and Turkowski P 1987 J. Math. Phys. 28 171-173
[13] Moody R V and Patera J 1991 J. Phys. A: Math. Gen. 25 2227-2257

Inst. Phys. Conf. Ser. No 173: Section 8
Paper presented at 24th Int. Coll. Group Theoretical Methods in Physics, Paris, France, 15–20 July 2002
©*2003 IOP Publishing Ltd*

Quantization on Compact Groups

G.Chadzitaskos and J.Tolar

Department of Physics,
Faculty of Nuclear Sciences and Physical Engineering,
Czech Technical University, Břehová 7,
CZ - 115 19 Prague
e-mail: chadzita@br.fjfi.cvut.cz, jiri.tolar@fjfi.cvut.cz

Abstract. Let a group manifold be a configuration space of a quantum system. Harmonic analysis of L^2 functions on a compact group allows simple definition of Wigner symbols on a "phase space" with discrete momenta. Wigner symbols of quantum observables are real functions on "phase space" and our approach leads to a non–commutative $\star$–product among them, i.e. deformation quantization.

1. Motivation

The Weyl quantization procedure is well known over flat configuration spaces R^n [1, 2, 3]. It can be formally presented via the Weyl–Wigner correspondence [4, 5] which, however, does not lend itself to general configuration manifolds.

In this contribution we are going to show that on the Wigner–symbol side, the Weyl–Wigner correspondence can be generalized to compact groups as configuration manifolds.

For simplicity consider a Hermitian operator $\hat{H}$ on a finite–dimensional complex Hilbert space with an orthonormal basis $\{|i\rangle\}$. Since the matrix $\langle i| \hat{H} |j\rangle$ is Hermitian, the sum of all elements of any antidiagonal is a real number. The antidiagonals consist of the elements $\langle (i+j)\mathrm{mod}M| \hat{H} |(i-j)\mathrm{mod}M\rangle$ for fixed i and all j, where M is the dimension of the Hilbert space. The finite Fourier transform of any antidiagonal is also a real number. Since the Fourier transform in quantum mechanics is a transformation between coordinates and momenta, the matrix of Fourier transformation of the antidiagonals is a real matrix and each element is labeled by the coordinate and the momentum.

Similarly, by applying simple formulae of harmonic analysis [6], Wigner symbols will be defined for any compact group as a configuration space.

2. Wigner symbol

- Let $\mathcal{M}$ be a compact group, quantum Hilbert space $L^2(\mathcal{M}, dx)$, where dx is the invariant measure normalized to unity.

- Let $\hat{H}$ be a selfadjoint integral operator acting on $L^2(\mathcal{M}, dx)$ with Hilbert–Schmidt kernel $H(x, y)$, i.e.

$$(\hat{H}\psi)(x) = \int_{\mathcal{M}} H(x, y)\psi(y)dy,$$

where

$$H(x, y) = \overline{H(y, x)}.$$

644

- Let $\{\pi_i(\mathcal{M}), i \in I\}$ be the set of all irreducible representations of $\mathcal{M}$. According to the Peter–Weyl theorem, any L^2 function on a compact group $\mathcal{M}$ admits a Fourier expansion into the *complete orthogonal basis* of all matrix elements $\{\phi_k(x) = D^i_{mn}(x), k = (i, m, n) \in U)\}$ of all irreducible representations $\pi_i(\mathcal{M})$. We assume that the basis $\{\phi_k(x)\}$ of the Hilbert space $\mathcal{H}$ is *normalized*.

- Let the operator $\hat{T}$ act on $L^2(\mathcal{M} \times \mathcal{M})$

$$\hat{T} : f(x, y) \rightarrow f(xy, xy^{-1}),$$

and let the inverse operator $\hat{T}^{-1}$ exist (this is the case e.g. for Abelian groups and for Lie groups for which the exponential map is onto).

Then the Wigner symbol of the operator $\hat{H}$ is a function on $\mathcal{M} \times U$ with the first variable in the group, $x \in \mathcal{M}$, and the second in the set of indices $k = (i, m, n) \in U$:

$$W_H(x, k) = \int_{\mathcal{M}} (\hat{T}(H(x, y)))\phi_k(y)dy.$$

It can be written

$$W_H(k, x) = \hat{\mathcal{F}}(\hat{T}(H(x, y)))$$

where $\hat{\mathcal{F}}$ is the Fourier transform of the second variable.

3. Weyl–Wigner Quantization

The Weyl–Wigner quantization is defined here by the non–commutative multiplication between the Wigner symbols — the $\star$–product. Assuming the existence of inverse operators $\hat{\mathcal{F}}^{-1}$ and $\hat{T}^{-1}$, the general scheme is: given the Wigner symbols W_F, W_G and W_{FG}, we define the $\star$–product

$$(W_F \star W_G)(x, k) =$$

$$= W_{FG}(x, k) = \hat{\mathcal{F}}_k(\hat{T}\langle x|\hat{F}\hat{G}|y\rangle) = \hat{\mathcal{F}}_k(\hat{T}(\int_{\mathcal{M}} dz \langle x|\hat{F}|z\rangle\langle z|\hat{G}|y\rangle)) =$$

$$= \hat{\mathcal{F}}_k(\hat{T}(\int_{\mathcal{M}} dz(\hat{T}^{-1}\hat{\mathcal{F}}^{-1}W_F)(x, z)(\hat{T}^{-1}\hat{\mathcal{F}}^{-1}W_G)(z, y))).$$

The form of the resolution of identity employed in the above formula follows from the assumed normalizations of the Haar measure and of the Fourier basis $\{\phi_k(x)\}$ which forms a complete orthonormal system in $L^2(\mathcal{M}, dx)$. In this way the quantization is converted into integration over the group manifold. Since the function $H(x, y)$ can be expanded in a double Fourier series

$$H(x, y) = \sum_{mn \in U} h_{m,n}\phi_m(x)\overline{\phi_n(y)},$$

where $h_{m,n} = \overline{h_{n,m}}$, one can use the orthogonality relations of the Fourier basis to simplify the relations. The results are best seen in the following examples.

4. Examples

4.1. Quantization on a periodic chain

The construction is based on the operator formulation of quantum mechanics on finite discrete space [1, 7, 8]. We assume that the number M of points of the chain is prime. It will guarantee the existence of the inverse operator $\hat{T}^{-1}$.

Let position coordinate q_i take one of M distinct values $q_i = i = 0, 1, \ldots, M - 1$. With each value of q_i a vector $|i\rangle$ of an orthonormal basis of M–dimensional Hilbert space $\mathcal{H}$ is connected. Then the *position operator* is defined by

$$\hat{Q} = \sum_{j=0}^{M-1} j|j\rangle\langle j|.$$

The eigenvectors $\{|i\rangle\}$ of $\hat{Q}$ form a basis of the Hilbert space $\mathcal{H}$ and i are the corresponding eigenvalues.

Using the Fourier basis $\phi_m(k) = \exp(\frac{2\pi i}{M}km)$ the conjugate *momentum operator* is defined by

$$\hat{P} = \sum_{k=0}^{M-1} k|k\rangle\langle k|,$$

where $|k\rangle$, $k = 0, \ldots, M - 1$, are obtained from the eigenvectors $|j\rangle$ by the discrete Fourier transform

$$|k\rangle = \frac{1}{\sqrt{M}} \sum_{j} e^{\frac{2\pi i}{M}kj}|j\rangle. \tag{1}$$

The Wigner symbol of a selfadjoint operator $\hat{H}$ is a real matrix

$$W_H(m, k) = \sum_{l=0}^{M-1} h_{l,k-l} e^{\frac{2\pi i}{M}m(2l-k)},$$

where the algebraic operations are done modulo M.

In order to get the $\star$–product, it is necessary to compute

$$(W_G \star W_H)(n, l) = W_{GH}(n, l) = \sum_{k,m} g_{m,k} h_{k,l-m} e^{\frac{2\pi i}{M}n(2m-l)}$$

with substitutions

$$g_{m,k} = \frac{1}{M} \sum_{n=0}^{M-1} W_G(n, m - k) e^{\frac{2\pi i}{M}n(m+k)}$$

and

$$h_{k,l-m} = \frac{1}{M} \sum_{n=0}^{M-1} W_H(n, k - l + m) e^{\frac{2\pi i}{M}n(m+l-k)}.$$

646

4.2. Quantization on a circle

An arbitrary function on a circle can be expanded in the Fourier series

$$f(x) = \sum_k f_k \phi_k(x) = \sum_k f_k e^{ikx},$$

where $k = 0, \pm 1, \pm 2, \ldots$ and $x \in (-\pi, \pi)$. The integral kernel corresponding to an operator $\hat{H}$ is

$$H(x, y) = \sum_{k,l} h_{k,l} e^{ikx} e^{-ily},$$

where $x, y \in (-\pi, \pi)$, $k, l = 0, \pm 1, \pm 2, \ldots$, and $h_{k,l} = \overline{h_{l,k}}$. The Wigner symbols of $\hat{H}$ and $\hat{G}$ are

$$W_H(x, k) = \sum_l h_{l,k-l} e^{ix(2l-k)}, \quad W_G(z, m) = \sum_n g_{n,m-n} e^{iz(2n-m)},$$

and the Wigner symbol corresponding to the product of the operators $\hat{H}$ and $\hat{G}$ is

$$W_{GH}(x, l) = \sum_{k,m} g_{m,k} h_{k,l-m} e^{ix(2m-l)}.$$

Let us now start with the symbols W_G and W_H. In order to determine the $\star$–product $W_G \star W_H$, we have to substitute the coefficients

$$h_{m,n} = \frac{1}{2\pi} \int_{-\pi}^{\pi} W_H(x, m-n) e^{ix(m+n)} dx, \quad g_{m,n} = \frac{1}{2\pi} \int_{-\pi}^{\pi} W_G(x, m-n) e^{ix(m+n)} dx,$$

with the result

$$(W_G \star W_H)(x, l) = W_{GH}(x, l) =$$

$$= \sum_{k,m} e^{ix(2m-l)} \frac{1}{2\pi} \int_{-\pi}^{\pi} dz \frac{1}{2\pi} \int_{-\pi}^{\pi} dy \, W_G(z, k-m) W_H(y, l-m-k) e^{iz(k+m)} e^{iy(l+k+m)}.$$

5. Conclusion

The problem of quantization on compact groups as configuration spaces was converted into the problem of determining the $\star$–product [9] of the Wigner symbols expressed in terms of integration over the group manifold. We have shown it for the examples of a periodic chain and a circle. Also the quantization on unitary groups $SU(n)$ is straightforward.

Acknowledgements

The authors wish to acknowledge support of the Ministry of Education of Czech Republic under the research project MSM210000018.

References

[1] Weyl, H. (1950): *Theory of Groups and Quantum Mechanics*. Dover, New York, 272–280.

[2] Moyal, J.E. (1949): Proc. Camb. Phil. Soc. **45**, 99.

[3] Wigner, E.P. (1932): Phys. Rev **40**, 749.

[4] Agarwal, G.S., Wolf, E. (1970): Phys. Rev. **D2**, 2161.

[5] Tolar, J. (1977): Quantization Methods. Lecture notes, Institut für Theoretische Physik der Technischen Universität, Clausthal.

[6] Barut, A.O., Raczka, R. (1977): *Theory of Group Representations and Applications*. PWN — Polish Scientific Publishers, Warszawa, 166–179.

[7] Schwinger, J. (1960): Proc. Nat. Acad. Sci. (US), **46**, reprinted in *Quantum Kinematics and Dynamics*, Benjamin, New York (1970).

[8] Šťovíček, P. and Tolar, J. (1984): Quantum mechanics in a discrete space-time. Rep. Math. Phys. **20**, 157-170.

[9] Bayen, F., Flato, M., Fronsdal, C., Lichnerowicz, A., Sternheimer, D. (1978): Deformation theory and quantization I, II, Ann. Phys. (N.Y.) **110**, 61, 111.

Inst. Phys. Conf. Ser. No 173: Section 8
Paper presented at 24th Int. Coll. Group Theoretical Methods in Physics, Paris, France, 15–20 July 2002
©2003 IOP Publishing Ltd

Special functions, raising and lowering operators

N Cotfas

Faculty of Physics, University of Bucharest, PO Box 76-54, Postal Office 76, Bucharest,
Romania, E-mail address: ncotfas@yahoo.com

Abstract. The Schrödinger equations which are exactly solvable in terms of associated special functions are directly related to some self-adjoint operators defined in the theory of hypergeometric type equations. The fundamental formulae occurring in a supersymmetric approach to these Hamiltonians are consequences of some formulae concerning the general theory of associated special functions. We use this connection in order to obtain a *general theory of Schrödinger equations exactly solvable in terms of associated special functions*, and to extend certain results known in the case of some particular potentials.

1. Introduction

It is well-known [2, 4] that, in the case of certain potentials, the Schrödinger equation is exactly solvable and its solutions can be expressed in terms of the so-called *associated special functions*. Our purpose is to present a general theory of these quantum systems. Our systematic study recovers a number of earlier results in a natural unified way and also leads to new findings.

The number of articles concerning exactly solvable quantum systems and related subjects is very large (see [2, 4, 5] and references therein). Our approach is based on the formalism of the factorization method [4] and on raising/lowering operators presented in general form (for the first time to our knowledge) by Jafarizadeh and Fakhri [5]. We re-obtain these operators in a simpler way, and use them in a rather different way. More details can be found in [3].

2. Orthogonal polynomials and associated special functions

Many problems in quantum mechanics and mathematical physics lead to equations of hypergeometric type

$$\sigma(s)y''(s) + \tau(s)y'(s) + \lambda y(s) = 0 \tag{1}$$

where $\sigma(s)$ and $\tau(s)$ are polynomials of at most second and first degree, respectively, and λ is a constant. This equation can be reduced to the self-adjoint form

$$[\sigma(s)\varrho(s)y'(s)]' + \lambda\varrho(s)y(s) = 0 \tag{2}$$

by choosing a function ϱ such that $[\sigma(s)\varrho(s)]' = \tau(s)\varrho(s)$. For $\lambda = \lambda_l = -\frac{1}{2}l(l-1)\sigma'' - l\tau'$ with $l \in \mathbb{N}$ there exists a polynomial Φ_l of degree l satisfying (1), that is,

$$\sigma(s)\Phi_l''(s) + \tau(s)\Phi_l'(s) + \lambda_l\Phi_l(s) = 0. \tag{3}$$

If there exists a finite or infinite interval (a, b) such that

$$\sigma(s)\varrho(s)s^k|_{s=a} = 0 \qquad \sigma(s)\varrho(s)s^k|_{s=b} = 0 \qquad \text{for all } k \in \mathbb{N} \tag{4}$$

and if $\sigma(s) > 0$, $\varrho(s) > 0$ for all $s \in (a, b)$, then the polynomials Φ_l are orthogonal with weight function $\varrho(s)$ in the interval (a, b). In this case Φ_l are known as *classical orthogonal polynomials* [7].

Consider a system of classical orthogonal polynomials, and let $\kappa(s) = \sqrt{\sigma(s)}$. By differentiating the equation (3) m times and multiplying it by $\kappa^m(s)$, we get for each $m \in \{0, 1, 2, ..., l\}$ the *associated differential equation* which can be written as $H_m \Phi_{l,m} = \lambda_l \Phi_{l,m}$, where

$$H_m = -\sigma(s)\frac{d^2}{ds^2} - \tau(s)\frac{d}{ds} + \frac{m(m-2)}{4}\frac{\sigma'^2(s)}{\sigma(s)} + \frac{m\tau(s)}{2}\frac{\sigma'(s)}{\sigma(s)} - \frac{1}{2}m(m-2)\sigma''(s) - m\tau'(s)$$

and $\Phi_{l,m}(s) = \kappa^m(s)\Phi_l^{(m)}(s)$ are known as the *associated special functions*. The set $\{\Phi_{m,m}, \Phi_{m+1,m}, \Phi_{m+2,m}, ...\}$ is an orthogonal sequence ([7], pag. 8) in the Hilbert space

$$\mathcal{H} = \left\{ \varphi : (a, b) \longrightarrow \mathbb{R} \ \middle| \ \int_a^b |\varphi(s)|^2 \varrho(s)ds < \infty \right\} \quad \text{with} \quad \langle \varphi, \psi \rangle = \int_a^b \varphi(s)\psi(s)\varrho(s)ds .$$

For each $m \in \mathbb{N}$, let $\mathcal{H}_m$ be the linear span of $\{\Phi_{m,m}, \Phi_{m+1,m}, \Phi_{m+2,m}, ...\}$. In the sequel we shall restrict us to the case when $\mathcal{H}_m$ is dense in $\mathcal{H}$ for all $m \in \mathbb{N}$. For this it is sufficient the interval (a, b) to be finite, but not necessary.

3. Raising and lowering operators. Factorizations for H_m

Lorente has shown recently [6] that a factorization of H_0 can be obtained by using the well-known three term recurrence relation satisfied by Φ_l and a consequence of Rodrigues formula. Following Lorente's idea we obtain a factorization of H_m by using the definition $\Phi_{l,m}(s) = \kappa^m(s)\Phi_l^{(m)}(s)$ and a three term recurrence relation.

Differentiating $\Phi_{l,m}(s) = \kappa^m(s)\Phi_l^{(m)}(s)$ we get the relation

$$\Phi_{l,m+1}(s) = \left(\kappa(s)\frac{d}{ds} - m\kappa'(s) \right) \Phi_{l,m}(s) \qquad \text{for all } m \in \{0, 1, ..., l-1\}. \tag{5}$$

If we differentiate (3) $m-1$ times and multiply the obtained relation by κ^{m-1} then we get for each $m \in \{1, 2, ..., l-1\}$ the three term recurrence relation

$$\Phi_{l,m+1}(s) + \left(\frac{\tau(s)}{\kappa(s)} + 2(m-1)\kappa'(s) \right) \Phi_{l,m}(s) + (\lambda_l - \lambda_{m-1})\Phi_{l,m-1}(s) = 0 \tag{6}$$

and the relation $\left(\frac{\tau(s)}{\kappa(s)} + 2(l-1)\kappa'(s) \right) \Phi_{l,l}(s) + (\lambda_l - \lambda_{l-1})\Phi_{l,l-1}(s) = 0$. A direct consequence of these formulae is the relation

$$(\lambda_l - \lambda_m)\Phi_{l,m}(s) = \left(-\kappa(s)\frac{d}{ds} - \frac{\tau(s)}{\kappa(s)} - (m-1)\kappa'(s) \right) \Phi_{l,m+1}(s) \tag{7}$$

satisfied for all $m \in \{0, 1, ..., l-1\}$.

The operators $A_m : \mathcal{H}_m \longrightarrow \mathcal{H}_{m+1}$ and $A_m^+ : \mathcal{H}_{m+1} \longrightarrow \mathcal{H}_m$ defined by

$$A_m = \kappa(s)\frac{d}{ds} - m\kappa'(s) \qquad A_m^+ = -\kappa(s)\frac{d}{ds} - \frac{\tau(s)}{\kappa(s)} - (m-1)\kappa'(s) \tag{8}$$

satisfy the relations $A_m \Phi_{l,m} = \Phi_{l,m+1}$ and $A_m^+ \Phi_{l,m+1} = (\lambda_l - \lambda_m)\Phi_{l,m}$ (see figure 1).

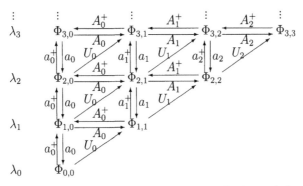

Figure 1. The functions $\Phi_{l,m}$ satisfy the relation $H_m\Phi_{l,m} = \lambda_l\Phi_{l,m}$, and are related (up to some multiplicative constants) through the operators A_m, A_m^+, a_m, a_m^+, U_m and $U_m^{-1} = U_m^+$.

Since $\sigma^m(s)\Phi_l^{(m)}(s)\Phi_k^{(m+1)}(s)$ is a polynomial and the function $\sigma(s)\varrho(s)$ satisfies (4), integrating by parts one obtains $\langle A_m\Phi_{l,m}, \Phi_{k,m+1}\rangle = \langle\Phi_{l,m}, A_m^+\Phi_{k,m+1}\rangle$, that is, the operators A_m and A_m^+ are mutually adjoint. From the relation

$$||\Phi_{l,m+1}||^2 = \langle\Phi_{l,m+1}, \Phi_{l,m+1}\rangle = \langle A_m\Phi_{l,m}, \Phi_{l,m+1}\rangle = \langle\Phi_{l,m}, A_m^+\Phi_{l,m+1}\rangle = (\lambda_l - \lambda_m)||\Phi_{l,m}||$$

it follows that $\lambda_l > \lambda_m$ for all $l > m$, and $||\Phi_{l,m+1}|| = \sqrt{\lambda_l - \lambda_m}\,||\Phi_{l,m}||$. This is possible only if $\sigma''(s) \leq 0$ and $\tau'(s) < 0$. Particularly, we have $\lambda_l \neq \lambda_k$ if and only if $l \neq k$.

The operators $H_m : \mathcal{H}_m \longrightarrow \mathcal{H}_m$ are self-adjoint, admit the factorizations

$$H_m - \lambda_m = A_m^+ A_m \qquad H_{m+1} - \lambda_m = A_m A_m^+ \tag{9}$$

and satisfy the intertwining relations $H_m A_m^+ = A_m^+ H_{m+1}$ and $A_m H_m = H_{m+1} A_m$.

4. Creation and annihilation operators. Coherent states

For each $m \in \mathbb{N}$, the sequence $\{|m, m >, |m + 1, m >, |m + 2, m >, ...\}$, where $|l, m >= \Phi_{l,m}/||\Phi_{l,m}||$ is an orthonormal basis of $\mathcal{H}$, and $U_m : \mathcal{H} \longrightarrow \mathcal{H}$, $U_m|l, m\rangle = |l + 1, m + 1\rangle$ is a unitary operator. The mutually adjoint operators (see figure 1) a_m, $a_m^+ : \mathcal{H}_m \longrightarrow \mathcal{H}_m$, $a_m = U_m^+ A_m$, $a_m^+ = A_m^+ U_m$ satisfy the relations

$$a_m|l, m\rangle = \sqrt{\lambda_l - \lambda_m}\,|l - 1, m\rangle \qquad a_m^+|l, m\rangle = \sqrt{\lambda_{l+1} - \lambda_m}\,|l + 1, m\rangle \tag{10}$$

and allow us to factorize H_m as $H_m - \lambda_m - a_m^+ a_m$.

The Lie algebra $\mathcal{L}_m$ generated by $\{a_m^+, a_m\}$ is isomorphic to $su(1, 1)$ if $\sigma'' < 0$, and it is isomorphic to the Heisenberg-Weyl algebra $h(2)$ if $\sigma'' = 0$.

Let $m \in \mathbb{N}$ be a fixed natural number, and let $|n\rangle = |m + n, m\rangle$, $e_n = \lambda_{m+n} - \lambda_m$, $\varepsilon_0 = 1$, $\varepsilon_n = e_1 e_2 ... e_n$. Since $0 = e_0 < e_1 < e_2 < ... < e_n < ...$ and

$$a_m|n\rangle = \sqrt{e_n}\,|n - 1\rangle \qquad a_m^+|n\rangle = \sqrt{e_{n+1}}\,|n + 1\rangle \qquad (H_m - \lambda_m)|n\rangle = e_n|n\rangle \tag{11}$$

we can define a system of coherent states by using the general setting presented in [1].

If $R = \limsup_{n\to\infty} \sqrt[n]{\varepsilon_n} \neq 0$ then we can define

$$|z\rangle = \frac{1}{N(|z|^2)}\sum_{n\geq 0}\frac{z^n}{\sqrt{\varepsilon_n}}|n\rangle \qquad \text{where} \qquad (N(|z|^2))^2 = \sum_{n=0}^{\infty}\frac{|z|^{2n}}{\varepsilon_n} \tag{12}$$

for any z in the open disk $C(0, R)$ of center 0 and radius R. We get in this way a continuous family $\{|z\rangle| \, z \in C(0, R)\}$ of normalized coherent states such that $a_m|z\rangle = z|z\rangle$.

5. Application to Schrödinger type operators

The problem of factorization of operators H_m is a very important one since it is directly related to the factorization of some Schrödinger type operators [2, 4]. If we use a change of variable $s = s(x)$ such that $ds/dx = \kappa(s(x))$ or $ds/dx = -\kappa(s(x))$ and define the new functions $\Psi_{l,m}(x) = \sqrt{\kappa(s(x))\,\varrho(s(x))}\,\Phi_{l,m}(s(x))$ then the relation $H_m\Phi_{l,m} = \lambda_l\Phi_{l,m}$ becomes an equation of Schrödinger type

$$-\frac{d^2}{dx^2}\Psi_{l,m}(x) + V_m(x)\Psi_{l,m}(x) = \lambda_l\Psi_{l,m}(x). \tag{13}$$

For example, by starting from the equation of Jacobi polynomials with $\alpha = \mu - 1/2$, $\beta = \eta - 1/2$, and using the change of variable $s(x) = \cos x$ we obtain the Schrödinger equation corresponding to the Pöschl-Teller potential [1]

$$V_0(x) = \frac{1}{4}\left[\frac{\mu(\mu-1)}{\cos^2(x/2)} + \frac{\eta(\eta-1)}{\sin^2(x/2)}\right] - \frac{(\mu+\eta)^2}{4}. \tag{14}$$

If we choose the change of variable $s = s(x)$ such that $ds/dx = \kappa(s(x))$, then the operators corresponding to A_m and A_m^+ are the adjoint conjugate operators

$$\begin{aligned}
\mathcal{A}_m &= [\kappa(s)\varrho(s)]^{1/2}A_m[\kappa(s)\varrho(s)]^{-1/2}|_{s=s(x)} = \frac{d}{dx} + W_m(x) \\
\mathcal{A}_m^+ &= [\kappa(s)\varrho(s)]^{1/2}A_m^+[\kappa(s)\varrho(s)]^{-1/2}|_{s=s(x)} = -\frac{d}{dx} + W_m(x)
\end{aligned} \tag{15}$$

where the *superpotential* $W_m(x)$ is given by the formula

$$W_m(x) = -\frac{\tau(s(x))}{2\kappa(s(x))} - \frac{2m-1}{2\kappa(s(x))}\frac{d}{dx}\kappa(s(x)). \tag{16}$$

From the relations satisfied by $\mathcal{A}_m$ and $\mathcal{A}_m^+$ we get the formulae

$$-\frac{d^2}{dx^2} + V_m(x) - \lambda_m = \mathcal{A}_m^+\mathcal{A}_m \qquad -\frac{d^2}{dx^2} + V_{m+1}(x) - \lambda_m = \mathcal{A}_m\mathcal{A}_m^+ \tag{17}$$

$$V_m(x) - \lambda_m = W_m^2(x) - \dot{W}_m(x) \qquad V_{m+1}(x) - \lambda_m = W_m^2(x) + \dot{W}_m(x) \tag{18}$$

$$W_m(x) = -\frac{\dot{\Psi}_{m,m}(x)}{\Psi_{m,m}(x)} \qquad V_m(x) = \frac{\ddot{\Psi}_{m,m}(x)}{\Psi_{m,m}(x)} + \lambda_m \tag{19}$$

where the dot sign means derivative with respect to x.

If we choose the change of variable $s = s(x)$ such that $ds/dx = -\kappa(s(x))$ then the corresponding formulae are very similar (only some signs are changed). In the case of Pöschl-Teller potential (14) we get $W_0(x) = \frac{1}{2}\left[\mu\cot\frac{x}{2} - \eta\tan\frac{x}{2}\right]$.

Acknowledgments

The author gratefully acknowledge the financial support of CNCSIS.

References

[1] Antoine J-P, Gazeau J-P, Monceau P, Klauder J R and Penson K A 2001 *J. Math. Phys.* **42** 2349-87
[2] Cooper F, Khare A and Sukhatme U 1995 *Phys. Rep.* **251** 267–385
[3] Cotfas N 2002 *Los Alamos preprint* quant-ph/0206129 (http://xxx.lanl.gov)
[4] Infeld L and Hull T E 1951 *Rev. Mod. Phys.* **23** 21–68
[5] Jafarizadeh M A and Fakhri H 1998 *Ann. Phys., NY* **262** 260–76
[6] Lorente M 2001 *J. Phys. A: Math. Gen.* **34** 569–88
[7] Nikiforov A F, Suslov S K and Uvarov V B 1991 *Classical Orthogonal Polynomials of a Discrete Variable* (Berlin: Springer) pp 2–17

Inst. Phys. Conf. Ser. No 173: Section 8
Paper presented at 24th Int. Coll. Group Theoretical Methods in Physics, Paris, France, 15–20 July 2002
©2003 IOP Publishing Ltd

All Positive Energy UIRs of D=6 Conformal Supersymmetry

V K Dobrev

Institute of Nuclear Research and Nuclear Energy, Bulgarian Academy of Sciences, 72 Tsarigradsko Chaussee, 1784 Sofia, Bulgaria

Abstract. We present the classification of the positive energy (lowest weight) unitary irreducible representations of the $D = 6$ superconformal algebras $osp(8^*/2N)$.

1. Introduction

Recently, there was renewed interest in superconformal field theories in arbitrary dimensions, in particular, due to their duality to AdS supergravities. Particularly important are those for $D \leq 6$ since in these cases the relevant superconformal algebras satisfy [1] the Haag-Lopuszanski-Sohnius theorem [2]. Until recently such classification was known only for the $D = 4$ superconformal algebras $su(2, 2/N)$, cf. [3] (for $N = 1$), [4]. Recently, the classification for $D = 3$ (for even N), $D = 5$, and $D = 6$ (for $N = 1, 2$) was given [5] but the results were conjectural and there was not enough detail in order to check these conjectures. On the other hand the applications of $D = 6$ unitary irreps require firmer theoretical basis. Among the many interesting applications we shall mention the analysis of OPEs and 1/2 BPS operators [6, 7, 8].

In particular, it is important that some general properties of abstract superconformal field theories can be obtained by using the BPS nature of a certain class of superconformal primary operators and the model independent nature of superconformal OPEs. In the classification of UIRs of superconformal algebras an important role is played by the representations with "quantized" conformal dimension since in the quantum field theory framework they correspond to operators with "protected" scaling dimension and therefore imply "non-renormalization theorems" at the quantum level.

Motivated by the above we decided to reexamine the list of UIRs of the $D = 6$ superconformal algebras in detail. More than that we treat the superalgebras $osp(8^*/2N)$ for arbitrary N. Thus, we give the final list of UIRs for $D = 6$. With this we also confirm all but one of the conjectures of [5] for $N = 1, 2$. Our main tool is the explicit construction of the norms. This, on the one hand, enables us to prove the unitarity list, and, on the other hand, enables us to give explicitly the states of the irreps. This contribution is based on [9].

2. Positive energy UIRs of D=6 conformal supersymmetry

The superconformal algebras in $D = 6$ are $\mathcal{G} = osp(8^*/2N)$ (real forms of $osp(8/2N) \cong D(4, N)$, [10]). We label their physically relevant representations by the signature:

$$\chi = [\, d \,;\, n_1 ,\, n_2 ,\, n_3 \,;\, a_1 ,\, ..., a_N \,] \tag{1}$$

where d is the conformal weight, n_1, n_2, n_3 are non-negative integers which are Dynkin labels of the finite-dimensional irreps of the $D = 6$ Lorentz algebra $so(5,1)$, and $a_1, ..., a_N$ are non-negative integers which are Dynkin labels of the finite-dimensional irreps of the internal (or R) symmetry algebra $usp(2N)$. The even subalgebra of $osp(8^*/2N)$ is the algebra $so^*(8) \oplus usp(2N)$, and $so^*(8) \cong so(6,2)$ is the $D = 6$ conformal algebra. The odd subspace of $osp(8^*/2N)$ is $8N$-dimensional.

In [9] we gave a constructive proof for the UIRs of $osp(8^*/2N)$ following the methods used for the $D = 4$ superconformal algebras $su(2, 2/N)$, cf. [4]. The main tool is an adaptation of the Shapovalov form to the Verma modules V^Λ over the complexification $\mathcal{G}^{\mathcal{C}} = osp(8/2N)$ of $\mathcal{G}$.

To introduce Verma modules we use the standard triangular decomposition:

$$\mathcal{G}^{\mathcal{C}} = \mathcal{G}^+ \oplus \mathcal{H} \oplus \mathcal{G}^- \tag{2}$$

where $\mathcal{G}^+$, $\mathcal{G}^-$, resp., are the subalgebras corresponding to the positive, negative, roots, resp., and $\mathcal{H}$ denotes the Cartan subalgebra.

We consider lowest weight Verma modules, so that $V^\Lambda \cong U(\mathcal{G}^+) \otimes v_0$, where $U(\mathcal{G}^+)$ is the universal enveloping algebra of $\mathcal{G}^+$, $\Lambda \in \mathcal{H}^*$ is the lowest weight, and v_0 is a lowest weight vector v_0 such that:

$$Z\, v_0 = 0, \quad Z \in \mathcal{G}^- ; \qquad H\, v_0 = \Lambda(H)\, v_0, \quad H \in \mathcal{H} \tag{3}$$

The lowest weight Λ is characterized by its values on the Cartan subalgebra $\mathcal{H}$. In order to have Λ corresponding to χ, one can choose a basis H_j in $\mathcal{H}$ so that to obtain the entries in the signature χ as the values $\Lambda(H_j)$.

The conditions of unitarity are intimately related with the conditions for reducibility of the Verma modules w.r.t. to the odd positive roots. Using [11] we find that the Verma module $V^{\Lambda(\chi)}$ is reducible if the conformal weight takes one of the following $8N$ values $d_{ij}^\pm$, $(i = 1, 2, 3, 4, j = 1, ..., N)$, corresponding to the odd roots $\alpha_{ij}^\pm$:

$$d = d_{1j}^\pm \doteq \tfrac{1}{2}(3n_1 + 2n_2 + n_3) + 2(3 - N) \mp 2(r_j + N - j + 1) \tag{4}$$

$$d = d_{2j}^\pm \doteq \tfrac{1}{2}(n_3 + 2n_2 - n_1) + 2(2 - N) \mp 2(r_j + N - j + 1) \tag{5}$$

$$d = d_{3j}^\pm \doteq \tfrac{1}{2}(n_3 - 2n_2 - n_1) + 2(1 - N) \mp 2(r_j + N - j + 1) \tag{6}$$

$$d = d_{4j}^\pm \doteq -\tfrac{1}{2}(n_1 + 2n_2 + 3n_3) - 2N \mp 2(r_j + N - j + 1) \tag{7}$$

We note that $d_{11}^- > d_{21}^- > d_{31}^- > d_{41}^-$. The value d_{11}^- is the biggest among all $d_{ij}^\pm$; it is called 'the first reduction point' in [3].

We suppose that we consider representations which are unitary when restricted to the even part $\mathcal{G}_{\bar{0}}$. This is justified a posteriori since the unitary bounds of the even part are weaker than the supersymmetric ones. Thus, we shall factorize the even part and we shall consider only the states created by the action of the odd generators, i.e., $\mathcal{F}_0 = \left(U(\mathcal{G}^+)/U(\mathcal{G}_{\bar{0}}^+) \right) v_0$. We introduce notation X_{ij}^+ for the odd generator corresponding to the positive root α_{ij}^-, and Y_{ij}^+ will correspond to α_{ij}^+. Since the odd generators are Grassmann there are only 2^{8N} states in $\mathcal{F}_0$ and choosing an ordering we give these states explicitly as follows:

$$\Psi_{\bar{\varepsilon}\bar{\nu}} = \left(\prod_{i=1}^4 (Y_{i1}^+)^{\varepsilon_{i1}} \right) \cdots \left(\prod_{i=1}^4 (Y_{iN}^+)^{\varepsilon_{iN}} \right) \times$$
$$\times \left(\prod_{i=1}^4 (X_{iN}^+)^{\nu_{iN}} \right) \cdots \left(\prod_{i=1}^4 (X_{i1}^+)^{\nu_{i1}} \right) v_0 \tag{8}$$

where $\varepsilon_{ij}, \nu_{ij} = 0, 1$, and $\bar{\varepsilon}, \bar{\nu}$, denote the set of all ε_{ij}, ν_{ij}, resp.

The main result is:

Theorem: All positive energy unitary irreducible representations of the conformal superalgebra $osp(8^*/2N)$ characterized by the signature χ in (1) are obtained for real d and are given in the following list:

$$d \geq d_{11}^- = \tfrac{1}{2}(3n_1 + 2n_2 + n_3) + 2r_1 + 6, \quad \text{no restrictions on } n_j \tag{9}$$

$$d = d_{21}^- = \tfrac{1}{2}(n_3 + 2n_2) + 2r_1 + 4, \quad n_1 = 0 \tag{10}$$

$$d = d_{31}^- = \tfrac{1}{2}n_3 + 2r_1 + 2, \quad n_1 = n_2 = 0 \tag{11}$$

$$d = d_{41}^- = 2r_1, \quad n_1 = n_2 = n_3 = 0. \tag{12}$$

Remark: For $N = 1, 2$ the Theorem was conjectured by Minwalla [5], except that he conjectured unitarity also for the open interval (d_{31}^-, d_{21}^-) with conditions on n_j as in (11). We should note that this conjecture could be reproduced neither by methods of conformal field theory [6], nor by the oscillator method [12] (cf. [5]), and thus was in doubt. To compare with the notations of [5] one should use the following substitutions: $n_1 = h_2 - h_3$, $n_2 = h_1 - h_2$, $n_3 = h_2 + h_3$, $r_1 = k$, and h_j are all integer or all half-integer. The fact that $n_j \geq 0$ for $j = 1, 2, 3$ translates into: $h_1 \geq h_2 \geq |h_3|$, i.e., the parameters h_j are of the type often used for representations of $so(2N)$ (though usually for $N \geq 4$). Note also that the statement of the Theorem is arranged in [5] according to the possible values of n_i first and then the possible values of d. To compare with the notation of [6] we use the substitution $(n_1, n_2, n_3) \rightarrow (J_3, J_2, J_1)$. Some UIRs at the four exceptional points d_{i1}^- were constructed in [13] by the oscillator method (some of these were identified with Cartan-type signatures like (1) in, e.g., [5], [7]).

The **Proof** of the Theorem requires to show that there is unitarity as claimed, i.e., that the conditions are *sufficient* and that there is no unitarity otherwise, i.e., that the conditions are *necessary*. For the sufficiency we need all norms, but for the necessity part we need only the knowledge of a few norms. We give the necessity part in a Proposition. We need notation for the following one-particle norm:

$$x = x(d) \equiv \| X_{11}^+ \|^2 = \tfrac{1}{2}(d - d_{11}^-) + 3 \tag{13}$$

Proposition: There is no unitarity if d is in any of the open intervals $(d_{j+1,1}^-, d_{j1}^-)$, $j = 1, 2, 3$, also if $d < d_{41}^-$, and also if $d = d_{j1}^-$, $j = 2, 3, 4$, and $n_1 > 0$, $n_1 + n_2 > 0$, $n_1 + n_2 + n_3 > 0$, resp.

Proof: a) Consider d in the open interval (d_{21}^-, d_{11}^-), which means that $3 > x > 2 - n_1$. Consider the norm:

$$\| X_{11}^+ X_{21}^+ X_{31}^+ X_{41}^+ v_0 \|^2 = (x - 3)(x + n_1 - 2)(x + n_1 + n_2 - 1)(x + n_1 + n_2 + n_3)$$

The first term is strictly negative, while the other three terms are strictly positive, independently of the values of n_i, i.e., the norm is negative.

b) Consider d in the semi-open interval $(d_{31}^-, d_{21}^-]$, which means that $2 \geq x + n_1 > 1 - n_2$. Consider the norm:

$$\| X_{11}^+ X_{31}^+ X_{41}^+ v_0 \|^2 = (x - 2)(x + n_1 + n_2 - 1)(x + n_1 + n_2 + n_3)$$

The first term is strictly negative, while the other two terms are strictly positive, independently of the values of n_i, (including the boundary case $x - 2 = -n_1$ since then $n_1 > 0$), i.e., the norm is negative.

c) Consider d in the semi-open interval $(d_{41}^-, d_{31}^-]$, which means that $1 \geq x + n_1 + n_2 > -n_3$. Consider the norm:

$$\| X_{11}^+ X_{41}^+ v_0 \|^2 = (x - 1)(x + n_1 + n_2 + n_3)$$

The first term is strictly negative, while the second is strictly positive, independently of the values of n_i, i.e., the norm is negative.

d) Consider d in the infinite interval $d \leq d_{41}^-$. Consider the norm of $X_{11}^+ v_0$ which is equal to $x \leq -(n_1 + n_2 + n_3)$. Thus, x is strictly negative, independently of the values of n_i (including the boundary case since then $n_1 + n_2 + n_3 > 0$).

Thus, the Proposition is proved.

3. Unitarity at the reduction points

At the four reduction points $d = d_{i1}^-$ the Verma modules are reducible and the UIRs are realized as the irreducible factor-modules $L_\Lambda = V^\Lambda / I^\Lambda$ where I^Λ is the maximal invariant submodule of the Verma module V^Λ. The corresponding oddly generated subspaces $\mathcal{F}$ have less states than $\mathcal{F}_0$. In this section we give the number of these states in the following table:

$$d = d_{11}^-$$

$a_j \neq 0,$	$j = 1, ..., N$	$|\mathcal{F}| = 15.2^{8N-4}$
$a_j = 0,$	$j = 1, ..., k < N$	$|\mathcal{F}| = 2^{8N-4-k}(2^{4+k} - 2^{k+1} + 1)$
$a_j = 0,$	$j = 1, ..., N$	$|\mathcal{F}| = 2^{6N-6}(2^{2N+6} - (2^{N+3} + 1)(2^N - 1))$

$$d = d_{21}^-, \quad n_1 = 0$$

$a_j \neq 0,$	$j = 1, ..., N$	$|\mathcal{F}| = 11.2^{8N-4}$
$a_1 = 0,$	$N = 1$	$|\mathcal{F}| = 168$

$$d = d_{31}^-, \quad n_1 = n_2 = 0$$

$a_j \neq 0,$	$j = 1, ..., N$	$|\mathcal{F}| = 5.2^{8N-4}$
$a_1 = 0,$	$N = 1$	$|\mathcal{F}| = 53$

$$d = d_{41}^-, \quad n_1 = n_2 = n_3 = 0$$

$a_j \neq 0,$	$j = 1, ..., N$	$|\mathcal{F}| = 2^{8N-4}$
$a_j = 0,$	$j = 1, ..., N$	$|\mathcal{F}| = 1$

References

[1] Nahm W, 1978 Nucl. Phys. **B135**, 149.

[2] Haag R, Lopuszanski J T and Sohnius M, 1975 Nucl. Phys. **B88**, 257.

[3] Flato M and Fronsdal C, 1984 Lett. Math. Phys. **8**, 159.

[4] Dobrev V K and Petkova V B, 1985 Lett. Math. Phys. **9**, 287 & Phys. Lett. **162B**, 127; 1987 Fortschr. d. Phys. **35**, 537; 1986 Proceedings, eds. A.O. Barut and H.-D. Doebner, Lecture Notes in Physics, vol. 261 (Springer:Berlin) p. 291 & p. 300.

[5] Minwalla S, 1998 Adv. Theor. Math. Phys. **2**, 781.

[6] Ferrara S and Sokatchev E, 2001 Int. J. Theor. Phys. **40**, 935.

[7] Eden B, Ferrara S and Sokatchev E, 2001 J. High En. Phys. 0111:020.

[8] Ferrara S and Sokatchev E, 2001, hep-th/0110174.

[9] Dobrev V K, 2002 J. Phys. **A35**, 7079-7100; hep-th/0201076.

[10] Kac V G, 1977 Adv. Math. **26**, 8-96; 1977 Comm. Math. Phys. **53**, 31-64.

[11] Kac V G, 1978 Lect. Notes in Math. **676** (Springer:Berlin) 597-626.

[12] Bars I and Gunaydin M, 1983 Comm. Math. Phys. **91**, 31.

[13] Gunaydin M, van Nieuwenhuizen P, Warner N P, 1985 Nucl. Phys. **B255**, 63.

Inst. Phys. Conf. Ser. No 173: Section 8
Paper presented at 24th Int. Coll. Group Theoretical Methods in Physics, Paris, France, 15–20 July 2002
©*2003 IOP Publishing Ltd*

Supersymmetry and Special Functions

Andrzej M Frydryszak

Institute of Theoretical Physics, University of Wrocław, Wrocław, Poland

Abstract. An implementation of the notion of special functions to the case of superanalysis and description of supersymmetric mechanical systems is discussed. In particular Grassmannian Hermite functions and Grassmannian Laguerre functions are briefly described.

1. Introduction

The aim this paper is to describe briefly the way in which superized versions of some special functions connected with supersymmetric quantum mechanics are introduced. Worth noting is the fact that, as in the conventional case, there are superized functional transformations relating them.

2. Generalized Hermite functions

2.1. Motivation from supermechanics

As in the conventional case Grassmannian counterparts of the Hermite polynomials appear as a result of quantization of the system generalizing the harmonic oscillator. There are two possibilities to write the supersymmetric generalization: even, and odd [1, 2]. To see them let us consider [3] $\mathbb{Z}_2$-graded configuration space M with $dim M = (N_0, N_1)$, the gradation mapping $J : M \longrightarrow M$, $M = M_0 + M_1$, $J(\overset{s}{\phi}) = (-1)^s \overset{s}{\phi}$, $\overset{s}{\phi} \in M_s$, $s = 1, 2$ and the $\mathbb{Z}_2$-graded metric G in M given by

$$<\overset{s}{\phi},\overset{s}{\phi}> = \sum_{i,j=1}^{N_s} \overset{s}{G}{}^{ij} \overset{s}{\phi}_i \overset{s}{\phi}_j, \qquad <\overset{s}{\phi},\overset{s'}{\phi}> - 0 \quad for \quad s \neq s' \quad \overset{s}{G}{}^{ij} = (-1)^s \overset{s}{G}{}^{ji}$$

(s is the parity of a homogeneous supervector). Trajectories in M are superfields with the following expansion

$$\overset{0}{\phi}_j(t, \vartheta) = x_j(t) + i\vartheta_\alpha x_j^\alpha(t) + \frac{1}{2}\vartheta^2 b_j(t), \quad \overset{1}{\phi}_j(t, \vartheta) = y_j(t) + \vartheta_\alpha y_j^\alpha(t) + \frac{1}{2}\vartheta^2 f_j(t)$$

An action S defines a supersymmetric system and has the form $S = \int L(D_\alpha\phi, \phi) dt d\vartheta_1 d\vartheta_2$. One can realize such a system in two natural ways: even and odd.

The even realization is typical and a simple example is provided by: Graded Superfield Oscillator (GSO)[4]. It is a supersymmetrization of bosonic harmonic oscillator and fermionic oscillator[5]. For $N_1 = 2k$ we have the following component action

$$S = \frac{1}{2} \int dt \overset{0}{G}{}^{ij} [(\dot{x}_i\dot{x}_j - b_ib_j) - (\delta_\beta^\alpha x_{\alpha i}\dot{x}_j^\beta) - \omega(x_ib_j + b_ix_j + i\epsilon_{\alpha\beta}x_i^\alpha x_j^\beta)$$

$$\frac{1}{2}\int dt \overset{1}{G}{}^{ij}[-(\dot{y}_i\dot{y}_j + f_if_j) + (\delta^\alpha_\beta y_{\alpha i}\dot{y}^\beta_j)] - \omega(y_if_j + f_iy_j + i\epsilon_{\alpha\beta}y^\alpha_i y^\beta_j). \qquad (1)$$

As a supersymmetric partner for bosonic and fermionic oscillators there are fermionic and bosonic (respectively) rotators. The canonical momenta have the same parity as conjugate coordinates and $\mathbb{Z}_2$-graded Poisson bracket $\{.,.\}_0$, is an even mapping. The phase space of this model has two sectors which we denote $P_{(0,0)}$ and $P_{(1,1)}$ with $(p,q) \in P_{(0,0)}$ and $(\Pi, \Theta) \in P_{(1,1)}$. The pair (p,q) describes even coordinates and their conjugate momenta and similarly (Π, Θ) denotes pairs of the odd coordinates and momenta.

A simple odd realization of such a system is given by the Odd Graded Superfield Oscillator (OGSO) [6]. Assuming that $N_1 = N_0$ one can write the component odd action as follows

$$S = \int dt \sum_{s=1,2} \overset{s}{G}{}^{ij}\{\frac{1}{2}[Tr\, \overset{s}{\delta}_q\,(\dot{x}_i\dot{y}_j - b_if_j)+$$

$$\frac{1}{2}(\overset{s}{\delta}{}^\alpha_{q\beta}x_{\alpha i}\dot{y}^\beta_j - \overset{s}{\delta}{}^\beta_{q\alpha}\dot{x}^\alpha_i y_{\beta j})] + (-1)^s\omega(x_if_j + i\epsilon_{\alpha\beta}x^\alpha_i y^\beta_j + b_iy_j)]\}. \qquad (2)$$

In this realization momenta have opposite parity with respect to the parity of conjugate coordinates. In the odd phase space there is defined an odd Poisson bracket (anti-bracket) $\{\cdot,\cdot\}_1$. It is an odd mapping with shifted grade properties. In analogy to the previous case we denote sectors in the phase space of this system as $P_{(0,1)}$ and $P_{(1,0)}$, where $(p, \Theta) \in P_{(0,1)}$ and $(\Pi, q) \subset P_{(1,0)}$. As before Greek letters denote odd entities.

Quantizing these systems, in the "oscillator" sectors one obtains generalized Hermite polynomials. For example for fermionic oscillator in $P_{(1,1)}$ one gets the following [5]

$$H = \frac{1}{2}(-\Pi\overset{1}{G}{}^{-1}\Pi + \Theta\overset{1}{G}\Theta); \qquad \Pi = \overset{1}{G}\Theta$$

$$a_\pm = \frac{1}{\sqrt{2}}(\Theta_i \mp i(\overset{1}{G}{}^{-1}\Pi)_i), \qquad \{a_{-i}, a_{-j}\} = \{a_{+i}, a_{+j}\} = 0, \qquad \{a_{-i}, a^{+j}\} = \delta^j_i$$

$$H = a^+a_- - \frac{1}{2}n, \qquad a_{-i}\psi_0 = 0, \qquad \psi_0 = e^{-\frac{1}{2}\Theta\overset{1}{G}\Theta}$$

$$\psi^{\alpha_1\alpha_2...\alpha_m} = a^{\alpha_1}_+ \ldots a^{\alpha_m}_+ \psi_0 \sim e^{\frac{1}{2}\Theta\overset{1}{G}\Theta}\partial^{\alpha_1} \ldots \partial^{\alpha_m} e^{-\frac{1}{2}\Theta\overset{1}{G}\Theta}$$

The family of Grassmannian functions obtained in this way $\{\psi_0, \psi^{\alpha_i}, \psi^{\alpha_i\alpha_j}, \ldots, \psi^{\alpha_1...\alpha_n}\}$ consists of the so called (even) Grassmannian Hermite polynomials.

2.2. Grassmannian Hermite functions

The Grassmannian Hermite polynomials, mentioned in previous section, can be discussed in more formal way [7], similar to the conventional one [8]. Namely, as a subset of general solutions of the equation

$$(A \cdot \partial^2 + B_\alpha\partial^\alpha + \Lambda)\Im(\Theta^{\alpha_1}, \Theta^{\alpha_2}, \ldots) = 0,$$

$\Im(\Theta^{\alpha_i})$ is a function generally possessing external indices. A, B_α and constant Λ have fixed parity i.e. $|A| = |B_\alpha| + 1 = |\Lambda|$. A is a multinomial in Θ of degree not greater then two and B_α are multinomials of constant degree, at most equal one. For a choice: A is a constant, $B_\alpha \sim \Theta_\alpha$ there is a set of solutions for the above equation called Grassmannian

Special Functions of type I. They fulfill Rodrigues-like formula, orthogonality relations etc. In particular Grassmannian Hermite polynomials have the form

$$h^{\alpha_n \dots \alpha_1} = K_n exp\{\Theta^2\}\partial^{\alpha_n} \dots \partial^{\alpha_1} exp\{-\Theta^2\}$$

For details cf. [7].

3. Generalized Laguerre functions

Generalized Laguerre functions can be obtained as Grassmannian Fourier-Wigner transform of the Grassmannian Hermite functions. To this end one can use the methods of the harmonic analysis on the Heisenberg [9, 10] group generalized to the case of even and odd phase superspace. It is interesting that Schrödinger quantization for the odd systems can be performed with the use of a generalization of complex numbers to the $\mathbb{Z}_2$-graded algebra with an odd multiplication and odd imaginary unit [11] (the odd Planck constant has been mentioned in [12]). For the even and odd systems we obtain the even and odd versions of Grassmannian special functions.

3.1. Even case

In this case we need generalization of the notion of Heisenberg group to the sector $P_{(1,1)}$ of the phase-superspace [6]. It is called the Fermionic Heisenberg group (the conventional Heisenberg group is related to the $P_{(0,0)}$ sector).

Let $\mathcal{P}_n$ be the free Q-module $\mathcal{P}_n = Q^{2n,1}$ with the fixed basis $\{e_i\}_{i=0}^{2n}$, $|e_0| = 0$, $|e_i| = 1$, $i = 1, 2, \dots, 2n$ (Q is the Banach -Grassmann algebra and $|\cdot|$ denotes Grassmannian parity of an element [14]). Moreover, let $B(\cdot, \cdot)$ be the graded symplectic even form defined on $Q^{2n,0}$ with values in Q. In our basis

$$v = \sum_{i=1}^{n} \Pi_i e_i + \sum_{i=1}^{n} \Theta_i e_{n+i}, \quad B(v, v') = -\sum_{i=1}^{n} (\Pi^i \Theta_i' + \Theta_i \Pi^{i'}), \tag{3}$$

where $|\Pi^i| = |\Theta_i| = 1$. Now we consider the module $\mathcal{P}$ with coordinates

$$(v, t) \equiv (\Pi, \Theta, t) = (\Pi^1, \dots, \Pi^n, \Theta_1, \dots, \Theta_n, t) \tag{4}$$

(where $t \in Q_0$) and with the following multiplication law

$$(v, t) \circ (v', t') = \left(v + v', t + t' + \frac{1}{2}B(v, v')\right). \tag{5}$$

$\mathcal{P}$ equipped with this multiplication forms a group. It is called the Fermionic Heisenberg group and denoted by FH_n.

It turns out that using generalization of the methods of harmonic analysis on Heisenberg group [9] to the Grassmannian fermionic case one can obtain in particular Grassmannian version of the Laguerre polynomials [6].

3.2. Odd case

To describe the Odd Heisenberg group [11] a new structure replacing the complex numbers (the algebra of oddons) has to be introduced. It provides the odd multiplication in the set of observables needed for the quantization of odd systems to preserve the graded structure of the Poisson-Buttin bracket on the level of quantum antibracket relations. We want to introduce it in such a way that the structure of the appropriate Hilbert Q-module will remain preserved. To this end we enlarge the basic Banach-Grassmann algebra Q [13, 14, 15] to the new structure: on the classical level to the $Q_{RO} \equiv Q_R \oplus \hat{1} Q_R$ and on the quantum level $Q_{CO} \equiv Q_C \oplus \hat{i} Q_C$. Q_R and Q_C are real and complex Banach-Grassmann algebras [11]. The $\hat{1}$ and $\hat{i}$ are new elements, both of odd Grassmannian parity and moreover fulfilling relations of the form $\hat{1}^2 = 1$ and $\hat{i}^2 = -1$. The Q_{RO} we shall refer to as the real oddons and Q_{CO} as the complex oddons, to stress that these algebras contain the odd imaginary unit and/or the odd unit. The Q_{CO} is not a graded commutative algebra.

To define the Odd Heisenberg group [11] let us consider as an extension of the phase space $P_{(0,1)}$ by the time dimension the free $Q^{\mathbb{RO}}$-module $T_n = Q_{\mathbb{RO}}^{n|n+1}$ with the basis $\{E_i, e_i, e_0\}_{i=1}^n$, where $|e_i| = |e_0| = 0$, $|E_i| = 1$; $i = 1, 2, \ldots, n$. Let $B(\cdot, \cdot)$ be the odd symplectic form defined on $Q_{\mathbb{RO}}^{n,n}$ with values in $Q_{\mathbb{RO}}$.

$$v = \sum_{i=1}^n p_i E_i + \sum_{i=1}^n \Theta_i e_i, \quad B(v, v') = \sum_{i=1}^n (p^i \Theta'_i - \Theta_i p'^i) \tag{6}$$

Now let OH_n be the set of vectors of the form

$$(v, \tau) = (p, \Theta, \tau) = (p^1, p^2, \ldots, p^n, \Theta_1, \Theta_2, \ldots, \Theta_n, \tau) \tag{7}$$

where $\tau = t \cdot \hat{1}$, $t \in Q_0^R$, $\Theta_i \in Q_1^R$, $p^i \in Q_0^{\mathbb{RO}}$. In the set OH_n we define the action in the following form

$$(v, \tau) \star (v', \tau') = (v + v', \tau + \tau' + \frac{1}{2} B(v, v')) \tag{8}$$

The $(OH_n, \star)$ is a group, we shall call it the Odd Heisenberg group. It the case of the Odd Heisenberg group we need generalization of the methods of harmonic analysis on Heisenberg group to the Grassmannian case extended by the oddons. Using odd multiplication provided in the new structure we can obtain in particular odd Grassmannian version of the Laguerre polynomials [11].

References

[1] Leites D A 1977 Dokl. Akad. Nauk. SSSR 236 804
[2] Leites D A Supplement 3 in Berezin F A and Shubin M A 1992 The Schrödinger Equation (Dordrecht: Kluwer Academic Publisher)
[3] Frydryszak A 1993 J. Phys. A: Math. and General 26 7227-7234
[4] Frydryszak A 1989 Lett. Math. Phys. 18 87-95
[5] Finkelstein R and Villasante M 1986 Phys. Rev. D 6 1666
[6] Frydryszak A 1992 Lett. Math. Phys. 26 105-114
[7] Frydryszak A 1990 Lett. Math. Phys. 20 159-163
[8] Nikiforov A and Ouvarov W 1983 Fonctions Speciales de la Physique Mathématique (Moscow: Editions Mir)
[9] Taylor M E 1986 Noncommutative Harmonic Analysis (AMS Providence: Rhode Island)
[10] Howe R 1980 Bull. Amer. Math. Soc.(N.S.) 3 821
[11] Frydryszak A 1998 Lett. Math. Phys. 44 89-97
[12] Volkov D V and Soroka V A 1988 Sov. J. Nucl. Phys 46 110
[13] Jadczyk A and Pilch K 1981 Commun. Math. Phys. 78 373
[14] Frydryszak A and Jakóbczyk L 1988 Lett. Math. Phys. 16 101-107
[15] Frank M 1996 Hilbert C^*-modules and related subjects - a guided reference overview funct-an/9605003

Inst. Phys. Conf. Ser. No 173: Section 8
Paper presented at 24th Int. Coll. Group Theoretical Methods in Physics, Paris, France, 15–20 July 2002
©2003 IOP Publishing Ltd

Quantization of Phases and Moduli in terms of the Group $SO^\uparrow(1,2)$

H A Kastrup
DESY, Theory Group
Notkestr. 85, D-22603 Hamburg, Germany
E-mail: hans.kastrup@desy.de

Abstract. The problem of quantizing phase and modulus associated with the polar coordinates of a 2-dimensional phase space dates back to the first years of quantum mechanics: The symplectic space $P = \{\varphi \in (-\pi, +\pi], p > 0\}$ has the global structure $S^1 \times R^+$ and cannot be quantized in the conventional naive way. The appropriate method is the group theoretical quantization which here leads to the group $SO(1,2)$ and the positive discrete series of its irreducible unitary representations. The basic classical observables $p \cos\varphi, p \sin\varphi$ and p correspond to the self-adjoint operators K_1, K_2 and K_3 of the $SO(1,2)$ Lie algebra. The approach provides appropriate quantum observables for the phase space P and a promising basis for the description of quantum optical structures

1. Introduction

The transformation

$$q(\varphi, I) = \sqrt{2I/\omega} \cos\varphi \ , \quad p(\varphi, I) = -\sqrt{2\omega I} \sin\varphi \ , \tag{1}$$

is locally canonical,

$$dq \wedge dp = d\varphi \wedge dI \ , \tag{2}$$

and transforms the Hamiltonian and the Poisson bracket relation

$$H(q,p) = \frac{1}{2}p^2 + \frac{1}{2}\omega^2 q^2 \ , \ (q,p) \in R^2 \ , \ \{q,p\} = 1 \ , \tag{3}$$

into

$$H = \omega I \ , \ \{q,p\} = \{\varphi, I\} = 1 \ . \tag{4}$$

The phase space

$$P = \{(\varphi \in (-\pi \ \pi], I > 0)\} \tag{5}$$

of the angle and action variables φ and I has the *global* structure

$$S^1 \times R^+ \ , \ R^+ = \{r \in R, \ r > 0\} \ , \tag{6}$$

which is quite different from the one we started from, namely R^2!
Dirac was the first to interpret [1] the polar decomposition of the complex amplitudes
($\bar{a}$ denotes the complex conjugate of a!)

$$a = \frac{1}{\sqrt{2}}(\sqrt{\omega}q + \frac{i}{\sqrt{\omega}}p) = I^{1/2}e^{-i\varphi} \ , \tag{7}$$

$$\bar{a} = \frac{1}{\sqrt{2}}(\sqrt{\omega}q - \frac{i}{\sqrt{\omega}}p) = I^{1/2}e^{i\varphi} \ , \tag{8}$$

$$I = \bar{a}a \ , \tag{9}$$

in terms of quantum mechanical operators $\hat{I} = N = a^+ a$ and $\hat{\varphi}$ by using a corresponding polar decomposition of the annihilation and creation operators

$$a = e^{-i\hat{\varphi}} \sqrt{N} = \sqrt{N+1} e^{-i\hat{\varphi}} , \quad a^+ = \sqrt{N} e^{i\hat{\varphi}} = e^{i\hat{\varphi}} \sqrt{N+1} , \tag{10}$$

with the commutator

$$[\hat{\varphi}, N] = i . \tag{11}$$

But even before the second of Dirac's just quoted papers appeared, London had realized [2] that the commutation relation (11) cannot hold, because

$$\langle n_2 | [\hat{\varphi}, \hat{N}] | n_1 \rangle = (n_1 - n_2) \langle n_2 | \hat{\varphi} | n_1 \rangle = i \delta_{n_2 n_1} ! \tag{12}$$

In a second paper [3] London suggested the use of the operators

$$E_- = a N^{-1/2} , \quad E_+ = N^{-1/2} a^+ \tag{13}$$

instead.

That proposal, however, was not used for about 40 years, but rediscovered by Susskind and Glogower in the mid-sixties [4] and worked out in more detail by Carruthers and Nieto [5]. The subject is still controversial to-day [6]

2. Group theoretical quantization

A new promising ansatz in order to quantize the phase space (5) consistently, is group theoretical quantization, beautifully reviewed by Isham [7]. In our special context the group $SO(1,2)$ is the relevant quantizing group [8], leading to a consistent quantum theory with possible applications in quantum optics [9]. Some of its essential features are the following: The basic classical "observables" on the phase space turn out to be the functions

$$h_1 = I \cos \varphi , \quad h_2 = -I \sin \varphi , \quad h_3 = I , \tag{14}$$

which obey the (Poisson) Lie algebra

$$\{h_3, h_1\} = -h_2 , \quad \{h_3, h_2\} = h_1 , \quad \{h_1, h_2\} = h_3 . \tag{15}$$

This is the Lie algebra of the group $SO(1,2)$ or one of its (infinitely) many covering groups, like, e.g. $SU(1,1)$.

Quantization is implemented by employing appropriate irreducible unitary representations of the group. The 3 self-adjoint generators K_j of the corresponding 1-parameter unitary subgroups are the quantized versions of the classical observables (14):

$$K_j = \hat{h}_j , \quad j = 1, 2, 3 , \tag{16}$$

which obey the commutation relations

$$[K_3, K_1] = i K_3 , \quad [K_3, K_2] = i K_2 , \quad [K_1, K_2] = -i K_3 . \tag{17}$$

The appropriate irreducible unitary representations are those of the positive discrete series which are characterized by the existence of a (ground) state $|k, 0\rangle$ for which

$$K_- |k, 0\rangle = 0 , \quad K_- = K_1 - i K_2 . \tag{18}$$

The generator K_3 of the compact subgroup has the spectrum

$$\mathrm{spec}(K_3) = \{k + n, k > 0, n = 0, 1, \ldots\} \tag{19}$$

The positive parameter k, which characterizes an irreducible unitary representation, can take the values $k = 1, 2, \ldots$ for the group $SO(1,2)$ and the values $k = 1/2, 1, 3/2, 2, \ldots$ for $SU(1,1)$. Taking the expectation values of the operators K_1 and K_2 with respect to the different coherent states associated with $SO(1,2)$ all of which are characterized by a complex number shows that these operators indeed "measure" the cos and the sin of the phase of those complex numbers.

3. Harmonic oscillator

One may illustrate many features of the generalized cos- and sin-operators K_1 and K_2 by means of the harmonic oscillator. A reason is that one can construct a realization of the $SU(1,1)$ Lie algebra generators from the oscillator annihilation and creation operators a and a^+:

$$K_- = \sqrt{N+2k}\,a\,, \quad K_+ = (K_-)^+\,, \quad K_3 = a^+a+k = N+k\,, \tag{20}$$

More interesting is the inverse:

Given the self-adjoint operators $K_\pm, K_3$ of a representation characterized by k, then one can define annihilation and creation operators - and therefore operators $\hat{q}$ and $\hat{p}$ - by solving the relations (20) for a and a^+!

Another point of interest is the following: For $k = 1/2$ we can identify the Hamiltonian H of the harmonic oscillator with the operator ωK_3. But for $k = 1/2$ the group $SU(1,1)$ has the following explicit irreducible unitary representation

$$(f_2, f_1) \quad = \frac{1}{2\pi} \int_0^{2\pi} d\varphi\, \bar{f}_2(\varphi)\, f_1(\varphi)\,, \tag{21}$$

$$|k = 1/2, n\rangle = e^{in\varphi}\,, n = 0, 1, \dots \,, \tag{22}$$

$$K_3 \quad = \frac{1}{i}\,\partial_\varphi + 1/2\,, \tag{23}$$

$$K_- \quad = e^{-i\varphi}\,\frac{1}{i}\,\partial_\varphi\,, \tag{24}$$

$$K_+ \quad = e^{i\varphi}\,(\frac{1}{i}\,\partial_\varphi + 1)\,, \tag{25}$$

$$H \quad = \omega K_3\,. \tag{26}$$

Thus, it is possible to describe the quantum physics of the harmonic oscillator in this Hilbert space. Any element $f(\varphi)$ may be expanded in a series

$$f(\varphi) = \sum_{n=0}^{\infty} c_n e^{in\varphi}\,. \tag{27}$$

The use of this Hilbert space also allows for a critical evaluation of the usual - somewhat controversial - notion of "phase states" [5, 6]

For the coherent states

$$|\alpha\rangle = e^{-|\alpha|^2/2} \sum_{n=0}^{\infty} \frac{\alpha^n}{\sqrt{n!}}\,|n\rangle\,, \quad \alpha = |\alpha|e^{i\beta}\,, \tag{28}$$

we get - using the relations (20) - the expectation values

$$\langle\alpha|K_1|\alpha\rangle = |\alpha|\cos\beta\,\langle\alpha|\sqrt{N+1}|\alpha\rangle\,, \tag{29}$$

$$\langle\alpha|K_2|\alpha\rangle = -|\alpha|\sin\beta\,\langle\alpha|\sqrt{N+1}|\alpha\rangle\,, \tag{30}$$

$$\langle\alpha|K_3|\alpha\rangle = |\alpha|^2 + 1/2\,, \tag{31}$$

where

$$\langle\alpha|\sqrt{N+1}|k,\alpha\rangle = e^{-|\alpha|^2} \sum_{n=0}^{\infty} \sqrt{n+1}\,\frac{|\alpha|^{2n}}{n!}\,. \tag{32}$$

We see here explicitly how the operators K_1 and K_2 measure the phase of the complex number α associated with the state $|\alpha\rangle$:

$$\tan\beta = -\frac{\langle K_2\rangle_\alpha}{\langle K_1\rangle_\alpha}\,. \tag{33}$$

The operators

$$K_1 = \frac{1}{2}(K_+ + K_-) = \cos\varphi \frac{1}{i}\partial_\varphi + \frac{1}{2}e^{i\varphi} , \qquad (34)$$

$$K_2 = \frac{1}{2i}(K_+ - K_-) = \sin\varphi \frac{1}{i}\partial_\varphi + \frac{1}{2i}e^{i\varphi} . \qquad (35)$$

have eigenvalues h_1 and h_2 and eigenfunctions

$$f_{h_1}(\varphi) = |2\cos\varphi|^{-1/2}|\tan(\varphi/2 + \pi/4)|^{ih_1} e^{-i\varphi/2} , \ h_1 \in R , \qquad (36)$$

$$f_{h_2}(\varphi) = |2\sin\varphi|^{-1/2}|\tan(\varphi/2)|^{ih_2} e^{-i\varphi/2} , \ h_2 \in R , \ \varphi \in (0, 2\pi) . \qquad (37)$$

Many more details and results will be contained in a forthcoming publication [10].

References

[1] Dirac P A M 1925 Proc. Royal Soc. London Ser. A 109 642-653;
    1927 Proc. Royal Soc. London Ser. A 114 243-265
[2] London F 1926 Zeitschr. f. Physik 37 915-925
[3] London F 1927 Zeitschr. f. Physik 40 193-210
[4] Susskind L and Glogower J 1964 Physics 1 49-61
[5] Carruthers P and Nieto M M 1968 Rev. Mod. Phys. 40 411-440
[6] See, e.g. Lynch R 1995 Phys. Reports 256 367-436; Peřinová V, Lukš A and Peřina J 1998 Phase in Optics (Singapore: World Scientific)
[7] Isham C J 1984 in Relativity, Groups and Topology II ed. by Dewitt B S and Stora R (Amsterdam: North-Holland) 1059-1290
[8] Loll R 1990 Phys. Rev. D 41 3785-3791; Bojowald M, Kastrup H A, Schramm F and Strobl T 2000 Phys. Rev. D 62 044026-1-26
[9] Kastrup H A 2001 e-print quant-ph/0109013
[10] Kastrup H A 2002 to appear

Inst. Phys. Conf. Ser. No 173: Section 8
Paper presented at 24th Int. Coll. Group Theoretical Methods in Physics, Paris, France, 15–20 July 2002
©*2003 IOP Publishing Ltd*

p-Mechanical Brackets and Method of Orbits

Vladimir V. Kisil ‡

School of Mathematics, University of Leeds, Leeds LS2 9JT, UK

E-mail: kisilv@amsta.leeds.ac.uk

Abstract. We use the orbit method of Kirillov to derive the *p*-mechanical brackets [8]. They generate the quantum (Moyal) and classic (Poisson) brackets on respective orbits corresponding to representations of the Heisenberg group. This highlights connections of *p*-mechanical brackets with deformation quantisation and Moyal's product. The *p*-brackets are invariant under automorphisms of the Heisenberg group, this leads to the symplectic invariance of quantum and classic mechanics.

AMS classification scheme numbers: 81R05, 81R15, 22E27, 22E70, 43A65

1. The Heisenberg Group and Its Representations

Let (s, x, y), where x, $y \in \mathbb{R}^n$ and $s \in \mathbb{R}$, be an element of the Heisenberg group $\mathbb{H}^n$ [3, 4]. The group law on $\mathbb{H}^n$ is given as follows:

$$(s, x, y) * (s', x', y') = (s + s' + \frac{1}{2}\omega(x, y; x', y'), x + x', y + y'), \qquad (1)$$

where the non-commutativity is solely due to ω—the *symplectic form* [1, § 37] on $\mathbb{R}^{2n}$:

$$\omega(x, y; x', y') = xy' - x'y. \qquad (2)$$

The Lie algebra $\mathfrak{h}^n$ of $\mathbb{H}^n$ is spanned by left-invariant vector fields

$$S = \frac{\partial}{\partial s}, \qquad X_j = \frac{\partial}{\partial x_j} - \frac{y_j}{2}\frac{\partial}{\partial s}, \qquad Y_j = \frac{\partial}{\partial y_j} + \frac{x_j}{2}\frac{\partial}{\partial s} \qquad (3)$$

on $\mathbb{H}^n$ with the Heisenberg *commutator relations* $[X_i, Y_j] = \delta_{i,j}S$ and all other commutators vanishing. The exponential map $\exp : \mathfrak{h}^n \to \mathbb{H}^n$ respecting the multiplication (1) and Heisenberg commutators is

$$\exp : sS + \sum_{k=1}^{n}(x_k X_k + y_k Y_k) \mapsto (s, x_1, \ldots, x_n, y_1, \ldots, y_n).$$

The adjoint representation $\mathrm{Ad} : \mathbb{H}^n \to \mathbb{H}^n$ given by $\mathrm{Ad}\,(g)h = g^{-1}hg$ fixes the unit $e \in \mathbb{H}^n$. The differential $\mathrm{ad} : \mathfrak{h}^n \to \mathfrak{h}^n$ of Ad at e is a linear map given by the Lie commutator: $\mathrm{ad}\,(A) : B \mapsto [B, A]$. The dual space $\mathfrak{h}^n_*$ to the Lie algebra $\mathfrak{h}^n$ is realised by the left invariant first order differential forms on $\mathbb{H}^n$. By the duality between $\mathfrak{h}^n$ and $\mathfrak{h}^n_*$ the map ad generates the *co-adjoint representation* [5, § 15.1] $\mathrm{ad}^* : \mathfrak{h}^n_* \to \mathfrak{h}^n_*$:

$$\mathrm{ad}^*(s, x, y) : (h, q, p) \mapsto (h, q + hy, p - hx), \quad \text{where } (s, x, y) \in \mathbb{H}^n \qquad (4)$$

‡ On leave from the Odessa University.

and $(h, q, p) \in \mathfrak{h}_n^*$ in bi-orthonormal coordinates to the exponential ones on $\mathfrak{h}^n$. There are two types of orbits in (4) for ad *—Euclidean spaces $\mathbb{R}^{2n}$ and single points:

$$\begin{aligned}
\mathcal{O}_h &= \{(h, q, p) : \text{ for a fixed } h \neq 0 \text{ and all } (q, p) \in \mathbb{R}^{2n}\}, &(5)\\
\mathcal{O}_{(q,p)} &= \{(0, q, p) : \text{ for a fixed } (q, p) \in \mathbb{R}^{2n}\}. &(6)
\end{aligned}$$

The *orbit method* of Kirillov [5, § 15] starts from the observation that the above orbits parametrise irreducible unitary representations of $\mathbb{H}^n$. All representations are *induced* [5, § 13] by a character $\chi_h(s, 0, 0) = e^{-ihs}$ of the centre of $\mathbb{H}^n$ generated by $(h, 0, 0) \in \mathfrak{h}_n^*$ and shifts (4) from the *left* on orbits. Using [5, § 13.2, Prob. 5] we get a neat formula, which (unlike many other in literature) respects all *physical units* [7]:

$$\rho_h(s, x, y) : f_h(q, p) \mapsto e^{ihs + \frac{i}{2}(qx + py)} f_h(q - hy, p + hx). \tag{7}$$

The derived representation $d\rho_h$ of $\mathfrak{h}^n$ defined on the vector fields (3) is:

$$d\rho_h(S) = ihI, \qquad d\rho_h(X_j) = h\partial_{p_j} + \tfrac{i}{2}q_j I, \qquad d\rho_h(Y_j) = -h\partial_{q_j} + \tfrac{i}{2}p_j I \tag{8}$$

Operators D_h^j, $1 \leq j \leq n$ representing vectors from the complexification of $\mathfrak{h}^n$:

$$D_h^j = d\rho_h(X_j - iY_j) = h(\partial_{p_j} + i\partial_{q_j}) + \tfrac{p_j + iq_j}{2}I = 2h\partial_{\bar{z}_j} + \tfrac{z_j}{2}I \tag{9}$$

where $z_j = p_j + iq_j$ are used to give the following classic result in terms of orbits:

Theorem 1 (Stone–von Neumann, cf. [5, § 18.4], [3, Chap. 1, § 5]) *All unitary irreducible representations of $\mathbb{H}^n$ are parametrised up to equivalence by two classes of orbits (5) and (6) of co-adjoint representation (4) in $\mathfrak{h}_n^*$:*

(i) The infinite dimensional representations by transformation ρ_h (7) for $h \neq 0$ in Fock [3, 4] space $F_2(\mathcal{O}_h) \subset L_2(\mathcal{O}_h)$ of null solutions to the operators D_h^j (9):

$$F_2(\mathcal{O}_h) = \{f_h(p, q) \in L_2(\mathcal{O}_h) \mid D_h^j f_h = 0, \ 1 \leq j \leq n\}. \tag{10}$$

(ii) The one-dimensional representations as multiplication by a constant on $\mathbb{C} = L_2(\mathcal{O}_{(q,p)})$ which drops out from (7) for $h = 0$:

$$\rho_{(q,p)}(s, x, y) : c \mapsto e^{\frac{i}{2}(qx + py)}c. \tag{11}$$

Note that $f_h(p, q)$ is in $F_2(\mathcal{O}_h)$ if and only if the function $f_h(z)e^{-|z|^2/(4h)}$, $z = p + iq$ is in the classical Segal–Bargmann space [3, 4], particularly is analytical in z. Furthermore the space $F_2(\mathcal{O}_h)$ is spanned by the Gaussian *vacuum vector* $e^{-(q^2 + p^2)/(4h)}$ and all *coherent states*, which are "shifts" of the vacuum vector by operators (7).

Commutative representations (11) are always neglected but their union naturally (see the appearance of Poisson brackets in (20)) acts as the classical *phase space*:

$$\mathcal{O}_0 = \bigcup_{(q,p) \in \mathbb{R}^{2n}} \mathcal{O}_{(q,p)}. \tag{12}$$

Furthermore the structure of orbits of $\mathfrak{h}_n^*$ echoes in Equation (21) and Proposition 2.

2. Convolution Algebra of $\mathbb{H}^n$ and Commutator

Using invariant measure dg the linear space $L_1(\mathbb{H}^n, dg)$ can be upgraded to an algebra with the convolution multiplication:

$$(k_1 * k_2)(g) = \int_{\mathbb{H}^n} k_1(g_1)\, k_2(g_1^{-1}g)\, dg_1 = \int_{\mathbb{H}^n} k_1(gg_1^{-1})\, k_2(g_1)\, dg_1. \tag{13}$$

Inner *derivations* D_k of $L_1(\mathbb{H}^n)$ are given by the *commutator*:

$$D_k : f \mapsto [k, f] = k * f - f * k = \int_{\mathbb{H}^n} k(g_1) \left(f(g_1^{-1}g) - f(gg_1^{-1}) \right) dg_1. \tag{14}$$

A unitary representation ρ_h of $\mathbb{H}^n$ extends to $L_1(\mathbb{H}^n, dg)$ by the formula:

$$[\rho_h(k)f](q, p) = \int_{\mathbb{R}^{2n}} \left(\int_{\mathbb{R}} k(s, x, y)e^{ihs} ds \right) e^{\frac{i}{2}(qx+py)} f(q - hy, p + hx) dx\, dy, \tag{15}$$

thus $\rho_h(k)$ for a fixed $h \neq 0$ depends only from $\hat{k}_s(h, x, y)$—the partial Fourier transform $s \to h$ of $k(s, x, y)$. Then the representation of the composition of two convolutions depends only from

$$(k' * k)\hat{_s}(h, x, y) = \int_{\mathbb{R}^{2n}} \hat{k}'_s(h, x', y')\hat{k}_s(h, x - x', y - y')e^{\frac{ih}{2}(xy'-yx')} dx'dy'.$$

The last expression for the full Fourier transforms of k' and k turn to be the *star product* known in deformation quantisation, cf. [9, (9)–(13)]. Consequently the representation of commutator (14) depends only from:

$$[k', k]\hat{_s} = 2i \int_{\mathbb{R}^{2n}} \hat{k}'_s(2h, x', y')\hat{k}_s(2h, x - x', y - y') \sin(\frac{h}{2}(xy' - yx')) dx'dy', \tag{16}$$

which turn to be exactly the "Moyal brackets" [9] for the full Fourier transforms of k' and k. Also the expression (16) vanishes for $h = 0$ as can be expected from the commutativity of representations (11).

3. *p*-Mechanical Brackets on $\mathbb{H}^n$

A right inverse operator $\mathcal{A}$ to the vector field S (3) on $\mathbb{H}^n$ defined by:

$$\mathcal{A}e^{ihs} = \begin{cases} \frac{1}{ih}e^{ihs}, & \text{if } h \neq 0, \\ s, & \text{if } h = 0; \end{cases} \tag{17}$$

extends by linearity to $L_1(\mathbb{H}^n)$. We introduce [8] a modified convolution operation $\star$ on $L_1(\mathbb{H}^n)$ and the associated modified commutator:

$$k_1 \star k_2 = k_1 * (\mathcal{A}k_2), \qquad \{[k_1, k_2]\} = k_1 \star k_2 - k_2 \star k_1. \tag{18}$$

Then from (15) one gets $\rho_h(\mathcal{A}k) = (ih)^{-1}\rho_h(k)$ for $h \neq 0$. Consequently the modification of (16) for $h \neq 0$ is only slightly different from the original one:

$$\{[k', k]\}\hat{_s} = \int_{\mathbb{R}^{2n}} \hat{k}'_s(2h, x', y')\hat{k}_s(2h, x - x', y - y')\frac{2}{h} \sin(\frac{h}{2}(xy' - yx')) dx'dy', \tag{19}$$

However the last expression for $h = 0$ is significantly distinct from the vanishing (16). From the natural assignment $\frac{2}{h} \sin(\frac{h}{2}(xy' - yx')) = xy' - yx'$ for $h = 0$ we get the Poisson brackets for the Fourier transforms of k' and k defined on $\mathcal{O}_0$ (12):

$$P_{(q,p)} \{[k', k]\} = \frac{\partial \hat{k}'}{\partial q} \frac{\partial \hat{k}}{\partial p} - \frac{\partial \hat{k}'}{\partial p} \frac{\partial \hat{k}}{\partial q} \tag{20}$$

Furthermore the dynamical equation based on the modified commutator (18) with a suitable Hamilton type function $H(s, x, y)$ for an observable $f(s, x, y)$ on $\mathbb{H}^n$

$$\dot{f} = \{[H, h]\} \text{ is reduced } \begin{cases} \text{by } \rho_h,\, h \neq 0 \text{ on } \mathcal{O}_h \text{ (5) to Moyal's equation [9, (8)];} \\ \text{by } \rho_{(q,p)} \text{ on } \mathcal{O}_0 \text{ (12) to Poisson's equation [1, § 39].} \end{cases} \tag{21}$$

The same connections are true for the solutions of these three equations, see [8] for the harmonic oscillator example. Rephrasing the title of [9] we could say that *quantum and classic mechanics live and work **together** on the Heisenberg group and are separated only in irreducible representations of* $\mathbb{H}^n$.

4. Symplectic Invariance of Mechanics from Automorphisms of $\mathbb{H}^n$

Let $A : \mathbb{R}^{2n} \to \mathbb{R}^{2n}$ be a linear *symplectomorphism*, i.e. a map preserving the symplectic form (2): $\omega\left(A(x,y); A(x',y')\right) = \omega(x,y;x',y')$. Then it is follows from (1) that the map $\alpha : \mathbb{H}^n \to \mathbb{H}^n$ such that $\alpha(s,x,y) = (s, A(x,y))$ is an automorphism of $\mathbb{H}^n$. Also α fixes the unit e of $\mathbb{H}^n$ and its differential $d\alpha : \mathfrak{h}^n \to \mathfrak{h}^n$ at e is given by the same matrix as α in exponential coordinates. By the duality we obtain adjoint map $d\alpha^* : \mathfrak{h}_n^* \to \mathfrak{h}_n^*$ defined by the expression

$$d\alpha^* : (h,q,p) \mapsto (h, A^t(q,p)), \tag{22}$$

where A^t is the transpose of A. Obviously $d\alpha^*$ preserves any orbit $\mathcal{O}_h$ (5) and maps the orbit $\mathcal{O}_{(q,p)}$ (6) to $\mathcal{O}_{A(q,p)}$. Because both representations ρ_h and $\rho_h \circ d\alpha^*$ for $h \neq 0$ correspond to the same orbit $\mathcal{O}_h$ they should be equivalent, the intertwining operators between them form the *metaplectic representation* [3, § 4.2] of the symplectic group.

For a symplectomorphism A of $\mathbb{R}^{2n}$ let us denote by α again the linear transformation of $L_1(\mathbb{H}^n)$ of the form $\alpha(k)(s,x,y) = k(s, A(x,y))$. Because $\det \alpha = 1$ the map α is an automorphism of the convolution algebra $L_1(\mathbb{H}^n)$ with the multiplication $*$ (13): $\alpha(k_1) * \alpha(k_2) = \alpha(k_1 * k_2)$. Moreover α commutes with the antiderivative $\mathcal{A}$ (17), thus α is an automorphism of $L_1(\mathbb{H}^n)$ with the modified multiplication $\star$ (18) as well: $\alpha(k_1) \star \alpha(k_2) = \alpha(k_1 \star k_2)$. Thus we have

Proposition 2 *The p-mechanical brackets are invariant under the symplectic automorphisms of* $\mathbb{H}^n$: $\{\!\{\alpha k_1, \alpha k_2\}\!\} = \alpha \{\!\{k_1, k_2\}\!\}$. *Consequently the dynamical equation (21) has symplectic symmetries which are reduced*

(i) by ρ_h, $h \neq 0$ on $\mathcal{O}_h$ (5) to metaplectic representation in quantum mechanics;

(ii) by $\rho_{(q,p)}$ on $\mathcal{O}_0$ (12) to symplectic symmetries of classic mechanics [1, § 38].

All orbits $\mathcal{O}_h$, $h \neq 0$ (5) possess the natural symplectic structure [5, § 15.1], which however degenerates on $\mathcal{O}_0$ (12). Symplectic automorphisms of $\mathbb{H}^n$ were used in [2] to obtain a non-trivial symplectic structure on $\mathcal{O}_0$ as well. Another application of symplectic symmetries is the description of the *Weyl quantisation* [3] as the *intertwining operator* [6] between classical and metaplectic representations.

[1] V. I. Arnol'd, *Mathematical methods of classical mechanics*, Graduate Texts in Mathematics, vol. 60, Springer-Verlag, New York, 1991, Translated from the 1974 Russian original by K. Vogtmann and A. Weinstein, Corrected reprint of the second (1989) edition. **MR** # 96c:70001

[2] Alastair Brodlie, *Symplectic structure on the family of single-point orbits of the Heisenberg group*, in preparation, 2002.

[3] Gerald B. Folland, *Harmonic analysis in phase space*, Annals of Mathematics Studies, vol. 122, Princeton University Press, Princeton, NJ, 1989. **MR** # 92k:22017

[4] Roger Howe, *Quantum mechanics and partial differential equations*, J. Funct. Anal. **38** (1980), no. 2, 188–254. **MR** # 83b:35166

[5] A. A. Kirillov, *Elements of the theory of representations*, Springer-Verlag, Berlin, 1976, Translated from the Russian by Edwin Hewitt, Grundlehren der Mathematischen Wissenschaften, Band 220. **MR** # 54 #447

[6] Vladimir V. Kisil, *Meeting Descartes and Klein somewhere in a noncommutative space*, Procedsings of ICMP2000, American Mathematical Society, 2002, E-print: arXiv:math-ph/0112059, p. 25.

[7] _____, *p-Mechanics respects physical units*, (2002), in preparation.

[8] _____, *Quantum and classic brackets*, Int. J. Theor. Phys. **41** (2002), no. 1, 63–77, E-print: arXiv:math-ph/0007030.

[9] Cosmas K. Zachos, *Deformation quantization: Quantum mechanics lives and works in phase-space*, Int. J. Mod. Phys. A **17** (2002), 297–316, E-print: `arXiv:hep-th/0110114`.

Inst. Phys. Conf. Ser. No 173: Section 8
Paper presented at 24th Int. Coll. Group Theoretical Methods in Physics, Paris, France, 15–20 July 2002
©2003 IOP Publishing Ltd

The Kähler form on the loop group and the Radul cocycle on pseudo-differential operators

Jean-Pierre Magnot

Laboratoire de Mathématiques Appliquées
Université Blaise Pascal (Clermont II)
Complexe Universitaire des Cézeaux
63177 Aubière Cedex, France.

Abstract.
    We show that the Kähler form of the loop group is the pull back of the Radul cocycle on formal pseudo-differential operators in a shorter way that in [1]. This new proof focus on symbolic calculus.

Introduction

The Kähler form ω of the loop group, defined by A.Pressley [7], has been computed in terms of a "two step" trace of a curvature by D.Freed [2], using Fredholm and Toeplitz operators and particular choices of orthonormal basis along which the trace was taken. One can first remark that D.Freed used the setting of pseudo-differential operators to conclude that the curvature of the loop group was Hilbert-Schmidt. Following this, we proved in a joint work with A. Cardona, C. Ducourtioux and S. Paycha [1] that ω is proportional to a pull back of the Radul cocycle, using the tool of renormalized traces. In that work, we interpreted the "two step" trace used by D.Freed and produced results that did not depend on any choice of L^2 orthomornal basis. Our proofs used direct computations of renormalized traces of pseudo-differential operators. In the present work, we provide a proof of this result avoiding Fredhom operators and renormalized traces. We focus instead on the structure of the algebra of formal classical symbols where only the Radul cocycle is defined.

1. Symbols of Pseudo-differential operators on a trivial bundle over S^1

For the definition of classical pseudo-differential operators acting on smooth sections of a vector bundle and their basic properties, we refer the reader to [3]. We use here the work of H. Widom [11] for the symbolic calculus on pseudo-differential operators.

    We identify the circle S^1 with the group $U(1) \subset \mathbb{C}$. A map $f : S^1 \to \mathbb{C}$ is interpreted as a 2π periodic map on $\mathbb{R}$. The cotangent bundle T^*S^1 is a trivial line bundle that we identify with $S^1 \times \mathbb{R}$. $T^*S^1 - S^1$ has two connected components and each connected component can be identified with $S^1 \times \mathbb{R}_+^*$. This enables us to identify T^*S^1 with $S^1 \times \mathbb{R}$. These identifications are fixed all along this work.

    Let V be a complex vector space of dimension n. We now work on the trivial bundle $E = S^1 \times V$. Let D be the Dirac operator on this bundle, $D = -iD_x$.

1.1. Some formal symbols of pseudo-differential operators

In this section, we recall the formal symbols that we use in the sequel. These symbols are mainly classical ones, even if a logarithmic (and hence non classical) symbol (see [5] for a definition) will play a crucial role later on. Recall that the formal symbol σ of a classical operator of order k can be written in terms of formal series $\sigma = \sum_{j=-\infty}^{k} \sigma_j$, where each σ_j is a smooth funtion $T^*S^1 - S^1 = S^1 \times \mathbb{R}^* \to M_n(\mathbb{C})$ such that, $\forall t > 0$ and $\forall (x, \xi) \in S^1 \times \mathbb{R}^*$, $\sigma_j(x, t\xi) = t^j \sigma_j(x, \xi)$.

Applying the results of [11] to formal symbols of pseudo-differential operators, we recover the well-known formula for the composition of two classical pseudo-differential operators, see e.g [3]:

Proposition 1 *Let D_x be the derivation of S^1 induced by the usual derivation on $\mathbb{R}$. Let D_ξ be the directional derivation with respect to the fibers of $T^*S^1 = S^1 \times \mathbb{R}$. Let A and B be two pseudo-differential operators acting on sections of $E = S^1 \times V$, and let σ and σ' be their formal symbols. The formal symbol σ'' of $A \circ B$ is given by*

$$\sigma'' = \sum_{\alpha \in \mathbb{N}} \frac{(-i)^\alpha}{\alpha!} D_\xi^\alpha \sigma D_x^\alpha \sigma'.$$

Proof : We apply [11], Proposition 3.6, taking ∇ as the flat derivation on T^*S^1, and the obvious projection $T^*S^1 \times S^1 = S^1 \times \mathbb{R} \times S^1 \to \mathbb{R}$ as linear function associated to ∇. $\square$

Let $\mathcal{F}Cl$ be the algebra of formal symbols of classical pseudo-differential operators acting on $C^\infty(S^1, V)$, and if $o \in \mathbb{Z}$, let $\mathcal{F}Cl^o$ be the vector subspace of $\mathcal{F}Cl$ made of symbols of order $\leq a$.

The operator $D = -iD_x$ splits $C^\infty(S^1, \mathbb{C}^n)$ into three spaces : its kernel (built of constant maps), the vector space E_+ spanned by eigenvectors related to positive eigenvalues, and the vector space E_- spanned by eigenvectors related to negative eigenvalues.

Let us restrict ourselves to $E_+ \oplus E_-$. The following elementary result will be useful for the sequel.

Lemma 1 *(i) $\sigma(D) = \xi$*

(ii) $\sigma(|D|) = |\xi|$

(iii) $\sigma(\epsilon) = \frac{\xi}{|\xi|}$, where $\epsilon = D|D|^{-1} = |D|^{-1}D$ is the sign of D.

(iv) Let p_{E_+} (resp. p_{E_-}) be the projection on E_+ (resp. E_-), then $\sigma(p_{E_+}) = \frac{1}{2}(Id + \frac{\xi}{|\xi|})$ and $\sigma(p_{E_-}) = \frac{1}{2}(Id - \frac{\xi}{|\xi|})$.

Proof : (i) We have $(-iD_x f)\widehat{}(\xi) = \xi \hat{f}(\xi)$. Hence, $\sigma(D) = \xi$.

(ii) If v is an eigenvector of D associated to the eigenvalue λ, v is also an eigenvector of $|D|$ associated to the eigenvalue $|\lambda|$. Let us compute $|D|(e^{ix \cdot \xi})$, where $x \mapsto e^{ix \cdot \xi}$ denotes the function defined in $]0, 2\pi[$ and extended to a 2π-periodic distribution on $\mathbb{R}$. This distribution on $\mathbb{R}$ defines a distribution on S^1, hence it makes sense to write $D_x(e^{ix \cdot \xi})$ and $|D|(e^{ix \cdot \xi})$. We have $D_x(e^{ix \cdot \xi}) = -i\xi e^{ix \cdot \xi}$. Hence, by easy computations in terms of Fourier series, $|D|(e^{ix \cdot \xi}) = |\xi| e^{ix \cdot \xi}$. Thus, using [11], $\sigma(|D|)(x, \xi) = e^{-ix \cdot \xi} |D|(e^{ix \cdot \xi}) = |\xi|$.

(iii) $\sigma(|D|^{-1})$ is the inverse of $\sigma(|D|)$ in the algebra of formal operators. Hence, using the fact that we have $D_x^\alpha |\xi| = 0$, applying Proposition 1, $1 = \sigma(|D|) \circ \sigma(|D|^{-1}) = \sum \frac{(-i)^\alpha}{\alpha!} D_x^\alpha \sigma(|D|) D_\xi^\alpha \sigma(|D|^{-1}) = \sum \frac{(-i)^\alpha}{\alpha!} D_x^\alpha |\xi| D_\xi^\alpha \sigma(|D|^{-1}) = |\xi| \sigma(|D|^{-1})$. Thus, $\sigma(|D|^{-1}) = \frac{1}{|\xi|}$. By the same techniques, $\sigma(\epsilon) = \frac{\xi}{|\xi|}$. (iv) is an easy consequence of (iii). $\square$

Let us now give an easy but very useful lemma:

Lemma 2 *Let $f : \mathbb{R}^* \to V$ be a 0-positively homogeneous function with values in a topological vector space V. Then, for any $n \in \mathbb{N}^*$, $f^{(n)} = 0$ where $f^{(n)}$ denotes the n-th derivative of f.*

Since a 0-positively homogeneous function is constant on $\mathbb{R}^+$ and on $\mathbb{R}^-$, the proof is obvious.

1.2. Splitting of the algebra of formal symbols

In this section, we define two ideals of the algebra $\mathcal{F}Cl$, that we call $\mathcal{F}Cl_+$ and $\mathcal{F}Cl_-$, such that $\mathcal{F}Cl = \mathcal{F}Cl_+ \oplus \mathcal{F}Cl_-$. This decomposition is explicit in [4], Section 4.4., p. 216, and we give an explicit description here.

Definition 1 *Let σ be a partial symbol of order o on E. Then, we define, for $\xi \in T^*S^1 - S^1$,*

$$\sigma_+(\xi) = \begin{cases} \sigma(\xi) & \text{if } \xi > 0 \\ 0 & \text{if } \xi < 0 \end{cases} \quad \text{and} \quad \sigma_-(\xi) = \begin{cases} 0 & \text{if } \xi > 0 \\ \sigma(\xi) & \text{if } \xi < 0 \end{cases}. \quad \text{We define } p_+(\sigma) = \sigma_+$$

and $p_-(\sigma) = \sigma_-$.

The maps $p_+ : \mathcal{F}Cl(S^1, E) \to \mathcal{F}Cl(S^1, E)$ and $p_- : \mathcal{F}Cl(S^1, E) \to \mathcal{F}Cl(S^1, E)$ are clearly algebra morphisms that leave the order invariant and are also projections.

Definition 2 *We define $\mathcal{F}Cl_+(S^1, E) = Im(p_+) = Ker(p_-)$ and $\mathcal{F}Cl_-(S^1, E) = Im(p_-) = Ker(p_+)$.*

Since p_+ is a projection, we have the splitting

$$\mathcal{F}Cl(S^1, E) = \mathcal{F}Cl_+(S^1, E) \oplus \mathcal{F}Cl_-(S^1, E).$$

Let us give another characterization of p_+ and p_-. Looking more precisely at the formal symbols of p_{E_+} and p_{E_-} computed in Lemma 1, we observe that $\sigma(p_{E_+}) = \begin{cases} 1 & \text{if } \xi > 0 \\ 0 & \text{if } \xi < 0 \end{cases}$

and $\sigma(p_{E_-}) = \begin{cases} 0 & \text{if } \xi > 0 \\ 1 & \text{if } \xi < 0 \end{cases}$. In particular, using lemma 2, we have that $D_x^\alpha \sigma(p_{E_+})$, $D_\xi^\alpha \sigma(p_{E_+})$, $D_x^\alpha \sigma(p_{E_-})$, $D_\xi^\alpha \sigma(p_{E_-})$ vanish for $\alpha > 0$. From this, we have the following result.

Proposition 2 *Let $A \in \mathcal{F}Cl(S^1, E)$. $p_+(A) = \sigma(p_{E_+}) \circ A = A \circ \sigma(p_{E_+})$ and $p_-(A) = \sigma(p_{E_-}) \circ A = A \circ \sigma(p_{E_-})$.*

Proof: The proof follows from the fact that $\sigma(p_{E_+}) \circ A = A.1_{\xi>0}$. □

1.3. The Wodzicki residue and the Radul cocycle

Let us now define a non trivial cocycle on the algebra of pseudo-differential operators on the circle, first described in [9], Section 3.4. For this, we need to **fix** a formal symbol $log(a)$ that is not classical but logarithmic. We assume that this symbol is given by

$$log(a) = \sigma^a + log|\xi|, \tag{1}$$

where σ^a is a classical symbol of order 0 such that $log(a)$ is the formal symbol of an operator of the type $log(Q)$, where Q is a classical elliptic positive operator of order 1. Notice that, for $o \in \mathbb{Z}$ and for $\sigma \in \mathcal{F}Cl^o$, $[\sigma, log(a)] \in \mathcal{F}Cl^o$. Recall that ([12], see e.g [4]) the Wodzicki residue (which is a trace on $\mathcal{F}Cl$) is given, for $\sigma \in \mathcal{F}Cl$, by

$$res(\sigma) = \frac{1}{2\pi} \sum_{\xi=-1;+1} \int_{S^1} tr(\sigma_{-1}(x, \xi))dx.$$

Unlike when the basis manifold is of higher dimension, res is not the only trace on $\mathcal{F}Cl$. Since $\sigma = \sigma_+ + \sigma_-$, we can also define $res_+(\sigma) = res(\sigma_+) = \frac{1}{2\pi} \int_{S^1} tr(\sigma_{-1}(x, 1))dx$ and $res_-(\sigma) = res(\sigma_-) = \frac{1}{2\pi} \int_{S^1} tr(\sigma_{-1}(x, -1))dx$, which gives two other traces on $\mathcal{F}Cl$ that are not proportional to res. We can now define the Radul cocycle.

Proposition 3 *[9] The map $c : (\sigma, \sigma') \mapsto res(\sigma[\sigma', log(a)])$ defines a non trivial cocycle on $\mathcal{F}Cl$, called the Radul cocycle.*

2. The Kähler form of the loop group and the Radul cocycle

2.1. The Kähler form on the loop group

This section is based on [8]. Let G be a compact Lie group with Lie algebra $\mathfrak{g}$ equipped with its bi-invariant metric $(.,.)_\mathfrak{g}$, embedded in the group $U(p)$ of unitary $p \times p$ matrices. Since G is embedded in $U(p)$, the bi-invariant metric on G is given by $\forall x, y \in \mathfrak{g}$, $(x, y)_\mathfrak{g} = -tr(ad_x ad_y)$.

We consider now the Lie group $C_b^\infty(S^1, G) = \{\gamma \in C^\infty(S^1, G) | \gamma(0) = Id\}$ of based loops on G. The Lie algebra of this Lie group is given by $C_b^\infty(S^1, \mathfrak{g}) = \{X \in C^\infty(S^1, \mathfrak{g}) | X(0) = 0\}$. The bracket on this Lie algebra is given pointwise by the bracket on $\mathfrak{g}$, and hence, for any $X \in C_b^\infty(S^1, \mathfrak{g})$, the Maurer-Cartan form ad_X is given pointwise by the Maurer-Cartan form on $\mathfrak{g}$. Thus, for any $X \in C_b^\infty(S^1, g)$, ad_X is a multiplication operator.

An interesting metric on $C_b^\infty(S^1, G)$ is the $H^{1/2}$ metric which is homogeneous Kähler [7].

Its symplectic form is left invariant, and is given on the Lie algebra by

$$\omega(X, Y) = \frac{1}{2\pi} \int_{S^1} (D_x X(x), Y(x))_\mathfrak{g} dx.$$

The associated almost complex structure is the scalar pseudo-differential operator $\frac{D_x}{|D|}$. The holomorphic bundle is given

by E_+, the eigenspace corresponding to the eigenvalue 1 of $J = \frac{D}{|D|}$ on $C_b^\infty(S^1, \mathfrak{g}_\mathbb{C})$.

Recall, from Section 1, that $p_{E_+} = \frac{1}{2}\left(1 + \frac{D}{|D|}\right)$ is a classical pseudo-differential operator, with formal symbol $\frac{1}{2}\left(1 + \frac{\xi}{|\xi|}\right)$.

3. The Kähler form as a pull back of a Radul cocycle

The following proposition was proved in [1], rephrasing the work of D.Freed in terms of renormalized traces. We give here another formulation and shorter proof than the one given in the previous reference, that does not use the notion of renormalized trace. First of all, we need an easy technical lemma :

Lemma 3 *Let* $o \in \mathbb{Z}$ *and* $\sigma \in \mathcal{F}Cl^0$. *Let* $log(a)$ *be a formal symbol as in Equation (1). Then, if* σ *or* $log(a)$ *takes values in scalar matrices,* $[\sigma, log(a)]$ *is of order* -1, *and its positively homogeneous part of order* -1 *equals to* $\frac{-i}{\xi} D_x \sigma_0(x, \xi)$.

Proof: We already know , [6] Lemma 2, that $[\sigma, log(a)]$ is of order -1. Let us now compute the partial symbol of order -1 using Proposition 1.

$$\{[\sigma, log(a)]\}_{-1} = [\sigma_{-1}, \sigma_0^a] + [\sigma_0, \sigma_{-1}^a]$$
$$- i\{D_\xi \sigma_0 D_x \sigma_0^a - D_\xi \sigma_0^a D_x \sigma_0\}$$
$$+ i D_\xi log|\xi| D_x \sigma_0.$$

The first line of the right hand side vanishes because one of the two symbols in each bracket takes values in scalar matrices. The second one vanishes because, by the lemma 2, $D_\xi \sigma_0(ad_Y) = D_\xi \sigma_0^a = 0$. Thus, $\{[\sigma, log(a)]\}_{-1} = i D_\xi log|\xi| D_x \sigma_0 = \frac{i}{\xi} D_x \sigma_0$. $\square$

Let us now consider $X \mapsto ad_X$ the adjoint representation of $C_b^\infty(S^1, g)$. ad_X is a multiplication operator, and hence has a formal symbol with only a partial symbol of order 0, that equals pointwise to ad_X. We identify ad_X with its formal symbol. The symbol of the holomorphic part of ad_X is $p_+(ad_X)$ by proposition 2. Then, we have the following :

Theorem 1 *If $\log(a)$ takes values in scalar matrices,*

$$i\omega(X,Y) = c\,(p_+(ad_X), p_+(ad_Y)) = \mathrm{res}_+\,(ad_X\,[ad_Y, \log(a)])\,,$$

where c is the Radul cocycle defined in Proposition 3.

Proof :

$$c\,(p_+(ad_X), p_+(ad_Y)) = \frac{1}{2\pi}\sum_{|\xi|=1}\int_{S^1} tr(p_+(ad_X)p_+(\{([ad_Y,\log(a)]\}_{-1}))$$

$$= \frac{1}{2\pi}\{\int_{S^1} tr((ad_X)\{[ad_Y,\log(a)]\}_{-1})\}_{\xi=1}$$

Let $Y = f(.) \otimes v \in C_b^\infty(\mathbb{R},\mathbb{C}) \otimes \mathfrak{g}_\mathbb{C}$, such that f is 2π-periodic. We remark that $D_x ad_Y = D_x f(x) \otimes ad_v = ad_{D_x Y}$. Hence, applying the last lemma,

$$c(p_+(ad_X), p_+(ad_Y)) = -\{\frac{1}{2\pi}\int_{S^1}(-tr(ad_X ad_{D_x Y}))\}\{\frac{i}{\xi}\}_{\xi=1}$$

$$= \frac{-i}{2\pi}\int_{S^1}(X, D_x Y)_\mathfrak{g} = i\omega(X,Y)\quad\square$$

References

[1] A.Cardona, C.Ducourtioux, J-P.Magnot, S.Paycha Weighted traces on pseudodifferential operators and geometry on loop groups, To appear in *Infin. Dimens. Anal. Quantum Probab. Relat. Top.*

[2] D.Freed The geometry of loop groups,*J. Diff. Geometry* **28**, 223-276 (1988)

[3] P.Gilkey *Invariance theory, the heat equation, and the Atiyah-Singer index theorem* Second edition. Studies in Advanced Mathematics CRC Press, Boca Raton, FL (1995)

[4] Ch.Kassel le residu non commutatif, *Séminaire Bourbaki* **708** , Asterisque , 199-229 (1989)

[5] M.Lesch On the non commutative residue for pseudo-differential operators with log-polyhomogeneous symbol *Ann. Global Anal. Geom.* **17** (1999), no 2, 151-187

[6] J-P.Magnot Pontrjagin and Chern forms on current spaces *Prépublications de l'université Blaise Pascal* 2002-01 (2002)

[7] A.Pressley The Energy flow on the loop space on a compact Lie group *J. London Math. Soc. (2)* , **26** (1982), 557-566

[8] A.Pressley, G.Segal *Loop Groups* Oxford Univ. Press (1988)

[9] A.O.Radul, Lie algebras of differential operators, their central extensions, and W-algebras, *Funct.Anal.Appl.* **25**, 25-39 (1991)

[10] R.T.Seeley, Complex powers of an elliptic operator *Proc. Symp. Pure Math.* **10** 288-307 (1967)

[11] H.Widom A complete symbolic calculus for pseudo-differential operators *Bull. Sc. Math. 2e serie* **104** p. 19-63 (1980)

[12] M.Wodzicki Local invariants in spectral asymetry *Inv. Math.* **75**, p. 143-178 (1984)

Inst. Phys. Conf. Ser. No 173: Section 8
Paper presented at 24th Int. Coll. Group Theoretical Methods in Physics, Paris, France, 15–20 July 2002
©2003 IOP Publishing Ltd

Cohomology and Bessel Functions theory

Mustapha Mekhfi

LSMC Laboratory, Dép. de Physique,
Université Es-senia 31100, Oran, Algérie

Abstract. The reduced Bessel function $\frac{J_n(r)}{r^n}$ and the exponential function $\exp(in\theta)$ have been associated with homotopic (n)loops living on the punctured plane as two independent realizations of the topological invariants of the punctured plane.Cohomological quantum mechanics has been the framework of such association .Old and new properties of Bessel functions (one new in this paper)are derived from simpler relationships between exponential functions as consequences of such association.
PACS:03.65.-w;02.30.Gp

1. Cohomological Quantum Mechanics

Let M be a compact manifold with local coordinates x^μ and let $H^r(M)$ be the rth de Rham cohomology group.Let $c_1,.........c_k$ be elements of the homology group $H_r(M)$ with k the rth Betti number and of the same class $[c_i] = [c_j]$.Then for any set of numbers $b_1,......b_k$ a corollary of the de Rham's theorem states that there exists a closed r-form ω such that

$$\int_{c_i} \omega = b_i \qquad 1 \le i \le k$$

The numbers b_i above are the periods of closed r-forms over cycles c_i.The main features of such actions is that there are defined on the product $H_r(M) \times H^r(M)$ and are therefore topological actions,that is invariant under any infinitesimal deformations which keeps the cycle within its homology class $H^r(M)$

$$\delta x^\mu = \epsilon^\mu$$

Bessel functions comes from considerations of the first non trivial action with period b_1 on the punctured plane .The topological ingredients are loops and 1-forms .The gauge fixed topological action associated with the punctured plane is ($\lambda \in R$ and $x^1 + ix^2 = r\exp(i\theta)$)

$$S_{GF} = \lambda \int_c d\theta + gauge\ fixed\ terms$$

A very practical way to gauge fix these actions is to use BRST symmetry.In the process of gauge fixing a function called prepotential V is introduced and should be selected as to properly define topological invariants.The correct choice is [1]

$$V = \lambda(\theta + \Phi(\theta))$$

Where the function $\Phi(\theta)$ is any function but periodic .The simplest case without $\Phi(\theta)$ has been selected by Baulieu and Rabinovici [2].Such truncated solution neither lead to a complete description of the invariants on the punctured plane nor to the theory of Bessel functions .

2. Bessel functions are realization of homotopic loops

The Hamiltonian associated with the gauge fixed action S_{GF} is

$$H = \frac{1}{2}\{Q, \bar{Q}\}$$

Where $p_i = -i\frac{\partial}{\partial x_i}$ and $\bar{\psi}_i = \frac{\partial}{\partial \psi_i}$ are canonical momenta for the coordinates x_i and the ghost fields ψ_i respectively with the generators $Q = \psi_i(p_i + i\frac{\partial V}{\partial x_i})$ and $\bar{Q} = \bar{\psi}_i(p_i - i\frac{\partial V}{\partial x_i})$ .Bessel functions will enter in play when we consider the eigenvalue problem associated to the Hamiltonian.Looking for a solution of the form $F(r)f(\theta)$ we get

$$\ddot{f}(\theta) + (\zeta^2 - \lambda^2 - W)f(\theta) = 0; \quad \ddot{F}(r) + \frac{1}{r}\dot{F} + (-\frac{\zeta^2}{r^2} + 2E)F = 0$$

Where ζ is the separation parameter.The r- component equation is the differential equation defining Bessel functions $J_\zeta(\sqrt{2Er})$,while the θ-component equation is a differential equation of Sturn-Liouville type.On dimensional grounds,one may select the following candidates,together with their hermitian conjugates .

$$\{Q, \epsilon^{ij} x_i \bar{\psi}_j\} = -i\partial_\theta + i\partial_\theta V + ghost\ terms\ dropped$$
$$\{Q, x_i \bar{\psi}_j\} = r\partial_r + ghost\ terms\ dropped$$

The first invariant is the effective winding number W (integer eigenvalues) corrected by the prepotential V.To define the second topological invariant one needs to interpret Bessel functions as describing homotopic loops .To this end we have to restrict ourselves to non interacting case i.e. $\lambda = 0; \zeta = n$ .In this case the wavefunction is as follows ($z \sim \sqrt{2Er}$)

$$J_n(z)\ \exp in\theta$$

This is not to be interpreted as in ordinary quantum mechanics (particles),that is as the probability amplitude to describe a point particle at position (r, θ) but as a set of functions describing homotopic loops.The exact interpretation is as follows:Since a homotopic loop on the punctured plane needs only one variable the θ variable for instance (any deformation δr within the equivalence class of the loop is irrelevant), we may say that the (n)loop is described by the function $\exp in\theta$ .What then is the meaning of the remaining r dependent function, which in this case is the Bessel function $J_n(z)$ .This function could be thought of as another independent function which describes the same (n) loop .This is possible apart from the fact that this function goes to zero as $r \rightarrow 0$ and therefore it cannot be associated with a loop "trapped" around the origin, which by definition should not vanish .It is the reduced Bessel $\frac{J_n(z)}{(z)^n}$ that is the appropriate function describing the (n)loop, as this function does not vanish at the origin, $\frac{J_n(z)}{(z)^n} \rightarrow 0$.We therefore consider that positively oriented loops ($n \succ 0$) are associated with the reduced Bessel function $\frac{J_\zeta(z)}{z^n}$. Negatively oriented loops ($n \prec 0$) on the other hand are associated with $z^n J_n(z)$.

Now that reduced Bessel functions are supposed to be yet another realization of homotopic loops(in the z variable),it then follows the existence of two operators .One is the analogous of the winding number operator $W_\theta = -i\frac{\partial}{\partial \theta}$ which we denote W_r and which turns out to have the form

$$W_r = -\frac{1}{2}(z\frac{d}{dz} + (\frac{d}{zdz})^{-1})$$

and it acts on the wavefunctions as follows $W_r \frac{J_n(z)}{z^n} = n \frac{J_n(z)}{z^n}$ and $W_r (z^n J_n(z)) = -n \ z^n J_n(z)$. The other one is the analogous of the fundamental group element defining the punctured plane $\Pi_\theta(m) = \exp im\theta$ but in the r variable. This operator needs not be constructed as it is available in the literature .It has the form

$$d_m = \frac{d}{zdz} \cdots \frac{d}{zdz} = (\frac{d}{zdz})^m \ ; m \in Z;$$

and acts as follows

$$(-1)^m d_m \frac{J_n(z)}{z^n} = \frac{J_{n+m}(z)}{z^{n+m}} \ ; \quad d_m \left(z^n J_n(z) \right) = z^{n-m} J_{n-m}(z); \ m \in N$$

We thus have two operators acting on reduced Bessel functions W_r and d_m and these are the analogous (r-components) of the winding number operator $W_\theta = -i\frac{d}{d\theta}$ and the fundamental group element $\Pi_\theta(m)$.

3. Exp^s vs Bessel^s

The remark we make toward finding new properties of Bessel functions (one in this paper) is that $\zeta = n$ (the separation parameter) is a common parameter to $\exp in\theta$ and to $\frac{J_n(r)}{r^n}$ and both describing the same object ,that is the (n)loop, hence we suggest and prove that :Certain specific relationships among exponentials will be transported to reduced Bessel functions via the following correspondence rules CR$^{'s}$

These CR$^{'s}$ are collected in the following tableau.

θ-components	r -components
Function	Function
$\exp in\theta \ ; -\pi \le \theta \le \pi$	$\frac{J_n(r)}{r^n}$
$1; (n=0)$	$J_0(r)$
Homotopic group elements	Homotopic group elements
$\Pi_\theta(m) = \exp(im\theta) \ ; m \in Z$	$\Pi_r(m) = (-1)^m (\frac{d}{rdr})^m \ ; m \in Z$
$1; (m=0)$	$1 \equiv d_{-|m|} d_{|m|} = d_{|m|} d_{-|m|}$
Winding number	Winding number
$W_\theta = -i\frac{d}{d\theta}$	$W_r = -\frac{1}{2}(r\frac{d}{dr} + (\frac{d}{rdr})^{-1})$

Few remarks are in order at this stage .First note that the topological invariant $r\frac{d}{dr}$ gets a meaning in the new framework.That is we have $r\frac{d}{dr} = -2W_r + \Pi_r(-1)$ .It is a combination of the winding number W_r and the homotopy group element $\Pi_r(-1)$ .Second, note the global restriction on the angle $-\pi \le \theta \le \pi$.This is because we identified the exponential with the whole loop hence the whole interval 2π.Relationships valid within truncated values of the angle or in another interval other that the above will not map to corresponding relationships in Bessel functions.To understand the restriction to the specific interval,consider the true identity

$$-i\theta = \sum_{m\in Z/0} (-1)^m \frac{\exp im\theta}{m} \ ; -\pi \prec \theta \prec \pi$$

It serves to define the interval of validity of the angle θ .Finally we may note that the periodicity of the exponential is not transported by CR^s as globality fixes the angle to be in a fixed interval.Applying the CR^s we get a series of additional correspondences listed in the following tableau.

θ-components	r -components
$\exp im\theta \exp in\theta = \exp i(n+m)\theta$	$(-1)^m(\frac{d}{rdr})^m\frac{J_n(r)}{r^n} = $
	$\frac{J_{n+m}(r)}{r^{n+m}}$
$1 \exp in\theta = \exp in\theta$	$(\frac{d^2}{dr^2} + (2n+1)\frac{d}{rdr} + 1)\frac{J_n(r)}{r^n} = 0$
$(i\theta)^n$; c number	$\frac{d^n}{d\lambda^n}\frac{J_\lambda(r)}{r^\lambda}\big|_{\lambda=0}$
$\exp i\lambda\theta$	$\frac{J_\lambda(r)}{r^\lambda}$
$-i\frac{d}{d\theta}\exp i\lambda\theta = \lambda \exp i\lambda\theta$	$r(J_{\lambda+1} + J_{\lambda-1}) = 2\lambda\, J_\lambda(r)$
The Unification Formula for Bessel Functions of Different Orders	
$\exp i(n+\lambda)\theta = $	$\frac{J_{n+\lambda}(r)}{r^{n+\lambda}} = $
$\exp(-\lambda\sum_{m\in Z/(0)}(-1)^m\frac{exp(im\theta)}{m}) \exp in\theta$	$\exp(-\lambda\sum_{m\in Z/(0)}\frac{1}{m}(\frac{d}{rdr})^m)\frac{J_n(r)}{r^n}$
etc	etc

4. New property of Bessel functions

The $CR^{'s}$ gave us among various known formulas a very specific and new formula unifying integer and real orders [3]

$$\frac{J_{n+\lambda}(z)}{z^{n+\lambda}} - \exp(-\lambda \sum_{m\in Z/(0)} \frac{1}{m}(\frac{d}{zdz})^m)\frac{J_n(z)}{z^n}$$

This formula has been tested using different methods.It has been shown to be true by direct analytical computations in [4] and by mapping the differential equation of integer order reduced Bessel function to the differential equation of reduced Bessel function of real order using the exponential operator above [5] .On the other hand such mapping has been successfully applied to Neumann '^{s} and Hankel functions as well, with a formula similar to the above and when applied to Polynomials such as Hermite and Laguerre it gives deformed versions of these polynomials which are generalization of the older one with certain shared properties .

5. Conclusions and outlook

The study of the classical topological action $S_{cl} = b_1$ on the punctured plane, where b_1 is the period of the unique one form on the punctured plane over the one- cycles (loops) [6] allowed us to probe all the topological invariants of the punctured plane.These are the winding number and the elements of the fundamental group and they came duplicated as independent realizations in the r and the θ variables.But the purpose of the study is to associate Bessel functions to the objects involved, namely homotopic loops on the punctured plane (Recall that in standard quantum mechanics the objects are particles) and then to infer old and possibly new properties of Bessel functions (one new in this paper) by using the topological properties of the punctured plane .As a consequence of this association,we predicted the existence of the operators d_m and W_r and how they should act on Bessel functions (while d_m were known in the literature we had to work out W_r) .Another consequence of this association is that there exists a correspondence rules $CR^{'s}$ between the exponential and reduced Bessel functions as these functions are nothing but independent realizations of the same homotopic loops in the θ and r variables respectively.It then follows that simple relationships between exponentials are translated into analogous relationships between reduced Bessel functions which are more involved to be guessed.The above analysis is however more promising if applied to more general actions b_i on manifolds of richer (co)homology ($H^r(M)$) $H_r(M)$ .In this more general case new topological invariants will show up together with the functions they act on and new $CR^{'s}$ between these new functions.

References

[1] M.Mekhfi :Mod.Phys.Lett A.11(1996)
[2] L.Baulieu and Rabinovici : Phys.lett.B.316 (1993)
[3] M.Mekhfi : International Journal of Theoretical Physics vol3.N8.(1996)
[4] M.Mekhfi : International Journal of Theoretical Physics vol39.N4.(2000)
[5] M.Mekhfi &M.Abdelouahab : Journal of dynamical systems and geometric theories, JDSG Vol.1,N.1,(2002)77-81
[6] For more details see M. Mekhfi : hep-th/0207256

Inst. Phys. Conf. Ser. No 173: Section 8
Paper presented at 24th Int. Coll. Group Theoretical Methods in Physics, Paris, France, 15–20 July 2002
©2003 IOP Publishing Ltd

Representations of Classical Lie Algebras from their Quantum Deformations

P. Moylan

The Pennsylvania State University
Abington College
Abington, Pennsylvania 19001 USA

Abstract. We make use of a well-know deformation of the Poincaré Lie algebra in $p + q + 1$ dimensions ($p + q > 0$) to construct the Poincaré Lie algebra out of the Lie algebras of the de Sitter and anti de Sitter groups, the generators of the Poincaré Lie algebra appearing as certain irrational functions of the generators of the de Sitter groups. We have obtained generalizations of this "anti-deformation" for the $SO(p + 2, q)$ and $SO(p + 1, q + 1)$ cases with arbitrary p and q. Similar results have been established for q deformations $U_q(so(p, q))$ with small p and q values. Combining known results on representations of $U_q(so(p, q))$ (for q both generic and a root of unity) with our "anti-deformation" formulae, we get representations of classical Lie algebras which depend upon the deformation parameter q. Explicit results are given for the simplest example (of type A_1) i.e. that associated with $U_q(so(2, 1))$.

1. Introduction.

We start with a well-known deformation [1], [2] of the Poincaré Lie algebra in $p + q + 1$ dimensions ($p + q > 0$), which is defined in terms of the generators $\mathbf{L}_{ij}$ of (pseudo) rotations and the translation generators $\mathbf{P}_i$ by the following:

$$\mathbf{L}_{ij} \;\; \rightarrow \;\; \mathbf{L}_{ij} \; , \tag{1.1a}$$

$$\mathbf{P}_i \;\; \rightarrow \;\; \mathbf{L}_{p+q+1,i}^{\pm} \;\; = \;\; \frac{i}{2\,Y}[\mathbf{Q}_2, \, \mathbf{P}_i] \; + \; \mathbf{P}_i \tag{1.1b±}$$

where $\mathbf{Q}_2 = \frac{1}{2} \sum_{i,\,j\,=\,0}^{p+q} \mathbf{L}_{ij}\,\mathbf{L}^{ji}$ is the second order Casimir operator of $SO_0(p + 1, q)$, and Y satisfies $Y^2 = \pm \sum_{i,\,j\,=\,0}^{p+q} \mathbf{P}_i\,\mathbf{P}^i$. ([,] denotes commutator.) Choice of the plus sign in this equation for Y^2 leads to the Lie algebra of $SO_0(p + 2, q)$ and the minus sign gives the commutation relations of $SO_0(p + 1, q + 1)$. Now Eqns. (1b±) may be considered as algebraic equations for the translation generators $\mathbf{P}_i$ of the Poincaré group, and we may attempt to solve these equations for the $\mathbf{P}_i$. The solution to this problem for $p = 0$, $q = 3$ and for the choice of eqn. (1.b−) has been given by us in [1]. The general solution for the case of eqn. (1.b+) ($p = 0$, $q = 3$) has been presented in [3]. We have also obtained a generalization of this "anti-deformation" to higher dimensions i.e. we have been able to solve eqns.(1.b±) for the $\mathbf{P}_i$ [3], but only by working in a particular class of irreducible representations, namely that which occurs in the decomposition of the left regular representation of $SO_0(p, q)$ groups on real hyperbolic spaces [4]. The proof of commutativity of the Poincaré translation generators for these higher dimensional cases makes use of an integral transform [4], which intertwines certain representations of $SO_0(p, q)$ induced from the maximal parabolic subgroup with representations which are restrictions of the $SO_0(p, q)$ left regular representation on eigenspaces of the Laplace-Beltrami operator on the hyperbolic space.

Here we report on some analogous findings for q-deformations of $so(p+1,q+1)$ algebras in lowest dimensions i.e. for $p+q+1 = 2,3$ and 4 [5] [6]. In particular, in the $p=1, q=0$ case, we start with the Euclidean group in two dimensions $\mathcal{E}(2)$, with generators $\mathbf{L}_{12}$ (rotation generator) and $\mathbf{P}_i$ $(i = 1,2)$ (translation generators), and define the following (c.f. [5]):

$$\tilde{\mathbf{L}}_{3i} = \left[\frac{\left([-i\mathbf{L}_{21}]_{\sqrt{q}}\right)^2}{[2]_{\sqrt{q}} Y}, \mathbf{P}_i \right] + \mathbf{P}_i \ , \ Y := \sqrt{\sum_{i=1}^{2} \mathbf{P}^i \mathbf{P}^i} \quad ([m]_q = \frac{q^{m/2} - q^{-m/2}}{q^{1/2} - q^{-1/2}}). \tag{1.2}$$

We readily obtain the "anti-deformation" by solving eqns. (1.2) for the $\mathbf{P}_i$. Our results are given below in section 2.

2. An Embedding of $\mathcal{E}(2)$ into a skew field extension of $U_q(so(2,1))$.

The q-deformation $U_q(so(3,\mathcal{C}))$ is defined as the associative algebra over $\mathcal{C}$ with generators $H, X^\pm$ and relations [5], [6]:

$$[H, X^\pm] = \pm 2X^\pm \ , \tag{2.1a}$$

$$[X^+, X^-] = [H]_q \ . \tag{2.1b}$$

Let I be the unit element in $U_q(so(3,\mathcal{C}))$, then the Casimir element of $U_q(so(3,\mathcal{C}))$ is

$$\Delta_q = X^+ X^- + ([\frac{1}{2}(H - I)]_q)^2 - \frac{1}{4} =$$

$$= X^- X^+ + ([\frac{1}{2}(H + I)]_q)^2 - \frac{1}{4} \ . \tag{2.2}$$

The real form $U_q(so(2,1))$ of $U_q(so(3,\mathcal{C}))$ is defined as follows. The generators of $U_q(so(2,1))$ are given by the following expressions:

$$\mathbf{L}_{32} = -\frac{i}{2}(X^+ - X^-) \ , \ \mathbf{L}_{13} = \frac{1}{2}(X^+ + X^-) \ , \ \mathbf{L}_{21} = \frac{i}{2} H \ . \tag{2.3}$$

Thus

$$X^\pm = \mathbf{L}_{13} \pm i\mathbf{L}_{32} \ . \tag{2.4}$$

The operators $i\mathbf{L}_{12}$, $i\mathbf{L}_{13}$, $i\mathbf{L}_{32}$ are preserved under the following antilinear anti-involution ω of $U_q(so(3,\mathcal{C}))$

$$\omega(H) = H \ , \ \omega(X^\pm) = -X^\mp \ . \tag{2.5}$$

For the coproduct on $U_q(so(3,\mathcal{C}))$ we take: [7]

$$\Delta(H) = H \otimes I + I \otimes H , \qquad \Delta(X^\pm) = X^\pm \otimes q^{\frac{H}{4}} + q^{-\frac{H}{4}} \otimes X^\pm . \tag{2.6}$$

The Lie algebra $\mathcal{E}(2)$ is the Lie algebra of the Euclidean group, $E(2)$, which is the semidirect product of $SO(2)$ with the group of translations of the plane, $\mathbb{R}^2$. A basis for the Lie algebra $\mathcal{E}(2)$ consists of the generator of rotations $\mathbf{L}_{12}$ and two commuting translation generators $\mathbf{P}_i$ $(i = 1, 2)$. They satisfy the following commutation relations:

$$[\mathbf{L}_{12}, \mathbf{P}_2] = \mathbf{P}_1 \ , \ [\mathbf{L}_{12}, \mathbf{P}_1] = -\mathbf{P}_2 \ , \tag{2.7a}$$

$$[\mathbf{P}_1, \mathbf{P}_2] = 0 \ . \tag{2.7b}$$

It is useful to work with the complexified translations generators, which are:

$$\mathbf{P}^{\pm} = -\mathbf{P}_1 \pm i\,\mathbf{P}_2 . \tag{2.8}$$

We also define as above

$$H = -2i\,\mathbf{L}_{21} . \tag{2.9}$$

Then using (2.7) we verify that

$$[H, \mathbf{P}^{\pm}] = \pm 2\mathbf{P}^{\pm} , \quad [\mathbf{P}^+, \mathbf{P}^-] = 0 . \tag{2.10}$$

We now solve eqns. (1.2) for the $\mathbf{P}_i$, our solution expresses the translation generators of $\mathcal{E}(2)$ as irrational functions of $U_q(so(2,1))$. Thus it gives an embedding of $\mathcal{E}(2)$ into an algebraic extension $K'(U_q(so(2,1)))$ of the skew field $K(U_q(so(2,1)))$ [8]. Explicitly the solution is given by:

$$\mathbf{P}_1 = D^{-1}\left(\{I - \frac{1}{2Y}\frac{[H]_q}{[H]_{\sqrt{q}}}\}\mathbf{L}_{31} + \frac{i[2]_{\sqrt{q}}}{2Y}[\frac{H}{2}]_q\mathbf{L}_{32} \right) , \tag{2.11a}$$

and

$$\mathbf{P}_2 = D^{-1}\left(\{I - \frac{1}{2Y}\frac{[H]_q}{[H]_{\sqrt{q}}}\}\mathbf{L}_{32} - \frac{i[2]_{\sqrt{q}}}{2Y}[\frac{H}{2}]_q\mathbf{L}_{31} \right) , \tag{2.11b}$$

where

$$D = -\frac{1}{4Y^2}\left\{ [H]_{\sqrt{q}}^2 - (\frac{[H]_q}{[H]_{\sqrt{q}}} - 2Y)^2 \right\} . \tag{2.12}$$

Furthermore

$$Y^2 = \Delta_q + \frac{1}{4}I . \tag{2.13}$$

One readily verifies that the $\mathbf{P}_i$ as defined by eqns. (2.11a) and (2.11b) satisfy the defining commutation relations for the translation generators of $\mathcal{E}(2)$, and verify that $Y^2 = \mathbf{P}^+\,\mathbf{P}^-$.

The embedding given by eqns. (1.2) extends to a homorphism τ from $K'(U_q(so(2,1)))$ to $K'(U(\mathcal{E}(2)))$ (an algebraic extension of the skew field of $U(\mathcal{E}(2))$. $(U(\mathcal{E}(2))$ is the enveloping algebra of $\mathcal{E}(2)$.) In fact, since $\mathbf{P}_i$ in (2.11a) and (2.11b) commute, it is easy to see that τ defined as $\tau(X^{\pm}) = \tilde{X}^{\pm}$ and $\tau(H) = H$ is an isomorphism. If we take the standard coproduct on $U(\mathcal{E}(2))$ [9] and call it $\tilde{\Delta}$, then one verifies that $\tau(\Delta(X^{\pm})) \neq \tilde{\Delta}(\tau(X^{\pm}))$ even for $q = 1$. However, we can treat the tensor product of representations as in [6] where we gave a description of $U_q(so(4,\mathbb{C}))$ similar to the above description of $U_q(so(3,\mathbb{C}))$. (It is well-known that $U_q(so(4,\mathbb{C}))$ is constructed out of two mutually commuting pairs of $U_q(so(3,\mathbb{C}))$ [10].) There we introduced two commuting pairs of translation operators defined on the tensor product representation of two representations of $U_q(so(3,\mathbb{C}))$. They were defined implicitly by equations similar to eqns. (1.2), and, as above for $U_q(so(3,\mathbb{C}))$, we were able to solve the equations for these four translation operators.

A few comments about the higher dimensional q deformed cases: the above remarks in the previous paragraph outline the main ideas of our generalization to $U_q(so(2,2))$ and $U_q(so(3,1))$. $(U_q(so(2,2))$ and $U_q(so(3,1))$ are real forms of $U_q(so(4,\mathbb{C}))$.) We have also obtained a description of the Rac representation of $U_q(so(3,2))$ [11] along these lines. This uses the fact that the Rac representation remains irreducible under $U_q(so(2,2))$.

3. Representations

For $\sigma \in \mathcal{C}$ and for any $q \in \mathcal{C}$ ($q \neq 0$ and not a root of unity) the following formulae define a representation $d\pi^{\sigma,\epsilon}$ of $U_q(so(3,\mathcal{C}))$ [12]: ($\epsilon = 0$ or $\frac{1}{2}$)

$$d\pi^{\sigma,\epsilon}(H)|m> = 2\,m|m> \quad , \quad d\pi^{\sigma,\epsilon}(X^{\pm})|m> = [-\sigma \pm m]_q|m \pm 1> \quad . \tag{3.1}$$

For $q^N \neq 1$ ($|q| = 1$): (1) $\sigma = i\rho - \frac{1}{2}$ ($\rho \in \mathbb{R}$) and the representation space $\mathcal{D}^{i\rho-1/2}$ is the linear span of the $|m>$ ($m = n + \epsilon$, $n = 0, \pm1, \pm2, \ldots$), and $d\pi^{\sigma,\epsilon}$ is the (infinitesimally unitarizable) principal series of $U_q(so(2,1))$; (2) $\sigma = \epsilon \mod(2)$ and $\sigma = \ell$ with $\ell < -\frac{1}{2}$ and a) the representation space $X_+^{-\ell,\epsilon}$ is the linear span of the above $|m>$ with $m > -\ell$, b) the representation space $X_-^{-\ell,\epsilon}$ is the linear span of the $|m>$ with $m < \ell$. $d\pi^{\sigma,\epsilon}$ acts irreducibly on $X_{\pm}^{-\ell,\epsilon}$. These give q deformed discrete series of $U_q(so(2,1))$.

For $q^M = 1$ ($M \in \mathbb{Z}$, $M > 2$), let $q = e^{\frac{2\pi i}{m}}$ and set $M = m$ for m odd, and set $M = \frac{m}{2}$ for m even. Define $\sigma = \frac{1}{2}(d-1) - \frac{1}{2}M$ ($d = 1, 2, \ldots M$) and let $V_d =$ linear span of the $|s_3 >$ ($s_3 = \sigma, \sigma - 1, \ldots \sigma - (d-1)$). The action $d\pi^{\sigma}$ of the basic generators H and $X^{\pm}$ on V_d is given by: [13]

$$d\pi^{\sigma}(H)|s_3> = -2s_3|s_3> \quad , \quad d\pi^{\sigma}(X^{\pm})|s_s> = [-\sigma \pm s_3]_q|s_3 \pm 1> \quad . \tag{3.2}$$

These finite dimensional highest weight modules are all infinitesimally unitary. For which of the above representations do eqns. (2.11) determine a representation of $\mathcal{E}(2)$ on the given representation space? The following theorem provides the answer to this question.

Theorem: For $q^N \neq 1$ we have representations of $\mathcal{E}(2)$ on $\mathcal{D}^{i\rho-1/2}$ and on $X_{\pm}^{-\ell,\epsilon}$ but the representation of $\mathcal{E}(2)$ is infinitesimally unitary only on $\mathcal{D}^{i\rho-1/2}$. For $q^N = 1$ ($N \in \mathbb{Z}$, $N > 2$) none of the representations $d\pi^{\sigma}$ lead to representations of $\mathcal{E}(2)$ on V_d.

The main ingredient in the proof of the theorem involves determining the action of the operator D of eqn. (2.12) on the given representation space, and, in particular, deciding whether zero lies in the resolvent set of the operator in its given action on the representation space.

References

[1] M. Havlíček, P. Moylan, *J. Math. Phys.*, **34**, 11, 5320-5332, (1993).
[2] A. O. Barut, A. Bohm, *Phys. Rev.*, **139**, 1107, (1965); J. Fang, C. Fronsdal, *Phys. Rev. D*, **22**, 1361 (1980); P. Šťovíček, *J. Math. Phys.*, **29**, 5320-5327, 1988.
[3] P. Moylan, *Tachyons and Representations of $SO_0(2,3)$*, 7th Wigner Symposium (2001).
[4] R.S. Strichartz, *J. Funct. Anal.*, **114**, 493-508, (1993).
[5] P. Moylan, *Czech. J. Phys.*, **47**, 1251-1258, (1998).
[6] P. Moylan, *Czech. J. Phys.*, **48**, 1457-1464, (1998).
[7] E.G. Kalnins, H. Manocha, W. Miller, Jr., *J. Math. Phys.*, **33**, (7), 2365, (1992).
[8] P. M. Cohn, *Skew Fields: The General Theory of Division Rings*, Encylopedia of Mathematics and Its Applications, **57**, Camb. Univ. Press, (1995).
[9] J. Dixmier, *Enveloping Algebras*, Graduate Studies in Mathematics, **11**, American Mathematical Society, North Holland Publ., Amsterdam, (1991).
[10] V.K. Dobrev, *Jour. Phys. A*, **26**, 1317 (1993).
[11] V.K. Dobrev, P. Moylan, *Phys. Lett. B*, **315**, 292, (1993).
[12] C. de Concini, V.G. Kac, in *Operator Algebras, Unitary Representations, Enveloping Algebras and Invariant Theory*, p. 471-506, Birkhauser, Boston, (1990).
[13] G. Keller, *Lett. Math. Phys.*, **21**, 273-286, (1991); H. Steinacker, *Comm. Math. Phys*, **192**, 687, (1998).

Inst. Phys. Conf. Ser. No 173: Section 8
Paper presented at 24th Int. Coll. Group Theoretical Methods in Physics, Paris, France, 15–20 July 2002
©*2003 IOP Publishing Ltd*

On the Symmetries of the 16 Genetic Code-Doublets

Tidjani Négadi

**Département de Physique, Faculté des Sciences, Université d'Oran,
31100, Es-Sénia, Oran, Algérie. E-mail: tidjani_negadi@yahoo. com**

Abstract. A unified classification of the Genetic Code-Doublets is presented (i) by using our recent *matrix* method, based on Neubert's Quartet Model of the Periodic System and (ii) by introducing a new set of transformations generating a Klein's 4-group, describing their symmetries and including the Rumer transformation. The straightforward extension to the 64 codons-system of the genetic table is briefly outlined.

1. Introduction

In this short communication, we present an algebraic (matrix) classification of the 16 Genetic Code-Doublets as well as a new set of transformations, constituting a Klein's 4-group, which describe their symmetries. Our approach is *self-contained* in the sense that we use the same formalism to describe the atoms, the molecular building blocks of living matter (the four bases U, C, A and G) and, finally, the 16 Doublets-System and the Genetic Code itself. More precisely, our classification is based on the Quartet Model of the Periodic System of the natural Elements (PSE), introduced by Neubert [1] in 1970, and constructed from special solutions of the Schrödinger equation of the hydrogen atom. Our original formulation of this model, into a matrix framework, gives it "life" in the sense that we can use the elementary mathematical operations of matrix algebra to construct new objects, starting with the atoms which are described by simple 2×2 matrices. In the second section, we give a brief resume of our matrix method which allows the building of a well defined base-matrix $\mathcal{B}$. The third section is devoted to the consideration of the symmetries of the 16 doublets $B_i B_{i+1}$ (i=1 or 2), classified in the direct product $\mathcal{B} \otimes \mathcal{B}$. We introduce three (commuting) transformations, R_i ($i = 1, 2, 3$) which, together with the identity transformation, constitute a Klein's 4-group capable of describing the symmetries associated to the "$B_1 B_2$-portion" of genetic code, in the line of thought of Rumer [2] and Danckwerts and Neubert [3], and leading to its Standard (Universal) form *as well as* the symmetries associated to the "$B_2 B_3$-portion", considered by Jiménez Montaño et al. [4], and leading to the more symmetrical Mitochondrial Code.

2. Atoms, molecules and bases as matrices

The basis for our matrix representation has its roots in the Quartet-Model of the Periodic System of the Elements, introduced by Neubert more than thirty years ago [1]. This model exploits a freedom in the choice of a dimensionless parameter, χ, multiplying or, more precisely, dividing the dimensionless radial variable of the radial Schrödinger equation of the hydrogen atom. There are two interesting series of states: $\chi=\frac{1}{2}$ for the non relativistic hydrogen atom and $\chi=1$ for the PSE because these latter states carry twice the orbital angular momentum and consequently incorporate automatically the Madelung Rule, the pillar of the electronic filling order in atomic systems (see Ref. [1]). Adding to the PSE states two kinds of spins, the ordinary spin s (=1/2) and the "topical spin" σ(=1/2), Neubert got the 4-shell classification of the PSE where 120 elements are arranged into 30 different Quartets with four atoms each. We have introduced a two-fold extension of this (static) model : (i) by assigning to each atom a matrix representation $\mathcal{E}^{(i,j)}$ and (ii) by defining an atomic number function, $\mathcal{Z}^{(i,j)}$, the (unique) associated numerical matrix element, giving precisely the atomic number of that atom [5]. Any atom could be written using one of the following four matrix forms

$$\mathcal{E}_{11}^{(i,j)} : \mathcal{Z}_{11}^{(i,j)} = Z_i + 2(j-1), \quad \mathcal{E}_{12}^{(i,j)} : \mathcal{Z}_{12}^{(i,j)} = \mathcal{Z}_{11}^{(i,j)} + 1, \tag{1}$$
$$\mathcal{E}_{21}^{(i,j)} : \mathcal{Z}_{21}^{(i,j)} = \mathcal{Z}_{11}^{(i,j)} + 2i^2, \quad \mathcal{E}_{22}^{(i,j)} : \mathcal{Z}_{22}^{(i,j)} = \mathcal{Z}_{21}^{(i,j)} + 1,$$

In each one of these 2×2 matrices, all the matrix elements are zero except at the indicated position. In Eq. (1), the upper indices, i (=1,2,3,4) and j (=1,2,. . . ,i^2) which specify a given Quartet of four atoms, are related to the quantum numbers n, l and m_l by i=n, and j=n^2-l(l+1)+m_l (m=-l, -l+1, . . . , +l) and Z_i (=(2/3)$i(i-1)(2i-1)+1$) is the key function which gives the atomic number of the element which begins the shell i. Note that half the atoms are represented by even matrices and the other half by odd matrices. As we shall deal, in the next section, only with the four atoms hydrogen, carbon, nitrogen and oxygen ‡, let us give them (compactly) as concrete examples. From Eq. (1), we have $H:=\mathcal{E}_{11}^{(1,1)}$, $C:=\mathcal{E}_{12}^{(2,1)}$, $N:=\mathcal{E}_{11}^{(2,2)}$ and $O:=\mathcal{E}_{12}^{(2,2)}$ with $\mathcal{Z}_{11}^{(1,1)}=1$, $\mathcal{Z}_{12}^{(2,1)}=6$, $\mathcal{Z}_{11}^{(2,2)}=7$ and $\mathcal{Z}_{12}^{(2,2)}=8$. We have found more convenient to display in place of the numerical matrix element $\mathcal{Z}$ (the atomic number) the corresponding symbol of the atom and to manipulate it exactly as a number, for the ease of reading. To each matrix, $\mathcal{E}$, is associated a *dual* matrix $\tilde{\mathcal{E}}=\eta\mathcal{E}\eta^{-1}$ where the matrix η is given by $\eta_{11}=\eta_{22}=0, \eta_{12}=\eta_{21}=1; \eta=\eta^{-1}=\eta^t$. The dual of a matrix is necessary when multiplication of matrices is used to construct a molecule; otherwise some products would vanish. As examples, the water molecule H_2O has the matrix expression HHO while the hydrogen cyanid molecule HCN writes $HC\tilde{N}$ (their dual forms are $\tilde{H}\tilde{H}\tilde{O}$ and $\tilde{H}\tilde{C}N$ respectively). Now, we consider the four special molecules uracil, cytosine, adenine and guanine. They are given, in a *unique* manner, by $U=(C\tilde{C})^2N^2H^4O\tilde{O}$, $C=(C\tilde{C})^2N^3H^5O$, $A=(C\tilde{C})^2CN^5\tilde{H}^5$ and $G=(C\tilde{C})^2CN^5\tilde{H}^5\tilde{O}$ and their duals (as for the atoms) are evaluated as $\tilde{U}=\eta U\eta^{-1}=(\tilde{C}C)^2\tilde{N}^2\tilde{H}^2\tilde{O}O$, etc. Visualizing, we have

$$U = \begin{pmatrix} U & 0 \\ 0 & 0 \end{pmatrix}, \quad C = \begin{pmatrix} 0 & C \\ 0 & 0 \end{pmatrix}, \quad A = \begin{pmatrix} 0 & A \\ 0 & 0 \end{pmatrix}, \quad G = \begin{pmatrix} G & 0 \\ 0 & 0 \end{pmatrix}. \tag{2}$$

The structure of these matrices is not casual. First, U and G are even while C and A are odd so this grouping into even, $\{U, G\}$, and odd, $\{C, A\}$, matrices corresponds naturally to the *keto-amino* grouping and paves the way for the expression of the Rumer symmetry (see section 3). Second, we have just the right things (see Eq. (2)) to build a (RNA) base-matrix (or its dual) in an economic manner. By adding the four bases in Eq . (2), using the duals of A and G, we obtain the *unique* definition (modulo the dual $\tilde{B}=\eta B\eta^{-1}$)

$$B := U + C + \tilde{A} + \tilde{G} = \begin{pmatrix} U & C \\ A & G \end{pmatrix}. \tag{3}$$

This nice and quick result is intimately linked to the detailed atomic composition of the bases and depends crucially and essentially on the number of carbon and oxygen atoms. To see this, let us consider, for example, thymine ($C_5N_2H_6O_2$) in place of uracil. In this case the matrix for thymine is odd so we would have three odd matrices and one even matrix and the construction of a (DNA) base-matrix would be computationally costly but not impossible (see at the end of the next section where a hint is given).

---

‡ These four atoms enter for exactly 100% of the composition of the nitrogenous bases (see above Eq. (2)) and more than 98% of the living matter.

3. The 16 Doublets and their symmetries

We are now in a position to consider the 16 doublets. In the preceeding section, we have used matrix multiplication to construct a given single object (molecule) but now, we are going to use the Kronecker (or direct) product to construct and classify the 16 doublets at the same time, in contrast with the constructions in [3,4] requiring more than one step to get the set of 16 doublets. From Eq. (3), we have

$$\mathcal{D}_1 : (\mathcal{D}_1)_{pq} := (\mathcal{B} \otimes \mathcal{B})_{js,kt} = \mathcal{B}_{jk}\mathcal{B}_{st} = \begin{pmatrix} UU & \underline{UC} & \underline{CU} & CC \\ UA & UG & CA & CG \\ AU & \underline{AC} & \underline{GU} & GC \\ AA & AG & GA & \underline{GG} \end{pmatrix} \quad (4.1)$$

where $p = 2(j-1) + s$, $q = 2(k-1) + t$ and $1 \le p \le 4, 1 \le q \le 4$. These are the 16 possible doublets, all in one matrix, $\mathcal{D}_1$. It is also useful to introduce the transformed form

$$\mathcal{D}_2 = \begin{pmatrix} 1 & 0 & 0 & 0 \\ 0 & 0 & 1 & 0 \\ 0 & 1 & 0 & 0 \\ 0 & 0 & 0 & 1 \end{pmatrix} \mathcal{D}_1 \begin{pmatrix} 1 & 0 & 0 & 0 \\ 0 & 0 & 1 & 0 \\ 0 & 1 & 0 & 0 \\ 0 & 0 & 0 & 1 \end{pmatrix}^{-1} = \begin{pmatrix} UU & \underline{CU} & \underline{UC} & CC \\ AU & \underline{GU} & \underline{AC} & GC \\ UA & CA & UG & CG \\ AA & GA & AG & \underline{GG} \end{pmatrix} \quad (4.2)$$

which could also be obtained from Eq. (4. 1) with the substitution $j \leftrightarrow s$, $k \leftrightarrow t$. $\mathcal{D}_1$, Eq. (4. 1), and $\mathcal{D}_2$, Eq. (4. 2), constitute our classification of the 16 doublets. In $\mathcal{D}_1$, the doublets sharing the same *first* base form *compact* clusters (the four quadrants or Quartets) whereas in $\mathcal{D}_2$, the doublets sharing the same *second* base form *compact* clusters. This partitioning is interesting and could be seen as another way to classify the doublets, in agreement with the Taylor-Coates suggestion, [6], that second-position bases connect amino acids that have similar properties while first-position bases connect amino acids from the same biosynthetic pathway. Let us note these Quartets "U_i", "C_i", "A_i" and "G_i" for $\mathcal{D}_i$, $i = 1, 2$. Now, there are two ways to look at the genetic code: (i) from the perspective of its $B_1 B_2$-portion (with $3'\text{-}B_1 B_2 B_3\text{-}5'$ the symbol of a codon) and this leads to the existence of the two sets $\mathcal{M}_1$, which contains doublets for which the third base has no influence on the coded amino acid, and $\mathcal{M}_2$ which contains doublets which do not code the amino acid uniquely but requires the knowledge of the third base ($\mathcal{M}_1$ is underlined in Eq s. (4. 1) and (4. 2)); this is the traditional point of view where $\mathcal{M}_1$ and $\mathcal{M}_1$ are *exchanged* by the transformation $U \leftrightarrow G, C \leftrightarrow A$, [2], or the α-transformation in [3] and (ii) from the perspective of its $B_2 B_3$-portion and this leads to the existence of two other sets $\mathcal{M}'_1$ which contains doublets ending in a *strong* base S (C/G) and $\mathcal{M}'_2$ which contains doublets ending in a *weak* base W (U/A). Jiménez Montaño and his collaborators, [4], have shown that this latter point of view leads to the more symmetrical Mitochondrial Code through the interesting decomposition of the 16 doublets set into $\mathcal{M}'_1$ and $\mathcal{M}'_2$ which are *exchanged* by the same transformation α of Ref. [3] belonging to the Klein 4-group as for $\mathcal{M}_1$ and $\mathcal{M}_2$, separating in this way the *"eukaryotes"* from the *"prokaryotes"* , as the former use frequently codons with $B_2 B_3$ in $\mathcal{M}'_1$ and the latter use frequently codons with $B_2 B_3$ in $\mathcal{M}'_2$. (It is clear that $\mathcal{D}_1$ or $\mathcal{D}_2$ could be used in the two views). Let us now introduce our three transformations

$$R_1 := \begin{pmatrix} 0 & 0 & 0 & 1 \\ 0 & 0 & 1 & 0 \\ 0 & 1 & 0 & 0 \\ 1 & 0 & 0 & 0 \end{pmatrix}, R_2 := \begin{pmatrix} 0 & 1 & 0 & 0 \\ 1 & 0 & 0 & 0 \\ 0 & 0 & 0 & 1 \\ 0 & 0 & 1 & 0 \end{pmatrix}, R_3 := \begin{pmatrix} 0 & 0 & 1 & 0 \\ 0 & 0 & 0 & 1 \\ 1 & 0 & 0 & 0 \\ 0 & 1 & 0 & 0 \end{pmatrix}, \quad (5)$$

which are defined by $R_1 = \eta \otimes \eta$, $R_2 = \eta \oplus \eta$ and $R_3 = R_1 R_2$, where η is defined in section 2. It is straightforward to see that they close under matrix multiplication, they are mutually commutative, idempotent ($R_i^2 = id_{4\times 4}, i = 1, 2, 3$), their own inverses and transpose so that, by adjoining the identity element, $id_{4 \times 4}$, these four matrices constitute an Abelian group, the Klein's 4-group. The action of R_1, R_2, and R_3 on $\mathcal{D}_2$, for example, is displayed below from left to right, respectively

$$\begin{pmatrix} GG & AG & GA & AA \\ CG & UG & CA & UA \\ GC & AC & GU & AU \\ CC & UC & CU & UU \end{pmatrix}, \begin{pmatrix} GU & AU & GC & AC \\ CU & UU & CC & UC \\ GA & AA & GG & AG \\ CA & UA & CG & UG \end{pmatrix}, \begin{pmatrix} UG & CG & UA & CA \\ AG & GG & AA & GA \\ UC & CC & UU & CU \\ AC & GC & AU & GU \end{pmatrix}$$

- R_1 performs a *three-fold exchange* at the same time: $\mathcal{M}_1 \leftrightarrow \mathcal{M}_2$, $\mathcal{M}'_1 \leftrightarrow \mathcal{M}'_2$ and "U_2" $\leftrightarrow$ "G_2", "C_2" $\leftrightarrow$ "A_2", with an overall Rumer-type exchange, $U \leftrightarrow G$ and $C \leftrightarrow A$, for the two bases in all the doublets.

- R_2 leaves "U_2", "C_2", "A_2" and "G_2" as well as $\mathcal{M}'_1$ and $\mathcal{M}'_2$ *globally* invariant, with an overall Rumer-type exchange, $U \leftrightarrow G$ and $C \leftrightarrow A$, for the *first* base in all the doublets.

- R_3 *exchanges* at the same time "U_2" and "G_2" *and* "A_2" and "C_2" on the one hand, and $\mathcal{M}'_1 \leftrightarrow \mathcal{M}'_2$ on the other, with an overall Rumer-type exchange, $U \leftrightarrow G$ and $C \leftrightarrow A$ for the *second* base in all the doublets.

In conclusion, we have constructed a unified classification of the genetic code-doublets and defined a set of transformations, constituting a Klein's 4-group, to describe their symmetries. The most important transformation is manifestely the (Rumer) transformation, R_1, as it exhibits *diversity* in its exchange properties. The action of these transformations on $\mathcal{D}_1$ is the same except for the inversion of the roles of R_2 and R_3 concerning $\mathcal{M}'_1$ and $\mathcal{M}'_2$ (putting i=1 in the Quartets and making the exchange *first* base $\leftrightarrow$ *second* base). Let us remark that besides the Rumer transformation ($U \leftrightarrow G$, $C \leftrightarrow A$) there exist two other ones $U \leftrightarrow C$, $A \leftrightarrow G$ and $U \leftrightarrow A$, $C \leftrightarrow G$. They correspond respectively to the *purine-pyrimidine* and *strong-weak* groupings. We are able to show that the construction of a base-matrix $\mathcal{B}$ reflecting these groupings and adapted to the study of the symmetries of the doublets, in the same way as Eq. (3) is adapted to the keto-amino grouping, is possible. It suffices to manipulate the definitions in Eq. (2), using essentially the Hadamard transformation. Finally, the above classification is easily extended to the 64 codons, 3'-$\mathcal{B}_1 \otimes \mathcal{B}_2 \otimes \mathcal{B}_3$-5', and consEq ently to the genetic code table. The corresponding symmetries are described by the eight-dimensional matrix versions of R_1, R_2 and R_3, constituting also a Klein's 4-group. (This work, I would tentatively consider as a "synthesis" of the (apparently unrelated) references [1] and [3], is dedicated to Karl-Dietrich Neubert.)

I would like to acknowledge Jean-Pierre Gazeau and Richard Kerner (Université Paris 7-Denis Diderot) for their kind invitation to Group 24.

4. References

[1] Neubert D 1970 Naturforsch. 210 25a
[2] Rumer Yu. B 1966 Proc. Acad. Sci. U. S. S. R. 167 1393
[3] Danckwerts H-J and Neubert D 1975 J. Mol. Evol. 5 327
[4] Jiménez Montaño M A de la Mora Basàñez C R and Pöschel T 1996 BioSystems 39 117
[5] Négadi T 2002 Int. J. Quantum Chemistry (to appear)
[6] Taylor F J R and Coates D 1989 Biosystems 22 177

Inst. Phys. Conf. Ser. No 173: Section 8
Paper presented at 24th Int. Coll. Group Theoretical Methods in Physics, Paris, France, 15–20 July 2002
©*2003 IOP Publishing Ltd*

A group theoretical approach to quantum theories on commutative and non-commutative spheres

K Odaka and S Sakoda

Department of Applied Physics, National Defence Academy, Yokosuka, JAPAN

Abstract. The space and the time are regarded as the non-commutative sphere and a c-number parameter, respectively, and quantum mechanics and a quantum field theory on such a space-time are constructed according to the group theoretical approach and we compare them to the commutative case. In the construction of the field theories it is important that the quantum fields transform covariantly under the symmetry transformation.

One of the simplest models for a non-commutative manifold is the non-commutative sphere $\hat{S}^2$ [1] and there are many works [2] on field theories on the compact space-time of $\hat{S}^2$. In these works the quantization of the fields is performed by using the path integrals. Meanwhile, we regard the space and the time as $\hat{S}^2$ and a c-number parameter, respectively, and construct quantum mechanics and a quantum field theory on such a space-time according to the group theoretical approach [3] [4], which is established for the Poincaré group. Quantum mechanics and quantum field theories are also constructed on the commutative sphere S^2 in order to compare their results.

The algebra of $\hat{S}^2$ is generated by $\hat{n}_i$ and the basic relations are

$$[\hat{n}_i, \hat{n}_j] = i\kappa\epsilon_{ijk}\hat{n}_k, \quad \sum_{i=1}^{3}\hat{n}_i\hat{n}_i = r^2 \tag{1}$$

where κ and r are real numbers and r means the radius. For simplicity, r is fixed to be 1. The derivative operator $\hat{l}_i$ for every $\hat{n}$ is defined by the following algebra

$$[\hat{l}_i, \hat{l}_j] = i\epsilon_{ijk}\hat{l}_k, \quad [\hat{l}_i, \hat{n}_j] = i\epsilon_{ijk}\hat{n}_k. \tag{2}$$

These algebra shift to the Lie algebra associated with the euclidean group E(3) (the group of rotations and translations) in the limit $\kappa \to 0$.

Now, it is well known that the coadjoint orbits of E(3) are precisely the phase space for a system whose configuration space is a sphere S^2 and therefore the group E(3) is a symmetry transformation one for the Wigner theorem [4]. Thus, the generators of E(3) are candidates for a preferred class of observables and we can quantize the system by using the irreducible unitary representations of E(3) obtained from the induced representation theory. These representations are realized on the Hilbert space $L^2(S^2, d\Omega(\theta\phi))$ and they are labeled by the irreducible representations s of the isotropy group U(1). The generators of the translation $\hat{x}_i$ and the rotation $\hat{l}_i$ have the form

$$\hat{x}_i\psi_s(\theta\phi) = x_i\psi_s(\theta\phi) \quad \vec{x} = (\sin\theta\cos\phi, \sin\theta\sin\phi, \cos\theta), \tag{3}$$

$$\begin{aligned}
\hat{l}_1\psi_s(\theta\phi) &= \{\bar{l}_1 + s\tfrac{\sin\theta\cos\phi}{1+\cos\theta}\}\psi_s(\theta\phi), \\
\hat{l}_2\psi_s(\theta\phi) &= \{\bar{l}_2 + s\tfrac{\sin\theta\sin\phi}{1+\cos\theta}\}\psi_s(\theta\phi), \\
\hat{l}_3\psi_s(\theta\phi) &= \{\bar{l}_3 + s\}\psi_s(\theta\phi),
\end{aligned} \tag{4}$$

where $\psi_s(\theta\phi) \in L^2(S^2, d\Omega(\theta\phi))$, $\bar{l}_1 = i(\sin\phi\frac{\partial}{\partial\theta} + \cot\theta\cos\phi\frac{\partial}{\partial\phi})$, $\bar{l}_2 = -i(\cos\phi\frac{\partial}{\partial\theta} - \cot\theta\sin\phi\frac{\partial}{\partial\phi})$ and $\bar{l}_3 = -i\frac{\partial}{\partial\phi}$. The quantization of the system is completed by translating the classical Hamiltonian which consists of the generators into the quantum one. Since we are interested in models which are invariant under the rotations, we take the Hamiltonian

$$\hat{H} = \sum_{i=1}^{3} \hat{l}_i^2. \tag{5}$$

Let's apply the above-mentioned quantization method to the non-commutative case. If we denote $\hat{m}_i = \hat{n}_i/\kappa$, $\hat{j}_i = \hat{l}_i - \hat{m}_i$, the algebra (1) and (2) are reduced to the Lie algebra $su(2) \oplus su(2)$

$$[\hat{m}_i, \hat{m}_j] = i\epsilon_{ijk}\hat{m}_k, \quad [\hat{j}_i, \hat{j}_j] = i\epsilon_{ijk}\hat{j}_k, \quad [\hat{j}_i, \hat{m}_j] = 0. \tag{6}$$

Therefore taking SU(2)⊗SU(2) as the symmetry transformation group, we can realize the phase space as its coadjoint orbit and the algebra of $\hat{S}^2$ is regarded as ones of the observables in quantum mechanics. Its unitary irreducible representations are constructed on the Hilbert space

$$\mathcal{H} = \{|m, m_3 > \otimes|j, j_3 >\} \quad (-m \le m_3 \le m, \ -j \le j_3 \le j) \tag{7}$$

where m and j are fixed. If $\kappa = 1/\sqrt{m(m+1)}$, the radius is independent of the choice of the representation and then we can take the commutative limit $m \to \infty$. We adopt the Hamiltonian (5) in analogy with the commutative case. The eigenstates of the Hamiltonian are

$$\mathcal{H} = \{|l, l_3 >\} = \{ \sum_{m_3+j_3=l_3} C_{m_3j_3}^{l,l_3}|m, m_3 > \otimes|j, j_3 >\} \tag{8}$$

where $C_{m_3j_3}^{l,l_3}$ is the Clebsch-Gordan coefficient and $|m - j| \le l \le m + j, \ -l \le l_3 \le l$.

Let's introduce the coordinates which correspond to the classical ones. They are defined by the parameters of the group. The group SU(2)⊗SU(2) may be the maximum symmetry transformation group but the Hamiltonian is not invariant under this transformation. Thus, we pay attention to the subgroup SU(2) under which the Hamiltonian is invariant and its coadjoint orbit is considered as the configuration space. Since the orbit is SU(2)/U(1), a point on the orbit corresponds to a group element $\sigma(\theta\phi) = e^{-i\phi\hat{l}_3}e^{-i\theta\hat{l}_2}e^{i\phi\hat{l}_3}$. The coherent state labeled by a point on the orbit is defined as

$$|\theta\phi, s > = \sum_A \frac{1}{\sqrt{N_l}}\hat{U}_l(\sigma(\theta\phi))|l, s > \quad (A : |j - m| \le l \le j + m) \tag{9}$$

where $\hat{U}_l(\sigma(\theta\phi))$ is a representation and $N_l = (2l + 1)/4\pi$. For a state $|ll_3 > \in \mathcal{H}$ the wave function on the orbit is given by

$$< l, l_3|\theta\phi, s > = \frac{1}{\sqrt{N_l}}e^{-il_3\phi}d_{l_3s}^l(\theta)e^{is\phi} \tag{10}$$

where $d_{l_3s}^l(\theta)$ is a rotational matrix and it satisfies

$$\int_{S^2} d\Omega e^{-il_3\phi}d_{l_3s}^l(\theta)e^{il_3'\phi}d_{l_3's}^{l'}(\theta) = N_l\delta_{ll'}\delta_{l_3l_3'}. \tag{11}$$

Thus, the set of the coherent states $\{|\theta\phi, s >\}$ is over-complete in the Hilbert space $\mathcal{H}$. Since the wave function $\psi_s(\theta\phi) = < \theta\phi, s|\psi >$ transforms in the same manner as the commutative case under the SU(2) transformation, the generators of the rotation $\hat{l}_i$ have the same form as (4).

Now, let's construct a field theory on $\hat{S}^2$ as many particle problem. Then, the creation $a^\dagger_{ll_3}(t)$ and the annihilation $a_{ll_3}(t)$ operators and the Fock space are introduced as follows,

$$[a_{ll_3}(t), a^\dagger_{l'l'_3}(t)]_\pm = \delta_{ll'}\delta_{l_3l'_3}, \quad [a^{(\dagger)}_{ll_3}(t), a^{(\dagger)}_{l'l'_3}(t)]_\pm = 0, \tag{12}$$

$$a_{ll_3}(t)|0> = 0 \quad a^\dagger_{ll_3}a^\dagger_{l'l'_3}\cdots|0> = |(ll_3),(l'l'_3),\cdots) \tag{13}$$

where - and + indicate boson and fermion, respectively. This particle picture is one for the particle which corresponds to the eigenstate of the Hamiltonian. Their time evolution is described by

$$i\frac{\partial}{\partial t}a_{ll_3}(t) = l(l+1)a_{ll_3}(t), \quad i\frac{\partial}{\partial t}a^\dagger_{ll_3}(t) = -l(l+1)a^\dagger_{ll_3}(t), \tag{14}$$

and the operators transform under the SU(2) transformation as

$$\hat{V}(g)a_{ll_3}(t)\hat{V}(g)^{-1} = \mathcal{D}^l_{l_3l'_3}(g^{-1})a_{ll'_3}(t),$$
$$\hat{V}(g)a^\dagger_{ll_3}(t)\hat{V}(g)^{-1} = \mathcal{D}^{l*}_{l_3l'_3}(g^{-1})a^\dagger_{ll'_3}(t) \tag{15}$$

where $\mathcal{D}^l_{l_3l'_3}(g^{-1})$ is an unitary matrix representing an element g.

For simplicity we take $s = 1/2$ and a field is defined by

$$\Phi(\theta\phi, t; 1/2) = \sum_A \sum_{l_3} < \theta\phi, 1/2|l, l_3 > a_{ll_3}(t). \tag{16}$$

This field transforms under the infinitesimal transformation $k = 1 - i\sum_i \xi_i \hat{l}_i$ as

$$\hat{V}(k)\Phi(\theta\phi, t; 1/2)\hat{V}(k)^{-1} =$$
$$\Phi(\theta\phi, t; 1/2) + i\sum_i \xi_i(\bar{l}_i + \frac{1}{2}f_i(\theta\phi))\Phi(\theta\phi, t; 1/2) \tag{17}$$

where $f_1(\theta\phi) = \sin\theta\cos\phi/(1 + \cos\theta)$, $f_2(\theta\phi) = \sin\theta\sin\phi/(1 + \cos\theta)$ and $f_3 = 1$.

Now, in the field theories we should construct the interaction Hamiltonian which is invariant under the symmetry transformation. Such an interaction Hamiltonian includes the Clebsch-Gordan coefficients to couple together the creation and annihilation operators and it is very complex. Then, it is rather easy to construct the Hamiltonian out of the fields which transform covariantly under the symmetry transformation [3]. The covariance means that each field is multiplied with a position-independent matrix under the transformation. Thus, we construct the covariant field from the field (16).

We introduce another particle picture $(b_{ll_3}(t), b^\dagger_{ll_3}(t))$ and field as

$$\Phi(\theta\phi, t, -1/2) = \sum_A \sum_{l_3} < \theta\phi, -1/2|l, l_3 > b_{ll_3}(t) \tag{18}$$

and a two component field is defined by

$$\Phi(\theta\phi, t) = \left(\begin{array}{c} \Phi(\theta\phi, t; 1/2) \\ \Phi(\theta\phi, t; -1/2) \end{array} \right). \tag{19}$$

This field transforms as

$$\hat{V}(k)\Phi(\theta\phi, t)\hat{V}(k)^{-1} = \Phi(\theta\phi, t) + i\sum_i \xi_i(\bar{l}_i + \frac{1}{2}\sigma_3 f_i(\theta\phi))\Phi(\theta\phi, t), \tag{20}$$

where σ_i is the Pauli matrix and then it is not covariant. Thus, the field $\Phi(\theta\phi, t)$ is transformed as

$$\Psi(\theta\phi, t) = M(\theta\phi)\Phi(\theta\phi, t), \tag{21}$$

where $M(\theta\phi)$ is a 2×2 matrix and its explicit form is found in [5] as

$$M(\theta\phi) = \exp\left(-i\frac{x_1\sigma_2 - x_2\sigma_1}{2\sqrt{1-x_3^2}}\tan^{-1}\frac{\sqrt{1-x_3^2}}{x_3}\right). \tag{22}$$

Since

$$M(\theta\phi)(\bar{l}_i + \frac{1}{2}\sigma_3 f_i(\theta\phi))M(\theta\phi)^{-1} = \bar{l}_i + \frac{1}{2}\sigma_i, \tag{23}$$

the covariance of the field $\Psi(\theta\phi, t)$ can be confirmed as follows,

$$\hat{V}(k)\Psi(\theta\phi, t)\hat{V}(k)^{-1} = \Psi(\theta\phi, t) + i\sum_i \xi_i(\bar{l}_i + \frac{1}{2}\sigma_i)\Psi(\theta\phi, t). \tag{24}$$

By using the covariant field the free Hamiltonian

$$H_0 = \sum_A \sum_{l_3}\left(l(l+1)a_{ll_3}^\dagger(t)a_{ll_3}(t) + l(l+1)b_{ll_3}^\dagger(t)b_{ll_3}(t)\right) \tag{25}$$

is rewritten as

$$H_0 = \int_{S^2} d\Omega\left(\Psi^\dagger(\theta\phi, t)\sum_i(\bar{l}_i + \frac{1}{2}\sigma_i)^2\Psi(\theta\phi, t)\right), \tag{26}$$

and the field equation is given by the covariant form

$$i\frac{\partial}{\partial t}\Psi(\theta\phi, t) = \sum_i\left(\bar{l}_i + \frac{1}{2}\sigma_i\right)^2\Psi(\theta\phi, t). \tag{27}$$

The covariant field can be splitted into two sections by the projection operator $(1 \pm \vec{x}\vec{\sigma})/2$ and each section corresponds to a particle picture. The commutation relation is given by the non-local form

$$[\Psi(\theta\phi, t),\ \Psi^\dagger(\theta'\phi', t)]_\pm =$$
$$M(\theta\phi)\begin{pmatrix} <\theta\phi, 1/2|\theta'\phi', 1/2> & 0 \\ 0 & <\theta\phi, -1/2|\theta'\phi', -1/2> \end{pmatrix}M(\theta'\phi')^{-1} \tag{28}$$

If we take the commutative limit $(A : 1/2 \leq l < \infty)$, it reduce to the local form

$$[\Psi_i(\theta\phi, t),\ \Psi_j^\dagger(\theta'\phi', t)]_\pm = \delta_{ij}\delta(\cos\theta - \cos\theta')\delta(\phi - \phi'). \tag{29}$$

The invariant interaction Hamiltonian is simply written, for example, as

$$H_I = \int_{S^2} d\Omega\sum_i :\left(\Psi^\dagger(\theta\phi, t)\sigma_i\Psi(\theta\phi, t)\Psi^\dagger(\theta\phi, t)\sigma_i\Psi(\theta\phi, t)\right): \tag{30}$$

where : : denotes the normal ordering.

References

[1] Madore J 1991 J. Math. Phys. 32 332-335
[2] See, for example, Prešnajder P 2000 J. Math. Phys. 41 2789-2804
[3] Isham C 1984 in Relativity, Groups and Topology II (Amsterdam: North-Holland)
[4] Weinberg S 1995 The Quantum Theory of Fields (New York: Cambridge University Press)
[5] Ohnuki Y and Kitakado S 1993 J. Math. Phys. 34 2827-2851

Inst. Phys. Conf. Ser. No 173: Section 8
Paper presented at 24th Int. Coll. Group Theoretical Methods in Physics, Paris, France, 15–20 July 2002
©2003 IOP Publishing Ltd

Milne phase for the Coulomb quantum problem related to Riemann's hypothesis

H C Rosu[1], J M Morán-Mirabal[1], M Planat[2]

Instituto Potosino de Investigación Científica y Tecnológica, Apdo Postal 3-74 Tangamanga, San Luis Potosí, MEXICO
Laboratoire de Physique et Métrologie des Oscillateurs du CNRS, 25044 Besançon Cedex, FRANCE

Abstract. We use the Milne phase function in the continuum part of the spectrum of the particular Coulomb problem that has been employed by Bhaduri, Khare, and Law as an equivalent physical way for calculating the density of zeros of the Riemann's function on the critical line. The Milne function seems to be a promising approximate method to calculate the density of prime numbers.

From a 1995 PRE paper of Bhaduri, Khare, and Law [1] one can obtain the following formula for the density of zeros of Riemann's zeta function on the critical line

$$n_Z(\epsilon) = -\frac{\ln \pi}{2\pi} + \frac{1}{2\pi}\text{Re}\left[\Psi\left(\frac{1}{4} + i\frac{\epsilon}{2}\right)\right], \tag{1}$$

where the digamma function is the logderivative of the gamma function, $\Psi(z) = \Gamma'(z)/\Gamma(z)$. Formula 16 of the same paper gives the phase shift of a repulsive Coulomb potential obtained from an inverted oscillator with a hard wall at the origin for the unconventional value of the partial wave number $l = -\frac{1}{4}$. Taking the derivative of that formula, one obtains

$$n_C(\epsilon) = -\frac{F'(\epsilon)}{2\pi} + \frac{1}{2\pi}\text{Re}\left[\Psi\left(\frac{1}{4} + i\frac{\epsilon}{2}\right)\right], \tag{2}$$

where

$$F(\epsilon) = \frac{\pi}{2} - \tan^{-1}(\text{cosech}\pi\epsilon). \tag{3}$$

Thus, under an appropriate shift, the two expressions given by (1) and (2) differ only by an exponentially small term as noted by the Indian team of authors.

Our aim in this work is to apply another technique based on the so-called Milne phase function for the calculation of the 'density of states' in the continuum of the same Coulomb problem. Indeed, being a phase, one might think a priori that Milne's function has something to do with the nontrivial zeros of the Riemann function. Not only this is a different procedure, but it might be a quite competitive approximation for n_Z. Previously, Korsch [2] applied the same method with very good results in the case of bound states for a few illustrative cases of elementary quantum mechanics. He also established that in this approach the density of quantum states is nowhere unique and not even necessarily positive (!?). For the Coloumb problem under focus here the nonuniqueness issue is actually an advantage because in a certain sense one can choose by trial and error a better Milne approximation for n_Z. The Milne function is expressed through the following formula

$$n_M(y, \epsilon) \equiv \frac{1}{\rho^2} = \frac{1}{\left[\alpha\phi_1(y, \epsilon) + \beta\phi_2(y, \epsilon)\right]^2 + \frac{1}{\alpha^2}\phi_2^2(y)}, \tag{4}$$

where ρ is the solution of the Pinney nonlinear equation and ϕ_1 and ϕ_2 are the two linear independent solutions of the repulsive Coulomb problem, i.e.

$$\phi_1(y, \epsilon) = \sin\left(ky - \frac{\epsilon}{2}\ln(2ky) - \frac{l_r\pi}{2} + \text{Arg}\Gamma\left(l_r + 1 + \frac{i\epsilon}{2}\right)\right), \tag{5}$$

and

$$\phi_2(y, \epsilon) = \cos\left(ky - \frac{\epsilon}{2}\ln(2ky) - \frac{l_r\pi}{2} + \text{Arg}\Gamma\left(l_r + 1 + \frac{i\epsilon}{2}\right)\right), \tag{6}$$

where, as we mentioned before, $l_r = -\frac{1}{4}$. The other symbols are as follows: $\epsilon = -\frac{E}{\hbar\omega}$ is a reduced spectral parameter, where E is the spectral parameter in the initial inverted oscillator problem and ω is the angular frequency of the oscillator; $k = \frac{m\omega}{2\hbar}$; $y = x^2$, where $x \geq 0$ is the oscillator coordinate. The superposition constants α and β are determined by arbitrary 'initial' conditions at some point y_0. We fix them in the most 'economic' way, which is at the point $y_0 = \frac{1}{2k}$ allowing one to eliminate the logarithm in the argument of the trigonometric functions. Moreover, we employ the Eliezer-Gray prescription, for details see the master thesis of Espinoza [3]. Thus

$$\alpha = \phi_1\left(\frac{1}{2k}\right) = \sin\left(\frac{1}{2} + \frac{\pi}{8} + \text{Arg}\Gamma\left(\frac{3}{4} + \frac{i\epsilon}{2}\right)\right), \tag{7}$$

$$\beta = \frac{d\phi_1}{dy}\Big|_{y=\frac{1}{2k}} = (1 - \epsilon)\cos\left(\frac{1}{2} + \frac{\pi}{8} + \text{Arg}\Gamma\left(\frac{3}{4} + \frac{i\epsilon}{2}\right)\right). \tag{8}$$

A three-dimensional plot of $n_M(y, \epsilon)$ is displayed in Fig. 1, where an expected oscillatory behaviour can be seen. Not only the procedure based on the Milne phase function could compete very well with other approximate methods for the density of prime numbers but there is a further advantage on which we briefly comment in the following. As is well known, the Milne phase function enters as a basic ingredient in the Ermakov-Lewis phase-amplitude approach for parametric oscillator problems (for recent applications see [4]). To transform the Coulomb problem at hand

$$\frac{d^2\phi}{dy^2} + \left[k^2 - \frac{k\epsilon}{y} + \frac{3}{16y^2}\right]\phi = 0 \tag{9}$$

into a parametric dynamical problem for a unit mass classical particle, one can use the well-known map to canonical classical variables $\phi = q$ and $\frac{d\phi}{dy} = p$ leading to

$$\dot{q} \equiv \frac{dq}{dy} = p \tag{10}$$

$$\dot{p} \equiv \frac{dp}{dy} = -\left[k^2 - \frac{k\epsilon}{y} + \frac{3}{16y^2}\right]q, \tag{11}$$

where the coordinate y plays the role of the classical Hamiltonian time. Various quantities, such as the Ermakov-Lewis invariant and geometrical angles can be calculated easily for this particularly interesting Coulomb problem (due to its connection with prime numbers). But we want to emphasize a different point here. In the parametric oscillator interpretation and of y as Hamiltonian time, one can use $n_M(y, \epsilon)$ as a direct tool for time-energy (and therefore time-imaginary axis) characterization of the density of prime numbers as can be inferred from Fig. 1.

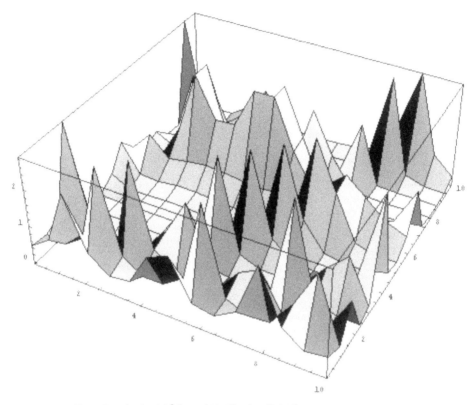

Figure 1. $n_M(y, \epsilon) = 1/\rho^2$ for $y \in [0.1, 10]$ and $\epsilon \in [0.1, 10]$.

References

[1] Bhaduri M T C Khare A and Law D B 1995 Phys. Rev. E52 486, chao-dyn/9406006
[2] Korsch H J 1985 Phys. Lett. A109 313
[3] Espinoza P B 2000 *Ermakov-Lewis dynamic invariants with some applications*, Master thesis supervised by Rosu H C, math-ph/0002005
[4] Rosu H C 2002 Physica Scripta 65 296, quant-ph/0104003; Rosu H C and Espinoza P 2001 Phys. Rev. E63 037603, physics/0004014; Hawkins R M and Lidsey J E 2002 Phys. Rev. D66 023523

Inst. Phys. Conf. Ser. No 173: Section 8
Paper presented at 24th Int. Coll. Group Theoretical Methods in Physics, Paris, France, 15–20 July 2002
©*2003 IOP Publishing Ltd*

Hierarchy of Gradings on $sl(3, \mathbf{C})$ and Summary of Fine Gradings on $sl(4, \mathbf{C})$

Milena Svobodová

Department of Mathematics, Faculty of Nuclear Sciences and Physical Engineering, Czech Technical University, Trojanova 13, 120 00 Prague, Czech Republic

Abstract. Fine gradings of a Lie algebra provide basic information about the structure of the algebra. Among others they can also be used for defining additive quantum numbers on the algebra. Starting from the fine gradings, one can arrive at the hierarchy of all gradings of the particular algebra by coarsening the fine ones. The coarsest grading - the top of the hierarchy - is the algebra itself. On $sl(3, \mathbf{C})$ we provide explicitly all the fine gradings (together with the additive quantum numbers) and then the whole hierarchy of gradings, that is 17 gradings altogether (4 of them fine). On $sl(4, \mathbf{C})$ we give just an overview of its eight fine gradings as well as the additive quantum numbers corresponding to each of the fine gradings.

1. Introduction

A **grading** of a Lie algebra L is a decomposition of L into a direct sum $\Gamma = \oplus_{j \in \mathcal{J}} L_j$ such that for all $j, k \in \mathcal{J}$ there exists $l \in \mathcal{J}$ fulfilling $[L_j, L_k] \subseteq L_l$. The index set $\mathcal{J}$ is not uniquely defined. But we are able to choose $\mathcal{J}$ as a subset of an Abelian group. The index set $\mathcal{J}$ is not necessarily closed under the operation '+' - if $j + k \notin \mathcal{J}$ then the subspaces L_j and L_k commute. If $j + k \in \mathcal{J}$ then $[L_j, L_k] \subseteq L_{j+k}$. Also the Abelian group into which the index set is embedded is not unique. We choose our Abelian groups as small as possible, i.e. finite. Therefore the indices are in general multi-component.

A grading $\tilde{\Gamma} = \oplus_{j \in \tilde{\mathcal{J}}} \tilde{L}_j$ is a **refinement** of the grading $\Gamma = \oplus_{j \in \mathcal{J}} L_j$ iff for each $j \in \tilde{\mathcal{J}}$ there exists $i \in \mathcal{J}$ such that $\tilde{L}_j \subseteq L_i$. If $|\tilde{\mathcal{J}}| > |\mathcal{J}|$ we say that $\tilde{\Gamma}$ is a **proper refinement** of Γ and we note this property symbolically by $\tilde{\Gamma} \prec \Gamma$. **Fine grading** is such a grading that cannot be properly refined.

A method how to find all fine gradings on classical simple Lie algebras is described in [1, 2]. Let us just point out here that it is based on the use of automorphisms - each grading subspace is an eigensubspace of certain set of automorphisms on the Lie algebra. Then the number of automorphisms used to split the algebra into its fine grading determines the number of components in the indices $j \in \mathcal{J}$.

If $\oplus_{j \in \mathcal{J}} L_j$ is a grading of L and g is an automorphism on L then $\oplus_{j \in \mathcal{J}} g L_j$ is also a grading of L, and $\oplus_{j \in \mathcal{J}} L_j$ and $\oplus_{j \in \mathcal{J}} g L_j$ are called **equivalent gradings**. Clearly when stating number of gradings or fine gradings of some algebra further in the text we mean the number of all non-equivalent (fine) gradings.

Let us demonstrate these terms on the well explored algebra $sl(2, \mathbf{C})$. This algebra has 4 gradings; two of them are fine (one toroidal - Cartan (Γ_1) and one consisting of linear hulls of Pauli matrices (Γ_2)).

$\Gamma_1 = L_0 \oplus L_1 \oplus L_2$, where $\qquad L_0 = \mathbf{C} \left(\begin{smallmatrix} 1 & 0 \\ 0 & -1 \end{smallmatrix} \right) \quad L_1 = \mathbf{C} \left(\begin{smallmatrix} 0 & 1 \\ 0 & 0 \end{smallmatrix} \right) \quad L_2 = \mathbf{C} \left(\begin{smallmatrix} 0 & 0 \\ 1 & 0 \end{smallmatrix} \right)$

$\Gamma_2 = L_{(0,1)} \oplus L_{(1,0)} \oplus L_{(1,1)}$, with $\quad L_{(0,1)} = \mathbf{C} \left(\begin{smallmatrix} 0 & 1 \\ 1 & 0 \end{smallmatrix} \right) \quad L_{(1,0)} = \mathbf{C} \left(\begin{smallmatrix} 1 & 0 \\ 0 & -1 \end{smallmatrix} \right) \quad L_{(1,1)} = \mathbf{C} \left(\begin{smallmatrix} 0 & 1 \\ -1 & 0 \end{smallmatrix} \right)$

Then there is the coarsest grading whose index set contains just one element, that is the algebra $sl(2, \mathbf{C})$ itself. The last grading Γ_0 is not fine. A non-fine grading always arises as a coarsening of some fine grading (i.e. suitable collecting of a few grading subspaces of the fine grading). Of course, it can occur that by coarsening of two different fine gradings one obtains the same non-fine grading. And this is just the case of Γ_0, which is a coarsening of both Γ_1 and Γ_2.

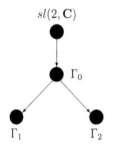

Here the hierarchy of gradings of $sl(2, \mathbf{C})$ is depicted. An arrow connects Γ (head) with $\tilde{\Gamma}$ (tail) when Γ is a direct refinement of $\tilde{\Gamma}$, i.e. $\Gamma \prec \tilde{\Gamma}$ and there exists no such a grading $\tilde{\tilde{\Gamma}}$ that $\Gamma \prec \tilde{\tilde{\Gamma}} \prec \tilde{\Gamma}$.

$\Gamma_0 = \tilde{L}_0 \oplus \tilde{L}_1$, where
$\tilde{L}_0 = L_0 = L_{(1,0)}, \tilde{L}_1 = L_1 \oplus L_2 = L_{(0,1)} \oplus L_{(1,1)}$

2. Hierarchy of Gradings on $sl(3, \mathbf{C})$

As well as in the case of $sl(2, \mathbf{C})$, we have one toroidal (Cartan) fine grading (Γ_1) and one fine grading of the Pauli type (Γ_2) on $sl(3, \mathbf{C})$. (Here we deal with generalized 3×3 Pauli matrices P and Q introduced in Table 2.) These two types of fine gradings appear in every Lie algebra $sl(n, \mathbf{C})$. More information about $n \times n$ Pauli matrices and the corresponding fine gradings of the Lie algebras $sl(n, \mathbf{C})$ can be found in [3].

$$
L_0 = \mathbf{C} \begin{pmatrix} 2 & 0 & 0 \\ 0 & -1 & 0 \\ 0 & 0 & -1 \end{pmatrix} + \mathbf{C} \begin{pmatrix} 0 & 0 & 0 \\ 0 & 1 & 0 \\ 0 & 0 & -1 \end{pmatrix} \quad L_1 = \mathbf{C} \begin{pmatrix} 0 & 0 & 0 \\ 0 & 0 & 0 \\ 1 & 0 & 0 \end{pmatrix} \quad L_2 = \mathbf{C} \begin{pmatrix} 0 & 1 & 0 \\ 0 & 0 & 0 \\ 0 & 0 & 0 \end{pmatrix}
$$
$$
L_3 = \mathbf{C} \begin{pmatrix} 0 & 0 & 0 \\ 0 & 0 & 0 \\ 0 & 1 & 0 \end{pmatrix} \quad L_4 = \mathbf{C} \begin{pmatrix} 0 & 0 & 0 \\ 0 & 0 & 1 \\ 0 & 0 & 0 \end{pmatrix} \quad L_5 = \mathbf{C} \begin{pmatrix} 0 & 0 & 0 \\ 1 & 0 & 0 \\ 0 & 0 & 0 \end{pmatrix} \quad L_6 = \mathbf{C} \begin{pmatrix} 0 & 0 & 1 \\ 0 & 0 & 0 \\ 0 & 0 & 0 \end{pmatrix}
$$

Table 1. **Cartan grading**. $\Gamma_1 = \oplus_{j \in \mathcal{J}} L_j$, the index set $\mathcal{J}$ is equal to the additive group $\mathbf{Z}_7$.

$$
L_{(j,k)} = \mathbf{C}(P^j Q^k) \text{ for all } j, k = 0, 1, 2, (j, k) \neq (0, 0)
$$
$$
P = \begin{pmatrix} 0 & 1 & 0 \\ 0 & 0 & 1 \\ 1 & 0 & 0 \end{pmatrix}, Q = \mathrm{diag}(1, \omega, \omega^2), \omega = \exp\left(\tfrac{2\pi i}{3}\right)
$$

Table 2. **Pauli grading**. $\Gamma_2 = \oplus_{(j,k) \in \mathcal{J}} L_{(j,k)}$, the index set $\mathcal{J}$ is a subset of the additive group $\mathbf{Z}_3 \times \mathbf{Z}_3$.

Moreover, we have another two fine gradings (Γ_3, Γ_4) on $sl(3, \mathbf{C})$, see Tables 3,4.

$$L_{(1,0,0)} = \mathbf{C}\begin{pmatrix} 1 & 0 & 0 \\ 0 & -1 & 0 \\ 0 & 0 & 0 \end{pmatrix} + \mathbf{C}\begin{pmatrix} 0 & 0 & 0 \\ 0 & 1 & 0 \\ 0 & 0 & -1 \end{pmatrix} \quad L_{(0,1,0)} = \mathbf{C}\begin{pmatrix} 0 & 0 & 1 \\ 0 & 0 & 0 \\ -1 & 0 & 0 \end{pmatrix} \quad L_{(1,1,0)} = \mathbf{C}\begin{pmatrix} 0 & 0 & 1 \\ 0 & 0 & 0 \\ 1 & 0 & 0 \end{pmatrix}$$

$$L_{(0,0,1)} = \mathbf{C}\begin{pmatrix} 0 & 1 & 0 \\ -1 & 0 & 0 \\ 0 & 0 & 0 \end{pmatrix} \quad L_{(1,0,1)} = \mathbf{C}\begin{pmatrix} 0 & 1 & 0 \\ 1 & 0 & 0 \\ 0 & 0 & 0 \end{pmatrix} \quad L_{(0,1,1)} = \mathbf{C}\begin{pmatrix} 0 & 0 & 0 \\ 0 & 0 & 1 \\ 0 & -1 & 0 \end{pmatrix} \quad L_{(1,1,1)} = \mathbf{C}\begin{pmatrix} 0 & 0 & 0 \\ 0 & 0 & 1 \\ 0 & 1 & 0 \end{pmatrix}$$

Table 3. $\Gamma_3 = \oplus_{(j,k,l) \in \mathcal{J}} L_{(j,k,l)}$, the index set $\mathcal{J}$ is a subset of the additive group $\mathbf{Z}_2 \times \mathbf{Z}_2 \times \mathbf{Z}_2$.

$$L_{(0,0)} = \mathbf{C}\begin{pmatrix} 0 & 0 & 0 \\ 0 & 1 & 0 \\ 0 & 0 & -1 \end{pmatrix} \quad L_{(0,1)} = \mathbf{C}\begin{pmatrix} 2 & 0 & 0 \\ 0 & -1 & 0 \\ 0 & 0 & -1 \end{pmatrix} \quad L_{(1,0)} = \mathbf{C}\begin{pmatrix} 0 & 1 & 0 \\ 0 & 0 & 0 \\ -1 & 0 & 0 \end{pmatrix} \quad L_{(1,1)} = \mathbf{C}\begin{pmatrix} 0 & 1 & 0 \\ 0 & 0 & 0 \\ 1 & 0 & 0 \end{pmatrix}$$

$$L_{(2,1)} = \mathbf{C}\begin{pmatrix} 0 & 0 & 0 \\ 0 & 0 & 0 \\ 0 & 1 & 0 \end{pmatrix} \quad L_{(3,1)} = \mathbf{C}\begin{pmatrix} 0 & 0 & 0 \\ 0 & 0 & 1 \\ 0 & 0 & 0 \end{pmatrix} \quad L_{(4,0)} = \mathbf{C}\begin{pmatrix} 0 & 0 & -1 \\ 1 & 0 & 0 \\ 0 & 0 & 0 \end{pmatrix} \quad L_{(4,1)} = \mathbf{C}\begin{pmatrix} 0 & 0 & 1 \\ 1 & 0 & 0 \\ 0 & 0 & 0 \end{pmatrix}$$

Table 4. $\Gamma_4 = \oplus_{(j,k) \in \mathcal{J}} L_{(j,k)}$, the index set $\mathcal{J}$ is a subset of the additive group $\mathbf{Z}_2 \times \mathbf{Z}_5$.

Altogether there exist 17 gradings of $sl(3, \mathbf{C})$. Their hierarchy is depicted on the following picture.

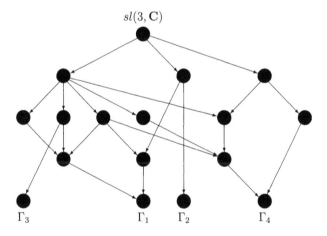

$sl(3, \mathbf{C})$

$\Gamma_3 \qquad \Gamma_1 \quad \Gamma_2 \qquad \Gamma_4$

3. Fine Gradings on $sl(4, \mathbf{C})$

This algebra already has eight fine gradings. (They can be found in [4].) Here we only present a brief overview of the structure of subspaces and additive groups. Again, of course, there are the two fine gradings that appear in each $sl(n, \mathbf{C})$, namely the Cartan grading (Γ_1) and the Pauli grading (Γ_2). Concerning the relation of equivalence, we have proceeded as follows: In the first place, we check the dimensions of grading subspaces. If they differ, the gradings are necessarily non-equivalent. In the second place, we compare the number of one-dimensional grading subspaces which consist of semisimple and nilpotent elements. Again, if these numbers do not coincide the gradings are not equivalent. Lastly, we count the number of commuting pairs of one-dimensional grading subspaces. By that characteristic we are able to distinguish those fine gradings that have not been proven non-equivalent by the previous two steps.

Fine gradings of sl$(4, \mathbf{C})$		
	Dimensions of grading subspaces	**Additive groups of indices**
Γ_1	1-dimensional:12, 3-dimensional:1	$\mathbf{Z}_{13}$
Γ_2	1-dimensional:15	$\mathbf{Z}_4 \times \mathbf{Z}_4$
Γ_3	1-dimensional:12, 3-dimensional:1	$\mathbf{Z}_2 \times \mathbf{Z}_2 \times \mathbf{Z}_2 \times \mathbf{Z}_2$
Γ_4	1-dimensional:13, 2-dimensional:1	$\mathbf{Z}_2 \times \mathbf{Z}_8$
Γ_5	1-dimensional:13, 2-dimensional:1	$\mathbf{Z}_2 \times \mathbf{Z}_9$
Γ_6	1-dimensional:15	$\mathbf{Z}_2 \times \mathbf{Z}_2 \times \mathbf{Z}_6$
Γ_7	1-dimensional:15	$\mathbf{Z}_2 \times \mathbf{Z}_2 \times \mathbf{Z}_2 \times \mathbf{Z}_2$
Γ_8	1-dimensional:13, 2-dimensional:1	$\mathbf{Z}_2 \times \mathbf{Z}_2 \times \mathbf{Z}_4$

To determine all gradings and their hierarchy on this algebra is an extremely complicated task since there is a huge number of them. It is just the same with nearly all the classical simple Lie algebras of rank higher than one.

To conclude with we provide the number of non-equivalent fine gradings for some low-rank Lie algebras. (See Table 5.) Some of the figures shown in this table are results from [2, 4, 5, 6, 7].

$sl(2, \mathbf{C}) \simeq sp(2, \mathbf{C})$	$sl(3, \mathbf{C})$	$sl(4, \mathbf{C}) \simeq o(6, \mathbf{C})$	$sl(5, \mathbf{C})$	$sl(6, \mathbf{C})$
2	4	8	5	11
$o(3, \mathbf{C})$	$o(4, \mathbf{C})$	$o(5, \mathbf{C}) \simeq sp(4, \mathbf{C})$	$o(7, \mathbf{C})$	$sp(6, \mathbf{C})$
2	4	3	4	3

Table 5. Number of fine gradings for some low-rank Lie algebras

References

[1] J. Patera, H. Zassenhaus, *On Lie gradings I*, Linear Algebra and its Applic. **112** (1989) 87-159

[2] M. Havlíček, J. Patera, and E. Pelantová, *On Lie gradings II,III*, Linear Algebra and its Applic. **277** (1998) 97-125, **314** (2000) 1-47

[3] J. Patera, H. Zassenhaus, *The Pauli matrices in n-dimensions and finest gradings of Lie algebras of type A_{n-1}*, J. Math. Phys. **29** (1988) 665-673

[4] J. Patera, E. Pelantová, and M. Svobodová, *The eight fine gradings of sl(4, C) and o(6, C)*, J. Math. Phys. **43** (2002).

[5] J. Patera, E. Pelantová, and M. Svobodová, *Fine gradings of o(5, C), sp(4, C) and of their real forms*, J. Math. Phys. **42** (2001) 3839-3853.

[6] M. A. Abdelmalek, X. Leng, J. Patera, P. Winternitz, *Grading refinements in the contractions of Lie algebras and their invariants*, J. Phys. A **29** (1996) 7519–7543.

[7] M. Havlíček, J. Patera, E. Pelantová, and J. Tolar, *On fine gradings and their symmetries*, Czechoslovak J. of Phys., **51**, no 4. (2001) 383-392.

Inst. Phys. Conf. Ser. No 173: Section 8
Paper presented at 24th Int. Coll. Group Theoretical Methods in Physics, Paris, France, 15–20 July 2002
©2003 IOP Publishing Ltd

Quadratic algebras in simple exclusion models

Reidun Twarock

Centre for Mathematical Science, City University, London EC1V 0HB, England

Abstract. The computation of steady state properties of simple exclusion models can be referred to the algebraic problem of finding representations for particular types of quadratic algebras. In the case of purely diffusive processes, these are diffusion algebras. We explain the connection of simple exclusion models and diffusion algebras and provide a state of the art account concerning their construction, classification and representation theory.

1. Introduction

Simple exclusion models are stochastic processes describing nearest-neighbour interactions of N species of particles on a one-dimensional lattice. The various applications of these models range from traffic flow models [1] to models for interface growth [2]. In general, the setting is as follows:

Consider a linear lattice or *chain* with L sites, where each site is either occupied by a particle or empty. We indicate the *type* or *species* of a particle as s_j, $j = 1, \ldots, N$, where N denotes the number of different species. A *configuration* at time t is specified by an L-tuple $(\tau_1(t), \ldots, \tau_L(t)) = \{\tau_i(t)\}$ where $\tau_i(t) = 0$ if site i is empty at time t and $\tau_i(t) = s_j$ if site i is occupied at time t by a particle of species s_j. We assume that the time evolution is purely diffusive, that is we exclude collisions or death and birth of particles.

As a first step, we formulate laws for the time evolution of individual particles. During each infinitesimal time interval dt, each particle moves with probability p_R to its right and with p_L to its left if the corresponding site is empty. Otherwise, it does not move. We distinguish the following scenarios: Symmetric exclusion models correspond to $p_R = p_L \neq 0$, asymmetric exclusion models to $p_R = 0$ and $p_L \neq 0$, or $p_L = 0$ and $p_R \neq 0$, and partially asymmetric exclusion models to $p_R \neq p_L$ with $p_R, p_L \neq 0$. At the boundaries of the chain, a particle may enter or leave with a given probability (open boundary case) or propagate periodically (closed boundary case). Based on these rules, we model the time evolution of the system as follows. Let $P_L(\tau_1, \tau_2, \ldots, \tau_L)$ denote the probability of finding the system in a configuration $(\tau_1, \tau_2, \ldots, \tau_L)$. Then we have

$$\frac{d}{dt} P_L(\tau_1, \tau_2, \ldots, \tau_L) = \sum_{\sigma_1} (\Gamma_1)_{\tau_1}^{\sigma_1} P_L(\sigma_1, \tau_2, \ldots, \tau_L)$$

$$+ \sum_{i=1}^{L-1} \sum_{\sigma_i, \sigma_{i+1}} \Gamma_{\tau_i, \tau_{i+1}}^{\sigma_i, \sigma_{i+1}} P_L(\tau_1, \ldots, \sigma_i, \sigma_{i+1}, \ldots, \tau_L) \qquad (1)$$

$$+ \sum_{\sigma_L} (\Gamma_L)_{\tau_L}^{\sigma_L} P_L(\tau_1, \ldots, \tau_{L-1}, \sigma_L)$$

where $(\Gamma_1)_{\tau_1}^{\sigma_1}$ and $(\Gamma_L)_{\tau_L}^{\sigma_L}$ encode the *transition probabilities* at the respective boundaries of the chain and the $\Gamma_{\tau_i, \tau_{i+1}}^{\sigma_i, \sigma_{i+1}}$ encode the transition probabilities in the middle of the chain, that is the probabilities for a transition of the configuration $(\ldots, \sigma_i, \sigma_{i+1}, \ldots)$ into the configuration $(\ldots, \tau_i, \tau_{i+1}, \ldots)$.

When the system reaches a steady state, one has

$$\frac{d}{dt} P_L(\tau_1, \tau_2, \ldots, \tau_L) = 0 \tag{2}$$

for the weights $P_L(\tau_1, \tau_2, \ldots, \tau_L)$ of each configuration in the steady state. In applications, one is usually interested in quantities that can be expressed in terms of the steady state weights, and it is thus important to device a strategy to calculate them.

2. Connection of the steady state weights with quadratic algebras

The correspondence between steady state weights and quadratic algebras is provided by the matrix product approach [3] which consists in the following ansatz. Associate with each species of particles s_j a generator D_j, $j = 1, \ldots, N$, and with each configuration a monomial in these generators, that is formally asign to a configuration $(\tau_1, \tau_2, \ldots, \tau_L)$ the monomial $D_{\tau_1} D_{\tau_2} \ldots D_{\tau_L}$, where $D_{\tau_k} = D_j$ if a particle of species s_j is located at τ_k.

Then one considers the following ansatz for the steady state weights. For a *closed chain*, corresponding to periodic boundary conditions, one uses $P_L^s(\tau_1, \tau_2, \ldots, \tau_L) = \mathrm{Tr}(D_{\tau_1} D_{\tau_2} \ldots D_{\tau_L})$ and for an *open chain* one uses $P_L^s(\tau_1, \tau_2, \ldots, \tau_L) = \langle 0| D_{\tau_1} D_{\tau_2} \ldots D_{\tau_L} |0\rangle$, where $\langle 0|$ and $|0\rangle$ are suitably chosen vectors.

Inserting this ansatz into the evolution equation (1) with purely diffusive motion (that is assuming that only the rates $g_{\alpha\beta} := \Gamma_{\alpha\beta}^{\beta\alpha}$ and $\Gamma_{\alpha\beta}^{\alpha\beta}$ are non-zero [4]) leads to the following quadratic relations as conditions on the generators D_j, $j = 1, \ldots, N$:

$$g_{\alpha\beta} D_\alpha D_\beta - g_{\beta\alpha} D_\beta D_\alpha = x_\alpha D_\beta - x_\beta D_\alpha \tag{3}$$

where $\alpha, \beta \in \{1, \ldots, N\}$, $g_{\alpha\beta} \in \mathbb{R}^{\geq 0}$ and $x_\alpha \in \mathbb{R}$. If representations can be found, then the steady state weights are obtainable via this ansatz.

3. Diffusion algebras

We focus on algebras with the PBW (Poincaré Birkhoff Witt) property: An algebra with generators $\{D_\alpha | \alpha = 1, \ldots, N\}$ has the *PBW property* if it admits a basis (PBW-basis) of ordered monomials of the form $D_{\alpha_1}^{k_1} D_{\alpha_2}^{k_2} \ldots D_{\alpha_n}^{k_n}$, $k_j \in \mathbb{N}^0$, with $\alpha_1 > \alpha_2 > \ldots > \alpha_n$.

Based on this we have the following terminology [5]: Those algebras with relations (3), that have the PBW property, are called *diffusion algebras*.

The PBW property implies restrictions on the rates $g_{\alpha\beta}$. To obtain an exhaustive list of diffusion algebras, one uses the Diamond lemma [6]. It implies that an algebra with a set of generators $\mathcal{D} := \{D_\alpha | \alpha = 1, \ldots, N\}$ and quadratic relations has the PBW property if for any subset $\mathcal{D}_s \subset \mathcal{D}$ the reductions $D_\alpha D_\beta D_\gamma \to D_\beta D_\alpha D_\gamma \to D_\beta D_\gamma D_\alpha \to D_\gamma D_\beta D_\alpha$ and $D_\alpha D_\beta D_\gamma \to D_\alpha D_\gamma D_\beta \to D_\gamma D_\alpha D_\beta \to D_\gamma D_\beta D_\alpha$, $\alpha, \beta, \gamma \in \mathcal{D}_s$, coincide when expressed in the PBW basis. This leads to the following strategy for the construction of diffusion algebras [7]: Determine all three generator algebras compatible with the requirements of the Diamond lemma, then determine all N generator algebras which have three generator subalgebras in this list of three generator algebras.

In particular, one thus obtains a complete list of diffusion algebras, which are organized in five families. Different algebras in the families are given in dependence on the choice of the decomposition of the index set $I_N = \{1, \ldots, N\}$ into ordered subsets I, S, T_a° $(a = 1, \ldots, M_T^\circ)$, $T_b^\bullet$ $(b = 1, \ldots, M_T^\bullet)$, and the choice of structure constants.

Using the following notation for *normal ordering* of the generators D_α and D_β, that is

$$: D_\alpha D_\beta : \quad := \begin{cases} D_\alpha D_\beta & \text{if} \quad \alpha < \beta \\ D_\beta D_\alpha & \text{if} \quad \beta < \alpha \end{cases} \tag{4}$$

one obtains the list of algebras below, which is exhaustive and contains all possible diffusion algebras with N generators [7]:

1. *Diffusion algebras of type* $A_I(I, S, T_a^\circ, T_b^\bullet)$, $N_I \geq 3$:

$$
\begin{aligned}
g[D_i, D_j] &= x_j D_i - x_i D_j, & \forall i, j \in I \\
g_s[D_s, D_i] &= x_i D_s, & \forall s \in S, \, i \in I \\
g_a^\circ : D_i D_t : &= -x_i D_t, & \forall a, \, t \in T_a^\circ, \, i \in I \\
g_b^+ D_i D_t &= -x_i D_t, & \forall b, \, t \in T_b^\bullet, \, i \in I : i < t \\
g_b^- D_t D_i &= x_i D_t, & \forall b, \, t \in T_b^\bullet, \, i \in I : i > t
\end{aligned}
$$

where $g, g_s, g_a^\circ, g_b^\pm \neq 0$, $x_i \neq 0$.

2. *Diffusion algebras of type* $A_{II}(I, T_a^\circ, T_b^\bullet)$, $N_I \geq 3$:

$$
\begin{aligned}
(g_i - g_j) D_i D_j &= x_j D_i - x_i D_j, & \forall i < j \in I \\
(g_i + g_a^\circ) : D_i D_t : &= -x_i D_t, & \forall a, \, t \in T_a^\circ, \, i \in I \\
(g_i + g_b^+) D_i D_t &= -x_i D_t, & \forall b, \, t \in T_b^\bullet, \, i \in I : i < t \\
(g_b^- - g_i) D_t D_i &= x_i D_t, & \forall b, \, t \in T_b^\bullet, \, i \in I : i > t
\end{aligned}
$$

where $x_i \neq 0$, $g_i \neq g_j$ for $i \neq j$ and $g_i \notin \{g_a^\circ, \mp g_b^\pm\}$.

3. *Diffusion algebras of type* $B(I = \{\mathbf{i}, \mathbf{j}\}, S, T_a^\circ, T_b^\bullet)$:

$$
\begin{aligned}
g D_{\mathbf{i}} D_{\mathbf{j}} - (g - \Lambda) D_{\mathbf{j}} D_{\mathbf{i}} &= x_{\mathbf{j}} D_{\mathbf{i}} - x_{\mathbf{i}} D_{\mathbf{j}} \\
g_s D_{\mathbf{i}} D_s - (g_s - \Lambda) D_s D_{\mathbf{i}} &= -x_{\mathbf{i}} D_s, & \forall s \in S \\
g_s D_s D_{\mathbf{j}} - (g_s - \Lambda) D_{\mathbf{j}} D_s &= x_{\mathbf{j}} D_s, & \forall s \in S \\
g_a^\circ : D_{\mathbf{i}} D_t : &= -x_{\mathbf{i}} D_t, & \forall t \in T_a^\circ, \\
(g_a^\circ - \Lambda) : D_{\mathbf{j}} D_t : &= -x_{\mathbf{j}} D_t, & \forall t \in T_a^\circ \\
g_b^+ D_{\mathbf{i}} D_t &= -x_{\mathbf{i}} D_t, & \forall t \in T_b^\bullet \\
g_b^- D_t D_{\mathbf{j}} &= x_{\mathbf{j}} D_t, & \forall t \in T_b^\bullet
\end{aligned}
$$

where $x_i \neq 0$, $g \neq 0$, $g_s \neq 0$ for all s and $g_s \neq \Lambda$ for s such that either $s < \mathbf{i}$ or $s > \mathbf{j}$, $g_a^\circ \notin \{0, \Lambda\}$ and $g_b^\pm \neq 0$.

4. *Diffusion algebras of type* $C(I = \{\mathbf{i}\}, R_a)$:

$$
g_r D_{\mathbf{i}} D_r - (g_r - \Lambda_a) D_r D_{\mathbf{i}} = -x_{\mathbf{i}} D_r, \quad \forall r \in R_a,
$$

where $x_i \neq 0$, $g_r \neq 0$ for $r < \mathbf{i}$ and $g_r \neq \Lambda_a$ for $r > \mathbf{i}$.

5. *Diffusion algebras of type* $D(R)$:

$$
D_r D_s - q_{sr} D_s D_r = 0, \quad \forall r < s \in R.
$$

Furthermore, in all five cases the commutation relations for the subset $S \cup T$, where $T := \{T_a^\circ, a = 1, \ldots, M_T^\circ\} \cup \{T_b^\bullet, b = 1, \ldots, M_T^\bullet\}$, are given by

$$
D_\alpha D_\beta - q_{\beta\alpha} D_\beta D_\alpha = 0
$$

where $q_{\alpha\beta} \neq 0$ for $\alpha < \beta$, and $q_{\alpha\beta} = 0$ for $\alpha < \beta$ if $\alpha \in S$ and $\beta \in T$.

Note that the splitting of the index set is such that the set I labels those generators D_i which have $x_i \neq 0$ in the relations (3), and such that T labels those generators which have only product relations, that is relations of the form (4) with all generators from the set I. The

706

set T is furthermore subdivided into sets $T_a^\bullet$ and T_a° according to their ordering with respect to the set I.

While being exhaustive, the list of diffusion algebras is repetitive. The issue of a classification of diffusion algebras is addressed in [8]. As criterion of equivalence one uses linear isomorphism: A *linear isomorphism* Φ is a bijective algebra homomorphism which on the set of generators D is given an affine mapping $\Phi : D_\alpha' \mapsto \sum_{\beta=1}^N a_{\alpha\beta} D_\beta + b_\alpha$ with $a_{\alpha\beta}, b_\alpha \in \mathbb{R}$. Simple examples of linear isomorphisms are the rescaling of generators, or the action of the symmetric group $\mathcal{S}_N$. However, while any type of rescaling is possible, care is needed when permuting generators due to the requirements on the mutual ordering of the generators imposed by the fact that $g_{\alpha\beta} \neq 0$ for $\alpha < \beta$. One shows that there exist members of different families in the construction scheme that are linearly isomorphic in the above sense [8]. In particular, in the classification scheme one obtains that diffusion algebras split into three families, which are called SYM, $ASYM$ and $PASYM$ according to their relations with simple exclusion models, that is they correspond to symmetric, partially asymmetric, or asymmetric exclusion models, respectively.

First results about the representation theory of diffusion algebras are given in [9]. One observes that the representation theory for different families in the classification scheme has different characteristic features: for example, while the representations for the SYM family are generically finite dimensional, they are generically infinite dimensional for the other two.

4. Discussion

The construction of diffusion algebras has led to a complete account of quadratic algebras related to models with purely diffusive motion. Several of these algebras have already been implemented in various applications (see [5] for references to explicit examples), and the mathematical analysis points to new possibilities for applications, especially in the case of the partially asymmetric exclusion models. Possible generalizations may include the analysis of more general dynamics including death and birth of particles, or generalizations to processes beyond two body interaction.

Acknowledgements

This contribution has been written during a visit to Uppsala University, Sweden, and financial support by The Swedish Foundation for International Cooperation in Research and Higher Education (STINT) is gratefully acknowledged. Furthermore, I would like to thank Uppsala University for their warm hospitality during my stay.

References

[1] Nagel K and Schreckenberg M 1992 J. Physique 12 2221
[2] Krug J and Spohn H 1991 in: Solids far from Equilibrium, ed. C. Godreche (Cam. Univ. Press)
[3] Derrida B, Evans M R, Hakim V and Pasquier V 1993 J. Phys. A: Math. Gen. 26 1493
[4] Arndt F, Heinzel T and Rittenberg V 1998 J. of Phys. A: Math. Gen. 31 833
[5] Isaev A P, Pyatov P N and Rittenberg V 2001 J. Phys. A: Math. Gen. 34 5815-5834
[6] Bergman G M 1978 Adv. in Math. 29 178-218
[7] Pyatov P N and Twarock R 2002 J. Math. Phys. 43 3268-3279
[8] Cox A, Martin P P and Twarock R, in preparation.
[9] Twarock R, in: Proceedings of Quantum Theory and Symmetries, Krakow 2001.

Inst. Phys. Conf. Ser. No 173: Section 8
Paper presented at 24th Int. Coll. Group Theoretical Methods in Physics, Paris, France, 15–20 July 2002
©2003 IOP Publishing Ltd

Finite su(2)-oscillator; applications to signal processing

Luis Edgar Vicent

Instituto de Matemáticas, UNAM.
Cuernavaca, México.

Abstract. The one-dimensional finite oscillator model satisfies that under commutator brackets, all operators close into the (dynamical) Lie algebra $su(2)$; therefore the eigenvalues of position, momentum and energy operators are discrete and the eigenvalue of a compact $u(1)$ central operator fixes the dimension of the $su(2)$ representation. In this work we review the one-dimensional finite oscillator model with $u(2)$ as the dynamical algebra where the wave functions, which involve Kravchuk polynomials, are discrete functions of angular momentum theory.

1. Introduction

1.1. Oscillator models

The harmonic oscillator is characterized by the following postulates:
i) There exist a position operator Q.
ii) There exist a self-adjoint and compact Hamiltonian operator H (generator of time evolution), satisfying

$$[H, Q] := -iP; \qquad [H, P] = iQ,$$

where the first(Hamilton-Lie) equation is purely geometrical (defining the momentum operator P) and the second describes the oscillator dynamics.
iii) Under commutation, the set $\{Q, P, H\}$ close into a Lie algebra:

$$[H[Q, P]] = 0 \quad \Longleftrightarrow \quad [Q, P] = iG(\hat{1}, H, C).$$

Here, the operator G chooses a model for the oscillator, being function of a central $\hat{1}$ operator, and/or H, and/or Casimir operators C. *E.g.* taking $G = \hat{1}$ (with $\hbar = 1$) we get the standard (continuous) quantum oscillator, whose algebra $\mathrm{Span}\{H, Q, P, \hat{1}\} \supset \mathrm{Span}\{Q, P, \hat{1}\}$, the Heisenbger-Weyl algebra.

1.2. The algebra $u(2)$

Here we consider the algebra $u(2) = u(1) \oplus su(2)$; the Casimir operator of $su(2)$ having as eigenvalues $j(j + 1)$; $j = 0, \frac{1}{2}, 1, \cdots$ (labelling the $(2j + 1)$-dimensional irreducible representation space of the algebra); the central algebra $u(1)$ being generated by E_j, a multiple of $\hat{1}$. Taking just $E_j := j\hat{1}$, then

$$-i[Q, P] = H - (j + \tfrac{1}{2})\hat{1},$$

fixing our model. If J_1, J_2, J_3 denote the well-known generators of angular momentum in $su(2)$, the (postulates of the) model of finite oscillator give then a new physical interpretation:

$$Q = J_1 \longleftrightarrow \text{position } q \in \{-j, -j+1, \cdots, j\},$$
$$-P = J_2 \longleftrightarrow \text{- momentum } p \in \{-j, -j+1, \cdots, j\},$$
$$H = J_3 + j + \tfrac{1}{2} \longleftrightarrow \text{Hamiltonian } H \in \{\tfrac{1}{2}, 3, \cdots, 2j + \tfrac{1}{2}\},$$

where $H - \frac{1}{2} = J_3 + j$ is the mode number $n \in \{0, 1, \cdots, N\}$, for $N := 2j$. The Lie commutation relations are the usual ones:

$$[J_j, J_k] = i\epsilon_{jkl}J_l, \quad \text{and} \quad [J_k, E_j] = 0.$$

Since $u(2)$ is compact, the oscillator model shall have discrete and finite values of position, momentum and energy mode: $q|_{-N/2}^{N/2}$, $p|_{-N/2}^{N/2}$ and $n|_0^N$, respectively.

2. Abstract, position and mode eigenbasis

Denoting by $\mu|_{-j}^j$ the eigenvalues of J_3, these are related to the mode number through $n = j + \mu$ (and $\mu = n - N/2$). Within the representation labelled by $j = N/2$, the abstract eigenbasis (independent of the model) is given by:

$$J_3|j, \mu\rangle_3 = \mu|j, \mu\rangle_3; \qquad \mu|_{-j}^j$$
$$\vec{J}^2|j, \mu\rangle_3 = j(j+1)|j, \mu\rangle_3$$

where $\vec{J}^2 = J_1^2 + J_2^2 + J_3^2$. Taking $j = N/2$, the finite oscillator is defined on $N + 1$ (equidistant) points: $\{-\frac{1}{2}N, -\frac{1}{2}N + 1, \cdots, \frac{1}{2}N\}$. Then, the Kronecker-position eigenbasis is given by

$$Q|N, q\rangle_1 = q|N, q\rangle_1; \qquad q|_{-N/2}^{N/2}$$
$$\vec{J}^2|N, q\rangle_1 = \tfrac{1}{2}N(\tfrac{1}{2}N + 1)|N, q\rangle_1,$$

and the oscillator also has $N + 1$ mode states:

$$H|N, n\rangle_H = n|N, n\rangle_H; \qquad n|_0^N$$
$$\vec{J}^2|N, n\rangle_H = \tfrac{1}{2}N(\tfrac{1}{2}N + 1)|N, n\rangle_H.$$

Then, the connection between this later (mode) eigenbasis and the former abstract one, is given by:

$$|j, \mu\rangle_3 := |2j, j + \mu\rangle_H$$

and

$$|N, n\rangle_H := |\tfrac{1}{2}N, n - j\rangle_3$$

whereas the 1- and 3- basis are interrelated by a rotation:

$$J_1 = e^{-i\frac{\pi}{2}J_2} J_3 e^{i\frac{\pi}{2}J_2} \quad \Rightarrow \quad |N, q\rangle_1 = e^{-i\frac{\pi}{2}J_2}|N, j + q\rangle_H.$$

It is worth to note that, using the $su(2)$ shift operators,

$$J_\pm := \tfrac{1}{\sqrt{2}}(J_1 \pm iJ_2)$$

we can reach any energy state starting from the lowest mode of the oscillator, $|j, -j\rangle_3 = |N, 0\rangle$:

$$J_+|N, n\rangle_H = \sqrt{\tfrac{1}{2}(n + 1)(N - n)}|N, n + 1\rangle_H$$
$$J_-|N, n\rangle_H = \sqrt{\tfrac{1}{2}n(N - n + 1)}|N, n - 1\rangle_H,$$

and we get

$$|N, n\rangle_H = \left[2^n \binom{N}{n}\right]^{-1/2} J_+^n|N, 0\rangle_H$$

3. Finite oscillator wavefunctions; Kravchuk functions

The overlap between the position and mode eigenbasis are the wavefunctions of the finite oscillator:

$$\phi_n^{(N)}(q) := {}_1\langle N, q | N, n\rangle_H, \qquad N = 2j, \quad n|_0^N, \quad q|_{-j}^j,$$
$$= {}_H\langle N, j + q | e^{i\frac{\pi}{2}J_2} | N, n\rangle_H.$$

Recalling the (abstract) definition of the d-Wigner functions,

$$d_{m,m'}^j(\beta) := {}_3(j, m | e^{-i\beta J_2} | j, m')_3 = d_{m',m}^j(\beta)$$

we note that the wavefunctions can then be written as

$$\phi_n^{(N)}(q) = d_{q,n-j}^j(-\pi/2)$$

and these particular case of d-functions admit an exact expression in terms of (symmetric) Kravchuk polynomials, known from angular momentum theory, *i.e.*

$$d_{n-j,q}^j(\pi/2) = \frac{(-1)^n}{2^j} \sqrt{\binom{2j}{n}\binom{2j}{j+q}} K_n(j + q; \tfrac{1}{2}, 2j),$$

where

$$K_n(x; \tfrac{1}{2}, N) := {}_2F_1(-n, \tfrac{1}{2}N - x; -N; 2) = K_x(n; \tfrac{1}{2}, N).$$

The $\phi_n^{(N)}(q)$ wavefunctions are orthogonal and complete over the $N + 1$ points, having as summation measure the binomial distribution.

In the $N \rightarrow \infty$ limit (at positions $x = q/\sqrt{N}$), as the number and density of points increase without bound the binomially-distributed factor of normalization becomes the decreasing Gaussian integration measure of the Hermite polynomials; Kravchuk functions converge to the well-known wave functions of quantum oscillator:

$$\lim_{N\to\infty} (\frac{N}{2})^{1/4} \phi_n^{(N)}(x\sqrt{\tfrac{N}{2}}) = \phi_n^{osc}(x) := \frac{e^{-x^2}}{\sqrt{\sqrt{\pi}2^n n!}} H_n(x).$$

4. Applications of the model: 1D-finite signals

$N + 1$-modal waveguides (optical fibers) can be modelled using this scheme; one-dimensional finite strings of data, obtained for instance placing a linear array of sensors perpendicularly to the fiber axis at the output of an optical fiber, can be Fourier(-Kravchuk) transformed and then decomposed into their mode components for further processing.

Here we shortly describe the fractional Fourier-Kravchuk transforms. The evolution of the finite oscillator is governed by the Hamiltonian operator H, rotating the $Q - P$ plane by the inner automorphism of the $su(2)$ algebra:

$$e^{-i\theta H} \begin{pmatrix} Q \\ P \end{pmatrix} e^{i\theta H} = \begin{pmatrix} \cos\theta & \sin\theta \\ -\sin\theta & \cos\theta \end{pmatrix} \begin{pmatrix} Q \\ P \end{pmatrix},$$

where the oscillator evolution angle θ defines the $so(2)$ cycle of fractional Fourier-Kravchuk transforms of power α through $\alpha = 2\theta/\pi$, and

$$\mathcal{K}^\alpha := e^{-i\frac{1}{2}\pi\alpha(J_3+j)} = e^{i\frac{1}{4}\pi\alpha} e^{-i\frac{1}{2}\pi\alpha H};$$

$$\mathcal{K}^\alpha |N,n\rangle_H = e^{-i\frac{1}{2}\pi n\alpha}|N,n\rangle_H,$$

having $\mathcal{K}^{\alpha_1}\mathcal{K}^{\alpha_2} = \mathcal{K}^{\alpha_1+\alpha_2}$ with $\mathcal{K}^0 = \hat{1}$.

Expressing wavefuncions $|\psi\rangle$ on the Kronecker eigenbasis by means of its expantion

$$|\psi\rangle = \sum_{q=-N/2}^{N/2} \psi^{(N)}(q)|N,q\rangle_1 := \sum_{q=-N/2}^{N/2} {}_1\langle N,q|\psi\rangle|N,q\rangle_1,$$

the Fourier-Kravchuk transform acts on the coefficients $\psi^{(N)}(q)$ through a 3-axis rotation unitary Kernel:

$$\mathcal{K}^\alpha : \psi^{(N)}(q) \mapsto \psi^{(N,\alpha)}(q) = \sum_{q'=-N/2}^{N/2} K_{q,q'}^{(N,\alpha)}\psi^{(N)}(q'),$$

with $\psi^{(N,0)(q)} = \psi^{(N)}(q)$ and where

$$
\begin{aligned}
K_{q,q'}^{(N,\alpha)} &:= {}_1\langle N,q|e^{-i\frac{1}{2}\pi(J_3+j)\alpha}|N,q'\rangle_1 \\
&= \sum_{n=0}^{N} {}_1\langle N,q|N,n\rangle_H e^{-i\frac{1}{2}\pi n\alpha} {}_H\langle N,n|N,q'\rangle_1 \\
&= e^{-i\frac{1}{4}\pi N\alpha}(-i)^{q-q'} d_{q,q'}^{N/2}(\tfrac{1}{2}\pi\alpha).
\end{aligned}
$$

References

[1] Arik M, Atakishiyev N M and Wolf K B 1999 J. Phys. A: Math. Gen. 32 L371-6
[2] Wigner E P 1950 Phys. Rev. 77 711-712
[3] Atakishiyev N M and Suslov S K 1991 Theor. Math. Phys. 85 1055-1062
[4] Atakishiyev N M and Wolf K W 1997 J. Opt. Soc. Am. 14 1467-1477
[5] Atakishiyev N M, Chumakov S M and Wolf K W 1998 J. Math. Phys. 39 6247-6261
[6] Atakishiyev N M, Pogosyan G S, Vicent L E and Wolf K W 2001 J. Phys. A: Math. Gen. 34 9381-9398
[7] Atakishiyev N M, Pogosyan G S, Vicent L E and Wolf K W 2001 J. Phys. A: Math. Gen. 34 9399-9415

Nonlinear Systems and their Symmetries

Inst. Phys. Conf. Ser. No 173: Section 9
Paper presented at 24th Int. Coll. Group Theoretical Methods in Physics, Paris, France, 15–20 July 2002
©2003 IOP Publishing Ltd

Infinite-dimensional Lie groups of symmetries of the ideal MHD equilibrium equations

O I Bogoyavlenskij

Department of Mathematics, Queen's University, Kingston, Canada

Abstract. A method for constructing ideal magnetohydrodynamics equilibria is introduced that consists of the application to any known MHD equilibrium the symmetry transforms [1]. The transforms form an infinite-dimensional abelian Lie groups of symmetries that depend upon the given MHD equilibria.

1. The symmetry transforms

In this paper, we introduce a method for constructing ideal MHD equilibria that is based on the symmetries of the MHD equilibrium equations discovered in [1]. Applying the symmetries to any known equilibrium, we obtain continuous families of new MHD equilibria. For example, by applying the method to the well-behaved axially and helically symmetric plasma equilibria [1, 2], we derive the MHD equilibria which are bounded in the whole Euclidean space R^3 and are rapidly decreasing as $x^2 + y^2 \longrightarrow \infty$.

The method is different from the method of Backlund transforms for the soliton equations such as the Korteweg - de Vries equation [3], the Kadomtsev - Petviashvili equation [4], the Sine - Gordon equation [5], etc. The method of Backlund transforms is based on the resolution of certain auxiliary differential equations which usually cannot be solved explicitly. All soliton equations depend only on a part of spatial variables (one for KdV and SG and two for KP). Unlike the Backlund transforms, the method of symmetry transforms produces new solutions in explicit algebraic form and is applicable to the equations in all three spatial variables x, y, z.

The system of magnetohydrodynamics equilibrium equations has the form

$$\rho(\mathbf{V} \cdot \mathrm{grad})\mathbf{V} + \frac{1}{\mu}\mathbf{B} \times \mathrm{curl}\,\mathbf{B} = -\,\mathrm{grad}\,P, \tag{1}$$

$$\mathrm{div}(\rho\mathbf{V}) = 0, \quad \mathrm{div}\,\mathbf{B} = 0, \quad \mathrm{curl}(\mathbf{V} \times \mathbf{B}) = 0, \tag{2}$$

where $\mathbf{B}$ is the magnetic vector field, μ is the constant magnetic permeability, $\mathbf{V}$ is the plasma velocity vector field, $\rho = \rho(\mathbf{x})$ is its density, P is the pressure.

We consider the incompressible plasma flows. The condition of incompressibility $\mathrm{div}\,\mathbf{V} = 0$ is widely used in the MHD literature [6,7]. For example, it is applicable with a high accuracy for the subsonic plasma flows with Mach number $M \ll 1$, $M^2 = V^2/(\gamma P/\rho)$. Then the continuity equation $\mathrm{div}(\rho\mathbf{V}) = 0$ implies $\mathbf{V} \cdot \mathrm{grad}\,\rho(\mathbf{x}) = 0$. Hence the plasma density $\rho(\mathbf{x})$ is constant on the plasma streamlines.

For the non-collinear vector fields $\mathbf{B}$ and $\mathbf{V}$, the third equation (2) implies the existence of either magnetic surfaces or a magnetic foliation. Indeed, in any simply connected domain $E \subset R^3$, $\mathbf{x} = (x, y, z)$, the equation $\mathrm{curl}(\mathbf{V} \times \mathbf{B}) = 0$ yields

$$\mathbf{V} \times \mathbf{B} = \mathrm{grad}\,\psi(\mathbf{x}), \quad \psi(\mathbf{x}) = \int_{\mathbf{x}_0}^{\mathbf{x}} (\mathbf{V} \times \mathbf{B}) \cdot \mathbf{ds}, \tag{3}$$

Hence the surfaces $\psi(\mathbf{x}) = \mathrm{const}$ are magnetic surfaces, because $\mathbf{B} \cdot \mathrm{grad}\,\psi = 0$, $\mathbf{V} \cdot \mathrm{grad}\,\psi = 0$ [6]. If the vector fields $\mathbf{B}$, $\mathbf{V}$ are defined only in some non-simply connected domain D (for example toroidal), then the function $\psi(\mathbf{x})$ (3) is multivalued in general. However, the differential form $d\psi$ is well-defined. Indeed, we have $d\psi(\mathbf{Y}(\mathbf{x})) = (\mathbf{V}(\mathbf{x}) \times \mathbf{B}(\mathbf{x})) \cdot \mathbf{Y}(\mathbf{x})$ for any tangent vector $\mathbf{Y}(\mathbf{x})$. The equation $d\psi(\mathbf{Y}(\mathbf{x})) = 0$ defines an integrable foliation in the domain D which is generated by the vector fields $\mathbf{B}$ and $\mathbf{V}$.

For toroidal domains D (which are the most important for the tokamak applications) the function $\psi(\mathbf{x})$ is defined up to a constant nI, where n is an integer and I is the integral (3) over the shortest non-contractable loop of the torus.

Any given smooth non-field-aligned MHD equilibrium $\mathbf{B}$, $\mathbf{V}$, ρ, P in R^3 ($\mathbf{B}$ and $\mathbf{V}$ are non-collinear) defines a distribution of magnetic surfaces $\psi(\mathbf{x}) = \mathrm{const}$ in R^3 [6]. Let E_m be the set of all incompressible equilibria that have the same magnetic surfaces as the given one. We introduce the transforms $T\colon E_m \longrightarrow E_m$ that depend on the arbitrary functions $a(\mathbf{x}) \neq 0$, $b(\mathbf{x})$, $c(\mathbf{x})$ that are constant on the magnetic surfaces $\psi(\mathbf{x}) = \mathrm{const}$. The $a(\mathbf{x})$, $b(\mathbf{x})$, $c(\mathbf{x})$ and $\rho(\mathbf{x})$ are either functions of $\psi(\mathbf{x})$ or, more generally, satisfy some functional equations $F(a(\mathbf{x}), \psi(\mathbf{x})) = 0$, $G(b(\mathbf{x}), \psi(\mathbf{x})) = 0$, $H(c(\mathbf{x}), \psi(\mathbf{x})) = 0$, $R(\rho(\mathbf{x}), \psi(\mathbf{x})) = 0$. For the MHD equilibria in a toroidal domain D (a tokamak), functions $a = a(\psi)$, $b = b(\psi)$, $c = c(\psi)$ and $\rho(\psi)$ are periodic with period I defined above.

The transforms T are defined by the formulae

$$\mathbf{B}_1 = b(\mathbf{x})\mathbf{B} + c(\mathbf{x})\sqrt{\mu\rho(\mathbf{x})}\,\mathbf{V}, \quad \mathbf{V}_1 = \frac{c(\mathbf{x})}{a(\mathbf{x})\sqrt{\mu\rho(\mathbf{x})}}\mathbf{B} + \frac{b(\mathbf{x})}{a(\mathbf{x})}\mathbf{V}, \qquad (4)$$

$$\rho_1(\mathbf{x}) = a^2(\mathbf{x})\rho(\mathbf{x}), \quad P_1 = CP + (C\mathbf{B}^2 - \mathbf{B}_1^2)/(2\mu), \quad b^2(\mathbf{x}) - c^2(\mathbf{x}) = C,$$

where $C \neq 0$ is a constant. Let us prove that transforms (4) define new solutions to the equilibrium equations (1) - (2). Let $h(\mathbf{x})$ be any function that is constant on the magnetic surfaces, for example $a(\mathbf{x})$, $b(\mathbf{x})$, $c(\mathbf{x})$, or $\rho(\mathbf{x})$. Hence we have

$$\mathbf{B} \cdot \mathrm{grad}\,h(\mathbf{x}) = 0, \quad \mathbf{V} \cdot \mathrm{grad}\,h(\mathbf{x}) = 0. \qquad (5)$$

Applying the classical identity $\mathbf{B} \times \mathrm{curl}\,\mathbf{B} = -(\mathbf{B} \cdot \mathrm{grad})\mathbf{B} + \mathrm{grad}(\mathbf{B}^2/2)$, we present the equation (1) in the form

$$\rho(\mathbf{V} \cdot \mathrm{grad})\mathbf{V} - (\mathbf{B} \cdot \mathrm{grad})\mathbf{B}/\mu + \mathrm{grad}(P + \mathbf{B}^2/(2\mu)) = 0. \qquad (6)$$

Using formulae (4) and (5), we get

$$\rho_1(\mathbf{V}_1 \cdot \mathrm{grad})\mathbf{V}_1 - (\mathbf{B}_1 \cdot \mathrm{grad})\mathbf{B}_1/\mu + \mathrm{grad}(P_1 + \mathbf{B}_1^2/(2\mu)) =$$

$$(b^2(\mathbf{x}) - c^2(\mathbf{x}))\left(\rho(\mathbf{V} \cdot \mathrm{grad})\mathbf{V} - (\mathbf{B} \cdot \mathrm{grad})\mathbf{B}/\mu + \mathrm{grad}(P + \mathbf{B}^2/(2\mu))\right) = 0.$$

Thus the functions ρ_1, $\mathbf{B}_1$, $\mathbf{V}_1$, P_1 satisfy equation (6) and therefore equation (1).

Equations $\mathrm{div}(\rho_1\mathbf{V}_1) = 0$ and $\mathrm{div}\,\mathbf{B}_1 = 0$ easily follow from equations (5) and $\mathrm{div}(\rho\mathbf{V}) = 0$, $\mathrm{div}\,\mathbf{B} = 0$. Substituting (4), we obtain

$$\mathrm{curl}(\mathbf{V}_1 \times \mathbf{B}_1) = \mathrm{curl}\left(\frac{C}{a(\mathbf{x})}\mathbf{V} \times \mathbf{B}\right) = \mathrm{grad}\,\frac{C}{a(\mathbf{x})} \times (\mathbf{V} \times \mathbf{B}) + \frac{C}{a(\mathbf{x})}\mathrm{curl}(\mathbf{V} \times \mathbf{B}).$$

Applying equations (5), we find that $\operatorname{grad}(C/a(\mathbf{x}))$ is collinear with the vector field $\mathbf{V} \times \mathbf{B}$. Hence $\operatorname{grad}(C/a(\mathbf{x})) \times (\mathbf{V} \times \mathbf{B}) = 0$. This fact and equation $\operatorname{curl}(\mathbf{V} \times \mathbf{B}) = 0$ proves that $\operatorname{curl}(\mathbf{V}_1 \times \mathbf{B}_1) = 0$. Hence formulae (4) for $C \neq 0$ define a new solution to equations (1) - (2).

For $C \neq 0$, the transform (4) is invertible:

$$C\mathbf{B} = b(\mathbf{x})\mathbf{B}_1 - c(\mathbf{x})\sqrt{\mu\rho_1(\mathbf{x})}\mathbf{V}_1, \quad C\mathbf{V} = \frac{-c(\mathbf{x})}{a_1(\mathbf{x})\sqrt{\mu\rho_1(\mathbf{x})}}\mathbf{B}_1 + \frac{b(\mathbf{x})}{a_1(\mathbf{x})}\mathbf{V}_1, \quad (7)$$

where $a_1(\mathbf{x}) = 1/a(\mathbf{x})$. We shall refer to the transforms T (4) as the symmetries of the MHD equilibrium equations (1) - (2).

For $C = 0$, or $b(\mathbf{x}) = \pm c(\mathbf{x})$, the transform (4) is not invertible and its range consists of the field-aligned solutions $\mathbf{B}_1 = \pm\sqrt{\mu\rho_1(\mathbf{x})}\mathbf{V}_1$, $P_1 + \mathbf{B}_1^2/(2\mu) = C_0 = \text{const}$, that are the known Chandrasekhar equipartition equilibria [7].

Remark 1. The new vector fields $\mathbf{B}_1$ and $\mathbf{V}_1$ (4) are linearly dependent on the original $\mathbf{B}$ and $\mathbf{V}$. Hence the new MHD equilibrium $\mathbf{B}_1$, $\mathbf{V}_1$, ρ_1, P_1 has the same magnetic surfaces as the original one $\mathbf{B}$, $\mathbf{V}$, ρ, P. This property shows that the symmetries (4) map the set E_m into itself.

2. The groups of symmetries

We consider the set G_m of all transforms (4) with $C \neq 0$ for which the smooth functions $a(\mathbf{x})$, $b(\mathbf{x})$ and $c(\mathbf{x})$ are constant on the magnetic surfaces for a given MHD equilibrium. Each transform (4) corresponds to a triple of functions (a, b, c) that satisfy the conditions

$$a(\mathbf{x}) \neq 0, \quad b^2(\mathbf{x}) - c^2(\mathbf{x}) \equiv \text{const} = C \neq 0. \qquad (8)$$

The domain E_m for these transforms consists of all divergence-free MHD equilibria that have the same magnetic surfaces as the given one. Remark 1 and the invertibility of the transforms (4) for $C \neq 0$ prove that the range of these transforms is the same as their domain, E_m. Hence the composition of the transforms is well defined. Let us show that the composition assigns on the set G_m the structure of an abelian group. Indeed, the composition of the transforms (4) is equivalent to the 3×3 matrix multiplication

$$
\begin{pmatrix} a_2 & 0 & 0 \\ 0 & b_2 & c_2\sqrt{\mu\rho_1} \\ 0 & \frac{c_2}{\sqrt{\mu\rho_2}} & b_2\sqrt{\frac{\rho_1}{\rho_2}} \end{pmatrix}
\times
\begin{pmatrix} a_1 & 0 & 0 \\ 0 & b_1 & c_1\sqrt{\mu\rho} \\ 0 & \frac{c_1}{\sqrt{\mu\rho_1}} & b_1\sqrt{\frac{\rho}{\rho_1}} \end{pmatrix}
=
\begin{pmatrix} a & 0 & 0 \\ 0 & b & c\sqrt{\mu\rho} \\ 0 & \frac{c}{\sqrt{\mu\rho_2}} & b\sqrt{\frac{\rho}{\rho_2}} \end{pmatrix},
$$

where $a = a_2a_1$, $b = b_2b_1 + c_2c_1$, $c = c_2b_1 + b_2c_1$. Hence the multiplication of the triples has the form

$$(a_2, b_2, c_2) \cdot (a_1, b_1, c_1) = (a_2a_1, b_2b_1 + c_2c_1, c_2b_1 + b_2c_1), \qquad (9)$$

that implies $C = b^2 - c^2 = C_2C_1 \neq 0$. The unit triple is $(1, 1, 0)$, the inverse transform (7) corresponds to the triple $(a, b, c)^{-1} = (a^{-1}, C^{-1}b, -C^{-1}c)$. It is evident that the multiplication (9) is commutative and associative. Hence the transforms (4) form an abelian group G_m.

The group G_m is infinite-dimensional and depends on the topology of the distribution of magnetic surfaces for the given MHD equilibrium because the functions $a(\mathbf{x})$, $b(\mathbf{x})$, $c(\mathbf{x})$ are constant on them.

To study the structure of the group G_m, we introduce the parametrization $a(\mathbf{x}) = \tau \exp \alpha(\mathbf{x})$, where $\alpha(\mathbf{x})$ is a smooth function that is constant on the magnetic surfaces and $\tau = \pm 1$. For $C = \sigma k^2$, $\sigma = \pm 1$, $k > 0$, the second equation (8) is resolved in the form: $\sigma = 1$: $b(\mathbf{x}) = \eta k \,\mathrm{ch}\,\beta(\mathbf{x})$, $c(\mathbf{x}) = \eta k \,\mathrm{sh}\,\beta(\mathbf{x})$; $\sigma = -1$: $b(\mathbf{x}) = \eta k \,\mathrm{sh}\,\beta(\mathbf{x})$, $c(\mathbf{x}) = \eta k \,\mathrm{ch}\,\beta(\mathbf{x})$, where $\eta = \pm 1$ and $\beta(\mathbf{x})$ is an arbitrary smooth function that is constant on the magnetic surfaces. Hence each transform (4) corresponds to a sextuple $(\alpha(\mathbf{x}), \beta(\mathbf{x}), k, \tau, \sigma, \eta)$. In view of the known identities for the hyperbolic functions $\mathrm{ch}\,t$ and $\mathrm{sh}\,t$, the multiplication (9) takes the simple form

$$(\alpha_1(\mathbf{x}), \beta_1(\mathbf{x}), k_1, \tau_1, \sigma_1, \eta_1) \cdot (\alpha_2(\mathbf{x}), \beta_2(\mathbf{x}), k_2, \tau_2, \sigma_2, \eta_2) =$$

$$(\alpha_1(\mathbf{x}) + \alpha_2(\mathbf{x}), \beta_1(\mathbf{x}) + \beta_2(\mathbf{x}), k_1 k_2, \tau_1 \tau_2, \sigma_1 \sigma_2, \eta_1 \eta_2).$$

Hence the group G_m is the direct sum

$$G_m = A_m \oplus A_m \oplus R^+ \oplus Z_2 \oplus Z_2 \oplus Z_2. \tag{10}$$

Here R^+ is the multiplicative group of positive numbers $k > 0$. The A_m is the additive abelian group of smooth functions in R^3 that are constant on the magnetic surfaces for the given MHD equilibrium. The group A_m is also a linear space and an associative algebra with respect to the multiplication of functions.

Remark 2. An additional algebraic structure. The group G_m has an additional algebraic structure: the G_m is a module over the associative algebra $A_m \oplus A_m$ with the multiplication defined by the multiplication of functions: $(\gamma(\mathbf{x}), \zeta(\mathbf{x})) \cdot (\alpha(\mathbf{x}), \beta(\mathbf{x}), k, \tau, \sigma, \eta) = (\gamma(\mathbf{x})\alpha(\mathbf{x}), \zeta(\mathbf{x})\beta(\mathbf{x}), k, \tau, \sigma, \eta)$. This operation results in the transform: $a(\mathbf{x}) \longrightarrow \tau |a(\mathbf{x})|^{\gamma(\mathbf{x})}$, $b(\mathbf{x}) = \eta k \,\mathrm{ch}\,\beta(\mathbf{x}) \longrightarrow \eta k \,\mathrm{ch}(\zeta(\mathbf{x})\beta(\mathbf{x}))$, $c(\mathbf{x}) = \eta k \,\mathrm{sh}\,\beta(\mathbf{x}) \longrightarrow \eta k \,\mathrm{sh}(\zeta(\mathbf{x})\beta(\mathbf{x}))$. Such an additional algebraic structure does not exist for the groups of symmetries of the soliton equations [3-5].

Remark 3. Subgroups of the group G_m. The group G_m (10) has eight periodic elements which form the subgroup $\Gamma_p = Z_2 \oplus Z_2 \oplus Z_2 \subset G_m$ and are defined by the sextuples $(0, 0, 1, \pm 1, \pm 1, \pm 1)$. All other elements are non-periodic. The subgroup Γ_p represents all eight components G_{mj}, $j = 1, \cdots, 8$, of the group G_m. The scale and reflection symmetries form the subgroup $\Gamma_{sr} = R \oplus R^+ \oplus Z_2 \oplus Z_2 \oplus Z_2 \subset G_m$ that consists of the sextuples $(r, 0, k, \tau, \sigma, \eta)$. The subgroup Γ_{sr} does not depend on the considered MHD equilibria and is the intersection of all infinite-dimensional groups G_m.

Acknowledgments. The work was supported by an Alexander von Humboldt Research Award of Germany and a Killam Research Fellowship of Canada.

References

[1] Bogoyavlenskij O I 2000 Phys. Rev. E 62 8616-8627
[2] Bogoyavlenskij O I 2000 Phys. Rev. Lett. 84 1914-1917
[3] Gardner G S, Greene J M, Kruskal M D and Miura R M 1967 Phys. Rev. Lett. 19 1095
[4] Kadomtsev B B and Petviashvili V P 1970 Sov. Phys. Dokl. 15 539-541
[5] Ablowitz M J, Kaup D J, Newell A C and Segur H 1973 Phys. Rev. Lett. 31 125-127
[6] Moffatt H K 1989 On the existence, structure and stability of MHD equilibrium states (Turbulence and nonlinear dynamics in MHD flows, Amsterdam: North-Holland) p 185
[7] Chandrasekhar S 1956 Proc. Natl. Acad. Sci. USA 42 273-276

Inst. Phys. Conf. Ser. No 173: Section 9
Paper presented at 24th Int. Coll. Group Theoretical Methods in Physics, Paris, France, 15–20 July 2002
©2003 IOP Publishing Ltd

Lie Symmetries of Nonlinear Reaction-Diffusion Systems with Variable Diffusivities

Roman Cherniha† and John R King‡

† Institute of Mathematics, Ukrainian National Academy of Sciences, Tereshchenkivs'ka Str. 3, Kyiv 01601, Ukraine
‡ Division of Theoretical Mechanics, Nottingham University University Park, Nottingham NG7 2RD, UK

E-mail: cherniha@imath.kiev.ua, john.king@nottingham.ac.uk

Abstract. The work is a natural continuation of the paper published in *J.Phys.A:Math.Gen.,* **33** *(2000), 267-282, 7839-41*, whereby a classical symmetry analysis is performed of a general system of two semilinear reaction-diffusion equations having variable diffusivities.

1. Introduction

In the papers [1] and [2], complete description of Lie symmetries is obtained for multidimensional semilinear systems of two reaction-diffusion (RD) equations of the form

$$\lambda_1 U_t = \Delta U + F(U, V),$$
$$\lambda_2 V_t = \Delta V + G(U, V), \tag{1}$$

where λ_1 and $\lambda_2 \in \mathbf{R}$, F and G are arbitrary smooth functions, $U = U(t, x)$, $V = V(t, x)$ are unknown functions of $n + 1$ variables t, $x = (x_1, \ldots, x_n)$, Δ is the Laplacian, and the t subscript to the functions U and V denotes differentiation with respect to this variable. Here we consider RD systems for the case $n = 1$ but with the variable diffusivities $D_1(U)$ and $D_2(V)$, i.e. reaction-diffusion systems of the form

$$U_t = (D_1(U)U_x)_x + F(U, V),$$
$$V_t = (D_2(V)V_x)_x + G(U, V), \tag{2}$$

where the x subscript to the functions U and V denotes differentiation with respect to this variable. It can be seen that RD system (2) for $D_k = \frac{1}{\lambda_k}$, $k = 1, 2$ takes the form (1) for $n = 1$ therefore we assume the vector $(D_1, D_2) \neq const$. Section 2 is devoted to search of all possible Lie symmetries which the system (2) can admit. The main results of this section are presented in the form of theorem 1. In section 3, discussion of the results obtained in this paper and other recently published papers devoted to search of Lie symmetries of RD systems is presented.

2. The main results

It is easily checked that RD system (2) is invariant under the operators

$$P_t = \frac{\partial}{\partial t}, \ P_x = \frac{\partial}{\partial x} \tag{3}$$

for arbitrary functions D_1, D_2, F and G. Following [3], this algebra is called *the trivial Lie algebra* of the system (2). Thus, we aim to find all pairs of functions (D_1, D_2) and (F, G)

that lead to extensions of the trivial Lie algebra (3) for system (2). It should be stressed that the most interesting case is that when at least one of the functions F and G is non-constant. Otherwise RD system is non-coupled, i.e. two independent equations with constant reaction terms are obtained therefore this case is omitted below. Now let us formulate a theorem which gives complete information on the classical symmetry of the system (2).

Theorem 1 *All possible maximal algebras of invariance (MAI) of the system (2) for any fixed pairs of functions $(D_1, D_2) \neq const$ and $(F, G) \neq const$ are presented in table 1. Any other system of the form (2) with non-trivial Lie symmetry is reduced by a local substitution of the form*

$$
\begin{aligned}
x &\to c_{00}x, \quad t \to c_{01}t + c_{02}\exp(c_{03}t), \\
U &\to c_1\exp(c_3t)U + c_5t + c_{10}, \\
V &\to c_2\exp(c_4t)V + c_6t + c_{20},
\end{aligned}
\tag{4}
$$

either to one of those given in table 1 or to a system of the form (2) for $(F, G) = const$ (the constants c with subscripts are determined by the form of the system in question).

Remark 1. In table 1, $d_1 \neq 0, d_2 \neq 0, \alpha_1 \neq 0, \alpha_2 \neq 0, \beta, \gamma, \lambda, \lambda_0 > 0$ are arbitrary constant; D_2, f and g are arbitrary smooth functions of the relevant arguments, while $P_0(t, x)$ and $P_\lambda(t, x)$ are arbitrary solutions of the linear equations $P_t = d_1 P_{xx}$ and $P_t = P_{xx} + \lambda P$, respectively. The proof of theorem 1 is based on the classical Lie scheme (see, e.g., [4], [5]) however it is very non-trivial and cumbersome because RD system (2) contains four arbitrary functions. We aim to present one in a forthcoming paper devoted to investigation of (2) in the multidimensional case ($n > 1$) together with a discussion of possible physical applications.

3. Discussion

It is worth commenting on systems and Lie algebras listed in table 1. It was established in [1] and [2] that there are a wide range of systems with non-trivial Lie symmetries among RD systems with constant diffusivities of the form (1). Moreover those have no analogues in the scalar case. In contrast to case of constant diffusivities, there are only 13 non-equivalent RD systems with variable diffusivities

and non-constant reaction term(s) that are invariant under non-trivial Lie algebras. Moreover the systems listed in the cases 6, 8–13 (see table 1) have direct analogues among single nonlinear reaction-diffusion equations (see [6], [3]). However, in contrast to the scalar case, all systems listed in table 1 (except for cases 11 and 13) contain at least two arbitrary functions. The most interesting of them are systems 9, 10 and 12 because the relevant MAI contain the classical three-dimensional algebras $sl(2, \mathbf{R})$ and $so(3)$ as subalgebras. Indeed, one can check that the operators P_x, D_3, K satisfy the commutation relations of the algebra $sl(2, \mathbf{R})$ while the triplets of operators $\sqrt{\frac{3}{4\lambda_0}}P_x, X_1, X_2$ and $\sqrt{\frac{3}{4\lambda_0}}P_x, X_3 + X_4, X_3 - X_4$ do the same for algebra $so(3)$.

Finally, it should be stressed that a complete description of Lie symmetries of a *nonlinear PDE system* containing *arbitrary functions of two or more dependent variables* is a quite difficult task. The pioneering works were published only a few years ago. To our knowledge the first one is the paper [7], where the authors have solved this problem in the case of a nonlinear system containing a two-dimensional PDE and an ODE, while the paper [1] is the first involving nonlinear systems of multidimensional PDEs. Note that Lie symmetries of RD systems with so called cross-diffusion terms were investigated in [8].

Table 1. Lie symmetries of system (2).

The form of system	Basic operators of MAI
1. $U_t = d_1 U_{xx} + U^{1+\beta} f(V),\ \beta \neq 0$ $V_t = (D_2(V)V_x)_x + U^\beta g(V)$	$P_t,\ P_x$ $D_1 = D_{00} - \frac{2}{\beta} U \partial_U$
2. $U_t = d_1 U_{xx} + \exp(\gamma U) f(V),\ \gamma \neq 0$ $V_t = (D_2(V)V_x)_x + \exp(\gamma U)g(V)$	$P_t,\ P_x$ $D_1 = D_{00} - \frac{2}{\gamma}\partial_U$
3. $U_t = d_1 U_{xx}$ $V_t = (D_2(V)V_x)_x + g(V)$	$P_t,\ P_x,\ U\partial_U$ $X_0^\infty = P_0(t,x)\partial_U$
4. $U_t = d_1 U_{xx} + \lambda U + f(V)$ $V_t = (D_2(V)V_x)_x + g(V)$	$P_t,\ P_x$ $X_\lambda^\infty = P_\lambda(t,x)\partial_U$
5. $U_t = d_1 U_{xx} + \gamma U \log U + f(V)$ $V_t = (D_2(V)V_x)_x + g(V)$	$P_t,\ P_x$ $Q_\gamma = \exp(\gamma t) U\partial_U$
6. $U_t = d_1 (U^{\alpha_1} U_x)_x + U^{\gamma_1} f(\omega)$ $V_t = d_2 (V^{\alpha_2} V_x)_x + V^{\gamma_2} g(\omega)$ $\omega = U^{-\alpha_1} V^{\alpha_2},\ \gamma_k = 1 + \alpha_k - \gamma\alpha_k,\ k = 1,2$	$P_t,\ P_x$ $D_1^\gamma = 2(\gamma - 1)t\partial_t + \gamma x \partial_x +$ $\frac{2}{\alpha_1} U\partial_U + \frac{2}{\alpha_2} V\partial_V$
7. $U_t = d_1 (U^{\alpha_1} U_x)_x + U^\gamma f(\omega)$ $V_t = d_2 (\exp(\alpha_2 V) V_x)_x + U^{\gamma - 1} g(\omega)$ $\omega = U^{-\alpha_1} \exp(\alpha_2 V),\ \gamma = 1 + \alpha_1 - \gamma\alpha_1$	$P_t,\ P_x$ $D_2^\gamma = 2(\gamma - 1)t\partial_t + \gamma x \partial_x +$ $\frac{2}{\alpha_1} U\partial_U + \frac{2}{\alpha_2}\partial_V$
8. $U_t = d_1 (\exp(\alpha_1 U) U_x)_x + \exp(\alpha_1(1 - \gamma)U)g(\omega)$ $V_t = d_2 (\exp(\alpha_2 V) V_x)_x + \exp(\alpha_2(1 - \gamma)V)g(\omega)$ $\omega = \alpha_1 U - \alpha_2 V$	$P_t,\ P_x$ $D_3^\gamma = 2(\gamma - 1)t\partial_t + \gamma x \partial_x +$ $\frac{2}{\alpha_1}\partial_U + \frac{2}{\alpha_2}\partial_V$
9. $U_t = d_1 (U^{-4/3} U_x)_x + U f(U/V)$ $V_t = d_2 (V^{-4/3} V_x)_x + V g(U/V)$	$P_t,\ P_x$ $D_3 = 2x\partial_x - 3(U\partial_U + V\partial_V)$ $K = x^2 \partial_x - 3x(U\partial_U + V\partial_V)$
10. $U_t = d_1 (U^{-4/3} U_x)_x + U f(U/V) - \lambda_0 d_1 U^{-1/3}$ $V_t = d_2 (V^{-4/3} V_x)_x + V g(U/V) - \lambda_0 d_2 V^{-1/3}$	$P_t,\ P_x,\ X_1 = \sin\left(\sqrt{\frac{4\lambda_0}{3}}x\right)\partial_x$ $-\sqrt{3\lambda_0}\cos\left(\sqrt{\frac{4\lambda_0}{3}}x\right)(U\partial_U + V\partial_V)$ $X_2 = \cos\left(\sqrt{\frac{4\lambda_0}{3}}x\right)\partial_x$ $+\sqrt{3\lambda_0}\sin\left(\sqrt{\frac{4\lambda_0}{3}}x\right)(U\partial_U + V\partial_V)$
11. $U_t = d_1 (U^{-4/3} U_x)_x - \lambda_0 d_1 U^{-1/3}$ $V_t = d_2 (V^{-4/3} V_x)_x - \lambda_0 d_2 V^{-1/3}$	$P_t,\ P_x,\ X_1,\ X_2$ $D_4 = 4t\partial_t + 3(U\partial_U + V\partial_V)$
12. $U_t = d_1 (U^{-4/3} U_x)_x + U f(U/V) + \lambda_0 d_1 U^{-1/3}$ $V_t = d_2 (V^{-4/3} V_x)_x + V g(U/V) + \lambda_0 d_2 V^{-1/3}$	$P_t,\ P_x,\ X_3 = \exp\left(\sqrt{\frac{4\lambda_0}{3}}x\right)(\partial_x$ $-\sqrt{3\lambda_0}(U\partial_U + V\partial_V))$ $X_4 = \exp\left(\sqrt{\frac{4\lambda_0}{3}}x\right)(\partial_x$ $+\sqrt{3\lambda_0}(U\partial_U + V\partial_V))$
13. $U_t = d_1 (U^{-4/3} U_x)_x + \lambda_0 d_1 U^{-1/3}$ $V_t = d_2 (V^{-4/3} V_x)_x + \lambda_0 d_2 V^{-1/3}$	$P_t,\ P_x$ X_3, X_4, D_4

720

Acknowledgments

The authors gratefully acknowledge the financial support of the Royal Society of London. The first author thanks Embassy of France in Ukraine for the financial support enabled him to participate at the GROUP-24 Colloquium, Paris, July 15-20, 2002.

[1] Cherniha R M and King J R 2000. *J. Phys. A: Math. Gen.* **33**, 267-282, 7839-41.
[2] Cherniha R and King J R 2003. *J.Phys.A: Math.Gen.* **36**, 405-425.
[3] Cherniha R M and Serov M I 1998. *Euro. J. Appl. Math.* **9**, 527–542.
[4] Ovsiannikov L V 1978. *The Group Analysis of Differential Equations.* Nauka, Moscow.
[5] Olver P 1986. *Applications of Lie Groups to Differential Equations.* Springer, Berlin.
[6] Dorodnitsyn V A 1982. *USSR Comput. Math. and Math. Phys.* **22**, 115–122.
[7] Torrisi M, Tracina R and Valenti A 1996. *J. Math. Phys.* **37**, 4758-4767.
[8] Nikitin A G and Wiltshire R J 2001. *J. Math. Phys.* **42**, 1666-1688.

Inst. Phys. Conf. Ser. No 173: Section 9
Paper presented at 24th Int. Coll. Group Theoretical Methods in Physics, Paris, France, 15–20 July 2002
©*2003 IOP Publishing Ltd*

On Geometric Aspects of Weierstrass Representations Associated with CP^N Harmonic Maps

A. M. Grundland$^{+\dagger}$ and W. J. Zakrzewski^{++}

$^+$ Centre de Recherches Mathématiques, Université de Montréal, C. P. 6128, Succ.
Centre-ville, Montréal, (QC) H3C 3J7, Canada
$^{++}$ Department of Mathematical Sciences,University of Durham,
Durham DH1 3LE, UK

Abstract. We introduce a Weierstrass-type system of equations corresponding to CP^N fields which generalise the systems, previously constructed, for CP^1 and CP^2. We use a set of conserved quantities for the CP^N model to suggest a possible geometrical interpretation of such maps.

It is easiest to define the CP^N σ models in terms of the Lagrangian density[1]

$$L = \tfrac{1}{4}(D_\mu z)^\dagger \cdot D_\mu z, \quad z^\dagger \cdot z = 1, \tag{1}$$

where z is a vector field of N components and

$$D_\mu = \partial_\mu - z\,(z^\dagger \cdot \partial_\mu z). \tag{2}$$

Here $\mu = 1, 2$, denotes x and y. Defining $z = \frac{f}{|f|}$ it is easy to check that the Euler-Lagrange equations for the vector f are

$$\left(1 - \frac{ff^\dagger}{|f|^2}\right)\left[\partial\bar{\partial}f - \partial f\frac{(f^\dagger \cdot \bar{\partial}f)}{|f|^2} - \bar{\partial}f\frac{(f^\dagger \cdot \partial f)}{|f|^2}\right] = 0, \tag{3}$$

where the derivatives are abbreviated $\partial = \partial/\partial\zeta$, $\bar{\partial} = \partial/\partial\bar{\zeta}$, $\zeta = x + iy$. As is well known [4] equations (3) can be written as a compatibility condition for a set of two linear spectral equations for a N component auxiliary vector Ψ

$$\partial\Psi = \frac{2}{1+\lambda}[\partial P, P]\,\Psi \qquad \bar{\partial}\Psi = \frac{2}{1-\lambda}[\bar{\partial}P, P]\,\Psi, \tag{4}$$

where the projector matrix is given by

$$P = \frac{1}{A}f f^\dagger, \qquad A = f^\dagger \cdot f \tag{5}$$

The compatibility conditions for (4) can be written in the form of a conservation law

$$\partial K + \bar{\partial}M = 0, \tag{6}$$

where the matrices K and L are given by

$$K_{ij} = \bar{f}_j\bar{\Phi}_i^2 - f_i\bar{\varphi}_j^2 \qquad M_{ij} = \bar{f}_j\varphi_i^2 - f_i\Phi_j^2, \tag{7}$$

where we have introduced the following notation

$$F_{ij} = f_i\,\partial f_j - f_j\partial f_i, \qquad G_{ij} = f_i\,\bar{\partial}f_j - f_j\bar{\partial}f_i. \tag{8}$$

† To whom correspondence should be addressed (grundlan@crm.umontreal.ca)

$$\varphi_i^2 = \frac{1}{A^2} \bar{f}_k F_{ki}, \qquad \Phi_i^2 = \frac{1}{A^2} f_k \bar{G}_{ki}. \tag{9}$$

Note that we have two constraints; namely,

$$\bar{f}_k \varphi_k^2 = 0, \qquad f_k \Phi_k^2 = 0 \tag{10}$$

which tell us that we have only $N - 1$ independent φ_i's ie in our further discussion we can take as independent $\varphi_2, ..., \varphi_N$ (and similarily for Φ_i). At the same time, using homogeneity of eq. (3), we can set, say, $f_1 = 1$ and so we end up with

$$\varphi_i^2 = \frac{1}{A^2} \left[(1 + f_k \bar{f}_k) \partial f_i - f_i (\bar{f}_k \partial f_k) \right], \tag{11}$$

where $A = 1 + |f_2|^2 + |f_3|^2 ... + |f_N|^2$ and all the sums over repeated indices run over $k = 2, ..., N$. This allows us to invert (11) and so express all ∂f_i in terms of φ_l's.

$$\partial f_i = A \left[\varphi_i^2 + f_i \bar{f}_k \varphi_k^2 \right]. \tag{12}$$

To introduce a generalised Weierstrass system we need a set of φ_i and ψ_i which generalise the φ and ψ of the CP^1 case and φ_i, ψ_i, $i = 1, 2$ of the CP^2 case [3]. We introduce new complex variables

$$\psi_i = f_i \bar{\varphi}_i. \qquad \text{no} \quad \text{summation} \tag{13}$$

Then to complete the generalisation of the Weierstrass system we need to prescribe $\bar{\partial}\varphi_i$ and $\partial\psi_i$. Note that as (no summation)

$$\partial\psi_i = \partial(f_i \bar{\varphi}_i) = \partial f_i \bar{\varphi}_i + f_i \left(\bar{\partial}\varphi_i \right)^\star \tag{14}$$

where $\star$, like $^-$, also denotes complex conjugation. So we need only to specify $\bar{\partial}\varphi_i$. To do this note that as

$$\varphi_i^2 = \frac{1}{A} \partial f_i - f_i \frac{\bar{f} \cdot \partial f}{A^2} \tag{15}$$

we have

$$\bar{\partial}\varphi_i^2 = -2\frac{\varphi_i^2}{A^3} \left(\bar{f}_k \bar{\partial} f_k + f_k \bar{\partial}\bar{f}_k \right)$$
$$+ \frac{1}{A^2} \left[(1 + |f|^2) \partial\bar{\partial} f_i + (\bar{f}_k \bar{\partial} f_k) \partial f_i - \bar{\partial} f_i (\bar{f}_k \partial f_k) - f_i (\bar{f}_k \partial\bar{\partial} f_k) \right] \tag{16}$$

However, the 2nd derivatives $\partial\bar{\partial} f_i$ can be eliminated by using the Euler-Lagrange equations (3) and we end up with

$$\bar{\partial}\varphi_i = -\frac{\varphi_i}{2A} (f_k \bar{\partial}\bar{f}_k). \tag{17}$$

Making use of equations (12) and (14) we obtain the modified Weierstrass system of equations for φ_i, ψ_i, where $i = 2, 3...N$ given by (no summation)

$$\bar{\partial}\varphi_i = -\varphi_i A (\psi \cdot \bar{\varphi}), \qquad \partial\psi_i = \varphi_i A |\varphi_i|^2, \qquad A = 1 + \sum_{k=1}^{N} \frac{|\psi_k|^2}{|\varphi_k|^2}. \tag{18}$$

From our construction it is clear that this system of equations is equivalent to the equations of the CP^N model (3). Next we define a set of real variables X_i, constructed out of our ψ_i's and φ_i's treating $X_i(\zeta, \bar{\zeta})$ as a map of R^2 into R^M, for some M so that we can discuss the geometry of these surfaces. To construct such coordinates it is convenient to exploit the conservation laws (6) for our system of equations. We note that we can drop the Φ terms in (7) and we still have our conservation laws; namely, we define

$$K'_{ij} = -f_i \bar{\varphi}_j^2, \qquad M'_{ij} = \varphi_i^2 \bar{f}_j \tag{19}$$

and then note that we still have conservation laws

$$\partial K' + \bar{\partial} M' = 0. \tag{20}$$

Note that as our conservation laws do not involve Φ_i then they can be written entirely in terms of Weierstrass variables φ_i and ψ_i. Next we consider ($l = 1, ..., N$ – no summation)

$$X_{ll} = \int_\gamma \bar{f}_l \varphi_l^2 \, d\zeta + \int_\gamma f_l \, \bar{\varphi}_l^2 \, d\bar{\zeta} = \int_\gamma \bar{\psi}_l \varphi_l \, d\zeta + \int_\gamma \psi_l \, \bar{\varphi}_l \, d\bar{\zeta}. \tag{21}$$

These quantities have been constructed from the diagonal entries of matrices M' and K'. Looking at the diagonal terms we note that $\sum_l X_{ll} = 0$; this follows from the tracelessness of matrices K' and M'. From the off-diagonal entries we construct

$$X_{lk} + iY_{lk} = \int_\gamma (a\bar{f}_l \varphi_k^2 + \bar{a}\bar{f}_k \varphi_l^2) d\zeta + \int_\gamma (\bar{a} f_l \, \bar{\varphi}_k^2 + a f_k \, \bar{\varphi}_l^2) d\bar{\zeta} \tag{22}$$

where $a = (1+i)/2$. In our expression we take all $l, k = 1, ..., N$ - and for $k = 1$ or $l = 1$ we can use our constraints (10) to rewrite all our expressions in terms of independent φ_i and ψ_i, $i = 2, ..., N$. Note that the conservation laws (20) guarantee that X_i do not depend on the choice of the contour γ (but only its endpoints). Next, following [2, 5] we introduce the metric

$$g_{\alpha,\beta} = \sum_{lk} \frac{\partial X_{lk}}{\partial \alpha} \frac{\partial X_{lk}}{\partial \beta}, \tag{23}$$

where α and β are ζ or $\bar{\zeta}$. We find that

$$g_{\zeta\zeta} = \left(\sum_i \bar{f}_i \varphi_i^2 \right)^2 = 0 \tag{24}$$

as this expression is exactly our constraint (10). Similarly $g_{\bar{\zeta}\bar{\zeta}} = 0$. The only nonzero term of the metric is

$$g_{\zeta\bar{\zeta}} = \left(1 + |f_2|^2 + ... + |f_N|^2 \right) \left[|\sum_k \bar{f}_k \varphi_k^2|^2 + |\varphi_2|^4 + ... + |\varphi_N|^4 \right]. \tag{25}$$

Of course, we can rewrite this expression to involve generalised Weierstrass variables φ_i and ψ_i by using (13). Note, however, that expressing all quantities in terms of f_i and ∂f_i, through (11), our expressions simplify further and we obtain

$$g_{\zeta\bar{\zeta}} = |Dz|^2, \qquad D = (D_1 - iD_2)/2 \tag{26}$$

where D_μ are covariant derivatives given by (2). Our work here generalises the approach of [6] and we intend to use it to look at other aspects of their work when generalised to CP^N.

1. Acknowledgements

Partial support for AMG's work was provided by a grant from NSERC of Canada and the Fonds FCAR du Quebec.

[1] see *e.g.* W.J. Zakrzewski, Low Dimensional Sigma Models, Adam Hilger, Bristol (1989)
[2] see *e.g.* B. Konopelchenko and G. Landolfi, Stud. Appl. Maths. **104**, 129–169 (1999);

[3] A.M. Grundland and W.J. Zakrzewski – The Weierstrass representation for surfaces immersed into R^8 and CP^2 maps, J. Math. Phys. 6, 43, 3352–3361 (2002).

[4] V.E. Zakharov and A.V. Mikhailov – Zh. Eksp. Theor. Fiz. **74**, 1953 (1978); (Sov. Phys. JETP **47**, 1017 (1979))

[5] P. Bracken, A.M. Grundland and L. Martina, J. Math. Phys. **40**, 3379–3403 (1999) and reference therein.

[6] B. Konopelchenko, Stud. Appl. Math. **96**, 9–51 (1996); B. Konopelchenko and G. Landolfi, J. Geo. and Phys. **29**, 319–333 (1999)

Inst. Phys. Conf. Ser. No 173: Section 9
Paper presented at 24th Int. Coll. Group Theoretical Methods in Physics, Paris, France, 15–20 July 2002
©2003 IOP Publishing Ltd

Integrability and exact solution of an electronic model with long range interactions

K.E. Hibberd

Departamento de Física Teórica, Universidad de Zaragoza, 50009, Zaragoza, Spain,
Email: keh@posta.unizar.es

J.R. Links

Centre for Mathematical Physics, The University of Queensland, 4072, Australia,
Email: jrl@maths.uq.edu.au

Abstract. We present an electronic model with long range interactions. Through the quantum inverse scattering method, integrability of the model is established using a one-parameter family of typical irreducible representations of $gl(2|1)$. The eigenvalues of the conserved operators are derived in terms of the Bethe ansatz, from which the energy eigenvalues of the Hamiltonian are obtained.

1. Introduction

The Quantum Inverse Scattering Method (QISM) [1] is one of the most powerful tools in the exact study of quantum systems. It can be applied in a number of contexts, including both one-dimensional systems with nearest neighbour interactions such as the Heisenberg [1] and Hubbard [2] models, and also for the analysis of models with long range interactions such as the Gaudin Hamiltonians [3] and extensions [4]. These latter constructions in particular have received renewed attention as it has been realised that the reduced BCS model, which was recently proposed to describe superconducting correlations in metallic grains of nanoscale dimensions [5], can be shown to be integrable through the use of Gaudin Hamiltonians in non-uniform external fields [6]. Formulating the reduced BCS model in the framework of the QISM reproduces the exact solution originally obtained by Richardson and Sherman [7], and opens the way for the calculation of form factors and correlation functions [8].

Motivated by this result, one can investigate to what extent the construction can be generalised to yield new classes of models, with some examples already given in [9]. Here, we will consider a case where an underlying *superalgebraic* structure (i.e., one with both bosonic and fermionic degrees of freedom) is employed to yield an electronic model. The supersymmetric formulation of integrable systems can be traced back to the work of Kulish [10], and recently supersymmetric Gaudin Hamiltonians have been analysed in detail in [11].

In this article we present a Hamiltonian derived through the QISM from a solution of the Yang-Baxter equation (YBE) associated with a typical irreducible representation of the Lie superalgebra $gl(2|1)$, which has the explicit form

$$H = \sum_{j}^{D} \epsilon_j n_j - g \sum_{j,k}^{D} \sum_{\sigma=\pm} Q_{j\sigma}^{\dagger} Q_{k\sigma}. \tag{1}$$

Above, the energy levels ϵ_j are two-fold degenerate, g is an arbitrary coupling parameter and D is the total number of distinct energy levels. Also, n_j is the fermion number operator for energy level ϵ_j and for the parameters α_j, $j = 1...D$ we define

$$Q_{j\sigma} = c_{j\sigma}\sqrt{\alpha_j + 1}.X_j^{n_{j,-\sigma}},$$

with $X_j = \sqrt{\alpha_j/(\alpha_j + 1)}$. The $c_\sigma, c_\sigma^\dagger$, $\sigma = \pm$, are two-fold degenerate Fermi annihilation and creation operators.

The Hamiltonian has a similar form to the reduced BCS model [5]. There, Cooper pairs are scattered into vacant energy levels while the one particle states are blocked from scattering. In the Hamiltonian above, there is correlated scattering depending on the occupation numbers. One of the features of this model is that the scattering couplings can be varied through the choice of the parameters α_i. Via the algebraic Bethe ansatz method and using the minimal typical representation of $gl(2|1)$, from which these free parameters arise, we establish the exact solvability of the model. Here we outline the necessary definitions and constructions, while full details will be presented elsewhere.

The Lie superalgebra $gl(2|1)$ has generators E_j^i, $i, j = 1, 2, 3$ with supercommutator relations

$$[E_j^i, E_l^k] = \delta_j^k E_l^i - (-1)^{([i]+[j])([k]+[l])}\delta_l^i E_j^k.$$

Above, the BBF grading $[1] = [2] = 0$, $[3] = 1$ is chosen and the elements are realised in terms of the Fermi operators through (cf. [12]) $E_2^1 = S^+ = c_+^\dagger c_-$, $E_1^2 = S^- = c_-^\dagger c_+$, $E_1^1 = \alpha$ $1 \mid n_+$, $E_2^2 = -\alpha - 1 + n_-$, $E_3^3 - 2\alpha + 2 - n$, $E_3^1 = Q_+^\dagger$, $E_3^2 = Q_-^\dagger$, $E_1^3 = Q_+$ and $E_2^3 = Q_-$, and we set $S^z = (n_+ - n_-)/2$. The Casimir invariant of the algebra, $C = \sum_{i,j=1}^3 E_j^i \otimes E_i^j(-1)^{[j]}$, which commutes with all the elements of $gl(2|1)$, will also be needed, and has the eigenvalue $\xi_C = -2\alpha(\alpha + 1)$ in the above representation. Below we let $V(\alpha)$ denote the four-dimensional module on which the representation acts, with the basis $|+-\rangle, |+\rangle, |-\rangle, |0\rangle$.

2. The Yang-Baxter equation and integrability

To construct the model, we use the supersymmetric formulation of the QISM [10]. We take the following solution of the YBE which acts on $W \otimes W \otimes V(\alpha)$, where W denotes the three-dimensional vector module of $gl(2|1)$,

$$R_{12}(u - v)L_{13}(u)L_{23}(v) = L_{23}(v)L_{13}(u)R_{12}(u - v) \tag{2}$$

with

$$R(u) = I \otimes I + \frac{\eta}{u}\sum_{m,n=1}^3 (-1)^{[n]}e_n^m \otimes e_m^n, \tag{3}$$

and the L-operator is given by

$$L(u) = I \otimes I + \frac{\eta}{u}\sum_{m,n=1}^3 (-1)^{[n]}e_n^m \otimes E_m^n. \tag{4}$$

The representations taken for the operators E_m^n are as stated above, the variable u represents the rapidity η is arbitrary and I is the identity operator.

By the usual procedure of the QISM, we define a transfer matrix acting on the D-fold tensor product space (for distinct α_i) $V(\alpha_1) \otimes V(\alpha_2) \otimes ... \otimes V(\alpha_D)$ via

$$t(u) = str_0 \left(G_0 L_{0D}(u - \epsilon_D)...L_{01}(u - \epsilon_1)\right),$$

which gives a mutually commuting family satisfying $[t(u), t(v)] = 0$. Above, str_0 denotes the supertrace taken over the auxiliary space labelled by 0 and G can be any matrix which satisfies $[R(u), G \otimes G] = 0$.

For the BBF grading we choose $G = \text{diag}(\exp(\beta\eta), \exp(\beta\eta), 1)$ and by employing the algebraic Bethe ansatz method the eigenvalues of the transfer matrix are found to be (cf. [10])

$$\Lambda(u) = \exp(\beta\eta) \prod_i^D \left(1 - \frac{\eta\alpha_i}{(u - \epsilon_i)}\right) \prod_j^P a(v_j - u)$$

$$+ \exp(\beta\eta) \prod_i^D \left(1 - \frac{\eta\alpha_i}{(u - \epsilon_i)}\right) \prod_j^P a(u - v_j) \prod_k^M a(\gamma_k - u)$$

$$- \prod_i^D \left(1 - \frac{2\eta\alpha_i}{(u - \epsilon_i)}\right) \prod_j^M a(\gamma_j - u), \tag{5}$$

where $a(u) = 1 + \eta/u$. The parameters v_i, w_j satisfy the Bethe ansatz equations

$$\prod_k^M a(\gamma_k - v_j) = -\prod_i^P \frac{a(v_i - v_j)}{a(v_j - v_i)}, \quad \prod_i^D \frac{\gamma_l - \epsilon_i - 2\eta\alpha_i}{\gamma_l - \epsilon_i - \eta\alpha_i} = \exp(\beta\eta) \prod_j^P a(\gamma_l - v_j).$$

We now introduce the operators

$$T_j = \lim_{u \to \epsilon_j} \frac{(u - \epsilon_j)}{\eta^2} t(u), \quad \text{which satisfy} \quad [T_j, T_k] = 0. \tag{6}$$

By taking the *quasi-classical* expansion $T_j = \tau_j + o(\eta)$, this leads to

$$\tau_j = -\beta\psi_j + \sum_{i \neq j}^D \frac{\theta_{ji}}{\epsilon_j - \epsilon_i}$$

where $\theta = \sum_{m,n}^3 E_m^n \otimes E_n^m (-1)^{[m]}$ and $\psi = E_3^3$. It is easily deduced that these operators satisfy $[\tau_j, \tau_k] = 0$.

Writing $K = \sum_{i,j}^D (S_i^+ S_j^- + S_i^- S_j^+ + 2S_i^z S_j^z)$, which satisfies $[K, \tau_j] = 0, \forall j$, we define the Hamiltonian as follows;

$$H = \frac{1}{2\beta^2} \sum_j^D (1 + 2\beta\epsilon_j)\tau_j + \frac{1}{4\beta^3} \sum_{j,k}^D \tau_j\tau_k + \frac{1}{2\beta} \sum_j^D C_j - \frac{K}{2\beta} + 2\sum_j^D \epsilon_j(\alpha_j + 1)$$

$$= \frac{1}{2\beta} \sum_j^D \sum_{k \neq j}^D \theta_{jk} - \frac{1}{2\beta} \sum_j^D (1 + 2\beta\epsilon_j)\psi_j + \frac{1}{4\beta} \sum_{i,j}^D \psi_i\psi_j - \frac{1}{\beta} \sum_j^D \alpha_j(\alpha_j + 1)$$

$$- \frac{K}{2\beta} + 2\sum_j^D \epsilon_j(\alpha_j + 1).$$

The term involving θ_{jk} may be simplified using the Casimir invariant and the commutation relations of the algebra $gl(2|1)$

$$\sum_j^D \sum_{k \neq j}^D \theta_{jk} = \sum_{k,j}^D \theta_{jk} - \sum_j^D C_j,$$

$$= K - \frac{1}{2} \sum_{j,k}^D \psi_j\psi_k + \sum_j^D \psi_j - 2\sum_{j,k}^D \sum_{\sigma=\pm} Q_{j\sigma}^+ Q_{k\sigma} + 2\sum_j^D \alpha_j(\alpha_j + 1).$$

728

For $g = 1/\beta$ we obtain the Hamiltonian (1), which establishes integrability since $[H, \tau_j] = 0, \forall j$.

From (5,6) we obtain the eigenvalues of τ_j for the BBF grading,

$$\lambda_j = -2\beta\alpha_j + \alpha_j \sum_i^M \frac{1}{\gamma_i - \epsilon_j} - 2\sum_{i \neq j}^D \frac{\alpha_j \alpha_i}{\epsilon_j - \epsilon_i}, \tag{7}$$

as the quasi-classical limit of the eigenvalues of the transfer matrix. The corresponding Bethe ansatz equations are

$$\beta + \sum_j^P \frac{1}{\gamma_l - v_j} = \sum_i^D \frac{\alpha_i}{\epsilon_i - \gamma_l}, \quad \sum_l^M \frac{1}{\gamma_l - v_j} = 2\sum_{i \neq j}^P \frac{1}{v_i - v_j}. \tag{8}$$

For a given solution of the Bethe ansatz equations we find that the number of electrons, $N = 2D - M, n_+ - n_- = M - 2P$ and the eigenvalue of K reads

$$\xi_K = \frac{1}{2}(M - 2P)(M - 2P + 2). \tag{9}$$

The energy eigenvalues can be computed using (7,8,9) and are given by

$$E = 2\sum_i^D \epsilon_j - \sum_l^M \gamma_l - 2g\sum_j^D \alpha_j - gM.$$

Similar results have been obtained for the FBB and BFB gradings, which will appear elsewhere.

Acknowledgements

We thank Petr Kulish for useful discussions. Jon Links acknowledges the Australian Research Council for financial support and Katrina Hibberd is supported by project number BFM2000-1057 from the Ministerio de Ciencia y Tecnologia, Spain.

References

[1] L.D. Faddeev, Int. J. Mod. Phys. **A10**, (1995) 1845.
[2] B.S. Shastry, Phys. Rev. Lett. **56**, (1986) 1529, 2453.
[3] M. Gaudin, J. Phys. (Paris) **37**, (1976) 1087.
[4] K. Hikami, P.P. Kulish and M. Wadati, J. Phys. Soc. Jpn. **61**, (1992) 3071.
[5] J. von Delft and D. C. Ralph, Phys. Rep. **345**, (2001) 61;
    J. von Delft, Ann. Phys. (Leipzig) **10**, (2001) 219.
[6] M.C. Cambiaggio, A.M.F. Rivas and M. Saraceno, Nucl. Phys. **A624**, (1997) 157.
[7] R.W. Richardson, Phys. Lett. **3**, (1963) 277; **5**, (1963) 82;
    R.W. Richardson and N. Sherman, Nucl. Phys. **52**, (1964) 221, 253.
[8] H.-Q. Zhou, J. Links, R.H. McKenzie and M.D. Gould, Phys. Rev. B **65**, (2002) 060502(R).
[9] J. Links, H.-Q. Zhou, M.D. Gould and R.H. McKenzie, J. Phys. A: Math. Gen. **35**, (2002) 6459;
    J. Links and K.E. Hibberd, Int. J. Mod. Phys. B **16**, (2002) 2009;
    J. Links, H.-Q. Zhou, R.H. McKenzie and M.D. Gould, Int. J. Mod. Phys. B, to appear;
    X.-W. Guan, A. Foerster, J. Links and H.-Q. Zhou, Nucl. Phys. B, to appear.
[10] P.P. Kulish, J. Sov. Math. **35**, (1986) 2648.
[11] P.P. Kulish and N. Manojlovic, Lett. Math. Phys. **55**, (2001) 77; J. Math. Phys. **42**, (2002) 64.
[12] A.J. Bracken, M.D. Gould. J.R. Links and Y.-Z. Zhang, Phys. Rev. Lett. **74**, (1995) 2768.

Inst. Phys. Conf. Ser. No 173: Section 9
Paper presented at 24th Int. Coll. Group Theoretical Methods in Physics, Paris, France, 15–20 July 2002
©2003 *IOP Publishing Ltd*

Conservation laws for fractional differential equations

M Klimek

Institute of Mathematics and Computer Science, Technical University of Częstochowa, Dąbrowskiego 73, 42-200 Częstochowa, Poland

E-mail: klimek@matinf.pcz.czest.pl

Abstract. The derivation of conservation laws for partial linear differential equations of fractional order containing Riemann-Liouville type derivatives is studied. Equations of this type are obtained in modern transport theory when underlying continuous time random process is characterized by diverging waiting time (see [1, 2, 3] and references therein). It is shown that the area where the stationarity-conservation law is fulfiled is restricted by the asymptotic properties of solutions. As an example fractional Rayleigh equation and its conservation law are discussed.

1. Introduction: Fractional derivatives and their properties

Let us recall the notion of the left Riemann-Liouville fractional derivative of order α. It is an integro-differential operator defined by the following formula for $m \leq Re\alpha < m + 1, t > 0$ [4]:

$$D_t^\alpha := \left(\frac{d}{dt}\right)^{m+1} \int_0^t \frac{f(s)}{(t-s)^{\alpha-m}} ds \tag{1}$$

In the process of derivation of conservation law the Leibniz's rule of the respective differential calculus is a crucial tool. It is rather complicated for fractional differential calculus and algebra of pointwise multiplication of functions [4] being valid in principle only in the subset of analytic functions. Thus we proposed [5] to study Leibniz's rule of the above fractional derivative in the algebra defined by Laplace convolution for which the following lemma is valid.

Lemma 1 *Let $m \leq Re\alpha < m + 1$ and the function g be piecewise continuous in $(0, +\infty)$. If the fractional derivative of function f does exist and the function itself fulfils the condition:*

$$\lim_{t\to 0^+} f^{(k)} * t^{m-\alpha} = 0 \tag{2}$$

for $k = 0, 1.., m$ and " $*$ " denoting the Laplace convolution then the fractional derivative acts as follows:

$$D_t^\alpha(f * g) = (D_t^\alpha f) * g \tag{3}$$

When both functions f and g obey the assumptions of the above lemma then we obtain an anlog of Leibniz's rule for derivatives of fractional order with $\beta \in [0, 1]$:

$$D_t^\alpha(f * g) = \beta(D_t^\alpha f) * g + (1 - \beta)f * (D_t^\alpha g) \tag{4}$$

The notion of fractional derivation is extended to partial fractional derivatives of Riemann-Liouville type defined by the formula [4]:

$$D_k^{\alpha_k} f(\vec{x}) := \frac{1}{\Gamma(m_k + 1 - \alpha_k)} \left(\partial^k\right)^{m_k+1} \int_0^{x_k} \frac{f(\vec{x} + (s - x_k)\vec{e}_k)}{(x_k - s)^{\alpha_k - m_k}} ds \tag{5}$$

where $m_k \leq Re\alpha_k < m_k + 1$.

The upper index in the formula denotes the fractional order of the partial derivative while the lower one says that it was taken with respect to coordinate x_k.

Let $x_1, ..., x_m$ be a subset of coordinates in our n-dimensional model for which the fractional partial derivatives appear in the equation. We define multiplication of functions of n-variables using the Laplace convolution in the form (we denote $(\vec{e}_l)_k = \delta_{lk}$):

$$f * g(\vec{x}) := \int_0^{x_1} ... \int_0^{x_m} f\left(\vec{x} - \sum_{l=1}^m s_l \vec{e}_l\right) g\left(\vec{x} + \sum_{l=1}^m (s_l - x_l)\vec{e}_l\right) ds_1...ds_m \tag{6}$$

The fractional Leibniz's rule is valid for functions fulfiling the respective assumptions concerning their behaviour at $x_k = 0$ with $\beta_k \in [0, 1]$ for $k = 1, ..., m$:

$$D_k^{\alpha_k} f * g = \beta_k (D_k^{\alpha_k} f) * g + (1 - \beta_k) f * D_k^{\alpha_k} g \tag{7}$$

The standard Leibniz's rule works for first order derivatives acting on coordinates $x_{m+1}, ..., x_n$:

$$\partial_j(f * g) = (\partial_j f) * g + f * \partial_j g \tag{8}$$

2. Stationarity - conservation laws for some fractional partial equations

The general equation contains the fractional and differential parts:

$$\Lambda(D, \partial)\phi = [\tilde{\Lambda}(D) + \Lambda(\partial)]\phi = \left(\sum_{k=1}^m \tilde{\Lambda}_k D_k^{\alpha_k} + \sum_{l=1}^N \Lambda_{\mu_1...\mu_l} \partial^{\mu_1}...\partial^{\mu_l} + \Lambda_0\right)\phi = 0 \tag{9}$$

We constructed stationarity-conservation laws for the homogenous form of the equation (9) remembering that the addition of the initial terms restricts only the area of application of the stationarity equation [5, 6]. We assumed that for given variables $x_1, ..., x_m$ the equation includes only fractional derivatives in $\tilde{\Lambda}(D)$ while for the remaining coordinates $x_{m+1}, ..., x_n$ only partial integer order derivatives appear in the operator $\Lambda(\partial)$. The coefficients Λ (we allow also matrices) are symmetric with respect to permutation of each set of indices $(\mu_1...\mu_l)$. They depend on coordinates and should obey the restriction in the form ($l = 1, ..., N$ $k, j = 1, ..., m$):

$$\partial^{\mu_1} \Lambda_{\mu_1...\mu_l} = 0 \qquad \partial^k \Lambda_{\mu_1...\mu_l} = 0 \qquad \partial^k \tilde{\Lambda}_j = 0 \tag{10}$$

Each direction of the space yields the component of the current which for coordinates $x_1, .., x_m$ is given by the $\tilde{\Gamma}$ operator of the form:

$$\tilde{\Gamma}_k = 2\tilde{\Lambda}_k \tag{11}$$

while for the part $j = m + 1, ..., n$ we obtain it using results of [7]:

$$\Gamma_j = \sum_{l=1}^{N-1} \sum_{k=1}^l \Lambda_{j\mu_1...\mu_l} (-\overleftarrow{\partial}^{\mu_1})...(-\overleftarrow{\partial}^{\mu_k})\partial^{\mu_{k+1}}...\partial^{\mu_l} \tag{12}$$

For an arbitrary pair of functions f and g for which we can use the fractional Leibniz's rule (7) the operators $\tilde{\Gamma}$ and Γ fulfil the equality:

$$\sum_{j=1}^m D_j^{\alpha_j} f * \tilde{\Gamma}_j g = -f * \tilde{\Lambda}(-\overleftarrow{D})g + f * \tilde{\Lambda}(D)g \tag{13}$$

$$\sum_{j=m+1}^n \partial^j f * \Gamma_j g = -f\Lambda(-\overleftarrow{\partial}) * g + f * \Lambda(\partial)g \tag{14}$$

and this property yields stationarity-conservation law for discussed fractional differential equation.

Proposition 1 *Let the function ϕ be a solution of the equation (9) and let ϕ' solve the conjugated equation:*

$$\phi' \Lambda(-\overleftarrow{D}, -\overleftarrow{\partial}) = \phi' [\tilde{\Lambda}(-\overleftarrow{D}) + \Lambda(-\overleftarrow{\partial})] = \tag{15}$$

$$\phi' \left(-\sum_{k=1}^{m} \tilde{\Lambda}_k \overleftarrow{D}_k^{-\alpha_k} + \sum_{l=1}^{N} \Lambda_{\mu_1 \dots \mu_l} (-\overleftarrow{\partial}^{\mu_1}) \dots (-\overleftarrow{\partial}^{\mu_l}) + \Lambda_0 \right) = 0$$

Then the current given by the components $(k = 1, ..., m \quad j = m + 1, ..., n)$:

$$J_k = \phi' * \tilde{\Gamma}_k \phi \qquad J_j = \phi' * \Gamma_j \phi \tag{16}$$

fulfils the stationarity-conservation equation:

$$\sum_{k=1}^{m} D_k^{\alpha_k} J_k + \sum_{j=m+1}^{n} \partial^j J_j = 0 \tag{17}$$

provided the solutions ϕ and ϕ' obey the conditions of lemma 1 in the neighbourhood of $x_k = 0 \quad k = 1, ..., m$.

The stationarity-conservation equation can be rewritten in the form of the standard conservation law for modified components of the above current ($m_k < \alpha_k < m_k + 1 \quad k = 1, ..., m \quad j = m + 1, ..., n$):

$$J_k' = (\partial^k)^{m_k} (J_k *_k \Phi_{\alpha_k - m_k}) \qquad J_j' = J_j \tag{18}$$

where the convolution $*_k$ is taken with respect to coordinate x_k:

$$f *_k g(\vec{x}) = \int_0^{x_k} f(\vec{x} - s\vec{e}_k) g(\vec{x} + (s - x_k)\vec{e}_k) ds_k \tag{19}$$

The new current J' obeys the conservation law:

$$\sum_{l=1}^{n} \partial^l J_l' = 0 \tag{20}$$

3. Stationary and conserved charges for mixed fractional-differential models

When the stationarity-conservation law for a given model is valid in the whole space of coordinates we can discuss the construction of conserved charges connected with corresponding conservation laws. Two cases were considered: when the time-derivative in the operator of the equation was a fractional one or when it was the partial derivative of the first order [5, 6]. For model with time-derivative of fractional order we obtain the charge Q which is a stationary function of order α_t

$$Q = \int_{R^{n-1}} d\vec{x} \ J_t(\vec{x}, t) \qquad D_t^{\alpha_t} Q = 0 \tag{21}$$

provided the respective boundary terms vanish. For components $J_j \quad j = m + 1, ..., n$ that means that they vanish at the infinity in the respective j- direction while the components $J_k \quad k = 2, .., m$ obey the corresponding asymptotic condition for $m_k < \alpha_k < m_k + 1$:

$$\lim_{|x_k| \to \infty} (\partial^k)^{m_k} (J_k *_k \Phi_{\alpha_k - m_k}) = 0 \tag{22}$$

The second case is the model with standard time-derivative. Then the charge Q is a strictly stationary function of time that means it is a true constant function:

$$Q = \int_{R^{n-1}} d\vec{x} \; J_t(\vec{x}, t) \qquad \partial^t Q = 0 \tag{23}$$

when the asymptotic conditions described above for respective components of the currents are fulfiled.

The derivation of the stationarity-conservation laws and corresponding stationary charges can be extended to the generalized equation containing both polynomial of fractional derivatives and polynomial of classical partial derivatives [6].

4. Example: conserved currents for the fractional Rayleigh equation

In the mentioned papers [5, 6] we have applied the presented procedure to fractional equations of anomalous diffusion and to fractional versions of Fokker-Planck, Klein-Kramers and Cattaneo equations. We add now the derivation of the stationarity-conservation law for the fractional Rayleigh equation [3]. In the nonhomogenous form it looks as follows with $k^* = k_B T/m$, $0 < \alpha < 1$:

$$[D_t^\alpha - \eta^* \partial^v v - \eta^* k^* (\partial^v)^2] P(v, t) = P_0(v) t^{-\alpha}/\Gamma(1 - \alpha) \tag{24}$$

$$\tilde{P}(v, t) \left[-\overset{\leftarrow}{D}_t^\alpha + \eta^* \overset{\leftarrow}{\partial}^v v - \eta^* k^* (\overset{\leftarrow}{\partial}^v)^2 \right] = \tilde{P}_0(v) t^{-\alpha}/\Gamma(1 - \alpha) \tag{25}$$

After modification of the probability density functions P and $\tilde{P}$

$$P(v, t) = exp(-mv^2/4k^*) W(v, t) \qquad \tilde{P}(v, t) = exp(mv^2/4k^*) \tilde{W}(v, t) \tag{26}$$

we arrive at the version of fractional Rayleigh equation and of its conjugation which obey the condition (10):

$$[D_t^\alpha - \eta^* k^* (\partial^v)^2 - V(v)] W(v, t) = W_0(v) t^{-\alpha}/\Gamma(1 - \alpha) \tag{27}$$

$$\tilde{W}(v, t) \left[-\overset{\leftarrow}{D}_t^\alpha - \eta^* k^* (\overset{\leftarrow}{\partial}^v)^2 - V(v) \right] = \tilde{W}_0(v) t^{-\alpha}/\Gamma(1 - \alpha) \tag{28}$$

$$V(v) = \eta^* (1/2 - v^2/4k^*) \tag{29}$$

Applying explicit formulae for the operator Γ (11,12) and for conserved currents (16) we obtain them in the form:

$$J_t = 2\tilde{P} *_t P \qquad J_v = \eta^* k^* \tilde{P} *_t \left(-\partial^v + \overset{\leftarrow}{\partial}^v \right) P + \eta^* v \tilde{P} *_t P \tag{30}$$

Then from proposition 1 we know that the following stationarity-conservation law is fulfiled:

$$D_t^\alpha J_t + \partial^v J_v = 0 \tag{31}$$

in the area where the initial conditions vanish $W_0(v) = \tilde{W}_0(v) = 0$ and the solutions W and $\tilde{W}$ obey the assumptions which allow the application of Leibniz's rule (7).

References

[1] 2000 *Applications of Fractional Calculus in Physics* Ed. Hilfer R (Singapore: World Scientific)
[2] Metzler R and Klafter J 2000 Phys. Rep. **339** 1-77
[3] Metzler R and Klafter J 2000 *J. Phys. Chem.* **B 104** 3851-3857
[4] Samko S G, Kilbas A A and Marichev O I 1993 *Fractional Derivatives and Integrals. Theory and Applications* (Amsterdam: Gordon and Breach)
[5] Klimek M 2001 *J. Phys. A: Math. Gen.* **34** 6167-6184
[6] Klimek M 2002 *J. Phys. A: Math. Gen.* **35** 6675-6693
[7] Klimek M 2002 *J. Math. Phys.* **43** 3610-3635

Inst. Phys. Conf. Ser. No 173: Section 9
Paper presented at 24th Int. Coll. Group Theoretical Methods in Physics, Paris, France, 15–20 July 2002
©2003 IOP Publishing Ltd

A New Powerful Method for Solving Non-linear Differential Equations: Burgers' Equation as an example

S. Otarod

Department of Physics Razi University, Kermanshah, Iran

Abstract. The method of separation of variables is used to solve Burger's equation in a special form and in it's most general form. In a new approach to the application of separation method we will show that this Method can be used effectively for solving nonlinear partial differential equations.

1. Introduction

The method of separation of variables is an old well known method. So far this method did not have a wide application in non-linear partial differential equations. The reason was that in most cases, by applying this technique, the variables will not be separated directly, and everyone will stop analysis immediately at this point. In fact the analysis has to be started exactly at this point. The following two simple examples (Burgers' Equation in a special form and in its most general form) prove this. Moreover we announce that the hydrodynamical equations, Sine- Gordon Equation and Fischer Equation can be solved by using this new method.

2. Example1: a simple form of Burgers' equation

Consider the following equation.

$$uu_x + u_y = 1 \tag{1}$$

This is a special form of Burger's equation. Here u is a function of x and y, that is $u = u(x, y)$. Suppose this function can be written in a separated form as,

$$u(x, y) = u(x)u(y) \tag{2}$$

$u(x)$ and $u(y)$ are two completely different functions and have nothing in common. Substituting equation (2) in Equation(1) will result in

$$u(y)^2 u(x)\frac{du(x)}{dx} + u(x)\frac{du(y)}{dy} = 1 \tag{3}$$

The variables have not been separated so far. Therefore one of the possible ways to separate the variables is to assume

$$\frac{du(x)}{dx} = 0 \tag{4}$$

as a result $u(x) = c$ and from equation (3) we will have $u(y) = \frac{1}{c}y + d$ and therefore

$$u_1(x, y) = y + cd \tag{5}$$

Moreover from the second assumption

$$\frac{du(y)}{dy} = 0 \tag{6}$$

we will come to

$$u_2(x, y) = \sqrt{2}\sqrt{x + \beta} \tag{7}$$

2.1. Solutions in non-separated form

So far two groups of solutions have been found. To find other solutions in non-separated form, we continue the analysis as follows;

We assume f to be an arbitrary function of $u_1(y)$ and $u_2(x)$ i.e, $f = f(u_1, u_2)$. Where, $u_1(y)$ and $u_2(x)$ are the solutions found above (equations (5) and (7)). Now we suppose that a group of solutions of Burgers' equations, can have the following form $U(x, y) = U(f(u_1, u_2)$ Substituting this in equation(1) will lead to;

$$U\frac{dU}{df}\frac{\partial f}{\partial u_2(x)}\frac{\partial u_2(x)}{\partial x} + \frac{dU}{df}\frac{\partial f}{\partial u_1(x)}\frac{\partial u_1(y)}{\partial y} = 1 \tag{8}$$

Since f is quite arbitrary, we assume;

$$\frac{\partial f}{\partial u_2(x)}\frac{\partial u_2(x)}{\partial x} = \gamma \tag{9}$$

and

$$\frac{\partial f}{\partial u_1(y)}\frac{\partial u_1(y)}{\partial y} = \lambda \tag{10}$$

Where γ and λ are two arbitrary constants. As a result, equation(8) will be written as

$$\gamma U\frac{dU}{df} + \lambda\frac{dU}{df} = 1 \tag{11}$$

Simultaneous solution of equations (9),(10) and (11) will results in;

$$U(x, y) = -A + \sqrt{2x + 2Ay + D} \tag{12}$$

where A and D are arbitrary constants. This result can be found in the reference (2).

3. Example 2: The most general form of Burger's equation

The ideas in the previous section help us to solve the most general form of Burgers' Equation i.e;

$$UU_x + U_y = \epsilon U_{xx} \tag{13}$$

Here we will not consider the trivial solutions in separated form and we will try to extract directly the general solutions in non-separated forms. For this purpose we suppose that this equation have a set of solutions as;

$$U(f(\chi(x), \psi(y))$$

Substituting this in Equation(13) will give;

$$U\frac{dU}{df}\frac{\partial f}{\partial \chi}\frac{\partial \chi}{\partial x} + \frac{dU}{df}\frac{\partial f}{\partial \psi}\frac{\partial \psi}{\partial y} = \epsilon\frac{\partial}{\partial x}[\frac{dU}{df}\frac{\partial f}{\partial \chi}\frac{\partial \chi}{\partial x}] \tag{14}$$

In this case too, the assumptions;

$$\frac{\partial f}{\partial \psi}\frac{\partial \psi}{\partial y} = \lambda \tag{15}$$

and

$$\frac{\partial f}{\partial \chi}\frac{\partial \chi}{\partial x} = \gamma \tag{16}$$

will lead us to;

$$\gamma U\frac{dU}{df} + \lambda\frac{dU}{df} = \epsilon\gamma^2\frac{d}{df}[\frac{dU}{df}] \tag{17}$$

Where, as before, γ and λ are two arbitrary constants. Here too, simultaneous solution of the above three equations will results in;

$$f = \gamma x + \lambda y \tag{18}$$

$$U = \frac{2\lambda e^{\frac{\lambda(f+\delta)}{\epsilon\gamma^2}}}{1 - \gamma e^{\frac{\lambda(f+\delta)}{\epsilon\gamma^2}}} \tag{19}$$

4. Discussion

Although the selected examples were very simple, they are very enlightening and they clearly show how much this method is effective for solving a vast variety of non-linear partial differential equations.

One point deserves to be mentioned here. What was represented above is not all we can say about the capability of this method. It seems that there are still more points that we can think about. For example, we are not sure that the above assumptions are the only possible assumptions that help us to solve Burger's equation and probably there is still more to be done. Therefore we can review the problem once more to see whether other assumptions, except those we have done before (equations (4),(6) and u=u(f)), will result in new solutions of Burgers equation.

Moreover, once more, we have to emphasize that hydrodynamical equations has been solved by Otarod and Ghanbari [1], and also Fischer Equation and Sine-Gordon equation have been solved by the author.

References

[1] S.Otarod and J.Ghanbari, 2000, The interstellar medium in M31 and M33 **232.**
    WE -Heraeus Seminar/E.M.Berkhuijsen et al. Aachen: Shaker 2000, 57.
[2] J.Kevorkian, *Partial Differential Equations, Analytical Solution Techniques.* 1990, Wadsworth,INC., Belmont, California 94002.

Inst. Phys. Conf. Ser. No 173: Section 9
Paper presented at 24th Int. Coll. Group Theoretical Methods in Physics, Paris, France, 15–20 July 2002
©2003 IOP Publishing Ltd

On Infinite Symmetries and Essential Conservation Laws for Navier-Stokes Equations

V. Rosenhaus

Department of Mathematics and Statistics, California State University, Chico, Chico, CA 95929, USA

Abstract. We discuss the relations between infinite symmetries and local conservation laws. We will introduce essential conservation laws for partial differential equations and demonstrate that known infinite series of conservation laws for the Navier-Stokes equations (associated with infinite symmetries) contain only one essential vector conservation law.

By a conservation law for a given differential equation $\omega(x, u, u_i, \ldots) = 0$ is usually meant a continuity equation

$$D_\mu K_\mu \doteq 0, \qquad \mu = 1, \ldots, n, \tag{1}$$

which is satisfied for any solutions of the equation ($\doteq$), see e.g. [1, 2]. The following cases give rise to *trivial conservation laws* [1]: 1) $D_\mu K_\mu = 0$. 2) $K_\mu \doteq 0$ $(\forall \mu = 1, \ldots, n)$. We will be interested in non-trivial conservation laws and further reduce the set of conservation laws to *essential conservation laws*. By an *essential* conservation law we will mean such non-trivial conservation law $D_\mu K_\mu \doteq 0$, which gives rise to a non-vanishing conserved quantity

$$D_t \iiint K_t dx dy dz \doteq 0, \qquad K_t \neq 0. \tag{2}$$

In the paper, we consider essential conservation laws for equations with infinite groups. For Lagrangian differential equations, the problem of the relationship between infinite symmetries, with arbitrary functions of not all base variables, and conservation laws was discussed in [3,4], and applications in [5, 6]. The approach is based on the Noether identity (see e.g. [7], or [8] for a version used here):

$$X_\alpha = \alpha E + \sum_{i=1}^{n} D_i R_{\alpha i}, \tag{3}$$

where E is the Euler-Lagrange operator

$$E = \frac{\partial}{\partial u} - D_i \frac{\partial}{\partial u_i} + \sum_{i \leq j} D_i D_j \frac{\partial}{\partial u_{ij}} + \cdots,$$

and

$$R_{\alpha i} = \alpha \frac{\partial}{\partial u_i} + \left\{ \sum_{k \geq i} (D_k \alpha) - \alpha \sum_{k \leq i} D_k \right\} \frac{\partial}{\partial u_{ik}} + \ldots$$

The application of the identity (3) to Lagrangian L gives $X_\alpha L = \alpha \omega + D_i(R_{\alpha i} L)$. Therefore, for a Noether symmetry transformation with the vector α, $X_\alpha L = D_i M_i$, we obtain

$$D_i(M_i - R_{\alpha i} L) = \alpha \omega. \tag{4}$$

In case of a finite Lie group, any Noether (variational) symmetry α leads to a corresponding conservation law; this is the statement of the First Noether Theorem [9]. According to the *Second Noether Theorem*, an infinite Noether symmetry

$$\alpha = ap(x) + b^i D_i p(x) + c^{ij} D_i D_j p(x) + \dots, \tag{5}$$

where *p(x)* is an arbitrary function of *all independent variables*, does not lead to conservation laws, but a certain relation between the original equations of the system:

$$a_b \omega^b - D_i(b_b^i \omega^b) + D_i D_j(c_b^{ij} \omega^b) + \dots = 0.$$

For the situation of infinite symmetry group with arbitrary functions of $k < n$ independent variables, or their combinations, it was demonstrated in [4], that infinite symmetries could lead only to a finite number of conservation laws. Particularly, Noether symmetries with *arbitrary* function of time $\gamma(t)$

$$\alpha = a\gamma(t) + b\gamma'(t) + c\gamma''(t) + \dots + h\gamma^{(l)}(t), \tag{6}$$

instead of conservation laws, lead to a set of additional constraints for the function u and its derivatives [3,4]. The satisfaction of the "strict" boundary conditions for some particular functions $\gamma(t)$

$$R_{\alpha 1} L\Big|_{x^1 \to \partial D} = R_{\alpha 2} L\Big|_{x^2 \to \partial D} = \dots = R_{\alpha(n-1)} L\Big|_{r^{n-1} \to \partial D} = 0. \tag{7}$$

is critical in order to avoid additional constraints and generate a finite number of essential conservation laws.

For equations without well-defined Lagrangian functions it is unclear weather the relations between symmetries and conservation laws, in general, exist. However, for differential systems for which $E(\beta_a w^a) \doteq 0$, with some β_a, it is possible to establish a correspondence between symmetries and conservation laws (through Noether identity [7]). If those systems also admit infinite symmetry algebras, we can apply the approach above. Let us demonstrate that for Navier-Stokes equations we will get a result similar to the case of Lagrangian systems: infinite symmetries lead to a finite number of essential conservation laws determined by boundary conditions. Consider Navier-Stokes equations

$$u_t^j - \nu u_{ii}^j + u^i u_i^j + p_j = 0, \qquad u_i^i = 0, \qquad i, j = 1, 2, 3 \tag{8}$$

where $\vec{u} = (u^1, u^2, u^3)$ is the velocity vector of an incompressible fluid of viscosity ν, and pressure p. The system (7) can be written in the form satisfying the condition $E(\omega^a) = 0$:

$$D_t u^j + D_i(-\nu u_i^j + u^i u^j + \delta_{ij} p) = 0, \qquad D_i u^i = 0. \qquad i, j = 1, 2, 3 \tag{9}$$

The symmetry algebra of Navier-Stokes equations includes a number of infinite Lie algebras [10-11]:

$$X_f = f\frac{\partial}{\partial x} + f'\frac{\partial}{\partial u^1} - xf''\frac{\partial}{\partial p}, \qquad X_g = g\frac{\partial}{\partial y} + g'\frac{\partial}{\partial u^2} - yg''\frac{\partial}{\partial p}$$

$$X_h = h\frac{\partial}{\partial z} + h'\frac{\partial}{\partial u^3} - zh''\frac{\partial}{\partial p}, \qquad X_\gamma = \gamma\frac{\partial}{\partial p}, \tag{10}$$

where $f(t), g(t), h(t), \gamma(t)$ are arbitrary functions. The following infinite series of conservation laws, associated with the symmetries (10) were reported in [12-13] (and later in [14]):

1. $D_t(qu^1) + D_1\left[q((u^1)^2 + p - \nu u_1^1) - q'xu^1\right] + D_2\left[q'(u^1u^2 - \nu u_2^1) - q'xu^2\right]$
$$+ D_3\left[q'(u^1u^3 - \nu u_3^1) - q'xu^3\right] \doteq 0, \tag{11}$$

where $q = q(t)$ is arbitrary. Note that the set (11) includes two more continuity equations obtained from (11) by cyclic permutation of indices. Conserved densities are

$$J^t = q(t)u^1 \quad (J^{ti} = q(t)u^i, \ i = 1,2,3.)$$

2. $D_t(Ku_1^1) + D_1\left[-K'u^iu_i^1 + Kp_1 - \nu K u_{ii}^1 + Ku_1^2\right] + D_2\left[K'u^2\right] + D_3\left[K'u^3\right] \doteq 0,$ (12)

and permutations, $K(t)$ is arbitrary. Conserved densities are

$$J^t = K(t)u_1^1, \quad (J_\mu^{ti} = K(t)u_\mu^i, \ i = 1,2,3, \ \mu = 1,2,3,4.)$$

3. Conserved densities are

$$J_\mu^{ti} = K(t)u_\mu^i - K', \ \mu = 1,2,3,4. \tag{13}$$

4. Conserved densities are

$$J_\mu^{ti} = Ku_\mu^i + 2tK' - 2tK'' - 3K', \quad i = 1,2,3, \ \mu = 1,2,3,4. \tag{14}$$

Let us show that the infinite series (11)-(14) lead to only one essential vector conservation law. The density (12), with appropriate boundary conditions (velocity vector vanishes on the boundary along with all its derivatives), gives rise to the conserved quantity of the form

$$D_t \iiint_D K(t)u_x^1 dxdydz = D_t \iint dydz K(t)u^1\Big|_{x\to\partial D} \doteq 0. \tag{15}$$

(Other components can be treated similarly). Equation (15) automatically holds for velocity components vanishing on the boundary, and does not give any conserved quantities. We can make the same conclusion about infinite series (13), since (13) is a combination of (12) and a trivial conservation law $D_t(K'(t)) - D_1(xK''(t)) = 0$. For the same reasons, equation (14) leads to no essential conservation laws, since (14) is a combination of (13) ($K \to K + 2tK'$) and a trivial conservation law $D_t(2tK'' + 3K') - D_1[x(2tK''' + 5K'')] = 0$.

Consider finally, an infinite series of continuity equations (11). Integrating over the whole space, and taking into account the following boundary conditions:

$$u^1, u_1^1, p, q'(t)xu^1 \underset{x\to\partial D}{\to} 0, \quad u^1, u^2, u_2^1, q'(t)xu^2 \underset{y\to\partial D}{\to} 0, \quad u^1, u^3, u_3^1, q'(t)xu^2 \underset{z\to\partial D}{\to} 0, \tag{16}$$

we get a conserved quantity

$$D_t \iiint_D q(t)u^1 dxdydz \doteq 0. \tag{17}$$

with arbitrary function $q(t)$). Let $\iiint_D u^1 dxdydz = m(t)$. We have $D_t(q(t)m(t)) \doteq 0$. Selecting $q(t) = 1$ we get $D_t m(t) \doteq 0$. Therefore,

$$q'(t)m(t) \doteq 0. \tag{18}$$

740

For arbitrary $q(t)$ the only solution of (18) is $m(t) \doteq 0$, or $\iiint_D u^1 dx dy dz \doteq 0$ which implies no conservation laws. Another solution is $q(t) = const$. In this case we obtain the momentum conservation law

$$D_t \iiint_D u^1 dx dy dz \doteq 0. \tag{19}$$

Conservation of other components of the momentum can be obtained similarly.

Let us note that the conservation law (19) follows directly from our approach. As we know, arbitrary functions of time lead to additional constraints, rather than conservation laws. In order to avoid additional constraints we have to select functions $q(t)$ such that the boundary conditions will be weaker than in the case of arbitrary $q(t)$. It follows immediately from (18) and (16), that the choice $q'(t) = 0$ leads to the weaker boundary conditions:

$$u^1, u_1^1, p \underset{x \to \partial D}{\to} 0, \quad u^1, u^2, u_2^1 \underset{y \to \partial D}{\to} 0, \quad u^1, u^3, u_3^1 \underset{z \to \partial D}{\to} 0, \tag{20}$$

and we get the conservation law (19).

References

[1] Olver, P.J. Applications of Lie Groups to Differential Equations (Springer, New York, 1986).
[2] Ibragimov, N.H. Transformation Groups Applied to Mathematical Physics. (Reidel, Boston, 1985).
[3] Rosenhaus, V. in Proc. Symmetry, Perturbation Theory, pp. 183-190, Cala Gonone, Italy, 2001.
[4] Rosenhaus, V. Infinite Symmetries and Conservation Laws, to be published in J. Math. Phys.
[5] Rosenhaus, V. On Conservation Laws and Boundary Conditions for "Short Waves" Equation, to be published in Rep. Math. Phys.
[6] Rosenhaus, V. On Conservation Laws for the Equation of Non-stationary Transonic Gas Flows, to be published in J. Dynamic Sys. Geom. Theor.
[7] Rosen, J. Some properties of the Euler-Lagrange operators, Preprint TAUP-269-72 (Tel-Aviv Univ., Tel-Aviv, 1972).
[8] Rosenhaus, V. and Katzin, G.H., J. Math. Phys. **35** (1994) 1998-2012.
[9] Noether, E., Nachr. König. Gessell. Wissen. Göttingen, Math.-Phys. Kl. (1918), pp. 235-257.
[10] Buchnev A.A. Dinamika Sploshnoi Sredy **7**, 212-214 (1971) (in Russian).
[11] Lloyd S.P. Acta Mech. **38**, (1981) 85-98
[12] Caviglia G., Int. J. Eng.Sci. **24** (1986) 1295-1302.
[13] Caviglia G., J. Math. Phys. **29** (1988) 812-816.
[14] Gusyatnikova V.N., Yamaguzhin V.A., Acta Appl. Math. **15** (1989) 65-81.

Inst. Phys. Conf. Ser. No 173: Section 9
Paper presented at 24th Int. Coll. Group Theoretical Methods in Physics, Paris, France, 15–20 July 2002
©*2003 IOP Publishing Ltd*

Green's function for dissipative quantum systems and its relation to nonlinear evolution equations

Dieter Schuch

Institut für Theoretische Physik, J. W. Goethe–Universität, Robert-Mayer-Strasse 8-10, D-60054 Frankfurt am Main, Germany

Abstract. Quantum systems interacting with a dissipative environment can effectively be described by a nonlinear modification of the Schrödinger equation. For some standard problems that possess analytical solutions, e.g. the free motion and the harmonic oscillator, also the nonlinear equation can be solved in a closed form and has Gaussian-shaped wave packet solutions. The equation determining the time-evolution of the wave-packet width is - already in the case of the usual conservative Schrödinger equation - nonlinear and provides the quantities that are necessary to construct the Green's function for the respective system. The Green's functions for the conservative and dissipative cases will be given explicitly in position and momentum space.

1. Introduction

Dissipation and irreversibility are common features of realistic physical systems but conventional quantum mechanics is based on a conservative, reversible Hamiltonian description of the system. There are several approaches to include dissipative effects, like friction forces, on the quantum level. It has been shown [1] that an effective description of a dissipative system interacting with some environment is possible in the framework of a pure state wave-function $\Psi(x, t)$ for the system alone, where Ψ obeys a nonlinear Schrödinger equation (NLSE) of the form

$$i\hbar\frac{\partial}{\partial t}\Psi_{NL} = \left\{ -\frac{\hbar^2}{2m}\frac{\partial^2}{\partial x^2} + V(x) + \gamma\frac{\hbar}{i}\left(ln\Psi_{NL} - < ln\Psi_{NL} >\right) \right\}\Psi_{NL} . \tag{1}$$

The real part of the logarithmic nonlinearity corresponds to a linear velocity-dependent frictional force and its imaginary part to a diffusion term that breaks the time-symmetry in the equation for the probability density $\rho = \Psi^*\Psi$ (for details see [1]). It is also possible to show [2] that this approach is physically equivalent to an approach by Caldirola [3] and Kanai [4] using explicitly time-dependent Hamiltonians which, itself, under certain conditions, has been proven by Sun and Yu [5] to be equivalent to the conventional system-plus-reservoir approach.

In the following, we will concentrate on the logarithmic NLSE, in particular, on those cases where the dissipative, as well as the corresponding conservative, systems have analytical solutions. Explicitly, the free motion and the harmonic oscillator in one dimension will be treated. However, extension to higher dimensions and the inclusion, e.g, of magnetic fields are possible. Green's functions for propagation in time will be determined for these systems, with and without dissipation, in position and momentum space and a comparision will show the similarities and differences in the various cases.

2. Green's functions for Gaussian WPs in position space

For Gaussian WPs it is always possible to construct a Green's function using a Gaussian-type ansatz. It shall be shown how the parameters that determine the time-dependence of the Green's function are connected with the dynamical variables of the WP solutions of the corresponding SE, as well as NLSE.

For this purpose, the WP can be written explicitly in the form

$$\Psi_{WP,L}(x,t) = N_L(t) \, exp\left\{ i \left[y(t)\tilde{x}^2 + \frac{<p>}{\hbar}\tilde{x} + K(t) \right] \right\} \tag{2}$$

where $\tilde{x} = x - <x> = x - \eta(t)$ (the explicit form of the normalization coefficient $N_L(t)$ and of the purely time-dependent term $K(t)$ is not relevant for the following discussion) shows that the maximum at position $<x>$ follows the classical trajectory η (the mean value is defined as $<...> = \int \Psi^* ... \Psi dx$). The WP-width, $\sqrt{<\tilde{x}^2>}$, (where $<\tilde{x}^2> = <x^2> - <x>^2$), is connected with the imaginary part of the complex coefficient $\tilde{x}^2$ in the exponent, $y(t)$, via $(2\hbar y_I/m) = \hbar/2m <\tilde{x}^2> = \beta(t)$.

Inserting WP (2) into the SE proves that $<x> = \eta(t)$ obeys the classical Newtonian equation for the corresponding point particle, e.g., for the harmonic oscillator: $\ddot{\eta} + \omega^2\eta = 0$ (for $V = 0$, in all following equations, the terms containing ω disappear).

To determine the time-dependence of the WP-width, the complex (quadratically) nonlinear equation of Riccati-type

$$\frac{2\hbar}{m}\dot{y} + \left(\frac{2\hbar}{m}y\right)^2 + \omega^2 = 0 \tag{3}$$

must be solved.

The WP solution (2), at time t, can also be obtained from an initial WP at, e.g., $t' = 0$, with the help of a Green's function, via $\Psi_{WP,L}(x,t) = \int dx' \, \Psi_{WP,L}(x',0) \, G_L(x,x',t,t' = 0)$.

For the Gaussian WP being considered, with an initial distribution

$$\Psi_{WP,L}(x',0) = \left(\frac{m\beta_0}{\pi\hbar}\right)^{1/4} exp\left\{ \frac{im}{2\hbar}\left[i\beta_0 x'^2 + 2\frac{p_0}{m}x' \right] \right\}, \tag{4}$$

where $\beta_0 = \hbar/2m <\tilde{x}^2>_0 = 1/\alpha_0^2$ and $p_0 = <p> (t = 0)$, the Green's function can be written as

$$G_L(x,x',t,0) = \left(\frac{m}{2\pi i\hbar\alpha_0\hat{z}}\right)^{1/2} exp\left\{ \frac{im}{2\hbar}\left[\frac{\dot{\hat{z}}}{\hat{z}}x^2 - 2\frac{x}{\hat{z}}\left(\frac{x'}{\alpha_0}\right) + \frac{\hat{u}}{\hat{z}}\left(\frac{x'}{\alpha_0}\right)^2 \right] \right\}. \tag{5}$$

Inserting the WP, expressed with the help of (4) and (5), into the SE, shows that the time-dependent parameters $\hat{z}(t)$ and $\hat{u}(t)$ fulfil the same equation of motion as the classical trajectory $\eta(t)$ and, in addition, are uniquely connected via the relation

$$\dot{\hat{z}}\hat{u} - \dot{\hat{u}}\hat{z} = 1. \tag{6}$$

From there, it follows that knowledge of $\hat{z}(t)$ allows one to determine $\hat{u}(t)$ by simple integration (for given initial conditions), $\hat{u} = -\hat{z}\int^t \hat{z}^{-2}dt'$. For comparison with the dissipative case and the momentum space representation, the quantity $K_L(x,x') = \Psi_{WP,L}(x',0) \, G_L(x,x',t,0)$ shall be introduced. The explicit form is given in Table 1, the coefficient $C_{...}(t)$ in front of the exponential function is only relevant for the

Table 1. Products of the Green's functions with the corresponding initial wave packets; $K_{L,NL}(...,...') = G_{L,NL}(...,...',t,0)\Psi(...',0)$.

$$K_L(x,x') = C_{L,x}\, exp\left\{\frac{im}{2\hbar}\left[\frac{\dot{\hat{z}}}{\hat{z}}x^2 - 2\left(\frac{x}{\hat{z}} - \frac{p_0\alpha_0}{m}\right)\left(\frac{x'}{\alpha_0}\right) + \left(\frac{\dot{\hat{u}}}{\hat{z}} + i\right)\left(\frac{x'}{\alpha_0}\right)^2\right]\right\}$$

$$K_{NL}(x,x') = C_{NL,x}\, F(t)\, exp\left\{\frac{im}{2\hbar}\left[\frac{\dot{\hat{z}}}{\hat{z}}x^2 - 2\left(\frac{x}{\hat{z}} - \frac{p_0\alpha_0}{m}\right)f_1\left(\frac{x'}{\alpha_0}\right) + \left(\frac{\dot{\hat{u}}}{\hat{z}} + i\right)f_2\left(\frac{x'}{\alpha_0}\right)^2\right]\right\}$$

$$K_L(p,p') = C_{L,p}\, exp\left\{\frac{-im}{2\hbar}\left[\frac{\dot{\hat{z}}}{\hat{z}}\frac{p^2}{m^2} - 2i\left(\frac{p}{m\hat{z}} - \frac{p_0}{\varepsilon_0 m}\right)\left(\frac{p'}{\varepsilon_0 m}\right) + \left(\frac{\dot{\hat{u}}}{\hat{z}} + i\right)\left(\frac{p'}{\varepsilon_0 m}\right)^2\right]\right\}$$

$$K_{NL}(p,p') = C_{NL,p}\, F(t)\, exp\left\{\frac{-im}{2\hbar}\left[\frac{\dot{\hat{z}}}{\hat{z}}\frac{p^2}{m^2} - 2i\left(\frac{p}{m\hat{z}} - \frac{p_0}{\varepsilon_0 m}\right)f_1\left(\frac{p'}{\varepsilon_0 m}\right) + \left(\frac{\dot{\hat{u}}}{\hat{z}} + i\right)f_2\left(\frac{p'}{\varepsilon_0 m}\right)^2\right]\right\}$$

normalization and shall not be given here explicitly. The last, necessary, step is to explicitly integrate $K_L(x,x')$, with respect to x', $\Psi_{WP,L}(x,t) = \int dx'\, K_L(x,x')$, to yield the WP solution in the form shown in Table 2, where the complex variable λ combines the two time-dependent parameters as $\lambda = \hat{u} + i\hat{z}$.

Comparison with the same WP, written in the form given in Eq.(2), shows that the relations $\hat{z}\alpha_0 p_0/m = <x> = \eta(t)$ and $\dot{\hat{z}}/\hat{z} - 1/\hat{z}\lambda = (2\hbar y/m)$ are valid. From there, it follows that - apart from a constant factor - $\hat{z}$ is the particle trajectory. Using relation (6), the latter relation can be written as $(2\hbar y/m) = \dot{\lambda}/\lambda$, which, when inserted into the Riccati Eq.(3), linearizes the equation to $\ddot{\lambda} + \omega^2\lambda = 0$.

For the log NLSE, Gaussian WP solutions also exist. Written in form(2), here, the maximum follows the classical trajectory, according to $\ddot{\eta} + \gamma\dot{\eta} + \omega^2\eta = 0$ and, in the Riccati equation, an additional linear term $\gamma(2\hbar y/m)$ occurs. Linearization can also be achieved here, via $(2\hbar y) = \dot{\tilde{\lambda}}/\tilde{\lambda}$, where $\tilde{\lambda} = \tilde{\hat{u}} + i\tilde{\hat{z}}$ and the quantities with a tilde are the original ones, multiplied by $e^{\gamma t/2}$ (see, e.g., [6]). Considering the same initial Gaussian distribution (4), the Green's function, for the dissipative case, can be expressed in terms of the parameters with tilde. In particular, the quantity $K_{NL}(x,x')$, given explicitly in Table 1, looks very similar to the one without dissipation. The term $F(t)$ provides only a purely time-dependent phase-factor and the time-dependent factors $f_1(t)$ and $f_2(t)$ must fulfil $f_1^2/f_2 = e^{\gamma t}$. Via integration, the WP can be obtained in a form similar to the conservative case (see Table 2). In this case, $\tilde{\hat{z}}\alpha_0 p_0/m = <x> = \eta(t)$ and $\dot{\tilde{\hat{z}}}/\tilde{\hat{z}} - e^{-\gamma t}/\tilde{\hat{z}}\lambda = \dot{\tilde{\lambda}}/\tilde{\lambda} = (2\hbar y/m)$, with $\dot{\tilde{\hat{z}}}\tilde{\hat{u}} - \dot{\tilde{\hat{u}}}\tilde{\hat{z}} = e^{\gamma t}$, are valid.

3. Green's functions for Gaussian WPs in momentum space

The solution in the momentum space representation, where $p_{op} = p$ and $x_{op} = -(\hbar/i)\partial/\partial p$, can be obtained via Fourier transformation of the solution in position space, $\Psi(p,t) = (2\pi\hbar)^{-1/2}\int dx\, \Psi(x,t)\, e^{ipx/\hbar}$. In particular, the Gaussian WP(2) takes the form

$$\Psi_{WP,L}(p,t) = \left(\frac{a}{\hbar}\right)^{1/2} N_L(t)\, exp\left\{-\frac{a}{2\hbar^2}\tilde{p}^2 - i\frac{<x>}{\hbar}\tilde{p} + i\left(K_L - \frac{<x><p>}{\hbar}\right)\right\}, \quad (7)$$

with $\tilde{p} = p - <p>$ and the complex quantity $a(t)$ that is related to $y(t)$ via $am/i\hbar = (2\hbar y/m)^{-1} = \lambda/\dot{\lambda}$. In particular, the real part of $a(t)$ is related to the momentum uncertainty via $a_R m/\hbar = m\hbar/2 <\tilde{p}^2>$ and $a(t)$ also fulfils a Riccati equation (see [7]).

However, inserting the Fourier-transformed WP in the dissipative case into the momentum space SE with the same additional logarithmic term does not solve this equation. This is connected with the fact that the mean value of the gradient of $ln\Psi$ in position space

Table 2. Wave packets represented with the help of the time-dependent parameters.

$$\Psi_{WP,L}(x,t) = \left(\frac{m}{\pi\hbar}\right)^{1/4}\left(\frac{1}{\lambda}\right)^{1/2}exp\left\{\frac{im}{2\hbar}\left[\frac{\dot{z}}{z}x^2 - \frac{1}{\dot{z}\lambda}\left(x - \frac{p_0\alpha_0}{m}\dot{z}\right)^2\right]\right\}$$

$$\Psi_{WP,NL}(x,t) = \left(\frac{m}{\pi\hbar}\right)^{1/4}\left(\frac{1}{\tilde{\lambda}e^{\gamma t/2}}\right)^{1/2}F(t)\,exp\left\{\frac{im}{2\hbar}\left[\frac{\dot{\tilde{z}}}{\tilde{z}}x^2 - \frac{e^{-\gamma t}}{\dot{\tilde{z}}\tilde{\lambda}}\left(x - \frac{p_0\alpha_0}{m}\dot{\tilde{z}}\right)^2\right]\right\}$$

$$\Psi_{WP,L}(p,t) = \left(\frac{1}{\pi\hbar m}\right)^{1/4}\left(\frac{i}{\lambda}\right)^{1/2}exp\left\{-\frac{im}{2\hbar}\left[\frac{\dot{z}}{z}\frac{p^2}{m^2} + \frac{1}{\dot{z}\lambda}\frac{\tilde{p}^2}{m^2}\right]\right\}$$

$$\Psi_{WP,NL}(p,t) = \left(\frac{m}{\pi\hbar m}\right)^{1/4}\left(\frac{i}{\tilde{\lambda}e^{\gamma t/2}}\right)^{1/2}F(t)exp\left\{-\frac{im}{2\hbar}\left[\frac{\dot{\tilde{z}}}{\tilde{z}}\frac{p^2}{m^2} + \frac{e^{-\gamma t}}{\dot{\tilde{z}}\tilde{\lambda}}\frac{\tilde{p}^2}{m^2}\right]\right\}$$

is proportional to velocity (like the frictional force) whereas, in momentum space, it is proportional to position which represents a different physical situation. However, it is possible to express the friction term with the help of $\tilde{\lambda}$ and the kinetic energy operator to yield a form of our NLSE that is applicable in position and momentum space (for details see [7]),

$$i\hbar\frac{\partial}{\partial t}\Psi_{NL}(p,t) = \left\{\left(1 + \gamma\frac{\tilde{z}}{\lambda}\right)\frac{p^2}{2m} - \frac{m}{2}\omega^2\hbar^2\frac{\partial^2}{\partial p^2} - \gamma\left(\frac{\tilde{\lambda}}{\tilde{z}}\right)\frac{<p^2>}{2m}\right\}\Psi_{NL}(p,t) . \tag{8}$$

The WP in form (7), inserted into this NLSE, again yields the dissipative classical equation of motion and the Riccati equation also obtains an additional linear term, $\gamma(am/i\hbar)$.

For the initial Gaussian distribution

$$\Psi_{WP,L}(p',0) = \left(\frac{b_0}{\pi\hbar m}\right)^{1/4}exp\left\{\frac{im}{2\hbar}\left[i\frac{b_0}{m}(p' - p_0)^2\right]\right\} \tag{9}$$

with $b_0 = m\hbar/2 < \tilde{p}^2 >_0 = 1/\varepsilon_0^2$, the WP can, again, be obtained via a Green's function according to $\Psi_{WP,L}(p,t) = \int dp' \, \Psi_{WP,L}(p',0) \, G_L(p,p',t,t'=0)$. In the conservative case, the Green's function and the quantity $K_L(p,p') = \Psi_L(p',0)G_L(p,p't,0)$ can again be expressed with the help of $\hat{u}$ and $\hat{z}$, in the dissipative case with $\hat{\tilde{u}}$ and $\hat{\tilde{z}}$, as shown in Table 1. After integration, this yields the WPs in the form given in Table 2.

4. Comparison and conclusions

Comparison of the quantities K in Table 1 (and, hence, the Green's functions which are simply obtained by dividing K by the initial WP) shows that transition from position to momentum space (in SE and NLSE) mainly requires the replacement of the quantities $\hat{u}$ and $\hat{z}$ without 'dot' (for time-derivative) by those with dot and vice-versa; the coefficient in front of the exponential function follows essentially from the normalization condition.

Transition from the SE to the NLSE (in position as well as momentum space) is achieved by replacing $\hat{u}$ and $\hat{z}$ by the corresponding quantities with tilde (i.e., multiplied by $e^{\gamma t/2}$) and introducing the time-dependent functions $F(t)$, $f_1(t)$ and $f_2(t)$.

Very similar relations apply to the WP solutions, expressed using the parameters $\hat{z}$ and λ. The time-dependence of the parameters can be obtained from the corresponding nonlinear Riccati equation, or its linearized complex Newtonian form.

5. References

[1] Schuch D, Chung K-M and Hartmann H 1983 *J. Math. Phys.* **24** 1652
    Schuch D, Chung K-M 1986 *Int. J. Quant. Chem.* **29** 1561

745

[2] Schuch D 1997 *Phys. Rev. A* **55** 935
    Schuch D 1999 *Int. J. Quant. Chem.* **72** 537
[3] Caldirola P 1941 *Nuovo Cimento* **18** 393
[4] Kanai E 1948 *Progr. Theor. Phys.* **3** 440
[5] Yu L H and Sun C P 1994 *Phys. Rev. A* **49** 592
    Sun C P and Yu L H 1995 *Phys. Rev. A* **51** 1845
[6] Schuch D 1989 *Int. J. Quant. Chem., Quant. Chem. Symp.* **23** 59
[7] Schuch D 1994 *Int. J. Quant. Chem., Quant. Chem. Symp.* **28** 251

Inst. Phys. Conf. Ser. No 173: Section 9
Paper presented at 24th Int. Coll. Group Theoretical Methods in Physics, Paris, France, 15–20 July 2002
©2003 IOP Publishing Ltd

On the Role of Weak Transversality in Lie Group Theory

P. Tempesta

Centre de Recherches Mathématiques, Université de Montréal, C. P. 6128, succ. Centre–ville, Montréal, Québec H3C 3J7, Canada
e-mail address: tempesta@CRM.UMontreal.ca

Abstract. The notion of weak transversality for systems of PDE's is introduced and its role in the Lie method of symmetry reduction is discussed.

1. Strong and weak transversality

The purpose of this article is to further develop, compare and apply alternative reduction methods based on Lie group theory [1]–[6]. We shall consider a system

$$\Delta_\nu \left(x, u^{(n)} \right) = 0, (\nu = 1, ..., m) \tag{1}$$

of m partial differential equations of order n, involving p independent variables $(x_1, x_2, ..., x_p)$ and q dependent variables $(u_1, u_2, ..., u_q)$. Suppose that we have found its Lie point symmetry group G. The corresponding symmetry algebra L has dimension r and has a basis realized by vector fields of the form

$$\mathbf{v}_a = \sum_{i=1}^p \xi_a^i \left(x, u \right) \frac{\partial}{\partial x^i} + \sum_{\alpha=1}^q \varphi_a^\alpha \left(x, u \right) \frac{\partial}{\partial u^\alpha}, \qquad a = 1, .., r = \dim G. \tag{2}$$

with the condition that their prolongation should annihilate the system on its solution set:

$$pr\, \mathbf{v}\, \Delta_\nu \left. \right|_{\Delta_\mu=0} = 0, \qquad \nu,\, \mu = 1, ..., m.$$

Let us now consider a subgroup $G_0 \subset G$ and its Lie algebra $\mathfrak{g}_0$. A solution $u = f(x)$ of the system (1) is G_0 invariant if its graph $\Gamma_f \sim \{x, f(x)\}$ is a G_0 invariant set: $g \cdot \Gamma_f = \Gamma_f$, $g \in G_0$. A G_0–invariant solution will satisfy the $q \times r_0$ characteristic equations

$$Q_a^\alpha = \varphi_a^\alpha \left(x, u \right) - \sum_{i=1}^p \xi_a^i \left(x, u \right) u_{x_i}^\alpha = 0, \quad \alpha = 1, ..., q, \quad a = 1, ..., r_i = \dim \mathfrak{g}_i.$$

Let us introduce the matrices Ξ_1 and Ξ_2 of the coefficients of the vector fields $\mathbf{v}_a$:

$$\Xi_1 = \left\{ \xi_a^i \left(x, u \right) \right\}, \qquad \Xi_2 = \left\{ \xi_a^i \left(x, u \right), \varphi_a^\alpha \left(x, u \right) \right\},$$

$$\Xi_1 \in \mathbb{R}^{r \times p}, \qquad \Xi_2 \in \mathbb{R}^{r \times (p+q)} \tag{3}$$

and the matrix $\{Q_a^\alpha\}$ of characteristics of the vector fields (2). Let us now consider a specific subalgebra $L_0 \subset L$. If the group G_0 acts regularly and transversally on $M \sim X \times U$ then

$$rank \left\{ \xi_a^i \left(x, u \right) \right\} = rank \left\{ \xi_a^i \left(x, u \right), \varphi_a^\alpha \left(x, u \right) \right\}. \tag{4}$$

This rank is equal to the dimension of the generic orbits of G_0 on M. If the transversality condition (4) is satisfied, at least locally, then Lie's classical reduction method is directly applicable: all dependent variables can be expressed in terms of invariants and a reduction to a system with q dependent variables and $p - s$ independent ones is immediate. We shall call the rank condition (4) "*strong transversality*". Quite recently [3] a method was proposed for obtaining group invariant solutions when equation (4) is not satisfied. The method of Ref. [3] can actually be simplified by introducing the concept of "*weak transversality*" [4].

Definition 1 *The local transversality condition will be said to be satisfied in the weak sense if it holds only on a subset $\widetilde{M} \subset M$, rather than on the entire space M:*

$$rank \left\{ \xi_a^i (x, u) \right\} |_{\widetilde{M}} = rank \left\{ \xi_a^i (x, u), \varphi_a^\alpha (x, u) \right\} |_{\widetilde{M}} \tag{5}$$

In other words, even if the transversality condition is not in general satisfied, there may exist a class S of functions $u = f(x)$ such that for them the condition (5) holds. The "weak transversality" method is quite simple, when applicable. It consists of several steps.

1. Determine the class S of functions $u = f(x)$ satisfying (5).
2. Require that the condition $rank\, Q = 0$ be satisfied on the set S.
3. Substitute the obtained expressions into the original system (1). By construction, the solutions, if they exist, will be G_0–invariant.

2. Weak transversality and invariant solutions.

Example 1. The isentropic compressible fluid model.

The equations describing the non–stationary isentropic flow of a compressible ideal fluid are

$$\overrightarrow{u_t} + \overrightarrow{u} \cdot \nabla \overrightarrow{u} + k\, a \nabla a = 0 \tag{6}$$

$$a_t + \overrightarrow{u} \cdot \nabla a + k^{-1} a \nabla \cdot \overrightarrow{u} = 0, \tag{7}$$

where $\overrightarrow{u} = u_1 (x, y, z, t), u_2 (x, y, z, t), u_3 (x, y, z, t)$ is the velocity field, $a = a(x, y, z, t)$ is the velocity of sound, related to the pressure p and the density ρ by the formula $a = \left(\frac{\gamma p}{\rho} \right)^{1/2}$, γ is the adiabatic exponent and $k = 2/(\gamma - 1)$. The full symmetry group G of eqs. (6)–(7) was derived in ref. [5]. Let us consider the subalgebra generated by

$$L_3 = y\partial_x - x\partial_y + u_2\partial_{u_1} - u_1\partial_{u_2}, \quad K_1 = t\partial_x + \partial_{u_1}, \quad K_2 = t\partial_y + \partial_{u_2},$$

$$D = x\partial_x + y\partial_y + z\partial_z + u_1\partial_{u_1} + u_2\partial_{u_2} + u_3\partial_{u_3} + a\,\partial_a.$$

It is immediate to observe that transversality is violated, because $rank\, \Xi_1 = 3$ and $rank\, \Xi_2 = 4$. If we force the matrix Ξ_2 to be of rank 3, then we get the following constraints

$$u_1 = \frac{x}{t}, \quad u_2 = \frac{y}{t}, \quad u_3 = u_3 (x, y, z, t), \quad a = a(x, y, z, t). \tag{8}$$

From the characteristic system $Q_a^\alpha (x, u^{(1)}) = 0$ we deduce $u_3 = z\, W(t), a = z\, A(t)$, where W and A are arbitrary functions of time. Now, substituting relations (8) into the system (6)–(7) we obtain

$$A(t) = \sqrt{-\frac{1}{k} (W^2 + W')} \tag{9}$$

and a second order ODE for W

$$W'' + 2 \left(2 + \frac{1}{k} \right) WW' + 2 \left(1 + \frac{1}{k} \right) W^3 + \frac{4}{k\, t} (W' + W^2) = 0 \tag{10}$$

In general, eq. (10) does not have the Painlevé property. For special values of the parameter k, namely $k = -1$ and $k = -2$, it does. In these cases it can be reduced to a canonical form (see [7], p. 334) via a linear transformation of the type $W = \alpha(t)\,U(z(t)) + \beta(t)$.

1. For $k = -1$ we have

$$W'' = -2WW' + p(t)\left(W' + W^2\right), \tag{11}$$

where $p(t) = 4/t$. This equation can be integrated and its solution, which is regular, is

$$W = \frac{c_1 t^2 \left(I_{-\frac{5}{6}}\left(\frac{c_1 t^3}{3}\right) + c_2 I_{\frac{5}{6}}\left(\frac{c_1 t^3}{3}\right)\right)}{I_{\frac{1}{6}}\left(\frac{c_1 t^3}{3}\right) + c_2 I_{-\frac{1}{6}}\left(\frac{c_1 t^3}{3}\right)}, \tag{12}$$

where c_1 and c_2 are constants and $I_n(x)$ is the modified Bessel function of the first kind. Correspondingly we find $A = c_1 t^2$ in eq. (9).

2. For $k = -2$ we have

$$W'' = -3WW' - W^3 + q(t)\left(W' + W^2\right), \tag{13}$$

where $q(t) = 2/t$. In this case, we can integrate and the general solution is:

$$W = \frac{4\,t^3 + c_1}{t^4 + c_1\,t + c_2}. \tag{14}$$

The solution of eq. (9) is

$$A = 2\sqrt{3}\sqrt{\frac{t^2}{t^4 + c_1 t + c_2}}, \tag{15}$$

where c_1, and c_2 are constants. The solutions for $k = -1$ and $k = -2$ represent nonscattering waves.

3. Partially invariant solutions and the transversality condition

In this section we will study the role of the local transversality condition (and in particular of the notion of weak transversality) in the theory of partially invariant solutions. Let G be a group, acting regularly with s–dimensional orbits. Under the hypothesis that $G(\Gamma_f)$ is a submanifold, we call the number $\delta = \dim G(\Gamma_f) - \dim \Gamma_f$ the defect δ of the function f with respect to G. The usual G–invariant functions correspond to the case $\delta = 0$. A function will be said to be *generic* if $\delta = m_0 = \min\{s, k - p\}$. When $0 < \delta < m_0$, the function f will be said to be partially invariant [2]. Let $\mathfrak{g}$ be a subalgebra of the symmetry algebra of a system Δ, and Q the corresponding characteristic matrix. Then $u = f(x)$ is a partially invariant solution of Δ with defect δ with respect to $\mathfrak{g}$ if and only if [2, 6] $rank\left(Q\left(x, u^{(1)}\right)\right) = \delta$. Let $u_0 = u_0(x_0)$ be a solution of Δ and G be an r–dimensional subgroup of the symmetry group of Δ, acting regularly on M, whose generators are given by (2). If the condition

$$rank\left(\xi_a^i(x_0, u_0)\right) < rank\left(\xi_a^i(x_0, u_0), \phi_a^\alpha(x_0, u_0)\right) \tag{16}$$

is satisfied, then $u_0 = f(x_0)$ is a partially invariant solution of Δ (or possibly a generic one).

Example 2. The Euler equations for an incompressible nonviscuous fluid in (3+1) dimensions:

$$\overrightarrow{u_t} + \overrightarrow{u}\cdot\nabla\overrightarrow{u} + \nabla p = 0,\ \nabla\cdot\overrightarrow{u} = 0, \tag{17}$$

admit the following subalgebra of symmetry generators

$$D = x\partial_x + y\partial_y + z\partial_z + 2t\partial_t - u_1\partial_{u_1} - u_2\partial_{u_2} - u_3\partial_{u_3} - 2p\partial_p,$$

$$L_3 = y\partial_x - x\partial_y + u_2\partial_{u_1} - u_1\partial_{u_2},$$

$$X = t^k \, \partial_x + k \, t^{k-1} \partial_{u_1} - k \, (k-1) \, t^{k-2} x \, \partial_p,$$

$$Y = t^k \, \partial_y + k \, t^{k-1} \partial_{u_2} - k \, (k-1) \, t^{k-2} y \, \partial_p, \tag{18}$$

which is a subalgebra of the Galilei–similitude algebra for a given $k \in \mathbb{R}$. The requirement of *weak transversality* implies

$$u_1 = k\frac{x}{t}, \qquad u_2 = k\frac{y}{t}, \qquad u_3 = u_3\,(x, y, z, t), \qquad p = p(x, y, z, t). \tag{19}$$

At this stage, we can choose to have not only group invariant but also partially invariant solutions. Indeed, substituting formulas (19) directly into the Euler equations (17) or the Navier–Stokes equations without requiring any invariance properties, we obtain the following partially invariant solution of the Euler equations:

$$u_1 = k\frac{x}{t}, \qquad u_2 = k\frac{y}{t}, \qquad u_3 = -\frac{2kz}{t} + x^2 F\left(tx^{-\frac{1}{k}}, \frac{y}{x}\right)$$

$$p = -\frac{k\,(k-1)\,(x^2 + y^2)}{2\,t^2} - \frac{k\,(2k+1)\,z^2}{t^2} + f\,(t), \tag{20}$$

where $\xi = tx^{-\frac{1}{k}}, \eta = \frac{y}{x}$ and $f\,(t)$ and $F\,(\xi, \eta)$ are arbitrary functions of their arguments. For a discussion of the irreducibility of this solution, see [4]. The same procedure can be applied to the Navier–Stokes equations.

4. Conclusions

The notion of weak transversality considerably extends the applicability of Lie's reduction method. It is useful in the determination of either group invariant or (irreducible) partially invariant solutions of systems of PDE's and can be used with other complementary methods to study physical problems. Missing at this stage are criteria which tell us which approach based on Lie group theory will be more fruitful.

Acknowledgements

This paper is based on my joint work in collaboration with A. M. Grundland and P. Winternitz.

[1] P. J. Olver, *Application of Lie groups to Differential Equations* (Springer–Verlag, New York, 1986).
[2] L. V. Ovsiannikov, *Group Properties of Differential Equations* (in russian), (Novosibirsk, 1962); *Group Analysis of Differential Equations* (Academic Press, 1982).
[3] I. M. Anderson, M. E. Fels and C. G. Torre, Group invariant solutions without transversality, Comm. Math. Phys. **212**, 653 (2000).
[4] A. M. Grundland, P. Tempesta and P. Winternitz, Weak Transversality and Partially invariant solutions, arXiv:math–ph/0206003 (2002).
[5] A. M. Grundland and L. Lalague, Invariant and partially–invariant solutions of the equations describing a non-stationary and isentropic flow for and ideal and compressible fluid in (3+1) dimensions, J. Phys. A: Math. Gen **29**, 1723 (1996).
[6] J. Ondich, The reducibility of partially invariant solutions of systems of partial differential equations, Euro. J. Appl. Math. **6**, 329 (1995).
[7] E. L. Ince, *Ordinary differential equations* (Dover, New York, 1956).

MINI-SYMPOSIUM

Quantum Group Theory and Integrable Systems

Invited Talks

Inst. Phys. Conf. Ser. No 173: Mini-symposium
Paper presented at 24th Int. Coll. Group Theoretical Methods in Physics, Paris, France, 15–20 July 2002

q-Painleve systems arising from a q-difference analogue of the KP hierarchy

M Noumi

Department of Mathematics - Kobe University Rokko, Kobe 657-8501 Japan

Abstract. Given a pair (M, N) of positive integers, we introduce a system of nonlinear q-difference equations of Painlevé type in M time variables, by similarity reduction from a q-analogue of the N-reduced KP hierarchy. This system has affine Weyl group symmetry of type $A_{M-1}^{(1)} \times A_{N-1}^{(1)}$; both the time evolution and the Bäcklund transformations are described in terms of a birational action of the affine Weyl group on the space of $M \times N$ matrices. This talk is based on a joint work with K.Kajiwara and Y.Yamada (nlin.SI/0106029, nlin.SI/0112045).

Inst. Phys. Conf. Ser. No 173: Mini-symposium
Paper presented at 24th Int. Coll. Group Theoretical Methods in Physics, Paris, France, 15–20 July 2002
©*2003 IOP Publishing Ltd*

On constructions of Fock Space

A Ram

Department of Mathematics University of Wisconsin-Madison 480 Lincoln Drive Madison, WI 53706 USA

Abstract.

The Fock space representation for quantum groups has taken a central role in the connection between representations of quantum groups, Hecke algebras, and geometry. Different ways of constructing the Fock space (the basic representation of the affine quantum group) lead to different points of view.

Inst. Phys. Conf. Ser. No 173: Mini-symposium
Paper presented at 24th Int. Coll. Group Theoretical Methods in Physics, Paris, France, 15–20 July 2002
©2003 IOP Publishing Ltd

Representations of quantum groups: constructions and characters from a Hopf algebraic point of view.

M Rosso

DMA ENS 45, rue d'Ulm - F 75230 Paris cedex 05 France

Abstract.

We give concrete realizations of finite dimensional representations of quantized enveloping algebras using quantized shuffles and some Hopf algebraic constructions. This leads to character formulas. This also applies to the root of unity case.

Inst. Phys. Conf. Ser. No 173: Mini-symposium
Paper presented at 24th Int. Coll. Group Theoretical Methods in Physics, Paris, France, 15–20 July 2002
©*2003 IOP Publishing Ltd*

On examples of locally compact quantum groups

Leonid Vainerman

Département de Mathématiques, Université de Caen, Campus II. Boulevard de Maréchal
Juin, B.P. 5186, F-14032 Caen Cedex, France; email:Leonid.Vainerman@math.unicaen.fr

Abstract. We give a short overview of some recent examples of locally compact quantum groups.

The first special class of objects which are now called locally compact (l.c.) quantum groups was introduced in 1961 by G.I. Kac in order to generalize classical results of harmonic analysis and duality on unimodular groups. Later on, this theory was extended by G.I. Kac-L. Vainerman and independently by M. Enock-J.-M. Schwartz who called the corresponding objects *Kac algebras*. A Kac algebra is a Hopf-von Neumann algebra with an involutive antipode and a (left) invariant weight - a Haar measure. Commutative and co-commutative Kac algebras can be identified respectively with l.c. groups and their duals, but already in the 1960's G.I. Kac and V.G. Paljutkin gave concrete examples of objects which were neither groups nor their duals - the first known examples of quantum groups. Our reference to the Kac algebra theory is [5].

Quantum groups [3] gave new important examples of Hopf ⋆-algebras obtained by deformation of universal enveloping of Lie algebras or of function algebras on Lie groups. Their antipodes were not involutive, so they where not Kac algebras, and a more general theory was needed. As a first step, compact and discrete quantum groups were constructed, see e.g. [18], [4]. Later on, a more general theory was proposed in [17].
    Already in the Kac algebra setting, it was known that all the information on the quantum group could be encoded in one object - a *multiplicative unitary* which was taken in [1] as a starting point of the theory.
    A general definition of a l.c. quantum group was proposed in [9], [10] along the lines of Kac algebras, but with a weaker set of axioms, and the theory extending that of Kac algebras was developed.
    A l.c. quantum group is a collection $\mathcal{G} = (A, \Delta, \phi, \psi)$, where A is either a C^*- or a von Neumann algebra equipped with a co-associative comultiplication $\Delta : A \to A \otimes A$ and two faithful semi-finite normal weights ϕ and ψ - right and left Haar measures. The antipode is not explicitly present in this definition, but can be constructed from the above data using the multiplicative unitary, canonically associated with $\mathcal{G}$. Kac algebras, compact and discrete quantum groups are special cases of a l.c. quantum group, and all important concrete examples of operator algebraic quantum groups fit into this framework. One can find an exposition of this theory in [11]. To simplify the notations, in what follows we denote a l.c. quantum group by (A, Δ); usually we deal with the case when A is a von Neumann algebra and $\Delta : A \to A \otimes A$ is a normal monomorphism of von Neumann algebras.
    Algebraically, examples of l.c. quantum groups are usually given in terms of generators of certain Hopf ⋆-algebras subject to some commutation relations. It is much harder to represent these generators as (often unbounded) operators acting on a Hilbert space and to give a meaning to the relations of commutation between them. Finally, it is even more difficult to associate an operator algebra with the above system of operators and commutation relations

and to construct comultiplication, antipode and invariant weights as applications related to this algebra. There is no general approach to these highly nontrivial problems, and one must design specific methods in each specific case. Recent examples of this type can be found in [19], [20]. Also in [8], a l.c. quantum group related to $SU_q(1,1)$ was constructed.

There are other examples of l.c. quantum groups which are easier to construct. For instance, given a non-commutative l.c. group G, one can replace the canonical co-commutative comultiplication Δ on the von Neumann algebra $\mathcal{L}(G)$ of the left regular representation of G with a new comultiplication of the form $\Delta_\Omega(\cdot) = \Omega\Delta(\cdot)\Omega^{-1}$, where Ω is an element from $\mathcal{L}(G) \otimes \mathcal{L}(G)$ such that Δ_Ω remains co-associative. This construction (called *twisting*) was developed purely algebraically by V.G. Drinfeld, and on an operator algebraic level in [6], [16] and [13], where numerous concrete examples were obtained as well. Note that in a sense dual approach was proposed by M. Rieffel.

Another construction has been developed in [7]. Given two finite groups, G_1 and G_2, viewed respectively as a co-commutative l.c. quantum group $(\mathcal{L}(G_1), \Delta_1)$ and a commutative l.c. quantum group $(L^\infty(G_2), \Delta_2)$, let us try to find a ring group (A, Δ) which makes the sequence

$$(L^\infty(G_2), \Delta_2) \to (A, \Delta) \to (\mathcal{L}(G_1), \Delta_1) \tag{1}$$

exact. G.I. Kac explained that: 1) (A, Δ) exists if and only if G_1 and G_2 are subgroups of a group G such that $G_1 \cap G_2 = \{e\}$ and $G = G_1 G_2$. Equivalently, G_1 and G_2 must act on each other (as on sets), and these actions must be compatible. 2) To get all possible (A, Δ) (they are called *extensions* of $(L^\infty(G_2), \Delta_2)$ by $(\mathcal{L}(G_1), \Delta_1)$), one must find all possible 2-cocycles for the above mentioned actions, compatible in certain sense. Under these conditions, [7] gives the explicit construction of (A, Δ) (the cocycle bicrossed product construction). Kac-Paljutkin examples mentioned above are exactly of this type. Later on, both algebraic and analytic aspects of this construction were intensively studied by S. Majid, see e.g. [12], who gave also a number of examples of operator algebraic quantum groups, some of them were not Kac algebras.

Very recently, the theory of extensions of the form (1), with l.c. G_1 and G_2, has been developed in [14]. If G_1 and G_2 are Lie groups, instead of the condition $G = G_1 G_2$, one should require $G_1 G_2$ to be an open dense subset of G, as suggested by S. Baaj-G. Skandalis. Then, for the corresponding *Lie algebras* we have $g = g_1 \oplus g_2$ — the direct sum of vector spaces. So, to construct examples of l.c. quantum groups, one can start with such a decomposition of Lie algebras and try to construct a corresponding pair of groups (G_1, G_2). This problem proved to be not so easy to resolve (typically, one must deal with non-connected Lie groups), and often it has no solution at all. In [15] the case of complex and real Lie groups G_1 and G_2 of low dimentions was studied in detail, a complete classification of the corresponding l.c. quantum groups with two or three generators was obtained, and all the ingredients of their structure were computed, as well as their infinitesimal objects (Hopf $\star$-algebras and Lie bialgebras).

Using the same construction for some l.c. groups coming from number theory, a completely different example of "non-regular" l.c. quantum group was constructed in [2].

1. References

[1] S. BAAJ & G. SKANDALIS, Unitaires multiplicatifs et dualité pour les produits croisés de C^*-algèbres. *Ann. Scient. Ec. Norm. Sup.*, 4e série, **26** (1993), 425–488.

[2] S. BAAJ, G. SKANDALIS & S. VAES, Non-semi-regular quantum groups coming from number theory. *Preprint* (2002).

[3] V.G. DRINFELD, Quantum groups. *Proceedings IC Berkeley* (1986), 798–820.

[4] E. EFFROS & Z.-J. RUAN, Discrete quantum groups : The Haar measure. *Int. J. Math.* **5**, No.5 (1994), 681–723.

[5] M. ENOCK & J-M. SCHWARTZ, Kac Algebras and Duality of Locally Compact Groups, Springer-Verlag, 1992.

[6] M. ENOCK & L. VAINERMAN, Deformation of a Kac Algebra by an Abelian Subgroup. *Comm. Math. Phys.* **66** (1993), 619–650.

[7] G.I. KAC, Extensions of groups to ring groups. *Math. USSR Sbornik* **5** (1968), 451–474.

[8] E. KOELINK & J. KUSTERMANS, A locally compact quantum group analogue of the normalizer of $SU(1,1)$ in $SL(2,C)$. *Preprint math.QA/0105117* (2001).

[9] J. KUSTERMANS & S. VAES, Locally compact quantum groups. *Ann. Scient. Ec. Norm. Sup.* **33 (6)** (2000), 837–934.

[10] J. KUSTERMANS & S. VAES, Locally compact quantum groups in the von Neumann algebraic setting. To appear in *Math. Scand* (2000).

[11] J. KUSTERMANS, S. VAES, L. VAINERMAN, A. VAN DAELE & S.L. WORONOWICZ, Locally Compact Quantum Groups. To appear in *Lecture Notes for the school on Noncommutative Geometry and Quantum Groups in Warsaw (17-29 September 2001). Banach Center Publications*.

[12] S. MAJID, Foundations of quantum group theory. Cambridge University Press (1995).

[13] D. NIKSHYCH, K_0-ring and twisting of finite dimentional Hopf algebras. *Comm. in Algebra* **26**, No. 1 (1998), 321–342.

[14] S. VAES & L. VAINERMAN, Extensions of locally compact quantum groups and the bicrossed product construction. *Adv. in Math.* **175**, No. 1 (2003), 1–101.

[15] S. VAES & L. VAINERMAN, On low-dimensional locally compact quantum groups. *Locally compact quantum groups and groupoids, IRMA Lectures in Mathematics and Theoretical Physics*, **2**. Editor L. Vainerman. Walter de Gruyter (2003), 127–187.

[16] L. VAINERMAN, 2-Cocycles and Twisting of Kac Algebras. *Comm. Math. Phys.* **191** (1998), 697–721.

[17] A. VAN DAELE, An algebraic framework for group duality. *Adv. in Math.* **140** (1998), 323–366.

[18] S.L. WORONOWICZ, Compact matrix pseudogroups. *Comm. Math. Phys.* **111** (1987), 613-665.

[19] S.L. WORONOWICZ, Quantum $'az + b'$ group on complex plane. *Int. J. Math.* **12**, No. 4 (2001), 461-503.

[20] S.L. WORONOWICZ & S. ZAKRZEVSKI, Quantum $'ax + b'$ group. *Preprint KMMF* (1999).

Inst. Phys. Conf. Ser. No 173: Mini-symposium
Paper presented at 24th Int. Coll. Group Theoretical Methods in Physics, Paris, France, 15–20 July 2002
©2003 IOP Publishing Ltd

Haar weight on some quantum groups

S. L. Woronowicz

Department of Mathematical Methods in Physics, Faculty of Physics, University of Warsaw,
Hoża 74, 00-682 Warszawa, Poland.
stanislaw.woronowicz@fuw.edu.pl

Abstract. We present a number of examples of locally compact quantum groups. These are quantum deformations of the group of affine transformations $\mathbb{R}$ ('$ax+b$' group) and $\mathbb{C}$ (Gz group). Starting from a modular multiplicative unitary W we find (under certain technical assumption) a simple formula expressing the (right) Haar weight on the quantum group associated with W. The formula works for quantum '$ax+b$' and '$az+b$' groups.

1. Introduction

It is difficult to overestimate the role of multiplicative unitaries in the present theory of locally compact quantum groups. The concept introduced by Baaj and Skandalis [1] is present in purely theoretical considerations in axiomatic formulation of the theory [2, 3, 4]. It is also very useful, when one considers particular examples of quantum groups (cf. examples presented is Section 4). Usually having the Haar weight h on a quantum group $G = (A, \Delta)$ one uses the GNS-construction to define a Hilbert space H and an embedding $A \subset B(H)$. Then the multiplicative unitary W is defined by taking the linear mapping:

$$A \otimes_{\text{alg}} A \ni a \otimes b \longmapsto \Delta(a)(I \otimes b) \in A \otimes A$$

and pushing it down to the level of $H \otimes H$. This is so called Kac-Takesaki operator [4]. On the other hand in a number of cases, a multiplicative unitary is found independently and the general theory [1, 9, 5] is used to construct the corresponding locally compact quantum group. This was the way used in recent works [11, 12, 6] devoted to quantum deformations of the groups of affine transformations of $\mathbb{R}$ and $\mathbb{C}$. In general the multiplicative unitary that we start with need not coincide with the Kac-Takesaki operator and the problem of existence of the Haar weight must be investigated.

To pass from a multiplicative unitary W to the corresponding quantum group one has to assume that W have certain properties. For example it is sufficient to assume manageability [9] or at least modularity [5].

The Haar weight on the quantum '$ax+b$' and '$az+b$' groups was found by Van Daele some time ago [8]. To this end he had to use very particular properties of the groups. In autumn 2002, I played with the quantum '$az+b$' group acting on the straight line girandole Γ (see formula (11) below). The aim was to show that a certain measure on Γ is relatively invariant. The problem reduced to a complicated formula relating the Fourier transform of a special function appearing in the theory with the holomorphic continuation of the function itself. To my surprise that was the same formula as the one used earlier to prove the modularity of the multiplicative unitary staying behind the quantum '$az+b$' group. It was a strong indication that the (relative) invariance of some measure on a homogenous space is closely related to the modularity of the corresponding multiplicative unitary.

Trying to explain this phenomenon when the homogenous space is the group itself (the case of proper homogenous spaces will be considered in a separate paper). I arrived to a simple formula describing a Haar weight on a quantum group coming from a modular multiplicative unitary. It is a special case of a formula found earier by Van Daele [8, 7]. The formula works if a certain technical assumption is satisfied. As a rule this is not the case for manageable multiplicative unitaries. Therefore the modularity introduced in [5] seems to be of real importance.

We shortly describe the content of the paper. In section 2 we recall the basic concepts: multiplicative unitaries, modularity and manageability and the passage from modular multiplicative unitaires to quantum groups. Next we explain what the Haar weight is. Then after a short discussion of the trace and related weights we formulate our main theorem. It contains an explicit formula introducing a faithful lower semicontinuous weight h on a quantum group of the kind mentioned above. If h is locally finite then according to the theorem, h is the Haar weight. Section 3 contains the proof of the formula. It is very straightforward and refers directly to the concept of the right invariance.

The examples are presented in Section 4. These are quantum '$ax+b$' and '$az+b$' groups with different values of the deformation parameter q. Only the result concerning '$az+b$' group with $\Im q \neq 0$ and $|q| \neq 1$ is new (the others were obtained earlier by Van Daele). The main concern was to show that in considered cases our technical assumption (local finiteness) is satisfied. Nevertheless we provided all the necessary information to have a complete description of the considered groups. In particular the modular multiplicative unitaries are presented in full detail including the definitions of the special functions used.

The paper uses the advanced concepts of the theory of C*-algebras and locally compact quantum groups in a very limited way. In the case of any problem concerning the notation we refer to our earlier papers.

2. Basic concepts and the main theorem

Let H be a separable Hilbert space. A unitary operator W acting on $H \otimes H$ is by definition a multiplicative unitary if (using the leg numbering notation) we have:

$$W_{23}W_{12} = W_{12}W_{13}W_{23}.$$

This is the famous pentagon equation of Baaj and Skandalis. An important property of multiplicative unitaries is *the modularity* introduced in [5]. We say a multiplicative unitary W acting on $H \otimes H$ is modular if there exist two strictly positive selfadjoint operators $\widehat{Q}$ and Q on H and a unitary $\widetilde{W} \in B(\overline{H} \otimes H)$ such that

$$W^*(\widehat{Q} \otimes Q)W = \widehat{Q} \otimes Q \qquad \text{and} \qquad (1)$$

$$(x \otimes y|W|z \otimes u) = \left(\overline{z} \otimes Qy\middle|\widetilde{W}\middle|\overline{x} \otimes Q^{-1}u\right) \qquad (2)$$

for all $x,z \in H$, $y \in D(Q)$ and $u \in D(Q^{-1})$. In this definition $\overline{H}$ denotes the Hilbert space complex conjugate to H and $H \ni z \mapsto \overline{z} \in \overline{H}$ is the canonical antiunitary identification.

Demanding in addition the equality $\widehat{Q} = Q$ we obtain a stronger condition called *the manageability*. The theory of manageable multiplicative unitaries is developed in [9]. It is shown there that they give rise to objects that in many respects behave like locally compact topological groups. For the purpose of this paper these objects are called (locally compact) quantum groups. In [5] we have shown that all the essential results of the theory of manageable multiplicative unitaries hold for modular ones.

For any $x, y \in H$, ω_{xy} will denote the linear functional on $B(H)$ defined by the formula:
$$\omega_{xy}(a) = (y|a|x)$$
for any $a \in B(H)$. Functionals ω_{xy} are normal. Moreover the set $B(H)_*$ of all normal functionals on $B(H)$ coincides with the norm closed linear span of all functionals of the form ω_{xy}, where $x, y \in H$. If $y = x$ then we write ω_x instead of ω_{xx}. Clearly ω_x is a positive linear functional on $B(H)$.

We have to devote a few lines to the trace and related weights on $B(H)$. The functionals ω_x may by applied to unbounded positive selfadjoint operators: If $T = S^*S$ where S is a closed operator, then by definition
$$\omega_x(T) = \begin{cases} \|Sx\|^2 & \text{if } x \in \mathscr{D}(S) \\ \infty & \text{otherwise} \end{cases}$$
One can easily show that the right hand side is independent of the choice of S in the formula $T = S^*S$.

Let S be a closed operator on H and $c \in B(H)$. We choose two orthonormal basis $e = (e_n)_{n=1,2,...}$ and $f = (f_m)_{m=1,2,...}$ on H. Assume that $e_n \in \mathscr{D}(S)$ for all $n = 1, 2, \ldots$. Then
$$\sum_n \omega_{S^*e_n}(c^*c) = \sum_m \omega_{c^*f_m}(S^*S). \tag{3}$$

Indeed both sides equal to $\sum_{nm} |(S^*e_n|c^*f_m)|^2$. The reader should notice that the left hand side of the above formula is independent of the basis f whereas the right hand side does not depend on e. Consequently both sides are independent of the choice of the bases. With a certain abuse of notation, for any positive $a \in B(H)$ we write
$$\sum_n \omega_{S^*e_n}(a) = \mathrm{Tr}(SaS^*).$$

Clearly the mapping $B(H)^+ \ni a \mapsto \mathrm{Tr}(SaS^*) \in \mathbb{R}_+ \cup \{+\infty\}$ is a normal semifinite weight on $B(H)$. In what follows, $S = \widehat{Q}$ will be strictly positive selfadjoint. One can easily show that $\mathrm{Tr}(\widehat{Q}c^*c\widehat{Q}) < \infty$ implies that the range of c^* is contained in the domain of $\widehat{Q}$: $c^*H \subset \mathscr{D}(\widehat{Q})$.

The way from a modular multiplicative unitary W to the corresponding quantum group $G = (A, \Delta)$ is short:
$$A = \{(\omega \otimes id)W : \omega \in B(H)_*\}^{\text{norm closure}} \tag{4}$$
and for any $a \in A$
$$\Delta(a) = W(a \otimes I)W^*. \tag{5}$$

It is known that A is a C^*-algebra and that $\Delta \in Mor(A, A \otimes A)$. Moreover $\Delta(a)(I \otimes b) \in A \otimes A$ for any $a, b \in A$. Using this fact one can show that for any $a \in A$ and any $\varphi \in A'$ (A' denotes the dual of A), the convolution product
$$\varphi * a = (id \otimes \varphi)\Delta(a)$$
belongs to A. If a and φ are positive then $\varphi * a$ is positive.

The (right) Haar weight on the quantum group $G = (A, \Delta)$ is by definition (see e.g. [4]), a lower semicontinuous faithful locally finite weight h on A such that
$$h(\varphi * a) = \varphi(I)h(a) \tag{6}$$
for any positive functional φ on A and any positive $a \in A$ such that $h(a) < \infty$. We recall that h is locally finite if the set $\{c \in A : h(c^*c) < \infty\}$ is dense in A. This concept is the C^*-version of the semifiniteness used for normal weights on von Neumann algebras.

The following theorem is a simplified version of a result of A. Van Daele [8, 7]:

Theorem 2.1 *Let $G = (A, \Delta)$ be the quantum group associated with a modular multiplicative unitary W and Q and $\widehat{Q}$ be strictly positive selfadjoint operators entering formulae (2) and (1). For any positive $a \in A$ we put*

$$h(a) = \mathrm{Tr}(\widehat{Q}a\widehat{Q}). \tag{7}$$

Clearly h is faithful lower semicontinuous weight h on A. Assume that h is locally finite. Then h is a (right) Haar weight on the quantum group G.

3. The proof.

This section is devoted to the proof of theorem 2.1. We have to show (6). To this end we rewrite (2) in a form involving $\widehat{Q}$.

Lemma 3.1 *Let $y, u \in H$ and $x, z \in D(\widehat{Q})$. Then*

$$\left(\widehat{Q}x \otimes y \left| W \right| z \otimes u\right) = \left(\overline{\widehat{Q}z} \otimes y \left| \widetilde{W} \right| \overline{x} \otimes u\right). \tag{8}$$

Proof. Assume for the moment that $y \in \mathscr{D}(\widehat{Q}^{-1})$ and $u \in \mathscr{D}(\widehat{Q})$. Inserting in (2), $\widehat{Q}z$, $Q^{-1}y$ and Qu instead of z, y and u respectively we obtain

$$\left(x \otimes Q^{-1}y \left| W \right| \widehat{Q}z \otimes Qu\right) = \left(\overline{\widehat{Q}z} \otimes y \left| \widetilde{W} \right| \overline{x} \otimes u\right).$$

Now, taking into account (1) we get (8). Using a simple continuity argument one can extent the validity of (8) to all $y, u \in H$.

The formula (8) is appropriate to prove the right-invariance (6) with $\varphi \in B(H)_*$. However to show (6) in full generality one has to strengthen the above lemma. The C*-algebra of all compact operators acting on H will be denoted by $\mathscr{K}(H)$. According to [5, Statement 2 of Theorem 2.3], the operator W belongs to the multiplier algebra of $\mathscr{K}(H) \otimes A$; $W \in M(\mathscr{K}(H) \otimes A)$. Therefore for any representation π of the C*-algebra A acting on a Hilbert space K, $\pi \in Mor(A, \mathscr{K}(K))$ and $(id \otimes \pi)W \in M(\mathscr{K}(H) \otimes \mathscr{K}(K)) = M(\mathscr{K}(H \otimes K)) = B(H \otimes K)$. In other words $(id \otimes \pi)W$ is a unitary operator acting on $H \otimes K$. It will be denoted by W_π:

$$W_\pi = (id \otimes \pi)W.$$

We shall use the following

Lemma 3.2 *There exists a unitary operator $\widetilde{W}_\pi$ acting on $\overline{H} \otimes K$ such that*

$$\left(\widehat{Q}x \otimes y \left| W_\pi \right| z \otimes u\right) = \left(\overline{\widehat{Q}z} \otimes y \left| \widetilde{W}_\pi \right| \overline{x} \otimes u\right) \tag{9}$$

for any $y, u \in K$ and $x, z \in D(\widehat{Q})$.

Proof. Using [5, Statement 6.(ii) of Theorem 2.3] one can easily show that $\widetilde{W} \in M(\mathscr{K}(\overline{H}) \otimes A)$. Therefore setting $\widetilde{W}_\pi = (id \otimes \pi)\widetilde{W}$ we obtain a unitary operator acting on $\overline{H} \otimes K$. Formula (8) shows that

$$(\omega_{z,\widehat{Q}x} \otimes id)W = (\omega_{\overline{x},\overline{\widehat{Q}z}} \otimes id)\widetilde{W}.$$

Applying π to both sides we obtain

$$(\omega_{z,\widehat{Q}x} \otimes id)W_\pi = (\omega_{\overline{x},\overline{\widehat{Q}z}} \otimes id)\widetilde{W}_\pi.$$

Therefore

$$(\omega_{z,\widehat{Q}x} \otimes \omega_{uy})W_\pi = (\omega_{\overline{x},\overline{Q}z} \otimes \omega_{uy})\widetilde{W}_\pi.$$

The last formula clearly coincides with (9).

Now the proof of (6) is a matter of simple computation. Let φ be a positive linear functional on A. Using the GNS construction we find a representation π of A acting on a Hilbert space K and a vector $x \in K$ such that

$$\varphi(a) = (x|\pi(a)|x) = \omega_x(\pi(a)),$$

for all $a \in A$. Then

$$\varphi * a = (id \otimes \omega_x)(id \otimes \pi)\Delta(a)$$
$$= (id \otimes \omega_x)(id \otimes \pi)(W(a \otimes I)W^*)$$
$$= (id \otimes \omega_x)(W_\pi(a \otimes I)W_\pi^*),$$

Assume that a is positive. Then $a = c^*c$, where $c \in A$. Let $(e_n)_{n=1,2,\dots}$ be an orthonormal basis in H such that $e_n \in \mathscr{D}(\widehat{Q})$ for all $n = 1,2,\dots$. Using (7) we obtain

$$h(\varphi * a) = \sum_n (\omega_{\widehat{Q}e_n} \otimes \omega_x)(W_\pi(c^*c \otimes I)W_\pi^*)$$
$$= \sum_n \left(\widehat{Q}e_n \otimes x \Big| W_\pi(c^*c \otimes I)W_\pi^* \Big| \widehat{Q}e_n \otimes x\right)$$
$$= \sum_n \left\| (c \otimes I)W_\pi^*(\widehat{Q}e_n \otimes x) \right\|^2,$$

To proceed with our computations we chose an orthonormal basis $(\varepsilon_k)_{k=1,2,\dots}$ in K. Then

$$h(\varphi * a) = \sum_{nmk} \left| \left(e_m \otimes \varepsilon_k \Big| (c \otimes I)W_\pi^* \Big| \widehat{Q}e_n \otimes x \right) \right|^2$$
$$= \sum_{nmk} \left| \left(\widehat{Q}e_n \otimes x \Big| W_\pi \Big| c^*e_m \otimes \varepsilon_k \right) \right|^2.$$

If $h(a) < \infty$ then $c^*e_m \in \mathscr{D}(\widehat{Q})$ for all $m = 1,2,\dots$. Inserting in (9), e_n, x, c^*e_m and ε_k instead of x, y, z and u respectively we obtain

$$\left(\widehat{Q}e_n \otimes x \Big| W_\pi \Big| c^*e_m \otimes \varepsilon_k\right) = \left(\overline{\widehat{Q}c^*e_m} \otimes x \Big| \widetilde{W}_\pi \Big| \overline{e_n} \otimes \varepsilon_k\right).$$

Therefore

$$h(\varphi * a) = \sum_{nmk} \left| \left(\overline{\widehat{Q}c^*e_m} \otimes x \Big| \widetilde{W}_\pi \Big| \overline{e_n} \otimes \varepsilon_k\right) \right|^2$$
$$= \sum_{nmk} \left| \left(\overline{e_n} \otimes \varepsilon_k \Big| \widetilde{W}_\pi^* \Big| \overline{\widehat{Q}c^*e_m} \otimes x\right) \right|^2$$
$$= \sum_m \left\| \widetilde{W}_\pi^* \left(\overline{\widehat{Q}c^*e_m} \otimes x\right) \right\|^2 = \sum_m \left\| \overline{\widehat{Q}c^*e_m} \otimes x \right\|^2$$
$$= \|x\|^2 \sum_m \left\| \widehat{Q}c^*e_m \right\|^2,$$

where in the forth step we used the unitarity of $\widetilde{W}_\pi$. On the other hand $\|x\|^2 = \varphi(I)$ and by formula (3):

$$\sum_m \left\| \widehat{Q}c^*e_m \right\|^2 = \sum_m \omega_{c^*e_m}(\widehat{Q}^2) = \sum_n \omega_{\widehat{Q}e_n}(a) = h(a).$$

Inserting these data into our computations we obtain (6). The proof of Theorem 2.1 is complete.

4. Examples

In this section we shall show that Theorem 2.1 allows us to reproduce in a simple way the results of Van Daele concerning the Haar measures on quantum '$ax+b$' and '$az+b$' groups. We start with '$az+b$' groups [6, 11]. These groups are quantum deformations of the group of affine transformations of complex plane $\mathbb{C}$.

Let Γ be a selfdual subgroup of the multiplicative group of non-zero complex numbers and α be so called Fresnel function on Γ. The particular form of Γ and α depends on the value of the deformation parameter q. The following value of q were considered so for:

1. q is real and belongs to the open interval $]0, 1[$. The quantum '$az+b$' group with this value of q is presented in [11, Appendix A]. In this case Γ is the system of concentric circles with radii forming the geometric progression:

$$\Gamma = \left\{ q^{i\theta+k} : \theta \in \mathbb{R}, k \in \mathbb{Z} \right\} \quad \text{and} \quad \alpha(q^{i\theta+k}) = q^{i\theta k}. \tag{10}$$

We say that Γ is a *target plate*.

2. $q = e^{\frac{2\pi i}{N}}$ where N is an even integer strictly larger than 2. The quantum '$az+b$' group with this value of q is presented in [11]. In this case Γ is the system of straight lines dividing $\mathbb{C}$ into N equal sectors:

$$\Gamma - \left\{ q^k r : k \in \mathbb{Z}, r > 0 \right\} \quad \text{and} \quad \alpha(q^k r) - e^{\frac{\pi i}{N} k^2} e^{\frac{N}{4\pi i} (\log r)^2}. \tag{11}$$

We say that Γ is a *straight line girandole*.

3. $q = e^{\frac{1}{\rho}}$ where $\rho = c + \frac{iN}{2\pi}$, $c < 0$, N is an even integer and $N \neq 0$. The quantum '$az+b$' group with this value of q is presented in [6]. In this case Γ is the system of N logarithmic spirals dividing $\mathbb{C}$ into equal sectors:

$$\Gamma = \left\{ e^{\frac{k+it}{\rho}} : k \in \mathbb{Z}, t \in \mathbb{R} \right\} \quad \text{and} \quad \alpha\left(e^{\frac{n+it}{\rho}} \right) = e^{i\Im \frac{(k+it)^2}{2\rho}}. \tag{12}$$

We say that Γ is a *logarithmic girandole*.

In what follows we shall treat all three cases simultaneously. For any $\gamma, \gamma' \in \Gamma$ we set

$$\chi(\gamma, \gamma') = \frac{\alpha(\gamma\gamma')}{\alpha(\gamma)\alpha(\gamma')}$$

It is easy to see that χ is a bicharacter on $\Gamma \times \Gamma$.

Let μ be the Haar measure on Γ and $H = L^2(\Gamma, \mu)$ be the corresponding Hilbert space. We chose the normalization of μ in such a way that the Fourier transform

$$(\mathscr{F}g)(\gamma) = \int_\Gamma \chi(\gamma, \gamma') g(\gamma') d\mu(\gamma')$$

is a unitary operator acting on H. We shall use two operators a and b acting on H. Operator b is the multiplication operator:

$$(bx)(\gamma) = \gamma x(\gamma).$$

Clearly b is a normal (unbounded) operator acting on H with $Spb = \overline{\Gamma}$, where $\overline{\Gamma} = \Gamma \cup \{0\}$ is the closure of Γ. The second operator $a = \mathscr{F} b \mathscr{F}^*$. Then for any $\gamma' \in \Gamma$ we have:

$$(\chi(a, \gamma')x)(\gamma) = x(\gamma\gamma').$$

Operators a and b satisfy the following commutation relations:

$$\chi(a,\gamma)\chi(b,\gamma') = \chi(\gamma,\gamma')\chi(b,\gamma')\chi(a,\gamma) \tag{13}$$

for any $\gamma, \gamma' \in \Gamma$. In brief (see [6, 11] for details):

$$ab = q^2 ba, \quad a^*b = ba^*.$$

The modular multiplicative unitary producing the quantum '$az+b$' group is given by

$$W = \mathbb{F}_q(b^{-1}a \otimes b)\chi(b^{-1} \otimes I, I \otimes a), \tag{14}$$

where $\mathbb{F}_q$ is a continuous function on $\overline{\Gamma}$ called *the quantum exponential function*. For $|q| < 1$ (cases 1 and 3) $\mathbb{F}_q$ is given by the formula:

$$\mathbb{F}_q(\gamma) = \prod_{k=0}^{\infty} \frac{1 + \overline{q^{2k}\gamma}}{1 + q^{2k}\gamma} \tag{15}$$

For $q = e^{\frac{2\pi i}{N}}$ (case 2), $\mathbb{F}_q(\gamma) = F_N(q^{-2}\gamma)$, where F_N is the function introduced in [11, formula 1.5]. We recall that

$$F_N(q^k r) = \begin{cases} \displaystyle\prod_{s=1}^{\frac{k}{2}} \left(\frac{1+q^{2s}r}{1+q^{-2s}r}\right) \frac{f_0(qr)}{1+r} & \text{for } k\text{ - even} \\[4mm] \displaystyle\prod_{s=0}^{\frac{k-1}{2}} \left(\frac{1+q^{2s+1}r}{1+q^{-2s-1}r}\right) f_0(r) & \text{for } k\text{ - odd} \end{cases}$$

where

$$f_0(z) = \exp\left\{\frac{1}{\pi i} \int_0^{\infty} \log(1 + a^{-\frac{N}{2}}) \frac{da}{a + z^{-1}}\right\}.$$

It is known that the operator W introduced by (14) is a modular multiplicative unitary with $Q = |a|$ and $\widehat{Q} = |b|$. The corresponding C*-algebra (4) is the crossed product algebra

$$A = \left\{g(a)f(b) : g \in C_\infty(\Gamma), f \in C_\infty(\overline{\Gamma})\right\}^{\text{norm closed linear envelope}}.$$

We recall that for any locally compact space Λ, $C_\infty(\Lambda)$ denotes the C*-algebra of all continuous functions vanishing at infinity on Λ. Consequently $C_\infty(\Gamma) = \{g \in C_\infty(\overline{\Gamma}) : g(0) = 0\}$.

For any positive $r \in A$ we set:

$$h(r) = \text{Tr}(|b| \, r \, |b|). \tag{16}$$

Let $c = g(a)f(b)$. One can verify that the operator $c\widehat{Q} = c|b|$ is an integral operator:

$$(c|b|x)(\gamma') = \int_\Gamma K_c(\gamma',\gamma)x(\gamma)d\mu(\gamma)$$

with the kernel $K_c(\gamma',\gamma) = (\mathscr{F}g)(\gamma'\gamma^{-1})f(\gamma)|\gamma|$. Therefore

$$\begin{aligned} h(c^*c) &= \int_{\Gamma \times \Gamma} |K_c(\gamma',\gamma)|^2 \, d\mu(\gamma')d\mu(\gamma) \\ &= \int_{\Gamma \times \Gamma} |(\mathscr{F}g)(\gamma'\gamma^{-1})|^2 |f(\gamma)|^2 \, |\gamma|^2 d\mu(\gamma')d\mu(\gamma) \\ &= \int_{\overline{\Gamma}} |g(\gamma)|^2 d\mu(\gamma) \int_{\overline{\Gamma}} |f(\gamma)|^2 \, |\gamma|^2 d\mu(\gamma), \end{aligned} \tag{17}$$

where in last step we used the right invariance of the Haar measure μ and the unitarity of $\mathscr{F}$. The reader should notice that the measure $d\gamma = |\gamma|^2 d\mu(\gamma)$ is locally finite on $\overline{\Gamma}$. Therefore the intersection $L^2(\overline{\Gamma}, d\gamma) \cap C_\infty(\overline{\Gamma})$ is dense in $C_\infty(\overline{\Gamma})$. On the other hand, obviously, $L^2(\Gamma, \mu) \cap C_\infty(\Gamma)$ is dense in $C_\infty(\Gamma)$. Formula (17) shows now that $\{c \in A : h(c^*c) < \infty\}$ is dense in A and using Theorem 2.1 we get

Theorem 4.1 *Formula* (16) *defines the Haar weight on the quantum 'az + b' group.*

Operator Q appearing in (2) is of great interest. It implements a one parameter group $(\tau_t)_{t \in \mathbb{R}}$ of *scaling automorphisms* of A: $\tau_t(a) = Q^{2it} a Q^{-2it}$. In our case $Q = |a|$ and $\widehat{Q} = |b|$. Using the commutation relations (13) one can show that $Q^{2it} \widehat{Q} Q^{-2it} = |q^{2it}| \widehat{Q}$. Using (7) we obtain the relative invariance of the Haar weight:

$$h \circ \tau_t = |q^{-4it}| \, h$$

for any $t \in \mathbb{R}$. If q is not real then the Haar weight is not invariant under scaling group. Instead we have the relative invariance (up to a positive factor). The relative invariance of the Haar weight appeared first in the paper of Kustermans and Vaes [2] as a consequence of their axioms. At that time all known quantum groups had the Haar weight invariant under scaling group. Alfons Van Daele was the first person, who noticed that this may not be true for quantum 'ax + b' and 'az + b' group. That was the main reason why he computed the Haar weights on these groups finding the first examples of the phenomenon foreseen by Kustermans and Vaes.

Remark 4.2 The infinite product (15) is not convergent for $q = e^{\frac{2\pi i}{N}}$ because the factors $\frac{1 + q^{2k}\overline{\gamma}}{1 + q^{2k}\gamma}$ as well as partial products $\prod_{\ell=0}^{k} \frac{1 + q^{2\ell}\overline{\gamma}}{1 + q^{2\ell}\gamma}$ are periodic in k. However if in this case for any real positive r, one put

$$\prod_{k=0}^{\infty} \frac{1 + q^{-2k-1}r}{1 + q^{2k+1}r} = f_0(r)$$

$$\prod_{k=0}^{\infty} \frac{1 + q^{-2k}r}{1 + q^{2k}r} = \frac{f_0(qr)}{1 + r}$$

then the formula (15) works also for $q = e^{\frac{2\pi i}{N}}$.

The rest of the section is devoted to the quantum 'ax + b' group [12]. This is a quantum deformation of the group of affine transformations of $\mathbb{R}$. It is constructed and investigated in [12]. For this group the deformation parameter $q^2 = e^{-i\hbar}$, where $\hbar = \frac{\pi}{2k+3}$ and $k = 0, 1, 2, \ldots$. Let $H = L^2(\mathbb{R})$. We consider three operators acting on H:

$$(sx)(t) = i\left(t\frac{d}{dt} + \frac{1}{2}\right)x(t),$$
$$(bx)(t) = \quad tx(t),$$
$$(\beta x)(t) = \quad x(-t).$$

Operators b, s and β are selfadjoint. Furthermore β is unitary. The operator a is related to s by the formula $a = e^{-\hbar s}$. Clearly a is selfadjoint and strictly positive. We have the formulae: $\beta a = a\beta$, $\beta b = -b\beta$ and $ab = q^2 ba$, where the last equality stands for Zakrzewski relation (cf. [10, Section 2]). To introduce the multiplicative unitary related to the quantum 'ax + b' group we shall use the appropriate special function $F_\hbar$. This is a function of two variables: the first one runs over $\mathbb{R}$, whereas the second one equals ± 1. It is given by the formula

$$
F_\hbar(r,\rho) = \begin{cases} V_\theta(\log r) & \text{for} \quad r > 0 \\[2mm] \left[1 + i\rho |r|^{\frac{\pi}{\hbar}}\right] V_\theta\left(\log |r| - \pi i\right) & \text{for} \quad r < 0. \end{cases}
$$

where $\theta = \frac{2\pi}{\hbar}$ and V_θ is the meromorphic function on $\mathbb{C}$ such that

$$
V_\theta(x) = \exp\left\{\frac{1}{2\pi i}\int_0^\infty \log(1 + t^{-\theta})\frac{dt}{t + e^{-x}}\right\}
$$

for all $x \in \mathbb{C}$ such that $|\Im x| < \pi$. With this notation

$$
W = F_\hbar\left(e^{i\hbar/2}b^{-1}a \otimes b, (-1)^k \beta \otimes \beta\right)^* (|b| \otimes I)^{I\otimes is}.
$$

It is known (cf. [5]) that W is a modular multiplicative unitary with $Q = a^{\frac{1}{2}}$ and $\widehat{Q} = |b|^{\frac{1}{2}}$. The corresponding C*-algebra (4) is given by

$$
A = \overline{\left\{g(s)\left(f_1(b) + \beta f_2(b)\right) : g, f_1, f_2 \in C_\infty(\mathbb{R}), \ f_2(0) = 0\right\}}^{\text{norm closed linear envelope}}.
$$

For any positive $r \in A$ we set:

$$
h(r) = \mathrm{Tr}\left(|b|^{\frac{1}{2}} r |b|^{\frac{1}{2}}\right). \tag{18}
$$

Let $c = g(s)\left(f_1(b) + \beta f_2(b)\right)$. One can verify that the operator $c\widehat{Q} = c|b|^{\frac{1}{2}}$ is an integral operator:

$$
(c|b|^{\frac{1}{2}}x)(t') = \int_\mathbb{R} K_c(t', t)x(t)dt
$$

with the kernel

$$
K_c(t', t) = \begin{cases} |t'|^{-\frac{1}{2}} \widetilde{g}(\frac{t'}{t}) f_1(t) & \text{for } \frac{t'}{t} > 0, \\[2mm] |t'|^{-\frac{1}{2}} \widetilde{g}(-\frac{t'}{t}) f_2(t) & \text{for } \frac{t'}{t} < 0, \end{cases}
$$

where

$$
\widetilde{g}(\rho) = \frac{1}{2\pi}\int_\mathbb{R} g(\tau)\rho^{-i\tau}d\tau
$$

for any $\rho > 0$. Therefore

$$
h(c^*c) = \int_{\mathbb{R}\times\mathbb{R}} |K_c(t', t)|^2 \, dt'dt
$$

$$
= \int_0^\infty |\widetilde{g}(\rho)|^2 \frac{d\rho}{\rho} \int_\mathbb{R} \left(|f_1(t)|^2 + |f_2(t)|^2\right)dt
$$

$$
= \frac{1}{2\pi}\int_\mathbb{R} |g(\tau)|^2 d\tau \int_\mathbb{R} \left(|f_1(t)|^2 + |f_2(t)|^2\right)dt.
$$

This formula shows that $\{c \in A : h(c^*c) < \infty\}$ is dense in A and using Theorem 2.1 we get

Theorem 4.3 *Formula* (18) *defines the Haar weight on the quantum 'ax + b' group.*

Also in this case the Haar weight is not invariant with respect to the scaling group. Remembering that $Q = a^{\frac{1}{2}}$ and $\widehat{Q} = |b|^{\frac{1}{2}}$ and using the Zakrzewski relation we obtain $Q^{2it}\widehat{Q}Q^{-2it} = e^{\frac{\hbar t}{2}}\widehat{Q}$ and

$$h \circ \tau_t = e^{-\hbar t} h$$

for any $t \in \mathbb{R}$.

[1] S. Baaj et G. Skandalis, *Unitaries multiplicatifs et dualité pour les produits croisé de C*-algèbres*, Ann. Sci. Ec. Norm. Sup., 4e série, **26** (1993), 425-488.

[2] J. Kustermans and S. Vaes: Locally compact quantum groups. *Annales Scientifiques de l'Ecole Normale Suprieure*. 33 (6) (2000), 837–934.

[3] T. Masuda and Y. Nakagami, *A von Neumann algebra framework for the duality of the quantum groups*, Publ. RIMS Kyoto Univ., **30** (1994), 799-850.

[4] T. Masuda, Y. Nakagami and S.L. Woronowicz, *A C*-algebraic framework for the quantum groups*, Dept. of Math. Methods in Phys., University of Warsaw, preprint (2003).

[5] P.M. Sołtan and S.L. Woronowicz, *A remark on manageable multiplicative unitaries*, Letters on Math. Phys., **57** (2001), 239 - 252.

[6] P.M. Sołtan, *New deformations of the group of affine transformations of the plane*, Doctor dissertation (in Polish), University of Warsaw (2003).

[7] S. Vaes and A. Van Daele: The Heisenberg commutation relations, commuting squares and the Haar measure on locally compact quantum groups. Proceedings of the OAMP Conference, Constantza, 2001

[8] A. Van Daele: The Haar measure on some locally compact quantum groups, Preprint of the K.U. Leuven (2000), arXiv:math.OA/0109004v1 (2001).

[9] S.L. Woronowicz, *From multiplicative unitaries to quantum groups*, International J. Math., **7** (1996), 127-149.

[10] S.L. Woronowicz, *Quantum exponential function*, Reviews in Mathematical Physics, Vol. **12**, No. 6 (2000), 873 – 920.

[11] S.L. Woronowicz, *Quantum 'az + b' group on complex plane*, International J. Math., **12** (2001), 461-503.

[12] S.L. Woronowicz and S. Zakrzewski, *Quantum 'ax + b' group*, Reviews in Mathematical Physics, **14**, Nos 7 & 8 (2002), 797 – 828.

MINI-SYMPOSIUM

Quantum Group Theory and Integrable Systems

Contributions

Inst. Phys. Conf. Ser. No 173: Mini-symposium
Paper presented at 24th Int. Coll. Group Theoretical Methods in Physics, Paris, France, 15–20 July 2002
©2003 IOP Publishing Ltd

RTT presentation of ortho-symplectic super Yangians. Quasi-Hopf Deformation

D. Arnaudon[a], J. Avan[b], N. Crampé[a], L. Frappat[a], É. Ragoucy[a]

[a] Laboratoire d'Annecy-le-Vieux de Physique Théorique LAPTH, CNRS, UMR 5108 associée à l'Université de Savoie, B.P. 110, F-74941 Annecy-le-Vieux Cedex, France
[b] Laboratoire de Physique Théorique et Hautes Énergies LPTHE, CNRS, UMR 7589, Universités Paris VI/VII, 4, place Jussieu, B.P. 126, F-75252 Paris Cedex 05, France

Abstract. We introduce a rational solution of the (super) Yang–Baxter equation. This solution is used to define the Yangian of ortho-symplectic superalgebras.

1. Introduction

The (super) Yangians $Y(\mathfrak{a})$ based on a simple Lie algebras $\mathfrak{a}$ are deformations of $\mathfrak{a}[u] = \mathfrak{a} \otimes \mathbb{C}[u]$. Their presentation within the RTT formalism [1] was not given explicitly in all the cases. In this presentation, the generators are gathered in

$$T(u) = \sum_{n \in \mathbb{Z}_{\geq 0}} T_{(n)} \, u^{-n} = \sum_{i,j=1} \sum_{n \in \mathbb{Z}_{\geq 0}} T^{ij}_{(n)} \, u^{-n} \, E_{ij} = \sum_{i,j=1} T^{ij}(u) \, E_{ij} \tag{1}$$

with $T^{ij}_{(0)} = \delta_{ij}$ and the relations are encoded in

$$R_{12}(u - v) \, T_1(u) \, T_2(v) = T_2(v) \, T_1(u) \, R_{12}(u - v) \, R_{12}(u) \tag{2}$$

where $R(u)$ satisfies the Yang–Baxter equation

$$R_{12}(u) \, R_{13}(u + v) \, R_{23}(v) = R_{23}(v) \, R_{13}(u + v) \, R_{12}(u); \, . \tag{3}$$

The expressions of the R-matrix for $Y(sl(N))$, $Y(sl(M|N))$, $Y(so(M))$, and $Y(sp(N))$ are

- For $Y(sl(N))$ and $Y(sl(M|N))$, $R(u) = \mathbb{I} + \dfrac{P}{u}$, where P is the (graded) permutation matrix.

- For $Y(so(M))$ $R(u) = \mathbb{I} + \dfrac{P}{u} - \dfrac{K}{u + \frac{M-2}{2}}$, where K is P partially transposed.

- For $Y(sp(N))$ (with N even) $R(u) = \mathbb{I} + \dfrac{P}{u} - \dfrac{K}{u + \frac{N+2}{2}}$, where K is now P partially transposed with some signs.

These matrices are rational solution of the Yang–Baxter equation.

2. Notations

We now consider $Y(osp(M|N))$, with N even. Let $\theta_0 = \pm 1$. As usual E_{ij} denotes the elementary matrix with entry 1 in row i and column j and zero elsewhere, the indices i and j running from 1 to $M + N$. We define the following data:

On orthogonal side:

The conjugate index $\bar{i}$ of $i \in \{1, ..., M\}$ is $\bar{i} = M + 1 - i$, its $\mathbb{Z}_2$ gradation and sign θ_i are $(-1)^{[i]} = \theta_0$ and $\theta_i = 1$.

On symplectic side:

The conjugate index $\bar{i}$ of $i \in \{M + 1, ..., M + N\}$ is $\bar{i} = 2M + N + 1 - i$, its $\mathbb{Z}_2$ gradation and sign θ_i are

$$(-1)^{[i]} = -\theta_0, \qquad \theta_i = \begin{cases} +1 & \text{for} & M + 1 \leq i \leq M + \frac{N}{2} \\ -1 & \text{for} & M + \frac{N}{2} + 1 \leq i \leq M + N \end{cases} \tag{4}$$

Transposition:

For $A = \sum_{ij} A^{ij} E_{ij}$, we define the transposition t by

$$A^t = \sum_{ij} (-1)^{[i][j]+[j]} \theta_i \theta_j \, A^{ij} \, E_{\bar{j}\bar{i}} = \sum_{ij} \left(A^{ij}\right)^t E_{ij} \,.$$

It satisfies $(A^t)^t = A$ and, for $\mathbb{C}$-valued matrices, $(AB)^t = B^t A^t$.

3. A rational solution of (super) Yang–Baxter equation

Let P be the (super)permutation operator (i.e. $X_{21} \equiv P X_{12} P$)

$$P = \sum_{i,j=1}^{M+N} (-1)^{[j]} E_{ij} \otimes E_{ji}$$

and K its transpose in the first space (in the above sense)

$$K \equiv P^{t_1} = \sum_{i,j=1}^{M+N} (-1)^{[i][j]} \theta_i \theta_j E_{\bar{j}\bar{i}} \otimes E_{ji} \,.$$

Define

$$R(u) = \mathbb{I} + \frac{P}{u} - \frac{K}{u + \kappa} \,.$$

The matrix $R(u)$ satisfies

$$R_{12}^{t_1}(-u - \kappa) = R_{12}(u), \qquad \text{(crossing symmetry)}$$
$$R_{12}(u) R_{12}(-u) = (1 - 1/u^2)\mathbb{I}, \qquad \text{(unitarity)}$$

provided that $2\kappa = (M - N - 2)\theta_0$.

The R-matrix $R(u)$ satisfies the super Yang–Baxter equation

$$R_{12}(u) R_{13}(u + v) R_{23}(v) = R_{23}(v) R_{13}(u + v) R_{12}(u)$$

for $2\kappa = (M - N - 2)\theta_0$, where the graded tensor product is understood.

With suitably corrected signs, it satisfies the ordinary Yang–Baxter equation.

4. Example : case of $osp(1|2)$

As an example, we give the R-matrix of $osp(1|2)$. The gradation is chosen for convenience to be $([1], [2], [3]) = (0, 1, 0)$ (we have exchanged, for this example only, the indices 1 and 2 with respect to the previous definitions). In addition, $(\theta_1, \theta_2, \theta_3) = (1, 1, -1)$ and $\kappa = \frac{3}{2}$. We get for the R-matrix

$$R_{osp(1|2)}(u) = \begin{pmatrix} \frac{u+1}{u} & & & & & & & \\ & 1 & & \frac{1}{u} & & & & \\ & & \frac{2u+1}{2u+3} & & \frac{-2}{2u+3} & & \frac{4u+3}{u(2u+3)} & \\ & \frac{1}{u} & & 1 & & & & \\ & & \frac{-2}{2u+3} & & -\frac{2u^2+3u-3}{u(2u+3)} & & \frac{2}{2u+3} & \\ & & & & & 1 & & \frac{1}{u} \\ & & \frac{4u+3}{u(2u+3)} & & \frac{2}{2u+3} & & \frac{2u+1}{2u+3} & \\ & & & & & \frac{1}{u} & & 1 \\ & & & & & & & \frac{u+1}{u} \end{pmatrix}$$

5. Definition of the (super) Yangians

We consider the Hopf (super)algebra $\mathcal{U}(R)$ generated by the operators $T^{ij}_{(n)}$, for $1 \leq i, j \leq M + N, n \in \mathbb{Z}_{\geq 0}$, with $T^{ij}_{(0)} = \delta_{ij}$, encapsulated into a $(M + N) \times (M + N)$ matrix

$$T(u) = \sum_{n \in \mathbb{Z}_{\geq 0}} T_{(n)} u^{-n} = \sum_{i,j=1}^{M+N} \sum_{n \in \mathbb{Z}_{\geq 0}} T^{ij}_{(n)} u^{-n} E_{ij} = \sum_{i,j=1}^{M+N} T^{ij}(u) E_{ij} .$$

One defines $\mathcal{U}(R)$ by imposing the following constraints (RTT relations) on $T(u)$

$$R_{12}(u - v) T_1(u) T_2(v) = T_2(v) T_1(u) R_{12}(u - v)$$

The Hopf algebra structure of $\mathcal{U}(R)$ is given by [1]

$$\Delta(T(u)) = T(u) \dot{\otimes} T(u) \qquad \text{i.e.} \quad \Delta(T^{ij}(u)) = \sum_{k=1}^{M+N} T^{ik}(u) \otimes T^{kj}(u)$$

$$S(T(u)) = T(u)^{-1} \quad ; \quad \epsilon(T(u)) = \mathbb{I}_{M+N}$$

Theorem [2]: *The operators generated by $C(u) = T^t(u - \kappa) T(u)$ lie in the centre of the algebra $\mathcal{U}(R)$ and $C(u) = c(u)\mathbb{I}$. Furthermore, $\Delta(c(u)) = c(u) \otimes c(u)$ and the two-sided ideal $\mathcal{I}$ generated by $C(u) - \mathbb{I}$ is also a coideal. The quotient $\mathcal{U}/\mathcal{I}$ is then a Hopf algebra.*

The Yangian of $osp(M|N)$ is defined to be $\mathcal{U}/\mathcal{I}$.

6. Double Yangians

As usual, the quantum double of $Y(osp(M|N))$ is defined by

$$R_{12}(u - v) L^{\pm}_1(u) L^{\pm}_2(v) = L^{\pm}_2(v) L^{\pm}_1(u) R_{12}(u - v)$$
$$R_{12}(u - v) L^{+}_1(u) L^{-}_2(v) = L^{-}_2(v) L^{+}_1(u) R_{12}(u - v)$$

with $L^+(u) \equiv T(u)$ and

$$L^+(u) = \sum_{n \in \mathbb{Z}_{\geq 0}} L_{(n)}\, u^{-n} \qquad\qquad L^-(u) = \sum_{n \in \mathbb{Z}_{\geq 0}} L_{(-n)}\, u^n$$

7. Deformed double Yangians

Starting from a centrally extended version of the above double Yangian, we obtain the R-matrix of the deformed double Yangian by a twisting procedure [3, 4] (using the same method as for the construction of dynamical quantum algebras [6])

$$\mathcal{R}_r(u) = \mathcal{F}_{21}(-u)\, \mathcal{R}(u)\, \mathcal{F}_{12}(u)^{-1}$$

where $\mathcal{F}$ is given by the infinite product (τ being a suitable automorphism)

$$\mathcal{F}_{21}(u) = \overset{\frown}{\prod_{n \geq 1}} \tau^n \left(\mathcal{R}_{12}^{-1}(u + nr) \right)$$

$\mathcal{F}$ hence satisfies the linear equation

$$\mathcal{F}_{12}(u + r) = \tau(\mathcal{F}_{12}(u))\, \mathcal{R}_{12}(u + r)\ .$$

The twist $\mathcal{F}$ also satisfies the shifted cocycle relation

$$\mathcal{F}_{12}(r)\, (\Delta \otimes \mathrm{id})\mathcal{F}(r) = \mathcal{F}_{23}(r + c^{(1)})\, (\mathrm{id} \otimes \Delta)\mathcal{F}(r)\ ,$$

so that the R-matrix of the deformed double Yangian satisfies the GNF equation [7, 8, 9]

$$\widetilde{\mathcal{R}}_{12}(r + c^{(3)})\, \widetilde{\mathcal{R}}_{13}(r)\, \widetilde{\mathcal{R}}_{23}(r + c^{(1)}) = \widetilde{\mathcal{R}}_{23}(r)\, \widetilde{\mathcal{R}}_{13}(r + c^{(2)})\, \widetilde{\mathcal{R}}_{12}(r)\ .$$

This yields a QTQHA (Quasi-triangular quasi-Hopf algebra) structure $\mathcal{D}Y_r(\mathfrak{g})$, the algebraic relations being given by

$$R_{12}(u_1 - u_2, r)\, L_1(u_1)\, L_2(u_2) = L_2(u_2)\, L_1(u_1)\, R_{12}(u_1 - u_2, r - c)$$

with $L(u)$ formal power series in u, i.e. $L(u) = \sum_{n \in \mathbb{Z}} L_n u^{-n}$.

Similarly, one can construct a dynamical double Yangian along the lines of [4]. For more details on this subject, see [5].

[1] L.D. Faddeev, N.Yu. Reshetikhin and L.A. Takhtajan, *Quantization of Lie groups and Lie algebras*, Leningrad Math. J. **1** (1990) 193.

[2] D. Arnaudon, J. Avan, N. Crampé, L. Frappat, E. Ragoucy, *R-matrix presentation for super Yangians* $Y(osp(m|2n))$, math.QA/0111325.

[3] D. Arnaudon, J. Avan, L. Frappat, E. Ragoucy, M. Rossi, *On the Quasi-Hopf structure of deformed double Yangians*, Lett. Math. Phys. **51** (2000) 193 and math.QA/0001034.

[4] D. Arnaudon, J. Avan, L. Frappat, and É. Ragoucy, *Yangian and quantum universal solutions of Gervais–Neveu–Felder equations*, Commun. Math. Phys., **226** (2002) 183 and math.QA/0104181.

[5] D. Arnaudon, J. Avan, L. Frappat, and E. Ragoucy. *Deformed double Yangians and quasi-Hopf algebras*. In G. Halbout, editor, *Deformation Quantization, 68ème rencontre entre Physiciens Théoriciens et Mathématiciens, Strasbourg 2001*, IRMA Lectures in Mathematics and Theoretical Physics 1, page 85, de Gruyter, (2002).

[6] D. Arnaudon, E. Buffenoir, E. Ragoucy, Ph. Roche, *Universal solutions of quantum dynamical Yang-Baxter equations*, Lett. Math. Phys. **44** (1998) 201 and q-alg/9712037.

[7] J.L. Gervais, A. Neveu, *Novel triangle relation and absence of tachyons in Liouville string field theory*, Nucl. Phys. **B238** (1984) 125.

[8] G. Felder, *Elliptic quantum groups*, Proc. ICMP Paris 1994, pp 211 and hep-th/9412207.

[9] O. Babelon, D. Bernard, E. Billey, *A quasi-Hopf algebra interpretation of quantum $3 - j$ and $6 - j$ symbols and difference equations*, Phys. Lett. **B375** (1996) 89.

Inst. Phys. Conf. Ser. No 173: Mini-symposium
Paper presented at 24th Int. Coll. Group Theoretical Methods in Physics, Paris, France, 15–20 July 2002
©2003 IOP Publishing Ltd

Integrable models with comodule algebra symmetry

Ángel Ballesteros[1], Fabio Musso[2] and Orlando Ragnisco[3]

[1]Departamento de Física, Universidad de Burgos, E-09001-Burgos, Spain
[2]SISSA, Via Beirut 2/4, Trieste, I-34013 Italy
[3]Dipartimento di Fisica, Terza Università di Roma, Via della Vasca Navale 84, I-00146-Roma, Italy

Abstract. A general method to construct classical and quantum N-dimensional Hamiltonian systems with comodule algebra symmetry is presented as a generalization of the dynamical coalgebra symmetry approach to integrable models. Several examples are given, and their Liouville integrability is discussed through the analysis of the Casimir elements and the representation theory of the underlying comodule algebra.

1. Introduction

It is well known that quantum groups were introduced as new symmetries connected with the integrability properties of different nonlinear quantum models [1]. In particular, quantum groups and algebras were used to obtain nontrivial solutions (quantum R matrices) of the Quantum Yang-Baxter Equation, which is the integrability condition for (1+1) quantum field theories [2] and is also connected with two dimensional models in lattice statistical physics [3], conformal field theory [4] and knot theory [5]. However, further examples of quantum algebras (non-cocommutative coalgebras) with no associated quantum R matrix were obtained (different examples of such non-coboundary deformations are given in [6, 7]). In this context, it seemed natural to study whether arbitrary coalgebras could be interpreted in general as symmetries of integrable Hamiltonian systems without making use of (either classical or quantum) Yang-Baxter equations.

The method introduced in [8] provides such a systematic construction of integrable systems by starting from a Hamiltonian defined on a given (either Poisson or non-commutative) coalgebra A with a number of Casimir operators/functions, and the coalgebra structure is used to obtain both the Hamiltonian on the space $A \otimes A \otimes \cdots^{(N)} \otimes A$ and the explicit form of the integrals of the motion. By following this procedure, different examples of completely integrable N-body Hamiltonians with coalgebra symmetry have been obtained. In particular, the Calogero-Gaudin (CG) Hamiltonian [9, 10] was proven to have $sl(2)$ coalgebra symmetry [11, 12] and its well-known constants of the motion were identified with the coproducts of the $sl(2)$ Casimir. On the other hand, the $N = 2$ rational Calogero-Moser Hamiltonian was also proven to share this $sl(2)$ coalgebra symmetry [13].

It is also important to remark that, given a Hamiltonian with certain coalgebra symmetry, any deformation of the underlying coalgebra provides an integrable deformation of the initial system. This property was shown for the first time by constructing an integrable deformation of the CG system through the quantum algebra $so_q(2, 1)$ [11]. An integrable deformation of a generalized Smorodinsky-Winternitz Hamiltonian was obtained in [15] by making use of the non-standard deformation $sl_h(2)$ [16]. Let us also mention another interesting example of coalgebra-invariant system linked to non-coboundary deformations: an analogue [8] of the Ruijsenaars-Schneider model [17], that was obtained by using the Poisson version of a quantum deformation of the (1+1) dimensional Poincaré algebra [18]. Other examples of this construction can be found in [13, 19, 20], and the superintegrability properties of the models with coalgebra symmetry will be analysed in [21].

In this contribution we present the generalization [22] of the abovementioned formalism by using the notion of comodule algebras (note that the possibility of using arbitrary Poisson maps in order to generate integrable systems was already pointed out in [23]). The comodule algebra approach refers to systems defined on $V \otimes H \otimes \cdots^{(N-1)} \otimes H$ where the algebra V has to be a comodule algebra of a given coalgebra H. In particular, it is remarkable that comodule algebras are relevant in the context of noncommutative geometry (if a given quantum space $\mathcal{Q}$ is covariant under the action of a quantum group G_q, the algebra $\mathcal{Q}$ is a G_q-comodule algebra [24]-[27]). We shall illustrate the method with three integrable models: a deformation of the classical harmonic oscillator, a model of interacting q-oscillators and a prototype of integrable Hamiltonian on a quantum space (see [22] for a detailed description of these and other examples). Remarkably enough, in all these models the algebraic superintegrability provided by the coalgebra is broken, and only complete integrability is guaranteed by the comodule algebra symmetry [21].

2. Integrable systems from comodule algebras

A (right) coaction of a Hopf algebra (H, Δ) on a vector space V is a linear map $\phi : V \rightarrow V \otimes H$ such that

$$(\phi \otimes id) \circ \phi = (id \otimes \Delta) \circ \phi \tag{1}$$

If V is an algebra, we shall say that V is a H-comodule algebra if the coaction ϕ is a homomorphism with respect to the product on the algebra

$$\phi(ab) = \phi(a)\,\phi(b) \qquad \forall a, b \in V.$$

Moreover, if $[\cdot, \cdot]$ is the Lie (Poisson) bracket on V, we shall have that V is a Lie (Poisson) H-comodule algebra if

$$\phi([a, b]_V) = [\phi(a), \phi(b)]_{V \otimes H} \qquad \forall a, b \in V$$

The coalgebra map can be iterated by using recursively (1) in order to construct the N-th coaction as a homomorphism that maps V within $V \otimes H \otimes \cdots^{(N-1)} \otimes H$. Let $\{X_1, \ldots, X_l\}$ be the generators of V and let C be a Casimir function/operator of V. It can be proven [22] that the Hamiltonian

$$H^{(N)} := \phi^{(N)}(\mathcal{H}(X_1, \ldots, X_l)) = \mathcal{H}(\phi^{(N)}(X_1), \ldots, \phi^{(N)}(X_l)),$$

together with the following iterated coactions of the Casimir are a set of N functions in involution

$$C^{(m)} := \phi^{(m)}(C(X_1, \ldots, X_l)) = C(\phi^{(m)}(X_1), \ldots, \phi^{(m)}(X_l)) \qquad (m = 2, \ldots, N).$$

3. Examples

3.1. An integrable deformation of the harmonic oscillator

We consider as the coalgebra H a Poisson analogue of a non-standard deformation of the Schrödinger algebra $\mathcal{S}_\sigma$ introduced in [28]. In this case, the V algebra is $gl(2)$ subalgebra of $\mathcal{S}_\sigma$ and the coaction map $\phi^{(2)} : gl(2) \rightarrow gl(2) \otimes \mathcal{S}_\sigma$ can be taken as the restriction (see [22])

$$\phi^{(2)}(X) := \Delta(X) \qquad X \in gl(2). \tag{2}$$

By using the coaction map (2) and an appropriate symplectic realization of both $gl(2)$ and $\mathcal{S}_\sigma$ we obtain the two-body Hamiltonian

$$
H^{(2)}_\sigma = \frac{1}{2}(p_1^2 + p_2^2) + \frac{q_2^2}{2} + \frac{q_1^2}{2(1 + \sigma\lambda_2 p_2)^2}
$$
$$
+ \sigma\lambda_2 \left(p_1^2 p_2 + \frac{q_2(\lambda_1^2 - 2q_1 p_1)}{2(1 + \sigma\lambda_2 p_2)} \right) + \sigma^2 \lambda_2^2 \left(\frac{1}{2} p_1^2 p_2^2 + \frac{(\lambda_1^2 - 2q_1 p_1)^2}{8(1 + \sigma\lambda_2 p_2)^2} \right),
$$

which is just an integrable deformation of the two-dimensional isotropic oscillator. The integral of motion is obtained as the coaction of the $gl(2)$ Casimir and reads

$$
C^{(2)}_\sigma = -\frac{\{2(p_2 q_1 - p_1 q_2) + \sigma p_1(2p_1 q_1 - 4p_2 q_2 - \lambda_1^2) - \sigma^2\lambda_2^2 p_1 p_2(-2p_1 q_1 + 2p_2 q_2 + \lambda_1^2)\}^2}{16(1 + \sigma\lambda_2 p_2)^2}
$$

As expected, the limit $\sigma \to 0$ of $C^{(2)}_\sigma$ is just $-(p_2 q_1 - p_1 q_2)^2/4$. Note that further iterations of the coaction map would provide the corresponding generalization in arbitrary dimension.

3.2. q-oscillator systems

Let us now consider as the dynamical algebra V the "q-oscillator algebra" A_q [29]-[31]. The coalgebra H will be the quantum algebra $su_q(2)$. Through the coaction $\phi^{(2)} : A_q \to A_q \otimes su_q(2)$ defined in [31] and by taking as Hamiltonian $H = A^+ A$ [31, 32], we obtain a two-body integrable Hamiltonian

$$
H^{(2)} = A^+ A\, q^{2J} + (q - q^{-1})\, q^{-2N} X_+ X_- + (q - q^{-1})^{\frac{1}{2}} q^{-N+J} \{q^{-1} A X_+ + q A^+ X_-\}.
$$

The coaction of the Casimir $\phi^{(2)}(C_q)$ yields the constant of the motion $C^{(2)}$:

$$
\phi^{(2)}(C_q) = H^{(2)} - \frac{q^{-2(N+J)} - 1}{q^{-2} - 1}.
$$

The k-dimensional model [22] is obtained through the k-th iteration of the coaction map:

$$
H^{(k)} = \phi^{(k)}(A^+ A) = (A^+ A)\,\Delta^{(k-1)}(q^{2J}) + (q - q^{-1})\,q^{-2N}\,\Delta^{(k-1)}(X_+ X_-)
$$
$$
+ (q - q^{-1})^{\frac{1}{2}} \{(q^{-N} A)\,\Delta^{(k-1)}(X_+ q^J) + (A^+ q^{-N})\,\Delta^{(k-1)}(q^J X_-)\},
$$

where $\Delta^{(k-1)}$ is the $(k-1)$-th coproduct in $su_q(2)$. This Hamiltonian is completely integrable, since it commutes with the m-th order Casimirs,

$$
C^{(m)} = H^{(m)} - \frac{1}{q^{-2} - 1}\left(q^{-2\phi^{(m)}(N)} - 1\right) \qquad \text{with} \quad m = 2, \ldots, k.
$$

Once again, by construction, all these integrals are in involution.

3.3. Integrable Hamiltonians on quantum spaces

The RE algebra $\mathcal{A}$ [33] provides a representative example of a four-dimensional comodule algebra with two Casimir elements. The RE algebra is a $GL_q(2)$-comodule algebra and, therefore, it can be interpreted as a $GL_q(2)$ quantum space. If we rewrite the generators $\{\alpha, \beta, \gamma, \delta\}$ of $\mathcal{A}$ as the entries of a matrix K, the k-th coaction map $\phi^{(k)} : \mathcal{A} \to GL_q(2) \otimes GL_q(2) \otimes \ldots^{(k-1)} \otimes GL_q(2) \otimes \mathcal{A}$ will be given by

$$
\phi^{(k)}(K) = \begin{pmatrix} \phi^{(k)}(\alpha) & \phi^{(k)}(\beta) \\ \phi^{(k)}(\gamma) & \phi^{(k)}(\delta) \end{pmatrix} = \left\{\prod_{l=1}^{k-1} T_{k-l}\right\} \cdot K \cdot \left\{\prod_{l=1}^{k-1} T_l^t\right\}
$$

where T_l is the l-th copy of the quantum group $GL_q(2)$. If we consider as the Hamiltonian any function of the type $H^{(m)} = H(\phi^{(m)}(\alpha), \phi^{(m)}(\beta), \phi^{(m)}(\gamma), \phi^{(m)}(\delta))$, its integrability will be guaranteed by the $2(m-1)$ coaction images of the casimir functions

$$c_1^{(k)} = \phi^{(k)}(c_1) = \left\{ \prod_{l=1}^{k-1} \det{}_q T_l \right\} c_1, \quad c_2^{(k)} = \phi^{(k)}(c_2) = \left\{ \prod_{l=1}^{k-1} (\det{}_q T_l)^2 \right\} c_2, \quad 2 \le k \le m$$

where the q-determinant on each $GL_q(2)$ copy is given by $\det_q T_l = a_l d_l - q b_l c_l$. Then,

$$\begin{aligned} \left[H^{(m)}, c_1^{(k)} \right] = 0 && \left[H^{(m)}, c_2^{(p)} \right] = 0 && 2 \le k, p \le m \\ \left[c_1^{(l)}, c_1^{(n)} \right] = 0 && \left[c_2^{(l)}, c_2^{(n)} \right] = 0 && 2 \le l < n \le m \\ \left[c_1^{(k)}, c_2^{(p)} \right] = 0 && k, p = 2, \ldots, m. \end{aligned}$$

Obviously, this procedure can be used to define integrable systems on other quantum spaces.

Acknowledgements. A.B. has been partially supported by MCyT (Project BFM2000-1055). O.R. has been partially supported by INFN and by MIUR (COFIN2001 "Geometry and Integrability"). Partial financial support from INFN-CICYT is also acknowledged.

References

[1] Reshetikhin N Y, Takhtadzhyan L A and Faddeev L D 1990 *Leningrad Math. J.* **1** 193
[2] Yang C N 1967 *Phys. Rev. Lett.* **19** 1312
[3] Baxter R J 1972 *Ann. Phys.* **70** 193
[4] Álvarez–Gaumé L, Gómez C and Sierra G 1989 *Phys. Lett.* **B220** 142
[5] Reshetikhin N Yu and Turaev V G 1990 *Commun. Math. Phys.* **127** 1
[6] Ballesteros A, Herranz F J and Parashar P 1997 *J. Phys. A: Math. Gen.* **30** L149
[7] Sobczyk J 1996 *J. Phys. A: Math. Gen.* **29** 2887
[8] Ballesteros A and Ragnisco O 1998 *J. Phys. A: Math. Gen.* **31** 3791
[9] Calogero F 1995 *Phys. Lett. A* **201** 306
[10] Gaudin M 1983 *La Fonction d'Onde de Bethe* Masson, Paris
[11] Ballesteros A, Corsetti M and Ragnisco O 1996 *Czech. J. Phys.* **46** 1153
[12] Karimipour V 1996 preprint hep-th/9602161
[13] Ballesteros A and Ragnisco O 2002 *J. Math. Phys* **43** 954
[14] Musso F and Ragnisco O 2000 *J. Math. Phys* **41** 7386
[15] Ballesteros A and Herranz F J 1999 *J. Phys. A: Math. Gen.* **32** 8851
[16] Ohn C 1992 *Lett. Math. Phys.* **25** 85
[17] Ruijsenaars S N M and Schneider H 1986 *Ann. Phys.* **170** 370
[18] Vaksman L and Korogodskii L I 1989 *Sov. Math. Dokl.* **39** 173
[19] Ballesteros A and Herranz F J 2001 *J. Nonlinear Math. Phys.* **8** Suppl. 18
[20] Ballesteros A 1999 *Int. J. Mod. Phys. B* **13** 2903
[21] Ballesteros A and Ragnisco O, in preparation
[22] Ballesteros A, Musso F and Ragnisco O 2002 *J. Phys. A: Math. Gen.* **35** 8197
[23] Grabowski J, Marmo G and Michor P W 1999 *Mod. Phys. Lett. A* **14** 2109
[24] Majid S 1990 *Int. J. Mod. Phys.* **A5** 1
[25] Majid S 1995 *Foundations of Quantum Group Theory*, Cambridge University Press
[26] Chari V and Pressley A 1994 *A Guide to Quantum Groups*, Cambridge University Press
[27] Chaichian M and Demichev A, 1996 *Introduction to Quantum Groups*, World Scientific
[28] Ballesteros A, Herranz F J, Negro J and Nieto L M 2000 *J. Phys. A: Math. Gen.* **33** 4859
[29] Biedenharn L C 1989 *J. Phys. A: Math. Gen.* **22** L873
[30] Macfarlane A J 1989 *J. Phys. A: Math. Gen.* **22** 4581
[31] Kulish P P 1993 *Theor. Math. Phys.* **94** 137
[32] Chaichian M and Kulish P P 1996 in *From Field Theory to Quantum Groups*, World Scientific
[33] Kulish P P and Sklyanin E K 1992 *J. Phys. A: Math. Gen.* **25** 5963

Inst. Phys. Conf. Ser. No 173: Mini-symposium
Paper presented at 24th Int. Coll. Group Theoretical Methods in Physics, Paris, France, 15–20 July 2002
©2003 IOP Publishing Ltd

Topological invariants of three-manifolds from $U_q(osp(1|2n))$

Sacha C Blumen‡

School of Mathematics and Statistics, University of Sydney, Sydney, NSW 2006, Australia

Abstract. We create Reshetikhin-Turaev topological invariants of closed orientable three-manifolds from the quantised superalgebra $U_q(osp(1|2n))$ at certain even roots of unity. To construct the invariants we develop tensor product theorems for finite dimensional modules of $U_q(osp(1|2n))$ at roots of unity.

1. Introduction

Topological invariants of closed orientable three-manifolds may be constructed from modular or quasimodular Hopf algebras [1, 2]. Reshetikhin and Turaev's construction using modular Hopf algebras relies upon several theorems relating framed links in S^3 to closed orientable three-manifolds. The Lickorish-Wallace theorem states that each framed link in S^3 determines a closed, orientable 3-manifold and that every such 3-manifold may be obtained by performing surgery upon a framed link in S^3. Kirby, and Fenn and Rourke showed that homeomorphism classes of closed orientable three-manifolds may be generated by performing surgery upon elements of equivalence classes of framed links in S^3, where the equivalence relations are generated by the Kirby moves. By taking such combinations of isotopy invariants of links in S^3 as to render them unchanged under the Kirby moves one obtains a topological invariant of 3-manifolds.

RT took invariants of isotopy derived from the quantised algebra $U_q(sl_2)$ at even roots of unity. Their method was adapted for the quantum algebras related to the A_n, B_n, C_n, D_n Lie algebras at all roots of unity [2, 3], and the exceptional quantum algebras and quantum superalgebras $U_q(osp(1|2))$ and $U_q(gl(2|1))$ at odd roots of unity [4]. Here we create invariants from $U_q(osp(1|2n))$ at $q = \exp(2\pi i/N)$ where $N \equiv 2 \mod 4$, $N \geq 6$.

Resethikhin-Turaev invariants can be constructed from a class of Hopf (super)algebras more general than (quasi)modular Hopf (super)algebras. For a Hopf (super)algebra A, invariants may be constructed if A has the following properties (where we take the quantum superdimension and quantum supertrace as appropriate if A is a quantised superalgebra)

(i) there exists a finite collection of mutually non-isomorphic left A-modules $\{V_\lambda\}_{\lambda \in I}$ for some index set I such that $\dim(V_\lambda) < \infty$ and $dim_q(V_\lambda) \neq 0$, $\forall \lambda \in I$,

(ii) for any finite collection of left A-modules $V_{\lambda_1}, V_{\lambda_2}, \ldots, V_{\lambda_s}$ where $\lambda_i \in I$ for all i,

$$V_{\lambda_1} \otimes V_{\lambda_2} \otimes \cdots \otimes V_{\lambda_s} = \mathcal{V} \oplus \mathcal{Z},$$

where $\mathcal{V} = \bigoplus_{\lambda \in I}(V_\lambda)^{\oplus m(\lambda)}$, $m(\lambda) \geq 0$ is the multiplicity of V_λ in the direct sum, and $\mathcal{Z}$ is a possibly empty left A-module with zero quantum dimension,

(iii) for each V_λ there is a dual module $(V_\lambda)^\dagger \cong (V_{\lambda^*})$ such that $\lambda^* \in I$ and there exists a distinguished module V_0 such that $(V_0)^\dagger \cong V_0$,

‡ sachab@maths.usyd.edu.au

(iv) the central element $\delta = v - \sum_{\lambda \in I} d_\lambda \chi_\lambda(v^{-1}) C_\lambda$ vanishes upon acting on any V_μ where $\mu \in I$. Here $\chi_\zeta(v) = q^{-(\zeta, \zeta + 2\rho)}$ and $C_\lambda = tr_\lambda[(id \otimes \pi)(1 \otimes q^{2h_\rho}) R^T R]$, where R is the universal R-matrix of A and $R^T = P \cdot R \cdot P$ where P is the (graded) permutation operator. $\{d_\lambda\}_{\lambda \in I}$ is a set of $\mathbb{C}$-valued constants such that at least one d_λ is non-zero, and

(v) the sum $z = \sum_{\lambda \in I} d_\lambda q^{-(\lambda.\lambda + 2\rho)} dim_q(V_\lambda)$ is non-zero.

Condition (iv) ensures that combinations of isotopy invariants of links are unchanged under some of the Kirby moves.

2. $U_q(osp(1|2n))$ at roots of unity and its finite dimensional modules

The quantised superalgebra $U_q(osp(1|2n))$ at roots of unity is not quasi-triangular. However, the quantised superalgebra now has a class of central elements which do not exist at generic q. These central elements generate an ideal of $U_q(osp(1|2n))$, which is also a two-sided co-ideal. The quotient of $U_q(osp(1|2n))$ by this ideal turns out to be a quasi-triangular Hopf superalgebra. We denote this algebra by $U_q^{(N)}(osp(1|2n))$, the details of which may be found in [6].

Associated with $osp(1|2n)$ is a vector space H^* over $\mathbb{C}$ with a basis $\{\varepsilon_i| 1 \leq i \leq n\}$ and a symmetric non-degenerate $\mathbb{C}$-bilinear form defined by $(\varepsilon_i, \varepsilon_j) = \delta_{i,j}$. The even (resp. odd) positive roots are $\Phi_0^+ = \{\varepsilon_i \pm \varepsilon_j, 2\varepsilon_k| 1 \leq i < j \leq n, 1 \leq k \leq n\}$ (resp. $\Phi_1^+ = \{\varepsilon_i| 1 < l \leq n\}$). Define $2\rho = \sum_{\alpha \in \Phi_0^+} \alpha - \sum_{\beta \in \Phi_1^+} \beta$, $\mathbb{Z} = \{\ldots, -1, 0, 1, \ldots\}$, $\mathbb{Z}_+ = \{0, 1, 2, \ldots\}$ and $\mathbb{Z}_n = \{0, 1, 2, \ldots, n-1\}$. Let $X \subset H^* = \sum_{i=1}^n \mathbb{Z}_+ \varepsilon_i$ and $X_N = X/NX$. Let $N' = N$ if N is odd, and $N/2$ if N is even.

Define $\phi_0 = \{\xi \in \Phi_0^+| \xi/2 \notin \Phi_0^+\}$, $\phi_1 = \Phi_1^+$, $\phi = \phi_0 \cup \phi_1$. For $U_q^{(N)}(osp(1|2n))$ where $n \geq 2$, $N \geq 3$ we define $\overline{\Lambda}_N^+ \subset X$ by $\overline{\Lambda}_N^+ = \left\{\lambda \in X| 0 \leq 2(\lambda + \rho, \tilde{\alpha}) \leq \tilde{N}, \forall \alpha \in \phi\right\}$ where $\tilde{\alpha} = 2\alpha/(\alpha, \alpha)$, $\tilde{N} = N$ if $N \equiv 2 \mod 4$, and $\tilde{\alpha} = \alpha$, $\tilde{N} = 2N'$ if $N \equiv 0, 1, 3 \mod 4$. For $U_q^{(N)}(osp(1|2))$ and $N \geq 3$ we define $\overline{\Lambda}_N^+ \subset X$ by $\overline{\Lambda}_N^+ = \{\lambda \in X| 0 \leq 2(\lambda + \rho, \alpha) \leq N''\}$, where $N'' = N/4$ if $N \equiv 2 \mod 4$ and $N'' = N'$ otherwise. For each $\overline{\Lambda}_N^+$ we define $\Lambda_N^+ \subset X$ identically except we replace $\leq$ with $<$ wherever appearing. Λ_N^+ plays the role of the index set I of the collection of $U_q^{(N)}(osp(1|2n))$ modules in the properties (i)–(v) above.

Let V be the fundamental module of $U_q^{(N)}(osp(1|2n))$ with highest weight ε_1. V is irreducible, $(2n + 1)$ dimensional and has the same structure as the fundamental module of $U_q(osp(1|2n))$ [7]. In [7] we prove the following lemmas and theorems.

Lemma 2.1 *Set $N \geq 4$ to be even. For each $\mu \in \overline{\Lambda}_N^+$ there exists a finite dimensional left $U_q^{(N)}(osp(1|2n))$ module V_μ with highest weight μ such that $V_\mu \subseteq V^{\otimes t}$ for some $t \geq 1$. The quantum superdimension of V_μ is $sdim_q(V_\mu) = \lim_{q^N \to 1}(sdim_q(V_\mu^{gen}))$ where V_μ^{gen} is the finite dimensional irreducible left $U_q(osp(1|2n))$ module with highest weight μ. $sdim_q(V_\mu) \neq 0$ for all $\mu \in \Lambda_N^+$ and $sdim_q(V_\mu) = 0$ for all $\mu \in \overline{\Lambda}_N^+ - \Lambda_N^+$.*

Let W be the Weyl group of $osp(1|2n)$ and τ be the maximal element of W. Then $-\tau(\lambda) \in \Lambda_N^+$, $\forall \lambda \in \Lambda_N^+$, which implies that for each V_λ there exists a dual module $(V_\lambda)^\dagger$ with highest weight an element of Λ_N^+.

Theorem 2.1 *Set $N \geq 4$ to be even, V to be the fundamental module of $U_q^{(N)}(osp(1|2n))$ and $\varepsilon_1 \in \Lambda_N^+$. Then for each $t \in \mathbb{Z}_+$, $V^{\otimes t} = \mathcal{V} \oplus \mathcal{Z}$, where $\mathcal{V} = \bigoplus_{\lambda \in \Lambda_N^+} (V_\lambda)^{\oplus m(\lambda)}$, $m(\lambda) \in \mathbb{Z}_+$*

and $\mathscr{Z}$ is a possibly empty direct sum of indecomposable left $U_q^{(N)}(osp(1|2n))$ modules, each of which has zero quantum superdimension.

Theorem 2.2 Set $N \geq 4$ to be even and let V_{λ_i} be finite dimensional left $U_q^{(N)}(osp(1|2n))$ modules such that $\lambda_i \in \Lambda_N^+$ for all i. Then

$$V_{\lambda_1} \otimes V_{\lambda_2} \otimes \cdots \otimes V_{\lambda_s} = \mathscr{V}' \oplus \mathscr{Z}',$$

where $s \geq 1$, and $\mathscr{V}'$ and $\mathscr{Z}'$ have the same form as $\mathscr{V}$ and $\mathscr{Z}$ respectively do in theorem 2.1.

Theorem 2.3 Set $N \geq 3$ to be odd, $\varepsilon_1 \in \Lambda_N^+$ and $t \in \mathbb{Z}_{N/2+1/2-n}$. Then $V^{\otimes t} = \overline{\mathscr{V}}$ where $\overline{\mathscr{V}}$ has the same form as $\mathscr{V}$ does in theorem 2.1.

3. Finding the set $\{d_\lambda\}_{\lambda \in \Lambda_N^+}$

Properties (i), (ii) and (iii) hold for $U_q^{(N)}(osp(1|2n))$ at even N. We consider condition (iv): there exists at least one set $\{d_\lambda\}_{\lambda \in \Lambda_N^+}$ of solutions to the equations

$$\chi_\mu(v) = \sum_{\lambda \in \Lambda_N^+} d_\lambda \chi_\lambda(v^{-1}) \chi_\mu(C_\lambda), \tag{1}$$

for all $\mu \in \Lambda_N^+$. Now at generic q, the eigenvalue of C_λ in an irreducible representation with highest weight $\mu \in \Lambda_N^+$ is given by $sch_\lambda(q^{2(\mu+\rho)}) = S_{\lambda,\mu}/Q_\mu$ where $S_{\lambda,\mu} = (-1)^{[\lambda]} \sum_{\sigma \in W} \varepsilon'(\sigma) q^{2(\lambda+\rho,\sigma(\mu+\rho))}$, $Q_\mu = \sum_{\sigma \in W} \varepsilon'(\sigma) q^{2(\rho,\sigma(\mu+\rho))}$, and where $\varepsilon'(\sigma) = -1$ if the number of components of σ that are reflections with respect to the elements of Φ_0 is odd and $\varepsilon'(\sigma) = 1$ otherwise. Our proofs for theorems 2.1 and 2.2 tell us that $sch_\lambda(q^{2(\mu+\rho)})$ is well behaved when q is taken to the Nth root of unity, and yields the desired $\chi_\mu(C_\lambda)$. To simplify finding $\{d_\lambda\}_{\lambda \in \Lambda_N^+}$ we initially consider $Q_\mu q^{-(\mu+2\rho,\mu)} = \sum_{\lambda \in \Lambda_N^+} d'_\lambda q^{(\lambda+2\rho,\lambda)} S'_{\lambda,\mu}$ where $d'_\lambda = (-1)^{[\lambda]} d_\lambda$ and $S'_{\lambda,\mu} = (-1)^{[\lambda]} S_{\lambda,\mu}$. To solve for the $\{d'_\lambda\}_{\lambda \in \Lambda_N^+}$ we consider $Q_\mu q^{-(\mu+2\rho,\mu)} = \sum_{\lambda \in X_N} x_\lambda q^{(\lambda+2\rho,\lambda)} S'_{\lambda,\mu}$ and set $x_\lambda = cq^{-(\lambda,2\rho)}$. We then have $Q_\mu q^{-(\mu+2\rho,\mu)} = \sum_{\sigma \in W} \varepsilon'(\sigma) \sum_{\lambda \in X_N} cq^{(\lambda,\lambda)} q^{2(\lambda+\rho,\sigma(\mu+\rho))}$.

Lemma 3.1 Set $N \equiv 0$ mod 4, $N \geq 4$. Then there does not exist any set $\{d_\lambda\}_{\lambda \in \Lambda_N^+}$ of solutions to equations (1).

We set $N \equiv 2$ mod 4, $N \geq 6$ in the remainder of this paper. So that x_λ is independent of μ we undertake the mapping $\lambda \to \sigma(\lambda) - \sigma(\mu)$ in the summation, which may be done as $\sigma(\lambda) - \sigma(\mu) \in X_N$ and the summation remains over X_N. We then obtain $Q_\mu q^{-(\mu,\mu+2\rho)} = \sum_{\lambda \in X_N} cq^{(\lambda,\lambda+2\rho)-(\mu,\mu+2\rho)} \sum_{\sigma \in W} \varepsilon'(\sigma) q^{2(\rho,\sigma(\mu+\rho))}$ from which $c^{-1} = \sum_{\lambda \in X_N} q^{(\lambda,\lambda+2\rho)}$ and $x_\lambda = q^{-(\lambda,2\rho)}/\sum_{\lambda \in X_N} q^{(\lambda,\lambda+2\rho)}$.

Now $x_{\lambda'} q^{(\lambda',\lambda'+2\rho)} S'_{\lambda',\mu} = x_\lambda q^{(\lambda,\lambda+2\rho)} S'_{\lambda,\mu}$ where $\lambda' = \lambda + N'\varepsilon_i$ for any ε_i. It follows that $\sum_{\lambda \in X_N} x_\lambda q^{(\lambda+2\rho,\lambda)} S'_{\lambda,\mu} = 2^n \sum_{\lambda \in X_{N/2}} x_\lambda q^{(\lambda+2\rho,\lambda)} S'_{\lambda,\mu}$.

Let $\overline{N} = N'$ and let $\overline{\Lambda}_{\overline{N}}^+$ be the fundamental domain for $X_{\overline{N}}$ under the action of the affine Weyl group $W_{\overline{N}}$ of $U_q^{(N)}(so_{2n+1})$ [3]. The affine Weyl groups $W_{N/2}$ of $U_q^{(N)}(osp(1|2n))$ and $U_q^{(N)}(so_{2n+1})$ are identical and as $\overline{\Lambda}_{\overline{N}}^+ = \overline{\Lambda}_N^+$ it follows that $\overline{\Lambda}_N^+$ is a fundamental domain for $X_{N/2}$ under the action of $W_{N/2}$.

Now $S'_{\sigma(\lambda+\rho)-\rho,\mu} = \varepsilon'(\sigma)S'_{\lambda,\mu}$ for any $\sigma \in W$. If $\lambda \in \overline{\Lambda}^+_N - \Lambda^+_N$ either there exists some $\sigma \in W$ such that $\sigma(\lambda+\rho) - \rho = \lambda$ and $\varepsilon'(\sigma) = -1$, or there exists some $w \in W$ such that $\varepsilon'(w) = 1$ and $\lambda = w(\lambda+\rho) - \rho + kN'\varepsilon_i$ for some ε_i, $k \in \mathbb{Z}$. As $S'_{\lambda',\mu} = -S'_{\lambda,\mu}$ where $\lambda' = \lambda + N'\varepsilon_i$, it follows that $S'_{\lambda,\mu} = 0$ for $\lambda \in \overline{\Lambda}^+_N - \Lambda^+_N$.

Then $2^n \sum_{\nu \in X_{N/2}} x_\nu q^{(\nu.\nu+2\rho)} S'_{\nu,\mu} = 2^n \sum_{\lambda \in \Lambda^+_N} \sum_{\sigma \in W} \varepsilon'(\sigma) x_{\sigma(\lambda+\rho)-\rho} q^{(\lambda.\lambda+2\rho)} S'_{\lambda,\mu}$. As $\sum_{\lambda \in \Lambda^+_N} d'_\lambda q^{(\lambda+2\rho,\lambda)} S'_{\lambda,\mu} = 2^n \sum_{\lambda \in \Lambda^+_N} \sum_{\sigma \in W} \varepsilon'(\sigma) x_{\sigma(\lambda+\rho)-\rho} q^{(\lambda.\lambda+2\rho)} S'_{\lambda,\mu}$, we obtain

$$d_\lambda = q^{(2\rho.\rho)} \gamma \, sdim_q(V_\lambda) / \sum_{\mu \in X_{N/2}} q^{(\mu.\mu+2\rho)},$$

where γ is the denominator in the expression of the quantum superdimension. Note that $d_{\lambda^*} = d_\lambda$ where $\lambda^* = -\tau(\lambda)$. The denominator of d_λ does not vanish: $\sum_{\mu \in X_{N/2}} q^{(\mu.\mu+2\rho)} = \sum_{\mu \in X_N} q^{(\mu.\mu+2\rho)} / 2^n$ and $\sum_{\mu \in X_N} q^{(\mu.\mu+2\rho)} = \prod_{k=0}^{n-1} G(N, 2k+1)$ where $G(N,m) = \sum_{i=0}^{N-1} q^{i(i+m)} = (1+i)\sqrt{N}/x^{m^2}$ and $x = \exp(\pi i/2N)$.

4. Constructing the invariant

The final matter we need to consider is condition (v): that $z = \sum_{\lambda \in \Lambda^+_N} d_\lambda q^{-(\lambda+2\rho,\lambda)} sdim_q(V_\lambda) \neq 0$. If $\{d_\lambda\}_{\lambda \in \Lambda^+_N}$ is a unique set of solutions to equations (1) then [2] implies that $z \neq 0$. Our set $\{d_\lambda\}_{\lambda \in \Lambda^+_N}$ is not necessarily unique however direct calculation gives $z \neq 0$.

Now it is a relatively simple matter to construct the invariants. Denote a framed link in S^3 by L and the 3-manifold it gives rise to by M_L. Let $\Sigma(L)$ stand for the Reshetikhin-Turaev functor applied to L (see [1, 3, 4]). Set A_L to be the linking matrix of L defined by: a_{ii} is the framing number of the ith component of L and a_{ij}, $i \neq j$ is the linking number between the ith and j^{th} components of L. Let $\sigma(A_L)$ be the number of nonpositive eigenvalues of A_L. Then

$$\mathscr{F}(M_L) = z^{-\sigma(A_L)} \sum(L)$$

is a topological invariant of M_L.

References

[1] Reshetikhin N Y and Turaev V G 1991 Invent. Math. 103 547-597
[2] Turaev V G and Wenzl H 1993 Int. J. Math. 4 no. 2 323-358
[3] Zhang R B 1996 Comm. Math. Phys. 182 no. 3 619-636
[4] Zhang R B 1994 Mod. Phys. Lett. A 9 no. 16 1453-1465, 1995 Rev. Math. Phys. 7 no. 5 809-831, 1997 Lett. Math. Phys. 41 no. 1 1-11
[5] Khoroshkin S M and Tolstoy V N 1991 Comm. Math. Phys. 141 no. 3 599-617
[6] Zhang R B 1992 J. Math. Phys. 33 no. 11 3918-3930
[7] Blumen S C "Tensor products of finite dimensional $U_q(osp(1|2n))$ modules at roots of unity", to appear.

Inst. Phys. Conf. Ser. No 173: Mini-symposium
Paper presented at 24th Int. Coll. Group Theoretical Methods in Physics, Paris, France, 15–20 July 2002
©2003 IOP Publishing Ltd

Quantum groups and interacting quantum fields

Christian Brouder

Laboratoire de Minéralogie-Cristallographie, CNRS UMR7590, Universités Paris 6 et 7, IPGP, 4 place Jussieu, 75252 Paris Cedex 05, France

Abstract. If C is a cocommutative coalgebra, a bialgebra structure can be given to the symmetric algebra $S(C)$. The symmetric product is twisted by a Laplace pairing and the twisted product of any number of elements of $S(C)$ is calculated explicitly. This is used to recover important identities in the quantum field theory of interacting scalar bosons.

1. Introduction

Quantum groups appear to be a powerful tool for quantum field calculations. First, Fauser pointed out a connection between Wick's theorem and the concept of Laplace pairing introduced by Rota and his school [1]. In reference [2], we showed that the Laplace pairing is a quantum group concept and we solved quantum field problems with quantum group tools. The quantum group approach to free scalar fields was presented in detail in [3]. Here we show that quantum groups can also deal with interacting fields. In the first part, a bialgebra $S(C)$ is built from a cocommutative coalgebra C, the symmetric product of $S(C)$ is twisted by a Laplace pairing and general identities are derived. In the second part, these identities are translated into the language of quantum field theory.

2. The abstract setting

If C is a cocommutative coalgebra with coproduct Δ' and counit ε', the symmetric algebra $S(C) = \bigoplus_{n=0}^{\infty} S^n(C)$ can be equipped with the structure of a bialgebra. The product of the bialgebra $S(C)$ is the symmetric product (denoted by $\cdot$) and its coproduct Δ is defined on $S^1(C) = C$ by $\Delta a = \Delta' a$ and extended to $S(C)$ by $\Delta 1 = 1 \otimes 1$ and $\Delta(u \cdot v) = \sum u_{(1)} \cdot v_{(1)} \otimes u_{(2)} \cdot v_{(2)}$. The elements of $S^n(C)$ are said to be of degree n. The counit ε of $S(C)$ is defined to be equal to ε' on $S^1(C) = C$ and extended to $S(C)$ by $\varepsilon(1) = 1$ and $\varepsilon(u \cdot v) = \varepsilon(u)\varepsilon(v)$. It can be checked recursively that Δ is coassociative and cocommutative and that $\sum \varepsilon(u_{(1)})u_{(2)} = \sum u_{(1)}\varepsilon(u_{(2)}) = u$. Thus, $S(C)$ is a commutative and cocommutative bialgebra which is graded as an algebra.

A Laplace pairing is a bilinear map $(|)$ from $S(C) \times S(C)$ to the complex numbers such that $(1|u) = (u|1) = \varepsilon(u)$, $(u \cdot v|w) = \sum(u|w_{(1)})(v|w_{(2)})$ and $(u|v \cdot w) = \sum(u_{(1)}|v)(u_{(2)}|w)$ for any u, v and w in $S(C)$. The powers Δ^k of the coproduct are defined by $\Delta^0 a = a$, $\Delta^1 a = \Delta a$ and $\Delta^{k+1}a = (\mathrm{Id} \otimes \ldots \otimes \mathrm{Id} \otimes \Delta)\Delta^k a$. Their action is denoted by $\Delta^k a = \sum a_{(1)} \otimes \ldots \otimes a_{(k+1)}$.

From the definition of the Laplace pairing and of the powers of the coproduct a straighforward recursive proof yields, for u^i and v^j in $S(C)$

$$(u^1 \cdot \ldots \cdot u^k|v^1 \cdot \ldots \cdot v^l) = \sum \prod_{i=1}^{k} \prod_{j=1}^{l} (u^i_{(j)}|v^j_{(i)}). \tag{1}$$

For example $(u \cdot v \cdot w|s \cdot t) = \sum(u_{(1)}|s_{(1)})(u_{(2)}|t_{(1)})(v_{(1)}|s_{(2)})(v_{(2)}|t_{(2)})(w_{(1)}|s_{(3)})(w_{(2)}|t_{(3)})$.

A Laplace pairing is entirely determined by its value on C. In other words, once we know $(a|b)$ for all a and b in C, equation (1) enables us to calculate the Laplace pairing on $S(C)$.

The Laplace pairing induces a twisted product $\circ$ on $S(C)$ by $u \circ v = \sum (u_{(1)}|v_{(1)}) u_{(2)} \cdot v_{(2)}$. By applying the counit to both sides of this equality we obtain the useful relation [3]

$$\varepsilon(u \circ v) = (u|v) \tag{2}$$

If we follow the proofs given in [3] we can easily show that the twisted product is associative, 1 is the unit of $\circ$, $(u \circ v|w) = (u|v \circ w)$ and $\Delta(u \circ v) = \sum u_{(1)} \circ v_{(1)} \otimes u_{(2)} \cdot v_{(2)}$. If we use the last identity recursively, we obtain for $u^1, \ldots, u^k$ in $S(C)$

$$\Delta(u^1 \circ \ldots \circ u^k) = \sum u^1_{(1)} \circ \ldots \circ u^k_{(1)} \otimes u^1_{(2)} \cdot \ldots \cdot u^k_{(2)}. \tag{3}$$

This leads us to the important relation

$$u^1 \circ \ldots \circ u^k = \sum \varepsilon(u^1_{(1)} \circ \ldots \circ u^k_{(1)}) u^1_{(2)} \cdot \ldots \cdot u^k_{(2)}. \tag{4}$$

To show (4) recursively, we observe that it is true for $k = 2$. We denote $U = u^1 \circ \ldots \circ u^k$ and we assume that the property is true up to the twisted product of k terms. Since, by definition, $U \circ v = \sum (U_{(1)}|v_{(1)}) U_{(2)} \cdot v_{(2)}$, equation (3) yields $U \circ v = \sum (u^1_{(1)} \circ \ldots \circ u^k_{(1)}|v_{(1)}) u^1_{(2)} \cdot \ldots \cdot u^k_{(2)} \cdot v_{(2)}$ and the result follows for the twisted product of $k + 1$ terms because of equation (2).

Finally, we shall prove the second important identity

$$\varepsilon(u^1 \circ \ldots \circ u^k) = \sum \prod_{i=1}^{k-1} \prod_{j=i+1}^{k} (u^i_{(j-1)}|u^j_{(i)}) = \sum \prod_{j>i} (u^i_{(j-1)}|u^j_{(i)}). \tag{5}$$

For $k = 2$, equation (5) is true because of equation (2). Assume that it is true up to k and denote $U = u^1 \circ \ldots \circ u^k$. From equation (2) and $U = \sum \varepsilon(U_{(1)}) U_{(2)}$ we find

$$\varepsilon(U \circ u^{k+1}) = \sum \varepsilon(U_{(1)})(U_{(2)}|u^{k+1}) = \sum \varepsilon(u^1_{(1)} \circ \ldots \circ u^k_{(1)})(u^1_{(2)} \cdot \ldots \cdot u^k_{(2)}|u^{k+1})$$

$$= \sum \varepsilon(u^1_{(1)} \circ \ldots \circ u^k_{(1)}) \prod_{n=1}^{k} (u^n_{(2)}|u^{k+1}_{(n)}),$$

where we used equations (3) and (1). Equation (5) is true up to k thus

$$\varepsilon(u^1 \circ \ldots \circ u^{k+1}) = \sum \prod_{i=1}^{k-1} \prod_{j=i+1}^{k} (u^i_{(j-1)}|u^j_{(i)}) \prod_{n=1}^{k} (u^n_{(k)}|u^{k+1}_{(n)})$$

$$= \sum \prod_{i=1}^{k-1} \prod_{j=i+1}^{k+1} (u^i_{(j-1)}|u^j_{(i)})(u^k_{(k)}|u^{k+1}_{(k)}) = \sum \prod_{i=1}^{k} \prod_{j=i+1}^{k+1} (u^i_{(j-1)}|u^j_{(i)})$$

and the identity is proved for the twisted product of $k + 1$ elements.

We considered the symmetric algebra $S(C)$, but the same results are obtained for the tensor algebra $T(C)$. A related construction was made by Hivert in [4].

3. Quantum fields

The previous construction is now applied to interacting quantum field theory. The scalar fields are defined by the usual formula [5]

$$\phi(x) = \int \frac{d\mathbf{k}}{(2\pi)^3 \sqrt{2\omega_k}} \left(e^{-ip \cdot x} a(\mathbf{k}) + e^{ip \cdot x} a^\dagger(\mathbf{k}) \right),$$

where $\omega_k = \sqrt{m^2 + |\mathbf{k}|^2}$, $p = (\omega_k, \mathbf{k})$, $a^\dagger(\mathbf{k})$ and $a(\mathbf{k})$ are the creation and annihilation operators acting on the symmetric Fock space of scalar particles. Interacting fields are products of fields at the same point. Thus, we define the powers of fields $\phi^n(x)$ as the normal product of n fields at x (i.e. $\phi^n(x) = :\phi(x)\ldots\phi(x):$). This definition is meaningful for all $n > 0$ and is extended to $n = 0$ by saying that $\phi^0(x)$ is the unit operator. In the following we shall consider the divided powers of fields defined by $\phi^{(n)}(x) = \phi^n(x)/n!$.

We consider the coalgebra $\mathscr{C}$ generated by $\phi^{(n)}(x)$, where x runs over spacetime and n goes from 0 to 3 for a ϕ^3 theory and from 0 to 4 for a ϕ^4 theory. We do not consider here the topology of this space. The coproduct of $\mathscr{C}$ is $\Delta\phi^{(n)}(x) = \sum_{k=0}^{n} \phi^{(k)}(x) \otimes \phi^{(n-k)}(x)$ and its counit is $\varepsilon(\phi^{(n)}(x)) = \delta_{n,0}$. Scalar fields are bosons, so we work with the symmetric algebra $S(\mathscr{C})$. The product of $S(\mathscr{C})$ is the normal product of operators, which is commutative and denoted by $:uv:$. Notice that the counit is equal to the expectation value over the vacuum: $\varepsilon(u) = \langle 0|u|0\rangle$.

In $S(\mathscr{C})$, the Laplace pairing is entirely determined by the value of $(\phi^{(n)}(x)|\phi^{(m)}(y))$, which is itself determined by the value of $(\phi(x)|\phi(y)) = G(x,y)$ if we consider $\phi^{(n)}(x)$ as a product of fields. More precisely $(\phi^{(n)}(x)|\phi^{(m)}(y)) = \delta_{n,m}G(x,y)^{(n)}$, where the right hand side is a divided power $G(x,y)^{(n)} = (1/n!)G(x,y)^n$. In general, $G(x,y)$ is a distribution. In quantum field theory we use two special cases: the Wightman function $G_+(x,y) = \langle 0|\phi(x)\phi(y)|0\rangle$ and the Feynman propagator $G_F(x,y) = \langle 0|T(\phi(x)\phi(y))|0\rangle$. As shown in [2], when the Laplace pairing is defined with $G(x,y) = G_+(x,y)$, the twisted product equals the operator product of fields. When it is defined with $G(x,y) = G_F(x,y)$ the twisted product equals the time-ordered product.

Notice that $\Delta^{k-1}\phi^{(n)}(x) = \sum \phi^{(m_1)}(x) \otimes \ldots \otimes \phi^{(m_k)}(x)$, with a sum over all nonnegative integers m_i such that $\sum_{i=1}^{k} m_i = n$. Thus, we can specialize equation (1) to our coalgebra $\mathscr{C}$

$$(:\phi^{(n_1)}(x_1)\ldots\phi^{(n_k)}(x_k):|:\phi^{(p_1)}(y_1)\ldots\phi^{(p_l)}(y_l):) = \sum_{M}\prod_{i=1}^{k}\prod_{j=1}^{l}G(x_i,y_j)^{(m_{ij})}, \quad (6)$$

where the sum is over all $k \times l$ matrices M of nonnegative integers m_{ij} such that $\sum_{j=1}^{l} m_{ij} = n_i$ and $\sum_{i=1}^{k} m_{ij} = p_j$. This formula was given in reference [6].

Equation (4) applied to $\mathscr{C}$ yields a classical result of quantum field theory

$$\phi^{(n_1)}(x_1)\circ\ldots\circ\phi^{(n_k)}(x_k) = \sum_{i_1=0}^{n_1}\cdots\sum_{i_k=0}^{n_k}\langle 0|\phi^{(i_1)}(x_1)\circ\ldots\circ\phi^{(i_k)}(x_k)|0\rangle$$
$$:\phi^{(n_1-i_1)}(x_1)\ldots\phi^{(n_k-i_k)}(x_k):. \quad (7)$$

This equation was published by Epstein and Glaser for the operator product and the time-ordered product [7]. It is now often used in the Epstein-Glaser approach to renormalisation (see e.g. [8]). Equation (4) is clearly more compact and also more general than equation (7): it is still valid if the elements of $\mathscr{C}$ (i.e. $\phi^{(n_i)}(x_i)$) are replaced by elements of $S(\mathscr{C})$.

Finally, if we specialize equation (5) to $\mathscr{C}$ we obtain

$$\langle 0|\phi^{(n_1)}(x_1)\circ\ldots\circ\phi^{(n_k)}(x_k)|0\rangle = \sum_{M}\prod_{i=1}^{k-1}\prod_{j=i+1}^{k}G(x_i,x_j)^{(m_{ij})}, \quad (8)$$

where the sum is over all symmetric $k \times k$ matrices M of nonnegative integers m_{ij} such that $\sum_{j=1}^{k} m_{ij} = n_j$ and $m_{ii} = 0$ for all i.

When the twisted product is the operator product, this expression was given by Brunetti, Fredenhagen and Köhler [9]. Notice that (8) was proved here with a few lines of algebra, whereas the quantum field proof is combinatorial. As remarked by Rota, a great virtue of Hopf algebras is to replace combinatorics by algebra.

When the twisted product is the time-ordered product, equation (8) has a diagrammatic interpretation. The diagrammatic calculation of $\langle 0|T(\phi^{(n_1)}(x_1)\ldots\phi^{(n_k)}(x_k))|0\rangle$ would be: draw all diagrams that have k vertices x_1 to x_k and for which each vertex x_i has n_i edges. There is a one to one correspondence between these diagrams and the matrices M satisfying the conditions stated above: m_{ij} is the number of edges linking vertices x_i and x_j. The condition $m_{ii} = 0$ means that there is no tadpoles. The graphs are not directed (i.e. the edges do not carry arrows) because the Feynman propagator $G_F(x,y)$ is symmetric (i.e. $G_F(y,x) = G_F(x,y)$).

4. Perspective

This paper shows that non trivial results of quantum field theory can be derived easily from a general quantum group construction. A word of caution must be added concerning equation (8). When the twisted product is the operator product, equation (8) is valid. It defines a state on $T(\mathscr{C})$ by $\omega(a_1 \otimes \ldots \otimes a_k) = \varepsilon(a_1 \circ \ldots \circ a_k)$ and the Laplace pairing is a positive semidefinite form on $T(C) \times T(C)$. However, when the twisted product is the time-ordered product, equation (8) is ill-defined because the powers $G_F(x,y)^n$ are singular products of distributions and renormalisation is necessary. The first step of a "quantum group" renormalisation of scalar field theories was done in [2]. It uses the fact that equation (7) is still valid in renormalised quantum field theory, so that the Laplace pairing must be replaced by a Sweedler's 2-cocycle [10] in the definition of a renormalised time-ordered product. To go further, we can implement renormalisation abstractly by starting from a bialgebra B and putting a bialgebra structure on the "squared" tensor algebra $T(T(B)^+)$ [4]. This construction is inspired by Pinter's approach to renormalization [11] and is related to the Faà di Bruno bialgebra of composition of series [10]. These results will be presented in a forthcoming publication.

5. Acknowledgements

I am very grateful to Robert Oeckl for his comments on this paper and to Alessandra Frabetti for her enthusiastic support. This is IPGP contribution #0000.

References

[1] Fauser B 2001 J. Phys. A: Math. Gen. 34 105-116
[2] Brouder Ch and Oeckl R 2002 hep-ph/0206054
[3] Brouder Ch and Oeckl R 2002 hep-th/0208118
[4] Hivert F 1999 Combinatoire des fonctions quasi-symétriques (Marne-la-Vallée University: Ph.D. thesis)
[5] Reed M and Simon B 1975 Methods of Modern Mathematical Physics. Vol. II (New York: Academic Press)
[6] Rota G-C and Stein J A 2001 Algebraic Combinatorics and Computer Science (Milano: Springer) 267-314
[7] Epstein H and Glaser V 1973 Ann. Inst. Henri Poincaré 19 211-295
[8] Hollands S and Wald R M 2002 gr-qc/0111108
[9] Brunetti R, Fredenhagen K and Köhler M 1996 Commun. Math. Phys. 180 633-652
[10] Majid S 1995 Foundations of Quantum Group Theory (Cambridge: Cambridge University Press)
[11] Pinter G 2000 Lett. Math. Phys. 54 227-33

Inst. Phys. Conf. Ser. No 173: Mini-symposium
Paper presented at 24th Int. Coll. Group Theoretical Methods in Physics, Paris, France, 15–20 July 2002
©*2003 IOP Publishing Ltd*

Dirac monopoles from the Matsumoto non-commutative spheres

Tomasz Brzeziński‡
Department of Mathematics, University of Wales Swansea, Swansea SA2 8PP, UK

Andrzej Sitarz
Institute of Physics, Jagiellonian University, Reymonta 4, 30059 Kraków, Poland

Abstract. It is shown that the non-commutative three-sphere introduced by Matsumoto is a total space of the quantum Hopf bundle over the classical two-sphere. A canonical connection is constructed, and is shown to coincide with the standard Dirac magnetic monopole.

1. Introduction

One of the first examples of a non-commutative three-sphere was constructed by Matsumoto in [11]. This example was overlooked for a considerable time, mainly due to a focused attention attracted by a plethora of examples of quantum spheres coming from quantum groups. The interest in Matsumoto's sphere was revived recently by general revival of interest in non-commutative spheres triggered by papers [5], [8] and [4], where the Matsumoto spheres were re-derived through the construction of projective modules with vanishing lower Chern classes and were shown to satisfy the axioms of a non-commutative manifold proposed by A. Connes.

In a follow-up [12] to his original paper, Matsumoto has also constructed a non-commutative Hopf fibering, thus making the first step toward developing gauge theory with non-commutative spheres. The aim of this note is to review Matsumoto's construction of non-commutative spheres, and then reformulate the Hopf fibering in terms of more modern language of quantum principal bundles as initiated in [3], in order to develop gauge theory on such spheres. In particular we prove that Matsumoto's Hopf fibering is a quantum principal bundle or a Hopf-Galois extension. We also construct a connection in this quantum principal bundle, and show that it coincides with the standard Dirac magnetic monopole. In other words we show that a classical physical particle can be equivalently described as a particle existing in a non-commutative world.

2. Review of Matsumoto's construction

The main idea of Matsumoto's construction is based on the deformation of the topological construction of the classical three-sphere via the *Heegaard splitting*. Classically, one starts with the solid two-torus $D^2 \times S^1$. The boundary of $D^2 \times S^1$ is the two-torus $S^1 \times S^1$. A loop on $S^1 \times S^1$ is called a *meridian* if it vanishes in the fundamental group of $D^2 \times S^1$, and it is called a *longitude* if it generates this group. The three-sphere S^3 can be obtained by gluing two copies of solid two-tori through a homeomorphism which exchanges a meridian

‡ TB thanks the Engineering and Physical Sciences Research Council for an Advanced Fellowship.

with a longitude on their common boundary. Algebraically, in terms of algebras of continuous functions this construction can be summarised as

$$C(S^3) = \{(a, b) \in C(D^2 \times S^1) \oplus C(S^1 \times D^2) \mid \rho(\pi(a)) = \pi(b)\},$$

where $\pi : C(D^2 \times S^1) \to C(S^1 \times S^1)$ is the canonical projection (restriction of the domain to the boundary), and $\rho : C(S^1 \times S^1) \to C(S^1 \times S^1)$ is an isomorphism corresponding to exchanging a meridian for a longitude.

Now, one can replace the algebra of functions on a solid torus, by a (non-commutative) C^*-algebra D_θ of functions on a non-commutative solid torus. This is generated by a normal operator y and a unitary v such that $vy = e^{2\pi i\theta}yv$, where θ is a non-zero real number. There is a C^*-algebra surjection π_θ from D_θ to the non-commutative torus A_θ. The latter is generated by unitaries U, V subject to the relations $VU = e^{2\pi i\theta}UV$. The surjection π_θ is given by $\pi_\theta(v) = V$, $\pi_\theta(y) = U$, and is the non-commutative version of the canonical projection π above. Next, take $D_{-\theta}$ generated by a normal operator x and a unitary u, and the corresponding non-commutative torus $A_{-\theta}$ generated by $\hat{U}, \hat{V}$. A C^*-algebra map

$$\rho_\theta : A_{-\theta} \to A_\theta, \qquad (\hat{U}, \hat{V}) \mapsto (V, U),$$

is an isomorphism, which corresponds to the map interchanging meridians with longitudes. The deformed algebra of functions on the three-sphere can be thus defined as

$$C_\theta(S^3) = \{(a, b) \in D_{-\theta} \oplus D_\theta \mid \rho_\theta(\pi_\theta(a)) = \pi(b)\}.$$

The algebra $C_\theta(S^3)$ can be seen as generated by two normal operators $S = (u, y)$, $T = (x, v)$ and relations $TS = \lambda ST$, $(1 - TT^*)(1 - S^*S) = 0$ and $\|S\| = \|T\| = 1$, where $\lambda = e^{2\pi i\theta}$. In [13], Matsumoto and Tomiyama observed that, equivalently, $C_\theta(S^3)$ is a C^*-algebra which has presentation with generators a, b and relations

$$aa^* = a^*a, \quad bb^* = b^*b, \quad ab = \lambda ba, \quad ab^* = \bar{\lambda}b^*a, \quad aa^* + bb^* = 1. \tag{1}$$

The presentation (1) makes it clear that $C_\theta(S^3)$ is a deformation of functions on the classical sphere, and it also allows one to identify the Matsumoto sphere with one of the spheres in [5].

3. Non-commutative Hopf bundle

There is an action of $U(1)$ on the deformed S^3, which algebraically is represented as a *coaction* of a Hopf algebra (quantum group) $C(U(1))$ of functions on $U(1)$ on $C_\theta(S^3)$. Explicitly, let $C(U(1))$ be generated by a unitary Z (i.e., $ZZ^* = Z^*Z = 1$). Then the coaction $\Delta_R : C_\theta(S^3) \to C_\theta(S^3) \otimes C(U(1))$ is defined as a $*$-algebra map given on generators by

$$\Delta_R(a) = a \otimes Z, \qquad \Delta_R(b) = b \otimes Z.$$

The quantum quotient space under this (co)action is defined as a *subalgebra of coinvariants*,

$$C_\theta(S^3/U(1)) = C_\theta(S^2) := \{f \in C_\theta(S^3) \mid \Delta_R(f) = f \otimes 1\}.$$

One easily computes that $C_\theta(S^2)$ is generated by $z = aa^*$, $x_+ = ba^*$, $x_- = ab^*$. This is a commutative algebra with an additional relation $z^2 + x_+x_- = z$, and thus coincides with the usual algebra of functions on the two-sphere, i.e., $C_\theta(S^2) = C(S^2)$.

Note that the coaction of $C(U(1))$ on $C_\theta(S^3))$ defines a $\mathbf{Z}$-grading on the latter. An element $f \in C_\theta(S^3)$ has degree $n = 0, 1, 2, \ldots$ if $\Delta_R(f) = f \otimes Z^n$, and has degree $-n$ if $\Delta_R(f) = f \otimes Z^{*n}$. Clearly $C(S^2)$ is a subalgebra of degree zero elements.

To make sure that the above construction defines a quantum principal bundle, one needs to check that the action represented by Δ_R is free. Algebraically, this means that we need to prove that the extension of algebras $C(S^2) \to C_\theta(S^3)$ is a *Hopf-Galois extension*, i.e., that the map can $: C_\theta(S^3) \otimes_{C(S^2)} C_\theta(S^3) \to C_\theta(S^3) \otimes C(U(1))$ given by $f \otimes g \mapsto f\Delta_R(g)$ is bijective. Consider a map $\chi : C_\theta(S^3) \otimes C(U(1)) \to C_\theta(S^3) \otimes_{C(S^2)} C_\theta(S^3)$,

$$\chi(f \otimes Z^n) = \sum_{m=0}^{n} \lambda^{-m(n-m)} \binom{n}{m} f a^{*n-m} b^{*m} \otimes a^{n-m} b^m,$$

$$\chi(f \otimes Z^{*n}) = \sum_{m=0}^{n} \lambda^{-m(n-m)} \binom{n}{m} f a^{n-m} b^m \otimes a^{*n-m} b^{*m},$$

for all $f \in C_\theta(S^3)$ and $n = 0, 1, 2, \ldots$ One easily checks that $\text{can} \circ \chi = \text{id}$. The proof that also $\chi \circ \text{can} = \text{id}$ is also based on direct calculation. To facilitate this calculation one needs to make a couple of observations. First, since both maps are obviously left $C_\theta(S^3)$-linear, suffices it to check the identity on elements of the form $1 \otimes f$. Second, note that the tensor product is over $C(S^2)$, hence one may restrict oneself to elements of type $f = a^m b^n$ with $m, n \geq 0$. Now the calculation is straightforward. Thus we conclude that χ is the inverse of can, i.e., we have defined a quantum $U(1)$-principal bundle over S^2 with the total space given by the Matsumoto's three-sphere. This quantum bundle is denoted by $(C(S^2) \subset C_\theta(S^3))^{C(U(1))}$.

4. Non-commutative setup for the Dirac monopole

Since the Matsumoto three-sphere is a total space of a quantum (non-commutative) principal bundle $(C(S^2) \subset C_\theta(S^3))^{C(U(1))}$, it constitutes a proper setup for studying connections (cf. [3]). In particular we can define a special type of connection, known as a *strong connection* [9]. There are various equivalent ways of defining such connections [7, Theorem 2.3]. One of the possibilities is to say that a strong connection in $(C(S^2) \subset C_\theta(S^3))^{C(U(1))}$, is a left $C(S^2)$-linear, right $C(U(1))$-colinear splitting $s : C_\theta(S^3) \to C(S^2) \otimes C_\theta(S^3)$ of the product $\mu : C(S^2) \otimes C_\theta(S^3) \to C_\theta(S^3)$ such that $s(1) = 1 \otimes 1$. Such an s can be defined as follows. Since $C_\theta(S^3)$ is a **Z**-graded algebra, suffices it to define s on homogeneous elements only. Let

$$s(f) = f \sum_{m=0}^{n} \binom{n}{m} \begin{cases} (a^{n-m} b^m)^* \otimes a^{n-m} b^m & \text{if } f \text{ has degree } n \geq 0 \\ a^{n-m} b^m \otimes (a^{n-m} b^m)^* & \text{if } f \text{ has degree } -n < 0 \end{cases}$$

Clearly s is left $C(S^2)$-linear, since $C(S^2)$ consists of all elements of degree 0. Obviously $s(1) = 1 \otimes 1$. The last equation in (1) implies that $\mu \circ s = \text{id}$, i.e., s is a splitting of the product. Finally, degree counting ensures that s commutes with right coactions, i.e., it is right colinear. Thus we have constructed a strong connection in $(C(S^2) \subset C_\theta(S^3))^{C(U(1))}$. Note that s is determined by a map $\ell : C(U(1)) \to C_\theta(S^3) \otimes C_\theta(S^3)$ given by

$$\ell(Z^n) = \sum_{m=0}^{n} \binom{n}{m} (a^{n-m} b^m)^* \otimes a^{n-m} b^m, \quad \ell(Z^{*n}) = \sum_{m=0}^{n} \binom{n}{m} a^{n-m} b^m \otimes (a^{n-m} b^m)^*.$$

This map, which should be noted satisfies a number of conditions such as bicovariance, can be taken for another definition of a strong connection. The reader interested in details of the equivalence of these notions of a connection to more familiar notion of a connection form or a gauge field is referred to [7] and, in greater generality, to [2].

Strong connections provide one with a contact point between geometry of quantum principal bundles and non-commutative geometry. In non-commutative geometry, one studies finitely generated projective modules as modules of sections on non-commutative vector bundles. In classical geometry every vector bundle can be viewed as a bundle associated to a principal bundle: the standard fibre is a representation space of the structure (gauge) group. Given a quantum principal bundle or a Hopf-Galois extension $(B \subset P)^H$ (H is a Hopf algebra, P is a right H-comodule algebra with the coaction Δ_R and B are coinvariants of P) and a right H-comodule (corepresentation) V one identifies sections of the associated bundle with the left B-module $\Gamma = \mathrm{Hom}^H(V, P)$ of right H-comodule maps, i.e., maps $\phi : V \to P$ such that $\Delta_R \circ \phi = (\phi \otimes \mathrm{id}) \circ \varrho^V$, where $\varrho^V : V \to V \otimes H$ is the coaction (cf. [1, Theorem 4.3]). By [7, Corollary 2.6] if H has a bijective antipode, $(B \subset P)^H$ admits a strong connection, and V is finite dimensional, then Γ is a finitely generated projective left B-module. Furthermore a strong connection induces a connection in module Γ, which coincides with the Grassmann or Levi-Civita connection. Such a connection characterises projective module and is uniquely (up to a gauge transformation) determined by a projector, i.e., a matrix e with entries from B such that $e^2 = e$ (cf. [6]).

In the case of deformed Hopf fibering $(C(S^2) \subset C_\theta(S^3))^{C(U(1))}$, we can take a one-dimensional corepresentation $V = \mathbf{C}$ of $C(U(1))$, with $\varrho^V(1) = 1 \otimes Z$. The resulting Γ is a module of sections of a line bundle, and there is an algorithm of how to read the projector out of the strong connection (cf. [2]). Explicitly, if we write $\ell(Z) = \sum_i l_i \otimes r_i$, then the entries of the projector are $e_{ij} = r_i l_j$. Thus the projector corresponding to s above comes out as

$$e = \begin{pmatrix} aa^* & ab^* \\ ba^* & bb^* \end{pmatrix} = \begin{pmatrix} z & x_- \\ x_+ & 1-z \end{pmatrix}.$$

This is precisely the projector which describes the well-known Dirac monopole [10].

5. Conclusion

In this short note we have shown that Matsumoto's non-commutative Hopf fibering fits perfectly into the framework of quantum principal bundles, i.e., it is a Hopf-Galois extension. Furthermore we have constructed a strong connection in this quantum principal bundle, and, by considering the associated line bundle, we have identified this connection with the Dirac magnetic monopole potential. By this means we have obtained a non-commutative geometric description of magnetic monopoles.

References

[1] Brzeziński T 1995 J. Geom. Phys. 20 349–370
[2] Brzeziński T and Hajac P 2001 Swansea Preprint MRR-01-15
[3] Brzeziński T and Majid S 1993 Commun. Math. Phys. 157 591–638 (Erratum: Commun. Math. Phys. 167 235 (1995))
[4] Connes A and Dubois-Violette M 2001 Preprint ArXiv:math.QA/0107070
[5] Connes A and Landi G 2001 Commun. Math. Phys. 221 141–159
[6] Cuntz J and Quillen D 1995 J. Am. Math. Soc. 8 251–289
[7] Dąbrowski L, Grosse H and Hajac P 2001 Commun. Math. Phys. 220 301–331
[8] Dąbrowski L, Landi G and Masuda T 2001 Commun. Math. Phys. 221 161–168
[9] Hajac P 1996 Commun. Math. Phys. 182 579–617
[10] Landi G 2000 Rev. Math. Phys. 12 1367–1390
[11] Matsumoto K 1991 Japan J. Math. 17 333–356
[12] Matsumoto K 1991 Yokohama Math. J. 38 103–111
[13] Matsumoto K and Tomiyama J 1992 J. Math. Soc. Japan 44 13–41

Inst. Phys. Conf. Ser. No 173: Mini-symposium
Paper presented at 24th Int. Coll. Group Theoretical Methods in Physics, Paris, France, 15–20 July 2002
©2003 IOP Publishing Ltd

Q-Boson Realization of the Quantum Algebras

Čestmír Burdík[1] and Ondřej Navrátil[2]
[1]Dep. of Mathematics, FNSPE-CTU, Trojanova 13, 120 00 Prague 2, Czech Republic
[2]Dep. of Applied Mathematics, FTS-CTU, Na Florenci 25, 110 00 Prague 1, Czech Republic

Abstract. We describe a construction of boson realizations of simple Lie algebras and quantum groups. The construction is demonstrated on the examples of simple Lie algebra $gl(n)$ and quantum algebra $U_q(A_n)$.

1. Boson realization in the Lie algebra case

Let X, Y be Lie algebras and $U(X), U(Y)$ their enveloping algebras. The homomorphism τ

$$\tau : X \to W_{2n} \otimes U(Y)$$

we will call the generalized boson realization.

$$[\tau(x_1), \tau(x_2)] = \tau([x_1, x_2]) \quad x_1, \, x_2 \in X.$$

The Weyl algebra W_{2n} is generated by n-pairs creation and annihilation operators for which

$$[a_i, a_j^+] = \delta_{ij} 1.$$

In 1985 we formulate [1] the method how to obtain such boson realization. The examples of Lie groups were studied in [2, 3].

Let me remember this method. Let $\mathcal{L}$ is simple Lie algebra with a system of simple roots $\Pi^+ = (\alpha_1, \ldots, \alpha_k)$. Any positive roots α we can write in the form

$$\alpha = \sum_{i=1}^{k} n_i \alpha_i \, ,$$

where $n_i \in 0, 1, 2, \ldots$

Let H_i, $(i = 1, \ldots, k)$, E_α and F_α be the Cartan generators and root vectors of the algebra with the following commutation relations

$$[H_i, E_\alpha] = \alpha(i) E_\alpha,$$
$$[E_\alpha, F_\alpha] = \alpha(i) H_i,$$
$$[E_\alpha, E_\beta] = N_{\alpha\beta} E_{\alpha+\beta},$$
$$[F_\alpha, F_\beta] = N_{\alpha\beta} F_{\alpha+\beta}.$$

For any simple root α_r we will denote $\Pi_r = \Pi^+ \setminus \{\alpha_r\}$ and Φ_r^+ the system of positive roots of the type

$$\alpha = \sum_{i=1, i \neq r}^{k} n_i \alpha_i \, ,$$

where $n_i \geq 0$. We will denote $\mathcal{L}_r$ the algebra generated by elements E_α, F_α, where $\alpha \in \Phi_r^+$, and H_i, $i = 1, 2, \ldots, k$. This subalgebra of the algebra $\mathcal{L}$ is reductive.

The positive roots β, which are not elements of Φ_r^+, this is

$$\beta = \sum_{i=1}^{k} n_i \alpha_i,$$

where $n_r > 0$, generate a nilpotent subalgebra $\mathcal{L}$ which we denote $\mathcal{N}_+$ and similarly F_β generate the algebra $\mathcal{N}_-$. We obtain by this way decomposition

$$\mathcal{L} = \mathcal{N}_+ \oplus \mathcal{L}_r \oplus \mathcal{N}_-.$$

The enveloping algebra $U(\mathcal{L})$ can be written as $U(\mathcal{L}) = U(\mathcal{N}_+) \cdot U(\mathcal{L}_r) \cdot U(\mathcal{N}_-)$.

Now let φ is a any representation of $\mathcal{L}_r$ on vector space V such that

$$\varphi(Z) = \varphi\left(\sum_{i=1}^{k} a_k H_k\right) = \lambda = \text{const.},$$

where Z ‡ is a element of the center of $\mathcal{L}_r$. Because $a_r \neq 0$, we can write $\varphi(H_r)$ by using $\varphi(H_i)$, $i \neq r$ and λ. The representation φ is given by two representations φ_1 and φ_2 of the algebras $\mathcal{L}_1$ and $\mathcal{L}_2$, connected with the systems of roots $\Pi_1 = \{\alpha_1, \ldots, \alpha_{r-1}\}$ and $\Pi_2 = \{\alpha_{r+1}, \ldots, \alpha_k\}$.

The representation φ we can expand to the algebra $\mathcal{L}_r \oplus \mathcal{N}_-$ if we put

$$\varphi(F_\beta)v = 0.$$

The subset $W \subset U(\mathcal{L}) \otimes V$ generated by the relations

$$xz \otimes v - x \otimes \varphi(z)v,$$

where $x \in U(\mathcal{L})$, $z \in U(\mathcal{L}_r)U(\mathcal{N})_-$ and $v \in V$ is evidently invariant space with respect to the left regular representation of $\mathcal{L}$ on $U(\mathcal{L}) \otimes V$ and we can define a factor representation on $\mathcal{L}$ on $(U(\mathcal{L}) \otimes V)/W$.

The main result of our paper is that for the good choice of the bases in $U(\mathcal{N}_+)$ we can rewrite this representation by using $p = \dim(\mathcal{N}_+)$ pairs of creation and annihilation operators

$$a^+ |n\rangle = |n+1\rangle \quad \text{and} \quad a |n\rangle = n |n-1\rangle.$$

The fact that the representation of W_{2p} is faithful and representation φ is any gives us that we can obtain by this way generalized boson realization.

$$\tau : \mathcal{L} \to W_{2p} \otimes U(\mathcal{L}_1 \oplus \mathcal{L}_2).$$

Example. Let E_{ij}, $i, j = 1, \ldots, n$, is base of the Lie algebra $gl(n)$. Simple Lie algebra $sl(n)$ we obtain, if we will consider E_{ij}, $i \neq j$, and $H_i = E_{i+1,i+1} - E_{i,i}$. Simple positive roots are $\alpha_i = E_{i,i+1}$, $i = 1, \ldots, n-1$.

If we take in general construction $\alpha_r = E_{r,r+1}$, we obtain that subalgebra $\mathcal{L}$ has basis $E_{i,j}$, $i, j \leq r$ or $i, j \geq r+1$, and subalgebras $\mathcal{N}_+$ have basis $E_{i,j}$, $i \leq r$, $j \geq r+1$. Therefore we need $p = r(n-r)$-pairs boson operators. The central element Z is in this case

$$Z = (n-r)\sum_{k=1}^{r-1} kH_k + r\sum_{k=1}^{n-r} kH_{n-k} = (r-n)\sum_{k=1}^{r} E_{k,k} + r\sum_{k=r+1}^{n} E_{k,k}.$$

‡ Such Z exist by the construction

Direct calculation gives the realization

$$\tau(E_{ij}) = a_{ij}^+, \qquad\qquad\qquad i \le r, j \ge r+1,$$

$$\tau(E_{ij}) = \sum_{k=r+1}^{n} a_{ik}^+ a_{jk} + E_{ij}, \qquad\qquad i, j \le r,$$

$$\tau(E_{ij}) = -\sum_{k=1}^{r} a_{kj}^+ a_{ki} + E_{ij}, \qquad\qquad i, j \ge r+1,$$

$$\tau(E_{ij}) = -\sum_{k=1}^{r}\sum_{l=r+1}^{n} a_{kl}^+ a_{ki} a_{jl} - \sum_{k=1}^{r} a_{ki} E_{kj} + \sum_{k=r+1}^{n} a_{jk} E_{ik}, \quad i \ge r+1, j \le r.$$

2. Boson realization for quantum algebras

The quantum algebras $U_q(\mathcal{L})$ were defined by Drinfeld and Jimbo [4, 5].

Let q be an independent variable, $\mathcal{A} = \mathcal{C}[q, q^{-1}]$ and $\mathcal{C}(q)$ is a division field of $\mathcal{A}$.

$$[n]_d = (q^{dn} - q^{-dn})/(q^d - q^{-d}) \in \mathcal{A},$$
$$[n]_d! = [n]_d \cdot [n-1]_d \cdot \ldots \cdot [1]_d$$

and

$$\begin{bmatrix} n \\ j \end{bmatrix}_d = \frac{[n]_d!}{[n-j]_d! \cdot [j]_d!}.$$

Let $\mathcal{L}$ is a simple Lie algebra with Cartan matrix $(a_{ij}) = 2(\alpha_i, \alpha_j)/(\alpha_i, \alpha_i)$, $i, j = 1, \ldots, k$. Let d_i be the smallest natural numbers such the matrix $(d_i a_{ij})$ is symmetric and positive.

Now the $\mathcal{C}(q)$-algebra $U_q(\mathcal{L})$ is defined by using E_i, F_i, K_i and K_i^{-1}, $i = 1, \ldots, k$, which satisfy the commutation relations

$$K_i K_j = K_j K_i, \qquad K_i K_i^{-1} = K_i^{-1} K_i = 1,$$

$$K_i E_j K_i^{-1} = q_i^{a_{ij}} E_j, \qquad K_i F_j K_i^{-1} = q_i^{-a_{ij}} F_j,$$

$$E_i F_j - F_j E_i = \delta_{ij} \frac{K_i - K_i^{-1}}{q_i - q_i^{-1}},$$

$$\sum_{s=0}^{1-a_{ij}} (-1)^s \begin{bmatrix} 1 - a_{ij} \\ s \end{bmatrix}_{d_i} E_i^{1-a_{ij}-s} E_j E_i^s = 0 \quad i \ne j,$$

$$\sum_{s=0}^{1-a_{ij}} (-1)^s \begin{bmatrix} 1 - a_{ij} \\ s \end{bmatrix}_{d_i} F_i^{1-a_{ij}-s} F_j F_i^s = 0 \quad i \ne j,$$

where $q_i = q^{d_i}$.

In the quantum algebra case it is not in general possible to express the element K_r with respect of the element of the center of the algebra $U_q(\mathcal{L}_r)$ and elements K_i, $i \ne r$. Therefore we must a representation φ of whole subalgebra $U_q(\mathcal{L}_r)$.

The second difference is that we will not use in quantum case the representation of Weyl algebra $\mathcal{W}$, but the representations of the algebra of "deformed" quantum oscillators

$$a^+ |n\rangle = |n+1\rangle \quad \text{and} \quad a |n\rangle = [n] |n-1\rangle.$$

798

If we define the operators q^x and q^{-x}

$$q^{\pm x}\,|n\rangle = q^{\pm n}\,|n\rangle\,,$$

we obtain the faithful representation of the q-deformed algebra $\mathcal{H}$, generated by elements a^+, a, $q^{\pm x}$, which fulfill the following relations [Biedenharn, Macfarlane, Hayashi][6, 7, 8].

$$aa^+ - q^{-1}a^+a = q^x, \qquad aa^+ - qa^+a = q^{-x},$$
$$q^xa^+q^{-x} = qa^+, \qquad q^xaq^{-x} = q^{-1}a,$$
$$q^xq^{-x} = q^{-x}q^x = 1\,.$$

Again similarly as in the Lie algebra case we obtain the generalized boson realization

$$\rho\,:\,U_q(\mathcal{L}) \to \mathcal{H}_p \otimes U_q(\mathcal{L}_r)\,,$$

where $\mathcal{H}_p$ is p-times tensor product of q-deformed oscillator algebras.

Example. Let $j = 2, \ldots, n$. Then the formulas

$$\rho(E_1) = a_1^+\,,$$
$$\rho(E_j) = a_{j-1}a_j^+ + q^{x_j - x_{j-1}} \otimes E_j\,,$$
$$\rho(K_1) = q^{x_1 + X_1^n} \otimes K_1\,,$$
$$\rho(K_j) = q^{x_j - x_{j-1}} \otimes K_j\,,$$
$$\rho(F_j) = a_j a_{j-1}^+ \otimes K_j^{-1} + F_j\,,$$
$$\rho(F_1) = -\frac{1}{q - q^{-1}}\left[q^{X_1^n} a_1 \otimes K_1 - q^{-X_1^n} a_1 \otimes K_1^{-1}\right] - \sum_{s=2}^{n} q^{X_s^n} a_s \otimes K_1 G_s\,,$$

where

$$G_2 = E_2, \quad G_k = E_k G_{k-1} - q^{-1}G_{k-1}E_k\,, \quad k = 3, \ldots, n\,,$$

and $q^{X_r^s} = q^{x_r + x_{r+1} + \ldots + x_s}$, give realization of the quantum algebra $U_q(A_n)$. The other quantum groups were studied in [9, 10, 11, 12, 13].

References

[1] Burdík Č 1985 *Realization of the real semisimple Lie Algebras: method of construction*, J. Phys. A: Math. Gen. 15 3101-3111

[2] Burdík Č 1986 *A new class of realizations of the Lie algebra* $sp(n, R)$, J. Phys. A: Math. Gen. 19 2465-2471

[3] Burdík Č 1988 *A new class of realizations of the Lie algebra* $so(q, 2n - q)$, J. Phys. A: Math. Gen. 21 289-295

[4] Drinfeld V G 1986 *Quantum Groups*, Proceeding of the ICM, Berkeley, 798-820

[5] Jimbo M 1985 *A q-difference analoque of* $U(g)$ *and the Yang-Baxter equation*, Lett. Math. Phys. 10 63-69

[6] Biedenharn L C M 1989 *The Quantum Group* $SU_q(2)$ *and q-analoque of the Boson Operators*, J. Phys. A: Math. Gen. 15 L837

[7] MacFarlane A J 1989 *On q-analoques of the quantum harmonic oscillator and the quantum group* $SU(2)_q$, J. Phys. A: Math. Gen. 15 4581 63-69

[8] Hyashi T 1990 *Q-Analogues of Clifford and Weyl Algebras - Spinor and Oscillator Representations of Quantum Envelopings Algebras*, Commun. Math. Phys. 127 129

[9] Burdík Č and Navrátil O 1990 *The boson realizations of the quantum group* $U_q(sl(2))$, J. Phys. A: Math. Gen. 23 L1205-L1208

[10] Burdík Č Černý L and Navrátil O 1993 *The boson realizations of the quantum group* $U_q(sl(n + 1, C))$, J. Phys. A: Math. Gen. 25 L83-L88

[11] Burdík Č and Navrátil O 1998 *The q-boson realizations of the quantum groups* $U_q(B_n)$, Czech. J. Phys. 48 1301-1306

[12] Burdík Č and Navrátil O 1999 *The q-boson realizations of the quantum groups* $U_q(D_n)$, J. Phys. A: Math. Gen 32 p. 6141

[13] Burdík Č and Navrátil O 1999 *The q-boson realizations of the quantum groups* $U_q(C_n)$, International Journal of Modern Physics A Vol 14 No. 28 p. 4491

Inst. Phys. Conf. Ser. No 173: Mini-symposium
Paper presented at 24th Int. Coll. Group Theoretical Methods in Physics, Paris, France, 15–20 July 2002
©2003 IOP Publishing Ltd

Dynamical systems associated to bicrossproducts

Oscar Arratia[1] and Mariano A del Olmo[2]

[1] Departamento de Matemática Aplicada a la Ingeniería, Universidad de Valladolid, E-47011, Valladolid, Spain

[2] Departamento de Física Teórica, Universidad de Valladolid, E-47011, Valladolid, Spain

Abstract. We present new aspects related to the structure of bicrossproduct shared by some quantum deformations of kinematical algebras. The actions associated with this bicrossproduct structure allow us to construct a nonlinear action over a new group related with the group of translations. We compute the flow linked to the action and the associated dynamical systems. These results can be applied in the theory of representations of quantum groups as well as in the theory of dynamical systems.

1. Introduction

Quantum groups have become a topic that reconciles modern physics and classical thoughts in the realm of noncommutative geometry. However, in some special situations like the one we describe in this paper the 'old' classical geometry allows us to develop part of the theory needed to make some physics using standard commutative tools. In fact, we have successfully constructed induced representations for several quantum kinematical algebras [1]–[3]. The particular feature shared by all the cases we have studied is a bicrossproduct structure [4] with one of the factors being commutative. In this respect we recall that the bicrossproduct construction is the quantum counterpart of the semidirect product of (Lie) groups which is the kind of structure exhibited by kinematical symmetries such as the Poincaré and Galilei groups. We will use an approach based on the inhomogeneous Cayley–Klein algebras in order to provide a unified discussion of the previously mentioned symmetries.

The paper is structured in the following way: section 2 makes a brief description of the Cayley–Klein algebras presenting their quantum deformations and bicrossproduct structure. In section 3 we describe the classical dynamical systems derived from the bicrossproduct structure and we discuss the appropriate way to integrate them in order to compute the corresponding one–parameter flows. Finally section 4 summarizes the main results obtained along the paper.

2. Quantum Cayley–Klein algebras

The Cayley–Klein pseudo-orthogonal algebras is a family of $(N+1)N/2$-dimensional real Lie algebras characterized by N real parameters $(\omega_1, \omega_2, \ldots, \omega_N)$ and denoted $\mathfrak{so}_{\omega_1,\omega_2,\ldots,\omega_N}(N+1)$ [5]. When all the ω_i's are different to zero the algebra $\mathfrak{so}_{\omega_1,\omega_2,\ldots,\omega_N}(N+1)$ is isomorphic to some of the pseudo-orthogonal algebras $\mathfrak{so}(p,q)$ with $p+q = N+1$ and $p \geq q > 0$. If some of the coefficients ω_i vanishes the corresponding algebra is inhomogeneous and can be obtained from $\mathfrak{so}(p,q)$ by means of a sequence of contractions. In the particular case of $\omega_1 = 0$, the algebras $\mathfrak{so}_{0,\omega_2,\ldots,\omega_N}(N+1)$ can be realized as algebras of groups of affine transformations on $\mathbb{R}^N$. In this case the first N generators are denoted by P_i stressing, in this way, its role as generators of translations. The remaining generators are denoted J_{ij}, and they originate compact and 'noncompact' rotations. These inhomogeneous algebras, denoted by $\text{iso}_{\omega_2,\ldots,\omega_N}(N)$, are characterized by the following nonvanishing commutators

$$[J_{ij}, J_{ik}] = \omega_{ij} J_{jk}, \qquad [J_{ij}, J_{jk}] = -J_{ik}, \qquad [J_{ik}, J_{jk}] = \omega_{jk} J_{ij},$$

$$[J_{ij}, P_i] = P_j, \qquad [J_{ij}, P_i] = -\omega_{ij} P_i, \qquad \omega_{ij} = \prod_{s=i+1}^{j} \omega_s, \quad 1 \leq i < j < k \leq N.$$

The standard deformation of $U(\text{iso}_{\omega_2,\omega_3,\omega_4}(4))$ was considered in [6] and the general case $U(\text{iso}_{\omega_2,\omega_3,\ldots,\omega_N}(N))$ was developed in [7]. In [8] it was proved that the standard quantum Hopf algebras $U_z(\text{iso}_{\omega_2,\omega_3,\ldots,\omega_N}(N))$ have a structure of bicrossproduct. Using a basis adapted to the bicrossproduct structure it can be shown that only the following commutators are different from the classical case

$$[J_{iN}, P_j] = \delta_{ij}\left(\frac{1 - e^{-2zP_N}}{2z} - \frac{z}{2}\sum_{s=1}^{N-1}\omega_{sN}P_s^2\right) + z\omega_{iN}P_iP_j. \tag{1}$$

The bicrossproduct structure $U_z(\text{iso}_{\omega_2,\omega_3,\ldots,\omega_N}(N)) = \mathcal{K} \rhd\!\!\blacktriangleleft \mathcal{L}$ is described by a right action of $\mathcal{K}$ over $\mathcal{L}$ and a left coaction of $\mathcal{L}$ over $\mathcal{K}$, where $\mathcal{K} = U(\text{so}_{\omega_2,\omega_3,\ldots,\omega_N}(N))$ and $\mathcal{L}$ the commutative Hopf subalgebra generated by $P_1, P_2, \ldots, P_N$. We are only interested in the action which is given by

$$P_i \lhd J_{jk} = [P_i, J_{jk}], \qquad j < k, \ i, j, k = 1, 2, \ldots, N.$$

3. One–parameter flows

In [8] $U_z(T_N)$ was considered as a noncommutative deformation of the Lie algebra of the group of translations of $\mathbb{R}^N$. However, since $U_z(T_N)$ is commutative we can consider it as a classical algebra of functions over a group. In this way we have the bicrossproduct decomposition

$$U_z(\text{iso}_{\omega_2,\omega_3,\ldots,\omega_N}(N)) = U(\text{so}_{\omega_2,\omega_3,\ldots,\omega_N}(N))\!\blacktriangleright\!\!\lhd F(T_{z,N}),$$

where $T_{z,N}$ is the space $\mathbb{R}^N$ equipped with the group composition law

$$(\alpha_1', \alpha_2', \ldots, \alpha_{N-1}', \alpha_N')(\alpha_1, \alpha_2, \ldots, \alpha_{N-1}, \alpha_N) =$$
$$(\alpha_1' + e^{-z\alpha_N'}\alpha_1, \alpha_2' + e^{-z\alpha_N'}\alpha_2, \ldots, \alpha_{N-1}' + e^{-z\alpha_N'}\alpha_{N-1}, \alpha_N' + \alpha_N). \tag{2}$$

The generators P_i of $U_z(T_N)$ give in this context a global chart over $T_{z,N}$,

$$P_i(\alpha) = \alpha_i, \qquad \alpha \in T_{z,N}.$$

The structure of $U(\text{so}_{\omega_2,\omega_3,\ldots,\omega_N}(N))$–module algebra of $F(T_{z,N})$ implies that an action of the group $SO_{\omega_2,\omega_3,\ldots,\omega_N}(N)$ on $T_{z,N}$ is defined. At the infinitesimal level this action is described by the vector fields

$$\hat{J}_{ij} = -P_j\frac{\partial}{\partial P_i} + \omega_{ij}P_i\frac{\partial}{\partial P_j},$$

$$\hat{J}_{iN} = \sum_{j=1}^{N-1}-\left[\delta_{ij}\left(\frac{1-e^{-2zP_N}}{2z} - \frac{z}{2}\sum_{s=1}^{N-1}\omega_{sN}P_s^2\right) + z\omega_{iN}P_iP_j\right]\frac{\partial}{\partial P_j} \tag{3}$$
$$+ \omega_{iN}P_i\frac{\partial}{\partial P_N}.$$

The integration of the vector fields $\hat{J}_{ij}$ when $i < j < N$ is immediate and gives the well known linear flows associated to simple compact or noncompact rotations in the ij plane. On the contrary the computation of the flows associated with the 'deformed' fields J_{iN} requires a detailed analysis of certain two dimensional autonomous systems to which we devote the rest of the paper. First of all note that the function

$$h_{\omega,z} = \sum_{j=1}^{N-1} \omega_{jN} P_j^2 e^{zP_N} + \frac{\cosh(zP_N) - 1}{\frac{z^2}{2}} \tag{4}$$

together with

$$h_{\omega,z}^{iN,k} = P_k e^{zP_N}, \qquad k \in \{1, 2, \ldots, N-1\} - \{i\} \tag{5}$$

form a complete set of invariants for $\hat{J}_{iN}$. On the other side, to obtain the integral curves $\hat{J}_{iN}$ it is necessary to solve the system of N differential equations

$$\dot{\alpha}_j = -z\omega_{iN}\alpha_i\alpha_j, \qquad\qquad j \neq i, N,$$
$$\dot{\alpha}_i = -\frac{1-e^{-2z\alpha_n}}{2z} + \frac{z}{2}\sum_{s=1}^{N-1}\omega_{sN}\alpha_s^2 - z\omega_{iN}\alpha_i^2, \tag{6}$$
$$\dot{\alpha}_N = \omega_{iN}\alpha_i.$$

The invariants $h_{\omega,z}^{iN,k}$ (5) allow us to remove $N-2$ degrees of freedom. From $h_{\omega,z}^{iN,k}(\alpha) = \alpha_k e^{z\alpha_N} = \beta_k$ we obtain $\alpha_k = \beta_k e^{-z\alpha_N}$, restricting the study of the N–dimensional system to the following family of 2–dimensional systems depending on the N parameters β_k, ω and z

$$\dot{\alpha}_i = -\left[\frac{1-e^{-2z\alpha_N}}{2z} - \frac{z}{2}\left(\sum_{s\neq i,N}\omega_{sN}\beta_s^2\right)e^{-2z\alpha_N} + \frac{z}{2}\omega_{iN}\alpha_i^2\right], \tag{7}$$
$$\dot{\alpha}_N = \omega_{iN}\alpha_i.$$

The case $z = 0$ is trivial since it corresponds to linear systems analogous to those of the fields $\hat{J}_{ij}$. For nonvanishing values of z the equations can be rescaled as follows

$$x(t) = z\alpha_i(t), \qquad y(t) = z\alpha_N(t).$$

Setting $a = \omega_{iN}$, $b - 1 = z^2\rho = z^2\sum_{s\neq i,N}\omega_{sN}\beta_s^2$ the above system (7) becomes

$$\dot{x} = -\frac{1}{2}ax^2 - \frac{1}{2} + \frac{1}{2}be^{-2y}, \qquad \dot{y} = ax. \tag{8}$$

In this form the limit $z \to 0$ cannot be studied, but in advantage it depends on only two parameters: a and b. The possibility of reabsortion of the parameter z is followed from the fact that all the Hopf algebras $U_z(\mathrm{iso}_{\omega_2,\omega_3,\ldots,\omega_N}(N))$ are isomorphic (for fixed values of the parameters ω_s) whenever z is nonzero. The function (4) gives rise to the following invariant of the system (8)

$$h_{a,b} = ax^2e^y + e^y + be^{-y}. \tag{9}$$

It is easy to check that (8) has fixed points only for positive values of b. In fact, there is only one hyperbolic (resp. elliptic) equilibrium point if in addition $a < 0$ (resp. $a > 0$). In the case $a = 0$ all the points with $y = \frac{1}{2}\ln b$ are fixed points.

Finally, using (9) and performing a quadrature, it is possible to give a closed expression for the flow associated to the system (8)

$$\Phi_{a,b}^t(x,y) = \left(\frac{(ax^2e^y - e^y + be^{-y})\,S_a(t) + (2xe^y)\,C_a(t)}{(ax^2e^y + e^y + be^{-y}) + (-ax^2e^y + e^y - be^{-y})\,C_a(t) + (2axe^y)\,S_a(t)},\right.$$
$$\left.\ln\frac{1}{2}\left[(ax^2e^y + e^y + be^{-y}) + (-ax^2e^y + e^y - be^{-y})\,C_a(t) + (2axe^y)\,S_a(t)\right]\right).$$

where

$$C_\omega(t) = \frac{e^{\sqrt{-\omega}t} + e^{-\sqrt{-\omega}t}}{2}, \qquad S_\omega(t) = \frac{e^{\sqrt{-\omega}t} - e^{-\sqrt{-\omega}t}}{2\sqrt{-\omega}}.$$

Note that the expression of $\Phi_{a,b}$ is valid for all values of the parameters a and b. Figure 1 shows the qualitative form of the integral curves in the most relevant cases.

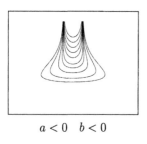
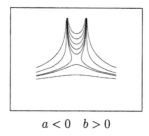
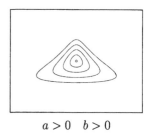

| $a < 0 \quad b < 0$ | $a < 0 \quad b > 0$ | $a > 0 \quad b > 0$ |

Figure 1. Qualitative aspect of the orbits associated to the flow $\Phi_{a,b}$.

4. Concluding remarks

In this work we have shown how to deal with quantum bicrossproduct algebras when one of their factors is commutative. In these cases classical vector fields arise. The difference between deformed and non–deformed actions is that linearity is not preserved in the deformed cases. However, the integrability of the deformed dynamical systems remains unchanged as we have seen explicitly in the case worked out.

Applications of the present work to the construction of induced representations can be found in [1]–[3]. A more detailed version of this work can be found in [9].

Acknowledgments

This work has been partially supported by DGES of the Ministerio de Educación y Cultura de España under Project PB98–0360, and the Junta de Castilla y León (Spain).

References

[1] Arratia O and del Olmo M A 2000 J. Math. Phys. 41 4817
[2] Arratia O and del Olmo M A 2001 Representations of quantum bicrossproduct algebras (preprint: UVA 2001; math.QA/0110275)
[3] Arratia O and del Olmo M A 2002 Induced Representations of Quantum Kinematical Algebras and Quantum Mechanics (math.QA/0202101; to be published in J. Phys. A: Math. Gen. **35** (2002)).
[4] Majid S 1990 J. Alg. 130 17; 1990 Isr. J. Math. 72 133; 1995 Foundations of quantum group theory (Cambridge Univ. Press: Cambridge)
[5] Herranz F J, de Montigny M, del Olmo M A and Santander M 1994 J. Phys. A: Math. Gen. 27 2515
[6] Ballesteros A, Herranz F J, del Olmo M A and Santander M 1994 J. Math. Phys. 35 4928
[7] Ballesteros A, Herranz F J, del Olmo M A and Santander M 1995 Lett. Math. Phys. 33 273
[8] de Azcárraga J A, del Olmo M A, Pérez-Bueno J C and Santander M 1997 J. Phys. A: Math. Gen. 30 3069
[9] Arratia O and del Olmo M A 2002 J. Phys. A: Math. Gen. 35 5347

Inst. Phys. Conf. Ser. No 173: Mini-symposium
Paper presented at 24th Int. Coll. Group Theoretical Methods in Physics, Paris, France, 15–20 July 2002
©2003 IOP Publishing Ltd

Combinatorics of the q-shuffle product and the $q$uon Weyl Algebra.

Gérard Duchamp†, Christophe Tollu‡, Frédéric Toumazet‡

† LIFAR, Université de Rouen, 76134 Mont Saint-Aignan Cedex, France – e-mail:
gerard.duchamp@univ-rouen.fr

‡ LIPN, CNRS and Université de Paris-Nord, 93430 Villetaneuse, France – e-mail:
christophe.tollu@lipn.univ-paris13.fr,
frederic.toumazet@lipn.univ-paris13.fr.

Research of all authors supported by a JemSTIC grant from the CNRS.

Abstract. We give two dual faithful representations of the multidimensional q-Weyl algebra using operators on q-shuffle algebras. The realizability problem is then naturally connected with the algebra of the symmetric group.

1. Introduction

Deformation theory allows to get at once relations within algebras especially when one introduces free parameters.

So, having a collection of operators $a_i^{\pm}$ (the superscripts $+$ and $-$ respectively stand for creation and for annihilation), there is a great temptation to interpolate between the boson and fermion statistics [4, 5]

$$a^- a^+ \pm a^+ a^- = 1$$

by resorting to the q-bracket product:

$$[a^-, a^+]_q = a^- a^+ - q\, a^+ a^- = 1$$

If one further asks for the qumutation of operators bearing distinct labels, one obtains the q-Weyl Algebra (see the precise definition below). Don Zagier proved the realizability of this algebra for specializations of q within $]-1, 1[$ (see [10]).

The present briefly expounds how operators on words can shed some light on this algebra by providing two faithful representations.

2. Two dual representations of the n-dimensional q-Weyl Algebra

Let $\mathcal{F} = (a_i^+; a_i^-)_{i \in I}$ be a family of (pairs of) symbols. The general q-Weyl algebra is defined by generators — the latter symbols — and relations

$$a_i^- a_j^+ - q\, a_j^+ a_i^- = \delta_{i,j} \qquad (i, j \in I). \tag{1}$$

This algebra will be denoted by $\mathcal{W}_{\mathcal{F}}(q)$. The above relations allow for a rearrangement of the monomials such that (for example) all creations appear on the left and all annihilations appear on the right (see prop. 1), which leaves no possibility for further moves. More precisely, one has:

Proposition 1 *For two sequences* $J = (j_1, j_2, \cdots j_r);\ K = (k_1, k_2, \cdots k_s)$ *let*

$$E_{J,K} = a_{j_1}^+ a_{j_2}^+ \cdots a_{j_r}^+ a_{k_1}^- a_{k_2}^- \cdots a_{k_s}^-$$

then the set $\{E_{J,K}\} \cup \{1_{\mathcal{W}_{\mathcal{F}}(q)}\}$, *where* J *and* K *describe all nonempty sequences of* I *and* $1_{\mathcal{W}_{\mathcal{F}}(q)}$ *denotes the unit of the algebra, is a basis of* $\mathcal{W}_{\mathcal{F}}(q)$.

Let us now recall some basic facts about words [1, 6] and the q-shuffle product [3, 8, 9] (see also [7] for the classical shuffle product).

Let A be an alphabet (a set of symbols), a word is any sequence of symbols $w = a_1 a_2 \cdots a_n$ (n is the length of w: $|w| = n$). The set of all words (including the empty sequence ϵ of length 0) is denoted by A^*. Concatenation naturally endows A^* with the structure of a monoid. The following concatenation

$$(a_1 a_2 \cdots a_r)(b_1 b_2 \cdots b_s)$$

results in the word (z_i) with $z_i = a_i$ for $1 \leq i \leq r$ and $z_i = b_{i-r}$ for $r + 1 \leq i \leq r + s$ (here ϵ is neutral). The C- (resp. $C[q]$-) algebra of this monoid is the free algebra of noncommutative polynomials denoted by $C\langle A \rangle$ (resp. $C[q]\langle A \rangle$).

There is [3] a unique graded law $\sqcup\!\sqcup_q$ satisfying the relations

$$\begin{cases} u \sqcup\!\sqcup_q \epsilon = \epsilon \sqcup\!\sqcup_q u = u \\ a^{-1}(u \sqcup\!\sqcup_q v) = \underbrace{(a^{-1}u) \sqcup\!\sqcup_q u + q^{|u|} u \sqcup\!\sqcup_q (a^{-1}v)}_{q\text{–derivative.}}, \end{cases}$$

We also define, for each symbol $a \in A$, an operator ∂_a on the free algebra which acts by derivations:

$$\partial_a(\epsilon) = 0;\ \partial_a(b) = \delta_{a,b};\ \partial_a(uv) = \partial_a(u) + q^{|u|} u \partial_a(v).$$

The blunt simplification by (the first) "a" can formally be defined by

$$a^{-1}(\epsilon) = 0;\ a^{-1}(bu) = \delta_{a,b} u.$$

We now are in a position to see that these operators are dual to each other. Moreover, they give two faithful representations of $\mathcal{W}_{\mathcal{F}}(q)$ on $C[q]\langle A \rangle$.

Consider the table

S_I	$a^{-1}(?)$	$a \sqcup\!\sqcup_q (?)$
S_{II}	$a(?)$	∂_a

then the elements of the (second and third) columns are mutually dual (for the standard scalar product defined on words by $\langle u | v \rangle = \delta_{u,v}$) and each line provides a faithful representation of $\mathcal{W}_{\mathcal{F}}(q)$.

Theorem 2 *(i)* $S_{I,II}$ *are two mutually dual faithful representations of the n-dimensional q-Weyl algebra.*

(ii) S_{II} *generalizes (for a collection of variables) the classical Bargmann-Fock representation.*

2.1. The realizability problem

The realizability problem can be plainly expressed as follows :
Is there a Hilbert space $\mathcal{H}$ and a faithful rep ρ of $\mathcal{W}_{\mathcal{F}}(q)$ ($q \in C$) on $\mathcal{H}$ such that, for each $i \in I$, $\rho(a_i^+)$ and $\rho(a_i^-)$ are mutually adjoint ?
The problem can be easily solved for $|I| = 1$ by taking the scalar product

$$\langle a^n | a^m \rangle = \delta_{n,m} [n]!_q$$

where $[k]_q = 1 + q + q^2 + \cdots q^{k-1}$ and $[n]!_q = [2]_q [3]_q \cdots [n]_q$. Then the classical orthonormal

basis $\left(\dfrac{a^n}{\sqrt{[n]!_q}} \right)_{n \in \mathbf{F}}$ can be straightforwardly derived [2].

The general case crucially involves the symmetric group. The analysis goes as follows: let $(\mathcal{H}, \rho)$ be any solution with vacuum vector $|0$, $i_1, i_2 \cdots i_n$ be distinct indices of I and $\sigma \in \mathbf{S}_n$. We denote, for short, the vector $\rho(a_{i_1}^+ a_{i_2}^+ \cdots a_{i_n}^+)|0$ by $(i_1 i_2 \cdots i_n)$; then

$$\langle (i_1 i_2 \cdots i_n) | (i_{\sigma(1)} i_{\sigma(2)} \cdots i_{\sigma(n)}) \rangle = q^{l(\sigma)}.$$

Thus, if the algebra is realizable, the element of $C[\mathbf{S}_n]$

$$U_n = \sum_{\sigma \in \mathbf{S}_n} q^{l(\sigma)} \sigma$$

is invertible. The converse implication also holds because [3]

$$a_1 a_2 \cdots a_n \cdot U_n = a_1 \sqcup\!\sqcup_q a_2 \sqcup\!\sqcup_q \cdots \sqcup\!\sqcup_q a_n.$$

3. Concluding remarks

Orthonormal bases of $S_{I,II}$ can be computed with the square root of the matrix of U_n. Let $\Omega_k = \sum_{l(\sigma)=k} \sigma$; one remarks that

$$U_n = \sum_{0 \le k \le \frac{n(n-1)}{2}} q^k \Omega_k = 1 + q \sum_{1 \le k \le \frac{n(n-1)}{2}} q^{k-1} \Omega_k = 1 + qU^+$$

so that the square root of U_n explicitly reads

$$\sqrt{U_n} = \sum_{k \ge 0} \binom{\frac{1}{2}}{k} q^k (U^+)^k$$

This computation, along with a study of the isometric algebra (i.e. generated by the elements Ω_k), will be the subject of a forthcoming paper.

[1] J. Berstel and C. Reutenauer, *Rational series and their languages*, EATCS Monographs on Theoretical Computer Science, Springer, Berlin, 1988.
[2] G. Duchamp and J. Katriel, *Ordering relations for q-boson operators, continued fractions techniques, and the q-CBH enigma*. Journal of Physics A **28** (1995) 7209–7225.
[3] G. Duchamp, A. Klyachko, D. Krob and J.-Y. Thibon, *Noncommutative symmetric functions III: Deformations of Cauchy and convolution algebras*, Discrete Mathematics and Theoretical Computer Science **1** (1997), 159–216.

806

[4] D.I Fivel, *Interpolation between Bose and Fermi statistics using generalized commutators*, Physical Review Letters, **65** (1990), 3361–3364.

[5] O.W. Greenberg, *Particles with small violations of Fermi or Bose statistics*, Physical Rev. D **43** (1991), 4111–4120.

[6] N. Lothaire, *Algebraic Combinatorics on words*, Cambridge University Press, Cambridge, 2001.

[7] C. Reutenauer, *Free Lie algebras*, London Mathematical Society Monographs, Clarendon Press, Oxford, 1993.

[8] M. Rosso, *Groupes quantiques et algèbres de battage quantiques*, C.R. Acad. Sci. Paris Ser. I **320** (1995), 145–148.

[9] V. Schechtman and A. Varchenko, *Quantum groups and homology of local systems*, in "Algebraic geometry and analytical geometry", ICM-90 Satellite Conference, 182–191, Springer, 1991.

[10] D. Zagier, *Realizability of a model in infinite statistics*, Commun. Math. Phys. **147** (1992), 199–210.

Inst. Phys. Conf. Ser. No 173: Mini-symposium
Paper presented at 24th Int. Coll. Group Theoretical Methods in Physics, Paris, France, 15–20 July 2002
©2003 IOP Publishing Ltd

Decoupling of Tensor factors in Cross Product and Braided Tensor Product Algebras

Gaetano Fiore

Dip. Matematica ed Applicazioni Università Federico II & INFN, Napoli, Italy

Abstract. We briefly review and illustrate our procedure to 'decouple' by transformation of generators: either a Hopf algebra H from a H-module algebra $\mathcal{A}_1$ in their cross-product $\mathcal{A}_1 \rtimes H$; or two (or more) H-module algebras $\mathcal{A}_1, \mathcal{A}_2$. These transformations are based on the existence of an algebra map $\mathcal{A}_1 \rtimes H \to \mathcal{A}_1$.

1. Decoupling of tensor factors in cross product algebras

Let H be a Hopf algebra over $\mathbf{C}$, say, $\mathcal{A}$ a unital (right, say) H-module algebra. We denote by $\triangleleft$ the right action, namely the bilinear map such that, for any $a, a' \in \mathcal{A}$ and $g, g' \in H$

$$\triangleleft : (a, g) \in \mathcal{A} \times H \to a \triangleleft g \in \mathcal{A},$$
$$a \triangleleft (gg') = (a \triangleleft g) \triangleleft g', \qquad (aa') \triangleleft g = (a \triangleleft g_{(1)}) (a' \triangleleft g_{(2)}). \tag{1}$$

We have used the Sweedler-type notation $\Delta(g) = g_{(1)} \otimes g_{(2)}$ for the coproduct Δ. The cross-product algebra $\mathcal{A} \rtimes H$ is $H \otimes \mathcal{A}$ as a vector space, and so we denote as usual $g \otimes a$ simply by ga; $H 1_{\mathcal{A}}$, $1_H \mathcal{A}$ are subalgebras isomorphich to $H, \mathcal{A}$, and so we omit to write either unit $1_{\mathcal{A}}, 1_H$ whenever multiplied by non-unit elements; for any $a \in \mathcal{A}$, $g \in H$ the product fulfills

$$ag = g_{(1)} (a \triangleleft g_{(2)}). \tag{2}$$

$\mathcal{A} \rtimes H$ is a H-module algebra itself, if we extend $\triangleleft$ on H as the adjoint action: $h \triangleleft g = Sg_{(1)} h g_{(2)}$. If H is a Hopf $*$-algebra, $\mathcal{A}$ a H-module $*$-algebra, then, as known, these two $*$-structures can be glued in a unique one to make $\mathcal{A} \rtimes H$ a $*$-algebra itself.

Theorem 1 *[6] Let H be a Hopf algebra, $\mathcal{A}$ a right H-module algebra. If there exists a "realization" $\tilde{\varphi}$ of $\mathcal{A} \rtimes H$ within $\mathcal{A}$ acting as the identity on $\mathcal{A}$, i.e. an algebra map*

$$\tilde{\varphi} : \mathcal{A} \rtimes H \to \mathcal{A}, \qquad \tilde{\varphi}(a) = a, \tag{3}$$

then $\tilde{\zeta}(g) := g_{(1)} \tilde{\varphi}(Sg_{(2)})$ defines an injective algebra map $H \to \mathcal{A} \rtimes H$ such that

$$[\tilde{\zeta}(g), \mathcal{A}] = 0 \tag{4}$$

for any $g \in H$. Moreover $\mathcal{A} \rtimes H = \tilde{\zeta}(H) \mathcal{A}$. Consequently, the center of the cross-product $\mathcal{A} \rtimes H$ is given by $\mathcal{Z}(\mathcal{A} \rtimes H) = \mathcal{Z}(\mathcal{A}) \tilde{\zeta}(\mathcal{Z}(H))$, and if $H_c, \mathcal{A}_c$ are Cartan subalgebras of H and $\mathcal{A}$ respectively, then $\mathcal{A}_c \tilde{\zeta}(H_c)$ is a Cartan subalgebra of $\mathcal{A} \rtimes H$. Finally, if $\tilde{\varphi} : \mathcal{A} \rtimes H \to \mathcal{A}$ is a $*$-algebra map, then also $\tilde{\zeta} : H \to \tilde{\mathcal{C}}$ is.

The equality $\mathcal{A} \rtimes H = \tilde{\zeta}(H) \mathcal{A}$ means that $\mathcal{A} \rtimes H$ is equal to a product $\mathcal{A} H'$, where $H' \equiv \tilde{\zeta}(H) \subset \mathcal{A} \rtimes H$ is a subalgebra isomorphic to H and *commuting* with $\mathcal{A}$, i.e. is isomorphic to the *ordinary* tensor product algebra $H \otimes \mathcal{A}$. In other words, if $\{a_I\}, \{g_J\}$ are resp. sets of generators of $\mathcal{A}, H$, then $\{a_I\} \cup \{\tilde{\zeta}(g_J)\}$ is a more manageable set of generators of $\mathcal{A} \bowtie H$ than $\{a_I\} \cup \{g_J\}$. The other statements allow to determine casimirs and complete sets of commuting observables, key ingredients to develop representation theory.

Well-known examples of maps (3) are the "vector field" or the "Jordan-Schwinger" realizations of the UEA $U\mathbf{g}$'s ($\mathbf{g}$ being a Lie algebra), where $\mathcal{A}$ is the Heisenberg algebra on a $\mathbf{g}$-covariant space or Clifford algebra, respectively. In the case of e.g. the Heisenberg algebra on the Euclidean space $\mathbf{R}^3$, the well-known realization of the three generators J^{ij}, ($i \neq j$) of $\mathbf{g} = so(3)$ as "vector fields" (namely homogeneous first order differential operators)

$$\tilde{\varphi}(J^{jk}) := x^j p^k - x^k p^j$$

gives nothing but the orbital angular momentum operator in 1-particle quantum mechanics ($x_i, p^i \equiv -i\partial^i$ denote the position and momentum components respectively). Then

$$\tilde{\zeta}(J^{jk}) = J^{jk} - \tilde{\varphi}(J^{jk}) = J^{jk} - (x^j p^k - x^k p^j).$$

gives the difference between the total and the orbital angular momentum, i.e. the "intrinsic" angular momentum (or "spin"), which indeed commutes with the x^i, p^i's. Maps $\tilde{\varphi}$ have been determined [2, 5] also for a number of $U_q\mathbf{g}$-covariant, i.e. quantum group covariant, deformed Heisenberg (or Clifford) algebras. So the theorem is immediately applicable to them.

No map $\tilde{\varphi}$ can exist if we take as $\mathcal{A}$ just the space on which the Heisenberg algebra is built (in the previous example $\mathbf{R}^3$), since one cannot realize the non abelian algebra $\mathcal{A} \bowtie H$ in terms of the abelian $\mathcal{A}$. Surprisingly, a map $\tilde{\varphi}$ may exist if we deform the algebras. For instance, in Ref. [8] a map $\tilde{\varphi}$ realizing $U_q so(3)$ has been determined for $\mathcal{A}$ the q-deformed fuzzy sphere $\hat{S}^2_{q,M}$. To treat other examples we generalize the previous results by weakening our assumptions. Namely, we require at least that H admits a Gauss decomposition

$$H = H^+ H^- = H^- H^+$$

into two Hopf subalgebras H^+, H^- for each of which analogous maps $\tilde{\varphi}^+, \tilde{\varphi}^-$ (3) coinciding on $H^+ \cap H^-$ exist. Then Theorem 1 will apply separately to $\mathcal{A} \bowtie H^+$ and $\mathcal{A} \bowtie H^-$. What about the whole $H \bowtie \mathcal{A}$?

Theorem 2 [6] *Under the above assumptions setting* $\tilde{\zeta}^{\pm}(g^{\pm}) := g^{\pm}_{(1)} \tilde{\varphi}^{\pm}(S g^{\pm}_{(2)})$ *(where* $g^{\pm} \in H^{\pm}$ *respectively) defines injective algebra maps* $H^{\pm} \to \mathcal{A} \bowtie H^{\pm}$ *such that* $[\tilde{\zeta}^{\pm}(g^{\pm}), \mathcal{A}] = 0$ *for any* $g^{\pm} \in H^{\pm}$. *Moreover*

$$\mathcal{A} \bowtie H = \tilde{\zeta}^+(H^+)\, \tilde{\zeta}^-(H^-)\, \mathcal{A} = \tilde{\zeta}^-(H^-)\, \tilde{\zeta}^+(H^+)\, \mathcal{A}. \tag{5}$$

Any $c \in \mathcal{Z}(\mathcal{A} \bowtie H)$ *can be expressed in the form*

$$c = \tilde{\zeta}^+\left(c^{(1)}\right) \tilde{\zeta}^-\left(c^{(2)}\right) c^{(3)}, \tag{6}$$

where $c^{(1)} \otimes c^{(2)} \otimes c^{(3)} \in H^+ \otimes H^- \otimes \mathcal{Z}(\mathcal{A})$ *and* $c^{(1)}c^{(2)} \otimes c^{(3)} \in \mathcal{Z}(H) \otimes \mathcal{Z}(\mathcal{A})$; *conversely any such* $c \in \mathcal{Z}(\mathcal{A} \bowtie H)$. *If* $H_c \subset H^+ \cap H^-$ *and* $\mathcal{A}_c$ *are Cartan subalgebras resp. of* H *and* $\mathcal{A}$, *then* $\mathcal{A}_c \tilde{\zeta}^+(H_c)$ $[\equiv \mathcal{A}_c \tilde{\zeta}^-(H_c)]$ *is a Cartan subalgebra of* $\mathcal{A} \bowtie H$.

If $\tilde{\varphi}^{\pm}$ *are* $*$-*algebra map, then also* $\tilde{\zeta}^{\pm} : H^{\pm} \to \tilde{C}$ *are. If* $\tilde{\varphi}^{\pm}(\alpha^*) = [\tilde{\varphi}^{\mp}(\alpha)]^*$ $\forall \alpha \in H^{\mp} \bowtie \mathcal{A}$, *then* $\tilde{\zeta}^{\pm}(g^{\mp*}) = [\tilde{\zeta}^{\mp}(g^{\mp})]^*$, *with* $g^{\mp} \in H^{\mp}$.

As a consequence of this theorem, $\forall g^+ \in H^+$, $g^- \in H^-$ $\exists c^{(1)} \otimes c^{(2)} \otimes c^{(3)} \in \mathcal{Z}(\mathcal{A}) \otimes H^- \otimes H^+$ (depending on g^+, g^-) such that

$$\tilde{\zeta}^+(g^+)\tilde{\zeta}^-(g^-) = c^{(1)}\tilde{\zeta}^-(c^{(2)})\tilde{\zeta}^+(c^{(3)}). \tag{7}$$

These will be the "commutation relations" between elements of $\zeta^+(H^+)$ and $\zeta^-(H^-)$. Their form will depend on the specific algebras considered.

As an application we consider now the pair $(H, \mathcal{A})$ with $H = U_q so(3)$ and $\mathcal{A} = \mathbf{R}_q^3$, the (algebra of functions on) the 3-dim quantum Euclidean space, whose generators we denote resp. by E^+, E^-, K, K^{-1} and p^+, p^0, p^-. In our present conventions

$$p^0 p^\pm = q^{\pm 1} p^\pm p^0 \qquad\qquad [p^+, p^-] = (1 - q^{-1}) p^0 p^0 \qquad\qquad (8)$$

$$K E^\pm = q^{\pm 1} E^\pm K \qquad\qquad [E^+, E^-]_{q^{-1}} = \frac{K^2 - 1}{q^2 - 1} \qquad\qquad (9)$$

$$
\begin{aligned}
&K p^0 = p^0 K, &&K p^\pm = q^{\mp 1} p^\pm K &&[p^0, E^\pm] = \mp p^\mp \\
&p^\pm E^\mp = q^{\pm 1} E^\mp p^\pm &&[p^+, E^+]_q = p^0 &&[p^-, E^-]_{q^{-1}} = -q^{-1} p^0
\end{aligned}
\qquad (10)
$$

$$\Delta(K) = K \otimes K \qquad\qquad \Delta(E^\pm) = E^\pm \otimes K + 1 \otimes E^\pm \qquad\qquad (11)$$

$$(p^0)^* = p^0, \quad (p^-)^* = p^+, \quad K^* = K, \quad (E^+)^* = E^-, \qquad\qquad (12)$$

where $[a, b]_w := ab - wba$. The first three relations give the algebra structure of $\mathbf{R}_q^3 \rtimes U_q so(3)$ (this underlies the quantum group of inhomogenous transformations of $\mathbf{R}_q^3$), (11) together with $\varepsilon(E^\pm) = 0, \varepsilon(K) = 1$ the coalgebra of H, (12) the $*$-structure corresponding to compact H and "real" $\mathbf{R}_q^3$ (this requires $q \in \mathbf{R}$). The element $P^2 := qp^+ p^- + p^0 p^0 + p^- p^+$ is central, positive definite under this $*$-structure and real under the other one (that requires $|q| = 1$). We enlarge the algebra by introducing also the square root and the inverse P, P^{-1}. Setting $P = 1$ we obtain the quantum Euclidean sphere S_q^2, and $S_q^2 \rtimes U_q so(3)$ can be interpreted as the algebra of observables of a quantum particle on S_q^2. The maps $\tilde\varphi^+, \tilde\varphi^-, \tilde\zeta^+, \tilde\zeta^-$, the algebra relations among the new generators $e^+ := \tilde\zeta^+(E^+)$, $e^- := \tilde\zeta^-(E^-)$, $k := \tilde\zeta^+(K) = \tilde\zeta^-(K)$, and the additional central element c of the form (6) are given by

$$\tilde\varphi^\pm(K) = \eta \frac{P}{p_0} \qquad \tilde\varphi^+(E^+) = \frac{1}{(q-1)p_0} p^- \qquad \tilde\varphi^-(E^-) = \frac{q}{(q-1)p_0} p^+ \qquad (13)$$

$$k = K \frac{\eta p_0}{P} \qquad e^+ = \frac{\eta}{P}\left[E^+ p^0 + \frac{qp^-}{1-q}\right] \qquad e^- = \frac{\eta}{P}\left[E^- p^0 + \frac{p^+}{1-q}\right] \qquad (14)$$

$$k e^\pm = q^{\pm 1} e^\pm k \qquad [e^+, e^-]_{q^{-1}} = \frac{k^2 + \eta^2}{q^2 - 1} \qquad c = \frac{e^+ e^- k^{-1}}{1+q} + \frac{k - qk^{-1}}{(1-q^2)^2}. \quad (15)$$

Here $\eta \in \mathbf{C}, \eta \neq 0$. (15) translates (7), and differs from (9)$_3$ by the presence at the rhs of the "central charge" $(\eta^2 + 1)/(q^2 - 1)$. Only for $\eta^2 = -1$ can the maps $\tilde\varphi^+, \tilde\varphi^-$ and $\tilde\zeta^+, \tilde\zeta^-$ be glued into maps $\tilde\varphi$ and $\tilde\zeta$ respectively; then the latter will be $*$-maps under the non-compact $*$-structure where $|q| = 1$, but not under (12). In order this to happen we need to take $\eta \in \mathbf{R}$. The $*$-representations of the e^+, e^-, k subalgebra for real q differ from the ones of $U_q su(2)$ in that they are lowest-weight but not highest-weight representations, or viceversa [4].

In Ref. [7, 6] we give maps $\tilde\varphi^\pm$ for the cross products $\mathbf{R}_q^N \rtimes U_q so(N)$ for all $N \geq 3$.

2. Decoupling of braided tensor products

As known, if H is a noncocommutative Hopf algebra (e.g. a quantum group $U_q\mathbf{g}$) and $\mathcal{A}_1, \mathcal{A}_2$ are two (unital) H-module algebras the tensor product algebra $\mathcal{A}_1 \otimes \mathcal{A}_2$ will not be in general a H-module algebra. If H is quasitriangular a H-module algebra can be obtained as the *braided* tensor product algebra $\mathcal{A}^+ := \mathcal{A}_1 \otimes^+ \mathcal{A}_2$, which is defined as follows: the vector space underlying the latter is still the tensor product of the vector spaces underlying $\mathcal{A}_1, \mathcal{A}_2$, and so we shall denote as usual $a_1 \otimes a_2$ simply by $a_1 a_2$; $\mathcal{A}_1 \mathbf{1}_{\mathcal{A}_2}, \mathbf{1}_{\mathcal{A}_1} \mathcal{A}_2$ are still subalgebras isomorphich to $\mathcal{A}_1, \mathcal{A}_2$, and so one can omit to write the units, except $1_{\mathcal{A}^+} \equiv 1_{\mathcal{A}_1} 1_{\mathcal{A}_2}$; but the "commutation relations" between $a_1 \in \mathcal{A}_1, a_2 \in \mathcal{A}_2$ are modified:

$$a_2 a_1 = (a_1 \vartriangleleft \mathcal{R}^{(1)}) (a_2 \vartriangleleft \mathcal{R}^{(2)}). \tag{16}$$

Here $\mathcal{R} \equiv \mathcal{R}^{(1)} \otimes \mathcal{R}^{(2)} \in H^+ \otimes H^-$ (again a summation symbol at the rhs has been suppressed) denotes the so-called universal R-matrix or quasitriangular structure of $H \equiv$ [3], and as before $H^\pm$ denote some positive and negative Borel Hopf subalgebras of H. If $H = U_q \mathbf{g}$, then in the limit $q \to 1$ H becomes the cocommutative Hopf algebra $U\mathbf{g}$ and $\mathcal{R} \to 1 \otimes 1$. As a consequence $a_2 a_1 \to a_1 a_2$ and thus $\mathcal{A}^+$ goes to the ordinary tensor product algebra. An alternative braided tensor product $\mathcal{A}^- = \mathcal{A}_1 \underline{\otimes}^- \mathcal{A}_2$. can be obtained by replacing in (16) $\mathcal{R}$ by $\mathcal{R}_{21}^{-1}$. This is equivalent to exchanging $\mathcal{A}_1$ with $\mathcal{A}_2$.

$\mathcal{A}^+$ (as well as $\mathcal{A}^-$) is a $*$-algebra if H is a Hopf $*$-algebra, $\mathcal{A}_1, \mathcal{A}_2$ are H-module $*$-algebras (we use the same symbol $*$ for the $*$-structure on all algebras $H, \mathcal{A}_1$, etc.), and $\mathcal{R}^* \equiv \mathcal{R}^{(1)*} \otimes \mathcal{R}^{(2)*} = \mathcal{R}^{-1}$. In the quantum group case this requires $|q| = 1$. Under the same assumptions also $\mathcal{A}_1 \rtimes H$ is a $*$-algebra.

If $\mathcal{A}_1, \mathcal{A}_2$ represent the algebras of observables of different two quantum systems, (16) will mean that in the composite system their degrees of freedom are "coupled" to each other. But again one can "decouple" them by a transformation of generators if there exists an algebra map $\tilde{\varphi}_1^+$, or an algebra map $\tilde{\varphi}_1^-$:

Theorem 3 *[7].* *Let $\{H, \mathcal{R}\}$ be a quasitriangular Hopf algebra and H^+, H^- be Hopf subalgebras of H such that $\mathcal{R} \in H^+ \otimes H^-$. Let $\mathcal{A}_1, \mathcal{A}_2$ be respectively a H^+- and a H^--module algebra, so that we can define $\mathcal{A}^+$ as in (16), and $\tilde{\varphi}_1^+$ be a map of the type (3), so that we can define the "unbraiding" map $\chi^+ : \mathcal{A}_2 \to \mathcal{A}^+$ by*

$$\chi^+(a_2) := \tilde{\varphi}_1^+(\mathcal{R}^{(1)}) (a_2 \vartriangleleft \mathcal{R}^{(2)}). \tag{17}$$

Then χ^+ is an injective algebra map such that

$$[\chi^+(a_2), \mathcal{A}_1] = 0, \tag{18}$$

namely the subalgebra $\tilde{\mathcal{A}}_2^+ := \chi^+(\mathcal{A}_2) \approx \mathcal{A}_2$ commutes with $\mathcal{A}_1$. Moreover $\mathcal{A}^+ = \mathcal{A}_1 \tilde{\mathcal{A}}_2^+$. Finally, if $\mathcal{R}^* = \mathcal{R}^{-1}$ and $\tilde{\varphi}_1^+$ is a $*$-algebra map then χ^+ is, and $\mathcal{A}_1, \tilde{\mathcal{A}}_2^+$ are closed under $*$. By replacing everywhere $\mathcal{R}$ by $\mathcal{R}_{21}^{-1}$ we obtain an analogous statement valid for $\mathcal{A}^-, \chi^-$.

Of course, we can use the above theorem iteratively to completely unbraid the braided tensor product algebra of an arbitrary number M of copies of $\mathcal{A}_1$. One can also combine the two methods illustrated here to decouple the tensor factors in 'mixed' tensor products such as

$$(\mathcal{A}_1 \rtimes H) \underline{\otimes}^\pm \mathcal{A}_2, \qquad \mathcal{A}_1 \underline{\otimes}^\pm (\mathcal{A}_2 \rtimes H), \qquad (\mathcal{A}_1 \underline{\otimes}^\pm \mathcal{A}_2) \rtimes H. \tag{19}$$

References

[1] Cerchiai B.L., Madore J., Schraml S. and J. Wess 2000, Eur. Phys. J. C 16, 169.
[2] Chu C. S. and Zumino B. 1995, Proceedings of ICGTMP XX, Toyonaka (Japan) 1994, q-alg/9502005.
[3] Drinfeld V. 1986, Proceedings of the International Congress of Mathematicians, Berkeley (USA) 1986, Vol. 1, 798.
[4] Fiore G 1995, J. Math. Phys. 36, 4363-4405; 1996, Int. J. Mod. Phys. A11, 863-886.
[5] Fiore G. 1995, Commun. Math. Phys. 169, 475-500.
[6] Fiore G. 2002, J. Phys. A: Math. Gen. 35, 657-678.
[7] Fiore G., Steinacker H. and Wess J. "Unbraiding the braided tensor product", math/0007174, to appear in J. Math. Phys.
[8] Grosse H., Madore J., Steinacker H. 2001, J.Geom.Phys. 38, 308-342

Inst. Phys. Conf. Ser. No 173: Mini-symposium
Paper presented at 24th Int. Coll. Group Theoretical Methods in Physics, Paris, France, 15–20 July 2002
©2003 IOP Publishing Ltd

Differential calculus on quantum spaces

Ulrich Hermisson

Institut für Mathematik und Informatik, Universität Greifswald,
Jahnstraße 15a, 17487 Greifswald, Germany

Abstract. While for any commutative algebra there is a canonically defined differential calculus, its generalization to noncommutative algebras with quantum group operation (quantum spaces) needs some consideration. This article introduces into a result that shows for the example of the quantum 2-sphere of Podleś that imposing just three basic conditions: the Leibniz rule, the compatibility with the quantum group operation (covariance in the sense of Woronowicz) and a dimension condition, leaves only a few cases of possible generalized differential calculi.

1. Introduction

Let $\mathscr{B}$ be an algebra over $\mathbf{C}$ (the complex numbers). A *derivation* of $\mathscr{B}$ is defined as a $\mathbf{C}$-linear map d from $\mathscr{B}$ into a $\mathscr{B}$-bimodule satisfying the Leibniz rule

$$\mathrm{d}(a\,b) = a\,\mathrm{d}b + \mathrm{d}a\,b \qquad \text{for all } a, b \in \mathscr{B}.$$

In this paper, $\mathrm{d}a\,b$ means $(\mathrm{d}a)\,b$. Example: For any $a \in \mathscr{B}$, $\mathrm{d}x := a\,x - x\,a$ defines a derivation of $\mathscr{B}$. Note that $\ker(\mathrm{d})$ is always a subalgebra of $\mathscr{B}$ and that d is $\ker(\mathrm{d})$-linear. If $\mathscr{B}$ denotes the smooth functions on a smooth manifold and d the exterior derivative, then $\dim_{\mathbf{C}}(\ker(\mathrm{d}))$ is the number of connected components of the manifold. We set $\Gamma(\mathrm{d}) = \mathrm{Lin}_{\mathbf{C}}\{a\,\mathrm{d}b \mid a, b \in \mathscr{B}\}$ (the $\mathbf{C}$-linear span), which is a $\mathscr{B}$-bimodule. We write $\mathrm{d}_1 \le \mathrm{d}_2$, if d_1 and d_2 are derivations of $\mathscr{B}$ and $\sum_i a_i \mathrm{d}_1 b_i = 0$ whenever $\sum_i a_i \mathrm{d}_2 b_i = 0$, and consider derivations $\mathrm{d}_1, \mathrm{d}_2$ of $\mathscr{B}$ identical, if $\mathrm{d}_1 \le \mathrm{d}_2$ and $\mathrm{d}_2 \le \mathrm{d}_1$. Thus, $\mathrm{d}_1 \le \mathrm{d}_2$ means that the $\mathbf{C}$-linear map $\Gamma(\mathrm{d}_2) \to \Gamma(\mathrm{d}_1) : a\,\mathrm{d}_2 b \mapsto a\,\mathrm{d}_1 b$ is well-defined. The set of derivations of $\mathscr{B}$ with $\le$ is a complete lattice. If $\mathscr{B}$ is a $*$-algebra, then $a\,\mathrm{d}^* b := \mathrm{d}(b^*)\,a^*$ defines an involution on this lattice. If $v : \mathscr{B}_1 \to \mathscr{B}_2$ is an algebra homomorphism and d_2 a derivation of $\mathscr{B}_2$, then $a\,\mathrm{d}_1 b := v(a)\,\mathrm{d}_2(v(b))$ defines a derivation $v(\mathrm{d}) := \mathrm{d}_1$ of $\mathscr{B}_1$.

The $\mathscr{B}$-bimodule generated by the symbols $\mathrm{d}b$, $b \in \mathscr{B}$, with the relations $\mathrm{d}c = 0$, $c \in \mathbf{C}$, $\mathrm{d}(a + b) = \mathrm{d}a + \mathrm{d}b$ and $\mathrm{d}(a\,b) = a\,\mathrm{d}b + \mathrm{d}a\,b$, $a, b \in \mathscr{B}$, defines the universal derivation of $\mathscr{B}$. If $\mathscr{B}$ is commutative, then adding the relation $\mathrm{d}b\,a = a\,\mathrm{d}b$ defines the commutative universal derivation. It coincides with the exterior derivative if $\mathscr{B}$ is the algebra of smooth functions on a smooth manifold. In this case, the vector fields are the $\mathscr{B}$-linear functionals on $\Gamma(\mathrm{d})$.

The goal is to generalize this commutative universal derivation to the noncommutative case, retaining only the Leibniz rule and the compatibility with a group operation, the group being generalized to a quantum group. The quantum group is specified as in [9] by its noncommutative function algebra, which is a Hopf algebra $\mathscr{A}$ over $\mathbf{C}$ with comultiplication Δ, counit ε and antipode S, cf. [10]. We set $\otimes = \otimes_{\mathbf{C}}$ and

$$a_{(1)} \otimes a_{(2)} = \Delta(a), \qquad a_{(1)} \otimes \cdots \otimes a_{(n+1)} = a_{(1)} \otimes \cdots \otimes a_{(n-1)} \otimes \Delta(a_{(n)})$$

for $n = 2, 3, \ldots$ (Sweedler's notation) and use the map $^+ : \mathscr{A} \to \mathscr{A}$ defined by $a^+ = a - \varepsilon(a)1$. We assume that $\mathscr{B}$ is a left $\mathscr{A}$-comodule algebra and use Δ and Sweedler's notation also for the comodule operation, i.e. $\Delta : \mathcal{B} \to \mathcal{A} \otimes \mathcal{B}$. We call a derivation d of $\mathcal{B}$ *equivariant* if and only if the $\mathbf{C}$-linear map

$$\Gamma(\mathrm{d}) \to \mathcal{A} \otimes \Gamma(\mathrm{d}) : a\,\mathrm{d}b \mapsto a_{(1)}\,b_{(1)} \otimes a_{(2)}\,\mathrm{d}b_{(2)}, \qquad a, b \in \mathcal{B},$$

is well-defined. An equivariant derivation is the same as a covariant first order differential calculus introduced in [12], [7]. The above comments on derivations can be transferred to the equivariant case in the natural way. The universal derivation and, for commutative $\mathcal{A}$, the commutative universal derivation of $\mathcal{B}$ are equivariant. Example: The smallest equivariant derivation $\lceil \mathrm{d} \rceil$ above a given derivation d is obtained as $a \lceil \mathrm{d} \rceil b := a_{(1)}\,b_{(1)} \otimes a_{(2)}\,\mathrm{d}b_{(2)}$.

The quantum group $\mathcal{A}$ generalizes the function algebra of a Lie group G, and the quantum space $\mathcal{B}$ that of a G-manifold M. The map $\Delta : \mathcal{B} \to \mathcal{A} \otimes \mathcal{B}$ generalizes the dual of the group operation $G \times M \to M$. We will mostly consider the particular case of a quantum homogeneous space by using a subalgebra $\mathcal{B}$ of $\mathcal{A}$ with the comultiplication as the comodule operation.

The quantum groups in [9] are *coquasitriangular*, which means that the Hopf algebra $\mathcal{A}$ is equipped with a $\mathbf{C}$-linear map $r : \mathcal{A} \otimes \mathcal{A} \to \mathbf{C}$ satisfying

$$r(a_{(1)} \otimes b_{(1)})\,a_{(2)}\,b_{(2)} = b_{(1)}\,a_{(1)}\,r(a_{(2)} \otimes b_{(2)}),$$
$$r(ab \otimes c) = r(a \otimes c_{(1)})\,r(b \otimes c_{(2)}), \qquad r(1 \otimes c) = \varepsilon(c),$$
$$r(a \otimes bc) = r(a_{(1)} \otimes c)\,r(a_{(2)} \otimes b), \qquad r(a \otimes 1) = \varepsilon(a)$$

for all $a, b, c \in \mathcal{A}$. This notion is the dual of a universal R matrix and generalizes commutativity, which is the special case $r = \varepsilon \otimes \varepsilon$.

2. Construction and Classification

See [2] and [3] for details of the following results.

Theorem 1 *Let $\mathcal{A}$ be a Hopf algebra and $\mathcal{B}$ a subalgebra of $\mathcal{A}$ with $\Delta(\mathcal{B}) \subseteq \mathcal{A} \otimes \mathcal{B}$. Let $e^1, \ldots, e^N \in \mathcal{B}$ be $\mathbf{C}$-linearly independent elements with $\Delta(e^i) = \sum_j \pi_j^i \otimes e^j$. For a derivation d of $\mathcal{B}$, if $\mathrm{d}e^1, \ldots, \mathrm{d}e^N$ is a basis of $\Gamma(\mathrm{d})$ as a left $\mathcal{B}$-module $(*)$, we define partial derivatives $\partial_i : \mathcal{B} \to \mathcal{B}$ by $\mathrm{d}a = \sum_i \partial_i(a)\,\mathrm{d}e^i$. Now, the map $\mathrm{d} \mapsto \{\varepsilon \circ \partial_1, \ldots, \varepsilon \circ \partial_N\}$ is a one-to-one correspondence between*

- *the equivariant derivations of $\mathcal{B}$ with the property $(*)$ and*

- *the sets $\{\chi_1, \ldots, \chi_N\}$ of linear functionals on $\mathcal{B}$ with the property*

$$\chi_i(e^j) = \delta_{ij}, \quad \chi_i(1) = 0, \quad \chi_i(ab) = \sum_j \chi_j(a)\,f_i^j(b) + \varepsilon(a)\,\chi_i(b) \quad and$$
$$\partial_i(\mathcal{B}) \subseteq \mathcal{B}, \quad where \quad \partial_i(a) = \sum_j a_{(1)}\,\chi_j(a_{(2)})\,S(\pi_i^j).$$

In particular, an equivariant derivation d of $\mathcal{B}$ with the property $(*)$ is uniquely determined by $\{\varepsilon \circ \partial_1, \ldots, \varepsilon \circ \partial_N\}$, its partial derivatives evaluated at the point ε.

Lemma 1 *Let $\mathcal{A}$ be a coquasitriangular Hopf algebra and $\mathcal{B}$ a left $\mathcal{A}$-comodule algebra. Let $b^1, \ldots, b^N \in \mathcal{B}$ be $\mathbf{C}$-linearly independent elements with $\Delta(b^i) = \sum_j \psi_j^i \otimes b^j$ and v a comodule algebra endomorphism of $\mathcal{B}$. Let Γ be the $\mathcal{B}$-bimodule generated by the symbols $\gamma_1, \ldots, \gamma_N$ with the relations $\gamma_i a = \sum_j r(\psi_i^j \otimes a_{(1)})\,v(a_{(2)})\,\gamma_j$, $a \in \mathcal{B}$, $i = 1, \ldots, N$. Then $\gamma_1, \ldots, \gamma_N$ is a basis of Γ as a left $\mathcal{B}$-module, and $\mathrm{d}a := \omega a - a\omega$ with $\omega = \sum_i b^i \gamma_i$ defines an equivariant derivation of $\mathcal{B}$. If v is bijective, then $a\gamma_j = \sum_i \gamma_i\,r(\psi_j^i \otimes S(a_{(1)}))\,v^{-1}(a_{(2)})$, and $\gamma_1, \ldots, \gamma_N$ is a basis of Γ as a right $\mathcal{B}$-module, too.*

Theorem 2 *Let $\mathcal{A}$ be a coquasitriangular Hopf algebra and $\mathcal{B}$ a subalgebra of $\mathcal{A}$ with $\Delta(\mathcal{B}) \subseteq \mathcal{A} \otimes \mathcal{B}$. We assume that if $b_{ij} \in \mathcal{B}$, $a_{ij} \in \mathcal{A}$ and $\sum_j b_{ij} a_{jk} = \delta_{ik} 1 = \sum_j a_{ij} b_{jk}$, then $a_{ij} \in \mathcal{B}$ for i, j, k in any finite set. Let $e^1, \dots, e^N \in \mathcal{B}$ be $\mathbf{C}$-linearly independent elements with $\Delta(e^i) = \sum_j \pi_j^i \otimes e^j$ and d the equivariant derivation of $\mathcal{B}$ from the preceding lemma. Let $\chi^m(a) = r(b^m \otimes v(a)) - \varepsilon(b^m)\varepsilon(a)$ and χ the matrix with the coefficients $\chi^{mn} = \chi^m(e^n)$. If χ is invertible, then $\Gamma(d) = \Gamma$, and $de^1, \dots, de^N$ is a basis of $\Gamma(d)$ as a left $\mathcal{B}$-module. The following formulae apply:*

$$da = \sum_i \partial_i(a) \, de^i \quad \text{with} \quad \partial_i(a) = \sum_{mn} a_{(1)} \chi^n(a_{(2)}) (\chi^{-1})_{mn} S(\pi_i^m) \in \mathcal{B},$$

$$\omega = \sum_i \omega_i \, de^i \quad \text{with} \quad \omega_i = \sum_{mn} \varepsilon(b^n) (\chi^{-1})_{mn} S(\pi_i^m) \in \mathcal{B}.$$

The foregoing construction applies for example to the quantizations of symmetric spaces by Noumi et al. [5].

Theorem 3 *Let $\mathcal{A}$ be a Hopf algebra and $\mathcal{B}$ a subalgebra of $\mathcal{A}$ with $\Delta(\mathcal{B}) \subseteq \mathcal{A} \otimes \mathcal{B}$.*

(i) Let $\mathcal{R}$ be a right ideal of $\mathcal{B}^+$ and $p : \mathcal{B}^+ \to \mathcal{B}^+/\mathcal{R}$ be the canonical projection. Then $a \, db := ab_{(1)} \otimes p(b_{(2)}^+)$ uniquely determines an equivariant derivation $d_{\mathcal{R}} := d$ of $\mathcal{B}$. Let $^{\overline{}} : \mathcal{A} \to \mathcal{A}/(\mathcal{B}^+\mathcal{A})$ be the canonical projection and $\overline{\Delta} := (^{\overline{}} \otimes \mathrm{id}) \circ \Delta$. Then $\lfloor \mathcal{R} \rfloor := \overline{\Delta}^{-1}(\overline{\mathcal{A}} \otimes \mathcal{R})$ is a right ideal of $\mathcal{B}^+$, $\lfloor \mathcal{R} \rfloor \subseteq \mathcal{R}$, such that $\overline{\Delta}(\lfloor \mathcal{R} \rfloor) \subseteq \overline{\mathcal{A}} \otimes \lfloor \mathcal{R} \rfloor$ and $d_{\mathcal{R}} = d_{\lfloor \mathcal{R} \rfloor}$.

(ii) Let d be a derivation of $\mathcal{B}$. Then $\mathcal{R}_d := \{ \sum_i \varepsilon(a_i) b_i^+ \mid \sum_i a_i \, db_i = 0 \}$ is a right ideal of $\mathcal{B}^+$. If d is equivariant, then $\overline{\Delta}(\mathcal{R}_d) \subseteq \overline{\mathcal{A}} \otimes \mathcal{R}_d$.

(iii) If $\mathcal{R}_1 \subseteq \mathcal{R}_2$, then $d_{\mathcal{R}_2} \leq d_{\mathcal{R}_1}$ for right ideals $\mathcal{R}_1$, $\mathcal{R}_2$ of $\mathcal{B}^+$. If $d_1 \leq d_2$, then $\mathcal{R}_{d_2} \subseteq \mathcal{R}_{d_1}$ for derivations d_1, d_2 of $\mathcal{B}$. Moreover, $\mathcal{R}_{d_{\mathcal{R}}} \subseteq \mathcal{R}$, if $\mathcal{R}$ is a right ideal of $\mathcal{B}^+$, and $d_{\mathcal{R}_d} \leq d$, if d is an equivariant derivation of $\mathcal{B}$.

In particular, we get the universal derivation $d_{\{0\}}$ and, if $\mathcal{A}$ is commutative, the commutative universal derivation $d_{(\mathcal{B}^+)^2}$. In line with the interpretation of d as the exterior derivative, $\mathcal{R}_d$ may be viewed as the functions vanishing together with their first derivatives in the point ε and correspondingly the $\mathbf{C}$-linear functionals on $\mathcal{B}^+/\mathcal{R}_d \cong \Gamma(d)/(\mathcal{B}^+\Gamma(d))$ as the tangent space in ε. We call its $\mathbf{C}$-dimension the left dimension of d at ε. The lattices of equivariant derivations of $\mathcal{B}$ (with maps $d_2 \to d_1$ for $d_1 \leq d_2$) and right ideals of $\mathcal{B}^+$ (with inclusion maps) are categories. In this language, $d \to \mathcal{R}_d$ is a left adjoint functor for the functor $\mathcal{R} \to d_{\mathcal{R}}$.

That a right $\mathcal{B}$-module $\mathcal{A}$ is *flat* means that for all left ideals $\mathcal{L}$ of $\mathcal{B}$ the surjective map $\mathcal{A} \otimes_{\mathcal{B}} \mathcal{L} \to \mathcal{A}\mathcal{L} : a \otimes_{\mathcal{B}} b \mapsto ab$ is also injective, and that it is *faithfully flat* means that, in addition, $\mathcal{A}\mathcal{L} \neq \mathcal{L}$. The case $\mathcal{B} = \mathcal{A}$ of the following result is due to Woronowicz [12].

Theorem 4 *Let $\mathcal{A}$ be a Hopf algebra and $\mathcal{B}$ a subalgebra of $\mathcal{A}$ with $\Delta(\mathcal{B}) \subseteq \mathcal{A} \otimes \mathcal{B}$. If $\mathcal{A}$ is faithfully flat as a right $\mathcal{B}$-module, then $\mathcal{R} \mapsto d_{\mathcal{R}}$ and $d \mapsto \mathcal{R}_d$ as in Theorem 3 establish a one-to-one correspondence between*

- *the right ideals $\mathcal{R}$ of $\mathcal{B}^+$ with $\overline{\Delta}(\mathcal{R}) \subseteq \overline{\mathcal{A}} \otimes \mathcal{R}$ and*

- *the equivariant derivations of $\mathcal{B}$.*

The proof relies on a theorem by Takeuchi [11] requiring the faithfully flat condition. Example: Let $\mathcal{A}$ be Sweedler's 4-dimensional complex Hopf algebra [10] which is generated as an algebra by the symbols g and y with the relations $g^2 = 1$, $y^2 = 0$ and $gy = -yg$, the comultiplication being given by $\Delta(g) = g \otimes g$ and $\Delta(y) = 1 \otimes y + y \otimes g$. This implies $\varepsilon(g) = 1$, $\varepsilon(y) = 0$ and $S(g) = g$, $S(y) = gy$. But, as a downside of simplicity, the only $*$-operation, up to isomorphisms, is given by $g^* = g$ and $y^* = y$, thus $y^*y = 0$, so $\mathcal{A}$ doesn't underlie a C^*-algebra.

For $\mathscr{B} = \mathscr{A}$, the case already treated in [12], the faithfully flat condition is trivially satisfied, and $\overline{\mathscr{A}} = 1$. Hence, the equivariant derivations correspond one-to-one to all right ideals of $\mathscr{A}^{+}$: $\mathrm{Lin}_{\mathbf{C}}\{y + g\,y\}$ and $\mathrm{Lin}_{\mathbf{C}}\{y - g\,y, c_1\,(g-1) + c_2\,y\}$ with $c_1, c_2 \in \mathbf{C}$. By the way, the derivations of $\mathscr{A}$ for which the linear map

$$\Gamma(\mathrm{d}) \to \mathscr{A} \otimes \Gamma(\mathrm{d}) \otimes \mathscr{A} : a\,\mathrm{d}b \mapsto a_{(1)}\,b_{(1)} \otimes a_{(2)}\,\mathrm{d}b_{(2)} \otimes a_{(3)}\,b_{(3)}$$

exists (i.e. the bicovariant first order differential calculi in the sense of [12]) correspond to the ad-invariant right ideals $\mathrm{Lin}_{\mathbf{C}}\{y + g\,y\}$ and $\mathrm{Lin}_{\mathbf{C}}\{y - g\,y, g - 1\}$, see [12].

Apart from $\mathscr{B} = 1$ and $\mathscr{B} = \mathscr{A}$, the only subalgebras of $\mathscr{A}$ with $\Delta(\mathscr{B}) \subseteq \mathscr{A} \otimes \mathscr{B}$ are $\mathscr{B} = \mathrm{Lin}_{\mathbf{C}}\{1, c_1\,g + c_2\,g\,y\}$ with $c_1, c_2 \in \mathbf{C}$, $(c_1, c_2) \neq (0, 0)$. In this case $\mathscr{A}$ is free as a right $\mathscr{B}$-module, since $\mathscr{A} = \mathscr{B} \oplus g\mathscr{B}$, if $c_2 \neq 0$, and $\mathscr{A} = \mathscr{B} \oplus y\mathscr{B}$, if $c_1 \neq 0$. In particular, $\mathscr{A}$ is faithfully flat as a right $\mathscr{B}$-module. Therefore, according to Theorem 4, $\mathrm{d}_{\{0\}}$ and $\mathrm{d}_{\mathscr{A}^{+}}$ are the only equivariant derivations of $\mathscr{B}$. Note that, though $\mathscr{B}$ is commutative, for $c_1 = 0$ the universal commutative derivation of $\mathscr{B}$ is not equivariant.

For the quantum 2-sphere of Podleś $\mathscr{B} = \mathcal{O}(S_{qc}^2)$, $\mathscr{A} = \mathcal{O}(\mathrm{SL}_q(2))$, we obtain the following result:

Theorem 5 *If the faithfully flat condition holds, that is, in this case, $c \neq -q^{2n}/(q^{2n} + 1)^2$ for $n = 0, 1, 2, \ldots$, see [4], then for $c = 0$ three 2-dimensional equivariant derivations of $\mathscr{B}$ exist, one exists for each of $c = \mp q/(\pm q + 1)^2$ and no one otherwise.*

If the faithfully flat condition is not satisfied, we get for $c = -q^4/(q^4 + 1)^2$ two 2-dimensional equivariant derivations constructed as in Theorem 3, but also examples which can not be constructed in this way.

References

[1] Apel J and Schmüdgen K: Classification of three-dimensional covariant differential calculi on Podles' quantum spheres 1994 Lett. Math. Phys. 32, 25-36

[2] Hermisson U: Construction of covariant differential calculi on quantum homogeneous spaces 1998 Lett. Math. Phys. 46, 313-322; math.QA/9806008

[3] Hermisson U: Derivations with quantum group action 2002 Commun. Algebra 30, 101-117; math.QA/0005106

[4] Müller E F and Schneider H-J: Quantum homogeneous spaces with faithfully flat module structures 1999 Israel J. Math. 111, 157-190

[5] Noumi M and Sugitani T: Quantum symmetric spaces and related q-orthogonal polynomials 1995 in: Group theoretical methods in physics (Singapore: World Scientific), 28-40

[6] Podleś P: Quantum spheres 1987 Lett. Math. Phys. 14, 193-202

[7] Podleś P: Differential calculus on quantum spheres 1989 Lett. Math. Phys. 18, 107-119

[8] Podleś P: The classification of differential structures on quantum 2-spheres 1992 Commun. Math. Phys. 150, 167-179

[9] Reshetikhin N Yu, Takhtadzhyan L A and Faddeev L D: Quantization of Lie groups and Lie algebras 1990 Leningrad Math. J. 1, 193-225

[10] Sweedler M E 1969 Hopf Algebras (New York: Benjamin)

[11] Takeuchi M: Relative Hopf modules—equivalences and freeness criteria 1979 J. Algebra 60, 452-471

[12] Woronowicz S L: Differential calculus on compact matrix pseudogroups (quantum groups) 1989 Commun. Math. Phys. 122, 125-170

Inst. Phys. Conf. Ser. No 173: Mini-symposium
Paper presented at 24th Int. Coll. Group Theoretical Methods in Physics, Paris, France, 15–20 July 2002
©2003 IOP Publishing Ltd

On generalized Freudenthal-Kantor triple systems and Yang-Baxter equations

Noriaki Kamiya[1] and Susumu Okubo[2]

[1] Center for Mathematical Sciences, University of Aizu, Aizuwakamatsu 965-8580, Japan
[2] Department of Physics and Astronomy, University of Rochester, Rochester, N.Y. 14627, U.S.A.

E-mail: kamiya@u-aizu.ac.jp and okubo@pas.rochester.edu

Abstract. The concept of Yang-Baxter equation has been mainly investigated from the point of view of a triple system, several triple systems are introduced in this paper and it seems that these triple systems are useful in mathematical physics.

1. Introduction

The concept of a triple system (= vector space equipped with a triple product $< xyz >$) plays an important role in the construction of simple Lie algebras or Lie superalgebras without the notion of root systems and Cartan matrix([2]-[6]). We have also given a construction of Jordan algebras or Jordan superalgebras by means of the concept of triple systems([10]).

On the other hand, it is interesting for us to study how the triple system behaves in the field of mathematical physics ([7]-[9]). In particular, we show that generalized Freudenthal-Kantor triple systems may be utilized to obtain a class of solutions of Yang-Baxter equations.

2. Freudenthal-Kantor Triple System

For $\varepsilon = \pm 1$ and $\delta = \pm 1$, a vector space $U(\varepsilon, \delta)$ over a field Φ with triple product $L(x, y)z := < xyz >$ is called a (ε, δ)-*Freudenthal-Kantor triple system* if

$$< abc >\in U(\varepsilon, \delta) \tag{2.0}$$

$$< ab < cde >>=<< abc > de > +\varepsilon < c < bad > e > + < cd < abe >>> \tag{2.1}$$

$$K(< abc >, d) + K(c, < abd >> +\delta K(a, K(c, d)b) = 0, \tag{2.2}$$

where $K(a, b)c =< acb > -\delta < bca >$.

In particular, it is said to be a *generalized Freudenthal-Kantor triple system* if the conditions (2.0) and (2.1) are satisfied.

We recall that a δ-Lie triple system that satisfies the following identities ([6]):

$$[xyz] = -\delta[yxz] \tag{2.3}$$

$$[xyz] + [yzx] + [zxy] = 0 \tag{2.4}$$

$$[L(x, y), L(z, w)] = L([xyz], w) + L(z, [xyw]) \tag{2.5}$$

where $L(x, y)z = [xyz]$; we often use the notation $[xyz]$ for the product of Lie triple systems instead of $< xyz >$, the traditional notation.

For $\delta = 1$ (resp. $\delta = -1$), we call it a Lie triple system (resp. anti-Lie triple systems).

Proposition 1 *Let $(U, [xyz])$ be a δ-Lie triple system. Then $(U, [xyz])$ has the structure of a generalized Freudenthal-Kantor triple system with $\varepsilon = -\delta$.*

Proposition 2 *Let $U(\varepsilon, \delta)$ be a (ε, δ)-Freudenthal-Kantor triple system. If P is a linear transformation of $U(\varepsilon, \delta)$ such that $P < xyz >=< PxPyPz >$ and $P^2 = -\varepsilon\delta Id$, then $(U(\varepsilon, \delta), [-, -, -])$ is a Lie triple system for the case of $\delta = 1$ and an anti-Lie triple system for the case of $\delta = -1$ with respect to the product*

$$[xyz] :=< xPyz > -\delta < yPxz > +\delta < xPzy > - < yPzx > .$$

Definition A generalized Freudenthal-Kantor triple system is called *quasiclassical* if there exists a nondegenerate bilinear form $< \cdot | \cdot >$ satisfying the following condition:
  (i) $< x|y >=< y|x >$,
  (ii) $< y| < xvu >>=< x| < yuv >> .$

3. Yang-Baxter Equation

If we have

$$R_{12}(\theta)R_{13}(\theta')R_{23}(\theta'') = R_{23}(\theta'')R_{13}(\theta')R_{12}(\theta) \tag{3.1}$$

for parameters θ, θ', and θ'' satisfying

$$\theta' - \theta + \theta'', \tag{3.2}$$

then the relation is called a *Yang–Baxter equation* (e.g., see [1],[7]). Although we can generalize our result to the case of super space, we will consider here only a nonsuper case for simplicity. Assume that V possesses a non-degenerate bilinear symmetric inner product $< .|. >$ so that we have $< y|x >=< x|y >$. We can then introduce two θ–dependent triple products $[x, y, x]_\theta$ and $[x, y, z]_\theta^*$ satisfying

$$(i) \quad < x|[y, u, v]_\theta >=< y|[x, v, u]_\theta^* > \tag{3.3}$$

$$(ii) \quad R(\theta)(x \otimes y) = \sum_{j=1}^{N}[e^j, y, x]_\theta^* \otimes e_j = \sum_{j=1}^{N} e_j \otimes [e^j, x, y]_\theta. \tag{3.4}$$

Here, e_j and $e^j (j = 1, 2, \ldots, N)$ are the basis and its dual basis of V, respectively. Then, the Yang-Baxter equation (hereafter abbreviated as YBE) can be rewritten as a triple product relation

$$\sum_{j=1}^{N}[v, [u, e_j, z]_{\theta'}, [e^j, x, y]_\theta]_{\theta''}^* = \sum_{j=1}^{N}[u, [v, e_j, x]_{\theta'}^*, [e^j, z, y]_{\theta''}^*]_\theta. \tag{3.5}$$

We are hereafter interested only in the case when we have

$$[x, y, z]_\theta^* = [x, y, z]_\theta \tag{3.6a}$$

or equivalently

$$< y|[x, v, u]_\theta >=< x|[y, u, v]_\theta > . \tag{3.6b}$$

Lemma 3 *A neccessary and sufficient condition to have*

$$[R_{ij}(\theta), R_{kl}(\theta')] = 0 \tag{3.7}$$

for all $i, j, k, l = 1, 2, 3$ is the validity of

$$[u, v, [x, y, z]_\theta]_{\theta'} = [x, y, [u, v, z]_{\theta'}]_\theta. \tag{3.8}$$

Proposition 4 *Let V be a Lie triple system equipped with a symmetric inner product $< \ | \ >$ satisfying $< y|[xvu] >=< x|[yuv] >$ and $[u, v, [x, y, z]] = [x, y, [u, v, z]]$. Then, the θ–dependent triple product*

$$[x, y, z]_\theta = f(\theta)[x, y, z] + g(\theta) < x|y > z$$

for arbitrary functions $f(\theta)$ and $g(\theta)$ of θ gives a solution of Eq.(3.7), and hence of YBE.

Example   Let V be a Lie triple system equiped with basis e, f, x, y such that non-zero Lie triple product and inner products are assumed to be given by $[xff] = -[fxf] = x, [yff] = -[fyf] = y, [xfy] = -[fxy] = e, [yfx] = -[fyx] = e$, and $< e|f >=< f|e >=< f|f >= 1, < x|y >=< y|x >= -1$. Then this triple system has a structure of abelian and quasi-classical generalized Freudenthal-Kantor triple system satisfying the assumption of Proposition 4.

References

M.Jimbo,(ed.), Yang–Baxter Equation in Integrable Systems, World Scientific Pub., Singapore, 1989.

N.Kamiya, A Structure Theory of Freudenthal–Kantor Triple Systems, J.Alg.110, (1987)107–123.

N.Kamiya, A Structure Theory of Freudenthal–Kantor Triple Systems II, Comm. Math.Univ.Sancti Pauli 38, (1989) 41 –60.

N.Kamiya, The Construction of All Simple Lie Algebras Over C from Balanced Freudenthal –Kantor Triple Systems, Contribution to General Algebra 7, (1991) 205 –213, Verlag Hölder–Pichler–Tempsky.

N.Kamiya, On Freudenthal–Kantor Triple Systems and Generalized Structurable Algebras, Mathematics and Its Applications 303, (1994) 198–203, Kluwer Academic Press.

N.Kamiya and S.Okubo, On δ-Lie Supertriple Systems Associated with (ε, δ)- Freudenthal-Kantor Supertriple Systems, Proc. Edinburgh Math. Soc. 43, (2000) 243-260.

S.Okubo,Triple Products and Yang–Baxter Equations I and II, J. Math. Phys. 34, (1993) 3273–3315.

S.Okubo, Super–triple Systems, Normal and Classical Yang–Baxter Equations, Mathematics and Its Applications 303, (1994) 300–308, Kluwer Academic Press.

S.Okubo, Introduction to Octonion and Non– Associative Algebras in Physics, Cambridge Univ. Press (1995).

S.Okubo and N.Kamiya, Jordan-Lie Superalgebras and Jordan-Lie Triple systems, J.Alg. 198,(1997) 388-411.

S.Okubo and N.Kamiya, Quasi-classical Lie Superalgebras and Lie Supertriple Systems, to appear, Comm. in Algebras,

S.Okubo and N.Kamiya, Constraction of Lie superalgebras D(2,1;α),G(3) and F(4) from Some Triple Systems, to appear, Proc. Edinburgh Math. Soc.

Inst. Phys. Conf. Ser. No 173: Mini-symposium
Paper presented at 24th Int. Coll. Group Theoretical Methods in Physics, Paris, France, 15–20 July 2002
©*2003 IOP Publishing Ltd*

The Jordanian Bicovariant Differential Calculus

L. Mesref
Department of Physics, Kaiserslautern University, 67653 kaiserslautern, Germany

Abstract. We show that the Woronowicz prescription using a bimodule constructed out of a tensorial product of a bimodule and its conjugate and a bi-coinvariant singlet leads to a trivial differential calculus in the case of the Jordanian group.

1. Introduction

It is by now well known that our naive conception of space-time as a collection of points equipped with suitable topological and metric structures at energies much below the Planck scale should be modified. One possible approach to the description of physical phenomena at small distance is based on non-commutative geometry of space-time [1]. In the quantum groups picture [2] the symmetry is described by noncommutative non-cocommutative $*$ Hopf algebra. The connection with noncommutative differential geometry has been made by Woronowicz [3] who introduced the theory of bicovariant differential calculus. This turns out to be the appropriate language to describe quantum gauge theories. In this letter we show that this method leads to a trivial calculus in the case of the Jordanian group $U_h(2)$.

2. The Jordanian Quantum Group $U_h(2)$

We recall that there are only two quantum group structures which admit a central determinant on the space of 2×2 matrices: $GL_q(2)$ [4] and $GL_h(2)$ [5] (the deformation of $M(2)$ was considered and named "Jordanian" by Manin [6]). The continuous parameter h was introduced by Zakrzewski [7].

Let $\mathcal{A}$ be the associative unital C-algebra generated by the linear transformations $M^n{}_m$ $(n, m = 1, 2)$

$$M^n{}_m = \begin{pmatrix} a & b \\ c & d \end{pmatrix}, \tag{1}$$

the elements a, b, c, d satisfying the relations

$$
\begin{aligned}
[a, c] &= hc^2, & [b, a] &= h\left(a^2 - D_h\right), \\
[d, c] &= hc^2, & [d, b] &= h\left(D_h - d^2\right), \\
[a, d] &= h(d - a)c, & [b, c] &= h(ac + cd) \\
D_h &= ad - cb - hcd,
\end{aligned}
\tag{2}
$$

The 2×2 matrix belonging to $U_h(2)$ preserves the nondegenerate bilinear form B_{nm} [9]

$$B_{nm} M^n_k M^m_l = D_h B_{kl}, \quad B^{nm} M^k_n M^l_m = D_h B^{kl}, \quad B_{kn} B^{nl} = \delta^l_k, \tag{3}$$

$$B_{nm} = \begin{pmatrix} 0 & -1 \\ 1 & h \end{pmatrix}, \quad B^{nm} = \begin{pmatrix} h & 1 \\ -1 & 0 \end{pmatrix}, \quad B^{nm} B_{nm} = -2. \tag{4}$$

3. $U_h(2)$ **Woronowicz Bicovariant Differential Calculus**

Zakrzewski [7] has applied the general construction of the Leningrad School [10] to the following R matrix which controls the noncommutativity of the elements M^n_m

$$R = \begin{pmatrix} 1 & -h & h & h^2 \\ 0 & 0 & 1 & -h \\ 0 & 1 & 0 & h \\ 0 & 0 & 0 & 1 \end{pmatrix}. \tag{5}$$

The R matrix becomes the permutation operator $R^{nm}_{kl} = \delta^n_l \delta^m_k$ in the classical limit $h = 0$.

The braiding R matrix satisfy the Yang-Baxter equation

$$R^{ij}_{pq} R^{pk}_{lr} R^{qr}_{mn} = R^{jk}_{pq} R^{ip}_{rm} R^{rq}_{lm}. \tag{6}$$

The noncommutativity of the elements M^n_m is expressed as

$$R^{pq}_{nm} M^n_k M^m_l = M^p_n M^q_m R^{nm}_{kl}. \tag{7}$$

With the nondegenerate bilinear form B, the R matrix has the form

$$R^{+nm}_{kl} = R^{nm}_{kl} = \delta^n_k \delta^m_l + B^{nm} B_{kl},$$
$$R^{-nm}_{kl} = R^{-1nm}_{kl} = R^{nm}_{kl}. \tag{8}$$

The R matrix satisfies the Hecke relations $R^{\pm 2} = 1$ and the relation

$$B_{nm} R^{an}_{kc} R^{cm}_{lb} = \delta^a_b B_{kl}. \tag{9}$$

Now, we are going to consider the bicovariant bimodule Γ over $U_h(2)$. Let θ^a be a right invariant basis of Γ_{inv}, the linear subspace of all right -invariant elements of Γ i.e. $\Delta_R(\theta^a) = \theta^a \otimes I$. In the $h = 0$ the right coaction Δ_R coincides with the pullback for 1-forms. The left action is defined as

$$\Delta_L(\theta^a) = M^a_b \otimes \theta^b. \tag{10}$$

In the Jordanian quantum case we have $\theta^a M^n_m \neq M^n_m \theta^a$ in general, the bimodule structure of Γ being non-trivial for $h \neq 0$. There exist linear functionals $f^a_b : Fun(U_h(2)) \to \mathcal{C}$ for these left invariant basis such that

$$\theta^a M^n_m = (M^n_m * f^a_b) \theta^b = (f^a_b \otimes id) \Delta(M^n_m) \theta^b = f^a_b(M^n_k) M^k_m \theta^b. \tag{11}$$

Using these conditions, we find from Eq.(3) and Eq.(9) $f^a_b(M^n_k) = (D_h)^{\frac{1}{2}} R^{an}_{kb}$. The linear functionals $\bar{f}^a_b$ are defined by

$$\bar{\theta}_b M^n_m = \left(M^n_m * \bar{f}^a_b \right) \bar{\theta}_a, \tag{12}$$

and

$$\bar{f}^a_b(S(M^n_m)) = (D_h)^{\frac{-1}{2}} R^{-an}_{mb}. \tag{13}$$

The representation with the upper index of $\bar{\theta}^a$ is defined by using the nondegenerate bilinear form B:

$$\bar{\theta}^b = \bar{\theta}_a B^{ab}. \tag{14}$$

This gives

$$\Delta_L(\bar{\theta}^a) = M^a_b \otimes \bar{\theta}^b, \tag{15}$$

which defines the new functionals $\tilde{f}^a_b$ corresponding to the basis $\bar{\theta}^a$

$$\tilde{f}^a_b = B_{bc} \bar{f}^c_d B^{da}. \tag{16}$$

We can deduce the relation between the left and the right multiplication for the basis $\theta_b^a = \theta^a \bar{\theta}^b$

$$\theta^{ab} M_m^n = \left(M_m^n * f_{Ad}{}^{ab}{}_{cd} \right) \theta^{cd} = f_{Ad}{}^{ab}{}_{cd} \left(M_k^n \right) M_m^k \theta^{cd}, \tag{17}$$

where

$$f_{Ad}{}^{ab}{}_{cd} = \tilde{f}_d^b * f_c^a. \tag{18}$$

The exterior derivative d is defined as

$$dM_m^n = \frac{1}{\mathcal{N}} [X, M_m^n]_- = \theta^{ab} \left(M_m^n * \chi_{ab} \right)$$

$$= \chi_{ab} \left(M_k^n \right) \theta^{ab} M_m^k, \tag{19}$$

where $X = B_{ab} \theta^{ab} = -\theta^{12} + \theta^{21} + h \theta^{22}$ is the singlet representation of θ^{ab} and is both left and right co-invariant, $\mathcal{N} \in \mathcal{C}$ is the normalization constant which we take purely imaginary $N^* = -N$ and χ_{ab} are the quantum analogue of right- invariant vector fields. Using (13),(16), (18)

$$dM_m^n = \frac{1}{\mathcal{N}} \left(B_{ab} \delta_m^k M_k^n \theta^{ab} - B_{cd} f_{Ad}{}^{cd}{}_{ab} \left(S \left(M_m^k \right) \right) M_k^n \theta^{ab} \right)$$

$$= \frac{1}{\mathcal{N}} \left(B_{ab} \delta_m^k - B_{cd} R^{ct}{}_{ma} R^{dk}{}_{tb} \right) M_k^n \theta^{ab} = 0. \tag{20}$$

We see that it is a trivial calculus, since $dM_m^n = 0$. To obtain nontrivial calculi we have followed, in a recent paper [11], the Karimipour [12] method for our $4D$ calculus and constructed a Jordanian trace. This trace has permitted us to define an invariant $U_h (2)$ Yang-Mills Lagrangian. The Jordanian BRST and anti-BRST transformations [13] can also be carried out and will be reported in a future work.

References

[1] A. Connes, Noncommutative Geometry, Academic Press, 1994.
[2] V. G. Drinfeld, Proc. Int. Congress of Mathematicians (Californian, 1986), p.798; L. D. Faddeev, N. Yu. Reshetikhin and L. A. Takhtajan, Algebra and Analysis 1, 178 (1989).
[3] S.L. Woronowicz, Commun. Math. Phys. 111, 613 (1987), ibid. 122, 125 (1989); U. Carow-Watamura, M. Schlieker, S. Watamura and W. Weich, Commun. Math. Phys. 142, 605 (1991); B. Jurco, Lett. Math. Phys. 22, 177 (1991).
[4] L.A. Takhtajan, in M. L. Ge and B.H. Zhao (eds), Introduction to Quantum Groups and Integrable Massive Models of Quantum Field Theory, World Scientific, Singapore, 1991.
[5] E.E. Demidov et al., Prog. Theor. Phys. Supp 102 , 203 (1990).
[6] Yu. I. Manin: Quantum groups and non-commutative geometry, Montreal University preprint, CRM-1561 (1988).
[7] S. Zakrzewski, Lett. Math. Phys. 22, 287 (1991).
[8] L.D. Faddeev, N. Yu. Reshetikhin and LA. Takhtajan, Alg. Anal. 1, 178 (1989) (in Russian); Algebraic Analysis vol. 1 (New York: Academic) pp 129-39 (1988).
[9] M. Dubois-Violette and G. Launer, Phys. Lett. B245, 175 (1990).
[10] L.D. Faddeev: Integrable Models in 1+1 -dimensionnal Quantum Fields theory (Les Houches Lectures 1982), Elsevier, Amsterdam.
[11] L. Mesref, The Jordanian U_h (2) Yang-Mills Theory, preprint KL-TH/02-01, to appear in Int. J. Mod. Phys. A.
[12] V. Karimipour, Lett. Math. Phys. 30, 87 (1994); Lett. Math. Phys. 35, 303 (1985).
[13] L. Mesref, Quantum BRST and anti-BRST invariance of the U_q (2) BF-Yang-Mills Theory: The quantum Batalin-Vilkovisky operator, preprint KL-TH/02-03, to appear in Int. J. Mod. Phys. A.

Inst. Phys. Conf. Ser. No 173: Mini-symposium
Paper presented at 24th Int. Coll. Group Theoretical Methods in Physics, Paris, France, 15–20 July 2002
©2003 IOP Publishing Ltd

Differential calculus on a novel cross-product quantum algebra

Deepak Parashar

Max Planck Institute for Mathematics in the Sciences, Inselstrasse 22-26, D - 04103 Leipzig, Germany, and
Department of Mathematics, University of Wales Swansea, Singleton Park, Swansea SA2 8PP, UK (from October 2002)
E-mail: Deepak.Parashar@mis.mpg.de

Abstract. We investigate the algebro-geometric structure of a novel two-parameter quantum deformation which exhibits the nature of a semidirect or cross-product algebra built upon $GL(2) \otimes GL(1)$, and is related to several other known examples of quantum groups. Following the R-matrix framework, we construct the $L^{\pm}$ functionals and address the problem of duality for this quantum group. This naturally leads to the construction of a bicovariant differential calculus that depends only on one deformation parameter, respects the cross-product structure and has interesting applications. The corresponding Jordanian and hybrid deformation is also explored.

1. The quantum algebra $\mathscr{A}_{r,s}$

The biparametric q-deformation $\mathscr{A}_{r,s}$ is defined [1] to be the semidirect or cross-product $GL_r(2) \underset{s}{\rtimes} \mathbb{C}[f, f^{-1}]$ built on the vector space $GL_r(2) \otimes \mathbb{C}[f, f^{-1}]$ where $GL_r(2) = \mathbb{C}[a,b,c,d]$ modulo the relations

$$
\begin{aligned}
ab &= r^{-1}ba, & bd &= r^{-1}db \\
ac &= r^{-1}ca, & cd &= r^{-1}dc \\
bc &= cb, & [a,d] &= (r^{-1} - r)bc
\end{aligned}
\tag{1}
$$

and $\mathbb{C}[f, f^{-1}]$ has the cross relations

$$
\begin{aligned}
af &= fa, & cf &= sfc \\
bf &= s^{-1}fb, & df &= fd
\end{aligned}
\tag{2}
$$

$\mathscr{A}_{r,s}$ can also be interpreted as a skew Laurent polynomial ring $GL_r[f, f^{-1}; \sigma]$ where σ is the automorphism given by the action of element f on $GL_r(2)$. If we let $A = GL_r(2)$ and $H = \mathbb{C}[f, f^{-1}]$, then A is a left H-module algebra and the action of f is given by

$$
f \triangleright a = a, \qquad f \triangleright b = sb, \qquad f \triangleright c = s^{-1}c, \qquad f \triangleright d = d
\tag{3}
$$

2. The dual algebra $\mathscr{U}_{r,s}$

Knowing properties of cross-product algebras [2, 3], we already know that the algebra dual to $\mathscr{A}_{r,s}$ would be the cross-coproduct coalgebra $\mathscr{U}_{r,s} = U_r(gl(2)) \underset{s}{\rtimes} \mathbb{C}[[\phi]]$ with ϕ as an element dual to f. As a vector space, the dual is $\mathscr{U}_{r,s} = U_r(gl(2)) \otimes U(u(1))$. Now, the duality relation between $\langle GL_r(2), U_r(gl(2)) \rangle$ is already well-known [4], while that between

$\langle \mathbb{C}[f,f^{-1}],U(u(1))\rangle$ is given by $\langle f,\phi\rangle = 1$, i.e., $U(u(1)) = \mathbb{C}[[\phi]]$. More precisely, we work algebraically with $\mathbb{C}[s^{\phi},s^{-\phi}]$ where $\langle f,s^{\phi}\rangle = s$. This induces duality on the vector space tensor products, the left action dualises to the left coaction, and this results in the dual algebra being a cross-coproduct $\mathscr{U}_{r,s} = U_r(gl(2)) \rtimes_s \mathbb{C}[[\phi]]$. Let us recall that $U_r(gl(2))$, the algebra dual to $GL_r(2)$, is isomorphic to the tensor product $U_r(sl(2)) \otimes \tilde{U}(u(1))$ where $U_r(sl(2))$ has the usual generators $\{H,X_{\pm}\}$ and $\tilde{U}(u(1)) = \mathbb{C}[[\xi]] = \mathbb{C}[r^{\xi},r^{-\xi}]$ with ξ central. Therefore, $\mathscr{U}_{r,s}$ is nothing but $U_r(sl(2))$ and two central generators ξ and ϕ, where ξ is the generating element of $\tilde{U}(u(1))$ and ϕ is the generating element of $U(u(1))$. Also note that s^{ϕ} is dually paired with the element f of $\mathscr{A}_{r,s}$.

3. R-matrix relations

In the quantum group language, $\mathscr{A}_{r,s}$ is understood as a novel Hopf algebra [5, 6] generated by $\{a,b,c,d,f\}$ arranged in the matrix form

$$T = \begin{pmatrix} f & 0 & 0 \\ 0 & a & b \\ 0 & c & d \end{pmatrix} \qquad (4)$$

with the labelling $0,1,2$, and $\{r,s\}$ are the two deformation parameters. The R-matrix

$$R = \begin{pmatrix} r & 0 & 0 & 0 \\ 0 & S^{-1} & 0 & 0 \\ 0 & \Lambda & S & 0 \\ 0 & 0 & 0 & R_r \end{pmatrix} \qquad (5)$$

is in block form, i.e., in the order $(00), (01), (02), (10), (20), (11), (12), (21), (22)$ (which is chosen in conjunction with the block form of the T-matrix) where

$$R_r = \begin{pmatrix} r & 0 & 0 & 0 \\ 0 & 1 & 0 & 0 \\ 0 & \lambda & 1 & 0 \\ 0 & 0 & 0 & r \end{pmatrix}; \qquad S = \begin{pmatrix} s & 0 \\ 0 & 1 \end{pmatrix}; \qquad \Lambda = \begin{pmatrix} \lambda & 0 \\ 0 & \lambda \end{pmatrix}; \qquad \lambda = r - r^{-1}$$

The RTT relations $RT_1T_2 = T_2T_1R$ (where $T_1 = T \otimes 1$ and $T_2 = 1 \otimes T$) then yield the commutation relations (1) and (2) between the generators. The Hopf algebra structure underlying $\mathscr{A}_{r,s}$ is $\Delta(T) = T \dot{\otimes} T$, $\varepsilon(T) = \mathbf{1}$. The Casimir operator $\delta = ad - r^{-1}bc$ is invertible and the antipode is

$$S(f) = f^{-1}, \quad S(a) = \delta^{-1}d, \quad S(b) = -\delta^{-1}rb, \quad S(c) = -\delta^{-1}r^{-1}c, \quad S(d) = \delta^{-1}a \quad (6)$$

The quantum determinant $\mathscr{D} = \delta f$ is group-like but not central. There are several interesting features of this deformation [5, 6, 1] which cannot be mentioned here. In the R-matrix formulation of matrix quantum groups, a basic step is to construct functionals (matrices) L^+ and L^- which are dual to the matrix of generators (4) in the fundamental representation. These functions are defined by their value on the matrix of generators T

$$\langle (L^{\pm})^a_b, T^c_d \rangle = (R^{\pm})^{ac}_{bd} \qquad (7)$$

where

$$(R^+)^{ac}_{bd} = c^+(R)^{ca}_{db} \qquad (8)$$

$$(R^-)^{ac}_{bd} = c^-(R^{-1})^{ac}_{bd} \qquad (9)$$

and c^+, c^- are free parameters. For $\mathscr{A}_{r,s}$ we make the following ansatz for the $L^\pm$ matrices:

$$L^+ = \begin{pmatrix} J & 0 & 0 \\ 0 & M & P \\ 0 & 0 & N \end{pmatrix} \quad \text{and} \quad L^- = \begin{pmatrix} J' & 0 & 0 \\ 0 & M' & 0 \\ 0 & Q & N' \end{pmatrix} \tag{10}$$

where

$$
\begin{aligned}
J &= s^{-\frac{1}{2}(F-A+D-1)} r^{\frac{1}{2}(F-A-D+1)}, & J' &= s^{-\frac{1}{2}(F-A+D-1)} r^{-\frac{1}{2}(F-A-D+1)} \\
M &= s^{-\frac{1}{2}(F-A-D+1)} r^{\frac{1}{2}(-F+A-D+1)}, & M' &= s^{-\frac{1}{2}(F-A-D+1)} r^{-\frac{1}{2}(-F+A-D+1)} \\
N &= s^{-\frac{1}{2}(F+A+D-1)} r^{\frac{1}{2}(-F-A+D+1)}, & N' &= s^{-\frac{1}{2}(F+A+D-1)} r^{-\frac{1}{2}(-F-A+D+1)} \\
P &= \lambda C, & Q &= -\lambda B
\end{aligned}
\tag{11}
$$

and $\{A,B,C,D,F\}$ is the set of generating elements of the dual algebra $\mathscr{U}_{r,s}$. This is consistent with the action on the generators of $\mathscr{A}_{r,s}$ and gives the correct duality pairings. The commutation algebra is given by the RLL relations $R_{12}L_2^\pm L_1^\pm = L_1^\pm L_2^\pm R_{12}$, $R_{12}L_2^+ L_1^- = L_1^- L_2^+ R_{12}$, where $L_1^\pm = L^\pm \otimes \mathbf{1}$, $L_2^\pm = \mathbf{1} \otimes L^\pm$, and R_{12} is the same as (5). Finally, we obtain a single-parameter deformation of $U(gl(2)) \otimes U(u(1))$ as an algebra. Including the coproduct, we again obtain $\mathscr{U}_{r,s}$ as a semidirect product $U_r(gl(2)) \rtimes_s U(u(1))$.

4. Differential calculus on $\mathscr{A}_{r,s}$

The R-matrix procedure [7] is known to provide a natural framework to construct differential calculus on matrix quantum groups. We note here that the $\mathscr{A}_{r,s}$ deformation is not a full matrix quantum group, but an appropriate quotient of one (of multiparameter q-deformed $GL(3)$, to be precise). Nevertheless, it turns out that the constructive differential calculus methods [8] work equally well for such quotients. The bimodule Γ (space of quantum one-forms ω) is characterised by the commutation relations between ω and $a \in \mathscr{A}$ ($\equiv \mathscr{A}_{r,s}$)

$$\omega a = (\mathbf{1} \otimes g)\Delta(a)\omega \tag{12}$$

and the linear functional $g \in \mathscr{A}'(= \mathrm{Hom}(\mathscr{A}, \mathbb{C}))$ is defined in terms of the $L^\pm$ matrices

$$g = S(L^+)L^- \tag{13}$$

Thus, in terms of components we have

$$\omega_{ij} a = [(\mathbf{1} \otimes S(l_{ki}^+)l_{jl}^-)\Delta(a)]\omega_{kl} \tag{14}$$

using $L^\pm = l_{ij}^\pm$ and $\omega = \omega_{ij}$ where $i,j = 1..3$. From these relations, one can obtain the commutation relations of all the left-invariant one-forms with the generating elements of $\mathscr{A}$. The left-invariant vector fields χ_{ij} on $\mathscr{A}$ are given by the expression

$$\chi_{ij} = S(l_{ik}^+)l_{kj}^- - \delta_{ij}\varepsilon \tag{15}$$

The vector fields act on the generating elements as

$$\chi_{ij} a = (S(l_{ik}^+)l_{kj}^- - \delta_{ij}\varepsilon)a \tag{16}$$

Furthermore, using the formula $\mathbf{d}a = \sum_i (\chi_i * a)\omega^i$, we obtain the action of the exterior derivatives ($\mathbf{d} : \mathscr{A} \longrightarrow \Gamma$):

$$\mathbf{d}a = (r^{-2} - 1)a\omega^1 - \lambda b\omega^+ \tag{17}$$

$$\mathbf{d}b = \lambda^2 b\omega^1 - \lambda a\omega^- + (r^{-2} - 1)b\omega^2 \tag{18}$$

$$\mathbf{d}c = (r^{-2} - 1)c\omega^1 - \lambda d\omega^+ \tag{19}$$

$$\mathbf{d}d = \lambda^2 d\omega^1 - \lambda c\omega^- + (r^{-2} - 1)d\omega^2 \tag{20}$$

$$\mathbf{d}f = (r^{-2} - 1)f\omega^0 \tag{21}$$

where $\omega^0 = \omega_{11}, \omega^1 = \omega_{22}, \omega^+ = \omega_{23}, \omega^- = \omega_{32}, \omega^2 = \omega_{33}$. $\mathbf{d}\mathscr{A}$ generates Γ as a left $\mathscr{A}$-module, and this defines a first-order differential calclulus $(\Gamma, \mathbf{d})$ on $\mathscr{A}_{rs}$. The calculus is bicovariant due to the coexistence of the left $(\Delta_L : \Gamma \longrightarrow \mathscr{A} \otimes \Gamma)$ and the right $(\Delta_R : \Gamma \longrightarrow \Gamma \otimes \mathscr{A})$ actions. Curiosly, using the Leibniz rule it can be checked that

$$\mathbf{d}(af - fa) = 0, \quad \mathbf{d}(cf - sfc) = 0, \quad \mathbf{d}(bf - s^{-1}fb) = 0, \quad \mathbf{d}(df - fd) = 0, \tag{22}$$

which is consistent with cross relations (2), and so the differential calculus also respects the cross-product structure of $\mathscr{A}_{rs}$.

5. The Jordanian deformation $\mathscr{A}_{m,k}$

$\mathscr{A}_{rs}$ can be contracted [6] (by means of singular limit of similarity transformations) to obtain a nonstandard or Jordanian analogue, say $\mathscr{A}_{m,k}$, with deformation parameters $\{m, k\}$ and the associated R-matrix is triangular. In analogy with $\mathscr{A}_{rs}$, $\mathscr{A}_{m,k}$ can also be considered as the semidirect or cross-product $GL_m(2) \rtimes_k \mathbb{C}[f, f^{-1}]$ built upon the vector space $GL_m(2) \otimes \mathbb{C}[f, f^{-1}]$, where $GL_m(2)$ is itself a Jordanian deformation of $GL(2)$. Thus, $\mathscr{A}_{m,k}$ can also be interpreted as a skew Laurent polynomial ring $GL_m[f, f^{-1}; \sigma]$ where σ is the automorphism given by the action of element f on $GL_m(2)$.

6. Conclusions

The $\mathscr{A}_{rs}$ and $\mathscr{A}_{m,k}$ deformations provide interesting new examples of cross-product quantum algebras, both of which have $GL(2) \otimes GL(1)$ as their classical limits. The differential calculus on $\mathscr{A}_{rs}$ also has an inherent cross-product structure, embeds the calculus on $GL_q(2)$ and is also related to the calculus on $GL_{p,q}(2)$. It would be interesting to investigate the calculus on the Jordanian $\mathscr{A}_{m,k}$, and on the *hybrid/intermetiate* [9] deformation obtained during the course of the contraction of $\mathscr{A}_{rs}$ to $\mathscr{A}_{m,k}$.

References

[1] Parashar D 2001 J. Math. Phys. 42 5431-5443
[2] Majid S 1995 Foundations of Quantum Group Theory (Cambridge: CUP)
[3] Klimyk A and Schmüdgen K 1997 Quantum Groups and Their Representations (Springer)
[4] Sudbery A 1990 Proc. Workshop on Quantum Groups, Argonne (edited by Curtright T, Fairlie D and Zachos C) pp. 33-51
[5] Basu-Mallick B 1994 hep-th/9402142
[6] Parashar D and McDermott R J 2000 J. Math. Phys. 41 2403-2416
[7] Faddeev L D, Reshetikhin N Y and Takhtajan L A 1990 Len. Math. J. 1 193-225
[8] Jurčo B 1991 Lett. Math. Phys. 22 177-186; 1994 preprint CERN-TH 9417/94
[9] Ballesteros A, Herranz F J and Parashar P 1999 J. Phys. A: Math. Gen. 32, 2369-2385

Inst. Phys. Conf. Ser. No 173: Mini-symposium
Paper presented at 24th Int. Coll. Group Theoretical Methods in Physics, Paris, France, 15–20 July 2002
©2003 IOP Publishing Ltd

q-Deformations of Classical (and Quantum) Integrable Systems Associated with Coalgebras

Ángel Ballesteros[1], Fabio Musso[2] and Orlando Ragnisco[3]

[1]Departamento de Física, Universidad de Burgos, E-09001-Burgos, Spain
[2]SISSA, Via Beirut 2/4, Trieste, I-34013 Italy
[3]Dipartimento di Fisica, Università di Roma TRE, Via della Vasca Navale 84, I-00146-Roma, Italy

Abstract.   The idea of constructing classical integrable systems with co-algebra symmetry is briefly reviewed. The undeformed and q-deformed $sl(2)$ cases are considered in detail, with special emphasis on the explicit solution of the equations of motion. Some open problems, related with supersymmetric extensions and to higher rank Lie algebras are also mentioned.

1. Introduction

In a number of recent papers ([1],[2],[3],[4], [5], [6]) the co-algebra approach for constructing classical and quantal integrable systems has been introduced, successfully applied to several examples, and generalized for instance to co-module algebras. In Section 1 we give a terse outline of the basic definitions and we present our working example, the $sl(2)$ Gaudin magnet, and its q-deformation. In Section 2, after recalling the results obtained in the quantum case, we will present the solution to the equations of motion in the classical case, both for the undeformed and the q-deformed magnet; surprisingly enough, it turns out that things are much simpler in the quantum than in the classical version: while the solution to the quantum q-deformed magnet is a straightforward generalization of that holding for the undeformed one, a similar assertion cannot be made for the q-deformed classical dynamics, described by a coupled system of nonlinear equations. In Section 3, a few words will be said on the supersymmetric extension(s) of the quantum $sl(2)$ model, as well as on the classical $sl(N)$ case.

2. Integrable systems from co-algebras

As is well known ([7]), a coalgebra A is a (unital, associative) algebra endowed with a co-product, i.e. a homomorphism $\Delta : A \to A \otimes A$ which is coassociative, i.e. $(id \otimes \Delta)\Delta(a) = (\Delta \otimes id)\Delta(a), \quad \forall a \in A$.

The main examples are:

- $U(g)$, the universal enveloping algebra of a simple Lie algebra g, with generators X_α, where:

$$\Delta(X_\alpha) := X_\alpha \otimes id + id \otimes X_\alpha \tag{1}$$

The coproduct extends to the whole $U(g)$ thanks to the homomorphism property:

$$\Delta(XY) := \Delta(X)\Delta(Y) \tag{2}$$

which obviously implies $\Delta([X, Y]) = [\Delta(X), \Delta(Y)]$.

- The second example is provided by a Poisson co-algebra. Let P being a Poisson algebra, i.e. a vector space endowed with a commutative multiplication and a Lie product $\{\cdot, \cdot\}$, coupled by the Leibnitz rule; then $P \otimes P$ is again a Poisson algebra with the bracket

$$\{a \otimes b, b \otimes c\}_{P \otimes P} := \{a, c\} \otimes bd + ac \otimes \{b, d\}$$

We say that the pair (P, Δ) is a Poisson coalgebra iff P is a Poisson algebra and Δ is a coassociative homomorphism between P and $P \otimes P$, namely:

$$\Delta(\{a, b\}_P) = \{\Delta(a), \Delta(b)\}_{P \otimes P}$$
$$\Delta(ab) = \Delta(a)\Delta(b)$$

Clearly, the simplest possible instance obtains by taking $g = sl(2)$ or its Poisson dual:

$$\{X_3, X_\pm\} = \pm X_\pm; \quad \{X_+, X_-\} = X_3$$

equipped with the primitive coproduct (1), and with the Casimir:

$$C = X_3^2 + X_+ X_- + X_- X_+$$

Another interesting example is provided by its Drinfeld-Jimbo deformation $sl_q(2)$ ([11]),([12]), and its Poisson "dual":

$$\{X_3, X_\pm\} = \pm X_\pm; \quad \{X_+, X_-\} = \frac{\sin(zX_3)}{z}$$

equipped with the q-deformed coproduct:

$$\Delta(X_3) = X_3 \otimes id + id \otimes X_3 \tag{3}$$
$$\Delta(X_\pm) = \exp(-\frac{1}{2}zX_3) \otimes X_\pm + X_\pm \otimes \exp(\frac{1}{2}zX_3) \tag{4}$$

and with the q-deformed Casimir:

$$C = \left(\frac{\sinh(\frac{z}{2}X_3)}{\frac{z}{2}}\right)^2 + X_+ X_- + X_- X_+$$

The following general Proposition can be easily proven:

Let A be a coalgebra with generators X_α and with Casimirs C_i. Then, the mth-coproducts of the Casimirs (Poisson-)commute among themselves and with the Nth-coproducts of the generators:

$$\left\{\Delta^{(m)}(C_i), \Delta^{(n)}(C_j)\right\} = 0 \quad m, n = 2, \ldots, N \quad \forall i, j; \tag{5}$$
$$\left\{\Delta^{(m)}(C_i), \Delta^{(N)}(X_\alpha)\right\} = 0 \quad m = 2, \ldots N \quad \forall \alpha.$$

The proof relies on the fact that the coproduct is a coassociative homomorphism, and consequently works in the classical as well as in the quantum context, *both* in the undeformed and in the q-deformed case.

Hence, for rank $r = 1$ Lie algebras (and their Poisson dual), complete integrability is established for any Hamiltonian being a smooth function of the Nth-coproduct of the Casimir and of the generators:

$$H = f(\Delta^{(N)}(C), \Delta^{(N)}(X_\alpha))$$

The most significant example is provided by the Gaudin Hamiltonian ([8]),([9])([10]):

$$H = uS_z^{(N)} + v \sum_{j,k=1}^{N} \vec{S}^j \cdot \vec{S}^k \tag{6}$$

where

$$\vec{S}^j = (S_x^j, S_y^j, S_z^j); \quad S_z^{(N)} = \Delta^{(n)}(S_z) = \sum_{j=1}^{N} S_z^j$$

and the spin variables $\vec{S}^j = (S_x^j, S_y^j, S_z^j)$ are related to the Cartan-Weyl generators X_α in the usual way.

A complete family of commuting conserved quantities is given by:

$$\{H, \Delta^{(m)}(C)\}, \quad m = 2, \cdots N$$

where, up to trivial numerical constants

$$\Delta^{(m)}(C) = \sum_{j,k=1}^{m} \vec{S}^j \cdot \vec{S}^k$$

Complete integrability is preserved by q-deformation; we give here the classical q-deformed version of the Gaudin Hamiltonian (6):

$$H_z = \mu\sigma_z^{N)} + v \sum_{j,k=1}^{N} [\vec{\sigma}^j \cdot \vec{\sigma}^k + \exp(z\beta_{ij})(\vec{\sigma}_x^j \cdot \vec{\sigma}_x^k + \vec{\sigma}_y^j \cdot \vec{\sigma}_y^k)] \tag{7}$$

where the "dynamical parameters" β_{ij} read:

$$\beta_{ij} = -\frac{\sigma_z^i - \sigma_z^j}{2} - \sum_{l=1}^{i-1} \sigma_z^l + \sum_{l=j+1}^{N} \sigma_z^l$$

and by $\vec{\sigma}$ we have denoted the deformed generators.

The Gaudin Hamiltonian as well as its q-deformation are actually superintegrable, both at the classical and at the quantum level. The reason for that is easily understood, observing that one can indeed construct the following two families of commuting integrals

$$(i) \ \Delta^{(m)}(C) \ (m = 2, \cdots, N); \quad (ii) \ \tilde{\Delta}^{(m)}(C) \ (m = 2, \cdots, N) \tag{8}$$

where in (i) the interaction takes place among the first $\binom{m}{2}$ pairs, and in (ii) among the last ones. Clearly, the two families share $\Delta^{(N)}(C)$, so that we get $2(N-1)$ conserved quantities.

3. Solution to the (quantum and) classical problem

The quantum problem has been explicitly solved both for $sl(2)$ and $sl_q(2)$, first in an infinite-dimensional representation and then in a finite-dimensional one (spin $j = 1/2$), by a technique which is reminiscent of the celebrated Bethe ansatz, and may be viewed as a simplified version of it. Referring for the details to [3], [5], [13], we recall here the following results:

- the spectrum of the commuting observables $\Delta^{(m)}(C)$ has been derived;
- the common set of eigenvectors of those commuting observables has been constructed, in two steps, in the so-called "lowest weight" representation. First, the kernel of $\Delta^{(N)}(X_-)$ (that we have called the "Bethe sea") has been fully characterized: the appropriate basis is built up thanks to a recursive formula that can be solved in terms of Γ functions (or of q-shifted factorials in the q-deformed case); the rest of the basis is then determined by acting on the "Bethe sea" by the raising operators X_+ and its coproducts.

As already mentioned in the introduction, no conceptual difference exists at the quantum level among $sl(2)$ and $sl_q(2)$, the solution procedure being exactly the same. Of course, the spectrum and the form of the eigenvectors are different, but from a mathematical point of view the only interesting feature is related to the onset of q-special functions in the deformed case. However, this picture is no longer valid at the classical level.

For the sake of simplicity, in the following we shall take as a Hamiltonian the N-body Casimir, $\Delta^{(N)}(C)$, both for $sl(2)$ and for $sl_q(2)$.

For $sl(2)$ the classical equations of motions are trivially solved. Indeed we have:

$$\frac{\partial X_3^{(j)}}{\partial t_N} := \left\{ X_3^{(j)}, \Delta^{(N)}(C) \right\} = 2\Delta^{(N)}(X_-)X_+^{(j)} - 2\Delta^{(N)}(X_+)X_-^{(j)}$$

$$\frac{\partial X_\pm^{(j)}}{\partial t_N} := \left\{ X_\pm^{(j)}, \Delta^{(N)}(C) \right\} = \pm 2[\Delta^{(N)}(X_\pm)X_3^{(j)} - \Delta^{(N)}(X_3)X_\pm^{(j)}] \qquad (9)$$

It is readily seen from (9) that one has separation of variables (indeed, the *same linear* time-evolution holds for each degree of freedom). Introducing the matrices:

$$L^{(j)} := \begin{pmatrix} X_3^{(j)} & \sqrt{2}X_-^{(j)} \\ \sqrt{2}X_+^{(j)} & -X_3^{(j)} \end{pmatrix};$$

$$M^{(N)} := \begin{pmatrix} \Delta^{(N)}(X_3) & \sqrt{2}\Delta^{(N)}(X_-) \\ \sqrt{2}\Delta^{(N)}(X_+) & -\Delta^{(N)}(X_3) \end{pmatrix}$$

eqns (9) can be cast in the simple "Lax" form:

$$\frac{\partial L^{(j)}}{\partial t_N} = \left[L^{(j)}, M^{(N)} \right] \qquad (10)$$

yielding

$$L^{(j)}(t_N) = \exp(-M^{(N)}t_N)L^{(j)}(0)\exp(M^{(N)}t_N) \qquad (11)$$

The motion is characterized by the eigenvalues of $M^{(N)}$, namely $\lambda_\pm = \pm\sqrt{\Delta^{(N)}(C)}$; accordingly, it is either hyperbolic or periodic. In the periodic case, the period is the same $\forall j$, and moreover it depends on the initial condition only through the Hamiltonian itself. It is worth noticing that instead of the "natural" single-particle coordinates $X^{(j)}$, one could have chosen the partial coproducts $\Delta^{(j)}(X_\alpha)$ getting the same time-evolution.

What about $sl_q(2)$? We know that the system remains superintegrable, and one is tempted to conjecture that again the essential features of the motion are encoded in the Hamiltonian itself, that is in the deformed Casimir. In particular, one would expect the existence of "separation" coordinates that for negative Hamiltonian undergo a periodic motion with the same period. Unfortunately at the moment we cannot give a complete answer to the above conjecture: indeed, while we have found "separation coordinates", we still lack a closed form solution of the equations of motion.

The q-deformed version of (9) read:

$$\frac{\partial X_3^{(j)}}{\partial t_N} := \left\{ X_3^{(j)}, \Delta^{(N)}(C) \right\} =$$
$$2\Delta^{(N)}(X_-)X_+^{(j)}E_z^{(j)} - 2\Delta^{(N)}(X_+)X_-^{(j)}E_z^{(j)} \tag{12}$$

$$\frac{\partial X_\pm^{(j)}}{\partial t_N} := \left\{ X_\pm^{(j)}, \Delta^{(N)}(C) \right\} = \pm 2[\Delta^{(N)}(X_\pm)E_z^{(j)}s_z(X_3^{(j)}) - s_z(\Delta^{(N)}(X_3))X_\pm^{(j)}]$$

$$\pm X_\pm^{(j)} \sum_{k=1}^N \text{sgn}(j-k)E_z^{(k)}[\Delta^{(N)}(X_+)X_-^{(k)} + \Delta^{(N)}(X_-)X_+^{(k)}] \tag{13}$$

where we have introduced the shorthand notations:

$$E_z^{(j)} := \exp[\frac{z}{2}\sum_{l=1}^N \text{sgn}(l-j)X_3^{(j)}]; \quad s_z(t) := \frac{\sinh(zt)}{z}$$

A careful inspection to (12),(13) suggests the use of the new "collective" variables:

$$S_3^{(j)} = \sum_{k=1}^j s_z(X_3^{(k)})E_z^{(k)}; \quad S_\pm = \sum_{k=1}^j X_\pm^{(k)}E_z^{(k)} \tag{14}$$

which evolve according to the system:

$$\frac{\partial S_3^{(j)}}{\partial \tau_N} = 4z S_3^{(j)}[K_- S_+^{(j)} - K_+ S_-^{(j)}] \tag{15}$$

$$\frac{\partial S_\pm^{(j)}}{\partial \tau_N} = \pm \frac{K_\pm}{z}[(S_3^{(j)})^2 - c_3] \pm 2z K_\mp(S_\pm^{(j)})^2 \mp S_\pm^{(j)}(K_3 + zK_+K_-) \tag{16}$$

In (15), (16) τ_N is the time labelling the flow generated by $\Delta^{(N)}(C)$, K_α denotes $\Delta^{(N)}(X_\alpha)$ and $c_3 = \exp(-z\Delta^{(N)}(X_3))$. Hence, we have succeded in "decoupling" the degrees of freedom, finding coordinates $S_\alpha^{(j)}$ obeying the same time-evolution, no matter what the value of j is, but the explicit solution of the above system with quadratic nonlinearity is postponed to a forthcoming paper.

4. Open problems and concluding remarks

There are a number of open problems that we are now actively working on, with encouraging preliminary results.

- One point concerns the generalization of the approach to higher rank Lie algebras, where the coproducts of the Casimirs and of the Cartan generators are generically not enough to ensure complete integrability. We are looking for a purely algebraic characterization of the "lacking" integrals, and we have solved the problem for $sl(3)$, where the dimension of a generic symplectic leaf is 6. So, we need $3N$ independent commuting functions arising from the Nth tensor product of $U(sl(3))$; an elementary calculation shows that from the coproducts of the Casimirs and of the other elements of the maximal abelian subalgebra of $U(sl(3))$ we get $2(N-1) + 3 = 2N + 1$ quantities. So the question is: where the remaining $N - 1$ quantities come from? The answer is the following ([14]): Let

$$C_3 = c_{jkl} X_j X_k X_l$$

be the cubic Casimir of $sl(3)$; then the "lacking" $N - 1$ integrals are given by:

$$I_m = c_{jkl} \Delta^{(m)}(X_j) \Delta^{(m-1)}(X_k) \Delta^{(m-1)}(X_l)$$

- The "coalgebra approach" can be extended to superalgebras. and we are now about to write down the relevant results for the quantum $Osp(1|2)$. We stress that, as well as in the "nonsupersymmetric case", the system constructed in this way differs from those already considered in the literature and solved through the Bethe ansatz ([15], [16]) .

- As is well known, a powerful geometric approach to integrability, at least for classical mechanical systems, is the so-called bi-hamiltonian approach. We can assert that so far we have a clear picture of the link between the two points of view in the simplest case, namely $sl(2)$. Unveiling the bihamiltonian structure of the q-deformed system does not seem to be an easy task, inasmuch as we don't know in that case any Lax representation.

Acknowledgements. A.B. has been partially supported by MCyT (Project BFM2000-1055). O.R. has been partially supported by INFN and by MIUR (COFIN2001 "Geometry and Integrability"). Partial financial support from INFN-CICYT is also acknowledged.

References

[1] Ballesteros A, Corsetti M and Ragnisco O 1996 *Czech. J. Phys.* **46** 1153
[2] Ballesteros A and Ragnisco O 1998 *J. Phys. A: Math. Gen.* **31** 3791
[3] Musso F and Ragnisco O 2000 *J. Math. Phys* **41** 7386
[4] Ballesteros A and Ragnisco O 2002 *J. Math. Phys* **43** 954
[5] Musso F and Ragnisco O 2001 *J. Phys. A: Math. Gen.* **34** 2625
[6] Ballesteros A, Musso F and Ragnisco O 2002 *J. Phys. A: Math. Gen.* **35** 8197
[7] Tjin T 1995 *J. Mod. Phys A* **203** 6175
[8] Gaudin M 1976, *J. de Physique* **37** 1087
[9] Gaudin M 1983 *La Fonction d'Onde de Bethe* Masson, Paris
[10] Sklianin E K *Lett. Math. Phys.* **47** 275
[11] Drinfeld VG 1986 *Quantum Groups* Proc. Int. Congress Math., MRSI Berkeley, 798
[12] Jimbo M 1985 *Lett. Math. Phys.* **10** 63
[13] Calogero F and Van Diejen J F 1996 *J. Math. Phys.* **37** 4243
[14] Musso F 2002, in preparation
[15] Kulish P P and Manojlovic N 2001 *J. Math. Phys* **42** 4757
[16] Brzezinski T and Macfarlane A J 1994 *J. Math. Phys.* **35** 3261

Inst. Phys. Conf. Ser. No 173: Mini-symposium
Paper presented at 24th Int. Coll. Group Theoretical Methods in Physics, Paris, France, 15–20 July 2002
©2003 IOP Publishing Ltd

Sum Rules for Codon Usage Frequencies

L. Frappat[a], A. Sciarrino[b] , P. Sorba[a]

[a] Laboratoire d'Annecy-le-Vieux de Physique Théorique LAPTH
Annecy-le-Vieux Cedex, France
[b] Dipartimento di Scienze Fisiche, Università di Napoli "Federico II"
and I.N.F.N., Sezione di Napoli, 80126 Napoli, Italy

Abstract. In the crystal basis model of the genetic code, in which the codons are assigned to irreducible representations of $U_{q \to 0}\left(sl(2)_H \oplus sl(2)_V\right)$, it is derived that the sum of the usage probabilities for codons in quartets with nucleotide C and A in 3rd position is independent of the biological species (for vertebrates). Comparison with experimental data shows the prediction is verified within 5 %.

The genetic code, i.e. the association between codons and amino acids (a.a.), is degenerate. Degeneracy refers to the fact that almost all the amino acids are encoded by multiple codons (synonymous codons). Degeneracy is found primarily in the third position of the codon, i.e. the nucleotide in the third position can change without changing the corresponding amino acid. For amino acids encoded by more than one codon, it is natural to ask if there are preferred codons. For longtime the common belief has been that there was no preference for the use of a particular codon, except for the enhancement due to the abundance of particular nucleotides. It is presently believed that the non uniform usage of synonymous codons is a widespread phenomenon. The main reasons for the codon usage biases are believed to be: the mutational biases, the translation efficiency, the natural selection and the abundance of the specific anticodons in the tRNA. The aim of this talk is not to compare the different proposed explanations, but to put in evidence a general pattern of the bias. It is experimentally observed that the pattern of codon usage varies between species and even among tissues within a species, see refs. [1, 2] which contain a large number of references to the original works on the subject. Most of the analyses of the codon usage frequencies have adressed to analyze the relative abundance of a specified codon in different genes of the same biological species or in the comparison of the relative abundance in the same gene for different biological species. Little attention has been paid to analyze codon usage frequency summed over the whole available sequences to infer global correlations between different biological species. Indeed a correlation between suitable ratios of codon usage frequencies has been remarked in [3] for biological species belonging to the vertebrate class and in [4] for biological organisms including plants. Such correlations fit well in a mathematical model of the genetic code, called crystal basis model, proposed by the authors in [5]. Moreover in [6] it has also been observed that the ratio of the previously defined quantities exhibits for vertebrates an almost universal behaviour, i.e. independent on the biological species and on the nature of the amino-acid, for the subset of the amino acids encoded by quartets or sextets. These remarks suggest the possible existence of a general bias in the codon usage frequency of a specific codon belonging to a quartet or quartet sub-part of a sextet, i.e. the four codons differing for the last nucleotide. In [7] this aspect has been investigated and a set of sum rules, which should be verified by all the biological species belonging to vertebrates, has been derived. Let us define the usage probability for the codon XZN ($X, Z, N \in \{A, C, G, U\}$) as

$$P(XZN) = \lim_{n_{tot} \to \infty} \frac{n_{XZN}}{n_{tot}} \qquad (1)$$

where n_{XZN} is the number of times the codons XZN has been used in the biosynthesis process of the corresponding amino-acid and n_{tot} is the total number of codons used to synthetise this amino-acid. In our model each codon XZN is described by a state belonging to an irreducible representation (irrep.), denoted $(J_H, J_V)^\xi$, of the algebra $U_q\big(sl(2)_H \oplus sl(2)_V\big)$ in the limit $q \to 0$ (so-called crystal basis), see Table 3, where the upper label ξ removes the degeneracy of the irreps.. In the crystal basis model, we write the usage probability as a function of the biological species (b.s.), of the particular amino-acid and of the labels $J_H, J_V, J_{H,3}, J_{V,3}$ describing the state XZN:

$$P(XZN) = P(b.s.; J_H, J_V, J_{H,3}, J_{V,3}) \qquad (2)$$

We make the hypothesis that the r.h.s. of eq.(2) can be written as the sum of a function ρ independent on the biological species and a b.s. depending function f_{bs}:

$$P(XZN) = \rho^{XZ}(J_H, J_V, J_{H,3}, J_{V,3}) + f_{bs}^{XZ}(J_H, J_V, J_{H,3}, J_{V,3}) \qquad (3)$$

We assume also:

$$f_{bs}^{XZ}(J_H, J_V, J_{H,3}, J_{V,3}) \approx F_{bs}^{XZ}(J_H; J_{H,3}) + G_{bs}^{XZ}(J_V; J_{V,3}) \qquad (4)$$

Let us analyse in the light of the above considerations the usage probability for the quartets Ala, Gly, Pro, Thr and Val and for the quartet sub-part of the sextets Arg (i.e. the codons of the form CGN), Leu (i.e. CUN) and Ser (i.e. UCN). For Thr, Pro, Ala and Ser we can write, using Table 3 and eqs. (2)-(4), with $N = A, C, G, U$:

$$P(NCC) + P(NCA) = \rho_{C+A}^{NC} + F_{bs}^{NC}(\tfrac{3}{2}; x) + G_{bs}^{NC}(\tfrac{3}{2}; y)$$
$$+ F_{bs}^{NC}(\tfrac{1}{2}; x') + G_{bs}^{NC}(\tfrac{1}{2}; y') \qquad (5)$$

where ρ_{C+A}^{NC} denotes the contribution of the universal functions (i.e. not depending on the biological species) and the labels x, y, x', y' depend on the nature of the first two nucleotides NC, see Table 3. We get:

$$P(NCG) + P(NCU) = \rho_{G+U}^{NC} + F_{bs}^{NC}(\tfrac{3}{2}; x) + G_{bs}^{NC}(\tfrac{3}{2}; y)$$
$$+ F_{bs}^{NC}(\tfrac{1}{2}; x') + G_{bs}^{NC}(\tfrac{1}{2}; y') \qquad (6)$$

It follows that the difference between eq. (5) and eq. (6) is a quantity independent of the biological species:

$$P(NCC) + P(NCA) - P(NCG) - P(NCU) = \rho_{C+A-G-U}^{NC} \qquad (7)$$

Analogous equations hold for Leu, Val, Arg and Gly. Since the probabilities for one quadruplet are normalised to one, from eq.(7) we deduce that, for all the eight amino acids, the sum of probabilities of codon usage for codons with nucleotide A and C (or U and G) in 3rd position is independent of the biological species:

$$P(XZC) + P(XZA) = \text{Const.} \quad (XZ = NC, CU, GU, CG, GG) \qquad (8)$$

In order to check our proposed sum rules eq.(8), we have considered the data for species from the GenBank (release 127.0 of Dec. 2001) for vertebrates with large codon statistics, see Table 1, where the number of codons is reported. In Table 2 we report the experimental data

Table 1. Data for vertebrates from GenBank Release 127.0 [15 December 2001]
1) Homo sapiens – 18611700; 2) Mus musculus – 8079821; 3) Rattus norvegicus – 3324518; 4) Gallus gallus – 1019029; 5) Xenopus laevis – 929562; 6) Bos taurus – 764195; 7) Danio rerio – 535583; 8) Oryctolagus cuniculus – 441547; 9) Macaca fascicularis – 403875; 10) Sus scrofa – 380357; 11) Canis familiaris – 229526; 12) Takifugu rubripes – 152479; 13) Ovis aries – 134027; 14) Oncorhynchus mykiss – 131431; 15) Cricetulus griseus – 109395; 16) Rattus sp. – 06164; 17) Pan troglodytes – 88272; 18) Oryzias latipes – 85610; 19) Macaca mulatta – 81673; 20) Felis cattus – 66930; 21) Equus caballus – 59932.

for the selected 21 vertebrates. The comparison shows that our predictions are verified within 3-6 % (for the species with highest statistics within about 3 %), which is an amazing result. In Table 4 we report the mean value and the standard deviation of the probability of usage of the codons XZN ($XZ = NC, CU, GU, CG, GG$) and of $P_{CA}(a.a.)$, $P_{CU}(a.a.)$ and $P_{CG}(a.a.)$ ($P_{CA}(a.a.) = P_C(a.a.) + P_A(a.a.)$) computed over all biological species given in Table 1. It can be remarked that these probabilities show a rather large spread which is

Table 2. Sum of usage probability of codons $P_{CA}(a.a.) \equiv P(XZC) + P(XZA)$. The number in the first column denotes the biological species (b.s.) of Table 1. The amino acid, encoded by the quartet XZN, are labelled by the standard letter.

b.s.	$P_{CA}(P)$	$P_{CA}(A)$	$P_{CA}(T)$	$P_{CA}(S)$	$P_{CA}(V)$	$P_{CA}(L)$	$P_{CA}(R)$	$P_{CA}(G)$
1	0.60	0.63	0.64	0.60	0.35	0.33	0.51	0.59
2	0.59	0.61	0.65	0.59	0.36	0.34	0.52	0.59
3	0.59	0.62	0.65	0.60	0.37	0.34	0.52	0.59
4	0.60	0.59	0.62	0.59	0.35	0.31	0.53	0.58
5	0.60	0.60	0.62	0.56	0.38	0.33	0.50	0.58
6	0.60	0.63	0.65	0.60	0.35	0.33	0.52	0.60
7	0.53	0.56	0.61	0.56	0.34	0.32	0.55	0.63
8	0.61	0.65	0.64	0.63	0.35	0.33	0.55	0.61
9	0.60	0.63	0.63	0.60	0.37	0.35	0.50	0.58
10	0.61	0.64	0.65	0.62	0.36	0.33	0.53	0.61
11	0.61	0.62	0.64	0.60	0.38	0.34	0.52	0.59
12	0.58	0.58	0.61	0.59	0.38	0.32	0.55	0.59
13	0.61	0.63	0.66	0.60	0.35	0.35	0.55	0.61
14	0.62	0.61	0.68	0.60	0.38	0.33	0.53	0.57
15	0.61	0.62	0.66	0.58	0.37	0.33	0.51	0.59
16	0.58	0.62	0.66	0.59	0.37	0.34	0.52	0.59
17	0.62	0.54	0.72	0.59	0.29	0.35	0.58	0.58
18	0.56	0.59	0.63	0.60	0.35	0.31	0.55	0.63
19	0.61	0.61	0.65	0.61	0.34	0.34	0.50	0.59
20	0.61	0.63	0.64	0.60	0.38	0.34	0.52	0.60
21	0.58	0.63	0.66	0.62	0.37	0.34	0.53	0.61

surprisingly reduced in the sum and that essentially two species: *Danio rerio* (zebrafish) and *Pan troglodytes* (chimpanze), differ sensibly from the average value for most amino acids. It is interesting a comparison between the theoretical previsions of our results for the correlation matrix Γ and the experimental values. Theoretical values for the entries of Γ:

$$\Gamma_{CA} = \Gamma_{GU} = -1$$
$$\Gamma_{CG} = \Gamma_{AU} = -\Gamma_{CU} = \Gamma_{AG} \tag{9}$$

Experimental values computed for the average value over the eight amino acids:

$$< \Gamma_{CA} >= -0.85 < \Gamma_{CG} >= 0.29 < \Gamma_{CU} >= -0.68$$
$$< \Gamma_{GU} >= -0.79 < \Gamma_{AU} >= 0.67 < \Gamma_{AG} >= -0.56 \tag{10}$$

The average values differ, within 20%, from the expected ones, except for Γ_{CG} which is almost half of the expected value. This feature may be due to the known suppression of the codons containing a dinucleotide CG.

In conclusion, in the crystal basis model of the genetic code, we have derived that the sum of the usage probabilities of two suitably choosen codons, in quartets and sextets, should approximatively be independent on the biological species for vertebrates. This prevision is in satisfactory agreement with the experimental data. One can state the above results saying that the sum of the probability of codon usage $XZC + XZA$ is not depending on the nature of the biological species, without any reference to the model. However a deeper analysis of Table 2 shows that P_{C+A} for Pro, Thr, Ala, Ser and Gly is of the order of 0.62, for Leu and Val of the order 0.35 while for Arg is of order 0.52. In the crystal basis model the roots, i.e. the dinucleotide formed by the first two nucleotides of the first 5 amino acids belong to the same irrep. (1,1), the roots of Leu and Val belong to the irrep. (0,1), while the root of Arg belongs to the irrep. (1,0). This is an interesting result, especially for Pro whose molecule has a different structure than the others amino acids (Pro has an imino group instead of an amino group).

Acknowledgments: Partially supported for one of us (A.S.) by the M.U.R.S.T. through National Research Project *SINTESI 2000*.

References
[1] Duret L., Mouchiroud D., Proc. Natl. Acad. Sci. USA **96** (1999) 4482.
[2] Kanaya S., Yamada Y., Kinoouchi M., Kudo Y. Ikemura T., J. Mol. Evol. **53** (2001) 290.
[3] Frappat L., Sciarrino A., Sorba P., Phys. Lett. A **259** (1999) 339, physics/9812041.
[4] Chiusano M.L., Frappat L., Sciarrino A., Sorba P., Europhys. Lett. **55** (2001) 287.
[5] Frappat L., Sciarrino A., Sorba P., Phys. Lett. A. **250** (1998) 214, physics/9801027.
[6] Frappat L., Sciarrino A., Sorba P., J. Biol. Phys. **27** (2001) 1, physics/0003037.
[7] Frappat L., Sciarrino A., Sorba P., *Sum rules for codon usage probabilities*, physics/0205013.

Table 3. The eukaryotic or standard code code.
The upper label denotes different irreducible representations.

codon	amino acid	J_H	J_V	$J_{3,H}$	$J_{3,V}$	codon	amino acid	J_H	J_V	$J_{H,3}$	$J_{V,3}$
CCC	Pro P	3/2	3/2	3/2	3/2	UCC	Ser S	3/2	3/2	1/2	3/2
CCU	Pro P	$(1/2$	$3/2)^1$	1/2	3/2	UCU	Ser S	$(1/2$	$3/2)^1$	$-1/2$	3/2
CCG	Pro P	$(3/2$	$1/2)^1$	3/2	1/2	UCG	Ser S	$(3/2$	$1/2)^1$	1/2	1/2
CCA	Pro P	$(1/2$	$1/2)^1$	1/2	1/2	UCA	Ser S	$(1/2$	$1/2)^1$	$-1/2$	1/2
CUC	Leu L	$(1/2$	$3/2)^2$	1/2	3/2	UUC	Phe F	3/2	3/2	$-1/2$	3/2
CUU	Leu L	$(1/2$	$3/2)^2$	$-1/2$	3/2	UUU	Phe F	3/2	3/2	$-3/2$	3/2
CUG	Leu L	$(1/2$	$1/2)^3$	1/2	1/2	UUG	Leu L	$(3/2$	$1/2)^1$	$-1/2$	1/2
CUA	Leu L	$(1/2$	$1/2)^3$	$-1/2$	1/2	UUA	Leu L	$(3/2$	$1/2)^1$	$-3/2$	1/2
CGC	Arg R	$(3/2$	$1/2)^2$	3/2	1/2	UGC	Cys C	$(3/2$	$1/2)^2$	1/2	1/2
CGU	Arg R	$(1/2$	$1/2)^2$	1/2	1/2	UGU	Cys C	$(1/2$	$1/2)^2$	$-1/2$	1/2
CGG	Arg R	$(3/2$	$1/2)^2$	3/2	$-1/2$	UGG	Trp W	$(3/2$	$1/2)^2$	1/2	$-1/2$
CGA	Arg R	$(1/2$	$1/2)^2$	1/2	$-1/2$	UGA	Ter	$(1/2$	$1/2)^2$	$-1/2$	$-1/2$
CAC	His H	$(1/2$	$1/2)^4$	1/2	1/2	UAC	Tyr Y	$(3/2$	$1/2)^2$	1/2	1/2
CAU	His H	$(1/2$	$1/2)^4$	$-1/2$	1/2	UAU	Tyr Y	$(3/2$	$1/2)^2$	$-3/2$	1/2
CAG	Gln Q	$(1/2$	$1/2)^4$	1/2	$-1/2$	UAG	Ter	$(3/2$	$1/2)^2$	$-1/2$	$-1/2$
CAA	Gln Q	$(1/2$	$1/2)^4$	$-1/2$	$-1/2$	UAA	Ter	$(3/2$	$1/2)^2$	$-3/2$	$-1/2$
GCC	Ala A	3/2	3/2	3/2	1/2	ACC	Thr T	3/2	3/2	1/2	1/2
GCU	Ala A	$(1/2$	$3/2)^1$	1/2	1/2	ACU	Thr T	$(1/2$	$3/2)^1$	$-1/2$	1/2
GCG	Ala A	$(3/2$	$1/2)^1$	3/2	$-1/2$	ACG	Thr T	$(3/2$	$1/2)^1$	1/2	$-1/2$
GCA	Ala A	$(1/2$	$1/2)^1$	1/2	$-1/2$	ACA	Thr T	$(1/2$	$1/2)^1$	$-1/2$	$-1/2$
GUC	Val V	$(1/2$	$3/2)^2$	1/2	1/2	AUC	Ile I	3/2	3/2	$-1/2$	1/2
GUU	Val V	$(1/2$	$3/2)^2$	$-1/2$	1/2	AUU	Ile I	3/2	3/2	$-3/2$	1/2
GUG	Val V	$(1/2$	$1/2)^3$	1/2	$-1/2$	AUG	Met M	$(3/2$	$1/2)^1$	$-1/2$	$-1/2$
GUA	Val V	$(1/2$	$1/2)^3$	$-1/2$	$-1/2$	AUA	Ile I	$(3/2$	$1/2)^1$	$-3/2$	$-1/2$
GGC	Gly G	3/2	3/2	3/2	$-1/2$	AGC	Ser S	3/2	3/2	1/2	$-1/2$
GGU	Gly G	$(1/2$	$3/2)^1$	1/2	$-1/2$	AGU	Ser S	$(1/2$	$3/2)^1$	$-1/2$	$-1/2$
GGG	Gly G	3/2	3/2	3/2	$-3/2$	AGG	Arg R	3/2	3/2	1/2	$-3/2$
GGA	Gly G	$(1/2$	$3/2)^1$	1/2	$-3/2$	AGA	Arg R	$(1/2$	$3/2)^1$	$-1/2$	$-3/2$
GAC	Asp D	$(1/2$	$3/2)^2$	1/2	$-1/2$	AAC	Asn N	3/2	3/2	$-1/2$	$-1/2$
GAU	Asp D	$(1/2$	$3/2)^2$	$-1/2$	$-1/2$	AAU	Asn N	3/2	3/2	$-3/2$	$-1/2$
GAG	Glu E	$(1/2$	$3/2)^2$	1/2	$-3/2$	AAG	Lys K	3/2	3/2	$-1/2$	$-3/2$
GAA	Glu E	$(1/2$	$3/2)^2$	$-1/2$	$-3/2$	AAA	Lys K	3/2	3/2	$-3/2$	$-3/2$

Table 4. Mean value, standard deviation for the probabilities $P(XZN)$ corresponding to the eight amino acids related to quartets or sextets for the choice of biological species of Table 1.

	$P_U(P)$	$P_C(P)$	$P_A(P)$	$P_G(P)$	$P_U(T)$	$P_C(T)$	$P_A(T)$	$P_G(T)$
$\bar{x}$	0.28	0.33	0.26	0.13	0.23	0.39	0.26	0.13
σ	0.028	0.043	0.034	0.028	0.030	0.050	0.034	0.027
	$P_U(A)$	$P_C(A)$	$P_A(A)$	$P_G(A)$	$P_U(S)$	$P_C(S)$	$P_A(S)$	$P_G(S)$
$\bar{x}$	0.27	0.40	0.21	0.12	0.30	0.38	0.22	0.10
σ	0.026	0.046	0.035	0.029	0.027	0.036	0.026	0.020
	$P_U(V)$	$P_C(V)$	$P_A(V)$	$P_G(V)$	$P_U(L)$	$P_C(L)$	$P_A(L)$	$P_G(L)$
$\bar{x}$	0.17	0.26	0.10	0.47	0.15	0.25	0.08	0.52
σ	0.036	0.023	0.026	0.045	0.034	0.018	0.017	0.035
	$P_U(R)$	$P_C(R)$	$P_A(R)$	$P_G(R)$	$P_U(G)$	$P_C(G)$	$P_A(G)$	$P_G(G)$
$\bar{x}$	0.16	0.34	0.18	0.31	0.17	0.33	0.26	0.23
σ	0.042	0.039	0.026	0.043	0.029	0.034	0.033	0.032
	$P_{CA}(P)$	$P_{CA}(A)$	$P_{CA}(T)$	$P_{CA}(S)$	$P_{CA}(V)$	$P_{CA}(L)$	$P_{CA}(R)$	$P_{CA}(G)$
$\bar{x}$	0.595	0.611	0.646	0.598	0.359	0.334	0.527	0.596
σ	0.020	0.027	0.024	0.016	0.020	0.012	0.020	0.015
	$P_{CU}(P)$	$P_{CU}(A)$	$P_{CU}(T)$	$P_{CU}(S)$	$P_{CU}(V)$	$P_{CU}(L)$	$P_{CU}(R)$	$P_{CU}(G)$
$\bar{x}$	0.613	0.672	0.614	0.687	0.430	0.401	0.506	0.507
σ	0.030	0.028	0.027	0.026	0.031	0.022	0.049	0.024
	$P_{CG}(P)$	$P_{CG}(A)$	$P_{CG}(T)$	$P_{CG}(S)$	$P_{CG}(V)$	$P_{CG}(L)$	$P_{CG}(R)$	$P_{CG}(G)$
$\bar{x}$	0.462	0.513	0.511	0.479	0.728	0.769	0.652	0.567
σ	0.058	0.058	0.063	0.046	0.056	0.048	0.055	0.058

Inst. Phys. Conf. Ser. No 173: Mini-symposium
Paper presented at 24th Int. Coll. Group Theoretical Methods in Physics, Paris, France, 15–20 July 2002
©2003 IOP Publishing Ltd

About rational-trigonometric deformation

V N Tolstoy

Institute of Nuclear Physics, Moscow State University, 119992 Moscow, Russia

Abstract. We consider a rational-trigonometric deformation in context of rational and trigonometric deformations. The simplest examples of these deformations are presented in different fields of mathematics. Rational-trigonometric differential Knizhnik-Zamolodchikov and dynamical equations are introduced.

1. Introduction

In classical mathematics there are three classes of meromorphic functions: rational, trigonometric, elliptic. According to these classes there are three types of quantum deformations: *rational, trigonometric, elliptic*. It turns out we can also introduce a *rational-trigonometric deformation*. All these deformations will be called standard. In this short and sketched paper we discussed the simplest examples of the standard deformations in the arithmetic (number theory), geometry, differential calculus, theory functions and Lie algebras. We also introduce rational-trigonometric differential Knizhnik-Zamolodchikov (KZ) and dynamical (DD) equations. The rational-trigonometric differential KZ equations are connected with a rational-trigonometric classical r-matrix [6] which is a sum the simplest rational and trigonometric r-matrices depending on spectral parameter. It turns out that the rational-trigonometric differential KZ (DD) equations are sums of the rational and trigonometric differential KZ (DD) equations.

2. Trigonometric, rational and rational-trigonometric deformations

2.1. Standard deformations of numbers

Let z be a complex number, $z \in \mathbb{C}$. a). The trigonometric deformation (or q-deformation) of z:

$$z \rightarrow (z)_q = \frac{1 - q^z}{1 - q}, \qquad (1)_q = 1, \qquad (0)_q = 0, \tag{1}$$

where q is a deformation parameter, $(z)_{q=1} = z$. This deformation is well-known. b). The rational deformation (or η-deformation) of z [7]:

$$z \rightarrow (z)_\eta = \frac{z}{1 + \eta(z - 1)}, \qquad (1)_\eta = 1, \qquad (0)_\eta = 0, \tag{2}$$

where η is a deformation parameter, $(z)_{\eta=0} = z$. c). The rational-trigonometric deformation (or (q, η)-deformation) of z [7]:

$$z \rightarrow (z)_{q\eta} = \frac{(z)_q}{1 + \eta(z - 1)_q}, \qquad (1)_{q\eta} = 1, \qquad (0)_{q\eta} = 0, \tag{3}$$

where q and η are deformation parameters, $(z)_{q=1,\eta} = (z)_\eta$, $(z)_{q,\eta=0} = (z)_q$. *Remark.* There is an elliptic deformation of z (for example, see [3])

2.2. Standard deformations of two-dimensional plane

Let x and y be two commuting coordinate variables, i.e.

$$xy - yx = 0 \qquad \text{(the } (x,y) - \text{plane)}. \tag{4}$$

a). The trigonometric deformation of the (x,y)-plane:

$$xy - qyx = 0 \qquad \text{(the Manin's plane)}. \tag{5}$$

b). The rational deformation of the (x,y)-plane:

$$xy - yx = \eta y^2 . \tag{6}$$

c). The rational-trigonometric deformation of the (x,y)-plane:

$$xy - qyx = \eta y^2 . \tag{7}$$

All these deformation of the (x,y)-plane are well-known. *Remark.* No elliptic deformation of the (x,y)-plane is known. It is an open problem.

2.3. Standard deformations of the differential calculus

Let $\partial_x := \frac{\partial}{\partial x}$ be the usual derivative. a). The trigonometric deformation of ∂_x:

$$\partial_x^{(q)} f(x)_q = \frac{f(qx) - f(x)}{qx - x} = \frac{f(qx) - f(x)}{x(q-1)} . \tag{8}$$

b). The rational deformation of ∂_x:

$$\partial_x^{(\eta)} f(x)_q = \frac{f(x+\eta) - f(x)}{x + \eta - x} = \frac{f(x+\eta) - f(x)}{\eta} . \tag{9}$$

The deformations a) and b) for ∂_x are well-known.

c). The rational-trigonometric deformation of ∂_x:

$$\partial_x^{(q\eta)} f(x)_q = \frac{f(qx+\eta) - f(x)}{qx + \eta - x} = \frac{f(qx+\eta) - f(x)}{x(q-1) + \eta} . \tag{10}$$

All these deformations of ∂_x have the following properties: (i) they are defined by the general formulas

$$\partial_x^{(df)} f(x) = \frac{\Delta^{(df)} f(x)}{\Delta^{(df)} x} = \frac{f(x') - f(x)}{x' - x} , \tag{11}$$

(ii) they satisfy the Leibniz rule

$$\partial_x^{(df)} f(x)\varphi(x) = \left(\partial_x^{(df)} f(x) \right) \varphi(x) + f(x') \left(\partial_x^{(df)} \varphi(x) \right) , \tag{12}$$

where $x' = qx$ for the q-deformation, $x' = x + \eta$ for the η-deformation and $x' = qx + \eta$ for the (q,η)-deformation. *Remark.* There is an explicite formula for the elliptic deformation of ∂_x.

2.4. Standard deformations of functions

Let $\exp(x)$ be the usual exponential of x:

$$\exp(x) = 1 + x + \frac{1}{2!}x^2 + \cdots + \frac{1}{n!}x^n + \cdots . \tag{13}$$

and let $F_{n,m}(x)$ be a standard hypergeometric series:

$$F_{n,m}(x) = \sum_{k=0}^{\infty} \frac{(a_1)_{(k)}(a_2)_{(k)}\cdots(a_n)_{(k)}}{k!(b_1)_{(k)}(b_2)_{(k)}\cdots(b_m)_{(k)}}x^k, \tag{14}$$

where

$$(a)_{(k)} = a(a+1)\cdots(a+k-1). \tag{15}$$

a). The trigonometric deformation of $\exp(x)$ and $F_{n,m}(x)$.

$$\exp_q(x) = 1 + x + \frac{1}{(2)_q!}x^2 + \cdots + \frac{1}{(n)_q!}x^n + \cdots , \tag{16}$$

where

$$(n)_q! = (1)_q(2)_q \cdots (n)_q . \tag{17}$$

If we replace the parameters a_i, b_j, and $k!$ in (14) by the q-analogs $(a_i)_q$, $(b_j)_q$ and $(k)_q!$ we obtain the basic hypergeometric (q-hypergeometric) series $F_{n,m}^{(q)}(z)$. b). The rational deformation of $\exp(x)$ and $F_{n,m}(x)$.

$$\exp_\eta(x) = 1 + x + \frac{1}{(2)_\eta!}x^2 + \cdots + \frac{1}{(n)_\eta!}x^n + \cdots , \tag{18}$$

where

$$(n)_\eta! = (1)_\eta(2)_\eta \cdots (n)_\eta . \tag{19}$$

If we replace the parameters a_i, b_j, and $k!$ in (14) by the η-analogs $(a_i)_\eta$, $(b_j)_\eta$ and $(k)_\eta!$ we obtain the basic hypergeometric (η-hypergeometric) series $F_{n,m}^{(\eta)}(z)$.

c). The rational-trigonometric deformation of $\exp(x)$ and $F_{n,m}(x)$.

$$\exp_{q\eta}(x) = 1 + x + \frac{1}{(2)_{q\eta}!}x^2 + \cdots + \frac{1}{(n)_{q\eta}!}x^n + \cdots , \tag{20}$$

where

$$(n)_{q\eta}! = (1)_{q\eta}(2)_{q\eta} \cdots (n)_{q\eta} . \tag{21}$$

The replacement of a_i, b_j, and $k!$ in (14) by the (q,η)-analogs $(a_i)_{q\eta}$, $(b_j)_{q\eta}$ and $(k)_{q\eta}!$ gives us the (q,η)-hypergeometric series

$$F_{n,m}^{(q,\eta)}(x) = \sum_{k=0}^{\infty} \frac{((a_1)_{q\eta})_{(k)}((a_2)_{q\eta})_{(k)}\cdots((a_n)_{q\eta})_{(k)}}{(k)_{q\eta}!((b_1)_{q\eta})_{(k)}((b_2)_{q\eta})_{(k)}\cdots((b_m)_{q\eta})_{(k)}}x^k, \tag{22}$$

where

$$((a)_{q\eta})_{(k)} = (a)_{q\eta}(a+1)_{q\eta}\cdots(a+k-1)_{q\eta} . \tag{23}$$

Setting here $q = 1$ we obtain the η-hypergeometric series $F_{n,m}^{(\eta)}(x)$. We can also introduce the η- and (q,η)- analogs of other special functions.

All deformed exponentials $\exp_q(x)$, $\exp_\eta(x)$ and $\exp_{q\eta}(x)$ can be obtained from the functional equation $f_1(x+y) = f_2(y)f_3(x)$ with regular functions $f_i(z)$ satisfying the initial conditions $f_i(0) = 1$, where the variables x and y satisfy the relations (5)–(7) (see [7]). *Remark.* Analogous formulas exist also for the elliptic case they can be found in [3].

2.5. Standard deformations of universal enveloping algebras

Let g be a finite-dimensional Lie algebra, $g[u, u^{-1}]$ be a loop algebra and $g[u]$ be a non-negative loop algebra over g. We denote by $U(g)$, $U(g[u])$ and $U(g[u, u^{-1}])$ their universal enveloping algebras.

a). The trigonometric deformation (q-deformation) of $U(g)$, $U(g[u])$ and $U(g[u, u^{-1}])$ are well-known. They are denoted by $U_q(g)$, $U_q(g[u])$ and $U_q(g[u, u^{-1}])$ [1].

b). The rational deformation of $U_q(g[u])$ and $U_q(g[u, u^{-1}])$ are well-known. They are Yangian $Y_\eta(g)$ [1] and its double $DY_\eta(g)$ [2].

c). The rational-trigonometric deformation of $U_q(g[u])$ and $U_q(g[u, u^{-1}])$ are Drinfeldian $D_{q\eta}(g)$ [6], [7] and its double $DD_{q\eta}(g)$. *Remark.* An elliptic deformation exists only for $U(sl_n[u])$. It is well-known too.

3. Rational-trigonometric differential Knizhnik-Zamolodchikov and dynamical equations

We consider the case $g = gl_M$ although results of this section are also valid for an arbitrary simple complex Lie algebra g. Let e_{ab}, $a, b = 1, \ldots, M$, be the standard Cartan-Weyl basis of gl_M: $[e_{ab}, e_{cd}] = \delta_{bc}e_{ad} - \delta_{ad}e_{cb}$. The element $C_2 := \sum_{a,b=1}^{N} e_{ab}e_{ba} \in U(gl_M)$ is a gl_M-scalar, i.e. $[C_2, x] = 0$ for any $x \in gl_M$, and it is called the second order Casimir element. The element $\Omega := \frac{1}{2}\big(\Delta(C_2) - C_2 \otimes \mathrm{id} - \mathrm{id} \otimes C_2\big) = \sum_{a,b=1}^{M} e_{ab} \otimes e_{ba} \subset U(gl_M) \otimes U(gl_M)$, where Δ is a trivial co-product $\Delta(x) = x \otimes \mathrm{id} + \mathrm{id} \otimes x$ ($\forall x \in gl_M$), is called the Casimir two-tensor. The two-tensor can be presented in the form $\Omega = \Omega^+ + \Omega^-$, where $\Omega^+ = \frac{1}{2}\sum_a e_{aa} \otimes e_{aa} + \sum_{1 \le a < b \le M} e_{ab} \otimes e_{ba}$ and $\Omega^- = \frac{1}{2}\sum_a e_{aa} \otimes e_{aa} + \sum_{1 \le a < b \le M} e_{ba} \otimes e_{ab}$. Note that $(\omega \otimes \omega)(\Omega^{\pm}) = \Omega^{\mp}$, where ω is the Cartan automorphism: $\omega(e_{ab}) = -e_{ba}$. For any $x \in U(gl_M)$ we set $x_{(i)} = \underbrace{\mathrm{id} \otimes \cdots \otimes \mathrm{id} \otimes \overset{i-th}{x} \otimes \mathrm{id} \otimes \cdots \otimes \mathrm{id}}_{N-times}$. We consider $U(gl_M)$ as a subalgebra of $(U(gl_M))^{\otimes N}$, the embedding $U(gl_M) \hookrightarrow (U(gl_M))^{\otimes N}$ being given by the N-fold co-product, that is $x \mapsto \Delta^N(x) = x_{(1)} + \ldots + x_{(N)}$ for any $x \in gl_M$. For a nonzero complex number κ we consider differential operators $\nabla_{z_1}^{(r)}, \ldots, \nabla_{z_N}^{(r)}$ and $\nabla_{z_1}^{(t)}, \ldots, \nabla_{z_N}^{(t)}$ with coefficient in $(U(gl_M))^{\otimes N}$ depending on complex variable $z_1, \ldots, z_N$ and $\lambda_1, \cdots, \lambda_M$ (see [4] and [5]):

$$\nabla_{z_i}^{(r)}(z; \lambda) = \kappa \frac{\partial}{\partial z_i} - \sum_{a=1}^{M} \lambda_a(e_{aa})_{(i)} - \sum_{\substack{j=1 \\ j \ne i}}^{N} \frac{\Omega_{(ij)}}{z_i - z_j}, \tag{24}$$

$$\nabla_{z_i}^{(t)}(z; \lambda) = \kappa z_i \frac{\partial}{\partial z_i} - \sum_{a=1}^{M}(\lambda_a - e_{aa})(e_{aa})_{(i)} - \sum_{\substack{j=1 \\ i \ne i}}^{N} \frac{z_i\Omega_{(ij)}^+ + z_j\Omega_{(ij)}^-}{z_i - z_j}. \tag{25}$$

The operators $\nabla_{z_1}^{(r)}, \ldots, \nabla_{z_N}^{(r)}$ (resp. $\nabla_{z_1}^{(t)}, \ldots, \nabla_{z_N}^{(t)}$) are called the rational (resp. trigonometric) Knizhnik-Zamolodchikov (KZ) operators. We set

$$\nabla_{z_i}^{(rt)}(z; \lambda) = \hbar \nabla_{z_i + \frac{\eta}{\hbar}}^{(t)}(z + \tfrac{\eta}{\hbar}; (1 + \tfrac{\eta}{\hbar})\lambda). \tag{26}$$

It is easy to see that

$$\nabla_{z_i}^{(rt)}(z; \lambda) = \hbar \nabla_{z_i}^{(t)}(z; \lambda) + \eta \nabla_{z_i}^{(r)}(z; \lambda). \tag{27}$$

The rational (trigonometric or rational-trigonometric) KZ equations is the system of the differential equations

$$\nabla_{z_i}^{(\alpha)}(z;\lambda)u(z;\lambda) = 0 , \qquad i = 1,\ldots,N; \quad \alpha = (r)\,((t)\text{ or }(rt)) \tag{28}$$

for a function $u(z;\lambda) := u(z_1,\ldots,z_N;\lambda_1,\cdots,\lambda_M)$ taking values in an N-fold tensor product of gl_M-modules.

We also consider the differential operators $D_{\lambda_1}^{(r)},\ldots,D_{\lambda_M}^{(r)}$ and $D_{\lambda_1}^{(t)},\ldots,D_{\lambda_M}^{(t)}$ with coefficient in $(U(gl_M))^{\otimes N}$ depending on complex variables $z_1,\ldots,z_N$ and $\lambda_1,\cdots,\lambda_M$ [4]:

$$D_{\lambda_a}^{(r)}(z;\lambda) = \kappa\frac{\partial}{\partial\lambda_a} - \sum_{i=1}^{N} z_i(e_{aa})_{(i)} - \sum_{\substack{b=1\\b\neq a}}^{M} \frac{e_{ab}e_{ba} - e_{aa}}{\lambda_a - \lambda_b} , \tag{29}$$

$$D_{\lambda_a}^{(t)}(z;\lambda) = \kappa\lambda_a\frac{\partial}{\partial\lambda_a} + \frac{a_{aa}^2}{2} - \sum_{i=1}^{N} z_i(e_{aa})_i -$$
$$- \sum_{b=1}^{M}\sum_{1\leq i<j\leq N}(e_{ab})_{(i)}(e_{ba})_{(j)} - \sum_{\substack{b=1\\b\neq a}}^{M}\frac{\lambda_b(e_{ab}e_{ba} - e_{aa})}{\lambda_a - \lambda_b} . \tag{30}$$

Remind that $e_{ab} = \sum_{i=1}^{M}(e_{ab})_{(i)}$. The operators $D_{z_1}^{(r)},\ldots D_{z_M}^{(r)}$ (resp. $D_{z_1}^{(t)},\ldots D_{z_M}^{(t)}$) are called the rational (resp. trigonometric) differential dynamic (DD) operators. We set

$$D_{\lambda_a}^{(rt)}(z;\lambda) := \hbar D_{\lambda_a+\frac{\eta}{\hbar}}^{(t)}((1+\tfrac{\eta}{\hbar})z;\lambda+\tfrac{\eta}{\hbar}) . \tag{31}$$

It is easy to see that

$$D_{z_i}^{(rt)}(z;\lambda) = \hbar D_{\lambda_a}^{(t)}(z;\lambda) + \eta D_{\lambda_a}^{(r)}(z;\lambda) . \tag{32}$$

The rational (trigonometric or rational-trigonometric) DD equations is the system of the differential equations

$$D_{\lambda_a}^{(\alpha)}(z;\lambda)u(z;\lambda) = 0 , \qquad a = 1,\ldots,M; \quad \alpha = (r)\,((t)\text{ or }(rt)) \tag{33}$$

for a function $u(z;\lambda) := u(z_1,\ldots,z_N;\lambda_1,\cdots,\lambda_M)$ taking values in an N-fold tensor product of gl_M-modules. From (27), (32) and Theorem 5.8 of the paper [4] for any $i = 1,\ldots,N$ and $a = 1,\ldots,M$ we have the duality:

$$\nabla_{z_i}^{(rt)}(z;\lambda)_M \simeq D_{z_i}^{(rt)}(\lambda;z)_N , \qquad D_{\lambda_a}^{(rt)}(z;\lambda)_M \simeq \nabla_{\lambda_a}^{(rt)}(\lambda;z)_N , \tag{34}$$

where the index M (N) is connected with the Lie algebra $gl_M(gl_N)$.

Acknowledgment. This work was supported by Russian Foundation for Fundamental Research, grant No. RFBR-02-01-00668, and CRDF RMI-2334-MO-02.

References

[1] Drinfeld V G 1987 Quantum groups Amer. Math. Soc. Providence 798-820.
[2] Khoroshkin S M and Tolstoy V N 1996 Lett. Math. Phys. 36 373–402
[3] Spiridonov V 2001 Uspehi Math. Nauk
[4] Tarasov V and Varchenko A 2001 arXiv:math.QA/0112005
[5] Tarasov V and Varchenko A 2001 Int. Math. Res. Notices 15 801-829 (arXiv:math.QA/0002132)
[6] Tolstoy V N 1997 Connection between Yangians and Quantum Affine Algebras Proc. the X-th Max Born Symposium Wroclaw 1996 PWN - Polish Sci. Publishers Warszawa 99-117
[7] Tolstoy V N (2001) Rational-trigonometric deformations Proc. Karpacz Winter School of Theor. Physics Poland 2001 AIP Conf.Proc. Vol 589 American Institute of Physics Melville New York 296–306

SATELLITE COLLOQUIUM

Coherent States, Wavelets and Signal Processing

COHERENT STATES, WAVELETS AND APPLICATIONS
G24 – LLN

A Satellite Workshop held at UCL, Louvain-la-Neuve (Belgium)
July 10–12, 2002

Although coherent states and wavelets have developed by now into full-fledged subdisciplines, with plenty of applications in physics, mathematics and engineering, both have deep roots in group theory,. The construction of coherent states was identified as a group-theoretical problem in the pioneering work of Perelomov and Gilmore, whereas wavelets are simply the coherent states associated to the affine group of the line (or the similitude group of the plane for the two-dimensional case). It was therefore appropriate to have both topics represented in the GROUP-24 Colloquium.

However, their importance warrants a special treatment, thus more speaking time than available in such a wide ranging and crowded conference. It was thus resolved to hold a dedicated satellite workshop the preceding week in Louvain-la-Neuve, Belgium. The meeting was attended by some 32 participants, most of them invited, and 21 talks were presented. The present volume contains, in a specific section at the end, most of these contributions, plus two extra ones, by M A Man'ko and H Ishi, actually given in Paris, but more appropriate in the context of the LLN workshop (M A Man'ko was in fact supposed to come to LLN, but could not make it). For the sake of clarity, the contributions have been divided into two sections. The first one contains talks of a more mathematical nature, devoted to various aspects of coherent states, quantization, frames or phase space methods. The second subset is devoted to wavelets proper, and essentially to their applications to fields as diverse as quantum field theory, analysis of DNA sequences, solar physics, texture analysis, image retrieval or statistical analysis. This diversity attests of the vitality of the wavelet community, and also fully justifies the idea of a separate meeting.

The LLN workshop was supported by the Physics Department and the Faculty of Science of the Université catholique de Louvain, and by the Fonds National de la Recherche Scientifique de Belgique (FNRS), under the activities of the Contact Group "Wavelets and Applications". We thank them all warmly.

The local Organizing Committee

Jean-Pierre Antoine (UCL, Louvain-la-Neuve), chairman
Françoise Bastin (ULg, Liège)
Christine De Mol (ULB, Brussels)
Dominique Lambert (FUNDP, Namur)

Part A

Coherent States and Phase Space Methods

Inst. Phys. Conf. Ser. No 173: Satellite colloquium
Paper presented at 24th Int. Coll. Group Theoretical Methods in Physics, Paris, France, 15–20 July 2002
©2003 IOP Publishing Ltd

Symplectic geometry of the Wigner transform on noncompact symmetric spaces

S Twareque Ali† and Marco Bertola‡

Department of Mathematics and Statistics, Concordia University,
7141 Sherbrooke St. West, Montréal, Québec, Canada H4B 1R6

Abstract. We introduce a notion of Wigner transform on the symmetric spaces $\mathcal{X} = SO_0(1, n)/SO(n)$ which satisfies the usual marginality and covariance properties. Recalling the notion of the Helgason dual of $\mathcal{X}$ and denoting it by Ξ we show that there exists a natural and canonical $SO_0(1, n)$-invariant symplectic structure and briefly describe the corresponding geometric quantization, thus showing that the Wigner transform maps states in $L^2(\mathcal{X})$ to functions on the phase space $\mathcal{X} \times \Xi$, yielding an intermediate position–momentum representation of the quantum mechanics on $\mathcal{X}$.

1. Introduction

The purpose of this short note is to introduce an integral transform that generalizes the Wigner transform used in Quantum Mechanics [6] on the standard phase space $T^*\mathbb{R}^n$. In some earlier work, a generalized Wigner function was constructed on coadjoint orbits of a locally compact group, using the Plancherel theorem for non-unimodular groups [1, 2]. The present approach is a generalization of a similar transform introduced in [5], where it was worked out for the simplest of noncompact symmetric spaces, the unit disk with the hyperbolic metric. Reasons of space do not allow us to treat the subject in full generality and hence we will restrict ourselves to the important class of rank-one, reduced, noncompact Riemannian symmetric spaces, that is, the n-dimensional hyperbolic spaces $\mathcal{X}$. A general treatment is possible for more general noncompact symmetric spaces and will be the topic of a forthcoming publication.

The rest of this paper is organized as follows: in Section 2 we set up the notation and recall the basic properties of the geometry of the symmetric space $\mathcal{X}$. In Section 3 we introduce the definition of the Wigner transform, drawing on similarities with the standard definition in [6]. In Section 4 we show how to define an invariant symplectic structure on $\mathcal{X} \times \Xi$ and in Section 5 we briefly indicate the corresponding geometric quantization. We conclude in Section 6 with a short summary and outlook on open problems and generalizations.

2. Preliminaries on the geometry of $\mathcal{X}$

In this section we set up the notation and recall some standard facts about the n-dimensional hyperboloid. We consider the n-dimensional manifold $\mathcal{X}$

$$\mathcal{X} := \left\{ X \in \mathbb{R}^{n+1}, \ X \cdot X := (X^0)^2 - \sum_{j=1}^{n} (X^i)^2 = 1, \ X^0 > 0 \right\}, \qquad (1)$$

† e-mail: stali@mathstat.concordia.ca
‡ e-mail: bertola@mathstat.concordia.ca

realized as an embedded submanifold in the flat Minkowski space $\mathbb{R}^{n+1}$ with standard metric $ds^2 = -dX^{0^2} + \sum_{j=1}^{n} dX^{j^2}$, where X^i are the canonical coordinates of $\mathbb{R}^{n+1}$. By abuse of notation we will denote with a lowercase x both a point of the manifold $\mathcal{X}$ and its coordinates in the ambient space of embedding, which allows us to write sums and differences of points of $\mathcal{X}$ in the sense of the sum or difference of the corresponding points in the ambient space. The manifold $\mathcal{X}$ is the "upper" unit hyperboloid in $\mathbb{R}^{n+1}$ and it inherits a Riemannian metric from the embedding in $\mathbb{R}^{n+1}$. The inherited metric has constant sectional curvature -1. Since the orthochronous Lorentz group in $n+1$ dimensions, $SO_0(1,n)$, acts transitively on $\mathcal{X}$ with isotropy subgroup isomorphic to $SO(n)$, the manifold can also be thought of as the coset space

$$\mathcal{X} \simeq SO_0(1,n)/SO(n) \,. \tag{2}$$

The manifold is not only a homogeneous space but also isotropic and in the mathematical literature this is called a (maximally) symmetric space. An equivalent definition is that for each point $x \in \mathcal{X}$ one can define an involutive isometry $\vartheta_x : \mathcal{X} \mapsto \mathcal{X}$, that is, a nontrivial isometry that squares to the identity and that acts as the geodesic reflection around x. In order to be completely explicit, we will spell out the formulae that realize these reflections.

First of all, we recall that the geodesic distance between two points $x, y \in \mathcal{X}$ is given by the formula

$$\cosh(d(x,y)) = x \cdot y \,, \qquad x \cdot x = y \cdot y = 1. \tag{3}$$

We define the *mid-point map* as the map $m : \mathcal{X} \times \mathcal{X} \to \mathcal{X}$ which associates to any pair of points $(x,y) \in \mathcal{X} \times \mathcal{X}$ the mid-point $m(x,y)$ along the geodesic joining them. It is straightforward to check that m is given by the $SO_0(1,n)$-equivariant map

$$m(x,y) := \frac{x+y}{\sqrt{2(1 + x \cdot y)}} \,, \quad x, y \in \mathbb{R}^{n+1}, \ x \cdot x = y \cdot y = 1. \tag{4}$$

Indeed $m(x,y) \cdot m(x,y) = 1$ and $m(x,y) \cdot y = m(x,y) \cdot x$, so that the geodesic distance of $m(x,y)$ from x is the same as its distance from y. The fact that it lies on the geodesic γ connecting x and y follows from the fact that γ is the intersection of $\mathcal{X}$ with the linear hyperplane containing both points in the ambient Minkowskian space. Then $m(x,y)$ also belongs to the same hyperplane and hence lies on γ.

The reflection around x is the map $\vartheta_x : \mathcal{X} \to \mathcal{X}$, which takes a point y and maps it to its opposite along the geodesic joining y to x. Explicitly,

$$\vartheta_x(y) = 2(x \cdot y)x - y \,. \tag{5}$$

The map $\vartheta_x(\cdot)$, for fixed x, is the involutive isometry around x which enters the definition of symmetric space. We clearly have the relation

$$m(\vartheta_x(y), y) = x \,, \quad m(x,y) = m(\vartheta_{m(x,y)}(x), \vartheta_{m(x,y)}(y)) = m(y,x) \,. \tag{6}$$

For fixed y, the map $\vartheta_\bullet(y) : \mathcal{X} \to \mathcal{X}$ is not an isometry: for later purposes, we compute the Jacobian of this diffeomorphism of $\mathcal{X}$.

Lemma 2.1 *We have the formula (valid for any integrable function φ)*

$$\int_\mathcal{X} \varphi(m(x,y))dx = \int_\mathcal{X} \varphi(m)J(m;y)dm \,, \tag{7}$$

where $J(x;y)$ is the inverse of the Jacobian of the mid-point map,

$$J(m;y) = (2(1 + x(m,y) \cdot y))^{\frac{n}{2}} = (2m \cdot y)^n \tag{8}$$

Proof Writing in ambient coordinates, we have the extended map

$$m := \mathbb{R}^{n+1} \times \mathbb{R}^{n+1} \to \mathcal{X} \hookrightarrow \mathbb{R}^{n+1} , \tag{9}$$

$$m(x, y) := \frac{x + y}{\sqrt{(x + y) \cdot (x + y)}} , \tag{10}$$

whose restriction to $\mathcal{X} \times \mathcal{X}$ coincides with the mid-point map. Differentiating,

$$J^\mu_\nu := \frac{\partial m^\mu}{\partial x^\nu} = \frac{1}{\sqrt{(x + y) \cdot (x + y)}} (\delta^\mu_\nu - F^\mu F_\nu) \tag{11}$$

$$F^\mu := \frac{x^\mu + y^\mu}{\sqrt{(x + y)^2}} . \tag{12}$$

Since J^μ_ν vanishes on the normal bundle to $\mathcal{X} \hookrightarrow \mathbb{R}^{n+1}$, its determinant (as a map from $T_x\mathcal{X}$ to $T_m\mathcal{X}$) is easily computed to be $[(x + y)^2]^{\frac{n}{2}}$. Expressing it in terms of m when $x, y \in \mathcal{X}$ (i.e. $x \cdot x = y \cdot y = 1$) we find

$$[(x + y)^2]^{\frac{n}{2}} \Big|_{x^2 = y^2 = 1} = 2\, m \cdot y \tag{13}$$

$$\det \left[\frac{\partial m^\mu}{\partial x^\nu} \right] = (2\, m \cdot y)^n . \qquad \square \tag{14}$$

2.1. Fourier–Helgason transform on $\mathcal{X}$

The main goal of this section is to recall the definition and properties of the Fourier–Helgason transform on $\mathcal{X}$; this integral transform is the precise analog of the Fourier-Plancherel transform on $\mathbb{R}^N$. It consists of an isometry between two Hilbert spaces

$$\mathcal{H} : L^2(\mathcal{X}, \mathrm{d}\mu) \longrightarrow L^2(\mathcal{L}, \mathrm{d}\rho) , \tag{15}$$

where the measure $\mathrm{d}\mu$ is the $SO_0(1, n)$ invariant measure on $\mathcal{X}$ and $L^2(\mathcal{L}, \mathrm{d}\rho)$ denotes the Hilbert space of sections of a line–bundle $\mathcal{L}$ (specified later) over another suitably defined manifold, which we call the Helgason-dual of $\mathcal{X}$ and denote by Ξ. As a set Ξ can be identified with the set of "horocycles" with oriented distance, but we will not need any of this generality for the purposes of this letter (see [3, 4] for the general abstract theory). Suffice it to say here that the Helgason-dual of $\mathbb{R}^N$ is just the dual space $\mathbb{R}^{N^\vee}$ and the transform would then coincide with the ordinary Fourier transform. It is useful to keep in mind this parallel when comparing the formulae.

Before introducing the Helgason transform we want to give a concrete realization of the dual space Ξ; in fact Ξ can be realized as the asymptotic projective null-cone times the positive real line

$$V_+ := \left\{ \xi \in \mathbb{R}^{n+1^\vee} : \xi \cdot \xi := -(\xi_0)^2 + \sum_{j=1}^n (\xi_j)^2 = 0 , \ \xi_0 > 0 \right\} \tag{16}$$

$$\Xi := \mathbb{R}_+ \times \mathbb{P}V_+ \tag{17}$$

$$p = (\nu, [\xi]) \in \Xi, \tag{18}$$

where $\mathbb{P}V_+$ denotes the projectivized forward cone (the set of "rays" on the cone); a convenient realization of $\mathbb{P}V_+$ makes it diffeomorphic to the $(n - 1)$-dimensional sphere as follows

$$\mathbb{P}V_+ \simeq \{ \vec{\xi} \in \mathbb{R}^n : \|\vec{\xi}\| = 1 \}$$

$$[\xi_0 : \xi_1 : \ldots : \xi_n] \mapsto \frac{1}{\xi_0} \vec{\xi} . \tag{19}$$

The Fourier-Helgason transform (FH for brevity), is defined in a similar way as the ordinary Fourier transform using the eigenfunctions of the invariant differential operator of second order, i.e., the Laplacian. In the present case the eigenfunctions of the (unique) invariant differential operator (the Laplacian) are

$$E_{\nu,\xi}(x) := (\xi \cdot x)^{-\frac{n-1}{2}-i\nu} , \qquad (20)$$

$$\xi \in \Xi^{\vee}, \ V_+ := \left\{ \xi \in \mathbb{R}^{n+1^{\vee}}, \,, \xi \cdot \xi = 0, \ \xi_0 > 0 \right\} . \qquad (21)$$

The first remark is that they are not defined on $\mathbb{R}_+ \times \mathbb{P}V_+$ but rather on $\mathbb{R}_+ \times V_+$; however the action of $\mathbb{R}_+^{\times}$ on V_+ just rescales these eigenfunctions by a factor which is constant in x. In other words, they are sections of an appropriate line bundle over Ξ which we denote by $\mathcal{L}$ and V_+ is thought of as the total space of the tautological $Gl_0(1; \mathbb{R})$–bundle over $\mathbb{P}V_+$. Secondly we note that the inner product $\xi \cdot x$ is positive on the product space $V_+ \times \mathcal{X}$, so that the complex exponential is uniquely defined.

Remark 2.2 *One can prove directly that they are eigenfunctions of the the Laplacian as follows. Write the d'Alembertian in $\mathbb{R}^{n+1}$ in the coordinates $(r, x) \in \mathbb{R}_+ \times \mathcal{X}$ which are valid within the forward light cone. If $X^0, ..., X^n$ are the flat coordinates, then we have $X = r\, x$ (a sort of "polar" decomposition) and then*

$$\Box = -\frac{\partial^2}{\partial z_0^2} + \sum_{j=1}^{n} \frac{\partial^2}{\partial z_j^2} = -\frac{\partial^2}{\partial r^2} - \frac{n}{r}\frac{\partial}{\partial r} + \frac{1}{r^2}\triangle_X . \qquad (22)$$

Using the homogeneity of the functions $E_{\xi,\nu}$ we obtain

$$0 = \Box(\xi \cdot z)^s = (\xi \cdot x)\left(-\frac{\partial^2}{\partial r^2} - \frac{n}{r}\frac{\partial}{\partial r} \right) r^s + r^{s-2}\triangle_X(\xi \cdot x)^s \qquad (23)$$

$$= ((-s(s-1) - ns) + \triangle_X(\xi \cdot x)^s)\, r^{s-2} \qquad (24)$$

Therefore the functions $(\xi \cdot x)^s$ are eigenfunctions of the Laplacian on $\mathcal{X}$ with eigenvalue $s(s-1) + ns = s(s+n-1)$. Inserting the value $s = -\frac{n-1}{2} + i\nu$ gives

$$\triangle_X E_{\xi,\nu}(x) = -\left(\frac{(n-1)^2}{4} + \nu^2 \right) E_{\xi,\nu}(x) . \qquad (25)$$

The FH transform $\mathcal{H}$ and its inverse $\mathcal{H}^{-1}$ are defined as

$$(\mathcal{H}F)(\nu, \xi) := \int_X (x \cdot \xi)^{\frac{1-n}{2}+i\nu} F(x)\, d\mu(x) , \qquad \forall F \in C_0^{\infty}(X), \qquad (26)$$

$$(\mathcal{H}^{-1}G)(x) := \int_{\imath\Xi} (x \cdot \xi)^{\frac{1-n}{2}-i\nu} G(\nu, \xi)\, d\rho(\nu, \xi) , \quad \forall G \in C_0^{\infty}(\mathcal{L}) \qquad (27)$$

where $C_0^{\infty}(\mathcal{L})$ denotes the space of compactly supported smooth sections of the line–bundle $\mathcal{L}$. The integration in (27) is performed along any smooth embedding $\imath\Xi$ into the total space of the (tautological) line-bundle $\mathcal{L}$ and the measure $d\rho$ is given by

$$d\rho(\nu, \xi) = \frac{d\nu}{|\mathbf{c}(\nu)|^2}\, d\sigma_0 . \qquad (28)$$

In this formula the $(n-1)$-form $d\sigma_0$ is defined on the null cone V_+ by

$$d\sigma_0 = \sum_{j=1}^{n}(-)^j \frac{\xi_j}{\xi_0} d\xi_1 \wedge \cdots \widehat{d\xi_j} \wedge \cdots \wedge d\xi_n , \qquad \xi_0 = \sqrt{\sum \xi_i^2} . \qquad (29)$$

This $(n-1)$–form is closed on V_+ and hence the integration is independent of the particular embedding of Ξ; moreover it scales with a power $n-1$ under dilations on the cone and hence the integral (27) is also scale-invariant and hence well defined for sections G of the line-bundle $\mathcal{L}$. For concreteness we fix such an embedding as follows

$$
\begin{aligned}
\imath : \Xi &\longrightarrow \mathbb{R}_+ \times V_+ \\
(\nu, [\xi_0 : \cdots : \xi_n]) &\longmapsto \left(\nu, \left(1, \tfrac{\xi_1}{\xi_0}, \ldots, \tfrac{\xi_n}{\xi_0}\right)\right)
\end{aligned} \tag{30}
$$

Then the form $\mathrm{d}\sigma_0$ restricted to this embedded sphere becomes just the ordinary volume form on the sphere. The measure $\mathrm{d}\rho$ involves also the *Harish–Chandra* c-function

$$
\mathbf{c}(\lambda) := \frac{1}{4} \frac{2^{n+i\lambda}\Gamma\left(\frac{n}{2}\right)\Gamma(i\lambda)}{\sqrt{\pi}\Gamma\left(\frac{n-1}{2} + i\lambda\right)} . \tag{31}
$$

Its definition in general and its properties can be found in [3, 4]. The factor $|\mathbf{c}(\lambda)|^2$ entering the inversion formula can be simplified to

$$
|\mathbf{c}(\lambda)|^{-2} = \frac{\lambda \sinh(\pi\lambda)\left|\Gamma\left(\frac{n-1}{2} + i\lambda\right)\right|^2}{4^{n-2}\left(\Gamma\left(\frac{n}{2}\right)\right)^2} \tag{32}
$$

Note that the transform $\mathcal{H}$ maps functions on $\mathcal{X}$ to sections of $\mathcal{L}$ and the "inverse"§ transform $\mathcal{H}^{-1}$ takes sections to functions. We then have

Proposition 2.3 (Theorem Ch. III in [4]) *The Fourier–Helgason transform defined in Eqs. (26,27) extends to an isometry of $L^2(\mathcal{X}, \mathrm{d}\mu)$ onto $L^2(\mathcal{L}, \mathrm{d}\rho)$ so that we have*

$$
\int_{\mathcal{X}} |F(x)|^2 \, \mathrm{d}\mu(x) = \int_{\imath\Xi} \left|\widehat{F}(\xi, \nu)\right|^2 \mathrm{d}\rho(\xi, \nu) \tag{33}
$$

3. Wigner Transform

We first recall that the usual Wigner transform is given by

$$
\mathcal{W}_K(x, \xi) := (2\pi)^{-n} \int_{\mathbb{R}^n} \mathrm{d}^n\eta \, e^{-i\eta\cdot\xi} K\left(x + \frac{1}{2}\eta, x - \frac{1}{2}\eta\right) . \tag{34}
$$

In order to justify the definition on the symmetric space $\mathcal{X}$ we perform the change of variables

$$
\eta' := x + \frac{1}{2}\eta \;\Rightarrow\; \eta = 2(\eta' - x) , \tag{35}
$$

so that the integral (34) becomes

$$
\mathcal{W}_K(x, \xi) = (\pi)^{-n} \int_{\mathbb{R}^n} \mathrm{d}^n\eta' \, e^{-i2(\eta'-x)\cdot\xi} K\left(\eta', 2x - \eta'\right) = \tag{36}
$$

$$
= (\pi)^{-n} \int_{\mathbb{R}^n} \mathrm{d}^n\eta \, e^{-i\eta\cdot\xi} e^{i(2x-\eta)\cdot\xi} K\left(\eta', 2x - \eta'\right) . \tag{37}
$$

Note that in the vector space $\mathbb{R}^n$ the map

$$
\vartheta_x(\eta) := 2x - \eta , \tag{38}
$$

is the reflection of η around the center x; this makes sense in any symmetric space. The basic idea is to replace the flat eigenfunctions, $e^{i\eta\cdot x}$, of the Laplacian by the corresponding eigenfunctions of the Laplacian on $\mathcal{X}$, i.e., $E_p(x)$ defined in eq. (21).

§ The name "inverse" is still unjustified since so far it is defined only for compactly supported sections. The justification for the symbol comes in the theorem below.

Definition 3.1 *For a smooth, compactly supported function* $K : X \times X \mapsto \mathbb{C}$ *we define its Wigner Transform*

$$\mathcal{W}_K(x,p) := (x \cdot \xi)^{n-1} \int_{\mathcal{X}} (x' \cdot \xi)^{\frac{1-n}{2}+i\nu} (\vartheta_x(x') \cdot \xi)^{\frac{1-n}{2}-i\nu} K(\vartheta_x(x'), x') J(x,x') d\mu(x')$$

$$= (x \cdot \xi)^{n-1} \int_{\mathcal{X}} (\vartheta_x(x') \cdot \xi)^{\frac{1-n}{2}+i\nu} (x' \cdot \xi)^{\frac{1-n}{2}-i\nu} K(x', \vartheta_x(x')) J(x,x') d\mu(x') , \quad (39)$$

$$J(x,x') := \det \partial_x \vartheta_x(x')^{-1} = (2x \cdot x')^n = (2 \cosh(d(x,x')))^n , \quad p := (\nu, [\xi]). \quad (40)$$

That the two expressions are the same is due to the fact that $\vartheta_x : \mathcal{X} \to \mathcal{X}$ *is an isometry.*

Note that the function W_K is defined on $\mathcal{X} \times \Xi$ since the expression on the RHS is invariant under rescaling of ξ.

It follows by simple manipulation of the integrals that this Wigner function fulfills the two *marginality* conditions. This means that if we chose a factorized $K(x,x') = \psi(x)\overline{\varphi}(x')$, we have

Proposition 3.2 (Marginality) *For* $K(x,x') = \psi(x)\overline{\varphi}(x')$ *we have*

$$\int_{\mathcal{X}} \mathcal{W}_{\psi \otimes \varphi^*}(x, (\xi,\nu))(x \cdot \xi)^{1-n} d\mu(x) = \widehat{\psi}(\xi,\nu)\widehat{\varphi}^*(\xi,\nu) \quad (41)$$

$$\int_{\Xi} \mathcal{W}_{\psi \otimes \varphi^*}(x, (\xi,\nu))(x \cdot \xi)^{1-n} d\rho(\xi,\nu) = \psi(x)\varphi(x)^* \quad (42)$$

The factor $(x \cdot \xi)^{n-1}$ in the definition of the transform is needed for invariance properties as the next proposition asserts and restores a section of $\mathcal{L} \otimes \overline{\mathcal{L}}$. Moreover, the reader is invited to check that the Wigner function is $SO_0(1,n)$ covariant.

Proposition 3.3 (Invariance) *For any* $g \in SO_0(1,n)$ *we have*

$$W_{gK}(x,p) = gW_K(x,p) =: W_K(gz, gp) \quad (43)$$

where, by definition

$$gp := (\nu, [\xi \cdot g^{-1}]) . \quad (44)$$

4. Symplectic Structure of $\mathcal{X} \times \Xi$

The identification of $\mathcal{X} \times \Xi$ with the cotangent bundle $T^*\mathcal{X}$ is tempting, but not obvious; following the parallel between the functions $E_p(x)$ and $\mathrm{e}^{i\eta \cdot x}$ one may be tempted to consider $p = (\nu, [\xi])$ as the canonical momentum by identifying $[\xi] \in PV_+$ with the direction of the wave-vector and ν with its magnitude. While it is clear that they play similar rôles, it is however obvious that the analogy cannot be that strict. It is also clear that whatever identification we are trying to achieve, it should be aimed at finding a suitable symplectic structure on $\mathcal{X} \times \Xi$, which would then allow a closer parallel between the standard Wigner transform and the present one. In fact, a key point in the formulation of the Wigner transform on $\mathbb{R}^N$ is that it maps states to functions on the phase space T^*R^N; whence the urge to endow $\mathcal{X} \times \Xi$ with a $SO_0(1,n)$–invariant symplectic structure.

We start with the trivial remark that, in the flat case, the symplectic form $\omega = \mathrm{d}p_i \wedge \mathrm{d}x^i$ is the differential of $\Theta = p_i \mathrm{d}x^i$ and it can be "invariantly" obtained by considering the plane waves $\mathrm{e}^{ip \cdot x}$ as follows

$$\Theta = \Im \left(\mathrm{d}_x \ln \left(\mathrm{e}^{ip \cdot x} \right) \right) = p \cdot \mathrm{d}x . \quad (45)$$

The idea is to replace the eigenfunctions of the flat Laplacian with the corresponding eigenfunctions of the hyperbolic Laplacian.

Definition 4.1 (Symplectic form on $\mathcal{X} \times \Xi$) *The following form defines a symplectic struc-*
ture on $\mathcal{X} \times \Xi$

$$\Theta := \Im\left(\mathrm{d}_x E_{\xi,\nu}(x)\right) = \Im\left(\mathrm{d}_x \ln\left(\left(\xi \cdot x\right)^{\frac{1-n}{2}+i\nu}\right)\right) = \frac{\nu}{\xi \cdot x}\,\xi \cdot \mathrm{d}x \qquad (46)$$

$$\omega := \mathrm{d}\Theta \qquad (47)$$

Notice that Θ and ω are invariant under rescaling of ξ and hence defined on $\mathcal{X} \times \Xi$.

By way of this definition of the symplectic potential and corresponding symplectic form we obtain a natural identification of $\mathcal{X} \times \Xi \simeq T^*\mathcal{X}$. More precisely we have the following Lemma.

Lemma 4.2 *The blow-up of the cotangent bundle $T^*\mathcal{X}$ along the zero section is identified with $\mathcal{X} \times \Xi$ via the map*

$$\begin{aligned}
\tau: \quad \mathcal{X} \times \Xi \quad &\longmapsto T^*\mathcal{X} \\
(x, (\nu, [\xi])) &\longmapsto \tau_x(\nu, [\xi]) = \left[\frac{\nu}{\xi \cdot x}\,\xi \cdot \mathrm{d}x, x\right] \in T_x^*\mathcal{X} \times \{x\} \subset T^*\mathcal{X}
\end{aligned} \qquad (48)$$

Proof We need to check the surjectivity of $\tau_x(\nu, [\xi])$ onto the cotangent space to $\mathcal{X}$ at the point x, whereas the injectivity is obvious.

Since $x \cdot x = 1$, we have $x \cdot \mathrm{d}x = 0$ and hence

$$\mathrm{d}x^0 = \frac{x_i \mathrm{d}x^i}{x_0}. \qquad (49)$$

Since the map τ_x is really defined on $\mathbb{P}V_+$, we can assume that $\xi_0 = 1$ in the formulæ to come. Therefore, using the (global) coordinates $x^1, \ldots, x^n$,

$$\frac{\nu}{\xi \cdot x}\,\xi \cdot \mathrm{d}x = \frac{\nu}{\xi \cdot x}\left(\mathrm{d}x^0 + \xi_i \mathrm{d}x^i\right) = \frac{\nu}{\xi \cdot x}\left(\xi_i + \frac{x_i}{x_0}\right)\mathrm{d}x^i. \qquad (50)$$

This map is clearly surjective, since $\vec{\xi}^2 = 1$ and $\left|\frac{\vec{x}}{x^0}\right|^2 < 1$. Indeed, solving for $\vec{\xi}$ and $\nu \in \mathbb{R}_+$ for any $\eta \in T_x^*\mathcal{X} \setminus \{0\}$, $\eta = \eta_i \mathrm{d}x^i$,

$$\frac{\nu}{\xi \cdot x}\left(\vec{\xi} + \frac{\vec{x}}{x_0}\right) = \vec{\eta} \qquad (51)$$

$$S^n \ni \vec{\xi} = \frac{1}{x^0 |\vec{\eta}|^2}\left(\vec{\eta} \cdot \vec{x} + \sqrt{|\vec{\eta}|^2 + (\vec{\eta} \cdot \vec{x})^2}\right)\vec{\eta} - \frac{1}{x^0}\vec{x} \qquad (52)$$

$$\nu = \vec{\eta} \cdot \vec{x} + \frac{|\vec{\eta}|^2}{\vec{\eta} \cdot \vec{x} + \sqrt{|\vec{\eta}|^2 + (\vec{\eta} \cdot \vec{x})^2}}. \qquad (53)$$

The inverse image of $0 \in T_x^*\mathcal{X}$ is clearly $\{0\} \times \mathbb{P}V_+ \simeq \{0\} \times S^n$, which constitutes the blow-up of the zero section of the cotangent bundle. $\qquad \square$

Note that the one form Θ, thought of as a one-form on the cotangent bundle, is clearly $SO_0(1, n)$ invariant and so is the symplectic form ω. We have proven

Theorem 4.3 *The pair $(\mathcal{X} \times \Xi, \omega)$ is naturally an $SO_0(1, n)$ symplectic manifold.*

5. Geometric Quantization

The symplectic manifold $(\mathcal{M}, \omega) := (\mathcal{X} \times \Xi, \omega)$ can be geometrically quantized. The quantization is here rather simple since $\mathcal{M}$ is essentially the (blow-up of the) cotangent bundle of $\mathcal{X}$. Since the symplectic form is defined through the symplectic potential

$$\omega := \mathrm{d}\Theta, \qquad \Theta := \frac{\nu}{\xi \cdot x}\,\xi \cdot \mathrm{d}x, \qquad (54)$$

Θ gives directly the connection of the prequantum bundle:

$$\nabla := d + i\Theta .\tag{55}$$

Notice that the integrality condition is empty for $\mathcal{X} = SO_0(n,1)/SO(n)$ since this space is retractable. The symplectic form reads

$$\omega := \frac{d\nu \wedge \xi \cdot dx}{\xi \cdot x} + \frac{\nu}{\xi \cdot x}d\xi_\mu \wedge dx^\mu - \frac{\nu}{(\xi \cdot x)^2}x^\rho \xi_\mu d\xi_\rho \wedge dx^\mu .\tag{56}$$

It is then clear that there are two obvious polarizations: one spanned by the ∂_{x^μ} and the other spanned by the tangent vectors to Ξ. Choosing the second gives the usual $L^2(\mathcal{X}, d\mu)$ for the Hilbert space of quantum states. The other polarization gives $L^2(\imath\Xi, d\rho)$ up to a gauge. Indeed choosing this latter polarization implies that the flat sections are

$$F(x; ([\xi], \nu)) := (\xi \cdot x)^{i\nu} f([\xi], \nu)\tag{57}$$

Since the Hermitian metric of the line bundle will kill the $U(1)$ factor $(\xi \cdot x)^{i\nu}$, the section F is equivalent to the function $f([\xi], \nu)$ which belongs to $L^2(\imath\Xi)$.

Equivalently, we can change the connection by subtracting the exact form

$$d \ln [(\xi \cdot x)^\nu] ,\tag{58}$$

(where the differential is taken w.r.t. both x and ξ, ν), and obtain a connection $\widetilde{\Theta}$ which has the same curvature (symplectic form). Then, the choice of the second polarization above gives directly $L^2(\imath\Xi)$.

6. Conclusion and outlook

In this contribution we have defined a Wigner transform that retains many of the properties of the standard Wigner transform on $\mathbb{R}^N$. We started from an interpretation of the latter in terms of the natural structure of a symmetric space of $\mathbb{R}^N$ and then extended these to the symmetric space $\mathcal{X} = SO_0(1,n)/SO(n)$. We mentioned earlier that this definition can be carried over to more general symmetric spaces and that this will be the subject of a future publication. The proposed definition is a generalization of the definition in [5] which is limited to the Poincaré disk. Moreover, we have constructed a natural invariant symplectic structure on $\mathcal{X} \times \Xi$ and briefly described the geometric quantization of this space, which makes it clear that the space Ξ is to be fully regarded as the "momentum" representation of the quantum mechanics on $\mathcal{X}$.

Two main questions remain open to further research. The first is to find a viable formula for the $\star$-product of Wigner functions. The second is to investigate wether it is possible to introduce a measure on $\mathcal{X} \times \Xi$ which makes W_K an element of an L^2 space and sets up an isometry between the Hilbert–Schmidt operators on $L^2(\mathcal{X}) \simeq L^2(\Xi)$ and $L^2(\mathcal{X} \times \Xi)$. We leave these interesting and difficult points to a subsequent publication.

References

[1] Ali S T, Führ H, and Krasowska A, Plancherel inversion as unified approach to wavelet transforms and Wigner functions, Annales H. Poincaré, to appear

[2] Ali S T, Atakishiyev N M, Chumakov S M, and Wolf K B 2000, The Wigner function for general Lie groups and the wavelet transform, Annales H. Poincaré 1 685–714

[3] Helgason S 1978, Differential Geometry, Lie Groups, and Symmetric Spaces (New York: Academic)

[4] Helgason S 1994, Geometric Analysis on Symmetric Spaces, Math. Surveys and Monographs, Vol. 39 (Providence, RI: Amer. Math. Soc.)

[5] Tate T 2001, Weyl calculus and Wigner transform on the Poincaré disk, in Noncommutative Differential Geometry and its Applications to Physics (Shonan, 1999), pp. 227–243, Math. Phys. Stud. vol. 23 (Dordrecht: Kluwer)

[6] Wigner E 1932, On the quantum correction for thermodynamic equilibrium, Phys. Rev. 40 749–759

Inst. Phys. Conf. Ser. No 173: Satellite colloquium
Paper presented at 24th Int. Coll. Group Theoretical Methods in Physics, Paris, France, 15–20 July 2002
©*2003 IOP Publishing Ltd*

Wavelet transform and ⋆ exponential

Michel Cahen and Simone Gutt†

Département de Mathématique, Campus Plaine CP218
Université Libre de Bruxelles,
bd du Triomphe, 1050 Bruxelles

Abstract. In this contribution, we relate the continuous wavelet transform to the ⋆ exponential. This is done for compact groups and for semi simple groups admitting a holomorphic discrete series, using Berezin dequantization prodecure.

1. Introduction

The continuous wavelet transform [1, 2] is closely related to Plancherel's decomposition theorem; it induces an association between some operators acting on a representation space of a Lie group and some functions on the group.

In the context of deformation quantisation [3, 4], the notion of ⋆ representation and ⋆ exponential have been introduced. The ⋆ exponential associates to some functions on a Lie group some distributions on a homogeneous symplectic or Poisson manifold (typically an orbit of the group in the dual of its Lie algebra).

The symbolic calculus introduced by Berezin [2] can be interpreted as a dequantisation procedure; it associates to some operators on a representation space, realised in terms of sections of a bundle over a homogeneous symplectic manifold, some functions on that manifold [6].

In this paper, we relate those three maps in the case of a compact group and for the holomorphic discrete series representations of a semisimple group.

2. Compact Lie groups

Let G be a connected, simply connected, compact Lie group; let $\mathfrak{g}$ denote its Lie algebra; let T be a maximal torus of G and $\mathfrak{T}$ its Lie algebra. Let C be a Weyl chambre of $i\mathfrak{T}$.

Let (V, π) be a unitary irreducible representation of G. Then there exists a dominant (relative to C) integral weight λ of G such that (up to equivalence) (V, π) is the highest weight representation associated to λ [7]. The weight λ exponentiates to a character e^λ of T.

The flag manifold G/T has a G invariant complex structure induced by the choice of C. The complex line bundle

$$L = G \times_{e^\lambda} \mathbb{C}$$

over G/T is a holomorphic line bundle and the theorem of Borel Weil Bott tells us that the space of holomorphic sections of this bundle L is an irreducible, unitary G module isomorphic to (V, π) [8].

---

† e-mail : mcahen@ulb.ac.be, sgutt@ulb.ac.be

Let $x \in G/T$ and let q be a point of L projecting on x, which does not belong to the 0-section. If $s : G/T \to L$ is a holomorphic section the evaluation at x of s is continuous; i.e. the function l_q defined by

$$l_q(s)q = s(x)$$

depends continuously on s [6]. Denote by $H^0(L)$ the space of holomorphic sections of L, endowed with the unitary structure :

$$< s, s' > = \int_{G/T} s(x)\overline{s'(x)} \, d\mu(x)$$

where $d\mu(x)$ is a G invariant measure on G/T. By Riesz's theorem, there exists an element $e_q \in H^0(L)$ such that

$$l_q(s) = < s, e_q > .$$

The section e_q is called a coherent state [6]. The following properties of coherent states are classical.

(i) if $c \in \mathbb{C}^*$, $e_{cq} = \bar{c}^{-1} e_q$

(ii) if $g \in G$, $g \cdot e_q = e_{gq}$

(The natural action of G on L is given by $g \cdot [g_1, z] = [gg_1, z]$).
Properties (i,ii) imply that the function $L \to \mathbb{R} : q \to < e_q, e_q > |q|^2$ is the pull back of a function $\epsilon : G/T \to \mathbb{R}$ and that this function is a constant.

Berezin's symbol map is a map

$$\wedge : V^* \otimes V (= \text{End} V) \longrightarrow L^2(G/T) \quad A \to \hat{A}$$

$$\hat{A}(x) = \frac{< Ae_q, e_q >}{< e_q, e_q >}$$

where q belongs to the fibre of L above x but not to the zero section of L. This function can be uniquely extended to a neighborhood of the diagonal in $G/T \times G/T$:

$$\hat{A}(x, y) = \frac{< Ae_q, e_{q'} >}{< e_q, e_{q'} >}$$

where q' belongs to the fibre of L above y but not to the zero section. Furthermore $\hat{A}$ is holomorphic in x and antiholomorphic in y.

On the other hand, we have a natural injection

$$V^* \otimes V \xrightarrow{i} L^2(G)$$

$$i(\underline{u} \otimes v)(g) = < v, \pi(g)u > \qquad u, v \in V.$$

This map is called by some authors the wavelet transform [2]. Finally we have an adapted Fourier transform $\mathcal{E} : L^2(G) \to L^2(G/T)$

$$\mathcal{E}(f) = \int_G f(g)\widehat{\pi(g)} dg$$

where dg is a biinvariant Haar measure on G [4].

Our first remark is that the wavelet transform and the adapted Fourier transform are related. More precisely

Proposition 2.1 *Let G be a compact, connected, simply connected Lie group and let (V, π) be an irreducible unitary representation of G. Let $L \to G/T$ (T = a maximal torus of G) be the holomorphic line bundle such that $H^0(L)$ (= space of holomorphic sections of L) is isomoprhic (as G module) to (V, π). Then the diagram*

$$V^* \otimes V \quad \xrightarrow{\;i\;} \quad L^2(G)$$

$$\wedge \downarrow \qquad \diagup \mathcal{E}$$

$$L^2(G/T)$$

is commutative.

Proof. Let $u, v \in V$ and let $f_{u,v} = i(\underline{u} \otimes v)$. Then

$$(\mathcal{E} f_{u,v})(x) = \int_G < v, \pi(g)u > \frac{< e_q, \pi(g^{-1})e_q >}{< e_q, e_q >} \, dg$$

$$= \frac{1}{\epsilon} \int_G < v, \pi(g)u > < e_q, \pi(g^{-1})e_q > |q|^2 \, dg.$$

Observe that if

$$L_{u,v,u',v'} \quad = \int_G < u, \pi(g)v > < u', \pi(g^{-1})v' > dg,$$

$$L_{\pi(g_1)u,v,u',v'} = \int_G < \pi(g_1)u, \pi(g)v > < u', \pi(g^{-1})v' > dg$$

$$= \int_G < u, \pi(g_1^{-1}g)v > < u', \pi(g^{-1})v' > dg$$

$$= L_{u,v,u',\pi(g_1^{-1})v'}.$$

Similarly, $L_{u,\pi(g_1)v,u',v'} = L_{u,v,\pi(g_1^{-1})u',v'}$. Hence there exists a constant C such that

$$L_{u,v,u',v'} = C < u, v' > < u', v >$$

and one checks that $C = \dfrac{\text{vol } G}{\dim V}$.

    Thus :

$$(\mathcal{E} f_{u,v})(x) = \frac{C}{\epsilon} |q^2| < v, e_q > < e_q, u >$$

$$= \frac{C}{\epsilon} |q|^2 < e_q, e_q > \widehat{\underline{u} \otimes v}(x) = C \widehat{\underline{u} \otimes v}(x).$$

Hence the proposition is proven, provided one chooses the normalizations so that $C = 1$.

    Let us indicate a few properties of the adapted Fourier transform which prove the interest of proposition 1.

Kirillov's character formula. For any $f \in L^2(G)$,

$$Tr\pi(f) = \epsilon \int_{G/T} \mathcal{E}(f) \, d\mu.$$

if $\pi(f) = \int_G f(g)\pi(g)\,dg$.

Proof. Let s_i ($i \le N = \dim V$) be an orthonormal basis of $H^0(L)$. Then

$$Tr\pi(g) = \sum_i < \pi(g)s_i, s_i >$$

$$= \int_{G/T} \sum_i (\pi(g)s_i)(x)\overline{s_i(x)}\,d\mu(x)$$

$$= \int_{G/T} \sum_i < s_i, \pi(g^{-1})e_q > < e_q, s_i > |q^2|\,d\mu(x)$$

$$= \int_{G/T} < e_q, \pi(g^{-1})e_q > |q|^2\,d\mu(x)$$

$$= \int_{G/T} \epsilon\widehat{\pi(g)}(x)\,d\mu(x)$$

Hence

$$Tr\pi(f) = Tr \int_G f(g)\pi(g)\,dg = \int_G \int_{G/T} \epsilon f(g)\widehat{\pi(g)}(x)\,d\mu(x)\,dg$$

$$= \epsilon \int_{G/T} \mathcal{E}(f)\,d\mu.$$

Kernel of $\mathcal{E}$. Let ψ be a matrix coefficient of a unitary irreducible representation (W, ν) of G which is not equivalent to (V, π). Then :

$$\mathcal{E}(\psi) = 0$$

Proof. We may assume $\psi(g) = < b, \nu(g)a >$ (for $a, b \in W$). Then

$$\mathcal{E}(\psi)(x) = \int_G < b, \nu(g)a > \frac{< \pi(g)e_q, e_q >}{< e_q, e_q >}\,dg = 0$$

as we have orthogonality in $L^2(G)$.

Link with "$\star$" product. A non commutative, associative product (denoted $\star$) may be defined on $\wedge(V^\star \otimes V)$ by

$$\wedge(A \circ B) = \wedge(A) \star \wedge(B)$$

Then we have

Proposition 2.2 *Let A, B belong to $V^\star \otimes V$; let $\times$ denote the standard convolution of functions on G. Then :*

$$\mathcal{E}(i(A) \times i(B)) = \widehat{B \circ A}.$$

Proof. Choose $A = \underline{u} \otimes v$ and $B = \underline{u'} \otimes v'$. Then :

$$(i(A) \times i(B))(g) = \int_G i(A)(g_1^{-1}g)i(B)(g_1)\,dg_1$$

$$= \int_G < v, \pi(g_1^{-1}g)u > < v', \pi(g_1)u' > dg_1$$

$$= C < v, u' > < v', \pi(g)u > .$$

Hence :

$$\mathcal{E}(i(A) \times i(B))(x) = C <v, u'> \int_G <v', \pi(g)u> \frac{<e_q, \pi(g^{-1})e_q>}{<e_q, e_q>} dg$$

$$= \frac{C|q|^2}{\epsilon} <v, u'> <v', e_q> <e_q, u>$$

$$= C \widehat{B \circ A}(x)$$

which gives the proposition when adjusting the normalisation.

3. Non compact semisimple Lie groups

Let G be a non compact, connected, linear, semi simple Lie group and let $\mathfrak{g}$ be its Lie algebra; let θ be a Cartan involution of $\mathfrak{g}$ and let $\mathfrak{g} = \mathfrak{K} \oplus \mathfrak{P}$ be the corresponding decomposition of $\mathfrak{g}$. Assume $\mathfrak{K}$ admits a one dimensional center C and that the centralizer of C in $\mathfrak{g}$ is $\mathfrak{K}$. Let K be the connected subgroup of G with algebra $\mathfrak{K}$; then K is a maximal compact subgroup of G and rank G = rank K. One knows that in this situation G admits holomorphic discrete series representations[9].

Let $\mathfrak{H}$ be a Cartan subalgebra of $\mathfrak{K}$; it contains C and is a Cartan subalgebra of $\mathfrak{g}$. Let Δ be the set of roots of $\mathfrak{g}^{\mathbb{C}}$ relative to $\mathfrak{H}^{\mathbb{C}}$; if $\mathfrak{g}^{\alpha}$ is a root space it belongs to $\mathfrak{K}^{\mathbb{C}}$ if α vanishes on C and it belongs to $\mathfrak{P}^{\mathbb{C}}$ otherwise. (If $\mathfrak{g}^{\alpha} \subset \mathfrak{K}^{\mathbb{C}}$, α is called a compact root; if $\mathfrak{g}^{\alpha} \subset \mathfrak{P}^{\mathbb{C}}$, α is called a non compact root). Choose an ordering of roots such that any non compact positive root is larger than any compact root. Denote by Δ_1 (resp. Δ_2) the set of compact (resp. non compact) roots; denote by

$$\mathfrak{P}_+ = \oplus_{\alpha \in \Delta_2^+} \mathfrak{g}^{\alpha} \qquad \mathfrak{P}_- = \oplus_{\alpha \in \Delta_2^+} \mathfrak{g}^{-\alpha}$$
$$\mathfrak{B} = \mathfrak{H}^{\mathbb{C}} \oplus_{\alpha \in \Delta} \mathfrak{g}^{-\alpha}$$

These are subalgebras of $\mathfrak{g}^{\mathbb{C}}$; $\mathfrak{P}_+$ and $\mathfrak{P}_-$ are abelian subalgebras. Let P_+, P_-, B denote the corresponding connected subgroups of $G^{\mathbb{C}}$.

Theorem 3.1 (Harish-Chandra) *The map $P_+ \times K^{\mathbb{C}} \times P_- \to G^{\mathbb{C}}$ given by multiplication is a holomorphic embedding into $G^{\mathbb{C}}$; the image is an open set of $G^{\mathbb{C}}$. The set GB is open in $G^{\mathbb{C}}$ and there exists an open bounded set $\Omega \subset P_+$ such that $GB = \Omega K^{\mathbb{C}} P_-$.*

A holomorphic discrete series representation is constructed as follows. Let $\lambda \neq 0$ be an element of $\mathfrak{H}^{\mathbb{C}*}$ such that $\lambda(h) \in 2\pi i \mathbb{Z}$ for any $h \in \mathfrak{H}$ which satisfies $\exp h = 1$. Then λ exponentiates to a character $\tilde{\lambda}$ of H (the connected subgroup of $G^{\mathbb{C}}$ with Lie algebra $\mathfrak{H}$). This character $\tilde{\lambda}$ extends to a character of B. Denote by $\Gamma_\lambda = \{\phi : GB \to \mathbb{C} \mid \phi$ is holomorphic and $\phi(xb) = \tilde{\lambda}(b^{-1})\phi(x)\}$, by V_λ the subspace of those functions such that

$$\int_G |\phi(g)|^2 dg < \infty$$

and by π_λ the representation of G on V_λ :

$$(\pi_\lambda(g)\phi)(g') = \phi(g^{-1}g').$$

Theorem 3.2 (Harish-Chandra) *Let $\delta = \dfrac{1}{2} \sum_{\alpha \in \Delta^+} \alpha$ and assume that for the scalar product induced on $i\mathfrak{H}^*$ by the Killing form of $\mathfrak{g}^{\mathbb{C}}$, $(\lambda + \delta, \alpha) < 0$ for all $\alpha \in \Delta_2^+$. Then (V_λ, π_λ) is a* *unitary irreducible representation of G and the matrix coefficient of π_λ are square integrable. Furthermore the representation of K with highest weight λ occurs in π_λ/K.*

The representation (V_λ, π_λ) is a holomorphic discrete series representation of G.

Let us mimic what we did in the compact case. Consider $\mathcal{A}_1 = \{$ space of finite rank-operators on $(V_\lambda, \pi_\lambda)\}$; define

$$i : \mathcal{A}_1 \longrightarrow L^2(G)$$

$$i(\underline{u} \otimes v)(y) = < v, \pi_\lambda(g)u > \qquad u, v \in V_\alpha$$

This is indeed in $L^2(G)$ as we have a square integrable representation.

If $\varphi \in V_\lambda$, the evaluation of φ at a point $y \in GB$ is continuous. Hence there exists an element $e_y \in V_\lambda$ such that :

$$< \varphi, e_y >= \varphi(y)$$

Such a function is called a coherent state. It depends antiholomorphic on y. It has the following properties

(i) $e_{yb} = \overline{\lambda(b^{-1})}e_y \qquad (b \in B)$

(ii) $e_{gy} = g \cdot e_y \qquad (g \in G)$

Define now a symbol map $\wedge : \mathcal{A}_1 \to C^0(GB/B)$ by :

$$(\widehat{\underline{u} \otimes v})(x) = \frac{< (\underline{u} \otimes v)e_y, e_y >}{< e_y, e_y >}$$

for any $y \in GB$ which projects on x. By property (i) this is well defined and :

$$(\widehat{\underline{u} \otimes v})(x) = \frac{< e_y, u > < v, e_y >}{< e_y, e_y >} = \frac{\bar{u}(y)v(y)}{< e_y, e_y >}$$

Finally define an adapted Fourier transform $\mathcal{E} : i(\mathcal{A}_1) \to C^0(GB/B)$ by :

$$\mathcal{E}(i(u \otimes v))(x) = \int_G < v, \pi_\lambda(g)u > < \pi_\lambda(g)e_y, e_y > \frac{1}{||e_y||^2} \, dg$$

$$= \int_G < v, \pi_\lambda(g)u > < \overline{e_y, \pi_\lambda(g)e_y} > \frac{1}{||e_y||^2} \, dg$$

$$= \frac{d_{\pi_\lambda}^{-1}}{||e_y||^2} < v, e_y > < e_y, u >$$

where d_{π_λ} is the so called formal degree of the representation. Hence we have :

Proposition 3.3 *Let (V_λ, π_λ) be a holomorphic discrete series representation of the non compact connected linear semi simple group G. Let $\mathcal{A}_1$ be the space of finite rank operators on (V_λ, π_λ). Then the diagram :*

$$\begin{array}{ccc} \mathcal{A}_1 & \xrightarrow{\ i\ } & L^2(G) \\ \wedge \downarrow & \swarrow \mathcal{E} & \\ C^0(GB/B) & & \end{array}$$

is commutative.
We simply mention the fact that proposition 2 remains valid in this framework.

4. Example of $SU(1,1)$

Let us examine what happens in the case of the rank 1 group $SU(1,1)$. Recall that

$$G = SU(1,1) = \left\{ A = \begin{pmatrix} \alpha & \beta \\ \bar\beta & \bar\alpha \end{pmatrix} \Big| \alpha, \beta \in \mathbb{C}, \det A = 1 \right\}$$

and its Lie algebra is given by:

$$\mathfrak{g} = su(1,1) = \left\{ x = \begin{pmatrix} ir & \gamma \\ \bar\gamma & -ir \end{pmatrix} \Big| r \in \mathbb{R}, \gamma \in \mathbb{C} \right\}$$

The Cartan involution $\theta X = -^\tau \bar X$; hence

$$\mathfrak{H} = \mathfrak{K} = \left\{ \begin{pmatrix} ir & 0 \\ 0 & -ir \end{pmatrix} \Big| r \in \mathbb{R} \right\} = \mathcal{C}. \qquad \mathfrak{P} = \left\{ Y = \begin{pmatrix} 0 & \gamma \\ \bar\gamma & 0 \end{pmatrix} \Big| \gamma \in \mathbb{C} \right\}$$

The centralizer of $\mathcal{C}$ in $\mathfrak{g}$ is $\mathfrak{K}$. Also $\mathfrak{g}^{\mathbb{C}} = sl(2,\mathbb{C})$. The root spaces are

$$\mathfrak{g}^\alpha = \left\{ \begin{pmatrix} 0 & \delta \\ 0 & 0 \end{pmatrix} \Big| \delta \in \mathbb{C} \right\} \qquad \mathfrak{g}^{-\alpha} = \left\{ \begin{pmatrix} 0 & 0 \\ \epsilon & 0 \end{pmatrix} \Big| \epsilon \in \mathbb{C} \right\}$$

A root will be positive if $\alpha(Z_1) > 0$ where $Z_1 (\in i\mathfrak{H}) = \begin{pmatrix} 1 & 0 \\ 0 & -1 \end{pmatrix}$; one has $\alpha(Z_1) = 2$.
Thus

$$\mathfrak{P}_+ = \mathfrak{g}^\alpha \qquad \mathfrak{P}_- = \mathfrak{g}^{-\alpha} \qquad \mathfrak{B} = \left\{ \begin{pmatrix} \beta & 0 \\ \epsilon & -\beta \end{pmatrix} \Big| \beta, \epsilon \in \mathbb{C} \right\}$$

and

$$P_+ = \left\{ \begin{pmatrix} 1 & \delta \\ 0 & 1 \end{pmatrix} \Big| \delta \in \mathbb{C} \right\} \qquad P_- = \left\{ \begin{pmatrix} 1 & 0 \\ \epsilon & 1 \end{pmatrix} \Big| \epsilon \in \mathbb{C} \right\}$$

$$B = \left\{ \begin{pmatrix} 1 & \delta \\ 0 & 1 \end{pmatrix} \Big| \beta \in \mathbb{C}^*, \gamma \in \mathbb{C} \right\}$$

The open set GB in $G^{\mathbb{C}} = SL(2,\mathbb{C})$ is given by $\left\{ \begin{pmatrix} a & b \\ c & d \end{pmatrix} \Big| ad - bc = 1, a \neq 0, |\frac{b}{d}| < 1 \right\}$
and GB/B is isomorphic to the unit disc in the complex plane. The weight λ considered to construct the discrete series is such that

$$p \leq -2$$

A point $y \in GB = \Omega K^{\mathbb{C}} P_-$ is labelled by 3 complex coordinates z ($|z| < 1$), a ($a \neq 0$), b. A point $u \in B$ is labelled by (a,b).
The character $\tilde\lambda$ of B is

$$\tilde\lambda(a,b) = a^p$$

where p is an integer ≤ -2. A holomorphic function $\tilde\psi(z,a,b)$ belongs to $\Gamma_{\tilde\lambda}$ provided

$$\tilde\psi(z, aa', b' + a'^2 b) = a(1-p)\tilde\psi(z,a,b)$$

which means that there exists a holomorphic function ψ on the unit disc such that

$$\tilde\psi(z,a,b) = a^{-p}\psi(z)$$

Restricted to G one has

$$\tilde{\psi}(z = \frac{\beta}{\bar{\alpha}}, a = \frac{1}{\alpha}, b = \frac{\bar{\beta}}{\bar{\alpha}}) = \bar{\alpha}^p \psi(z)$$

$$|\tilde{\psi}(g)|^2 = |\psi(z)|^2 \frac{1}{(1 - z\bar{z})^p} \qquad g = \begin{pmatrix} \alpha & \beta \\ \bar{\beta} & \bar{\alpha} \end{pmatrix}$$

Thus $V_{\tilde{\lambda}} = \left\{ \psi : D \to \mathbb{C} \middle| \text{ holomorphic s.t. } \int_D \frac{i|\psi(z)|^2 \, dz \wedge d\bar{z}}{(1 - z\bar{z})^p} < \infty \right\}$ and the representation is given by

$$(g \cdot \psi)z = (-\bar{\beta}z + \alpha)^p \psi(\frac{\bar{\alpha}z - \beta}{-\bar{\beta}z + \alpha}) \qquad g = \begin{pmatrix} \alpha & \beta \\ \bar{\beta} & \bar{\alpha} \end{pmatrix}$$

The group $SU(1,1)$ contains subgroups isomorphic to the affine group of the line; for example the subgroup

$$H = \left\{ \begin{pmatrix} cht + iq & sht - iq \\ sht + it & cht - iq \end{pmatrix} \middle| t, q \in \mathbb{R}. \right\}$$

The restriction of the above discrete series representation to H gives formulas worth analysing.

5. Aknowledgment

The first author had many nice and constructive conversations with Twareque Ali and his interest in these matters was stimulated by him.

References

[1] Fabec R and Olafsson G 2001, The continuous wavelet transform and symmetric spaces, preprint math.FA/011100
[2] Ali S T, Atakishiyev N, Chumakov S, and Wolf K B 2000, The Wigner function for general Lie groups and the wavelet transform, preprint
[3] Bayen F, Flato M, Fronsdal C, Lichnerowicz A, and Sternheimer D 1977, Quantum mechanics as a deformation of classical mechanics, Lett. Math. Phys. 1 521–530
[4] Fronsdal C 1978, Some ideas about quantization, Reports Math. Phys. 15 111–145
[5] Berezin F A 1975, General concept of quantization, Commun. Math. Phys. 40 153–174
[6] Cahen M, Gutt S, and Rawnsley J 1990, Quantization of Kähler manifolds I, J. Geom. Phys. 7 45–62
[7] Wallach N 1973, Harmonic Analysis on Homogeneous Spaces, p 93 (......)
[8] Wallach N 1973, Harmonic Analysis on Homogeneous Spaces, p 156 (......)
[9] Knapp A 1986, Representation Theory of Semisimple Groups. An Overview Based on Examples (Princeton: Princeton Univ. Press, Math Series 36)

Inst. Phys. Conf. Ser. No 173: Satellite colloquium
Paper presented at 24th Int. Coll. Group Theoretical Methods in Physics, Paris, France, 15–20 July 2002
©*2003 IOP Publishing Ltd*

Phase space techniques for the study of generalized Anti-Wick operators

Elena Cordero †

Department of Mathematics, University of Torino, Italy

Abstract. We give a generalization of pseudodifferential operators with Anti-Wick symbols in the modulation spaces and depending on two different window functions. We establish sufficient and necessary boundedness and Schatten class properties in dependence on the interplay between the operator symbol and the windows. Phase space analysis is used both to define the Anti-Wick operator and to study the related regularity properties: the necessary results provided seem to reveal that this approach is optimal.

1. Introduction

The classical Anti-Wick correspondence has been introduced as a quantization rule in physics and since then variations have occurred in various fields of mathematics under names such as Toeplitz operators or wave packets [4, 7]. Recently they have come again into focus in applied mathematics and signal analysis as localization operators and as short-time Fourier transform multipliers [5, 6, 11, 14]. The Anti-Wick correspondence is a mapping

$$a \rightarrow A_a$$

from a symbol a defined on the phase space $\mathbb{R}^{2d}$ to an operator A_a acting on a subspace of $L^2(\mathbb{R}^d)$. The class of symbols depends on the applications: in PDE or in physics smooth symbols have been used (belonging to Hörmander or Shubin classes [12]), while in the context of signal processing rougher symbols are needed. Next, we note that the classical coherent states (translations and modulations of the Gaussian function) occur implicitly in the definition of A_a, however, in the application to localization other type of windows have turned out to be useful.

Accordingly, we will consider symbols that may be tempered distributions and replace the Gaussian windows with two arbitrary windows φ_1, φ_2 in suitable function spaces. In this way we obtain a generalization of the classical definition of Anti-Wick operator, and this can be written conveniently by means of standard phase space representations. For the precise formulation, we define the operators of translation and modulation by

$$T_x f(t) = f(t - x) \quad \text{and} \quad M_\omega f(t) = e^{2\pi i \omega \cdot t} f(t) \,,$$

and let $V_g f$ be the short-time Fourier transform (STFT) of f with respect to the window g

$$V_g f(x, \omega) = \langle f, M_\omega T_x g \rangle = \int_{\mathbb{R}^d} f(t) \, \overline{g(t - x)} \, e^{-2\pi i \omega \cdot t} \, dt \,, \qquad x, \omega, t \in \mathbb{R}^d \,,$$

† e-mail: cordero@dm.unito.it

whenever the integral or the inner product exist, e.g., for $(f, g) \in L^2(\mathbb{R}^d) \times L^2(\mathbb{R}^d)$ or $(f, g) \in \mathcal{S}'(\mathbb{R}^d) \times \mathcal{S}(\mathbb{R}^d)$. Then the Anti-Wick operator $A_a^{\varphi_1, \varphi_2}$ with symbol a and windows φ_1, φ_2 is defined to be

$$A_a^{\varphi_1, \varphi_2} f(t) = \int_{\mathbb{R}^{2d}} a(x, \omega) V_{\varphi_1} f(x, \omega) M_\omega T_x \varphi_2(t) \, dx d\omega$$

whenever this vector-valued integral makes sense.

If $\varphi_1(t) = \varphi_2(t) = e^{-\pi t^2}$, then $A_a = A_a^{\varphi_1, \varphi_2}$ is the classical Anti-Wick operator [1].

To prove our results we have heavily exploited the weak definition of the previous integral, that is

$$\langle A_a^{\varphi_1, \varphi_2} f, g \rangle = \int a(x, \omega) \, V_{\varphi_1} f(x, \omega) \, \langle M_\omega T_x \varphi_2, g \rangle \, dx d\omega$$

$$= \langle a, \overline{V_{\varphi_1} f} \, V_{\varphi_2} g \rangle \qquad \text{for } f, g \in \mathcal{S}(\mathbb{R}^d) \,.$$

Here the brackets $\langle \cdot, \cdot \rangle$ express the duality on a suitable pair of dual spaces $B' \times B$ and extend the inner product on $L^2(\mathbb{R}^{2d})$.

In order to measure the phase space distribution of symbols and so-called windows, we use the modulation spaces. For their basic properties we refer, for instance, to [8, Ch. 11-13] and the original literature quoted there.

Given a fixed nonzero window $g \in \mathcal{S}(\mathbb{R}^d)$, a nonnegative, even weight function m on $\mathbb{R}^{2d}$ of polynomial growth, and $1 \le p, q \le \infty$, the *modulation space* $M_m^{p,q}(\mathbb{R}^d)$ consists of all tempered distributions $f \in \mathcal{S}'(\mathbb{R}^d)$ such that $V_g f \in L_m^{p,q}(\mathbb{R}^{2d})$ (weighted mixed-norm spaces). The norm on $M_m^{p,q}$ is

$$\|f\|_{M_m^{p,q}} = \|V_g f\|_{L_m^{p,q}} = \left(\int_{\mathbb{R}^d} \left(\int_{\mathbb{R}^d} |V_g f(x, \omega)|^p m(x, \omega)^p \, dx \right)^{q/p} d\omega \right)^{1/p} \,.$$

If $p = q$, we write M_m^p instead of $M_m^{p,p}$, and if $m(z) \equiv 1$ on $\mathbb{R}^{2d}$, then we write $M^{p,q}$ and M^p for $M_m^{p,q}$ and $M_m^{p,p}$.

$M_m^{p,q}(\mathbb{R}^d)$ is a Banach space whose definition is *independent of the choice of the window* $g \in \mathcal{S}(\mathbb{R}^d)$. Moreover, for so-called moderate weights, and $g \in \mathcal{S}(\mathbb{R}^d) \setminus \{0\}$, then $\|V_g f\|_{L_m^{p,q}}$ is an equivalent norm for $M_m^{p,q}(\mathbb{R}^d)$ (see [8, Thm. 11.3.7]). We will use the polynomial weights defined by

$$v_s(z) = v_s(x, \omega) = \langle z \rangle^s = (1 + x^2 + \omega^2)^{s/2}, \quad \tau_s(z) = \tau_s(x, \omega) = \langle \omega \rangle^s = (1 + \omega^2)^{s/2},$$

where $z = (x, \omega) \in \mathbb{R}^{2d}$ and $s \in \mathbb{R}$.

Among the modulation spaces one can recognize well-known function spaces:

(i) $L^2(\mathbb{R}^d) = M^2(\mathbb{R}^d)$.

(ii) Sobolev spaces: $H^s(\mathbb{R}^d) = \{f : \hat{f}(\omega)\langle \omega \rangle^s \in L^2(\mathbb{R}^d)\} = M_{\tau_s}^2(\mathbb{R}^d)$.

(iv) Shubin-Sobolev spaces [12, 2]: $Q_s(\mathbb{R}^d) = L_s^2(\mathbb{R}^d) \cap H^s(\mathbb{R}^d) = M_{v_s}^2(\mathbb{R}^d)$.

(v) Feichtinger's algebra: $M^1(\mathbb{R}^d) = S_0(\mathbb{R}^d)$.

(vi) The Schwartz class [10]: $\mathcal{S}(\mathbb{R}^d) = \bigcap_{s \ge 0} M_{v_s}^\infty(\mathbb{R}^d)$.

(vii) The space of tempered distributions [10]: $\mathcal{S}'(\mathbb{R}^d) = \bigcup_{s \ge 0} M_{1/v_s}^\infty(\mathbb{R}^d)$.

2. Sufficient and necessary regularity conditions

In view of the definition of generalized Anti-Wick operators, it is natural to expect that certain norms imposed on the STFT of the symbol a and the windows φ_1, φ_2 (in other words, modulation spaces) will play a role in the formulations of the results. We present some of our results which underline the importance of modulation spaces and the associated phase space methods. For details and proofs, we refer to [3].

Recall that the singular values $\{s_k(L)\}_{k=1}^{\infty}$ of a compact operator $L \in B(L^2(\mathbb{R}^d))$ are the eigenvalues of the positive self-adjoint operator $\sqrt{L^*L}$. For $1 \leq p < \infty$, the Schatten class S_p is the space of all compact operators whose singular values lie in l^p. For consistency, we define $S_\infty := B(L^2(\mathbb{R}^d))$ for the space of bounded operators on $L^2(\mathbb{R}^d)$.

We shall use the notation $A \lesssim B$ to indicate $A \leq cB$ for a suitable constant $c > 0$.

Theorem 2.1 (Sufficient conditions) *Let $s \geq 0$, then the following results hold:*

(i) If $a \in M_{1/\tau_s}^{\infty}(\mathbb{R}^{2d})$, $\varphi_1, \varphi_2 \in M_{v_s}^1(\mathbb{R}^d)$, then $A_a^{\varphi_1, \varphi_2}$ is bounded on $M^{p,q}(\mathbb{R}^d)$ for all $1 \leq p, q \leq \infty$, and the operator norm satisfies the uniform estimate

$$\|A_a^{\varphi_1, \varphi_2}\|_{op} \lesssim \|a\|_{M_{1/\tau_s}^{\infty}} \|\varphi_1\|_{M_{v_s}^1} \|\varphi_2\|_{M_{v_s}^1}.$$

In particular, $A_a^{\varphi_1, \varphi_2}$ is bounded on $L^2(\mathbb{R}^d)$.

(ii) If $1 \leq p \leq 2$, $a \in M_{1/\tau_s}^{p,\infty}(\mathbb{R}^{2d})$ $\varphi_1 \in M_{v_s}^1(\mathbb{R}^d)$, $\varphi_2 \in M_{v_s}^p(\mathbb{R}^d)$ then $A_a^{\varphi_1, \varphi_2} \in S_p$, with

$$\|A_a^{\varphi_1, \varphi_2}\|_{S_p} \lesssim \|a\|_{M_{1/\tau_s}^{p,\infty}} \|\varphi_1\|_{M_{v_s}^1} \|\varphi_2\|_{M_{v_s}^p}.$$

(iii) If $2 \leq p \leq \infty$, $a \in M_{1/\tau_s}^{p,\infty}(\mathbb{R}^{2d})$ $\varphi_1 \in M_{v_s}^1(\mathbb{R}^d)$, $\varphi_2 \in M_{v_s}^{p'}$ then $A_a^{\varphi_1, \varphi_2} \in S_p$, with

$$\|A_a^{\varphi_1, \varphi_2}\|_{S_p} \lesssim \|a\|_{M_{1/\tau_s}^{p,\infty}} \|\varphi_1\|_{M_{v_s}^1} \|\varphi_2\|_{M_{v_s}^{p'}}.$$

(iv) Any distribution $a \in \mathcal{E}'(\mathbb{R}^d)$ with compact support can be represented as

$$a = \sum_{|\alpha| \leq m} \partial^\alpha f_\alpha$$

for compactly supported continuous functions f_α on $\mathbb{R}^{2d}$ [13, p. 263, Cor. 2]. If $\varphi_1, \varphi_2 \in M_{v_m}^1(\mathbb{R}^d)$, then $a \in M_{1/\tau_m}^{1,\infty}(\mathbb{R}^d)$ and $A_a^{\varphi_1, \varphi_2}$ is a trace class operator.

As a consequence of the embedding of the Potential spaces into modulation spaces (if $1 \leq p \leq \infty$, $s \in \mathbb{R}$, then $W_s^p(\mathbb{R}^d) \hookrightarrow M_{\tau_s}^{p,\infty}(\mathbb{R}^d)$, [3]), we derive a result obtained in [2, Thm. 4.7].

Corollary 2.2 *Let $1 \leq p \leq \infty$, $a \in W_{-s}^p(\mathbb{R}^{2d})$ for some $s \geq 0$, and $\psi_1, \varphi_2 \in M_{v_s}^1(\mathbb{R}^d)$. Then*

$$\|A_a^{\varphi_1, \varphi_2}\|_{S_p} \lesssim \|a\|_{W_{-s}^p} \|\varphi_1\|_{M_{v_s}^1} \|\varphi_2\|_{M_{v_s}^1}.$$

Let us give a brief outline of the arguments and methods used to prove the Anti-Wick sufficient properties of Theorem 2.1.

(i) Connection between Anti-Wick operator and Weyl calculus [2]. Namely, if $W(\varphi_2, \varphi_1)$ is the cross-Wigner distribution of φ_2, φ_1, i.e.

$$W(\varphi_2, \varphi_1)(x, \omega) = \int_{\mathbb{R}^d} e^{-2\pi i \omega t} \varphi_2(x + \frac{t}{2}) \overline{\varphi_1(x - \frac{t}{2})} \, dt,$$

and L_σ is the Weyl transform with Weyl symbol σ (see [7, 8, 14]). Then we have [7, 2]

$$A_a^{\varphi_1, \varphi_2} = L_{a*W(\varphi_2, \varphi_1)}.$$

(ii) We then exploit boundedness properties for the Weyl transform with a symbol σ in a modulation space. For instance, if $\sigma \in M^{\infty,1}$, then L_σ is bounded on $L^2(\mathbb{R}^d)$ and $\|L_\sigma\|_{op} \lesssim \|\sigma\|_{M^{\infty,1}}$. For more details, we refer to [8, 9]

(iii) Finally, we use convolution relations for modulation spaces and estimates for the cross-Wigner distribution (see [8, 3]).

Theorem 2.1 is optimal in the following sense.

Theorem 2.3 (Necessary boundedness and Hilbert-Schmidt conditions) *Let $a \in \mathcal{S}'(\mathbb{R}^{2d})$ and $s \geq 0$.*
(i) If there exists a constant $C = C(a) > 0$ depending only on a such that

$$\|A_a^{\varphi_1,\varphi_2}\|_{S_\infty} \leq C \|\varphi_1\|_{M_{v_s}^1} \|\varphi_2\|_{M_{v_s}^1}$$

for all $\varphi_1, \varphi_2 \in \mathcal{S}(\mathbb{R}^d)$, then $a \in M_{1/\tau_s}^\infty$.
(ii) Likewise, if

$$\|A_a^{\varphi_1,\varphi_2}\|_{S_2} \leq C \|\varphi_1\|_{M^1} \|\varphi_2\|_{M^1},$$

for all $\varphi_1, \varphi_2 \in \mathcal{S}(\mathbb{R}^d)$, then $a \in M^{2,\infty}$.

In view of the previous result we believe that modulation spaces are the appropriate function classes for the study of Anti-Wick operators.

Acknowledgment: This note reports a joint work [3] with Karlheinz Gröchenig, University of Connecticut, Storrs, USA.

References

[1] Berezin F A 1971, Wick and anti-Wick symbols of operators, Mat. Sb. (N.S.) , 86(128): 578–610
[2] Boggiatto P, Cordero E, and Gröchenig K 2002, Generalized anti-Wick operators with symbols in distributional Sobolev spaces, preprint
[3] Cordero E and Gröchenig K 2002, Time-frequency analysis of Gabor localization operators, preprint
[4] Córdoba A and Fefferman C 1978, Wave packets and Fourier integral operators, Comm. Partial Diff. Eq. 3 979–1005
[5] Daubechies I 1988, Time-frequency localization operators: a geometric phase space approach, IEEE Trans. Inform. Theory 34 605–612
[6] H G Feichtinger and Nowak K 2002, A first survey of Gabor multipliers, in Feichtinger H G and Strohmer T S (eds), Advances in Gabor Analysis (Boston: Birkhäuser)
[7] Folland G B 1989, Harmonic Analysis in Phase Space (Princeton, NJ: Princeton Univ. Press)
[8] Gröchenig K 2001, Foundations of Time-Frequency Analysis (Boston: Birkhäuser)
[9] Gröchenig K and Heil C 1999, Modulation spaces and pseudodifferential operators, Integral Equ. Oper. Theory 34 439–457
[10] Gröchenig K and Zimmermann G 2001, Hardy's theorem and the short-time Fourier transform of Schwartz functions, J. London Math. Soc. 63 205–214
[11] Ramanathan J and Topiwala P 1993, Time-frequency localization via the Weyl correspondence, SIAM J. Math. Anal. 24 1378–1393
[12] Shubin M A 2001, Pseudodifferential Operators and Spectral Theory, Second edition, translated from the 1978 Russian original by Stig I. Andersson (Berlin: Springer)
[13] Trèves F 1967, Topological Vector Spaces, Distributions and Kernels (New York: Academic)
[14] Wong M W 2002, Wavelets Transforms and Localization Operators, Operator Theory: Advances and Applications, vol.136 (Boston: Birkhäuser)

Inst. Phys. Conf. Ser. No 173: Satellite colloquium
Paper presented at 24th Int. Coll. Group Theoretical Methods in Physics, Paris, France, 15–20 July 2002
©2003 IOP Publishing Ltd

Admissible vectors and traces on the commuting algebra

Hartmut Führ

Institute for Biomathematics and Biometry, GSF National Research Center for Environment and Health, Ingolstädter Landstraße 1, D-85764 Neuherberg

Abstract. Given a representation of a unimodular locally compact group, we discuss criteria for associated coherent state expansions in terms of the commuting algebra. It turns out that for those representations that admit such expansions there exists a unique finite trace on the commuting algebra such that the admissible vectors are precisely the tracial vectors for that trace. This observation is immediate from the definition of the group Hilbert algebra and its associated trace. The trace criterion allows to discuss admissibility in terms of the central decomposition of the regular representation. In particular, we present a new proof of the admissibility criteria derived for the type I case. In addition we derive admissibility criteria which generalize the Wexler-Raz biorthogonality relations characterizing dual windows for Weyl-Heisenberg frames.

1. Introduction

Given a representation $(\pi, \mathcal{H}_\pi)$ of a unimodular, separable locally compact group G, we want to discuss the existence and characterization of vectors giving rise to coherent state expansions on $\mathcal{H}_\pi$. For this purpose, a vector $\eta \in \mathcal{H}_\pi$ is called **bounded** if the coefficient operator

$$V_\eta : \mathcal{H}_\pi \to \mathrm{L}^2(G) \ , \quad (V_\eta \varphi)(x) = \langle \varphi, \pi(x)\eta \rangle$$

is a bounded map. We are interested in inverting this operator, hence the following notion is natural: A pair of bounded vectors (η, ψ) is called **admissible** if $V_\eta^* V_\phi$ is the identity operator on $\mathcal{H}_\pi$. Note that this property gives rise to the weak-sense inversion formula

$$z = \int_G \langle z, \pi(x)\eta \rangle \, \pi(x)\psi \, d\mu_G(x) \ ,$$

which can be read as a continuous expansion of z in terms of the orbit $\pi(G)\psi \subset \mathcal{H}_\pi$. Identities of this type are known as **coherent state expansion** in mathematical physics. A single vector η is called admissible if (η, η) is an admissible pair. It is obvious from the definition that (η, ψ) is admissible iff (ψ, η) is. In such a case η is called the **dual vector** of ψ.

Admissible vectors were first discussed almost exclusively in connection with irreducible, so-called *discrete series* or *square-integrable* representations [10]. The existence of admissible vectors for these representations is a fairly straightforward consequence of Schur's Lemma, and slightly more complicated in the nonunimodular case. Recently exhaustive criteria for the existence and characterization of admissible vectors were established for the case that the regular representation λ_G of G is type I, using the Plancherel formula of the group. However, if we want to include discrete groups in this general discussion, the type I restriction is rather too rigid. Indeed, a discrete group G has a type I regular representation iff G itself is type I [11], and the latter is only the case if G is a finite extension of an abelian normal subgroup [14].

Let us now give a short survey of the paper. It initiated from the idea to replace the decomposition into irreducibles by the central decomposition of λ_G, and to try to come up with criteria using the latter. This however requires an understanding of how admissible vectors are recognised in terms of their image under the central decomposition. The main result of this paper, Theorem 2.2, provides the key to this problem, by relating admissibility to the natural trace on the von Neumann algebra $VN_r(G)$. Since the trace decomposes, the problem of characterizing admissible pairs can be translated to the characterization of tracial pairs for the fibre von Neumann algebras of the central decomposition (Proposition 3.1). As an application of this result we obtain a characterization of admissible pairs in the case that λ_G is type I (Theorem 3.2). A slightly weaker version of this result had been proved, by somewhat different arguments, in an earlier paper ([7, Theorem 1.6]). In the final section we sketch how the trace criterion gives rise to admissibility criteria in terms of certain orthogonality relations. As a special case we obtain the Wexler-Raz biorthogonality relations in Gabor analysis.

2. The group Hilbert algebra, traces and admissible pairs

Throughout the paper, G denotes a separable unimodular locally compact group, and $(\pi, \mathcal{H}_\pi)$ a (unitary, strongly continuous) representation of G. λ_G is the **left regular representation**, acting on $L^2(G)$ by $(\lambda_G(x)f)(y) = f(x^{-1}y)$. We denote the commuting algebra of λ_G as $VN_r(G)$, and the bicommutant by $VN_l(G)$ (the right/left group von Neumann algebras).

Our definitions and notations regarding the group Hilbert algebra are taken from [5]. In order to define the group Hilbert algebra, we let for $f, g \in L^2(G)$, and additionally either f or g in $C_c(G)$,

$$\mathcal{U}_f(g) = g * f \ .$$

The (full) group Hilbert algebra consists of all $f \in L^2(G)$ for which $\mathcal{U}_f$ extends to a bounded operator. Note that these operators then lie in $VN_r(G)$. Writing $f^*(x) = \overline{f(x^{-1})}$, we note that $V_f = \mathcal{U}_{f*} = \mathcal{U}_f^*$. Hence the bounded vectors are precisely the elements of the full Hilbert algebra.

Let us recall the definition of a trace. Given a von Neumann algebra $\mathcal{A}$, we let $\mathcal{A}^+$ denote the cone of positive elements. A mapping $tr : \mathcal{A}^+ \to \mathbb{R}^+ \cup \infty$ is called a **trace** if it satisfies the following two properties:

- $tr(S + \alpha T) = tr(S) + \alpha tr(T)$, for all $S, T \in \mathcal{A}^+$, $\alpha \in \mathbb{R}^+$. Note the conventions $\alpha\infty = \infty$ for $\alpha > 0$ and $0\infty = 0$.
- $tr(UTU^*) = tr(T)$, for all $T \in \mathcal{A}^+$ and all unitary $U \in \mathcal{A}$.

Further relevant properties, that traces may or may not have, are faithfulness, normality and semifiniteness; we refer the reader to [6] for the definitions. tr is **finite** if $tr(\mathrm{Id}_\mathcal{H}) = 1$. A trace tr uniquely extends to linear functional on the two-sided ideal

$$\mathfrak{M}_{tr} = \{S \in \mathcal{A} : tr(|S|) < \infty\} \ ,$$

for finite traces this is obviously $\mathcal{A}$ itself. We will denote the extension by tr as well. If the trace is normal, the associated linear functional is ultra-weakly continuous [6, III.6, Proposition 1].

The group Hilbert algebra induces a faithful, normal and semifinite trace on $VN_r(G)^+$, by letting for $T \in VN_r(G)^+$

$$tr(T) = \begin{cases} \|f\|^2 & T = \mathcal{U}_f^*\mathcal{U}_f \text{ for a bounded vector } f \\ \infty & \text{otherwise} \end{cases}$$

We note that for bounded vectors $f, g \in L^2(G)$, $V_f^*V_g$ is in $\mathfrak{M}_{tr}$, with

$$tr(V_f^*V_g) = \langle f, g \rangle \tag{59}$$

For our arguments it will sometimes be convenient to assume that $\mathcal{H}_\pi = \mathcal{H} \subset L^2(G)$ is a closed leftinvariant subspace, on which π acts by left translation. This is not a restriction, thanks to the following lemma which collects a few facts about admissible pairs. We expect most of these statements to be widely known.

Lemma 2.1 (a) *For any bounded vector η, V_η intertwines π with λ_G.*

(b) *If π has an admissible pair (η, ψ), then both V_η and V_ψ are topological embeddings into $L^2(G)$. Conversely, given a bounded vector η such that V_η is a topological embedding, a dual vector for η is given by $\psi = (V_\eta^* V_\eta)^{-1}\eta$. ψ is the unique dual window with minimal norm.*

(c) *There exists an admissible pair (η, ψ) iff there exists an admissible vector.*

(d) *If there exists an admissible pair, π is unitarily equivalent to a subrepresentation of λ_G.*

Proof. Part (a) is immediate, and (d) then follows from (c). The first statement of (b) is obvious. For the existence of a dual vector, we observe that $S = V_\eta^* V_\eta$ is a strictly positive operator with bounded inverse. Hence $V_{S^{-1}\eta} = V_\eta \circ S^{-1}$ is bounded, i.e., $S^{-1}\psi$ is a bounded vector. The computation

$$V_\psi^* V_\eta = V_{S^{-1}\eta}^* V_\eta = S^{-1} V_\eta^* V_\eta = \mathrm{Id}_{\mathcal{H}_\pi} \ ,$$

shows that ψ is a dual vector. A similar calculation shows $V_{S^{-1/2}\eta}^* V_{S^{-1/2}\eta} = \mathrm{Id}_{\mathcal{H}_\pi}$, i.e., (c).

Hence, for the proof of minimality of $\|\psi\|$ (which is the only thing left to show), we may assume that $\mathcal{H}_\pi = \mathcal{H} \subset L^2(G)$, and that π is left translation on $\mathcal{H}$. Then the commuting algebra $\pi(G)'$ is readily identified as the **reduced von Neumann algebra** $\{pTp : T \in VN_r(G)\} \subset VN_r(G)$, where p denotes the projection onto $\mathcal{H}$.

The set of all dual windows is an affine subspace, since the difference of two dual windows is in the linear subspace

$$W = \{x \in \mathcal{H}_\pi \text{ bounded vector} : V_x^* V_\eta = 0\} \ .$$

Hence a dual window of minimal norm is necessarily unique. Now (59) entails for $x \in W$

$$\langle x, \psi \rangle = tr(V_x^* V_\psi) = tr(V_x^* V_{S^{-1}\eta}) = tr(V_x^* V_\eta S^{-1}) = tr(0) = 0 \ ,$$

hence $\psi \perp W$, and $\|\psi\|$ is indeed minimal. $\qquad\square$

The characterization of admissible vectors in terms of the trace requires one more piece of notation: Given a particular trace tr on a von Neumann algebra $\mathcal{A}$, we call a pair of elements (η, ψ) of the underlying Hilbert space **tracial** if

$$\forall T \in \mathcal{A}^+ : tr(T) - \langle T\eta, \psi \rangle \ .$$

Theorem 2.2 *Let $\mathcal{H} \subset L^2(G)$ be a closed, leftinvariant subspace, with associated leftinvariant projection p, and let π denote the restriction of λ_G to $\mathcal{H}$.*

(a) *There exists an admissible pair for $\mathcal{H}$ iff $tr(p) < \infty$.*

(b) *For all pairs $(\eta, \psi) \in \mathcal{H} \times \mathcal{H}$ of bounded vectors: (η, ψ) is admissible iff (η, ψ) is tracial for $\pi(G)'$.*

Proof. For part (a), first assume that there exists an admissible vector η. Hence $p = V_\eta^* V_\eta = \mathcal{U}_\eta^* \mathcal{U}_\eta^*$, and thus $tr(p) = \|\eta\|^2 < \infty$.

Conversely, if $tr(p) < \infty$, then $p = \mathcal{U}_\eta^* \mathcal{U}_\eta^*$, for some bounded vector $\eta \in L^2(G)$. Now the computation

$$V_{p\eta}^* V_{p\eta} = p V_\eta^* V_\eta p = p$$

shows that $p\eta \in \mathcal{H}$ is admissible.

For (b), we compute, for any $T = V_g^* V_g$ with g bounded, and for and any pair (η, ψ) of bounded vectors,

$$\begin{aligned}
\langle T\eta, \psi \rangle &= \langle \eta * g^* * g, \psi \rangle \\
&= \langle \eta * g^*, \phi * g^* \rangle \\
&= \langle g * \eta^*, g * \phi^* \rangle \\
&= \langle g, g * \phi^* * \eta \rangle \\
&= \langle g, (V_\eta^* V_\phi) g \rangle \ .
\end{aligned}$$

Hence, assuming that (η, ϕ) are admissible, we obtain

$$\langle T\eta, \phi \rangle = \|g\|^2 = tr(T) \ ,$$

as desired. Conversely, assuming traciality of (η, ϕ), the above calculation yields

$$\|g\|^2 = tr(T) = \langle T\eta, \psi \rangle = \langle g, V_\eta^* V_\phi g \rangle \ ,$$

for all bounded vectors. By polarization this leads to

$$\langle h, g \rangle = \langle h, V_\eta^* V_\phi g \rangle \ ,$$

for all bounded vectors h, g, and since these are total, $V_\eta^* V_\phi = \mathrm{Id}_{\mathcal{H}}$ follows. $\qquad\square$

REMARK: The equivalent conditions from part (a) imply in particular that $\pi(G)'$ is a finite von Neumann algebra. However, finiteness of $\pi(G)'$ is not sufficient, as the following equivalences show:

$$V N_r(G) \text{ is finite } \Leftrightarrow G \text{ is an SIN-group} \tag{60}$$

$$tr(\mathrm{Id}_{L^2(G)}) < \infty \Leftrightarrow G \text{ is discrete} \tag{61}$$

Here (60) is [5, 13.10.5], whereas (61) follows combining Theorem 2.2 (a) with [7, Proposition 0.4]. Recall that SIN-groups are defined by having a conjugation-invariant neighborhood-base at unity. Clearly this class comprises the locally compact abelian groups, hence for any nondiscrete LCA group $V N_r(G)$ is finite, but $tr(\mathrm{Id}_{L^2(G)}) = \infty$.

3. Application to the central decomposition

In this section we consider the central decomposition of the regular representation and its use for the characterization of admissible pairs. In particular we recover the characterization obtained in [7] for the case that λ_G is type I. The following facts concerning the central decomposition of λ_G can be found in [5, 18.7.7,18.7.8].

Let $\check{G}$ denote the space of quasi-equivalence classes of factor representations of G, endowed with the natural Borel structure. Then there exists a standard positive measure ν_G on $\check{G}$, and a measurable field of factor representations $\rho_\sigma \in \sigma$, for ν_G-almost every σ, such that

$$\lambda_G \simeq \int_{\check{G}}^{\oplus} \rho_\sigma d\nu_G(\sigma) \ .$$

The operator effecting the unitary equivalence is called the **Plancherel transform**. Moreover, the direct integral provides a decomposition of $V N_r(G)$ and the natural trace:

$$V N_r(G) = \int_{\check{G}}^{\oplus} \mathcal{A}_\sigma d\nu_G(\sigma)$$

for a measurable family of von Neumann algebras $\mathcal{A}_\sigma$ on $\mathcal{H}_\sigma$, as well as

$$tr(T) = \int_{\check{G}} tr_\sigma(T_\sigma) d\nu_G(\sigma) \ ,$$

when $(T_\sigma)_{\sigma\in\hat{G}}$ denotes the operator field corresponding to T under the central decomposition, and tr_σ is a faithful normal, semifinite trace on the factor $\mathcal{A}_\sigma$, which exists for ν_G-almost every σ. In particular, ν_G-almost every $\mathcal{A}_\sigma$ is of type I or II.

Now admissibility is easily translated to traciality in the fibres:

Proposition 3.1 *Let π denote the restriction of λ_G to a closed, leftinvariant subspace $\mathcal{H} \subset \mathrm{L}^2(G)$. Let P denote the projection onto $\mathcal{H}$, then P decomposes into a measurable field of projections $\widehat{P}_\sigma$, and $\pi(G)'$ decomposes under the central decomposition into the von Neumann algebras $\mathcal{C}_\sigma = \widehat{P}_\sigma \mathcal{A}_\sigma \widehat{P}_\sigma$.*

(a) Let $\eta, \psi \in \mathcal{H}$, and denote the respective Plancherel transforms by $(\widehat{\eta}_\sigma)_{\sigma\in\hat{G}}, (\widehat{\psi}_\sigma)_{\sigma\in\hat{G}}$. Then

$$(\eta, \psi) \text{ is admissible for } \mathcal{H} \Leftrightarrow (\widehat{\eta}_\sigma, \widehat{\psi}_\sigma) \text{ is tracial for } \mathcal{C}_\sigma \quad (\nu_G a.e.)$$

(b) $\mathcal{H}$ has an admissible pair of vectors iff $\int_{\hat{G}} tr(\widehat{P}_\sigma) d\nu_G(\sigma) < \infty$. In particular, almost all $\mathcal{C}_\sigma$ are finite von Neumann algebras.

The (potential) use of the proposition consists in the fact that we only need to characterize tracial pairs for factor representations. Unfortunately, we are not aware of any explicit criteria for tracial vectors associated to type II factors. For type I factors, though, they are easily derived, as the proof of the next theorem shows.

Note that if λ_G is type I, the fibre spaces in the central decomposition are just the Hilbert-Schmidt spaces $\mathcal{B}_2(\mathcal{H}_\sigma)$, where σ runs through the unitary dual, and λ_G decomposes into left action on $\mathcal{B}_2(\mathcal{H}_\sigma)$ via σ [5, 18.8].

Theorem 3.2 *Let G be unimodular with λ_G type I. Let $\mathcal{H} \subset \mathrm{L}^2(G)$ be a leftinvariant subspace. Then there exists a measurable field of projections $\widehat{P}_\sigma$ on $\mathcal{H}_\sigma$ such that*

$$P \simeq \int_{\hat{G}}^{\oplus} 1 \otimes \widehat{P}_\sigma \, d\nu_G(\sigma) \ .$$

(a) (η, ψ) is admissible $\Leftrightarrow$ for ν_G-almost every $\sigma \in \hat{G} : \widehat{\psi}(\sigma)^* \widehat{\eta}(\sigma) = \widehat{P}_\sigma$.

(b) There exists an admissible vector for $\mathcal{H}$ iff

$$\nu_\mathcal{H} = \int_{\hat{G}} \mathrm{rank}(\widehat{P}_\sigma) d\nu_G(\sigma) < \infty.$$

Proof. The existence of the $\widehat{P}_\sigma$ follows from the type I property. In the following it is convenient to use tensor-product notation for rank-one operators, i.e., $x \otimes y$ denotes the operator $z \mapsto \langle z, y \rangle x$. Given a fixed $\sigma \in \hat{G}$, the elements of $\mathcal{K} = \mathcal{B}_2(\mathcal{H}_\sigma) \circ \widehat{P}_\sigma$ can be written uniquely as $\eta = \sum_{i\in I} \eta_i \otimes e_i$, where $(e_i)_{i\in I}$ is a fixed orthogonal basis of $\widehat{P}(\mathcal{H}_\sigma)$. It follows that $\mathcal{K}$ is conveniently identified with $\mathcal{H}_\sigma \otimes \ell^2(I)$. In this identification the left action of σ on $\mathcal{K}$ becomes $\sigma \otimes 1$. Moreover, the commuting algebra is easily identified with $1 \otimes \mathcal{B}_2(\ell^2(I))$, and its trace is the usual operator trace.

A weak-operator dense subspace of $\mathcal{B}(\ell^2(I))$ is spanned by the operators $e_{i,k} = \delta_i \otimes \delta_k$, where $\delta_i \in \ell^2(I)$ denotes the usual Kronecker-δ concentrated at i. Now, given $\widehat{\eta}_\sigma = \sum_{i\in I} \eta_i \otimes e_i$ and $\widehat{\psi}_\sigma = \sum_{i\in I} \psi_i \otimes e_i$, we compute

$$tr(e_{i,k}) = \delta_{i,k}$$

and

$$\langle (1 \otimes e_{i,k})\eta, \psi \rangle = \langle \eta_i, \psi_k \rangle \ ,$$

whence we obtain the following traciality condition

$$(\widehat{\eta}_\sigma, \widehat{\psi}_\sigma) \text{ tracial} \iff \forall i, k \ : \ \langle \eta_i, \psi_k \rangle = \delta_{i,k}$$

$$\iff \left(\sum_{i \in I} \psi_i \otimes e_i \right)^* \left(\sum_{i \in I} \eta_i \otimes e_i \right) = \widehat{P}_\sigma \ ,$$

which proves part (a). Part (b) follows easily from (a), see [7]. $\qquad\square$

We wish to point out that the admissibility criteria, however abstract they may appear, have been made explicit for certain classes of representations, in particular for multiplicity-free representations. See [9] for a discussion of quasiregular representations of semidirect product groups, and [8] for a treatment of Weyl-Heisenberg frames with integer sampling ratio.

Another interesting class of representations are the factor subrepresentations of the regular representation, i.e., the atoms in the central decomposition, and the elements of their quasi-equivalence classes. These representations were already considered in [13], though not with a view to constructing admissible vectors.

Corollary 3.3 *Let π be a factor representation.*

(a) *π has admissible vectors iff π is equivalent to a subrepresentation of λ_G, and $\pi(G)'$ is a finite von Neumann algebra. In particular, π has either type I or II, and there exists a faithful, finite, normal trace tr on $\pi(G)'$, unique up to normalization.*

(b) *The trace on $\pi(G)'$ can be normalized in such a way that the following equivalence holds:*

$$(\eta, \psi) \text{ is admissible} \Leftrightarrow (\eta, \psi) \text{ is tracial} \ .$$

4. Checking admissibility using biorthogonality relations

While the discussion of the type I case shows that the characterization of admissible vectors via the trace on the commutant can be used to some effect, in the general case the merits are much less obvious. In this section we sketch a procedure to arrive at more concrete necessary and sufficient conditions for admissible pairs, in terms of certain scalar products. We will then demonstrate that the Wexler-Raz biorthogonality relations are a special instance of this approach.

For the formulation of the admissibility conditions, we require

- A family $(T_i)_{i \in I} \subset \pi(G)'$ spanning a weak-operator dense subspace of $\pi(G)'$. Recall that the density requirement means that for each $S \in \pi(G)'$ there exists a net $(S_j)_{j \in J}$ in the span such that for all pairs $y, z \in \mathcal{H}_\pi$ we have $\langle S_j y, z \rangle \to \langle S y, z \rangle$.
- An admissible pair (η_0, ψ_0).

Then for a pair of bounded vectors (η, ψ) we have the following equivalence:

$$(\eta, \psi) \text{ is admissible} \iff \forall i \in I \ : \ \langle T_i \eta, \psi \rangle = \langle T_i \eta_0, \psi_0 \rangle \ . \tag{62}$$

The proof of the condition is immediate from the assumptions and Theorem 2.2. The criterion is explicit as soon as the T_i and the admissible pair (η_0, ψ_0) are known explicitly. Clearly, generators are preferable which provide particularly simple relations.

Let us now elaborate on Weyl-Heisenberg frames and the associated admissibility criteria. Weyl-Heisenberg frames are obtained by picking a window function $\eta \in L^2(\mathbb{R})$ and translating it along a time-frequency lattice Γ. The shifts are described in terms of the operators

$$T_x : f \mapsto f(\cdot - x) \ , \quad M_\omega : f \mapsto e^{2\pi i \omega \cdot} f \ .$$

For the following we fix $\alpha, \beta > 0$. In the following we assume $\alpha\beta \leq 1$, which is a well-known necessary and sufficient condition for the existence of Weyl-Heisenberg frames. The sufficiency is proved by the admissible vector η_0 given below, for necessity confer, among others, [1, 2, 12]. Given $\eta \in L^2(\mathbb{R})$, we wish to decide whether the family

$$\{M_{\alpha m} T_{\beta n} \eta : m, n \in \mathbb{Z}\}$$

constitutes a frame of $L^2(\mathbb{R})$. Recall that the latter property means that the coefficient map

$$T_{f;\alpha,\beta} : f \mapsto (\langle f, M_{\alpha m} T_{\beta n} \eta \rangle)_{m,n \in \mathbb{Z}}$$

defines a topological embedding $L^2(\mathbb{R}) \hookrightarrow \ell^2(\mathbb{Z}^2)$. As in the proof of Lemma 2.1 we see that f generates a Weyl-Heisenberg frame iff there exists a **dual window** g generating a Weyl-Heisenberg frame and satisfying $T_{g;\alpha,1}^* T_{f;\alpha,\beta} = \mathrm{Id}$.

Since the time-frequency shifts $M_{\alpha m} T_n$ do not constitute a group of operators, the group-theoretic interpretation of the problem requires a slight detour in the form of the next lemma. Note that $\mathbb{T}$ denotes the set of complex numbers with modulus one.

Lemma 4.1 *Define the group $G = \mathbb{Z} \times \mathbb{Z} \times \mathbb{T}$, with group law*

$$(m, n, z)(m', n', z') = (m + m', n + n', zz'e^{-2\pi i \alpha\beta m'n}) .$$

G acts on $L^2(\mathbb{R})$ via the representation

$$\pi(m, n, z) = M_{\alpha m} T_{\beta n} z .$$

For all $(f, g) \in L^2(\mathbb{R})$ with $T_{f;\alpha,\beta}, T_{g;\alpha,\beta}$ bounded, f generates a Weyl-Heisenberg frame with dual window g iff (f, g) is an admissible pair for π.

Proof. The statements concerning G and π are immediate from the definitions. For the last statement, observe that

$$V_g^* V_f h = \int_{\mathbb{T}} \sum_{m,k \in \mathbb{Z}} \langle h, \pi(m, k, z) f \rangle \, \pi(m, k, z) g \, dz$$

$$= \int_{\mathbb{T}} \sum_{m,k \in \mathbb{Z}} \langle h, \pi(m, k, 0) f \rangle \, \pi(m, k, 0) g \, dz$$

$$= T_{g;\alpha,\beta}^* T_{f;\alpha,\beta} .$$

$\square$

REMARK: The representation π is type I iff $\alpha\beta$ is rational. For the only-if part confer [1], Remark 2., whereas the if-part follows from the fact that the group itself is type I if $\alpha\beta$ is rational (a straightforward application of Mackey's theory). In the case where $1/(\alpha\beta) \in \mathbb{Z}$, there exist admissibility criteria which employ the so-called Zak transform; here the representation is even multiplicity-free. See [8] for an interpretation of the Zak transform criterion in the light of Theorem 3.2

Following the general procedure sketched above, we now observe that

- $\eta_0 = \sqrt{\alpha}\chi_{[0,\beta)}$ is an admissible vector [3].
- The **commuting lattice**

$$\Lambda_c = \{M_{m/\beta} T_{n/\alpha} : m, n \in \mathbb{Z}\}$$

generates a weak-operator dense subspace of $\pi(G)'$ ([4, Appendix 6.1]).

Hence, after verifying that

$$\langle M_{m/\beta} T_{n/\alpha} \eta_0, \eta_0 \rangle = \alpha\beta \delta_{m,0} \delta_{n,0} ,$$

we obtain the **Wexler-Raz biorthogonality relations** as a special case of (62):

Corollary 4.2 *Let g, γ be such that $T_{g;\alpha,\beta}, T_{\gamma;\alpha,\beta}$ are bounded. Then γ is a dual window for g iff*

$$\langle M_{m/\beta} T_{n/\alpha} \gamma, g \rangle = \alpha\beta \delta_{m,0} \delta_{n,0} \ . \tag{63}$$

REMARK: A more general "Wexler-Raz-relation" is

$$T_{f;\alpha,\beta}^* T_{g;\alpha,\beta} h = \frac{1}{\alpha\beta} T_{h;1/\beta,1/\alpha}^* T_{g;1/\beta,1/\alpha} f \tag{64}$$

proved for suitable f, g, h in [4]. (64) is easily seen to imply (63). It is not clear whether (64) has a counterpart in the general setting.

Concluding remarks

Von Neumann algebra techniques have been used previously for establishing criteria for the existence of cyclic and/or admissible vectors, see for instance [1, 12]. In particular the coupling constant has proved to be a powerful tool for existence results, see [2, 12]. However, these techniques seem to be of limited use for the explicit construction of admissible vectors. By contrast, this paper aims at providing criteria for these vectors, though it is clear that much remains to be done to make these criteria work. The authors of [4] used the trace on the commuting algebra in the Weyl-Heisenberg frame context, but did not point out the close connection to admissibility.

References

[1] Baggett L 1990, Processing a radar signal and representations of the discrete Heisenberg group, Coll. Math. 60/61 195-203

[2] Bekka M B 2002, Square integrable representations, von Neumann algebras and an application to Gabor analysis, Preprint

[3] Daubechies I and Grossmann A 1986, Painless nonorthogonal expansions, J. Math. Phys. 27 1271-1283

[4] Daubechies I, Landau H J, and Landau Z 1995, Gabor time-frequency lattices and the Wexler-Raz identity, J. Fourier Anal. Appl. 1 437-478

[5] Dixmier J 1977, C*-Algebras (Amsterdam: North Holland)

[6] Dixmier J 1981, Von Neumann Algebras (Amsterdam: North Holland)

[7] Führ H 2002, Admissible vectors for the regular representation, Proc. Amer. Math. Soc. 130 2959-2970

[8] Führ H 2002, Plancherel transform criteria for Weyl-Heisenberg frames with integer oversampling, submitted, available as math.FA/0206309

[9] Führ H and M. Mayer M 2002, Continuous wavelet transform from semidirect products: Cyclic representations and Plancherel measure, J. Fourier Anal. Appl., to appear

[10] Grossmann A, Morlet J, and Paul T 1985, Transforms associated to square integrable group representations I: General results, J. Math. Phys. 26 2473-2479

[11] Kaniuth E 1969, Der Typ der regulären Darstellung diskreter Gruppen, Math. Ann. 182 334-339

[12] Rieffel M 1981, Von Neumann algebras associated to pairs of lattices in Lie groups, Math. Ann. 257 403-418

[13] Rosenberg J 1978, Square-integrable factor representations of locally compact groups, Trans. Amer. Math. Soc. 261 1-33

[14] Thoma E 1968, Eine Charakterisierung diskreter Gruppen vom Typ I, Invent. Math. 6 190-196

Inst. Phys. Conf. Ser. No 173: Satellite colloquium
Paper presented at 24th Int. Coll. Group Theoretical Methods in Physics, Paris, France, 15–20 July 2002
©2003 IOP Publishing Ltd

Localization of frames

Karlheinz Gröchenig ‡
Department of Mathematics, The University of Connecticut, Storrs, CT 06269-3009, USA

Abstract. We discuss stable, overcomplete, non-orthogonal phase-space expansions with respect to generalized coherent states. The key concepts are Weyl-Heisenberg frames and quantitative measures of phase-space localization. On an abstract level, we introduce and investigate the localization of general frames.

1. Introduction

Frames are an important tool for the construction of stable non-orthogonal series expansions with respect to overcomplete systems in a Hilbert space. They have become immensely useful in many applications involving redundant data sets. By definition, frames are a concept related to Hilbert space, but their usefulness in applications stems from additional properties. For instance, "good" wavelet frames can be used to characterize the membership of distributions in Besov-Triebel-Lizorkin spaces [7]; good Gabor frames (often called Weyl-Heisenberg frames) can be used to characterize the phase space distribution of functions [5]; frames consisting of reproducing kernels in certain Hilbert spaces encode pointwise information and yield sampling theorems [1].

These applications go "beyond Hilbert space". The goal is to recognize the finer properties of functions by means of the magnitudes of the frame coefficients. Such properties, typically the smoothness and decay of functions or their phase-space localization, are measured by Banach space norms. Consequently, in studying such properties with frames, the emphasis shifts from the Hilbert space to an associated family of Banach spaces, and we seek to characterize these Banach spaces by the values of the frame coefficients.

In this note, we try to understand which special properties of functions can be characterized by their frame coefficients. Formulated differently, our goal is determine those additional properties that make a frame useful. The key idea is a new, abstract concept for the localization of frames. First we will discuss the localization of Weyl-Heisenberg frames, i.e., discrete sets of (generalized) coherent states; in this case, the abstract concept coincides with the intuitive concept of phase-space localization of quantum mechanical states. Based on this example, we then introduce the localization of abstract frames and develop briefly an axiomatic theory of localized frames. We will formulate some of their most important properties and mention a few applications.

For details, proofs, and extended references we refer to [9].

Notation. A set $\mathcal{E} = \{e_x : x \in \mathcal{X}\}$ in a Hilbert space $\mathcal{H}$ is called a frame if there exist constants $A, B > 0$ such that for all $f \in \mathcal{H}$

$$A\|f\|^2 \leqslant \sum_{x \in \mathcal{X}} |\langle f, e_x \rangle|^2 \leqslant B\|f\|^2 . \tag{1}$$

‡ e-mail: groch@math.uconn.edu

The frame operator imitates an orthogonal expansions and is defined as

$$Sf := S_{\mathcal{E}}f = \sum_{x \in \mathcal{X}} \langle f, e_x \rangle e_x \,. \tag{2}$$

Note that $\mathcal{E}$ is a frame if and only if S is invertible on $\mathcal{H}$.

2. Weyl-Heisenberg frames and phase-space localization

Fix a non-zero function $g \in L^2(\mathbb{R}^d)$ and consider the time-frequency shift or phase space shift defined by

$$\pi(z)g(t) = M_\omega T_x g(t) = e^{2\pi i \omega \cdot t} g(t - x) \,.$$

In quantum mechanics g is a state, $z = (x, \omega) \in \mathbb{R}^{2d}$ is a point in phase-space, $x, t \in \mathbb{R}^d$ are position variables and $\omega \in \mathbb{R}^d$ the momentum variable. In time-frequency analysis g is called a "window", $z \in \mathbb{R}^{2d}$ is a point in the time-frequency plane, x, t are time variables, and $\omega \in \mathbb{R}^d$ is the frequency. For details, proofs, and extended references on time-frequency analysis we refer to [8].

Definition 1 Given lattice constants $\alpha, \beta > 0$, the countable collection of phase-space shifts

$$\mathcal{G}(g, \alpha, \beta) = \{ M_{\beta l} T_{\alpha k} g : k, l \in \mathbb{Z}^d \}$$

is called a *Weyl-Heisenberg system (Gabor system)*. If $\mathcal{G}(g, \alpha, \beta)$ is a frame, we call it a *Weyl-Heisenberg frame (Gabor frame)*. This means that there exist $A, B > 0$ such that

$$A\|f\|_2^2 \leqslant \sum_{k,l \in \mathbb{Z}^d} |\langle f, M_{\beta l} T_{\alpha k} g \rangle|^2 \leqslant B\|f\|_2^2 \tag{3}$$

for all $f \in L^2(\mathbb{R}^d)$. The associated frame operator (the Gabor frame operator) is defined as

$$Sf := S_{g,\alpha,\beta}f = \sum_{k,l \in \mathbb{Z}^d} \langle f, M_{\beta l} T_{\alpha k} g \rangle M_{\beta l} T_{\alpha k} g \,. \tag{4}$$

REMARKS: 1. If $\mathcal{G}(g, \alpha, \beta)$ is a frame, then

$$\gamma = S^{-1}g$$

is well-defined and is called the *dual window*. The Weyl-Heisenberg system $\mathcal{G}(\gamma, \alpha, \beta)$ is again a frame [8, Prop. 5.2.1] and is called the (canonical) *dual (Weyl-Heisenberg) frame*.

2. It is easy to see that S commutes with all phase-space shifts $M_{\beta l} T_{\alpha k}, k, l \in \mathbb{Z}^d$, therefore $S^{-1}(M_{\beta l} T_{\alpha k} g) = M_{\beta l} T_{\alpha k} S^{-1} g = M_{\beta l} T_{\alpha k} \gamma$. By factoring the identity operator, we then obtain two related expansions involving the Weyl-Heisenberg frame $\mathcal{G}(g, \alpha, \beta)$:

$$f = S^{-1}Sf = \sum_{k,l \in \mathbb{Z}^d} \langle f, M_{\beta l} T_{\alpha k} g \rangle M_{\beta l} T_{\alpha k} \gamma \tag{5}$$

$$= SS^{-1}f = \sum_{k,l \in \mathbb{Z}^d} \langle f, M_{\beta l} T_{\alpha k} \gamma \rangle M_{\beta l} T_{\alpha k} g \tag{6}$$

Here (5) is a reconstruction of f from the frame coefficients $\langle f, M_{\beta l} T_{\alpha k} g \rangle$, whereas (6) is a (non-orthogonal) series expansion with respect to the set $\mathcal{G}(g, \alpha, \beta)$. Both series converge unconditionally in $L^2(\mathbb{R}^d)$ [8].

The coefficients $\langle f, M_{\beta l} T_{\alpha k} g \rangle$ and $\langle f, M_{\beta l} T_{\alpha k} \gamma \rangle$ are interpreted as the phase-space content of the state f in the phase-space cell centered at $(\alpha k, \beta l) \in \mathbb{R}^{2d}$. In time-frequency analysis they measure the amplitude of the frequency band near βk at time αk.

To see how well this interpretation works, we consider three extreme examples.

Example I. Let $g = \chi_{[0,1]} \in L^2(\mathbb{R})$ and choose $\alpha \leqslant 1, \beta \leqslant 1$. Then $\mathcal{G}(g, \alpha, \beta)$ is frame for $L^2(\mathbb{R})$ [2, 8] and by (5) and (6) there exists a dual window $\gamma \in L^2(\mathbb{R}^d)$ such that

$$f = \sum_{k,l \in \mathbb{Z}} \langle f, M_{\beta l} T_{\alpha k} \chi \rangle M_{\beta l} T_{\alpha k} \gamma . \tag{7}$$

This series expansion is perfectly adequate to describe the L^2-behavior of f, and even its decay, since χ is well-localized in position. However, the coefficients $\langle f, M_{\beta l} T_{\alpha k} \chi \rangle$ fail to provide useful information about the momentum distribution of f. For example, it is impossible to characterize a Schwartz function solely by the size of these coefficients. The reasons for this failure are twofold: firstly, $\hat{\chi}$ decays like $|\omega|^{-1}$ and is thus badly localized in momentum space. As a consequence, the dual window γ cannot be in the Schwartz class $\mathcal{S}$ and so the series (7) does not converge in $\mathcal{S}$. Secondly, we may interpret the sum over $l \in \mathbb{Z}$ as the Fourier series of the restriction $f \cdot T_{\alpha k} \chi$. The cut-off with a characteristic function introduces artificial discontinuities, thus $(\langle f, M_{\beta l} T_{\alpha k} \chi \rangle : l \in \mathbb{Z}\}) \notin \ell^1$ in general.

Example II. Next consider the Gaussian window $\phi(t) = e^{-\pi t^2}$ and the associated classical coherent states. By a fundamental theorem of Seip-Wallsten [15], $\mathcal{G}(\phi, \alpha, \beta)$ *is frame for $L^2(\mathbb{R})$ if and only if $\alpha\beta < 1$.*

In this case, the Schwartz class $\mathcal{S}$ possesses a natural characterization in terms of the Gabor coefficients [5]: a function $f \in L^2(\mathbb{R})$ belongs to $\mathcal{S}(\mathbb{R})$ if and only if

$$|\langle f, M_{\beta l} T_{\alpha k} \phi \rangle| = \mathcal{O}(1 + |k| + |l|)^{-N}) \qquad \text{for all } N \geqslant 0 . \tag{8}$$

Why does the phase-space characterization of Schwartz functions work in this case? The answer lies in a deep theorem of Janssen [13]: *If $g \in \mathcal{S}$, then the (canoncial) dual window γ is also in $\mathcal{S}$.* Based on Janssen's Theorem, it can be shown that the partial sums of the reconstruction series

$$f = \sum_{k,l \in \mathbb{Z}} \langle f, M_{\beta l} T_{\alpha k} \phi \rangle M_{\beta l} T_{\alpha k} \gamma \tag{9}$$

converge in $\mathcal{S}$. In this example, our intuition about phase-space localization is correct.

Intuitively, the difference between these two examples is easy to understand. The rectangular function $\chi_{[0,1]}$ lacks localization in phase space,
expansions we need sets of well localized coherent states.

Example III. It is well-known that the set $\mathcal{G}(\phi, 1, 1) = \{M_l T_k \phi : k, l \in \mathbb{Z}\}$ is complete in $L^2(\mathbb{R})$, however, it is *not* a frame. Although series expansions of the form (9) exist, they are not stable. Even for $f \in \mathcal{S}$, the partial sums converge only in the sense of tempered distributions, but in general not even in $L^2(\mathbb{R})$ [12]. even in $L^2(\mathbb{R})$, let alone in $\mathcal{S}$. Therefore the discrete set of coherent states $\mathcal{G}(\phi, 1, 1)$ on the lattice $\mathbb{Z} \times \mathbb{Z}$ is not useful in applications. Although $\mathcal{G}(\phi, 1, 1)$ is well localized in phase-space, it does not possess a dual frame with similar localization properties.

3. Measures of phase-space concentration and modulation spaces

To quantify the phase-space localization of a state, we use a suitable phase-space representation of a function or distribution f. For our purpose the most convenient phase-space distribution is the short-time Fourier transform.

Definition 2 Fix a nonzero function $g \in \mathcal{S}(\mathbb{R}^d)$. Then the *short-time Fourier transform* (STFT) of a function $f \in L^2(\mathbb{R}^d)$ or a tempered distribution $f \in \mathcal{S}'(\mathbb{R}^d)$ with respect to g is defined as

$$V_g f(x, \omega) = \langle f, M_\omega T_x g \rangle = \int_{\mathbb{R}^d} f(t) g(t - x) \, e^{2\pi i \omega \cdot t} \, dt. \tag{10}$$

Here $V_g f(x, \omega)$ is interpreted as the phase-space content of f in a neighborhood of the point (x, ω) in phase-space. Of course, it depends also on the window g, but the decay behavior of $V_g f$ is usually independent of the particular choice of g (see below).

With slightly different normalizations the STFT is also known under names such as coherent state transform, ambiguity function, or (cross) Wigner distribution, see [8, Ch. 3,4].

We can now introduce measures for the phase-space localization of a function by imposing integrability and decay conditions on $V_g f$. We use (mixed) L^p-norms to obtain Banach space norms, and suitable weight functions $m \geq 0$ to fine-tune the decay of f in phase-space $\mathbb{R}^{2d}$.

For simplicity we will only use symmetric weight functions of polynomial growth satisfying one of the following conditions:

$$m(z_1 + z_2) \leq C \, (1 + |z_1|)^s \, m(z_2) \qquad \text{for some } s \geq 0 \tag{11}$$

$$m(z_1 + z_2) \leq C \, e^{A|z_1|^c} \, m(z_2), \qquad \text{for some } 0 \leq c < 1. \tag{12}$$

for all $z_1, z_2 \in \mathbb{R}^{2d}$. In the former case we call m s-moderate, in the latter case m is a sub-exponential weight.

We now introduce phase-space localization as follows.

Definition 3 Fix a nonzero $g \in \mathcal{S}(\mathbb{R}^d)$ and a weight satisfying (11) or (12). Then the *modulation space* $M_m^{p,q}$ consists of all tempered distributions $f \in \mathcal{S}'(\mathbb{R}^d)$, for which the norm

$$\|f\|_{M_m^{p,q}} := \left(\int_{\mathbb{R}^d} \left(\int_{\mathbb{R}^d} |V_g f(x, \omega)|^p \, m(x, \omega)^p \, dx \right)^{q/p} d\omega \right)^{1/q} \tag{13}$$

is finite. If $p = q$, we write $M_m^{p,p} = M_m^p$, and if $m \equiv 1$, we write $M^{p,q} = M_m^{p,q}$.

REMARKS: 1. It can be shown that $M_m^{p,q}$ is a Banach space and that the definition is independent of the choice of the window g in (13), see [8], Thms. 11.3.5 and 11.3.7.

2. Modulation spaces have been introduced and investigated by H. G. Feichtinger in the 1980's in analogy to the Besov spaces [4]. Later they were recognized to be the appropriate function spaces for phase-space analysis or time-frequency analysis, and now they occur in many applications, for instance, in the formulation of uncertainty principles, the construction of phase-space expansions, non-linear approximation with Weyl-Heisenberg frames, or as symbol classes for pseudodifferential operators. For references and precise details we refer to [8] and the original literature.

4. Window design

The three examples discussed above suggest that for useful phase-space expansions it is imperative to use a *pair of dual windows* (g, γ) with simultaneous, good phase-space localization. The construction of suitable pairs of dual windows is known as the problem of *window design* and has become the subject of intensive research. Here we formulate the strongest result known so far [5, 10].

Theorem 1 ([10]) *Assume that (a) $\mathcal{G}(g, \alpha, \beta)$ is a Weyl-Heisenberg frame for $L^2(\mathbb{R}^d)$ and that (b) $g \in M_v^1$ where either $v(z) = (1+|z|)^s$ for $s \geqslant 0$ or $v(z) = e^{A|z|^c}$ for $A > 0$ and $0 \leqslant c < 1$. Then the Gabor frame operator S is invertible simultaneously on all modulation spaces $M_m^{p,q}$ where m satisfies either (11) or (12). Consequently $\gamma = S^{-1}g \in M_v^1$ as well.*

Theorem 1 says that both the window g and the dual window γ possess the same phase-space localization. According to the theorem it suffices to choose a *single* window g with suitable phase-space localization and then verify that the Weyl-Heisenberg set $\mathcal{G}(g, \alpha, \beta)$ is a frame for $L^2(\mathbb{R}^d)$. Then the dual window $\gamma = S^{-1}g$ possesses automatically the same phase-space localization.

Theorem 1 is fairly deep. The proof requires tools from the theory of symmetric Banach algebras, the analysis of convolution operators on a certain Heisenberg group, and a theorem of Ludwig about the symmetry of nilpotent groups. See [10] for details.

5. Time-frequency analysis of distributions

Next we investigate the phase-space expansions with "good" windows. As a consequence of Theorem 1 we can characterize the phase-space localization of tempered distributions entirely by the magnitude of the Weyl-Heisenberg frame coefficients [5, 8]. In the following v is either $v(z) = (1 + |z|)^s$ for $s \geqslant 0$ or $v(z) = e^{A|z|^c}$ for $A > 0$ and $0 \leqslant c < 1$, and m satisfies (11) or (12).

Theorem 2 *Assume that (i) $\mathcal{G}(g, \alpha, \beta)$ is frame for $L^2(\mathbb{R}^d)$ and that (ii) $g \in M_v^1$ (consequently also $\gamma \in M_v^1$). Then the following are equivalent:*

(a) $f \in M_m^{p,q}$.

(b) $\left(\sum_{l \in \mathbb{Z}^d} \left(\sum_{k \in \mathbb{Z}^d} |\langle f, M_{\beta l} T_{\alpha k} g \rangle|^p m(\alpha k, \beta l)^p \right)^{q/p} \right)^{1/q} < \infty$.

(c) $\left(\sum_{l \in \mathbb{Z}^d} \left(\sum_{k \in \mathbb{Z}^d} |\langle f, M_{\beta l} T_{\alpha k} \gamma \rangle|^p m(\alpha k, \beta l)^p \right)^{q/p} \right)^{1/q} < \infty$.

In this case the frame expansion $f = \sum_{k,l \in \mathbb{Z}^d} \langle f, M_{\beta l} T_{\alpha k} g \rangle M_{\beta l} T_{\alpha k} \gamma$ converges unconditionally in $M_m^{p,q}$ for $1 \leqslant p, q < \infty$ (and weakly if $p = \infty$ or $q = \infty$).

In some situations it may be useful to work with non-uniform Weyl-Heisenberg frames. In this case we consider an arbitrary discrete set $\mathcal{Z} = \{z_j = (x_j, \omega_j) \in \mathbb{R}^{2d}\}$ in phase-space and the corresponding *non-uniform Weyl-Heisenberg system* $\{\pi(z_j)g = M_{\omega_j} T_{x_j} g : j \in J\}$.

If $\{\pi(z_j)g : j \in J\}$ is frame for $L^2(\mathbb{R}^d)$, then by frame theory we know that

$$f \in L^2(\mathbb{R}^d) \iff \sum_{j \in J} |\langle f, \pi(z_j)y \rangle|^2 < \infty.$$

But again the question arises how other properties of functions can be characterized by means of the coefficients $\langle f, \pi(z_j)g \rangle, j \in J$, in particular the phase-space localization of f. For instance, it is not even clear how a function $f \in \mathcal{S}$ or $f \in M^1$ can be characterized by the set of coefficients $\langle f, \pi(z_j)g \rangle, j \in J$.

To answer questions of this type, we again need more information about the dual frame $\{\tilde{e}_j := S^{-1}(\pi(z_j)g) : j \in J\}$ and we need to understand the convergence of frame expansion

$$f = \sum_{j \in J} \langle f, \pi(z_j)g \rangle \tilde{e}_j = S^{-1}\left(\sum_{j \in J} \langle f, \pi(z_j)g \rangle \pi(z_j)g \right). \tag{14}$$

However, since $\mathcal{Z}$ is not a lattice, the frame operator no longer commutes with the phase-space shifts $\pi(z_j)$ and lacks some important invariance properties. So the structure of the dual frame is much more complicated. The treatment of non-uniform Weyl-Heisenberg frames is therefore considered a very difficult problem.

6. Axiomatic theory of frame localization

To obtain a characterization of phase-space localization with non-uniform Weyl-Heisenberg frames in the style of Theorem 2, we now digress for a moment and turn to abstract frames and their localization. Inspired by the example of non-uniform Weyl-Heisenberg frames, we have developed a concept of localization for arbitrary frames. In the following we consider index sets $\mathcal{N}, \mathcal{X} \subseteq \mathbb{R}^d$. We assume that $\mathcal{N}$ indexes an orthonormal basis $\{g_n : n \in \mathcal{N}\}$ of a Hilbert space $\mathcal{H}$.

The theory of localized frames involves several elements: (a) a class of Banach spaces associated with $\mathcal{H}$ and a fixed orthonormal basis $\{g_n\}$, (b) a suitable notion of localization, and (c) the analysis of the frame operator on the associated Banach spaces.

In the following m is again either an s-moderate or a sub-exponential weight function, i.e., m satisfies either (11) or (12).

Definition 4 Assume that $\ell_m^p(\mathcal{N}) \subseteq \ell^2(\mathcal{N})$. Then $\mathcal{H}_m^p$ is the subspace of $\mathcal{H}$ defined by

$$\mathcal{H}_m^p = \{f \in \mathcal{H} : f = \sum_{n \in \mathcal{N}} c_n g_n \text{for } c \in \ell_m^p(\mathcal{N})\} \tag{15}$$

with norm $\|f\|_{\mathcal{H}_m^p} = \|c\|_{\ell_m^p}$.

For several concrete orthonormal bases, the Banach spaces $\mathcal{H}_m^p$ turn out to be well known function spaces. Choosing a Wilson basis for $L^2(\mathbb{R}^d)$ of exponential decay in phase-space, we have $\mathcal{H}_m^p = M_m^p$ for $1 \leqslant p \leqslant 2$ [6]. On the other hand, if we choose a wavelet basis for $L^2(\mathbb{R}^d)$, then the $\mathcal{H}_m^p$ coincide with certain Besov spaces [14].

Now we introduce the key concept.

Definition 5 A frame $\mathcal{E} = \{e_x : x \in \mathcal{X}\}$ of $\mathcal{H}$ is s-*localized* with respect to the orthonormal basis $\{g_n\}$, if

$$|\langle e_x, g_n \rangle| \leqslant C(1 + |x - n|)^{-s} \tag{16}$$

for all $n \in \mathcal{N}$ and $x \in \mathcal{X}$.

Similarly, $\mathcal{E}$ is *exponentially localized*, if for some $\alpha > 0$

$$|\langle e_x, g_n \rangle| \leqslant C e^{-\alpha |x - n|}$$

for all $n \in \mathcal{N}$ and $x \in \mathcal{X}$.

Whereas the frame operator of a general frame is defined only on the Hilbert space $\mathcal{H}$ and usually carries no information beyond $\mathcal{H}$, the frame operator of localized frames is well-defined on the associated Banach spaces $\mathcal{H}_m^p$. The next result states some properties of localized frames and their frame operator.

Proposition 3 *Given $1 \leqslant p \leqslant \infty$, $s \geqslant 0, \epsilon > 0$ and an s-moderate weight m on $\mathbb{R}^d$.*
If $\mathcal{E}$ is $(s + d + \epsilon)$-localized frame for $\mathcal{H}$, then the following properties hold:
(a) The coefficient operator $f \to (\langle f, e_x \rangle)_{x \in \mathcal{X}}$ is bounded from $\mathcal{H}_m^p$ to $\ell_m^p(\mathcal{X})$.
(b) The synthesis operator $c \to \sum_{x \in \mathcal{X}} c_x e_x$ is bounded from $\ell_m^p(\mathcal{X})$ to $\mathcal{H}_m^p$.
(c) The frame operator $S = \sum_{x \in \mathcal{X}} \langle f, e_x \rangle e_x$ maps $\mathcal{H}_m^p$ into $\mathcal{H}_m^p$, and the series converges unconditionally for $1 \leqslant p < \infty$ (and weakly for $p = \infty$).
If $\mathcal{E}$ is exponentially localized, then (a) — (c) hold for all sub-exponential weights.

Our main theorem is analogous to Theorem 1 for Weyl-Heisenberg frames and states that the dual frame $\{\widetilde{e}_x := S^{-1} e_x, x \in \mathcal{X}\}$ inherits its localization properties from the original frame. Recall that the constant d is the dimension of the index sets $\mathcal{X}$ and $\mathcal{N} \subseteq \mathbb{R}^d$.

Theorem 4 *Assume that $\mathcal{E} = \{e_x : x \in \mathcal{X}\} \subseteq \mathcal{H}$ is an $(s + d + \epsilon)$-localized frame with respect to some orthonormal basis $\{g_n\}$ for some $\epsilon > 0$.*

(a) Then frame operator S is invertible simultaneously on all Banach spaces $\mathcal{H}_m^p$, where $1 \leqslant p \leqslant \infty$ and m is an s-moderate weight.

(b) The dual frame $\{\widetilde{e}_x = S^{-1}e_x : x \in \mathcal{X}\}$ is also $(s + d + \epsilon)$-localized, and the elements $\widetilde{e}_x$ satisfy the estimate

$$|\langle \widetilde{e}_x, g_n \rangle| \leqslant C'(1 + |n - x|)^{-s-d-\epsilon} \tag{17}$$

uniformly for all $x \in \mathcal{X}$ and $n \in \mathcal{N}$.

(c) The frame expansion

$$f = \sum_{x \in \mathcal{X}} \langle f, e_x \rangle \widetilde{e}_x = \sum_{x \in \mathcal{X}} \langle f, \widetilde{e}_x \rangle e_x \tag{18}$$

converges unconditionally in $\mathcal{H}_m^p$ for $1 \leqslant p < \infty$.

(d) The elements in $\mathcal{H}_m^p$ can be characterized by the magnitude of the frame coefficients, and we have the norm equivalence

$$\|f\|_{\mathcal{H}_m^p} \asymp \left(\sum_{x \in \mathcal{X}} |\langle f, e_x \rangle|^p m(x)^p \right)^{1/p} \asymp \left(\sum_{x \in \mathcal{X}} |\langle f, \widetilde{e}_x \rangle|^p m(x)^p \right)^{1/p}. \tag{19}$$

In its essence, Theorem 4 states that the *dual frame possesses same localization properties as original frame*. An analogous result holds for exponentially localized frames.

In contrast to Theorem 1 there is no group theoretic structure in the background. This makes the proof of Theorem 4 much more difficult. The main tool to prove the estimates (17) is an important theorem of Jaffard about off-diagonal decay of inverse matrices [11].

7. Non-uniform Weyl-Heisenberg frames revisited

Applying Theorem 4 to the case of non-uniform Weyl-Heisenberg frames discussed earlier, we can now answer the question of how to measure phase-space localization with non-uniform Weyl-Heisenberg frames. In this case, the underlying orthonormal basis is a Wilson basis [3], and the abstract localization property (16) can be shown to coincide with a modulation space norm. With some work, one can then prove the following result. (In the statement we use a metric $d(z, w)$ on $\mathbb{R}^d$ that is related to the Euclidean metric, but that incorporates the symmetry properties of a Wilson basis. We omit the definition.)

Theorem 5 *Let $\mathcal{Z} = \{z_j : j \in J\}$ be a separated set in $\mathbb{R}^{2d}$ and let $\{\pi(z_j)g : j \in J\}$ be a non-uniform Weyl-Heisenberg frame for $L^2(\mathbb{R}^d)$ satisfying the localization estimate*

$$|V_g g(z)| \leqslant C(1 + |z|)^{-s-2d-\epsilon}, \qquad \text{for some } \epsilon > 0. \tag{20}$$

Then the frame operator $Sf = \sum_{j \in J} \langle f, \pi(z_j)g \rangle \pi(z_j)g$ is invertible simultaneously on all M_m^p for each $1 \leqslant p \leqslant \infty$ and all s-moderate weights m.

The dual frame $\widetilde{e}_{z_j} = S^{-1}(\pi(z_j)g)$ satisfies the localization estimates

$$|V_g \widetilde{e}_{z_j}(w)| \leqslant C'(1 + d(w - z_j))^{-s-2d-\epsilon} \quad j \in J.$$

The frame expansions

$$f = \sum_{j \in J} \langle f, \pi(z_j)g \rangle \widetilde{e}_{z_j} = \sum_{j \in J} \langle f, \widetilde{e}_{z_j} \rangle \pi(z_j)g$$

converge unconditionally in the modulation spaces M_m^p for $1 \leqslant p < \infty$, and

$$A\|f\|_{M_m^p} \leqslant \left(\sum_{j \in J} |\langle f, \pi(z_j)g \rangle|^p m(z)|^p \right)^{1/p} \leqslant B\|f\|_{M_m^p}, \quad f \in M_m^p.$$

In particular, $f \in \mathcal{S}(\mathbb{R}^d)$ if and only if $|\langle f, \pi(z_j)g \rangle| = \mathcal{O}(|z_j|^{-N})$ for all $N \geqslant 0$.

Similar conclusions hold for exponentially localized Weyl-Heisenberg frames, i.e., assuming $|V_g g(z)| \leqslant C^{-\alpha|z|}$.

References

[1] Aldroubi A and Gröchenig K 2001, Nonuniform sampling and reconstruction in shift-invariant spaces, SIAM Rev. 43(4) 585–620

[2] Daubechies I, Grossmann A, and Meyer Y 1986, Painless nonorthogonal expansions, J. Math. Phys. 27 1271–1283

[3] Daubechies I, Jaffard S, and Journé J-L 1991, A simple Wilson orthonormal basis with exponential decay, SIAM J. Math. Anal. 22(554–573

[4] Feichtinger H G (1989), Atomic characterizations of modulation spaces through Gabor-type representations, Proc. Conf. Constructive Function Theory, Edmonton, July 1986, Rocky Mount. J. Math. , 19 113–126

[5] Feichtinger HG and Gröchenig K 1997, Gabor frames and time-frequency analysis of distributions, J. Functional Anal. 146 464–495

[6] Feichtinger HG, Gröchenig K, and Walnut D 1992, Wilson bases and modulation spaces, Math. Nachr. 155 7–17

[7] Frazier M and Jawerth B 1990, A discrete transform and decompositions of distribution spaces, J. Functional Anal. 93 34–170

[8] Gröchenig K 2001, Foundations of Time-Frequency Analysis (Boston: Birkhäuser)

[9] Gröchenig K 2002, Localization of frames, Banach frames, and the invertibility of the frame operator, Preprint

[10] Gröchenig K and Leinert M 2001, Wiener's lemma for twisted convolution and Gabor frames, Preprint

[11] Jaffard S 1990, Propriétés des matrices "bien localisées" près de leur diagonale et quelques applications, Ann. Inst. H. Poincaré Anal. Non Linéaire 7 461–476

[12] Janssen AJEM 1981, Gabor representation of generalized functions, J. Math. Anal. Appl. 83 377–394

[13] Janssen AJEM 1995, Duality and biorthogonality for Weyl-Heisenberg frames, J. Fourier Anal. Appl. 1 403–436

[14] Meyer Y 1990, Ondelettes et opérateurs. I Ondelettes. (Paris: Hermann)

[15] Seip K and Wallstén R 1992, Density theorems for sampling and interpolation in the Bargmann-Fock space. I , J. Reine Angew. Math. 429 107–113

Inst. Phys. Conf. Ser. No 173: Satellite colloquium
Paper presented at 24th Int. Coll. Group Theoretical Methods in Physics, Paris, France, 15–20 July 2002
©2003 IOP Publishing Ltd

Lattices of coherent states and square integrability

A L Hohouéto[†], T Kengatharam[†], S T Ali[†], and J-P Antoine[‡] [†]

[†] Department of Mathematics and Statistics, Concordia University, Montréal, Canada
[‡] Institut de Physique Théorique, Université Catholique de Louvain,
Louvain-La-Neuve, Belgium

Abstract. We develop a method of discretization of the continuous theory of coherent states on a general semidirect product Lie group. The group is assumed to have a unitary representation which is square integrable on some homogeneous space. We show also that the existence of a discrete frame of coherent states in the carrier space of a unitary representation of such a group implies the square integrability of this representation on the label space.

1. Introduction

Let V be a n-dimensional real vector space, $S \subset GL(V)$ be a semisimple connected Lie group acting on V, and $G = V \rtimes S$ the resulting semidirect product group with the group law

$$(x, s) \cdot (x', s') = (x + s \cdot x', ss'), \quad (x, s), (x', s') \in G \quad (1)$$

$s \cdot x$ denoting the action of S on V. Wavelet groups are of this form, with $V = \mathbb{R}^n$ as the group of translations, $S = \mathbb{R}_+ \times H$, where $\mathbb{R}_+$ is the groups of dilations and $H \subset GL(n, \mathbb{R})$ is the group of rotations. The relativity groups are of the same form, but without a subgroup of dilations. In the case of wavelet groups, the relationship between the existence of a discrete frame of wavelets (coherent states) in the carrier space $\mathfrak{H}$ of a unitarily induced representation U of G and the square integrability of U on G is known [1, 2]. In this paper, we extend Aniello *et al.*'s results in order to encompass the case of the relativity groups. Now, the square integrability of U does no hold on the group any more, but is to be taken in a broad sense, that is, modulo an appropriate subgroup H_0 (or, what is the same, modulo a section $\sigma : \Gamma = G/H_0 \to G$) [3]. Thus, in Section 2, we show that, for a suitably chosen subgroup H_0 and a unitarily induced representation of G which is square integrable on Γ, there always exists in $\mathfrak{H}$ a discrete frame of coherent states labeled by (elements of) Γ. This proceeds from a discretization (sampling) of the continuous theory of coherent states of a general semidirect product group [3]. We give an explicit construction of such a frame in the simple case of the Euclidean group in two dimensions. In Section 3, we address the converse problem and show that the converse also occurs as well. We even point out that, for a certain class of sections, the existence of a discrete frame of coherent states in $\mathfrak{H}$ ensures the existence of a continuous one.

2. Sampling the continuous theory

Let V^* be the dual of V, k_0 be a fixed element of V^*, S_0 be its stabilizer under the dual action $S \times V^* \to V^*$ defined by

$$(s[k], x) = (k, s^{-1} \cdot x), \quad \forall s \in S, x \in V, k \in V^*, \quad (2)$$

[†] e-mail: al_hohoueto@yahoo.fr, santhar@vax2.concordia.ca, stali@mathstat.concordia.ca,
antoine@fyma.ucl.ac.be

and $\mathcal{O}^* \simeq S/S_0$ its orbit under the dual action. We will assume that $\mathcal{O}^*$ is m-dimensional as manifold ($m \leq n$). Set $V_{k_0}^* = T_{k_0}\mathcal{O}^*$ the tangent space of $\mathcal{O}^*$ at k_0 and consider N_0 the annihilator of V_0^* in V. $H_0 = N_0 \rtimes S_0$ is a closed subgroup of G; therefore, we can form the homogeneous space $\Gamma = G/H_0$, which has a coadjoint orbit structure. $V_0 = T_{k_0}^*\mathcal{O}^*$ denoting the cotangent space of $\mathcal{O}^*$ at k_0, Γ is isomorphic to $V_0 \times \mathcal{O}^*$ as Borel space, while as symplectic manifold it is isomorphic to the cotangent bundle $T^*\mathcal{O}^* = \bigcup_{k \in \mathcal{O}^*} T_k^*\mathcal{O}^*$. The construction of coherent states exploits the structure of the principal fiber bundle $\pi : G \to \Gamma$, and, then, relies on the choice of suitable (Borel) sections $\sigma : \mathcal{O}^* \to S$ and $\sigma_{pr} : \Gamma \to G$. For σ, we choose a smooth section on some dense open set $\mathcal{O} \subset \mathcal{O}^*$. The base section σ_{pr} is defined, for $(q, p) \in \Gamma = V_0 \times \mathcal{O}^*$, by

$$\sigma_{pr}(q, p) = (\sigma(p)q, \sigma(p)). \tag{3}$$

The admissible sections $\widehat{\sigma} : \Gamma \to G$ are of the form

$$\widehat{\sigma}(q, p) = \sigma_{pr}(q, p)(n(q, p), s_0(p)) = (\widehat{q}, \Lambda(p)), \tag{4}$$

where $n(q, p) = \theta(p)q + \varphi(p)$, $s_0 : \mathcal{O}^* \to S_0$, $\theta : \mathcal{O}^* \to \mathcal{L}(V)$, $\varphi : \mathcal{O}^* \to N_0$ are smooth functions, and $\mathrm{Ker}\,\theta = \mathrm{Ran}\,\theta = N_0$. Because of its irrelevance in the theory (it corresponds essentially to a change of gauge), φ is set to zero. The space Γ is endowed with an invariant measure μ of the form

$$d\mu(q, p) = \rho(p)dqd\nu(p), \quad \rho(p) = \frac{f(p)}{m(p)}, \quad f, m \text{ positive and } C^\infty, \tag{5}$$

where dq is the Lebesgue measure on V_0 and $d\nu$ is the Haar measure on $\mathcal{O}^*$. A local chart of $\mathcal{O}^*$ can be used to make the functions ρ smooth and nonzero almost everywhere (with respect to $d\nu$) on $\mathcal{O}^*$. Then, considering the induced representation $U = \mathrm{Ind}_{H_0}^G \chi \otimes L$, where χ is a unitary character of V_0 and L a unitary irreducible representation of S_0 in the N-dimensional Hilbert space $\mathfrak{K}$, choosing a set of linearly independent vectors $\{\eta^j ; j = 1, \ldots, N\}$ in the carrier space $\mathfrak{H} = \mathfrak{K} \otimes L^2(\mathcal{O}^*, d\nu)$ of U such that the projection operator $\mathbb{F} = \sum_{j=1}^n |\eta^j\rangle\langle\eta^j|$ is invariant under S_0, it can be shown (under some further admissibility condition on $\widehat{\sigma}$ and some regularity and support conditions on the η^js) that U is square integrable and the operator

$$A_{\widehat{\sigma}} = \int_\Gamma U(\widehat{\sigma}(q, p))\mathbb{F}U(\widehat{\sigma}(q, p))^\dagger d\mu(q, p) \tag{6}$$

defines a rank-N frame (that is, it is self-adjoint, positive, bounded, invertible with a bounded inverse) if and only if there exists two reals $0 < A \leq B < \infty$ such that

$$A \leq (2\pi)^n \sum_{j=1}^N \int_{\mathcal{O}^*} \|\eta^j(\sigma(p)^{-1}[k])\|_{\mathfrak{K}}^2 \frac{m(\sigma(p)^{-1}[k])}{|\mathcal{I}(p, k)|}\rho(p)d\nu(p) \leq B. \tag{7}$$

$\mathcal{I}(p, k)$ is the determinant of the Jacobian of the adjoint map $F(p)^* = (\mathbb{I} + \theta(p))^*$ restricted to O_{k_0}, (O_{k_0}, ψ) being some suitably chosen local chart of $\mathcal{O}^*$ at k_0. For more details, see [3].

The discretization of the continuous theory is based on a result due to Borel [4]. By this result, there exists in S a uniform lattice $\mathcal{X}_I = \{s_\tau; \tau \in I \subset \mathbb{N}\}$, that is, a discrete co-compact subgroup. Therefore, there is a compact set $\mathcal{Q} \subset \mathcal{O}^*$ such that $\mathcal{O}^* = \bigcup_{\tau \in I} s_\tau[\mathcal{Q}]$. The pair $(\mathcal{X}_I, \mathcal{Q})$ is called a *frame generator* [5]. In fact, using the compactness of $\mathcal{Q}$, it can be shown [1] that, for any open set $\Omega \subset \mathcal{O}^*$, there is a natural integer N_0 such that

$$\mathcal{O}^* = \bigcup_{\tau \in I} \bigcup_{\nu=0}^{N_0} s_\tau s_\nu[\Omega]. \tag{8}$$

Consider next, as in the continuous case, a set of η^js with supports $K_j \subset O_{k_0}$ such that

- $k_0 \in \overset{\circ}{K}_j$ for all j,
- $s_0(p)[K_j] \subset O_{k_0}$, for all p,
- $\psi(F(p)^* s_0(p)[K_j]) = R_j$ is a regular hyperparallelepiped in $\mathbb{R}^m$ for all p,

and set $\Omega = \bigcap_{j=1}^N \overset{\circ}{K}_j$ and $K = \bigcup_{j=1}^N K_j$. A finite open covering of K can be constructed such that $K \subset \bigcup_{n \in \Theta} \Lambda(p_n)^{-1}[\Omega]$ ($\Theta \subset \mathbb{N}$) and the open sets $\Lambda(p_n)^{-1}[\Omega]$ are pairwise disjoint. Take, for each j,

$$\eta^j = f^j \otimes \eta \chi_{K_j}, \tag{9}$$

where $\{f^j, j = 1, \ldots, N\}$ is an orthonormal basis of $\mathfrak{K}$, η is a continuous complex-valued function in $L^2(\mathcal{O}^*, d\nu)$, and χ_{K_j} is the characteristic function of K_j. Denoting by $R_{j,\kappa}, \kappa = 1, \ldots, m$, the edges of R_j, using an admissible affine section $\hat{\sigma}$, and defining, for $l = (l_\kappa)$ in $\mathbb{Z}^m$,

$$q_l^j = (q_{l_\kappa}^j)_{\kappa=1}^m, \quad \text{with} \quad q_{l_\kappa}^j = [\text{measure}(R_{j,\kappa})]^{-1} 2\pi l_\kappa,$$
$$\mu^j = \text{volume}(R_j),$$

it can be shown that, for $\tau \in I$, $\nu = 0, \ldots, N_0$, $n \in \Theta$, $l \in \mathbb{Z}^m$ and $j = 1, \ldots, N$, the vectors

$$\eta_{\tau,\nu,n,l}^j = U(s_\tau s_\nu \Lambda(p_n) \hat{q}_{n,l}^j, s_\tau s_\nu \Lambda(p_n)) \eta^j, \tag{10}$$

form a discrete frame in $\mathfrak{H}$, that is, there exists A, B real such that

$$0 < A\|\phi\|_{\mathfrak{H}}^2 \leqslant \sum_{j=1}^N \sum_{\tau \in I} \sum_{\nu=0}^{N_0} \sum_{n \in \Theta} \sum_{l \in \mathbb{Z}^m} |\langle \eta_{\tau,\nu,n,l}^j | \phi \rangle_{\mathfrak{H}}|^2 \leqslant B\|\phi\|_{\mathfrak{H}}^2 < \infty, \tag{11}$$

for all ϕ in $\mathfrak{H}$. See [6] for a proof.

Let us apply this construction to the Euclidean group in two dimensions $E(2) = \mathbb{R}^2 \rtimes SO(2)$ used as a pedagogical example. Details on the continuous theory of coherent states of that group can be found in [7]. The elements of the group are of the form $(x, r_\theta) \simeq (x, \theta)$, $x \in \mathbb{R}^2$, $\theta \in [0, 2\pi)$, where r_θ denotes the rotation in the plane by the angle θ. The orbits of the elements of $\mathbb{R}^2$ (taken as the dual of $\mathbb{R}^2$) are circles. We fix the element $k_0 = (1, 0)$ in $\mathbb{R}^2$ and take the unit circle S^1 as its orbit. Its stabilizer is $S_0 = \{\mathbb{I}_2\}$. Under the usual parametrization of the unit circle $\psi(p) = (\cos p, \sin p)$, $p \in (-\pi, \pi)$ and the coordinatization

$$(q, p) \in \Gamma \simeq \mathbb{R} \times S^1 \simeq T^* S^1 = \{(q, p) \in \mathbb{R} \times S^1 : q_1 p_1 + q_2 p_2 = 0\}, \tag{12}$$

the section $\hat{\sigma}$ is defined by

$$\hat{\sigma}(q, p) = ((-q \sin p, q \cos p), p). \tag{13}$$

The representation of $E(2)$ is given by

$$(U(x, \alpha)\phi)(\theta) = \exp[i\langle x, \psi(\theta)\rangle]\phi(\theta - \alpha), \quad \phi \in L^2(S^1, d\theta). \tag{14}$$

Setting, for $d \in \mathbb{N}$, $\mathbb{N}_d = \{0, 1, \ldots, d\}$ and using the uniform lattice

$$\mathcal{X}_N = \{r_{2\pi\tau/N}, \tau \in \mathbb{N}_{N-1}\} \subset SO(2) \tag{15}$$

for a given $N \in \mathbb{N}^*$, we split S^1 as

$$S^1 = \bigcup_{\tau=0}^{N-1} \bigcup_{\nu=0,1} r_{2\pi\tau/N} r_{\pi\nu/N}[(-\pi/2, \pi/2)]. \tag{16}$$

886

For $\theta \in S^1$, $n \in \mathbb{N}_{2N+1}$ and $m \in \mathbb{N}_2$ and $l \in \mathbb{Z}$, the computations [6] yield frame vectors of the form

$$\eta_{n,m,l}(\theta) = \exp\left[2i\pi l \sin\left(\theta - \frac{(n-m)\pi}{N}\right)\right]\eta\left(\theta - \frac{(n-1)\pi}{N}\right), \tag{17}$$

where η is an even function of $L^2(S^1, d\theta)$, which has its support in the open interval $(-\pi/2, \pi/2)$ and verifies the admissibility condition

$$\int_{-\pi/2}^{\pi/2} \frac{|\eta(\theta)|^2}{\cos\theta} d\theta < \infty. \tag{18}$$

A similar construction has been done in [8, 9] for the Poincaré group in 1+3 dimensions $\mathcal{P}_+^\uparrow(1,3) = \mathbb{R}_{1,3}^4 \rtimes \mathrm{SL}(2, \mathbb{C})$.

3. Discrete frames and square integrability

To establish the converse of the result above, the group theoretical setting is the same. In addition, assume that U is a unitary representation of G, and there exists a lattice $\{q_l, \ l \in \mathbb{Z}^m\}$ in V_0, an at most countable discrete subset $\{p_n, \ n \in J \subset \mathbb{N}\} \subset \mathcal{O}^*$ containing k_0, and a set of linearly independent vectors $\{\eta^j; \ j = 1, \dots, N\} \subset \mathfrak{H}$ with supports in O_{k_0} such that the set $\{\eta_{n,l}^j = U(\sigma_{pr}(q_l, p_n))\eta^j \mid n \in J; l \in \mathbb{Z}^m; j = 1, \dots, N\}$ is a frame. Therefore, using results of functional analysis, it is not difficult to show that, for all ϕ in $\mathfrak{H}$, the map

$$\Gamma \ni (q, p) \mapsto \sum_{j=1}^N |\langle U(\sigma_{pr}(q, p))\eta^j | \phi\rangle_{\mathfrak{H}}|^2 \tag{19}$$

is in $L^1(\Gamma, d\mu)$. We do not have yet the square integrability, but this follows if we assume in addition that the representation U is irreducible. Furthermore, if we consider affine admissible sections such that $n(q, p) = n(p)$, $\forall q \in V_0$, the operator $A_{\hat{\sigma}}$ is a multiple of the identity in $\mathfrak{H}$. This is equivalent to saying that there exists in $\mathfrak{H}$ a continuous tight frame.

References

[1] Aniello P, Cassinelli G, De Vito E, and Levrero A 1998, Wavelet transforms and discrete frames associated to semidirect products, J. Math. Phys. 39 3965-3973
[2] Aniello A, Cassinelli G, De Vito E, and Levrero A 2001, On discrete frames associated with semidirect products, J. Fourier Anal. Appl. 7 199-206
[3] Ali S T, Antoine J-P, and Gazeau J-P 2000 Coherent States, Wavelets and their Generalizations (New York: Springer-Verlag)
[4] Raghunathan M S 1982, Discrete Subgroups of Lie Groups (Berlin: Springer-Verlag)
[5] Bernier D and Taylor K F 1996, Wavelets from square integrable representations, SIAM J. Math. Anal. 27 594-608
[6] Hohouéto A L, Kengatharam T, Ali S T, and Antoine J-P 2003, Coherent states lattices and square integrability of representations, Preprint UCL-IPT–03–07, Louvain-la-Neuve and Concordia University
[7] De Bièvre S 1989, Coherent states over symplectic homogeneous spaces, J. Math. Phys. 30 1401-1407
[8] Hohouéto A L 1999, On some discrete relativistic frames, PhD. Thesis, Université Nationale du Bénin, Abomey-Calavi.
[9] Antoine J-P and Hohouéto A L 2002, Discrete frames of Poincaré coherent states in 1+3 dimensions, J. Fourier Anal. Appl. 9 141–172

Inst. Phys. Conf. Ser. No 173: Satellite colloquium
Paper presented at 24th Int. Coll. Group Theoretical Methods in Physics, Paris, France, 15–20 July 2002
©2003 IOP Publishing Ltd

Unified view of tomographic and other transforms in signal analysis

Margarita A Man'ko

P.N. Lebedev Physical Institute, Moscow, Russia

Abstract. Unified construction of tomograms, wavelets and quasidistributions is discussed. Relation of the construction to Klauder's coherent states (continuous representation) is elucidated. Tomographic representation of nonlinear equations is presented.

1. Introduction

The continuous representation (coherent states and generalized coherent states) was introduced by Klauder in [1]. Coherent states are widely used in quantum physics (see, for example, [2, 3]) and in signal analysis [4]. The construction of coherent states [1] is connected with a choice of fiducial (reference) vector and the action on the vector by an appropriate operator which depends on continuous parameters. On the other hand, recently it was shown [5] that the integral transforms used in signal analysis can be treated in view of a unified approach (for review of the approach, see [6]). Our aim is to discuss the unified approach and find its relation to the Klauder's coherent states. Another goal is to review an extension of the tomographic transform approach to analyze properties of nonlinear equations first suggested in [7] and developed in [8].

2. Wavelet-like transforms, quasidistributions, and tomograms

Following [5] we present a unified general construction of three types of transforms used in signal analysis. The first class consists of wavelet-type transforms, the second of quasidistributions, and in the third class the tomographic transforms are. Quasidistributions are transforms like the Wigner–Ville one. Husimi–Kano positive quasidistributions will be also discussed.

In quantum mechanics, quasidistributions describe a quantum state in terms of phase-space quasiprobability densities. In signal analysis, quasidistributions describe the structure of analytic signals in the time–frequency plane. There also exist quasidistributions characterizing the signal structure in the time–scale plane. We refer to quasiprobability densities because the corresponding functions are not conventional probabilities, being either complex or nonpositive. The corresponding observables do not commute and the uncertainty relation prevents the existence of a joint probability distribution function for noncommuting observables.

The general setting for our construction is as follows [5].

Signals $f(t)$ are considered to be vectors $\mid f\rangle$. With α being a set of parameters, $\{U(\alpha)\}$ is a family of operators. In many cases, the family of operators $\{U(\alpha)\}$ generates a Lie group. However, this is not a necessary condition for the consistency of the formalism provided the completeness conditions discussed below are satisfied.

In this setting, three types of transforms are defined. Consider a reference vector $\mid h\rangle$ chosen in such a way that out of the set $\{U(\alpha) \mid h\rangle = \mid h\rangle\}$ a complete set of vectors can be chosen to serve as a basis. Completeness relation for the vectors $\mid h, \alpha\rangle$ means existence of measure $d\mu(\alpha)$ in the unity operator decomposition

$$\int \mid h, \alpha\rangle\langle h, \alpha \mid d\mu(\alpha) = \hat{1}.$$

Two of the transforms considered are given by scalar products

$$W_f^{(h)}(\alpha) = \langle U(\alpha) h \mid f\rangle, \qquad Q_f(\alpha) = \langle U(\alpha) f \mid f\rangle. \tag{1}$$

If $U(\alpha)$ is operator of unitary irreducible representation of some Lie group, transform (1) is matix element of the irreducible representation. If one compares the wavelet-type transform in (1) and the definition of coherents states done in [1], their identity is obvious.

We will denote the transforms of the $W_f^{(h)}$-type as *wavelet-type* transforms and those of the Q_f-type as *quasidistribution* transforms.

In general, if $U(\alpha)$ are unitary operators, there are self-adjoint operators $B(\alpha)$ such that

$$W_f^{(h)}(\alpha) = \langle h \mid e^{iB(\alpha)} \mid f\rangle, \qquad Q_f^{(B)}(\alpha) = \langle f \mid e^{iB(\alpha)} \mid f\rangle. \tag{2}$$

In this case, because $B(\alpha)$ has a real valued spectrum, another transform may be defined by means of Dirac (or Kronecker) delta function

$$M_f^{(B)}(X) = \langle f \mid \delta\Big(B(\alpha) - X\Big) \mid f\rangle. \tag{3}$$

Equation (3) defines what we call the tomographic transform of analytic signal or *tomogram*. In contrast to the quasiprobabilities, the transform $M_f^{(B)}(X)$ is positive and it can be correctly interpreted as a probability distribution.

For a normalized vector $\mid f\rangle$, $\langle f \mid f\rangle = 1$, the tomogram is a normalized function $\int M_f^{(B)}(X) \, dX = 1$ and, therefore, it may be interpreted as a probability distribution for the random variable X corresponding to the observable defined by the operator $B(\alpha)$.

In view of the geometric meaning of the construction, the three classes of transforms are mutually related

$$M_f^{(B)}(X) = \frac{1}{2\pi} \int Q_f^{(kB)}(\alpha) \, e^{-ikX} \, dk, \qquad Q_f^{(B)}(\alpha) = \int M_f^{(B/p)}(X) \, e^{ipX} \, dX. \tag{4}$$

Wavelet-type transforms, quasidistributions, and tomograms are related by the formulas

$$Q_f^{(B)}(\alpha) = W_f^{(f)}(\alpha), \tag{5}$$

$$W_f^{(h)}(\alpha) = \frac{1}{4} \int e^{iX} \left[M_{f_1}^{(B)}(X) - iM_{f_2}^{(B)}(X) - M_{f_3}^{(B)}(X) + iM_{f_4}^{(B)}(X) \right] dX, \tag{6}$$

where

$$\mid f_1\rangle = \mid h\rangle + \mid f\rangle, \qquad \mid f_3\rangle = \mid h\rangle - \mid f\rangle,$$
$$\mid f_2\rangle = \mid h\rangle + i \mid f\rangle, \qquad \mid f_4\rangle = \mid h\rangle - i \mid f\rangle.$$

Another important case concerns operators $U(\alpha)$, which can be represented in the form $U(\alpha) = e^{ib(\alpha)} P_h e^{-ib(\alpha)}$, with P_h being a projector on a reference vector $\mid h\rangle$. This creates a quasidistribution of the Husimi–Kano type $H_f^{(b)}(\alpha) = \langle f \mid U(\alpha) \mid f\rangle$.

Let us discuss some examples of the general construction.

We take the operator $B(\alpha)$, $\alpha = (\mu, \nu)$ in the form $B^{(S)}(\alpha) = \mu\hat{t} + \nu\hat{\omega}$, with $\hat{\omega} = -i\,\partial/\partial t$.

For the tomogram, that is,

$$M_f^{(S)}(X, \mu, \nu) = \langle f \mid \delta\left(\mu\hat{t} + \nu\hat{\omega} - X\right) \mid f \rangle, \tag{7}$$

using a Fourier transform representation of the delta-operator, one obtains [9]

$$M_f^{(S)}(X, \mu, \nu) = \frac{1}{2\pi|\nu|}\left|\int \exp\left[\frac{i\mu t^2}{2\nu} - \frac{itX}{\nu}\right] f(t)\,dt\right|^2. \tag{8}$$

The tomogram (8) is also related to the Wigner–Ville quasidistribution $WV(\tau, \omega)$ by

$$M_f^{(S)}(X, \mu, \nu) = \int \exp\left[-ik(X - \mu\omega - \nu\tau)\right] WV(\tau, \omega)\,\frac{dk\,d\omega\,d\tau}{(2\pi)^2}. \tag{9}$$

The Wigner–Ville quasidistribution is given by the following formula, where we introduce $\alpha = (\tau, \Omega)$,

$$WV(\tau, \Omega) = \int f\left(\tau + \frac{u}{2}\right) f^*\left(\tau - \frac{u}{2}\right) e^{-i\Omega u}\,du. \tag{10}$$

3. Nonlinear equations in tomographic and Weyl–Wigner–Moyal representations

Let us consider the following generalized nonlinear Schrödinger equation:

$$i\frac{\partial\psi}{\partial s} = -\frac{1}{2}\frac{\partial^2\psi}{\partial x^2} + U\left[|\psi|^2\right]\psi, \tag{11}$$

where s and x are the time-like and space-like variables and $\psi = \psi(x, s)$ is a complex wave function describing the system's evolution in the configuration space; $U = U\left[|\psi|^2\right]$ is an arbitrary real functional of $|\psi|^2$.

Let us introduce a notation for tomogram related to the analytic signal $\psi(x, s)$ by (8) with $t \to y$, $f(t) \to \psi(y, s)$, i.e., $M_f^{(s)}(X, \mu, \nu) \to w(X, \mu, \nu, s)$, where we denote the tomogram as $w(X, \mu, \nu, s)$. Since (8) implies the change of variables in (11), one obtains the tomographic form of the nonlinear equation under consideration [8]:

$$\frac{\partial w(X, \mu, \nu, s)}{\partial s} + \mu\frac{\partial w(X, \mu, \nu, s)}{\partial \nu} - 2\,\mathrm{Im}\,U\left\{\int w(y, \mu', 0, s)\right.$$

$$\left. \times \exp\left[i\left(y + \mu'\left[\left(\frac{\partial}{\partial X}\right)^{-1}\frac{\partial}{\partial \mu} - \frac{i}{2}\nu\frac{\partial}{\partial X}\right]\right)\right]\frac{dy\,d\mu'}{2\pi}\right\} w(X, \mu, \nu, s) = 0. \tag{12}$$

It has been recently shown that equation (11) with $U[|\psi|^2] = q_0|\psi|^{2\beta}$, for $q_0 < 0$ and any real positive value of β, has the following envelope soliton-like solutions [12]:

$$\Psi(x, s) = \left[\frac{|E|(1 + \beta)}{|q_0|}\right]^{1/2\beta} \mathrm{sech}^{1/\beta}\left[\beta\sqrt{2|E|}\,\xi\right] \exp\left\{i\left[V_0 x - \left(E + \frac{V_0^2}{2}\right)s\right]\right\}, \tag{13}$$

where the real numbers V_0 and E are arbitrary and negative, respectively, and still $\xi = x - V_0 s$ (V_0 is the soliton velocity).

The tomogram of the soliton solution of generalized nonlinear Schrödinger equation is given by the formula ($V_0 = 0$)

$$w(X, \mu, \nu) = \frac{1}{2\pi|\nu|}\left|\left[\frac{|E|(1 + \beta)}{|q_0|}\right]^{1/\beta}\int \mathrm{sech}^{1/\beta}\left[\beta\sqrt{2|E|}\,y\right]\exp\left(\frac{i\mu}{2\nu}y^2 - \frac{iXy}{\nu}\right)dy\right|^2 \tag{14}$$

890

and it does not depend on the time-like variable s.

Numerical calculations and 3D plots along with density plots of both Wigner function and tomorgams of bright solitons for varios values of free parameters are presented in [7, 8].

4. Conclusion

The main result in our work is the formulation of an unified view for some linear and nonlinear transforms through the operator formulation developed in section 2. The formulation emphasizes the basic unity of these transforms, which are related by explicit formulas. Nevertheless, for each particular application, one type of transform may be more convenient than the others. The relation of wavelet transforms to the coherent states introduced as continuous representation scheme [1] is also the result of this work.

The other result of the work is demonstration of the possibility to construct the tomographic representation for soliton solutions of nonlinear equations.

Acknowledgment

This study was supported by the Russian Foundation for Basic Research under Projects Nos. 01-00-16516 and 01-02-17745.

References

[1] Klauder JR 1963, Continuous-representation theory. II. Generalized relation between quantum and classical dynamics, J. Math. Phys. 4 1058–1073
[2] Klauder JR and Skagerstam B-S 1985, Coherent States: Applications in Physics and Mathematical Physics (Singapore: World Scientific)
[3] Gazeau J-P and Klauder JR 1999, Coherent states for systems with discrete and continuous spectrum, J. Phys. A: Math. Gen. 32 123–132
[4] Aslaksen EW and Klauder JR 1968, Unitary representations of the affine group J. Math. Phys. 206–211; — 1969, Continuous representation theory using the affine group, J. Math. Phys. 10 2267–2275
[5] Man'ko MA, Man'ko VI, and Mendes RV 2001, Tomograms and other transforms: a unified view, J. Phys. A: Math. Gen. 34 8321–8332
[6] Man'ko MA 2001, Tomograms, wavelets, and quasidistributions in the geometric picture, J. Russ. Laser Res. 22 505–533
[7] De Nicola S, Fedele R, Man'ko M, and Man'ko V 2002, Wigner picture and tomographic representation of envelope solitons, Proc. Intern. Workshop 'Nonlinear Physics: Theory and Experiment. II' (Gallipoli, Lecce, Italy, 27 June – 6 July 2002) eds Ablowitz M, Boiti M, Pempinelli F, and Prinari B (Singapore: World Scientific)
[8] De Nicola S, Fedele R, Man'ko M, and Man'ko V 2002, Tomography of solitons J. Opt. B: Quantum Semiclass. Opt. (submitted)
[9] Man'ko VI and Mendes RV 1999, Noncommutative time–frequency tomography, Phys. Lett. A 263 53–59
[10] Dodonov VV and Man'ko VI 1989, Invariants and Evolution of Nonstationary Quantum Systems, Proc. PN Lebedev Physical Institute Vol 183 (New York: Nova Science)
[11] Fedele R and Schamel H 2002, Solitary waves in the Madelung's fluid: Connection between the nonlinear Schrodinger equation and the Korteweg-de Vries equation, European J. Phys. B 27 313–320
[12] Fedele R 2002, Envelope solitons versus solitons, Phys. Scr. 65 502–508

Part B

Wavelets and their Applications

Inst. Phys. Conf. Ser. No 173: Satellite colloquium
Paper presented at 24th Int. Coll. Group Theoretical Methods in Physics, Paris, France, 15–20 July 2002
©2003 *IOP Publishing Ltd*

Wavelet based regularization for Euclidean field theory

M V Altaisky†

Joint Institute for Nuclear Research, Dubna, 141980, Russia; and
Space Research Institute, Moscow, 117997, Russia

Abstract. It is shown that Euclidean field theory with polynomial interaction, can be regularized using the wavelet representation of the fields. The connections between wavelet based regularization and stochastic quantization are considered.

1. Introduction

The connections between quantum field theory and stochastic differential equations have been calling constant attention for quite a long time [1, 2]. We know that stochastic processes often possess self-similarity. The renormalization procedure used in quantum field theory is also based on self-similarity. So, it is natural to use for the regularization of field theories the wavelet transform (WT), the decomposition with respect to the representation of the affine group. In this paper, we present two ways of regularization. First, the direct substitution of WT of the fields into the action functional leads to a field theory with scale-dependent coupling constants. Second, the WT, substituted into the Parisi-Wu stochastic quantization scheme [3], provides a stochastic regularization with no extra vertexes introduced into the theory.

2. Scalar field theory on affine group

The Euclidean field theory is defined on $\boldsymbol{R}^d$ by the generating functional

$$W_F[J] = \mathcal{N} \int \mathcal{D}\phi \exp\left[-S[\phi(x)] + \int d^dx J(x)\phi(x)\right], \qquad (1)$$

where $S[\phi]$ is the Euclidean action. In the simplest case of a scalar field with the fourth power interaction, one has

$$S[\phi] = \int d^nx \frac{1}{2}(\partial_\mu\phi)^2 + \frac{m^2}{2}\phi^2 + \frac{\lambda}{4!}\phi^4(x). \qquad (2)$$

The ϕ^4 theory is often referred to as the Ginsburg-Landau model for its ferromagnetic applications. The ϕ^3 theory is also a useful model.

The perturbation expansion generated by the functional (1) is usually evaluated in k-space. The reformulation of the theory from the coordinate (x) to momentum (k) representation is a particular case of decomposition of a function with respect to the representation of a Lie group G. $G : x' = x + b$ for the case of the Fourier transform, but other groups may be used as well. For a locally compact Lie group G acting transitively on

† e-mail: altaisky@mx.iki.rssi.ru

the Hilbert space $\mathcal{H}$, it is possible to decompose state vectors with respect to a representation of G [8, 9]

$$|\phi\rangle = C_\psi^{-1} \int_G U(g)|\psi\rangle d\mu(g)\langle\psi|U(g)|\phi\rangle. \tag{3}$$

The constant C_ψ is determined by the norm of the action of $U(g)$ on the fiducial vector $\psi \in \mathcal{H}$,

$$C_\psi = \|\psi\|^{-2} \int_{g\in G} |\langle\psi|U(g)|\psi\rangle|^2 d\mu(g);$$

$d\mu(g)$ is the left-invariant measure on G.

Using the decomposition (3), it is possible to define a field theory on a non-abelian Lie group. Let us consider the fourth power interaction model

$$\int V(x_1, x_2, x_3, x_4)\phi(x_1)\phi(x_2)\phi(x_3)\phi(x_4)dx_1 dx_2 dx_3 dx_4.$$

Using the notation $U(g)|\psi\rangle \equiv |g, \psi\rangle$, $\langle\phi|g, \psi\rangle \equiv \phi(g)$, $\langle g_1, \psi|D|g_2, \psi\rangle \equiv D(g_1, g_2)$, we can rewrite the generating functional (1) in the form

$$\begin{aligned}
W_G[J] = \int \mathcal{D}\phi(g) \exp\Bigg(&-\frac{1}{2} \int_G \phi(g_1)D(g_1, g_2)\phi(g_2)d\mu(g_1)d\mu(g_2) \\
&-\frac{\lambda}{4!} \int_G V(g_1, g_2, g_3, g_4)\phi(g_1)\phi(g_2)\phi(g_3)\phi(g_4)d\mu(g_1)d\mu(g_2)d\mu(g_3)d\mu(g_4) \\
&+\int_G J(g)\phi(g)d\mu(g)\Bigg),
\end{aligned} \tag{4}$$

where $V(g_1, g_2, g_3, g_4)$ is the result of the transform $\phi(g) := \int \overline{U(g)\psi(x)}\phi(x)dx$ applied to $V(x_1, x_2, x_3, x_4)$ in all arguments x_1, x_2, x_3, x_4.

Let us turn to the particular case of the affine group.

$$x' = ax + b, \quad U(g)\psi(x) = a^{-d/2}\psi((x-b)/a)), \quad x, x', b \in \mathbf{R}^d. \tag{5}$$

The scalar field $\phi(x)$ in the action $S[\phi]$ can be written in the form of a wavelet decomposition

$$\begin{aligned}
\phi(x) &= C_\psi^{-1} \int \frac{1}{a^{d/2}}\psi\left(\frac{x-b}{a}\right)\phi_a(b)\frac{dadb}{a^{d+1}}, \\
\phi_a(b) &= \int \frac{1}{a^{d/2}}\bar{\psi}\left(\frac{x-b}{a}\right)\phi(x)d^dx.
\end{aligned} \tag{6}$$

In the scale-momentum (a, k) representation, the matrix element of the free field inverse propagator

$$\langle a_1, b_1; \psi|D|a_2, b_2; \psi\rangle = \int \frac{d^n k}{(2\pi)^n}e^{ik(b_1-b_2)}D(a_1, a_2, k)$$

has the form

$$D(a_1, a_2, k) = a_1^{d/2}\overline{\hat{\psi}(a_1 k)}(k^2 + m^2)a_2^{d/2}\hat{\psi}(a_2 k). \tag{7}$$

The field theory (4) with the propagator $D^{-1}(a_1, a_2, k)$ gives the standard Feynman diagram technique, but with an extra wavelet factor $a^{d/2}\hat{\psi}(ak)$ on each line and the integrations over the measure $d\mu(a, k) = \frac{d^d k}{(2\pi)^d}\frac{da}{a^{d+1}}$ instead of $\frac{d^d k}{(2\pi)^d}$.

Recalling the power law dependence of the coupling constants on the cutoff momentum resulting from the Wilson expansion, we can define a scalar field model on the affine group, with the coupling constant depending on scale. Thus, the fourth power interaction can be written as

$$V[\phi] = \int \frac{\lambda(a)}{4!} \phi_a^4(b) d\mu(a, b), \quad \lambda(a) \sim a^\nu. \tag{8}$$

The one-loop order contribution to the Green function G_2 in the theory with interaction (8) can be evaluated [11] by integration over $z = ak$:

$$\int \frac{a^\nu a^d |\hat{\psi}(ak)|^2}{k^2 + m^2} \frac{d^d k}{(2\pi)^d} \frac{da}{a^{n+1}} = \int \frac{d^d k}{(2\pi)^d} \frac{C_\psi^{(\nu)} k^{-\nu}}{k^2 + m^2}, C_\psi^{(\nu)} = \int |\hat{\psi}(z)|^2 \frac{dz}{z^{1-\nu}}. \tag{9}$$

Therefore, there are no UV divergences for $\nu > n - 2$. This is a kind of asymptotically free theory which is hardly appropriate, say, to spin systems. What is required to get a finite theory is an interaction vanishing outside a given range of scales. Such a model is presented in the next section by means of the stochastic quantization framework.

3. Stochastic quantization with wavelets

Let us remind the basic ideas of the stochastic quantization [3, 14, 12, 13]. Let $S[\phi]$ be the action of the field $\phi(x)$. Instead of calculating the physical Green functions, it is possible to introduce the "extra-time" variable τ: $\phi(x) \to \phi(x, \tau)$ and evaluate the moments $\langle \phi(x_1, \tau_1) \dots \phi(x_m, \tau_m) \rangle_\eta$ by averaging over the random process $\phi(x, \tau, \cdot)$ governed by the Langevin equation with the Gaussian random force

$$\dot{\phi}(x, \tau) + \frac{\sigma^2}{2} \frac{\delta S}{\delta \phi(x, \tau)} = \eta(x, \tau), \langle \eta(x, \tau) \eta(x', \tau') \rangle = \sigma^2 \delta(x - x') \delta(\tau - \tau'). \tag{10}$$

The physical Green functions are then obtained by taking the limit $\tau_1 = \dots = \tau_m = T \to \infty$.

The stochastic quantization procedure has been considered as a possible candidate for the regularization of gauge theories, for it respects local gauge symmetries in a natural way. However a δ-correlated Gaussian random noise in the Langevin equation still yields sharp singularities in the perturbation theory. For this reason, a number of modifications based on the noise regularization $\eta(x, \tau) \to \int dy R_{xy}(\partial^2) \eta(y, \tau)$ have been proposed [4, 7, 6].

In this paper, following [15], we start with the random processes defined directly in wavelet space. The use of the wavelet coefficients instead of the original stochastic processes provides an extra analytical flexibility of the method: there exist more than one set of random functions $W(a, b, \cdot)$ the images of which have coinciding correlation functions. It is easy to check that the random process generated by wavelet coefficients with the correlation function $\langle \widehat{W}(a_1, \mathbf{k}_1) \widehat{W}(a_2, \mathbf{k}_2) \rangle = C_\psi (2\pi)^d \delta^d(\mathbf{k}_1 + \mathbf{k}_2) a_1^{d+1} \delta(a_1 - a_2) D_0$ has the same correlation function as the white noise [17].

As an example, let us consider the Kardar-Parisi-Zhang equation [16]:

$$\dot{Z} - \nu \Delta Z = \frac{\lambda}{2} (\nabla Z)^2 + \eta. \tag{11}$$

Substituting the wavelet transform

$$Z(x) = C_\psi^{-1} \int \exp(\imath(\mathbf{k}\mathbf{x} - k_0 t)) a^{\frac{d}{2}} \hat{\psi}(ak) \hat{Z}(a, k) \frac{d^{d+1}k}{(2\pi)^{d+1}} \frac{da}{a^{d+1}}$$

into (11), with the random force of the form

$$\langle \hat{\eta}(a_1, k_1) \hat{\eta}(a_2, k_2) \rangle = C_\psi (2\pi)^{d+1} \delta^{d+1}(k_1 + k_2) a_1^{d+1} \delta(a_1 - a_2) D(a_2, k_2), \tag{12}$$

leads to the integral equation

$$(-\imath\omega + \nu\boldsymbol{k}^2)\hat{Z}(a,k) = \eta(a,k) - \tfrac{\lambda}{2}a^{\frac{d}{2}}\overline{\hat{\psi}(a\boldsymbol{k})}C_\psi^{-2}\int(a_1 a_2)^{\frac{d}{2}}\hat{\psi}(a_1\boldsymbol{k}_1)\hat{\psi}(a_2(\boldsymbol{k}-\boldsymbol{k}_1))$$

$$\boldsymbol{k}_1(\boldsymbol{k}-\boldsymbol{k}_1)\hat{Z}(a_1,k_1)\hat{Z}(a_2,k-k_1)\frac{d^{d+1}k_1}{(2\pi)^{d+1}}\frac{da_1}{a_1^{d+1}}\frac{da_2}{a_2^{d+1}}.$$

From this the one-loop contribution the stochastic Green function follows:

$$G(k) = G_0(k) - \lambda^2 G_0^2(k)\int\frac{d^{d+1}k_1}{(2\pi)^{d+1}}\Delta(k_1)$$

$$\boldsymbol{k}_1(\boldsymbol{k}-\boldsymbol{k}_1)|G_0(k_1)|^2\boldsymbol{k}\boldsymbol{k}_1 G_0(k-k_1) + O(\lambda^4), \tag{13}$$

where $G_0^{-1}(k) = -\imath\omega + \nu\boldsymbol{k}^2$. The difference from the standard approach [16] is in the appearance of the effective force correlator

$$\Delta(k) \equiv C_\psi^{-1}\int\frac{da}{a}|\hat{\psi}(a\boldsymbol{k})|^2 D(a,\boldsymbol{k}), \tag{14}$$

which has the meaning of the effective force averaged over all scales.

Let us consider a single-band forcing [17] $D(a,\boldsymbol{k}) = \delta(a-a_0)D(\boldsymbol{k})$ and the "Mexican hat" wavelet $\hat{\psi}(k) = (2\pi)^{d/2}(-\imath\boldsymbol{k})^2\exp(-\boldsymbol{k}^2/2)$. In the leading order in small parameter $x = |\boldsymbol{k}|/|\boldsymbol{k}_1| \ll 1$, the contribution to the stochastic Green function is:

$$G(k) = G_0(k) + \lambda^2 G_0^2(k)\frac{S_d}{(2\pi)^d}\frac{a_0^3 k^2}{\nu^2}\frac{d-2}{8d}\int_0^\infty D(\boldsymbol{q})e^{-a_0^2 q^2}q^{d+1}dq + O(\lambda^4).$$

4. Langevin equation for the ϕ^3 theory with scale-dependent noise

Let us apply the same scale-dependent noise (12) to the Langevin equation for ϕ^3 theory. The standard procedure of the stochastic quantization then comes from the Langevin equation [6]

$$\dot{\phi}(x,\tau) + \left[-\Delta\phi + m^2\phi + \frac{\lambda}{2!}\phi^2\right] = \eta(x,\tau). \tag{15}$$

Applying the wavelet transform to this equation, we get

$$(-\imath\omega + \boldsymbol{k}^2 + m^2)\hat{\phi}(a,k) = \hat{\eta}(a,k) - \tfrac{\lambda}{2}a^{\frac{d}{2}}\overline{\hat{\psi}(a\boldsymbol{k})}C_\psi^{-2}\int(a_1 a_2)^{\frac{d}{2}}\hat{\psi}(a_1\boldsymbol{k}_1)$$

$$\times\hat{\psi}(a_2(\boldsymbol{k}-\boldsymbol{k}_1))\hat{\phi}(a_1,k_1)\hat{\phi}(a_2,k-k_1)\frac{d^{d+1}k_1}{(2\pi)^{d+1}}\frac{da_1}{a_1^{d+1}}\frac{da_2}{a_2^{d+1}}. \tag{16}$$

Iterating the integral equation (16), we yield the correction to the stochastic Green function

$$G(k) = G_0(k) + \lambda^2 G_0^2(k)\int\frac{d^{d+1}q}{(2\pi)^{d+1}}\Delta(q)|G_0(q)|^2 G_0(k-q) + O(\lambda^4). \tag{17}$$

The analytical expressions for the stochastic Green functions, can be obtained in the $\omega \to 0$ limit. As an example, we take the $D(a,\boldsymbol{q}) = \delta(a-a_0)D(\boldsymbol{q})$ for ϕ^3 theory and the Mexican hat wavelet. The one loop contribution to the stochastic Green function $G(k) = G_0(k) + G_0^2\lambda^2 I_3^2 + O(\lambda^4)$ is

$$\lim_{\omega\to 0}I_3^2 = \int\frac{d^d q}{(2\pi)^d}\Delta(\boldsymbol{q})\frac{1}{2(q^2+m^2)}\cdot\frac{1}{q^2+(\boldsymbol{k}-\boldsymbol{q})^2+2m^2}.$$

The same procedure can be applied in higher loops. As it can be seen, for constant or compactly supported $D(\boldsymbol{q})$ all integrals are finite due to the exponential factor coming from wavelet ψ.

Acknowledgement

The author is grateful to Profs. H.Hüffel and V.B.Priezzhev for useful discussions.

References

[1] Nelson E 1985, Quantum Fluctuations (Princeton, NJ : Princeton University Press)
[2] Glim G and Jaffe A 1981, Quantum Physics (New York: Springer)
[3] Parisi G and Wu Y-S 1981, Scientica Sinica 24
[4] Breit J, Gupta S, and Zaks A 1984, Stochastic quantization and regularization, Nucl. Phys. B 233 61–68
[5] Bern Z, M Halpern M, Sadun L, and Taubes C 1985, Continuum regularization of QCD, Phys. Lett. B 165 151–156
[6] Iengo R and Pugnetti S 1988, Stochastic quantization, non-Markovian reqularization and renormalization, Nucl. Phys. B 300 128–142
[7] Bern Z, Halpern M, Sadun L, and Taubes C 1987, Continuum regularization of quantum field theory (i). Scalar prototype. Nucl. Phys. B 284 1–91
[8] Carey A 1976, Square-integrable representations of non-unimodular groups. Bull. Austr. Math. Soc. 15 1–12
[9] Duflo M and Moore C1976, On regular representations of nonunimodular locally compact group, J. Funct. Anal. 21 209–243
[10] Chui C 1991, An introduction to wavelets (New York: Academic)
[11] Altaisky M 2001, ϕ^4 field theory on a Lie group, in Sidharth B G and Altaisky M V (eds.), Frontiers of Fundamental Physics 4, pp.124–128 (New York: Kluwer and Plenum)
[12] Damgaard P and Hüffel H 41987, Phys. Rep. 152 227–248
[13] Namiki M 1992, Stochastic Quantization (Heidelberg: Springer)
[14] Zinn-Justin J 1986, Renormalization and stochastic quantization, Nucl. Phys. B 275(FS17) 135–159
[15] Altaisky M V 1999, Scale-dependent function in statistical hydrodynamics: a functional analysis point of view, European J. Phys. B 8(4) 613–617
[16] Kardar M, Parisi G, and Zhang Y-C 1986, Dynamic scaling of growing interfaces, Phys. Rev. Lett. 56 889–892
[17] Altaisky M V 2002, Langevin equation with scale-dependent noise. Communication E5-2002-35, Joint Institute For Nuclear Research, Dubna

Inst. Phys. Conf. Ser. No 173: Satellite colloquium
Paper presented at 24th Int. Coll. Group Theoretical Methods in Physics, Paris, France, 15–20 July 2002
©2003 IOP Publishing Ltd

Measuring a curvature radius with directional wavelets

J-P Antoine† and L Jacques‡

Institut de Physique Théorique, Université catholique de Louvain,
B-1348 Louvain-la-Neuve, Belgium

Abstract. We present in this paper a new technique based on the Continuous Wavelet Transfrom (CWT) for estimating the curvature radius of contours delimiting simple objects in images. This method exploits the link which exists between the angular response of directional wavelets and the curvature radius of the analyzed curves.

1. Introduction

There is currently a high interest in image processing in obtaining a good representation of the curves that define shapes (contours). Each element of answer, even partial, to this important issue may help one to find new ways to represent information in natural images.

In this paper, after a brief review of the CWT, we show how the 2-dimensional continuous wavelet transform based on a conical directional wavelet [2] is able to extract information on the curvature radius of contours delimiting objects. The main idea is to compare the angular response of the CWT energy on straight lines with its response on curves.

2. The Continuous Wavelet Transform and conical wavelets

The two-dimensional continuous wavelet transform of an image $I \in L^2(\mathbb{R}^2)$ is defined by the scalar product of I, in the sense of $L^2(\mathbb{R}^2)$, with a probe function ψ, i.e.,

$$W_I(\vec{b}, a, \theta) = \int_{\mathbb{R}^2} d^2\vec{x}\, \overline{\psi_{\vec{b}\,a\,\theta}(\vec{x})}\, I(\vec{x})\,, \tag{1}$$

where $\vec{x} = (x, y)$, and the overbar denotes the usual complex conjugation [1]. The function $\psi_{\vec{b}\,a\,\theta}$ is a copy of ψ translated by the position $\vec{b}$, dilated by a factor a, and oriented in the direction θ, that is,

$$\psi_{\vec{b}\,a\,\theta}(\vec{x}) = a^{-1}\psi(r_\theta^{-1}\frac{\vec{x} - \vec{b}}{a}), \tag{2}$$

with r_θ the usual 2×2 rotation matrix of angle θ. Notice that the wavelets $\psi_{\vec{b}\,a\,\theta}$ are $L^2(\mathbb{R}^2)$ normalized. The function ψ must be well localized in position and in frequency. It has also to satisfy an *admissibility* requirement which, under some regularity assumptions, is equivalent to the zero average of ψ.

Let us remark finally that the formula (1) can be rewritten in frequency space as

$$W_I(\vec{b}, a, \theta) = \int_{\mathbb{R}^2} d^2\vec{k}\, a\, \overline{\hat{\psi}(ar_\theta^{-1}\vec{k})}\, \hat{I}(\vec{k})\, e^{i\vec{k}\cdot\vec{b}}, \tag{3}$$

† e-mail: antoine@fyma.ucl.ac.be
‡ e-mail: ljacques@fyma.ucl.ac.be

using the usual Plancherel formula.

Conical wavelets were first introduced in [2], more precisely the particular example called *Cauchy* wavelet. They are 2-D wavelets strictly supported in a convex cone in frequency space. This property makes these functions well adapted to the detection of very oriented features in images, such as straight lines. Our favorite type is the Gaussian conical wavelet, given in frequency space by

$$\widehat{\psi}_m^{\varphi}(\vec{k}) = \begin{cases} (\vec{k} \cdot \vec{e}_{\varphi - \frac{\pi}{2}})^m \, (\vec{k} \cdot \vec{e}_{\frac{\pi}{2} - \varphi})^m \, e^{-\frac{12}{T}|\vec{k}|^2}, & \text{if } \vec{k} \in \mathcal{C}(\varphi) \\ 0, & \text{otherwise,} \end{cases} \tag{4}$$

where $\mathcal{C}(\varphi)$ is the frequency cone of half-aperture φ defined by

$$\mathcal{C}(\varphi) = \{\vec{k} \in \mathbb{R}^2 : |\arg \vec{k}| \leqslant \varphi\}, \tag{5}$$

$\vec{e}_\nu = (\cos \nu, \sin \nu)$ and m represents the *moments* of the wavelet on the boundaries of the cone.

3. Angular selectivity

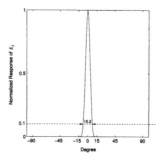

Figure 1. Analysis of a straight line: wavelet energy $E_I = |W_I|^2$ as a function of θ.

The conical wavelet (4) is well adapted to the selection of very oriented features in images. To understand this fact, let us analyze an infinite vertical straight line defined by $I(\vec{r}) = \delta(x)$, where δ is the Dirac distribution. We easily compute that $\widehat{I}(\vec{k}) = 2\pi\delta(k_y)$. Then, if we fix $\vec{b}$ on the origin, we see from (3) that the response of W_I will be important when the frequency line intersects the support of $\widehat{\psi}(ar_\theta^{-1}\vec{k})$. This induces an angular response of width equal to the aperture of the cone defining the wavelet, namely 2φ, and determines the *angular selectivity* of ψ. This phenomenon is illustrated in Figure 1 by the analysis of an infinite straight line with a conical wavelet of parameters $\varphi = 15°$ and $m = 4$. , The wavelet *energy* $E_I = |W_I|^2$ is drawn as a function of θ and is different from zero on a interval equal to $30°$. If we place a threshold T at 10% of the maximum of E_I, one can show that the latter is larger than T on an interval of size $\Delta_l(T) = 15.2°$.

4. Analysis of a disk: calibration of the method

Consider now a disk of radius R centered on the origin. What happens to the angular response of the CWT on a point $\vec{R} = (R, 0)$ of the circle defining this disk ? First, because the wavelet has a zero average, it sees only the circular edge of this shape. Next, in comparison with the

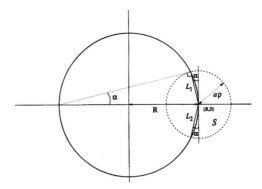

Figure 2. Graphical explanation of the angular behavior of the CWT on the edge of a disk of radius R.

straight line analysis, the angular response of the CWT is modified in function of the radius R. Let us explain this behavior. Given $S = \bigcup_\theta \operatorname{supp} \psi_{\vec{R}a\theta}$, the union for a fixed scale a of all the (numerical) supports of $\psi_{\vec{R}a\theta}$ (by numerical support, we mean the region where the function is numerically non-negligible). S is clearly a disk centered on $\vec{R}$ with a size proportional to a, but depending on the wavelet parameters m and φ. That is, $S = B(\vec{R}, a\rho(\varphi, m))$ for a certain function $\rho : S_1 \times \mathbb{N} \to \mathbb{R}$, and with $B(\vec{c}, R)$ the ball of radius R centered on $\vec{c}$.

On the point $\vec{R}$, the edge of the disk may be locally approximated by two straight lines: L_1 and L_2. These ones are oriented respectively in directions $\frac{\pi}{2} + \alpha$ and $-\frac{\pi}{2} - \alpha$, with $\alpha = \arcsin(\frac{a\rho}{2R})$.

We have seen in the preceeding section that, given a threshold T in $[0, 1]$, a straight line is detected by E_l on an angular interval of length $\Delta_l(T)$. Therefore, the portion of the disk contained in S and approximated by L_1 and L_2, will be "seen" above the threshold T for $\theta \in [-\alpha - \frac{1}{2}\Delta_l(T), \alpha + \frac{1}{2}\Delta_l(T)]$, that is, on an angular interval of size

$$\Delta_c(T) \simeq \Delta_l(T) + 2\arcsin\left(\frac{a\rho}{2R}\right) \tag{6}$$

$$\simeq \Delta_l(T) + \frac{a\rho}{R}, \tag{7}$$

The last equality holds only for $\frac{a}{R}$ sufficiently small, i.e., for small scales or for large radii. Therefore, the radius of the disk can be established from the width $\Delta_c(T)$ with

$$R \simeq \frac{a\rho}{\Delta_c(T) - \Delta_l(T)}. \tag{8}$$

Even if the expression (8) is very approximative, it shows there exists a one-to-one relation between the radius R of the disk and the angular width $\Delta_c(T)$. This is in agreement with our computations presented in Fig. 3 for a conical wavelet ($a = 6, \varphi = 15°$) and disks of radius ranging from 10 to 250 pixels in a 256×256-size image. Using this curve, we fit a third degree polynomial in $\kappa \equiv (\Delta_c - \Delta_l)^{-1}$ to the determination of R, i.e.,we determine the coefficients p_j in the expression

$$R = p_3\kappa^3 + p_2\kappa^2 + p_1\kappa + p_0 + \epsilon. \tag{9}$$

The result is $p_3 = 3.4026 \, 10^{-1}$, $p_2 = -6.6601$, $p_1 = 6.6203 \, 10^1$ and $p_0 = -1.0779 \, 10^1$ with an error $|\epsilon| < 0.9931$. Notice, however, that these values are only valid for $T = 10\%$, $a = 6$, $\varphi = 15°$ and $m = 4$.

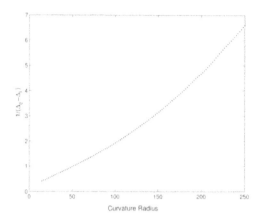

Figure 3. Angular width ($T = 10\%$) vs. radius for the conical wavelet ($a = 6, \varphi = 15°$)

5. Analysis of the Archimedes spiral

Figure 4. Archimedes's spiral on a 256×256 grid for $c = 128/6\pi$

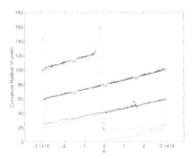

Figure 5. Estimated curvature radius of the Archimedes spiral.

Up to now, we have analyzed only a disk, for which the curvature radius is constant over the whole circular edge. Let us now test our method on a shape whose curvature radius varies, namely, the Archimedes spiral (Fig. 4). The latter is defined, in polar coordinates, by $r(\theta) = c\theta$ for the (extended) angle $\theta \in \mathbb{R}$ and the constant $c \in \mathbb{R}$. The curvature radius R of this spiral follows the rule

$$R(\theta) = c\,\frac{(1 + \theta^2)^{3/2}}{(2 + \theta^2)}. \tag{10}$$

Using the calibration (9) of the last section, we have measured κ on a spiral defined on a 256×256 grid for $c = 128/6\pi$ and estimated R with the same conical wavelet. The results are presented in Fig. 5. The plain line is the theoretical curvature radius and each dot corresponds to a point of the image grid passed through by the spiral. Except for the extremities of the curve where the corresponding singularities disturbs the approximation, the estimated curvature radius is relatively close to the theoretical one. Indeed, the absolute error is almost everywhere inferior to 5 pixels for $\theta \in [4.58, 16.75]$.

6. Analysis of a gravitational lens

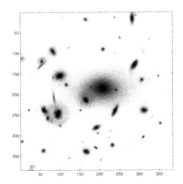

Figure 6. The cluster Abell 2218.

To test our technique on a more natural example, we choose an astronomical image of arcs produced by a gravitational lens present in the cluster Abell 2218 [4] (Fig. 6). The more intense arcs are relatively well defined in this image.

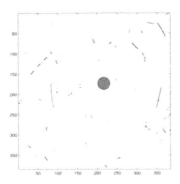

Figure 7. Edges for which R is greater than 50 pixels.

We first compute all the edges in the image with the Canny algorithm [3], a classical method of image processing. Next, for each point of these edges, we estimate the curvature radius according to (9) with the same conical wavelet as before. Fig. 7 displays the points for which the curvature radius is greater than 50 pixels. We see that most of the gravitational arcs are detected by this filtering. Finally, we select one arc in this image (inside the dotted area of Fig. 7) and plot the curvature radius as a function of a path joining the extremities of this arc (Fig. 8). The average curvature radius is about 78 pixels. This points more or less to the center of the cluster (the gray disk in Fig. 7), which is in agreement whith astronomical interpretation. However, due to the intensity variation on the arc and to the presence of noise in this image, it is clear that the error is large on this curvature radius estimation.

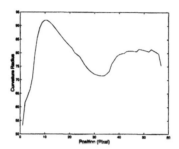

Figure 8. Curvature radius of one gravitional candidate.

7. Conclusion

We have shown that the width of the angular response of a directional continuous wavelet analysis is directly linked to the curvature radius of the edges defining shapes. In relatively simple images, the latter can thus be relatively well approximated. However, the technique used is far to be resistant to intensity variation and noise in images. In addition, we do not know at this moment if a more precise rule than the calibration (9) is available for an arbitrary fixed scale a.

References

[1] Antoine J-P, Murenzi R, and Vandergheynst P 1996, Two-dimensional directional wavelets in image processing, Int. J. of Imaging Systems and Technology, 7 152-165
[2] Antoine J-P, Murenzi R, and Vandergheynst P 1999, Directional wavelets revisited: Cauchy wavelets and symmetry detection in patterns, Appl. Comput. Harmon. Anal., 6 314-345
[3] Canny J 1986, A computational approach to edge detection, IEEE Trans. Pattern Anal. Machine Intell., 8 679-698
[4] NASA 2000, Hubble opens its eye on the Universe and captures a cosmic magnifying glass. http://hubblesite.org/newscenter/archive/2000/07/image/b

Inst. Phys. Conf. Ser. No 173: Satellite colloquium
Paper presented at 24th Int. Coll. Group Theoretical Methods in Physics, Paris, France, 15–20 July 2002
©2003 IOP Publishing Ltd

Extracting structural and dynamical informations from wavelet-based analysis of DNA sequences

A Arnéodo[(1)]†, **B Audit**[(2)], **C Vaillant**[(1)], **Y d'Aubenton-Carafa**[(3)], and **C Thermes**[(3)]

(1) Centre de Recherche Paul Pascal, avenue Schweitzer, 33600 Pessac, France.
(2) Computational Genomics Group, EMBL-European Bioinformatics Institute, Wellcome Trust Genome Campus, Cambridge CB10 1SD, UK.
(3) Centre de Génétique Moléculaire du CNRS, Allée de la Terrasse, 91198 Gif-sur-Yvette, France

Abstract. The packaging of the eucaryotic genomic DNA involves the wrapping around the histone proteins [1] followed by the successive foldings of higher order structured nucleoprotein complexes [2]. The bending properties of DNA play an essential role in these compaction processes [3, 4]. This hierarchically organized pathway is likely to be reflected in the fractal behavior of DNA bending signals in eucaryotic genomes, but the challenge is to somehow extract this structural information by a clever reading of the DNA sequences. We show that when using an adapted mathematical tool, the "wavelet transform microscope" [5, 6], to explore the fluctuations of bending profiles, one reveals a characteristic scale of 100-200bp that separates two different regimes of (long-range) power-law correlations (PLC) that are common to eucaryotic as well as eubacterial and archaeal genomes. The same analysis of the DNA text yields strikingly similar results to those obtained with bending profiles, and this for all three kingdoms. In the small-scale regime, PLC are observed in eucaryotic genomes, in nuclear replicating DNA viruses and in archaeal genomes, which contrasts with their total absence in the genomes of eubacteria and their viruses, thus indicating that small-scale PLC are likely to be related to the mechanisms underlying the wrapping of DNA around histone proteins. These results together with the observation of PLC between particular sequence motifs known to participate in the formation of nucleosomes (e.g. AA dinucleotides) show that the $10 - 200$ bp PLC provide a very efficient diagnostic of the nucleosomal structure and this in coding as well as in noncoding regions [7, 8]. We discuss possible interpretations of these PLC in terms of the physical mechanisms that might govern the positioning and dynamics of the nucleosomes along the DNA chain through cooperative processes [8]. We further speculate that the large-scale PLC are the signature of the higher-order structure and dynamics of chromatin.

The availability of fully sequenced genomes offers the possibility to study the scale-invariance properties of DNA sequences on a wide range of scales extending from tens to thousands of nucleotides. Actually, scale invariance measurement enables us to evidence particular correlation structures between distant nucleotides or groups of nucleotides. During the past few years, there has been intense discussion about the existence, the nature and the origin of long-range correlations in genomic sequences [9, 10, 11, 12]. If it is now well admitted that long-range correlations do exist in DNA sequences [6, 11, 13], their biological interpretation is still debated [9, 10, 11, 12, 13, 14, 15, 16, 17]. Most of the models proposed so far are based on the genome plasticity and are supported by the reported absence of power-law correlations (PLC) in coding DNA sequences [5, 6, 13, 18]. In a previous work [17], from a systematic analysis of human exons, CDS's and introns, we have found that PLC are not only present in non-coding sequences but also in coding regions somehow hidden in their inner codon structure. Here we report the results of a recent study [7, 8] that

† Present address: Ecole Normale Supérieure de Lyon, 46, allée d'Italie, F-69364 Lyon Cedex 07 France

906

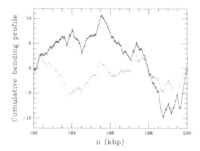

Figure 1. Cumulative bending profiles for a human DNA fragment (chromosome 21, positions 192 kb to 200 kb). Abscissa is the position on the sequence; the curves are cumulative representations of the PNuc (black) and DNase (grey) codings (in order to facilitate the comparison, the mean drift of the curves has been eliminated) (from [8]).

demonstrate that the long-range correlations observed in DNA sequences are more likely the signature of the hierarchical structural organization of chromatin. In contrast to previous interpretations, we propose some understanding of these correlations as a necessity for chromosome condensation-decondensation processes in relation with DNA replication, gene expression and cell division [8].

A major problem of fractal analysis applied to DNA sequences is that these display a mosaic structure which is characterized by "patches" resulting from compositional biases with an excess of one type of nucleotide. When mapping DNA sequences to numerical sequences using the "DNA walk" representation, these patches appear as trends in the DNA walk landscapes that are likely to break the scale invariance [9, 10, 11, 12, 13, 18]. In previous works [5, 6], we have emphasized the wavelet transform (WT) as a well suited technique to overcome this difficulty. By considering analyzing wavelets that make the "WT microscope" blind to low frequency trends, any bias in the DNA walk can be removed and the existence of PLC associated with specific scale invariance properties can be revealed accurately. When exploring sequences selected from the human genome, we have found that the fluctuations in the patchy landscapes of both coding and non-coding DNA walks are monofractal with Gaussian statistics in the small-scale range, which justifies the use of a single exponent H usually called the Hurst or roughness exponent [5, 6]. H values larger than the uncorrelated random walk value $H = 1/2$, correspond to the existence of long-range correlations that we will refer to as "persistence". To estimate this exponent, we just have to investigate the behavior across scales (a) of the root mean square (r.m.s.) fluctuations of wavelet coefficients:

$$\sigma(a) \propto a^{H}. \tag{1}$$

Wavelet coefficients actually reflect the local variation (over size a) of the concentration of nucleotides. Persistence ($H > 1/2$), therefore means that these concentrations fluctuate more smoothly (over short distances) than for uncorrelated sequences, but in the same time with a larger amplitude (over large distances) around the mean value [8].

Here we summarize the results of a comparative analysis of the persistence properties for both DNA texts and DNA bending profiles of various eucaryotic, eubacterial and archaeal genomes [7, 8]. To study the DNA texts, we construct "DNA walks" according to the binary coding method extensively used by Voss [11]; this method decomposes the nucleotide sequence into four sequences corresponding to A, C, T or G (coding with 1 at the nucleotide position and 0 at other positions). To construct DNA bending profiles that account for the fluctuations of the local double helix curvature, we use the trinucleotide model proposed in [19] (here called Pnuc) and which was deduced from experimentally determined nucleosome positioning. To test that our results do not come out from a simple recoding of the DNA text, we use also the trinucleotide coding table defined in [20] (here called DNase) and which is based on sensitivity of DNA fragments to DNase I and which more likely codes for the DNA local flexibility properties. As illustrated in Fig. 1, it is worth noting that the differences between the two tables are clearly sufficient to produce, at least at first sight, significantly different DNA bending profiles.

The first completely sequenced eucaryotic genome *Saccharomyces cerevisiae (S.c.)* provides an opportunity to perform a comparative wavelet analysis of the scaling properties displayed by each chromosome. When looking at the global estimate of $\sigma(a)$ over the DNA walks corresponding to "A" in each of the 16 yeast chromosomes shown in Fig. 2(a), one sees that all present superimposable behavior, with notably the same characteristic scale that separates two different scaling regimes. At small scales, $20 \lesssim a \lesssim 200$ (expressed in nucleotide units), PLC are observed as characterized by $H = 0.59 \pm 0.02$, a mean value which is significantly larger than $1/2$. At large scales, $200 \lesssim a$, stronger PLC with $H = 0.82 \pm 0.01$ become dominant with a cutoff around 10000bp (a number by no means accurate) above which uncorrelated behavior is observed. A similar wavelet analysis of the bending profiles of the yeast chromosomes obtained when using the Pnuc coding table (Fig. 2(a)) reveals striking similarities with the curves resulting from the DNA walk analysis, in both the small-scale and the large-scale regimes. These observations are not simply due to a "recoding" of the DNA sequences since when using the DNase coding table, one notices a significant weakening of the H exponent observed in the large-scale regime ($H \simeq 0.6$). The existence of these two scaling regimes is confirmed in Fig. 3(a), where the probability density functions (pdfs) of wavelet coefficient values of the yeast Pnuc bending profiles computed at different scales are shown to collapse on a single curve, as predicted by the self-similarity relationship [5, 6]

$$a^H \rho_a(a^H T) = \rho(T), \tag{2}$$

provided one uses the scaling exponent value $H = 0.60$ in the scale range $10 \lesssim a \lesssim 100$ and $H = 0.75$ in the scale range $200 \lesssim a \lesssim 1000$. In the small-scale regime, the pdfs are very well approximated by Gaussian distributions. In the large-scale regime, the pdfs have stretched exponential-like tails. The fact that the self-similarity relationship (2) is satisfied in both regimes corroborates the monofractal nature of the roughness fluctuations of the yeast bending profiles. Similar quantitative results are obtained for the corresponding DNA texts [7, 8]. Let us emphasize that we have also examined a number of eucaryotic DNA sequences from different organisms (human, rodent, avian, plant and insect) and that we have observed the same characteristic features as those obtained in Fig. 2(a) for *S.c.* (see Fig. 2(c) for the human chromosome 21 and Table 1 of [8]). Note that the characteristic scale found for higher eucaryotes is slightly smaller $a^* \simeq 100 - 140 \, bp$ than for *S.c.* and that the cross-over between the two PLC regimes is remarkably robust for the four "A", "C", "G" and "T" DNA walks. In a work under progress, we are investigating the possible departure from Gaussian statistics in the small-scale regime when increasing the $(G + C)$ content of the considered DNA sequence.

The striking overall similarity of the results obtained with these different eucaryotic genomes prompted us to also examine the scale invariance properties of bacterial genomes [7, 8]. In Fig. 2(b) are reported the results obtained for *Escherichia coli* which are quite typical of what we have observed with other eubacterial genomes. Again, there exists a well defined characteristic scale $a^* \simeq 200$bp that delimits a transition to very strong PLC with $H = 0.80 \pm 0.05$ at large scales. Let us point out that as for *S.c.* (Fig. 2(a)), if

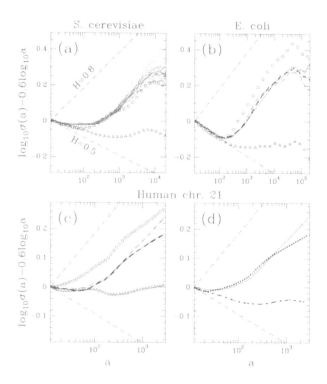

Figure 2. Global estimate of the r.m.s of WT coefficients: $\log_{10} \sigma(a) - 0.6 \log_{10} a$ is plotted versus $\log_{10} a$; the dashed lines corresponding to uncorrelated ($H = 1/2$) and strongly correlated ($H = 0.80$) regimes are drawn to guide the eyes. (Some horizontal line in this logarithmic representation will correspond to $H = 0.6$). The analyzing wavelet is the "Mexican hat" wavelet [6]. (a) *S. cerevisiae:* "A" DNA walks of the 16 *S. cerevisiae* chromosomes (——) and of the corresponding bending profiles obtained with the Pnuc (○) and DNase (△) coding tables when averaged over the 16 chromosomes. (b) *Escherichia coli:* "A" (grey ——), "T" (grey – – –), "G" (black ——) and "C" (black – – –) DNA walks and the corresponding Pnuc (○) and DNase (△) bending profiles. (c) *Human chromosome 21:* "A", "C", "G" and "T" DNA walks and corresponding Pnuc and DNase bending profiles; same symbols as in (b). (d) *Human chromosome 21:* comparative analysis of DNA walks for all adenines (——), adenines part of a dinucleotide AA (···) and isolated adenines not part of a dinucleotide AA (– – –) (from [7]).

one uses the DNase table for human (Fig. 2(c)) and *E. coli* (Fig. 2(b)) sequences, one no longer observes the strong PLC as obtained with the Pnuc table. In Fig. 3(b) are reported the wavelet coefficient pdfs of the *E. coli* Pnuc bending profile that corroborate the existence of a cross-over scale between two different monofractal scaling regimes characterized by $H = 0.50 \pm 0.02$ and $H = 0.80 \pm 0.05$ respectively. In order to examine if these properties actually extend homogeneously over the whole genomes, $\sigma(a)$ was calculated over a window of width $l = 2000$, sliding along the bending profiles. The results reported in Fig. 4 for *Yeast*, a human contig and *E. coli* confirm the existence of a characteristic scale $a^* \simeq 100 - 200\mathrm{bp}$ which seems to be robust all along the corresponding DNA molecules and this for all investigated genomes in the three kingdoms. Note that analogous results are

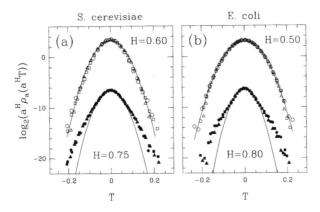

Figure 3. Probability distribution functions of wavelet coefficient values of Pnuc bending profiles. The analyzing wavelet is the Mexican hat [6]. (a) *Saccharomyces cerevisiae*: $\log_2(a^H \rho_a(a^H T))$ vs T for the set of scales $a = 12$ ($\triangle$), 24 ($\square$), 48 ($\circ$), 192 ($\blacktriangle$), 384 ($\blacksquare$), and 768 ($\bullet$) in nucleotide units; $H = 0.60$ ($H = 0.75$) in the small (large) scale regime. (b) *Escherichia coli*: same as in (a) but with $H = 0.50$ ($H = 0.80$) in the small (large) scale regime (from [7]).

obtained for the four mononucleotide DNA walks as for the bending profiles [7, 8].

There exists however an important difference between eucaryotic and eubacterial genomes: no PLC are observed for the latter in the small-scale regime where uncorrelated Brownian motion-like behavior with $H = 1/2$ is observed (Figs. 2(b) and 3(b)). As discussed in previous works [5, 6, 9, 10, 13, 18], separate analyses of coding and non-coding eucaryotic DNA walks actually show that introns display PLC (with a mean H value of 0.60 ± 0.02) in the small-scale regime, while exons have no such correlations. At this point, it may seem that PLC are inherent to non-coding sequences only, but that is not the case. As shown in Fig. 5 for *Archaeoglobus fulgidus*, the wavelet investigation of five archaeal genomes (which are mostly coding) also reveals the presence of small-scale PLC as observed in eucaryotic genomes, although somewhat less pronounced [8]. Note that the strong large-scale PLC are present in all eubacterial, archaeabacterial and eucaryotic genomes.

What mechanism or phenomenon might explain the small-scale PLC in eucaryotic genomes? Their total absence in eubacterial genomes raises the possibility that they could be related to certain nucleotide arrangements in the 150 *bp* long DNA regions which are wrapped around histone proteins to form the eucaryotic nucleosome [1, 2]. Indeed, eubacterial genomic DNA is associated with histone-like proteins (e.g. HU), but no nucleosome-type structure has been detected in these organisms [21]. Along this line, the observation of small-scale PLC in archaeal genomes is consistent with the presence in archaebacteria of structures similar to the eucaryotic nucleosomes [22]. This analysis has also been extended to viral genomes. Small-scale PLC are clearly detected in most eucaryotic viral double-strand DNA genomes as shown for *Epstein-Barr* virus in Fig. 5. This further supports the hypothesis of nucleosome-based PLC since nucleosomes are present on double-strand DNA viruses [23]. The Poxviridae, which are the only animal DNA viruses replicating in the cytoplasm of their host cells, code for an eubacterial-type of histone-like protein [24], and no PLC are found in this scale range as shown in Fig. 5 for *Melanoplus sanguinipes* virus [8]. This observation is consistent with our hypothesis and suggests that the genomic DNA of these viruses is

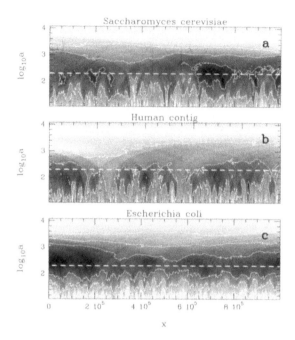

Figure 4. Space-scale wavelet like representation (x and a are expressed in nucleotide units) of the local estimate of the r.m.s $\sigma(a, x)$ of the WT coefficients of the "A" DNA walk. $\sigma(a)$ is computed over a window of width $l = 2000$, sliding along the first 10^6bp of the *yeast* chromosome IV (a), a human contig (b) and of the *Escherichia coli* genome (c). $\log_{10} \sigma(a) - 2/3 \log_{10} a$ is coded using 128 grey levels from black (min) to white (max). The horizontal white dashed line marks the scale $a^* = 200$bp where some minimum is observed consistently along the entire genomes as a separation between two different monofractal scaling regimes (see text). Note that in the human contig, the actual characteristic scale seems closer to $a^* = 150$ *bp*.

submitted to packaging processes different from the other animal viruses. Other classes of virus genomes like the single and double-strand RNA viruses (to the exception of the retroviruses) are very unlikely associated to nucleosomes. In all cases except retroviruses, we observe a total absence of small-scale PLC as shown in Fig. 5(c). In the case of retroviruses, it is known that the integrated viral DNA is associated to nucleosomes in the cell nucleus [25]; we clearly confirm in Fig. 5(c) the presence of small-scale PLC ($H \simeq 0.57 \pm 0.02$). Finally, bacteriophage genomes do not present any small-scale PLC (Fig. 5 for *T4* bacteriophage and [8]) as already observed for their eubacterial hosts. This wavelet based fractal analysis of viral and cellular genomes of all three kingdoms sustains the fact that small-scale PLC are a signature of nucleosomal DNA [7, 8].

To further investigate this PLC nucleosomal diagnostic, we ask whether particular dinucleotides which are known to participate to the positioning and formation of nucleosomes [26] (e.g. AA dinucleotides) would carry PLC specifically associated to eucaryotic genomes. This can be examined if one performs the analysis of different DNA walks generated with (i) all adenines, (ii) only adenines that are part of a dinucleotide AA and (iii) isolated adenines that are not part of a dinucleotide AA. The analysis of human

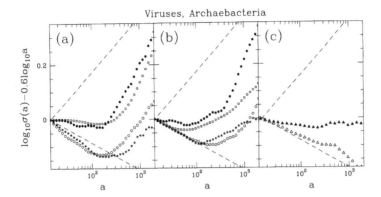

Figure 5. Global estimate of the r.m.s of WT coefficients of (a) "A" DNA walks and (b), (c) bending profiles. The various symbols correspond to the following genomes: *Archaeglobus fulgidus* (squares), *Epstein-Barr virus* (dots), *Melanoplus sanguinipes entomapoxvirus* (circles), *T4 bacteriophage* (stars), average over 21 single-strand RNA viruses (triangles) and 17 retroviruses (black triangles). Same representation as in Fig. 2.

(Fig. 2(d)) and eucaryotic (Table 1 in [8]) shows that the "isolated A" DNA walk exhibits a clear weakening of the PLC properties at small scale, while the "AA" DNA walk accounts for a major part of the observed PLC on the "A" DNA walk, which confirms the nucleosomal signature of small-scale PLC. Note that this observation is an additional illustration of the fact that recoding does not trivially conserve the correlation law [8].

Several studies have established the presence in genomic sequences of DNA motifs related to bending properties. A 10.2 base periodicity has been observed using either Fourier [27] or correlation function [28] analysis, specifically in eucaryotic genomes where it has been interpreted in relation to nucleosomal structures. However there is a fundamental difference between this nucleosome diagnostic based on periodicity and our analysis based on scale invariance properties which strongly suggests that the mechanisms underlying the nucleosomal structure of eucaryotic genomes are multi-scale phenomena that actually involve the whole set of scales in the $1 - 200$ bp range. In this respect, periodicity (which concerns 5% of sequences that present affinities for the histone octamer that are significantly larger than average [29]) and scale invariance (which concerns 95% of bulk genomic DNA sequences that have an affinity for the histone octamer similar to that of random sequences [29]) should not be considered as opposed to each other but rather complementary. In [8], we have proposed the following dynamical understanding of the observed small scale PLC. In contrast to the tight histone binding obtained with an adequate periodic distribution of bending sites, PLC would facilitate the positioning of the histone core throughout a major part of the genome. If one considers the translational positioning of nucleosomes as a mechanism of diffusion along the DNA chain and if we assume that, once the histone core is bound to DNA, the distribution of bending sites has a direct consequence on this diffusion process, then the PLC are likely to allow nucleosome mobility along DNA to proceed with an average displacement (after a given number of elementary steps) larger than with uncorrelated sequences. The persistent nature of the scale-invariant spatial organization of bending sites would be selected in order to favour the overall dynamic of compaction of nucleosomes by enabling them to explore larger segments of DNA. In other words, nucleosomes would require less energy

912

for similar amplitude of displacements. Persistence therefore offers some understanding of the modest free-energie of nucleosome formation observed for most DNA sequences, which also facilitates the translational mobility and thus the propensity of nucleosomes to be dynamical structures. Such properties could then favour an optimal compromise between DNA compaction and accessibility constraints. These hypotheses constitute new directions for the study of the effects of the small-scale PLC on the structural, mechanical and dynamical properties of DNA in chromatin.

The interpretation of the large scale PLC observed in the DNA bending profiles as well as in the DNA walks is an open problem. As suggested in [29], the signals involved in nucleosome formation may act collectively over large distances to the packing of nucleosomal arrays (10 nm filament) into high-order chromatin structures (30 nm fiber) [1, 2, 30]. Since DNA bending sites are key elements for nucleosomal structures, the detailed investigation of large-scale PLC (in the $200 - 5000$ bp range) observed in eucaryotic bending profiles should shed light on the compaction mechanisms at work in the hierarchical formation and dynamics of chromatin. An important clue provided by our studies is that similar long distance correlations in bending profiles are also observed in eubacteria (Figs. 2(b) and 3(b)) and archaebacteria (Fig. 5) [8]. Actually, all chromosomes are submitted to condensation-decondensation processes (in relation with DNA replication, gene expression, ...) which might result in common dynamical and structural properties. A deep understanding of the large-scale PLC and their interpretation in terms of these constraints remain challenging questions requiring further investigation.

This research was supported by the GIP GREG (project "Motifs dans les Séquences") and by the Ministère de l'Education Nationale, de l'Enseignement Supérieur, de la Recherche et de l'Insertion Professionnelle ACC-SV (project "Génétique et Environnement") and the Action BioInformatique (CNRS, 2000). BA acknowledges the support from the European Community through a Marie Curie Fellowship (contract: HPMF-CT-2001-01321).

References

[1] Luger K, Mäder AW, Richmond RK, Sargent DF, and Richmond TJ, 1997, Nature 389 (6648), 251–260
[2] van Holde K, 1989, Chromatin (New York: Springer)
[3] Drew HR and Travers AA 1985, J. Mol. Biol. 186 773–790
[4] Yao J, Lowary PT, and Widom J 1990, Proc. Natl. Acad. Sci. USA 87 7603–7607
[5] Arneodo A, Bacry E, Graves PV and Muzy J-F 1995, Phys. Rev. Lett. 74 3293–3296
[6] Arneodo A, d'Aubenton-Carafa Y, Bacry E, Graves PV, Muzy J-F, and Thermes C 1996, Physica D 96 291–320
[7] Audit B, Thermes C, Vaillant ., d'Aubenton-Carafa Y, Muzy J-F, and Arneodo A 2001, Phys. Rev. Lett. 86 2471–2474
[8] Audit B, Vaillant C, Arneodo A, d'Aubenton-Carafa Y, and Thermes C 2002, J. Mol. Biol. 316 903–918
[9] Stanley HE, Buldyrev SV, Goldberger AL, Havlin S, Ossadnik SM, Peng C-K, and Simons M 1993 Fractals 1 (3), 283–301
[10] Li W, Marr TG, and Kaneko , 1994, Physica D 75 392–416
[11] Voss RF 1994, Fractals 2 (1) 1–6
[12] Karlin S and Brendel V 1993, Science 259 677–679
[13] Buldyrev SV, Goldberger AL, Havlin S, Mantegna RN, Matsa ME, Peng C-K, Simons M, and Stanley HE 1995, Phys. Rev. E 51 5084–5091
[14] Li W., 1992, Int. J. Bifurc. Chaos 2 (1), 137–154.

[15] Buldyrev SV, Goldberger AL, Havlin S, Stanley HE, Stanley MHR, and Simons M 1993, Biophys. J. 65 2673–2679

[16] Herzel H, Trifonov EN, Weiss O, and Grosse I 1998, Physica A 249, 449

[17] Arneodo A, d'Aubenton-Carafa Y, Audit B, Bacry E, Muzy J-F, and Thermes C 1998, Eur. Phys. J. B 1 259–263

[18] Peng C-K, Buldyrev SV, Goldberger AL, Havlin S, Sciortino F, Simons M, and Stanley HE 1992, Nature 356 168–170

[19] Goodsell DS and Dickerson RE 1994, Nucl. Acids Res. 22 5497–5503

[20] Brukner I, Sanchez R, Suck D, and Pongor S 1995, J. Biomol. Struct. Dynam. 13 309–317

[21] Murphy LD and Zimmerman SB 1997, J. Struct. Biol. 119 336–346

[22] Reeve JN, Sandman K, and Daniels CJ 1997, Cell 89, 999–1002

[23] Challberg MO and Kelly TJ 1989, Annu. Rev. Biochem. 58 671–717

[24] Borca MV, Irusta PM, Kutish GF, Carillo C, Afonso CL, Burrage AT, Neilan JG, and Rock DL 1996, Arch. Virol. 141 301–313

[25] Stanfield-Oakley SA and Griffith JD 1996, J. Mol. Biol. 256 503–516

[26] Thaström A, Lowary PT, Widlund HR, Cao H, Kubista M, and Widom J 1999, J. Mol. Biol. 288 213–219

[27] Widom J 1996, J. Mol. Biol. 259 579–588

[28] Herzel H, Weiss O, and Trifonov EN 1999, Bioinformatics 15 187–193

[29] Lowary PT and Widom J 1997, Proc. Natl. Acad. Sci. USA 94 1183–1188

[30] Polach KJ and Widom J 1995, J. Mol. Biol. 254 130–149

Inst. Phys. Conf. Ser. No 173: Satellite colloquium
Paper presented at 24th Int. Coll. Group Theoretical Methods in Physics, Paris, France, 15–20 July 2002
©2003 IOP Publishing Ltd

Deficient splines wavelets

F Bastin† and P Laubin

University of Liège, Institute of Mathematics, B37,
4000 Liège, Belgium

1. Introduction

Splines are intensively used in different domains (numerical analysis, approximation theory,...) It is also well known that wavelets which are spline functions have been constructed some years ago; let us quote the works of Chui and Wang [2], Mallat [6], Meyer [7].

In some problems of numerical analysis (see for example [3], [8]), deficient splines are prefered to classical splines. Then a natural question arises in the context of wavelets: is it possible to construct bases wavelets which are splines of that kind? This problem is treated in a very general aspect in the papers [4, 5].

Here we present a constructive and direct way to obtain a mutiresolution analysis generated by deficient splines which are piecewise polynomials of degree 5, regularity 3, and have a compact support. Then, by a natural procedure, we obtain wavelets which are also functions of that type.

2. The results

Let us denote by V_0 the following set of quintic splines

$$V_0 := \{f \in L_2(\mathbb{R}) : \ f|_{[k,k+1]} = P_k^{(5)}, k \in \mathbb{Z} \text{ and } f \in C_3(\mathbb{R})\}.$$

Looking for $f \in V_0$ with support $[0,3]$ (a smaller interval would not give anything), we are led to a homogenous linear system of 18 unknowns and 16 equations; this suggests that two scaling functions will be needed to generate V_0.

Proposition 2.1 *A function f with support $[0,3]$ belongs to V_0 if and only if*

$$f(x) = \begin{cases} nx^4 + ax^5 & \text{if } x \in [0,1] \\ bx^5 + cx^4 + dx^3 + ex^2 + fx + g & \text{if } x \in [1,2] \\ h(3-x)^4 + j(3-x)^5 & \text{if } x \in [2,3] \\ 0 & \text{if } x < 0 \text{ or } x > 3 \end{cases}$$

with

$$n = -\tfrac{4}{5}c - \tfrac{3}{10}d \qquad a = \tfrac{7}{15}c + \tfrac{8}{45}d \qquad b = -\tfrac{19}{75}c - \tfrac{19}{450}d$$

$$e = -\tfrac{-18}{5}c - \tfrac{13}{5}d \qquad f = \tfrac{18}{5}c + \tfrac{21}{10}d \qquad g = -\tfrac{27}{25}c - \tfrac{29}{50}d$$

$$h = -c - \tfrac{1}{3}d \qquad j = \tfrac{46}{75}c + \tfrac{91}{450}d$$

† e-mail· F.Bastin@ulg.ac.be

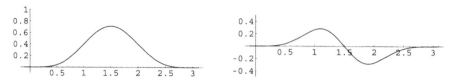

Figure 1. The scaling functions φ_s, φ_a

Theorem 2.2 *The following functions φ_a and φ_s*

$$\varphi_a(x) = \begin{cases} x^4 - \frac{11}{15}x^5 & \text{if } x \in [0,1] \\ -\frac{9}{8}(x - \frac{3}{2}) + 3(x - \frac{3}{2})^3 - \frac{38}{15}(x - \frac{3}{2})^5 & \text{if } x \in [1,2] \\ -(3-x)^4 + \frac{11}{15}(3-x)^5 & \text{if } x \in [2,3] \\ 0 & \text{if } x < 0 \text{ or } x > 3 \end{cases}$$

$$\varphi_s(x) = \begin{cases} x^4 - \frac{3}{5}x^5 & \text{if } x \in [0,1] \\ \frac{57}{80} - \frac{3}{2}(x - \frac{3}{2})^2 + (x - \frac{3}{2})^4 & \text{if } x \in [1,2] \\ (3-x)^4 - \frac{3}{5}(3-x)^5 & \text{if } x \in [2,3] \\ 0 & \text{if } x < 0 \text{ or } x > 3 \end{cases}$$

are respectively antisymmetric and symmetric with respect to $\frac{3}{2}$ and the family

$$\{\varphi_a(.-k),\ k \in \mathbb{Z}\} \cup \{\varphi_s(.-k),\ k \in \mathbb{Z}\}$$

constitutes a Riesz basis of V_0.

The result is obtained using Fourier techniques for the computations. The functions φ_s, φ_a are plotted in Figure 1.

To get the Riesz condition, we first show that each of the families $\{\varphi_a(.-k),\ k \in \mathbb{Z}\}$ and $\{\varphi_s(.-k),\ k \in \mathbb{Z}\}$ satisfies this condition. Then, using the fact that the $L^2(\mathbb{R})$-norm of functions of the linear hull of the union of the families can be written as the $L^2([0,1])$-norm of functions which belongs to a linear space of finite dimension, and the fact that, separately, the families satisfies the Riesz condition, we get the result.

To obtain that these families generate V_0, we just solve the equations obtained when we write down the problem.

Now, we construct the multiresolution analysis. For every $j \in \mathbb{Z}$ we define

$$V_j = \{f \in L^2(\mathbb{R})\ :\ f(2^{-j}.) \in V_0\}.$$

Proposition 2.3 *We have $V_j \subset V_{j+1}$ for every j and*

$$\bigcap_{j \in \mathbb{Z}} V_j = \{0\}, \quad \overline{\bigcup_{j \in \mathbb{Z}} V_j} = L^2(\mathbb{R}).$$

Moreover, the functions φ_a, φ_s satisfy the following scaling relation

$$\begin{pmatrix} \widehat{\varphi_s}(2\xi) \\ \widehat{\varphi_a}(2\xi) \end{pmatrix} = M_0(\xi) \begin{pmatrix} \widehat{\varphi_s}(\xi) \\ \widehat{\varphi_a}(\xi) \end{pmatrix}$$

where $M_0(\xi)$ is the matrix (called filter matrix)

$$M_0(\xi) = \frac{e^{-3i\xi/2}}{64} \begin{pmatrix} 51\cos(\frac{\xi}{2}) + 13\cos(\frac{3\xi}{2}) & -9i(\sin(\frac{\xi}{2}) + \sin(\frac{3\xi}{2})) \\ i(11\sin(\frac{3\xi}{2}) + 21\sin(\frac{\xi}{2})) & -7\cos(\frac{3\xi}{2}) + 9\cos(\frac{\xi}{2}) \end{pmatrix}.$$

Finally, using standard procedure, we get wavelets. We denote by W_0 the orthogonal complement of V_0 in V_1.

First we define the matrix $W(\xi)$ as follows

$$W(\xi) = \begin{pmatrix} \omega_s(\xi) & \omega_m(\xi) \\ \overline{\omega_m(\xi)} & \omega_a(\xi) \end{pmatrix}$$

with

$$\omega_a(\xi) = \sum_{l=-\infty}^{+\infty} |\widehat{\varphi_a}(\xi + 2l\pi)|^2 = \frac{23247 - 21362\cos\xi - 385\cos(2\xi)}{311850}$$

$$\omega_s(\xi) = \sum_{l=-\infty}^{+\infty} |\widehat{\varphi_s}(\xi + 2l\pi)|^2 = \frac{14445 + 7678\cos\xi + 53\cos(2\xi)}{34650}$$

$$\omega_m(\xi) = \sum_{l=-\infty}^{+\infty} \widehat{\varphi_s}(\xi + 2l\pi)\overline{\widehat{\varphi_a}(\xi + 2l\pi)} = -\frac{i}{51975}\sin\xi\,(6910 + 193\cos\xi).$$

Proposition 2.4 *A function f belongs to W_0 if and only if there exist 2π-periodic functions $p, q \in L^2_{loc}$, such that*

$$\widehat{f}(2\xi) = p(\xi)\widehat{\varphi_s}(\xi) + q(\xi)\widehat{\varphi_a}(\xi)$$

and

$$\overline{M_0(\xi)}W(\xi)\begin{pmatrix} p(\xi) \\ q(\xi) \end{pmatrix} + \overline{M_0(\xi + \pi)}W(\xi + \pi)\begin{pmatrix} p(\xi + \pi) \\ q(\xi + \pi) \end{pmatrix} = 0 \text{ a.e.}$$

We explicitly solve this matrix equation and find

Proposition 2.5 *There exist symmetric and antisymmetric solutions with support in $[0, 5]$.*

Introduce the notations

$$\psi_s(2\xi) = p_s(\xi)\overline{\widehat{\varphi_a}(\xi)} + q_s(\xi)\widehat{\varphi_a}(\xi)$$

$$\psi_a(2\xi) = p_a(\xi)\widehat{\varphi_s}(\xi) + q_a(\xi)\widehat{\varphi_a}(\xi).$$

Then we have (with the natural definition of M_1)

$$\begin{pmatrix} \widehat{\psi_s}(2\xi) \\ \widehat{\psi_a}(2\xi) \end{pmatrix} = M_1(\xi)\begin{pmatrix} \widehat{\varphi_s}(\xi) \\ \widehat{\varphi_a}(\xi) \end{pmatrix}.$$

The functions ψ_s, ψ_a are shown in Figure 2 (up to a multiplicative constant).

Proposition 2.6 *The family*

$$\{\psi_a(. - k) : k \in \mathbb{Z}\} \cup \{\psi_s(. - k) : k \in \mathbb{Z}\}$$

forms a Riesz basis of W_0.

918

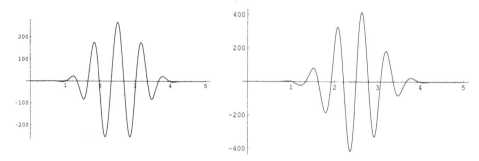

Figure 2. The wavelets ψ_s, ψ_a

To obtain this result, we proceed as follows, using essentially techniques of Goodman and Lee [4, 5]. Define

$$W_\psi(\xi) = \begin{pmatrix} \omega_{\psi_s}(\xi) & \omega_{\psi_s,\psi_a}(\xi) \\ \omega_{\psi_s,\psi_a}(\xi) & \omega_{\psi_a}(\xi) \end{pmatrix}$$

where

$$\omega_{\psi_a}(\xi) = \sum_{l=-\infty}^{+\infty} |\widehat{\psi_a}(\xi + 2l\pi)|^2,$$

$$\omega_{\psi_s}(\xi) = \sum_{l=-\infty}^{+\infty} |\widehat{\psi_s}(\xi + 2l\pi)|^2,$$

$$\omega_{\psi_s,\psi_a}(\xi) = \sum_{l=-\infty}^{+\infty} \widehat{\psi_s}(\xi + 2l\pi)\overline{\widehat{\psi_a}(\xi + 2l\pi)}.$$

The Riesz condition is satisfied if and only if there are $A, B > 0$ such that

$$A \leqslant \lambda_1(\xi), \lambda_2(\xi) \leqslant B$$

where $\lambda_i(\xi)$ are the eigenvalues of $W_\psi(\xi)$.

The functions form a basis if and only if for every ξ, the matrix

$$\begin{pmatrix} \overline{M_0(\xi)} & \overline{M_0(\xi + \pi)} \\ \overline{M_1(\xi)} & \overline{M_1(\xi + \pi)} \end{pmatrix} \begin{pmatrix} W(\xi) & 0 \\ 0 & W(\xi + \pi) \end{pmatrix}$$

is not singular.

The properties above are easily obtained using the three relations

$$W(2\xi) = M_0(\xi)W(\xi)M_0^*(\xi) + M_0(\xi + \pi)W(\xi + \pi)M_0^*(\xi + \pi)$$
$$0 = M_1(\xi)W(\xi)M_0^*(\xi) + M_1(\xi + \pi)W(\xi + \pi)M_0^*(\xi + \pi)$$
$$W_\psi(2\xi) = M_1(\xi)W(\xi)M_1^*(\xi) + M_1(\xi + \pi)W(\xi + \pi)M_1^*(\xi + \pi)$$

which result from the previous constructions and the properties of multiresolution analysis and functions in W_0.

Theorem 2.7 *It follows that the functions*

$$2^{j/2}\psi_s(2^j x - k), \quad 2^{j/2}\psi_a(2^j x - k) \quad (j, k \in \mathbb{Z})$$

form a Riesz basis of compactly supported deficient splines of $L^2(\mathbb{R})$ with symmetry properties.

To conclude, let us mention a result of Gilson, Faure, and Laubin [3] concerning an application of splines (deficient splines) in approximation theory of singular functions.

For $q \geq 1$, let $S_{3,q}^N$ the space of classical cubic splines on the interval $[0, 1]$ with respect to the subdivision $(j/N)^q$, $j = 0, \ldots, N$.

Proposition 2.8 *If $[\alpha_0, \alpha_1] \subset \mathbb{R}$ with $\alpha_0 > -1/2$ and $q(1 + 2\alpha_0) > 8$, then there exists $C > 0$ such that*

$$\inf_{u \in S_{3,q}^N} \|u - x^\alpha\|_{L^2(]0,1[)} \leqslant \frac{C}{N^4}$$

for every $N \in \mathbb{N}$ and $\alpha \in [\alpha_0, \alpha_1]$.
This result extends to deficient splines of odd degree and to Sobolev spaces.

References

[1] Bastin F and Laubin P 2002, Quintic deficient spline wavelets, Bull. Soc. Royale des Sc. de Liège, to appear

[2] Chui C K and Wang J Z 1992, On compactly supported spline wavelets and a duality principle, Trans. Amer. math. Soc. 330 903-915

[3] Gilson O, Faure O,and Laubin P, Quasi-optimal convergence using interpolation by non-uniform deficient splines, submitted for publication

[4] Goodman T N T, and Lee S L 1994, Wavelet of multiplicity r, Trans. Amer. Math. Soc. 342 307-324

[5] Goodman T N T, Lee S L, and Tang W S 1993, Wavelets in wandering subspaces, Trans. Amer. Math. Soc. 338 639-654

[6] Mallat S 1989, Multiresolution approximations and wavelet orthonormal bases of $L^2(\mathbb{R})$, Trans. Amer. Math. Soc. 315 69-88

[7] Meyer Y 1990, Ondelettes et opérateurs, I,II,III (Paris: Hermann)

[8] Rana S S and Dubey Y P 1997, Best error bounds of deficient quintic splines interpolation, Indian J. Pure Appl. Math. 28 1337-1344

Inst. Phys. Conf. Ser. No 173: Satellite colloquium
Paper presented at 24th Int. Coll. Group Theoretical Methods in Physics, Paris, France, 15–20 July 2002
©2003 IOP Publishing Ltd

A general recurrence relation between the moments of a scaling function

F Bastin and S Nicolay

Institute of Mathematics, University of Liège, Belgium

Abstract. Under natural and weak hypotheses, we prove a reproducing formula for polynomials. Then we obtain a new recurrence relation between the moments of a scaling function and a new exact formula for the computation of moments of even order.

1. Introduction

In [7], W. Sweldens and R. Piessens present the following result:

$$M_2 = (M_1)^2$$

where

$$M_j = \int_{\mathbb{R}} x^j \varphi(x) \, dx$$

are the moments of a compactly supported scaling function associated to a multiresolution analysis and in case the associated wavelet has at least three vanishing moments. Then, considering the shifted moments, they cancel the first and the second error terms in approximations and obtain an interesting quadrature formula. Their result about moments comes from a reproducing formula for polynomials.

A general result leading to a formula of that kind has been obtained by Y. Meyer in [6] but under rather strong regularity hypothesis.

In [1], under natural hypothesis and Strang-Fix conditions on a function φ (not necessarily a scaling function), we prove the reproducing formula for polynomials with absolute uniform convergence on compact sets and obtain the unicity of the coefficients. The proof we give does not follow the lines of the one of Y. Meyer and only use trigonometric Fourier series. Moreover, our result leads to relations showing that moments M_j of even order can be expressed in terms of a linear combination of products of moments of smaller order, with coefficients directly computable. In particular, we obtain $M_2 = (M_1)^2$.

Recurrence relations to compute the moments or approximations of them can be found in [2],[7]. These relations involve approximations or computation of auxiliary numbers related to the specific property of scaling functions. Here, we present relations leading to the exact computation of moments of even order using only combinatory coefficients.

In what follows, the set of natural numbers greater or equal to 0 (resp. strictly greater than 0) is denoted $\mathbb{N}$ (resp. $\mathbb{N}_0$) and the set of all integers (resp. all integers not equal to 0) is denoted $\mathbb{Z}$ (resp. $\mathbb{Z}_0$).

We also use the following notation $C_m^n = \frac{m!}{n! \, (m-n)!}$ where $m, n \in \mathbb{N}$, $m \geqslant n$.

2. Reproducing formula

Here is the result concerning the reproducing formula (see [1]).

Proposition 2.1 *Let φ be a function defined on $\mathbb{R}$ satisfying*

$$|\varphi(x)| \leqslant \frac{C}{(1+|x|)^{m+1+\varepsilon}}$$

for some $m \in \mathbb{N}_0, C, \varepsilon > 0$ and such that the functions $\varphi(.-k), k \in \mathbb{Z}$ satisfy

$$\int_{\mathbb{R}} \varphi(x-k)\,\varphi(x-j)\,dx = \delta_{kj}, \; j,k \in \mathbb{Z}.$$

If in addition φ is such that $M_0 = \widehat{\varphi}(0) = 1$ and satisfy the Strang-Fix conditions

$$D^j\widehat{\varphi}(2k\pi) = 0 \; for \; k \in \mathbb{Z}_0, 1 \leqslant j \leqslant m,$$

then for every $j = 0, \ldots, m$, there is a unique sequence $(a_k^{(j)})_{k \in \mathbb{Z}}$ such that

$$x^j = \sum_{k \in \mathbb{Z}} a_k^{(j)} \varphi(x-k) \quad ae$$

where the serie is absolutely and uniformly convergent on every compact subset of $\mathbb{R}$ and where $a_k^{(j)}$ is a polynomial of degree j in the variable k. These coefficients are

$$a_k^{(j)} = \int_{\mathbb{R}} x^j \varphi(x-k)\,dx, \quad j = 0, \ldots, m, \; k \in \mathbb{Z}.$$

In particular we have

$$a_0^{(j)} = M_j.$$

Another expression of the polynomials $a_k^{(j)}$ is obtained below. The proof can be found in [1]; it uses the previous result and recurrence technique. This expression leads to new relations between moments.

We use some definitions and notations: for $j, l \in \mathbb{N}_0$, we define

$$K_l(j) = \{(i_1, \ldots, i_l) \in \mathbb{N}_0^l : \sum_{k=1}^{l} i_k = j\}$$

and

$$K(j) = \bigcup_{l=1}^{j} K_l(j).$$

For $(i_1, \ldots, i_l) \in K(j)$, we write $i \in K(j)$. For $j \geqslant i_1 + \ldots + i_l$ we define

$$F_j(i_1, \ldots, i_l) = F_j(i) = (-1)^{i_1+1} \cdots (-1)^{i_l+1} C_j^{i_1} C_{j-i_1}^{i_2} \cdots C_{j-\sum_{k=1}^{l-1} i_k}^{i_l} M_{i_1} \ldots M$$

where

$$M_j = \int_{\mathbb{R}} x^j \varphi(x)\,dx = a_0^{(j)}.$$

For $j \in \mathbb{N}$, we also set

$$\sum_{i \in K(0)} F_j(i) = 1. \qquad (*)$$

Proposition 2.2 *Under the same hypothesis as in Proposition 2.1 and using the notations introduced above, we have the following relations*

$$a_k^{(j)} = \sum_{l=0}^{j} k^l \sum_{i \in K(j-l)} F_j(i), \quad k \in \mathbb{Z}, j = 1, \dots, m. \tag{1}$$

3. Relation between moments

We can deduce from the previous relations that the moments of even order can be expressed in terms of a linear combination of products of moments of smaller order in which the coefficients are of type C_m^l.

Corollary 3.1 *Under the same hypothesis as in Proposition 2.1 and using the same notations, we have*

$$M_j = \sum_{i \in K(j)} F_j(i) = \sum_{i \in \cup_{l=1}^{j} K_l(j)} F_j(i), \quad j = 1, \dots, m.$$

In particular, if j is even, we have

$$2M_j = \sum_{i \in \cup_{l=2}^{j} K_l(j)} F_j(i)$$

Proof. It suffices to take $k = 0$ in the relations (1) giving $a_k^{(j)}$ in the previous proposition. For j even, we have

$$F_j(j) = (-1)^{1+j} M_j = -M_j$$

hence the conclusion.

As example, we obtain

$$K_2(2) = \{(1,1)\}, \quad F_2((1,1)) = 2$$

hence

$$2M_2 = 2(M_1)^2;$$

in the same way

$$M_4 = 3(M_1)^4 + 4M_1 M_3,$$
$$M_6 = 45(M_1)^6 - 60(M_1)^3 M_3 + 6M_1 M_5 + 10(M_3)^2.$$

4. A numerical use

Of course, the previous relations can be used for numerical applications.
Following the ideas of [2], for a filter of the form

$$m_0(\omega) = 2^{-1/2} \sum_{k=1}^{2M} h_k \exp(i(k-1)\omega),$$

we can approximate the m-th moment with the relation

$$\mathcal{M}_m^{(r)} = i^{-m} [D^m \prod_{j=1}^{r} m_0(2^{-j}\omega)]_{\omega=0}. \tag{2}$$

Here, r represents the number of factors in the approximation. From (2), we can obtain the recurrence formula

$$\mathcal{M}_m^{(r+1)} = \sum_{k=0}^{m} C_m^k 2^{-rk} \mathcal{M}_{m-k}^{(r)} \mathcal{M}_k^{(1)}.$$

When m is even, we can replace this approximation with our formula where the preceding moments, which have been estimated, are involved. So, in this case, we replace r evaluations by 1.

References

[1] Bastin F and Nicolay S A note on moments of scaling functions, to appear in Rocky Mountain Journal of Mathematics

[2] Beylkin G, Coifman R, and Rokhlin V 1993, Fast wavelets transforms and numerical algorithms I, Communications on Pure and Applied Mathematics 44 141–183

[3] Daubcchies I 1992, Ten Lectures on Wavelets, (Philadelphia: SIAM, CBMS-NSF 61)

[4] Daubechies I 1993, Orthonormal compactly supported wavelets II. Variations on a theme, Siam J. Math. Anal. 24 499–519

[5] Fix G and Strang G 1969, Fourier analysis of the finite element method in Ritz-Galerkin theory, Studies in applied mathematics, Vol. XVIII 265–274

[6] Meyer Y 1990, Ondelettes et opérateurs, I (Paris: Hermann)

[7] Sweldens W and Piessens R 1992, Calculation of the wavelet decomposition using quadrature formulae, CWI Quarterly 5 33–52

Inst. Phys. Conf. Ser. No 173: Satellite colloquium
Paper presented at 24th Int. Coll. Group Theoretical Methods in Physics, Paris, France, 15–20 July 2002
©*2003 IOP Publishing Ltd*

Atom modulus quantization for matching pursuits

Christophe De Vleeschouwer

Belgian NFS researcher, Laboratoire de télécommunication
Université catholique de Louvain, Belgium

Abstract. This paper considers the selection and quantization of matching pursuits coefficients. It demonstrates that an optimal rate-distortion trade-off is achieved by selecting the atoms up to a dead-zone threshold, and by defining the modulus quantizer in terms of that threshold. In doing so, it takes into account quantization error re-injection resulting from inserting the modulus quantizer inside the MP atom computation loop. *In-loop* quantization results in a non-uniform optimal entropy constrained quantizer.

1. Introduction

Matching pursuit (MP) is a greedy and iterative approximation algorithm that generates sparse representation of a signal with respect to an overcomplete set of basis functions. The MP expansion is defined in terms of index, sign, and modulus of a subset of basis functions. The paper primarily deals with the modulus quantization. The study relies on two main assumptions. Firstly, linear reconstruction is assumed because complexity considerations dictate such reconstruction in most applications. Secondly, independent scalar modulus quantization is considered. A previous work has already addressed the MP atom modulus quantization issue in this context [1]. Our work not only provides an analytical derivation of the empirical results in [1], but also refines and completes them. Section 2 shows that an optimal MP expansion must stop as soon as the moduli of the extracted atoms become smaller than a threshold. Section 3 designs the optimal quantizer by taking into account the quantization error re-injection resulting from insertion of the modulus quantizer inside the MP atom computation loop (for improved coding efficiency). Section 4 measures the impact of our work.

2. Expansion in a rate/distortion framework

MP is an iterative expansion process. In a coding context, a critical question is when to stop the process. This section shows that an expansion is optimal in the rate/distortion (R/D) sense if it captures all and only all atoms larger than a threshold. Let a dictionary $\mathcal{D} = \{\varphi_k\}_{k=1}^{S} \in R^M$ be a frame such that $||\varphi_k|| = 1$ for all k. Given a source vector $x \in R^M$, MP approximates x by a linear combination of elements of $\mathcal{D}$, or atoms. At step i, the algorithm selects the dictionary function φ_{k_i} that maximizes $|< \varphi_{k_i}, R_i x >|$ and generates the residue for the next iteration, i.e. $R_{i+1}x = R_i x - < \varphi_{k_i}, R_i x > \varphi_{k_i}$. In the initial step, $R_0 x = x$. After n steps, the approximate signal is $x \approx \sum_{i=0}^{n-1}(s_i \cdot \alpha_i \cdot \varphi_{k_i})$, where $\alpha_i = | < \varphi_{k_i}, R_i x > |$ and $s_i = sign(< \varphi_{k_i}, R_i x >)$. To code the expansion, i.e. the index, sign, and modulus of the coefficients, it is desirable to achieve the lowest distortion D at the lowest number of bits R. For a given Lagrangian multiplier λ, the optimal trade-off is the one that minimizes the Lagrangian cost function $\mathcal{L}(\lambda) = D + \lambda R$. For an MP expansion, define D_i to be the distortion or MP residue energy after i atoms have been selected and

$\Delta D_i = D_i - D_{i-1}$ to be the decrease in residual energy due to the ith atom. As atoms are roughly selected in decreasing order of magnitude, we have $\Delta D_i < \Delta D_{i+1}$. Moreover, as MP representations are sparse, the number of bits devoted to an additional atom does not rapidly decrease with the iteration number. So, the incremental contribution $\Delta D + \lambda \Delta R$ of successive atoms to $\mathcal{L}(\lambda)$ can be assumed to be monotically increasing. This ensures the convexity of the (R, D) curve drawn along the expansion process. It also makes the selection of an additional atom worthwhile only until $\Delta D + \lambda \Delta R = 0$, defining a stopping criterion for the MP expansion. Defining m to be the modulus of an additional atom and $\xi_{Q(.)}(m)$ to be its quantization error for a quantization method defined by $Q(.)$, the decrease ΔD of the MP residue energy is $\Delta D = -\left(m^2 - \xi_{Q(.)}^2(m)\right)$. Moreover, the incremental rate ΔR can be divided in two parts: the first one $R_{Q(.)}(m)$ corresponds to the quantized atom modulus; the second one ΔR_{index} corresponds to the rate for the sign and index of the additional atom, and is not directly affected by the quantization method. As a consequence, the stopping criterion can be formulated in terms of the Lagrange multiplier λ, and of the threshold modulus of the last selected atom Θ,

$$\lambda = \frac{-\Delta D}{\Delta R} = \frac{\Theta^2 - \xi_{Q(.)}^2(\Theta)}{\Delta R_{index} + R_{Q(.)}(\Theta)} \tag{1}$$

To achieve a rate constrained optimal representation, atoms have to be selected until no atom larger than Θ can be found on any part of

3. Atom modulus quantizer design

At constant number of atoms, a coarser modulus quantization increases the distortion D but decreases the bit budget R of the MP expansion. The goal of this section is to find the quantizer that minimizes the Lagrangian cost function $\mathcal{L}(\lambda) = D + \lambda R$ for a given multiplier λ. We consider independent scalar quantization of atoms, which is appropriate when atoms are coded in a random order of modulus.

3.1. In-loop quantization of MP atom modulus

To improve coding efficiency, the quantization error is re-injected in the MP expansion process so that it can be corrected by subsequent MP iterations. Here, we analyze the evolution of the energy of the MP residue along the quantized expansion process in order to understand the impact of the re-injection on the distortion of the quantized moduli. As in Section 2, at step i, the algorithm selects the dictionary function φ_{k_i} that best matches $R_i x$. If $\hat{\alpha}_i$ is defined to be the quantized value of $\alpha_i = <R_i x, \varphi_{k_i}>$, then the residue based on the quantized modulus is given by $R_{i+1} x = R_i x - \hat{\alpha}_i \varphi_{k_i}$. Defining σ_i^2 as the quantization distortion on α_i, and observing that $R_i x - \alpha_i \varphi_{k_i}$ is orthogonal to φ_{k_i}, with $||.||$ referring to the quadratic norm in R^M, we have $||R_{i+1} x||^2 = ||R_i x||^2 - \alpha_i^2 + \sigma_i^2$. Without loss of generality, we can assume that at step i a fraction γ_i of the residual signal energy is captured by the MP expansion, i.e. $\alpha_i^2 = \gamma_i ||R_i x||^2$. Then, we have

$$||R_{n+1} x||^2 = \Pi_{i=0}^n (1 - \gamma_i) ||R_0 x||^2 + \sum_{i=0}^n \left(\Pi_{j=i+1}^n (1 - \gamma_j)\right) \sigma_i^2 \tag{2}$$

The factor $\left(\Pi_{j=i+1}^n (1 - \gamma_j)\right)$ is due to re-injection of the quantization error into the MP expansion loop.

3.2. Uniform quantization

For a uniform quantizer, by considering small bin width Ω, the quantization distortion for the ith Θ-subtracted modulus, i.e. σ_i^2 in (2), is independant of i and can be denoted by $D_{M_\Theta}(\Omega) \approx \Omega^2/12$ [2]. The distortion D_{UQ} resulting from the *in-loop* uniform quantization of N atoms is the second term of the sum in (2), i.e.

$$D_{UQ}(\Omega) = \sum_{i=0}^{N-1} \left(\Pi_{j=i+1}^{N-1}(1 - \gamma_j) \right) D_{M_\Theta}(\Omega) \tag{3}$$

The entropy H_{UQ} for N atoms is N times the Θ-subtracted modulus entropy, denoted by $H_{M_\Theta}(\Omega)$ and approximated in [2] as $H_{M_\Theta}(\Omega) \approx -\int f_{M_\Theta}(x) \log_2 f_{M_\Theta}(x) dx - \log_2 \Omega$. Given λ, the R/D optimal quantizer stepsize Ω solves

$$\lambda = -\frac{\partial D_{UQ}}{\partial H_{UQ}} = -\frac{\sum_{i=0}^{N-1} \left(\Pi_{j=i+1}^{N-1}(1 - \gamma_j) \right)}{N} \cdot \frac{\partial D_{M_\Theta}}{\partial H_{M_\Theta}} = \beta(N) \cdot \frac{\ln(2).\Omega^2}{6} \tag{4}$$

Equation (4) specifies the relationship between λ and Ω assuming $\beta(N) = \sum_{i=0}^{N-1} \left(\Pi_{j=i+1}^{N-1}(1 - \gamma_j) \right) / N$ is known. The multiplication by a factor $\beta(N)$ that is smaller than one is due to re-injection. To estimate $\beta(N)$, we note that all γ_i are much smaller than one, and approximate them by their mean value $\bar{\gamma}$. Then,

$$\beta(N) \approx \frac{\sum_{i=0}^{N-1}(1 - \sum_{j=i+1}^{N-1}\gamma_i)}{N} \approx (1 - N\bar{\gamma}/2) \approx \sqrt{(1 - N\bar{\gamma})} \tag{5}$$

In (2), by neglecting the quantization distortion in comparison with the energy of the residual signal, we also have

$$||R_N x||^2 \approx \Pi_{i=0}^{N-1}(1 - \gamma_i)||R_0 x||^2 \approx (1 - N\bar{\gamma})||R_0 x||^2 \tag{6}$$

From (5) and (6), $\beta(N)$ is estimated in terms of the ratio between the initial and final MP residual energy, i.e., $\beta(N) \approx \sqrt{||R_N x||^2/||R_0 x||^2}$. Since an optimal expansion selects the atoms up to a threshold Θ, $||R_N x||^2$ is also the energy E_Θ after all atoms larger than Θ have been selected. In that sense, $\beta(N)$ is a function of Θ and we have

$$\beta(N) \approx \sqrt{E_\Theta/||R_0 x||^2} \triangleq \beta'(\Theta) \tag{7}$$

In practice, experiments have shown that for an arbitrary signal, $\beta'(\Theta)$ can be estimated from an initial expansion that is computed without quantization. For video coding applications, $\beta'(\Theta)$ can be estimated from the expansion obtained on the previous frame [3].

threshold Θ, and the quantizer stepsize Ω. We begin by noting that $\xi_{Q(.)}(\Theta)$ is equal to $\Omega/2$. Experimentaly we have also observed that the cost in bits of the last selected atom, i.e. $R_{last} = \Delta R_{index} + R_{Q(.)}(\Theta)$

$$\frac{\Theta^2 - \Omega^2/4}{R_{last}} = \beta'(\Theta) \cdot \ln(2).\Omega^2/6 \tag{8}$$

For video coding, $R_{last} \in [17\dots19]$, and hence

$$\frac{\Omega}{\Theta} \approx 0.66 \cdot \sqrt{\beta'(\Theta)^{-1}} \tag{9}$$

3.3. Non-uniform quantization

In the previous paragraph we have shown that re-injection affects the optimal uniform quantizer stepsize. Now we consider the design of the entropy constrained quantizer in presence of re-injection but in absence of uniformity constraint. Non-uniformity is justified by the observation that atoms are roughly selected in decreasing order of magnitude. As a consequence, the initial atoms, which have more chance to be corrected by subsequent MP iterations, are also the largest ones. This suggests increasing the quantization stepsize with the atom magnitude. Designing the R/D optimal non-uniform quantizer consists of fixing its bin boundaries, or equivalently the sequence of stepsizes $\{\Omega_i\}_{i \geq 0}$, so that all bins have the same incremental benefit in distortion for a given incremental cost in rate, the ratio between them being defined by the Lagrangian multiplier λ. We propose a recursive approach: at each step, given the lower boundary of a quantizer bin, the upper boundary is determined. From Section 2, the lower boundary of the first bin is the stopping threshold Θ. Let us now consider the selection of the upper boundary of the ith quantizer bin, assuming its lower boundary is known. Let n_1 denote the hypothetical iteration index for which the atom modulus α_{n_1} equals the lower boundary of the ith quantizer bin. For any hypothetical iteration index $n_2 < n_1$, define α_{n_2} to be the selected atom modulus. As atoms are selected in decreasing order of magnitude, $\alpha_{n_2} > \alpha_{n_1}$ and the set of atoms selected between the n_2^{th} and n_1^{th} iterations of the MP expansion belong to $[\alpha_{n_1}, \alpha_{n_2}[$. Without loss of generality our problem is to find α_{n_2}, or equivalently n_2, so that the best Lagrangian R/D trade-off is achieved for the quantization of the $(n_1 - n_2)$ atoms belonging to $[\alpha_{n_1}, \alpha_{n_2}[$. Letting N be the hypothetical total number of atoms selected by the MP expansion up to the Θ threshold, the distortion $D_i(N, n_1, n_2)$ due to the quantization of the $(n_1 - n_2)$ atoms belonging to the ith bin $[\alpha_{n_1}, \alpha_{n_2}[$ with a stepsize $\Omega_i = \alpha_{n_2} - \alpha_{n_1}$ is an immediate consequence of (2) and can be written

$$D_i(N, n_1, n_2) = \sum_{k=n_2+1}^{n_1} \left(\Pi_{j=k+1}^{N-1}(1 - \gamma_j) \right) \cdot D_{M\Theta}(\Omega_i) \tag{10}$$

Similar to (4), given λ, the optimal i^{th} bin quantizer stepsize Ω_i solves

$$\lambda = -\frac{\sum_{k=n_2+1}^{n_1} \left(\Pi_{j=k+1}^{N-1}(1 - \gamma_j) \right)}{n_1 - n_2} \cdot \frac{\partial D_{M\Theta}}{\partial H_{M\Theta}} = \beta_i(N, n_1, n_2) \cdot \frac{\ln(2).\Omega_i^2}{6} \tag{11}$$

In (11), Ω_i is the width of the ith bin. Given the lower bound of the bin, it defines the upper bound as a function of λ. For small γ_i, and following the developments similar to the ones in (5) and (6), $\beta_i(N, n_1, n_2)$ can be approximated as

$$\beta_i(N, n_1, n_2) \approx \frac{||R_N x||^2}{||R_{n_1} x||^2} \cdot \sqrt{\frac{||R_{n_1} x||^2}{||R_{n_2} x||^2}} = \frac{||R_N x||^2}{\sqrt{||R_{n_1} x||^2 \cdot ||R_{n_2} x||^2}} \tag{12}$$

This factor depends on the residual signal energies after respectively n_1, n_2 and N MP iterations, with $N > n_1 > n_2$. Actually, $||R_{n_1} x||^2$ and $||R_{n_2} x||^2$ are measured once the atoms larger than the lower and higher boundary of the ith bin have been selected respectively. $||R_N x||^2$ measures the energy of the final residue, i.e. once all atoms larger than the stopping threshold Θ have been selected. Formally, let θ_i be the lower boundary of the ith bin, i.e. $\theta_i = \Theta + \sum_{l=1}^{i-1} \Omega_l^{j+1}$. Define E_θ to be the energy of the residue after all atoms larger than θ have been selected. So, (12) can be written as

$$\beta_i(N, n_1, n_2) \approx \frac{E_\Theta}{\sqrt{E_{\theta_i} \cdot E_{\theta_{i+1}}}} \triangleq \beta_i'(\Theta, \theta_i, \theta_{i+1}) \triangleq \beta_i'(\Theta) \tag{13}$$

For notation convenience, $\beta_i'(\Theta, \theta_i, \theta_{i+1})$ is simply refered to as $\beta_i'(\Theta)$ in the following. Similar to arguments used in deriving (9), (1) can be combined with (11) to arrive at

$$\frac{\Omega_i}{\Theta} \approx 0.66 \cdot \sqrt{\beta_i'(\Theta)^{-1}} \tag{14}$$

We note that $\beta_i'(\Theta)$ decreases as i increases, which indicates that the stepsize Ω_i increases with the quantizer bin index i. This is to be expected as quantization error of large moduli atoms in early MP iterations is corrected by subsequent iterations. The energy values needed to estimate $\{\beta_i'(\Theta)\}_{i>0}$ in (13) are derived either from a non-quantized expansion of the current signal or from the previous frame of the video sequence, thus avoiding excessive computation. This computation is performed by accumulating the atom contribution to the residue energy decrease on a quantizer bin basis [3]. For the kth atom, this contribution is $\alpha_k^2 - (\alpha_k - \hat{\alpha}_k)^2$.

4. Application to video coding

In Fig.1, three uniform and one non-uniform quantizers are compared. All uniform quantizer stepsizes have been selected according to equation (9), i.e. $\Omega/\Theta = 0.66\sqrt{\beta'(\Theta)^{-1}}$. For both UQ, IS and UQ, the quantization error re-injection is neglected, i.e. $\beta' = 1$. UQ, IS is the approach used in [1]. It stops the search as soon as an atom smaller than the threshold is found in one part of the frame. On the contrary, UQ conforms to the stopping criteria expressed in Section 2. The third uniform quantizer ULQ considers the quantization error re-injection and sets β' according to (7). The non-uniform quantizer, $NULQ$, is designed based on (14) and (13). As expected, $NULQ$ outperforms all uniform quantizers. However, at low bitrates the improvement is negligible. This is because most of the bits are devoted to motion vector coding, and few atoms are encoded. On the contrary, at high bitrates, the non-uniform quantizer results in 0.5 to 2 dBs improvement over [1].

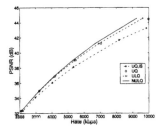

Figure 1. R/D curves for Mobile CIF video sequence encoded at 30 fps.

References

[1] Neff R and Zakho A 2000, Modulus quantization for MP video coding, IEEE Transactions on Circuits and Systems for Video Technology, 10 895-912

[2] Gray R M and Neuhoff D L 1998, Quantization, IEEE Transactions on Information Theory, 44 1-63

[3] De Vleeschouwer C and Zakhor A, In-loop atom modulus quantization for MP and its application to video coding, IEEE Transactions on Image Processing, submitted.

Inst. Phys. Conf. Ser. No 173: Satellite colloquium
Paper presented at 24th Int. Coll. Group Theoretical Methods in Physics, Paris, France, 15–20 July 2002
©*2003 IOP Publishing Ltd*

Multiscale observations of the solar atmosphere

J-F Hochedez and S Gissot

Departement of Solar Physics, Royal Observatory of Belgium, B-1180 Bruxelles, Belgium

L Jacques and J-P Antoine

Institut de Physique Théorique, Université catholique de Louvain,
B-1348 Louvain-la-Neuve, Belgium

Abstract. The observations of the Extreme ultraviolet Imaging Telescope (EIT) contain various small features of the solar coronal regions: EUV bright points, ephemeral regions, brightenings, loops segments, cosmic ray hits, etc. In this work we present an automated method extracting small objects. It is based on the continuous 2D Mexican Hat Wavelet transform. We study global and long-term statistical properties of the small features. Their distribution will be used for solar cycle studies and prediction of solar events (Space Weather applications).

1. Introduction

In the highly structured Solar Corona, several kind of processes are observed at different scales. The complex coronal dynamics is usually depicted as "solar activity on all scales" [2, 3]. Their connection entails a lot of physics, and remains challenging for modelling. Small scales studies are also necessary to Space Weather applications like flares or Coronal Mass Ejection (CME) predictions, and for the comprehension of the solar cycle (e.g. ephemeral regions rate evolution [4]). Although the quest for instruments with higher resolution and cadence is going on, like the SDO or Solar Orbiter missions, it is still of interest to push the limits of the existing datasets. The Extreme ultraviolet Imaging Telescope (EIT) of SoHO has accumulated a long-term image record of the Sun atmosphere over the first half of the solar cycle 23. In this work we present a new technique of small object extraction, based on the Continuous Wavelet Transform (CWT). After describing the basic tools of the method in Sections 2 and 3, the principle of the extraction of the small objects will be discussed in Section 4. Applications to the long-term study of the solar cycle will also be presented.

2. 2D CWT of simple objects

The 2D wavelet transform of an image considered as a function $f \in L^2(\mathbb{R}^2)$ is [1]

$$\mathcal{W}_f(\vec{b}, a) \;=\; \int_{\mathbb{R}^2} d\vec{x}\, f(\vec{x})\, \psi^*_{\vec{b},a}(\vec{x}), \tag{1}$$

where * denotes the complex conjugation, and $\psi_{\vec{b},a}$ is the shifted and dilated wavelet ψ given, in $L^1(\mathbb{R}^2)$ normalization, by

$$\psi_{\vec{b},a}(\vec{x}) \;=\; \frac{1}{a^2}\psi\!\left(\frac{\vec{x} - \vec{b}}{a}\right), \tag{2}$$

for $\vec{b} \in \mathbb{R}^2$ and $a \in \mathbb{R}^*_+$.

The wavelet coefficient maxima are found where the position and scale of the wavelet are closest to those of local features. We use the 2D Mexican Hat wavelet (MH):

$$\psi(\vec{x}) \;=\; (2 - |\vec{x}|^2)e^{-|\vec{x}|^2/2} \tag{3}$$

Like the wavelet itself, the Fourier transform of the MH is real and its analytical definition is:

$$\hat{\psi}(\vec{k}) \;=\; 2\pi|\vec{k}|^2 e^{-\vec{k}^2/2} \tag{4}$$

It has been shown that the MH wavelet has an optimal detectivity for Gaussian peaks on 1/f noise [5]. As in 1D, the Mexican Hat is known to have 2 vanishing moments so that any area of an image that can be approximated by a polynomial of degree 0 or 1 has vanishing wavelet coefficients. This filtering property allows to better detect singularities, e.g. spikes and peaks, and to a lesser extent elongated ridges such as magnetic loops.

3. Wavelet spectrum of an individual EIT image

The CWT is related to the energy of an image f by the Plancherel formula:

$$\|f\|_2^2 \;=\; \int_{\mathbb{R}^2} d^2\vec{x}\, |f(\vec{x})|^2 \;=\; \frac{1}{C_\psi} \int_{\mathbb{R}^2 \times \mathbb{R}_+^*} d^2\vec{b}\, \frac{da}{a}\, |\mathcal{W}_f(\vec{b}, a)|^2 \tag{5}$$

where

$$C_\psi \;=\; \int_{\mathbb{R}^2} d^2\vec{k}\, \frac{|\hat{\psi}(\vec{k})|^2}{\|\vec{k}\|^2}. \tag{6}$$

The quantity $|\mathcal{W}_f(\vec{b}, a)|^2$ may be interpreted as an energy density. Integrating it over the domain that we are studying, that is, the solar disk denoted by $\Omega_\odot$, we obtain:

$$\mu(a) \;=\; \iint_{\Omega_\odot} |\mathcal{W}_f(\vec{b}, a)|^2\, d^2\vec{b} \tag{7}$$

$\mu(a)$, the scale measure (also called the wavelet spectrum), which represents the energy integrated over the area of study. From a statistical point of view, the scale measure is the variance of the CWT over the area of interest at a scale a. We will see in the next section how to use it for thresholding our CWT coefficients.

The dataset was made of EIT images over the SOHO mission. Each of the four wavelengths were studied at a 5-days cadence. The earlier observation occured in January 1996, and the last in May 2001. In three of the four lines (or wavelength channel) the scale measure was found to follow a power law, whose exponent is almost systematically 1 [6]. The fourth one corresponds to the 304 channel. Its structure is the chromospheric network made of convection cells called "super granules", and characterized by its self-similar properties. The shape of the scale-measure is a trace of this property.

These results are shown in Figure 1; their interpretation is under way.

4. Extraction and dataset of small features

We develop here an automated method for the extraction of "small" objects in EIT images. We use the scale measure for thresholding the CWT, so that at any scale we keep only significant wavelet coefficients. We apply a 3σ statistical rule. We then obtain a thresholded set of points in the space-scale volume. In the second step of the small features extraction, we gather the selected points of the precedent step into connected regions. Thus we get a set of connected regions in the 3D space: each region is interpreted as a small object in the EIT image. For

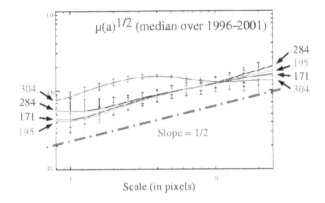

Figure 1. Square root of the scale-measure $\mu(a)^{1/2}$ averaged over half a solar cycle (minimum roughly around 1996 to maximum reached in 2001).

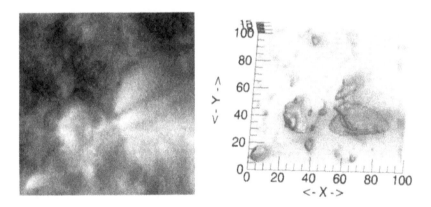

Figure 2. Representation of the connected regions (right) found for a rectangular region extracted from an EIT image (left).

each region of the dataset, the method provides caracteristic parameters. An example of such connected regions is shown in Figure 2.

In this figure small regions are lying over bigger regions. Those features are not detected by other methods. To finally retrieve the small features we have to impose bounds to the maximal scale allowed. The minimum scale is imposed by mathematical restrictions, whereas we impose an arbitrary maximum scale in order that the largest features be excluded from our dataset. The dataset is built by selecting the small features in a 3D space-scale volume. This volume is delimited spatially by the solar disk, and along the scale axis by the minimum and maximum scale formerly described: $\Omega_\odot \times [a_{min}, a_{max}]$.

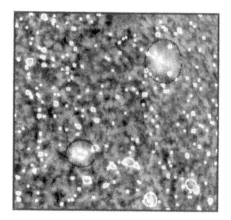

Figure 3. Small features detection applied on a 304-wavelength image.

We can summarize the two steps of the small features detection as follows:

Step 1: for each scale $a \in [a_{min}, a_{max}]$, threshold the CWT over the whole area of interest, namely, $\Omega_\odot$, and obtain the set A_f of the pixels of interest of f at scale a:

$$A_f(a) = \{\vec{b} \in \Omega_\odot : \mathcal{W}_f(\vec{b}, a) > 3\mu(a)\} \qquad (8)$$

We define the space-scale volume D_f by the union:

$$B_f = \bigcup_{a \in [a_{\min}, a_{\max}]} A_f(a), \qquad (9)$$

Step 2: select the connected regions of B_f and collect them with caracteristic parameters in the final dataset C_f. Each element of C_f is then considered as a feature.

All the selected features from this process are represented in the real EIT image. We draw the contour of the blob that we obtain (contour of the shape that contains the projection on the 2D image of this 3D object). We then get an exhaustive dataset of small features covering the solar disk (Figure 3). For each of these objects, the dataset provides precise informations: location, optimal scale, peak and background intensities.

Compared to classical methods based on thresholding, this detection is not as sensitive to the background luminosity of these objects. Meoreover, the surface of the active region increases at maximum of activity. Indeed, the method shall be robust and stable over the whole solar cycle.

5. Conclusion

The method developed here is a robust tool for building a long-term dataset of small features covering the solar disk in the EIT images. It is now possible to look for long-term trends of the distribution of small features: it has already been shown [7] that the small objects number follows a law that is the inverse of the curve of solar activity (a minimum for small objects number corresponds to a maximum for solar activity and vice-versa). Other such fundamental results are still to be found. Indeed several improvements of the method have yet to be performed. In particular, the use of directional wavelets should allow us to refine the search and identification of specific solar objects like coronal loops. The parameters that describe each object of the dataset, like the optimal scale, can be used to find other statistical trends of these objects. The use of the CWT for coronal images is a promising way to understand the physical link between phenomena at various scales.

References

[1] Antoine J-P, Carrette P, Murenzi R, and Piette B 1993, Image analysis with two-dimensional continuous wavelet transform, Signal Proc. 31 241-272

[2] Švestka Z 1999, Solar activity on all scales, in 8th SOHO Workshop: Plasma Dynamics and Diagnostics in the Solar Transition Region and Corona, Vial J-C and Kaldeich-Schümann B (eds), ESA Special Publications 446, 8 53

[3] Clette F, Poletto G, and Švestka Z 1999, Working Group 6: Activity on all Scales, in 8th SOHO Workshop: Plasma Dynamics and Diagnostics in the Solar Transition Region and Corona, Vial J-C and Kaldeich-Schümann B (eds), ESA Special Publications 446, 8 103

[4] Hochedez J-F, Clette F, Verwichte E, Berghmans D, and Cugnon P 2000, Mid-Term Variations in the Extreme UV Corona: The EIT/SOHO Perspective, in Proc. 1st Solar & Space Weather Euroconference, "The Solar Cycle and Terrestrial Climate", Santa Cruz de Tenerife, Spain (ESA SP-463, December 2000)

[5] Sanz J L, Herranz D, and Martínez-Gónzalez E 2001, Optimal detection of sources on a homogeneous and isotropic background, Astrophys. J. 552 484-492

[6] Hochedez J-F, Jacques L, Verwichte E, Berghmans D, Wauters L, Clette F, and Cugnon P 2002, Multiscale activity observed by EIT/SoHO, in Proc. Second Solar Cycle and Space Weather Euroconference, Sept. 2001, Vico Equense, Italy. Sawaya-Lacoste H (ed) (ESA SP-477, Noordwijk) 115 - 118

[7] Hochedez J-F, Jacques L, Zukhov A, Clette F, and Antoine J-P 2002, Small features in the EIT-SOHO images, Proc. SOHO-11 Symposium "From Solar Min to Max: Half a Solar Cycle with SOHO", March 2002, Davos (CH), ESA SP-508 295–298

Inst. Phys. Conf. Ser. No 173: Satellite colloquium
Paper presented at 24th Int. Coll. Group Theoretical Methods in Physics, Paris, France, 15–20 July 2002
©2003 IOP Publishing Ltd

Wavelet transform associated to a homogeneous Siegel domain

Hideyuki Ishi†

Department of Mathematical Sciences, Yokohama City University, Seto 22-2, Yokohama 236-0027, Japan

Abstract. Let $\mathcal{D}$ be a homogeneous Siegel domain of non-tube type and G a completely solvable Lie group acting on $\mathcal{D}$ simply transitively as affine transformations. After giving an irreducible decomposition of the unitary representation of G defined on the L^2-function space over the Shilov boundary of $\mathcal{D}$, we consider the continuous wavelet transform associated to each irreducible subrepresentation.

1. Introduction

Let $\mathcal{D}$ be a homogeneous Siegel domain on which a completely solvable Lie group G acts simply transitively as affine transformations (see [9]), and Σ the Shilov boundary of $\mathcal{D}$. The simplest example is the case that $\mathcal{D}$ is the upper half plane $\{ z \in \mathbf{C} \,;\, \Im z > 0 \}$ with $\Sigma = \mathbf{R}$ (the real axis), and G is the $ax + b$ group. We can define naturally a unitary representation π of G on the Hilbert space $L^2(\Sigma)$. Then $L^2(\Sigma)$ is decomposed into the direct sum of irreducible subspaces, and all the subrepresentations of G are square integrable (see Theorem 2). We introduce the continuous wavelet transform associated to such irreducible representations. This consideration has been carried out in the case that $\mathcal{D}$ is of rank 1 by Liu-Peng [7], and in the case that $\mathcal{D}$ is a symmetric tube domain by Liu [6].

Let L_0 be an irreducible subspace of $L^2(\Sigma)$, and $\phi \in L_0$ an admissible vector, that is, an element of L_0 for which

$$c_\phi := \int_G |(\phi|\pi(g)\phi)|^2 dg < \infty.$$

Then, following the general theory [4] (see also [1]), we define the wavelet transform $W_\phi : L_0 \to L^2(G)$ by

$$W_\phi f(g) := c_\phi^{-1/2}(f|\pi(g)\phi) \qquad (f \in L^2(W),\ g \in G).$$

This W_ϕ is an isometry and an intertwining operator from π to the left regular representation of G. Moreover we have the inverse formula

$$f = c_\phi^{-1/2} \int_G W_\phi f(g)\, \pi(g)\phi\, dg$$

(in the weak topology), which means that any vector f in L_0 is expressed as a 'continuous linear combination' of the family of vectors $\{\pi(g)\phi\}_{g \in G}$, and the function $W_\phi f$ gives the coefficients of the resolution.

† e-mail: hideyuki@yokohama-cu.ac.jp

In this article, we shall describe an irreducible decomposition of the function space $L^2(\Sigma)$ explicitly and give a formula for the calculation of the constant c_ϕ. If $\mathcal{D}$ is a tube domain, G is the semidirect product of a linear group with a vector group [9] , so that one can solve the problem by applying the argument of [1, section 9] (see also [6]).Thus we shall restrict our consideration to the case of non-tube type in what follows.

2. Siegel domains of non-tube type

Let $\nu_0, \nu_1, \ldots, \nu_r$ be numbers such that $n = \nu_0 + \nu_1 + \ldots + \nu_r$, and put $n' := \nu_1 + \ldots + \nu_r$. For real vector spaces $V_{lk} \subset \mathrm{M}(\nu_l, \nu_k; \mathbf{R})$ $(1 \leqslant k < l \leqslant r)$ and complex vector spaces $W_l \subset \mathrm{M}(\nu_l, \nu_0; \mathbf{C})$ $(l = 1, \ldots, r)$ satisfying the conditions

(a) $A \in V_{lk}, B \in V_{ki} \Rightarrow AB \in V_{li}$ $(1 \leqslant i < k < l \leqslant r)$,
(a') $A \in V_{lk}, U \in W_k \Rightarrow AU \in W_l$ $(1 \leqslant k < l \leqslant r)$,
(b) $A \in V_{li}, B \in V_{ki} \Rightarrow A^{\mathrm{t}}B \in V_{lk}$ $(1 \leqslant i < k < l \leqslant r)$,
(b') $U \in W_l, U' \in W_k \Rightarrow U^{\mathrm{t}}\bar{U}' \in (V_{lk})_{\mathbf{C}}$ $(1 \leqslant k < l \leqslant r)$,
(c) $A \in V_{lk} \Rightarrow A^{\mathrm{t}}A \in \mathbf{R}I_{\nu_l}$ $(1 \leqslant k < l \leqslant r)$,
(c') $U \in W_l \Rightarrow U^{\mathrm{t}}\bar{U} + \bar{U}^{\mathrm{t}}U \in \mathbf{R}I_{\nu_l}$ $(l = 1, \ldots, r)$,

we define $V \subset \mathrm{Sym}(n', \mathbf{R})$ and $W \subset \mathrm{M}(n', \nu_0; \mathbf{R})$ as follows:

$$V := \left\{ x = \begin{pmatrix} X_{11} & {}^{\mathrm{t}}X_{21} & \cdots & {}^{\mathrm{t}}X_{r1} \\ X_{21} & X_{22} & & {}^{\mathrm{t}}X_{r2} \\ \vdots & & \ddots & \\ X_{r1} & X_{r2} & & X_{rr} \end{pmatrix} ; \begin{array}{l} X_{kk} = x_{kk}I_{\nu_k}, \ x_{kk} \in \mathbf{R} \quad (k = 1, \ldots, r) \\ X_{lk} \in V_{lk} \qquad\qquad (1 \leqslant k < l \leqslant r) \end{array} \right\},$$

$$W := \left\{ u = \begin{pmatrix} U_1 \\ U_2 \\ \vdots \\ U_r \end{pmatrix} ; U_l \in W_l \ (l = 1, \ldots, r) \right\}.$$

Thanks to (b') and (c') , we can define a Hermitian map $Q : W \times W \to V_{\mathbf{C}}$ by $Q(u, u') := (u^{\mathrm{t}}\bar{u}' + \bar{u}'^{\mathrm{t}}u)/4$ $(u, u' \in W)$. Let Ω be an open convex cone $\{ x \in V ; x \gg 0 \}$ in the vector space V. Our Siegel domain $\mathcal{D} = \mathcal{D}(\Omega, Q)$ is the complex domain in $V_{\mathbf{C}} \times W$ given by

$$\mathcal{D}(\Omega, Q) := \{ (z, u) \in V_{\mathbf{C}} \times W ; \Im z - Q(u, u) \in \Omega \} .$$

Next we give an affine transformation group on $\mathcal{D}(\Omega, Q)$. Put

$$H := \left\{ T = \begin{pmatrix} T_{11} & 0 & \cdots & 0 \\ T_{21} & T_{22} & & 0 \\ \vdots & & \ddots & \\ T_{r1} & T_{r2} & & T_{rr} \end{pmatrix} ; \begin{array}{l} T_{kk} = t_{kk}I_{\nu_k}, \ t_{kk} > 0 \quad (k = 1, \ldots, r) \\ T_{lk} \in V_{lk} \qquad\qquad (1 \leqslant k < l \leqslant r) \end{array} \right\}.$$

Then H forms a Lie subgroup of $\mathrm{GL}(n', \mathbf{R})$ owing to the condition (a). Let G be the subgroup of $\mathrm{Sp}(2n, \mathbf{R})$ given by

$$\left\{ g(x, u, T) := \begin{pmatrix} I_{\nu_0} & 0 & \Re^{\mathrm{t}}u \\ \Im u & I_{n'} & \Re u & x + \Im(u^{\mathrm{t}}\bar{u})/2 \\ & & I_{\nu_0} & \\ & & -\Im^{\mathrm{t}}u & I_{n'} \end{pmatrix} \begin{pmatrix} I_{\nu_0} & & & \\ & T & & \\ & & I_{\nu_0} & \\ & & & {}^{\mathrm{t}}T^{-1} \end{pmatrix} ; \begin{array}{l} x \in V \\ u \in W \\ T \in H \end{array} \right\}.$$

This G acts on $\mathcal{D}(\Omega, Q)$ simply transitively by

$$g(x, u, T) \cdot (z_0, u_0) := (Tz_0{}^tT + x + 2iQ(Tu_0, u) + iQ(u, u), Tu_0 + u).$$

Indeed, the group H acts on the cone Ω simply transitively by $\Omega \ni x \mapsto Tx{}^tT \in \Omega$ $(T \in H)$.

Theorem 1 *All homogeneous Siegel domains are realized as $\mathcal{D}(\Omega, Q)$ or tube domains $\mathcal{D}(\Omega) = V + i\Omega \subset V_{\mathbb{C}}$. Therefore, all bounded homogeneous domains are biholomorphic to these domains.*

The Shilov boundary Σ of $\mathcal{D}(\Omega, Q)$ is described as $\Sigma = \{(x + iQ(u, u), u)\,;\, x \in V, u \in W\}$, and we define the L^2-space over Σ by $L^2(\Sigma) := \{f : \Sigma \to \mathbb{C}\,;\, \|f\|^2 := \int_V \int_W |f(x + iQ(u, u), u)|^2\, dm(u)\, dm(x) < \infty\}$. We set $n_{lk} := \dim V_{lk}$ $(1 \leqslant k < l \leqslant r)$, $m_k := \dim W_k$ and $d_k := 1 + (\sum_{i<k} n_{ki} + \sum_{l>k} n_{lk})/2$ $(k = 1, \ldots, r)$. Putting $\chi_\Sigma(g) := (t_{11})^{d_1+m_1}(t_{22})^{d_2+m_2} \cdots (t_{rr})^{d_r+m_r}$ for $g = g(x, u, T) \in G$, we define a unitary representation $(\pi, L^2(\Sigma))$ by

$$\pi(g)f(p_0) := \chi_\Sigma(g)^{-1}f(g^{-1} \cdot p_0) \qquad (p_0 = (x_0 + iQ(u_0, u_0), u_0) \in \Sigma).$$

3. Irreducible decomposition of $(\pi, L^2(\Sigma))$

Denote by V^* the dual vector space of V, and by a^* the contragredient action of H on V:

$$\langle x, a^*(T)\xi \rangle := \langle T^{-1}x{}^tT^{-1}, \xi \rangle \qquad (x \in V, \xi \in V^*, T \in H).$$

For $\varepsilon = (\varepsilon_1, \varepsilon_2, \ldots, \varepsilon_r) \in \{-1, 1\}^r$, we define a linear form E_ε^* by $\langle x, E_\varepsilon^* \rangle := \sum_{k=1}^r \varepsilon_k x_{kk}$ $(x \in V)$, and put

$$\mathcal{O}_\varepsilon^* := a^*(H)E_\varepsilon^* \subset V^*.$$

Then each orbit $\mathcal{O}_\varepsilon^*$ is open, and the disjoint union $\bigsqcup_{\varepsilon \in \{-1,1\}^r} \mathcal{O}_\varepsilon^*$ is dense in V^* [3, section 6]. For a function $f \in L^2(\Sigma)$, we denote by $\hat{f}$ the partial Fourier transform of f with respect to the variable $x \in V$:

$$\hat{f}(\xi, u) := (2\pi)^{-(\dim V)/2} \int_V e^{-i\langle x, \xi \rangle} f(x + iQ(u, u), u)\, dm(x) \qquad (\xi \in V^*,\ u \in W).$$

Then we have $\hat{f}(\xi, \cdot) \in L^2(W)$ for almost all $\xi \in V^*$, and the Plancherel formula yields $\|f\|^2 = \int_{V^*} \|\hat{f}(\xi, \cdot)\|_{L^2(W)}^2\, dm(\xi)$. Take an orthonormal basis $\{\mathbf{e}_{k,i}\}_{i=1,\ldots,m_k}$ of the complex vector space W_k with respect to the inner product $(U|U') := \operatorname{tr} U^t\bar{U}'$ $(U, U' \in W_k)$. Let Λ be the index set $\{\lambda = (k, i)\,;\, 1 \leqslant k \leqslant r,\ 1 \leqslant i \leqslant m_k\}$, and denote by $\{u_\lambda\}_{\lambda \in \Lambda}$ the coordinate of $u \in W$ with respect to the basis $\{\mathbf{e}_\lambda\}_{\lambda \in \Lambda}$ of W. For $\varepsilon \in \{-1, 1\}^r$ and $\lambda = (k, i) \in \Lambda$, let $D_{(\varepsilon),\lambda}$ be the differential operator on W given by

$$D_{(\varepsilon),\lambda} := \begin{cases} \left(\dfrac{u_\lambda}{2} - \dfrac{\partial}{\partial \bar{u}_\lambda}\right)\left(\dfrac{\bar{u}_\lambda}{2} + \dfrac{\partial}{\partial u_\lambda}\right) & (\varepsilon_k = -1), \\[2ex] \left(\dfrac{\bar{u}_\lambda}{2} - \dfrac{\partial}{\partial u_\lambda}\right)\left(\dfrac{u_\lambda}{2} + \dfrac{\partial}{\partial \bar{u}_\lambda}\right) & (\varepsilon_k = 1), \end{cases}$$

and for $\xi = a^*(T)E_\varepsilon^* \in \mathcal{O}_\varepsilon^*$, set $D_{\xi,\lambda} := \sigma_T \circ D_{(\varepsilon),\lambda} \circ \sigma_T^{-1}$, where σ_T is the operator on $L^2(W)$ given by $\sigma_T \varphi(u) := \varphi(T^{-1}u)$ $(\varphi \in L^2(W))$. For $\alpha = (\alpha_\lambda) \in \mathbf{Z}_{\geqslant 0}^M$ $(M := \dim_{\mathbb{C}} W = \sharp\Lambda)$, we define

$$\mathcal{L}_{\xi,\alpha} := \{\varphi \in L^2(W)\,;\, D_{\xi,\lambda}\varphi = \alpha_\lambda\varphi\ (\forall \lambda \in \Lambda)\},$$

$$L^2_{\varepsilon,\alpha}(\Sigma) := \left\{ f \in L^2(\Sigma); \begin{array}{ll} \hat{f}(\xi,\cdot) \in \mathcal{L}_{\xi,\alpha} & (\text{a.a. } \xi \in \mathcal{O}^*_\varepsilon) \\ \hat{f}(\xi,\cdot) = 0 & (\text{a.a. } \xi \notin \mathcal{O}^*_\varepsilon) \end{array} \right\}.$$

Then we have a unitary isomorphism $L^2_{\varepsilon,\alpha}(\Sigma) \ni f \mapsto \hat{f} \in \int^{\oplus}_{\mathcal{O}^*_\varepsilon} \mathcal{L}_{\xi,\alpha}\, dm(\xi)$.

Theorem 2 *One has an irreducible decomposition of $(\pi, L^2(\Sigma))$ as*

$$L^2(\Sigma) = \sum_{\varepsilon \in \{-1,1\}^r} \sum_{\alpha \in \mathbf{Z}^M_{\geq 0}} L^2_{\varepsilon,\alpha}(\Sigma),$$

and all irreducible subrepresentations are square integrable. Two subrepresentations $(\pi, L^2_{\varepsilon,\alpha}(\Sigma))$ and $(\pi, L^2_{\varepsilon',\alpha'}(\Sigma))$ are equivalent if and only if $\varepsilon = \varepsilon'$.

Let $\Upsilon^*_\varepsilon : \mathcal{O}^*_\varepsilon \to \mathbf{R}$ be a positive smooth function defined by

$$\Upsilon^*_\varepsilon(a^*(T)E^*_\varepsilon) := (t_{11})^{2s_1}(t_{22})^{2s_2}\cdots(t_{rr})^{2s_r} \qquad (T \in H)$$

with $s_k := 1 + \sum_{i<k} n_{ki}$ $(k = 1,\ldots,r)$.

Theorem 3 *For $\phi \in L^2_{\varepsilon,\alpha}(\Sigma)$, one has*

$$c_\phi = \int_{\mathcal{O}^*_\varepsilon} \|\hat{\phi}(\xi,\cdot)\|^2_{\mathcal{L}_{\xi,\alpha}} \Upsilon^*_\varepsilon(\xi)\, dm(\xi).$$

Indeed, ϕ is admissible if and only if the integral of the right hand side converges.

4. Concluding discussion

Let $N(Q)$ be the subgroup $\{ g(x,u,I_n); x \in V, u \in W \}$ of G. This $N(Q)$ is a 2-step nilpotent Lie group, called *the generalized Heisenberg group*. We have a diffeomorphism $N(Q) \ni g(x,u,I_n) \mapsto (x + iQ(u,u), u) \in \Sigma$, so that the Hilbert space $L^2(\Sigma)$ is naturally identified with $L^2(N(Q))$. Indeed, the partial Fourier transform can be regarded as a variation of the operator-valued Fourier transform for $N(Q)$ [8]. On the other hand, our solvable group G is the semidirect product of $N(Q)$ with H. This point of view suggests a generalization of the formula in Theorem 3 to a framework of wavelet transform for the semidirect product group of a unimodular (not necessarily commutative) Lie group N_0 with a subgroup H_0 of the automorphism group on N_0 [5].

References

[1] Ali S T, Antoine J-P, and Gazeau J-P 2000, Coherent states, wavelets and their generalizations, (New York: Springer)
[2] Duflo M and Moore C C 1976, On the regular representation of a nonunimodular locally compact group, J. Funct. Anal. 21 209–243
[3] Gindikin S G 1964, Analysis in homogeneous domains, Russian Math. Surveys 19 1–89
[4] Grossmann A, Morlet J, and Paul T 1985, Transforms associated to square integrable group representations I: General results, J. Math. Phys. 26 2473–2479
[5] Ishi H 2003, Wavelet transforms associated to semidirect product groups, in preparation.
[6] Liu H 1998, Wavelet transform and symmetric tube domains, J. Lie Theory 8 351–366
[7] Liu H and Peng L 1997, Admissible wavelets associated with the Heisenberg group, Pacific J. Math. 180 101–123

[8] Ogden R D and Vági S 1979, Harmonic Analysis of a nilpotent group and function theory on Siegel domains of type II, Adv. in Math. 33 31–92

[9] Piatetskii-Shapiro I I 1969, Automorphic functions and the geometry of classical domains (New York: Gordon and Breach)

Inst. Phys. Conf. Ser. No 173: Satellite colloquium
Paper presented at 24th Int. Coll. Group Theoretical Methods in Physics, Paris, France, 15–20 July 2002
©2003 IOP Publishing Ltd

Modeling 2D+1 textures

G Menegaz and S Valaeys

Audiovisual Communications Laboratory, Swiss Federal Institute of Technology

Abstract. We propose a novel wavelet based modeling technique for 2D+1 textures, i.e. static textures shot by a moving camera. The correct *perception* of motion is preserved by keeping unchanged the the temporal correlation between subsequent images, or frames. Global motion estimation is used to determine the movement of the camera and to identify the overlapping region between two successive texture images. Such an information is then exploited for the generation of the texture movies. The proposed method for synthesizing 2D+1 textures is able to emulate any piece-wise linear trajectory and is real-time on PIII processors.

1. Introduction

Dynamic textures are usually meant as multi-dimensional stochastic processes exhibiting some stationarity over time [1]. Some examples are smoke, waves and foliage. This can be regarded as a generalization of the bi-dimensional case, where temporal evolution is a feature of the global stochastic process [1, 2].

The novelty of our contribution is that we address the problem of modeling a different class of dynamic textures, for which the motion is not an intrinsic property of the considered process, but the result of a continuous change of the viewpoint. We aim at modeling the motion features as perceived by a moving observer. To make the distinction with respect of the 3D dynamic processes mentioned above, we call the considered class *2D+1 Texture Movies (TM)*. In this case, the key point is the preservation of the temporal correlation between subsequent images, or frames. We consider here the case of a static texture - the grass - shot by a moving camera, and generalize the DWT based Multiresolution Probabilistic Texture Modeling (MPTM) technique [3] to such a dynamic texture. Probabilistic modeling of static textures aims at generating a new image from a sample texture, such that it is *sufficiently different* from the original yet appears to be generated by the same underlying stochastic process. The goal of the proposed algorithm is to generalize such an idea to the generation of a progressively "growing" texture, where the direction and speed of growth is given *a-priori* by a predefined motion model. More specifically, we focus here on piece-wise linear trajectories. In this case, the main issue is the preservation of the perception of motion, namely the preservation of those visual features. Noteworthy, the trivial juxtaposition of temporally subsequent patches respectively sampled from successive frames is not a solution. The aliasing phenomena due to the sampling as well as the mismatch between the sampling grids associated to two successive frames would result in a discontinuity along the boundary.

This paper is organized as follows. Sec. 2 describes the 2D+1 texture model. Sec. 3 illustrates the movement-simulating algorithm. Results are discussed in Sec. 4 and Sec. 5 derives conclusions.

2. Modeling 2D+1 texture movies

Let Ω be the infinite lattice, and let Ω_t be the domain which is observed at time t, i.e. the spatial support associated with the observation at a given instant in time. Let then $I(\Omega_t)$ be the observation at time t and $\tilde{I}(\Omega_t)$ be the synthetic counterpart. Clearly:

$$\Omega_t \subset \Omega \qquad \forall t \tag{1}$$

Accordingly, Ω_{t_1} and Ω_{t_2} denote the domains covered by the observations at times t_1 and t_2, respectively. The specificity of the proposed approach is that it provides a solution to the following problem: *Given two sub-lattices Ω_{t_1} and Ω_{t_2} such that:*

$$\Omega_{t_1} \cap \Omega_{t_2} = \Omega_{\Delta t} \neq \emptyset, \tag{2}$$

generate a synthetic texture over Ω_{t_2} by growing it from the seed already present on $\Omega_{\Delta t}$ such that the impression of visual continuity is preserved.

If the two sets were disjoint, then the independent generation of the texture over the two domains would have been adequate. Conversely, where there is an overlap between the two domains, the independent generation of the texture would produce an apparent edge at the boundary or, equivalently, a flickering on the representation as a temporal sequence which destroys the continuity of the visual flow.

The key feature of the proposed model is the ability to synthesize a textures $\tilde{I}(\Omega_{t_2})$ over the domain Ω_{t_2} by growing the texture over $\overline{\Omega}_{\Delta t} = \Omega_{t_2} \backslash \Omega_{\Delta t}$ but keeping unchanged the texture already present over $\Omega_{\Delta t}$ and avoiding discontinuities along the boundary. The previous discussion holds unchanged also when the observations are themselves realizations of the stochastic process represented by the considered model for static textures. In this case, the following relation holds:

$$\tilde{I}(\Omega_{t+\Delta t}) = \tilde{I}(\Omega_t) \oplus \tilde{I}(\overline{\Omega}_{\Delta t}) \tag{3}$$

where $\tilde{I}(\Omega_{t+\Delta t})$ is the synthetic texture simulating the observation at time $t+\Delta t$, $\tilde{I}(\Omega_{\Delta t})$ is the texture seed, and the operator $\oplus$ indicates the juxtaposition of the textures stated. The spatial position of $\Omega_{t+\Delta t}$ can be easily recovered from the underlying motion model. Let $x, y \in \mathbb{R}$ be the spatial coordinates of the upper left corner of Ω_t and let h and w, with $h, w \in \mathbb{R}_*^+$, be the height, respectively the width, of the spatial domain Ω_t, assumed to be of rectangular shape. Given the estimated speed $\vec{v} = (v_x, v_y)$ at which the viewpoint moves, it is straightforward to derive the position of the domain $\Omega_{t+\Delta t}$ concerned by the observation at time $t + \Delta t$ as the one whose upper left corner has coordinates:

$$\begin{aligned} x + \Delta x &= x + v_x \cdot \Delta t \\ y + \Delta y &= y + v_y \cdot \Delta t \end{aligned} \tag{4}$$

Therefore, one can easily identify $\Omega_{\Delta t}$ and $\overline{\Omega}_{\Delta t}$.

3. Generalizing the DTW-MPTM to 2D+1 textures

The proposed method is s generalization of the DWT-MPTM to 2D+1 textures. It consists in synthesizing a texture area larger than the video frame size, preserving the texture over $\Omega_{\Delta t}$ while generating a limited amount of new texture, only when necessary, to cover $\overline{\Omega}_{\Delta t}$ avoiding discontinuities. It is worth pointing out that the straightforward solution of synthesizing each frame independently with the DWT-MPTM is not suitable because it creates a disjointed succession of rapid texture changes that fails to generate an impression of movement. One also quickly comes to the conclusion that a cut-and-paste approach at image level, in which the common section is correctly displaced and remaining empty parts of the frame are filled with newly synthesized patches of texture, creates unacceptable discontinuities. Another trivial solution would be to synthesize a much larger texture area than the frame size and to select the covered domain to be part of the frame according to the camera movement. This method is however suffers of some shortcomings. First, the required size of the synthetic texture should be known *a-priori*. Moreover, large amounts of texture could be produced without ever being needed.

A way to answer those concerns is to work in feature space. Although the DWT used for compression purposes is in general not shift-covariant, covariance properties hold for translations in transform space which correspond to translations at image level that can be broken down in horizontal and vertical shifts of $k \cdot 2^N$ and $h \cdot 2^N$ pixels, respectively, where $k, h \in \mathbb{Z}$ and N is the number of decomposition levels of the DWT. Working in feature space, we are consequently able to generate from a synthetic image S the following set:

$$\Gamma = \{S_{(k,h)}, k, h \in \mathbb{Z} | S_{(k,h)} = T_{(k,h)}S\} \tag{5}$$

where $T_{(k,h)}$ is the translation operator that applied to S produces a shift of $k \cdot 2^N$ and $h \cdot 2^N$ units in the horizontal and vertical directions, respectively. As the translation from S to $S_{(k,l)}$ takes in fact place in feature space, the remaining empty parts of $S_{(k,l)}$ can be filled by applying the DWT-MPTM algorithm locally without creating discontinuities. Any random translation can be obtained by extending the size of S so as to add to it a border of 2^N pixels on all sides. Accordingly, simulating a random translation is a two-step process: obtaining the correct $S_{(k,h)}$; selecting the correct area of $S_{(k,h)}$ which corresponds to the video frame. An example is shown in fig. 1. Let p and q, with $p, q \in \{0, 2 \cdot 2^N\}$ be the width in pixels of the border zone respectively in the horizontal and vertical direction of movement. To generate a horizontal, respectively vertical, movement of m, respectively n, pixels at image level, with $m \geq p$ corresponding to Δx and $n \geq q$ corresponding to Δy in Sec. 2, the correct $S_{(k,h)}$ is chosen so that:

$$k = \min_{\tilde{k}}\{(p + \tilde{k} \cdot 2^N) \geq m\} \tag{6}$$

$$h = \min_{\tilde{h}}\{(q + \tilde{h} \cdot 2^N) \geq n\} \tag{7}$$

The window of visibility is then correctly positioned inside $S_{(k,h)}$, namely:

$$new\_p = p + k \cdot 2^N - m \tag{8}$$
$$new\_q = q + h \cdot 2^N - n \tag{9}$$

4. Results and discussion

The performance of the proposed system has been evaluated in terms of *preservation of the perceptual features*. Before tackling this subject, it is important to mention that despite the great amount of research devoted to the identification of the *perceptual features* which determine texture perception, the problem is still unsolved. Two main guidelines can be identified. The first is based on the assumption that there exists a set of statistics which is *necessary and sufficient* to identify a *texture class*. Under such an hypothesis, a pair of textures sharing those statistics are *perceptually equivalent* [4, 8]. The problem is faced in an information theoretic manner, and leads to the definition of models based on statistical parameters. The way such parameters map to the hypothesized necessary and sufficient set is still unknown. The second consists in looking at the problem in a different perspective and aims at identifying and characterizing the *visual mechanisms* which are responsible for texture

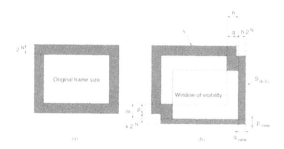

Figure 1. Simulation of movement. The size of S is bigger than that of the original. When the added border is not enough to simulate the required movement, a shifted version of S is created. The window of visibility is then moved inside $S_{(k,h)}$ to reproduce the correct movement.

perception (referring to particular tasks like analysis and discrimination) on a psychophysical or, more in general, neuro-physiological basis [5, 6, 7]. The focus is on the *local* parameters that are relevant for the analysis of the visual stimulus with regard to the considered task, as opposed to the analysis by synthesis approach followed in the other case. The problem is very complex and further investigation is needed to understand and model the involved visual processes. In this contribution, we do not put forward a general theory for texture perception neither a golden rule for the evaluation of a modeling technique. In our opinion, the visual mechanisms which subserve these phenomena need to be investigated further before being able to formulate a general theory. Instead, we have focused on a particular case - driven by an application - and we have faced it in an empirical way leading to what can be considered a "first order" solution. The identification of the features which determine the classification of the texture as belonging to a given class as well as the impression of motion will be an essential part of our future research.

The evaluation of the ability of any texture model, either static or dynamic, to reproduce the perceptual features of the original texture implies *ad-hoc* subjective tests respecting the paradigm set by psychophysics. However, as mentioned above, such a task was beyond the scope of this contribution. Nevertheless, some informal subjective tests involving non trained people of the laboratory revealed that the majority of the subjects were not able to distinguish between the original and synthetic samples.

5. Conclusion

We propose a novel generative model for 2D+1 textures, suitable for model-based coding of video. The integration of the motion information within the DWT-MPTM algorithm for static textures results in an dynamic generative model able to synthesize any 2D+1 texture movie with any piece-wise linear trajectory. A texture seed is extracted from a frame of the original sequence and is used as model for synthesizing the other frames. The motion vector field is estimated at any frame and is used to constrain the generating process such that the correct temporal correlation between the images is preserved. Among the issues deserving further investigation are the emulation of other camera functions like zooming and rotation as well as the rendering of perspective.

References

[1] Soatto S, Doretto G and Wu Y N 2001, Dynamic textures, Proc. of the International Conference on Computer Vision (ICCV)

[2] Bar-Joseph Z, El-Yaniv R, Lischinski D and Werman M 2001, Proc. IEEE Visualization, 403-410

[3] Menegaz G 2001, DWT-based non-parametric texture modeling, Proc. of the International Conference on Image Processing (ICIP), Thessaloniki, Greece, Oct. 2001

[4] Simoncelli E P and Olshausen B A 2001 Natural image statistics and neural representation, Ann. Rev. Neuroscience 24 1193-1215

[5] Landy M S 1996, Texture perception, Encyclopedia of Neuroscience (Amsterdam: Elsevier)

[6] Wolfson S S and Landy M S 1998, Examining edge- and region-based texture analysis mechanisms, Vision Research 38 439-446

[7] Li Z 2001, Modeling pre-attentive stereo grouping by intracortical interactions in early visual cortex, J. Vision, 1 ....

[8] Portilla J and Simoncelli E P 2000, A parametric texture model based on joint statistics of wavelet coefficients, Int. J. Computer Vision 40 49-71

Inst. Phys. Conf. Ser. No 173: Satellite colloquium
Paper presented at 24th Int. Coll. Group Theoretical Methods in Physics, Paris, France, 15–20 July 2002
©2003 IOP Publishing Ltd

An application of space-adaptive lifting: Image retrieval

P J Oonincx

Royal Netherlands Naval College, P.O. Box 10000, 1780 CA Den Helder, The Netherlands

Abstract. We show that adaptive lifting can be a useful tool for image retrieval systems that are based on shape detection and multiresolution structures of objects in a database against a background of texture. To measure the performance of our approach, feature vectors are computed based on moment invariants of lifting coefficients produced by the adaptive lifting scheme and retrieval rates are obtained by measuring distances between these vectors.

1. Introduction

We consider the problem of retrieving similar images of one single object from a database of objects. The images we are looking at are grayscale images of selected objects against a background of texture. Within the database one single object may appear in may different fashions in the images: translated, rotated etc. Furthermore, backgrounds will vary.

One of the most classical approaches to this problem is to use a filtering process that isolates the object and to identify the object with a vector. The entries of such vector are scalars obtained from using invariant measures on the object. Here we focus on the filtering. A possible choice is to use the wavelet transform as a preprocessing step. In this paper we show how new adaptive wavelet approaches may be used for even improving existing wavelet approaches. Simulation results on a database give a 'proof of principle' of the approach we use.

2. Space-adaptive lifting

The lifting scheme is a method for constructing wavelet transforms, see e.g. [1], that are not necessarily based on dilates and translates of one wavelet function. For a general introduction of the lifting scheme we would like to refer to [2]. Here we only discuss a special 2D case, which is used in our approach. This case uses a quincunx grid to project the pixels of an image on and is also referred to as Red-Black transform (RBT), see [3].

The first step of the RBT is to split all pixels of a given image into two sets, labeled as red and black pixels. This is done by projecting the image I onto a checkerboard yielding pixels on the red spots (I_r) and pixels on the black spots (I_b) of the checkerboard. By interpolating the 'red' data I_r a prediction $P(I_r)$ for I_b is obtained. Subtracting this estimate from the data I_b yields a detailed image $I_d = I_b - P(I_r)$. So for smooth images the error in the estimate $I_b - P(I_r)$ will generally be small and therefore I_d will not contain much information. However, I_d will attain high values at places where sharp transitions in the image I can be noted. The final step of the RBT consists of updating I_r by filtering I_d ($U(I_d)$) and adding it to I_r. This yields a coarse scale approximation $I_a = I_r + U(I_d)$.

When using the lifting scheme, the filters are chosen in a fixed fashion. In our application it is desirable to use small filters along sharp transitions in the image to guarantee large values for I_d along the details of the object of interest and to use large filters at the background of I to cancel out contributions to I_d resulting from the texture at the background.

These considerations lead to the idea of using different filters for different parts of the image. The image itself should indicate by local behaviour information whether a high or low order prediction filter should be used. Such an approach is commonly referred to as an adaptive approach. In this paper we follow the approach proposed by Baraniuk et al. in [4], called the space-adaptive approach. This approach follows the scheme as shown in Fig. 1.

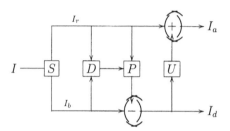

Figure 1. Generating coefficients via adaptive lifting: S splits the data into red/black labeled pixels, D brings in adaptivty.

As we see in this figure a decision operator $D(I_r, I_b)$ indicates the choice for P. This prediction filter P may vary from high to low order filters, depending on the regularity of the image locally. For the update operator, we choose the update filter that corresponds to the prediction filter with lowest order out of all possible to be chosen P. In particular we made the following choices for D, P and U. Based on a set of increasing quantizing levels applied on the relative local variance [5] the decision operator chooses for P filtering with respectively 2nd, 4th, 6th or 8th order Neville filters on a quincunx grid, see [6]. Consequently, U is a filtering operation using a 2nd order Neville filter on the quincunx grid.

3. Moment invariants

As already discussed in the introduction we propose to bring in space-adaptive lifting as a preprocessing tool for a certain type of image retrieval. Therefore, to compare its performance in relation to other classical filtering preprocessing tools we have to incorporate the scheme within a retrieval algorithm. For this we use the detail coefficients as input for a set of measures that are invariant under several affine transforms of the data, like translations, rotations and reflections. These commonly used invariants have been described already in 1962 by Hu [7]. The starting point for these invariants are $(p + q)$th order central moments μ_{pq} given by

$$\mu_{pq}(f) = \iint_{\text{R R}} (x - x_c)^p (y - y_c)^q f(x, y) \, d(x - x_c) \, d(y - y_c), \tag{1}$$

with x_c and y_c the coordinates of the center of mass.

A set of invariant measures can be constructed by considering homogeneous polynomials of these moments μ_{pq} up to a certain order. Here we use Hu's approach of constructing 7 invariant measure $J_1, \ldots, J_7$ by using moments up to order 3 ($0 \leqslant p + q \leqslant 3$), see [7]. As an example, one of these invariants is $\mu_{20}(f) + \mu_{02}(f)$, representing the variance of f.

At this point the invariants by Hu are still not consistent to uniform dilations (zooming in/out), neither do all invariants change evenly when the pixel values of an image are scaled (as a result of different lightning conditions). To overcome these problems we changed the invariants slightly by introducing the similitude invariance and the so-called homogeneity condition [5]. These normalisations lead to new central moments ν_{pq} with 7 corresponding invariants.

4. The testing algorithm

For measuring the performance of the adaptive approach, we computed detailed data I_d of images I using adaptive lifting at 4 levels, i.e., the approximated data I_a were used as input data to get details at the second level $(I_a)_b - P((I_a)_r)$ and a new approximate to be used again in the scheme and so on. Next, at each level j a set of coefficients W_j, that are considered to be related to the object in the image, is selected by thresholding. These coefficients are used to compute $\eta_{pq}^j = \mu_{pq}^j / |W_j|$ by converting the integrals in (1) into Riemann sums and by using the magnitude of the detailed data $|I_d|$ at each level for the function f. Finally, new normalised moments ν_{pq}^j are computed out of η_{pq}^j as discussed before. In our approach the root mean square (RMS) is used as a threshold value for the gradient of the detail data in order to find W_j, i.e.,

$$I_d^j(n,m) \in W_j \iff |\nabla I_d^j(n,m)| \geq \sqrt{\sum_{n=1}^{N} \sum_{m=1}^{M} |\nabla I_d^j(n,m)|^2 / NM}, \qquad (2)$$

with N and M respectively the number of rows and columns in the matrix (image detail) I_d^j. The gradients are approximated numerically.

Resuming, at each of the 4 levels k an image I is mapped to a 7-dimensional vector $J = (J_1^k, \ldots, J_7^k)$. Equivalently, when the process has been completed a 28-dimensional vector can be associated to the image I, namely

$$\tilde{J} = (J_1^1, \ldots, J_7^1, J_1^2, \ldots, J_7^2, J_1^3, \ldots, J_7^3, J_1^4, \ldots, J_7^4),$$

which can be seen as a kind of fingerprint of a certain image I. To identify in a given database images representing the same objects, we compute for all images in the database a fingerprint $\tilde{J}$ and measure all mutual Euclidean distances in $\mathbf{R}^{28}$. Small distances are related to similar images while large distances are associated with images of different type of objects.

5. Simulation results

Simulation results have been obtained by constructing a database of 64 images. These images are divided into 8 classes, each one consisting of images of one single object, see Fig. 2, but subjected to translation over various distances, rotations over various angles, reflections and scaling. Moreover, the images were pasted on an arbitrarily chosen (out of 4) wooden background texture, see [5].

For the simulation each image was used as a query to retrieve the other 7 relevant ones. The effectiveness of our approach (solid line) is shown in Fig. 3 with both the ideal case (crosses) and the case in which the lifting scheme with a fixed prediction filter was used (dotted line). In this figure the best possible performance using a fixed chosen filter (8th order) has been depicted. Distances among images are measured by weighted Euclidean distances between the feature vectors $\tilde{J}$. The average number of correctly retrieved images given a query image has been plotted against different number of allowed top retrievals. As we can see, retrieval rates increase by 5-10% by using an adaptive approach in our test case.

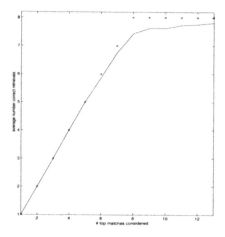

Figure 2. Object library of 8 different images to be transformed to 64 images.

Figure 3. Retrieval performance of adapted (solid) and non-adapted (dotted) approaches.

References

[1] Daubechies I and Sweldens W 1998, Factoring wavelet transforms into lifting steps, J. Fourier Anal. Appl. 4(3) 345–267

[2] Sweldens W 1997, The lifting scheme: A construction of second generation wavelets, SIAM J. Math. Anal. 29(2) 511–546

[3] Uytterhoeven G and Bultheel A 1997, The red-black wavelet transform, TW Report 271, Dept. Comp. Sc., Katholieke Universiteit Leuven

[4] Baraniuk R, Claypoole R, Davis G and Sweldens W, Nonlinear wavelet transforms for image compression via lifting, IEEE Trans. Imag. Proc., to appear

[5] Oonincx P J and de Zeeuw P M 2002, An image retrieval system based on adaptive wavelet lifting, CWI Report PNA-R0208

[6] Kovacevic J and Sweldens W 2000, Wavelet families of increasing order in arbitrary dimensions, IEEE Trans. Imag. Proc. 9(3) 480–496

[7] Hu M 1962, Visual pattern recognition by moment invariants, IRE Trans. Inf. Th. IT-8 179–187

Inst. Phys. Conf. Ser. No 173: Satellite colloquium
Paper presented at 24th Int. Coll. Group Theoretical Methods in Physics, Paris, France, 15–20 July 2002
©*2003 IOP Publishing Ltd*

Hybrid audio models and transform coding

B Torrésani

Laboratoire d'Analyse, Topologie et Probabilités, CMI, Université de Provence,
31 rue Joliot-Curie, 13453 Marseille Cedex 13, France

Abstract. This short note reports on recent results obtained by L. Daudet, S. Molla and the author on the application of hybrid, structured, non-linear coding technique for audiophonic signal compression. It presents the main ideas developed in this program, and provides a few references to the literature.

Digital audiophonic signal encoding has become an important issue in many application areas. Among the most popular approaches, transform coding (see for example [6, 11]) has received particular attention during the recent decades, as the rapid development of hardware and the discovery of novel mathematical approximation techniques has made it particularly efficient. The main characteristics of transform coding is that it starts with an expansion into a suitably chosen orthonormal basis of the spaces of signals under consideration. The (*functional*) approximation part comes from the fact that only a subset of the coefficients are retained, and encoded (after quantization, which is the second point where approximation comes into play).

Even though much may be done at the level of quantization, we shall focus here on the problem of functional approximation, which may thought in two different ways:

- *Linear approximation:* the retained coefficients correspond to a fixed set of basis functions. The basis has to be chosen accordingly, and one generally uses an orthonormal basis which approximates well the Karhunen-Loève basis of the signal (modeled as a random signal), under some additional constraints (essentially, the existence of efficient algorithms). This approach is the conventional one.

- *Non-linear approximation:* the retained coefficients are selected adaptively, for example as the set of coefficients such that the corresponding truncated expansion minimizes a given *distorsion* some norm of the approximation error. It is easy to see that a so-defined non-linear approximation automatically outperforms the linear one (using the same basis and the same number of retained coefficients) in terms of distorsion, but introduces an extra costs in terms of encoding: the retained set of basis functions being not fixed *a priori*, the corresponding information (the *addresses* of retained coefficients) has to be encoded as well, which introduces an extra cost. The latter has to be taken into account for a fair comparison of the encoding techniques.

In order to cope with the problem of address encoding for non-linear approximation, an extra concept has to be introduced: the concept of *structured* approximation. The latter stems from the fast that for given classes of signal (and/or functional spaces), and accordingly chosen orthonormal bases, the significant coefficients have a natural tendancy to *cluster* around some given types of structures (lines, trees,...) in their index space. Exploiting such information yields substantial gains in the addresses coding, and thus in the performances of encoders. In addition, this may also improve the quality of signal modeling, as we shall see on more specific examples below.

The present coding scheme is based upon structured non-linear approximations, with an additional non-linearity, introduced for the following reason. Audio signals (like other signal classes, images,...) may be thought as *compound objects*, containing significantly different features: mainly *tonals* (usually termed *partials* in the audio and speech literature), *transients*, and additional *residual* components. The adequate orthonormal bases for transform coding these components (at least the first two) are significantly different, and we propose to model the signal as a superimposition of two different components, each being transform coded with an adequate basis.

To our knowledge, our approach is fundamentally new, in that it does not rely on a prior segmentation of the signal into different components (see [5, 10]): superimposition is prefered to segmentation, which is from our point of view a more accurate model.

In order to get estimates for rate and distortion, signal models have to be considered. Random signal models are generally prefered. For modeling transient signals within the hybrid scheme, Hidden Markov Wavelet Trees (HMWT, see [1]) are particularly well adapted. Corresponding estimation algorithms (namely, Baum-Welsh's EM algorithm, and Viterbi's algorithm, see [9]) provide good estimates for transient structures. Similar ideas may also be developed for tonal features.

Preliminary results validating the hybrid coder approach have been presented in [2, 4]. The adaptation of HMWT models to transient signal coding has been presented in [3, 7]. Further problems, including the "transient/tonal balance" (i.e. the adaptive dynamic bit allocation to these two components), quantization issues, rate-distorsion estimates will be discussed in a forthcoming publication [8].

References

[1] Baraniuk R 1999, Optimal tree approximation using wavelets. In Aldroubi A J and Unser M (eds.), Wavelet Applications in Signal Processing, VII 196–207 (Bellingham, WA: SPIE)

[2] Daudet L 2000, Représentations structurelles de signaux audiophoniques. Méthodes hybrides pour des applications à la compression, PhD thesis, Université de Provence

[3] Daudet L, Molla S and Torrésani B 2001, Transient modeling and encoding using trees of wavelet coefficients, in "GRETSI'01", Toulouse, September 2001

[4] Daudet L and Torrésani B 2002, Hybrid models for audio signals encoding, Sig. Proc. 16 793-810

[5] Levine S 1998, Audio Representations for Data Compression and Compressed Domain Processing, PhD thesis, Stanford University

[6] Mallat S 1998, A wavelet tour on signal processing, (San Diego: Academic Press)

[7] Molla S and Torrésani B. 2002, Hidden Markov trees of wavelet coefficients for transient detection in audiophonic signals, Proceedings of the conference "Self-Similarity and Applications", Clermont, May 2002, Benassi A, editor

[8] Molla S and Torrésani B 2002, An hybrid audio scheme using Hidden Markov Models, in preparation

[9] Rabiner L R 1989, A tutorial on hidden markov models and selected applications in speech recognition, Proceedings of the IEEE, 77 257–286

[10] Verma T, Levine S, and Meng T 1997, Transient modeling synthesis: a flexible analysis/synthesis tool for transient signals, in "Proc. of the International Computer Music Conference", Greece

[11] Vetterli M and Kovacevic J 1996, Wavelets and SubBand Coding (Englewood Cliffs, NJ: Prentice Hall)

Inst. Phys. Conf. Ser. No 173: Satellite colloquium
Paper presented at 24th Int. Coll. Group Theoretical Methods in Physics, Paris, France, 15–20 July 2002
©*2003 IOP Publishing Ltd*

A wavelet-based model for forecasting non-stationary processes

S Van Bellegem, P Fryźlewicz and R von Sachs

FNRS Research Fellow, Institut de statistique, Université catholique de Louvain, Belgium
Graduate Student, Department of Mathematics, University of Bristol, UK
Professor, Institut de statistique, Université catholique de Louvain, Belgium

Abstract. In this article, we discuss recent results on modelling and forecasting covariance non-stationary stochastic processes using non-decimated wavelets.

1. Introduction

In this article, we are concerned with data generated by a univariate, discrete-time stochastic process. We focus on the analysis of its covariance structure, and therefore we assume that the process is zero-mean. Zero-mean processes arise, for example, when the global trend has been removed from the data. Trend estimation is a well studied problem, and some recent methods use wavelets.

The *autocovariance function* of a process X_t is denoted by $c_X(r, s) := \mathrm{Cov}(X_r, X_s)$, and, for *stationary processes*, it depends on the distance between r and s only, i.e. $c_X(r, s) = c_X(|r - s|)$. As it does not depend on any reference point in time, we say that the autocovariance function of a stationary process is homogeneous over time. All stationary processes X_t have the following Fourier representation:

$$X_t = \int_{(-\pi,\pi]} A(\omega) \exp(i\omega t) dZ(\omega), \quad t \in Z, \tag{1}$$

where $A(\omega)$ is the amplitude, and $Z(\omega)$ is a stochastic process with orthonormal increments [1]. Correspondingly, under mild conditions, the autocovariance function can be expressed as $c_X(\tau) = \int_{-\pi}^{\pi} f_X(\omega) \exp(i\omega\tau) d\omega$, where f_X is the *spectral density* of X_t [1].

The assumption of stationarity leads to an elegant theory from the point of view of both estimation and forecasting [1]. However, various studies based on statistical tests of stationarity have shown that many observed processes have a non-homogeneous autocovariance (spectral) structure [2, 3, 4]. Examples of such *non-stationary* processes abound e.g. in econometrics (returns on stock indices), biomedical statistics (electrocardiograms), meteorology (wind speed), and many other fields. The important question of how to model and forecast non-stationary processes arises, and one of the main motivations behind using wavelets here is that, being well-localised in both time and frequency, they have the potential to naturally handle phenomena whose spectral characteristics change over time.

2. The class of locally stationary wavelet processes and the wavelet spectrum

We now recall a definition of a class of zero-mean nonstationary processes built of non-decimated discrete wavelets [5], rather than harmonics $\exp(i\omega t)$ like in (1).

Definition 1 ([6]) *A triangular stochastic array* $X_{t,T}$ $(t = 0, \ldots, T-1, T > 0)$ *is in the class of* locally stationary wavelet (LSW) *processes if there exists a mean-square representation*

$$X_{t,T} = \sum_{j=-\lceil \log_2 T \rceil}^{-1} \sum_{k=-\infty}^{\infty} w_{j,k;T}\, \psi_{jk}(t)\, \xi_{jk}, \tag{2}$$

where j and k are scale and location parameters, respectively, $w_{j,k;T}$ are real constants, $\{\psi_{jk}(t)\}_{jk}$ is a non-decimated family of discrete compactly supported wavelets [5], ξ_{jk} is an orthonormal sequence of identically distributed zero mean random variables, and for each $j \leq -1$, there exists a Lipschitz-continuous function $W_j(z)$ on $[0, 1)$ such that

- $\sum_{j=-\infty}^{-1} W_j(z)^2 < \infty$ *uniformly in* $z \in [0, 1)$;
- *the Lipschitz constants L_j satisfy* $\sum_{j=-\infty}^{-1} 2^{-j} L_j < \infty$;
- *there exists a sequence of constants C_j satisfying* $\sum_{j=-\infty}^{-1} C_j < \infty$, *such that, for all T,*

$$\sup_{k=0,\ldots,T-1} |w_{j,k;T} - W_j(k/T)| \leq C_j/T.$$

By analogy with Dahlhaus [7], the time-varying quantity $W_j(z)$ is defined in *rescaled time* $z = t/T \in [0, 1)$. As the non-decimated wavelet system is overcomplete, the "amplitudes" $w_{j,k;T}^2$ are not uniquely defined and therefore not identifiable. However, due to the regularity of $W_j(z)$ in the rescaled time, the *wavelet spectrum* of $X_{t,T}$, defined by $S_j(z) := W_j(z)^2$, is unique. $S_j(z)$ measures the power of the process at a particular scale j and location z, and can be estimated by means of asymptotically unbiased multiscale estimators [6]. For stationary processes, $S_j(z)$ is independent of z for all j. Here, $j = -1$ denotes the finest scale.

As recalled in the Introduction, the autocovariance function of a stationary process is the Fourier transform of its spectral density. The next results shows an analogous link between the autocovariance function of an LSW process $X_{t,T}$, defined by $c_T(z, \tau) = \mathrm{Cov}(X_{[zT],T}, X_{[zT]+\tau,T})$, and its wavelet spectrum $S_j(z)$. We first define the *autocorrelation wavelets* $\Psi_j(\tau) = \sum_k \psi_{jk}(0)\psi_{jk}(\tau)$. They are symmetric with respect to τ and satisfy $\Psi_j(0) = 1$ for all scales. Also, like wavelets themselves, they enjoy good localisation properties.

Proposition 1 ([6]) *Under the assumptions of Definition 1,* $\|c_T - c\|_{L_\infty} = O(T^{-1})$, *where*

$$c(z, \tau) = \sum_{j=-\infty}^{-1} S_j(z)\Psi_j(\tau). \tag{3}$$

Function $c(z, \tau)$ is called the *local autocovariance function* (LACV) of $X_{t,T}$. Formula (3) is a *multiscale representation of the nonstationary autocovariance function* $c(z, \tau)$. The representation is unique because the set $\{\Psi_j\}_j$ is linearly independent, [6]. By way of example, it can be proved [8] that $\sum_{j=-\infty}^{-1} 2^j \Psi_j(\tau) = \delta_0(\tau)$ holds, which implies, in particular, that the wavelet spectrum of (stationary) white noise is proportional to $S_j(z) = 2^j$.

3. Forecasting

We now want to forecast an LSW process h steps ahead, basing on t observations $X_{0,T}, \ldots, X_{t-1,T}$. We set $T = t + h$ and consider a linear predictor $\hat{X}_{t+h-1,T} = \sum_{s=0}^{t-1} b_{t-1-s;T}^{(t,h)} X_{s,T}$, where the coefficients $b_{t-1-s}^{(t,h)}$ minimise the mean-square prediction error $\mathrm{E}(\hat{X}_{t+h-1,T} - X_{t+h-1,T})^2$.

Proposition 2 ([8]) *Assume $h = o(T)$ and let the length of ψ_j be ℓ_j. Assume also:*

$$\sum_{\tau=0}^{\infty} \sup_z |c(z,\tau)| < \infty \qquad \sum_{j<0}(C_j + \ell_j L_j)\ell_j < \infty \qquad \operatorname{ess\,inf}_{z,\omega} \sum_{j<0} S_j(z)|\hat{\psi}_j(\omega)|^2 > 0,$$

where $\hat{\psi}_j(\omega) = \sum_{s=-\infty}^{\infty} \psi_{j0}(s)\exp(i\omega s)$. The mean-square prediction error can be written as

$$\tilde{b}^{(t,h)\,\prime} B^{(t+h;T)}\tilde{b}^{(t,h)} \left(1 + o_T(1)\right), \tag{4}$$

where $B_{m,n}^{(t+h;T)} = c(\frac{n+m}{2T}, n - m)$, $n, m = 0, \ldots, t - 1 + h$, and $\tilde{b}^{(t,h)}$ is the vector $(b_{t-1}^{(t,h)}, \ldots, b_0^{(t,h)}, b_{-1}^{(t,h)}, \ldots, b_{-h}^{(t,h)})'$ with $b_{-1}^{(t,h)}, \ldots, b_{-h+1}^{(t,h)} = 0$ and $b_{-h}^{(t,h)} = -1$.

In theory, the coefficients of $\tilde{b}^{(t,h)}$ (and therefore $b^{(t,h)}$) are now found by minimising (4). In practice, however, the elements of $B^{(t+h;T)}$ need to be estimated. It is easily observed [8] that Formula (3) suggests the following multiscale estimator of the LACV:

$$\hat{c}\left(\frac{k}{T}, \tau\right) = \sum_{j,m=-[\log_2 T]}^{-1} A_{mj}^{-1}\left(\sum_{s=0}^{t-1} X_{s,T}\psi_{mk}(s)\right)^2 \Psi_j(\tau), \quad k = 0, \ldots, t-1, \tag{5}$$

where $A_{mj} = \sum_\tau \Psi_m(\tau)\Psi_j(\tau)$. The above estimator is asymptotically unbiased [8]. Also, it can be shown by simulation, and by exact calculation in some particular cases, that the above estimator enjoys better Mean-Square Error properties than other commonly used (asymptotically) unbiased estimators of local covariance. However, it is not consistent (its variance does not go to zero with T, [8]) and therefore has to be smoothed using e.g. a Gaussian kernel smoother with bandwidth g. Moreover, in (4), we also require the values of $c(k/T, \tau)$ for $k = t, \ldots, t+h-1$. Motivated by its slow evolution, we extrapolate the LACV by one-sided smoothing of the estimated autocovariances with the same bandwidth g. The next section discusses a data-driven method for choosing the smoothing parameter g.

4. The forecasting algorithm

As mentioned before, we estimate the entries of $B^{(t+h;T)}$ using (5) smoothed over k/T to achieve consistency. For simplicity, we choose the same bandwidth g for all τ. In this section and next, we only consider one-step-ahead forecasts.

Also, in practice, we only incorporate p past observations into the predictor, instead of the whole history of the process, which corresponds to only considering the bottom-right corner of $B^{(t+h;T)}$. The motivation behind this is that the "clipped" predictor often performs much better in practice, while also being less computationally expensive.

We select (g, p) by *adaptive forecasting*, i.e. we gradually update (g, p) according to the success of prediction. We first move backwards by s observations and choose the initial parameters (g_0, p_0) for forecasting $X_{t-s,T}$. Next, we forecast $X_{t-s,T}$ using not only (g_0, p_0) but also the 8 neighbouring pairs $(g_0 + \delta\epsilon_g, p_0 + \epsilon_p)$, for $\epsilon_g, \epsilon_p \in \{-1, 0, 1\}$ and δ fixed. As we already know the true value of $X_{t-s,T}$, we compare the 9 forecasts using a pre-selected criterion, and update (g, p) to be equal to the pair which gave the best forecast. We now use this updated pair, as well as its 8 neighbours, to forecast $X_{t-s+1,T}$, and continue in the same manner until we reach $X_{t-1,T}$. The updated pair (g_1, p_1) is used to perform the actual prediction, and can itself be updated later if we wish to forecast $X_{t,T}, X_{t+1,T}, \ldots$

Various criteria can be used to compare the quality of the pairs of parameters at each step. Denote by $\hat{X}_{t-i,T}(g, p)$ the predictor of $X_{t-i,T}$ computed using pair (g, p), and by $P_{t-i,T}(g, p)$ the length of the corresponding prediction interval. In [8], we propose the following criterion:

choose the pair which minimises $|X_{t-i,T} - \hat{X}_{t-i,T}(g,p)|/\mathrm{P}_{t-i,T}(g,p)$. We suggest that the length s of the "training segment" be chosen in such a way that $X_{t-s-p}, \ldots, X_{t-1}$ does not contain any apparent breakpoints observed after a visual inspection. To avoid dependence on the initial values (p_0, g_0), the algorithm can be iterated a few times along the training segment, e.g. until at least 95% of the observations fall within their 95% prediction intervals.

5. Example

We apply the algorithm to forecast the wind anomaly index (denoted here by X_t), plotted on <http://www.stats.bris.ac.uk/~mapzf/g24/wind.html>. This time series has been studied in meteorology in order to understand the El Niño effect in a specific region of the Pacific. The length of the series is $T = 910$. By trial and error, we have found that the pair $(g_0, p_0) = (70/910, 7)$ produces accurate one-step-ahead forecasts for many segments of the series. However, the results for the segment $X_{801}, \ldots, X_{840}$ using (g_0, p_0) are extremely bad (only 5% of the observations fall within the 95% prediction intervals). Suppose that we want to forecast $X_{841}, X_{842}, \ldots$. As we have no reason to believe that (g_0, p_0) has a chance of performing well here, we run the algorithm of the previous section with $s = 40$, $\delta = 1/910$, and the criterion described in the previous section. Three iterations along the training segment (with the "new" (g_0, p_0) always being set to the (g_1, p_1) obtained in the previous iteration) are sufficient to obtain 95% coverage of the 95% prediction intervals. The updated parameters are $(g_1, p_1) = (114/910, 14)$. Indeed, they do an excellent job in forecasting $X_{841}, \ldots, X_{848}$ one-step-ahead: over this segment, all the true values of X_t lie within their prediction intervals, and the main gain from using our procedure here is that the prediction intervals themselves are more than 15% narrower than those obtained from optimally fitted stationary ARMA models. A similar effect has been observed at several other points of the series. For other examples and more details, the reader is referred to [8].

References

[1] Brillinger D 1975, Time Series – Data Analysis and Theory (Austin, TX : Holt, Rinehart and Winston)

[2] von Sachs R and Neumann MH 2000, A wavelet-based test of stationarity, J. Time Ser. Anal. 21 597-613

[3] Fryźlewicz P 2002, Modelling and forecasting financial log-returns as locally stationary processes. Research Report, University of Bristol, Department of Mathematics

[4] Van Bellegem S and von Sachs R 2002, Forecasting economic time series using models of nonstationarity, Discussion paper No.0227, Université catholique de Louvain, Institut de statistique

[5] Nason G P and Silverman B 1995, The stationary wavelet transform and some statistical applications, in: Antoniadis A and Oppenheim G (eds.), Wavelets in Statistics, pp. 271-300 (New-York: Springer)

[6] Nason G P, von Sachs R, and Kroisandt G 2000, Wavelet processes and adaptive estimation of the evolutionary wavelet spectrum, J. Roy. Statist. Soc. Ser. B 62 271-292

[7] Dahlhaus R 1997, Fitting time series models to nonstationary processes, Ann. Statist. 25 1-37

[8] Fryźlewicz P, Van Bellegem S, and von Sachs R 2002, Forecasting non-stationary time series by wavelet process modelling, Discussion paper No.0208, Université catholique de Louvain, Institut de statistique

Louvain-la-Neuve Satellite Workshop
List of Participants

Ali S T, Concordia U, Montreal (Canada)
Altaisky M, JINR, Dubna (Russia)
Andrle M, Univ. Paris 7 – Denis Diderot (France)
Antoine J-P, UCL, Louvain-la-Neuve (Belgium)
Arnéodo A, CRPP, Pessac (France)

Bastin F, Univ. Liège (Belgium)

Cahen M, ULB, Bruxelles (Belgium)
Charoing V, Univ. Paris IX Dauphine (France)
Cordero E, Univ. Torino (Italy)

De Mol C, ULB, Bruxelles (Belgium)
De Vleeschouwer C, UCL, Louvain-la-Neuve (Belgium)
Demanet L, Caltech, Pasadena (USA)
Divorra Escoda O, EPFL, Lausanne (Switzerland)

Figueras i Ventura R-M, EPFL, Lausanne (Switzerland)
Führ H, GSF, Neuherberg (Germany)

Gissot S, Royal Observatory, Brussels (Belgium)
Gröchenig K, Univ. Connecticut, Storrs (USA)

Hochedez J-F, Royal Observatory, Brussels (Belgium)
Hohouéto A L, Concordia University, Montreal (Canada)

Jacques L, UCL, Louvain-la-Neuve (Belgium)
Jansen M, TU Eindhoven (The Netherlands)

Lambert D, FUNDP, Namur (Belgium)

Macq B, UCL, Louvain-la-Neuve (Belgium)
Menegaz G, EPFL, Lausanne (Switzerland)

Nicolay S, Univ. Liège (Belgium)

Oonincx P, Royal Netherlands Naval College, Den Helder (The Netherlands)

Rivoldini A, Royal Observatory, Brussels (Belgium)

Torrésani B, Univ. Provence, Marseille (France)

Van Bellegem S, UCL, Louvain-la-Neuve (Belgium)
Verhoeven O, Royal Observatory, Brussels (Belgium)

Group 24 in Paris
List of Participants

Aldaya, Victor, Spain
Aldrovandi, Ruben, Brazil
Ali, Syad Twareque, Canada
Alodjants, Alexander, Russia
Altaisky, Mikhail, Russia
Andrle, Miroslav, Czech Republic
Angelova, Maia, UK
Antoine, Jean-Pierre, Belgium
Arnaudon, Daniel, France
Aroyo, Mois, Spain
Aschieri, Paolo, Germany
Atakishiyev, Natig, Mexico
Auffret-Babak, Isabelle, UK
Authier, Evelyne, France

Baake, Michael, Germany
Bacry, Henry, France
Baláži, Peter, Czech Republic
Ballesteros, Ángel, Spain
Barbachoux, Cécile, France
Barrio, Rafael, Mexico
Bartlett, Stephen, Australia
Batista, Eliezer, UK
Bazunova, Nadezda, Estonia
Beige, Almut, Germany
Bender, Carl, USA
Bergeron, Herve, France
Besprosvany, Jaime, Mexico
Birman, Joseph L, USA
Blohmann, Christian, Germany
Blumen, Sacha, Australia
Bogoyavlenskij, Oleg, Canada
Böhm, Arno, USA
Bonatsos, Dennis, Greece
Bonechi, Francesco, Italy
Boroczky, Karoly, Hungary
Borowiec, Andrzej, Poland
Boya, Luis J, Spain
Boyle, L Laurence, UK
Brouder, Christian, France

Brzeziński, Tomasz, UK
Burdik, Čestmir, Czech Republic

Calixto, Manuel, Spain
Campoamor-Stursberg, Rutwig, Spain
Carbone, Alessandra, France
Cartier, Pierre, France
Carvalho Abreu, Everton Murilo, Brasil
Cassinelli, Giovanni, Italy
Castagnoli, Giuseppe, Italy
Catto, Sultan, USA
Celeghini, Enrico, Italy
Chadzitaskos, Goce, Czech Republic
Chakrabarti, Amitabha, France
Chavleishvili, Michael, Russia
Cherniha, Roman, Ukraine
Cho, Yongmin, South Korea
Chrúsciński, Dariusz, Poland
Coquereaux, Robert, France
Coste, Antoine, France
Cotfas, Nicolae, Romania

Daher, Naoum, France
de Guise, Hubert, Canada
de Montigny, Marc, Canada
de Souza Dutra, Alvaro, Brazil
de Vega, Hector, France
del Olmo, Mariano A, Spain
Demiralp, Ersan, Turkey
Dobrev, Vladimir, Bulgaria
Doebner, Heinz-Dietrich, Germany
Drigo Filho, Elso, Brazil
Duchamp, Gérard, France
Durand, Stéphane, Canada

Elkharrat, Avi, France
Engliš, Miroslav, Czech Republic

Finkelstein, David, USA
Fiore, Gaetano, Italy

Florek, Wojciech, Poland
Fowler, Patrick W, UK
Franceschelli, Sara, France
Françoise, Jean-Pierre, France
Frenkel, Edward, USA
Frydryszak, Andrzey, Poland
Fu, Hongchen, UK

Gadjiev, Bahruz, Russia
Gal'tsov, Dimitri, Russia
Garidi, Tarik, France
Gazeau, Jean-Pierre, France
Ge, Mo-Lin, RP China
George, Nathan, USA
Gilmore, Robert, USA
Giraud-Heraud, Yannick, France
Gnerlich, Ingrid, USA
Goldin, Gerald, USA
Gomes, Jose F, Brazil
Gourevitch, Dimitri, France
Gracia-Bondía, José, Costa Rica
Grimm, Uwe, UK
Grundland, Michel, Canada
Gueorguiev, Vesselin, USA
Gueorguieva, Ana, Bulgaria
Guerrero, Julio, Spain
Güler, Yurdahan, Turkey

Haouchine, Mustapha, Algeria
Henkel, Malte, France
Hennig, Joerg, Germany
Hermisson, Ulrich, Germany
Herranz, Francisco J, Spain
Herrera, Alfredo, Mexico
Hibberd, Katrina, Spain
Hlavaty, Ladislav, Czech Republic
Ho, Choon-Lin, Taiwan
Hohouéto, A Lionel, Canada
Horzela, Andrzej, Poland
Hounkonnou, M Norbert, Benin
Hsiao, Pai, France
Huguet, Eric, France
Hussin, Véronique, Canada

Iamaleev, Robert, Mexico
Ilisca, Ernest, France
Ishi, Hideyuki, Japan

Jacobsen, Jesper, France
Jakobsen, Hans Plesner, Denmark
Joyce, William, New Zealand
Jurčo, Branislan, Germany

Kamiya, Noriaki, Japan
Kapuscik, Edward, Poland
Karkar, Hassane, France
Kastrup, Hans, Germany
Katriel, Jacob, Israel
Kerner, Richard, France
Kijowski, Jerzy, Poland
Kim, Young S, USA
Kisil, Vladimir, UK
Klauder, John, USA
Klimek, Malgorzata, Poland
Klimov, Andrei, Mexico
Kramer, Peter, Germany
Kulish, Petr P, Russia

Lagarias, Jeffrey, USA
Lévai, Geza, Hungary
Levi, Decio, Italy
Lévy, Jean-Claude, France
Lévy-Leblond, Jean-Marc, France
Lin, Florence J, USA
Links, Jon, Australia
Lipkin, Harry, Israel
Lorente, Miguel, Spain
Lukierski, Jerzy, Poland

Magnot, Jean-Pierre, France
Man'ko, Margarita, Russia
Man'ko, Olga, Russia
Man'ko, Vladimir, Russia
Mann, Ady, Israel
Mansouri, Freydoon, USA
Masakova, Zuzana, Czech Republic
Matthes, Rainer, Germany
Mebarki, Noureddine, Algeria
Mekhfi, Mustapha, Algeria
Mensky, Michael, Russia
Mercier, Daniel, France
Mesref, Lahouari, Germany
Métens, Stéphane, France
Mir-Kasimov, Rufat, Russia
Monastyrsky, Michael, Russia

Moody, Robert, Canada
Moshinsky, Marcos, Mexico
Moskaliuk, Stepan, Austria
Mosseri, Rémy, France
Mostafazadeh, Ali, Turkey
Mourad, Jihad, France
Moylan, Patrick, USA
Mudrov, Andrey, Israel

Naudts, Jan, Belgium
Naumis, Gerardo, Mexico
Navrátil, Ondres, Czech Republic
Ne'eman, Yuval, Israel
Négadi, Tidjani, Algeria
Nikitin, Anatoliy, Ukraine
Noumi, Masatoshi, Japan
Nowicki, Anatol, Poland
Numa, Edgar, France

O'Raifeartaigh, Cormac, Ireland
Odaka, Kazuhiko, Japan
Okubo, Susumu, USA
Olkhov, Oleg, Russia
Omote, Minoru, Japan
Otarod, Saeed, Iran

Pan, Feng, PR China
Papadopolos, Zorka, Germany
Parashar, Deepak, Germany
Patera, Jiri, Canada
Penson, Karol, France
Picco, Marco, France
Pogosyan, George, Russia
Pop, Iulia, Sweden
Popov, Todor, Bulgaria
Proskurin, Denis, Russia
Pujol, Pierre, France
Pusz, Wieslaw, Poland

Quesne, Christiane, Belgium

Radford, Chris, Australia
Radosz, Andrzej, Poland
Ragnisco, Orlando, Italy
Rahula, Maido, Estonia
Ram, Arun, USA
Ramond, Pierre, USA

Rassat, André, France
Rausch de Traubenberg, Michel, France
Redlich, Krzysztof, Switzerland
Reis, Nuno, France
Ricotta, Regina, Brazil
Riquer, Veronica, Mexico
Rivero, Alejandro, Spain
Rosenhaus, Vladimir, USA
Rosso, Marc, France
Rosu, Haret, Mexico
Roubtsov, Vladimir, France

Sakoda, Seiji, Japan
Sanchez, Norma G, France
Sartori, Gianfranco, Italy
Saveliev, Vladimir, Kazakhstan
Schmidt, Heinz-Jürgen, Germany
Schuch, Dieter, Germany
Schweigert, Christoph, France
Sciarrino, Antonio, Italy
Serié, Emmanuel, France
Shima, Kazunari, Japan
Snobl, Libor, Czech Republic
Sogami, Ikuo, Japan
Solomon, Allan, UK
Soltan, Piotr, Poland
Sorace, Emanuele, Italy
Sorba, Paul, France
Spector, Nissan, France
Steinacker, Harold, Germany
Sternheimer, Daniel, France
Sutcliffe, Paul, UK
Suzuki, Osamu, Japan
Svobodová, Milena, Czech Republic
Sylos Labini, Francesco, France

Tafel, Jacek, Poland
Takemura, Kouichi, Japan
Takizawa, Michiru, Australia
Tarlini, Marco, Italy
Tempesta, Piergiulio, Canada
Thibon, Jean-Yves, France
Tolar, Jiri, Czech Republic
Tollu, Christophe, France
Tolstoy, Valery, Russia
Toumazet, François, France
Tounsi, Ahmed, France

Twarock, Reidun, UK

Vainerman, Leonid, Germany
Van der Jeugt, Joris, Belgium
Verger-Gaugry, Jean-Louis, France
Vicent, Luis Edgar, Mexico
Vinet, Luc, Canada
Vitiello, Giuseppe, Italy
Vorov, Oleg, France
Vourdas, Apostolos, UK

Walker, Michael, South Korea
Wang, Li, France
Weigt, Gerhard, Germany
Winternitz, Pavel, Canada

Wohlgenannt, Michael, Germany
Wolf, Kurt, Mexico
Woronowicz, Stanislaw, Poland
Wybourne, Brian, Poland

Xu, Xiaoping, PR China

Yang, Wen-Li, Germany

Zamiralov, Valeri, Russia
Zeiner, Peter, Austria
Zhang, Hechun, The Netherlands
Zhilinskii, Boris, France
Znojil, Miloslav, Czech Republic

Author Index